MOLECULAR EPIDEMIOLOGY: PRINCIPLES AND PRACTICES

EDITED BY NATHANIEL ROTHMAN, PIERRE HAINAUT, PAUL SCHULTE, MARTYN SMITH, PAOLO BOFFETTA AND FREDERICA PERERA

INTERNATIONAL AGENCY FOR RESEARCH ON CANCER
LYON, FRANCE
2011

Published by the International Agency for Research on Cancer,
150 cours Albert-Thomas, 69372 Lyon Cedex 08, France

©International Agency for Research on Cancer, 2011

Distributed by
WHO Press, World Health Organization, 20 Avenue Appia, 1211 Geneva 27, Switzerland
(tel: +41 22 791 3264; fax: +41 22 791 4857; email: bookorders@who.int)

Publications of the World Health Organization enjoy copyright protection in accordance with the provisions of Protocol 2 of the Universal Copyright Convention. All rights reserved.

The designations employed and the presentation of the material in this publication do not imply the expression of any opinion whatsoever on the part of the Secretariat of the World Health Organization concerning the legal status of any country, territory, city, or area or of its authorities, or concerning the delimitation of its frontiers or boundaries.

The mention of specific companies or of certain manufacturers' products does not imply that they are endorsed or recommended by the World Health Organization in preference to others of a similar nature that are not mentioned. Errors and omissions excepted, the names of proprietary products are distinguished by initial capital letters.

The authors alone are responsible for the views expressed in this publication.

The International Agency for Research on Cancer welcomes requests for permission to reproduce or translate its publications, in part or in full. Requests for permission to reproduce or translate IARC publications – whether for sale or for noncommercial distribution – should be addressed to the IARC Communications Group, at: com@iarc.fr.

IARC Library Cataloguing in Publication Data

Molecular epidemiology: Principles and practices / edited by Nathaniel Rothman ... [et al.]

(IARC Scientific Publications ; 163)

1. Biological Markers 2. Disease Susceptibility 3. Molecular Epidemiology -- methods
I. Rothman, Nathaniel II. Series

ISBN 978-92-832-2163-0 (NLM Classification: QH 506)
ISSN 0300-5085

Table of contents

Foreword v
Preface vii

Unit 1
Contextual framework for molecular epidemiology

Chapter 1 1
Molecular epidemiology: Linking molecular scale insights to population impacts

Chapter 2 9
Ethical issues in molecular epidemiologic research

Unit 2
Biomarkers: practical aspects

Chapter 3 23
Biological sample collection, processing, storage and information management

Chapter 4 43
Physical/chemical/immunologic analytical methods

Chapter 5 63
Assessment of genetic damage in healthy and diseased tissue

Chapter 6 99
Basic principles and laboratory analysis of genetic variation

Chapter 7 121
Platforms for biomarker analysis using high-throughput approaches in genomics, transcriptomics, proteomics, metabolomics, and bioinformatics

Chapter 8 143
Measurement error in biomarkers: sources, assessment, and impact on studies

Unit 3
Assessing exposure to the environment

Chapter 9 163
Environmental and occupational toxicants

Chapter 10 175
Infectious agents

Chapter 11 189
Dietary intake and nutritional status

Chapter 12 199
Assessment of the hormonal milieu

Chapter 13 215
Evaluation of immune responses

Unit 4
Integration of biomarkers into epidemiology study designs

Chapter 14 241
Population-based study designs in molecular epidemiology

Chapter 15 261
Family-based designs

Chapter 16 281
Analysis of epidemiologic studies of genetic effects and gene-environment interactions

Chapter 17 303
Biomarkers in clinical medicine

Chapter 18 323
Combining molecular and genetic data from different sources

Unit 5
Application of biomarkers to disease

Chapter 19 337
Cancer

Chapter 20 363
Coronary heart disease

Chapter 21 387
Work-related lung diseases

Chapter 22 407
Neurodegenerative diseases

Chapter 23 421
Infectious diseases

Chapter 24 441
Obesity

Chapter 25 453
Disorders of reproduction

Chapter 26 475
Studies in children

Chapter 27 493
Future perspectives on molecular epidemiology

Authors' list 501

Index 505

Foreword

Molecular epidemiology provides an exciting set of opportunities to contribute to the evidence base for the prevention of chronic diseases in the coming decades. In the mid- to late-1980s, the emergence of the polymerase chain reaction resulted in a step-change in the ability to investigate genetic polymorphisms and disease risk. This area was further transformed by the Human Genome Project and the widespread application of genome-wide association studies to large, multicentre case–control studies. Nevertheless, I believe the best is still to come from molecular epidemiology. This assertion is based on a combination of advances in understanding molecular mechanisms underlying disease development, powerful new laboratory technologies to interrogate patterns of gene, protein and metabolite levels, and their potential application to biobank specimens associated with large-scale population-based cohort studies.

There are risks to the fulfilment of this promise. First, the exquisite tools to study genetic susceptibility are as yet unmatched by tools of equal power to evaluate the environmental (non-genetic) basis of disease; without a balance between the genome and the exposome, their interplay in the causation of chronic diseases cannot be fully elucidated. Second, a systematic investment by research organizations and funders is needed in the type of translational research that draws advances in mechanistic knowledge and the associated technologies into epidemiology; interdisciplinary research across the laboratory sciences, epidemiology, clinical research, biostatistics and bioinformatics has never been more important.

This IARC publication, prepared by experts in the field, is a timely and valuable foundation for the future. It emphasizes the development and validation of appropriate methodology. It highlights the flow of knowledge from mechanisms of disease development, through the derivation of biomarkers, to their application in epidemiological studies. It illustrates the benefits of mentally crossing disease boundaries when considering the origins and consequences of underlying pathological processes. It stimulates inter-disciplinary thinking and orientates the laboratory towards public health.

The book spans great scale, highlighting at one end of the spectrum the increasing requirement to handle and interpret through computational means tens of millions of biomarker data points on tens of thousands of subjects while, at the other end, being attentive to the ethical questions affecting the individuals contributing to research through donation of their time, information and biological samples. If molecular epidemiology is to truly contribute to relieving the ever-increasing burden of chronic disease it will need excellent communication not only to the scientific audience that is the target of this book, but the people worldwide who are the subject of its investigations and concerns.

Christopher P. Wild
Director, International Agency for Research on Cancer

Major advances in our understanding of the origins and natural history of several chronic diseases have come from epidemiologic and laboratory research over the past 1–2 decades. While this knowledge has provided new opportunities for disease prevention and control, we are still limited by an incomplete grasp of causal mechanisms, which hold the key to further progress in preventive medicine and public health. However, recent conceptual breakthroughs in genomic and molecular sciences have fuelled optimism that the incorporation of innovative high-throughput technologies into robust epidemiologic designs will further dissect the genetic and environmental components underpinning complex diseases such as cancer, and thereby inform new clinical and public health interventions.

At this critical moment in the evolution of molecular epidemiology, the editors of this volume have enlisted scientific leaders in the field to review the major concepts, methods and tools of this interdisciplinary approach. The chapters summarize recent progress that has been made for several diseases and traits through molecular epidemiology, while suggesting promising directions for further discovery. Special attention is given to the process of selecting, validating and integrating molecular and biochemical biomarkers that sharpen our measures of causal factors and mechanisms, as well as disease outcomes, through epidemiologic research. The success of molecular epidemiology is due in no small part to advances in statistical methods and bioinformatics, as illustrated by the discovery of heritable mechanisms for many diseases and traits generated recently by large-scale genome-wide association studies.

As a fast-breaking interdisciplinary approach, molecular epidemiology faces formidable challenges, but the dividends are likely to increase by an order of magnitude as the next-generation "omics" technologies become available for epidemiologic application. With the evidence in this volume as a starting point, the stage is set for basic, clinical and population scientists to accelerate collaborative efforts that will contribute new biological insights and augment strategies for preventing and controlling disease on a global scale.

Joseph F. Fraumeni, Jr.
Director, Division of Cancer Epidemiology and Genetics, National Cancer Institute

Preface

We are pleased to present our textbook *Molecular Epidemiology: Principles and Practices*. As noted in prefaces by Christopher Wild and Joseph Fraumeni, Jr., this is an extremely exciting time in molecular epidemiology. Advanced tools and platforms have facilitated new efforts to be launched that are enabling a broad approach to studying the impact of a wide range of environmental exposures, broadly defined, and the inherited contribution to disease. These platforms are undergoing rapid evolution in the areas of exposure assessment and genomics and promise further advances in the near future. At the same time, there exist fundamental and basic principles of epidemiological study design: biologic sample collection, processing, and storage; and analysis of biological samples to ensure that reliable and accurate data are generated. The goal of our book is to provide a broad overview of these fundamental principles and their application to a wide range of diseases to help build a foundation that will allow the reader to appreciate, interpret and utilize these new technologies as they arise in the coming years.

We envision this collection of chapters as an orientation to the exciting opportunities that exist in molecular epidemiology. We also hope it will motivate readers to translate this information and harness these tools in meaningful ways that have a positive impact at the broadest public health level as well as at the personalized level. As noted in Chapter 1, "Knowledge is the basis for action."

The text is meant for graduate and post-graduate students in public health and the biologic sciences, as well as seasoned practitioners interested in the striking advances that have occurred in molecular epidemiology in recent years. The book represents an update and extension of its forerunner, by the same title, published in 1993 by Frederica Perera and Paul Schulte. In that ground-breaking effort, a broad approach was taken that included a discussion of the full range of biologic markers available to investigators carrying out molecular epidemiologic research and how these tools had been and could be applied to a wide range of diseases. In the current text, we have continued and expanded upon this approach.

The book begins with providing a contextual framework for molecular epidemiology focusing on both historical and ethical components of molecular epidemiology research. It then discusses practical aspects of using biomarkers including collection, processing and storage of biologic samples; the major types of biologic markers used in molecular epidemiology research; and measurement error. Next, examples are provided of biomarkers used in characterizing exposure to environmental and occupational toxins and infectious agents, and to assessing nutritional and hormonal status and the immune response. The integration and analysis of biomarkers in a spectrum of study designs, including population- and family-based studies and clinical trials, is presented, as well as a discussion of approaches to summarizing data across studies. Examples of the application of biomarkers to the study of several major diseases and conditions are given, including cancer, coronary heart disease, lung disease, neurodegenerative disease, infectious disease, reproductive disorders and obesity. Also discussed is the conduct of molecular epidemiology studies in children. The book concludes with a discussion of future directions in molecular epidemiologic research.

Finally, we sincerely thank the chapter authors and co-authors who made this book possible. Also, we would like to acknowledge the critical support of Jennifer Donaldson, the project manager and technical editor, without whom this book could not have been brought to fruition, and the support of IARC's publication staff, in particular John Daniel, Nicolas Gaudin and Sylvia Moutinho.

UNIT 1.
CONTEXTUAL FRAMEWORK FOR MOLECULAR EPIDEMIOLOGY

CHAPTER 1.

Molecular epidemiology: Linking molecular scale insights to population impacts

Paul A. Schulte, Nathaniel Rothman, Pierre Hainaut, Martyn T. Smith, Paolo Boffetta, and Frederica P. Perera

Summary

In a broad sense, molecular epidemiology is the axis that unites insights at the molecular level and understanding of disease at the population level. It is also a partnership between epidemiologists and laboratory scientists in which investigations are conducted using the principles of both disciplines. A key trait of molecular epidemiology is to evaluate and establish the relationship between a biomarker and important exogenous and endogenous exposures, susceptibility, or disease, providing understanding that can be used in future research and public health and clinical practice. When potential solutions or interventions are identified, molecular epidemiology is also useful in developing and conducting clinical and intervention trials. It can then contribute to the translation of biomedical research into practical public health and clinical applications by addressing the medical and population implications of molecular phenomena in terms of reducing risk of disease. This chapter summarizes the contributions and research endeavours of molecular epidemiology and how they link with public health initiatives and clinical practice.

Introduction

This is a unique and exciting period in the health sciences. For the first time, it is possible to look at both nature and nurture with sophisticated and molecular-level tools (1–12). The promise of using these and other tools to prevent, control, and treat chronic and infectious diseases stimulates the imagination and creativity of medical and health scientists and practitioners. The challenge is to effectively apply these tools, and knowledge from genetics, exposure assessment, population health and medicine, to health problems that afflict 21st century people. The means of meeting that challenge is the widespread conduct of molecular epidemiologic research. Driven by discoveries of basic biological phenomena at the molecular and genetic levels, molecular epidemiology is able to translate discovery of essential scientific knowledge into determination and quantification of hazards and risks, and then to investigate useful approaches for prevention, control, and treatment of disease and dysfunction (9,12–15).

To fully appreciate the potential contributions of molecular epidemiology, it is important to understand how it fits into the context of epidemiology and public health. Molecular epidemiology is a partnership between epidemiologists and laboratory scientists that conducts investigations using the principles of both laboratory and population research (1,2,16). This is a message that merits restatement as powerful genetic and analytic technologies become available to epidemiologists. Historically, molecular epidemiology was derived from those disciplines that made contributions to relating biological measurements to health and disease (1,2). These include bacteriology, immunology and infectious disease epidemiology; pathology and clinical chemistry; carcinogenesis and oncology; occupational medicine and toxicology; cardiovascular disease epidemiology; genetics, molecular biology, and genetic epidemiology; and traditional epidemiology and biostatistics. The term "molecular epidemiology" was first used in the infectious disease literature by Kilbourne to describe the "molecular determinants of epidemiologic events" (17). In 1977, Higginson used the term in the context of pathology in a paper entitled "The role of the pathologist in environmental medicine and public health" (18). Lower's landmark 1979 publication introduced genetic effect modifiers and brought attention to the importance of external exposure, individual susceptibility and biologic markers in terms of phenotype (19). In a seminal paper in 1982, Perera and Weinstein coined the term "molecular cancer epidemiology" and first proposed a formal and comprehensive framework for the use of biomarkers of internal dose, biologically effective dose, early biologic effect and susceptibility within a molecular epidemiological framework (2). In 1987, the National Research Council (NRC) adopted this basic conceptual framework for molecular epidemiology and subsequently published a series of reports on biological markers (20–22). In the 1980s through the mid-1990s, a series of important papers and books were published describing the evolution and progress of molecular epidemiology (1,17,23–36). More recently, the changing face of epidemiology in the genomics and epigenetic eras has been described (9,12,36–39).

In the past, molecular epidemiology was sometimes viewed as one of epidemiology's many subspecialties. Some subspecialties focus on the disease type (e.g. chronic, infectious, reproductive or cardiovascular), some on the origin of the hazard (e.g. occupational, environmental or nutritional), and still others focus on the approach to the disease (e.g. clinical, serological or analytical). Viewed in this context, molecular epidemiology may best fit into the category of subspecialty defined by the approach that is applicable to all of these areas. Molecular epidemiology is the use of all types of biological markers in the investigation of the cause, distribution, prevention and treatment of disease, in which biological markers are used to represent exposures, intervening factors, susceptibility, intermediate pathological events, preclinical and clinical disease, or prognosis.

More broadly, molecular epidemiology can be viewed as a hub that links various aspects of health research. Even the term molecular epidemiology is a linking term which brings together molecular-level thinking and population-level understanding. These insights can be useful in characterizing a health problem, conducting mechanistic research (at the molecular and population levels), understanding the solutions, and contributing to the clinical and public health practice. These four functions and the research that supports them are illustrated in Figure 1.1.

Characterizing a public health problem

Surveillance, the sentinel activity of public health and clinical practice, is the ongoing collection, analysis and interpretation of data on rates and trends of disease, injury, death and hazards. Molecular epidemiology plays an important role in surveillance by identifying the frequency of biological markers of exposure, disease or susceptibility in various population groups and in monitoring trends of biomarkers over time. Examples are population monitoring of blood lead concentrations, neonatal screening for genetic disease, and molecular typing of viruses in a geographical area. The validation of those biomarkers and the analysis of the data involve molecular epidemiologic skills and knowledge. Increasingly, biological specimen banks are being used as public health surveillance systems (40) and can play an important role in etiologic research (41).

Mechanistic research

Establishing the relationship between a biomarker and exposure, disease or susceptibility is the hallmark of molecular epidemiology, and leads to developing the knowledge that will eventually be used in further research and in clinical and public health practice.

Figure 1.1. Molecular epidemiology can serve as a hub for other components of health research and practice. Adapted from (74).

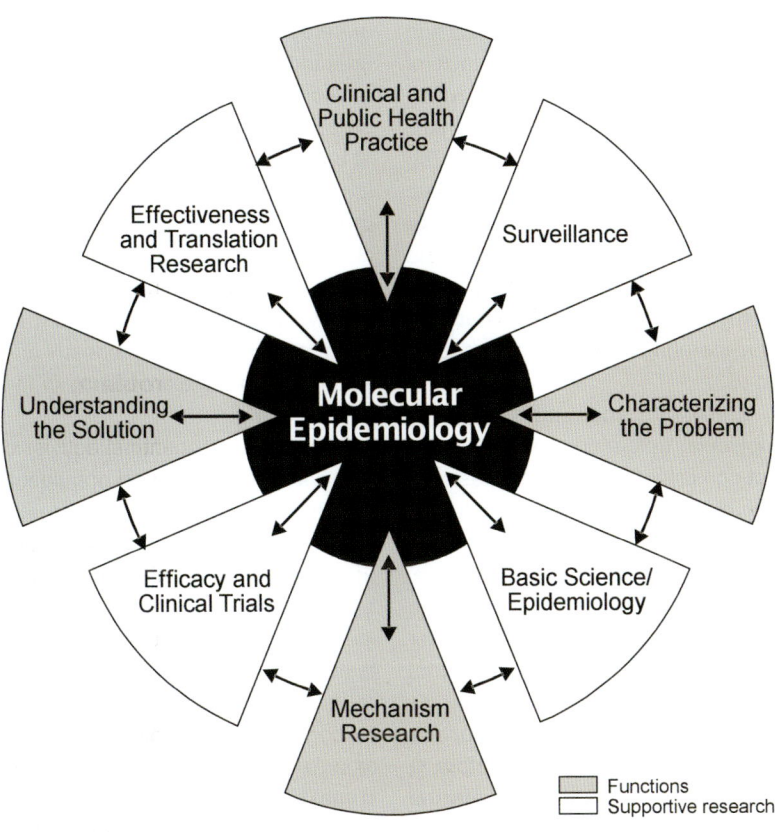

To achieve this progression, there is a need for parallel laboratory and population research to understand the mechanisms through which environmental exposures interact with host susceptibility factors to increase the risk of disease. The key mechanisms can then be blocked by interventions, such as exposure reduction, behaviour modification, chemoprevention or prophylaxis.

Understanding the solutions

When potential solutions or interventions are identified, molecular epidemiological knowledge is useful in the development and conduct of clinical and intervention trials, and monitoring the efficacy of policy interventions. Following assessment in trials, there is a need for research on the translation of findings to clinical and public health practitioners. This involves identifying the potential uses of the findings, the plan for communicating and disseminating this information, and ways to measure the impact of their use. Epidemiologists have a long history of providing the evidence base for demonstrating the efficacy and effectiveness of clinical and population interventions moved into practice (42). Molecular epidemiologic knowledge can be used in impact assessments to determine changes in incidence of the biomarkers as surrogates for disease or as indicators of disease risk.

Clinical and public health practice

Translation of biomedical research to useful clinical and public health applications is clearly a major challenge (15,42–44). Molecular epidemiologists can accept that challenge and contribute to the translation of knowledge from research endeavours. This entails a more expansive view of molecular epidemiology beyond a tool in etiologic research to a discipline that addresses the medical and population ramifications of molecular phenomena in terms of reducing risk of disease (45). Translation is a multifaceted process that has been described as involving four phases: 1) discovery to candidate health application; 2) health application to evidence-based practice guidelines; 3) practice guidelines to health practice; and 4) practice to population health impact (44).

At times, molecular epidemiology has been portrayed as a reductionist approach that merely identifies molecular risk factors and indicators in individuals. However, molecular epidemiology is first and foremost a means to gain sufficient biological understanding at the molecular and biochemical level of the process of disease causation to protect public health. From its outset, molecular epidemiology has had the vision that biological marker data can be used to prevent or reduce morbidity and mortality (1,2,21,22,46). Consequently, molecular epidemiology is the means to obtain molecular- and biochemical-level understanding in a population context.

The term molecular epidemiology is compelling. It inspires the scientific imagination, compelling thinking of incorporating the new resolving powers of molecular biology, genetics and analytical chemistry into epidemiology, and it stimulates hypothesis development and testing over a broader range of genetic and environmental factors. The term also focuses on the population distributions and implications of molecular events.

On the face of it, the fact that molecular epidemiology is focused both on biological processes in individuals and their distribution in populations makes the term sound contradictory (47). Yet this tension between identifying causal pathways at an individual biological level and understanding the causes of disease in populations has always been present in epidemiology. This seemingly contradictory nature of the molecular epidemiological endeavour may be most familiar as articulated by Geoffrey Rose, in that epidemiologists' efforts are concerned with unraveling both the determinants of individual cases and the determinants of incidence rates (48). Although this tension may be exemplified by molecular epidemiology, there is nothing inherent in the actions of molecular epidemiologists *per se* that limits the utility of their activities for public health. Of greater importance is that this tension itself, this struggle to reconcile two seemingly contradictory objectives, has been productive and inspiring (49). In this vein, some have argued that the integration of genomics into epidemiology can been seen as a further challenge to epidemiologists to take seriously the contextual factors that bear on biological processes (37,50).

In short, the relevance and usefulness of molecular epidemiologic research to public health depends on how successfully practitioners address challenges that face epidemiology and research in general. These issues—lack of biological realism or theoretical basis for research, lack of consistency in results, and worse still, in some cases lack of scientific rigor—are threats of which all epidemiology, indeed all scientific research, must be wary (51,52). Too often the attempt to substantiate molecular epidemiologic results by post-hoc searching through the scientific literature has led to finding biologic information that is not truly corroborating but only appears to be so (53).

Similarly, the criticism that molecular epidemiologic results are not consistent between studies, and are even sometimes contradictory, is partly due to the media and public misinterpretation of the nature of scientific investigations, but it is also partly due to the failure of molecular epidemiologists to say loud and clear that their studies must be repeated and confirmed in various populations and settings before a causal link can be strongly inferred (54,55). This is especially true when strong causal claims are made following small studies.

To continue to serve as a hub for health research, molecular epidemiology will need to continue expanding its contribution to surveillance, mechanistic research, efficacy trials, translational research and health policy. Critical for this holistic approach is the ability to assemble and communicate information, and, ultimately, evidence to decision-makers, medical and health professionals, and the public. This will involve fostering an evidence-based approach to research and adopting vigorous and stringent criteria for systematic integration of confirmation from many disciplines (e.g. genomics, biochemistry, exposure assessment, pathology, medicine and public health)(43). Specifically, this expansive view means not only thinking of causal mechanisms and being problem-oriented, but also being solution-oriented. How can the findings of molecular epidemiologic research be used to address a problem both at the patient and population levels? It is critical to focus on credibility, rather than statistical significance of research findings; encourage rigorous replication, not just discovery; and build public trust by communicating results honestly and acknowledging the limitations of the evidence (43).

If molecular epidemiology is to make a major impact on population health, it must have a global focus as well as a local one. Too often, the findings of research on genetic biomarkers have been seen as leading to expensive sophisticated tests and treatments for a few rather than for the many. Molecular epidemiologic researchers need to be aware of this concern in the context of their work. The result should be research and strategies to help develop affordable population-wide tools for combating common diseases (56).

Nomenclature, taxonomics and approaches

Other disciplines and terms overlap with molecular epidemiology. The terms genomics, population genomics, population genetics, and human genome epidemiology all can involve molecular epidemiologic approaches. Critical in all of these approaches is the use of valid epidemiological study designs, methodologies, and perspectives with valid and reliable indicators of susceptibility, genotype and phenotypes.

Another term that merits discussion and definition is "biomarker." The term biologic marker, or biomarker, is broadly defined to include any type of measurement made in a biologic sample and includes measurements of exogenous and endogenous exposures, as well as any phenomena in biologic systems at the biochemical, molecular, genetic, immunologic or physiologic level (1,20).

Historically, biomarkers have been used for many decades in etiologic and clinical research, beginning with seminal studies of infectious diseases followed by research on chronic diseases, such as cardiovascular disease (1,57–59). Over time, an appreciation of the heterogeneity in biomarkers developed with regard to the different aspects of the disease process reflected by them. Emerging from the seminal works in the 1980s and 1990s, three types of biomarkers were defined: biomarkers of exposure/dose (internal and biologically effective dose), biomarkers of effect (generally indicators of damage, alteration in homeostatic mechanisms, molecular or biochemical dysfunction, preclinical effects of early disease, and clinically apparent disease), and biomarkers of susceptibility (either inherited or acquired) (2,20–22,29,30). These have been linked in a continuum that is applicable to many exposure-disease relationships and have been further characterized with regard to the advantages and limitations of their application within the spectrum of epidemiologic studies (1,2,33,39).

The discovery of new biomarkers for medical, environmental and epidemiologic research is of growing importance. The global biomarkers market is projected to reach about $20.5 billion by 2014 (60). Increasingly, there are developments in a broad range of areas that include: biomarkers as tools in decision-making, regulation, diagnostics, personalized medicine, therapeutics, pharmacology, public and environmental health, and as dependent and independent variables in molecular epidemiologic research.

Implicit in biomarker-based research is the collection of biological specimens from individuals within an epidemiologic framework, analysis of those specimens and the amassing of the results in databases. The emergence of large-scale networks, multicentre collaborations and formal consortia has increasingly been observed and has been advanced as an approach to complex disease research efforts (12,61–63). Although there is a strong rationale for using consortia for exploring the role of environmental exposures and genetic variants in disease, this does not mean that smaller, single investigative approaches are without merit. Such studies still may provide useful leads, hypotheses, mechanistic insights and identification of risk factors; they are also helpful for validation of biomarkers. Nonetheless, large-scale consortia provide a powerful approach to achieve adequate statistical power (particularly in studies of individual genetic variants and gene-environment interactions) to identify effects and avoid false-positive reports and to address complex research problems (64–69). One unique, global collaboration, the Human Genome Epidemiology Network (HuGE Net), combines the traditional methodology of population-level investigation with molecular and genetic epidemiology data. HuGE Net is focused on the post-gene discovery phase and interpretation of epidemiologic information on human genes for the purpose of health promotion (70,71). This is one example of the convergence of classical and molecular epidemiology applications in a practical approach for disease prevention.

On the horizon

The great investment in biomedical research made in the past 50 years could yield many benefits in the next 50 years if the results of that research can be used and translated into practical advances (see the following chapters that discuss these advances). The skills, tools and insights of molecular epidemiology can contribute to that effort. Knowledge is the basis of action. Serving as the linking hub for laboratory and population research, molecular epidemiology can help translate it to practice. To do this, there will be a need to maintain current trends in the discipline and establish new ones. Continuation of the trend towards large-scale networks and biobanks, use of bioinformatics, and attention to individual and collective ethical issues will serve to move the field forward. But more powerful effects will be achieved by incorporating epigenetic and biological systems theory in research, expanding skill sets and professional knowledge to complement translation research and risk communication, and by fostering public health perspectives (35,72,73). A broad population-wide vision for using biological markers is required to leverage the power of molecular scale insight to give beneficial macro-scale impacts on public health.

Disclaimer: The findings and conclusions in this chapter are those of the author and do not necessarily represent the views of the Centers for Disease Control and Prevention.

References

1. Schulte PA, Perera FP, editors. Molecular epidemiology principles and practices. San Diego (CA): Academic Press Inc; 1993.

2. Perera FP, Weinstein IB (1982). Molecular epidemiology and carcinogen-DNA adduct detection: new approaches to studies of human cancer causation. *J Chronic Dis*, 35:581–600.doi:10.1016/0021-9681(82)90078-9 PMID:6282919

3. Khoury MJ (1999). Human genome epidemiology: translating advances in human genetics into population-based data for medicine and public health. *Genet Med*, 1:71–73. PMID:11336455

4. Pittinger CA, Brennan TH, Badger DA et al. (2003). Aligning chemical assessment tools across the hazard-risk continuum. *Risk Anal*, 23:529–535.doi:10.1111/1539-6924.00333 PMID:12836845

5. Waters MD, Olden K, Tennant RW (2003). Toxicogenomic approach for assessing toxicant-related disease. *Mutat Res*, 544:415–424.doi:10.1016/j.mrrev.2003.06.014 PMID:14644344

6. Weis BK, Balshaw D, Barr JR et al. (2005). Personalized exposure assessment: promising approaches for human environmental health research. *Environ Health Perspect*, 113:840–848.doi:10.1289/ehp.7651 PMID:16002370

7. National Research Council. Human biomonitoring for environmental chemicals. Washington (DC): The National Academies Press; 2006.

8. Ward MH, Wartenberg D (2006). Invited commentary: on the road to improved exposure assessment using geographic information systems. *Am J Epidemiol*, 164:208–211.doi:10.1093/aje/kwj183 PMID:16707652

9. Wild C, Vineis P, Garte S, editors. Molecular epidemiology of chronic diseases. Chichester (England): John Wiley & Sons; 2008.

10. Jayjock MA, Chaisson CF, Arnold S, Dederick EJ (2007). Modeling framework for human exposure assessment. *J Expo Sci Environ Epidemiol*, 17 Suppl 1;S81–S89. doi:10.1038/sj.jes.7500580 PMID:17505502

11. National Research Council. Applications of toxicogenomic technologies to predictive toxicology and risk assessment. Washington (DC): The National Academies Press; 2007.

12. Vineis P, Perera FP (2007). Molecular epidemiology and biomarkers in etiologic cancer research: the new in light of the old. *Cancer Epidemiol Biomarkers Prev*, 16:1954–1965.doi:10.1158/1055-9965.EPI-07-0457 PMID:17932342

13. Perera FP (2000). Molecular epidemiology: on the path to prevention? *J Natl Cancer Inst*, 92:602–612.

14. Riley LW, editor. Molecular epidemiology of infectious diseases: principles and practices. Washington (DC): ASM Press; 2004.

15. Schulte PA (2005). The use of biomarkers in surveillance, medical screening, and intervention. *Mutat Res*, 592:155–163. PMID:16051280

16. McMichael AJ (1994). Invited commentary–"molecular epidemiology": new pathway or new travelling companion? *Am J Epidemiol*, 140:1–11. PMID:8017398

17. Kilbourne ED (1973). The molecular epidemiology of influenza. *J Infect Dis*, 127:478–487.doi:10.1093/infdis/127.4.478 PMID:4121053

18. Higginson J (1977). The role of the pathologist in environmental medicine and public health. *Am J Pathol*, 86:460–484. PMID:836677

19. Lower GM Jr, Nilsson T, Nelson CE et al. (1979). N-acetyltransferase phenotype and risk in urinary bladder cancer: approaches in molecular epidemiology. Preliminary results in Sweden and Denmark. *Environ Health Perspect*, 29:71–79.doi:10.2307/3429048 PMID:510245

20. National Research Council (1987). Biological markers in environmental health research. *Environ Health Perspect*, 74:3–9. doi:10.1289/ehp.87743 PMID:3691432

21. National Research Council. Biologic markers in pulmonary toxicology. Washington (DC): National Academy Press; 1989.

22. National Research Council. Biologic markers in reproductive toxicology. Washington (DC): National Academy Press; 1989.

23. Perera FP, Poirier MC, Yuspa SH et al. (1982). A pilot project in molecular cancer epidemiology: determination of benzo[a]pyrene–DNA adducts in animal and human tissues by immunoassays. *Carcinogenesis*, 3:1405–1410.doi:10.1093/carcin/3.12.1405 PMID:6295657

24. Miller JR (1983). International Commission for Protection against Environmental Mutagens and Carcinogens. ICPEMC working paper 5/4. Perspectives in mutation epidemiology: 4. General principles and considerations. *Mutat Res*, 114:425–447. PMID:6835244

25. Tannenbaum SR, Skipper PL (1984). Biological aspects to the evaluation of risk: dosimetry of carcinogens in man. *Fund Appl Toxicol*, 4:S367–S373.doi:10.1016/0272-0590(84)90264-1 PMID:6745554

26. Harris CC, Vahakangus K, Autrup H et al. Biochemical and molecular epidemiology of human cancer risk. In: Scarpelli D, Craighead J, Kaufman N, editors. The pathologist and the environment. Baltimore (MD): Williams and Wilkins; 1985. p. 140–67.

27. Wogan GN, Gorelick NJ (1985). Chemical and biochemical dosimetry of exposure to genotoxic chemicals. *Environ Health Perspect*, 62:5–18. PMID:4085448

28. Perera FP (1987). Molecular cancer epidemiology: a new tool in cancer prevention. *J Natl Cancer Inst*, 78:887–898. PMID:3471998

29. Hulka BS, Wilcosky TC (1988). Biological markers in epidemiologic research. *Arch Environ Health*, 43:83–89.doi:10.1080/00039896.1988.9935831 PMID:3377561

30. Schulte PA (1989). A conceptual framework for the validation and use of biologic markers. *Environ Res*, 48:129–144.doi:10.1016/S0013-9351(89)80029-5 PMID:2647488

31. Hulka BS, Wilcosky TC, Griffith JD. Biological markers in epidemiology. New York (NY); Oxford University Press; 1990.

32. Groopman JD, Skipper PL, editors. Molecular dosimetry and human cancer: analytical, epidemiological, and social considerations. Boca Raton (FL): CRC Press; 1991.

33. Rothman N, Stewart WF, Schulte PA (1995). Incorporating biomarkers into cancer epidemiology: a matrix of biomarker and study design categories. *Cancer Epidemiol Biomarkers Prev*, 4:301–311. PMID:7655323

34. Carrington M, Hoelzel R, editors. Molecular epidemiology. New York (NY): Oxford University Press; 2001.

35. Wild CP (2005). Complementing the genome with an "exposome": the outstanding challenge of environmental exposure measurement in molecular epidemiology. *Cancer Epidemiol Biomarkers Prev*, 14:1847–1850.doi:10.1158/1055-9965.EPI-05-0456 PMID:16103423

36. Rebbeck TR, Ambrosone CB, Shields PG, editors. Molecular epidemiology: applications in cancer and other human diseases. New York (NY): Informa Healthcare; 2008.

37. Millikan R (2002). The changing face of epidemiology in the genomics era. *Epidemiology*, 13:472–480.doi:10.1097/00001648-200207000-00017 PMID:12094104

38. Jablonka E (2004). Epigenetic epidemiology. *Int J Epidemiol*, 33:929–935. doi:10.1093/ije/dyh231 PMID:15166187

39. Garcia-Closas M, Vermeulen R, Sherman ME *et al*. Application of biomarkers. In Schottenfeld D, Fraumeni JF Jr, editors. Cancer epidemiology and prevention. 3rd ed. New York (NY): Oxford University Press; 2006. p.70–88.

40. Brand AM, Probst-Hensch NM (2007). Biobanking for epidemiological research and public health. *Pathobiology*, 74:227–238. doi:10.1159/000104450 PMID:17709965

41. Pukkala E, Andersen A, Berglund G *et al*. (2007). Nordic biological specimen banks as basis for studies of cancer causes and control–more than 2 million sample donors, 25 million person years and 100,000 prospective cancers. *Acta Oncol*, 46:286–307.doi:10.1080/02841860701203545 PMID:17450464

42. Hiatt R, Samet J, Ness RB; American College of Epidemiology Policy Committee (2006). The role of the epidemiologist in clinical and translational science. *Ann Epidemiol*, 16:409–410.doi:10.1016/j.annepidem.2006.02.002 PMID:16647631

43. Ioannidis JPA (2006). Evolution and translation of research findings: from bench to where? *PLoS Clin Trials*, 1:e36.doi:10.1371/journal.pctr.0010036 PMID:17111044

44. Khoury MJ, Gwinn M, Yoon PW *et al*. (2007). The continuum of translation research in genomic medicine: how can we accelerate the appropriate integration of human genome discoveries into health care and disease prevention? *Genet Med*, 9:665–674.doi:10.1097/GIM.0b013e31815699d0 PMID:18073579

45. Grodsky J (2005). Genetics and environmental law: redefining public health. *Calif Law Rev*, 93:171–270.

46. Harris CC (1993). p53: at the crossroads of molecular carcinogenesis and risk assessment. *Science*, 262:1980–1981.doi:10.1126/science.8266092 PMID:8266092

47. Castiel LD (1999). Apocalypse...now? Molecular epidemiology, predictive genetic tests, and social communication of genetic contents. *Cad Saude Publica*, 15 Suppl 1;S73–S89.doi:10.1590/S0102-311X1999000500009 PMID:10089550

48. Rose G (2001). Sick individuals and sick populations. *Int J Epidemiol*, 30:427–432, discussion 433–434.doi:10.1093/ije/30.3.427 PMID:11416056

49. Parodi A, Neasham D, Vineis P (2006). Environment, population, and biology: a short history of modern epidemiology. *Perspect Biol Med*, 49:357–368.doi:10.1353/pbm.2006.0044 PMID:16960306

50. Shostak S (2003). Locating gene-environment interaction: at the intersections of genetics and public health. *Soc Sci Med*, 56:2327–2342.doi:10.1016/S0277-9536(02)00231-9 PMID:12719185

51. Taubes G (1995). Epidemiology faces its limits. *Science*, 269:164–169.doi:10.1126/science.7618077 PMID:7618077

52. Gori GB (1998). Presentation: epidemiology and public health: is a new paradigm needed or a new ethics? *J Clin Epidemiol*, 5:637–641.

53. Muscat JE (1996). Epidemiological reasoning and biological rationale. *Biomarkers*, 1:144–145 doi:10.3109/13547509609088683.

54. Gallo V, Egger M, McCormack V *et al*. (2011). STrengthening the Reporting of OBservational studies in Epidemiology - Molecular Epidemiology (STROBE-ME): An Extension of the STROBE Statement. *PLoS Med*, 8:e1001117.doi:10.1371/journal.pmed.1001117 PMID:22039356

55. Ioannidis JPA (2007). Genetic and molecular epidemiology. *J Epidemiol Community Health*, 61:757–758.doi:10.1136/jech.2006.059055 PMID:17699527

56. Brand A (2005). Public health and genetics–a dangerous combination? *Eur J Public Health*, 15:114–116.doi:10.1093/eurpub/cki090 PMID:15941755

57. Kannel WB, McGee D, Gordon T (1976). A general cardiovascular risk profile: the Framingham Study. *Am J Cardiol*, 38:46–51. doi:10.1016/0002-9149(76)90061-8 PMID:132862

58. Epstein FH, Napier JA, Block WD *et al*. (1970). The Tecumseh study. *Arch Environ Health*, 21:402–407. PMID:5504438

59. Lipid Research Clinics Program (1984). The Lipid Research Clinics Coronary Primary Prevention Trial results. *JAMA*, 251:351–364. doi:10.1001/jama.251.3.351 PMID:6361299

60. Markets and Markets (2009). Available from URL: http://www.marketsandmarkets.com/PressReleases/global-biomarker-market-worth-US-20.5-billion-by-2014.asp.

61. Boffetta P, Armstrong B, Linet M *et al*. (2007). Consortia in cancer epidemiology: lessons from InterLymph. *Cancer Epidemiol Biomarkers Prev*, 16:197–199.doi:10.1158/1055-9965.EPI-06-0786 PMID:17301250

62. Hunter DJ, Thomas G, Hoover RN, Chanock SJ (2007). Scanning the horizon: what is the future of genome-wide association studies in accelerating discoveries in cancer etiology and prevention? *Cancer Causes Control*, 18:479–484. doi:10.1007/s10552-007-0118-y PMID:17440825

63. Seminara D, Khoury MJ, O'Brien TR *et al*.; Human Genome Epidemiology Network; Network of Investigator Networks (2007). The emergence of networks in human genome epidemiology: challenges and opportunities. *Epidemiology*, 18:1–8.doi:10.1097/01.ede.0000249540.17855.b7 PMID:17179752

64. Epidemiology and Genetics Research Program (EGRP). Supported Epidemiology Consortia. Available from URL: http://epi.grants.cancer.gov/Consortia/.

65. Psaty BM, O'Donnell CJ, Gudnason V *et al*.; CHARGE Consortium (2009). Cohorts for Heart and Aging Research in Genomic Epidemiology (CHARGE) Consortium: Design of prospective meta-analyses of genome-wide association studies from 5 cohorts. *Circ Cardiovasc Genet*, 2:73–80. doi:10.1161/CIRCGENETICS.108.829747 PMID:20031568

66. Chinese SARS Molecular Epidemiology Consortium (2004). Molecular evolution of the SARS coronavirus during the course of the SARS epidemic in China. *Science*, 303:1666–1669.doi:10.1126/science.1092002 PMID:14752165

67. Truong T, Sauter W, McKay JD *et al*.; EPIC-lung (2010). International Lung Cancer Consortium: coordinated association study of 10 potential lung cancer susceptibility variants. *Carcinogenesis*, 31:625–633. PMID: 20106900.

68. Furberg H, Kim YJ, Dackor J *et al*.; Tobacco and Genetics Consortium (2010). Genome-wide meta-analyses identify multiple loci associated with smoking behavior. *Nat Genet*, 42:441–447.doi:10.1038/ng.571 PMID: 20418890

69. Tuberculosis Epidemiologic Studies Consortium (TBESC) Centers for Disease Control and Prevention. Available from URL: http://www.cdc.gov/tb/topic/research/TBESC/.

70. Khoury MJ (1999). Human genome epidemiology: translating advances in human genetics into population-based data for medicine and public health. *Genet Med*, 1:71–73. PMID:11336455

71. Khoury MJ. Human Genome Epidemiology Network (HuGENet™). Available from URL: http://www.cdc.gov/genomics/hugenet.

72. Sutherland JE, Costa M (2003). Epigenetics and the environment. *Ann N Y Acad Sci*, 983:151–160.doi:10.1111/j.1749-6632.2003.tb05970.x PMID:12724220

73. Wade PA, Archer TK (2006). Epigenetics: environmental instructions for the genome. *Environ Health Perspect*, 114:A140–A141. doi:10.1289/ehp.114-a140 PMID:16507439

74. Narayan KM, Benjamin E, Gregg EW *et al*. (2004). Diabetes translation research: where are we and where do we want to be? *Ann Intern Med*, 140:958–963. PMID:15172921

UNIT 1.
CONTEXTUAL FRAMEWORK FOR MOLECULAR EPIDEMIOLOGY

CHAPTER 2.

Ethical issues in molecular epidemiologic research

Paul A. Schulte and Andrea Smith

Introduction

Contemporary and future molecular epidemiologic research will be conducted against a backdrop of massive biological databases, comprehensive and longitudinal electronic medical records, large medical care expenditures, aging populations, emerging infectious diseases in some countries, and global climate change. These conditions will influence the ethical issues that arise in molecular epidemiologic research. Will these issues differ from epidemiologic or scientific research in general? Some of the issues will be unique to molecular epidemiology, and others will be relevant to all research. If the conduct of molecular epidemiology is to contribute to medical and public health research and have a positive impact, there is a need for investigators to be aware of and adhere to the generally accepted ethical principles discussed in this chapter. Further, it is important to realize that data that will be made available in the future from new genomic technology will continue to pose challenges to the ethical conduct of molecular epidemiologic research. Therefore, researchers will need to be aware of the dynamic nature of guidelines and regulations.

Distinctive ethical issues in molecular epidemiology

Three key features of molecular epidemiology form the basis for the distinctive ethical issues unique to the field. First and foremost is that molecular epidemiology relies on the collection of biologic specimens and the identification and use of biological markers derived from those specimens (1,2). The second feature is that many of the biological markers pertain to inherited genetic information. While similar to other biomedical information, genetic information is often perceived (rightly or wrongly) as being more powerful and sensitive, a perception reflected in the widespread use of the metaphor of genes as the blueprint for what makes us human (3). Moreover, critical in molecular epidemiologic research is the emerging capability to efficiently sequence nearly the entire genome, as well as the availability of information in public databases, most of which are restricted to bona fide researchers who gain formal permission (2,4,5). Lastly, molecular epidemiology continually involves the application of new technologies and methodologies whose validity and reliability are in the process of being established. Together these three features trigger the need for molecular epidemiologists to consider and address specific ethical issues in addition to the more generic ones typical of epidemiological studies (2,6–13). Epidemiology, as a population science, observes the characteristics of individual research participants to understand disease at the level of the population. As a result, the ethical concerns generated in the field are two-fold: there are those that pertain to interaction with individual study participants, and those that are concerned with populations. This means that molecular epidemiologists need to reflect upon ethical issues beyond those encountered in any particular study. The broader issues to be considered include how to distribute the scientific and social benefits of molecular epidemiologic research, particularly research that involves genomic data and addresses various social, political and scientific questions related to collective, as well as individual, rights (14–16).

Clearly, these are questions not answerable by molecular epidemiologists alone, and require the input and involvement of various other disciplines. Yet, important for molecular epidemiologists to bear in mind is the larger context in which their work is situated, and to build dialogue across disciplines in an effort to contribute to these larger issues. A review of ethical issues follows, primarily as they relate to the molecular epidemiologic research process, and a discussion on how they arise in: 1) the development of the study protocol, 2) obtaining participation and informed consent, 3) maintaining privacy of subjects and confidentiality of data, 4) interpreting and communicating test and study results, and 5) avoiding inappropriate inferences and actions (or lack of appropriate actions) based on study results. Wherever relevant, we point towards the broader population health ethics involved in molecular epidemiology, acknowledging that these discussions are merely introductory and far from exhaustive.

Most of the health research, including molecular epidemiologic research, conducted in the United States is regulated by the Common Rule (45 CFR Part 46, subpart A). The Common Rule pertains to individually identifiable data and does not apply to research conducted on specimens or health records that are not individually identifiable (12). Overlapping some aspects of the Common Rule is the Privacy Rule of the Health Insurance Portability and Accountability Act (HIPAA) (45 CFR Parts 160, 164). They both cover large, academic medical centre institutions, but differ on such issues as reviews preparatory to research, research involving health records of deceased individuals, and revocations of consents and authorizations (17).

The other major regulatory feature of research is the Institutional Review Board (IRB). IRBs review protocols for human subject research as defined by the Common Rule. They also are charged with addressing the ethical aspects of the increasing volume and variation of genetic molecular epidemiologic studies (2,13). These boards face significant challenges, as currently in many cases there is no general agreement on the ethical aspects of issues that arise. Nonetheless, as described in this chapter, there are some established principles and experiences and practices that can fill this gap.

Development of the study protocol

Ethics are an intrinsic aspect of the framing of the research question and in the selection of methods to carry out any study. The decision to use or focus on molecular biomarkers in a study can itself raise ethical issues. A starting point for considering the appropriateness of molecular biomarkers is whether or not the research question being addressed is of public health importance (18). If the answer is no, then the use of scarce resources to develop, validate or apply a biological marker can be wasteful and inefficient, and detract from efforts to address other public health issues of greater urgency. Ethically, molecular epidemiologic research should identify driving scientific and public health questions that cannot be answered by some other more accessible and less costly approach. Given the resource-intensive nature of biobanking and molecular technologies, the use of biomarkers within epidemiologic research should be done judiciously. Like all research, studies that propose to use biomarkers must ground their decisions in the available empirical evidence and sound scientific reasoning. In the genomic era, vast amounts of biological data are generated using technologies that simultaneously process hundreds of genes within hundreds of samples. Even in a small epidemiological study, such as one with 100 cases and 100 controls, investigators can easily obtain genetic and epigenomic data involving millions of variables for each participant (although such studies are likely to be both underpowered and likely to produce large numbers of false-positive findings unless they have replication efforts built into them). Bioinformatic approaches are needed to sort through such data sets and the literature. Ideally, such approaches are first conducted in iterative processes using existing databases before the initiation of a new study. This detailed preparation provides a rationale for the study design and focuses the scope of the research question.

One set of ethical concerns relevant to protocol development involves whether the investigator has any interests that conflict with the ultimate aim or potential outcomes of the research. Ideally, investigators should be involved in research to seek the prevention of disease through free inquiry and the pursuit of knowledge. Conflicting interests may lead investigators (consciously or not) to make choices about study design that could introduce biases, yielding results that deviate from less biased approaches. To foster a transparent and accountable process through peer review and other mechanisms, it is important that investigators acknowledge and identify their conflict of interest to their collaborators, research participants and other stakeholders. Not only do conflicts of interest jeopardize the validity and utility of any particular study, they also bear on the health research enterprise as a whole, since the ramifications of failing to disclose them can damage the public's trust in and support of science (20). The issue of conflict of interest is particularly acute in research using genetic material, due to the push by academic and research institutions (and commercial collaborators) to seek intellectual property rights, and other avenues of commercialization, of their research (13).

Turning to more methodological issues, the decision on where to conduct a molecular epidemiologic study, and on whom, should also be scrutinized with ethical considerations such as equity, justice and autonomy kept in mind. In light of these principles, many decisions relating to sample design that initially seem of little ethical consequence, gain stature. For example, how well the sample population reflects the target population is a matter that bears on both scientific validity and moral concerns. Within molecular epidemiologic research, an additional issue includes whether it is the responsibility of investigators to attempt to obtain ethnic, racial or social class diversity in studies. This question extends into the avoidance of socio-genetic marginalization, that is, the isolation of social groups and individuals as a consequence of discrimination on the basis of genetic information (22). In a similar vein, should one assess whether various ethnic groups are provided similar opportunities to be in a database? If not, characterization in a database can make one ethnic group appear more or less susceptible than another ethnic group lacking the same opportunity for characterization. Other questions about sample selection that should be taken into account are whether the sample is representative in terms of genetic and ethnic factors, as well as various other host or environmental factors of the study's target population. However, there is a cost associated with representativeness—loss of power and the need to adjust for confounding factors. Small groups included to make samples more representative may be subject to statistical power limitations and, for studies on restricted budgets, may decrease the ability of the study to accomplish its primary aims. At the same time, power issues can be surmounted in part if data are collected in a way that is consistent with previous studies that have included multiple ethnic populations, and if plans for pooling data with other studies are made, preferably early in the study design phase.

Molecular epidemiologic study design and analysis also can affect whether the research contributes to public health. The promise of genome-wide association and other genetic susceptibility studies, in terms of prevention and public health, may not be realized if a study is designed to minimize observing the effect of environment and lifestyle factors. To take full public health advantage of such research, environmental exposures, quantified by state-of-the-art exposure assessment methods when feasible, must be considered in the design, particularly in the selection of study populations and in the analysis (23). Such an approach may involve using analytical techniques that do not require relying on either significant main genetic or environmental effects as a threshold for investigating gene-environment interactions.

A particularly sticky issue relating to study design is the premature use of biological markers as variables in research before they have been validated (10,24); there are many examples of premature use in commerce (25). Validation is not an all-or-none state, but rather a process that is informed by continued research and investigation. Critical in any definition of validation is the extent to which the biomarker actually represents what it is intended to represent (1,26). The use of biomarkers that have not been validated for the purpose for which they are being used can lead to false or misleading findings, which may harm participants, groups or communities. For transitional studies in which the characteristics of a marker are being determined, and for which there are clearly no associated clinical findings, prognostic significance, or clear meaning, the needs of study participants may be different from those in studies with established biomarkers. In the case where a biomarker has a known association with a disease outcome (or exposure or susceptibility) and holds implications for individual risk, interventions such as medical screening, biological monitoring, or diagnostic evaluation may be appropriate follow-up measures.

Furthermore, ethical issues may arise during the design phase of a study protocol from a researcher's failure to anticipate how to respond to the distributional extremes in biomarker assay results (6). Possible responses may include repeat testing, risk communications counselling or clinical surveillance. With genetic markers of susceptibility, it may be important to consider the impact of the research not only on individual participants, but also on their families, given that knowing something about an individual's genes possibly means knowing something about their past, present and future family's genetic constitution.

Recruiting participants and informed consent

When recruiting potential research participants, a core ethical issue in molecular epidemiologic research is respect for individuals, which is upheld by ensuring their autonomy. This means that potential research subjects should be viewed and treated as self-ruling and able to voluntarily participate in and withdraw from research without coercion or prejudice. Autonomy also implies that those who are not capable of self-determination, such as children, are to be protected from exploitation and harm (27). Potential participants need to be informed of a broad range of information (e.g. purpose of the study, its duration, identity of the investigators and sponsors, ownership and other uses of specimens, the methods and procedures to be used, and all potential risks and benefits of participating in the study), some of which are unique to molecular epidemiology (6,28). The investment in population-based field studies to obtain biologic specimens and covariate information is generally quite large, making it cost-effective to collect and bank DNA and other biological materials for current and future research. Moreover, the number of biological specimen banks is growing, and as a result the nature of future research might not be known at the time of specimen collection (29). Accurately depicting the purpose of a molecular epidemiologic study can be difficult for the investigator, because there may be a multiplicity of purposes, some intended, others not even yet envisioned. At issue is how one should solicit consent for future use of specimens, and what to tell potential participants about this.

Future use of specimens requires additional procedures for obtaining consent (30). Some have proposed that informed consent for future use is best acquired by enabling participants to specify the research areas to which they sanction, or to permit them to give blanket approval, which informs them of the intention of banking specimens and their subsequent use for a wide range of research purposes (31,32). While such procedures clearly allow the maximum scientific benefit and potential public health impact to be obtained from such biobanks, they could be considered to deviate in important ways from the general standards of informed consent. In soliciting blanket consent for future use, investigators are generally unable to provide research participants specific and accurate information as to all the purposes of the study (as they are yet unknown); thus, the attendant potential harms and benefits of participation are not fully fleshed out. The resulting scenario is that the informed consent reflects a "potential" informed consent, not one in which research participants are fully informed and then knowingly choose to be involved (13). This appears to stand in contrast to the principles of informed consent as laid out in ethical codes of medical research, such as the Declaration of Helsinki (33). The evolution of technologies used in molecular epidemiologic research has pushed IRBs to consider how ethical codes apply. This is illustrated in the development of a large number of prospective cohort studies worldwide and the guidelines pertaining to them, such as the United Kingdom Biobank Ethics and Governance Framework and the independent advisory council formed to oversee the Biobank's activities (34–36). After careful consideration and review by IRBs, informed consent procedures have been developed that accomplish the dual purposes of protecting the rights of individual participants while also providing the opportunity for the maximum public health benefit from the substantial resources needed to establish and maintain such prospective studies.

Molecular epidemiologic studies have generally used a large number of biological markers analysed in specimens collected directly from research participants enrolled into formal case-control and prospective cohort studies. Increasingly, though, the source of the specimens may not be from participants directly, but from biobanks where specimens were collected before the development of a given study, and possibly even for a different purpose. Given this trend, it is important that the informed consent process address intellectual property rights and state who maintains ownership of the collected specimens (2). There are various issues that pertain to ownership or custodianship of biospecimens. Generally, however, there do not appear to be laws or regulations that directly address them. Nonetheless, participants have a right to know what future uses their specimens may be considered for. There also could be special concerns about future use of specimens among indigenous people or various 'island' populations that need to be considered (37,38). Overall, molecular epidemiologists involved with biobanks and surveillance efforts should think about both individual and collective rights and interests in creating or assessing such databases for public health research.

While procedures for dealing with biorepositories in the future can be established, what about the millions of human specimens currently in storage collected from a wide variety of formal and less-formal study designs, and obtained from study participants over several decades during which standards of informed consent and IRB review have undergone continuing evolution? These are highly valuable resources but ones where procedures and practices may not necessarily conform to current standards. For example, can a participant whose specimens are in a biorepository decide to discontinue participation and not have their samples continue to be used? General practice and a recent court case ascribe ownership to the institution that maintains the repository. However, this interpretation excludes the input of the research participant. A stewardship model has been described that respects a research participant's request to terminate participation in a DNA biorepository by destroying remaining DNA instead of continuing use of the specimen, as is a common response (28). The American College of Epidemiology has espoused four useful principles regarding the handling of biospecimens: (1) custodianship should encourage openness of scientific inquiry and maximize biospecimen use and sharing so as to exploit the full potential to promote health; (2) the privacy of participants must be protected and informed consent must provide provisions for unanticipated biospecimen use; (3) the intellectual investment of investigators involved in the creation of a biorepository is often substantial and should be respected; and (4) sharing of specimens needs to protect proprietary information and to address the concerns of third-party funders (39). While these principles are a good foundation, they do not specifically address the research participant except in the area of privacy. There also is the need to consider control of human specimens in terms of respect for persons and autonomy (28).

The issue of future use of specimens is more complex with larger studies involving whole-genome analyses. One problem in obtaining consent for future use of specimens is the apparent discrepancies between implementation of the Common Rule (45 CFR Subpart A) and the Privacy Rule of the Health Insurance Portability and Accountability Act (HIPAA). The Common Rule allows patients to consent to unspecified future research, whereas the HIPAA Rule requires that each authorization by a patient for release of protected health information include a specific research purpose (2,40,41). As noted by Vaught *et al.* (2007): "Because support of future research is a major purpose of biospecimen resources, this lack of harmony among federal regulations has had a significant effect on and created a great deal of confusion within the biospecimen community."

Until recently, there was little or no available guidance for addressing informed consent issues in population-based studies of low penetrance gene variants (42,43). Most existing guidance pertains to single genes of high penetrance that are investigated in family studies. Yet the risks and benefits of population-based research involving low penetrance gene variants are substantially different from those associated with family-based genetic epidemiologic research (44). When obtaining informed consent, these differences become particularly meaningful: "Recommendations developed for family-based research are not well suited for most population-based research because they generally fail to distinguish between studies expected to reveal clinically relevant information about participants and studies expected to have meaningful public health implications but involving few physical, psychological, or social risks for individual participants" (42). Further recommendations for obtaining informed consent have been developed by a US Centers for Disease Control and Prevention (CDC) workgroup that considered integrating genetic variation in population-based research (42). The workgroup provided a useful outline of the content, language and considerations for an informed consent document. Much of the language in these consent materials addresses the important distinction between genetic research expected to reveal clinically relevant information about individual participants, and that which is not. It is anticipated that the majority of population-based genetic research will not identify clinically relevant information. Thus, the workgroup did not recommend informing participants of individual results in these types of studies. However, they did note that the dividing line between low and high penetrance is difficult to define, since there is a spectrum of genetic variants with differing effect sizes. They therefore recommended "...when the risks identified are both valid and associated with proven intervention for risk reduction, disclosure may be appropriate" (42). A broader discussion of communicating test and study results follows in the next section.

Maintaining privacy of subjects and confidentiality of data

Molecular epidemiologic research participants explicitly agree to cooperate in a specified study when they consent to provide specimens and corollary demographic and risk factor information. Such participation generally does not include or imply consent to the distribution of the data in any way that identifies them individually to any other party, such as government agencies, employers, unions, insurers, credit agencies or lawyers. Such confidentiality and anonymity is premised on the ethical concept of respect for persons. Dissemination or revelation of results beyond the explicit purposes for which specimens were collected intrudes on subjects' privacy. Inadvertent labelling of a subject as "abnormal" or as "in the extremes of a distribution of biomarker assay results" could have a potentially deleterious impact on the person's ability to obtain insurance, a job, or credit, and can also affect the person socially or psychologically. Thus, as Nelkin and Tancredi noted, some union representatives are concerned that workers who participate in genetic research or screening will bear a genetic "scarlet letter" and that they will become "lepers" or genetic untouchables (45). The psychological impact of such stigmatization is virtually unknown.

Molecular epidemiology investigators must maintain the confidentiality of biomarker data because of the potential for misuse or abuse leading to discrimination, labelling and stigmatization (3,6,7). This can be increasingly difficult because ownership of stored specimens may be in question, and various investigators may request the use of them for research, litigation or commercial enterprise. In some cases, where specimens are identifiable or are capable of being linked to databases where identification is possible, it may be difficult to assure confidentiality. Informatics and the ability to link disparate databases are progressing at a rapid pace. In some countries, there may be a need for further legislation to prohibit unauthorized access to, or use of, specimen results. The Genetic Information Nondiscrimination Act (GINA) of 2008 was enacted to prohibit the use of genetic information in hiring or providing insurance. Nonetheless, the challenge to investigators will be to assure the rights of study participants while providing for a broad range of research opportunities.

As noted earlier, the regulation of privacy issues in the United States is addressed by the Federal Rule on the Protection of Human Subjects (the Common Rule), and, since 2003, the Privacy Rule of HIPAA. The lack of harmonization of these rules has been reported to "...create confusion, frustration, and misunderstanding by researchers, research subjects, and institutional review boards ... [Nonetheless] both rules seek to strike a reasonable balance between individuals' interests in privacy, autonomy, and well-being with the societal interest in promoting ethical scientific research" (17). The investigators concluded that the two rules should be revised to promote consistency and maximize privacy protections while minimizing the burdens on researchers.

The issue of identifiability of biological specimens (i.e. the linking of a specimen with its originator's identity) that arises with the advent of large-scale research platforms that assemble, organize, and store data and sometimes specimens, and make them available to researchers, has been thoughtfully addressed (46). At issue is the ease with which individuals can be identified from DNA or genomic data. Individual identifiability from a database "...should not be overstated, as it takes competence, perhaps a laboratory equipped for the purpose, computational power perhaps linking to other data, and determined efforts." (46). Nonetheless, identification is increasingly possible as the collection of biospecimens that can be used for matching grows and becomes more widely accessible. It has been demonstrated that an individual can be uniquely identified with high certainty with access to several hundred single-nucleotide polymorphisms (SNP) from that person (4,47).

The advent of the genome-wide association studies (GWAS), which genotype thousands of SNPs in large populations, have generated a series of questions concerning the practice of making summary data publicly available. This is due to the development of methods that use genotype frequencies and an individual's genotype profile generated elsewhere to infer whether the individual or a close relative participated in the study set (48,49). For published GWAS, the probability of inferring membership in a study is substantially decreased when less than 5000 SNPs are examined. Consequently, it is important for researchers to protect subject participation while making data available to bona fide researchers who provide sufficient and binding institutional support for protecting the confidentiality of IRB-approved research.

There is a need for proper balance between encouraging molecular epidemiologic research on genomic specimens and protecting the privacy and confidentiality of research participants. Figure 2.1 illustrates the flow of data that arises from these platforms. Among the design and governance issues are whether, and how, to de-identify the data, and at what stage to conduct scientific and ethical reviews (46).

The ultimate question is whether a completely open-access model is defensible when different amounts of genomic data are present and potentially unique to an individual to allow for identification. Clearly, in the spirit of medical research and privacy laws and ethics, there is a need for controlled access models for these types of data sets, or else consent documents need to make clear the lack of complete confidentiality that may arise from publicly accessible databases.

Interpreting and communicating test and study results

Molecular epidemiology research yields both individual test (assay) results and study results, and research participants may want or have a right to both (6,50). However, increasingly, the bioethics literature also has recognized a counter-right of informational privacy, that is, the right not to know about certain information about oneself (12,51). Providing test or study results, genetic or otherwise, requires more than merely sending results to participants, it also involves interpreting the results (52); this responsibility ultimately rests with the investigator. Some IRBs require investigators to provide individual test results to subjects as well as overall study results, while others may advise or forbid them not to communicate results of assays that have no clinical relevance (27,42). Even though participants are told that tests may be purely for research purposes and have no clinical value, they may still ultimately want to know if they are "all right." Investigators face difficult ethical issues in interpreting test and study results, and in deciding when biomarkers indicate an early warning where preventive steps should be taken.

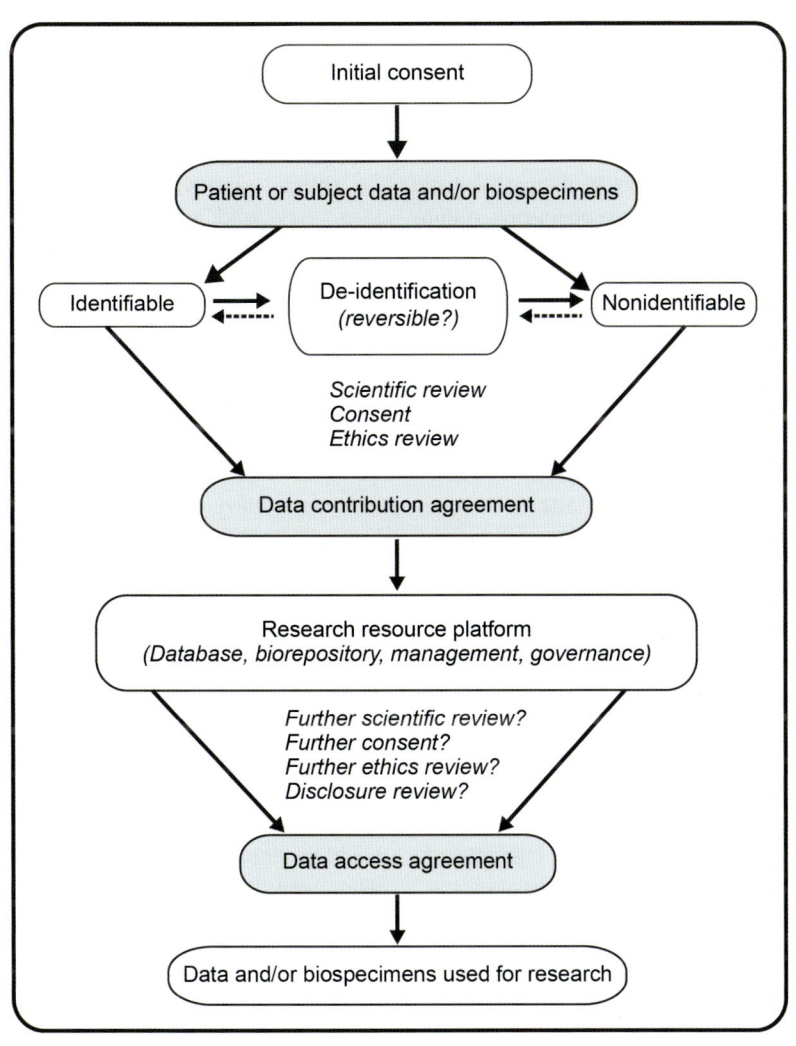

Figure 2.1. Steps in the protection of the identity of research subjects in large-scale databases and projects (46). Reprinted with permission from AAAS.

Prevention actions may include efforts to control exposures (in occupational or environmental settings), the need for subsequent testing, ongoing monitoring, or simply, and often most importantly, counselling and a demonstration of caring (6). Reporting molecular epidemiologic test results to study participants, particularly those involving genetic information, involves among other issues, defining the concept of clinical utility. Clinical utility is generally based on three criteria: (1) clinical validity (the association between the test result and a health condition or risk); (2) the likelihood of a clinical effective outcome; and (3) the value of the outcome to the individual (26,53).

The interpretation of biomarker data is a complex matter. For example, in cross-sectional studies of populations with occupational or environmental exposure and biomarkers of early biological effect, biomarkers will not be indicators of risk *per se*, but of exposure, susceptibility given exposure, or biological changes that could be homeostatic responses to an exposure (6,54). The investigator needs to sort out these changes against a background of extensive intraindividual and interindividual variability in biomarkers. It is also important to note that such studies are not usually those designed for the purpose of identifying risk and should not be construed as such. Current technological capabilities offer investigators and practitioners the opportunity to utilize techniques with heightened sensitivity for detecting changes at cellular and molecular levels and for detecting exposures to minute amounts of a xenobiotic (12). Yet at the same time, at these levels, inherited and acquired host factors and other confounding factors can be strong causes of wide variability in biomarker results unrelated to the exposure or risk factor of interest. Moreover, when multiple biomarkers are to be assessed, researchers have a responsibility to consider whether issues of multiple comparisons can lead to inappropriate selection of significance levels (6). Associations with biomarkers not included in original hypotheses should be evaluated at more rigorous levels of statistical significance with built-in replication strategies, and subsequent interpretations should be considered in that light. This is particularly the case with the development of "omic" platforms that have facilitated the use of critically important agnostic approaches that produce thousands to now millions of biomarker variables.

In general, the accurate interpretation and communication of genetic information is quite challenging due to its probabilistic character and the pleiotropic nature of genes. Moreover, the potential impact of genetic information on family relationships, reproduction, and personal integrity can further complicate its interpretation (53,55).

Using genetic and epigenetic information for public health purposes requires that variation in the population be accurately described and categorized, and that the concept of "abnormal" be thought of more in terms of susceptibility than deterministically; hence, the appropriate interpretation of biomarkers is one, which is probabilistic (56). Lloyd (1998) concluded that "…public and scientific misconceptions of susceptibility are probably one of the most prominent problems facing those interested in the development of genetic medicine." The same can be said for molecular epidemiology as well. For public health purposes, there is a need to define concepts (e.g. susceptibility) on a population level (18).

Another area of interpretation that is problematic is what is called individual risk assessment. Generally speaking, epidemiological studies (with or without biomarkers) yield group results. The disease risk pertains to the group as a whole and not necessarily to individual members of the group, although it is possible to compute an individualistic risk using a risk function equation (57). However, if the marker being used has not been validated for disease, the calculation of an individual's risk will be meaningless. Thus far, for the current generation of biomarkers used in chronic disease research, there are a small number of markers (such as a few genetic mutations linked to high risk of disease in cancer family syndromes) for which an individual probabilistic risk can be estimated based on the biomarker.

These vagaries of biomarker data may lead an investigator to conclude that a particular biomarker is of uncertain meaning with regard to risk. Nonetheless, investigators have an obligation to accurately portray the degree of uncertainty in test and study results. There is a range of opinions about communicating results of biomarker tests on individuals or groups if there is no clinical meaning, such as usually occurs in transitional studies to validate markers and in population-based genetic research. Some believe that autonomy of participants is not honoured if they do not receive results, while others believe that the information communicated by results has no meaning for participants and indeed could be detrimental (52). While the latter view has the appearance of being paternalistic, as it decides what is good for the participant without seeking the opinion or decision of the participant, it may also be viewed as "doing no harm" (6). Such an interpretation is premised on the notion that providing results lacking any clinical, prognostic, or other use may elevate the risk of harm to participants by creating opportunities for undue anxiety, stress, alarm and unnecessary medical testing. However, recent evidence suggests that most research participants want results provided to them, and that the risk of anxiety may be less than originally estimated (58,59). Nonetheless, individuals may have a right not to know certain information that might be very sensitive and troubling to them. Increasingly, molecular epidemiologists may also be dealing with epigenetic data, which may be far more complex and difficult to interpret than biomarker data currently under investigation (60–62).

The communication of the results of biologic tests (particularly genetic tests) in molecular epidemiologic studies is still a difficult area. While generally the literature identifies adherence to the principles of autonomy (beneficence, respect for persons, reciprocity, and justice), the actual ways to do that are still subject to interpretation and opinion. It is clear that the approach taken concerning communicating results should be made explicit in the informed consent process. However, there are differing opinions on whether, or to what extent, test results should be communicated to study participants. On one extreme, some argue for full disclosure of genetic information, while others argue for balance of benefit and harm and that disclosure should be limited to certain situations. US federal regulations regarding biomedical research have been characterized as not providing clear guidance on this matter (52).

Timeliness of communication of results is also important to consider. This particularly becomes an issue when results indicate an action that could reduce exposure or risk, or affect timely treatment. As discussed above, situations exist where additional support to participants may be warranted. Evaluating the impact of notifying research participants of results may not need to be a routine matter, but since the consequences of notification cannot always be anticipated, it may be useful to provide the opportunity for participants to obtain more information or provide feedback about the results (6).

Avoiding inappropriate actions based on study results

Molecular epidemiologic investigators must concern themselves with how study results are incorporated into epidemiologic knowledge and public health practice. In some sense, the results of molecular epidemiologic studies of biomarkers of susceptibility are particularly at risk of being misunderstood or abused (6,45,52,55,56,63–65). For example, many common low penetrance gene variants, some of which require specific environmental exposures to increase risk of disease, do not provide unambiguous information. Yet various groups in society may start using such genotype information as if it represented diagnoses rather than risk factors (66). The consequences of such misinterpretation and application of biomarker results can include discrimination, labelling and stigmatization of subjects. Moreover, the deleterious effects of the inappropriate application of results can extend to family members, communities, ethnic groups, and other social groups as well. Unfortunately, there is a paucity of research about the negative repercussions of molecular epidemiologic research findings on participants, family members, communities and society. There is the widely expressed concern that genetic biomarkers can be used in ways that are discriminating and unjust, but there is little published evidence (22,45,67). Similarly, this concern has also been voiced with epigenetic data (68).

To facilitate the appropriate use of study results as much as possible, investigators should assure their quality. Methodological considerations in study design bear directly on the kind and strength of the inferences that can be drawn (e.g. increasing generalizability of study results through sample selection, and achieving appropriate statistical power with large enough sample sizes). This in turn affects what evidence can be provided from any particular study and what prevention or interventions can be envisioned. Inappropriate actions can thus inadvertently occur when interventions (or lack thereof) are based on results from a study that used biased or inappropriate methods. Some aspects of the research process provide investigators greater control over ensuring the appropriate application of findings; namely by strengthening the study's internal and external validity (such as in regards to study design and selection of research participants). Other dimensions are less in the control of investigators, such as public perception, media coverage, and the application of the results in the policy arena. The importance of the availability of all relevant evidence becomes apparent here as well (69). Timely publication of negative results is also crucial, for they contribute to the evidence on a particular biomarker and help to define the uncertainty accompanying a particular finding.

Molecular epidemiology holds promise for our ability to identify changes earlier in the natural history of a disease that may be amenable to intervention, leading to prevention of clinical disease or a better prognosis. This contribution is not without potential ethical issues. Premature marketing or use of tests is one problematic area that results from an inappropriate assessment of whether biomarkers or molecular tests have been validated for the specific use intended (25,26,70,71).

Inappropriate action also includes the lack of action, such as where there is some evidence from molecular epidemiologic research that indicates the need for preventive measures, and none are taken. There is increasing concern that public health practice has failed to take action on preliminary findings on the basis of uncertainty in the evidence. Delays in recognizing risks from past exposures, and acting on the findings, such as for cigarette smoking and exposure to asbestos and benzene, are failures that were not only scientific but ethical, since they resulted in preventable harm to exposed populations (72). One explanation offered for such delay is the absence of adequate proof or evidence of the certainty of a causal relationship. Such a position reflects an unwillingness to accept what may appear to be a preponderance of evidence as a trigger for public health actions even if there are some uncertainties (73).

The precautionary principle, a contemporary re-definition of Bradford Hill's case for action, provides a common sense rule for doing good by preventing harm to public health from delay: when in doubt about the presence of a hazard, there should be no doubt about its prevention or removal (70). It shifts the burden of proof from showing presence of risk to showing absence of risk and aims to do good by preventing harm, subsuming the upstream strategies of the Driving Forces Pressure Stress Exposure Effect Action (DPSEEA) model and downstream strategies from molecular epidemiology for detection and prevention of risk (74). It has emerged because of ethical concerns about delays in detection of risks to human health and the environment, and serves to emphasize epidemiology's classic role for early detection and prevention. At the same time, such precautionary strategies can have significant unintended consequences that also must be considered (71,75). Further, the translation of epidemiologic findings into public health policy generally involves multiple parties with various vested interests. The arena is complex: the role in this arena of those who carry out molecular epidemiologic research is not altogether clear, and there is a concern that the perception of an investigator's ability to carry out objective research could potentially be compromised through advocacy.

In keeping with the wider field of epidemiology, it is important that molecular epidemiology strive towards disease detection and prevention in populations. A concern has been expressed that when a public health problem is reduced to the level of the individual, such as with molecular biomarkers, then so too shall the intervention lie at the individual level (76). In some instances, this may be perfectly appropriate, yet in others, it may lead to the non-individual level factors (such as ecological chemical exposures) that gave rise to the public health problem in the first place and allow it to persist unabated (77). Inappropriate action could result from appropriate research. While there is no clear path to follow to those studies that will be beneficial and to avoid those that will not, considering why and how a particular research question is being asked, and what truly is the best manner in which to answer it, may aid molecular epidemiology in a balancing act between a high-risk approach and population-wide applicability of findings.

The results of molecular epidemiologic research may be used to support regulation or litigation. For regulatory agencies, there is a need to balance the risk of premature use of inadequately validated data with the harm from unduly delaying the use of relevant data from overly cautious policies (12). Critical in assessing the validity of molecular epidemiologic research for regulation or litigation will be whether the studies are of sufficient size and methodologic quality, and whether the findings have been replicated or corroborated. However, the ability of molecular epidemiologic research to provide evidence of toxicant-induced injuries, long before any clinical symptoms emerge, could profoundly affect how regulation is conceived to protect the public from environmental risks (78).

Sharing the benefits of molecular epidemiologic research: Public health ethics

In addition to the scientific benefits of sharing genomic and molecular epidemiologic data, there are also social and ethical issues. Fourteen stakeholder groups (many of which are outside the scientific community) have been identified who have at least eight different perspectives on the question of donor privacy and scientific efficiency (16). The researchers conclude that, at present, society lacks the sophisticated ethical or policy framework to simultaneously weigh multiple perspectives and interests. More broadly, the benefits of molecular epidemiologic research involving genes may not be equally shared among poorer people in developed countries or among developing countries (21,79–81). The responsibilities of molecular epidemiologists to share the benefits of their research are generally viewed as limited. With that said, there is a need for molecular epidemiologists to consider broader questions, such as under what general conditions genome-based knowledge in molecular epidemiology could be further used in public health.

Beyond the need for molecular epidemiologists to address the rights of individuals is the need to consider broader questions, such as clarifying the general conditions under which molecular epidemiological research findings will contribute to public health in a wide-ranging way. Population-based data on genome-disease and genome-environment interactions are the primary point for assessing the added value of genome-based information for all health interventions in different health care settings. This includes the integration of genome-based information into existing population-based surveillance systems, and the use of large-scale biobanks to quantify disease incidence in various populations and subpopulations, as well as to understand their natural histories of disease through risk factors including genome-environment interactions (15). Making such potential benefits of molecular epidemiology manifest requires paying particular attention to the public health-specific ethical, legal and social implications of such research (15,77,82).

Whole-genome research

A core element of molecular epidemiologic research is the ability to utilize whole-genome and related "omic" technologies (see Chapters 6 and 7), because of the considerable cost and effort directed at conducting large studies. The area of whole-genome research is in its formative stage. The initial recommendations have been formulated to protect the confidentiality of participants and, at the same time, make the data available to researchers who propose projects and adhere to strict guidelines for protection of the data sets and participants. To this end, the US National Institutes of Health (NIH) has made available GWAS data to researchers through a registered access process using the database of genotypes and phenotypes (dbGaP) resource of the National Center for Biotechnology Information (NCBI) (83). The procedure requires institutional support for a faculty member (from a university, organization, or commercial entity) to access GWAS genotype data under agreed-upon conditions (Table 2.1). Sign-off by the sponsoring institution must guarantee the security and validity of the proposed analyses according to the precepts of the Trans-NIH GWAS Sharing Policy along with subsequent updates (84). The policy addresses issues of data sharing and availability of data sets. Moreover, guidelines have been proposed for issues of informed consent prospectively, and review of older studies for use in GWAS studies. This includes explicit assent from the overseeing

Table 2.1. Requirements for conducting genome-wide association studies

The NIH GWAS certificate expects that a Principal Investigator (PI) and their institution certify the following:
The data submission is consistent with all applicable laws and regulations, as well as institutional policies;
The appropriate research uses of the data and the uses that are explicitly excluded by the informed consent documents are delineated;
The identities of research participants will not be disclosed to the NIH GWAS data repository;

An IRB and/or Privacy Board, as applicable, has reviewed and verified that:
  • The submission of data to the NIH GWAS repository and subsequent sharing for research purposes are consistent with the informed consent of study participants from whom the data were obtained;
  • The investigator's plan for de-identifying data sets is consistent with the standards outlined in the policy;
  • It has considered the risks to individuals, their families, and groups or populations associated with data submitted to the NIH GWAS data repository; and
  • The genotype and phenotype data to be submitted were collected in a manner consistent with 45 C.F.R Part 46.

After publication, a full GWAS data set, stripped of all identifiers and with limited covariate data (e.g. case-control status, study or geographic entity, age group, sex, and broad racial and ethnic groups), is transferred to a Data Access Committee (DAC), according to the trans-NIH GWAS data posting policy of January 25, 2008 (84). All investigators, regardless of whether or not they are PIs on the GWAS or external to the project, who desire access to the individual level genotype data with limited covariate data can obtain access by submitting a secured application proposal to a certified DAC. Access to the data through the DAC requires the use of an ERA number, registration with the NIH, support of an investigator's institution (signing official), IT security program including use of a controlled-access and secure site, and a Data Use Certificate and modified SF-424 form. Proposal application forms are completed and sent to the DAC, which is composed of NIH officials who make the final decision regarding access to the data.

IRB that the conduct and availability of the GWAS study are consistent with the informed consent signed by the participants. NIH and other large funding organizations, such as the Wellcome Trust in the United Kingdom, have mandated that funded GWAS studies be made available through the above described registered access process.

Conclusion

Relevance and rigor of molecular epidemiologic research is essential for enlightened public health policy and practice. Such research cannot be used effectively as the basis of public health policy if it lacks respect for people or contains flawed science. Moreover, if there is to be robust participation in research, participants must be motivated and assured that the research is conducted within a strong ethical framework (5,28). Consideration of the ethical issues in molecular epidemiological research should lead to maintaining the relevance and rigor of the discipline and ensure that the contributions it makes will be of great value.

Disclaimer: The findings and conclusions in this chapter are those of the author and do not necessarily represent the views of the Centers for Disease Control and Prevention.

References

1. Schulte PA, Perera FP, editors. Molecular epidemiology: principles and practices. San Diego (CA): Academic Press; 1993.

2. Vaught JB, Lockhart N, Thiel KS, Schneider JA (2007). Ethical, legal, and policy issues: dominating the biospecimen discussion. *Cancer Epidemiol Biomarkers Prev,* 16:2521–2523.doi:10.1158/1055-9965.EPI-07-2758 PMID:18086753

3. Nelkin D, Lindee MS. The DNA mystique: the gene as a cultural icon. New York (NY): W.H. Freeman; 1995.

4. McGuire AL, Gibbs RA (2006). Genetics. No longer de-identified. *Science,* 312:370–371. doi:10.1126/science.1125339 PMID:16627725

5. Caulfield T, McGuire AL, Cho M *et al.* (2008). Research ethics recommendations for whole-genome research: consensus statement. *PLoS Biol,* 6:e73.doi:10.1371/journal.pbio.0060073 PMID:18366258

6. Schulte PA, Hunter D, Rothman N (1997). Ethical and social issues in the use of biomarkers in epidemiological research. Lyon: IARC Scientific Publication; (142):313–318. PMID:9354930

7. Soskolne CL (1997). Ethical, social, and legal issues surrounding studies of susceptible populations and individuals. *Environ Health Perspect,* 105 Suppl 4;837–841.doi:10.2307/3433291 PMID:9255569

8. Hainaut P, Vähäkangas K. Genetic analysis of metabolic polymorphisms in molecular epidemiological studies: social and ethical implications. In: Vineis P, Malats N, Lang M *et al.*, editors. Metabolic polymorphisms and susceptibility to cancer. Lyon: IARC Scientific Publication; 1999(148). p. 395–402.

9. American College of Epidemiology Ethics Guidelines (2000). *Ann Epidemiol,* 10:487–497.doi:10.1016/S1047-2797(00)90000-0 PMID:11188987

10. International Programme on Chemical Safety. Biomarkers in risk assessment: validity and validation. Environmental health criteria 222. Geneva, Switzerland: World Health Organization; 2001.

11. Sharp RR, Zigas PH. Ethical and legal considerations in biological markers research. In: Wilson SH, Suk WA, editors. Biomarkers of environmentally associated disease: technologies, concepts, perspectives. Boca Raton (FL): CRC Press LLC; 2002. p. 17–26.

12. National Research Council. Applications of toxicogenomics technologies to predictive toxicology and risk assessment. Washington (DC): National Academies Press; 2007.

13. Vähäkangas K (2004). Ethical aspects of molecular epidemiology of cancer. *Carcinogenesis,* 25:465–471.doi:10.1093/carcin/bgh043 PMID:14656936

14. Williams G (2005). Bioethics and large-scale biobanking: individualistic ethics and collective projects. Genomics. *Soc Policy,* 1:50–66.

15. Brand AM, Probst-Hensch NM (2007). Biobanking for epidemiological research and public health. *Pathobiology,* 74:227–238.doi:10.1159/000104450 PMID:17709965

16. Foster MW, Sharp RR (2007). Share and share alike: deciding how to distribute the scientific and social benefits of genomic data. *Nat Rev Genet,* 8:633–639.doi:10.1038/nrg2124 PMID:17607307

17. Rothstein MA (2005). Currents in contemporary ethics. Research privacy under HIPAA and the common rule. *J Law Med Ethics,* 33:154–159.doi:10.1111/j.1748-720X.2005.tb00217.x PMID:15934672

18. Millikan R (2002). The changing face of epidemiology in the genomics era. *Epidemiology,* 13:472–480.doi:10.1097/00001648-200207000-00017 PMID:12094104

19. Barnes MR, Gray IC, editors. Bioinformatics for geneticists. West Sussex, England: John Wiley & Sons, Ltd.; 2003.

20. Resnik DB (2004). Disclosing conflicts of interest to research subjects: an ethical and legal analysis. *Account Res,* 11:141–159. PMID:15675055

21. Serrano LaVertu D, Linares AM (1990). Ethical principles of biomedical research on human subjects: their application and limitations in Latin America and the Caribbean. *Bull Pan Am Health Organ,* 24:469–479. PMID:2073561

22. Sleeboom M. Socio-genetic marginalization in Asia: a plea for a comparative approach to the relationship between genomics, governance, and social-genetic identity. In: Arnason G, Nordal S, Arnason V, editors. Blood and data: ethical, legal and social aspects of human genetic databases. Reykjavik, Iceland: University of Iceland Press; 2004. p. 39–44.

23. Le Marchand L, Wilkens LR (2008). Design considerations for genomic association studies: importance of gene-environment interactions. *Cancer Epidemiol Biomarkers Prev,* 17:263–267.doi:10.1158/1055-9965.EPI-07-0402 PMID:18268108

24. Schulte PA, Talaska G (1995). Validity criteria for the use of biological markers of exposure to chemical agents in environmental epidemiology. *Toxicology*, 101:73–88. doi:10.1016/0300-483X(95)03020-G PMID:7631325

25. Vineis P, Christiani DC (2004). Genetic testing for sale. *Epidemiology*, 15:3–5. doi:10.1097/01.ede.0000101961.86080.f8 PMID:14712140

26. Schulte PA (2005). The use of biomarkers in surveillance, medical screening, and intervention. *Mutat Res*, 592:155–163. PMID:16051280

27. Weed DL, McKeown RE (2001). Ethics in epidemiology and public health I. Technical terms. *J Epidemiol Community Health*, 55:855–857.doi:10.1136/jech.55.12.855 PMID:11707476

28. Dressler LG (2007). Biospecimen "ownership": counterpoint. *Cancer Epidemiol Biomarkers Prev*, 16:190–191. doi:10.1158/1055-9965.EPI-06-1004 PMID:17301248

29. Goodman GE, Thornquist MD, Edelstein C, Omenn GS (2006). Biorepositories: let's not lose what we have so carefully gathered! *Cancer Epidemiol Biomarkers Prev*, 15:599–601.doi:10.1158/1055-9965.EPI-05-0873 PMID:16614097

30. Maschke KJ (2006). Alternative consent approaches for biobank research. *Lancet Oncol*, 7:193–194.doi:10.1016/S1470-2045(06)70590-3 PMID:16510329

31. Knoppers BM (2004). Biobanks: simplifying consent. *Nat Rev Genet*, 5:485. doi:10.1038/nrg1396 PMID:15243995

32. Hansson MG, Dillner J, Bartram CR et al. (2006). Should donors be allowed to give broad consent to future biobank research? *Lancet Oncol*, 7:266–269.doi:10.1016/S1470-2045(06)70618-0 PMID:16510336

33. Rickham PP (1964). Human experimentation: code of ethics of the World Medical Association. *Br Med J*, 2:177.doi:10.1136/bmj.2.5402.177 PMID:14150898

34. Biobank UK. Ethics and governance framework. Available from URL: http://www.ukbiobank.ac.uk/ethics/intro.php.

35. UK Biobank. Ethics and Governance Council (EGC). Available from URL: http://www.egcukbiobank.org.uk/.

36. Laurie G (2009). Role of the UK biobank ethics and governance council. Lancet, 374:1676.doi:10.1016/S0140-6736(09)61989-9 PMID:19914512

37. Knoppers BM, Hirtle M, Lormeau S et al. (1998). Control of DNA samples and information. *Genomics*, 50:385–401. doi:10.1006/geno.1998.5287 PMID:9676435

38. Burhansstipanov L, Bemis L, Kaur JS, Bemis G (2005). Sample genetic policy language for research conducted with native communities. *J Cancer Educ*, 20 Suppl;52–57. doi:10.1207/s15430154jce2001s_12 PMID:15916522

39. Ness RB; American College of Epidemiology Policy Committee (2007). Biospecimen "ownership": point. *Cancer Epidemiol Biomarkers Prev*, 16:188–189. doi:10.1158/1055-9965.EPI-06-1011 PMID:17301247

40. Bankhead C (2004). Privacy regulations have mixed impact on cancer research community. *J Natl Cancer Inst*, 96:1738–1740. PMID:15572753

41. Nosowsky R, Giordano TJ (2006). The Health Insurance Portability and Accountability Act of 1996 (HIPAA) privacy rule: implications for clinical research. *Annu Rev Med*, 57:575–590.doi:10.1146/annurev.med.57.121304.131257 PMID:16409167

42. Beskow LM, Burke W, Merz JF et al. (2001). Informed consent for population-based research involving genetics. *JAMA*, 286:2315–2321.doi:10.1001/jama.286.18.2315 PMID:11710898

43. Schulte PA. Interpretations of genetic data for medical and public health uses. In: Arnason G, Nordal S, Arnason V, editors. Blood and data: ethical, legal and social aspects of human genetic databases. Reykjavik, Iceland: University of Iceland Press; 2004. p. 277–282.

44. Clayton EW, Steinberg KK, Khoury MJ et al. (1995). Informed consent for genetic research on stored tissue samples. *JAMA*, 274:1786–1792.doi:10.1001/jama.274.22.1786 PMID:7500511

45. Nelkin D, Tancredi L. Dangerous diagnostics: the social power of biological information. New York (NY): Basic Books; 1989.

46. Lowrance WW, Collins FS (2007). Ethics. Identifiability in genomic research. *Science*, 317:600–602.doi:10.1126/science.1147699 PMID:17673640

47. Lin Z, Owen AB, Altman RB (2004). Genetics. Genomic research and human subject privacy. *Science*, 305:183.doi:10.1126/science.1095019 PMID:15247459

48. Homer N, Szelinger S, Redman M et al. (2008). Resolving individuals contributing trace amounts of DNA to highly complex mixtures using high-density SNP genotyping microarrays. *PLoS Genet*, 4:e1000167. doi:10.1371/journal.pgen.1000167 PMID:18769715

49. Jacobs KB, Yeager M, Wacholder S et al. (2009). A new statistic and its power to infer membership in a genome-wide association study using genotype frequencies. *Nat Genet*, 41:1253–1257.doi:10.1038/ng.455 PMID:19801980

50. Schulte PA, Singal M (1989). Interpretation and communication of the results of medical field investigations. *J Occup Med*, 31:589–594.doi:10.1097/00043764-198907000-00009 PMID:2769455

51. Chadwick RF (2004). The right not to know: a challenge for accurate self-assessment. *Philos Psychiatry Psychol*, 11:299–301 doi:10.1353/ppp.2005.0005.

52. Ravitsky V, Wilfond BS (2006). Disclosing individual genetic results to research participants. *Am J Bioeth*, 6:8–17.doi:10.1080/15265160600934772 PMID:17085395

53. Grosse SD, Khoury MJ (2006). What is the clinical utility of genetic testing? *Genet Med*, 8:448–450.doi:10.1097/01.gim.0000227935.26763.c6 PMID:16845278

54. Ashford NA (1994). Monitoring the worker and the community for chemical exposure and disease: legal and ethical considerations in the US. *Clin Chem*, 40:1426–1437. PMID:8013132

55. Schulte PA (2004). Some implications of genetic biomarkers in occupational epidemiology and practice. *Scand J Work Environ Health*, 30:71–79. PMID:15018031

56. Lloyd EA. Normality and variation: the human genome project and the ideal human type. In: Hull DL, Ruge M, editors. The philosophy of biology. New York/Oxford: Oxford University Press; 1998.

57. Truett J, Cornfield J, Kannel W (1967). A multivariate analysis of the risk of coronary heart disease in Framingham. *J Chronic Dis*, 20:511–524.doi:10.1016/0021-9681(67)90082-3 PMID:6028270

58. Stolt UG, Liss PE, Svensson T, Ludvigsson J (2002). Attitudes to bioethical issues: a case study of a screening project. *Soc Sci Med*, 54:1333–1344.doi:10.1016/S0277-9536(01)00099-5 PMID:12058850

59. Partridge AH, Wong JS, Knudsen K et al. (2005). Offering participants results of a clinical trial: sharing results of a negative study. *Lancet*, 365:963–964.doi:10.1016/S0140-6736(05)71085-0 PMID:15766998

60. Sutherland JE, Costa M (2003). Epigenetics and the environment. *Ann N Y Acad Sci*, 983:151–160.doi:10.1111/j.1749-6632.2003.tb05970.x PMID:12724220

61. Jablonka E (2004). Epigenetic epidemiology. *Int J Epidemiol*, 33:929–935. doi:10.1093/ije/dyh231 PMID:15166187

62. Weinhold B (2006). Epigenetics: the science of change. *Environ Health Perspect*, 114:A160–A167.doi:10.1289/ehp.114-a160 PMID:16507447

63. Ashford NA (1986). Policy considerations for human monitoring in the workplace. *J Occup Med*, 28:563–568.doi:10.1097/00043764-198608000-00007 PMID:3746474

64. Wagener DK (1995). Ethical considerations in the design and execution of the National and Hispanic Health and Nutrition Examination Survey (HANES). *Environ Health Perspect*, 103 Suppl 3;75–80.doi:10.2307/3432564 PMID:7635116

65. Vineis P, Schulte PA, McMichael AJ (2001). Misconceptions about the use of genetic tests in populations. *Lancet*, 357:709–712.doi:10.1016/S0140-6736(00)04136-2 PMID:11247571

66. Schulte PA, Sweeney MH (1995). Ethical considerations, confidentiality issues, rights of human subjects, and uses of monitoring data in research and regulation. *Environ Health Perspect*, 103 Suppl 3;69–74. doi:10.2307/3432563 PMID:7635115

67. Sharp RR, Foster MW (2006). Clinical utility and full disclosure of genetic results to research participants. *Am J Bioeth*, 6:42–44, author reply W10-2. doi:10.1080/15265160600938443 PMID:17085408

68. Rothstein MA. Exposed today, grandchildren pay. Seventh Annual Rabbi Seymour Siegel Memorial Lecture in Ethics. Duke Law School, February 26, 2008. Available from URL: http://www.law.duke.edu/news/story?id=1013&u=11.

69. Gallo V, Egger M, McCormack V et al. (2011). STrengthening the Reporting of OBservational studies in Epidemiology - Molecular Epidemiology (STROBE-ME): An Extension of the STROBE Statement. *PLoS Med*, 8:e1001117.doi:10.1371/journal. pmed.1001117 PMID:22039356

70. Richter ED, Laster R, Soskolne C (2005). The precautionary principle, epidemiology and the ethics of delay. *J Hum Ecol Risk Assess*, 11:17–27 doi:10.1080/10807030590919864.

71. Weed DL (2004). Precaution, prevention, and public health ethics. *J Med Philos*, 29:313–332. doi:10.1080/03605310490500527 PMID:15512975

72. Davis D. The secret history of the war on cancer. New York (NY): Basic Books; 2007.

73. Michaels D. Doubt is their product: how industry's assault on science threatens your health. New York (NY): Oxford University Press; 2008.

74. World Health Organization. Development of environment and health indicators for European Union countries: results of a pilot study. Bonn, Germany: World Health Organization Regional Office for Europe; 2004. Available from URL: http://www.euro.who.int/document/E85061.pdf.

75. Goldstein BD, Carruth RS (2005). Implications of the precautionary principle: is it a threat to science? *Hum Ecol Risk Assess*, 11:209–219 doi:10.1080/10807030590920033.

76. Pearce N (1996). Traditional epidemiology, modern epidemiology, and public health. *Am J Public Health*, 86:678–683.doi:10.2105/AJPH.86.5.678 PMID:8629719

77. Robert JS, Smith A (2004). Toxic ethics: environmental genomics and the health of populations. *Bioethics*, 18:493–514. doi:10.1111/j.1467-8519.2004.00413.x PMID:15580721

78. Grodsky JA (2005). Genetics and environmental law: redefining public health. *Calif Law Rev*, 93:171–270.

79. Berg K (2001). The ethics of benefit sharing. *Clin Genet*, 59:240–243.doi:10.1034/j.1399-0004.2001.590404.x PMID:11298678

80. Sheremeta L (2003). Population genetic studies: is there an emerging legal obligation to share benefits? *Health Law Rev*, 12:36–38. PMID:15742495

81. Sheremeta L, Knoppers BM (2003). Beyond the rhetoric: population genetics and benefit-sharing. *Health Law J*, 11:89–117. PMID:15600070

82. Porter J, Ogden J, Pronyk P (1999). Infectious disease policy: towards the production of health. *Health Policy Plan*, 14:322–328.doi:10.1093/heapol/14.4.322 PMID:10787648

83. National Center for Biotechnology Information. Database of genotypes and phenotypes (dbGaP). Available from URL: http://www.ncbi.nlm.nih.gov/gap?db=gap.

84. Policy for sharing of data obtained in NIH supported or conducted genome-wide association studies (GWAS). Notice number: NOT-OD-07–088. Available from URL: http://grants.nih.gov/grants/guide/notice-files/not-od-07-088.html.

UNIT 2.
BIOMARKERS: PRACTICAL ASPECTS

CHAPTER 3.

Biological sample collection, processing, storage and information management

Jimmie B. Vaught and Marianne K. Henderson

Summary

The collection, processing and storage of biological samples occur in the larger context of organizations known as biological resource centres or biospecimen resources. Biological resource centres are (1,2) service providers and repositories of living cells, as well as genomes of organisms, archived cells and tissues, and information relating to these materials. The US National Cancer Institute (3) defines a biospecimen resource as a "… collection of human specimens and associated data for research purposes, the physical entity where the collection is stored, and all relevant processes and policies." The complexities involved in proper sample management policies and procedures are often underestimated. Prior to initiating a study that will involve the collection of biological samples, many decisions need to be made that will affect the quality of the samples and the outcome of the study. The appropriate sample type(s) needs to be chosen. The processing protocol that will result in samples of suitable quality for the intended laboratory analyses must be selected from among various possible protocols. Consideration must be given to the proper storage conditions to maintain sample quality until analyses are completed. All of these activities must be monitored and controlled by appropriate sample tracking and laboratory informatics systems. A comprehensive quality management system, with standard operating procedures and other appropriate controls, is necessary to assure that biological samples are of consistent quality and right for the intended analyses and study goals.

Introduction

Although biological specimens have been collected for use in a variety of molecular epidemiology, clinical trial and basic research studies for many years, it has only recently been recognized that the protocols and practices involved in collecting, processing and storing specimens actually comprise "biospecimen science." As a result, many organizations (Appendix 3.1) have engaged in producing guidelines and best practices for these endeavours, now known as biological resource centres or biospecimen resources.

Appendix 3.1. Existing guidelines and best practices for biorepositories

Title	Authors/Origin	Reference/Link
Tissue Banking for Biomedical Research	National Cancer Centre/Singapore	http://www.bioethics-singapore.org/uploadfile/52533%20PMHT%20AppendixB-Dr%20Kon.pdf
Biorepository Protocols	Australian Biospecimen Network, ABN/Australia	http://www.abrn.net/
Biological Resource Centres: Underpinning the Future of Life Sciences and Biotechnology	OECD/International	http://www.oecd.org/dataoecd/55/48/2487422.pdf
European Human Frozen Tumor Tissue Bank – TUBAFROST	TUBAFROST/The Netherlands	http://www.tubafrost.org/
Human Tissue and Biological Samples for use in Research: Operational and Ethical Guidelines	MRC/UK	http://www.mrc.ac.uk/Utilities/Documentrecord/index.htm?d=MRC002420
Best Practices for Repositories I: Collection, Storage, and Retrieval of Human Biological Materials for Research	International Society for Biological & Environmental Repositories/USA	Cell Preserv Technol 2008;6:3-58
First-Generation Guidelines for NCI-Supported Biospecimen Resources	NCI/USA	http://biospecimens.cancer.gov/bestpractices
UN Recommendations on the Transport of Dangerous Goods. Model Regulations	UN Economic Commission for Europe, UNECE/International	http://www.unece.org/trans/danger/publi/unrec/rev13/13files_e.html
Specimen Collection, Preparation, and Handling	LabCorp/International	http://www.labcorp.com/datasets/labcorp/html/frontm_group/frontm/section/speccol.htm

Several organizations have published guidelines and best practices relevant to the discussion in this chapter (54). This table is adapted from the IARC publication *International Network of Biological Resource Centres for Cancer Research: Recommendations on Common Minimal Technical Standards* (2).

These terms reflect the fact that specimen management takes place in an environment that includes a wide range of policies concerning the specimens and data, as well as the physical structure, the biorepository. Biological resource centres are engaged in many activities beyond storage, such as acquiring, processing (e.g. aliquoting, DNA extraction) and distributing biological materials. The practices and policies that have been organized into formal documents testify to the importance of following proper steps that will result in the highest quality specimens for research purposes. The use of proper procedures to produce biological specimens of the appropriate quality, as well as the collection of relevant clinical, epidemiologic and quality control data, gives the biospecimens their value in research.

Context and public health significance

Biological specimens (or biospecimens), such as blood, urine, saliva, and many other types, are collected for a variety of reasons, for normal patient monitoring and care as well as for basic, clinical and epidemiologic research studies. Many medical advances, including studies of heart disease, AIDS and cancer, have resulted from preliminary developmental studies that have relied on access to and proper use of the appropriate biospecimens. The sources of biospecimens for these studies have been varied, as has their quality (1–4).

For molecular epidemiology studies, the ultimate success of a study depends on reliable laboratory analyses of these specimens. In order for laboratory analyses to be reliable, the collection, processing and storage of specimens must be performed under strictly controlled procedures. As the sensitivity and specificity of analytic techniques have increased to an extraordinary degree in recent years (see Chapter 4), it has become even more important to assure that biospecimens are of the highest quality. In addition, from the point in time that the specimens are collected until laboratory results are analysed and reported, all of the relevant information concerning the specimen, as well as data concerning the study participant and laboratory analyses, must be properly stored in interoperable information management systems. This could mean multiple systems or multiple databases interconnected in a single system. All of these steps must be performed under a well-planned quality assurance programme, and according to relevant legal and ethical standards (discussed in Chapter 2).

Examples/case studies

Prior to initiating a study that involves specimen collection, several key points must be considered. The answers to these questions will be important in determining whether the appropriate materials, equipment and procedures are in place:
• What are the goals of the study?
• What laboratory analyses will be needed to accomplish the study goals?
• What type of biospecimens will be necessary for the intended laboratory analyses?
• How many specimens will be collected? If necessary, a biostatistician should be consulted to assist in determining the number required to achieve statistical significance.
• What volume or size will be required for each specimen to assure that it is adequate for the intended analyses? Will it be necessary to store smaller volumes in aliquots for future unplanned use to avoid thawing a larger aliquot? For example, it is important to consider that new technologies have resulted in more sensitive analytical techniques to apply to older samples (see also Chapters 4 and 7), or older samples may become sources of information to study the natural history of a seemingly 'new' disease.
• What quality standards do the specimens need to meet for valid laboratory analyses? Have such quality measures been validated?
• Have specimen collection, processing and storage protocols been standardized and validated in pilot studies?
• If the specimens will be stored for some period of time before analysis, has the stability of the intended biomarker, or other analyte, been determined for the planned storage conditions?
• Will specimens need to be shipped to distant locations for analysis? If so, have packaging and shipping protocols been validated to assure the stability and safety of the specimens and personnel who will handle them?
• Have all other logistical issues been resolved, such as proper coding, labelling and identifying the types of storage vessels?
• What data will be collected with the sample and the study, and is an appropriate informatics system available to collect and process this information?
• Have all appropriate informed consent, privacy and other ethical and legal rules and regulations been reviewed and adhered to in the study planning?
• Are funding and other resources for the proposed study's specimen collection adequate? Will it be necessary to consider lower cost alternate methodologies?

If there is a significant amount of uncertainty in answering the above questions, then additional thought and planning will be needed before beginning the study. For example, before initiating the collection of blood and urine from 500 000 study participants in 2007, the United Kingdom Biobank conducted a series of sample processing validation studies (5–7). These studies showed the effects of sample processing delays, as well as storage conditions, on the results of the wide variety of assays to be conducted on samples that will be collected over a four-year period, but may be used for studies for 20 years or more. The long-term success of such a large and costly project depends on this careful approach to planning the most efficient specimen collection and processing to maintain the stability of the resulting sample aliquots, which are expected to number approximately 15 000 000 (5).

Among the issues outlined above, cost is a major consideration, especially when designing a study that will include a large collection of biospecimens. Often the costs of collecting, processing and storing biospecimens are not well understood or estimated before starting a study. The design and operation of the physical biorepository also needs to be well thought out. Baird and Frome (8) have outlined the major elements of cost and design for a large biorepository. It is also important to plan biospecimen collections with careful attention to the costs of analyses and storage, especially if long-term storage will be necessary. For example, if a study requires only nanogram quantities of DNA for genotyping purposes, one should consider collecting small amounts of blood or saliva on filter cards, instead of a large volume of blood that will yield hundreds of micrograms of DNA and incur larger processing and storage costs. Other alternate processing and storage approaches that may result in cost savings are considered in the Specimen collection section.

As shown in Figure 3.1, specimen collection, processing and storage are components of a series of steps that are used in any study involving

Figure 3.1. The lifecycle of biospecimens in biological resource centres. Used with permission from (2).

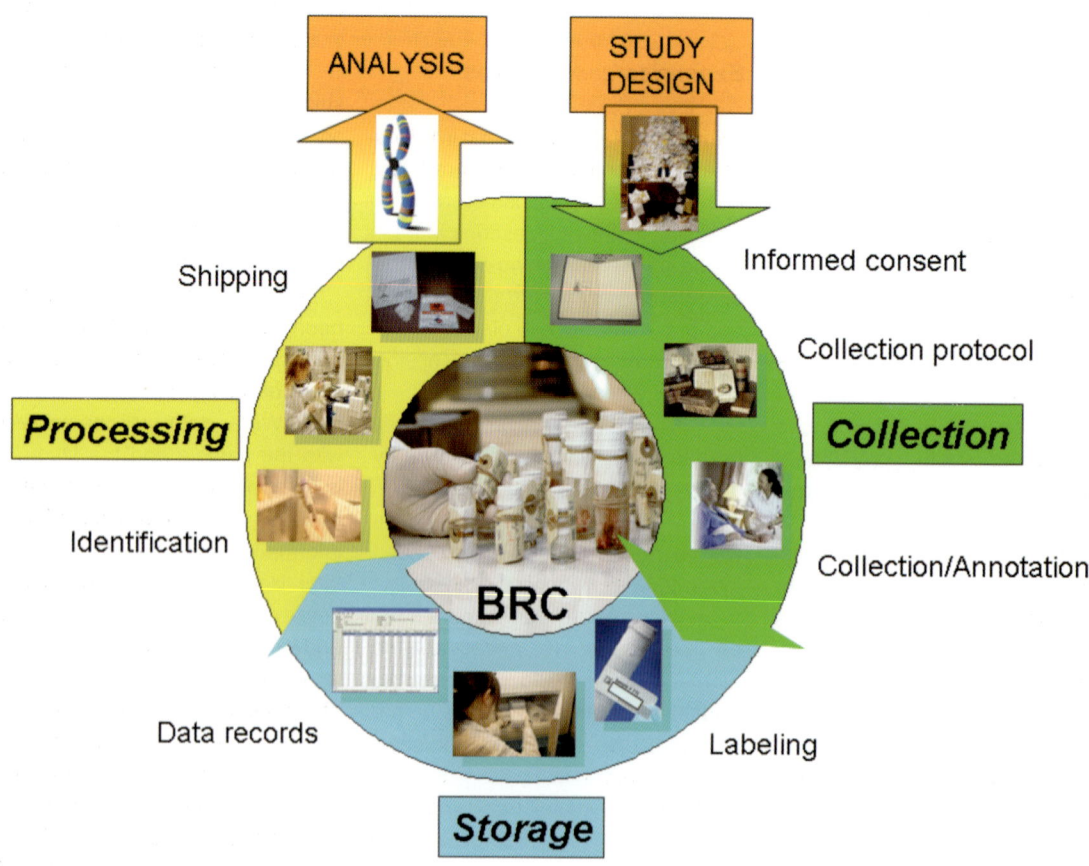

the collection of biospecimens. Each of these steps is discussed in turn in the following sections.

Specimen collection

Specimen types

A wide variety of specimen types may be collected for storage, and in many molecular epidemiology studies more than one of the following (discussed in detail) may be necessary, depending on the study goals (2,3). Additional collection, processing and storage guidance can be found in the International Society for Biological and Environmental Repositories (ISBER), National Cancer Institute (NCI) and International Agency for Research on Cancer (IARC) documents (2–4).
• Blood and blood fractions (plasma, serum, buffy coat, red blood cells)
• Tissue (from surgery, autopsy, transplant)
• Urine
• Saliva/buccal cells

Many other types of specimens may be collected, depending on availability and study goals, for example:
• Placental tissue, meconium, cord blood
• Bone marrow
• Breast milk
• Bronchoalveolar lavage
• Cell lines
• Exhaled air
• Feces
• Fluids from cytology (ascites, pleural fluid, synovial fluid, etc.)
• Hair
• Nail clippings
• Semen

Each of these specimen types should be collected, processed, and stored under conditions that preserve their stability with respect to the intended future analyses.

Of particular interest for molecular epidemiology studies are those specimen types that can be collected most conveniently and efficiently, and at the lowest cost for large population-based studies. The most common specimen types collected for these studies are discussed in the following sections: blood, tissue, urine and saliva.

Collection procedures

Collection procedures will vary according to specimen type and the intended analyses, but all procedures should be carefully designed and documented. It is normally a good practice to perform pilot studies to validate new specimen collection methods and protocols (4). The discussion in this section focuses on the specimens most commonly collected for molecular epidemiology studies. Additional information and collection protocols may be found in several references (2–4). Also see Chapter 12, Table 12.2 for additional information about specimen types collected for epidemiologic studies, and their advantages and disadvantages.

Blood collection

Collection of blood specimens (9) should be carried out by trained phlebotomists to avoid causing study participant discomfort, or compromising the quality or quantity of the sample. Standard protocols recommended by well-established organizations should be used.

An evacuated tube system (e.g. Becton-Dickenson Vacutainer®) with interchangeable glass or plastic tubes is commonly used to collect blood. The tubes, some with additives appropriate to a specific application, are differentiated by their colour-coded stoppers. Blood collection tubes should be drawn in a specific order to avoid cross-contamination of additives (10,11) (also see Chapter 12).

As shown in Table 3.1, blood is often fractionated before being analysed or stored (10,11). Fractionation of blood results in the following components:
• Mononuclear leukocytes (peripheral blood mononuclear cells, PBMCs) are the only cell type in blood that can be maintained in a viable state.
• Neutrophils (the most abundant type of granulocytes) are also nucleated and another source of DNA.
• Erythrocytes can be used to study adducts of haemoglobin.
• Plasma is obtained from an anticoagulated blood sample by separating out the cellular components.

Serum isolation requires no anticoagulants. To reduce contamination, serum should be separated from other blood components as soon as possible. Serum allows for improved analyses of antibodies, nutrients, lipids and lipoproteins. Either serum or plasma may be used for proteomic analyses, although according to recent Human Proteome Organization (HUPO) guidelines there are advantages and disadvantages in the use of either specimen (12). For studies intended to investigate the broadest array of proteins and peptides, plasma is the better choice, as the process of blood coagulation results in the loss of many proteins. Some differences in endogenous hormone analytical results have been found between serum and plasma, but as noted in Chapter 12, both are acceptable as specimens for such analyses.

Depending on the intended laboratory analyses, blood should be collected anticoagulated (consisting of plasma, buffy coat and red blood cells) or coagulated

Table 3.1. General guidelines for blood collection and processing

Blood Fraction	Collection additive	Preferred uses	Limitations/problems
Whole Blood	Anticoagulant (ACD, heparin, EDTA); protease inhibitor for proteomics	Genomics studies; Source of DNA, RNA	Anticoagulant effects need to be considered
Buffy Coat	Anticoagulant	DNA extraction; source of lymphocytes, cell lines as unlimited DNA source	Limited yield if blood not properly processed. As a source of DNA, whole blood collection is generally more economical
Serum	None	Proteomics; Source of DNA; Multiple analytes	DNA yield low (nanograms) but suitable for genomics applications
Plasma	Anticoagulant, possibly protease inhibitor	Proteomics (preferred sample)	DNA yield low (nanograms) but suitable for genomics applications.
		Source of DNA, multiple analytes	Analytical results may differ in serum and plasma.
Blood Clot	None	Source of DNA	Extraction difficult, costly

(consisting of serum and red blood cell clot) (9). There are several types of anticoagulants which need to be chosen carefully to avoid problems with certain laboratory applications (9,11). Other special collection tubes such as Serum Separator Tubes® and Cell Preparation Tubes® (SST, CPT, Becton-Dickenson) allow for more convenient separation of blood fractions, but some problems have been encountered in their use (9). Special collection tubes with protease inhibitors have been developed, which preserve proteins for proteomics analyses (9,12). The analysis of trace metals in blood also requires caution, as they may be present in the evacuated collection tubes. Lot-to-lot variation in the quality of collection tubes is also a potential source of spurious laboratory results.

There is no fixed time period that can be recommended for collecting and processing blood. However, depending on the intended analyses, the stability of blood with respect to various laboratory analyses may be affected or controlled as follows (9,11):

• Anticoagulants used in blood collection, as described above and in Table 3.1.
• Stabilizing agents are necessary to preserve some analytes, and should be included in the collection device or added as soon as possible after collection.
• The time elapsed between blood collection or removal from a storage unit and subsequent processing may be important, depending on the intended analyses. See the United Kingdom Biobank validation study summary for examples of such effects (summarized in reference 7).
• The temperatures at which blood specimens are processed and stored may be important, depending on the intended analyses (13).
• Thaw/refreeze cycles should generally be avoided due to the potential for instability of some analytes. However, thaw/refreeze effects are not well documented for all analytes and may need to be evaluated through pilot tests (13).
• Enzymatic degradation affects many biochemical markers. RNA and proteins are particularly susceptible to this and require special procedures to maintain their integrity during collection and processing. The addition of commercially available RNase inhibitors preserves RNA integrity.
• Special collection systems (ex. PAX DNA® Blood Collection System by PreAnalytiX®) allow for the collection, shipping, and short-term storage of blood at room temperature, and for subsequent extraction of DNA according to a single-tube protocol (14).

Tissue collection

The primary sources of tissues for research are biopsy, surgery and autopsy. As noted in the ISBER Best Practices and IARC Biological Resource Centre Guidelines (2,4), tissues must be collected under strict ethical and legal guidelines, and the collection of samples for research must never compromise the diagnostic integrity of a specimen. Generally it is preferable for a trained pathologist to be involved in the actual procurement of the tissue specimen during a surgical or autopsy procedure.

Other important considerations in collecting tissue are (adapted from ISBER Best Practices and IARC Biological Resource Centre Guidelines (2,4):

• Timing. In general, it is important to minimize the time between collection and stabilization and processing of tissue specimens. This time will vary according to the intended use, since different biomolecules degrade at different rates. The effects of collection timing on tissue and macromolecule preservation have not been well studied. The best approach is to collect, stabilize (freezing or fixing) and process tissue specimens as rapidly as possible. It is recommended that surgical or biopsy specimens be preserved within 1 hour (or less if possible) of excision; however, tissue subject to a delay up to two hours should still be collected (15). Detailed records of the timing of events from excision to fixation or freezing should be kept. Tissue banking staff must be present in pathology to freeze or fix the tissue as quickly as possible. Tissues must be snap frozen either directly or enclosed in a container immersed in the freezing medium (e.g. precooled isopentane). Liquid nitrogen is not recommended as a suitable freezing medium for direct snap freezing, due to the potential formation of cryo-artefacts. When dry ice or liquid nitrogen are not readily available, tissue collection into RNAlater® (16) may be a good alternative, provided that the tissue is not required for diagnostic purposes and permission is given by the pathologist.

• Surgical specimens. Remnant samples may be collected from diagnostic procedures or, with proper IRB approval, specimens may be resected specifically for research. Depending on the intended use, specimens may be transported or frozen immediately. Samples requiring snap freezing can be frozen in a Dewar flask of liquid nitrogen or on dry ice at the time of collection. Otherwise, it is recommended that samples be transported in saline on wet ice to the repository or laboratory for additional processing.

• Autopsy specimens. It is important to know the time interval between death and collection and processing of the specimen, as specimens may degrade quickly after death. Autopsy procedures may yield "normal" tissues (i.e. normal lung), or large quantities of a specimen that would not otherwise be available from surgical procedures. Tissue specimens collected at autopsy should be appropriately labelled as to the organ site, tissue type, and time of resection, and then immediately placed in a container of saline on wet ice for transport to the tissue repository for processing.

• Transplant tissue and organs that are inappropriate for transplant may sometimes be made available for research. Often transplant tissue is of a higher quality than either surgical or autopsy specimens, due to the special efforts made to preserve the integrity of the transplant organs.

Tissue fixation

Formalin- or alcohol-fixation and paraffin embedding may be used to preserve tissues at relatively low cost when adequate freezing procedures and storage facilities are not available (2). Formalin-fixation is also the standard practice for preservation of tissues collected during surgery or autopsy. Fixed paraffin blocks may be stored in light- and humidity-controlled facilities at room temperature (18–22°C). Formalin-fixed tissues may be used for DNA extraction. The DNA is usually fragmented but remains suitable for PCR-based analysis of short DNA fragments.

Due to degradation issues, formalin-fixed, paraffin-embedded tissues are of limited use as a source of RNA. However, RNAlater® (16) is a commercial aqueous, non-toxic tissue storage reagent that rapidly permeates tissues to stabilize and protect cellular RNA and eliminates the need to immediately freeze or otherwise stabilize tissue samples. Tissue samples can be harvested and submerged in RNAlater® for storage for specific periods without jeopardizing the quality or quantity of RNA extracted at a later time or date. However, specimens processed in RNAlater® cannot be further used for histomorphopathological analyses.

Alternatives to formalin fixation include ethanol, Optimal Cutting Temperature (OCT) media, methacarn, and Carnoy's solution, among others. To achieve an acceptable balance between the preservation of tissue morphology and nucleic acid integrity, it may be necessary to alter fixation methodology to achieve a study's goals. Several studies have explored the effects of the above standard fixatives, as well as newer ones for special applications (17–20). Although formalin-fixation remains the standard tissue preservation method, these alternatives should be considered for special research applications that require the preservation of particular macromolecules or morphological features.

Urine collection
(see also Chapter 12)

Many analytes, such as steroid hormones, pesticides and a wide variety of drugs and their metabolites, can be measured in urine for molecular epidemiology studies (11), making it a convenient specimen for a variety of studies. Urine collection can performed under several conditions, depending on the study design and analytical goals (4,11):

- First morning. Collected immediately upon rising in the morning, recommended for analytes requiring concentration for detection in laboratory assays.
- Random urine specimens are appropriate for drug monitoring and cytology studies.
- Fractional specimens. The study participant fasts after the last evening meal, and the second morning urine is collected. These specimens are used to compare urine analyte levels with their concentrations in blood.
- Timed urine collections (e.g. 12 and 24 hour) are used to allow comparisons of excretion patterns.

Urine collections should be maintained on ice or refrigerated for the duration of the collection. Collection vessels are generally larger than for other liquid specimens, and may range from 50 to 3000 mL. Depending on the analyte to be measured, a preservative may be needed. The type of preservative may differ according to test methodologies, time delay, and transport conditions. EDTA and sodium metabisulfite are examples of preservatives commonly used in urine collections (11).

Saliva/buccal cell collection

Saliva, with exfoliated buccal cells, is an excellent source of DNA for genetic studies (21). Self-collection of buccal cells is a safe, convenient method that can be used to reduce the cost of specimen collection and is often preferred over blood collection by study participants (discussed in Chapter 12). Several methods have been developed for collecting buccal cells, including swabs, cytobrushes and a mouthwash protocol. The mouthwash protocol has been successfully used in large population-based studies and has been shown to yield DNA of good quality and quantity for genetic analyses (21). However there are limitations to buccal cell DNA, as described below.

New methods are being developed for saliva collection. One such method has been developed by DNAGenotek (22). A proprietary reagent, Oragene, preserves saliva (and DNA) at room temperature. The method has been successfully used in epidemiologic studies (23). The yield and quality of DNA from the Oragene collection is similar to that for the mouthwash method.

Collection of blood, saliva on treated cards

New technologies, such as whole-genome amplification methods to increase genomic DNA yields, and the high cost of collecting and processing blood or mouthwash samples, have led to renewed consideration of treated filter paper cards as a method to collect DNA from blood (24) and buccal swabs (25) (also discussed in Chapter 12). Filter paper cards have been pre-treated to retard bacterial growth, inhibit nuclease activity, and release DNA during processing (26). The cards may be easier to use in paediatric and elderly populations to collect specimens, and can be mailed in an envelope with a desiccant at a nominal cost.

Blood collected on filter cards is well established as a source of DNA for genetic studies, as well as for a variety of other research and clinical applications. The US Centers for Disease Control and Prevention (CDC) uses blood spot cards in its nationwide neonatal screening programme (24). The US Armed Forces collect blood spot cards from all service members and stores them for possible identification purposes, as well as research and clinical purposes. DNA can be easily extracted from blood spots in amounts more than sufficient for genetic studies. This process has been automated, especially for forensic applications (27).

In addition to standard filter cards, new technologies for dry-state specimen collection have been developed. GenVault (28) uses small elements of treated filter paper in 384-well plates for storage of blood, DNA, plasma and serum specimens at room temperature. DNA and protein can be eluted from the elements by relatively straightforward methods.

Preserving specimen stability during collection

As noted above for tissue biospecimens, the elapsed time for collection, and between collection and stabilization, should be minimized, and the tissue temperature should be reduced as soon as possible after collection. This is especially important if freezing is the stabilization endpoint. If fixation is the stabilization endpoint, control of processing time between maximum and minimum durations may be required. Rapid processing may not be as critical for other types of biospecimens, such as blood. Optimal processing times vary depending on the analysis method for which a biospecimen is used.

Biorepositories should use the processing method that preserves the greatest number of analytes. The best scheme to preserve analytes is to divide specimens into aliquots or fractions of appropriate size or volume and/or preserve them by multiple processing methods.

Specimen processing

Specimens are processed according to the study design and the methods most appropriate for preserving the analytes of interest. For a particular specimen type and analysis, several processing methods may be appropriate. The IARC standards (2) list some of the more routine processing protocols. The general guidelines in this section outline some of the important considerations when choosing processing methods for specimens most commonly collected for molecular epidemiology studies. Additional issues concerning the processing and analyses of specimens for proteomic, metabolomic, physical, chemical, and immunologic applications are discussed in Chapters 4 and 7.

Blood – separation into fractions (e.g. plasma, serum, buffy coat, red blood cells)

The processing method used for blood specimens depends on the laboratory analyses to be performed. Cryopreservation is a cost-effective way of preserving viable lymphocytes for subsequent recovery of DNA, or for Epstein-Barr Virus (EBV) transformation to create lymphoblastoid cell lines as a source of unlimited amounts of DNA (29). Cryopreservation typically involves the use of a cryoprotectant, such as dimethyl sulfoxide (DMSO). However, commercial cryoprotectants that are less toxic have been developed (30). Whole blood may also be cryopreserved as an efficient and cost-effective approach to centralized processing and storage of viable cells in large-scale epidemiological studies (29).

Tissue – processing after surgery, autopsy

Specimens resected specifically for research may be either processed in the operating room or pathology suite, shortly after the time of collection, or may be transported to the repository for processing, depending upon the requirements of the specific protocol. Additional details are discussed above, and may also be found in the ISBER and IARC Guidelines (2,4).

Urine

Processing of urine before storage is fairly straightforward. The primary decision is the size of the aliquots to be stored and is based on the expected analyses. If the analytes are stable to thaw/refreeze cycles then larger aliquots can be stored.

Saliva/buccal cell processing from mouthwash protocol specimens

Buccal cells collected using the mouthwash protocol (21) are processed by centrifugation of the cell suspension, resuspension in a buffer, and either processed immediately or frozen for future use. Usually, additional processing involves DNA extraction. Note that a special consideration in processing buccal cell DNA is the high percentage of bacterial DNA present in these specimens, which requires special quantitation by real-time PCR.

DNA extraction

DNA extraction methodology is well established for a variety of specimen types, including whole blood, blood fractions, buccal cells, fresh and frozen tissues, and paraffin tissue blocks (31). The gold standard for DNA extraction is generally considered to be phenolchloroform extraction, but other standard methods that are more efficient, less expensive, and that utilize less toxic chemicals provide similar yields and DNA of similar molecular weight. Companies such as Gentra and Qiagen have collected DNA stability data of over 12 years' duration (32).

Techniques for measuring the quality and quantity of DNA range from absorbance at 260nm and 280nm, to fluorescence methods, to real-time PCR for detection of less than 25 picograms DNA. The A260/A280 ratio is a rough measure of DNA purity and protein contamination. Additional methods of measuring DNA quality include gel electrophoresis. The accuracy of DNA quantitation by these methods can vary widely and can affect the quality of downstream genomic analyses. Genomic assays may be very sensitive to the quantity of DNA. A study by the US National Institute of Standards and Technology found a great deal of variability between various methods and among laboratories participating in a DNA quantitation study (33). Great care must be taken to assure that DNA concentration is accurately measured before use in any assay, especially PCR-based genomic applications that require precise quantities of DNA.

RNA is less stable than DNA and is more difficult to extract intact. However, special methods and reagents have been developed that allow for preservation of RNA in blood and other specimens, as noted in the discussion of tissue fixation.

Saliva or blood collected on treated paper cards is available, for example, from Whatman® for laboratory applications. Enough DNA can be obtained from a 2mm punch of a paper card for about 500 single nucleotide polymorphism (SNP) genotypes. The extraction of DNA from blood spot cards can be automated as noted above (27).

Table 3.2 summarizes source material for nucleic acid extraction, and some of the procedural and methodological issues encountered with each specimen type.

Aliquoting

Dividing specimens into smaller sample aliquots is usually necessary to preserve them in volumes useful for routine analyses. The aliquoting protocol should be designed only to store the number of aliquots necessary for the intended analyses, plus additional long-term archival samples that will be available for unforeseen uses. In developing an aliquoting protocol, the consequences of repeated thawing and refreezing cycles should be considered. Although many analytes, such as steroid hormones (discussed in Chapter 12), are stable, other analyses may be affected by one or more thaw-freeze cycles (2,3).

Automated systems for specimen processing

Automated systems have been developed for specimen processing, and several of these systems are useful in processing specimens for molecular epidemiology studies. Generally automation is most applicable to DNA extraction and specimen aliquoting.

For DNA extraction several automated systems are available, depending on the specimen type and volume. For blood specimens, and other blood fractions and suspensions of buccal cells up to 10 mL, the Gentra AutoPure is one of the preferred systems (32). The AutoPure has been validated for use with plasma, serum, buffy coat, buccal cell and other cell suspensions. For smaller samples, in the volume range of 50 uL to 1 mL, the Qiagen EZ-1 and M-48 systems are available (32). Other commercial and custom systems have been developed for specialized automated applications.

The other major biorepository activity that is amenable to automation is aliquoting. DNA in solution, as well as for example serum and plasma, must be stored in volumes suitable for downstream laboratory analyses. If standard collection and storage vessels are used, and a standard aliquoting protocol can be developed, then aliquoting can be automated. An example of a system for automated aliquoting is from TECAN (34).

Storage

Depending on the intended laboratory analyses, and other considerations, specimens and their aliquots may be stored under

Table 3.2. Common DNA sources and extraction issues

Specimen source	Collection method	Extraction method	DNA yield	Advantages	Challenges
Whole Blood	Evacuated tube with anticoagulant	Manual or automated	100s of micrograms	High yield, minimal processing	Refusal to participate
Blood -Buffy Coat	Processing of anti-coagulated blood	Manual or automated (with some processing)	100s of micrograms	High yield, minimal storage volume	Variable yield and quality of buffy coat cellular material
Blood - Plasma, Serum	Processing of blood, with or without anticoagulant	Manual or automated	Nanograms	Good use of samples collected for other purposes	Low yield
Saliva	Mouthwash, Oragene	Manual or automated	10-50 micrograms	High compliance rate	Bacterial DNA
Blood clot	Evacuated tube, no anticoagulant	Manual (special processing necessary)	Variable	Good use of 'extra' samples	Extractions expensive,
DNA fragmented	None	Source of DNA	Extraction difficult, costly		
Tissue – Fresh or Frozen	Surgery, autopsy	Manual	Variable	Most appropriate sample for some studies	DNA fragmented, RNA quality low
Paraffin Embedded Tissue	Tissue sections from surgery, autopsy	Manual	Variable	Easily stored	DNA fragmented, RNA quality low

a variety of conditions as shown in Table 3.3. Most common specimens such as plasma, serum or DNA may be securely stored in mechanical freezers at −80 °C. However, lymphocytes, or other cellular specimens, should be stored in the vapour phase of liquid nitrogen at −150 °C or lower, when long-term cellular viability is necessary. Other storage conditions that are optimal for the preservation of specimen stability should be considered, for example for endogenous hormones, as discussed in Chapter 12. Although generally not necessary in terms of sample and analyte stability, storage in the liquid phase of a liquid nitrogen tank at −196 °C is an excellent option. Although thorough cost analyses have not been performed, it is generally accepted that over the long term, liquid nitrogen freezers are less expensive to maintain than mechanical freezers, due to lower electrical requirements for the equipment and less need to cool the equipment space. In addition, liquid nitrogen freezers are less susceptible to mechanical failure and can withstand power outages for long periods with no temperature deviations.

In situations where freezer systems may not be available, a lower-cost option is collection of saliva or blood spots on filter cards and storage at room temperature. Below are some general storage considerations (1,2,4):

• Adequate back-up storage capacity for low temperature units should be maintained. The power supply must be connected to a back-up generator system that immediately provides power during an electrical outage. Standard operating procedures and techniques for rapidly transferring material to back-up units during such emergencies should be documented.

• Where liquid nitrogen freezers are used, an adequate supply of liquid nitrogen must be maintained. Vapour phase liquid nitrogen storage is preferred over liquid phase storage, where cross-contamination of specimens may occur. Cryovials must be capable of withstanding liquid nitrogen temperatures. Screw cap vials that will not leak are necessary. A good storage container in liquid phase nitrogen is the CryoBio Systems plastic straw (35).

• Alarm systems should be in place to monitor the temperature of mechanical freezers, or in the case of liquid nitrogen freezers, the liquid nitrogen level and temperature.

• Dry ice is frequently used as a refrigerant for shipping and emergency back-up for mechanical freezers.

• A system for maintenance and repair of storage equipment, support systems and facilities should be in place.

• All equipment should be validated before use, or following

Table 3.3. General specimen storage guidelines

Temperature in °C	Preservation method	Recommended for
+18 to +20	Room temperature	Slides, tissue blocks
0 to +4	Refrigerator	Processing fresh specimens
−0.5 to −27	Freezer	Short-term DNA stability
−27 to −40	Freezer	DNA stability
−40 to −80	Freezer	DNA/RNA stability
−80 to −130	Freezer	Recommended for urine, blood, blood fractions (plasma, serum etc)
−130 to −150	Liquid nitrogen vapour	Recommended for storage of tissues, preservation of cellular viability
−196	Liquid nitrogen liquid phase	Storage of living cells

Adapted from (2).

repairs that affect the instrument's accuracy or other capabilities.

- Labels for storage vessels must be capable of withstanding the required storage conditions, i.e. the label material must not deteriorate and printing must be readable or scanable after long-term storage.

Automated freezer systems

Automated freezer systems are available for convenient storage and retrieval of samples. Commercial automated freezer systems include a custom system built for ARUP Laboratories (36) and systems developed by REMP (37). Generally automated systems are developed for storage at −80 °C, although some liquid nitrogen systems are available.

Automation is most useful for studies and facilities that are focused on one or a few specimen types that will be collected in large numbers and processed and stored in a systematic way. If samples can be stored in microplates (for example, 96- or 384-well), then automated storage and retrieval systems should be considered. However, due to the wide variety of specimen types and processing methods used in molecular epidemiology studies, it is often difficult to justify expensive automated storage and retrieval systems.

Storage system maintenance

Freezers and other storage equipment should be validated and maintained according to the manufacturer's recommendations. In addition, the biorepository should develop additional protocols to assure that equipment functions properly (3,4). A preventive maintenance programme should be in place, with maintenance performed at regularly established intervals.

Special procedures should be developed to assure that freezers are properly validated, in terms of maintaining their optimal temperatures, during initial installation and at regular intervals. As noted in the ISBER best practices: "...any device that provides a readout, data, or has a meter movement, is considered an instrument, and requires calibration." (4).

Freezer temperature monitoring

Freezer temperatures must be continuously monitored to assure proper storage conditions for samples. For mechanical freezers (−20° to −80 °C), temperatures are displayed on each freezer. For small biorepositories, regular (twice daily) manual logging of temperatures may be adequate. However, larger biorepositories should have additional automated systems for remote monitoring of temperatures to efficiently respond to malfunctions (4).

Liquid nitrogen freezers require monitoring of both temperature and liquid nitrogen levels. Temperature monitoring is performed as for mechanical freezers. Liquid nitrogen levels should be recorded manually, on a regular basis, with a stick to assure that normal levels (usually 8–10 cm) are maintained. It is possible for liquid nitrogen freezers to overfill, which is detrimental to samples. Automated systems should be used that can detect and sound alarms for levels of liquid nitrogen that are either too low or too high.

Information management

Driven by advances in molecular technologies, including genomics and proteomics, information management is critical to the molecular epidemiology research enterprise (38). Collation and analysis of the data associated with the collected specimens that support biomedical research require robust interoperability to allow maximum usage of the collections (3,4). Information management and analysis tools across the spectrum of biomedical research are challenged to provide high performance, scalability and user-friendly interfaces. Also, as data sharing and collaboration between global investigators increases, secure interfaces for data transfer among institutions is paramount.

To manage the vast amounts of data in a variety of formats and environments, robust, flexible and extensible informatics systems are required (38). Too often, initial research plans do not include a well-thought-out approach to handle the results of an investigation. Deliberate planning for data management is far less costly and time consuming compared with ad hoc efforts that occur post-collection. A plan for the various disparate data types and formats should be included with special considerations for multisite collection protocols. A major part of the integrated informatics system for molecular epidemiology is support for biospecimen collection, shipping, processing, storage, inventory and retrieval processes.

Specimen tracking

Today, biospecimen collections are documented and tracked by many forms of data management tools, spanning from laboratory notebooks for a few hundred sample vials to real-time, multiuser software implementations, which support collections with millions of vials. Clearly, there is a need for automated information systems, but the level of informatics sophistication needed for a collection is limited by the availability of funding. In addition, it is incumbent upon the custodian of human biospecimens to adhere to ethical standards to protect and use the samples (3,4). Documentation of the study protocol number and the informed consent for the study subject should be easily linked back to the biospecimen to guarantee that the specific use of the specimens has been verified before distribution. Information technology software for specimen tracking features secure, validated environments that adhere to ethical practices. As more and more collections are shared among investigators all over the world, information on patient/subject consent, sample collection techniques and processes, and annotation of the sample must be easily retrievable, exportable, and traceable through time.

Biorepository information systems should support inventory functions by tracking all phases of sample acquisition, processing, handling, quality control and distribution from collection site (patient/subject) to utilization (researcher) (3,4). The inventory tracking should include significant events, such as thaws, loss, depletion and destruction of specimens, whether intentional or accidental. Restocking of returned, unused samples from the researcher, if allowed per protocol, must also be documented. Current guidelines for biorepository information systems recommend the use of electronic (linear or two-dimensional) labels or barcodes to document and associate a unique identification number to the samples. No identifying information about the specimen should be encoded as part of the identifier (3,4). The system should also be able to track any pre-existing, external biospecimen identifiers, such as vial type, and notations from hand-written vial labels. Standard operating procedures for the development of identifiers should be maintained with the system and updated to include all labelling paradigms used in the repository.

Bar code scanning technologies have become faster and more accurate in recent years. There are several varieties of software solutions to generate bar codes, from stand-alone programs to those embedded within other applications. Bar code printing options are recommended based on the volume of labels being printed. For high-volume label printing, thermal transfer or direct thermal bar code printers are the instruments of choice (39). When choosing a device, the conditions under which the scanner will be used, the frequency of use, the type of bar code (linear or 2-D), and the distance from which the scanning will be performed should be considered (39). Cost considerations may influence the selection of the bar code scanning technology employed by the biospecimen resource (4).

Biorepository information systems can report available space in the repository and assign and reserve space for incoming specimens. The location of a specimen should be tracked, but should not be used as part of the identifier naming convention, as locations of specimens may change in time.

The user interface of the system must provide tools to search the inventory based on various specimen characteristics, as well as support the requisition of samples to use in research studies. Query interfaces should be easy to navigate by experienced and inexperienced users. Standard and customizable queries are available in all commercially available systems, although ease of use varies. Many of the currently available biorepository inventory systems include web-based access portals to make the systems easier to deploy and navigate.

Informatics system security

The size and scale of the informatics needs of the molecular epidemiology group will determine if the biorepository information system should include the subjects' demographic and study annotation, or whether these data can be held within another database. Robust biorepository management systems provide controlled user access for system security (39). The system should include role-based security for all repository staff, study coordinators and scientists with a need to access the biospecimens inventory. If the study annotation is held within the same data system, security measures should be enacted to protect the subjects' personal health information (PHI) from disclosure to unauthorized users of the data. Regulations governing the protection of individual identifying information vary from country to country, so it is important to reference the guidelines for the specific locations of study and analysis in the study planning process (4).

If the biospecimen inventory is physically separated from the study annotation, these systems should be designed to interoperate and easily link the full study data, to maximize the ability to mine and analyse the data. If the links between systems are unstructured, the result can be an extraordinarily challenging, expensive and time-consuming effort to produce scientific findings from the study.

The system security architecture for information systems can be two- or three-tiered, depending on the separation of the user interface client (tier one) from the application server (tier two), and then optionally from the data storage (tier three). Three-tiered systems are more flexible and scalable for groups that have large concurrent user needs with heavy data load requirements (39).

Inventory control

"Inventory control starts with an understanding of the conditions under which errors occur and ends with error-resistant processes, intelligent use of technology, a well-trained and highly motivated workforce, and an ongoing process of continuous improvement" (40).

Inventory controls for biorepository management systems include the creation and storage of audit trails to track data history, data verification routines to assure data quality, and process tracking to assure the integrity of the sample data (39). The audit trails will include any changes/additions/deletions of data identifying the user that made the modifications. The system should have the ability to generate configurable reports and data files to provide the most complete information on the specimen. Inventory controls should include complete documentation of the information management system, updated standard operating procedures for the biorepository processes, security measures, and on-going training for those who access the data system (4).

Specimen annotation

The recognized value of molecular epidemiology studies is the collection of appropriate amounts of data, that when combined with the study subject's specimens and laboratory analyses, can be used to study the environmental and genetic causes of disease. It is important to be able to maintain tight integration of the demographic and clinical annotation of biospecimens, whether the data resides within the same data system or in physically distinct systems. Some study collections may include data-use agreements that require specimens to be de-identified before release from the biorepository for analysis. During the study planning process, the rules that govern specimen access are key factors when considering the use of pre-collected biospecimens in a study (41).

The goals of each molecular epidemiology study will determine the specific clinical annotation that should be maintained. Discussions are ongoing across the international biomedical community to provide guidelines for minimal clinical annotation for various study types (2–4,42), to facilitate data pooling of studies across common research areas. The cohort, case–control, and family-based consortia will benefit from the comparison and harmonization of their study data elements and definitions, and this will allow faster mining to detect underlying patterns across their combined data sets.

System interoperability

Epidemiologists are employing newer genomic technologies within studies, which have resulted in exponentially larger data sets. Legacy databases, however, that were functional with smaller data sets and do not communicate with other systems, may need to be replaced or modified. Large data management challenges require the integration of heterogeneous data and tools in a scalable, high-performance system. These systems can manage vast quantities of data, and provide tools for query and analysis in a secure collaborative environment. Efforts to provide interoperability across many institutions and tools based on grid computing are ongoing. Grid technology can be viewed as an extension or application of the internet framework to create a more generic resource-sharing context (43). Cloud computing is a newer delivery model for large, hosted datacentres which offers various computational and data access on an as-needed, "utility company" model over the internet. It typically involves the provision of dynamically scalable and often virtualized resources, thus avoiding the capital expenditure for purchase and maintenance of infrastructure at each bioresource centre location (44).

Whether the study data is housed within one central data system or in a federated, grid or cloud framework, interoperability is essential for the analysis of the data and the publishing of results. Efficient electronic data exchange or sharing between interoperable systems is based on shared common data element (CDE) definitions (45). When combining data from systems that do not share CDEs, mapping of the data to a shared set of elements is required. Often, these mapping efforts are labour-intensive and can result in a loss of information, as local CDEs are fit into exchangeable definitions. It is possible that small differences in the way questions and responses are worded or presented in epidemiology survey instruments can lead to significant (potentially unrecognized) differences in interpretation. The goal of developing CDEs is to enable semantic interoperability—the ability to represent information precisely enough that it may pass between humans and electronic representations precisely without requiring absolute central control of data systems or external human expertise (38). Semantic interoperability is a key component to speed data pooling efforts across epidemiologic studies to replicate and validate study findings.

Informatics at the US National Cancer Institute

Biomedical informatics systems are evolving as the technology becomes available to "personalize medicine" for each patient. Towards this end, the NCI Center for Bioinformatics has begun the development of the cancer Biomedical Informatics Grid or caBIGTM (45). This is a voluntary network or grid connecting individuals and institutions to enable the sharing of biomedical data and tools, with a goal of creating a World Wide Web of cancer research. The focus is to speed the delivery of innovative approaches for the prevention and treatment of cancer. The infrastructure and tools created by caBIGTM should have broad utility outside the cancer community. An integral part of the caBIGTM plan is the cancer data standards repository (caDSR) that will be used to build and maintain a repository of CDEs for standardization of terms and data storage practices. Tools for many aspects of biomedical research are becoming available on the caGrid.

Information management systems from the US National Cancer Institute and Centers for Disease Control and Prevention

Several organizations and companies around the US and the world are creating solutions to address the information management challenges presented by molecular epidemiology studies. Informatics activities at the National Cancer Institute, Office of Biorepository and Biospecimen Research (NCI, OBBR) have focused on creating recommendations for best practices associated with biorepository data systems, and the minimal clinical data set that should accompany all NCI-funded specimen collections (3). ISBER is focusing on the creation of best practices for biorepository management data systems. This will foster the development of worldwide standardized methods for collection, long-term storage, retrieval and distribution of specimens that will enable their future use (4).

There is a large variety of highly sophisticated, off-the-shelf, open source, and/or custom software applications for biorepository information management (e.g. http://www.isber.org/ims-products.html). Specific needs of the biorepository and the available funding will help guide the selection of the system employed. One highly-focused custom system serves the CASPIRTM (US Centers for Disease Control and Prevention-ATSDR (Agency for Toxic Substances and Disease Registry) Specimen Packaging, Inventory, and Repository) biorepository (46). CASPIR is a central facility to store biological and environmental biospecimens that the CDC-ATSDR began to develop in 1995. The mission of this biorepository is "… to provide storage for valuable, mostly human, biological samples that have been collected from CDC and ATSDR diagnostic studies, epidemiologic outbreaks, and research studies for possible future use." It has a storage capacity of more than six million biospecimens and is managed through customized data management software called the Archival Specimen Tracking and Retrieval Operations (ASTROTM) system.

The custom BioSpecimen Inventory System-II (BSI-II) was initially developed on contract for the NCI's Division of Cancer Epidemiology and Genetics to support their large biospecimen inventory from hundreds of molecular epidemiology studies (39). The BSI-II is flexible, extensible, and is currently storing data associated with more than 10 million specimens in storage across several contract repositories. The NCI's caBIGTM project has developed an open-source, modular caTissue Suite tool set for biospecimen inventory management, tracking, and annotation. This software permits users to enter and retrieve data concerning the collection, storage, quality assurance, and distribution of biospecimens (47).

Additional issues

Although the issues discussed in the previous sections are critical to the successful collection and preservation of biospecimens, there are other important considerations, concerning the control of specimen quality, as well as the safety and security of personnel and facilities, that are equally important.

Quality assurance and quality control

A Quality Management System (QMS) is an essential element of biospecimen management (3,4). The key to an effective QMS is the development and adherence to Standard Operating Procedures (SOPs). SOPs should guide the collection, processing, storage and equipment maintenance processes described in this chapter. Biorepository staff should be trained to adhere to all relevant quality systems and SOPs. Additional elements that are important for a

QMS include: appropriate security systems, computerized inventory and specimen quality tracking systems, and a facility disaster plan (4).

Several formal quality programs are appropriate for a specimen QMS, including current Good Manufacturing Practices (cGMP) and International Organization for Standardization (ISO) (48) certification. cGMP certification is used in the USA to maintain quality standards that are appropriate for Food and Drug Administration inspection of laboratories and biorepositories that process and store specimens for clinical applications. For research biorepositories, ISO certification, in general, is more appropriate for organizations that will be collaborating with international partners, and wish to assure that they are operating under a common set of recognized international standards. Both cGMP and ISO require extensive documentation of the sources, quality and performance of materials, equipment, and procedures.

Safety in the laboratory and biorepository

Laboratories and biorepositories should assume that all human biospecimens are potentially infective and biohazardous. A predictable, small percentage of biospecimens will pose a risk to the biorepository workers who process them. All biospecimens should be treated as biohazards (49). In addition to taking biosafety precautions, biorepositories should adhere to key principles of general laboratory safety.

In the United States, the Occupational Safety and Health Administration (OSHA) regulations (50) require that appropriate vaccinations be offered to all personnel who may be potentially exposed to human blood, body fluids and tissues, or other potentially infectious materials. Biorepository work practices should be based on universal precautions similar to those used in laboratories and clinical settings. Good general laboratory work practices are outlined by Grizzle and Fredenburgh (49). The CDC/NIH booklet *Biosafety in Microbiological and Biomedical Laboratories* (51) outlines general biosafety guidelines. All biorepositories that handle human biospecimens should operate under the OSHA (or similar) blood-borne pathogens standards and develop an exposure control plan.

In addition to biosafety, biorepositories should follow strict general safety regulations and procedures regarding chemical, electrical, fire, physical and radiological safety (3,4,50).

The use of liquid nitrogen poses unique safety problems that are not usually noted in laboratory safety documentation. With a liquid temperature of −196 °C, flesh freezes almost instantly if it comes in direct contact with the liquid. Both face and eye protections are required. Oxygen level sensors should always be employed, since oxygen deprivation is a serious hazard in the event of a liquid nitrogen leak.

Proper packaging and shipping

Depending on whether they are known to contain infectious agents, and the intended analyses, specimen shipments may be regulated as infectious substances or as diagnostic specimens. To properly classify the specimens to be included in a shipment, consult references provided in the ISBER Best Practices (4) and by the International Air Transport Association (52).

Specimens are often exposed to temperature fluctuations during transit. The required shipping temperature depends on the intended analyses (3,4). Packaging materials and equipment are available to preserve specimens under ambient, refrigerated and frozen conditions, including liquid nitrogen dry shippers that can preserve specimens frozen at or below −150 °C for up to several weeks (3,4). Devices are available to monitor temperature trends during shipment, either by recording temperatures precisely at certain time intervals, or by changing colour if a certain temperature is exceeded during shipment.

Security systems for biospecimen facilities

Due to the irreplaceable nature of many specimens collected for molecular epidemiology studies, it is critical to protect them from destruction due to electrical outages, equipment failures, and similar problems. The most important systems to have in place are electrical back-up generators and equipment alarms (4).

Generators should be available to provide electrical service to all freezers and any other critical equipment immediately upon the loss of general electric service to the facility. They should be maintained in good working order and started on a regular basis to assure that they are functioning properly (4). The appropriate fuel should be in adequate supply for up to three days of electrical outage during an emergency situation.

Alarm systems should be provided in specimen storage areas to alert the staff when a freezer or other equipment is malfunctioning.

They should be designed to automatically (for example, by cell phone or paging device) notify biorepository staff and other appropriate facilities maintenance personnel during non-working hours. Procedures should be in place to immediately respond to such equipment emergencies, and to either move the specimens to a functioning back-up freezer, or take other appropriate action to preserve their integrity.

In general, these measures should be part of a broader disaster response plan that is designed to protect personnel as well as specimens (4).

Future directions and challenges

Specimen management under adverse or low-resource conditions

In general, the methods, equipment and supplies described in this chapter are practices that should be adopted under the conditions found in developed countries. However, it is not always possible in some developing countries with fewer resources to have access to liquid nitrogen or mechanical ultra-low freezers, for example, or even electricity in some situations. These special circumstances need to be carefully considered before specimen collection is initiated. Some of the materials described in other sections of this chapter may be useful. For example, if extreme temperatures with little or no local refrigeration is an issue, then blood or saliva can be collected on filter cards and shipped and stored at ambient temperature. Blood can also be collected and shipped at ambient temperature using the PaxGene® collection tubes. Tissues can be fixed in formalin and embedded in paraffin blocks for low-cost storage and transport. If possible, given local conditions, "cool packs" and other supplies can be provided from a central coordinating centre and used to transport specimens at refrigerated temperatures. Note that any such procedures that deviate from documented best practices must be validated in a preliminary pilot study before full-scale adoption.

A specific example of working under such conditions is the Costa Rica HPV Vaccine Trial conducted by the US NCI in collaboration with the Fundacion Inciensa (53). Given the conditions under which specimens had to be collected in Costa Rica, the following factors were considered and accounted for:

• Bad road conditions increase shipment time and specimen shaking. Road conditions change from the dry to rainy season every year, and affect access to some communities.

• Liquid nitrogen may be hard to find in some countries, but not impossible. For example, Nicaragua does not produce any gases, but has hospitals and factories that require oxygen and liquid nitrogen, so oxygen is imported from Costa Rica.

• The cost of liquid nitrogen, equipment and reagents are generally higher in Central America than in developed countries, and in some cases, dealers for a particular country are regional. For example, a particular product produced in the USA may have to be acquired from a Mexican dealer that represents that product for Mexico and Central America.

• In some countries the power supply may be regulated and/or in poor condition. If possible, a back-up power supply should be provided or alternate storage methods should be considered.

• High temperature and humidity during the day are common conditions that may require special shipping containers, such as coolers with cold packs.

• Permits for importation and exportation of human-derived substances and repository operation permits must be obtained before starting operations. Policies and procedures will vary according to the country of origin and the destination.

• Laboratory equipment and reagents may have to be imported, which will require a variable time for customs and regulatory issues or the delivery time policy of the local or international dealer. Because of this, inventory management must be highly coordinated to account for potential delays.

Alternate collection technologies

In addition to dry-state collection and storage on treated cards, other special collection and storage systems have been developed that are beginning to be used in population-based studies. These approaches, mentioned briefly in other sections of this chapter, may gain more widespread use in studies that require the collection of large numbers of specimens that will need purified DNA as the analytical derivative. Some examples are:

• Oragene, developed by DNAGenotek (22). Oragene is a reagent used for saliva collection. The reagent saliva mixture is stable at room temperature. DNA can be readily extracted either by using the company's manual procedure or an automated procedure, such as the Gentra AutoPure. At least one large epidemiology study, performed by the Karolinska Institute, has had success with this protocol (23).

• GenVault (28) has developed a small cellulose element, based on the Whatman treated card, that can be used in a 384-well microplate format to store DNA and other samples in the dry-state. DNA can be eluted from the elements using a simple protocol, and adequate amounts (up to 200 nanograms) of DNA can be extracted from each element, making this a convenient system for long-term economical archiving of DNA.

Biospecimen ethical, legal and policy issues

The ethical, legal and policy aspects of biospecimen collection are as complex, if not more so, than the technical matters outlined in this chapter. The following are some of the issues that have not been fully resolved in the international community:

• *Informed consent.* Formats and details vary greatly among institutions. Policies for handling of biospecimens after withdrawal of consent are not well defined.

• *Ownership.* It is often unclear who 'owns' biospecimens once they have been donated for research. Court cases in the USA have ruled that the study participant does not have any ownership rights after donating a specimen for research. The NCI Best Practices (3) uses the term "custodianship" to reflect the need for a biospecimen resource to develop a plan for long-term care of biospecimens.

• *Specimen and data access.* Biospecimen resources should have clear rules for outside access to specimens and collected data (3).

• *Privacy protection.* Study participants need to be assured that their identity will be protected, with respect to use of specimens they have donated and any resulting data. Privacy regulations are in place for this purpose (3). Due to advances in genomic technologies, it is becoming increasingly difficult to guarantee the protection of an individual's identity.

• *Intellectual property.* Inventions and data arising from research using annotated biospecimens may have commercial value. Institutions should have clear intellectual property guidelines, and use material transfer agreements to assure that the sharing of specimens and data are well controlled. The final disposition of specimens and data should be understood before initiating a transfer.

In summary, the issues surrounding the use of biospecimens in research are complex and must be approached with attention to the many technical factors that may affect the quality of the specimens. In addition, it is important to recognize that the quality of biospecimens is enhanced by the collection and proper control of various types of data. Finally, many issues discussed in this chapter are subject to strict local and national policies and regulations concerning privacy and informed consent.

References

1. OECD Best Practice Guidelines for Biological Resource Centres. Available from URL: http://www.oecd.org/dataoecd/7/13/38777417.pdf.

2. World Health Organization, International Agency for Research on Cancer. Common minimal technical standards and protocols for biological resource centers dedicated to cancer research. Available from URL: http://www.iarc.fr/en/publications/pdfs-online/wrk/wrk2/Standards_ProtocolsBRC.pdf.

3. National Cancer Institute, Office of Biorepositories and Biospecimen Research. NCI best practices for biospecimen resources. Available from URL: http://biospecimens.cancer.gov/bestpractices.

4. Campbell JD, Skubitz APN, Somiari SB et al. (2008). International Society for Biological and Environmental Repositories (ISBER). 2008 Best practices for repositories: collection, storage, retrieval and distribution of biological materials for research. *Cell Preserv Technol*, 6:3–58.

5. Manolio TA (2008). Biorepositories–at the bleeding edge. *Int J Epidemiol*, 37:231–233. doi:10.1093/ije/dym282 PMID:18381397

6. Elliott P, Peakman TC; UK Biobank (2008). The UK Biobank sample handling and storage protocol for the collection, processing and archiving of human blood and urine. *Int J Epidemiol*, 37:234–244.doi:10.1093/ije/dym276 PMID:18381398

7. Peakman TC, Elliott P (2008). The UK Biobank sample handling and storage validation studies. *Int J Epidemiol*, 37 Suppl 1;i2–i6.doi:10.1093/ije/dyn019 PMID:18381389

8. Baird PM, Frome RJ (2005). Large-scale repository design. *Cell Preserv Technol*, 3:256–266 doi:10.1089/cpt.2005.3.256.

9. Vaught JB (2006). Blood collection, shipment, processing, and storage. *Cancer Epidemiol Biomarkers Prev*, 15:1582–1584. doi:10.1158/1055-9965.EPI-06-0630 PMID:16985016

10. BD Diagnostics. Available from URL: http://www.bd.com/vacutainer/pdfs/plus_plastic_tubes_wallchart_orderofdraw_VS5729.pdf.

11. Landi MT, Caporaso NE. Sample collection, processing, and storage. In: Toniolo P, Boffeta P, Shuker DEG et al., editors. Applications of biomarkers in cancer epidemiology. Lyon: IARC Scientific Publication; 1997. p. 223–236.

12. Rai AJ, Gelfand CA, Haywood BC et al. (2005). HUPO plasma proteome project specimen collection and handling: towards the standardization of parameters for plasma proteome samples. *Proteomics*, 5:3262–3277. doi:10.1002/pmic.200401245 PMID:16052621

13. Jackson C, Best N, Elliott P (2008). UK Biobank Pilot Study: stability of haematological and clinical chemistry analytes. *Int J Epidemiol*, 37 Suppl 1;i16–i22. doi:10.1093/ije/dym280 PMID:18381388

14. PreAnalytix Blood DNA system. Available from URL: http://www.preanalytix.com/DNA.asp.

15. Eiseman E, Bloom G, Brower J et al. Case studies of existing human tissue repositories. Santa Monica (CA): RAND Science and Technology; 2003.

16. Applied Biosystems RNAlater tissue collection: RNA stabilization solution. Available from URL: https://products.appliedbiosystems.com/ab/en/US/adirect/ab?cmd=catNavigate2&catID=603386.

17. Stanta G, Mucelli SP, Petrera F et al. (2006). A novel fixative improves opportunities of nucleic acids and proteomic analysis in human archive's tissues. *Diagn Mol Pathol*, 15:115–123.doi:10.1097/00019606-200606000-00009 PMID:16778593

18. Vincek V, Nassiri M, Nadji M, Morales AR (2003). A tissue fixative that protects macromolecules (DNA, RNA, and protein) and histomorphology in clinical samples. *Lab Invest*, 83:1427–1435.doi:10.1097/01.LAB.0000090154.55436.D1 PMID:14563944

19. Cox ML, Schray CL, Luster CN et al. (2006). Assessment of fixatives, fixation, and tissue processing on morphology and RNA integrity. *Exp Mol Pathol*, 80:183–191.doi:10.1016/j.yexmp.2005.10.002 PMID:16332367

20. Olert J, Wiedorn KH, Goldmann T et al. (2001). HOPE fixation: a novel fixing method and paraffin-embedding technique for human soft tissues. *Pathol Res Pract*, 197:823–826. doi:10.1078/0344-0338-00166 PMID:11795830

21. García-Closas M, Egan KM, Abruzzo J et al. (2001). Collection of genomic DNA from adults in epidemiological studies by buccal cytobrush and mouthwash. *Cancer Epidemiol Biomarkers Prev*, 10:687–696. PMID:11401920

22. DNAGenotek. Oragene DNA stabilization system. Available from URL: http://www.dnagenotek.com.

23. Rylander-Rudqvist T, Håkansson N, Tybring G, Wolk A (2006). Quality and quantity of saliva DNA obtained from the self-administered oragene method–a pilot study on the cohort of Swedish men. *Cancer Epidemiol Biomarkers Prev*, 15:1742–1745. doi:10.1158/1055-9965.EPI-05-0706 PMID:16985039

24. Mei JV, Alexander JR, Adam BW, Hannon WH (2001). Use of filter paper for the collection and analysis of human whole blood specimens. *J Nutr*, 131 Suppl;1631S–1636S. PMID:11340130

25. Sigurdson AJ, Ha M, Cosentino M et al. (2006). Long-term storage and recovery of buccal cell DNA from treated cards. *Cancer Epidemiol Biomarkers Prev*, 15:385–388. doi:10.1158/1055-9965.EPI-05-0662 PMID:16492933

26. Whatman. Filter paper DNA isolation. Available from URL: http://www.whatman.com/NucleicAcidandProteinSamplePreparation.aspx.

27. Tack LC, Thomas M, Reich K et al. (2005). Automated forensic DNA purification optimized for FTA card punches and identifier STR-based PCR analysis. *J Assoc Lab Autom*, 10:231–236 doi:10.1016/j.jala.2005.04.004.

28. GenVault. DNA isolation. Available from URL: http://www.genvault.com.

29. Hayes RB, Smith CO, Huang WY et al. (2002). Whole blood cryopreservation in epidemiological studies. *Cancer Epidemiol Biomarkers Prev*, 11:1496–1498. PMID:12433734

30. Biolife Solutions. Available from URL: http://www.biolifesolutions.com/.

31. Santella RM (2006). Approaches to DNA/RNA Extraction and whole genome amplification. *Cancer Epidemiol Biomarkers Prev*, 15:1585–1587.doi:10.1158/1055-9965.EPI-06-0631 PMID:16985017

32. Qiagen. DNA extraction. Available from URL: http://www1.qiagen.com/Products/DNA.aspx.

33. Kline MC, Duewer DL, Redman JW, Butler JM (2005). Results from the NIST 2004 DNA quantitation study. *J Forensic Sci*, 50:570–578. doi:10.1520/JFS2004357 PMID:15932088

34. TECAN. Liquid handling and robotics. Available from URL: http://www.tecan.com/page/content/index.asp?MenuID=1&ID=2&Menu=1&Item=21.1.

35. Cryo Biol System. Available from URL: http://www.cryobiosystem-imv.com/.

36. ARUP. Automated storage and retrieval system. Available from URL: http://www.aruplab.com/LaboratoryExpertise/AutomationInitiative/as_rs.jsp.

37. REMP. Automated freezer systems. Available from URL: http://www.remp.com/.

38. Henderson MK, Mohla C, Jacobs KB, Vaught J (2005). Challenges of scientific data management for large epidemiologic studies. *Cell Preserv Technol,* 3:49–53 doi:10.1089/cpt.2005.3.49.

39. Biological Specimen Inventory System BSI-II. Available from URL: http://bsi-ii.com/.

40. Piasecki DJ. Inventory accuracy: people, processes and technology. Pleasant Prairie (WI): Ops Publishing; 2003.

41. Policy for Sharing of Data Obtained in NIH Supported or Conducted Genome-Wide Association Studies (GWAS). Available from URL: http://grants.nih.gov/grants/guide/notice-files/not-od-07-088.html.

42. Lopez AD, Mathers CD, Ezzati M *et al.* (editors). Global burden of disease and risk factors. Oxford University Press and the World Bank; 2006.

43. Grid Computing. Definition from English Wikipedia. Available from URL: http://en.wikipedia.org/wiki/Open_Grid_Forum.

44. Bernstein D, Ludvigson E, Sankar K *et al.* Blueprint for the intercloud: protocols and formats for cloud computing interoperability. Fourth International Conference on Internet and Web Applications and Services. *IEEE Computer Soc* 2009:328–336.

45. Information about the NCI Cancer Bioinformatics Grid (caBIG). Available from URL: https://cabig.nci.nih.gov/.

46. Gunter EW (1997). Biological and environmental specimen banking at the Centers for Disease Control and Prevention. *Chemosphere,* 34:1945–1953.doi:10.1016/S0045-6535(97)00056-8 PMID:9159897

47. NCI. caBIG's caTissue Suite tools. Available from URL: https://cabig.nci.nih.gov/tools/catissuesuite.

48. International Organization for Standardization. Available from URL: http://www.iso.org/iso/home.htm.

49. Grizzle WE, Fredenburgh J (2001). Avoiding biohazards in medical, veterinary and research laboratories. *Biotech Histochem,* 76:183–206. PMID:11549131

50. Occupational Safety and Health Administration. Hazardous and toxic substances. Available from URL: http://www.osha.gov/SLTC/hazardoustoxicsubstances/standards.html.

51. U.S. Department of Health and Human Services. Centers for Disease Control and Prevention. National Institutes of Health. Biosafety in microbiological and biomedical laboratories. 4th ed. Washington (DC): U.S. Government Printing Office; 1999. Available from URL: http://www.cdc.gov/od/ohs/biosfty/bmbl4/bmbl4toc.htm.

52. International Air Transport Association. Infectious substances shipping guidelines. 7th ed. Montreal, Canada: International Air Transport Association; 2010. Available from URL: http://iatabooks.com/.

53. Cortés B, Schiffman M, Herrero R *et al.* (2010). Establishment and operation of a biorepository for molecular epidemiologic studies in Costa Rica. *Cancer Epidemiol Biomarkers Prev,* 19:916–922 doi:10.1158/1055-9965.EPI-10-0066 PMID:20332271

54. Vaught JB, Caboux E, Hainaut P (2010). International efforts to develop biospecimen best practices. *Cancer Epidemiol Biomarkers Prev,* 19:912–915.doi:10.1158/1055-9965.EPI-10-0058 PMID:20233852

UNIT 2.
BIOMARKERS: PRACTICAL ASPECTS

CHAPTER 4.

Physical/chemical/immunologic analytical methods

Jia-Sheng Wang and John D. Groopman

Summary

Biomarkers can be used to measure the presence of a wide variety of parent compounds and metabolites in body fluids and excreta, and serve as biomarkers of internal dose. Chemical-macromolecular adducts formed in blood and tissue or excreted in urine serve as biomarkers of exposure as well, and in many instances reflect both exposure and additional relevant biological processes. An assortment of analytical techniques have been developed to identify and measure parent compounds, metabolites, chemical-DNA and protein adducts. This chapter will discuss many analytical techniques that measure biomarkers in molecular epidemiologic studies, including biological, physical, chemical and immunological methods.

Introduction

Over the past 25 years, the development, validation and application of molecular biomarkers that reflect events from environmental exposure to the formation of clinical disease (e.g. cancer) has rapidly expanded our knowledge of the mechanisms of pathogenesis and provided opportunities for devising improved tools for disease treatment and prevention. Molecular epidemiology and its evolving paradigm refers to the use of biomarkers in epidemiological research (i.e. incorporation of molecular, cellular and other biochemical measurements into epidemiological studies of the etiology, prevention, and control of health risks faced by human populations)(1–4) (Figure 4.1). The application of validated biomarkers to traditionally descriptive epidemiological studies helps to: delineate the continuum of events between an exposure and resulting disease; identify smaller exposures to specific xenobiotics; indicate earlier events in the natural history of diseases and reduce misclassification of dependent and independent variables; enhance individual and group risk monitoring and assessments; and reveal toxicologic mechanisms by which an exposure and a disease are related (5,6). A unique feature of molecular epidemiologic studies is the interdisciplinary collaboration between population and field scientists and laboratory scientists from various disciplines, such as epidemiology, toxicology, molecular biology, genetics, immunology,

Figure 4.1. Molecular epidemiology paradigm

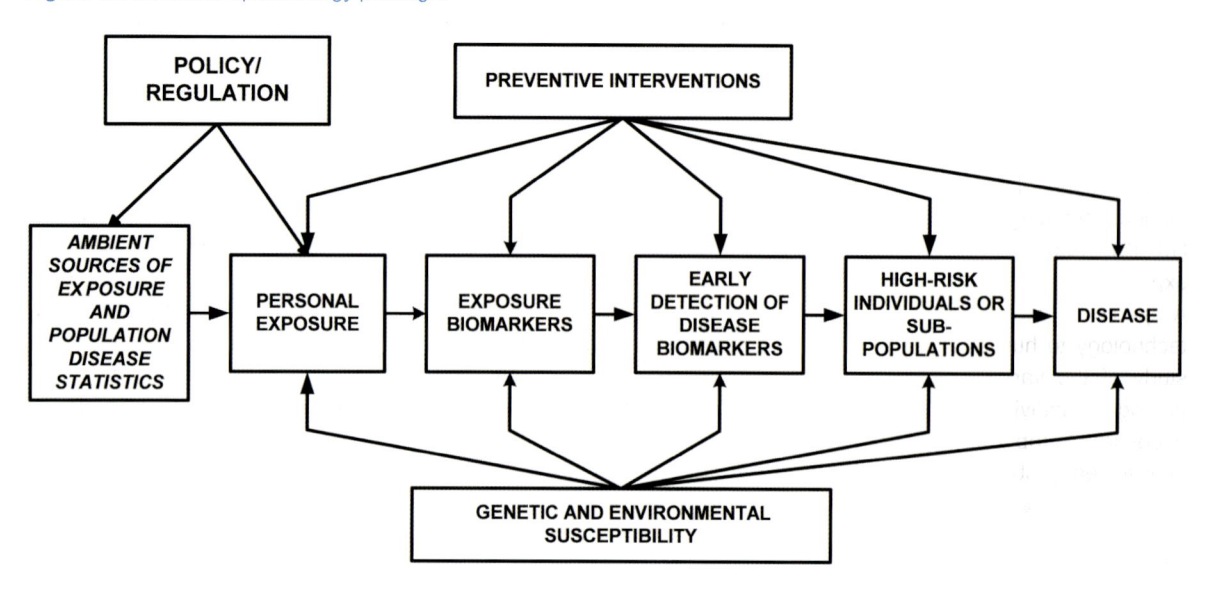

```
                POLICY/                    PREVENTIVE INTERVENTIONS
              REGULATION
                  │                    ┌────────┬────────┬────────┐
         ┌────────┼────────┐           │        │        │        │
         ▼        ▼        ▼           ▼        ▼        ▼        ▼
    ┌─────────┐  ┌──────────┐  ┌──────────┐  ┌──────────┐  ┌──────────┐  ┌──────────┐
    │ AMBIENT │  │          │  │          │  │  EARLY   │  │HIGH-RISK │  │          │
    │SOURCES OF│ │ PERSONAL │  │ EXPOSURE │  │DETECTION OF│ │INDIVIDUALS│ │ DISEASE  │
    │EXPOSURE │→ │ EXPOSURE │→ │BIOMARKERS│→ │ DISEASE  │→ │ OR SUB-  │→ │          │
    │  AND    │  │          │  │          │  │BIOMARKERS│  │POPULATIONS│ │          │
    │POPULATION│ │          │  │          │  │          │  │          │  │          │
    │ DISEASE │  │          │  │          │  │          │  │          │  │          │
    │STATISTICS│ │          │  │          │  │          │  │          │  │          │
    └─────────┘  └──────────┘  └──────────┘  └──────────┘  └──────────┘  └──────────┘
                       ▲             ▲             ▲            ▲             ▲
                       └─────────────┴─────────────┼────────────┴─────────────┘
                                                   │
                                    ┌──────────────────────────────┐
                                    │ GENETIC AND ENVIRONMENTAL    │
                                    │      SUSCEPTIBILITY          │
                                    └──────────────────────────────┘
```

MARKERS OF EXPOSURE ━━━━━━━━━━━━━━━━━━━━━━━━━━▶ MARKERS OF RISK

biochemistry, pathology and analytical chemistry. The analytic measurement of biomarkers is critical to molecular epidemiologic studies and requires special attention to the collection, handling and storage of biologic specimens, as well as development and validation of analytical methods (7).

Biomarker paradigm and validation strategies

As adapted from Perera and Weinstein (1) and the Committee on Biological Markers of the National Research Council (8), the development of disease as a result of exposure to an environmental agent or other toxicant is multistage: it starts with exposure and progresses to internal dose (e.g. deposited body dose), biologically effective dose (e.g. dose at the site of toxic action), early biological effect (e.g. at the subcellular level), altered structure or function (e.g. subclinical changes) and finally to clinical disease (Figure 4.1). Any step in this process may be modified by host-susceptibility factors including genetic traits and effect modifiers (e.g. diet or environmental exposures). Therefore, biomarkers are indicators of events for physiologic, cellular, subcellular and molecular alterations in the multistage development of specific diseases (9).

Molecular biomarkers are typically used as indicators of exposure, effect or susceptibility. A biomarker of exposure refers to measurement of the specific agent of interest, its metabolite(s), or its specific interactive products in a body compartment or fluid, which indicates the presence (and magnitude) of current and past exposure. A biomarker of effect indicates the presence (and magnitude) of a biological response to exposure to an environmental agent. Such a biomarker may be an endogenous component, a measure of the functional capacity of the system, or an altered state recognized as impairment or disease. A biomarker of susceptibility is an indicator or a metric of an inherent or acquired ability of an organism to respond to the challenge of exposure to a specific xenobiotic substance or other toxicant. Such a biomarker may be the unusual presence or absence of an endogenous component, or an abnormal functional response to an administered challenge (9). Molecular epidemiology and molecular dosimetry thus have great utility in addressing the relationships between exposure to environmental agents and development of clinical diseases, and in identifying those individuals at high risk for the diseases (2,6,10). Collectively, these data also help to inform the risk assessment process, where regulations can be tested against biological measurements of exposure to determine the efficacy of policies.

The development and application of molecular biomarkers

for environmental chemical agents should be based upon specific knowledge of their metabolism, interactive product formation and general mechanisms of action (11,12). Examples in the field are studies on the relationships between tobacco smoking and lung cancer (13–16) and between aflatoxin exposure and liver cancer (17,18). A specific application of biomarker technology to human cancer is the study of the variation in response among individuals following exposures to tobacco. For example, even in heavy tobacco smokers, less than 15% of these exposed people develop lung cancer (19); thus, intrinsic susceptibility factors must affect the time course of disease development and eventual outcome. The identification of those at highest risk for developing cancers should be facilitated by biomarker studies.

Extensive efforts have been made to identify these high-risk individuals using various genetic and metabolic susceptibility markers (e.g. measurement of polymorphism of genotype and phenotype of various enzymes involved in transformative metabolic reactions of certain known carcinogens and the DNA repair process)(20–23). Although this strategy has not yet proven to be broadly applicable to many other human diseases, progress is being made for many types of cancers (24).

The validation of any biomarker-effect link requires sequential or parallel experimental and human studies (12). Following the development of the hypothesis of an exposure-disease linkage, there is the need to develop the analytical methodology necessary to measure these biological markers in human and experimental samples. Conceptually, as shown in Figure 4.2, an appropriate animal model is used to determine the associative or causal role of the biomarker in the disease or effect pathway, and to establish relations between dose and response. The putative biomarker can then be validated in pilot human studies, where sensitivity, specificity, accuracy and reliability parameters can be established. Data obtained in these studies can then be used to assess intraindividual or interindividual variability, background levels, relationship of the biomarker to external dose or to disease status, as well as feasibility for use in larger population-based studies. It is important to establish a connection between the biomarker and exposure, effect or susceptibility. To fully interpret the information that

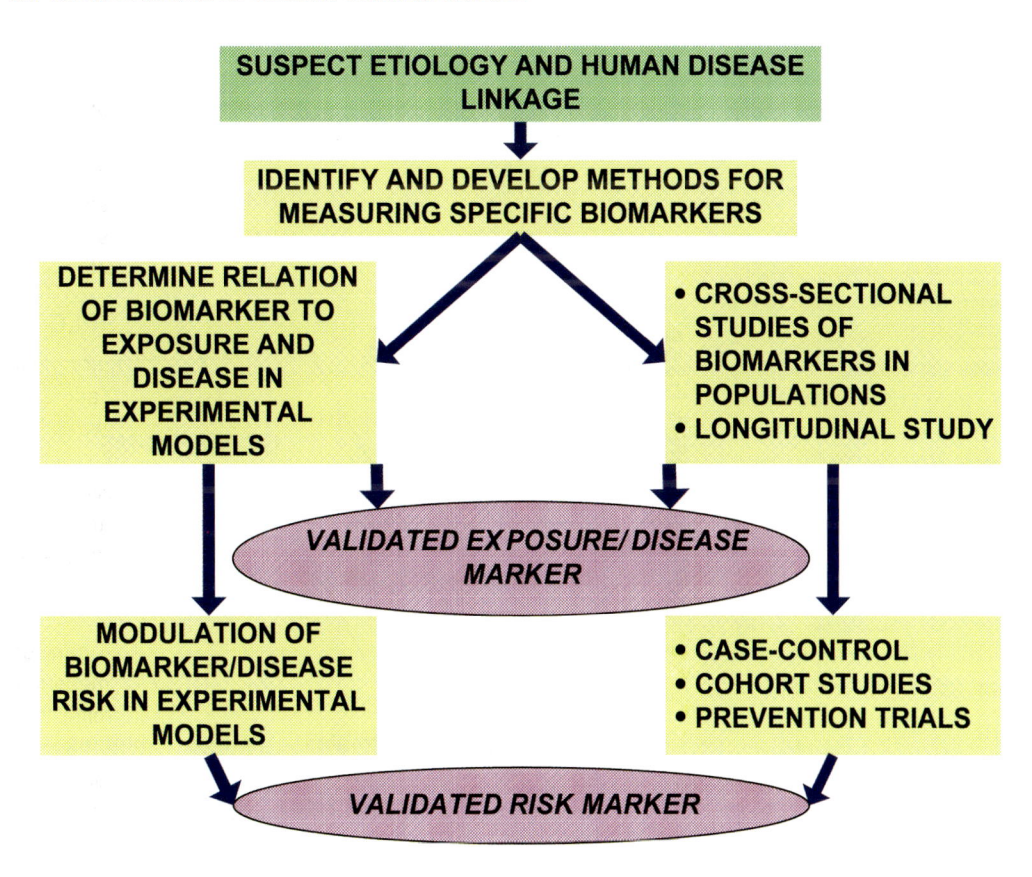

Figure 4.2. Validation scheme for molecular biomarker research

the biomarker provides, prospective epidemiological studies may be necessary to demonstrate the role that the biomarker plays in the overall pathogenesis of the disease or effect. To date, few biomarkers have been rigorously validated using this entire process.

Techniques and strategies for measuring parent compounds and metabolites

Many analytical techniques, including biological, physical, chemical, and immunological methods, have been developed and standardized by regulating agencies (e.g. US Environmental Protection Agency) for biohazard identification and risk assessment in various exogenous settings, such as environmental and occupational arenas. For establishment of a quantitative relationship with exposure, analytical methods have been extended to measure these parent compounds in biological samples; they could serve very well as a biomarker of exposure. A good example is the measurement of various heavy metals (e.g. lead, arsenic, cadmium and mercury) in human biospecimens, such as urine, blood, hair and tissues (25). In general, most heavy metal measurements have been in environmental samples and technologies including atomic absorption (AA) (26), inductively coupled plasma-optical emission spectrometer (ICP-OES) (27) and inductively coupled plasma-mass spectrometer (ICP-MS)(28). In addition to the measurement of environmental sources of heavy metals, methods such as X-ray fluorescence (XRF) permit the assessment of body burden of lead through its deposition in bone (29). Of importance to molecular epidemiology is the need to speciate the heavy metal compound, since toxicity can vary by both charge state (e.g. tri- and penta-valent state of arsenic) (30) and methylation (e.g. methyl mercury) (31). Measurement of parent organic compounds in biological samples, although still in practice, is less favoured because most organic toxic/carcinogenic compounds undergo metabolism and exert their toxicologic/carcinogenic effects through metabolic activation (22). Collectively, the problem faced in these investigations is the relationship between exposure and dose. Unlike most organic compounds, metals' measurements often reflect both exposure and dose. Figure 4.3 illustrates a nomograph relating the increase in analytical sensitivity by various instrument methods of the past 10–15 years. As these technologies have advanced, the number of non-detects in human epidemiological studies has dramatically decreased.

Incorporation of biomarkers in molecular epidemiology can provide critical information on interindividual variation in dose–response of environmental agents, the central focus in risk assessment. While biomarkers have been most widely applied in the area of clinical pharmacology, they are increasingly being used to document interindividual variation in the context of the much lower exposures found in the environment (32,33). There are many examples of variation in the pharmacologic action of drugs in people; in general, the response varies about 10-fold in the general population for most pharmaceutics. At issue with the extrapolation of these findings to environmentally occurring toxicants is the recognition that a limited variance in response is selected for during drug development, whereas in an environmental setting, there may be a much wider range of toxicologic outcomes. This is further complicated by the age, gender and health status of the general population when compared to people being treated

Figure 4.3. Nomograph for analytical methods

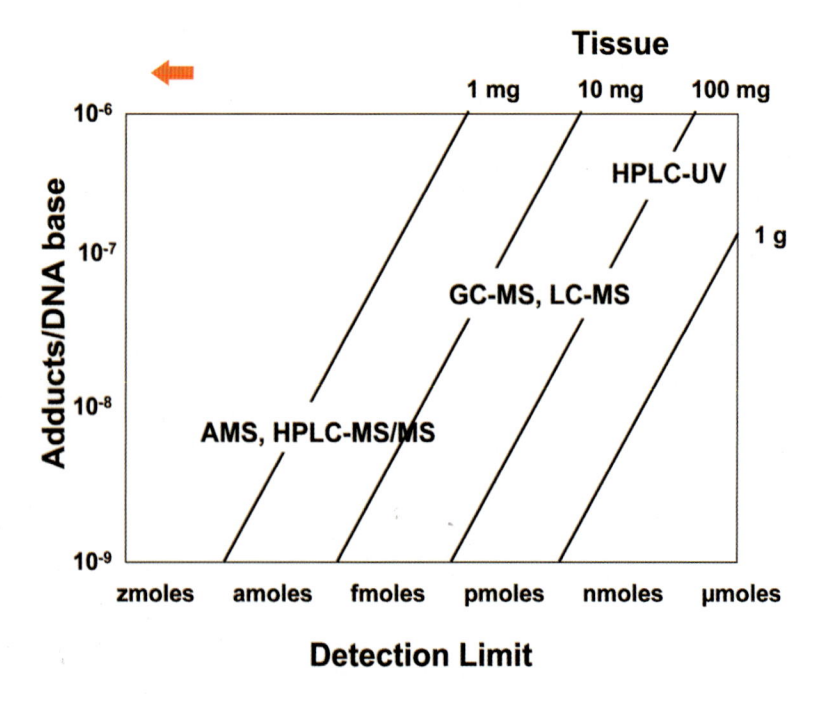

for a clinically diagnosed disease. Further, exposures to drugs tend to be of shorter duration then the lifetime exposures to environmental compounds that can have wide day-to-day variations in dose. While the ability of environmental exposures to reach the level where K_m of an enzyme becomes limiting is debatable, the effective concentration of an agent might well exceed this level at the cellular site of action (34).

A very significant consideration in the design of the molecular epidemiologic investigation is the balance between the analytical sensitivity of the chosen method and the operational sample throughput per day. Our infrastructure capacity to collect, annotate and store large numbers of samples far outpaces the number of samples that can be analysed. Most environmental and occupational toxicants of concern exist at trace levels in human samples. Thus, while the biological impact may be substantial, both the sample size available and analytical technology employed conspire to reduce the number of tests that can be run in a given day. Further, as analytical sensitivity increases, the contribution of noise to an analysis also increases, necessitating more extensive clean-up methods to maintain an appropriate signal-to-noise ratio. For most quantitative analyses, reliable measurements can be made only with signal-to-noise ratios that exceed 3:1. Unfortunately, judgement about the amount of material that will be used for an analysis results in many samples being at this borderline level of detection. Since the background from a urine (or other biological) sample can be very heterogeneous from person-to-person, many pilot studies have inadvertently generated false-positives when they do not employ multiple confirmatory analytical techniques and other quality control methods. Each sample matrix poses varied challenges for clean-up. Thus, as the complexity of sample clean-up increases, the effective number of samples that can be analysed in a given day diminishes. The largest automated, robotically driven clean-up strategy, The National Report on Human Exposure to Environmental Chemicals (http://www.cdc.gov/exposurereport/), is based upon the NHANES survey samples. This repository has been used to explore many environmental exposures to low molecular weight chemicals (35,36). However, most individual laboratories lack this type of infrastructure for large-scale sample preparation for chemical analysis; hence, these operational considerations impinge upon the size and scope of the molecular epidemiology study.

A variety of metabolites of toxicants/carcinogens found in body fluids and excreta (e.g. blood, urine, feces, hair and milk) have the potential for use as biomarkers of internal dose. These measures could provide information about the actual concentration of toxicants/carcinogens that have been absorbed and distributed in the body. Measurement of these metabolites has been incorporated into several human epidemiologic studies. For example, excretion of aflatoxin M_1 (AFM$_1$), one of the major metabolites of aflatoxin B_1 (AFB$_1$), has been used as a biomarker for the evaluation of human exposure to aflatoxin and was found to be associated with the risk of liver cancer (37,38). Reflecting its complex pharmacokinetics, this metabolite has also been measured in both human urine and milk samples (39). Specific metabolites of one of the tobacco-specific nitrosamines, 4-(methylnitrosamino)-1-(3-pyridyl)-1-butanone (NNK), a potent chemical carcinogen, have been detected and quantified in the urine of smokers; these metabolites were not found in the urine of non-smokers (40). Intraindividual and interindividual variations in these metabolites of NNK in smokers' urine were noted and might prove to be important in disease risk (41). Two metabolites of NNK, 4-(methylnitrosamino)-1-(3-pyridyl)-1-butanol (NNAL) and NNAL-β-O-D-glucosiduronic acid (NNAL-Gluc), excreted in the urine of smokeless tobacco users, were found to be associated with the presence of oral leukoplakia (42). Other examples of internal dose markers include the measurement of blood and serum levels of DDE (1,1-dichloro-2,2-bis(p-chlorophenyl)-ethylene), the major metabolite of DDT (2,2-bis(p-chlorophenyl)-1,1,1-trichloroethane), which have been used as biomarkers in breast cancer studies in women (43,44).

An example of the use of agent-specific biomarkers to identify etiologic factors in human cancer is the study of aristolochic acid (45,46). This compound has been associated with Balken endemic nephropathy (BEN), a chronic renal tubulointerstitial disease that often is accompanied by upper urinary tract urothelial cancer. Using ^{32}P-postlabelling/PAGE and authentic standards, both adenine and guanine aristolactam DNA adducts were detected in the renal cortex of patients with BEN, but not in patients with other chronic renal diseases. In addition, urothelial cancer tissue was obtained from residents of endemic villages with upper urinary tract malignancies. The AmpliChip p53 microarray was then used to sequence exons 2–11 of the *p53* gene where 19 base substitutions were identified. Mutations at A:T pairs accounted for 89% of all *p53* mutations, with 78% of these being A:T→T:A transversions. It was concluded that DNA adducts derived from aristolochic acid are present in renal tissues of patients with documented BEN. These adducts can be detected in transitional cell cancers, and A:T→T:A transversions dominate the *p53* mutational spectrum in the upper urinary tract malignancies found in this population.

The process of toxicant/carcinogen metabolism produces a variety of reactive electrophilic intermediates that constitute biologically effective forms of the ultimate chemicals, such as aflatoxin B_1-8,9-epoxide, benzo(a)pyrene-7,8-diol-9,10-epoxide (BPDE), and N-acetoxy-2-acetylaminofluorene (47–49). The actual concentration of the products formed by these ultimate forms of toxicants/carcinogens in the human body or target tissues should serve as biomarkers for biologically effective dose. Great efforts have been made over many years to develop methods to identify and detect these active metabolic products. These ultimate toxicants/carcinogens can be directly monitored in various *in vitro* models and *in vivo* animal systems by a variety of sensitive analytical techniques. These approaches are often difficult to apply in human populations simply because of the extremely short half-lives of these ultimate toxicants/carcinogens, and the high background of interfering substances. Alternative methods for measuring the formation of DNA and protein adducts in human blood and tissues include measuring their further metabolites, such as diols, conjugates (including glucuronide and mercapturic acids) or nucleic acid base adducts in human urine.

As discussed previously, clean-up methods are required to lower the noise and enhance the signal for the detection of any of the chemical-specific biomarkers in biological samples. Many of the studies to measure these biomarkers employ single or multiple chromatographic step(s) to facilitate biomarker detection. In a preparative mode, this chromatography usually consists of high-capacity chromatographic columns using reverse phase, normal phase and ion-exchange resins. These columns are generally gravity or low-pressure devices and provide a crude first stage enrichment method. Analytical chromatography is used as a low-capacity, but highly-selective, technology to separate many of the compounds in a complex mixture. Liquid chromatography is the most common mode for separation, but many compounds, if they are intrinsically or derivatisable to a volatile agent, can be used in gas chromatography (GC). GC has a much higher capacity to resolve different chemical species. All chromatography methods are coupled with a form of spectroscopy for selective detection. For example, GC and high performance liquid chromatography (HPLC) coupled online with UV, fluorescence and electron-capture detectors and mass spectrometry (MS), are necessary techniques in these studies. In general, modern analytical instrumentation can lead to a limit of detection of chemical biomarkers in the femptomole (fmol) to low picomole (pmol) range. As a caution, many studies report sensitivity and limits of detection using standards; however, the operational limits of detection are between 10 to 100-fold higher when measuring real biological samples.

Immunoassays, such as radioimmunoassay (RIA), enzyme-linked immunosorbent assay (ELISA) and immunoaffinity chromatography (IAC), are also used in biomarker analysis (50,51) for both preparative methods and analytical measurement. Further, several investigations have employed hybrid methods using HPLC separation with the collection of fractions followed by analysis using immunoassays (52). An obvious advantage of clean-up using chromatography before an immunoassay is the removal of potential cross-reactive materials that contribute to a false-positive result. It is also important to note that any clean-up before an analytical measurement lowers the number of samples per day and potentially introduces other artefacts, such as cross-contamination of samples and stability of the biomarker.

Internal standard development is an area of considerable importance that has received far less attention than it should. All quantitative measurements require the use of an internal standard to account for sample-to-sample recovery variations. In the case of mass spectrometry, internal standards generally employ an isotopically labelled material that is physicochemically identical to the chemical that is being measured. Obtaining such isotopically labelled materials does require chemical synthesis (if they are not commercially available), which has impeded the application of internal standards in many studies. In the case of immunoassays internal standards pose a different challenge, since the addition of an internal standard that is recognized by an antibody results in a positive value contribution. The dynamic range is usually less than 100 in immunoassays; therefore great care must be taken to spike a sample with an internal standard to obtain a valid result (53). In contrast, for example, most chromatographic methods result in dynamic ranges of analyses that can be over a 10 000-fold range of levels.

Techniques for measuring DNA adducts

The metabolically activated ultimate form of carcinogens can covalently interact with cellular DNA, which is a critical step in the process of carcinogenesis (23,54–56). Measurement of carcinogen-DNA adducts has an important role in human biomonitoring and molecular epidemiologic studies, as they are specific biomarkers that provide a way to measure human exposure to chemical carcinogens and provide information about a specific dose to a carcinogen target site (DNA). Moreover, it has been possible to establish a correlation between tumour incidence and exposure by measuring the level of these adducts (see (57) as an example).

Many different analytical techniques have been developed to identify and measure carcinogen-DNA adducts, including: immunoassays: ELISA, RIA, IAC, and immunohistochemical staining assay (IHC); radiometric postlabelling methods: ^{32}P-post-labelling; and various physicochemical methods: GC, HPLC, GC-MS, LC-MS, electrochemical detection (ECD), fluorescence and phosphorescence spectroscopy, or a combination of these methods (4,10,20,50,58–63). Capillary electrophoresis and other new separation techniques have improved the sensitivity and specificity of these methods. Nuclear magnetic resonance (NMR) spectrometry has also been used to determine stereospecificity and three-dimensional structure (64,65).

The ^{32}P-post-labelling assay, which radioactively labels adducts digested from sample DNA, has been widely applied because of its high sensitivity and the requirement for only microgram amounts of DNA. This assay has been especially useful for detection of adducts in single exposure experimental systems and as a means of elucidating the metabolic activation of previously uninvestigated potential carcinogens. ^{32}P-post-labelling can give an impression of total adduct burden, but it is rarely possible to quantify specific adducts accurately in human samples. Advances may lie in the use of better chemical standards, more advanced preparative techniques, and in connection with MS techniques (3,22). Carcinogen-DNA adduct detection by fluorescence has been applied to compounds that lead to either highly fluorescent products or adducts that can subsequently be derived to highly fluorescent chemical species. Physicochemical methods, including MS, offer the advantage of high chemical specificity. Major improvements in sensitivity have allowed the measurement of increasingly smaller amounts of adducted species in biological matrices. The sensitivities of individual methods vary and often depend on the amount of DNA that can be analysed. Detection limits for quantitative assays are typically in the range of one adduct in 10^7 or 10^9 nucleotides. However, accelerator mass spectrometry (AMS), which is highly sophisticated and involved in use of low levels of ^{3}H- or ^{14}C-labelled compound, has a detection limit of one adduct in 10^{12} nucleotides (66,67). A recent application of this technology has been in the identification of the fate of a variety of alkylanilines in experimental models (68). In this investigation, the ^{14}C-labelled 2,6-dimethyl- (2,6-DMA), 3,5-dimethyl- (3,5-DMA), and 3-ethylaniline (3-EA) compounds, associated with human bladder cancer (69), were administered to C57BL/6 mice, which were subsequently sacrificed 2, 4, 8, 16 and 24 hours post-dosing. Bladder, colon, kidney, liver, lung and pancreas were harvested from each animal, and DNA was isolated from each tissue. Adducts were detectable in the bladder and liver DNA samples from every animal at every time point, at levels that ranged from three per 10^9 to 1.5 per 10^7 nucleotides. Adduct levels were highest in animals given 3,5-DMA and lowest in those given 3-EA. Taken together, the results strongly suggest that these three alkylanilines are metabolized *in vivo* to electrophilic intermediates that covalently bind to DNA, and that adducts are formed in the DNA of bladder, which is a putative target organ for these alkylanilines (68).

Many analytical techniques have been used to measure composite and specific DNA adducts in cellular DNA isolated from peripheral lymphocytes, bladder, breast, lung and colonic tissues, as well as excreted DNA adducts in urine (3,50,60). These techniques have also been applied in the clinical setting to examine carcinogen-macromolecular adducts of people undergoing chemotherapy with alkylating agents, in an attempt to associate adduct levels with clinical outcome (70,71). Recently, these methods have also been applied to human clinical trials to validate various intervention tools for the assessment of chemopreventive agents in modulating various intermediate biomarkers (17,72).

Measurement of DNA adducts for studying complex mixtures of carcinogen exposure

Many studies have used DNA adducts to assess potential sources of carcinogen exposure. One classic study has examined a spectrum of molecular biomarkers to assess human exposure to complex mixtures of environmental pollution in Poland

(73). Measurement of genotoxic damage in peripheral blood samples from residents of high-exposure regions indicated that environmental pollution is associated with significant increases in carcinogen-DNA adducts (polynuclear aromatic hydrocarbon (PAH)-DNA and other aromatic adducts), sister chromatic exchanges, chromosomal aberrations and frequency of increased *ras* oncogene expression. The presence of aromatic adducts on DNA was found to be significantly correlated with chromosomal mutation, providing a possible link between environmental exposure and genetic alterations relevant to disease.

Tobacco smoke, the primary cause of lung cancer, contains several types of known carcinogens. The most abundant of these are PAHs, arylamines and the tobacco-specific nitrosamines, including the lung-specific carcinogen NNK. These carcinogens are metabolically activated to reactive species which form specific DNA adducts. Smokers are usually found to have significantly elevated levels of aromatic and/or hydrophobic adducts as compared with non-smokers, and some studies found that DNA-adduct levels are linearly related to total smoking exposure (74). One investigation measured the level of bulky, hydrophobic DNA adducts in lung parenchyma of smokers and ex-smokers by the ^{32}P-postlabelling method. Smokers had five-fold higher levels of DNA adducts than did ex-smokers. A positive linear correlation between bulky adduct levels and *CYP1A1* (AHH) activity was found in smokers. A statistically significant correlation was determined comparing pulmonary microsomal AHH activity and the level of BPDE-DNA adducts ($r = 0.91$; $P < 0.01$) (71). Additionally, BPDE-DNA adducts have been detected in oral mucosa cells of smokers and non-smokers. Levels of DNA damage were elevated in each of 16 smokers compared to 16 age-, race-, and sex-matched non-smokers. There was about a three-fold range between smokers and non-smokers (75).

Measurement of carcinogen-DNA adducts in target tissues can offer useful information related to the mechanism of carcinogenesis; however, the limitation of availability of these specimens in humans is an impediment to extensive studies. An alternative method is to use the DNA isolated from peripheral white blood cells (WBC). One example is the detection of PAH-DNA adducts in specific subsets of WBC. It was observed that DNA combined from lymphocyte and monocyte fractions of smokers, exhibited detectable levels of DNA adducts with a mean of 4.38 ± 4.29 adducts/10^8 nucleotides, while the corresponding values were $1.35 \pm 0.78/10^8$ ($P < 0.001$) in non-smokers (76). The elevated levels of PAH-DNA adducts in DNA obtained from WBC of smokers compared to non-smokers suggested that only certain subsets of WBC are a valid, readily accessible source for monitoring genotoxicity from cigarette smoke.

The decline of PAH-DNA and 4-aminobiphenyl-haemoglobin (4-ABP-Hb) adducts in peripheral blood following smoking cessation in serial samples from 40 heavy smokers (≥ 1 pack/day for ≥ 1 year) was described (77). The substantial reduction (50–75%) of PAH-DNA and 4-ABP-Hb adduct levels after quitting indicates that they are reflective of smoking exposure, which is essential information in the validation of biomarkers (77). The estimated half-life of the PAH-DNA adducts in leukocytes was 9–13 weeks; for 4-ABP-Hb adducts, it was 7–9 weeks. Women had higher levels of 4-ABP-Hb adducts at baseline and after smoking cessation. These results show that PAH-DNA and 4-ABP-Hb adducts can be useful as intermediate biomarkers by verifying smoking cessation and possibly identifying persons who are at increased risk of cancer from exposure to cigarette smoke, due to high levels of carcinogen binding.

In other reports, anti-BPDE-DNA adducts were detected in four of seven colon mucosa samples, but not in any of 11 human pancreas samples from smokers and non-smokers. Adduct levels in human colon samples varied between 0.2–1.0 adducts/10^8 nucleotides (78). DNA adducts have also been detected in biopsy samples of human urinary bladder tissue. Total PAH-DNA adduct levels, and the average levels of several specific adducts, were significantly elevated in samples from current smokers, compared to never-smokers and ex-smokers who had abstained from smoking for at least five years (79). Putative aromatic amine adducts were detected, one of which displayed chromatographic behaviour identical to the predominant adduct induced by the human urinary bladder carcinogen, 4-ABP, which is present in cigarette smoke. Immunohistochemical quantitation of 4-ABP-DNA adducts and *p53* nuclear overexpression in T1 bladder cancer of smokers and non-smokers was described (80). Mean relative staining intensity for 4-ABP-DNA adducts was significantly higher in current smokers compared to non-smokers. There was a linear relationship between mean level of relative staining and number of cigarettes smoked, with lower levels in the 1–19 cig/day group, compared to the 20–40 and the >40 cig/day groups. Nuclear overexpression of *p53* was observed in 27 (59%) of the 45 stage T1 tumours analysed. Nuclear staining of *p53* was correlated with smoking status, cig/day, and 4-ABP-DNA adducts. In another study, 4-ABP-DNA adducts in 11 human lung and eight urinary bladder mucosa specimens were analysed by alkaline hydrolysis and negative chemical ionization GC-MS. Adduct levels were found to be 0.32–49.5 adducts/10^8 nucleotides in the lung and 0.32–3.94 adducts/10^8 nucleotides in the bladder samples (81).

Carcinogen-DNA adducts in human breast tissue samples have been reported (82). A total of 31 breast tissue samples, which included tumour and tumour-adjacent tissues from 15 women with breast cancer and normal tissue samples from four women undergoing breast reduction, were analysed. Among the breast cancer cases, the mean aromatic/hydrophobic-DNA adduct level assayed was 5.3 ± 2.4 adducts/10^8 nucleotides, compared to 2.3 ± 1.5/10^8 nucleotides from the non-cancer patients. Five of 15 tissues from the cases displayed a pattern of adducts associated with tobacco smoke exposure; all of these positive samples were from current smokers. Tissue samples from the eight non-smoking cases did not exhibit this pattern. This study indicated that biomarkers may be useful in investigating specific environmental exposures that could contribute to breast cancer. In another study, BPDE-DNA and other adducts were also found in the smooth muscle layer of atherosclerotic lesions in abdominal aorta specimens by various analytical methods (83).

Alkylating agents, such as N-nitroso compounds, are potential human carcinogens. Humans are known to be exposed to N-nitrosoamines from diet, workplace, cigarette smoke, and through endogenous formation. These compounds alkylate DNA, leading to formation of various types of DNA adducts, such as 7-alkyl-2'-deoxyguanosine (dG) (e.g. 7-methyl-dGp and 7-ethyl-dGp). Several investigations have focused on the levels of 7-methyl-dG adducts in human lung tissue, where higher levels have been found in smokers compared to non-smokers (84–87). Separately, 7-methyl-dG levels in lung tissues have been associated with cytochrome P4502D6 and 2E1 genetic polymorphisms (84). One study analysed N^7-alkylguanine adduct levels in DNA in a group of 46 patients with larynx tumours by the ^{32}P-postlabelling method. The average level of N^7-alkylguanines was 26.2/10^7 nucleotides in tumour cells, 22.7/10^7 in non-tumour cells, and 13.1/10^7 in blood leukocytes. Males and smokers had significantly higher levels of adducts than females and non-smokers (88). In another study, 7-alkyl-2'-deoxyguanosine adducts were measured in eight separate lung segments of 10 autopsy specimens (89). Levels of 7-methyl-dGp were detected in all eight samples (ranging from 0.3–11.5 adducts/10^7 dG; mean 2.5 ± 2.3). In all but five of the samples, 7-ethyl-dGp levels were detected (ranging from <0.1–7.1 adducts/10^7 dG; mean 1.6 ± 1.7). 7-methyl-dG levels were approximately 1.5-fold higher than 7-ethyl-dG, and were positively correlated with each other in most individuals. There was no consistent pattern of adduct distribution in the different lung lobar segments (90).

Measurement of DNA adducts for studying occupational carcinogen exposure

Occupational exposure to many chemical carcinogens has been well documented. Since the initial reports in aniline dye workers, many laboratories have applied technologies for measuring these carcinogens and their metabolites to confirm exposure in the workplace (56). The occurrence of DNA adducts in exfoliated urothelial cells of a worker exposed to the aromatic amine 4,4'-methylene-bis(2-chloroaniline) (MOCA), an agent that induces lung and liver tumours in rodents and urinary bladder tumours in dogs, was reported (91). ^{32}P-postlabelling analysis revealed the presence of a single, major DNA adduct that cochromatographed with the major N-hydroxy-MOCA-DNA adduct, N-(deoxyadenosin-8-yl)-4-amino-3-chlorobenzyl alcohol, formed *in vitro*. PAH-DNA adducts in WBC and 1-hydroxypyrene in urine were examined in a group of 105 workers from a primary aluminum plant with different PAH exposures (92). Exposure was measured by personal monitoring and ranged from 0.4–150 fg/m^3. High exposure to PAH in the work atmosphere was associated with increased concentration of 1-hydroxypyrene in the urine. PAH-DNA adducts were detected in 93% of the worker samples. Workers with a high PAH exposure had significantly higher adduct levels than did those with a low PAH exposure. A good correlation was found between PAH exposure and the average PAH-adduct values in blood. A statistically significant correlation was also found between the average adduct values and the concentration of 1-hydroxypyrene in the urine of smokers.

The lymphocyte bulky PAH-DNA adduct levels have been examined in workers exposed to ambient air pollution (93). A significantly higher adduct level was found in bus drivers working in central Copenhagen compared with those driving in suburban areas. The urban drivers had higher adduct levels than rural controls. There was no observation of significant influence on adduct level by potential confounders including smoking and diet, *GSTM1*, or *NAT2*. A separate study measured BPDE-DNA adducts in 39 coke oven workers (exposed to PAH) and 39 non-exposed controls (each group consisting of smokers and non-smokers) (94). Adducts were detected in 51% of workers and in 18% of controls. The mean level in workers (15.7×10^8 nucleotides) was 15 times higher than in non-exposed controls. Although large interindividual variations were noted, smoking workers had 3.5 times more adducts than non-smokers.

Measurement of DNA adducts in excretion for studying carcinogen exposure

In addition to monitoring carcinogen-DNA adducts *in situ* in DNA, the excised products of these adducts can be determined in urine samples. These urinary biomarkers have been especially amenable to comprehensive validation studies (95). One example is the examination of the dose-dependent excretion of urinary aflatoxin biomarkers in the rat following a single exposure to AFB_1 (96). The relationship between AFB_1 dose and the excretion of the major nucleic acid adduct, AFB_1-N^7-Guanine (AFB-N^7-Gua), over the initial 24-hour period following exposure, showed an excellent linear correlation between dose and excretion in urine. In contrast, other oxidative metabolites, such as aflatoxin P_1 (AFP_1), revealed no linear excretion characteristic.

One approach for the development and validation of aflatoxin adduct biomarkers has entailed parallel experiments in animal models with the systematic evaluation of these molecular biomarkers in humans. The urinary excretion of aflatoxin metabolites in an area of China with a high incidence of liver cancer was studied (97). Total 24-hour urine samples were collected and analysed by an IAC-HPLC analysis to determine individual aflatoxins in the urine samples. The aflatoxins most commonly detected were AFB-N^7-Gua, AFM_1, AFP_1 and AFB_1; however, only AFB-N^7-Gua and AFM_1 showed a dose-dependent relationship between aflatoxin intake and urinary levels, which indicates that these two metabolites might be useful biomarkers of exposure. Interestingly, these studies also demonstrated that the kinetics of formation and excretion of AFB-N^7-Gua in urine is almost identical in the F344 rat and humans, thereby enhancing the value of rodent studies for assessing risk to humans.

Modification of N-3 at the position of adenine is a major route of DNA adduct formation for many alkylating carcinogens. The resulting 3-alkyldeoxyadenosines are unstable, rapidly depurinating to give the corresponding 3-alkyladenines (3-alkAde) that are excreted in urine. These can then be quantitated by immunochemical and/or GC-MS methods. In a study of cancer patients receiving methylnitrosourea (MNU) as part of combination chemotherapy, 24-hour urine samples were collected. An analysis of urinary 3-MeAde in smokers showed increased excretion of this biomarker. Overall, a dose-dependent excretion of 3-MeAde was observed (98).

Measurement for endogenous DNA damage

Endogenous oxidative DNA damage may play an important role in the formation of chronic degenerative diseases, including cancer (99). Among the many oxidatively damaged DNA bases formed, 8-hydroxy-2'-deoxyguanosine, or 8-oxo-7, 8-dihydro-2'-deoxyguanosine (oxo8dG) is a lesion that can be sensitively measured. Several techniques have been developed and applied to determine this damage product in biofluids and tissue samples from animals and humans. These methods include HPLC-EC, GC/MS, immunoassay, fluorescence postlabelling, ^{3}H-postlabelling and ^{32}P-postlabelling (100). IA column methods have also been described for the analysis of oxidative damage products of nucleic acids excreted in urine (101). Quantitative analysis of these adducts in the urine of rats fed a nucleic acid-free diet suggests that 8-oxo-7, 8-dihydroguanine is the principal repair product from oxo8dG in DNA. In addition to these reports, excretion of oxidative DNA damage products in urine has also been correlated with dietary antioxidant consumption in humans (102,103). Thus, these markers may eventually be used to assess protection status as well as risk of disease in people.

Malonaldehyde (MA) is the major reactive aldehyde resulting from the peroxidation of polyunsaturated fatty acids (PUFA) constituents of biological membranes, and is a by-product of prostaglandin biosynthesis (104). MA has been proven to be carcinogenic in rats, mutagenic in several bacterial and mammalian mutation assays, and readily reacts with DNA to produce several adducts. Relatively high endogenous levels of MA-DNA adducts have been detected in healthy individuals (1–10 adducts/10^7 nucleotides). Thus, MA is considered an important endogenously produced genotoxic agent that may contribute to the development of some human cancers, particularly to the carcinogenic effects associated with high dietary fat intake. Recent reviews on this field can be found in (105,106).

The effect of dietary fatty acid composition on the endogenous formation of MA-DNA adducts was investigated in a group of 59 healthy men and women (107). They were initially fed a milk fat-based diet (rich in saturated fatty acids) for two weeks to induce a homogeneous dietary background. Following this period, the subjects were randomly divided into two subgroups: 30 people were given a sunflower oil-based (SO) diet (rich in polyunsaturated fatty acid), and the remaining 29 people were fed a low erucic acid rapeseed oil-based (RO) diet (rich in monounsaturated fatty acid) for 25 days. At the end of the study, the fatty acid composition of plasma lipids and the level of MA-DNA adduct in total WBC were determined. A higher concentration of PUFA in plasma triglycerides and higher levels of MA-DNA adducts were found in the subjects of the SO diet group as compared with those in the RO diet group. The average adduct level (7.4 ± 8.7 adducts/10^7 nucleotides) in the SO group was 3.6-fold higher than that in the RO group, although large interindividual variation was noted.

Malondialdehyde (MDA)-DNA adducts were analysed in surgical specimens of normal breast tissues of 51 breast cancer patients, while normal breast tissue samples from 28 non-cancer patients served as controls (108). Two previously characterized MDA-deoxyadenosine (dA) and one MDA-deoxyguanosine (dG) adducts were detected in all tissue samples examined. Normal breast tissues from cancer patients exhibited significantly higher levels than those found in non-cancer controls. Ten of the 51 cancer patients and one of the 28 controls were found to contain the MDA-DNA adducts at the level of >1/10^7 nucleotides. Age and body mass did not significantly influence the levels of these adducts. However, the presence of a previously detected BP-DNA adduct in the breast tissues was associated with higher levels of the MDA-dA adducts in cancer patients. Interestingly, the level of MA-dA adducts was significantly lower in smokers and ex-smokers compared to non-smokers. Tumor tissues (n = 11) also displayed significantly lower levels of MA adducts than their corresponding normal adjacent tissues. These results suggest that lipid peroxidation products can accumulate in human breast tissues and reach relatively high levels in the breast tissues of women with breast cancer.

Estrogen is also known to be a major risk factor in breast cancer, and its biological effects are mediated by both receptor and metabolism (109). Estrogen can form DNA adducts, which have been measured in human samples (110). Formation of the 4-hydroxyestradiol-*N7*-guanine (4-OHE2-*N7*-guanine) adduct from the reaction of estradiol-3,4-quinone with DNA and its detection *in vivo*, has been established. Therefore, the development and application of methods to measure estrogen-guanine adducts will, in the future, explore the biological implications of these compounds to determine their contribution to estrogen toxicology.

Techniques for measuring carcinogen-protein adducts

Formation of carcinogen-protein adducts is considered to be a valuable surrogate for DNA adducts, since many chemical carcinogens bind to both DNA and protein in blood with similar dose–response kinetics (3,111). Haemoglobin and serum albumin are the proteins of choice, although efforts have been made to validate histone and collagen adducts, because they are readily accessible, more abundant than DNA, and have known rates of turnover. The lifespan of haemoglobin is ~60 days in rodents and 120 days in humans, and the half-life of serum albumin in humans is 23 days. Because protein adducts are stable and are not removed by active repair processes, they constitute a much more precise dosimetry tool when compared with DNA adducts. Interaction of a carcinogen with a protein typically occurs by substitution at a nucleophilic amino acid. For alkylating agents, the most commonly substituted amino acid is cysteine, but modifications for other carcinogens have been reported at lysine, aspartate, glutamate, tryptophan, histidine and valine (3,112).

Formation of haemoglobin or serum albumin adducts has been reported in experimental animals and humans for many categories of carcinogens, including AFB$_1$, aromatic amines, B(a)P, benzene, dimethylnitrosamine, ethylene oxide, 2-amino-3-methylimidazo(4,5-f) quinoline, methylmethane sulfonate, NNK, propylene oxide, styrene, and workplace and medicinal (psoriasis) PAHs (3,50,113). Techniques for measuring carcinogen-protein adducts include immunoassays (ELISA, RIA, and IAC) and analytical chemical methods (GC, GC-MS, HPLC, LC-MS, and AMS). Several combinative methods, such as IAC-HPLC with fluorescent detection and isotope dilution MS, have been applied to measure protein adducts (114). Sensitivity of these methods typically can be within the pmol and fmol range. For detection of haemoglobin or albumin adducts in humans, samples must be enriched for adducts, or adducts must be removed from the protein, before analysis (3,60,111). This is accomplished by either chemical or enzymatic release of the adduct or carcinogen from the protein or digestion of the protein into peptides and amino acids. Solvent extraction or IAC purification may then be used for partial purification before undergoing analysis with GC-MS, HPLC, or LC-MS.

Measurement of haemoglobin adducts for studying carcinogen exposure

A wide variety of aromatic amines and PAHs have been found to bind at high levels to haemoglobin (115). A brief summary showing the range of the different chemical haemoglobin adducts that have been detected in human non-smokers is found in Table 4.1. One carcinogen-Hb adduct that has been well characterized is formed by the potent urinary bladder carcinogen 4-ABP, and has been reported in human blood specimens (113). It has been concluded that 4-ABP-Hb adduct is closely associated with three major risk factors for bladder cancer: cigarette smoking, the type of tobacco smoked and acetylator phenotype. The relation between exposure to environmental tobacco smoke (ETS) and levels of 4-ABP-Hb adducts in non-smoking pregnant women compared to adduct levels in those women who smoked during pregnancy has been reported (116). A questionnaire on smoking and exposure to ETS was administered to pregnant women. Samples of maternal blood and cord blood were collected during delivery and analysed for 4-ABP-Hb adducts by GC-MS. The mean adduct level in smokers was approximately nine-fold higher than that in non-smokers. Among non-smokers, the levels of 4-ABP-Hb adducts increased with increasing ETS level. This relationship between ETS exposure and 4-ABP-Hb adduct levels supports the concept that ETS is a probable hazard during pregnancy.

Metabolic polymorphism, both in *NAT* and in *CYP1A2*, is also expected to affect the formation of 4-ABP-DNA- and Hb-adducts. Levels of DNA adducts in bladder cells and 4-ABP-Hb adducts in 79 individuals, together with the acetylator phenotype and genotype, were determined (117). Among the slow acetylators, levels of 4-ABP-Hb adducts were significantly higher compared to those present in rapid acetylators. This study indicated that clearance of low-dose carcinogens is decreased in the slow acetylator phenotype. Since the highest levels of adducts were found in individuals with rapid N-oxidation (*CYP1A2*) and slow N-acetylation (*NAT2*) phenotype, determination of phenotypes and genotypes may provide a better prediction and assessment of human cancer risk.

It was found that mean 3- and 4-ABP-Hb adduct levels in 151 subjects were statistically significantly higher in cigarette smokers compared to non-smokers, and that the level increased with increasing number of cigarettes smoked per day (118). Again, slow acetylators consistently exhibited higher mean levels of ABP-Hb adducts compared to rapid acetylators. The mean level of 4-ABP-Hb adduct was higher in subjects possessing the *GSTM1*-null versus *GSTM1*-non-null genotype (46.5 versus 36.0 pg/g Hb; P = 0.037). In another study, a polymorphic distribution of the *CYP1A2* and *NAT2* phenotypes was examined in relation to ABP-Hb adduct formation in 97 healthy males (119). Rapid oxidizers and subjects with the combined slow acetylator-rapid oxidizer phenotype showed the highest ABP-Hb adduct levels at

Table 4.1. Haemoglobin adducts in human non-smokers

Compound	fmole/g Haemoglobin
HPB from NNK	29.3 ± 25.9
2-Aminonapthalene	40 ± 20
4-ethylaniline	99 ± 10
2,6-dimethylaniline	157 ± 50
4-Aminobiphenyl	166 ± 77
3,5-dimethylaniline	220 ± 20
o-Toluidine	320 ± 90
p-Toluidine	640 ± 370
m-Toluidine	6400 ± 1900
N-(2-carbamoylethyl)valine	19000 ± 12000
Aniline	41000 ± 22000
N-(2-Hydroxyethyl)valine	58000 ± 25000

Table compiled from (69,115,120).

a low smoking dose. However, in a subset of 45 available samples, no association was seen between the ABP-Hb adduct levels and *GSTM1* genotype.

A wide variety of aromatic amines and PAHs have been found to bind at high levels to haemoglobin (115). Tobacco-specific nitrosamine binding to haemoglobin from pyridyloxobutylation has been detected at 29.3 ± 25.9 fmole/g haemoglobin (54). 2-Aminonapthalene, 4-ethylaniline, 2,6-dimethylaniline, 4-Aminobiphenyl, 3,5-dimethylaniline, o-Toluidine, p-Toluidine, m-Toluidine, N-(2-carbamoylethyl)valine, Aniline, and N-(2-Hydroxyethyl)valine have been measured at 40 ± 20, 99 ± 10, 157 ± 50, 166 ± 77, 220 ± 20, 320 ± 90, 640 ± 370, 6400 ± 1900, 19 000 ± 12 000, 41 000 ± 22 000, and 58 000 ± 25 000 fmole/g haemoglobin, respectively (69,115,120). One of the carcinogen-Hb adducts that has been well characterized is formed by 4-ABP, the potent urinary bladder carcinogen; 4-ABP-Hb adducts in human blood specimens have been reported (113). The results indicate that the 4-ABP-Hb adduct is closely associated with three major risk factors for bladder cancer: cigarette smoking, the type of tobacco smoked, and acetylator phenotype.

The role of aromatic amines in the development of bladder cancer in non-smokers in Los Angeles, USA was explored in a population-based case–control study involving 298 case subjects with bladder cancer and 308 controls. To assess arylamine exposure, levels of arylamine-haemoglobin adducts of nine selected alkylanilines (2,3-dimethylaniline (2,3-DMA), 2,4-DMA, 2,5-DMA, 2,6-DMA, 3,4-DMA, 3,5-DMA, 2-ethylaniline (2-EA), 3-EA, and 4-EA) were measured in peripheral blood collected from study subjects. Levels of all arylamine-haemoglobin adducts, with the exception of 2,6-DMA, were higher in smokers than in non-smokers, and levels of all arylamine-haemoglobin adducts were higher in cases than in controls. Arylamine-haemoglobin adducts of 2,6-DMA, 3,5-DMA, and 3-EA were all independently statistically significantly (all $P < 0.001$) associated with bladder cancer risk after adjusting for cigarette smoking at the time of blood collection, lifetime smoking history and other potential risk factors. These adducts were also independently associated with bladder cancer risk when only non-smokers at the time of blood draw were considered (highest quartile versus lowest quartile: 2,6-DMA, relative risk (RR) of bladder cancer = 8.1, 95% confidence interval (CI) = 3.6–18.0; 3,5-DMA, RR = 2.7, 95% CI = 1.2–6.0; 3-EA, RR = 4.3, 95% CI = 1.6–11.6). Thus, diverse arylamine exposures are strongly associated with bladder cancer risk among non-smokers (69).

Measurement of albumin adducts for studying carcinogen exposure

In addition to carcinogen-Hb adducts, carcinogen-albumin adducts have also been investigated, particularly for AFB_1 exposure (113,121). There are four analytical techniques currently available for measuring AFB_1-albumin adducts in human blood: ELISA, RIA, IAC-HPLC with fluorescence detection, and isotope dilution MS (IDMS) (50,114). Using RIA, levels of aflatoxin-serum albumin adducts in serum samples from residents of Guangxi, China were monitored; a highly significant association between AFB_1-albumin adduct level and AFB_1 intake was found (122). Further, it was calculated that about 2% of the ingested AFB_1 became covalently bound to serum albumin, a value very similar to that observed when rats were administered AFB_1. When the data for AFB-N^7-Gua adduct excretion in urine and serum albumin were compared, a statistically significant relationship was seen with a correlation coefficient of 0.73 (123). Using ELISA, AFB_1-albumin adducts in human sera from several regions of the world were investigated (124). It was found that 12–100% of serum samples from children and adults of various African countries contained AFB_1-albumin adducts, with levels up to 350 pg AFB_1-lysine equivalent/mg albumin. In studies conducted in the Gambia, West Africa, a strong dose–response relationship between aflatoxin exposure and AFB_1-albumin adducts was also seen (125), similar to that previously reported in China (122). From a practical perspective pertinent to epidemiologic studies, the measurement of serum AFB_1-albumin adduct offers a rapid, facile approach that can be used to screen very large numbers of people (17).

Three methods for AFB_1-albumin adduct measurement were compared using serum samples from highly exposed residents of Qidong (n=88), Fushui (n=65), and Chongming Island (n=115), China, and The Gambia (n=29). Although the average levels of AFB_1-albumin adducts were similar among these regions, great individual variations were found, as evidenced by the detectable level ranging from 0.124–25.925 pmol aflatoxin/mg albumin in the 297 samples. High correlations across the methods (r = 0.80–0.90) were obtained by comparing samples with each of the different analytical methods. Moreover, some of these samples can be stored for over 10 years with insignificant losses of albumin adduct levels (126). The ELISA and IDMS methods were compared in measurement of 20 human serum samples collected in Guinea, West Africa; high correlation between these two methods was found (r = 0.856, $P < 0.0001$) (127). In an experimental study, the level of AFB_1-albumin adducts formed as a function of a single dose of AFB_1 in rodents was compared to data from humans exposed to AFB_1. This comparison yielded a value for the three rat strains (Fischer 344, Wistar, and Sprague-Dawley) of between 0.3–0.51 pg AFB_1-lysine/mg albumin/1 μg AFB_1/kg body weight and a value for the mouse (C57BL) of <0.025. The best estimate for people from the Gambia and southern China was 1.56 pg/mg albumin for the same exposure. These data suggest that humans exposed to AFB_1 form amounts of albumin adducts closer to those observed in AFB_1-sensitive species and 1–2 orders of magnitude higher levels than the AFB_1-resistant species (128).

Measurement of DNA and protein adducts for studying cancer risk

DNA and protein adducts not only serve as biomarkers for exposure, but as biomarkers for cancer risk. A nested case-control study initiated in 1986 in Shanghai, examined the relationship between biomarkers for aflatoxin and hepatitis B virus (HBV) and the development of liver cancer (129,130). In this study, over 18 000 urine samples were collected from healthy males between the ages of 45 and 64. In the subsequent seven years, 50 of these individuals developed liver cancer. The urine samples for cases were age-matched and residence-matched with controls and analysed for both aflatoxin biomarkers and HBsAg status. A highly significant increase in the RR of 3.5 was observed for those liver cancer cases where urinary aflatoxin biomarkers (AFB-N^7-Gua and other AFB_1 metabolites) were detected. The RR for people who tested positive for the HBsAg was about eight, but individuals with both urinary aflatoxin biomarkers and positive HBsAg status had a RR for developing liver cancer of 57. These results show, for the first time, a relationship between the presence of carcinogen-specific biomarkers and cancer risk. Moreover, these findings provided the first demonstration of a multiplicative interaction between these two major risk factors for liver cancer. Further, when individual aflatoxin metabolites were stratified for liver cancer outcome, the presence of the AFB-N^7-Gua in urine always resulted in a two- to three-fold elevation in risk of developing liver cancer (130).

A case–control study measured BPDE-DNA adducts in DNA samples from WBC of lung cancer patients and healthy controls. High levels of adducts were found in WBC from lung cancer patients, with a range of 65–533 adducts/10^8 nucleotides. In WBC-DNA samples from healthy controls (smokers, non-smokers), the presence of adducts was detected only in smokers, but at a lower level than in lung cancer patients (74). PAH-DNA adducts in peripheral leukocytes were investigated from 119 non-small cell lung cancer patients and 98 controls (131). Among them, 31 cases had adduct measurements in leukocytes, lung tumour and non-tumour specimens collected at surgery, and 34 had paired leukocyte and tumour specimens. After adjustment for age, gender, ethnicity, season and smoking, DNA adducts in leukocytes were significantly higher in cases than controls; the odds ratio was 7.7 (95% CI = 1.7–34; $P<0.01$). DNA adducts in leukocytes were increased significantly in smokers and ex-smokers compared to non-smokers among cases and controls (separately and combined) after adjusting for age, gender, ethnicity, and season.

Aflatoxin-albumin adducts have been used as biomarkers for liver cancer risk. A nested case-control study carried out in Taiwan, China, followed a cohort of 8068 men for three years (132,133). Twenty-seven cases of hepatocellular carcinoma (HCC) were identified and matched with 120 healthy controls. Serum samples were analysed for AFB_1-albumin adducts by ELISA. The proportion of subjects with a detectable serum AFB_1-albumin adduct level was higher for HCC cases (74%) than matched controls (66%), giving an odds ratio of 1.5. There was a statistically significant association between detectable level of AFB_1-albumin adduct and HCC risk among men younger than 52 years old, showing a multivariate-adjusted odds ratio of 5.3, although no association was observed between AFB_1-albumin adduct level and HBsAg carrier status. Another prospective, nested case–control study (134) was carried out in Qidong, China in 1991. Serum samples from 804 healthy HBsAg-positive individuals (728 male, 76 female) aged 30–65 were obtained and stored frozen. Between the years 1993–95, 38 individuals developed liver cancer. The serum samples for 34 of these cases were matched by age, gender, residence and time of sampling to 170 controls. Serum AFB_1-albumin adduct levels were determined by RIA. The RR for HCC among AFB_1-albumin positive individuals was 2.4 (95% CI = 1.2–4.7).

Summary and perspectives for the future

Over the past 25 years, the development and application of

molecular biomarkers reflecting events from exposure to formation of clinical disease has rapidly expanded our knowledge of the mechanisms of disease pathogenesis. These biomarkers will have increasing potential for early detection, treatment, interventions and prevention. Biomarkers derived from toxicant/carcinogen metabolism include a variety of parent compounds and metabolites in body fluids and excreta, which serve as biomarkers of internal dose. Carcinogen-macromolecular adducts, such as DNA and protein adducts formed in blood and tissue or excreted in urine, serve as biomarkers for: exposure to the complex mixture of occupational carcinogens; biologically effective dose or early biological effect, to measure the actual dose to the carcinogen target site; and for risk assessment between carcinogen exposure and eventual cancer formation.

Many different analytical techniques have been developed to identify and measure parent compounds, metabolites, carcinogen-DNA and -protein adducts. These include immunoassays (ELISA, RIA, IHC, and IA), radiometric postlabelling methods (^{32}P-post-labelling) and various physicochemical methods (GC, HPLC, GC-MS, LC-MS, ECD, fluorescence and phosphorescence spectroscopy, or a combination of these methods). Capillary electrophoresis and other new separation techniques have improved sensitivity and specificity of these strategies. NMR spectrometry has also been used to determine stereospecificity and three-dimensional structure. Molecular epidemiologic studies that employ carcinogen-macromolecular adduct measurements are likely to be widely applied in the future. They have the potential to generate hypotheses regarding underlying basic biologic mechanisms that subsequently can be tested in the laboratory. The complex nature of gene–environment and chemical–biological interactions will be better understood through the use of advanced techniques, such as rapidly developing metabolomics and proteomics, which include NMR-MS and matrix-assisted laser desorption/ionization (MALDI)-MS, and incorporation of validated biomarkers into large-scale studies. In the future, integration of data for these biomarkers, together with other environmental and host susceptibility factors in molecular epidemiologic studies of human cancer, will assist in the elucidation of human cancer risk.

The molecular epidemiology investigations of aflatoxins probably represent one of the most extensive data sets in the field, and may serve as a template for future studies of other environmental agents. The development of aflatoxin biomarkers has been based upon the knowledge of their biochemistry and toxicology gleaned from both experimental and human studies. These biomarkers have subsequently been used in experimental models to provide data on the modulation of these markers under different situations of disease risk. This systematic approach provides encouragement for preventive interventions and should serve as a template for the development, validation and application of other biomarkers to cancer or other chronic diseases.

Acknowledgements

This work was supported in part by grants P01 ES006052 and P30 ES003819 from the US Public Health Service.

References

1. Perera FP, Weinstein IB (1982). Molecular epidemiology and carcinogen-DNA adduct detection: new approaches to studies of human cancer causation. *J Chronic Dis*, 35:581–600.doi:10.1016/0021-9681(82)90078-9 PMID:6282919

2. Hulka BS (1991). ASPO Distinguished Achievement Award Lecture. Epidemiological studies using biological markers: issues for epidemiologists. *Cancer Epidemiol Biomarkers Prev*, 1:13–19. PMID:1845163

3. Poirier MC, Santella RM, Weston A (2000). Carcinogen macromolecular adducts and their measurement. *Carcinogenesis*, 21:353–359.doi:10.1093/carcin/21.3.353 PMID:10688855

4. Wogan GN (1992). Molecular epidemiology in cancer risk assessment and prevention: recent progress and avenues for future research. *Environ Health Perspect*, 98:167–178. PMID:1486846

5. Schulte PA. A conceptual and historical framework for molecular epidemiology. In: Schulte PA, Perera FP, editors. Molecular epidemiology: principles and practices. San Diego (CA): Academic Press Inc; 1993. p. 3–44.

6. Perera FP (2000). Molecular epidemiology: on the path to prevention? *J Natl Cancer Inst*, 92:602–612.doi:10.1093/jnci/92.8.602 PMID:10772677

7. Hattis D, Silver K. Use of biomarkers in risk assessment. In: Schulte PA, Perera FP, editors. Molecular epidemiology: principles and practices. San Diego (CA): Academic Press Inc; 1993. p. 251–273.

8. Committee on Biological Markers of the National Research Council (1987). Biological markers in environmental health research. *Environ Health Perspect*, 74:3–9. PMID:3691432

9. Wang JS, Links JM, Groopman JD. Molecular epidemiology and biomarkers. In: Choy WN, editor. Genetic toxicology and cancer risk assessment. New York (NY): Marcel Dekker; 2001. p. 269–296.

10. Wogan GN (1989). Markers of exposure to carcinogens. *Environ Health Perspect*, 81:9–17. PMID:2667991

11. Groopman JD, Kensler TW (1993). Molecular biomarkers for human chemical carcinogen exposures. *Chem Res Toxicol*, 6:764–770.doi:10.1021/tx00036a004 PMID:8117914

12. Groopman JD, Kensler TW (1999). The light at the end of the tunnel for chemical-specific biomarkers: daylight or headlight? *Carcinogenesis*, 20:1–11.doi:10.1093/carcin/20.1.1 PMID:9934843

13. Hecht SS (1999). Tobacco smoke carcinogens and lung cancer. *J Natl Cancer Inst*, 91:1194–1210.doi:10.1093/jnci/91.14.1194 PMID:10413421

14. Hecht SS, Carmella SG, Foiles PG, Murphy SE (1994). Biomarkers for human uptake and metabolic activation of tobacco-specific nitrosamines. *Cancer Res*, 54 Suppl;1912s–1917s. PMID:8137311

15. Tang D, Phillips DH, Stampfer M et al. (2001). Association between carcinogen-DNA adducts in white blood cells and subsequent risk of lung cancer in the Physicians' Health Study. *Cancer Res*, 61:6708–6712. PMID:11559540

16. Hecht SS (2003). Tobacco carcinogens, their biomarkers and tobacco-induced cancer. *Nat Rev Cancer*, 3:733–744.doi:10.1038/nrc1190 PMID:14570033

17. Groopman JD, Wogan GN, Roebuck BD, Kensler TW (1994). Molecular biomarkers for aflatoxins and their application to human cancer prevention. *Cancer Res*, 54 Suppl;1907s–1911s. PMID:8137310

18. Kensler TW, Qian GS, Chen JG, Groopman JD (2003). Translational strategies for cancer prevention in liver. *Nat Rev Cancer*, 3:321–329.doi:10.1038/nrc1076 PMID:12724730

19. Mattson ME, Pollack ES, Cullen JW (1987). What are the odds that smoking will kill you? *Am J Public Health*, 77:425–431. doi:10.2105/AJPH.77.4.425 PMID:3826460

20. Kaderlik RK, Lin DX, Lang NP, Kadlubar FF (1992). Advantages and limitations of laboratory methods for measurement of carcinogen-DNA adducts for epidemiological studies. *Toxicol Lett*, 64–65:469–475.

21. Sugimura T, Inoue R, Ohgaki H, et al. (1995). Genetic polymorphisms and susceptibility to cancer development. *Pharmacogenetics*, 5:S161–S165.

22. Guengerich FP (2000). Metabolism of chemical carcinogens. *Carcinogenesis*, 21:345–351.doi:10.1093/carcin/21.3.345 PMID:10688854

23. Perera FP, Poirier MC, Yuspa SH et al. (1982). A pilot project in molecular cancer epidemiology: determination of benzo[a]pyrene–DNA adducts in animal and human tissues by immunoassays. *Carcinogenesis*, 3:1405–1410.doi:10.1093/carcin/3.12.1405 PMID:6295657

24. Wogan GN, Hecht SS, Felton JS et al. (2004). Environmental and chemical carcinogenesis. *Semin Cancer Biol*, 14:473–486.doi:10.1016/j.semcancer.2004.06.010 PMID:15489140

25. Perera FP, Weinstein IB (2000). Molecular epidemiology: recent advances and future directions. *Carcinogenesis*, 21:517–524. doi:10.1093/carcin/21.3.517 PMID:10688872

26. Nordberg GF, Jin T, Hong F et al. (2005). Biomarkers of cadmium and arsenic interactions. *Toxicol Appl Pharmacol*, 206:191–197.doi:10.1016/j.taap.2004.11.028 PMID:15967208

27. Savory J, Herman MM (1999). Advances in instrumental methods for the measurement and speciation of trace metals. *Ann Clin Lab Sci*, 29:118–126. PMID:10219699

28. Sengoku S, Wagatsuma K (2006). Comparative studies of spectrochemical characteristics between axial and radial observations in inductively coupled plasma optical emission spectrometry. *Anal Sci*, 22:245–248.doi:10.2116/analsci.22.245 PMID:16512416

29. Lobiński R, Schaumlöffel D, Szpunar J (2006). Mass spectrometry in bioinorganic analytical chemistry. *Mass Spectrom Rev*, 25:255–289.doi:10.1002/mas.20069 PMID:16273552

30. Hu H, Milder FL, Burger DE (1989). X-ray fluorescence: issues surrounding the application of a new tool for measuring burden of lead. *Environ Res*, 49:295–317.doi:10.1016/S0013-9351(89)80074-X PMID:2753011

31. Aposhian HV, Aposhian MM (2006). Arsenic toxicology: five questions. *Chem Res Toxicol*, 19:1–15.doi:10.1021/tx050106d PMID:16411650

32. Dopp E, Hartmann LM, Florea AM et al. (2004). Environmental distribution, analysis, and toxicity of organometal(loid) compounds. *Crit Rev Toxicol*, 34:301–333.doi:10.1080/10408440490270160 PMID:15239389

33. Guengerich FP (2008). Cytochrome p450 and chemical toxicology. *Chem Res Toxicol*, 21:70–83.doi:10.1021/tx700079z PMID:18052394

34. Guengerich FP (2000). Metabolism of chemical carcinogens. *Carcinogenesis*, 21:345–351.doi:10.1093/carcin/21.3.345 PMID:10688854

35. Liebler DC, Guengerich FP (2005). Elucidating mechanisms of drug-induced toxicity. *Nat Rev Drug Discov*, 4:410–420. doi:10.1038/nrd1720 PMID:15864270

36. Calafat AM, Kuklenyik Z, Reidy JA et al. (2007). Serum concentrations of 11 polyfluoroalkyl compounds in the U.S. population: data from the National Health and Nutrition Examination Survey (NHANES). *Environ Sci Technol*, 41:2237–2242. doi:10.1021/es062686m PMID:17438769

37. Muntner P, Menke A, DeSalvo KB et al. (2005). Continued decline in blood lead levels among adults in the United States: the National Health and Nutrition Examination Surveys. Arch Intern Med, 165:2155–2161.doi:10.1001/archinte.165.18.2155 PMID:16217007

38. Groopman JD, Donahue PR, Zhu JQ et al. (1985). Aflatoxin metabolism in humans: detection of metabolites and nucleic acid adducts in urine by affinity chromatography. Proc Natl Acad Sci USA, 82:6492–6496. doi:10.1073/pnas.82.19.6492 PMID:3931076

39. Zhu JQ, Zhang LS, Hu X et al. (1987). Correlation of dietary aflatoxin B1 levels with excretion of aflatoxin M1 in human urine. Cancer Res, 47:1848–1852. PMID:3102051

40. Zarba A, Wild CP, Hall AJ et al. (1992). Aflatoxin M1 in human breast milk from The Gambia, west Africa, quantified by combined monoclonal antibody immunoaffinity chromatography and HPLC. Carcinogenesis, 13:891–894.doi:10.1093/carcin/13.5.891 PMID:1587004

41. Carmella SG, Akerkar S, Hecht SS (1993). Metabolites of the tobacco-specific nitrosamine 4-(methylnitrosamino)-1-(3-pyridyl)-1-butanone in smokers' urine. Cancer Res, 53:721–724. PMID:8428352

42. Carmella SG, Akerkar SA, Richie JP Jr, Hecht SS (1995). Intraindividual and interindividual differences in metabolites of the tobacco-specific lung carcinogen 4-(methylnitrosamino)-1-(3-pyridyl)-1-butanone (NNK) in smokers' urine. Cancer Epidemiol Biomarkers Prev, 4:635–642. PMID:8547830

43. Kresty LA, Carmella SG, Borukhova A et al. (1996). Metabolites of a tobacco-specific nitrosamine, 4-(methylnitrosamino)-1-(3-pyridyl)-1-butanone (NNK), in the urine of smokeless tobacco users: relationship between urinary biomarkers and oral leukoplakia. Cancer Epidemiol Biomarkers Prev, 5:521–525. PMID:8827356

44. Wolff MS, Toniolo PG, Lee EW et al. (1993). Blood levels of organochlorine residues and risk of breast cancer. J Natl Cancer Inst, 85:648–652.doi:10.1093/jnci/85.8.648 PMID:8468722

45. Krieger N, Wolff MS, Hiatt RA et al. (1994). Breast cancer and serum organochlorines: a prospective study among white, black, and Asian women. J Natl Cancer Inst, 86:589–599.doi:10.1093/jnci/86.8.589 PMID:8145274

46. Nedelko T, Arlt VM, Phillips DH, Hollstein M (2009). TP53 mutation signature supports involvement of aristolochic acid in the aetiology of endemic nephropathy-associated tumours. Int J Cancer, 124:987–990.doi:10.1002/ijc.24006 PMID:19030178

47. Grollman AP, Jelaković B (2007). Role of environmental toxins in endemic (Balkan) nephropathy. October 2006, Zagreb, Croatia. J Am Soc Nephrol, 18:2817–2823.doi:10.1681/ASN.2007050537 PMID:17942951

48. Essigmann JM, Croy RG, Nadzan AM et al. (1977). Structural identification of the major DNA adduct formed by aflatoxin B1 in vitro. Proc Natl Acad Sci USA, 74:1870–1874. doi:10.1073/pnas.74.5.1870 PMID:266709

49. Conney AH (1982). Induction of microsomal enzymes by foreign chemicals and carcinogenesis by polycyclic aromatic hydrocarbons: G. H. A. Clowes Memorial Lecture. Cancer Res, 42:4875–4917. PMID:6814745

50. Miller EC (1978). Some current perspectives on chemical carcinogenesis in humans and experimental animals: Presidential Address. Cancer Res, 38:1479–1496. PMID:348302

51. Wang JS, Groopman JD. Biomarkers for carcinogen exposure: tumor initiation. In: Wallace K, Puga A, editors. Molecular biology of the toxic response. Washington (DC): Taylor & Francis; 1998. p. 145–66.

52. Groopman JD, Skipper PL, editors. Molecular dosimetry and human cancer: analytical, epidemiological and social consideration. Boca Raton (FL): CRC Press, Inc; 1991.

53. Wild CP, Chapot B, Scherer E et al. Application of antibody methods to the detection of aflatoxin in human body fluids. In: Bartsch H, Hemminki K, O'Neill K, editors. Methods for detecting DNA damaging agents in humans: applications in cancer epidemiology and prevention. Lyon: IARC Scientific Publication; 1988(89). p. 67–74.

54. Gange SJ, Muñoz A, Wang JS, Groopman JD (1996). Variability of molecular biomarker measurements from nonlinear calibration curves. Cancer Epidemiol Biomarkers Prev, 5:57–61. PMID:8770468

55. Harris CC (1991). Chemical and physical carcinogenesis: advances and perspectives for the 1990s. Cancer Res, 51 Suppl;5023s–5044s. PMID:1884379

56. Dipple A (1995). DNA adducts of chemical carcinogens. Carcinogenesis, 16:437–441. doi:10.1093/carcin/16.3.437 PMID:7697795

57. Pitot HC 3rd, Dragan Y. Chemical carcinogenesis. In: Klaassen CD, editor. Casarett and Doull's toxicology: the basic science of poisons. 5th ed. New York: McGraw-Hill; 1996. p. 201–267.

58. Hecht SS (1998). Biochemistry, biology, and carcinogenicity of tobacco-specific N-nitrosamines. Chem Res Toxicol, 11:559–603.doi:10.1021/tx980005y PMID:9625726

59. Weston A (1993). Physical methods for the detection of carcinogen-DNA adducts in humans. Mutat Res, 288:19–29. PMID:7686262

60. Strickland PT, Routledge MN, Dipple A (1993). Methodologies for measuring carcinogen adducts in humans. Cancer Epidemiol Biomarkers Prev, 2:607–619. PMID:8268781

61. Santella RM (1999). Immunological methods for detection of carcinogen-DNA damage in humans. Cancer Epidemiol Biomarkers Prev, 8:733–739. PMID:10498391

62. Randerath K, Reddy MV, Gupta RC (1981). 32P-labeling test for DNA damage. Proc Natl Acad Sci USA, 78:6126–6129.doi:10.1073/pnas.78.10.6126 PMID:7031643

63. Randerath K, Randerath E (1994). 32P-postlabeling methods for DNA adduct detection: overview and critical evaluation. Drug Metab Rev, 26:67–85.doi:10.3109/03602539409029785 PMID:8082582

64. Phillips DH, Farmer PB, Beland FA et al. (2000). Methods of DNA adduct determination and their application to testing compounds for genotoxicity. Environ Mol Mutagen, 35:222–233.doi:10.1002/(SICI)1098-2280(2000)35:3<222::AID-EM9>3.0.CO;2-E PMID:10737957

65. Patel DJ, Mao B, Gu Z et al. (1998). Nuclear magnetic resonance solution structures of covalent aromatic amine-DNA adducts and their mutagenic relevance. Chem Res Toxicol, 11:391–407.doi:10.1021/tx9702143 PMID:9585469

66. Geacintov NE, Cosman M, Hingerty BE et al. (1997). NMR solution structures of stereoisometric covalent polycyclic aromatic carcinogen-DNA adduct: principles, patterns, and diversity. Chem Res Toxicol, 10:111–146. doi:10.1021/tx9601418 PMID:9049424

67. Dingley KH, Curtis KD, Nowell S et al. (1999). DNA and protein adduct formation in the colon and blood of humans after exposure to a dietary-relevant dose of 2-amino-1-methyl-6-phenylimidazo[4,5-b]pyridine. Cancer Epidemiol Biomarkers Prev, 8:507–512. PMID:10385140

68. Vogel JS, Turteltaub KW, Finkel R, Nelson DE (1995). Accelerator mass spectrometry. Anal Chem, 67:353A–359A.doi:10.1021/ac00107a001 PMID:9306729

69. Skipper PL, Trudel LJ, Kensler TW et al. (2006). DNA adduct formation by 2,6-dimethyl-, 3,5-dimethyl-, and 3-ethylaniline in vivo in mice. Chem Res Toxicol, 19:1086–1090.doi:10.1021/tx060082q PMID:16918249

70. Gan J, Skipper PL, Gago-Dominguez M et al. (2004). Alkylaniline-hemoglobin adducts and risk of non-smoking-related bladder cancer. J Natl Cancer Inst, 96:1425–1431. doi:10.1093/jnci/djh274 PMID:15467031

71. Poirier MC, Beland FA (1992). DNA adduct measurements and tumor incidence during chronic carcinogen exposure in animal models: implications for DNA adduct-based human cancer risk assessment. Chem Res Toxicol, 5:749–755.doi:10.1021/tx00030a003 PMID:1489923

72. Poirier MC, Reed E, Litterst CL et al. (1992). Persistence of platinum-ammine-DNA adducts in gonads and kidneys of rats and multiple tissues from cancer patients. Cancer Res, 52:149–153. PMID:1727376

73. Kensler TW, Egner PA, Wang JB et al. (2004). Chemoprevention of hepatocellular carcinoma in aflatoxin endemic areas. Gastroenterology, 127 Suppl 1;S310–S318. doi:10.1053/j.gastro.2004.09.046 PMID:15508099

74. Perera FP, Hemminki K, Gryzbowska E et al. (1992). Molecular and genetic damage in humans from environmental pollution in Poland. Nature, 360:256–258.doi:10.1038/360256a0 PMID:1436106

75. Bartsch H (1996). DNA adducts in human carcinogenesis: etiological relevance and structure-activity relationship. Mutat Res, 340:67–79. PMID:8692183

76. Zhang YJ, Hsu TM, Santella RM (1995). Immunoperoxidase detection of polycyclic aromatic hydrocarbon-DNA adducts in oral mucosa cells of smokers and nonsmokers. Cancer Epidemiol Biomarkers Prev, 4:133–138. PMID:7537994

77. Santella RM, Grinberg-Funes RA, Young TL et al. (1992). Cigarette smoking related polycyclic aromatic hydrocarbon-DNA adducts in peripheral mononuclear cells. Carcinogenesis, 13:2041–2045.doi:10.1093/carcin/13.11.2041 PMID:1423873

78. Mooney LA, Santella RM, Covey L et al. (1995). Decline of DNA damage and other biomarkers in peripheral blood following smoking cessation. Cancer Epidemiol Biomarkers Prev, 4:627–634. PMID:8547829

79. Alexandrov K, Rojas M, Kadlubar FF et al. (1996). Evidence of anti-benzo[a]pyrene diolepoxide-DNA adduct formation in human colon mucosa. Carcinogenesis, 17:2081–2083.doi:10.1093/carcin/17.9.2081 PMID:8824539

80. Talaska G, al-Juburi AZ, Kadlubar FF (1991). Smoking related carcinogen-DNA adducts in biopsy samples of human urinary bladder: identification of N-(deoxyguanosin-8-yl)-4-aminobiphenyl as a major adduct. Proc Natl Acad Sci USA, 88:5350–5354. doi:10.1073/pnas.88.12.5350 PMID:2052611

81. Curigliano G, Zhang YJ, Wang LY et al. (1996). Immunohistochemical quantitation of 4-aminobiphenyl-DNA adducts and p53 nuclear overexpression in T1 bladder cancer of smokers and nonsmokers. Carcinogenesis, 17:911–916.doi:10.1093/carcin/17.5.911 PMID:8640937

82. Lin D, Lay JO Jr, Bryant MS et al. (1994). Analysis of 4-aminobiphenyl-DNA adducts in human urinary bladder and lung by alkaline hydrolysis and negative ion gas chromatography-mass spectrometry. Environ Health Perspect, 102 Suppl 6;11–16.doi:10.2307/3432144 PMID:7889831

83. Perera FP, Estabrook A, Hewer A et al. (1995). Carcinogen-DNA adducts in human breast tissue. Cancer Epidemiol Biomarkers Prev, 4:233–238. PMID:7606197

84. Izzotti A, De Flora S, Petrilli GL et al. (1995). Cancer biomarkers in human atherosclerotic lesions: detection of DNA adducts. Cancer Epidemiol Biomarkers Prev, 4:105–110. PMID:7742716

85. Kato S, Onda M, Matsukura N et al. (1995). Cytochrome P4502E1 (CYP2E1) genetic polymorphism in a case-control study of gastric cancer and liver disease. Pharmacogenetics, 5 Spec No;S141–S144.doi:10.1097/00008571-199512001-00016 PMID:7581484

86. Shields PG, Povey AC, Wilson VL et al. (1990). Combined high-performance liquid chromatography/32P-postlabeling assay of N7-methyldeoxyguanosine. Cancer Res, 50:6580–6584. PMID:2208119

87. Mustonen R, Schoket B, Hemminki K (1993). Smoking-related DNA adducts: 32P-postlabeling analysis of 7-methylguanine in human bronchial and lymphocyte DNA. Carcinogenesis, 14:151–154.doi:10.1093/carcin/14.1.151 PMID:8425264

88. Kato S, Petruzzelli S, Bowman ED et al. (1993). 7-Alkyldeoxyguanosine adduct detection by two-step HPLC and the 32P-postlabeling assay. Carcinogenesis, 14:545–550. doi:10.1093/carcin/14.4.545 PMID:8386066

89. Szyfter K, Hemminki K, Szyfter W et al. (1996). Tobacco smoke-associated N7-alkylguanine in DNA of larynx tissue and leucocytes. Carcinogenesis, 17:501–506.doi:10.1093/carcin/17.3.501 PMID:8631136

90. Kato S, Bowman ED, Harrington AM et al. (1995). Human lung carcinogen-DNA adduct levels mediated by genetic polymorphisms in vivo. J Natl Cancer Inst, 87:902–907. doi:10.1093/jnci/87.12.902 PMID:7666479

91. Blömeke B, Greenblatt MJ, Doan VD et al. (1996). Distribution of 7-alkyl-2′-deoxyguanosine adduct levels in human lung. Carcinogenesis, 17:741–748.doi:10.1093/carcin/17.4.741 PMID:8625485

92. Kaderlik KR, Talaska G, DeBord DG et al. (1993). 4,4′-Methylene-bis(2-chloroaniline)-DNA adduct analysis in human exfoliated urothelial cells by 32P-postlabeling. Cancer Epidemiol Biomarkers Prev, 2:63–69. PMID:8420614

93. van Schooten FJ, Jongeneelen FJ, Hillebrand MJ et al. (1995). Polycyclic aromatic hydrocarbon-DNA adducts in white blood cell DNA and 1-hydroxypyrene in the urine from aluminum workers: relation with job category and synergistic effect of smoking. Cancer Epidemiol Biomarkers Prev, 4:69–77. PMID:7894326

94. Nielsen PS, de Pater N, Okkels H, Autrup H (1996). Environmental air pollution and DNA adducts in Copenhagen bus drivers— effect of GSTM1 and NAT2 genotypes on adduct levels. Carcinogenesis, 17:1021–1027. doi:10.1093/carcin/17.5.1021 PMID:8640907

95. Rojas M, Alexandrov K, Auburtin G et al. (1995). Anti-benzo[a]pyrene diolepoxide-DNA adduct levels in peripheral mononuclear cells from coke oven workers and the enhancing effect of smoking. Carcinogenesis, 16:1373–1376.doi:10.1093/carcin/16.6.1373 PMID:7788857

96. Shuker DE, Farmer PB (1992). Relevance of urinary DNA adducts as markers of carcinogen exposure. Chem Res Toxicol, 5:450–460. doi:10.1021/tx00028a001 PMID:1391611

97. Groopman JD, Roebuck BD, Kensler TW (1992). Molecular dosimetry of aflatoxin DNA adducts in humans and experimental rat models. Prog Clin Biol Res, 374:139–155. PMID:1320272

98. Groopman JD, Zhu JQ, Donahue PR et al. (1992). Molecular dosimetry of urinary aflatoxin-DNA adducts in people living in Guangxi Autonomous Region, People's Republic of China. Cancer Res, 52:45–52. PMID:1727385

99. Prevost V, Shuker DE, Friesen MD et al. (1993). Immunoaffinity purification and gas chromatography-mass spectrometric quantification of 3-alkyladenines in urine: metabolism studies and basal excretion levels in man. Carcinogenesis, 14:199–204.doi:10.1093/carcin/14.2.199 PMID:8435861

100. Ames BN, Gold LS (1991). Endogenous mutagens and the causes of aging and cancer. Mutat Res, 250:3–16. PMID:1944345

101. Marnett LJ, Burcham PC (1993). Endogenous DNA adducts: potential and paradox. Chem Res Toxicol, 6:771–785.doi:10.1021/tx00036a005 PMID:8117915

102. Park EM, Shigenaga MK, Degan P et al. (1992). Assay of excised oxidative DNA lesions: isolation of 8-oxoguanine and its nucleoside derivatives from biological fluids with a monoclonal antibody column. Proc Natl Acad Sci USA, 89:3375–3379.doi:10.1073/pnas.89.8.3375 PMID:1565629

103. Simic MG, Bergtold DS (1991). Dietary modulation of DNA damage in human. Mutat Res, 250:17–24. PMID:1944333

104. Simic MG (1992). Urinary biomarkers and the rate of DNA damage in carcinogenesis and anticarcinogenesis. Mutat Res, 267:277–290. PMID:1376430

105. Vaca CE, Wilhelm J, Harms-Ringdahl M (1988). Interaction of lipid peroxidation products with DNA. A review. Mutat Res, 195:137–149. PMID:3277035

106. Marnett LJ, Riggins JN, West JD (2003). Endogenous generation of reactive oxidants and electrophiles and their reactions with DNA and protein. J Clin Invest, 111:583–593. PMID:12618510

107. Marnett LJ (2002). Oxy radicals, lipid peroxidation and DNA damage. Toxicology, 181-182:219–222.doi:10.1016/S0300-483X(02)00448-1 PMID:12505314

108. Fang JL, Vaca CE, Valsta LM, Mutanen M (1996). Determination of DNA adducts of malonaldehyde in humans: effects of dietary fatty acid composition. Carcinogenesis, 17:1035–1040.doi:10.1093/carcin/17.5.1035 PMID:8640909

109. Wang M, Dhingra K, Hittelman WN et al. (1996). Lipid peroxidation-induced putative malondialdehyde-DNA adducts in human breast tissues. Cancer Epidemiol Biomarkers Prev, 5:705–710. PMID:8877062

110. Yager JD, Davidson NE (2006). Estrogen carcinogenesis in breast cancer. N Engl J Med, 354:270–282.doi:10.1056/NEJMra050776 PMID:16421368

111. Rogan EG, Cavalieri EL (2004). Estrogen metabolites, conjugates, and DNA adducts: possible biomarkers for risk of breast, prostate, and other human cancers. *Adv Clin Chem*, 38:135–149.doi:10.1016/S0065-2423(04)38005-4 PMID:15521191

112. Skipper PL, Peng X, Soohoo CK, Tannenbaum SR (1994). Protein adducts as biomarkers of human carcinogen exposure. *Drug Metab Rev*, 26:111–124.doi:10.3109/03602539409029787 PMID:8082561

113. Skipper PL, Tannenbaum SR (1990). Protein adducts in the molecular dosimetry of chemical carcinogens. *Carcinogenesis*, 11:507–518.doi:10.1093/carcin/11.4.507 PMID:2182215

114. Skipper PL, Tannenbaum SR (1994). Molecular dosimetry of aromatic amines in human populations. *Environ Health Perspect*, 102 Suppl 6;17–21.doi:10.2307/3432145 PMID:7889842

115. McCoy LF, Scholl PF, Schleicher RL et al. (2005). Analysis of aflatoxin B1-lysine adduct in serum using isotope-dilution liquid chromatography/tandem mass spectrometry. *Rapid Commun Mass Spectrom*, 19:2203–2210.doi:10.1002/rcm.2045 PMID:16015671

116. Tannenbaum SR (1990). Hemoglobin-carcinogen adducts as molecular biomarkers in epidemiology. *Princess Takamatsu Symp*, 21:351–360. PMID:2134688

117. Hammond SK, Coghlin J, Gann PH et al. (1993). Relationship between environmental tobacco smoke exposure and carcinogen-hemoglobin adduct levels in nonsmokers. *J Natl Cancer Inst*, 85:474–478.doi:10.1093/jnci/85.6.474 PMID:8445675

118. Vineis P, Bartsch H, Caporaso NE et al. (1994). Genetically based N-acetyltransferase metabolic polymorphism and low-level environmental exposure to carcinogens. *Nature*, 369:154–156.doi:10.1038/369154a0 PMID:7909916

119. Yu MC, Ross RK, Chan KK et al. (1995). Glutathione S-transferase M1 genotype affects aminobiphenyl-hemoglobin adduct levels in white, black and Asian smokers and nonsmokers. *Cancer Epidemiol Biomarkers Prev*, 4:861–864. PMID:8634658

120. Landi MT, Zocchetti C, Bernucci I et al. (1996). Cytochrome P4501A2: enzyme induction and genetic control in determining 4-aminobiphenyl-hemoglobin adduct levels. *Cancer Epidemiol Biomarkers Prev*, 5:693–698. PMID:8877060

121. Schettgen T, Rossbach B, Kütting B et al. (2004). Determination of haemoglobin adducts of acrylamide and glycidamide in smoking and non-smoking persons of the general population. *Int J Hyg Environ Health*, 207:531–539.doi:10.1078/1438-4639-00324 PMID:15729833

122. Sabbioni G, Skipper PL, Büchi G, Tannenbaum SR (1987). Isolation and characterization of the major serum albumin adduct formed by aflatoxin B1 in vivo in rats. *Carcinogenesis*, 8:819–824.doi:10.1093/carcin/8.6.819 PMID:3111739

123. Gan LS, Skipper PL, Peng XC et al. (1988). Serum albumin adducts in the molecular epidemiology of aflatoxin carcinogenesis: correlation with aflatoxin B1 intake and urinary excretion of aflatoxin M1. *Carcinogenesis*, 9:1323–1325.doi:10.1093/carcin/9.7.1323 PMID:3133131

124. Groopman JD, Hasler JA, Trudel LJ et al. (1992). Molecular dosimetry in rat urine of aflatoxin-N7-guanine and other aflatoxin metabolites by multiple monoclonal antibody affinity chromatography and immunoaffinity/high performance liquid chromatography. *Cancer Res*, 52:267–274. PMID:1728400

125. Wild CP, Jiang YZ, Allen SJ et al. (1990). Aflatoxin-albumin adducts in human sera from different regions of the world. *Carcinogenesis*, 11:2271–2274.doi:10.1093/carcin/11.12.2271 PMID:2265478

126. Wild CP, Hudson GJ, Sabbioni G et al. (1992). Dietary intake of aflatoxins and the level of albumin-bound aflatoxin in peripheral blood in The Gambia, West Africa. *Cancer Epidemiol Biomarkers Prev*, 1:229–234. PMID:1339083

127. Wang LY, Hatch M, Chen CJ et al. (1996). Aflatoxin exposure and risk of hepatocellular carcinoma in Taiwan. *Int J Cancer*, 67:620–625.doi:10.1002/(SICI)1097-0215(19960904)67:5<620::AID-IJC5>3.0.CO;2-W PMID:8782648

128. Scholl PF, Turner PC, Sutcliffe AE et al. (2006). Quantitative comparison of aflatoxin B1 serum albumin adducts in humans by isotope dilution mass spectrometry and ELISA. *Cancer Epidemiol Biomarkers Prev*, 15:823–826.doi:10.1158/1055-9965.EPI-05-0890 PMID:16614131

129. Wild CP, Hasegawa R, Barraud L et al. (1996). Aflatoxin-albumin adducts: a basis for comparative carcinogenesis between animals and humans. *Cancer Epidemiol Biomarkers Prev*, 5:179–189. PMID:8833618

130. Ross RK, Yuan JM, Yu MC et al. (1992). Urinary aflatoxin biomarkers and risk of hepatocellular carcinoma. *Lancet*, 339:943–946.doi:10.1016/0140-6736(92)91528-G PMID:1348796

131. Qian GS, Ross RK, Yu MC et al. (1994). A follow-up study of urinary markers of aflatoxin exposure and liver cancer risk in Shanghai, People's Republic of China. *Cancer Epidemiol Biomarkers Prev*, 3:3–10. PMID:8118382

132. Tang D, Santella RM, Blackwood AM et al. (1995). A molecular epidemiological case-control study of lung cancer. *Cancer Epidemiol Biomarkers Prev*, 4:341–346. PMID:7655328

133. Wang LY, Hatch M, Chen CJ et al. (1996). Aflatoxin exposure and risk of hepatocellular carcinoma in Taiwan. *Int J Cancer*, 67:620–625.doi:10.1002/(SICI)1097-0215(19960904)67:5<620::AID-IJC5>3.0.CO;2-W PMID:8782648

134. Kuang SY, Frang X, Lu PX et al. (1996). Aflatoxin-albumin adducts and risk for hepatocellular carcinoma in residents of Qidong, People's Republic of China. [abstract]. *Proc Am Assoc Cancer Res*, 371:A1216.

UNIT 2.

BIOMARKERS: PRACTICAL ASPECTS

CHAPTER 5.

Assessment of genetic damage in healthy and diseased tissue

Joe Shuga, Pierre Hainaut, and Martyn T. Smith

Summary

DNA, along with other cellular components, is under constant attack by chemical, physical, and infectious agents present in the human environment, as well as by reactive metabolites generated by physiological processes. Mutations occur as the consequence of this damage, but may also be caused by improper DNA repair of alterations occurring during normal DNA replication and transcription. Genetic damage can occur at the level of the gene (e.g. point mutations, insertions, and deletions) or at the level of the chromosome (e.g. aneuploidy, translocations). Further, mutations can also take place in mitochondrial DNA. Another form of DNA modification is epigenetic methylation of CpG islands, which affects the dynamics of chromatin as well as the expression of a large panel of genes.

Recent technical advances have improved the capacity to detect and quantify genetic and epigenetic changes. This chapter summarizes current knowledge on mechanisms of DNA damage and mutagenesis, laying out the concepts for interpreting mutations as biomarkers in investigating the causes and consequences of cancer. It also outlines both established and novel methods for detecting genetic and epigenetic changes in normal and diseased tissues, and then discusses their application in the realm of molecular epidemiology.

Introduction

The sequencing of the human genome has established the existence of about 22 000 protein-coding genes (1,2). Together these protein-coding genes only comprise 2–3% of the total genome, which amounts to approximately 3.25×10^9 nucleotide base pairs. The great majority of DNA is actually not protein-coding and instead consists of regulatory sequences, sequences encoding regulatory and metabolic RNAs, and repetitive sequences. All sequences are assembled and replicated according to specific base pairing to form the double helix, which is packed with proteins into a structure called chromatin that forms chromosomes. Cells also contain non-genomic DNA: the mitochondrial genome, which is circular and composed of

16.6x10³ base pairs, is present in the cytoplasm at a copy number of ~10²-10⁴ per cell. Both genomic and mitochondrial DNA undergo structural alterations associated with disease (3–9).

Structural alterations in DNA occur through changes in DNA base pairing, as well as in its supra-molecular chromatin and chromosome organization (Figure 5.1). Base pairing changes are known as 'mutations,' while changes that do not modify the base pairing content of DNA but affect its expression, processing, metabolism, and stability are known as 'epigenetic changes.' The nature and type of mutations can vary by several orders of magnitude, from single base pair mutations to deletions or duplications encompassing whole chromosomes. Such changes are the causal defects of many diseases. Cancers, in particular, develop as the consequence of accumulated genetic and epigenetic changes that affect the expression and activity of selected sets of genes, providing cells with selective growth advantages on the path to malignancy (10–12).

In recent years, technical advances have improved the capacity to detect and quantify genetic changes, paving the way for novel methods for early detection of mutations and for better understanding the mechanisms that have caused their formation. This chapter outlines several of these methods and examines their application in the area of molecular epidemiology and early detection of disease (Figure 5.2). It also provides a brief presentation of the more established methods available for detecting and measuring mutations in normal and diseased tissues. This information is presented in the context of current knowledge on mechanisms of mutagenesis, laying out the key concepts for interpreting the significance of mutations as biomarkers in investigating the causes and consequences of cancer.

Mechanisms of mutagenesis

Mutations as biomarkers of early effects of carcinogens

In their seminal 1953 paper, Watson and Crick made one of the most famous understatements in biology: "It has not escaped our notice that the specific pairing we have postulated immediately suggests a possible copying mechanism for the genetic material." Since then, DNA replication mechanisms and their associated repair systems have developed into a prolific field of research. Many human diseases, such as cancer, neurodegenerative, inflammatory, or autoimmune diseases, can be described as a disruption in the balance between correct and incorrect DNA synthesis (13).

Figure 5.1. DNA, chromatin structures, and chromosomes. The architecture of the genetic material, from DNA double-helix to packed chromosomes, is represented in relation with the level at which different forms of genetic modifications may occur (mutations, epigenetic changes, chromosomal aberrations)

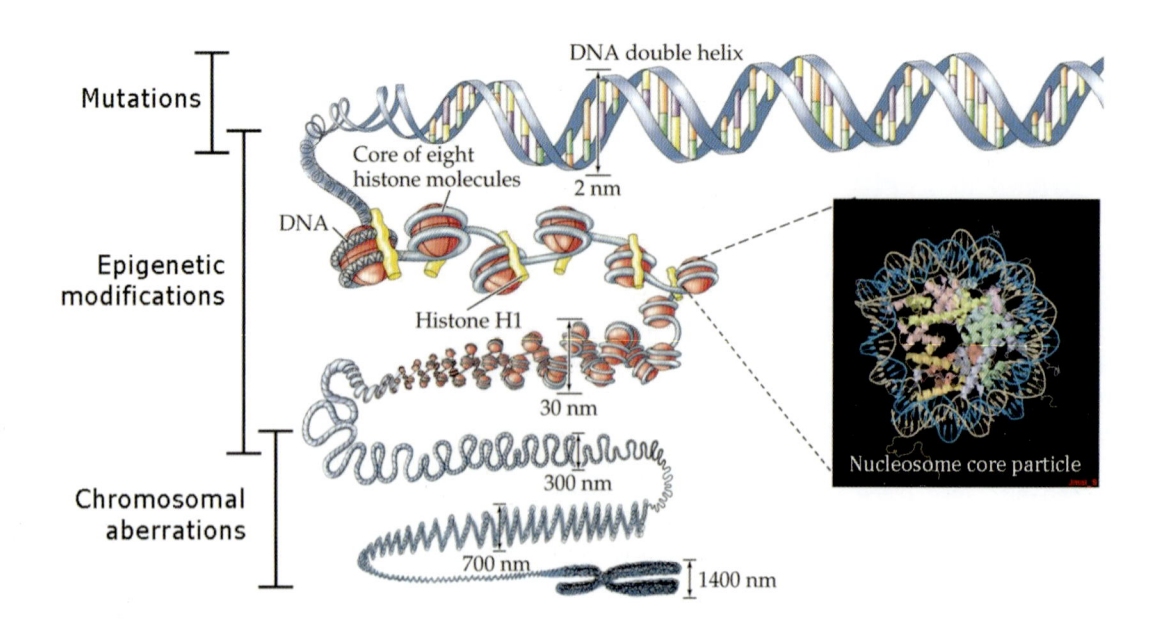

DNA, along with other cellular components, is under constant attack by reactive metabolites generated by physiological processes, as well as by chemical, physical, or infectious agents present in the human environment. It is estimated that each individual human cell can undergo damage to its DNA at a rate of up to 10^6 molecular lesions per day (14). This extensive DNA damage is compensated for and corrected by DNA repair systems. Thus, contrary to common perception, DNA is far from being carved in stone for eternity: its structure is highly variable, ever changing, and stabilized only by active biological processes that maintain the fidelity of DNA replication. Failure to detect, process, or repair DNA damage in an appropriate way leads to mutations.

Mutations occurring in the germline may be passed from one generation to the other and may form the underlying cause of inherited diseases. These germline mutations are present in the genome of every cell of the resulting offspring, even those cells and tissues that do not express a phenotypic defect caused by mutation. Genetic changes can also be acquired by a somatic cell after conception; such mutations are not transmitted from one generation to the other. However, these acquired mutations are transmitted to all cells descended from the original cell that underwent the mutation, giving rise to a clone (colony) of cells carrying the mutation as a marker and possibly as a phenotypic trait. This is particularly spectacular in the case of cancer, which results from the proliferation of a single or a small number of clone(s) having acquired a selective growth advantage as the result of mutation. Cancer involves deep modifications of the cell genome through multiple steps of somatic mutations (15,16).

Genetic damage can occur at the level of the gene (e.g. point mutations, insertions, and deletions) or at the level of the chromosome (e.g. aneuploidy, translocations). Historically, studies on genetic and genomic damage have tended to measure mutations in surrogate genes, such as hypoxanthine phosphoribosyltransferase (*HPRT*) and glycophorin A (*GPA*) (17), or to use cytogenetics to assess changes in chromosome structure and number, such as classical and banded chromosomal aberrations, sister chromatid exchanges, and micronucleus formation (18–21). These biomarkers have been shown to be associated with a wide range of carcinogenic exposures (22–26). However, mutations in surrogate genes are of limited value as biomarkers of early effect, since they are not on the causal pathway of disease.

During the past three decades, several hundred genes have been identified as recurrent sites for genetic or genomic damage in cancer cells. These genes provide

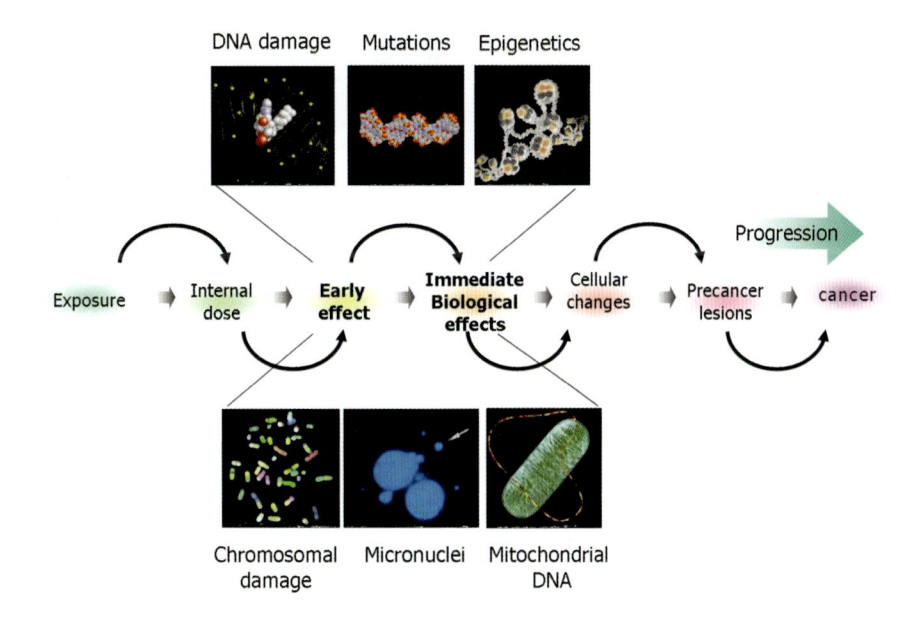

Figure 5.2. Scope of the chapter: biomarkers of early effects. This scheme shows sequential steps in the processes by which environmental exposures may deregulate genetic programmes, thus leading to cancer. This chapter focuses on the detection of genetic changes that are biomarkers of early effects of DNA damaging processes, and, in particular, on those biomarkers that are parts of the molecular pathways of disease causation. These biomarkers include DNA damage, mutations in genomic and mitochondrial DNA, chromosomal aberrations, epigenetic changes, and formation of micronuclei

a wide spectrum of biomarkers to detect early mutational and chromosomal effects of carcinogenic exposure in humans (27). These novel biomarkers measure changes frequently observed among cancer patients, including point mutations in genes such as *TP53*, ras, *BRCA1/2*, *HER1/2*, altered gene methylation, aneuploidy (chromosome loss or gain – including monosomy 7 and trisomy 8), and specific chromosome rearrangements such as translocations. Such changes are readily detectable in cancer cells. However, to exploit their value as biomarkers of early effect, they must be applicable to study individuals who may be at risk, but who do not yet have cancer. Such studies require detecting genetic changes that occur in single cells or a small number of cells that are morphologically undistinguishable from 'normal' cells. This is now feasible using cutting edge technologies such as real-time quantitative polymerase chain reaction (PCR), fluorescence in situ hybridization (FISH) analysis, and genotypic selection methods which introduce new levels of sensitivity and specificity. Such biomarkers are useful in epidemiological studies of environmentally induced cancers which have long latency periods, as well as providing early detection for those individuals at risk.

Sources of DNA damage

The elucidation of the human genome sequence has made it possible to identify genetic alterations in cancers on an unprecedented scale. Comprehensive analysis of the coding sequence of 13 023 genes in breast and colorectal cancers revealed that individual tumours accumulate, on average, approximately 90 mutant genes. However, only a subset (about 11 per tumour on average) appears to be mutated at a significant frequency and may be considered as potential 'drivers' of the neoplastic process. Most other mutations appear to be 'passengers,' occurring as a consequence of the genetic instability of cancer cells (28,29). Currently, the list of genes affected by potential driver mutations includes about 300 candidates. An exhaustive discussion of each mutation known to be associated with cancer is beyond the scope of this chapter. The Human Gene Mutation Database (HGMD) compiles a list of mutations in the coding regions of genes that are known to cause genetic defects (30–35) (http://www.hgmd.cf.ac.uk/ac/index.php). Single base pair substitutions account for about 50% of all mutations in the HGMD and include different subtypes (e.g. transitions or transversions) depending upon the nature of the base change. Other common changes include deletions, insertions, duplications, inversions, and alterations of unstable repeated sequences (Figure 5.3).

Epidemiological studies have demonstrated the links between carcinogen exposure and cancer in human populations. The type, route, and amount of exposure can

Figure 5.3. Distribution of mutation types in the HGMD database. The proportion of different types of mutations in the HGMD database is shown. A: Types of mutations. Indels: combined insertions and deletions. B: Types of base changes among missense mutations. Note that the two types of transitions are more common than the four types of transversions. Mutations at CpG sites represent about 65% of all G:C to A:T transitions. Source: http://lisntweb.swan.ac.uk/cmgt/index.htm

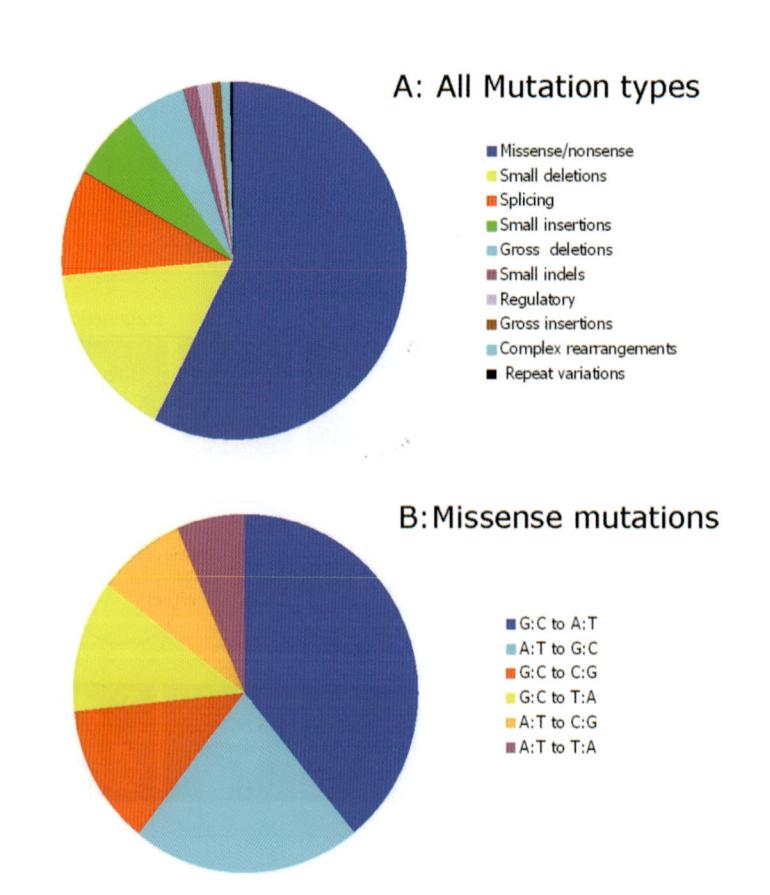

determine the type of cancer. It also influences the type of genetic, genomic, and epigenetic alterations, leading in some instances to genetic changes that are 'signature' of specific environmental carcinogens. A typical example of such a 'carcinogen fingerprint' is C to T transition mutations at hotspot dipyrimidines sites in *TP53* caused by ultraviolet radiation (36). However, such an unequivocal mutation pattern is the exception. In most instances, mutation patterns are complex, reflecting the diversity of exposures and mechanisms involved in carcinogenesis (37).

Many factors and agents can produce DNA damage leading to mutations. Highly reactive molecules, such as oxygen and nitrogen radicals, are produced as by-products of physiological and pathological processes. DNA binding compounds can also form as the result of the enzymatic transformation of exogenous compounds, a process called carcinogen activation. These reactive products can induce covalent or non-covalent anomalies in DNA, resulting in various forms of base damage, single- or double-strand cuts, nicks, and gaps, and crosslinks (both intrastrand and interstrand) (38). Different forms of damage elicit distinct DNA repair reactions. The main forms of base damage are oxidized, reduced, and fragmented bases, as well as covalent adducts of small chemical groups (e.g. alkyl adducts) or large compounds (the so-called 'bulky adducts' induced by metabolites of polycyclic aromatic hydrocarbons, arylamines, or mycotoxins). Imperfect repair of these lesions induces irreversible changes in the DNA base pairing. Carcinogen DNA fingerprints arise when a particular type of base pair change is frequently observed following exposure to a specific type of carcinogen. Table 5.1 shows a list of some chemicals that induce defined types of DNA lesions and describes the major types of mutations that result from these lesions in experimental systems.

Mutagenesis induced by exposure to carcinogens

Many carcinogens are lipophilic compounds that cross plasma membranes to accumulate in the

Table 5.1. DNA fingerprints of some exogenous and endogenous DNA-damaging agents

Site of pre-mutagenic lesion	Mutagen	Main mutations	Possible *TP53* fingerprint in:
*N*7-G	AFB1	GC > TA	Hepatocellular carcinoma
*N*2-G	B[a]P-7,8-diol-9,10-epoxide (BPDE)	GC > TA	Lung cancer, smokers
*O*6-G	*N*-Methyl-*N*-nitrosourea	GC > AT	Oral, esophageal cancer?
*O*6-G	NNK	GC > AT	Lung cancer?
C8-G	1-Nitrosopyrene	GC > AT, GC > TA	?
C8-G	4-Aminobiphenyl	GC > TA	Bladder cancer
C8-G	2-AAF	GC > TA	Bladder cancer?
C8-G	PhIP	GC > TA	?
8-oxo-G	Oxidative agents	GC > TA	Many cancers, including lung
1,*N*2-G	Malondialdehyde	GC > TA, GC > AT	?
*N*6-A	Stryene oxide	AT > CG	?
*N*6-A	Benzo[c]phenanthrene diol epoxides	AT > TA, AT > GC	Lung, esophageal cancer?
*N*6-A	BPDE	AT > GC	Lung cancer
3,*N*4-C	Vinyl chloride	GC > AT	Angiosarcoma of the liver
5-OH-C, 5-OH-U, uridine glycol	Oxidative agents	GC > AT	?
N3-U	Propylene oxide	GC > AT	?
Pyrimidine dimers	UV	CC > TT tandem,	
GC > AT	Non-melanoma skin cancer		
Apurinic	Depurinating agents	GC > TA, AT > TA	?

? – No clear mutation fingerprint identified so far
Table compiled from (235) and (236)

Unit 2 • Chapter 5. Assessment of genetic damage in healthy and diseased tissue cytoplasm and the nucleus. To neutralize their immediate, toxic effects, cells mobilize complex enzymatic machineries acting as a first line of defence against DNA damage. Cytochrome P (CYP) 450 enzymes initiate a cascade of metabolic detoxification reactions by catalysing the addition of an oxygen atom to the carcinogen, increasing its solubility in water (Figure 5.4) (39,40). This process is amplified by conjugation enzymes, such as glutathione S-transferase, converting the oxygenated carcinogen to a soluble compound which is eliminated from the cell. These efficient detoxification reactions provide a first line of protection against the toxicity of chemicals (41). However, the reactive, water-soluble compounds formed during this process often contain an electrophilic (electron-deficient) centre that can react with DNA bases at specific N and O positions resulting in the formation of covalent DNA adducts (Figure 5.4) (42–45).

The second line of defence is to remove damage through DNA repair proteins and pathways (46–48). The nucleotide excision repair pathway eliminates intra- and interstrand DNA crosslinks as well as bulky DNA adducts. Base excision repair (BER) eliminates and corrects bases damaged by small chemical groups (oxidized or methylated bases) or those fragmented by ionizing radiation or chemical oxidation. The frequent, miscoding, methylated base O^6-methylguanine is repaired through a specialized mechanism using the enzyme O^6-methylguanine DNA methyltransferase (49–51). Repair mechanisms involve steps of damage removal (e.g. by DNA glycosylases in BER) followed by base incorporation reactions mediated by polymerases. Furthermore, some DNA lesions are not repaired at the same rate on both strands of the double helix. The transcribed strand is preferentially repaired during transcription-coupled repair, generating strand asymmetry in the distribution of some mutations induced by exogenous carcinogens, such as in cigarette smoke (Figure 5.5). In addition, repair is dependent upon sequence context. For example, in *TP53*, there is evidence that repair is slower at some of the major mutation hotspots than at other positions (52,53).

Figure 5.4. Carcinogen metabolism, DNA damage, and mutations: the example of aflatoxins. Aflatoxin is a widespread contaminant of the staple diet in tropical areas. This mycotoxin is metabolized in the liver to form epoxides that bind covalently to guanine at codon 249 in TP53. There is a synergistic effect between aflatoxin and chronic HBV infection in inducing a specific mutation which is found in about 50% of hepatocellular carcinomas in large regions of Africa and South-East Asia

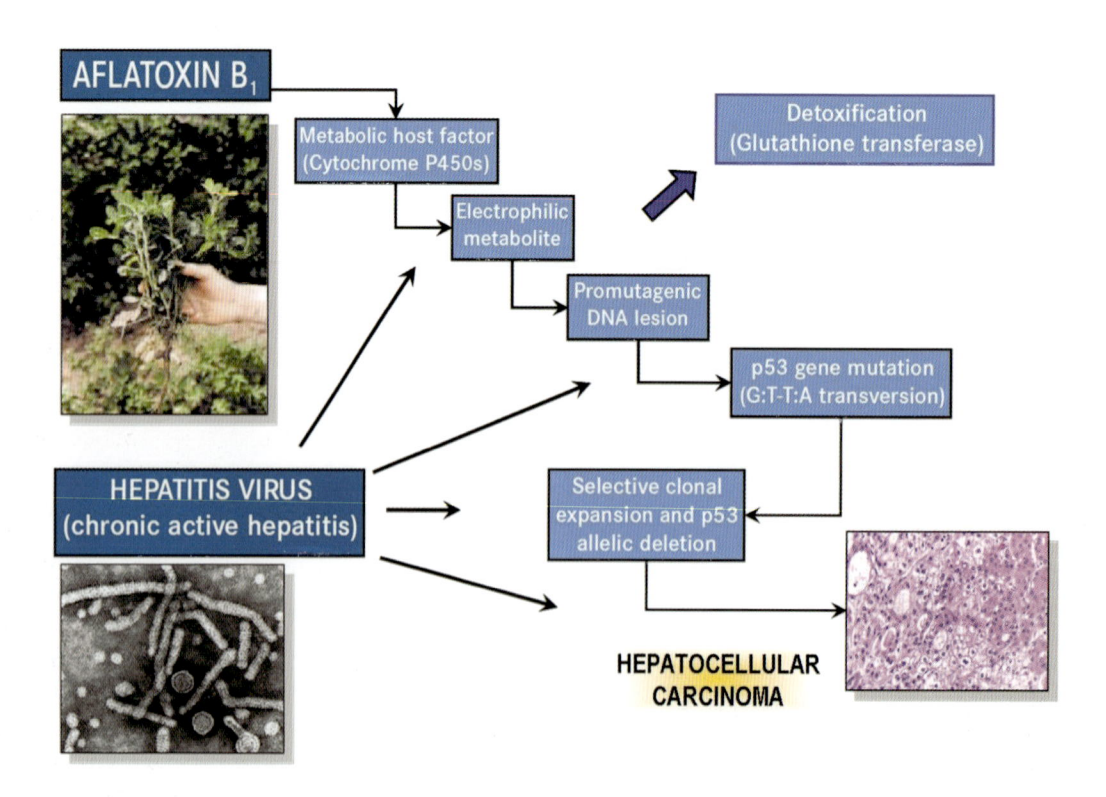

A third line of defence against DNA damage is provided by the cellular response to incomplete or imperfect DNA repair, which triggers suppressive mechanisms that prevent DNA replication. A key response in this process is stabilization and activation of the *p53* protein, which induces either apoptosis or cell-cycle arrest, terminal differentiation, and senescence (54), thereby permanently deleting damaged cells from the pool of cells capable of replicating their DNA. Cells that escape these mechanisms and proceed through replication undergo a replicative block due to the stalling of DNA polymerases at the site of a persistent lesion. To bypass this block, cells have evolved low-fidelity polymerases which resolve the lesion but are also error prone, often incorporating the wrong base at the site of damage (13,55). Mutations arise when DNA adducts are bypassed incorrectly by these low-fidelity DNA polymerases. The variable fidelity of DNA copying mechanisms is one of the main molecular mechanisms of evolution: high-fidelity DNA synthesis prevents mutations and maintains stable genetic information over many generations, while low-fidelity DNA synthesis serves to generate diversity, leading to advantages for some individuals in a population subjected to selection pressures.

Figure 5.5. Distribution and strand bias of TP53 mutations in lung cancers in relation with tobacco smoke. The proportion of each base change in lung cancers is shown for smokers and non-smokers. Differences between two symmetric base changes demonstrate a strand bias (e.g. G:T versus C:A). Note the strand bias for transversions (G:C to T:A and A:T to C:G) in smokers. In non-smokers, these mutations are less frequent and do not show a strand bias. Strand bias is indicative of DNA damage by bulky adducts that stall polymerase and trigger transcription-coupled repair on the transcribed strand. Source: (54) with permission of Oxford University Press.

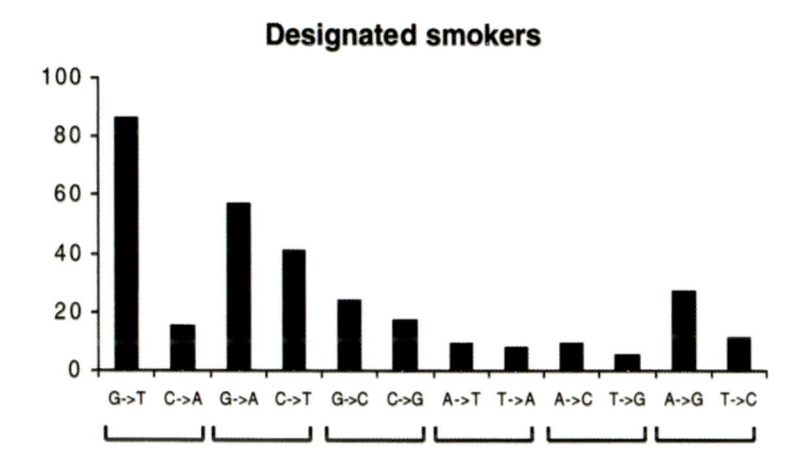

Spontaneous mutations

Many mutations occur without the involvement of exogenous DNA damaging agents. There are four main types of damage to DNA due to endogenous cellular processes: (1) base oxidation (e.g. 8-oxo-7,8-dihydroguanine (8-oxoG) and generation of single or double DNA strand breaks by reactive oxygen species); (2) base alkylation (e.g. methylation, such as formation of methylguanine); (3) hydrolysis (e.g. deamination, depurination, and depyrimidination); and (4) mismatch (due to DNA replication in which the wrong DNA base is incorporated into a newly synthesized DNA strand). It is estimated that spontaneous mutations occur in the coding regions of mammalian genomes at a rate of about 2.2×10^{-9} per base pair per year (56). This rate is similar among different genes, but is extremely variable at different base pairs. In particular, the CpG dinucleotides can mutate at a rate 10 times higher than other nucleotides, generating transitions (57,58). About 3–5% of cytosines at CpG dinucleotides in the human genome are methylated at position 5' by a post-replicative mechanism catalysed by DNA methyltransferase. The 5-methylcytosine (5mC) is less stable than cytosine and undergoes spontaneous deamination into thymine at a rate five times higher than the unmethylated base. The instability of CpG dinucleotides has led to their negative selection and subsequent loss during evolution. CpG dinucleotides represent less than 1% of the genome, one-seventh of their expected frequency assuming an equal proportion of all dinucleotide motifs, yet they are the site of roughly 25% of all known mutations in human disease that are listed in the HGMD database. In *TP53*'s DNA binding domain, which is a major site for mutations linked to cancer, there are 22 CpG dinucleotides located within 600 bp of coding sequence. Transitions at these CpG dinucleotides represent about 25% of all reported mutations, with a range from about 15% in lung cancers of smokers (in which many mutations are caused by tobacco carcinogens rather than by spontaneous mechanisms) to close to 50% in adenocarcinomas of the gastro-digestive tract (Figure 5.6) (59). Deamination of 5mC is enhanced by oxygen and nitrogen radicals, leading to a higher load of these mutations in cancers occurring within the context of inflammatory precursor lesions, such as Barrett's mucosa or ulcerative colitis (60,61).

Small insertions and deletions arise during replication through a mechanism known as the slipped-mispairing model (32). In this model, nucleotide skipping and/or misincorporation results from transient misalignment of the primer to the template due to the looping out of a base (or a short stretch of bases) from the template. This phenomenon preferentially occurs within runs of identical bases or in regions containing repetitive DNA sequences. Increased length of monotonic runs correlates with increased frequency of insertion/deletion events (62).

Mutation patterns in relation to cancer risk factors

The 'mutation pattern' concept is of central importance in assessing the value of mutations as biomarkers of early effects of carcinogens. Assessing a mutation pattern relies on six critical points: (1) type of mutation; (2) nucleotide change(s); (3) sequence context; (4) strand distribution; (5) occurrence of the mutation at a position of known structure or function (e.g. mutations in exons, introns, at mRNA splice junctions, or other structures involved in mRNA processing, within promoter regions, etc.); and (6) consequence of the mutation on the gene structure and its coding potential (e.g. silent, missense, nonsense, mutations affecting exon processing, or expression levels) (63). A mutation pattern occurs when there is a significant difference in any of these elements, or combination thereof, between a set of 'test' mutations (e.g. mutations identified in a particular type of disease and exposure, such as lung cancers of smokers) and a set of 'reference' mutations (e.g. lung cancers of never-smokers).

The formation of a mutation pattern can be seen as the result of a complex process of mutation selection through a succession of filters (Table 5.2). The first filter consists of the chemical properties of the carcinogen and of its cellular activation process. Carcinogens can damage DNA in specific ways, generating lesions that reflect the chemistry of DNA damage (64). Base position, accessibility, and

Figure 5.6. Patterns of TP53 mutations in lung cancers of smokers, colorectal cancer, and adenocarcinoma of the oesophagus. The proportion of different mutation types is shown in lung cancers of smokers and in two cancers which commonly develop in an inflammatory context (e.g. colitis in colon cancer, Barrett's mucosa in oesophagus). CpG transitions are more frequent in colon and esophageal ADC as compared to lung/smokers. In contrast, transversions are more common in lung/smokers than in colon and esophageal ADC. Data from IARC TP53 database, release R12 (http://www-p53.iarc.fr)

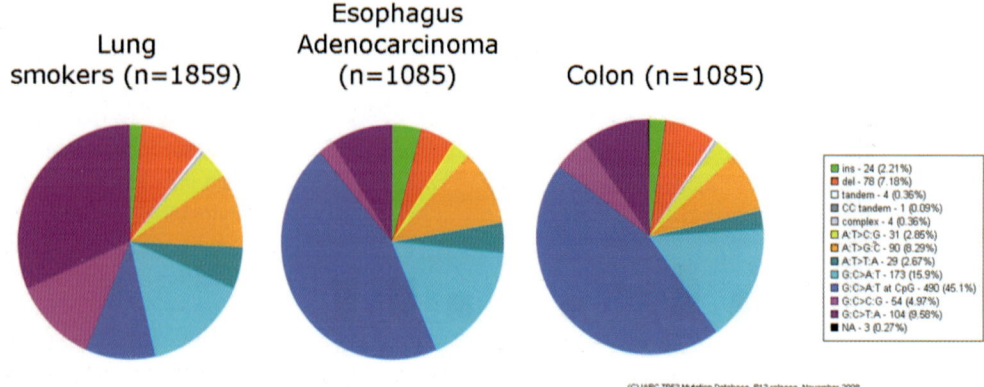

Table 5.2. Formation of a mutation pattern through a succession of 'filters:'*
The example of benzo(a)pyrene from tobacco smoke

Exposure	Tobacco smoke contains over 60 substances classified as carcinogenic to humans by IARC, including 1 to 40 ng Benzo[a]Pyrene (B[a]P)/cigarette		
	Filter	**Example**	**Type of Lesion**
Filter 1	Chemistry of DNA damage	B[a]P is metabolized by CYP450 to generate BPDE that binds on the N2 position of guanine	BPDE-N^2-dG adduct
Filter 2	Base position and sequence context	Adducts preferentially form at G adjacent to methylated cytosines at mCpG sites	Major adducts at codons 156, 157, 245, 248, 273
Filter 3	DNA repair	Transcription-coupled repair preferentially removes lesions on the TS	Strand bias with persistence of adducts on G on the NTS
Filter 4	DNA replication	Lesion bypass of an adducted template by Pol η mis-incorporates A instead of C; replication results in substitution of G to T opposite to misincorporated A	Formation of G to T transversions
Filter 5	Protein filter	Only mutations that inactivate *p53* protein contribute to the clonal expansion of cancer cells and are detectable in cancer lesions	Selection of mutations at codons 157, 245, 248, 273; counter-selection of mutation at codon 156, which is silent.
Mutation pattern in cancer	Excess of G to T transversions on the NTS at specific codons in lung cancers of smokers		

*Refers to specific criteria that can influence mutation pattern formation.
BPDE, B[a]P-7,8-diol-9,10-epoxide; TS, transcribed DNA strand; NTS, non-transcribed strand
Table compiled from (235)

sequence context are important factors that determine the type of DNA damage, forming a second filter (63). The third filter consists of DNA repair, which removes the majority of lesions, but does so in a selective manner such that all types of lesions are not eliminated with the same efficiency (46). A strand bias towards preferential repair of the transcribed strand is suggestive of selective removal of bulky, polymerase-blocking lesions during transcription (65). DNA replication and polymerase fidelity constitute the fourth filter (13). The final filter is the biological selection process that chooses cells with mutations that confer a selective advantage on the path to neoplastic transformation (66,67).

Until recently, most of our knowledge on mutation patterns in cancer was based on studies of a handful of genes frequently mutated in human cancers, including members of the *ras* family and *TP53*. Mutations in *K-ras* occur in up to 20–40% of common cancers, such as breast, colon, and pancreas and adenocarcinoma of the lung. The most common mutation is at codon 12, effectively limiting the spectrum of the mutation pattern to three different bases. *TP53* in contrast, offers a wider target for assessing mutation patterns since most mutations fall within a domain that spans about 600 nucleotides. Moreover, over 75% of all mutations are point mutations, providing a good representation of many different types of base changes. Figure 5.7 summarizes some of the most characteristic mutation patterns identified by sequencing *TP53*. Current efforts aimed at large-scale, high-throughput sequencing of tumour DNA are producing a wealth of mutation data that essentially recapitulate the mutation patterns observed in *TP53*. An interesting difference is the higher prevalence of the rare G to C transversions in breast cancer, identified by large-scale sequencing, as compared to *TP53* sequencing. This observation suggests that some unidentified carcinogen causing such mutations may be involved in mutagenesis leading to breast cancer (68).

Mutation databases provide a repository and quick access for published mutation data, with annotations that allow users to select

Figure 5.7. Examples of mutation patterns and carcinogen fingerprints in TP53. Three well described 'carcinogen fingerprints' are represented. Solar UV induces a characteristic DNA lesion, dipyrimidine dimer, leading to CC to TT transitions at adjacent cytosines in non-melanoma skin cancer. Aflatoxin metabolites form adducts on the N7 position of guanine at the third base of codon 249 in TP53, leading to AGG to AGT transversion mutations at that codon in hepatocellular carcinoma. Polycyclic aromatic hydrocarbons from tobacco smoke induce adduct formation on N2 position of several guanines, leading to frequent transversions at several codons in lung cancers of smokers. Source: (70) with kind permission from Springer.

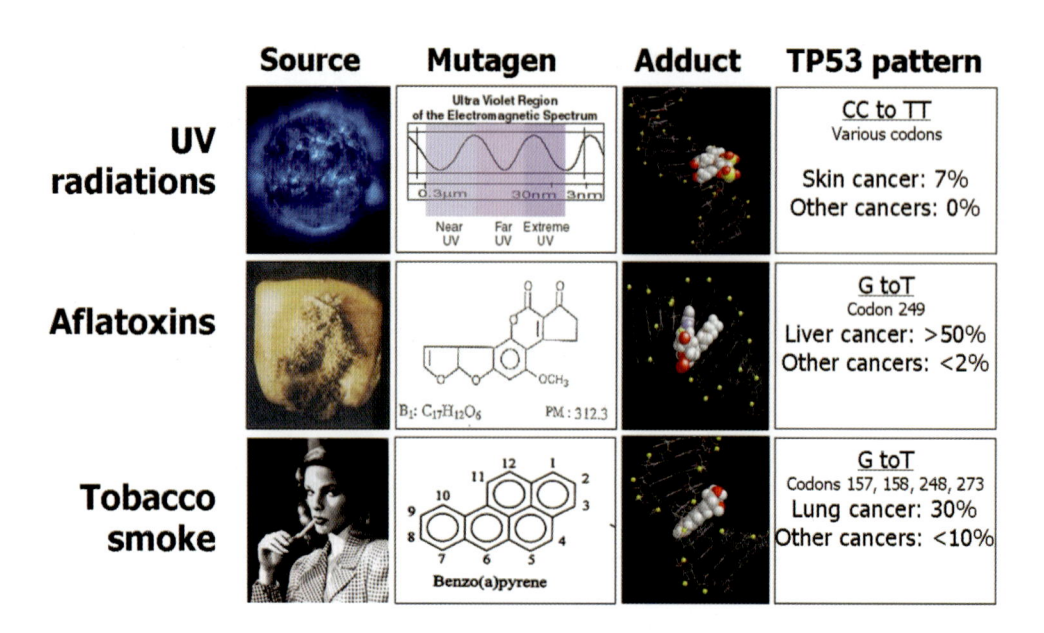

	Source	Mutagen	Adduct	TP53 pattern
UV radiations		Ultra Violet Region		CC to TT Various codons Skin cancer: 7% Other cancers: 0%
Aflatoxins		$B_1: C_{17}H_{12}O_6$ PM: 312.3		G to T Codon 249 Liver cancer: >50% Other cancers: <2%
Tobacco smoke		Benzo(a)pyrene		G to T Codons 157, 158, 248, 273 Lung cancer: 30% Other cancers: <10%

reference data sets and compare mutation patterns. Examples of such databases are given in Table 5.3. However, these databases are subject to many biases since they compile data from studies that differ in size, methods, design, case selection criteria, and annotations, and are prone to publication bias (reviewed in (69)).

Multistep carcinogenesis

Neoplasia is a multistep process. Experimental studies have demonstrated that the tumorigenic conversion of normal human fibroblasts requires the concerted disruption of several signalling pathways. The number and sequence of genetic changes required for neoplastic transformations varies according to species and according to cell type within species (70). Moreover, particular genes, chromosomal regions, and entire chromosomes are vulnerable to mutation at variable points in carcinogenesis (16). This suggests that certain mutations play a role in the ability of a cell to survive and continue to the next step of this multistep process and determine what the next mutation will be. These mutations, particularly early events, may provide markers indicative of genetic damage and potential cancer risk.

Since much of cancer research depends on backtracking from tumour tissue, it is difficult to assess the time point at which one mutation arose relative to another. Comparison between 'early stage' versus 'late stage' lesions does not entirely eliminate this difficulty, because lesions deemed 'early stage' are not necessarily the temporal predecessors of those deemed 'late stage.' However, models of temporal sequence of genetic events have been developed and have provided valuable information on the clonal and genetic progression of cancer. The archetype of these models was developed for colon cancer (16) (Figure 5.8). This model takes advantage of the fact that colon cancer has distinct morphological stages that define a pathological progression sequence, from polyp to adenoma and then carcinoma. By assessing the predominant mutations in each morphological stage, it has been possible to identify sequential genetic changes that underpin the morphological changes. From normal tissue, the model proposes that cells acquire one mutation after another, beginning with the loss of a key gene involved in cell proliferation (activated protein C (*APC*), detectable in benign polyps), aberrant methylation, further mutation of oncogenes (*K-ras*, detectable in many adenomas), and

Table 5.3. List and web links of selected mutation databases

Database Name	Content and Scope	Web Link
Catalogue of Somatic Mutations in Cancer (COSMIC)	Global catalogue of somatic mutations in 4773 cancer related genes; contains over 70 000 mutations	http://www.sanger.ac.uk/genetics/CGP/cosmic/
IARC *TP53* mutation database	Comprehensive database of *TP53* mutations in human tissues; contains over 25 000 entries	http://www-p53.iarc.fr/
The Human Genome Variation Society (HGVS)	The most comprehensive list of single-locus mutation databases and a portal to access them	http://www.hgvs.org/dblist/dblist.html
Mitochondrial Mutations (MITOMAP)	Compendium of polymorphisms and mutations of the human mitochondrial DNA	http://www.mitomap.org/
Mitelman Database of Chromosome Aberrations in Cancer	Chromosomal aberrations in relation to tumor characteristics, based either on individual cases or associations	http://cgap.nci.nih.gov/Chromosomes/Mitelman
The Mammalian Gene Mutation database (MGMD)	Searchable database of published mutagen-induced gene mutations in mammalian tissues	http://lisntweb.swan.ac.uk/cmgt/index.htm
Genetic Alterations in Cancer (GAC)	Comprehensive collection of data compiled from studies reported in the published literature on genetic alterations in tumors associated with exposure to specific chemical, physical, or biological agents that can be linked to genes implicated in the development of cancers	http://www.niehs.nih.gov/research/resources/databases/gac/index.cfm

finally loss of *DCC* and *TP53*, which are frequent events in carcinomas and may push the cell over the malignant cancer threshold. This concept has been expanded to other cancer types (Figure 5.8). It is important to note the differences in occurrence of mutations in each cancer type. For example, *TP53* mutations are believed to be early or predisposing events in astrocytoma and breast cancers, but are proposed to be later events in colon carcinogenesis. However, this order is not invariant and accumulation of mutations is the key factor in progression towards malignancy.

Mutations in mitochondrial DNA

The mitochondrial genome is ~16.6x10^3 base pairs in length, exists at a copy number of ~10^2-10^4 per cell, and is densely packed in protein coding sequence (~93% is used to encode 37 genes). Functional changes in mitochondria were associated with cancer as early as 1956, when Warburg proposed that irreversible damage to the respiration was a necessary first step in carcinogenesis (71–73). Changes in mitochondrial DNA (mtDNA), specifically, were associated with cancer as early as 1967, when a series of reports showed that the frequency of aberrations (in this case, multiple copy-length circular molecules) in mtDNA was increased in the leukocytes of granulocytic leukaemia patients (9,74–76). The presence of mtDNA mutations was reported in seven out of 10 colorectal cancer cell lines examined, with a predominance of mitochondria containing multiple copies of the mtDNA (7). It was then demonstrated that the mtDNA mutations were somatic, since they were found in the primary tumours from which the cell lines were derived, but not in normal tissues from the donors. In many cases, the mutations were homoplasmic, meaning that a single mutated mitochondrion had selectively proliferated over all others in a single cell (7).

Recent reports have documented somatic mtDNA mutations in tumours of the bladder, breast, prostate, head and neck, lung, liver, kidney, brain, stomach, pancreas and in the haematologic malignancies leukaemia and lymphoma. These findings support the notion that mtDNA mutations contribute to tumour growth (4,5,77–88). Furthermore, the copy number for mtDNA was recently found to be significantly increased in workers exposed to high levels (> 10ppm) of benzene, a carcinogen that causes leukaemia (89,90). It is not yet understood how mutations in mtDNA accumulate within tumours, but both theoretical and empirical approaches have suggested that

Figure 5.8. Four multistep models of carcinogenesis. A: Vogelstein and Kinzler model of mutation accumulation pattern in colon cancer (16). B: Cavanee and White model of astrocytoma progressing to secondary glioblastoma (73). C: Theoretical model of therapy-induced leukaemia. D: A simplified version of the Beckmann and Niederacher model of multistep carcinogenesis in breast cancer (74). These models are only intended to provide a rough overview of how these cancers may progress during typical carcinogenesis, and it should be kept in mind that these cancers may arise via different paths in individuals

A. Colon Cancer

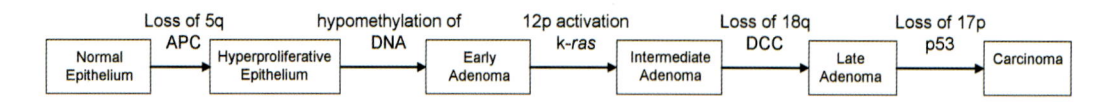

Normal Epithelium → (Loss of 5q, APC) → Hyperproliferative Epithelium → (hypomethylation of DNA) → Early Adenoma → (12p activation, k-*ras*) → Intermediate Adenoma → (Loss of 18q, DCC) → Late Adenoma → (Loss of 17p, p53) → Carcinoma

B. Astrocytoma

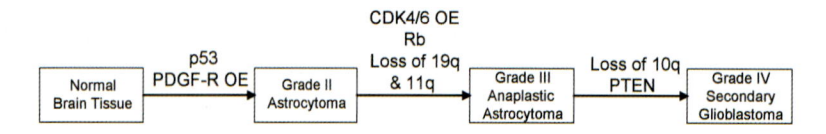

Normal Brain Tissue → (p53, PDGF-R OE) → Grade II Astrocytoma → (CDK4/6 OE, Rb, Loss of 19q & 11q) → Grade III Anaplastic Astrocytoma → (Loss of 10q, PTEN) → Grade IV Secondary Glioblastoma

C. Therapy Induced Leukemia

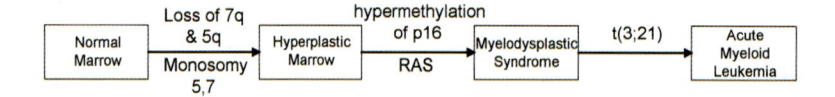

Normal Marrow → (Loss of 7q & 5q, Monosomy 5,7) → Hyperplastic Marrow → (hypermethylation of p16, RAS) → Myelodysplastic Syndrome → (t(3;21)) → Acute Myeloid Leukemia

D. Breast Cancer

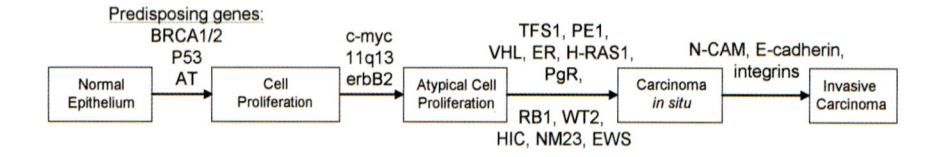

Normal Epithelium → (Predisposing genes: BRCA1/2, P53, AT) → Cell Proliferation → (c-myc, 11q13, erbB2) → Atypical Cell Proliferation → (TFS1, PE1, VHL, ER, H-RAS1, PgR, RB1, WT2, HIC, NM23, EWS) → Carcinoma *in situ* → (N-CAM, E-cadherin, integrins) → Invasive Carcinoma

---

they accumulate without selection (91,92).

A recent study used cytoplasmic hybrid technology to demonstrate that the metastatic potential of tumour cells was enhanced by mtDNA mutations associated with the overproduction of reactive oxygen species (3). However, many of the mutations reported so far are not associated with a detectable mitochondrial defect (7). Although it is still unclear how mutations in mtDNA contribute to carcinogenesis, these mutations are significant biomarkers in detecting tumour recurrence and in assessing genotoxic damage (5,93).

Detection of mutations

Many standard methods are available for detecting mutations in normal or diseased tissue samples; reviewing them is beyond the scope of this chapter. They differ by their sensitivity, scope (one or multiple genes), and by whether the detection aims to identify mutations at specific base positions or to screen large DNA fragments to detect mutations at any possible position within that fragment (Table 5.4). Independent of the technique used, the modern methodological cornerstones of mutation detection are PCR and DNA sequencing. This section briefly discusses the basic requirements for detecting somatic mutations, focusing on detecting low levels of mutant DNA in non-diseased or surrogate samples.

Table 5.4. Comparison of sensitivity of selected mutation detection methods

Technology	Detection limit (% mutant DNA)
RFLP	3-6
mEPCR	0.1
SOMA	<1
APEX*	3-6
DHPLC*	3-12
TTGE	10
Direct Sequencing	25

RFLP, restriction fragment length polymorphism; mEPCR, membrane expression of endothelial protein C receptor; SOMA, short oligonucleotide mass analysis; APEX, arrayed primer extension; DHPLC, denaturing high performance liquid chromatography; TTGE, temporal temperature gradient gel electrophoresis
Table compiled from (111) and (104)

Obtaining high-quality DNA

DNA is a robust molecule retrievable from biological materials stored in a wide range of conditions. However, DNA is sensitive to modifications by oxidation from prolonged contact with air, exposure to light (UV, in particular), enzymes, and by reaction with fixatives used in pathology. RNA may also be used as starting material for mutation detection. It is the recommended source for screening based on functional assays in which RNA is used to generate cDNA and express the protein *in vitro*, or when mutation detection is specifically aimed at identifying mutants with splicing defects. However, RNA is much more labile and unstable than DNA and is extremely sensitive to RNase present in biological materials.

The first challenge is to process and preserve specimens in a way that is compatible with obtaining good quality nucleic acids. Fresh frozen material is the best source and is mandatory for RNA. However, in many studies, the most routinely available material is tissue fixed in buffered formalin or alcohol and embedded in paraffin. Alcohol is preferable to formalin, as the latter induces the formation of covalent protein and DNA adducts. Other fixatives, such as alcoholic Bouin's, should be avoided since they contain chemicals that inhibit PCR. Fixed and embedded material yields DNA that is generally degraded by fragmentation and chemical modification. Though damage increases with overfixation, underfixation is also a problem as DNA may become degraded by chemical or enzymatic reactions. DNA fragmentation effectively limits the length of PCR-amplifiable fragments to 300–500 bp and DNA base modifications increase the risk of mutation artefacts during PCR. Despite these limitations, formalin-fixed tissue has been routinely used to detect mutations by PCR-based assays (94). The risk of artefactual mutation detection may be kept low by using strict laboratory protocols and mutation confirmation strategies (see below).

The second problem is to extract DNA from cells relevant for mutation detection analysis. Many tissue specimens obtained by resection or biopsy contain cells other than those suspected to contain mutations (e.g. stromal cells, blood vessels, infiltrated inflammatory cells, etc.) that are present in solid tumours. Tumors are heterogeneous in their cellular composition and contain areas of different stage, grade, or morphological differentiation. Surrogate specimens used as a source of cancer cells, such as sputum or exfoliated cells, may contain significant amounts of DNA of bacterial origin. The use of methods to enrich the specimen for DNA extraction in DNA from the appropriate source is recommended. With tissue sections, this may be achieved by prior assessment by a pathologist and delineation of areas of material to extract, either by marking segments of tissues on companion histological slides, or by using laser-guided microdissection to retrieve specific groups of cells from a histological section. In the case of haematologic malignancies, knowledge of cell surface protein characteristics can often be used along with antibodies and selection techniques (e.g. immunomagnetic or flow cytometric) to separate cancerous cells from normal blood cells.

PCR sensitivity and specificity

DNA extraction is easy to perform using standard protocols and commercially available kits. Controlling DNA quality by physical methods (spectrometry, fluorimetry, or gel analysis) is not mandatory since the 'gold standard' is suitability for PCR. The quality, amount, and specificity of PCR products should be systematically checked before analysis by sequencing or other mutation detection method.

It should be kept in mind that PCR generates miscoding artefacts. First, commercial polymerases used in PCR generate random incorporation errors at rates between less than

2×10^{-6} and 8.4×10^{-5} (95). Second, PCR is prone to contamination by adventitious material. In laboratories where the same gene(s) are routinely analysed, contamination by aerosols of PCR products is a serious problem which can be overcome with rigorous laboratory procedures (96), as well as the use of dUTP and uracil glycosylase in PCR reactions to prevent carryover contamination. The best practice is to compartmentalize the various steps of the protocol in different laboratory locations (DNA extraction, assembly of PCR reactions, and performance of PCR itself) with one-way circulation patterns to make sure those final PCR products never come into contact with biological materials for extraction.

The sensitivity of mutation detection depends on the PCR strategy (see below), the sequence context, and the methodology for mutation detection. In principle, the sensitivity of mutation detection is expressed as the minimal percentage of mutant material detectable in a background of wild-type material (Table 5.4). It is mandatory to confirm detected mutations in a second analysis performed using a batch of DNA independently extracted from the same specimen (not a second aliquot from the same extraction). It should be noted that up to 10% of mutations included in mutation databases may be false mutations resulting from fixation and/or PCR artefacts. The need for replication has a bearing on the cost of mutation detection studies.

Detection of point mutations in non-diseased tissues

Mutations occur in non-cancer cells at rates that are increased by exposure to carcinogens. They can also be detected in bodily fluids or exfoliated cells that contain only a small fraction of tumour-derived material, thus providing a means for detection of subclinical disease. The proportion of mutant DNA in such samples is too low for detection using conventional methods. However, mutations may be detectable using methods that are several orders of magnitude more sensitive than conventional ones.

The most remarkable studies on mutation detection in non-diseased tissues have been conducted on *TP53* mutations in skin (97,98) and on *TP53* mutations in non-cancerous liver, colon, and lung (99–101). In normal skin, detection is facilitated by the fact that the epithelium is made of juxtaposed patches of cells originating from single progenitors. Since many missense *TP53* mutations induce protein stabilization, immunohistochemistry can be used to detect patches of cells with *p53* accumulation, which are then microdissected and analysed by PCR/sequencing (97,98). Mutational signatures of solar UV have been detected in the DNA of normal skin of sun-exposed subjects (102). Studies in the liver, colon, and lung have used a sensitive genotypic assay (103). This method is based on the cloning of PCR products of the mutant allele into phage lambda followed by plaque assay and oligonucleotide hybridization to quantitate mutant PCR products. Its sensitivity is of one mutant DNA copy cell in about 10 million cells. Results have demonstrated increased mutation loads in the liver of patients with Wilson disease, in the colon of patients with ulcerative colitis (two oxyradical overload cancer precursor diseases) (99,101), and in the normal lung of heavy smokers without clinical evidence of cancer (100). However, these methods are labour-intensive and expensive, limiting their application in molecular epidemiology.

Detection of mutations in surrogate samples

Identifying cancer-related mutations in tissues other than cancer is a major goal for studies aimed at assessing the impact of environmental exposures, as well as developing molecular-based methods for early detection of cancer. This has led to the development of methods to detect mutations in exfoliated cells or DNA retrieved from bodily fluids or secretions (Table 5.5). Recent developments on the use of circulating free DNA (CFDNA) isolated from plasma or serum, provide a good example of the problems and challenges posed by mutation assessment in surrogate samples (104). The plasma of all subjects contains minute amounts of free DNA that occurs as a by-product of normal cell turnover in solid tissues. This DNA is unstable and does not accumulate at levels above one to 10 ng per ml. In patients with various cancers, inflammatory or autoimmune diseases, however, increased tissue destruction and cell turnover in the lesion results in abnormally high levels of CFDNA in the plasma/serum. It was estimated that for a patient with a tumour load of 100 g in size ($\sim3\times10^{10}$ cancer cells), up to 3.3% of the tumour DNA entered the circulation every day (105). Various types of DNA alterations have been reported in CFDNA, including point mutations, DNA hypermethylation, microsatellite instability, and losses of heterozygosity (LOH) in patients with many different types of cancer. In most cases, these alterations were identical to the ones detected in the patient's tumour tissue,

Table 5.5. Genetic and epigenetic changes both in tumors and matched CFDNA detected with different methods

Type of alteration	Cancer site	Method	Tumor	CFDNA	Reference
Point Mutation					
TP53	HNSCC	AS-PCR	11 (18%)	2/11 (18%)	(237)
K-ras2	Colorectal	PCR	35/135	29/35 (83%)	(238)
Hypermethylation					
CDKN2a	Esophagus	MSP	31/38 (82%)	7/31 (23%)	(239)
APC	Lung	RT-MSP	95/99 (96%)	42/89 (47%)	(240)
Genomic Instability					
LOH	Melanoma	Fluorescent PCR	34/40 (85%)	21/34 (62%)	(241)

CFDNA, circulating free DNA; CDKN2a, cyclin-dependent kinase inhibitor 2A; APC, adenomatous polyposis coli; RT-MSP, real-time methylated-specific PCR

supporting the tumoural origin of altered CFDNA. Thus, CFDNA may provide a very valuable source of genetic material as a surrogate for molecular analysis of cancer and pre-cancer patients, for detecting somatic alterations when biopsies are not available, and for accessing small amounts of tumour DNA when the exact position of a suspected primary lesion is not clearly defined. The fact that CFDNA can be obtained without invasive or painful procedures makes it particularly suitable for studies in a population-based context.

High-sensitivity detection of point mutations

The main problem for detecting mutation in non-diseased tissues or in surrogate samples is that the mutation is present in only a small fraction of the total DNA, a level too low for detection by conventional sequencing.

This section briefly describes recent high-throughput assays suitable for detection of low levels of mutant DNA in a background of wild-type DNA. The first four of these methods all have the same limitation: they require the prior knowledge of the exact position and type of the mutation, and are therefore limited to the detection of mutation hotspots. For detection of mutations at unselected sites, most studies have used pre-screening methods, such as single strand conformation polymorphism (SSCP), denaturing high performance liquid chromatography (DHPLC), denaturing gradient gel electrophoresis (DGGE), or related techniques (106–108). Several protocols are available to retrieve and re-amplify mutant DNA after pre-screening (either by excision of shifted bands detected by temporal temperature gradient gel electrophoresis (TTGE) or SSCP, or by collection of shifted peaks in DHPLC). This re-amplified, mutant-enriched material can then be analysed by direct sequencing. This approach has been successfully used to detect mutant TP53 DNA in the plasma of healthy subjects recruited in a prospective study (Figure 5.9) (109). A new method, arrayed primer extension (APEX), allows the detection of 'unknown' mutations within a given sequence (110–112). However, this assay is still in development and its suitability for large-scale studies remains to be demonstrated. Finally, this section concludes with a brief description of current high-throughput sequencing efforts and their perspectives for application in molecular epidemiology.

Mutation-enriched PCR

Mutation-enriched PCR (ME-PCR) is the most widely used procedure for genotypic selection of mutant DNA. It is based on restriction digestion using enzymes that cleave DNA at sites that are modified by mutations. This method selectively cleaves wild-type sequences, thus providing enrichment in mutant sequences. Two versions of this type of assay have been commonly used: restriction site mutation (RSM) is based on digestion before PCR amplification (113), and restriction fragment length polymorphism (RFLP) is based on restriction

Figure 5.9. Detection of low levels of KRAS (codon 12) mutation or of TP53 mutation in circulating free plasma DNA (CFDNA) of healthy subjects. KRAS2 (A) and TP53 (B) mutation detection in CFDNA. A: Detection of mutations in codon 12 of KRAS2 by ME-PCR (involving two consecutive RFLP analyses for enrichment of the mutant DNA). After MvaI digestion, the mutant PCR product (MT; white arrow) is excised, amplified, and sequenced. Black arrow – wild-type PCR product (WT). B: Detection of TP53 mutation at codon 282. Mutations in exons 5 to 9 are analysed by DHPLC. Samples with abnormal DHPLC chromatograms are sequenced from an independent PCR product. If the mutation is not detected by sequencing, a new PCR product is analysed by TTGE. Homoduplex products are excised from the TTGE gel, reamplified, and sequenced. Gray arrow – mutant-wild-type heteroduplexes; white arrow – mutant homoduplexes; black arrow – wild-type homoduplexes; white star – mutant control heteroduplexes (top two bands) and homoduplexes. Source: (113). Adapted and reprinted by permission from the American Association for Cancer Research.

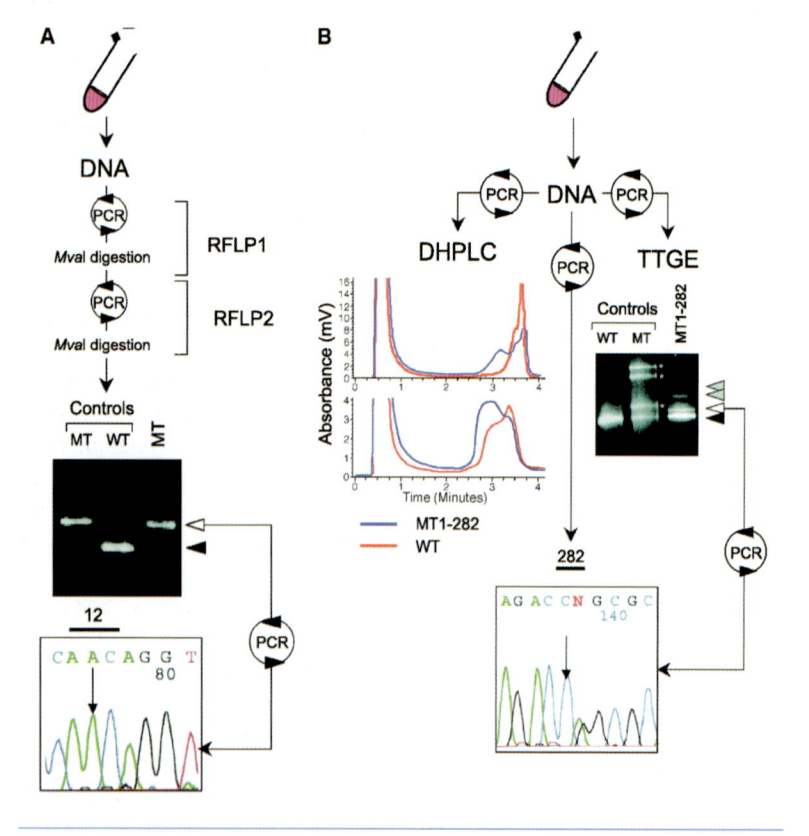

digestion after PCR (114). In both instances, mutations are identified by sequencing of digestion-resistant PCR products. A modified assay has been developed to detect mutations at DNA positions that do not fall within restriction sites. This assay uses two consecutive rounds of PCR to introduce a synthetic restriction site in the wild-type allele, thus generating a PCR product amenable to restriction. This method has been successfully applied to detect *K-ras* mutations in CFDNA present in the plasma of healthy subjects before diagnosis of cancer (Figure 5.9) (109).

Allele-specific PCR

Allele-specific PCR (AS-PCR) is based on the use of PCR primers that preferentially anneal with mutant DNA. The PCR products are then analysed using conventional methods (e.g. SSCP plus sequencing). The use of AS-PCR results in considerable improvement in sensitivity over conventional methods. AS-PCR analysis of *TP53* mutations resulted in the detection of mutated cells accounting for 0.01–1% of cells, sensitive enough to detect rare *TP53* mutations as early biomarkers of relapse in acute myelogenous leukaemia (AML) and acute lymphocytic leukaemia (ALL) (115,116). AS-PCR may be combined with PCR methods using fluorescent probes (the so-called 'Taqman' method) to detect rare mutations in a semiquantitative manner. A variant of AS-PCR that targets mutational hotspots in the *TP53* gene has been developed (117). This method combines PCR-SSCP with sequence-specific clamping by peptide nucleic acids (PNAs). PNAs are designed to preferentially bind to wild-type DNA, and not extend, thereby blocking amplification of wild-type DNA to yield a mutant enriched sample.

Combined Mut-Ex and allele-specific competitive blocker PCR

By combining two previously published methods (118,119), the Mut-Ex + allele-specific competitive blocker PCR (ACB-PCR technique provides one of the most sensitive genotypic selection methods (120). This assay begins with the denaturation of a heterogeneous sample of mutant and wild-type double stranded DNA. When reannealing, four types of DNA duplexes may be formed: the two homoduplexes of either wild-type or mutant DNA, and two types of heteroduplexes containing a mutant strand annealed to a wild-type strand. The proportion of each duplex depends upon the ratio of mutant to wild-type DNA in the sample. MutS, a thermostable mismatch repair

Figure 5.10. Short oligonucleotide mass analysis (SOMA) of TP53 R249S mutations. A: Principle of SOMA. DNA is amplified by PCR using primers that introduce a site for BpmI, a restriction enzyme that cleaves DNA away from its recognition site. Short oligonucleotides (8-mers) are generated by digestion, purified by HPLC, and analysed by electrospray mass spectrometry. B: Mass spectrum of the sense strand of the wild-type 8-mer (top spectrum) and of its breakdown products (bottom spectrum). Inset: expected mass of breakdown products. Presence of a specific species (framed) identifies the wild-type sequence (with G at third position of codon 249)

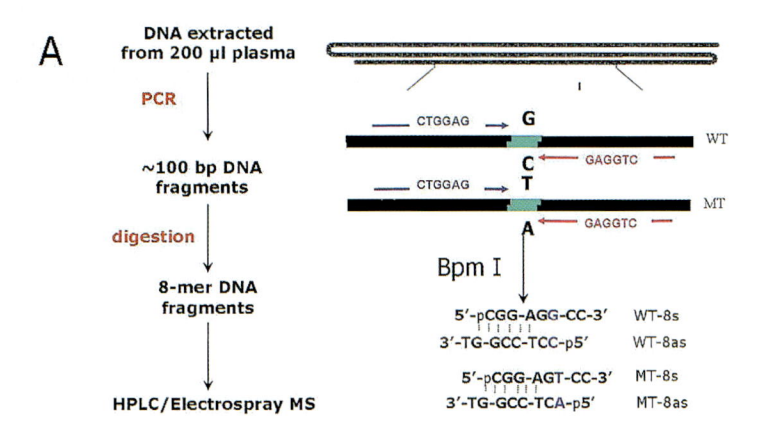

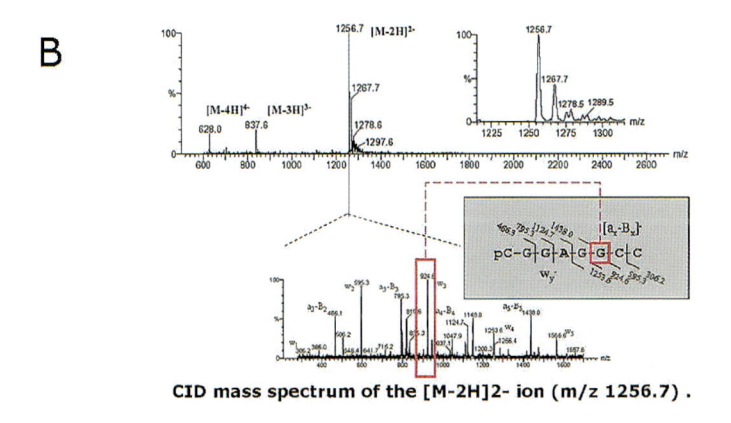

Short-oligonucleotide mass analysis

Short oligonucleotide mass analysis (SOMA) is a technique by which small sequences of mutated and wild-type DNA, produced by PCR amplification and restriction digestion, are characterized by high performance liquid chromatography (HPLC)-electrospray ionization tandem mass spectrometry (ESI-MS/MS) (121–123). DNA is amplified using primers that introduce restriction sites for enzymes that cleave DNA at positions away from their binding sites, such as BpmI. Short DNA fragments spanning the mutation site (seven to 15 base pair oligomers) are then produced by restriction digestion and separated by HPLC before ESI-MS/MS. The first MS analysis distinguishes the four single-stranded oligonucleotides corresponding to sense and antisense, wild-type, and mutant DNA. The second MS analyses oligonucleotide fragmentation products and detects mass fragments characterizing the mutated base (Figure 5.10). The use of an internal standard plasmid alongside test DNA allows the precise quantitation of mutant and wild-type sequences, which can be expressed in absolute copy numbers. This method has been applied to detection of *K-ras* and *TP53* mutations in the plasma DNA and tissues of healthy subjects and cancer patients (124,125). Quantitation of mutant CFDNA by SOMA in a case-control study of liver cancer in The Gambia (West Africa), has shown that *TP53* gene serine 249 mutation median levels were higher in hepatocellular carcinoma cases (2.8×10^3 copies/mL, range: 5×10^2-1.1×10^4) compared with median levels in cirrhotic patients and healthy controls (5×10^2 copies/mL, range: 5×10^2-2.6×10^3

protein, binds to the mispaired sequence of the heteroduplex and protects a short sequence of mutant DNA from digestion by the 3'-5' exonuclease activity of T7 DNA polymerase, whereas the wild-type DNA is digested. This Mut-Ex step results in a 10^3-fold enrichment of mutant alleles. The next step utilizes an additional selection technique, allele specific competitive blocking. This genotypic selection uses allele-specific primers to amplify mutant DNA, combined with blocker primer which preferentially anneals to the wild-type sequence. The blocker primer is modified with a 3'-dideoxyguanosine residue preventing extension. The combination of Mut-Ex and ACB-PCR results in the preferential amplification of the mutant allele, with a sensitivity of as few as one mutant allele per 10^7 copies of the wild-type allele.

and 5x10² copies/mL, range: 5x10²-2x10³ respectively) (124). This highly powerful method is rapid and amenable to scaling-up, making it one of the most powerful approaches for mutation detection in a large series of specimens.

BEAMing

BEAMing is an original method aimed at one-to-one conversion of a population of DNA fragments into a population of beads that can be counted. It derives its name from its principal components: beads, emulsion, amplification, and magnetics (Figure 5.11). First, PCR is used to amplify target DNA using primers that contain a sequence tag. Second, PCR products are mixed with beads coupled to an oligonucleotide that anneals with the tag. This mixture is emulsified to facilitate the reaction of individual PCR products with individual beads. Third, the DNA immobilized on the beads is denatured, hybridized with primers that anneal just upstream of the mutation site, and then a single nucleotide primer extension reaction is carried out using four fluorescently labelled nucleotide terminators. Flow cytometry is next used to rapidly measure the fluorescence of individual beads. The nature of the base changes is given by the fluorescence of the incorporated nucleotide. Counting fluorescent beads provides a precise estimate of the number of wild-type or mutant DNA copies, and allows quantitation of mutant and wild-type frequencies even when they are present at ratios less than 1:10 000. This method has been used to quantify mutant *APC* in the circulating plasma DNA of patients with colorectal cancer (105,126,127). Quantitation of mutant *APC* in the plasma of patients with advanced colorectal cancer detected on average 5.3x10³ (11.1%) copies/mL of mutant APC (range: 9.08x10² (1.9%)-1.2x10⁴ (27%)) (105).

Arrayed primer extension

Arrayed primer extension (APEX) is a genotyping and resequencing technology that allows the scanning of mutations over large regions of DNA. It combines the advantages of Sanger dideoxy sequencing with the high-throughput potential of the microarray format (Figure 5.12). A DNA sample is amplified, fragmented enzymatically, and annealed to arrayed 25-mer oligonucleotides that cover the sequence of interest. Each oligonucleotide hybridizes one base downstream of the preceding one, with their 3' ends one base upstream of the base to be identified. Once hybridized, they serve as primers for template-dependent DNA polymerase extension reactions by using four fluorescently labelled dideoxynucleotides. Each base is probed with two primers: one for the sense and another for the antisense strand. Image analysis and interpretation of fluorescence signals at each position provides a read-out of the sequence. This method has been adapted for the detection of *TP53* mutations in DNA isolated from plasma or from solid tumours, with a sensitivity of 0.1–5%, depending upon the sequence context and the nature of the mutation (which is higher than sequencing or conventional oligonucleotide hybridization arrays). Whether this method will prove robust enough for large-scale studies using non-diseased tissues or surrogate samples remains to be assessed.

High-throughput sequencing

The rapid development of long-range sequencing technologies makes it possible to comprehensively sequence the coding regions of the human genome. The cost and time

Figure 5.11. High-sensitivity, single DNA template mutation detection using BEAMing: APC mutations in circulating free plasma DNA. A: Extended beads were prepared attaching single PCR products to single beads in an emulsion mixture. B: Single base extensions were performed on the extended beads using four different fluorescent nucleotides. Normal DNA sequences contained a G at the queried position; mutant sequences contained an A. Source: (109). Copyright 2005 National Academy of Sciences, U.S.A.

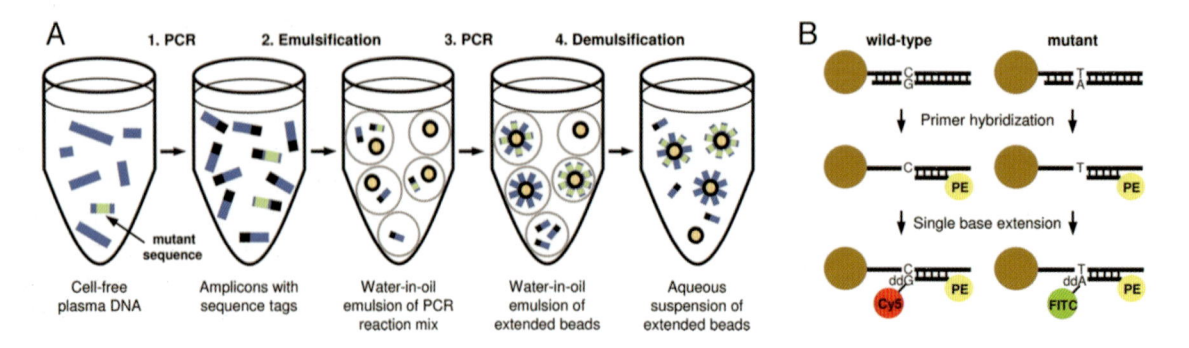

Figure 5.12. Detection of missense mutations in TP53 using arrayed primer extension (APEX). DNA is amplified by PCR and fragmented before hybridization to arrayed oligonucleotide that anneal with a sequence just 1 base upstream of the position of interest. Single base extension is then performed on the arrays using four different fluorescent nucleotides. The incorporated base is detected by acquisition and analysis of fluorescence data at each position of the array. Source: AsperBio (http://www.asperbio.com/)

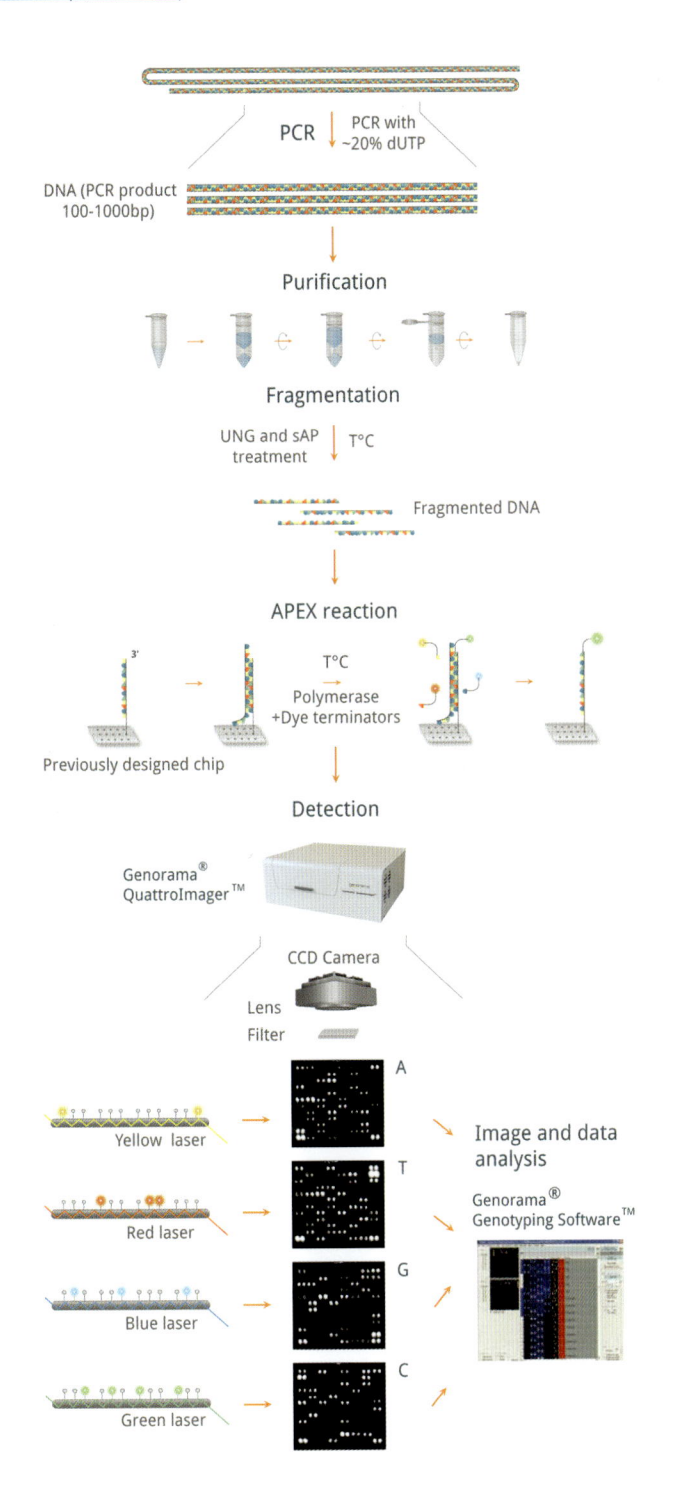

required to conduct high-throughput sequencing is decreasing at an extremely rapid pace, one that is reminiscent of the dramatic reduction in costs and increase in performance of microprocessors in the eighties and nineties. Within a few years, it is possible that high-throughput sequencing will largely replace current approaches for genome-wide analysis of multiple polymorphisms. The International Cancer Genome Consortium (http://www.icgc.org) has been organized to launch and coordinate several large-scale sequencing research projects with the primary goal of generating comprehensive catalogues of genomic abnormalities (somatic mutations, abnormal expression of genes, epigenetic modifications) in tumours. Collectively, high-throughput sequencing studies will generate a wealth of novel information on patterns of mutations in cancer. However, current technologies lack the sensitivity needed to detect somatic mutations in non-cancer tissues. Moreover, many of the mutations detected in such large-scale sequencing efforts appear to have no direct role in the development of cancer and simply happened to mutate as passengers in the tumour. Distinguishing 'drivers' from 'passengers,' and interpreting the significance of the latter as biomarkers of processes involved in mutagenesis, will require intensive research efforts.

Detection of genetic damage at the chromosome level

Chromosome aberrations encompass all types of changes in chromosome structure and number and have been shown to be involved in the development of cancer (e.g. leukemias and lymphomas (128,129). The most common numerical changes (resulting in aneuploid cells) are the loss (monosomy) or gain (trisomy) of one chromosome; less frequent types include the loss of both copies or the gain of more than one copy of a chromosome. Structural changes include translocations, inversions, breaks, and deletions. Chromosome loss can lead to the loss of tumour suppressor genes, while chromosome gain can lead to increased oncogene expression. Further, chromosome translocations, or other types of chromosome rearrangements, may lead to the formation of fusion genes with oncogenic properties.

Conventional cytogenetics

Chromosome aberrations are the only cytogenetic endpoint that has been shown to predict cancer risk (19,130), particularly in haematologic malignancies (Figure 5.13) (131). They may thus represent a promising early effect biomarker of carcinogen exposure. However, classical aberrations measure overall chromosome damage rather than specific events on the causal pathways of particular diseases. This decreases their specificity as biomarkers of exposure-related diseases, making it necessary to screen large populations or examine many cells from each subject to attain sufficient statistical power.

Many specific chromosome aberrations have been identified using classic karyotyping among patients with clinical syndromes. For example, an extra copy of chromosome 21 is routinely detected among children born with Down syndrome. As a result, classic karyotyping has become a widely used clinical diagnostic tool for many diseases, including leukaemia. However, classic cytogenetic techniques have several drawbacks for the detection of chromosome-specific aneusomy and rearrangements: the cells must be cultured to generate metaphase spreads, only a limited number (25–100) of scoreable cells can be examined, and recognition of specific chromosomes is problematic. The use of fluorescence in situ hybridization overcomes these problems.

Figure 5.13. Chromosome aberrations in haematological malignancies. In acute myeloid leukaemia (AML), loss of part or all of chromosomes 5 and 7 is a common event, along with trisomy of chromosome 8 and various specific translocations and inversions including inv(16), t(8;21), t(9;22), t(15;17), and t(11q23) (107). In acute lymphocytic leukaemia (ALL), particularly in childhood ALL, the translocation t(12;21) is common (~25%). In non-Hodgkin's lymphomas the translocation t(14;18) is frequently found (> 70%) in follicular lymphoma (98,132). Therefore, the detection of these changes at the chromosomal level could be very important in predicting risk of these diseases

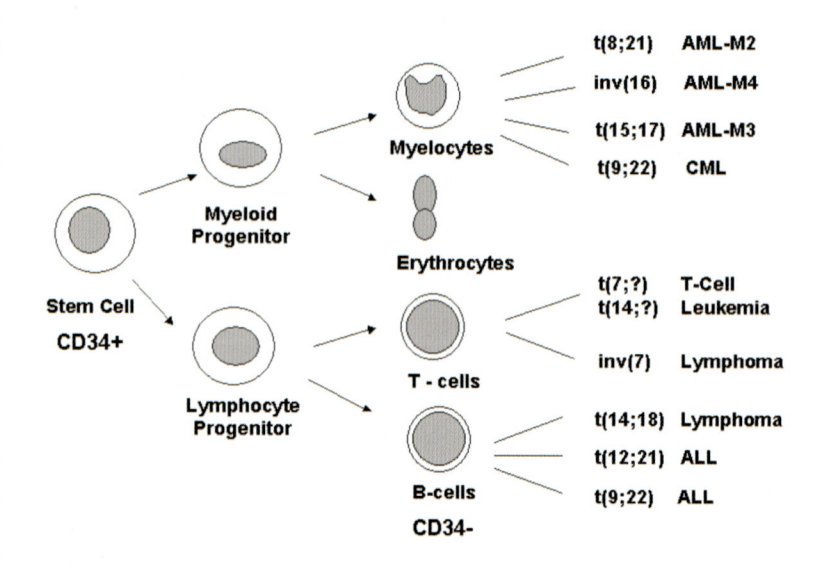

Molecular cytogenetics

Fluorescence in situ hybridization

Fluorescence in situ hybridization (FISH) has several advantages over conventional cytogenetics, including selectivity of specific DNA probes, multiple colour labelling, sensitivity of detection, and speed of microscopic analysis (132,133). Interphase FISH, in particular, offers several advantages over classical cytogenetics (134). First, it allows analysis of non-dividing cells. Second, a much larger number of cells, at least 10^3 or more, may be analysed. Third, the detection of aneuploidy is facilitated by simply counting the number of labelled regions corresponding to a particular chromosome within the nucleus. In contrast, metaphase FISH can readily detect structural rearrangements in addition to aneuploidy. The use of metaphase FISH makes it possible to directly compare and reconcile interphase FISH and conventional cytogenetics. Several studies have determined that FISH is more sensitive and convenient than classical cytogenetics, thus appearing to be the more suitable method for large-scale population biomonitoring (135–137). FISH is now widely used in the analysis of chromosomal changes in human cancers (e.g. leukemias) and in prenatal diagnostics (133,138).

FISH has been extensively used to analyse chromosomal damage induced by exposure to ionizing radiation (139,140), and has also been gradually applied in evaluating genetic damage in cancer cases and in exposed populations (141–143). For example, a specialized FISH assay used for radiation research in humans and experimental animals has been developed (144–146). This assay uses single-colour FISH for painting the chromosome pairs 1, 2, and 4 (or 3, 5 and 6) the same colour, which allows for the detection of both the numerical and structural chromosome aberrations among these painted chromosomes, and the structural rearrangements between these and other untargeted chromosomes. Since radiation is thought to cause equal levels of damage across all chromosomes (147), and chromosomes 1 through 6 (the largest chromosomes) make up 40% of the genome (148), measurement of damage in these large chromosomes may be extrapolated to the whole genome (144). This may not be true for chemical exposures, as certain chemicals may have selective or preferential effects on certain chromosomes (149), as observed for epoxide metabolites of 1,3-butadiene (150). Indeed, the hypothesis of equal levels of damage across the genome may not hold true for low doses of radiation, as inversion of chromosome 10 has been shown to be highly sensitive to low intensity radiation exposure (151). Interestingly inv(10) rearranges the *RET* gene and is associated with thyroid cancer, potentially caused by linear energy transfer radiation.

Current studies employ FISH to examine the cytogenetic changes in human blood cells caused by exposure to the established leukemogen, benzene. Pilot studies in highly exposed workers from China and controls have analysed five chromosomes by metaphase FISH, demonstrating striking, dose-dependent increases in monosomy and trisomy in some chromosomes, as well as several common structural changes (149,152). In particular, loss and long arm deletion of chromosomes 5 and 7, two of the most common cytogenetic changes in therapy- and chemical-related leukaemia, were significantly increased in benzene-exposed workers over controls (149).

While FISH can be used to measure both structural and numerical chromosome aberrations and is a powerful tool in molecular epidemiology, its sensitivity is limited to 1 in 10^3-10^4 cells and it is relatively expensive because of the high cost of probes. This makes it difficult to use FISH to detect rare translocations between multiple chromosomes, such as t(21q22) and t(11q23). The PCR technique allows much more sensitive detection of these types of changes and is also less expensive in comparison with FISH.

Other molecular cytogenetic methods

Novel cytogenetic methods have been recently developed, such as comparative genome hybridization (CGH), spectral karyotyping (SKY), and colour banding. CGH involves the comparison of total DNA extracted from normal and cancerous cells to detect specific gains or losses in genetic material associated with cancer (139). Initially developed using metaphase spreads as a template for hybridization, CGH is now commonly performed on cDNA or oligonucleotide microarrays representative of the whole genome. This method is now widely used to identify variations in copy numbers in tumour DNA, but has not been applied to the analysis of damage induced by environmental exposures in non-diseased human tissues (although some experimental studies in animals and cell lines have shown changes induced by carcinogens).

The SKY method involves painting each of the 24 different chromosomes a different colour using four or five fluorophores with combined binary ratio labelling, which allows the entire karyotype to be screened for chromosome aberrations (153). Since the human eye cannot effectively distinguish the 24 colors, this method requires the use of an automated imaging system. In colour banding, which is based on traditional banding techniques, each chromosome is labelled by subregional DNA probes in different colors, resulting in an unique 'chromosome bar code' (154). This method allows the rapid identification of chromosomes and chromosome rearrangements. These techniques, however, are relatively new and have not been employed as widely or extensively as FISH.

Measurement of chromosome rearrangements by PCR

Chromosome translocations produce novel fusion genes or products that can be detected at the DNA or RNA level by PCR or reverse-transcriptase PCR (RT–PCR), as well as by FISH. PCR holds several advantages over FISH, including the ability to detect very rare events (1 copy/10^{6-7} cells versus 1/10^{3-4} cells by FISH), and the ability to study large numbers of people easily and at low cost. However, the high sensitivity of PCR makes it prone to false-positive results caused by sample contamination (see above). The use of exonuclease-dependent real-time PCR ('TaqMan' technology, now generally called real-time PCR) allows for the absolute number of novel sequences to be quantified in a cell population without the need for gel electrophoresis. While no methods yet exist which employ PCR to measure rare aneuploidies or genome-wide structural damage, real-time and conventional PCR techniques, which measure specific chromosome rearrangements, have become available. RT–PCR has previously been used to detect translocations including t(14;18), t(8;21), t(9;22), and t(4;11). Using these techniques, t(9;22) and t(14;18) have been detected in unexposed individuals of different ages and in smokers (155–157). Both translocations were found to increase with age and the t(14;18) translocation was increased in cigarette smokers (158). Studies showing detectable t(8;21) by RT–PCR in an otherwise healthy benzene-exposed worker (152), demonstrate the potential of RT–PCR for monitoring specific aberrations in populations exposed to suspected or established leukemogens. Because many of these translocations have multiple breakpoints or translocation partners, multiplex assays have been developed to detect multiple or unknown rearrangements.

Principle of real-time PCR

Real-time PCR is comparable to conventional PCR in that it uses sense and antisense primers to frame and amplify a targeted sequence of DNA. However, real-time PCR employs an additional, non-extendable oligonucleotide probe, which is positioned between the two primers during the annealing phase of amplification (Figure 5.14) (159). The oligonucleotide probe is labelled with a fluorescent reporter dye (e.g. FAM (6-carboxyfluorescein)) at the 5' end and a quencher fluorescent dye (e.g. TAMRA (6-carboxytetramethylrhodamine)) at the 3' end. When the probe is intact, fluorescence resonance energy transfer to TAMRA quenches the FAM emission. During the extension phase of amplification, the Taq polymerase extends the primer to the region of the probe, at which point the 5' exonuclease property of Taq cleaves the reporter dye from the probe. This results in an increase in fluorescent signal that is proportional to the amount of amplification product, measured in real time by appropriate florescence detection systems. After each cycle, the fluorescence signal is measured resulting in an amplification plot. The cycle number in which the fluorescence crosses a defined threshold, Ct, is inversely proportional to the number of copies of DNA templates in the PCR. Cts of positive control samples are used to generate a standard curve to calculate copy numbers in unknown samples. Methods for the quantitative detection of translocations using the above TaqMan technology have recently been reported (160–165).

Measurement of t(14;18) associated with lymphocytic leukaemia and lymphoma

The t(14;18) translocation induces Bcl-2 protein overexpression in lymphocytic leukaemia and lymphoma. It may be caused by illegitimate V(D)J recombination in pre-B-cells as a result of aberrant immunoglobulin gene rearrangement (166,167), although some recent studies have concluded that most breaks on

Figure 5.14. Principle of real-time PCR. Source: (164). Copyright Elsevier (1999).

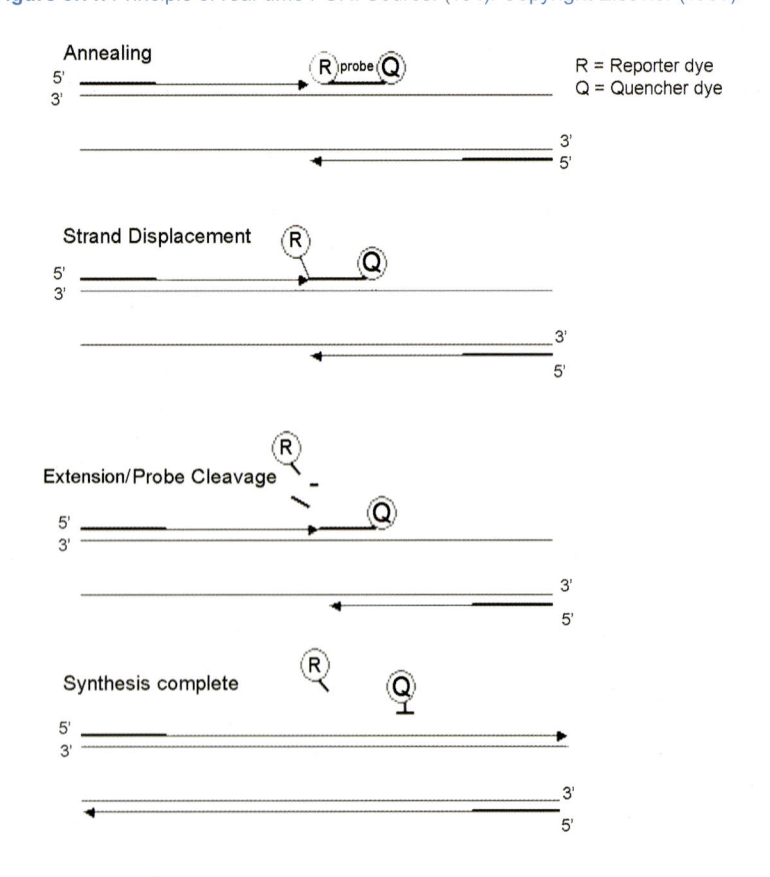

chromosome 18 are independent of V(D)J recombinase activity (94,168). This translocation was first identified at low levels in normal, healthy individuals (157). They subsequently showed that the incidence of t(14;18) increased with age and was higher in the blood of smokers (158). Recently, novel quantitative PCR procedures that measure very low levels of t(14;18) have been described (163,169). Rearrangements were detected at the Bcl-2 major breakpoint region in 36% of lymphoma cases, and a 98% concordance between real-time PCR and conventional PCR was found (163). In addition, using serial dilution it was demonstrated that real-time PCR was 100-fold more sensitive than conventional PCR. Bcl-2/JH fusion sequences were consistently detected when diluted 105-fold with normal genomic DNA. Others confirmed the sensitivity of this assay and concluded that the detection of single genome copies is possible if a stochastic multiple tube approach is taken (169).

Measurement of t(8;21) associated with acute myeloid leukaemia and myelodysplasia

The t(8;21) translocation results in the fusion of the *ETO* gene (8q22) with the *AML1* gene (21q22) and is one of the most frequent abnormalities observed in AML. The presence of t(8;21) is associated with high complete remission and survival rates (170), suggesting that the levels of the translocation may be predictive of relapse. A real-time RT-PCR method to detect *AML1/ETO* fusion transcript in patients with AML was developed (171). Each patient showed 10^3 copies of *AML1/ETO* transcript at diagnosis and a 2–4-log decrease in copy numbers following successful induction chemotherapy. In one patient, relapse was predicted by high copy number immediately after induction chemotherapy, which continued to increase during initial remission. These results suggest the t(8;21) translocation is detectable at low levels and may be a valuable biomarker of early effect or potential relapse.

Measurement of t(9;22) associated with leukaemia

A real-time RT–PCR method has been developed for the detection of the t(9;22) translocation (172), which is common in chronic myelogenous leukaemia (CML). This translocation results in the fusion of the *ABL* gene, an oncogene, with the *BCR* gene (173). The fusion gene product is expressed in malignant cells. By performing serial dilutions of a positive control diluted in wild-type RNA, a sensitivity of 10^{-5} was achieved, which is comparable to conventional PCR methods.

Micronuclei as indicators of chromosome damage in humans

Micronuclei are small cytoplasmic fragments of nuclear membrane-encapsulated DNA that are excluded from daughter nuclei during telophase after a clastogenic or aneugenic event. An increased frequency of micronuclei in reticulocytes (nascent red blood cells) has long been used as an index of acute mutagenic exposure in animals (174–178). Micronucleated (MN) reticulocytes are easily detected using DNA stains in both the peripheral blood and bone marrow of rodents following recent mutagenic exposure. Close to 70% of known human carcinogens are detected using this *in vivo* MN assay. Accordingly, this animal test is widely used as a pre-clinical cytogenetic assay for genotoxicity (21,179-181).

Reticulocytes are produced from rapidly dividing progenitors in the bone marrow and are well suited for rapid assessment of MN frequency because they, like mature red blood cells, have extruded their nucleus. However, the spleen filters micronucleated red cells out of circulation, so that erythroid micronuclei (known to haematologists as Howell-Jolly bodies) are found less frequently in the peripheral blood than in the marrow, and disappear from both of these compartments 2–3 days after aberrations occur. Furthermore, once an MN is formed, the genomic lesion can contribute to cell death, and, even if the cell is still able to proliferate, the MN is not replicated and will therefore be diluted in the progeny that are formed. Thus, the presence of micronuclei is mostly used as a biomarker of either recently occurring chromosome aberrations or of an increased tendency for these events to occur (indicated by relatively high baseline levels of MN formation or by an increased response upon exposure).

Since human marrow samples are not routinely collected when monitoring for exposure effects or cancer risk in healthy populations, reticulocytes are not used to evaluate MN formation in human tissues. Efforts to use micronuclei as mutation biomarkers in human tissues have instead mainly focused on cultured human lymphocytes or exfoliated cells (182). The MN assay in cultured human lymphocytes (the cytokinesis-block micronucleus (CBMN) assay) employs cytochalasin-B *in vitro* to interrupt cell division after the first telophase, allowing the visualization of daughter nuclei and excluded micronuclei (182–184). Cells that have undergone mitosis (a necessary step in MN formation) are identified by their binucleate appearance in the CBMN assay; cytochalasin-B prevents the dilution of micronuclei in the culture population by prohibiting proliferation. Although the CBMN assay is conducted on lymphocytes *in vitro*, it has been shown to be an accurate indicator of both carcinogenic exposure in the lymphocyte donor and increased cancer risk. For example, the CBMN assay has been used to establish that spontaneous MN generation increases with age of the lymphocyte donor (185,186), and that MN frequency is increased in the lymphocytes of nurses handling cytotoxic drugs (187), workers exposed to chlorinated solvents (188), and mortician students exposed to formaldehyde (189). Furthermore, a case-control study using the CBMN assay demonstrated that micronucleus frequency is increased in the lymphocytes of smokers with lung cancer (190). A nested case-control study recently showed that micronucleus frequency assessed in peripheral blood lymphocytes of disease-free subjects is a good predictor of cancer death risk (191).

The measurement of micronuclei in epithelial cells has also been used as a biomarker for monitoring DNA damage and is perhaps the least invasive method available for measuring DNA damage in humans (192). The MN assay in buccal cells was first proposed in 1983 (193), and has since been used to demonstrate that increased MN frequencies correlate with occupational exposures, lifestyle factors, dietary deficiencies, and various disease states. In early studies, exfoliated buccal mucosa cells were used to test for the genotoxic effects of betel nuts and quids of chewing tobacco (194,195). These studies often showed that higher MN frequencies were observed at the site where the quid or tobacco mixture was kept, compared to the opposite side. Other studies have demonstrated that increased MN frequency in buccal cells is significantly increased in people exposed to arsenic (196,197), formaldehyde (189), and smokeless tobacco (198). Furthermore, arsenic exposure has been shown to modulate the MN frequency in exfoliated bladder cells (199,200). As with the CBMN assay in lymphocytes, an increased MN frequency in epithelial cells has also been linked to cancers such as oral submucous fibrosis, oral carcinoma, and breast cancer (201–203). Therefore, although micronuclei are formed from a variety of chromosomal aberrations, only some of which are on the causal pathway to disease, MN frequency in cultured lymphocytes, red cells, and epithelial tissue remains a useful biomarker for evaluating exposure and cancer risk.

Detection of epigenetic changes

In addition to the wide range of genetic damage involved in carcinogenesis, epigenetic mechanisms (e.g. methylation of CpG islands in DNA) have gained attention as key players in certain cancer types. Although tumorigenesis is accompanied by global hypomethylation of the genome, some particular regions may become hypermethylated. Approximately 70% of known human promoter regions are found within CpG islands (204), and methylation of the cytosine within these CpG regions modulates expression by silencing genes. DNA methylation is normally controlled by the activity of the DNA methyltransferase family of enzymes, but it is also known that changes in local DNA structure, environmental exposure (e.g. nickel, plutonium, polycyclic aromatic hydrocarbons), and microsatellite instability can contribute to aberrant promoter methylation. Methylation patterns are reset during embryogenesis, although specific methylation patterns may be heritable through imprinting. Changes in the promoter methylation of a large series have been linked to aging and cancer; methods for measuring these epigenetic changes are critical to understanding the genetic basis of disease. Information on the diversity of methylated genes in cancer is available at the Methylation Cancer Database (MethCancerDB), which collects data on aberrant CpG methylation in human tumours and currently contains information on over 2000 genes (http://www.methcancerdb.net/methcancerdb/home.seam).

Aberrant methylation may result in changes in the dysregulation of every type of cell processes involved in carcinogenesis. Furthermore, changes in methylation patterns may occur synergistically with mutations or chromosomal aberrations. For example, in leukaemia and lymphoma, translocations cause the formation of novel fusion genes that produce excessive growth (128,129). Other genes undergo transcriptional silencing by methylation, which causes aberrant cell cycle control (205). Aberrant methylation and transcriptional silencing appears to be an early event in both solid tumours, including lung (206), colon (207), hepatocellular (208), and bladder (209), as well as haematologic malignancies (205).

Because of their widespread presence in the genome, their diversity, relevance to disease, and potential responsiveness to environmental changes, changes in methylation patterns are ideal candidate early effect biomarkers. Several different methods have been developed to detect aberrant methylation of genes, but either methylation-sensitive restriction enzyme digests of DNA or bisulfite conversion of DNA is at the core of most. So far, however, the application of these methods to non-diseased human tissues is still in its infancy.

Methylation-sensitive digestion

Restriction enzymes that have CpG sites in their recognition sequence and are sensitive to methylation status can be used to differentially digest DNA before using other techniques, such as PCR or Southern blotting, to analyse the resulting digest. It is preferable to use enzymes that have at least two CpG sites in their recognition sequence (e.g. NotI, SacII, EagI, and BsHII) to increase the specificity of the method. Southern blotting after digest is the gold standard for measuring aberrant methylation. There are very few examples of a Southern blot analysis that has not been confirmed by other techniques. However, the downside of Southern blot analysis is that it is time consuming and requires a large amount of high-quality DNA, which precludes the use of paraffin-embedded tissue. Methylation-specific restriction followed by PCR using primers that frame the digestion site can give false-positives due to incomplete digestion. Restriction landmark genome scanning is a whole-genome approach that relies on methylation-sensitive restriction digest, followed by radiolabelling the resulting fragments and 2D electrophoresis to resolve differences in the digest (210). Methylated CpG island amplification followed by microarray analysis is another modern technique, built on methylation-sensitive restriction enzymes, that facilitates high-throughput detection of aberrant methylation (211,212).

Bisulfite conversion assays

The basis for another class of methylation assays is bisulfite conversion of DNA. While normal cytosine is converted to uracil upon treatment with bisulfite, methylated cytosine is unaffected. Since uracil is read by polymerases as thymine, bisulfite conversion can be used to introduce polymorphisms into a sequence based on methylation status (Figure 5.15). After bisulfite conversion, methylation-specific PCR, which targets the created polymorphisms with primer(s) that overlap with the modified site at their 3' ends, can be used to distinguish between the two variants. Alternatively, standard PCR framing the modified site can be used to provide amplicon for SNP analysis using a variety of techniques including direct sequencing, pyrosequencing, methylation-sensitive single-strand conformational analysis, high-resolution melting analysis, methylation-sensitive single nucleotide primer extension, hybridization to arrayed oligonucleotides, or base-specific cleavage/MALDI-TOF. In a recent study, bisulfite conversion and pyrosequencing have been used to analyse the methylation patterns of several genes in lung cancer tissues from smokers, ex-smokers, and life-time never smokers, demonstrating substantial differences according to histology and to smoking history (213). A limitation of bisulfite conversion techniques is that only single-stranded DNA is susceptible to bisulfite, which requires that the DNA is denatured during analysis. Thus these techniques require optimization of temperature and salt concentrations during conversion to achieve a high degree of sensitivity and specificity. In addition, conditions of bisulfite conversion can lead to degradation of DNA, specifically depurination and random strand breaks.

Among aberrantly methylated genes in cancer, the tumour suppressor gene p16^{INK4a} has been

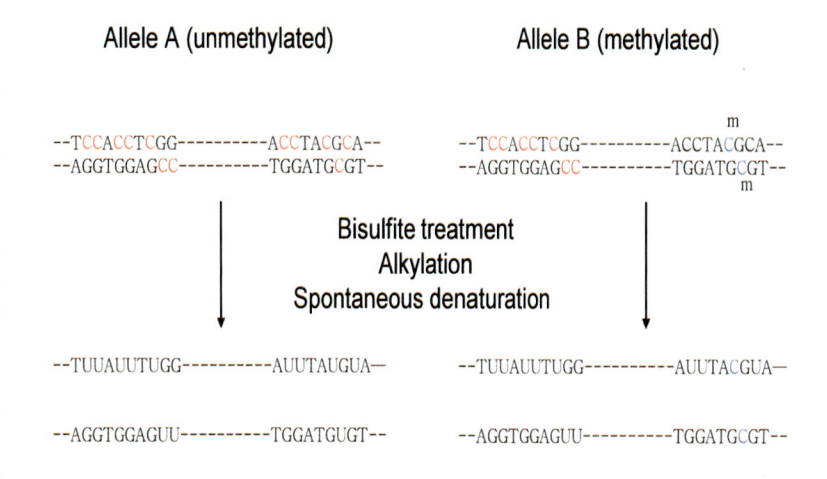

Figure 5.15. Detection and quantification of cytosine methylation using bisulfite conversion assay. Bisulfite conversion of a DNA sequence. Nucleotides in red are unmethylated cytosines that will be converted to uracils by bisulfite, while blue nucleotides are 5-methylcytosines resistant to conversion

one of the most extensively studied. This gene is a key component in the G1/S cell cycle checkpoint and has been shown to be involved in almost every type of solid cancer and in leukaemia. The INK4a locus is of special interest in cancer as it contains an alternative reading frame (ARF), encoding a different protein which also exerts tumour suppressive activities. Each ARF has its own promoters, which may be simultaneously or differentially methylated (214). A real-time methylation-specific PCR protocol has been developed for p16^{INK4a} and applied to bone marrow samples of patients with multiple myeloma (215). This method showed high concordance with conventional methods, plus the added sensitivity and specificity of the real-time technology. In addition, researchers correlated methylation status with p16 mRNA expression and observed that transcription was inversely correlated with methylation status. Other recent publications have found that promoter hypermethylation of p16 is associated with poor prognosis in recurrent early-stage hepatocellular carcinoma, and that aberrant p16 methylation is associated with nasopharyngeal carcinoma (216,217). As with other real-time methods, this application shows great potential for future studies involving methylation of key genes in carcinogenesis, as well as other biological processes.

Emerging technologies for measuring mutations in single cells

The ultimate goal of studies on genetic damage is to detect alterations at the level of a single cell, making it possible to capture modifications that precede and initiate pathological processes. Recently, several research groups have developed microfabricated genetic analysis systems based on performing PCR reactions in nanolitre volumes, either on a wafer or in emulsion, to separate individual templates for analysis and sequencing. These developments promise to vastly increase the throughput and sensitivity of long established methods for mutation detection (218–228). One such development is the BEAMing technique described earlier (126).

Microfabricated capillary array electrophoresis

Recently, a 96-channel microfabricated capillary array electrophoresis device was developed and applied (229) for high-throughput genotyping of the methylenetetrahydrofolate reductase gene (MTHFR), which has been shown to be predictive of increased risk of leukaemia (230). The microfabricated 96-channel capillary electrophoresis system used a radial channel format coupled with a four colour radial confocal fluorescence scanner. The chip was formed by microfabricating the capillary electrophoresis channels on one wafer and then bonding it to a drilled blank wafer. Samples were then introduced into reservoirs on the perimeter, electrophoresed into the channel intersection, and separated on a gel in under two minutes. This device was capable of rapid (< two minutes) two colour genotyping of 96 MTHFR allelic variants (229).

Single cell analysis using laboratory-on-a-chip technologies

Microfluidic and laboratory-on-a-chip technologies have advanced to the point that single cell genetic analysis is feasible on a high-throughput scale. The key to the concept of single cell genetic analysis is the idea of a PCR colony or 'polony' (231). The basic idea is the dilution of a PCR template into the single molecule limit followed by PCR amplification of this template in a format that limits product diffusion, either by placing the amplification in a gel or by defining a picolitre volumetric element (232,233). The amplification in a restricted volume proceeds to produce a colony of up to ~10^8 amplicons that can be sequenced by sequential extension and fluorescence or by pyrosequencing. In the case of the method used by the commercial company 454 Life Sciences, one of the primers is covalently linked to a bead that is statistically distributed and trapped in the picolitre volume reactor. This ensures that each bead will yield progeny from only one template. The bead further facilitates the retention of the PCR product as one goes through the steps of pyrosequencing. This concept has been extended to single cell genomics (231), which demonstrates the isolation of DNA from a single bacterial cell followed by multiple displacement amplification of the cellular DNA, producing enough product for cloning and sequencing from a single cell. Other work has demonstrated 'digital PCR,' where PCR template targets are diluted at the statistical limit into picolitre volume reactors in a microfabricated poly (dimethylsiloxane) (PDMS) structure (234). The observation of distinct stochastic product production allows the genetic characterization at the single molecule level. Thus, it is evident that there has been much success in the performance of single genomic copy amplification and PCR in picolitre-sized volumes followed by genetic analysis.

Microdroplet generation and single template PCR

Sample preparation and analysis is typically performed at the microlitre or millilitre scale, because of the limitations of manual and robotic sample transport and measuring technologies, as well as detection technologies. However, new technologies exploit microfabricated microfluidics for sample transport and manipulation and operate in tandem with miniaturized and sensitive detectors. The marriage of these two technological advances will dramatically impact the performance of clinical and point-of-care devices and is a critical part of single cell genetic analysis. Devices and methods, including the use of specific DNA probe-based affinity gels or beads for target capture and purification in DNA sequencing and DNA computing applications, have also advanced for performing sample preparation (218,219,221). A current version of one of these devices is shown in Figure 5.16 (228). This microdroplet generating nozzle forms uniform microdroplets containing PCR reagents, microbeads, and single cell templates. These methods have been used to sequence plasmid control templates and to detect amplicons of the gyrB and GAPDH control genes in single *E. coli* and lymphocyte cells, respectively (218,228).

Figure 5.16. Microfluidic device for single copy DNA template amplification. Single copy genetic amplification. A: target DNA or cells and beads are mixed with the PCR reagent (blue) at very dilute concentrations and pumped through a microfabricated droplet generator. Monodisperse nanolitre volume droplets of the PCR reagent are formed in a carrier oil (yellow) at the cross-injector and routed into a tube for temperature cycling. The number of droplets containing a single bead and a single target DNA/cell is controlled by varying their concentrations in the PCR solution and controlling the droplet volume. B: Each functional PCR mix droplet contains a bead covalently labelled with the reverse primer, dye labelled forward primer, and a single target copy. Subsequent steps of PCR generate dye labelled double stranded product on the bead surface. Following emulsion PCR, the droplets are broken and the beads are analysed by flow cytometry to quantify the bound clonal amplified product. C: Microdroplet generation for controlled formation of nanolitre PCR droplets: layout of device, showing the PCR solution inlet, the two oil inlets, and the droplet outlet ports (red). A three layer (glass-PDMS-glass) pneumatically controlled micropump is integrated on-chip to deliver PCR reagent containing dilute 34μm beads and template. The manifold layer (blue) controls valve actuation, and the via hole connects the glass-PDMS hybrid channel (green) to the thermally bonded all-glass channel and cross-injector (black). Etch depth, 100μm. D: Optical micrograph of droplet generation at the cross-injector. Droplets are generated at a frequency of 5.7Hz with a combined oil flow rate of 2.2μL/min and a PCR solution flow rate of 0.8μL/min. For this experiment, average bead concentration was 130 beads/μL (0.33 bead/droplet). E: Optical micrograph of droplets with a predictable stochastic distribution of beads. Adapted from (233), Figures 1 and 2. Copyright (2008) American Chemical Society.

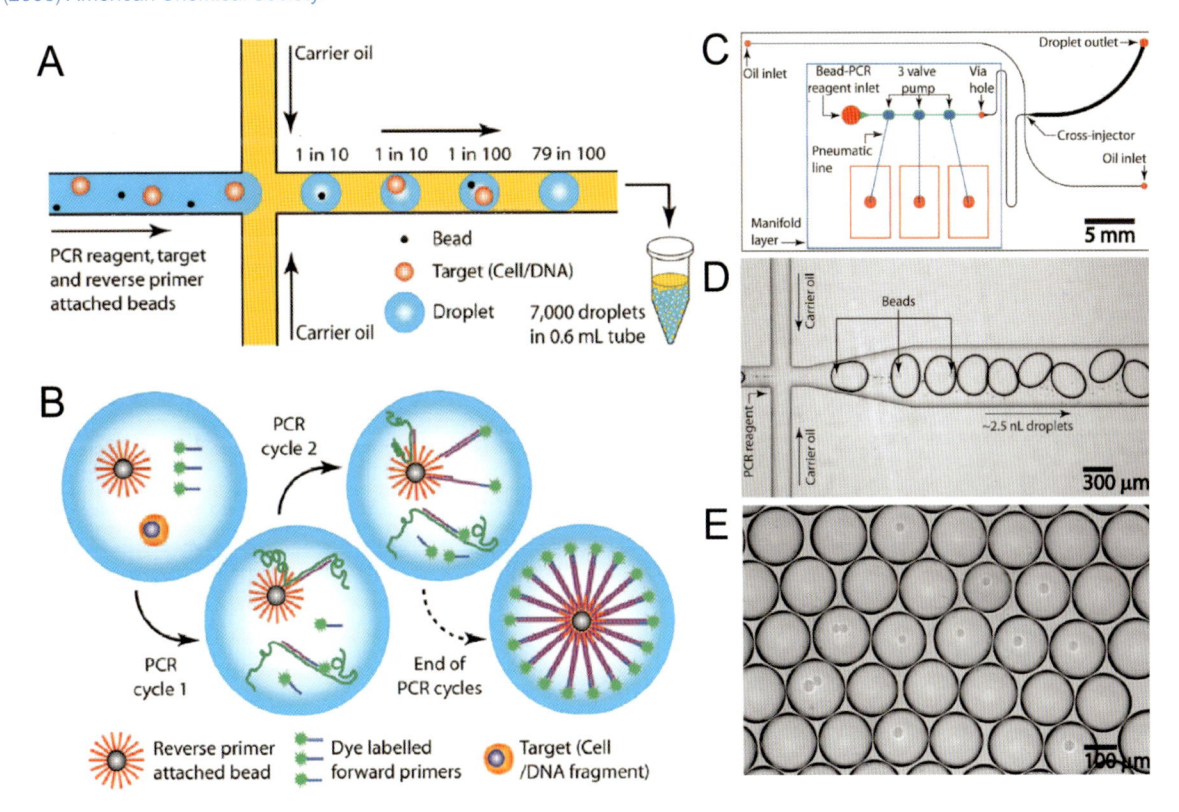

Conclusions

A new generation of candidate biomarkers of early effect in carcinogenesis is now available. Their validation in translation studies is currently a major focus in molecular epidemiology. Several of these makers are now available for application in large-scale studies on human populations. These methods utilize the latest advances in molecular cytogenetics and PCR allowing for genetic or epigenetic changes to be detected and measured in cancer-related genes and in specific regions of chromosomes that are rearranged, lost, or amplified in carcinogenesis. The combination of these methods with microfluidics and advanced fluorescence detection systems opens a wide horizon for innovative technologies and the development of novel laboratory instruments.

These new early effect biomarkers are on the causal pathway to disease and, as such, should have important application in predictive clinical tests of cancer risk. In addition, the high sensitivity of these assays will allow the detection of genetic damage in normal, healthy individuals, either as the result of ongoing, endogenous DNA damaging processes, or as the result of environmental exposure to chemical or physical carcinogens and infectious agents. These developments will have a considerable impact on our understanding of gene-environment interactions and of the early molecular steps of disease. However, their true value will only be assessed by their application into clinical trials and prospective epidemiological studies.

The implementation of these new biomarkers has important implications on the design, cost, conduct, and analysis of molecular epidemiology studies. Two particularly important aspects must be underlined. First, the cornerstone of such studies is the development of high-quality biobanks that include sophisticated systems for specimen collection, storage, and processing. Critical in this process is the definition of protocols for pre-analytical processing of the samples (from collection to storage and from storage to the bench). Heterogeneity and lack of quality control in these protocols are currently the main obstacles to the application of any type of novel molecular biomarker. Second, it should be anticipated that implementing such biomarkers will lead us to reconsider some statistical aspects of study design. The major problem in this respect will be to develop studies with sufficient power to distinguish between 'passenger' and 'driver' genetic changes. The use of highly sensitive methods at the single cell level will inevitably generate an unprecedented level of heterogeneity in current data on the human genome. Interpreting and mastering this heterogeneity is an important challenge which will be critical for the cost-effective implementation of molecular biomarkers.

References

1. Human Genome Sequencing Consortium I; International Human Genome Sequencing Consortium (2004). Finishing the euchromatic sequence of the human genome. *Nature*, 431:931–945.doi:10.1038/nature03001 PMID:15496913

2. Curwen V, Eyras E, Andrews TD et al. (2004). The Ensembl automatic gene annotation system. *Genome Res*, 14:942–950.doi:10.1101/gr.1858004 PMID:15123590

3. Ishikawa K, Takenaga K, Akimoto M et al. (2008). ROS-generating mitochondrial DNA mutations can regulate tumor cell metastasis. *Science*, 320:661–664.doi:10.1126/science.1156906 PMID:18388260

4. Czarnecka AM, Golik P, Bartnik E (2006). Mitochondrial DNA mutations in human neoplasia. *J Appl Genet*, 47:67–78.doi:10.1007/BF03194602 PMID:16424612

5. Fliss MS, Usadel H, Caballero OL et al. (2000). Facile detection of mitochondrial DNA mutations in tumors and bodily fluids. *Science*, 287:2017–2019.doi:10.1126/science.287.5460.2017 PMID:10720328

6. Penta JS, Johnson FM, Wachsman JT, Copeland WC (2001). Mitochondrial DNA in human malignancy. *Mutat Res*, 488:119–133.doi:10.1016/S1383-5742(01)00053-9 PMID:11344040

7. Polyak K, Li Y, Zhu H et al. (1998). Somatic mutations of the mitochondrial genome in human colorectal tumours. *Nat Genet*, 20:291–293.doi:10.1038/3108 PMID:9806551

8. Taylor RW, Turnbull DM (2005). Mitochondrial DNA mutations in human disease. *Nat Rev Genet*, 6:389–402.doi:10.1038/nrg1606 PMID:15861210

9. Clayton DA, Vinograd J (1967). Circular dimer and catenate forms of mitochondrial DNA in human leukaemic leucocytes. *Nature*, 216:652–657.doi:10.1038/216652a0 PMID:6082459

10. Chin L, Gray JW (2008). Translating insights from the cancer genome into clinical practice. *Nature*, 452:553–563.doi:10.1038/nature06914 PMID:18385729

11. Hanahan D, Weinberg RA (2000). The hallmarks of cancer. *Cell*, 100:57–70.doi:10.1016/S0092-8674(00)81683-9 PMID:10647931

12. Herceg Z, Hainaut P (2007). Genetic and epigenetic alterations as biomarkers for cancer detection, diagnosis and prognosis. *Mol Oncol*, 1:26–41.doi:10.1016/j.molonc.2007.01.004 PMID:19383285

13. Kunkel TA (2004). DNA replication fidelity. *J Biol Chem*, 279:16895–16898.doi:10.1074/jbc.R400006200 PMID:14988392

14. Lodish HF. Molecular cell biology. 5th ed. New York: W.H. Freeman and Company; 2003.

15. Lengauer C, Kinzler KW, Vogelstein B (1998). Genetic instabilities in human cancers. *Nature*, 396:643–649.doi:10.1038/25292 PMID:9872311

16. Vogelstein B, Kinzler KW (1993). The multistep nature of cancer. *Trends Genet*, 9:138–141.doi:10.1016/0168-9525(93)90209-Z PMID:8516849

17. Albertini RJ, Hayes RB (1997). Somatic cell mutations in cancer epidemiology. Lyon: IARC Scientific Publication; (142):159–184. PMID:9354918

18. Hagmar L, Bonassi S, Strömberg U et al. (1998). Cancer predictive value of cytogenetic markers used in occupational health surveillance programs. *Recent Results Cancer Res*, 154:177–184. PMID:10026999

19. Hagmar L, Brøgger A, Hansteen IL et al. (1994). Cancer risk in humans predicted by increased levels of chromosomal aberrations in lymphocytes: Nordic study group on the health risk of chromosome damage. *Cancer Res*, 54:2919–2922. PMID:8187078

20. Sorsa M, Wilbourn J, Vainio H (1992). Human cytogenetic damage as a predictor of cancer risk. Lyon: IARC Scientific Publication; (116):543–554. PMID:1428097

21. Kirkland DJ, editor. Basic mutagenicity tests: UKEMS recommended procedures (UKEMS Sub-Committee on Guidelines for Mutagenicity Testing. Report. Part 1 Revised). Cambridge (England); New York (NY): Cambridge University Press; 1990.

22. Holland N, Bolognesi C, Kirsch-Volders M et al. (2008). The micronucleus assay in human buccal cells as a tool for biomonitoring DNA damage: the HUMN project perspective on current status and knowledge gaps. *Mutat Res*, 659:93–108.doi:10.1016/j.mrrev.2008.03.007 PMID:18514568

23. Iarmarcovai G, Bonassi S, Botta A et al. (2008). Genetic polymorphisms and micronucleus formation: a review of the literature. *Mutat Res*, 658:215–233.doi:10.1016/j.mrrev.2007.10.001 PMID:18037339

24. Majer BJ, Laky B, Knasmüller S, Kassie F (2001). Use of the micronucleus assay with exfoliated epithelial cells as a biomarker for monitoring individuals at elevated risk of genetic damage and in chemoprevention trials. *Mutat Res*, 489:147–172.doi:10.1016/S1383-5742(01)00068-0 PMID:11741033

25. Natarajan AT, Palitti F (2008). DNA repair and chromosomal alterations. *Mutat Res*, 657:3–7. PMID:18801460

26. Wilson DM 3rd, Thompson LH (2007). Molecular mechanisms of sister-chromatid exchange. *Mutat Res*, 616:11–23. PMID:17157333

27. Toniolo P, Boffetta P, Shuker DEG et al. (editors). Application of biomarkers in cancer epidemiology. Lyon: International Agency for Research on Cancer; 1997.

28. Sjöblom T (2008). Systematic analyses of the cancer genome: lessons learned from sequencing most of the annotated human protein-coding genes. *Curr Opin Oncol*, 20:66–71.doi:10.1097/CCO.0b013e3282f31108 PMID:18043258

29. Sjöblom T, Jones S, Wood LD et al. (2006). The consensus coding sequences of human breast and colorectal cancers. *Science*, 314:268–274.doi:10.1126/science.1133427 PMID:16959974

30. Cooper DN, Ball EV, Krawczak M (1998). The human gene mutation database. *Nucleic Acids Res*, 26:285–287.doi:10.1093/nar/26.1.285 PMID:9399854

31. Krawczak M, Cooper DN (1997). The human gene mutation database. *Trends Genet*, 13:121–122.doi:10.1016/S0168-9525(97)01068-8 PMID:9066272

32. Cooper DN, Krawczak M. Human gene mutation. Human molecular genetics series. Oxford, UK: Bios Scientific Publishers; 1993.

33. Antonarakis SE, Krawczak M, Cooper DN (2000). Disease-causing mutations in the human genome. *Eur J Pediatr*, 159 Suppl 3;S173–S178.doi:10.1007/PL00014395 PMID:11216894

34. Stenson PD, Ball E, Howells K et al. (2008). Human Gene Mutation Database: towards a comprehensive central mutation database. *J Med Genet*, 45:124–126.doi:10.1136/jmg.2007.055210 PMID:18245393

35. Stenson PD, Ball EV, Mort M et al. (2003). Human Gene Mutation Database (HGMD): 2003 update. *Hum Mutat*, 21:577–581. doi:10.1002/humu.10212 PMID:12754702

36. Besaratinia A, Pfeifer GP (2006). Investigating human cancer etiology by DNA lesion footprinting and mutagenicity analysis. *Carcinogenesis*, 27:1526–1537.doi:10.1093/carcin/bgi311 PMID:16344267

37. Hainaut P, Hollstein M (2000). p53 and human cancer: the first ten thousand mutations. *Adv Cancer Res*, 77:81–137.doi: 10.1016/S0065-230X(08)60785-X PMID:10549356

38. Wogan GN, Hecht SS, Felton JS et al. (2004). Environmental and chemical carcinogenesis. *Semin Cancer Biol*, 14:473–486.doi:10.1016/j.semcancer.2004.06.010 PMID:15489140

39. Guengerich FP (2000). Metabolism of chemical carcinogens. *Carcinogenesis*, 21:345–351.doi:10.1093/carcin/21.3.345 PMID:10688854

40. Guengerich FP, Parikh A, Yun CH et al. (2000). What makes P450s work? Searches for answers with known and new P450s. *Drug Metab Rev*, 32:267–281.doi:10.1081/DMR-100102334 PMID:11139129

41. Burchell B, Coughtrie MW (1997). Genetic and environmental factors associated with variation of human xenobiotic glucuronidation and sulfation. *Environ Health Perspect*, 105 Suppl 4;739–747.doi:10.2307/3433277 PMID:9255555

42. Guengerich FP (2003). Activation of dihaloalkanes by thiol-dependent mechanisms. *J Biochem Mol Biol*, 36:20–27. PMID:12542971

43. Guengerich FP (2003). Cytochrome P450 oxidations in the generation of reactive electrophiles: epoxidation and related reactions. *Arch Biochem Biophys*, 409:59–71. doi:10.1016/S0003-9861(02)00415-0 PMID: 12464245

44. Guengerich FP (2003). Cytochromes P450, drugs, and diseases. *Mol Interv*, 3:194–204.doi:10.1124/mi.3.4.194 PMID:14993447

45. Miller JA (1994). Recent studies on the metabolic activation of chemical carcinogens. *Cancer Res*, 54 Suppl;1879s–1881s. PMID:8137303

46. Hanawalt PC, Ford JM, Lloyd DR (2003). Functional characterization of global genomic DNA repair and its implications for cancer. *Mutat Res*, 544:107–114.doi:10.1016/j.mrrev.2003.06.002 PMID:14644313

47. Hoeijmakers JH (2001). DNA repair mechanisms. *Maturitas*, 38:17–22, discussion 22–23.doi:10.1016/S0378-5122(00)00188-2 PMID:11311581

48. Hoeijmakers JH (2001). Genome maintenance mechanisms for preventing cancer. *Nature*, 411:366–374.doi:10.1038/35077232 PMID:11357144

49. Myrnes B, Giercksky KE, Krokan H (1982). Repair of O6-methyl-guanine residues in DNA takes place by a similar mechanism in extracts from HeLa cells, human liver, and rat liver. *J Cell Biochem*, 20:381–392.doi:10.1002/jcb.240200408 PMID:7183679

50. Samson L, Cairns J (1977). A new pathway for DNA repair in Escherichia coli. *Nature*, 267:281–283.doi:10.1038/267281a0 PMID:325420

51. Teo IA, Karran P (1982). Excision of O6-methylguanine from DNA by human fibroblasts determined by a sensitive competition method. *Carcinogenesis*, 3:923–928.doi:10.1093/carcin/3.8.923 PMID:7127672

52. Denissenko MF, Pao A, Pfeifer GP, Tang M (1998). Slow repair of bulky DNA adducts along the nontranscribed strand of the human p53 gene may explain the strand bias of transversion mutations in cancers. *Oncogene*, 16:1241–1247.doi:10.1038/sj.onc.1201647 PMID:9546425

53. Tornaletti S, Pfeifer GP (1994). Slow repair of pyrimidine dimers at p53 mutation hotspots in skin cancer. *Science*, 263:1436–1438. doi:10.1126/science.8128225 PMID:8128225

54. Hainaut P, Pfeifer GP (2001). Patterns of p53 G–>T transversions in lung cancers reflect the primary mutagenic signature of DNA-damage by tobacco smoke. *Carcinogenesis*, 22:367–374.doi:10.1093/carcin/22.3.367 PMID:11238174

55. Pluquet O, Hainaut P (2001). Genotoxic and non-genotoxic pathways of p53 induction. *Cancer Lett*, 174:1–15.doi:10.1016/S0304-3835(01)00698-X PMID:11675147

56. Livneh Z (2001). DNA damage control by novel DNA polymerases: translesion replication and mutagenesis. *J Biol Chem*, 276:25639–25642.doi:10.1074/jbc.R100019200 PMID:11371576

57. Kumar S, Subramanian S (2002). Mutation rates in mammalian genomes. *Proc Natl Acad Sci USA*, 99:803–808.doi:10.1073/pnas.022629899 PMID:11792858

58. Holliday R, Grigg GW (1993). DNA methylation and mutation. *Mutat Res*, 285:61–67. PMID:7678134

59. Jones PA, Rideout WM 3rd, Shen JC et al. (1992). Methylation, mutation and cancer. *Bioessays*, 14:33–36.doi:10.1002/bies.950140107 PMID:1546979

60. Olivier M, Hussain SP, Caron de Fromentel C et al. (2004). TP53 mutation spectra and load: a tool for generating hypotheses on the etiology of cancer. Lyon: IARC Scientific Publication; (157):247–270. PMID:15055300

61. Hussain SP, Hofseth LJ, Harris CC (2003). Radical causes of cancer. *Nat Rev Cancer*, 3:276–285.doi:10.1038/nrc1046 PMID:12671666

62. Schmutte C, Yang AS, Nguyen TT et al. (1996). Mechanisms for the involvement of DNA methylation in colon carcinogenesis. *Cancer Res*, 56:2375–2381. PMID:8625314

63. Greenblatt MS, Grollman AP, Harris CC (1996). Deletions and insertions in the p53 tumor suppressor gene in human cancers: confirmation of the DNA polymerase slippage/misalignment model. *Cancer Res*, 56:2130–2136. PMID:8616861

64. Antonarakis SE, Lyle R, Deutsch S, Reymond A (2002). Chromosome 21: a small land of fascinating disorders with unknown pathophysiology. *Int J Dev Biol*, 46:89–96. PMID:11902692

65. Essigmann JM, Wood ML (1993). The relationship between the chemical structures and mutagenic specificities of the DNA lesions formed by chemical and physical mutagens. *Toxicol Lett*, 67:29–39.doi:10.1016/0378-4274(93)90044-X PMID:8451766

66. Vrieling H, van Zeeland AA, Mullenders LH (1998). Transcription coupled repair and its impact on mutagenesis. *Mutat Res*, 400:135–142. PMID:9685614

67. Dogliotti E, Hainaut P, Hernandez T et al. (1998). Mutation spectra resulting from carcinogenic exposure: from model systems to cancer-related genes. *Recent Results Cancer Res*, 154:97–124. PMID:10026995

68. Hollstein M, Hergenhahn M, Yang Q et al. (1999). New approaches to understanding p53 gene tumor mutation spectra. *Mutat Res*, 431:199–209. PMID:10635987

69. Pfeifer GP, Besaratinia A (2009). Mutational spectra of human cancer. *Hum Genet*, 125:493–506.doi:10.1007/s00439-009-0657-2 PMID:19308457

70. Shi H, Le Calvez F, Olivier M, Hainaut P. Patterns of TP53 mutations in human cancer: interplay between mutagenesis, DNA repair and selection. In Hainaut P, Wiman KG, editors. 25 years of p53 research. The Netherlands: Springer; 2005. p. 293–319.

71. Hernandez-Boussard T, Montesano R, Hainaut P (1999). Sources of bias in the detection and reporting of p53 mutations in human cancer: analysis of the IARC p53 mutation database. *Genet Anal*, 14:229–233. PMID:10084119

72. Rangarajan A, Hong SJ, Gifford A, Weinberg RA (2004). Species- and cell type-specific requirements for cellular transformation. *Cancer Cell*, 6:171–183. doi:10.1016/j.ccr.2004.07.009 PMID:15324700

73. Furnari FB, Fenton T, Bachoo RM et al. (2007). Malignant astrocytic glioma: genetics, biology, and paths to treatment. *Genes Dev*, 21:2683–2710.doi:10.1101/gad.1596707 PMID:17974913

74. Beckmann MW, Niederacher D, Schnürch HG et al. (1997). Multistep carcinogenesis of breast cancer and tumour heterogeneity. *J Mol Med*, 75:429–439.doi:10.1007/s001090050128 PMID:9231883

75. Warburg O (1956). On respiratory impairment in cancer cells. *Science*, 124:269–270. PMID:13351639

76. Warburg O (1956). On the origin of cancer cells. *Science*, 123:309–314.doi:10.1126/science.123.3191.309 PMID:13298683

77. Weinhouse S, Warburg O, Burk D, Schade AL (1956). On respiratory impairment in cancer cells. *Science*, 124:267–272.doi:10.1126/science.124.3215.267 PMID:13351638

78. Clayton DA, Davis RW, Vinograd J (1970). Homology and structural relationships between the dimeric and monomeric circular forms of mitochondrial DNA from human leukemic leukocytes. *J Mol Biol*, 47:137–144.doi:10.1016/0022-2836(70)90335-9 PMID:5265062

79. Clayton DA, Smith CA, Jordan JM et al. (1968). Occurrence of complex mitochondrial DNA in normal tissues. *Nature*, 220:976–979. doi:10.1038/220976a0 PMID:5701854

80. Clayton DA, Vinograd J (1969). Complex mitochondrial DNA in leukemic and normal human myeloid cells. *Proc Natl Acad Sci USA*, 62:1077–1084.doi:10.1073/pnas.62.4.1077 PMID:5256408

81. He L, Luo L, Proctor SJ et al. (2003). Somatic mitochondrial DNA mutations in adult-onset leukaemia. Leukemia, 17:2487–2491. doi:10.1038/sj.leu.2403146 PMID:14523470

82. Jerónimo C, Nomoto S, Caballero OL et al. (2001). Mitochondrial mutations in early stage prostate cancer and bodily fluids. Oncogene, 20:5195–5198.doi:10.1038/sj.onc.1204646 PMID:11526508

83. Jones JB, Song JJ, Hempen PM et al. (2001). Detection of mitochondrial DNA mutations in pancreatic cancer offers a "mass"-ive advantage over detection of nuclear DNA mutations. Cancer Res, 61:1299–1304. PMID:11245424

84. Kirches E, Krause G, Warich-Kirches M et al. (2001). High frequency of mitochondrial DNA mutations in glioblastoma multiforme identified by direct sequence comparison to blood samples. Int J Cancer, 93:534–538. doi:10.1002/ijc.1375 PMID:11477557

85. Bianchi MS, Bianchi NO, Bailliet G (1995). Mitochondrial DNA mutations in normal and tumor tissues from breast cancer patients. Cytogenet Cell Genet, 71:99–103. doi:10.1159/000134072 PMID:7606938

86. Horton TM, Petros JA, Heddi A et al. (1996). Novel mitochondrial DNA deletion found in a renal cell carcinoma. Genes Chromosomes Cancer, 15:95–101.doi:10.1002/(SICI)1098-2264(199602)15:2<95::AID-GCC3>3.0.CO;2-Z PMID:8834172

87. LaBiche RA, Yoshida M, Gallick GE et al. (1988). Gene expression and tumor cell escape from host effector mechanisms in murine large cell lymphoma. J Cell Biochem, 36:393–403.doi:10.1002/jcb.240360408 PMID:3379107

88. Luciaková K, Kuzela S (1992). Increased steady-state levels of several mitochondrial and nuclear gene transcripts in rat hepatoma with a low content of mitochondria. Eur J Biochem, 205:1187–1193. doi:10.1111/j.1432-1033.1992.tb16889.x PMID:1374334

89. Tamura G, Nishizuka S, Maesawa C et al. (1999). Mutations in mitochondrial control region DNA in gastric tumours of Japanese patients. Eur J Cancer, 35:316–319.doi:10.1016/S0959-8049(98)00360-8 PMID:10448277

90. Gallardo ME, Moreno-Loshuertos R, López C et al. (2006). m.6267G>A: a recurrent mutation in the human mitochondrial DNA that reduces cytochrome c oxidase activity and is associated with tumors. Hum Mutat, 27:575–582.doi:10.1002/humu.20338 PMID:16671096

91. Petros JA, Baumann AK, Ruiz-Pesini E et al. (2005). mtDNA mutations increase tumorigenicity in prostate cancer. Proc Natl Acad Sci USA, 102:719–724.doi:10.1073/pnas.0408894102 PMID:15647368

92. Shidara Y, Yamagata K, Kanamori T et al. (2005). Positive contribution of pathogenic mutations in the mitochondrial genome to the promotion of cancer by prevention from apoptosis. Cancer Res, 65:1655–1663.doi:10.1158/0008-5472.CAN-04-2012 PMID:15753359

93. Shen M, Zhang L, Bonner MR et al. (2008). Association between mitochondrial DNA copy number, blood cell counts, and occupational benzene exposure. Environ Mol Mutagen, 49:453–457.doi:10.1002/em.20402 PMID:18481315

94. Glass DC, Gray CN, Jolley DJ et al. (2003). Leukemia risk associated with low-level benzene exposure. Epidemiology, 14:569–577.doi:10.1097/01.ede.0000082001.05563.e0 PMID:14501272

95. Coller HA, Khrapko K, Bodyak ND et al. (2001). High frequency of homoplasmic mitochondrial DNA mutations in human tumors can be explained without selection. Nat Genet, 28:147–150.doi:10.1038/88859 PMID:11381261

96. Taylor RW, Barron MJ, Borthwick GM et al. (2003). Mitochondrial DNA mutations in human colonic crypt stem cells. J Clin Invest, 112:1351–1360. PMID:14597761

97. Wardell TM, Ferguson E, Chinnery PF et al. (2003). Changes in the human mitochondrial genome after treatment of malignant disease. Mutat Res, 525:19–27. PMID:12650902

98. Albinger-Hegyi A, Hochreutener B, Abdou MT et al. (2002). High frequency of t(14;18)-translocation breakpoints outside of major breakpoint and minor cluster regions in follicular lymphomas: improved polymerase chain reaction protocols for their detection. Am J Pathol, 160:823–832.doi:10.1016/S0002-9440(10)64905-X PMID:11891181

99. Flaman JM, Frebourg T, Moreau V et al. (1994). A rapid PCR fidelity assay. Nucleic Acids Res, 22:3259–3260.doi:10.1093/nar/22.15.3259 PMID:8065949

100. Kwok S, Higuchi R (1989). Avoiding false positives with PCR. Nature, 339:237–238. doi:10.1038/339237a0 PMID:2716852

101. Ling G, Persson A, Berne B et al. (2001). Persistent p53 mutations in single cells from normal human skin. Am J Pathol, 159:1247–1253.doi:10.1016/S0002-9440(10)62511-4 PMID:11583952

102. Williams C, Pontén F, Ahmadian A et al. (1998). Clones of normal keratinocytes and a variety of simultaneously present epidermal neoplastic lesions contain a multitude of p53 gene mutations in a xeroderma pigmentosum patient. Cancer Res, 58:2449–2455. PMID:9622088

103. Hussain SP, Amstad P, Raja K et al. (2000). Increased p53 mutation load in noncancerous colon tissue from ulcerative colitis: a cancer-prone chronic inflammatory disease. Cancer Res, 60:3333–3337. PMID:10910033

104. Hussain SP, Amstad P, Raja K et al. (2001). Mutability of p53 hotspot codons to benzo(a)pyrene diol epoxide (BPDE) and the frequency of p53 mutations in nontumorous human lung. Cancer Res, 61:6350–6355. PMID:11522624

105. Hussain SP, Raja K, Amstad PA et al. (2000). Increased p53 mutation load in nontumorous human liver of wilson disease and hemochromatosis: oxyradical overload diseases. Proc Natl Acad Sci USA, 97:12770–12775.doi:10.1073/pnas.220416097 PMID:11050162

106. Ouhtit A, Nakazawa H, Armstrong BK et al. (1998). UV-radiation-specific p53 mutation frequency in normal skin as a predictor of risk of basal cell carcinoma. J Natl Cancer Inst, 90:523–531.doi:10.1093/jnci/90.7.523 PMID:9539248

107. Chiocca SM, Sandy MS, Cerutti PA (1992). Genotypic analysis of N-ethyl-N-nitrosourea-induced mutations by Taq I restriction fragment length polymorphism/polymerase chain reaction in the c-H-ras1 gene. Proc Natl Acad Sci USA, 89:5331–5335. doi:10.1073/pnas.89.12.5331 PMID:1351680

108. Gormally E, Caboux E, Vineis P, Hainaut P (2007). Circulating free DNA in plasma or serum as biomarker of carcinogenesis: practical aspects and biological significance. Mutat Res, 635:105–117.doi:10.1016/j.mrrev.2006.11.002 PMID:17257890

109. Diehl F, Li M, Dressman D et al. (2005). Detection and quantification of mutations in the plasma of patients with colorectal tumors. Proc Natl Acad Sci USA, 102:16368–16373. doi:10.1073/pnas.0507904102 PMID:16258065

110. Welsh JA, Castrén K, Vähäkangas KH (1997). Single-strand conformation polymorphism analysis to detect p53 mutations: characterization and development of controls. Clin Chem, 43:2251–2255. PMID:9439440

111. Yamanoshita O, Kubota T, Hou J et al. (2005). DHPLC is superior to SSCP in screening p53 mutations in esophageal cancer tissues. Int J Cancer, 114:74–79.doi:10.1002/ijc.20712 PMID:15523690

112. Børresen-Dale AL, Hovig E, Smith-Sørensen B (2001). Detection of mutations by denaturing gradient gel electrophoresis. Curr Protoc Hum Genet, Chapter 7:Unit 7 5.

113. Gormally E, Vineis P, Matullo G et al. (2006). TP53 and KRAS2 mutations in plasma DNA of healthy subjects and subsequent cancer occurrence: a prospective study. Cancer Res, 66:6871–6876.doi:10.1158/0008-5472.CAN-05-4556 PMID:16818665

114. Kurg A, Tõnisson N, Georgiou I et al. (2000). Arrayed primer extension: solid-phase four-color DNA resequencing and mutation detection technology. Genet Test, 4:1–7.doi:10.1089/109065700316408 PMID:10794354

115. Le Calvez F, Ahman A, Tõnisson N et al. (2005). Arrayed primer extension resequencing of mutations in the TP53 tumor suppressor gene: comparison with denaturing HPLC and direct sequencing. *Clin Chem*, 51:1284–1287.doi:10.1373/clinchem.2005.048348 PMID:15976115

116. Tõnisson N, Zernant J, Kurg A et al. (2002). Evaluating the arrayed primer extension resequencing assay of TP53 tumor suppressor gene. *Proc Natl Acad Sci USA*, 99:5503–5508.doi:10.1073/pnas.082100599 PMID:11960007

117. Jenkins GJ, Doak SH, Griffiths AP et al. (2003). Early p53 mutations in nondysplastic Barrett's tissue detected by the restriction site mutation (RSM) methodology. *Br J Cancer*, 88:1271–1276.doi:10.1038/sj.bjc.6600891 PMID:12698195

118. Kirk GD, Camus-Randon AM, Mendy M et al. (2000). Ser-249 p53 mutations in plasma DNA of patients with hepatocellular carcinoma from The Gambia. *J Natl Cancer Inst*, 92:148–153.doi:10.1093/jnci/92.2.148 PMID:10639517

119. Wada H, Asada M, Nakazawa S et al. (1994). Clonal expansion of p53 mutant cells in leukemia progression in vitro. *Leukemia*, 8:53–59. PMID:8289498

120. Zhu YM, Foroni L, McQuaker IG et al. (1999). Mechanisms of relapse in acute leukaemia: involvement of p53 mutated subclones in disease progression in acute lymphoblastic leukaemia. *Br J Cancer*, 79:1151–1157.doi:10.1038/sj.bjc.6690183 PMID:10098750

121. Behn M, Schuermann M (1998). Sensitive detection of p53 gene mutations by a 'mutant enriched' PCR-SSCP technique. *Nucleic Acids Res*, 26:1356–1358.doi:10.1093/nar/26.5.1356 PMID:9469850

122. Parsons BL, Heflich RH (1998). Detection of a mouse H-ras codon 61 mutation using a modified allele-specific competitive blocker PCR genotypic selection method. *Mutagenesis*, 13:581–588.doi:10.1093/mutage/13.6.581 PMID:9862188

123. Parsons BL, Heflich RH (1997). Evaluation of MutS as a tool for direct measurement of point mutations in genomic DNA. *Mutat Res*, 374:277–285. PMID:9100851

124. Parsons BL, Heflich RH (1998). Detection of basepair substitution mutation at a frequency of 1 x 10(-7) by combining two genotypic selection methods, MutEx enrichment and allele-specific competitive blocker PCR. *Environ Mol Mutagen*, 32:200–211.doi:10.1002/(SICI)1098-2280(1998)32:3<200::AID-EM2>3.0.CO;2-O PMID:9814434

125. Jackson PE, Qian GS, Friesen MD et al. (2001). Specific p53 mutations detected in plasma and tumors of hepatocellular carcinoma patients by electrospray ionization mass spectrometry. *Cancer Res*, 61:33–35. PMID:11196182

126. Laken SJ, Jackson PE, Kinzler KW et al. (1998). Genotyping by mass spectrometric analysis of short DNA fragments. *Nat Biotechnol*, 16:1352–1356.doi:10.1038/4333 PMID:9853618

127. Qian GS, Kuang SY, He X et al. (2002). Sensitivity of electrospray ionization mass spectrometry detection of codon 249 mutations in the p53 gene compared with RFLP. *Cancer Epidemiol Biomarkers Prev*, 11:1126–1129. PMID:12376521

128. Lleonart ME, Kirk GD, Villar S et al. (2005). Quantitative analysis of plasma TP53 249Ser-mutated DNA by electrospray ionization mass spectrometry. *Cancer Epidemiol Biomarkers Prev*, 14:2956–2962.doi:10.1158/1055-9965.EPI-05-0612 PMID:16365016

129. Szymañska K, Chen JG, Cui Y et al. (2009). TP53 R249S mutations, exposure to aflatoxin, and occurrence of hepatocellular carcinoma in a cohort of chronic hepatitis B virus carriers from Qidong, China. *Cancer Epidemiol Biomarkers Prev*, 18:1638–1643. doi:10.1158/1055-9965.EPI-08-1102 PMID:19366907

130. Diehl F, Li M, He Y et al. (2006). BEAMing: single-molecule PCR on microparticles in water-in-oil emulsions. *Nat Methods*, 3:551–559.doi:10.1038/nmeth898 PMID:16791214

131. Diehl F, Schmidt K, Choti MA et al. (2008). Circulating mutant DNA to assess tumor dynamics. *Nat Med*, 14:985–990.doi:10.1038/nm.1789 PMID:18670422

132. Ong ST, Le Beau MM (1998). Chromosomal abnormalities and molecular genetics of non-Hodgkin's lymphoma. *Semin Oncol*, 25:447–460. PMID:9728595

133. Rowley JD (1998). The critical role of chromosome translocations in human leukemias. *Annu Rev Genet*, 32:495–519. doi:10.1146/annurev.genet.32.1.495 PMID:9928489

134. Hagmar L, Bonassi S, Strömberg U et al. (1998). Chromosomal aberrations in lymphocytes predict human cancer: a report from the European Study Group on Cytogenetic Biomarkers and Health (ESCH). *Cancer Res*, 58:4117–4121. PMID:9751622

135. Bonassi S, Abbondandolo A, Camurri L et al. (1995). Are chromosome aberrations in circulating lymphocytes predictive of future cancer onset in humans? Preliminary results of an Italian cohort study. *Cancer Genet Cytogenet*, 79:133–135.doi:10.1016/0165-4608(94)00131-T PMID:7889505

136. Pedersen-Bjergaard J, Pedersen M, Roulston D, Philip P (1995). Different genetic pathways in leukemogenesis for patients presenting with therapy-related myelodysplasia and therapy-related acute myeloid leukemia. *Blood*, 86:3542–3552. PMID:7579462

137. Eastmond DA, Pinkel D (1990). Detection of aneuploidy and aneuploidy-inducing agents in human lymphocytes using fluorescence in situ hybridization with chromosome-specific DNA probes. *Mutat Res*, 234:303–318. PMID:2215545

138. Gray JW, Pinkel D (1992). Molecular cytogenetics in human cancer diagnosis. *Cancer*,69Suppl;1536–1542.doi:10.1002/1097-0142(19920315)69:6+<1536::AID-CNCR2820691306>3.0.CO;2-J PMID:1540892

139. Eastmond DA, Schuler M, Rupa DS (1995). Advantages and limitations of using fluorescence in situ hybridization for the detection of aneuploidy in interphase human cells. *Mutat Res*, 348:153–162.doi:10.1016/0165-7992(95)90003-9 PMID:85 44867

140. Kadam P, Umerani A, Srivastava A et al. (1993). Combination of classical and interphase cytogenetics to investigate the biology of myeloid disorders: detection of masked monosomy 7 in AML. *Leuk Res*, 17:365–374.doi:10.1016/0145-2126(93)90025-G PMID:8487586

141. Kibbelaar RE, Mulder JW, Dreef EJ et al. (1993). Detection of monosomy 7 and trisomy 8 in myeloid neoplasia: a comparison of banding and fluorescence in situ hybridization. *Blood*, 82:904–913. PMID:8338953

142. Poddighe PJ, Moesker O, Smeets D et al. (1991). Interphase cytogenetics of hematological cancer: comparison of classical karyotyping and in situ hybridization using a panel of eleven chromosome specific DNA probes. *Cancer Res*, 51:1959–1967. PMID:2004382

143. Cohen MM, Rosenblum-Vos LS, Prabhakar G (1993). Human cytogenetics. A current overview. *Am J Dis Child*, 147:1159–1166. PMID:8237909

144. Gray JW, Pinkel D, Brown JM (1994). Fluorescence in situ hybridization in cancer and radiation biology. *Radiat Res*, 137:275–289.doi:10.2307/3578700 PMID:8146269

145. Tucker JD, Senft JR (1994). Analysis of naturally occurring and radiation-induced breakpoint locations in human chromosomes 1, 2 and 4. *Radiat Res*, 140:31–36.doi:10.2307/3578565 PMID:7938452

146. Dulout FN, Grillo CA, Seoane AI et al. (1996). Chromosomal aberrations in peripheral blood lymphocytes from native Andean women and children from northwestern Argentina exposed to arsenic in drinking water. *Mutat Res*, 370:151–158. PMID:8917661

147. Rupa DS, Hasegawa L, Eastmond DA (1995). Detection of chromosomal breakage in the 1cen-1q12 region of interphase human lymphocytes using multicolor fluorescence in situ hybridization with tandem DNA probes. *Cancer Res*, 55:640–645. PMID:7834635

148. Zhang L, Rothman N, Wang Y et al. (1996). Interphase cytogenetics of workers exposed to benzene. *Environ Health Perspect*, 104 Suppl 6;1325–1329.doi:10.2307/3433184 PMID:9118914

149. Matsumoto K, Ramsey MJ, Nelson DO, Tucker JD (1998). Persistence of radiation-induced translocations in human peripheral blood determined by chromosome painting. *Radiat Res*, 149:602–613.doi:10.2307/3579907 PMID:9611099

150. Tucker JD, Breneman JW, Briner JF et al. (1997). Persistence of radiation-induced translocations in rat peripheral blood determined by chromosome painting. *Environ Mol Mutagen*, 30:264–272.doi:10.1002/(SICI)1098-2280(1997)30:3<264::AID-EM4>3.0.CO;2-L PMID:9366904

151. Tucker JD, Tawn EJ, Holdsworth D et al. (1997). Biological dosimetry of radiation workers at the Sellafield nuclear facility. *Radiat Res*, 148:216–226.doi:10.2307/3579605 PMID:9291352

152. Sachs RK, Chen AM, Brenner DJ (1997). Review: proximity effects in the production of chromosome aberrations by ionizing radiation. *Int J Radiat Biol*, 71:1–19. doi:10.1080/095530097144364 PMID:9020958

153. Morton NE (1991). Parameters of the human genome. *Proc Natl Acad Sci USA*, 88:7474–7476.doi:10.1073/pnas.88.17.7474 PMID:1881886

154. Zhang L, Rothman N, Wang Y et al. (1998). Increased aneusomy and long arm deletion of chromosomes 5 and 7 in the lymphocytes of Chinese workers exposed to benzene. *Carcinogenesis*, 19:1955–1961. doi:10.1093/carcin/19.11.1955 PMID:9855009

155. Xi L, Zhang L, Wang Y, Smith MT (1997). Induction of chromosome-specific aneuploidy and micronuclei in human lymphocytes by metabolites of 1,3-butadiene. *Carcinogenesis*, 18:1687–1693.doi:10.1093/carcin/18.9.1687 PMID:9328162

156. Scarpato R, Lori A, Panasiuk G, Barale R (1997). FISH analysis of translocations in lymphocytes of children exposed to the Chernobyl fallout: preferential involvement of chromosome 10. *Cytogenet Cell Genet*, 79:153–156.doi:10.1159/000134708 PMID:9533038

157. Smith MT, Zhang L, Wang Y et al. (1998). Increased translocations and aneusomy in chromosomes 8 and 21 among workers exposed to benzene. *Cancer Res*, 58:2176–2181. PMID:9605763

158. Schröck E, du Manoir S, Veldman T et al. (1996). Multicolor spectral karyotyping of human chromosomes. *Science*, 273:494–497.doi:10.1126/science.273.5274.494 PMID:8662537

159. Dantonio PD, Meredith-Molloy N, Hagopian WA et al. (2010). Proficiency testing of human leukocyte antigen-DR and human leukocyte antigen-DQ genetic risk assessment for type 1 diabetes using dried blood spots. *J Diabetes Sci Technol*, 4:929–941. PMID:20663459

160. Biernaux C, Loos M, Sels A et al. (1995). Detection of major bcr-abl gene expression at a very low level in blood cells of some healthy individuals. *Blood*, 86:3118–3122. PMID:7579406

161. Fuscoe JC, Setzer RW, Collard DD, Moore MM (1996). Quantification of t(14;18) in the lymphocytes of healthy adult humans as a possible biomarker for environmental exposures to carcinogens. *Carcinogenesis*, 17:1013–1020.doi:10.1093/carcin/17.5.1013 PMID:8640906

162. Liu Y, Hernandez AM, Shibata D, Cortopassi GA (1994). BCL2 translocation frequency rises with age in humans. *Proc Natl Acad Sci USA*, 91:8910–8914.doi:10.1073/pnas.91.19.8910 PMID:8090743

163. Bell DA, Liu Y, Cortopassi GA (1995). Occurrence of bcl-2 oncogene translocation with increased frequency in the peripheral blood of heavy smokers. *J Natl Cancer Inst*, 87:223–224.doi:10.1093/jnci/87.3.223 PMID:7707410

164. Innis MA, Gelfand DH, Sninsky JJ, editors. PCR applications: protocols for functional genomics. San Diego (CA): Academic Press; 1999.

165. Bories D, Dumont V, Belhadj K, et al. (1998). Real-time quantitative RT-PCR monitoring of chronic myelogenous leukemia. *Blood*, 92(10) (Suppl 1 Part 1–2):73A.

166. Krauter J, Wattjes MP, Nagel S, et al. (1998).AML1/MTG8 real time RT-PCR for the detection of minimal residual disease in patients with t(8;21)-positive AML. *Blood*, 92(10)(Suppl 1 Part 1–2):74A–75A.

167. Krauter J, Wattjes MP, Nagel S et al. (1999). Real-time RT-PCR for the detection and quantification of AML1/MTG8 fusion transcripts in t(8;21)-positive AML patients. *Br J Haematol*, 107:80–85.doi:10.1046/j.1365-2141.1999.01674.x PMID:10520027

168. Luthra R, McBride JA, Cabanillas F, Sarris A (1998). Novel 5′ exonuclease-based real-time PCR assay for the detection of t(14;18)(q32;q21) in patients with follicular lymphoma. *Am J Pathol*, 153:63–68.doi:10.1016/S0002-9440(10)65546-0 PMID:9665466

169. Preudhomme C, Merlat A, Roumier C, Duflos-Grardel N, Jouet JP, Cosson A, et al. Detection of BCR-ABL transcripts in chronic myeloid leukemia (CML) using a novel 'real time' quantitative RT-PCR assay: A report on 15 patients. *Blood* 1998;92(10)(Suppl 1 Part 1–2):93A.

170. Saal RJ, Gill DS, Cobcroft RG, Marlton P (1998). Quantitation of AML-ETO transcripts using real time PCR on the ABI7700 sequence detection system. *Blood*, 92(10) (Suppl 1 Part 1–2):74A.

171. Cayuela JM, Gardie B, Sigaux F (1997). Disruption of the multiple tumor suppressor gene MTS1/p16(INK4a)/CDKN2 by illegitimate V(D)J recombinase activity in T-cell acute lymphoblastic leukemias. *Blood*, 90:3720–3726. PMID:9345058

172. Schlissel MS, Kaffer CR, Curry JD (2006). Leukemia and lymphoma: a cost of doing business for adaptive immunity. *Genes Dev*, 20:1539–1544.doi:10.1101/gad.1446506 PMID:16778072

173. Aster JC, Longtine JA (2002). Detection of BCL2 rearrangements in follicular lymphoma. *Am J Pathol*, 160:759–763.doi:10.1016/S0002-9440(10)64897-3 PMID:11891173

174. Dölken L, Schüler F, Dölken G (1998). Quantitative detection of t(14;18)-positive cells by real-time quantitative PCR using fluorogenic probes. *Biotechniques*, 25:1058–1064. PMID:9863062

175. Satake N, Maseki N, Kozu T et al. (1995). Disappearance of AML1-MTG8(ETO) fusion transcript in acute myeloid leukaemia patients with t(8;21) in long-term remission. *Br J Haematol*, 91:892–898.doi: 10.1111/j.1365-2141.1995.tb05406.x PMID:8547135

176. Marcucci G, Livak KJ, Bi W et al. (1998). Detection of minimal residual disease in patients with AML1/ETO-associated acute myeloid leukemia using a novel quantitative reverse transcription polymerase chain reaction assay. *Leukemia*, 12:1482–1489. doi:10.1038/sj.leu.2401128 PMID:9737700

177. Mensink E, van de Locht A, Schattenberg A et al. (1998). Quantitation of minimal residual disease in Philadelphia chromosome positive chronic myeloid leukaemia patients using real-time quantitative RT-PCR. *Br J Haematol*, 102:768–774.doi:10.1046/j.1365-2141.1998.00823.x PMID:9722305

178. Heisterkamp N, Stephenson JR, Groffen J et al. (1983). Localization of the c-ab1 oncogene adjacent to a translocation break point in chronic myelocytic leukaemia. *Nature*, 306:239–242.doi:10.1038/306239a0 PMID:6316147

179. Heddle JA (1973). A rapid in vivo test for chromosomal damage. *Mutat Res*, 18:187–190. PMID:4351282

180. Matter B, Schmid W (1971). Trenimon-induced chromosomal damage in bone-marrow cells of six mammalian species, evaluated by the micronucleus test. *Mutat Res*, 12:417–425. PMID:4999599

181. Schmid W (1973). Chemical mutagen testing on in vivo somatic mammalian cells. *Agents Actions*, 3:77–85.doi:10.1007/BF01986538 PMID:4125456

182. Schmid W (1975). The micronucleus test. *Mutat Res*, 31:9–15. PMID:48190

183. Heddle JA, Cimino MC, Hayashi M et al. (1991). Micronuclei as an index of cytogenetic damage: past, present, and future. *Environ Mol Mutagen*, 18:277–291.doi:10.1002/em.2850180414 PMID:1748091

184. Hayashi M, Morita T, Kodama Y et al. (1990). The micronucleus assay with mouse peripheral blood reticulocytes using acridine orange-coated slides. *Mutat Res*, 245:245–249.doi:10.1016/0165-7992(90)90153-B PMID:1702516

185. MacGregor JT, Wehr CM, Gould DH (1980). Clastogen-induced micronuclei in peripheral blood erythrocytes: the basis of an improved micronucleus test. *Environ Mutagen*, 2:509–514.doi:10.1002/em.2860020408 PMID:6796407

186. Morita T, Asano N, Awogi T et al.; Collaborative Study of the Micronucleus Group Test. Mammalian Mutagenicity Study Group (1997). Evaluation of the rodent micronucleus assay in the screening of IARC carcinogens (groups 1, 2A and 2B) the summary report of the 6th collaborative study by CSGMT/JEMS MMS. Mutat Res, 389:3–122. PMID:9062586

187. Fenech M, Holland N, Chang WP et al. (1999). The HUman MicroNucleus Project–An international collaborative study on the use of the micronucleus technique for measuring DNA damage in humans. Mutat Res, 428:271–283. PMID:10517999

188. Fenech M (2000). The in vitro micronucleus technique. Mutat Res, 455:81–95. PMID:11113469

189. Fenech M, Morley AA (1985). Measurement of micronuclei in lymphocytes. Mutat Res, 147:29–36. PMID:3974610

190. Bolognesi C, Abbondandolo A, Barale R et al. (1997). Age-related increase of baseline frequencies of sister chromatid exchanges, chromosome aberrations, and micronuclei in human lymphocytes. Cancer Epidemiol Biomarkers Prev, 6:249–256. PMID:9107430

191. Fenech M, Morley AA (1986). Cytokinesis-block micronucleus method in human lymphocytes: effect of in vivo ageing and low dose X-irradiation. Mutat Res, 161:193–198. PMID:3724773

192. Anwar WA, Salama SI, el Serafy MM et al. (1994). Chromosomal aberrations and micronucleus frequency in nurses occupationally exposed to cytotoxic drugs. Mutagenesis, 9:315–317.doi:10.1093/mutage/9.4.315 PMID:7968572

193. da Silva Augusto LG, Lieber SR, Ruiz MA, de Souza CA (1997). Micronucleus monitoring to assess human occupational exposure to organochlorides. Environ Mol Mutagen, 29:46–52.doi:10.1002/(SICI)1098-2280(1997)29:1<46::AID-EM6>3.0.CO;2-B PMID:9020306

194. Suruda A, Schulte PA, Boeniger M et al. (1993). Cytogenetic effects of formaldehyde exposure in students of mortuary science. Cancer Epidemiol Biomarkers Prev, 2:453–460. PMID:8220090

195. Cheng TJ, Christiani DC, Xu X et al. (1996). Increased micronucleus frequency in lymphocytes from smokers with lung cancer. Mutat Res, 349:43–50. PMID:8569791

196. Murgia E, Ballardin M, Bonassi S et al. (2008). Validation of micronuclei frequency in peripheral blood lymphocytes as early cancer risk biomarker in a nested case-control study. Mutat Res, 639:27–34. PMID:18155071

197. Holland N, Bolognesi C, Kirsch-Volders M et al. (2008). The micronucleus assay in human buccal cells as a tool for biomonitoring DNA damage: the HUMN project perspective on current status and knowledge gaps. Mutat Res, 659:93–108.doi:10.1016/j.mrrev.2008.03.007 PMID:18514568

198. Stich HF, San RH, Rosin MP (1983). Adaptation of the DNA-repair and micronucleus tests to human cell suspensions and exfoliated cells. Ann N Y Acad Sci, 407:93–105.doi:10.1111/j.1749-6632. 1983.tb47816.x PMID:63 49490

199. Stich HF, Rosin MP (1983). Quantitating the synergistic effect of smoking and alcohol consumption with the micronucleus test on human buccal mucosa cells. Int J Cancer, 31:305–308.doi:10.1002/ijc.2910310309 PMID:6826255

200. Stich HF, Stich W, Parida BB (1982). Elevated frequency of micronucleated cells in the buccal mucosa of individuals at high risk for oral cancer: betel quid chewers. Cancer Lett, 17:125–134.doi:10.1016/0304-3835(82) 90024-6 PMID:6187434

201. Basu A, Mahata J, Roy AK et al. (2002). Enhanced frequency of micronuclei in individuals exposed to arsenic through drinking water in West Bengal, India. Mutat Res, 516:29–40. PMID:11943608

202. Vuyyuri SB, Ishaq M, Kuppala D et al. (2006). Evaluation of micronucleus frequencies and DNA damage in glass workers exposed to arsenic. Environ Mol Mutagen, 47:562–570.doi:10.1002/em.20229 PMID:16795086

203. Das RK, Dash BC (1992). Genotoxicity of 'gudakhu', a tobacco preparation. II. In habitual users. Food Chem Toxicol, 30:1045–1049.doi:10.1016/0278-6915(92)90115-2 PMID:1473798

204. Moore LE, Smith AH, Hopenhayn-Rich C et al. (1997). Decrease in bladder cell micronucleus prevalence after intervention to lower the concentration of arsenic in drinking water. Cancer Epidemiol Biomarkers Prev, 6:1051–1056. PMID:9419402

205. Moore LE, Smith AH, Hopenhayn-Rich C et al. (1997). Micronuclei in exfoliated bladder cells among individuals chronically exposed to arsenic in drinking water. Cancer Epidemiol Biomarkers Prev, 6:31–36. PMID:8993795

206. Desai SS, Ghaisas SD, Jakhi SD, Bhide SV (1996). Cytogenetic damage in exfoliated oral mucosal cells and circulating lymphocytes of patients suffering from precancerous oral lesions. Cancer Lett, 109:9–14.doi:10.1016/S0304-3835(96)04390-X PMID:9020897

207. Rajeswari N, Ahuja YR, Malini U et al. (2000). Risk assessment in first degree female relatives of breast cancer patients using the alkaline Comet assay. Carcinogenesis, 21:557–561.doi:10.1093/carcin/21.4.557 PMID:10753185

208. Ramirez A, Saldanha PH (2002). Micronucleus investigation of alcoholic patients with oral carcinomas. Genet Mol Res, 1:246–260. PMID:14963832

209. Saxonov S, Berg P, Brutlag DL (2006). A genome-wide analysis of CpG dinucleotides in the human genome distinguishes two distinct classes of promoters. Proc Natl Acad Sci USA, 103:1412–1417.doi:10.1073/pnas.0510310103 PMID:16432200

210. Issa JP, Baylin SB, Herman JG (1997). DNA methylation changes in hematologic malignancies: biologic and clinical implications. Leukemia, 11 Suppl 1;S7–S11. PMID:9130685

211. Belinsky SA, Nikula KJ, Palmisano WA et al. (1998). Aberrant methylation of p16(INK4a) is an early event in lung cancer and a potential biomarker for early diagnosis. Proc Natl Acad Sci USA, 95:11891–11896.doi:10.1073/pnas.95.20.11891 PMID:9751761

212. Hsieh CJ, Klump B, Holzmann K et al. (1998). Hypermethylation of the p16INK4a promoter in colectomy specimens of patients with long-standing and extensive ulcerative colitis. Cancer Res, 58:3942–3945. PMID:97 31506

213. Kanai Y, Ushijima S, Tsuda H et al. (1996). Aberrant DNA methylation on chromosome 16 is an early event in hepatocarcinogenesis. Jpn J Cancer Res, 87:1210–1217. PMID:9045955

214. Jones PA, Gonzalgo ML, Tsutsumi M, Bender CM (1998). DNA methylation in bladder cancer. Eur Urol, 33 Suppl 4;7–8. doi:10.1159/000052251 PMID:9615197

215. Hayashizaki Y, Hirotsune S, Okazaki Y et al. (1993). Restriction landmark genomic scanning method and its various applications. Electrophoresis, 14:251–258.doi:10.1002/elps.1150140145 PMID:8388788

216. Estécio MR, Yan PS, Ibrahim AE et al. (2007). High-throughput methylation profiling by MCA coupled to CpG island microarray. Genome Res, 17:1529–1536.doi:10.1101/gr.6417007 PMID:17785535

217. Toyota M, Issa JP (2002). Methylated CpG island amplification for methylation analysis and cloning differentially methylated sequences. Methods Mol Biol, 200:101–110. PMID:11951646

218. Vaissière T, Hung RJ, Zaridze D et al. (2009). Quantitative analysis of DNA methylation profiles in lung cancer identifies aberrant DNA methylation of specific genes and its association with gender and cancer risk factors. Cancer Res, 69:243–252. doi:10.1158/0008-5472.CAN-08-2489 PMID: 19118009

219. Esteller M, Tortola S, Toyota M et al. (2000). Hypermethylation-associated inactivation of p14(ARF) is independent of p16(INK4a) methylation and p53 mutational status. Cancer Res, 60:129–133. PMID:10646864

220. Lo YM, Wong IH, Zhang J et al. (1999). Quantitative analysis of aberrant p16 methylation using real-time quantitative methylation-specific polymerase chain reaction. Cancer Res, 59:3899–3903. PMID: 10463578

221. Ko E, Kim Y, Kim SJ et al. (2008). Promoter hypermethylation of the p16 gene is associated with poor prognosis in recurrent early-stage hepatocellular carcinoma. Cancer Epidemiol Biomarkers Prev, 17:2260–2267.doi:10.1158/1055-9965.EPI-08-0236 PMID:18723830

222. Ayadi W, Karray-Hakim H, Khabir A et al. (2008). Aberrant methylation of p16, DLEC1, BLU and E-cadherin gene promoters in nasopharyngeal carcinoma biopsies from Tunisian patients. *Anticancer Res*, 28(4B):2161–2167. PMID:18751390

223. Blazej RG, Kumaresan P, Cronier SA, Mathies RA (2007). Inline injection microdevice for attomole-scale sanger DNA sequencing. *Anal Chem*, 79:4499–4506. doi:10.1021/ac070126f PMID:17497827

224. Blazej RG, Kumaresan P, Mathies RA (2006). Microfabricated bioprocessor for integrated nanoliter-scale Sanger DNA sequencing. *Proc Natl Acad Sci USA*, 103:7240–7245.doi:10.1073/pnas.0602476103 PMID:16648246

225. Emrich CA, Tian H, Medintz IL, Mathies RA (2002). Microfabricated 384-lane capillary array electrophoresis bioanalyzer for ultrahigh-throughput genetic analysis. *Anal Chem*, 74:5076–5083.doi:10.1021/ac020236g PMID:12380833

226. Grover WH, Mathies RA (2005). An integrated microfluidic processor for single nucleotide polymorphism-based DNA computing. *Lab Chip*, 5:1033–1040.doi:10.1039/b505840f PMID:16175257

227. Harrison DJ, Fluri K, Seiler K et al. (1993). Micromachining a miniaturized capillary electrophoresis-based chemical analysis system on a chip. *Science*, 261:895–897. doi:10.1126/science.261.5123.895 PMID:17783736

228. Lagally ET, Emrich CA, Mathies RA (2001). Fully integrated PCR-capillary electrophoresis microsystem for DNA analysis. *Lab Chip*, 1:102–107.doi:10.1039/b109031n PMID:15100868

229. Lagally ET, Medintz I, Mathies RA (2001). Single-molecule DNA amplification and analysis in an integrated microfluidic device. *Anal Chem*, 73:565–570.doi:10.1021/ac001026b PMID:11217764

230. Lagally ET, Scherer JR, Blazej RG et al. (2004). Integrated portable genetic analysis microsystem for pathogen/infectious disease detection. *Anal Chem*, 76:3162–3170.doi:10.1021/ac035310p PMID:15167797

231. Paegel BM, Emrich CA, Wedemayer GJ et al. (2002). High throughput DNA sequencing with a microfabricated 96-lane capillary array electrophoresis bioprocessor. *Proc Natl Acad Sci USA*, 99:574–579.doi:10.1073/pnas.012608699 PMID:11792836

232. Toriello NM, Liu CN, Mathies RA (2006). Multichannel reverse transcription-polymerase chain reaction microdevice for rapid gene expression and biomarker analysis. *Anal Chem*, 78:7997–8003. doi:10.1021/ac061058k PMID:17134132

233. Kumaresan P, Yang CJ, Cronier SA et al. (2008). High-throughput single copy DNA amplification and cell analysis in engineered nanoliter droplets. *Anal Chem*, 80:3522–3529.doi:10.1021/ac800327d PMID:18410131

234. Shi Y, Simpson PC, Scherer JR et al. (1999). Radial capillary array electrophoresis microplate and scanner for high-performance nucleic acid analysis. *Anal Chem*, 71:5354–5361.doi:10.1021/ac990518p PMID:10596215

235. Skibola CF, Smith MT, Kane E et al. (1999). Polymorphisms in the methylenetetrahydrofolate reductase gene are associated with susceptibility to acute leukemia in adults. *Proc Natl Acad Sci USA*, 96:12810–12815.doi:10.1073/pnas.96.22.12810 PMID:10536004

236. Zhang K, Martiny AC, Reppas NB et al. (2006). Sequencing genomes from single cells by polymerase cloning. *Nat Biotechnol*, 24:680–686.doi:10.1038/nbt1214 PMID:16732271

237. Margulies M, Egholm M, Altman WE et al. (2005). Genome sequencing in microfabricated high-density picolitre reactors. *Nature*, 437:376–380. PMID:16056220

238. Shendure J, Porreca GJ, Reppas NB et al. (2005). Accurate multiplex polony sequencing of an evolved bacterial genome. *Science*, 309:1728–1732.doi:10.1126/science.1117389 PMID:16081699

239. Ottesen EA, Hong JW, Quake SR, Leadbetter JR (2006). Microfluidic digital PCR enables multigene analysis of individual environmental bacteria. *Science*, 314:1464–1467.doi:10.1126/science.1131370 PMID:17138901

240. Hemminki K, Thilly WG (2004). Implications of results of molecular epidemiology on DNA adducts, their repair and mutations for mechanisms of human cancer. Lyon: IARC Scientific Publication; (157):217–235. PMID: 15055298

241. Coulet F, Blons H, Cabelguenne A et al. (2000). Detection of plasma tumor DNA in head and neck squamous cell carcinoma by microsatellite typing and p53 mutation analysis. *Cancer Res*, 60:707–711. PMID:10676657

242. Kopreski MS, Benko FA, Kwee C et al. (1997). Detection of mutant K-ras DNA in plasma or serum of patients with colorectal cancer. *Br J Cancer*, 76:1293–1299.doi:10.1038/bjc.1997.551 PMID:9374374

243. Hibi K, Taguchi M, Nakayama H et al. (2001). Molecular detection of p16 promoter methylation in the serum of patients with esophageal squamous cell carcinoma. *Clin Cancer Res*, 7:3135–3138. PMID:11595706

244. Usadel H, Brabender J, Danenberg KD et al. (2002). Quantitative adenomatous polyposis coli promoter methylation analysis in tumor tissue, serum, and plasma DNA of patients with lung cancer. *Cancer Res*, 62:371–375. PMID:11809682

245. Fujiwara Y, Chi DD, Wang H et al. (1999). Plasma DNA microsatellites as tumor-specific markers and indicators of tumor progression in melanoma patients. *Cancer Res*, 59:1567–1571. PMID:10197630

UNIT 2.

BIOMARKERS: PRACTICAL ASPECTS

CHAPTER 6.

Basic principles and laboratory analysis of genetic variation

Jesus Gonzalez-Bosquet and Stephen J. Chanock

Summary

With the draft of the human genome and advances in technology, the approach toward mapping complex diseases and traits has changed. Human genetics has evolved into the study of the genome as a complex structure harbouring clues for multifaceted disease risk with the majority still unknown. The discovery of new candidate regions by genome-wide association studies (GWAS) has changed strategies for the study of genetic predisposition. More genome-wide, "agnostic" approaches, with increasing numbers of participants from high-quality epidemiological studies are for the first time replicating results in different settings. However, new-found regions (which become the new candidate "genes") require extensive follow-up and investigation of their functional significance. Understanding the true effect of genetic variability on the risk of complex diseases is paramount. The importance of designing high-quality studies to assess environmental contributions, as well as the interactions between genes and exposures, cannot be stressed enough. This chapter will address the basic issues of genetic variation, including population genetics, as well as analytical platforms and tools needed to investigate the contribution of genetics to human diseases and traits.

Introduction

New advances in microchip technologies and informatics allow geneticists to look across the genome agnostically using dense data sets with billions of data points. These developments have transformed the field, moving it away from the pursuit of hypothesis-driven, limited candidate studies to large-scale scans across the genome. Together these developments have spurred a dramatic increase in the discovery of genetic variants associated with or linked to human diseases and traits, many through genome-wide association studies (GWAS) (1). Already over 7400 novel regions of the genome have been associated with more than 75 human diseases or traits in large-scale GWAS (2). Each region now represents a new candidate "region" that harbours putative genes, which will require extensive mapping of the variants to explore the genomic architecture of the region and its contribution to human diseases and traits. The return to exploring candidate regions differs from the old approach of nominating favoured genes, because it is driven by findings that reach conclusive thresholds based on more rigorous statistical considerations.

While there is ample opportunity to survey thousands of genetic variants, often well chosen and based on an emerging understanding of the structure of genetic variation and its patterns of inheritance, the ability to analyse the interaction between genetic variants and the environment has lagged. This is mainly because the measurement tools for the latter have not undergone the transformative shift observed in assessing genetic variation. The integration of environmental exposure with genetic factors should provide insights into disease mechanisms and outcomes. Eventually these insights will be applied to treatment or preventive measures that are best suited for the individual (known as personalized medicine). Individualization of treatments based on the greatest likelihood for efficacy, while minimizing (or avoiding) deleterious toxicities, represents a long-term goal, but one that is in the distant future. While the opportunity to begin to develop evidence-based individualized therapeutics, also known as pharmacogenomics, is promising, its realization will require a nuanced understanding of the contribution of genetic variation to complex diseases.

This chapter will address the basic issues of genetic variation, including population genetics as well as analytical platforms and tools needed to investigate the contribution of genetics to human diseases and traits.

The scope of genetic variation

The spectrum of human genetic variation is enormous with respect to both the types of genetic variation and the sheer magnitude of the number of variants in any given genome. Even though two genomes are estimated to differ by less than 0.5%, there are still several million differences; the majority are vestigial, but a small proportion probably contribute to disease risk. The most common type of variation is a single nucleotide base change, followed by small insertions or deletions in sequence. Progressively larger structural alterations and copy number variants are fewer in absolute number, but perhaps affect more bases (Figure 6.1). So far, available technologies have accelerated the discovery and characterization of diversity in the human genome. In the first wave of annotation, common variants have

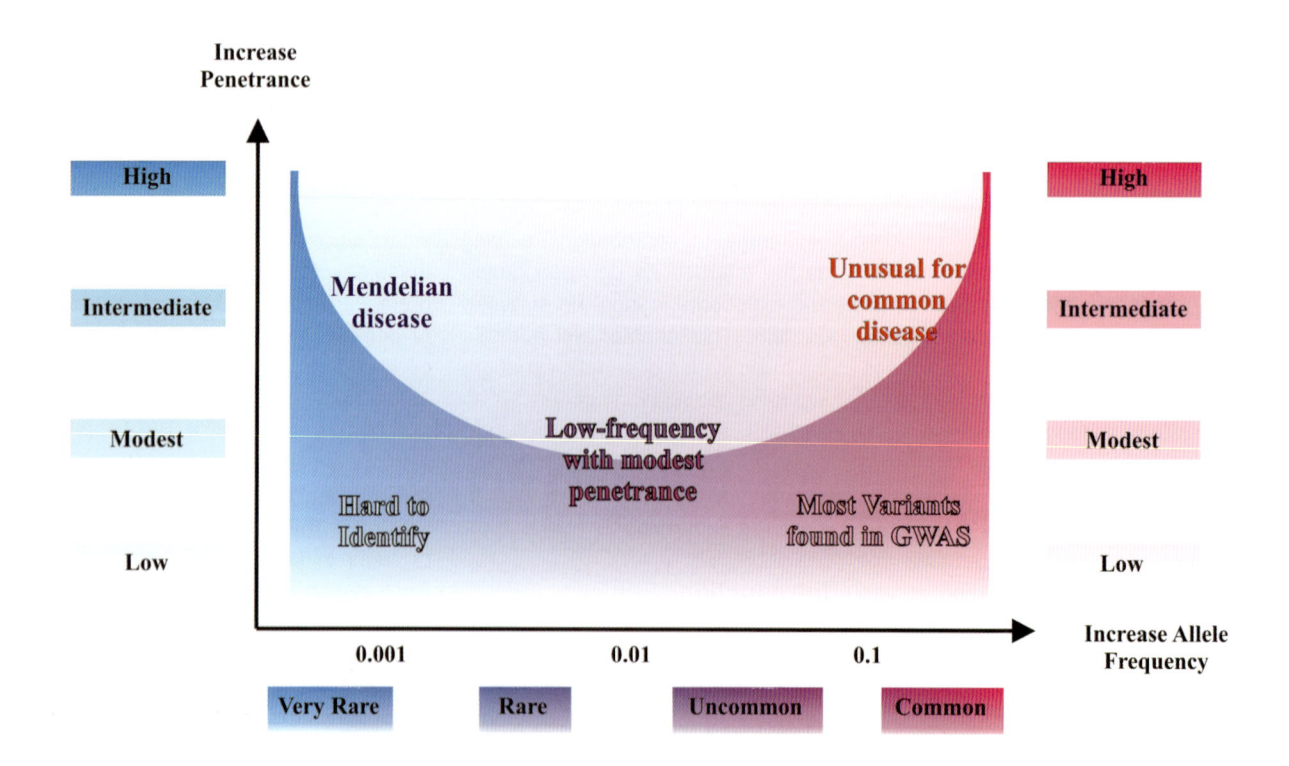

Figure 6.1. Genetic variant frequencies and estimated effect size for genetic contribution

been described, many of which are universal to all populations. The ability to ascertain estimates for lower frequency variants is dependent upon the number of subjects surveyed, as well as the population genetic history of the subjects used for discovery. New sequencing technologies, referred to as next-generation sequencing, allow for the ability to catalogue variants with lower frequencies and will certainly shift the paradigms further. Generally, the interrogation of genetic variation continues to reveal greater complexity in different human populations, which manifests as differences in frequencies of variants.

Single-nucleotide polymorphisms (SNPs)

The most common sequence variation in the genome, the single-nucleotide polymorphism (SNP), is the stable substitution of a single base, which by definition is observed in at least 1% of a population. Though this definition has been useful for cataloging genetic variation, the advent of next-generation sequencing technology has revealed the sheer breadth of variations in different populations with estimated frequencies well below 1%. Still, for the purpose of current applications of genetic variation, the SNP is the most commonly annotated variant. The minor allele frequency (MAF) is designated for the lower allele frequency observed at a locus in one particular population, but often there can be major differences in estimated MAFs between populations with distinct histories. The literature suggests that there are more than perhaps 15 million SNPs with a MAF greater than 1% (3–5), and 10 million SNPs with a MAF greater than 10% (3,6,7); however recent large-scale sequencing efforts, such as the 1000 Genomes Project, indicate these estimates are low (http://www.1000genomes.org/). There are estimated to be a greater number of SNPs with lower MAFs and, unlike common SNPs, the majority may be population-specific (Figure 6.2). The majority of common SNPs, with a MAF greater than 15–20%, are widespread in human populations (8,9). Only a small subset of high-frequency SNPs (less than 10%) appear to be found in a single population, again suggesting the universal ancestry of common SNPs (9).

Previously in candidate gene approach studies, SNPs in coding regions were often selected on the basis of an *in silico* predicted effect, but with little supporting biological evidence. The attempt to classify coding variants, known as a coding SNP (cSNP), has focused on the predicted effect on the actual coding sequence. The majority of cSNPs

Figure 6.2. Estimated number of SNPs in the human genome in relation with their minor allele frequency (MAF). Source: (5). Reprinted by permission from Macmillan Publishers Ltd: Nature Genetics, copyright (2003).

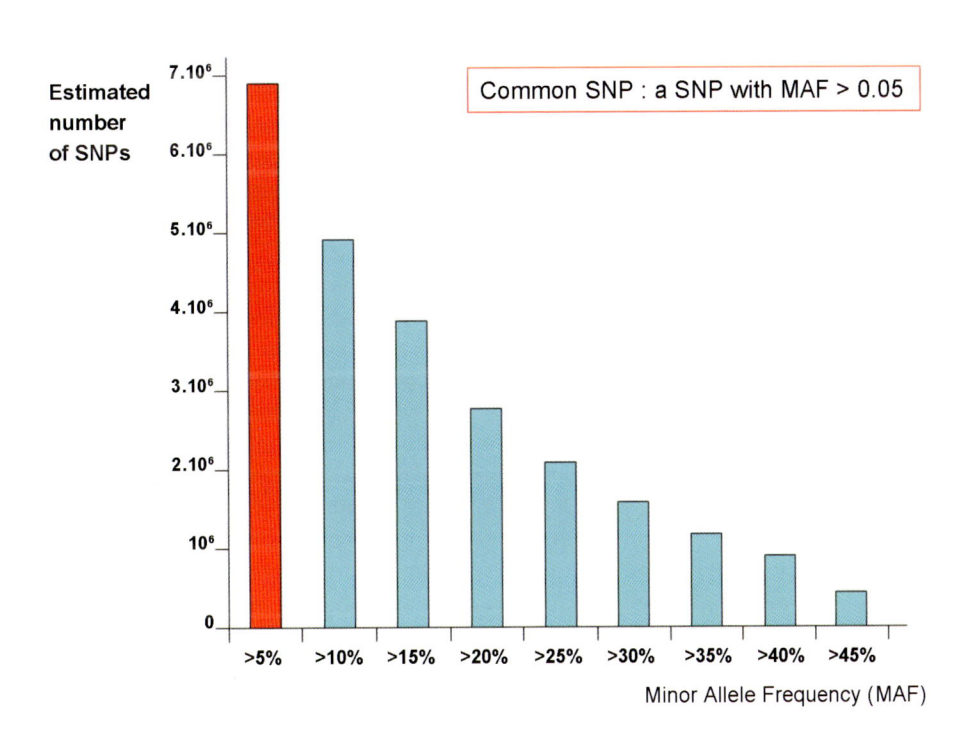

do not alter the predicted amino acid and are known as synonymous SNPs. However, a subset of variants are predicted to shift the amino acid and are known as non-synonymous coding SNPs. Though this subset was initially of great interest, very few non-synonymous coding SNPs have actually been conclusively associated with human diseases or traits, and even fewer have corroborative biological data to provide plausibility for the association (10,11). Nonetheless, the analysis of synonymous and non-synonymous SNPs has been quite informative for evolutionary studies (12,13).

There has been considerable effort to calculate the effect of a non-synonymous cSNP in conformational protein changes. A proliferation of prediction software has been created (e.g. Protein Data Bank (http://www.rcsb.org/pdb) and Swiss-Model (http://swissmodel.expasy.org//SWISS-MODEL.html)). Though new models and algorithms claim improved reliability for predicting deleterious changes in protein structure (14–16), without corroborative laboratory data the findings are merely *in silico* observations. Overall, between 50 000 and 250 000 SNPs could be functional, non-synonymous coding variants, or regulators of gene expression or splicing (10,11). It is likely that a subset of non-synonymous cSNPs contribute to regulatory differences in expression or genetic pathways (17–19), but most SNPs appear not to be functional and have been maintained on the backbone of an inherited block of DNA through generations. Subsets of SNPs that alter regulation or expression of a gene, called regulatory SNPs (rSNPs), are difficult to predict with high efficiency and most likely will be categorized on the basis of large-scale surveys of cell lines, as well as laboratory data.

Nearly half of the more than 10 million human SNPs in the international public database for SNPs, or dbSNP (http://www.ncbi.nih.gov/SNP/), have been validated with genotyping assays by the SNP Consortium and the International HapMap Project (8,20). Until recently, only a small percentage had been verified by sequencing, but with the advent of the 1000 Genomes Project, nearly all common (MAF >10%) and uncommon (MAF between 1 and 10%) variants should be confirmed by next generation sequence technology (21,22). In the current build, roughly one sixth of the variants in dbSNP are probably monoallelic, due to errors in either genotyping or, more likely, sequencing (23,24). In general, the reported SNPs have been biased towards high-frequency variants in populations of European ancestry.

Currently, the catalogue of uncommon variation, namely SNPs with MAFs under 1%, is incomplete. However, the 1000 Genomes Project is expected to generate a thorough catalogue of variants with greater than 1% MAF. The contribution of uncommon variants (MAF between 1% and 10%) represents an untapped portion of the genomic architecture. It will require either larger studies to provide sufficient power to detect association, or new design strategies to discover and characterize uncommon and rare variants (25,26). Rare or uncommon variants have been shown to be informative in the extremes of mapping human traits, such as with cholesterol levels (27). Rare variants or mutations can explain a proportion of the strong familial component of complex diseases, as well as the classical Mendelian inheritance of single or oligogenic diseases. These highly penetrant disease mutations are catalogued in a public database, the Online Mendelian Inheritance in Man (OMIM) (http://www.ncbi.nlm.nih.gov/omim/).

The correlation of common genetic variants

Most SNPs are not inherited independently but in blocks, resulting in sets of SNPs being transmitted together between generations (4,28,29). These blocks are defined by linkage disequilibrium (LD), which estimates the correlation between SNPs on shared chromosomes passed down from ancestral chromosomes. LD is defined as the non-random association of alleles at different loci (30). Initially, each SNP is a single mutation that has taken hold and become fixed in a population, either as a consequence of direct selection or because it is close enough on the chromosome to be included within a block of a shared segment. Individual SNPs that are strongly associated with each other are said to be in LD, although this correlation could be eroded over time by recombination (exchange of genetic material) during meiosis (31). Haplotypes are defined as sets of SNPs, and other genetic polymorphisms (larger in size), on chromosomal segments that are in strong LD.

There are several ways to determine haplotypes from genotypes; this is commonly referred to as resolving haplotype phase. The offspring haplotype phase can be determined if the parental genotypes are known or directly with biochemical methods (30). Based on the assumption that haplotypes are randomly joined into genotypes, phasing can be estimated using one of several statistical methods that can account for the ambiguity of unobserved haplotypes (30). Different methods have been developed to estimate haplotypes from unphased multilocus genotype data in unrelated individuals; the underlying principles are based on models that incorporate either a maximum likelihood (32), parsimony (33), combinational theory (34) or *a priori* distribution derived from coalescent theory (35). This last method is the basis for the phase reconstruction software PHASE (35,36), which has performed favourably in simulation studies (37), and its modified version designed for larger data sets, fastPHASE (22). Reconstructed haplotypes from unrelated individuals and LD structure have been used to study genomic association to complex traits (17,29). In fact, some research suggests that haplotypes would be better suited for candidate studies because of a perceived statistical advantage over the single-locus LD mapping (38–40), but the recent success with GWAS suggests otherwise.

The concept of LD also permits investigators to look at a set of SNPs and determine proxies for other untested SNPs (or tagSNPs) (41,42). This indirect approach is predicated on finding markers only, relegating the search for causal or functional variants to later work (Figure 6.3). Several approaches optimize the number of surrogate SNPs needed to account for untested variants, such as the "greedy algorithm." The latter estimates highly correlated SNPs, primarily on the basis of the MAF, to create heuristic bins of tagged SNPs. Thus, tagSNPs represent proxies for additional, highly correlated SNPs with comparable allele frequency and distribution in the population of interest. In a sense, tagSNPs are used to mark common haplotypes in the region (Tagger, embedded in Haploview software (http://www.broad.mit.edu/mpg/haploview) and TagZilla (http://tagzilla.nci.nih.gov)) (43). Consequently, the indirect approach of using a limited set of tagSNPs as a proxy of a LD block has emerged as the preferred approach, used by both GWAS and candidate gene studies (44).

Structural polymorphisms

Structural variations in the genome may be either cytologically visible or, more commonly, submicroscopic variants that can range in size from a few base pairs to thousands (45,46). These can include deletions, insertions and duplications collectively known as copy number variations (CNVs), as well as less-frequent inversions and

Figure 6.3. SNP selection strategy. A. SNP selection through haplotype blocks, based on the concept of linkage disequilibrium (LD). D' is a measure of LD between SNPs, represented in the figure through a heat map from white (low D') to red (high D"). A haplotype (represented by a dark triangle in the figure) is a set of SNPs in strong LD, or high D'. "TagSNPs" are proxies for other SNPs in the same haplotype. This is the so-called indirect approach (147). B. Selection of SNPs based on r2, another measure of LD. This method creates groups with similar LD (r2) into 'bins.' In the figure each spot represents a SNP, and those with similar r2 are included in the grey blocks or 'bins.' 'TagSNPs' are proxies for all these loci included in each 'bin' with comparable LD (148)

A.

Haplotype blocks: based on D' values for linkage disequilibrium (LD)

B.

Grouping of SNPs into bins based on r^2

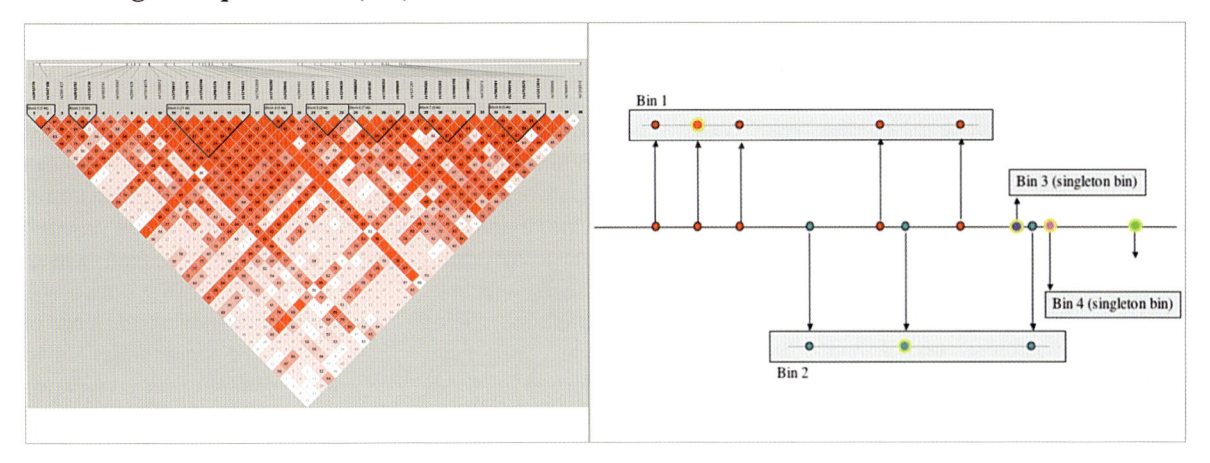

translocations (Figure 6.4) (47,48). Several of the inversions can be quite large, such as the 3.5 Mb on chromosome 17 seen in as much as 20% of the European population (49). On the other hand, insertion/deletions as small as two base pairs can be observed. Although structural variants in some genomic regions have no obvious phenotypic consequence (50–52), CNVs have been shown to influence gene dosage in select circumstances. Consequently, many have pursued CNVs because of the potential contribution of high estimated effects for complex diseases, either alone or in combination with other factors (53). Some observations, either by the failure to assemble the draft genome sequence or by actual experimentation, estimate the segmental duplicated genomic sequence could involve between 5–10% of the genome (51,54,55). Other clues come from the recognition that a notable number of SNPs failed the quality control metrics in the International HapMap Project; these were later determined to reside in regions now known to be enriched for CNVs (7,45,55–57). Current surveys suggest that CNVs are less common than previously reported (58), and many are infrequent (59). It is also notable that over three fourths of common CNVs are in LD with common SNPs (59).

Coordinated efforts are underway to establish a comprehensive catalogue of CNVs, such as the Database of Genomic Variants (http://projects.tcag.ca/variation/) (46,60) and the Human Genome Structural Variation Project (http://humanparalogy.gs.washington.edu/structuralvariation/). Recently, there have been several international efforts to establish standards for identification, validation and reporting of CNVs (46). The availability of several microarray platforms that can detect quantitative imbalances has accelerated CNV discovery, but there are still substantive technical challenges due to the breadth of polymorphic differences for which analyses are particularly unstable. New emerging algorithms should streamline moderate- to high-throughput, cost-effective methods to scan the genome for CNVs, as well as inversions or translocations based on stable sequence assemblies (59–64). Advances in techniques have improved determination of common CNVs, such as tiling arrays (which cover the genome through partial overlapping (tile-like) sets of fixed oligonucleotides), paired-end sequencing (sequence analysis of both ends of a larger fragment to improve alignment), and new dense SNP genotyping platforms based on probe intensity (e.g. Illumina and Affymetrix).

Figure 6.4. Spectrum of genomic variation. Challenges and standards in integrating surveys of structural variation: The range of genetic variation that must be taken into account when designing and analyzing genotype studies (46). The figure represents the whole spectrum of human genetic variation, from the molecular level with DNA sequence variation, exemplified by SNPs, to structural variation, a broad category that includes variations from 2 bp to whole chromosomal variations. The focus of recent genetic studies has been the subgroup in the midrange (with strong highlighting). These forms of variation have been studied with molecular methods to cytogenetic approaches. Reprinted by permission from Macmillan Publishers Ltd: Nature Genetics, copyright (2007).

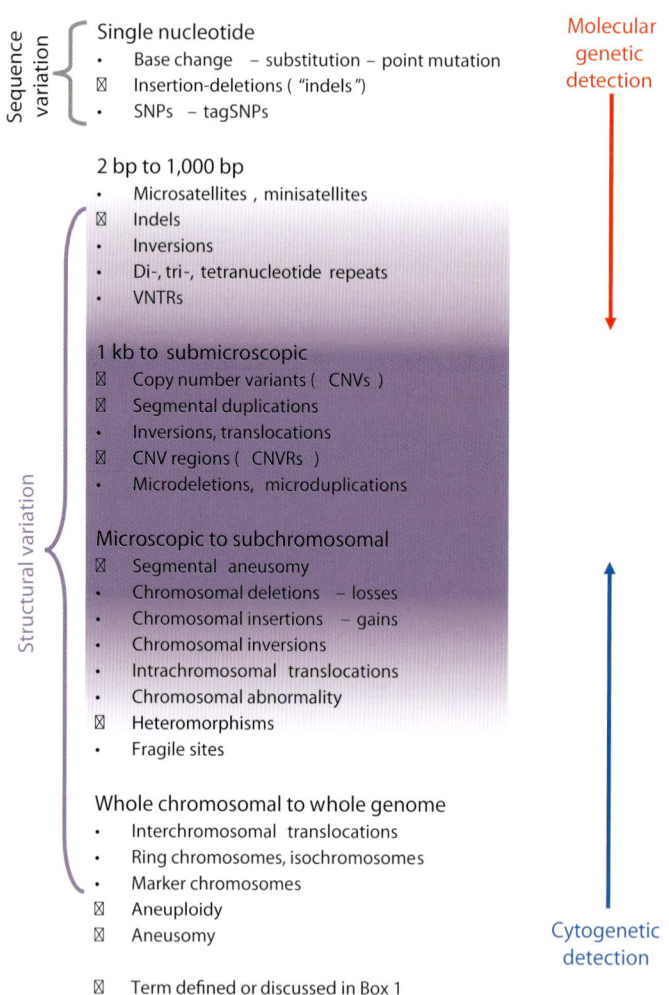

Sequence variation
Single nucleotide
- Base change – substitution – point mutation
- Insertion-deletions ("indels")
- SNPs – tagSNPs

Structural variation

2 bp to 1,000 bp
- Microsatellites, minisatellites
- Indels
- Inversions
- Di-, tri-, tetranucleotide repeats
- VNTRs

1 kb to submicroscopic
- Copy number variants (CNVs)
- Segmental duplications
- Inversions, translocations
- CNV regions (CNVRs)
- Microdeletions, microduplications

Microscopic to subchromosomal
- Segmental aneusomy
- Chromosomal deletions – losses
- Chromosomal insertions – gains
- Chromosomal inversions
- Intrachromosomal translocations
- Chromosomal abnormality
- Heteromorphisms
- Fragile sites

Whole chromosomal to whole genome
- Interchromosomal translocations
- Ring chromosomes, isochromosomes
- Marker chromosomes
- Aneuploidy
- Aneusomy

Term defined or discussed in Box 1

Molecular genetic detection

Cytogenetic detection

Short tandem repeats (STRs) represent a class of polymorphisms that occur when a pattern of two or more nucleotides are repeated in certain areas of the genome. Previously known as microsatellites, they were frequently employed to conduct linkage studies in potentially informative pedigrees. The patterns can range in length from 2–10 base pairs (usually tetra- or penta-nucleotide repeats) and are typically located in non-coding regions. Since longer repeat sequences can be susceptible to artefactual errors in genotyping accuracy, particularly related to problems of PCR amplification, the industry standard for both genetic analysis and forensic application is 4–5 base pair (bp) repeat units. Shorter repeat sequences (e.g. 2 or 3 bp) tend to suffer from artefacts, such as stutter and preferential amplification (65–67). By genotyping a sufficient number of STR loci, it is possible to generate a unique genetic profile of an individual.

Population genetics

The field of population genetics has advanced rapidly and emerged as central to the investigation of genetics and complex diseases. Overall, the discipline of population genetics seeks to characterize the genetic composition of biological populations, as well as the changes in genetic composition that occur from environmental and migratory factors, including natural selection. To draw conclusions about the likely patterns of genetic variation in actual populations, population geneticists develop abstract mathematical models of gene frequency dynamics and test these conclusions against empirical data. Some of the more robust concepts in population genetic analysis that are applied in disease mapping are discussed below.

Fitness for Hardy–Weinberg proportion

The fitness for Hardy–Weinberg proportion, an important tool for understanding population structure, examines the distribution of the allelic and genotypic frequencies. Though theoretical, it states that if certain assumptions are met, genotype and allele frequencies can be estimated from one generation to the next. The derivation of the Hardy–Weinberg principle for a single locus assumes: a randomly mating population; an infinitely large population, or a population size large enough that random fluctuations in allele and genotype frequencies are small; no mutation; no migration; and no fitness differences among genotypes. When all of these assumptions are met, Hardy–Weinberg Equilibrium (HWE) is established and four important conclusions can be drawn: 1) allele frequencies do not change from one generation to the next; 2) genotype frequencies can be inferred from allele frequencies; 3) only one generation is required to go from non-equilibrium to equilibrium; and 4) once the system is in HWE, it stays in HWE (68). Also, if these conditions are met, the genotypic and allelic frequencies of the offspring generation will be related by the following simple equations. For a trait in the population with two alleles (A_1 and A_2), if the A_1 allele frequency in the population is p, and the A_2 allele frequency is $q = (1-p)$, then expected genotype proportions (f) under HWP are:

$f(A_1A_1) = p^2$, $f(A_1A_2) = 2pq$, $f(A_2A_2) = q^2$

Random mating, or the absence of a genotypic correlation between mating partners, will generate a distribution of observed genotypes that should not deviate significantly from the expected proportions (Hardy–Weinberg Proportions (HWP)). This is predicated on Mendel's law of segregation, and, assuming the absence of selection, all parents contribute equal numbers of gametes to the pool. The HWE principles can be applied to family-based and case–control data to detect genotyping error, population stratification and association.

A violation of any of the above assumptions can produce deviation from HWE, which may include mating behaviour, population size and migration patterns. For example,

Table 6.1. Issues for generation of final, publication-grade build of high-density genotype data

- Eliminate samples with low completion rates (< 90%)
- Remove SNP assays with low call rates (< 90%)
- Determination of fitness for Hardy–Weinberg proportion
- Compare expected duplicates
- Investigate unexpected duplicates
- Assess concordance between duplicates
- Search for cryptic relatedness between subjects
- Assessment of population substructure (after filtering 1st degree relatives)
- Determine admixture with STRUCTURE analysis
- Estimate population stratification (principal component analysis)
- Assess genotype calling algorithm
- Validate significant genotype calls with second technology

systematic inbreeding will increase levels of homozygosity across the genome, as will small population sizes (68). Having more than one random mating population in a sample may also cause deviations from HWE, as well as mating with certain phenotypes (known as assortative mating), which will increase homozygosity as well. Small population size causes allele frequencies to drift from one generation to the next. In many cases, the deviations are also a screen for performance of the genotype technology, because a disproportionate number of heterozygotes or homozygotes could represent systematic errors in genotyping.

One of the most common reasons for not using data in association studies is presumed genotyping error. Many types of errors in genotyping can cause deviations from HWE; therefore tests for both assay specificity and deviation from HWE have been proposed to minimize the genotype error rate and thereby improve data quality (69,70). Deviation from HWE resulting from allelic drop-out, where some alleles are insufficiently amplified, can cause an excess of homozygotes and increase false-negative or false-positive results (71). However, caution should be exercised in association studies before removing data because of HWE deviations. If there is a systematic HWE deviation in both cases and controls, it may be easier to determine a genotyping error if both deviations occurred in the same direction (72). Non-systematic error is more problematic and should trigger a review of standard operating procedures for biospecimen handling, as well as an assessment of all information workflow. If the error is recognized, re-genotyping of the faulty samples might eliminate the problem. The power to detect deviations due to genotyping error under most modes of inheritance has been found to be very small (73). Even the deviation created by neighbouring SNPs, which diminish the performance of genotyping assays, does not produce a large enough deviation from HWE to be detected (74).

In GWAS, it is likely that hundreds if not thousands of markers will deviate from HWE. Understanding why and how HWE testing would help in the process of disease-gene discovery is becoming more important as the number of SNPs included in these studies increases into the hundreds of thousands (75). The control observed genotype frequencies are tested against control expected genotype frequencies to determine if there may be genotyping error (68).

Spectrum of differences in population substructure

The age of GWAS has generated sufficiently large data sets that can determine the degree of differences in underlying population substructure, also known as population stratification. An examination of thousands of markers not in LD permits investigators to assess the extent of admixture and exclude individuals who are outliers for the association analysis.

Classically, population stratification is present when there is a measurable difference in the distribution of alleles between subgroups that have different population histories. There are examples of this in older case–control studies where the cases and controls have been drawn from different populations. It is also possible to have stratification between cases and controls based on differences in exposures, as well in the distribution of common SNP markers (76). The ability to detect stratification with any marker or set of markers may also vary depending on the allele frequency in each subgroup (68).

In general, an assessment of the underlying structure can be estimated using standard algorithms to identify distinct populations (77). The most commonly used approach is implemented in the STRUCTURE program. This uses multilocus genotype data to examine population structure by attempting to separate subjects into groups (defined as k populations) and determining the distribution of shared alleles.

As the ability to understand population stratification (or differences between cases and controls due to systematic ancestral differences) has improved, several methods have been developed to study and account for these types of systematic study population structures. One approach commonly used for the correction of population stratification is to adjust simultaneously for a fixed number of top-ranked principal components resulting from a principal component analysis (PCA) (76). It is critical to look for underlying subgroups in stratified samples by testing sets of genetic markers not linked to the phenotype, and then adjust for inflation due to stratification (76,78,79). An alternative approach is to use a structured association method in association mapping, permitting case–control analysis in the context of known differences in population structure.

In select circumstances, in which the epidemiologic data suggest major differences between populations, it is possible to conduct mapping by admixture of linkage disequilibrium (MALD). This capitalizes on the concept of admixture, which is the genetic mix of two or more distinct populations. It relies on the differences in allele frequencies between populations to guide the search to focus on changes in the genome rather than a specific gene(s). So far it has been successful in mapping a key prostate cancer region on 8q24 and a type of end-stage renal disease that is more common in individuals of African American background (80–82).

Selection

Population geneticists often define evolution as a change in a population's genetic composition over time. The four factors that can bring about such a change are natural selection, mutation, random genetic drift, and migration into or out of the population. More controversial is a possibility of changes in the mating pattern, which some consider not to be part of classical evolutionary change. Natural selection occurs when some genotypic variants in a population enjoy a survival or reproduction advantage over others. Although the concept that natural selection favours the survival of individuals with a fitness advantage now almost seems intuitive, it was largely opposed when introduced by Darwin (83). Under Mendelian inheritance and with random mating, genotype frequencies after one generation do not change; the determinant of whether the allele will spread in the population is the fitness of heterozygotes versus that of wild-type homozygotes.

Mutation is the primary source of genetic variation driving differences within a population and thus preventing homogeneity. Although mutations that occur in the genome are initially thought to be random, the distribution of biologically significant mutations that cause diversity appears to be non-random (84,85). Gene function, gene structure and the roles of genes and gene products in genetic networks can influence whether particular mutations will contribute to advantageous phenotypic changes. Some mutations generate specific phenotypic changes, whereas pleiotropic mutations alter several seemingly unrelated traits. Mutations with pleiotropic effects will rarely change all phenotypic traits in a favourable way, and, in some instances, may even reduce fitness (86). The same mutation in a different genetic background may produce a different phenotypic effect because of interactions between alleles, under the phenomenon called epistasis. Also, populations exposed to repeated environmental changes may present with different genetic changes that produce a range of phenotypes suited to the environmental conditions, namely phenotypic plasticity.

Initially, when the environment favours a phenotype that is largely different from the average one in a population, mutations that cause this phenotypic change towards the new optimum are favoured (called strength of selection). Population size and history also influence genetic evolution. A small population size can accentuate the effects of random sampling of alleles, so-called genetic drift. In small populations, genetic drift will allow deleterious alleles to occasionally increase in frequency (84).

Random genetic drift refers to the chance fluctuations in gene frequency that arise in finite populations; it can be thought of as a type of "sampling error." In many evolutionary models, the population is assumed to be infinite or very large to avoid chance fluctuations. This assumption is often not realistic, and species with historically low effective population sizes, such as humans, show evidence for reduced variability and effectiveness of selection in comparison with other species (87,88). In the era of multispecies comparisons of genome sequences and GWAS, it is critical to assess the evolutionary role of genetic drift and its interactions with mutation, migration, recombination and selection. Therefore, population size plays a central part in modern studies of molecular evolution and variation (88).

One of the most influential variables for human genetic variation is geographic location, with genetic differentiation between populations increasing with geographic distance and genetic diversity decreasing with distance from Africa. Populations of African ancestry have the greatest diversity, resulting in shorter segments of LD (89–93). Modern population genetics estimates that the ancestral human population originated in Africa and radiated outward to other continental locations, both within Africa and elsewhere.

Alleles under positive selection can increase in prevalence in a population and leave distinctive signatures, or patterns of genetic variation, in DNA sequences. These can be identified by comparison with the background distribution of genetic variation, primarily evolved under neutrality (94). In some cases, these signatures, or differences in allele frequencies between populations, reflect major regional selective pressures like infectious diseases (e.g. malaria), environmental stresses (e.g. temperature), or dietary factors (e.g. milk consumption) (13,95,96).

When immigrants with a different genetic makeup enter a new population, the population's genetic composition will be altered. The evolutionary importance of migration stems from the fact that many species are composed of several distinct subpopulations, largely isolated from each other but connected by occasional migration. Migration between subpopulations gives rise to gene flow, limiting the extent to which subpopulations can diverge from each other genetically.

Laboratory analysis of human genetic variation

Genotype analysis

Genotyping is used to interrogate specific, unique loci in the genome following DNA amplification by polymerase chain reaction (PCR). One of the challenges of genotype analysis is that each allele in the genome must be assayed individually, unlike surveys of gene expression that can use a common signature (the polyA tail) to capture a high percentage of mRNA at once. An assay must be robust and reproducible in exceeding a sufficient threshold for detection. Even though amplification protocols are highly reliable, error can be introduced for SNP detection, particularly if there are neighbouring SNPs that alter allele-specific binding of probes or if local genomic sequence is enriched for guanine-cytosine (GC) content (Figure 6.5) (97,98). The presence of duplicates of part of the sequence (CNV), either in the segment amplified or neighbouring the SNPs, can undermine the fidelity of the assay, sometimes providing bias in allele calling (55). Based on the amplification of local sequence surrounding the SNP of interest, redundant sequences are amplified, either locally or elsewhere in the genome, and the fidelity of the polymorphisms between these different segments is undermined, as was observed in the International HapMap Project (45,56).

Initially, restriction fragment length polymorphism (RFLP) assays were used to identify patterns of DNA broken into pieces by restriction enzymes. The size of the fragments was used to develop a footprint of the region of interest (99). RFLP analysis is laborious and error-prone, and thus has been largely abandoned for probe intensity and microchip technologies that can be easily scaled and reliably performed. Examples of these are differential hybridization, primer extension, ligation reactions and allele-specific probe cleavage, all of which interrogate one SNP

Figure 6.5. Fidelity of the genotyping assay: error could be introduced in SNP detection. For example, the presence of a neighboring SNP under both TaqMan® (TM) probes (left panel) may alter allele-specific binding and bias the allele call (right panel).

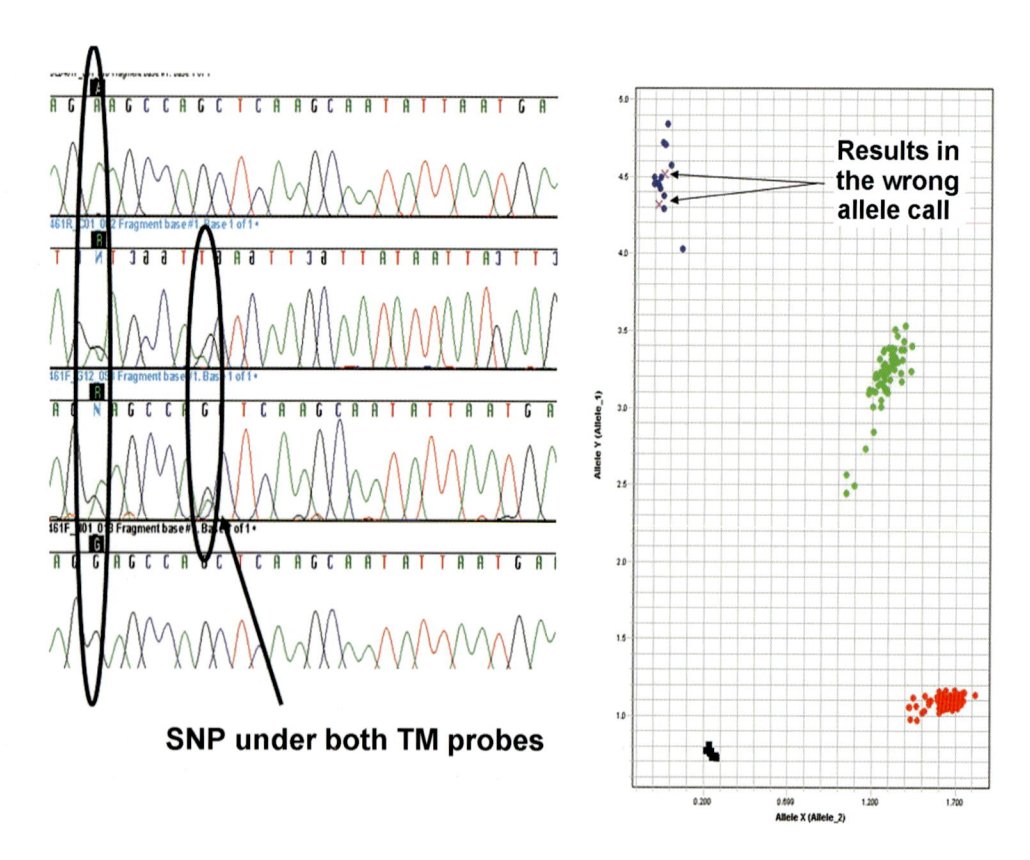

at a time. Occasionally, RFLPs are required to interrogate a region with high degrees of paralogy.

Low-density genotyping

The most commonly used technique for single SNP assays is the TaqMan® SNP genotyping assay (Applied Biosystems). It is a PCR-based assay designed to interrogate a single SNP that uses two locus-specific PCR primers and two allele-specific labelled probes (100). The 5′ exonuclease property of Taq polymerase is capitalized for detection of base-matching at a specific site. Attached to the 5′ end of each probe is an allele-specific reporter dye: each allele has a corresponding dye, which provides a benchmark for the ratio of the dyes as a reflection of the allele distributions. On the 3′ end of each probe is a single universal quencher dye, which prevents the excitation and emission of the reporter dyes. During PCR amplification, the two PCR primers anneal to the template DNA. The detection probes anneal specifically to the complementary sequence between the forward and reverse primer sites. During the elongation step of each cycle, the Taq polymerase comes in contact from the 5′ end with the reporter dye. Capitalising on the exonuclease property of Taq polymerase, the reporter dye is released from the probe and the fluorescence is released (i.e. no longer quenched by the quencher dye). In addition, the probe itself is digested by the Taq polymerase. After multiple cycles of PCR (that reach saturation for copying both alleles), fluorescence is detected for the two reporter dyes using an ABI 7900HT Sequence Detection System.

Careful attention must be paid to the unique flanking sequences to avoid overlap with adjacent, neighbouring SNPs or insertion/deletions. The throughput is moderate for single-plex TaqMan, but new miniaturization technologies have improved the efficiency of moderate-scale genotyping studies using either the Fluidigm® or BioTrove platforms (101,102).

Multiplexing has increased the technical capacity to interrogate large, predetermined, fixed sets of SNPs. The cost of high-density SNP platforms and the necessity for large-scale follow-up studies have incentivized the development of methodologies for selective replication efforts. The technologies that have been developed for these replication studies are based on direct oligonucleotide hybridization with probe fluorescence detection, the single-base sequencing method, or chip-based mass spectrometry (i.e. based on matrix-assisted laser desorption/ionization time-of-flight (MALDI-TOF)) (103). Matrix-assisted laser desorption/ionization (MALDI) enables analysis of biomolecules by ionization usually triggered by a laser beam. A matrix is used to protect the biomolecule from destruction; it can be multiplexed to perform roughly 30 SNP assays at one time.

High-density SNP detection

The first generation of custom bead-array technology by Illumina® enables custom detection of more than 1500 SNPs with excellent performance, and analysis of high-quality DNA generated by whole-genome amplification assays (104,105). This system combines high-multiplexing in a multisample array format, well suited for custom genotype analysis of samples. Though best used with native DNA, it can analyse whole-genome amplified DNA, but at a price of distortions of heterozygosity for roughly 5% of the SNPs.

The newer system of Illumina, known as the Infinium® Assay, features single-tube preparation of DNA followed by whole-genome amplification before genotyping thousands of unique SNPs. Hybridization to bead-bound 50mer oligomers is followed by single-base extension, which incorporates a labelled nucleotide for assay detection. This technology can be used to design custom sets of SNPs (between 7600 and 60 000 bead types) with high efficiency (106). It is the backbone of the fixed content chips, which have increased in size and coverage of the common SNPs in the genome. This began with the HumanHap300 and its complementary HumanHap 240, through to the HumanHap500, Human Hap610 and HumanHap 660w. The Infinium HD (high-density) series followed with the Human1M-Duo BeadChips, which has over 10^6 SNPs to be genotyped, primarily chosen as tagSNPs from HapMap II (8). The increasing content of the chips also provides an opportunity to detect a larger subset of the common CNVs. However, algorithms for detection of CNVs continue to evolve and should improve in the coming years.

The Affymetrix microchip system is based on an assay known as the whole-genome sampling analysis (WGSA) for highly multiplexed SNP genotyping (107). This method amplifies the human genome with a single primer amplification reaction using restriction enzyme-digested, adaptor-ligated human genomic DNA. After fragmentation, sequential labelling and hybridization of the targets is required before analysing the fragments on a microchip. The initial GeneChip® Human Mapping 500K Array spaced SNP markers by physical proximity, but the new Genome-Wide Human SNP Array 6.0 provides a denser set of SNPs (over 900 000), as well as probes that monitor common CNVs across the genome. The distribution of restriction enzyme sites in select regions of the genome does not permit assays across the full genome, limiting the coverage somewhat. The primary debate over the choice of platforms is the coverage of known SNPs in HapMap Stage 2: the SNPs selected for the Illumina platform have been primarily chosen according to the aggressive tag strategy, whereas the first-generation Affymetrix chips provided spaced coverage based on the physical map of the genome, but with higher density. The coverage of the latter has improved.

Methodological issues in GWAS genotyping

High-throughput genotyping facilities require sophisticated robotics for efficient laboratory flow and sample handling, as well as dedicated computational hardware and software able to effectively process both the quantity and complexity of the data. Despite the fact that new technologies and platforms has decreased the nominal price per genotype assayed, the pricing must also take into account the need to study duplicates and samples that must be redone due to technical inadequacies determined in the quality control assessment (see below).

Since replication is a central requirement to protect against the flurry of false-positives observed in GWAS, follow-up studies are needed to verify the results and thus justify the considerable effort required to investigate novel regions. To this end, custom panels are needed to explore regions at the same time that loci are analysed over sufficiently large data sets, so that genome-wide significance can be conclusively established (106,108). Normally, custom panels are more expensive and usually created for a single study (109). In this regard, scalability to meet the requirements of validation studies represents one of the biggest challenges in the design of these studies (110).

Important components of the optimization process include both a Laboratory Information Management System (LIMS) and robotic automation that accurately track and handle samples for efficient workflow management. Because of the high cost of these platforms, the hardware used for sample processing, and the software integrating both, there is little flexibility in choosing individual SNPs to be included within the already-designed, commercially available whole-genome scans.

Two high-density genotyping platforms, Affymetrix and Illumina®, achieve calling capabilities of between 500 000 and 2.5 million SNPs, as well as probe content to interrogate CNVs. Both platforms need between 400–800 ng of total high-quality DNA (usually at 50 ng/µl) for the assay, but because of the dead-space of the robotics (which can be 35% of the required amount for the assay) over 1 ug is required. Issues common to both platforms are the difficulties in assaying SNPs that reside close together (within 60 or fewer nucleotides), which, as previously mentioned, is inherent in this type of genotyping detection. Denser sets of SNPs on commercial platforms have increased coverage, but not always for all populations.

Coverage based on the HapMap II set of SNPs with minor allele frequencies greater than 5%, is one of the main factors driving the choice of platform (8,43). Figure 6.6 illustrates the minimum LD for any SNP assay assessed by the coefficient of correlation, r^2 (a measure of LD), for 2-SNP comparison. The closer the value is to 1, the stronger the correlation, and if the value is estimated to be 1.0, then both loci segregate together. New approaches are being developed to account for the complexity of LD patterns in distinct populations, such as multimarker strategies that have been proposed for analysing more complicated loci (111,112).

Sequence analysis

Until recently, DNA sequence analysis by capillary electrophoresis has been the platform of choice for medium- and small-scale projects, displacing the Sanger sequencing protocols that used gels or polymers as separation media for the fluorescently labelled DNA fragments (113). The advent of the 96-capillary 3730/3730 xl DNA Analyser (Applied Biosystems) was the central catalyst in the generation of the first draft sequence of the human genome (114).

Dideoxy sequencing is based on the principle of terminating DNA synthesis by incorporation of the dideoxy nucleotide terminator on the complementary strand of DNA fragments. The generated library of various length fragments can be assembled to read the specific DNA sequence. Sequencing-by-synthesis is based upon the principle of pyrophosphate release by nucleotide incorporation along the complementary strand of DNA to the varied-length template. As with dideoxy sequencing, the library of generated fragment lengths can be assembled into a specific DNA sequence. An amplification step by PCR is required, and thus has an intrinsic error below 0.3% (small but predictable) (115).

Efficient removal of unincorporated dye terminators is necessary before running samples on a capillary electrophoresis in

Figure 6.6. Genotyping platforms coverage of HapMap II SNPs. SNP coverage is plotted against LD measured by r2, or coefficient of correlation, for SNP-SNP comparison. Panels: A. HapMap CEU population: CEPH (Utah residents with ancestry from northern and western Europe USAB); B. HapMap YRI population: Yoruba in Ibadan, Nigeria; C. HapMap JPT population: Japanese in Tokyo, Japan, and CHB population: Han Chinese in Beijing, China.

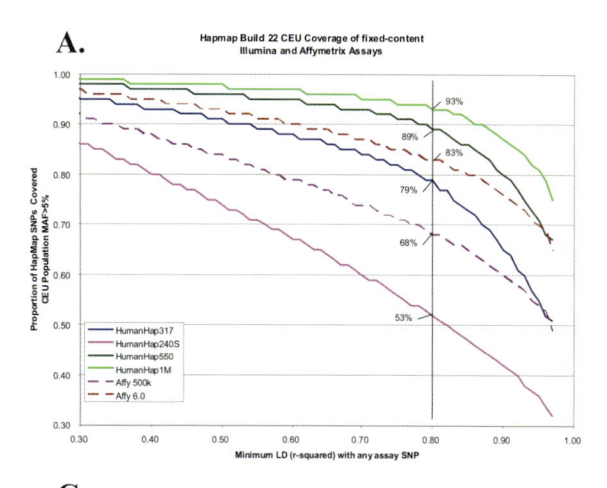

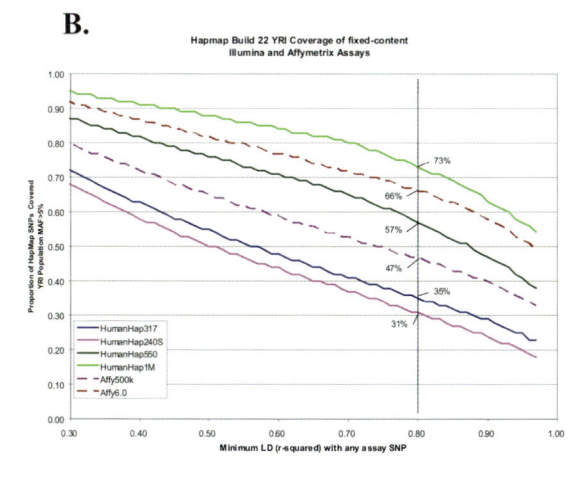

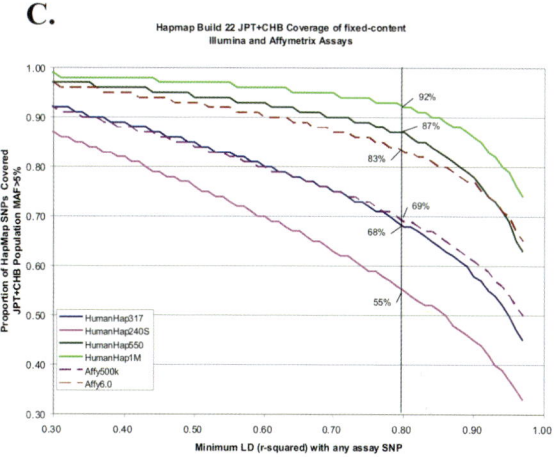

Genotyping platforms coverage of HapMap II SNPs in relation with LD, measured by r^2

which an electrical field is applied. This allows negatively-charged DNA fragments to move through the polymer towards the positive electrode. Standard software collects raw data files and translates the collected colour data images into consecutive nucleotide base calls.

Next-generation DNA sequencing

Next-generation sequencers have been developed to process millions of sequence reads in parallel rather than in batches of 96 at a time, setting them apart from conventional capillary-based sequencing. These techniques provide high speed and high-throughput from amplified single DNA fragments, avoiding the need for cloning of DNA fragments. Therefore, with minimal input of DNA, the sequencer produces libraries of shorter length reads of between 35–400 bp, depending on the platform, compared to those of capillary sequencers (650–800 bp). A limiting factor is the elevated cost for generating the sequence with high-throughput. There is a need to develop software applications and more efficient computer algorithms to analyse the increasing amount of data generated by these systems (113). Because of their novelty, the accuracy and associated quality of sequencing reads must be further validated, but the high number of reads provides increased coverage of each base position (25). The major challenge of the next-generation sequencing is the informatics of the dense data sets, which requires archiving and storing dense data sets that must be assembled to determine accurate reads. In this regard, error rates for next-generation sequencing runs and assembly constitute a new set of problems, particularly since the quantum increase in data makes their inspection more daunting.

The Roche/454 GS-FLX technology works on the principle of pyrosequencing, which uses pyrophosphate molecules released on nucleotide incorporation by DNA polymerase to fuel a downstream set of reactions that ultimately produces light from the cleavage of oxyluciferin by luciferase (116). The DNA strands of the library are amplified en masse by emulsion PCR (117) on the surfaces of hundreds of thousands of agarose beads. Each agarose bead surface contains up to 1 million copies of the original annealed DNA fragment to produce a detectable signal from the sequencing reaction. Imaging of the light flashes from luciferase activity records which templates are adding that particular nucleotide; the light emitted is directly proportional to the amount of a particular nucleotide incorporated. The current 454 instrument, the GS-FLX, produces an average read length of 400 bp per sample (per bead), with a combined throughput of ~100–150 Mb of sequence data per run. By contrast, a single ABI 3730 programmed to sequence 24 × 96 well plates per day produces ~440 kb of sequence data in 7 hours, with an average read length of 650 bp per sample (25).

The Illumina Genome Analyser is based on the concept of sequencing by synthesis (Solexa® Sequencing technology) to produce sequence reads of 35–150 bp from tens of millions of surface-amplified DNA fragments simultaneously (118). A mixture of single-stranded, adaptor oligo-ligated DNA fragments is incubated and amplified with four differentially-labelled fluorescent nucleotides. Each base incorporation cycle is followed by an imaging step that identifies it and by a chemical step that removes the fluorescent group. At the end of the sequencing run (~4 days), the sequence of each cluster is computed and subjected to quality control. A typical run yields ~40–50 million such sequences.

The Applied Biosystems SOLiD sequencer uses a unique sequencing process catalysed by DNA ligase. A SOLiD (Sequencing by Oligo Ligation and Detection) run requires days, and produces 3–4 Gb of sequence data with an average read length of approximately 50 bp (119). The specific process couples oligo adaptor-linked DNA fragments with 1-μm magnetic beads that are decorated with complementary oligos, and amplifies each bead-DNA complex by emulsion PCR. A SOLiD sequencing by ligation first anneals a universal sequencing primer, then goes through subsequent ligation of the appropriate labelled 8mer, followed by detection at each cycle by fluorescent readout. The unique attribute of this system is that an extra quality check of read accuracy is enabled that facilitates the discrimination of base calling errors from true polymorphisms or insertion/deletion (indel) events, the so-called "2 base encoding" (25).

The third generation of sequencing technologies is in development and should be available in the coming years. For example, single molecule sequencing is based on novel chemistry that enables direct measurement of billions of strands of DNA. The detection system measures incorporated bases on individual strands and thus avoids the requirement of amplification, which is subject to biases and errors.

Applications of high-throughput DNA sequencing

A major focus of this new technology is to rapidly and comprehensively catalogue human genetic variation, particularly common and uncommon genetic polymorphisms (e.g. SNPs and insertion/deletions). Since GWAS have relied on the genotyping of common alleles to discover novel associations with diseases' risks (120), follow-up of regions of association identified by GWAS is important to characterize common and uncommon variants which might be better markers (or even candidates) for further functional studies. Since GWAS point to new candidate regions, the detailed fine mapping of a region necessitates the generation of a comprehensive set of common and uncommon variants. Already there are select examples of next-generation sequencing analysis applied to regions to determine new variants for follow-up association testing, such as for regions 8q24 associated with prostate and colon cancer and 10q11.2 (containing the *MSMB* gene) associated with prostate cancer (121,122). Eventually the 1000 Genomes Project should provide a suitable map to begin to choose variants in a region of interest. Characterizing all common variants previous to a fine-mapping process has two major benefits: all common genetic variants are represented using a tagSNP approach; and the correlations among all genetic variants are known, which provides advantages in functional variant detection (121,122).

Since large-scale sequencing across the genome is still several years away, attention has focused on targeted sequencing of regions of high interest, such as those defined by GWAS or linkage studies, and, more recently, the opportunity to sequence across the exome (e.g. more than 180 000 known exons in the genome). Several different technologies have been developed to capture target sequence, either through liquid phase (e.g. biotinylated solution capture probes with long range or micro-droplet solution technique), or tiled arrays that contain probes that enrich for capture of DNA for sequence analysis. Each of the next-generation sequencing technologies have been successfully used with one or more target capture technologies. For instance, using the NimbleGen solution-based capture technique, the *KLK3* locus, recently identified as a signal for prostate cancer and prostate serum antigen levels, was resequenced to comprehensively catalogue all common variants for follow-up genotype and functional analyses (123–125). Recently, sequencing across the exome after enrichment with tiled arrays has successfully been used to identify high-penetrant mutations in the coding regions in individuals with Mendelian disorders (126). Exome sequencing represents the first step towards examining the portion of the genome that is easily interpretable, namely changes in coding structure. It requires careful annotation and analytical structures, however, to sift through the thousands of rare variants in unique individuals.

The sequencing of the first human genomes has underscored the challenge of unraveling the physical map, particularly in some regions of great redundancy and/or complexity; moreover, it illustrates the daunting problem of assembly (127,128). More genomes need to be sequenced to establish a reliable reference standard for the analysis of human genomic variations. The current reference is an amalgam of several genomes, thus the ability to unravel variation is particularly difficult. Two new developments should address this issue: sequencing with greater coverage, which diminishes the false-positives and -negatives of sequence determination, and an increase in read length, which will permit phasing of genomes.

The ambitious effort from an international research consortium, namely the 1000 Genomes Project, "…will involve sequencing the genomes of at least a thousand people from around the world to create the most detailed and medically useful picture to date of human genetic variation" (http://www.1000genomes.org/). The goal is to create a detailed map of human genetic variation relevant at or above the level of a frequency of 0.5–1% across the genome (113). By optimizing technology, costs will continue to fall enabling greater scope of study at a lower price. Reduction to affordable levels, targeted for the US$1000 range for an entire human genome sequence, offers the promise of personal genomics. There are still formidable barriers, however, with respect to informatics, storage and the ethical and social dilemmas posed by such analyses.

Next-generation sequencing technologies have already been applied to complementary fields of investigation in genetics. The intent has been to characterize a complex sample with a mixture of nucleic acids through their sequence without prior knowledge of it, in contrast to the probe hybridization used by the original SAGE technique (129,130). Thus, it is possible to characterize the sequence of mRNAs, methylated DNA, DNA or RNA regions bound by certain proteins, and other DNA or RNA regions involved in gene expression and regulation (113). Recent examples are its application to transcriptome profiling in stem cells (119); to whole transcriptome shotgun sequencing, or RNA-Seq, study into alternative splicing in human cells (131); and the identification of mammalian DNA sequences bound by transcription factors *in vivo*, by combining chromatin immunoprecipitation (ChIP) with parallel sequencing (ChIP-Seq) (132).

The Human Microbiome Project (HMP) (http://nihroadmap.nih.gov/hmp/) integrates genomics and metagenomics in an effort to characterize the genome sequences of organisms inhabiting a common environment (133). By understanding genomics, metagenomics and their relations, the HMP seeks to determine whether individuals share a core human microbiome and whether changes in the human microbiome can be correlated with changes in human health (134).

Quality control in the laboratory

The advent of new technologies and workflow paradigms required for high-throughput genotyping and sequencing has changed the nature of laboratory work in genetics. The bulk of the work has been shifted to high-throughput analyses, where so much data is processed in such a short time that the older shibboleths of quality control have been shed for more efficient approaches, which seek to identify potential errors in a high-volume workflow.

The efficient and meticulous sample handling process must begin at the moment of receipt of germline DNA for genotyping or sequencing. Close coordination between the laboratory performing the extraction and the biorepository storing the DNA samples is optimal and protects against handling and biorepository errors, an underappreciated problem.

Standard operating procedures (SOPs) for the process should be created and reviewed regularly for improvements and quality control purposes.

DNA quantification is not an exact science. Due to technical and workflow issues, it is actually quite difficult to reproducibly quantify DNA (135). Several different techniques can be used to measure DNA, but each one has limitations and, in some workflows, different applications of preparing for low- or high-throughput genetic analyses. Quantification methods should be chosen for specific genotype/ sequence platforms. The most commonly used techniques are spectrophotometric measurement of DNA optical density by PicoGreen (Turner BioSystems) analysis, NanoDrop spectrophotometer (NanoDrop Technologies), or by real-time PCR analysis using a standardized TaqMan™ assay (136). Real-time PCR can provide a preliminary test for sample quality as it relates to robust analysis in a high-throughput laboratory, but performance still must be gauged with specific technologies. Spectrophotometry and the PicoGreen assay measure total DNA present, regardless of source or quality, whereas a real-time PCR assay measures the total amplifiable human DNA. DNA quantitation by real-time PCR is particularly helpful for assessing the contribution of non-human DNA to samples collected from buccal swabs, cytobrush samples or other non-blood sources. Minor but real differences between techniques reflect dissimilarities in the ratio of single- and double-stranded DNA, critical for analysis using diverse technologies.

Because of the high volume of activities in high-throughput genotyping/sequencing facilities, unique genetic profiles of samples can be useful for quality assessment and control in the workflow. Many laboratories have incorporated into the upfront analysis a set of SNPs or a forensic panel of 15 small tandem repeats and amelogenin, also known as the AmpFLSTR Identifier assay (Applied Biosystems). The fingerprinting can be helpful to sleuth problems and identify contaminated samples before costly analysis. Furthermore, the results can serve as a proxy for the viability of the DNA and its success on the high-performance genotyping or sequencing technologies. Certainly, high failure rates indicate poor performance. The profiles can be used to match known duplicates and identify unexpected duplicates, which in turn stimulates close inspection of both biorepository issues and workflow in the laboratory (e.g. errors with plates or reagents).

For the conduct of many molecular epidemiology studies, sample availability has been a limiting factor. Naturally, there has been intense interest in the whole-genome amplification (WGA) technology to provide sufficient amounts of DNA for analysis. Thus, varying results reflect not only differences in the protocols and reagents, but the samples themselves. The quality of DNA used to amplify across the genome affects the success and fidelity of the process. WGA can generate large quantities of DNA for genotype assays, but approximately 5% of the genome is not faithfully reproduced, particularly regions with high GC content or near telomeres. Thus, the results of analyses of these regions should be cautiously interpreted. While the temptation to use WGA DNA in GWAS is great, the results so far have not been encouraging. Currently, there are two approaches that have been commercially optimized. These include a type of multiple displacement amplification (MDA) with the high-performance bacteriophage φ 29 DNA polymerase, which uses degenerate hexamers or generation of libraries of 200–2000 base pair fragments created by random chemical cleavage of genomic DNA. Ligation of adaptor sequences to both ends and PCR amplification is required. Quantities can vary greatly based on input DNA, but under optimal conditions an enrichment of 10 000-fold can be expected.

The rolling circle amplification (RCA) technique is an enzymatic process mediated by DNA polymerases. Long single-stranded DNA molecules are synthesized on a short circular template by using a single DNA primer. RCAs generate a large-scale DNA template with the advantage of not requiring a thermal cycling instrument (137,138). Differential success has been observed with whole blood, dried blood, buccal cell swabs, cultured cells and buffy coat cells. Intriguingly, WGA of water control specimens generates a small, monoallelic signal, which can be called as a single allele, thus underscoring the value of rigorous controls (139). Still, more laboratories have chosen MDA for whole-genome amplification (140).

The utility of duplicates drawn from the same sample remains a central theme of laboratory quality control, but with the advent of high-throughput laboratories the purpose has shifted slightly. Still duplicate testing is useful to detect problems with sample quality, prior storage, and informatic issues in sample management. In some cases, it can also reveal rare individuals enrolled in more than one study. Reproducibility of assays is key, and with the whole-genome-scan chips surpasses 99.8% concordance.

Errors in genotyping, mainly due to loss of one of the heterozygous alleles, occur in well below 1% of samples; therefore, when the rate creeps above 1%, close inspection of the process should be undertaken. If SOPs are followed closely, completion rates should be greater than 95% for most studies, but may be slightly lower depending on the quality of genomic DNA. Completion rates below 90% should raise substantive concern about technical or analytical problems. In GWAS studies, it is recommended that a second technology, such as TaqMan or sequencing, be performed to verify the accuracy and establish concordance (120). Errors with fitness for Hardy–Weinberg proportion (Hardy–Weinberg equilibrium (HWE) testing) can catch major genotype errors, but should probably not be used as a stringent threshold for excluding SNPs from analysis.

Bioinformatics

Large-scale genotyping and sequence analysis has shifted the burden of informatics towards high-performance tools that manage the computational and bioinformatic workflow needed to manipulate high-density data sets. The required tasks, archiving, analysis and access are destined to grow exponentially as studies are designed with increasing numbers of participants and larger and more complex variants to be interrogated. Accordingly, the efficiency of the laboratory flow is based on a high-throughput pipeline for both genetic analysis and informatic handling of the data sets. Major steps in the process include the choice of markers and platforms together with a sophisticated quality control process. Highly trained personnel are needed to effectively coordinate the flow of information. Central to the success of a laboratory is the functioning of a Laboratory Information Management System (LIMS), which is required to track samples, assays, reagents, equipment, robotics and processes through the entire workflow. The LIMS captures the movement of information from receipt of samples through the analytical steps and into the quality control regime required to provide a final, stable data set, linking the results of experimental data to *in silico* information via relational databases. Annotation of the genome is needed to provide clear points of reference for the genomic coordinates for the genotype and sequence assays. Careful quality control and quality assurance checks within the LIMS software, particularly with real-time monitoring, are needed to maintain assay reproducibility and reliable data flow.

The increasing number of loci explored by new platforms, as well as the quantum increases in the increments in study size, has forced major changes in laboratory data storage and management. Laboratory systems should be able to routinely process, monitor and assess quality control of large amounts of data (10^6–10^9 data points) generated by these studies. The increasing need for processing power mandates the use of scalable computational systems capable of parallel computing, with software applications specially designed for this multiprocessor environment and readily upgradable.

Suites of publicly available tools (e.g. PLINK (http://pngu.mgh.harvard.edu/~purcell/plink/summary.shtml) (141) and Genotype Library and Utilities (GLU) (http://cgf.nci.nih.gov/glu/docs)) have been developed for archiving and management of dense data sets, such as those encountered in GWAS. PLINK, now in version 1.06, is a free, open-source whole-genome association suite that focuses on the analysis of large-scale genotype/phenotype data, but lacks support for study design and planning, genotype generation, or CNV calling. Its integration with Haploview (http://www.broadinstitute.org/haploview/haploview) allows visualization, annotation and representation of some of the results. GLU (version 1.0) is also a suite created to manage, analyse and report high-throughput SNP genotype data (http://code.google.com/p/glu-genetics). GLU was created to address the need for new and scalable computational approaches, as well as storage, management, quality control and genetic analysis. It is a framework and software package designed with a set of powerful tools that can scale to effectively handle trillions of genotypes. The integration of GLU with a robust and fast SNP tagging tool, like TagZilla, increases its functionality and allows LD estimation and computation of MAF, HWE, and proportions (http://tagzilla.nci.nih.gov/).

Conclusions and future directions

Knowledge acquired by the draft of the human genome and its annotation, and advances in technology, have changed the approach towards mapping complex diseases and traits. Once oriented to the study of candidate genes and/or mutations, human genetics has evolved into the study of the genome as a complex structure harbouring clues for multifaceted disease risk; some known, but the majority unknown. The discovery of new candidate regions by GWAS has forced rethinking previous strategies for the study of genetic predisposition. More agnostic approaches, genome-wide, with increasing numbers of participants from high-quality epidemiological studies are, for the first time, replicating results in different settings. But new-found candidate regions lead to extensive follow-up and confirmation of their functional significance. Understanding the true effect of genetic variability on the risk of complex diseases is paramount, but also important is the design of high-quality studies to assess environmental contributions, as well as the interactions between genes and exposures.

If accurate measures of environmental factors must be addressed, increased efforts are needed in the study of the biological relevance of the regions already discovered. To date, there are a few examples where biological functional basis has been associated with a candidate region discovered via GWAS. Also, the gap between new-found genomic regions and their biological interpretation could become greater with the introduction of new resequencing technology, which is capable of interrogating more numbers of less frequent loci. New challenges arise with new technologies. High-throughput resequencing must standardise its technical protocols, quality control, calling algorithm and interpretation. Only deep resequencing of high numbers of individuals will create quality databases capable of testing rare variants in the population. Until these steps are readily available for new technologies, broad implementation will not be possible.

The new approach to the genomic study of complex diseases has resulted in a more ambitious "team" science, in which resources and study populations are pooled to identify novel genetic markers (Cf. Figure 6.1). In this regard, GWAS study thousands of the most common genetic variants across the genome (SNPs), without any prior hypothesis, conception or what is being defined as an agnostic manner. This initial phase requires adequately powered follow-up studies for replication that is central to the search for moderate- to high-frequency low-penetrance variants associated with human diseases and traits (120,142). Teams of scientists with specific responsibilities in each step of the process are necessary to ensure quality control and stable analytical results as part of the effort to map complex human diseases and traits.

Previously, family linkage studies have been used to identify rare genetic variants with high-penetrance susceptibility genes (143,144), but failed to be informative on more common genetic variants with low to moderate effect (145). With the advent of next-generation sequencing technologies and the discovery of many rare and uncommon variants, family studies will be required to assist in defining the most notable variants for follow-up studies. In this regard, family studies should prove invaluable in mapping many complex diseases, as well as the highly penetrant Mendelian disorders.

Based on the preliminary data published as a result of GWAS, it is not currently possible to draw final conclusions concerning the valid risk assessment of complex diseases. Education of both the public and scientific media is necessary to affect a rational approach towards implementing any risk reduction policies. These new challenges for public health officials will require careful attention to the ethical, moral and social implications of dense genomic data sets to assure the public, and the participants in the current studies, that confidentiality is protected (26).

Consortial efforts to describe human variation have focused on the description and characterization of three continental populations pursued by the International HapMap Project (http://www.hapmap.org). But using GWAS, other consortia and interest groups have focused on a more disease-specific approach that has resulted in the discovery of over 200 novel loci associated with human diseases/traits (2,8,20,57). Though the majority of the association studies to date have used high-throughput genotyping technology, new programs in comprehensive resequencing analysis would unveil an even greater catalogue of uncommon variants (http://www.1000genomes.org/).

In concert with the assessment of germline genetic variation, other programs are underway to characterize functional annotation through gene expression analysis. The ENCODE Project (ENCyclopedia Of DNA Elements) seeks to define functional elements (http://www.genome.gov/10005107) (25), and The Cancer Genome Atlas (TCGA) examines both somatic and germline alterations in select cancers (146). Together, these new developments promise to accelerate the discovery and characterization of novel genomic mechanisms in human diseases and traits.

References

1. Hunter DJ, Khoury MJ, Drazen JM (2008). Letting the genome out of the bottle–will we get our wish? *N Engl J Med*, 358:105–107. doi:10.1056/NEJMp0708162 PMID:18184955

2. Manolio TA, Brooks LD, Collins FS (2008). A HapMap harvest of insights into the genetics of common disease. *J Clin Invest*, 118:1590–1605.doi:10.1172/JCI34772 PMID:18451988

3. Kruglyak L, Nickerson DA (2001). Variation is the spice of life. *Nat Genet*, 27:234–236. doi:10.1038/85776 PMID:11242096

4. Reich DE, Cargill M, Bolk S et al. (2001). Linkage disequilibrium in the human genome. *Nature*, 411:199–204.doi:10.1038/35075590 PMID:11346797

5. Reich DE, Gabriel SB, Altshuler D (2003). Quality and completeness of SNP databases. *Nat Genet*, 33:457–458.doi:10.1038/ng1133 PMID:12652301

6. Lander ES, Linton LM, Birren B et al.; International Human Genome Sequencing Consortium (2001). Initial sequencing and analysis of the human genome. *Nature*, 409:860–921.doi:10.1038/35057062 PMID:11237011

7. Venter JC, Adams MD, Myers EW et al. (2001). The sequence of the human genome. *Science*, 291:1304–1351.doi:10.1126/science.1058040 PMID:11181995

8. Frazer KA, Ballinger DG, Cox DR et al.; International HapMap Consortium (2007). A second generation human haplotype map of over 3.1 million SNPs. *Nature*, 449:851–861. doi:10.1038/nature06258 PMID:17943122

9. Hinds DA, Stuve LL, Nilsen GB et al. (2005). Whole-genome patterns of common DNA variation in three human populations. *Science*, 307:1072–1079.doi:10.1126/science.1105436 PMID:15718463

10. Chanock SJ (2001). Candidate genes and single nucleotide polymorphisms (SNPs) in the study of human disease. *Dis Markers*, 17:89–98. PMID:11673655

11. Risch NJ (2000). Searching for genetic determinants in the new millennium. *Nature*, 405:847–856.doi:10.1038/35015718 PMID:10866211

12. Hughes AL, Packer B, Welch R et al. (2003). Widespread purifying selection at polymorphic sites in human protein-coding loci. *Proc Natl Acad Sci USA*, 100:15754–15757.doi:10.1073/pnas.2536718100 PMID:14660790

13. Hughes AL, Packer B, Welch R et al. (2005). Effects of natural selection on interpopulation divergence at polymorphic sites in human protein-coding Loci. *Genetics*, 170:1181–1187.doi:10.1534/genetics.104.037077 PMID:15911586

14. Miklós I, Novák A, Dombai B, Hein J (2008). How reliably can we predict the reliability of protein structure predictions? *BMC Bioinformatics*, 9:137.doi:10.1186/1471-2105-9-137 PMID:18315874

15. Edwards YJ, Cottage A (2003). Bioinformatics methods to predict protein structure and function. A practical approach. *Mol Biotechnol*, 23:139–166.doi:10.1385/MB:23:2:139 PMID:12632698

16. Heringa J (2000). Computational methods for protein secondary structure prediction using multiple sequence alignments. *Curr Protein Pept Sci*, 1:273–301.doi:10.2174/1389203003381324 PMID:12369910

17. Erichsen HC, Chanock SJ (2004). SNPs in cancer research and treatment. *Br J Cancer*, 90:747–751.doi:10.1038/sj.bjc.6601574 PMID:14970847

18. Cargill M, Altshuler D, Ireland J et al. (1999). Characterization of single-nucleotide polymorphisms in coding regions of human genes. *Nat Genet*, 22:231–238.doi:10.1038/10290 PMID:10391209

19. Stephens JC, Schneider JA, Tanguay DA et al. (2001). Haplotype variation and linkage disequilibrium in 313 human genes. *Science*, 293:489–493.doi:10.1126/science.1059431 PMID:11452081

20. International HapMap Consortium (2003). The International HapMap Project. *Nature*, 426:789–796. doi:10.1038/nature02168 PMID:14685227

21. Packer BR, Yeager M, Burdett L et al. (2006). SNP500Cancer: a public resource for sequence validation, assay development, and frequency analysis for genetic variation in candidate genes. *Nucleic Acids Res*, 34 Database issue;D617–D621.doi:10.1093/nar/gkj151 PMID:16381944

22. Stephens M, Sloan JS, Robertson PD et al. (2006). Automating sequence-based detection and genotyping of SNPs from diploid samples. *Nat Genet*, 38:375–381.doi:10.1038/ng1746 PMID:16493422

23. Marth G, Schuler G, Yeh R et al. (2003). Sequence variations in the public human genome data reflect a bottlenecked population history. *Proc Natl Acad Sci USA*, 100:376–381.doi:10.1073/pnas.222673099 PMID:12502794

24. Marth GT, Korf I, Yandell MD et al. (1999). A general approach to single-nucleotide polymorphism discovery. *Nat Genet*, 23:452–456.doi:10.1038/70570 PMID:10581034

25. Mardis ER (2008). The impact of next-generation sequencing technology on genetics. *Trends Genet*, 24:133–141.doi:10.1016/j.tig.2007.12.007 PMID:18262675

26. Birney E, Stamatoyannopoulos JA, Dutta A et al.; ENCODE Project Consortium; NISC Comparative Sequencing Program; Baylor College of Medicine Human Genome Sequencing Center; Washington University Genome Sequencing Center; Broad Institute; Children's Hospital Oakland Research Institute (2007). Identification and analysis of functional elements in 1% of the human genome by the ENCODE pilot project. *Nature*, 447:799–816.doi:10.1038/nature05874 PMID:17571346

27. Romeo S, Pennacchio LA, Fu Y et al. (2007). Population-based resequencing of ANGPTL4 uncovers variations that reduce triglycerides and increase HDL. *Nat Genet*, 39:513–516.doi:10.1038/ng1984 PMID:17322881

28. Bonnen PE, Wang PJ, Kimmel M et al. (2002). Haplotype and linkage disequilibrium architecture for human cancer-associated genes. *Genome Res*, 12:1846–1853.doi:10.1101/gr.483802 PMID:12466288

29. Sabeti PC, Reich DE, Higgins JM et al. (2002). Detecting recent positive selection in the human genome from haplotype structure. *Nature*, 419:832–837.doi:10.1038/nature01140 PMID:12397357

30. Slatkin M (2008). Linkage disequilibrium–understanding the evolutionary past and mapping the medical future. *Nat Rev Genet*, 9:477–485.doi:10.1038/nrg2361 PMID:18427557

31. Orr N, Chanock SJ (2008). Common genetic variation and human disease. *Adv Genet*, 62:1–32.doi:10.1016/S0065-2660(08)00601-9 PMID:19010252

32. Hill WG (1974). Estimation of linkage disequilibrium in randomly mating populations. *Heredity*, 33:229–239.doi:10.1038/hdy.1974.89 PMID:4531429

33. Clark AG (1990). Inference of haplotypes from PCR-amplified samples of diploid populations. *Mol Biol Evol*, 7:111–122. PMID:2108305

34. Eskin E, Halperin E, Karp RM (2003). Efficient reconstruction of haplotype structure via perfect phylogeny. *J Bioinform Comput Biol*, 1:1–20.doi:10.1142/S0219720003000174 PMID:15290779

35. Stephens M, Smith NJ, Donnelly P (2001). A new statistical method for haplotype reconstruction from population data. *Am J Hum Genet*, 68:978–989.doi:10.1086/319501 PMID:11254454

36. Stephens M, Donnelly P (2003). A comparison of bayesian methods for haplotype reconstruction from population genotype data. *Am J Hum Genet*, 73:1162–1169.doi:10.1086/379378 PMID:14574645

37. Marchini J, Cutler D, Patterson N et al.; International HapMap Consortium (2006). A comparison of phasing algorithms for trios and unrelated individuals. Am J Hum Genet, 78:437–450.doi:10.1086/500808 PMID:16465620

38. Akey J, Jin L, Xiong M (2001). Haplotypes vs single marker linkage disequilibrium tests: what do we gain? Eur J Hum Genet, 9:291–300. doi:10.1038/sj.ejhg.5200619 PMID:11313774

39. Schaid DJ (2004). Evaluating associations of haplotypes with traits. Genet Epidemiol, 27:348–364.doi:10.1002/gepi.20037 PMID:15543638

40. Tan Q, Christiansen L, Christensen K et al. (2005). Haplotype association analysis of human disease traits using genotype data of unrelated individuals. Genet Res, 86:223–231.doi:10.1017/S0016672305007792 PMID:16454861

41. Cardon LR, Abecasis GR (2003). Using haplotype blocks to map human complex trait loci. Trends Genet, 19:135–140.doi:10.1016/S0168-9525(03)00022-2 PMID:12615007

42. Johnson GC, Esposito L, Barratt BJ et al. (2001). Haplotype tagging for the identification of common disease genes. Nat Genet, 29:233–237.doi:10.1038/ng1001-233 PMID:11586306

43. Barrett JC, Fry B, Maller J, Daly MJ (2005). Haploview: analysis and visualization of LD and haplotype maps. Bioinformatics, 21:263–265.doi:10.1093/bioinformatics/bth457 PMID:15297300

44. Stram DO, Haiman CA, Hirschhorn JN et al. (2003). Choosing haplotype-tagging SNPS based on unphased genotype data using a preliminary sample of unrelated subjects with an example from the Multiethnic Cohort Study. Hum Hered, 55:27–36.doi:10.1159/000071807 PMID:12890923

45. McCarroll SA, Altshuler DM (2007). Copy-number variation and association studies of human disease. Nat Genet, 39 Suppl;S37–S42.doi:10.1038/ng2080 PMID:17597780

46. Scherer SW, Lee C, Birney E et al. (2007). Challenges and standards in integrating surveys of structural variation. Nat Genet, 39 Suppl;S7–S15.doi:10.1038/ng2093 PMID:17597783

47. Kidd JM, Cooper GM, Donahue WF et al. (2008). Mapping and sequencing of structural variation from eight human genomes. Nature, 453:56–64.doi:10.1038/nature06862 PMID:18451855

48. Feuk L, Carson AR, Scherer SW (2006). Structural variation in the human genome. Nat Rev Genet, 7:85–97.doi:10.1038/nrg1767 PMID:16418744

49. Stefansson H, Helgason A, Thorleifsson G et al. (2005). A common inversion under selection in Europeans. Nat Genet, 37:129–137.doi:10.1038/ng1508 PMID:15654335

50. Sharp AJ, Locke DP, McGrath SD et al. (2005). Segmental duplications and copy-number variation in the human genome. Am J Hum Genet, 77:78–88.doi:10.1086/431652 PMID:15918152

51. Sebat J, Lakshmi B, Troge J et al. (2004). Large-scale copy number polymorphism in the human genome. Science, 305:525–528. doi:10.1126/science.1098918 PMID:15273396

52. Iafrate AJ, Feuk L, Rivera MN et al. (2004). Detection of large-scale variation in the human genome. Nat Genet, 36:949–951.doi:10.1038/ng1416 PMID:15286789

53. Inoue K, Lupski JR (2002). Molecular mechanisms for genomic disorders. Annu Rev Genomics Hum Genet, 3:199–242.doi:10.1146/annurev.genom.3.032802.120023 PMID:12142364

54. Bailey JA, Yavor AM, Massa HF et al. (2001). Segmental duplications: organization and impact within the current human genome project assembly. Genome Res, 11:1005–1017. doi:10.1101/gr.GR-1871R PMID:11381028

55. Bailey JA, Gu Z, Clark RA et al. (2002). Recent segmental duplications in the human genome. Science, 297:1003–1007.doi:10.1126/science.1072047 PMID:12169732

56. Freeman JL, Perry GH, Feuk L et al. (2006). Copy number variation: new insights in genome diversity. Genome Res, 16:949–961. doi:10.1101/gr.3677206 PMID:16809666

57. International HapMap Consortium (2005). A haplotype map of the human genome. Nature, 437:1299–1320.doi:10.1038/nature04226 PMID:16255080

58. Buckley PG, Mantripragada KK, Piotrowski A et al. (2005). Copy-number polymorphisms: mining the tip of an iceberg. Trends Genet, 21:315–317.doi:10.1016/j.tig.2005.04.007 PMID:15922827

59. McCarroll SA, Kuruvilla FG, Korn JM et al. (2008). Integrated detection and population-genetic analysis of SNPs and copy number variation. Nat Genet, 40:1166–1174. doi:10.1038/ng.238 PMID:18776908

60. Khaja R, Zhang J, MacDonald JR et al. (2006). Genome assembly comparison identifies structural variants in the human genome. Nat Genet, 38:1413–1418.doi:10.1038/ng1921 PMID:17115057

61. Istrail S, Sutton GG, Florea L et al. (2004). Whole-genome shotgun assembly and comparison of human genome assemblies. Proc Natl Acad Sci USA, 101:1916–1921. doi:10.1073/pnas.0307971100 PMID:14769938

62. Cooper GM, Zerr T, Kidd JM et al. (2008). Systematic assessment of copy number variant detection via genome-wide SNP genotyping. Nat Genet, 40:1199–1203.doi:10.1038/ng.236 PMID:18776910

63. Korn JM, Kuruvilla FG, McCarroll SA et al. (2008). Integrated genotype calling and association analysis of SNPs, common copy number polymorphisms and rare CNVs. Nat Genet, 40:1253–1260.doi:10.1038/ng.237 PMID:18776909

64. Barnes C, Plagnol V, Fitzgerald T et al. (2008). A robust statistical method for case-control association testing with copy number variation. Nat Genet, 40:1245–1252. doi:10.1038/ng.206 PMID:18776912

65. Ballantyne KN, van Oorschot RAH, Mitchell RJ (2007). Comparison of two whole genome amplification methods for STR genotyping of LCN and degraded DNA samples. Forensic Sci Int, 166:35–41.doi:10.1016/j.forsciint.2006.03.022 PMID:16687226

66. Goellner GM, Tester D, Thibodeau S et al. (1997). Different mechanisms underlie DNA instability in Huntington disease and colorectal cancer. Am J Hum Genet, 60:879–890. PMID:9106534

67. Roewer L, Krawczak M, Willuweit S et al. (2001). Online reference database of European Y-chromosomal short tandem repeat (STR) haplotypes. Forensic Sci Int, 118:106–113. doi:10.1016/S0379-0738(00)00478-3 PMID:11311820

68. Ryckman K, Williams SM. Calculation and use of the Hardy-Weinberg model in association studies. Curr Protoc Hum Genet 2008;Chapter 1:Unit 1.18.

69. Hosking L, Lumsden S, Lewis K et al. (2004). Detection of genotyping errors by Hardy-Weinberg equilibrium testing. Eur J Hum Genet, 12:395–399.doi:10.1038/sj.ejhg.5201164 PMID:14872201

70. Gomes I, Collins A, Lonjou C et al. (1999). Hardy-Weinberg quality control. Ann Hum Genet, 63:535–538.doi:10.1046/j.1469-1809.1999.6360535.x PMID:11246455

71. Akey JM, Zhang K, Xiong M et al. (2001). The effect that genotyping errors have on the robustness of common linkage-disequilibrium measures. Am J Hum Genet, 68:1447–1456. doi:10.1086/320607 PMID:11359212

72. Wittke-Thompson JK, Pluzhnikov A, Cox NJ (2005). Rational inferences about departures from Hardy-Weinberg equilibrium. Am J Hum Genet, 76:967–986.doi:10.1086/430507 PMID:15834813

73. Leal SM (2005). Detection of genotyping errors and pseudo-SNPs via deviations from Hardy-Weinberg equilibrium. Genet Epidemiol, 29:204–214.doi:10.1002/gepi.20086 PMID:16080207

74. Cox DG, Kraft P (2006). Quantification of the power of Hardy-Weinberg equilibrium testing to detect genotyping error. Hum Hered, 61:10–14. doi:10.1159/000091787 PMID:16514241

75. Hirschhorn JN (2005). Genetic approaches to studying common diseases and complex traits. Pediatr Res, 57:74R–77R.doi:10.1203/01.PDR.0000159574.98964.87 PMID:15817501

76. Yu K, Wang Z, Li Q et al. (2008). Population substructure and control selection in genome-wide association studies. PLoS One, 3:e2551.doi:10.1371/journal.pone.0002551 PMID:18596976

77. Falush D, Stephens M, Pritchard JK (2003). Inference of population structure using multilocus genotype data: linked loci and correlated allele frequencies. *Genetics*, 164:1567–1587. PMID:12930761

78. Devlin B, Roeder K (1999). Genomic control for association studies. *Biometrics*, 55:997–1004.doi:10.1111/j.0006-341X.1999.00997.x PMID:11315092

79. Pritchard JK, Rosenberg NA (1999). Use of unlinked genetic markers to detect population stratification in association studies. *Am J Hum Genet*, 65:220–228.doi:10.1086/302449 PMID:10364535

80. Freedman ML, Haiman CA, Patterson N et al. (2006). Admixture mapping identifies 8q24 as a prostate cancer risk locus in African-American men. *Proc Natl Acad Sci USA*, 103:14068–14073.doi:10.1073/pnas.0605832103 PMID:16945910

81. Kao WH, Klag MJ, Meoni LA et al.; Family Investigation of Nephropathy and Diabetes Research Group (2008). MYH9 is associated with nondiabetic end-stage renal disease in African Americans. *Nat Genet*, 40:1185–1192. doi:10.1038/ng.232 PMID:18794854

82. Kopp JB, Smith MW, Nelson GW et al. (2008). MYH9 is a major-effect risk gene for focal segmental glomerulosclerosis. *Nat Genet*, 40:1175–1184.doi:10.1038/ng.226 PMID:18794856

83. Hurst LD (2009). Fundamental concepts in genetics: genetics and the understanding of selection. *Nat Rev Genet*, 10:83–93. doi:10.1038/nrg2506 PMID:19119264

84. Stern DL, Orgogozo V (2009). Is genetic evolution predictable? *Science*, 323:746–751. doi:10.1126/science.1158997 PMID:19197055

85. Stern DL, Orgogozo V (2008). The loci of evolution: how predictable is genetic evolution? *Evolution*, 62:2155–2177. doi:10.1111/j.1558-5646.2008.00450.x PMID:18616572

86. Cooper TF, Ostrowski EA, Travisano M (2007). A negative relationship between mutation pleiotropy and fitness effect in yeast. *Evolution*, 61:1495–1499.doi:10.1111/j.1558-5646.2007.00109.x PMID:17542856

87. Eyre-Walker A, Keightley PD, Smith NGC, Gaffney D (2002). Quantifying the slightly deleterious mutation model of molecular evolution. *Mol Biol Evol*, 19:2142–2149. PMID:12446806

88. Charlesworth B (2009). Fundamental concepts in genetics: effective population size and patterns of molecular evolution and variation. *Nat Rev Genet*, 10:195–205. doi:10.1038/nrg2526 PMID:19204717

89. Relethford JH (2004). Global patterns of isolation by distance based on genetic and morphological data. *Hum Biol*, 76:499–513. doi:10.1353/hub.2004.0060 PMID:15754968

90. Ramachandran S, Deshpande O, Roseman CC et al. (2005). Support from the relationship of genetic and geographic distance in human populations for a serial founder effect originating in Africa. *Proc Natl Acad Sci USA*, 102:15942–15947.doi:10.1073/pnas.0507611102 PMID:16243969

91. Li JZ, Absher DM, Tang H et al. (2008). Worldwide human relationships inferred from genome-wide patterns of variation. *Science*, 319:1100–1104.doi:10.1126/science.1153717 PMID:18292342

92. Rosenberg NA, Pritchard JK, Weber JL et al. (2002). Genetic structure of human populations. *Science*, 298:2381–2385.doi: 10.1126/science.1078311 PMID:12493913

93. Romero IG, Manica A, Goudet J et al. (2009). How accurate is the current picture of human genetic variation? *Heredity*, 102:120–126.doi:10.1038/hdy.2008.89 PMID:18766200

94. Sabeti PC, Schaffner SF, Fry B et al. (2006). Positive natural selection in the human lineage. *Science*, 312:1614–1620.doi:10.1126/science.1124309 PMID:16778047

95. Sabeti PC, Varilly P, Fry B et al.; International HapMap Consortium (2007). Genome-wide detection and characterization of positive selection in human populations. *Nature*, 449:913–918.doi:10.1038/nature06250 PMID:17943131

96. Tishkoff SA, Reed FA, Ranciaro A et al. (2007). Convergent adaptation of human lactase persistence in Africa and Europe. *Nat Genet*, 39:31–40.doi:10.1038/ng1946 PMID:17159977

97. Pompanon F, Bonin A, Bellemain E, Taberlet P (2005). Genotyping errors: causes, consequences and solutions. *Nat Rev Genet*, 6:847–859.doi:10.1038/nrg1707 PMID:16304600

98. Packer BR, Yeager M, Staats B et al. (2004). SNP500Cancer: a public resource for sequence validation and assay development for genetic variation in candidate genes. *Nucleic Acids Res*, 32 Database issue;D528–D532.doi:10.1093/nar/gkh005 PMID:14681474

99. Saiki RK, Scharf S, Faloona F et al. (1985). Enzymatic amplification of beta-globin genomic sequences and restriction site analysis for diagnosis of sickle cell anemia. *Science*, 230:1350–1354.doi:10.1126/science.2999980 PMID:2999980

100. Livak KJ, Marmaro J, Todd JA (1995). Towards fully automated genome-wide polymorphism screening. *Nat Genet*, 9:341–342.doi:10.1038/ng0495-341 PMID:7795635

101. Brenan CJ (2002). DNA-based molecular lithography for nanoscale fabrication. *IEEE Eng Med Biol Mag*, 21:164.doi:10.1109/MEMB.2002.1175178 PMID:12613226

102. Frederickson RM (2002). Fluidigm. Biochips get indoor plumbing. *Chem Biol*, 9:1161–1162. PMID:12445764

103. Sun X, Ding H, Hung K, Guo B (2000). A new MALDI-TOF based mini-sequencing assay for genotyping of SNPS. *Nucleic Acids Res*, 28:E68.doi:10.1093/nar/28.12.e68 PMID:10871391

104. Cunningham JM, Sellers TA, Schildkraut JM et al. (2008). Performance of amplified DNA in an Illumina GoldenGate BeadArray assay. *Cancer Epidemiol Biomarkers Prev*, 17:1781–1789.doi:10.1158/1055-9965.EPI-07-2849 PMID:18628432

105. Berthier-Schaad Y, Kao WH, Coresh J et al. (2007). Reliability of high-throughput genotyping of whole genome amplified DNA in SNP genotyping studies. *Electrophoresis*, 28:2812–2817.doi:10.1002/elps.200600674 PMID:17702060

106. Thomas G, Jacobs KB, Yeager M et al. (2008). Multiple loci identified in a genome-wide association study of prostate cancer. *Nat Genet*, 40:310–315.doi:10.1038/ng.91 PMID:18264096

107. Matsuzaki H, Loi H, Dong S et al. (2004). Parallel genotyping of over 10,000 SNPs using a one-primer assay on a high-density oligonucleotide array. *Genome Res*, 14:414–425.doi:10.1101/gr.2014904 PMID:14993208

108. Al Olama AA, Kote-Jarai Z, Giles GG et al.; UK Genetic Prostate Cancer Study Collaborators/British Association of Urological Surgeons' Section of Oncology; UK Prostate testing for cancer and Treatment study (ProtecT Study) Collaborators (2009). Multiple loci on 8q24 associated with prostate cancer susceptibility. *Nat Genet*, 41:1058–1060.doi:10.1038/ng.452 PMID:19767752

109. Barrett JC, Cardon LR (2006). Evaluating coverage of genome-wide association studies. *Nat Genet*, 38:659–662.doi:10.1038/ng1801 PMID:16715099

110. McCarthy MI, Abecasis GR, Cardon LR et al. (2008). Genome-wide association studies for complex traits: consensus, uncertainty and challenges. *Nat Rev Genet*, 9:356–369. doi:10.1038/nrg2344 PMID:18398418

111. Kim Y, Feng S, Zeng ZB (2008). Measuring and partitioning the high-order linkage disequilibrium by multiple order Markov chains. *Genet Epidemiol*, 32:301–312.doi:10.1002/gepi.20305 PMID:18330903

112. Schaid DJ (2004). Genetic epidemiology and haplotypes. *Genet Epidemiol*, 27:317–320.doi:10.1002/gepi.20046 PMID:15543637

113. Ansorge WJ (2009). Next-generation DNA sequencing techniques. *N Biotechnol*, 25:195–203.doi:10.1016/j.nbt.2008.12.009 PMID:19429539

114. Collins FS, Morgan M, Patrinos A (2003). The Human Genome Project: lessons from large-scale biology. *Science*, 300:286–290. doi:10.1126/science.1084564 PMID:12690187

115. Sanger F, Nicklen S, Coulson AR (1977). DNA sequencing with chain-terminating inhibitors. *Proc Natl Acad Sci USA*, 74:5463–5467.doi:10.1073/pnas.74.12.5463 PMID:271968

116. Margulies M, Egholm M, Altman WE et al. (2005). Genome sequencing in microfabricated high-density picolitre reactors. Nature, 437:376–380. PMID:16056220

117. Dressman D, Yan H, Traverso G et al. (2003). Transforming single DNA molecules into fluorescent magnetic particles for detection and enumeration of genetic variations. Proc Natl Acad Sci USA, 100:8817–8822.doi:10.1073/pnas.1133470100 PMID:12857956

118. Van Tassell CP, Smith TPL, Matukumalli LK et al. (2008). SNP discovery and allele frequency estimation by deep sequencing of reduced representation libraries. Nat Methods, 5:247–252.doi:10.1038/nmeth.1185 PMID:18297082

119. Cloonan N, Forrest ARR, Kolle G et al. (2008). Stem cell transcriptome profiling via massive-scale mRNA sequencing. Nat Methods, 5:613–619.doi:10.1038/nmeth.1223 PMID:18516046

120. Chanock SJ, Manolio T, Boehnke M et al.; NCI-NHGRI Working Group on Replication in Association Studies (2007). Replicating genotype-phenotype associations. Nature, 447:655–660.doi:10.1038/447655a PMID:17554299

121. Yeager M, Deng Z, Boland J et al. (2009). Comprehensive resequence analysis of a 97 kb region of chromosome 10q11.2 containing the MSMB gene associated with prostate cancer. Hum Genet, 126:743–750.doi:10.1007/s00439-009-0723-9 PMID:19644707122.

122. Yeager M, Xiao N, Hayes RB et al. (2008). Comprehensive resequence analysis of a 136 kb region of human chromosome 8q24 associated with prostate and colon cancers. Hum Genet, 124:161–170.doi:10.1007/s00439-008-0535-3 PMID:18704501

123. Parikh H, Deng Z, Yeager M et al. (2010). A comprehensive resequence analysis of the KLK15-KLK3-KLK2 locus on chromosome 19q13.33. Hum Genet, 127:91–99.doi:10.1007/s00439-009-0751-5 PMID:19823874

124. Ahn J, Berndt SI, Wacholder S et al. (2008). Variation in KLK genes, prostate-specific antigen and risk of prostate cancer. Nat Genet, 40:1032–1034, author reply 1035–1036.doi:10.1038/ng0908-1032 PMID:19165914

125. Eeles RA, Kote-Jarai Z, Giles GG et al.; UK Genetic Prostate Cancer Study Collaborators; British Association of Urological Surgeons' Section of Oncology; UK ProtecT Study Collaborators (2008). Multiple newly identified loci associated with prostate cancer susceptibility. Nat Genet, 40:316–321. doi:10.1038/ng.90 PMID:18264097

126. Ng SB, Turner EH, Robertson PD et al. (2009). Targeted capture and massively parallel sequencing of 12 human exomes. Nature, 461:272–276.doi:10.1038/nature08250 PMID:19684571

127. Levy S, Sutton G, Ng PC et al. (2007). The diploid genome sequence of an individual human. PLoS Biol, 5:e254.doi:10.1371/journal.pbio.0050254 PMID:17803354

128. Wheeler DA, Srinivasan M, Egholm M et al. (2008). The complete genome of an individual by massively parallel DNA sequencing. Nature, 452:872–876.doi:10.1038/nature06884 PMID:18421352

129. Mortazavi A, Williams BA, McCue K et al. (2008). Mapping and quantifying mammalian transcriptomes by RNA-Seq. Nat Methods, 5:621–628.doi:10.1038/nmeth.1226 PMID:18516045

130. Velculescu VE, Zhang L, Vogelstein B, Kinzler KW (1995). Serial analysis of gene expression. Science, 270:484–487. doi:10.1126/science.270.5235.484 PMID:7570003

131. Sultan M, Schulz MH, Richard H et al. (2008). A global view of gene activity and alternative splicing by deep sequencing of the human transcriptome. Science, 321:956–960. doi:10.1126/science.1160342 PMID:18599741

132. Robertson G, Hirst M, Bainbridge M et al. (2007). Genome-wide profiles of STAT1 DNA association using chromatin immunoprecipitation and massively parallel sequencing. Nat Methods, 4:651–657.doi:10.1038/nmeth1068 PMID:17558387

133. Hugenholtz P, Tyson GW (2008). Microbiology: metagenomics. Nature, 455:481–483.doi:10.1038/455481a PMID:18818648

134. Turnbaugh PJ, Ley RE, Hamady M et al. (2007). The human microbiome project. Nature, 449:804–810.doi:10.1038/nature06244 PMID:17943116

135. Bergen AW, Qi Y, Haque KA et al. (2005). Effects of DNA mass on multiple displacement whole genome amplification and genotyping performance. BMC Biotechnol, 5:24.doi:10.1186/1472-6750-5-24 PMID:16168060

136. Haque KA, Pfeiffer RM, Beerman MB et al. (2003). Performance of high-throughput DNA quantification methods. BMC Biotechnol, 3:20.doi:10.1186/1472-6750-3-20 PMID:14583097

137. Fire A, Xu SQ (1995). Rolling replication of short DNA circles. Proc Natl Acad Sci USA, 92:4641–4645.doi:10.1073/pnas.92.10.4641 PMID:7753856

138. Dean FB, Nelson JR, Giesler TL, Lasken RS (2001). Rapid amplification of plasmid and phage DNA using Phi 29 DNA polymerase and multiply-primed rolling circle amplification. Genome Res, 11:1095–1099.doi:10.1101/gr.180501 PMID:11381035

139. Bergen AW, Haque KA, Qi Y et al. (2005). Comparison of yield and genotyping performance of multiple displacement amplification and OmniPlex whole genome amplified DNA generated from multiple DNA sources. Hum Mutat, 26:262–270.doi:10.1002/humu.20213 PMID:16086324

140. Lasken RS (2009). Genomic DNA amplification by the multiple displacement amplification (MDA) method. Biochem Soc Trans, 37:450–453.doi:10.1042/BST0370450 PMID:19290880

141. Purcell S, Neale B, Todd-Brown K et al. (2007). PLINK: a tool set for whole-genome association and population-based linkage analyses. Am J Hum Genet, 81:559–575. doi:10.1086/519795 PMID:17701901

142. Hunter DJ, Thomas G, Hoover RN, Chanock SJ (2007). Scanning the horizon: what is the future of genome-wide association studies in accelerating discoveries in cancer etiology and prevention? Cancer Causes Control, 18:479–484.doi:10.1007/s10552-007-0118-y PMID:17440825

143. Hall JM, Lee MK, Newman B et al. (1990). Linkage of early-onset familial breast cancer to chromosome 17q21. Science, 250:1684–1689.doi:10.1126/science.2270482 PMID:2270482

144. Wooster R, Bignell G, Lancaster J et al. (1995). Identification of the breast cancer susceptibility gene BRCA2. Nature, 378:789–792.doi:10.1038/378789a0 PMID:8524414

145. Stratton MR, Rahman N (2008). The emerging landscape of breast cancer susceptibility. Nat Genet, 40:17–22.doi:10.1038/ng.2007.53 PMID:18163131

146. McLendon R, Friedman A, Bigner D et al.; Cancer Genome Atlas Research Network (2008). Comprehensive genomic characterization defines human glioblastoma genes and core pathways. Nature, 455:1061–1068.doi:10.1038/nature07385 PMID:18772890

147. Gabriel SB, Schaffner SF, Nguyen H et al. (2002). The structure of haplotype blocks in the human genome. Science, 296:2225–2229.doi:10.1126/science.1069424 PMID:12029063

148. Carlson CS, Eberle MA, Rieder MJ et al. (2004). Selecting a maximally informative set of single-nucleotide polymorphisms for association analyses using linkage disequilibrium. Am J Hum Genet, 74:106–120. doi:10.1086/381000 PMID:14681826

UNIT 2.
BIOMARKERS: PRACTICAL ASPECTS

CHAPTER 7.

Platforms for biomarker analysis using high-throughput approaches in genomics, transcriptomics, proteomics, metabolomics, and bioinformatics

B. Alex Merrick, Robert E. London, Pierre R. Bushel, Sherry F. Grissom, and Richard S. Paules

Summary

Global biological responses that reflect disease or exposure biology are kinetic and highly dynamic phenomena. While high-throughput DNA sequencing continues to drive genomics, the possibility of more broadly measuring changes in gene expression has been a recent development manifested by a diversity of technical platforms. Such technologies measure transcripts, proteins and small biological molecules, or metabolites, and respectively define the fields of transcriptomics, proteomics and metabolomics that can be performed at a cell-, tissue-, or organism-wide basis. Bioinformatics is the discipline that derives knowledge from the large quantity and diversity of biological, genetic, genomic and gene expression data by integrating computer science, mathematics, statistics and graphic arts. Gene, protein and metabolite expression profiles can be thought of as snapshots of the current, poorly-mapped molecular landscape. The ultimate aim of genomic platforms is to fully map this landscape to more completely describe all of the biological interactions within a living system, during disease and toxicity, and define the behaviour and relationships of all the components of a biological system. The development of databases and knowledge bases will support the integration of data from multiple domains, as well as computational modelling. This chapter will describe the technical platform methods involving DNA sequencing, mass spectrometry, nuclear magnetic resonance combined with separation systems, and bioinformatics to derive genomic and gene expression data and include the relevant bioinformatic tools for analysis. These genomic, or omics platforms should have wide application to epidemiological studies.

Introduction

The sequencing of the human genome stands as one of the major scientific achievements of the twentieth century. It embodies a defining moment in modern biology by which most high-throughput technologies are compared for size scope, and complexity. Beginning in 1990, it took roughly a decade for the first draft of the human genome to be completed. By 2003, about 99% of all gene-containing regions were described, numbering about 20 500 genes (1), although some regions of the genome, such as centromeres, telomeres and gene deserts, continue to undergo characterization and study. Data from the human genome project has provided a generalized human map of the three billion nucleotides comprising the DNA of a few human subjects. However, studies on the variations (polymorphisms) in human DNA sequences are currently underway; samples from 270 individuals of multiethnic backgrounds are being used in a consortium called the International HapMap Project for haplotype mapping (http://www.hapmap.org) (2). The goal is to identify the patterns of single nucleotide polymorphism (SNP) groups, called haplotypes or haps, among individual human beings. In addition, interpretation of the human genome has been greatly enhanced by the DNA sequencing of many other genomes that allow comparison of genetic organization, evolution and function. Nearly 300 genomes have been completely sequenced and range from unicellular organisms, like *E. coli* and *S. cervisiae*, to model invertebrate organisms, such as *Drosophila melanogaster* and *C. elegans*, to several mammalian species for which the completed and ongoing genome projects are all available online (3).

Although the conception of the idea for sequencing the human genome is relatively recent, the project could not have occurred without the preceding decades of biological and technological developments. Particularly noteworthy of these contributions are early cytogenetics and chromosomal studies at the beginning of the twentieth century by Morgan and colleagues, the discovery of DNA structure by Watson and Crick in 1953, DNA cloning in 1973 by Berg and Cohen, the DNA sequencing reaction in 1975 by Sanger, reverse transcriptase in 1970 and restriction endonucleases in 1971, and the polymerase chain reaction (PCR) by Mullis in 1983 (4). A new century now begins with an era of "omics," those fields describing a multitude of genomic functions aimed at further deciphering the biological meaning of sequences in the human genome.

The purpose of this chapter is to describe the technical platforms in genomics, transcriptomics, proteomics, metabolomics and bioinformatics that could be useful in epidemiologic studies. These analytical platforms favour high sample throughput and generation of large data sets.

Omes and omics

Gene expression constantly changes during health, adaptation, toxicity, disease and aging. While the genetic blueprint of an individual is relatively static, the various levels of gene expression to form and operate a complex organism are dynamically regulated, structurally complex and spatially determined. At any point in time, only a portion of a genome is expressed in specific cells and tissues. At the mRNA level of gene expression, the transcriptome represents all genes transcribed at any one moment, and the proteome is the complement of proteins making up cells and tissues. Small molecules and metabolites comprise the metabolome. The global study of each gene expression level is suffixed with "omics," such as transcriptomics, proteomics and metabolomics. Figure 7.1 suggests a sequence of gene expression based on genomic DNA sequences that are dynamically reflected in changes of transcripts, proteins and metabolites. Each level of gene expression (represented by upward, curved, dotted lines) has the opportunity to feed back and influence other levels reflective of highly integrated, multicellular processes in cells, tissues and organisms. Studies of each gene expression area utilize very different technical platforms to maximize large scale coverage of the transcriptome, proteome and metabolome. Technical platforms may involve mass parallel analysis using robotics, miniaturization, automation and computer processing. The integration of the many levels of gene expression is often referred to as systems biology by bioinformatics or computational biology. Bioinformatics represents an applied field of mathematics to biochemistry and molecular biology using statistics, computer science and artificial intelligence to design algorithms to derive biological meaning from gene expression data.

Figure 7.1. The interdependence of the cellular metabolome, proteome, transcriptome, and genome. Each type of characterization provides a functional indication of the activity of the proceeding set of molecules (solid lines). Conversely, there will be some degree of feedback regulation built into the system (dotted lines).

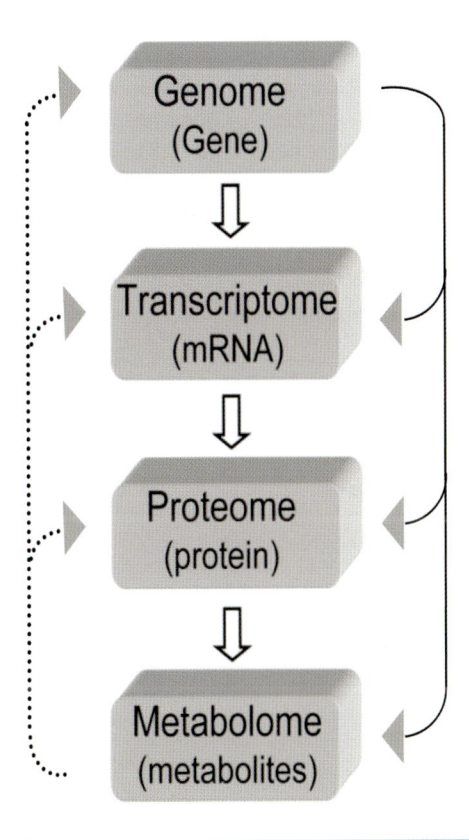

Genome (Gene) → Transcriptome (mRNA) → Proteome (protein) → Metabolome (metabolites)

Genomics

Chromosomal abnormalities are responsible for many developmental defects and malignancies, and include rearrangements in genomic DNA or changes in copy number, such as deletions, duplications and amplifications. Identification of genomic changes and mutations that underlie disease rely on comparisons of DNA sequences between affected and unaffected individuals. Finding disease-causing chromosomal abnormalities by genomic analysis is confounded by the fact that many sequence polymorphisms are functionally irrelevant and produce no observable biological consequence. Detection of many disease-causing mutations, such as those in *p53* in the Li-Fraumeni syndrome (5) and ATM in ataxia telangiectasia (6), have been found by sectional resequencing of genomic DNA or PCR-amplified DNA or RNA. However, de novo sequencing of individuals using automated Sanger-based capillary electrophoresis systems has so far been practical for only small regions of the human genome-containing candidate genes.

Recent advances in nucleic acid sequencing technologies using massive parallel sequencing, called next-generation sequencing, now allow sequencing of much larger genomic intervals (7). Sequencing of entire genomes can take place within a matter of several weeks, in a comprehensive search for chromosomal aberrations and mutations that affect phenotype. DNA sequencing does have inherent advantages in achieving single-base resolution and importantly for *de novo* analysis of samples without the prior knowledge of existing DNA sequence required for fabricated sequence platforms (8). New sequencing technologies that are high-throughput and low-cost while maintaining high accuracy and completeness are in continued development (9). New platforms often integrate real-time (RT) PCR and may incorporate microelectrophoresis, sequencing by hybridization, mass spectrometry, high-density oligonucleotide arrays, or incorporation of nanopore technology to bring within sight the goal of routine human genome sequencing (10) for personalized medicine (11).

There are several alternatives to whole-genome assessment of chromosomal abnormalities that do not involve traditional DNA sequencing. Alternative platforms involve selective genotyping of very focused genomic loci (short sections of DNA) that are potentially related to disease susceptibility. Haplotype mapping data have been extremely useful in providing candidate genomic regions with high polymorphic variation. Hundreds of thousands of loci can be very rapidly genotyped using BeadArray platforms, a technology based upon direct hybridization of whole-genome-amplified (WGA) genomic DNA to BeadArrays of locus-specific, 50mer oligonucleotide sequences (12). As such, genome-wide association studies (GWAS) comprise an important evolving field in genetic epidemiology in which more than 450 GWAS have been published, and the associations of greater than 2000 single nucleotide polymorphisms (SNPs) or genetic loci have been reported so far (13). Equally as important are high-density oligonucleotide microarrays for SNP detection in linkage analysis of susceptibility genes often used in cancer studies (14) or pharmacogenomics (15). In addition, fluorescence in situ hybridization (FISH) provides a visual map to examine all the chromosomes of a patient for abnormalities using fluorescent probes for specific genes or whole chromosome probes (16). Although high-throughput sequencing methodologies have been developed to accommodate the demand for sequence output, they consume large amounts of a valuable and potentially limiting genomic DNA. WGA can potentially remove DNA as a limiting factor for genomic analyses (17) that include multiple displacement amplification (MDA), primer extension preamplification (PEP), and degenerate oligonucleotide primed PCR (DOP) (18). However, genomic amplification technologies generate a certain level of replication error in sequence, which should be considered during verification studies. In summary, bead or chip DNA arrays or FISH platforms exemplify whole-genome technologies for high-throughput, compared to DNA sequencing for detection of genetic variation that can be linked to disease-based foci, protein, biomarkers and pharmacogenomic responses.

Transcriptomics

Transcriptomics studies the full, global complement of mRNA molecules expressed in cells and tissues. Some 20 500 genes are present within the human genome, of which about 10–15 000 are expressed at any one time in any particular tissue (19). Many expressed genes are necessary to perform basic functions of the cell, regardless of cell type or tissue, but a proportion of the expressed genes contribute to a cell's unique phenotype and specialized functions. Beginning around 1989, DNA microarrays, consisting of thousands of high-density cDNAs or oligonucleotides on support surfaces (called chips), were introduced and have evolved into powerful and versatile platforms for transcriptomic analysis (20–23). Spotted microarrays, either cDNAs or oligonucleotides, have been used extensively since the late 1990s, particularly by academia-based research scientists (see Figure 7.2). Commercial oligonucleotide arrays provide highly reproducible platforms representing the entire genome. Oligonucleotides from 25–70 bp in length are arrayed by either spotting pre-synthesized oligonucleotides directly onto glass, chemically synthesizing directly onto glass substrates (e.g. Agilent Technologies, Inc.),

Figure 7.2. Platforms for gene expression in transcriptomics, proteomics and metabolomics. Transcriptomic platforms are cDNA or oligonucleotides bound to glass slides or microbeads for analysis of mRNA. Metabolomic platforms are nuclear magnetic resonance (NMR) or mass spectrometry (MS) instruments for small biologic molecules or metabolites. Proteomic platforms can be gel-based or liquid chromatography-based (e.g. linear column gradients or multidimensional chromatography (MuDPIT)) for separation of proteins before identification (ID) by mass spectrometry. Use of stable isotopes greatly facilitates protein quantitation (ICAT, isotope coded affinity tags; iTRAQ, isobaric tags for relative and absolute quantitation; SILAC, stable isotope labelling by amino acids in culture), while non-isotopic or label-free methods can also be used that include spectral counting and ion precursor signal measurement. Retentate chromatography mass spectrometry (i.e. surface enhanced laser detection ionization (SELDI)) has been used for rapid profiling of biofluid samples using chemically reactive surfaces for separation and MALDI for generating protein mass spectra. Alternatives for MS-based proteomics involve affinity arrays, such as antibody microarrays or fluorescently tagged antibody bound bead suspensions (i.e. Luminex technology).

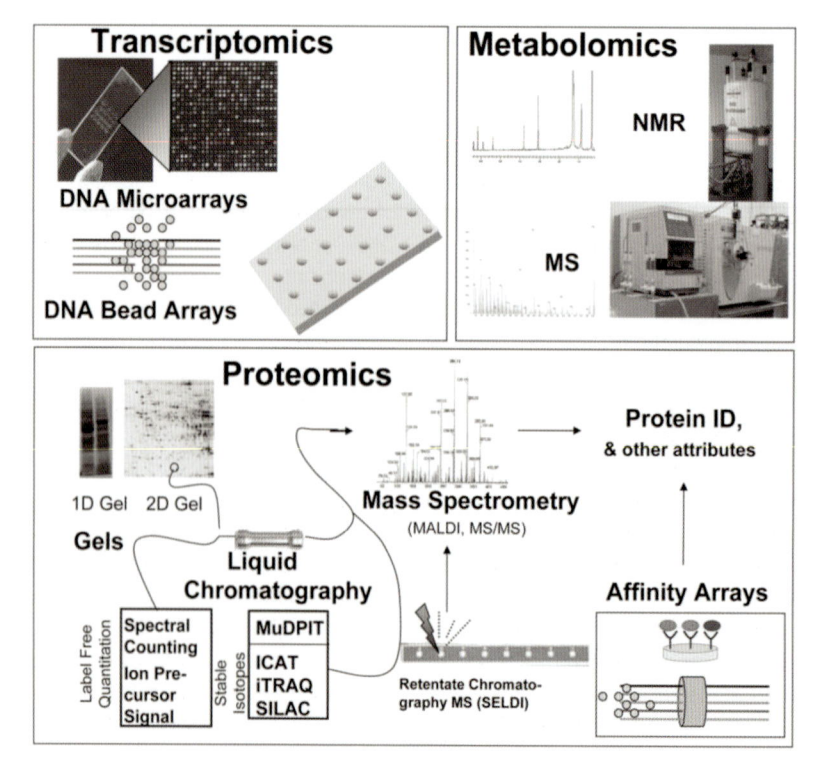

or by synthesizing directly onto quartz wafers by photolithographic technology (e.g. Affymetrix, Inc.) (24). In addition, oligonucleotides can be covalently linked with microbeads that can then be used in a 96-well microtiter dish format or on a glass substrate (e.g. Illumina, Inc.). With ever increasing technological advances, microarrays have progressed from chips with only several hundred probes to modern DNA chips reflecting expression of thousands, to even millions, of features per array.

The strength of any gene expression analysis, and the ability to determine expression profiles as potential biomarkers of exposure or effect, is dependent on proper experimental design and careful execution to minimize sources of variance and error and maximize useful biological information. Thus it is critical, in any microarray experiment, to design proper controls to include with samples of biological interest. Proper sampling and integrity of the RNA obtained from those samples are vital in determining the success of the analysis. Once RNA is isolated, proper labelling, hybridization, washing and scanning can dramatically influence the integrity of the resulting data. Since transcript expression is generally expressed in relative terms of a fold change of a particular gene expressed in one sample relative to a value in a control or normal sample, it is critical that the investigator structure the experiment in such a way as to minimize variations other than the one variable to be tested. Fortunately, commercial microarray providers have added increasingly more stringent quality control measurements in the production facilities. This has resulted in high reproducibility and low variation in microarrays coming from a manufacturer. Investigators have been allowed to shift resources away from multiple analyses of single samples to focusing on expanding the numbers of experimental samples. In turn, this has resulted in significant improvements in the confidence of results from microarray experiments. In fact, several large consortium efforts have demonstrated that comparable biological affects could be revealed in carefully controlled experiments in which multiple commercial microarray platforms, as well as rigorously quality-controlled spotted cDNA arrays, were used to analyse the same biological material (25–27).

Additional technological improvements have allowed for the reduction in the starting amounts of mRNA required in the labelling processes for the commercial platforms. Most labelling protocols used a single, PCR-based linear amplification of sample mRNA, which is used to incorporate a nucleotide conjugated with a fluorescent dye, biotin, or some other chemical modification. This amplification step has reduced the starting material for a sample to be analysed to only a few micrograms of total mRNA or less. Furthermore, protocols have been developed for additional rounds of PCR-based amplification of starting mRNA samples that make it possible to analyse very small quantities in the range of nanograms, and even picograms, of mRNA. These developments have facilitated gene expression profiling of samples derived from laser capture microdissection (LCM), for example, as well as biopsy samples, and other clinically derived samples that are limited in quantity. In addition, recent technological developments, particularly using bead-based microarrays (e.g. Illumina BeadChip), have opened up the possibility of using formalin-fixed, paraffin-embedded material for gene expression analysis.

An accessible, biological material fluid of principal interest to several clinical research scientists is blood. Many researchers are interested in testing the utility of gene expression profiling of peripheral blood leukocytes to generate biomarkers as surrogates for other tissues or organs affected in disease or injury processes (28). The utility of this approach has been demonstrated in studies of inflammatory responses and diseases in both animal models and humans, of neurological disorders, of angiomyolipoma (AML) and renal cell cancers, and of cardiac injury (29–36). Recent studies have used gene expression analysis of blood samples to generate molecular profiles as biomarkers of exposure and exposure-induced injury to arsenic, benzene, tobacco smoke and hepatotoxic levels of acetaminophen (37–41).

A common application in transcriptomics, useful in epidemiology studies, is to compare transcript outputs between normal and diseased tissues in what has been termed transcript profiling or expression profiling. Transcript expression studies can query all known or predicted genes in an organism, providing an abundance of information that represents a snapshot of the expression status of a tissue at any given time. One can gain considerable insight into molecular mechanisms from properly structured microarray experiments, both on the level of individual genes and on the level of biological pathways and processes. The potential for mRNA degradation makes expression profiling most applicable to freshly isolated tissues, cultured cells or flash-frozen tissue sections, but not paraffin-embedded tissue. While microarray approaches can be used to interrogate the entire genome on a single microarray chip, focused arrays representing distinct gene subsets have been used to focus upon changes in specific pathways or processes. These include both glass slide-based microarrays (e.g. the National Institute of Environmental Health Sciences' Human ToxChip (42)), and PCR-based gene expression analyses (e.g. SuperArrays).

DNA arrays can also reflect epigenetic effects upon gene expression. Epigenetics is defined as heritable changes in gene expression that are not due to DNA sequence alterations. Methylation is the most common epigenetic change and is detected by bisulfite conversion, methylation-sensitive restriction enzymes, methyl-binding proteins, and anti-methylcytosine antibodies. Combining these techniques with DNA microarrays and high-throughput sequencing has made the mapping of DNA methylation feasible on a genome-wide scale. Genomic DNA methylation occurs particularly at cytosines in clusters of cytosine-guanine dinucleotides, or CpG islands (p is the phosphodiester bond between C and G bases). Methylation of CpG islands in promoter regions frequently results in gene silencing, which normally occurs during development (43), but is often observed as an early alteration in some cancers by causing inactivation of tumour suppressors genes, such as von Hippel-Lindau disease (VHL), inhibitor of cyclin-dependent kinase 4a (p16^{INK4a}), and breast cancer gene 1 (*BRCA1*) (44).

DNA microarrays have also been developed for expression beyond profiling. In addition to SNP and comparative genomic hybridization (CGH) applications mentioned in the previous section, genome-wide localization of transcription factor binding sites can be accomplished by chromatin immunoprecipitation (ChIP) analysed on a microarray chip that forms the so-called ChIP-on-chip technique (45). The method can be innovatively combined with different types of DNA arrays, such as SNP chips, to form "ChIP-on-SNP" (46). The future for array technologies will also bring about a revolution in clinical DNA diagnostics (47), develop pharmaceuticals in pharmacogenomics (48), and personalized medicine (49).

Proteomics

The field for describing protein expression on a global scale is proteomics, which aims to detail the structure and functions of all proteins in an organism over time. The wide application of proteomics has generated great interest in many established disciplines of exposure biology and medicine, including the field of epidemiology (50,51). Chemical or toxicant exposure can bind to or modify proteins, produce changes in protein expression, and dysregulate critical biological pathways and processes that lead to toxicity and disease, which in theory should be detectable by proteomic analysis. Primary aims in proteomic analysis are the discovery of key modified proteins, the determination of affected pathways, and the development of biomarkers for association with and eventual prediction of disease.

The complexity of a proteome, represented by the total protein expression of a specific cell, organ, tissue or biofluid, presents numerous challenges for comprehensive analysis. Proteins are more complex than nucleic acids, and therefore proteomic analysis involves measurement of just some of the many attributes of proteins during any single expression analysis (52). Proteins exhibit many attributes of interest to biomarker development in epidemiology studies, including determination of protein sequence identity, quantity, post-translational modifications (PTM), protein–protein interactions, structure and function. Some of the challenges in proteomic analysis include: defining the identities and quantities of an entire proteome in a particular spatial location, such as serum or subcellular structures like mitochondria; the existence of multiple protein forms and complexes; the evolving structural and functional annotations of the human and rodent proteomes; and integration of proteomics data with transcriptomics or other expression data. Primary aims of proteomic analysis are to achieve maximal proteome identification, quantitative high-throughput protein measurement, timely analysis, and discovery-oriented platforms. Proteomic platforms represent combinations of technologies that describe protein attributes by the separation, quantitation and identification of all proteins in a biological sample. Proteomic analysis includes four broad categories of proteomic platforms: mass spectrometry has played a central role in proteomic platform development in large part because of its sensitive and versatile ability to identify proteins; the ability to separate proteins greatly determines the designation of platform type by gel-based separation or liquid chromatophic separation linked to mass spectrometry (53); solid phase adsorption, based on partitioning of peptides and proteins due to specific chemical properties, has been exploited in rententate chromatography combined with mass spectrometry; and finally, affinity chromatography, which sorts and identifies proteins in one reaction, is exemplified by use of antibodies in various formats (54). The following proteomic platforms represent some of the primary technologies being used for separating, identifying, and quantifying proteins during toxicoproteomic studies (Cf. Figure 7.2).

2D PAGE and DIGE

Two-dimensional polyacrylamide gel electrophoresis (2D PAGE) systems have been combined with mass spectrometry in an established and adaptable platform. Since 1975, 2D PAGE has been the most commonly used proteomic platform to separate and comparatively quantitate protein samples (55). Current state-of-the-art 2D gels use immobilized pH gradient (IPG) gels to separate proteins by charge. They are then resolved by mass spectrometry using sodium dodecyl sulfate (SDS) gel electrophoresis for effective separation of complex protein samples in μg to mg quantities. Either visible stains, such as Coomassie Blue or silver, or fluorescent staining are used for sensitive protein detection. After electronic alignment (registration) of stained proteins in 2D gels by image analysis software, intensities of identical protein spots are compared among treatment groups and a ratio (fold change) is calculated for each protein. A relatively new variation of the 2D PAGE technique, difference gel electrophoresis (DIGE), allows an investigator to measure three samples per gel that have been labelled with Cy2, Cy3 and Cy5 fluorescent dyes, which reduces some of the error associated with electronic registration during multiple gel alignment. This strategy allows for direct comparison of samples on one gel for better reproducibility and quantitation than conventional image analysis for comparison of multiple 2D gels. Thus, separation of proteins by 2D gels, using single stains or multiple fluors (i.e. DIGE), can be combined with mass spectrometry for ready protein identification to form a versatile and discovery-oriented platform for use in proteomic studies (56). In addition, some protein samples are sufficiently limited in their protein content that a simple size separation (one dimension) by SDS–PAGE can be used to identify protein bands of interest by mass spectrometry in 1D-Gel-MS.

Multidimensional LC-MS/MS

Proteomic platforms incorporating liquid chromatography (LC) as the primary means of separation (versus gel-based separations) have become the preferred means of analysis. There are many different types of LC separations often termed "multidimensional." In this proteomic platform, LC is used to separate protein digests by exploiting different biophysical properties of proteins before identification by tandem mass spectrometry (MS/MS). One of the most notable multidimensional LC-MS/MS platforms is Multidimensional Protein Identification Technology (MuDPIT). MuDPIT attempts to identify all proteins in a sample by two-dimensional separation of protein digests by charge (strong anion exchange matrix) and hydrophobicity (C18 column) with online LC immediately before entry into a tandem mass spectrometer (MS/MS) for protein identification (57). The platform has also been called shotgun proteomics, as entire protein lysates are trypsin digested into thousands of peptide fragments without any prior fractionation before separation and identification. Advantages of this newer platform are the potential for detection and identification of low abundance proteins that may not be observed in gel staining-based methods. One drawback is that LC-MS/MS platforms, like MuDPIT, are only semiquantitative and somewhat low-throughput in capacity.

The issue of protein quantitation in proteomics is an important one, since changes in protein expression may be a matter of altering existing gene and protein expression rather than turning them on (induction) or off (repression), which makes quantitation of proteins crucial in normal and diseased or control and experimental states. The use of stable isotopes, as detailed in the section below, takes advantage of the high resolution power of mass spectrometers to discriminate protein samples stable-isotopically labelled for quantitative comparison. However, rapid developments are being made using "label-free" approaches to quantitation if sample mass spectral data are sufficiently detailed (58). The two main approaches used are ion precursor signal intensities and spectral counting. Though they deliver relative sample quantitation, versus absolute protein measurement, they are simpler and less costly than stable isotopic methods, but not as precise. Label-free protein quantitation methods should be considered at the outset of designing LC-MS/MS proteomic studies and weighed against the considerable advantages and choices of stable isotopic approaches.

Stable isotope LC-MS/MS platforms: ICAT, iTRAQ and SILAC

A primary goal of proteomics is to comprehensively analyse all proteins in a sample, or as many as possible. However, the prospect of quantifying protein levels for comparison among protein samples has been a difficult aspect of proteomic analysis. Protein quantitation can be considered either in relative terms as a proportion of treatment (test) samples compared to control samples, or in absolute terms as the number of molecules (moles) or concentration (molarity). Internal standards are useful, but not realistic, for complex protein samples of unknown composition in most proteomic studies. Since many proteomic platforms are based in mass spectrometry, comparison of intensity signals seems the most direct means for comparative measurements; however, intensities are subject to many interfering factors. Quantitation by mass spectrometry has generally been regarded as semiquantitative under the best of circumstances.

The use of stable isotopes for tagging proteins has made great strides in proteomics for determining the relative amounts of proteins among samples (59). Stable isotopes of an element differ in mass due to the number of neutrons, but have the same elemental and chemical characteristics as the element. Stable isotopes are not radioactive. Common stable elements and their stable isotopes are ^1H and ^2H; ^{12}C and ^{13}C; ^{14}N and ^{15}N; ^{16}O and ^{18}O; ^{32}S and ^{34}S. A unique feature of high-resolution mass spectrometers is the ability to finely distinguish between small differences in mass, even to the point of resolving the relative abundance of stable isotopes in otherwise identical samples. Several proteomic platforms for protein quantitation and identification have been built around the use of isotopic tagging of proteins (isotope coded affinity tagging (ICAT), peptides (isotope tag for relative and absolute quantitation (iTRAQ), or metabolic incorporation of isotopically tagged amino acids in cell culture (stable isotope labelling with amino acids in cell culture (SILAC)). The applications of stable isotopes in proteomics have been recently reviewed for their sensitive detection of proteins in a quantitative and comparative fashion (60). As mentioned above, continuing improvements in spectral counting for use in LC-MS/MS platforms should have wide utility as a versatile, isotope-free method of protein quantitation when stable isotope use is not feasible (61).

SELDI-TOF mass spectrometry

Retentate chromatography-mass spectrometry (RC-MS) is a high-throughput proteomic platform that creates a laser-based mass spectrum (based on matrix-assisted laser desorption ionization time-of-flight (MALDI-TOF) mass spectrometry) from a chemically-absorptive surface. The principle of this approach is the adsorptive retention of a subset of sample proteins on a thin, chromatographic support (i.e. hydrophobic, normal phase, weak cation exchange, strong anion exchange or immobilized metal affinity supports). The absorptive surfaces are placed on thin metal chips which can be inserted into a MALDI-type mass spectrometer. The laser rapidly desorbs proteins from each sample on a metal chip to create a mass spectrum profile. RC-MS can be performed upon any protein sample, but thus far this platform has found greatest utility in the analysis of serum and plasma for disease biomarker discovery (62). The lead commercial platform of RC-MS proteomic platforms is the surface-enhanced laser desorption ionization time-of-flight mass spectrometry (SELDI-TOF-MS) instrument (63). Analysis of samples is relatively rapid (100/day), and only a few µl of sample is necessary. A downside is that protein identification of peaks is not readily accomplished without additional conventional separation and analysis.

Antibody arrays

Protein microarrays represent a promising new proteomic tool that closely emulates the design for parallel analysis of DNA microarray technology (64). Protein microarray formats can be divided into two types of multianalyte sensing formats: forward phase arrays and reverse phase arrays. In the forward phase array format, the analyte is captured from the solution phase by different capture molecules, such as an antibody immobilized on a substratum (i.e. glass slides). In contrast, the reverse array format immobilizes the individual test samples in each array spot so that, for example, hundreds of different patient blood or tissue samples are arrayed and probed with one detection protein, such as an antibody and a single analyte endpoint are measured for comparison across multiple samples. Such microarray studies have been carried out for metastatic ovarian cancer (65).

Many different types of capture molecules can be arrayed, including peptides (i.e. peptide substrates for kinases on phosphorylation arrays), proteins (protein–protein interaction arrays) and oligonucleotides (i.e. transcription factor binding arrays to oligonucleotides), but the most prevalent are antibody arrays in the forward phase format. Antibody arrays can directly separate proteins from complex biological fluids like plasma, serum or cell lysates by affinity binding to specific antigenic sites on target proteins.

Generally, current commercial antibody array platforms fall into three classes based on the targeted proteins: cytokine/chemokine arrays, cellular function protein arrays and cell signalling arrays. Although not all proteins for any given cell type or biofluid (i.e. blood, serum, plasma, urine, cerebral spinal fluid) are currently represented on antibody arrays, they do provide a rapid screen for protein alterations that may be relevant to tissue injury or disease (54). Antibodies can be placed in ordered array on glass slides or on a fluorescent microbead format (i.e. Luminex technology) for multiplexed separation, identification and quantitation (66). For example, in a study investigating the chemotherapeutic and radiotherapy of patients with rectal cancer, 40 tumour samples were analysed by DNA microarray and plasma samples were analysed by antibody (Luminex) bead microarray platforms. Using a kernel-based method with Least Squares Support Vector Machines to predict rectal cancer regression grade, investigators found that combining and integrating of microarray and proteomics data improved predictive power leading to the best model based on five genes and 10 proteins with an accuracy of 91.7%, sensitivity of 96.2% and specificity of 80% (67). In a different approach, a molecular epidemiological study used SELDI-TOF MS for *in vivo* studies of humans exposed to benzene. By using two sets of 10 exposed and 10 unexposed subjects, researchers identified with chemically-reactive surfaces and validated with antibody-coated chips three differentially expressed proteins in the serum of benzene-exposed individuals, two of which were identified as PF4 and CTAP-III, both members of the CXC-chemokine family (68). The same can be done with peptides (instead of antibodies) as the affinity ligand. This method was applied in the development of two diagnostic antibodies against avian influenza detection for epidemiologic studies, in which the epitopes of two monoclonal antibodies (mAbs) against avian influenza nucleoprotein (NP) were found using truncated NP recombinant proteins and peptide array techniques (69).

Future developments in proteomics will see incorporation of more sophisticated methods of quantitation in proteomic analysis (70), combining higher data density LC-MS/MS platforms with stable isotope labelled peptides, spectral counting, and parallel use of complementary proteomic platforms, such as tissue arrays (71). Study designs that remove abundant proteins from biofluids, enrich subcellular structures, and include cell-specific isolation from heterogeneous tissues will greatly increase differential expression capabilities. Advancement in mechanistic insights and biomarker development using proteomics will be furthered by completely defining plasma (serum) proteome and circulating microparticles in humans and rodent species as accessible biofluids (72). Some of the representative biomarkers and patterns of protein and message expression are shown in Table 7.1. Reviews on using proteomics to develop biomarkers and further mechanistic insights have been published (73–75).

Metabolomics

Analogous to the genomic characterization of cellular DNA and the proteomic characterization of the set of proteins expressed at a given time, cells also can be

Table 7.1. Recently identified biomarkers and signatures of toxicity using transcriptomics and proteomics

Biomarker/Signature	Identification	Condition	Reference No.
KIM-1	DNA microarray	Renal toxicity	(145)
Adipsin	DNA microarray	GI toxicity, functional gamma secretase inhibitors (FGSIs)	(146)
CXC-chemokines	SELDI	Benzene exposure	(68)
Troponin I,T	2D gel-MS	Myocardial ischemia, infarction	(147)
Aminopeptidase-P Annexin A1	2D gel-MS	Radioimmunotherapy	(148)
12 Lipid-gene signature	DNA microarray	Drug-induced phospholipidosis	(149)
Multigene blood signature	DNA microarray	Systemic sepsis	(32)

characterized in terms of the set of low-molecular-weight metabolites (typically < 1500 D) that comprise the cellular "metabolome." The cellular metabolome provides a functional readout of the cellular proteome (Cf. Figure 7.1). Although the analysis of homogeneous cell populations in culture, receiving identical nutrients and oxygenation, and exposed to the same levels of excreted waste products, represents the most ideal system for metabolomic characterization, the approach has been extended to the analysis of extracts and fluids derived from higher organisms. Urinary and blood metabolites have been among the most frequent targets for metabolomic characterization, but analyses of other fluids, such as cerebrospinal fluid (CSF), bronchoalveolar lavage fluid (76) and saliva (77), and of cellular extracts also have been performed. Typical ^{1}H NMR spectra illustrating the different metabolite composition of urine, blood, and saliva are shown in Figure 7.3. Currently, metabolomic characterization is being used for a wide range of objectives in human nutrition and toxicology, and for the development of pharmaceuticals and agricultural products. The underlying objectives of these studies include: discovery of metabolite signatures as prognostic indicators, diagnostics or biomarkers of disease states; establishing toxicological markers for drug development and environmental toxicology; understanding mechanisms of metabolic diseases; and correlation of metabolite phenotypes (metabotype) with genotype and environmental input (e.g. nutrition).

The screening of neonates for genetic disorders in intermediary metabolism is an application that predates the more recent interest in metabolomic characterization. The recent reviews of analytical approaches for clinical diagnosis of metabolic disorders (78,79) summarize methods for metabolite analysis and provide good examples of the application of MS to metabolomic analysis (Cf. Figure 7.2). Mass spectrometric analysis typically requires preparation of the metabolic components using either gas chromatography (GC) after chemical derivatization, or LC, with the newer method of ultra-performance liquid chromatography (UPLC) increasingly used. The use of capillary electrophoresis (CE) coupled to MS also has shown some promise. It was reported that a combination of approaches for metabolite extraction produced over 10 000 unique metabolite features, indicating both the complexity of the human metabolome and the potential of metabolomics in biomarker discovery (80).

Other more specialized techniques, such as Fourier transform infrared (FTIR) spectroscopy and arrayed electrochemical detection, have been used in some cases (81,82). The main limitation of FTIR is the low level of detailed molecular identification that can be achieved. In one study, MS was also employed for metabolite identification (82).

The extensive application of nuclear magnetic resonance (NMR) for metabolic characterization of urine and other body fluids has been developed primarily by Nicholson and coworkers (83–85). The primary advantage of the NMR approach (Cf. Figure 7.2) is the lack of required preparatory separations, which in turn leads to an unbiased and potentially quantitative measure of the constituents of the sample. Alternatively, while in principle it

Figure 7.3. Typical 1H NMR spectra illustrating the metabolite composition of urine (A), saliva (B) and plasma (C). Used with permission from (77).

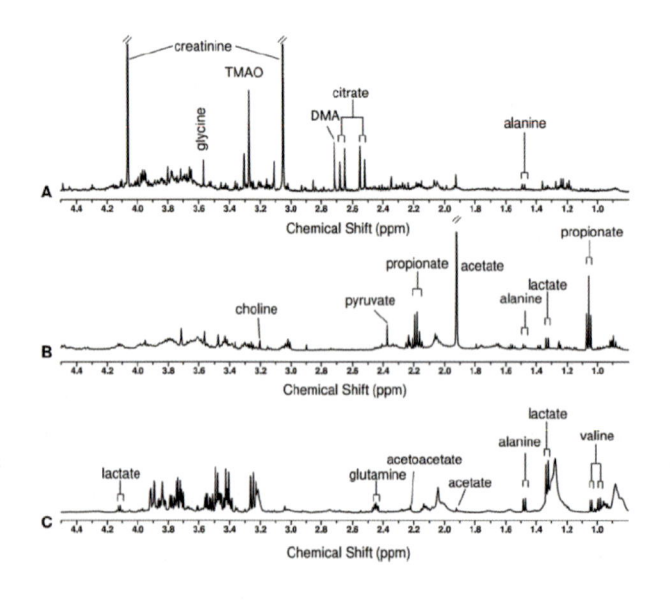

TMAO, trimethylamine oxide; DMA, dimethylamine.

is possible to obtain a molecular mass corresponding to each unfragmented metabolite observed in a mass spectrum, in an NMR spectrum, molecular information is distributed among the resonances of a compound, and the position of these resonances can depend critically on pH, salt concentration, divalent ions and other physical parameters. Additional caveats discussed in various reviews include loss of more volatile metabolites, metabolite contributions derived from intestinal bacteria (86), potential bacterial growth in stored fluids, and other factors (84,87). Some have proposed using dimethylamine and its nitroso metabolite as biomarkers for small bowel bacterial overgrowth (86), while others have suggested monitoring 4-hydroxyphenylacetate as a potential screening method for small bowel disease and bacterial overgrowth syndromes (88). Significant levels of ethanol in plasma or urinary samples generally indicate either bacterial contamination of the sample or small bowel bacterial overgrowth (87). Such conditions typically accompany renal failure or other serious illnesses.

In studies of chemical toxins or pharmaceutics, the xenobiotic and its metabolites and conjugates typically constitute an important source of variation of the metabolome. While of interest from a metabolic perspective, these compounds are not directly indicative of organ toxicity or therapeutic response, so that in general these metabolites are not relevant to the study. One approach to dealing with this issue involves stable isotope labelling of the compounds under study, so that the compound and metabolites derived from it will exhibit characteristic features in the NMR or mass spectrum. Alternatively, if the test compound and all of its metabolites and conjugates can be identified, these can simply be ignored or eliminated from the analysis. One general source of toxicity resulting from the administration of high levels of test compounds is the depletion of sulfur-containing amino acids that results from the excretion of glutathione and cysteine conjugates. Thus, it was reported that rats receiving high levels of acetaminophen excreted significant amounts of pyroglutamic acid. This effect of excess acetaminophen was prevented/reversed by supplementation with methionine (89). The glutathione analogue ophthalmic acid recently has been found to accumulate after high dosage with acetaminophen, and may also function as a biomarker for oxidative stress and glutathione depletion (90).

Various multivariate analyses of metabolite composition have been applied to detect differences among subject groups, such as those receiving different treatments or different chemical exposures. This type of analysis, termed "metabonomics" by the Nicholson group, most frequently utilizes principal component analysis (PCA) of the spectral data to reveal clustering behaviour that differentiates treated from control subjects. The clusters of data points in PC plots reveal the uniformity of the control and treated groups, as well as the extent to which the treated group yields a distinct metabolic phenotype. Since the axes of the PC plot are dependent on the data set and do not correspond to independent variables, the ability of different laboratories to utilize the published information is limited. Interestingly, a recent study that evaluated statistical methodology for the analysis of gene expression data found that the use of PCA to reveal clustering behaviour generally degrades cluster quality, and concluded that PCA was only useful in special cases (91). Hence, there is a critical need for the identification of metabolic biomarkers that provide universally quantifiable indications of organ function and toxicity.

Another critical issue related to the identification of biomarkers is the need to separate acute and chronic effects of illnesses or toxins. It is not unusual for a particular metabolite to become elevated during the acute phase of a toxic response, but to become depressed as the chronic effects become significant. Alternatively, in the absence of chronic effects, the metabolite level may return to pretreatment values. For this reason, plots of data obtained at different times after dosage, or trajectory plots for individual subjects, can provide critical information on the time-dependent response. Several of the issues discussed above are illustrated in studies identifying the association of the oxidative stress biomarker 8-oxoguanosine with Parkinson's disease (92). In this study, elevation of the 8-oxoguanosine level was observed in cerebrospinal fluid (CSF), but not in the serum of patients with Parkinson. Further, there was a significant negative correlation between the level of the biomarker and the duration of the disease. Finally, as indicated in Table 7.2, 8-oxoguanosine is also elevated in other conditions, e.g. amyotrophic lateral sclerosis (93). Another important limitation on the identification of some biomarkers relates to the chemical reactivity, which can deplete the free metabolite pool and lead to heterogeneous adduct formation and difficulties of detection. Homocysteine, which has long been linked to cardiovascular disease, provides one example (94).

Much of the early NMR work evaluating the effects of various toxins noted changes in the levels of tricarboxylic acid cycle (TCA) metabolites and other abundant molecules that may be present in the nutrient source and that are not organ-specific (85). More recently, several more specific metabolomic biomarkers have been correlated with various diseases or treatments (Table 7.2). Notably, most of the biomarkers that have been related to metastatic growth are proteins, but new metabolomic markers continue to be developed. The metabolite 12(S)-hydroxyeicosatetraenoic acid (12(S)-HETE) has been demonstrated to play a pivotal role in experimental melanoma invasion and metastasis, suggesting that 12-lipoxygenase expression may be important in early human melanoma carcinogenesis (95–97). Changes in phosphate-containing metabolites and in phospholipid composition have been correlated with tumour stage and metastatic spread. In studies of extracts from tumour cell lines, elevations in phosphorylcholine or other membrane-related phosphomonoesters have frequently been observed, and suggested to be correlated with metastatic potential (98,99). Increases have been observed in aspartyl-4-phosphate

Table 7.2. Biomarkers recently identified using metabolomics

Biomarker	Sample Analysis	Condition	Reference No.
NAN; 2-PY	NMR – human and rodent urine	Type 2 diabetes mellitus	(110)
NAN; 2-PY	NMR – rat urine	peroxosome proliferation	(111,112)
ADMA	MS – human blood plasma	Renal failure; atherosclerosis	(108,109)
Ophthalmic acid	MS – mouse serum, liver extract	Acetaminophen-induced hepatotoxicity	(90)
Pyroglutamic acid	NMR – rat urine	APAP-induced deficiency of sulfur- amino acids	(89)
3-nitrotyrosine	HPLC - human CSF	Amyotrophic lateral sclerosis	(150)
8-oxoguanosine	HPLC – human CSF	Alzheimer's disease	(93)
8-oxoguanosine	HPLC – human CSF	Parkinson's disease	(92)
Modified nucleosides	LC-IT-MS of human urine	Breast cancer	(102)
12(S)-HETE[a]	HPLC - tumor cell extracts	Human melanoma	(95-97)
Aspartyl-4-phosphate	DESI-MS/NMR – murine urine	Lung cancer/ tumour growth	(100)
Phosphorylcholine	31P NMR – cell extracts	Breast cancer cell extracts	(98)
Depressed lysophosphatidyl choline levels	31P NMR – blood plasma	Renal cell carcinoma	(101)
Elevated xanthine, hypoxanthine, urate	GC-MS – human urine	Lesch-Nyhan syndrome	(103)
Glc-Gal-pyridinoline	HPLC - human urine	Synovial degradation – RA	(104-106)
4-hydroxyphenyl acetate	GC-LC – human urine	SBBO	(88)
Dimethylamine, nitrosodimethylamine	GC – human serum; GC – whole blood	SBBO	(86)

[a] In most reported studies, concentration or enzymatic activity of 12-lipoxygenase, rather than 12(S)-HETE, has been determined.
2-PY, N-methyl-2-pyridone-5-carboxamide; ADMA, NG,NG-dimethylarginine; APAP, Acetaminophen; DESI, Desorption electrospray ionization; GC-LC, Gas chromatography-liquid chromatography; GC-MS, Gas chromatography-mass spectrometry; HPLC, High performance liquid chromatography; LC-IT-MS, Liquid chromatography-ion trap-mass spectrometry; NAN, N-methylnicotinamide; NMR, Nuclear magnetic resonance; RA, rheumatoid arthritis; SBBO, small bowel bacterial overgrowth.

that may correlate with tumour growth (100). Phosphorus-31 NMR studies of blood plasma derived from patients with advanced renal cell carcinoma have been found to exhibit depressed levels of lysophosphatidylcholine (101). A significantly improved discrimination of breast cancer patients based on metabolomic analysis of modified nucleosides present in urine was recently reported (102).

A metabolomic approach using a combination of gas chromatography and MS has identified elevations in the levels of the metabolites xanthine, hypoxanthine, urate and guanine in patients with Lesch-Nyhan syndrome (103). The urinary cross link product, Glc-Gal-pyridinoline (104), has been identified as a biomarker for synovial degradation observed in osteoarthritis (105); treatment with ibuprofen lowers the level of this excreted metabolite (106). Endogenously formed N^G,N^G-dimethylarginine, also referred to as asymmetric dimethylarginine (ADMA), is a potent inhibitor of nitric oxide synthase (107). Plasma levels increase as a consequence of renal failure (108), and ADMA has been identified as a biomarker for atherosclerosis (109). Metabolomic analyses have identified several pyridine derivatives in urine from diabetic rats (110), and the same derivatives have shown up as biomarkers of peroxisome proliferation (Table 7.2) (111,112). The presence of these compounds indicates a perturbation of the tryptophan-nicotinamide adenine dinucleotide (NAD) pathway.

As is typical for new technologies, some reports of putative biomarkers have proven controversial. Early identification of the partly-characterized metabolites CFSUM1 and CFSUM2, associated with chronic fatigue syndrome, were subsequently demonstrated to arise from incompletely derivatized pyroglutamic acid and serine (113,114), and the quantitative abnormalities of these metabolites in urine from patients with chronic fatigue syndrome/myalgic encephalomyelitis has been reported to be artefactual. Early analyses supporting the use of ^1H NMR of blood sera to diagnose coronary artery disease (115) have subsequently been found to be more equivocal than originally suggested and to compare unfavourably with angiography-based diagnosis (116).

As more specific biomarkers are identified, the power of this approach will continue to evolve, providing useful diagnostic information for pathological, environmental, toxicological, pharmaceutical and nutritional research, as well as enhancing the value of metabolomic analysis for basic research into mechanisms of toxicity. Future developments in mass spectrometry platforms in metabolomics will increase the detectable coverage of the metabolome in clinical specimens and experimental species, and permit better identification of metabolites in the process of converting raw data to biological knowledge (117).

Bioinformatics

The wealth of data generated through high-throughput omics approaches has become increasingly complex and too vast for conventional biomarker analysis strategies. Bioinformatics has played a crucial role in biomarker discovery and validation (Figure 7.4). It is a multidisciplinary field involving biology, computer science, mathematics and statistics to derive knowledge from biological, genomics and genetic data (118,119). Database systems, computational algorithms, statistical models, data mining methods and other analytical tools are typically employed in a bioinformatics framework to effectively manage, analyse and summarize the plethora of data. For example, proteomics approaches, such as SELDI-TOF and mass spectrometry in conjunction with bioinformatics tools, have greatly facilitated the discovery of new and better serum biomarkers to detect cancer (120).

The bioinformatics processes to translate omics data into clinically useful biomarkers can comprise a myriad of steps, beginning with initial analysis of the data to validation of the biomarkers (121). This multistep process typically involves discovery, data integration, predictive modeling, and delivery of the biomarkers to the clinic in a format to facilitate implementation (122). Bioinformatic analyses may take different approaches with several checkpoints along the way. A flowchart is described for the application of bioinformatics strategies to improve the identification of candidate biomarkers from cancer genome-wide expression analyses (123). The process proceeds with acquisition of gene expression data from cancer tissues, followed by the identification of candidate genes as biomarkers. The next step entails meta-mining public cancer data sets of the same type of pathophysiology to reduce the biomarker false-positive rate. The last step before use of the biomarkers in clinical trials involves validation of the candidates by RT–PCR, ELISA, tissue arrays, immunohistochemistry, and other types of bioassays.

What are these bioinformatics tools and processes and how are they used to aid in the discovery and validation of disease biomarkers? It is helpful to appreciate that a useful biomarker must be objective, highly accurate and very reliable in determining disease states and

Figure 7.4. Sequential scheme for the integration of omics with bioinformatics in the biomarker discovery process

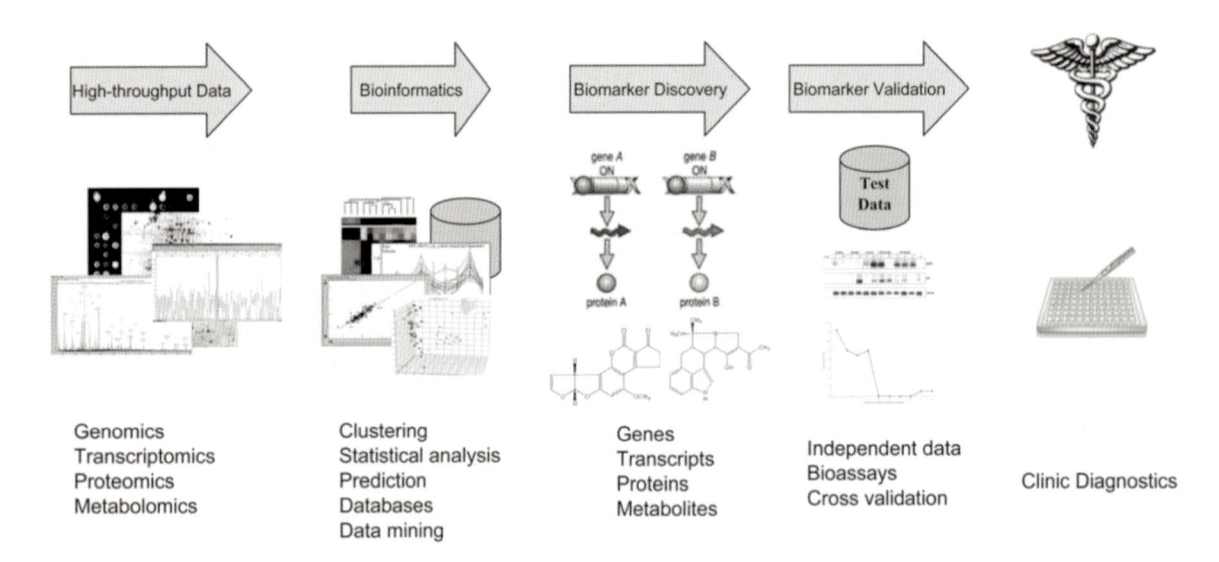

assessing risk. In other words, they must generalize well (i.e. extend) to broad cases, exhibit significance in their reporting, and be precise in their utility. Unfortunately, omics data possesses systematic variation due to the experimental error in the data acquisition process (124). There are technical limitations in the sensitivity of detecting biomarkers that are lowly expressed or non-abundant; however, biomarkers do not need to be highly expressed or in large abundance. Thus, the challenge in biomarker identification is successfully mining omics data with inherent error to find the features that reliably, accurately and objectively relate to the pathophysiology of a disease (125).

Data normalization and dimension reduction (data condensing) techniques have been widely used to preprocess omics data before biomarker discovery. Standard ways of dealing with data normalization have been adopted for omics data. Robust Multiarray Average (RMA), loess and quantile normalization methods have seemed to pass the test of time (126–128). A systematic variation normalization (SVN) approach was developed specifically to remove systematic error from microarray gene expression data (129). Baseline subtraction, signal smoothing and normalization methodologies were employed in the preprocessing of mass spectrometry proteomics data to reduce the noise and to make the analysis of spectral data comparable (130). In general, once the omics data is made unbiased and adjusted to make fair comparisons across samples, all the data can be used to mine for biomarkers or be filtered first to remove uninformative or redundant information. Some believe that the inclusion of uninformative features in the biomarker selection process will severely degrade the performance of the predictor model (131). Thus, it has been suggested to remove variables that do not contribute to a biological response of interest before the selection of biomarkers. Filtering of the data can be based on a signal or relative level, fold change, a confidence level, standard deviation from the mean of the distribution, P-values, mutual information, full or partial correlations, or more elaborate methods. The underlying omics data should be rich enough to be narrowed down to a core set of features that best represent the biology of the system for biomarker selection. Caution must be taken with respect to the preprocessing of omics data for biomarker identification, as the use of a particular combination of methods will surely add variability to the set of indicators selected. In other words, the preprocessing and filtering steps are sources of variability in and of themselves. The US Food and Drug Administration-led MicroArray Quality Control Phase II (MAQC-II) consortium set out to address this by trying to understand the limitations of various bioinformatics data analysis methods in developing and validating microarray-based predictive models, and determining if best practices for development and validation of predictive models based on microarray gene expression and genotyping data can be derived for biomarker discovery and personalized medicine (132).

The development and utilization of classification and prediction methods for analysis of omics data have transcended the process for identifying biomarkers from molecular signatures. One of the most widely used methods for classification is clustering. The process works by using a dissimilarity measure for the feature profiles in the omics data to iteratively form groups of samples that are tightly clustered. Features that cluster well together and can distinguish between groups of samples that differ in pathophysiology are considered to be potential biomarkers. Clustering and an F-test-like score based on within- and between-sample gene expression variance measures were used effectively to identify an intrinsic gene subset (i.e. a molecular portrait) that has a high predictive score for human breast tumours (133). More sophisticated bioinformatics methods have been developed to identify potential biomarkers. A hybrid approach was developed based on the genetic algorithm (GA) and k-nearest neighbours (KNN) classifier that is capable of identifying gene and protein molecular signatures of diseases based on microarray and proteomics data, respectively (134,135). The GA serves as a search tool to choose small subsets of predictors, whereas the KNN functions as a non-parametric (no distribution model assumed) pattern recognition method to evaluate the discriminative ability of the subsets. More recently, a hybrid approach for biomarker discovery from microarray gene expression data was developed to distinguish between types of cancer (136). This approach is based on Fisher's ratio (a measure for the linear discriminative power of variables) to select features "wrapped" with a classifier (hence, the procedure is called FR-Wrapper) to perform predictions. With these hybrid approaches, the two main objectives in biomarker discovery are met: 1) the identification of a small set of relevant indicators with minimum redundancy, and 2) the validation of the predictors using a classifier and cross-validation strategy. To balance false-positives and false-negatives in the selection of biomarkers, a clever method was proposed to use common peaks in mass spectrometry data as the predictive indicators (137). The procedure applies AdaBoost (a form of ensemble classifier training) to perform the classification and to select the informative common peaks.

Bioinformatics approaches to discover biomarkers can take on more sophisticated implementations. For instance, a dependence (interaction) network modeling scheme was suggested for identifying biomarkers from groups of genes or proteins (138). Very clear differences were observed in the dependence networks for cancer and non-cancer samples. On the other hand, a gene selection algorithm was used based on Gaussian processes to discover consistent gene expression patterns associated with ordinal clinical phenotypes (139). The method was able to identify subsets of genes as potential biomarkers for colon and prostate cancers. The integration of time-course microarray gene expression data with cytotoxicity measurements, by way of a partial least squares objective criterion, has been shown to be useful for identifying biomarkers in primary rat hepatocytes exposed to cadmium (140). The approach demonstrates the value of integrating omics data with associated biological data to glean more information about the biomarker's diagnostic utility. More recently, a bioinformatics approach was introduced that takes into account the inherent correlation of genes when using gene expression data for biomarker discovery (141).

Finally, a host of techniques and software to integrate omics data are summarized, to shed additional light on the complex molecular interactions that take place on a systems biology level (142).

Repositories of omics data from various studies that can be queried may bring about improved means for detecting biomarkers of a clinical process or phenotype than one which is isolated or from a small group of data sets (143). This realization has motivated the generators of omics data to store them in repositories for meta-mining purposes. Figure 7.5 represents a brief list of some of the publicly accessible databases that store, distribute and permit querying of omics data (a more comprehensive list is presented in (142)). A plasma proteome database at the Institute of Bioinformatics that stores comparisons of human plasma protein concentration levels along with their isoforms in normal and disease states should be useful for discovery of novel biomarkers (144). Another database, ONCOMINE, stores a collection of curated cancer gene expression profiles integrated with a therapeutic target database and biological resources, such as Gene Ontology, so that the data can be mined for putative biomarkers (123).

The use of bioinformatics tools has increased mechanistic understanding and development of biomarkers in the analysis of massive genomics, proteomics and metabolomics data (Cf. Figure 7.4). Bioinformatics techniques will continue to be useful in organizing and extracting candidate biomarkers for chemical exposures and disease for epidemiology, clinical and experimental studies. However, mere access to sophisticated bioinformatics tools will be insufficient to grapple with the identification of biomarkers from omics data (Figure 7.5). An ongoing and vigorous debate has emerged over the use and reproducibility of bioinformatics approaches and omics data for biomarker discovery and clinical applications (145,146). Clearly there is a need for rigorous quality control in the field of bioinformatics for the use of omics type data in clinical, diagnostic and regulatory settings (Lyle Burgoon, personal communication). A fundamental understanding of the inherent problems and issues with omics data, and knowing how, where and when to apply which type of bioinformatics approach, are essential to effectively translating omics biomarkers into clinically useful diagnostic tools and epidemiological markers.

Figure 7.5. Online bioinformatics resources for various omics fields, including genomics, transcriptomics, proteomics, and metabolomics

Data Type	Resource	Host	Description	URL
Genomics	Genomes OnLine Database (GOLD)	Genomesonline.org	Completed and ongoing genome projects	http://www.genomesonline.org/index
	GenBank	NIH-NCBI	Genetic sequence database	http://www.ncbi.nlm.nih.gov/Genbank
	EMBL-Bank	EMBL-EBI	Nucleotide sequence database	http://www.ebi.ac.uk/embl/
Transcriptomics	Gene Expression Omnibus (GEO)	NIH-NCBI	Gene expression/molecular abundance repository supporting MIAME-compliant data	http://www.ncbi.nlm.nih.gov/geo
	ArrayExpress	EMBL-EBI	Repository for MIAME-compliant microarray data	http://www.ebi.ac.uk/arrayexpress
	ArrayTrack™	FDA-NCTR	An integrated solution for managing, analysing and interpreting MIAME-compliant microarray gene expression data from pharmaco- and toxicogenomics studies	http://www.fda.gov/ScienceResearch/BioinformaticsTools/Arraytrack/default.htm
	Stanford Microarray Database (SMD)	Stanford University	Microarray-based MIAME-compliant gene expression	http://genome-www.stanford.edu/microarray
	ONCOMINE	University of Michigan	Database for mining and examining gene expression in cancer	http://www.oncomine.org
	Environment, Drugs and Gene Expression (EDGE)	University of Wisconsin-Madison	Toxicology-related gene expression data for mapping transcriptional changes from chemical exposure	http://edge.oncology.wisc.edu/
Proteomics	World-2DPAGE	Expert Protein Analysis System (ExPASy)	A dynamic portal to query simultaneously worldwide proteomics databases	http://ca.expasy.org/world-2dpage
	Plasma Protein Database	Pandey Lab and Institute of Bioinformatics	A comprehensive resource for all human plasma proteins along with their isoforms	http://www.plasmaproteomedatabase.org

Data Type	Resource	Host	Description	URL
Metabolomics	Open Proteomics Database (OPD)	University of Texas at Austin	For storing and disseminating mass spectrometry-based proteomics data	http://apropos.icmb.utexas.edu/OPD
	METLIN Metabolite Database	The Scripps Research Institute	A repository for mass spectral metabolite data	http://metlin.scripps.edu
	Human Metabolome Database (HMDB)	Genome Alberta and Genome Canada	Database of small metabolites found in the human body containing or linking chemical, clinical and molecular biology/biochemistry data	http://www.hmdb.ca/
Integrated	Chemical Effects in Biological Systems (CEBS)	NIH-NIEHS	Integrates study design, clinical pathology and histopathology, gene expression and proteomics data from all studies and enables discrimination of critical study factors	http://cebs.niehs.nih.gov/
	PharmGKB	Stanford	Collects, encodes and disseminates knowledge about the impact of human genetic variations on drug responses. Provides curated primary genotype and phenotype data, annotated gene variants, and gene-drug-disease relationships via literature review, and summarizes important PGx genes and drug pathways	http://www.pharmgkb.org/
	Kyoto encyclopaedia of genes and geomics pathways (KEGG)	Kyoto University, Japan	KEGG is an integrated database that relates genes to metabolic pathways; outlines functional relationships among genes; disease-associated genes; pharmaceuticals-targeted genes	http://www.genome.jp/kegg

References

1. Mundy C (2001). The human genome project: a historical perspective. *Pharmacogenomics*, 2:37–49.doi:10.1517/14622416.2.1.37 PMID:11258196

2. Thorisson GA, Smith AV, Krishnan L, Stein LD (2005). The International HapMap Project Web site. *Genome Res*, 15:1592–1593. doi:10.1101/gr.4413105 PMID:16251469

3. Liolios K, Tavernarakis N, Hugenholtz P, Kyrpides NC (2006). The Genomes On Line Database (GOLD) v.2: a monitor of genome projects worldwide. *Nucleic Acids Res*, 34 Database issue;D332–D334.doi:10.1093/nar/gkj145 PMID:16381880

4. Csako G (2006). Present and future of rapid and/or high-throughput methods for nucleic acid testing. *Clin Chim Acta*, 363:6–31.doi:10.1016/j.cccn.2005.07.009 PMID:161027 38

5. Royds JA, Iacopetta B (2006). p53 and disease: when the guardian angel fails. *Cell Death Differ*, 13:1017–1026.doi:10.1038/sj.cdd.4401913 PMID:16557268

6. Taylor AM, Byrd PJ (2005). Molecular pathology of ataxia telangiectasia. *J Clin Pathol*, 58:1009–1015.doi:10.1136/jcp.2005.026062 PMID:16189143

7. Nowrousian M (2010). Next-generation sequencing techniques for eukaryotic microorganisms: sequencing-based solutions to biological problems. *Eukaryot Cell*, 9:1300–1310.doi:10.1128/EC.00123-10 PMID:20601439

8. Rando OJ (2007). Chromatin structure in the genomics era. *Trends Genet*, 23:67–73. doi:10.1016/j.tig.2006.12.002 PMID:17188397

9. Metzker ML (2010). Sequencing technologies - the next generation. *Nat Rev Genet*, 11:31–46.doi:10.1038/nrg2626 PMID:19997069

10. Bentley DR (2006). Whole-genome re-sequencing. *Curr Opin Genet Dev*, 16:545–552.doi:10.1016/j.gde.2006.10.009 PMID:17055251

11. Bates S (2010). Progress towards personalized medicine. *Drug Discov Today*, 15:115–120.doi:10.1016/j.drudis.2009.11.001 PMID:19914397

12. Steemers FJ, Gunderson KL (2007). Whole genome genotyping technologies on the BeadArray platform. *Biotechnol J*, 2:41–49. doi:10.1002/biot.200600213 PMID:17225249

13. Ku CS, Loy EY, Pawitan Y, Chia KS (2010). The pursuit of genome-wide association studies: where are we now? *J Hum Genet*, 55:195–206.doi:10.1038/jhg.2010.19 PMID:20300123

14. Middeldorp A, Jagmohan-Changur S, Helmer Q et al. (2007). A procedure for the detection of linkage with high density SNP arrays in a large pedigree with colorectal cancer. *BMC Cancer*, 7:6.doi:10.1186/1471-2407-7-6 PMID:17222328

15. Giacomini KM, Brett CM, Altman RB et al.; Pharmacogenetics Research Network (2007). The pharmacogenetics research network: from SNP discovery to clinical drug response. *Clin Pharmacol Ther*, 81:328–345. doi:10.1038/sj.clpt.6100087 PMID:17339863

16. Mateuca R, Lombaert N, Aka PV et al. (2006). Chromosomal changes: induction, detection methods and applicability in human biomonitoring. *Biochimie*, 88:1515–1531.doi:10.1016/j.biochi.2006.07.004 PMID:16919864

17. Gunderson KL, Steemers FJ, Ren H et al. (2006). Whole-genome genotyping. *Methods Enzymol*, 410:359–376.doi:10.1016/S0076-6879(06)10017-8 PMID:16938560

18. Pinard R, de Winter A, Sarkis GJ et al. (2006). Assessment of whole genome amplification-induced bias through high-throughput, massively parallel whole genome sequencing. *BMC Genomics*, 7:216. doi:10.1186/1471-2164-7-216 PMID:16928277

19. Jongeneel CV, Iseli C, Stevenson BJ et al. (2003). Comprehensive sampling of gene expression in human cell lines with massively parallel signature sequencing. *Proc Natl Acad Sci USA*, 100:4702–4705.doi:10.1073/pnas.0831040100 PMID:12671075

20. Brown PO, Botstein D (1999). Exploring the new world of the genome with DNA microarrays. *Nat Genet*, 21 Suppl;33–37.doi:10.1038/4462 PMID:9915498

21. Davis TN (2004). Protein localization in proteomics. *Curr Opin Chem Biol*, 8:49–53. doi:10.1016/j.cbpa.2003.11.003 PMID:15036156

22. Lockhart DJ, Dong H, Byrne MC et al. (1996). Expression monitoring by hybridization to high-density oligonucleotide arrays. *Nat Biotechnol*, 14:1675–1680.doi:10.1038/nbt1296-1675 PMID:9634850

23. Schena M, Shalon D, Davis RW, Brown PO (1995). Quantitative monitoring of gene expression patterns with a complementary DNA microarray. *Science*, 270:467–470. doi:10.1126/science.270.5235.467 PMID:7569999

24. Fodor SP, Read JL, Pirrung MC et al. (1991). Light-directed, spatially addressable parallel chemical synthesis. *Science*, 251:767–773.doi:10.1126/science.1990438 PMID:1990438

25. Bammler T, Beyer RP, Bhattacharya S et al.; Members of the Toxicogenomics Research Consortium (2005). Standardizing global gene expression analysis between laboratories and across platforms. *Nat Methods*, 2:351–356.doi:10.1038/nmeth0605-477a PMID:15846362

26. Irizarry RA, Warren D, Spencer F et al. (2005). Multiple-laboratory comparison of microarray platforms. *Nat Methods*, 2:345–350.doi:10.1038/nmeth756 PMID:15846361

27. Ulrich RG, Rockett JC, Gibson GG, Pettit SD (2004). Overview of an interlaboratory collaboration on evaluating the effects of model hepatotoxicants on hepatic gene expression. *Environ Health Perspect*, 112:423–427. doi:10.1289/ehp.6675 PMID:15033591

28. Whitney AR, Diehn M, Popper SJ et al. (2003). Individuality and variation in gene expression patterns in human blood. *Proc Natl Acad Sci USA*, 100:1896–1901.doi:10.1073/pnas.252784499 PMID:12578971

29. Bennett L, Palucka AK, Arce E et al. (2003). Interferon and granulopoiesis signatures in systemic lupus erythematosus blood. *J Exp Med*, 197:711–723.doi:10.1084/jem.20021553 PMID:12642603

30. Borovecki F, Lovrecic L, Zhou J et al. (2005). Genome-wide expression profiling of human blood reveals biomarkers for Huntington's disease. *Proc Natl Acad Sci USA*, 102:11023–11028.doi:10.1073/pnas.0504921102 PMID:16043692

31. Burczynski ME, Twine NC, Dukart G et al. (2005). Transcriptional profiles in peripheral blood mononuclear cells prognostic of clinical outcomes in patients with advanced renal cell carcinoma. *Clin Cancer Res*, 11:1181–1189. PMID:15709187

32. Fannin RD, Auman JT, Bruno ME et al. (2005). Differential gene expression profiling in whole blood during acute systemic inflammation in lipopolysaccharide-treated rats. *Physiol Genomics*, 21:92–104. doi:10.1152/physiolgenomics.00190.2004 PMID:15781589

33. Heller RA, Schena M, Chai A et al. (1997). Discovery and analysis of inflammatory disease-related genes using cDNA microarrays. *Proc Natl Acad Sci USA*, 94:2150–2155.doi:10.1073/pnas.94.6.2150 PMID:9122163

34. Liew CC, Dzau VJ (2004). Molecular genetics and genomics of heart failure. *Nat Rev Genet*, 5:811–825.doi:10.1038/nrg1470 PMID:15520791

35. Sullivan PF, Fan C, Perou CM (2006). Evaluating the comparability of gene expression in blood and brain. *Am J Med Genet B Neuropsychiatr Genet*, 141B:261–268. doi:10.1002/ajmg.b.30272 PMID:16526044

36. Tang Y, Lu A, Aronow BJ, Sharp FR (2001). Blood genomic responses differ after stroke, seizures, hypoglycemia, and hypoxia: blood genomic fingerprints of disease. *Ann Neurol*, 50:699–707.doi:10.1002/ana.10042 PMID:11761467

37. Argos M, Kibriya MG, Parvez F et al. (2006). Gene expression profiles in peripheral lymphocytes by arsenic exposure and skin lesion status in a Bangladeshi population. *Cancer Epidemiol Biomarkers Prev*, 15:1367–1375.doi:10.1158/1055-9965.EPI-06-0106 PMID:16835338

38. Bushel PR, Heinloth AN, Li J et al. (2007). Blood gene expression signatures predict exposure levels. *Proc Natl Acad Sci USA*, 104:18211–18216.doi:10.1073/pnas.0706987104 PMID:17984051

39. Forrest MS, Lan Q, Hubbard AE et al. (2005). Discovery of novel biomarkers by microarray analysis of peripheral blood mononuclear cell gene expression in benzene-exposed workers. *Environ Health Perspect*, 113:801–807.doi:10.1289/ehp.7635 PMID:15929907

40. Lampe JW, Stepaniants SB, Mao M et al. (2004). Signatures of environmental exposures using peripheral leukocyte gene expression: tobacco smoke. *Cancer Epidemiol Biomarkers Prev*, 13:445–453. PMID:15006922

41. Fannin RD, Russo M, O'Connell TM et al. (2010). Acetaminophen dosing of humans results in blood transcriptome and metabolome changes consistent with impaired oxidative phosphorylation. *Hepatology*, 51:227–236. PMID:19918972

42. Afshari CA, Nuwaysir EF, Barrett JC (1999). Application of complementary DNA microarray technology to carcinogen identification, toxicology, and drug safety evaluation. *Cancer Res*, 59:4759–4760. PMID:10519378

43. Zilberman D, Henikoff S (2007). Genome-wide analysis of DNA methylation patterns. *Development*, 134:3959–3965.doi:10.1242/dev.001131 PMID:17928417

44. Esteller M (2008). Epigenetics in cancer. *N Engl J Med*, 358:1148–1159.doi:10.1056/NEJMra072067 PMID:18337604

45. Hoheisel JD (2006). Microarray technology: beyond transcript profiling and genotype analysis. *Nat Rev Genet*, 7:200–210.doi:10.1038/nrg1809 PMID:16485019

46. McCann JA, Muro EM, Palmer C et al. (2007). ChIP on SNP-chip for genome-wide analysis of human histone H4 hyperacetylation. *BMC Genomics*, 8:322.doi:10.1186/1471-2164-8-322 PMID:17868463

47. Beaudet AL, Belmont JW (2008). Array-based DNA diagnostics: let the revolution begin. *Annu Rev Med*, 59:113–129. doi:10.1146/annurev.med.59.012907.101800 PMID:17961075

48. Hardiman G (2008). Applications of microarrays and biochips in pharmacogenomics. *Methods Mol Biol*, 448: 21–30.doi:10.1007/978-1-59745-205-2_2 PMID:18370228

49. Allison M (2008). Is personalized medicine finally arriving? *Nat Biotechnol*, 26:509–517. doi:10.1038/nbt0508-509 PMID:18464779

50. Vineis P, Perera FP (2007). Molecular epidemiology and biomarkers in etiologic cancer research: the new in light of the old. *Cancer Epidemiol Biomarkers Prev*, 16:1954–1965.doi:10.1158/1055-9965.EPI-07-0457 PMID:17932342

51. Wild CP (2009). Environmental exposure measurement in cancer epidemiology. *Mutagenesis*, 24:117–125.doi:10.1093/mutage/gen061 PMID:19033256

52. Wetmore BA, Merrick BA (2004). Toxicoproteomics: proteomics applied to toxicology and pathology. *Toxicol Pathol*, 32:619–642.doi:10.1080/01926230490518244 PMID:15580702

53. Kline KG, Sussman MR (2010). Protein quantitation using isotope-assisted mass spectrometry. *Annu Rev Biophys*, 39:291–308. doi:10.1146/annurev.biophys.093008.131339 PMID:20462376

54. Sanchez-Carbayo M (2010). Antibody array-based technologies for cancer protein profiling and functional proteomic analyses using serum and tissue specimens. *Tumour Biol*, 31:103–112.doi:10.1007/s13277-009-0014-z PMID:20358423

55. Righetti PG, Castagna A, Antonucci F et al. (2004). Critical survey of quantitative proteomics in two-dimensional electrophoretic approaches. *J Chromatogr A*, 1051:3–17.doi:10.1016/j.chroma.2004.05.106 PMID:15532550

56. Yates JR 3rd (2004). Mass spectral analysis in proteomics. *Annu Rev Biophys Biomol Struct*, 33:297–316.doi:10.1146/annurev.biophys.33.111502.082538 PMID: 15139815

57. Macdonald N, Chevalier S, Tonge R et al. (2001). Quantitative proteomic analysis of mouse liver response to the peroxisome proliferator diethylhexylphthalate (DEHP). *Arch Toxicol*, 75:415–424.doi:10.1007/s002040100259 PMID:11693183

58. Neubert H, Bonnert TP, Rumpel K et al. (2008). Label-free detection of differential protein expression by LC/MALDI mass spectrometry. *J Proteome Res*, 7:2270–2279. doi:10.1021/pr700705u PMID:18412385

59. Ong SE, Pandey A (2001). An evaluation of the use of two-dimensional gel electrophoresis in proteomics. *Biomol Eng*, 18:195–205.doi:10.1016/S1389-0344(01)00095-8 PMID:11911086

60. Turner SM (2006). Stable isotopes, mass spectrometry, and molecular fluxes: applications to toxicology. *J Pharmacol Toxicol Methods*, 53:75–85.doi:10.1016/j.vascn.2005.08.001 PMID:16213756

61. Mueller LN, Brusniak MY, Mani DR, Aebersold R (2008). An assessment of software solutions for the analysis of mass spectrometry based quantitative proteomics data. *J Proteome Res*, 7:51–61.doi:10.1021/pr700758r PMID:18173218

62. Petricoin EF, Liotta LA (2004). SELDI-TOF-based serum proteomic pattern diagnostics for early detection of cancer. *Curr Opin Biotechnol*, 15:24–30.doi:10.1016/j.copbio.2004.01.005 PMID:15102462

63. Engwegen JY, Gast MC, Schellens JH, Beijnen JH (2006). Clinical proteomics: searching for better tumour markers with SELDI-TOF mass spectrometry. *Trends Pharmacol Sci*, 27:251–259.doi:10.1016/j.tips.2006.03.003 PMID:16600386

64. Cutler P (2003). Protein arrays: the current state-of-the-art. *Proteomics*, 3:3–18. doi:10.1002/pmic.200390007 PMID:12548629

65. Sheehan KM, Calvert VS, Kay EW et al. (2005). Use of reverse phase protein microarrays and reference standard development for molecular network analysis of metastatic ovarian carcinoma. *Mol Cell Proteomics*, 4:346–355.doi:10.1074/mcp.T500003-MCP200 PMID:15671044

66. Schwenk JM, Lindberg J, Sundberg M et al. (2007). Determination of binding specificities in highly multiplexed bead-based assays for antibody proteomics. *Mol Cell Proteomics*, 6:125–132. PMID:17060675

67. Daemen A, Gevaert O, De Bie T et al. (2008). Integrating microarray and proteomics data to predict the response on cetuximab in patients with rectal cancer. *Pac Symp Biocomput*, 166–177. PMID:18229684

68. Vermeulen R, Lan Q, Zhang L et al. (2005). Decreased levels of CXC-chemokines in serum of benzene-exposed workers identified by array-based proteomics. *Proc Natl Acad Sci USA*, 102:17041–17046.doi:10.1073/pnas.0508573102 PMID:16286641

69. Yang M, Berhane Y, Salo T et al. (2008). Development and application of monoclonal antibodies against avian influenza virus nucleoprotein. *J Virol Methods*, 147:265–274. doi:10.1016/j.jviromet.2007.09.016 PMID:18006085

70. Cho WC (2007). Proteomics technologies and challenges. *Genomics Proteomics Bioinformatics*, 5:77–85.doi:10.1016/S1672-0229(07)60018-7 PMID:17893073

71. Chung JY, Braunschweig T, Tuttle K, Hewitt SM (2007). Tissue microarrays as a platform for proteomic investigation. *J Mol Histol*, 38:123–128.doi:10.1007/s10735-006-9049-2 PMID:16953460

72. Merrick BA (2008). The plasma proteome, adductome and idiosyncratic toxicity in toxicoproteomics research. *Brief Funct Genomic Proteomic*, 7:35–49.doi:10.1093/bfgp/eln004 PMID:18270218

73. Kumar S, Mohan A, Guleria R (2006). Biomarkers in cancer screening, research and detection: present and future: a review. *Biomarkers*, 11:385–405.doi:10.1080/13547500600775011 PMID:16966157

74. Lemley KV (2007). An introduction to biomarkers: *applications to chronic kidney disease. Pediatr Nephrol,* 22:1849–1859. doi:10.1007/s00467-007-0455-9 PMID:17394023

75. Merrick BA, Bruno ME (2004). Genomic and proteomic profiling for biomarkers and signature profiles of toxicity. *Curr Opin Mol Ther,* 6:600–607. PMID:15663324

76. Azmi J, Connelly J, Holmes E *et al.* (2005). Characterization of the biochemical effects of 1-nitronaphthalene in rats using global metabolic profiling by NMR spectroscopy and pattern recognition. *Biomarkers,* 10:401–416. doi:10.1080/13547500500309259 PMID:16308265

77. Walsh MC, Brennan L, Malthouse JP *et al.* (2006). Effect of acute dietary standardization on the urinary, plasma, and salivary metabolomic profiles of healthy humans. *Am J Clin Nutr,* 84:531–539. PMID:16960166

78. Chace DH, Kalas TA, Naylor EW (2002). The application of tandem mass spectrometry to neonatal screening for inherited disorders of intermediary metabolism. *Annu Rev Genomics Hum Genet,* 3:17–45.doi: 10.1146/annurev.genom.3.022502.103213 PMID:12142359

79. Sim KG, Hammond J, Wilcken B (2002). Strategies for the diagnosis of mitochondrial fatty acid beta-oxidation disorders. *Clin Chim Acta,* 323:37–58.doi:10.1016/S0009-8981(02)00182-1 PMID:12135806

80. Want EJ, O'Maille G, Smith CA *et al.* (2006). Solvent-dependent metabolite distribution, clustering, and protein extraction for serum profiling with mass spectrometry. *Anal Chem,* 78:743–752.doi:10.1021/ac051312t PMID:16448047

81. Gamache PH, Meyer DF, Granger MC, Acworth IN (2004). Metabolomic applications of electrochemistry/mass spectrometry. *J Am Soc Mass Spectrom,* 15:1717–1726. doi:10.1016/j.jasms.2004.08.016 PMID:15589749

82. Kaderbhai NN, Broadhurst DI, Ellis DI *et al.* (2003). Functional genomics via metabolic footprinting: monitoring metabolite secretion by Escherichia coli tryptophan metabolism mutants using FT-IR and direct injection electrospray mass spectrometry. *Comp Funct Genomics,* 4:376–391.doi:10.1002/cfg.302 PMID:18629082

83. Lindon JC, Holmes E, Nicholson JK (2006). Metabonomics techniques and applications to pharmaceutical research & development. *Pharm Res,* 23:1075–1088.doi:10.1007/s11095-006-0025-z PMID:16715371

84. Nicholson JK, Connelly J, Lindon JC, Holmes E (2002). Metabonomics: a platform for studying drug toxicity and gene function. *Nat Rev Drug Discov,* 1:153–161.doi:10.1038/nrd728 PMID:12120097

85. Shockcor JP, Holmes E (2002). Metabonomic applications in toxicity screening and disease diagnosis. *Curr Top Med Chem,* 2:35–51.doi:10.2174/1568026023394498 PMID:11899064

86. Dunn S, Simenhoff M, Ahmed K *et al.* (1998). Effect of oral administration of freeze-dried Lactobacillus acidophilus on small bowel bacterial overgrowth in patients with end stage kidney disease: reducing uremic toxins and improving nutrition. *Int Dairy J,* 8:545–553 doi:10.1016/S0958-6946(98)00081-8.

87. London R, Houck D. Introduction to metabolomics and metabolic profiling. In: Hamadeh HK, Afshari CA, editors. Toxicogenomics: principles and applications. Hoboken (NJ): Wiley-LSS; 2004. p. 299–340.

88. Chalmers RA, Valman HB, Liberman MM (1979). Measurement of 4-hydroxyphenylacetic aciduria as a screening test for small-bowel disease. *Clin Chem,* 25:1791–1794. PMID:476929

89. Ghauri FY, McLean AE, Beales D *et al.* (1993). Induction of 5-oxoprolinuria in the rat following chronic feeding with N-acetyl 4-aminophenol (paracetamol). *Biochem Pharmacol,* 46:953–957.doi:10.1016/0006-2952(93)90506-R PMID:8373447

90. Soga T, Baran R, Suematsu M *et al.* (2006). Differential metabolomics reveals ophthalmic acid as an oxidative stress biomarker indicating hepatic glutathione consumption. *J Biol Chem,* 281:16768–16776.doi:10.1074/jbc.M601876200 PMID:16608839

91. Yeung KY, Ruzzo WL (2001). Principal component analysis for clustering gene expression data. *Bioinformatics,* 17:763–774. doi:10.1093/bioinformatics/17.9.763 PMID:11590094

92. Abe T, Isobe C, Murata T *et al.* (2003). Alteration of 8-hydroxyguanosine concentrations in the cerebrospinal fluid and serum from patients with Parkinson's disease. *Neurosci Lett,* 336:105–108.doi:10.1016/S0304-3940(02)01259-4 PMID:12499051

93. Abe T, Tohgi H, Isobe C *et al.* (2002). Remarkable increase in the concentration of 8-hydroxyguanosine in cerebrospinal fluid from patients with Alzheimer's disease. *J Neurosci Res,* 70:447–450.doi:10.1002/jnr.10349 PMID:12391605

94. Nekrassova O, Lawrence NS, Compton RG (2003). Analytical determination of homocysteine: a review. *Talanta,* 60:1085–1095.doi:10.1016/S0039-9140(03)00173-5 PMID:18969134

95. Chen YQ, Duniec ZM, Liu B *et al.* (1994). Endogenous 12(S)-HETE production by tumor cells and its role in metastasis. *Cancer Res,* 54:1574–1579. PMID:7511046

96. Liu B, Marnett LJ, Chaudhary A *et al.* (1994). Biosynthesis of 12(S)-hydroxyeicosatetraenoic acid by B16 amelanotic melanoma cells is a determinant of their metastatic potential. *Lab Invest,* 70:314–323. PMID:8145526

97. Winer I, Normolle DP, Shureiqi I *et al.* (2002). Expression of 12-lipoxygenase as a biomarker for melanoma carcinogenesis. *Melanoma Res,* 12:429–434.doi:10.1097/00008390-200209000-00003 PMID:12394183

98. Aiken NR, Gillies RJ (1996). Phosphomonoester metabolism as a function of cell proliferative status and exogenous precursors. *Anticancer Res,* 16 3B;1393–1397. PMID:8694507

99. Singer S, Souza K, Thilly WG (1995). Pyruvate utilization, phosphocholine and adenosine triphosphate (ATP) are markers of human breast tumor progression: a 31P- and 13C-nuclear magnetic resonance (NMR) spectroscopy study. *Cancer Res,* 55:5140–5145. PMID:7585561

100. Chen H, Pan Z, Talaty N *et al.* (2006). Combining desorption electrospray ionization mass spectrometry and nuclear magnetic resonance for differential metabolomics without sample preparation. *Rapid Commun Mass Spectrom,* 20:1577–1584.doi:10.1002/rcm.2474 PMID:16628593

101. Süllentrop F, Moka D, Neubauer S *et al.* (2002). 31P NMR spectroscopy of blood plasma: determination and quantification of phospholipid classes in patients with renal cell carcinoma. *NMR Biomed,* 15:60–68. doi:10.1002/nbm.758 PMID:11840554

102. Frickenschmidt A, Frohlich H, Bullinger D *et al.* (2008). Metabonomics in cancer diagnosis: mass spectrometry-based profiling of urinary nucleosides from breast cancer patients. *Biomarkers,* 13:435–449. doi:10.1080/13547500802012858 PMID:18484357

103. Ohdoi C, Nyhan WL, Kuhara T (2003). Chemical diagnosis of Lesch-Nyhan syndrome using gas chromatography-mass spectrometry detection. *J Chromatogr B Analyt Technol Biomed Life Sci,* 792:123–130. doi:10.1016/S1570-0232(03)00277-0 PMID:12829005

104. Seibel MJ, Gartenberg F, Silverberg SJ *et al.* (1992). Urinary hydroxypyridinium cross-links of collagen in primary hyperparathyroidism. *J Clin Endocrinol Metab,* 74:481–486.doi:10.1210/jc.74.3.481 PMID:1740480

105. Gineyts E, Garnero P, Delmas PD (2001). Urinary excretion of glucosyl-galactosyl pyridinoline: a specific biochemical marker of synovium degradation. *Rheumatology (Oxford),* 40:315–323.doi:10.1093/rheumatology/40.3.315 PMID:11285380

106. Gineyts E, Mo JA, Ko A *et al.* (2004). Effects of ibuprofen on molecular markers of cartilage and synovium turnover in patients with knee osteoarthritis. *Ann Rheum Dis,* 63:857–861.doi:10.1136/ard.2003.007302 PMID:15194584

107. MacAllister RJ, Whitley GSJ, Vallance P (1994). Effects of guanidino and uremic compounds on nitric oxide pathways. *Kidney Int,* 45:737–742.doi:10.1038/ki.1994.98 PMID:7515129

108. Zoccali C, Bode-Böger SM, Mallamaci F *et al.* (2001). Plasma concentration of asymmetrical dimethylarginine and mortality in patients with end-stage renal disease: a prospective study. *Lancet,* 358:2113–2117.doi:10.1016/S0140-6736(01)07217-8 PMID:11784625

109. Miyazaki H, Matsuoka H, Cooke JP et al. (1999). Endogenous nitric oxide synthase inhibitor: a novel marker of atherosclerosis. Circulation, 99:1141–1146. PMID:10069780

110. Salek RM, Maguire ML, Bentley E et al. (2007). A metabolomic comparison of urinary changes in type 2 diabetes in mouse, rat, and human. Physiol Genomics, 29:99–108. PMID:17190852

111. Ringeissen S, Connor SC, Brown HR et al. (2003). Potential urinary and plasma biomarkers of peroxisome proliferation in the rat: identification of N-methylnicotinamide and N-methyl-4-pyridone-3-carboxamide by 1H nuclear magnetic resonance and high performance liquid chromatography. Biomarkers, 8:240–271.doi:10.1080/1354750031000149124 PMID:12944176

112. Connor SC, Hodson MP, Ringeissen S et al. (2004). Development of a multivariate statistical model to predict peroxisome proliferation in the rat, based on urinary 1H-NMR spectral patterns. Biomarkers, 9:364–385.doi:10.1080/13547500400006005 PMID:15764299

113. Chalmers RA, Jones MG, Goodwin CS, Amjad S (2006). CFSUM1 and CFSUM2 in urine from patients with chronic fatigue syndrome are methodological artefacts. Clin Chim Acta, 364:148–158.doi:10.1016/j.cccn. 2005.05.036 PMID:16095585

114. Jones MG, Cooper E, Amjad S et al. (2005). Urinary and plasma organic acids and amino acids in chronic fatigue syndrome. Clin Chim Acta, 361:150–158.doi:10.1016/j.cccn. 2005.05.023 PMID:15992788

115. Brindle JT, Antti H, Holmes E et al. (2002). Rapid and noninvasive diagnosis of the presence and severity of coronary heart disease using 1H-NMR-based metabonomics. Nat Med, 8:1439–1445.doi:10.1038/nm802 PMID:12447357

116. Kirschenlohr HL, Griffin JL, Clarke SC et al. (2006). Proton NMR analysis of plasma is a weak predictor of coronary artery disease. Nat Med, 12:705–710.doi:10.1038/nm1432 PMID:16732278

117. Dunn WB (2008). Current trends and future requirements for the mass spectrometric investigation of microbial, mammalian and plant metabolomes. Phys Biol, 5:011001.doi:10.1088/1478-3975/5/1/011001 PMID:18367780

118. Luscombe NM, Greenbaum D, Gerstein M (2001). What is bioinformatics? A proposed definition and overview of the field. Methods Inf Med, 40:346–358. PMID:11552348

119. Mount DW, Pandey R (2005). Using bioinformatics and genome analysis for new therapeutic interventions. Mol Cancer Ther, 4:1636–1643.doi:10.1158/1535-7163.MCT-05-0150 PMID:16227414

120. Li J, Zhang Z, Rosenzweig J et al. (2002). Proteomics and bioinformatics approaches for identification of serum biomarkers to detect breast cancer. Clin Chem, 48:1296–1304. PMID:12142387

121. Azuaje F. Bioinformatics and biomarker discovery: "omic" data analysis for personalized medicine. Hoboken (NJ): Wiley-Blackwell; 2010.

122. Ginsburg GS, Haga SB (2006). Translating genomic biomarkers into clinically useful diagnostics. Expert Rev Mol Diagn, 6:179–191.doi:10.1586/14737159.6.2.179 PMID:16512778

123. Rhodes DR, Yu J, Shanker K et al. (2004). ONCOMINE: a cancer microarray database and integrated data-mining platform. Neoplasia, 6:1–6. PMID:15068665

124. Baggerly KA, Coombes KR, Morris JS (2005). Bias, randomization, and ovarian proteomic data: a reply to "producers and consumers.". Cancer Inform, 1:9–14.

125. Liu J, Zheng S, Yu JK et al. (2005). Serum protein fingerprinting coupled with artificial neural network distinguishes glioma from healthy population or brain benign tumor. J Zhejiang Univ Sci B, 6:4–10.doi:10.1631/jzus.2005.B0004 PMID:15593384

126. Yang YH, Dudoit S, Luu P et al. (2002). Normalization for cDNA microarray data: a robust composite method addressing single and multiple slide systematic variation. Nucleic Acids Res, 30:e15.doi:10.1093/nar/30.4.e15 PMID:11842121

127. Irizarry RA, Hobbs B, Collin F et al. (2003). Exploration, normalization, and summaries of high density oligonucleotide array probe level data. Biostatistics, 4:249–264.doi:10.1093/biostatistics/4.2.249 PMID:12925520

128. Bolstad BM, Irizarry RA, Astrand M, Speed TP (2003). A comparison of normalization methods for high density oligonucleotide array data based on variance and bias. Bioinformatics, 19:185–193. doi:10.1093/bioinformatics/19.2.185 PMID:12538238

129. Chou JW, Paules RS, Bushel PR (2005). Systematic variation normalization in microarray data to get gene expression comparison unbiased. J Bioinform Comput Biol, 3:225–241.doi:10.1142/S0219720005001028 PMID:15852502

130. Petricoin EF 3rd, Ardekani AM, Hitt BA et al. (2002). Use of proteomic patterns in serum to identify ovarian cancer. Lancet, 359:572–577.doi:10.1016/S0140-6736(02)07746-2 PMID:11867112

131. Tan CS, Ploner A, Quandt A et al. (2006). Finding regions of significance in SELDI measurements for identifying protein biomarkers. Bioinformatics, 22:1515–1523. doi:10.1093/bioinformatics/btl106 PMID:16567365

132. Shi L, Campbell G, Jones WD et al.; MAQC Consortium (2010). The MicroArray Quality Control (MAQC)-II study of common practices for the development and validation of microarray-based predictive models. Nat Biotechnol, 28:827–838.doi:10.1038/nbt.1665 PMID:20676074

133. Perou CM, Sørlie T, Eisen MB et al. (2000). Molecular portraits of human breast tumours. Nature, 406:747–752.doi:10.1038/35021093 PMID:10963602

134. Li L, Umbach DM, Terry P, Taylor JA (2004). Application of the GA/KNN method to SELDI proteomics data. Bioinformatics, 20:1638–1640.doi:10.1093/bioinformatics/bth098 PMID:14962943

135. Li L, Weinberg CR, Darden TA, Pedersen LG (2001). Gene selection for sample classification based on gene expression data: study of sensitivity to choice of parameters of the GA/KNN method. Bioinformatics, 17:1131–1142.doi:10.1093/bioinformatics/17.12.1131 PMID:11751221

136. Peng Y (2006). A novel ensemble machine learning for robust microarray data classification. Comput Biol Med, 36:553–573. doi:10.1016/j.compbiomed.2005.04.001 PMID:15978569

137. Fushiki T, Fujisawa H, Eguchi S (2006). Identification of biomarkers from mass spectrometry data using a "common" peak approach. BMC Bioinformatics, 7:358.doi:10.1186/1471-2105-7-358 PMID:16869977

138. Qiu P, Wang ZJ, Liu KJ et al. (2007). Dependence network modeling for biomarker identification. Bioinformatics, 23:198–206. doi:10.1093/bioinformatics/btl553 PMID:17077095

139. Chu W, Ghahramani Z, Falciani F, Wild DL (2005). Biomarker discovery in microarray gene expression data with Gaussian processes. Bioinformatics, 21:3385–3393. doi:10.1093/bioinformatics/bti526 PMID:15937031

140. Tan Y, Shi L, Hussain SM et al. (2006). Integrating time-course microarray gene expression profiles with cytotoxicity for identification of biomarkers in primary rat hepatocytes exposed to cadmium. Bioinformatics, 22:77–87.doi:10.1093/bioinformatics/bti737 PMID:16249259

141. Zuber V, Strimmer K (2009). Gene ranking and biomarker discovery under correlation. Bioinformatics, 25:2700–2707.doi:10.1093/bioinformatics/btp460 PMID:19648135

142. Joyce AR, Palsson BO (2006). The model organism as a system: integrating 'omics' data sets. Nat Rev Mol Cell Biol, 7:198–210. doi:10.1038/nrm1857 PMID:16496022

143. Waters M, Stasiewicz S, Merrick BA et al. (2008). CEBS–Chemical Effects in Biological Systems: a public data repository integrating study design and toxicity data with microarray and proteomics data. Nucleic Acids Res, 36 Database;D892–D900.doi:10.1093/nar/gkm755 PMID:17962311

144. Muthusamy B, Hanumanthu G, Suresh S et al. (2005). Plasma Proteome Database as a resource for proteomics research. Proteomics, 5:3531–3536.doi:10.1002/pmic.200401335 PMID:16041672

145. Amin RP, Vickers AE, Sistare F et al. (2004). Identification of putative gene based markers of renal toxicity. *Environ Health Perspect,* 112:465–479.doi:10.1289/ehp.6683 PMID:15033597

146. Searfoss GH, Jordan WH, Calligaro DO et al. (2003). Adipsin, a biomarker of gastrointestinal toxicity mediated by a functional gamma-secretase inhibitor. *J Biol Chem,* 278:46107–46116.doi:10.1074/jbc.M307757200 PMID:12949072

147. McDonough JL, Arrell DK, Van Eyk JE (1999). Troponin I degradation and covalent complex formation accompanies myocardial ischemia/reperfusion injury. *Circ Res,* 84:9–20. PMID:9915770

148. Oh P, Li Y, Yu J et al. (2004). Subtractive proteomic mapping of the endothelial surface in lung and solid tumours for tissue-specific therapy. *Nature,* 429:629–635.doi:10.1038/nature02580 PMID:15190345

149. Sawada H, Takami K, Asahi S (2005). A toxicogenomic approach to drug-induced phospholipidosis: analysis of its induction mechanism and establishment of a novel in vitro screening system. *Toxicol Sci,* 83:282–292.doi:10.1093/toxsci/kfh264 PMID: 15342952

150. Tohgi H, Abe T, Yamazaki K et al. (1999). Remarkable increase in cerebrospinal fluid 3-nitrotyrosine in patients with sporadic amyotrophic lateral sclerosis. *Ann Neurol,* 46:129–131.doi:10.1002/1531-8249(199907)46:1<129::AID-ANA21>3.0.CO;2-Y PMID:10401792

UNIT 2.
BIOMARKERS: PRACTICAL ASPECTS

CHAPTER 8.

Measurement error in biomarkers: Sources, assessment, and impact on studies*

Emily White

*Parts of this chapter appear in White E. Effects of biomarker measurement error on epidemiological studies. In: Toniolo P, Boffetta P, Shuker DEG, Rothman N, Hulka B, Pearce N, editors. *Applications of biomarkers in cancer epidemiology*. Lyon, IARC Scientific Publication; 1997. p. 73–94

Summary

Measurement error in a biomarker refers to the error of a biomarker measure applied in a specific way to a specific population, versus the true (etiologic) exposure. In epidemiologic studies, this error includes not only laboratory error, but also errors (variations) introduced during specimen collection and storage, and due to day-to-day, month-to-month, and year-to-year within-subject variability of the biomarker. Validity and reliability studies that aim to assess the degree of biomarker error for use of a specific biomarker in epidemiologic studies must be properly designed to measure *all* of these sources of error. Validity studies compare the biomarker to be used in an epidemiologic study to a perfect measure in a group of subjects. The parameters used to quantify the error in a binary marker are sensitivity and specificity. For continuous biomarkers, the parameters used are bias (the mean difference between the biomarker and the true exposure) and the validity coefficient (correlation of the biomarker with the true exposure). Often a perfect measure of the exposure is not available, so reliability (repeatability) studies are conducted. These are analysed using kappa for binary biomarkers and the intraclass correlation coefficient for continuous biomarkers. Equations are given which use these parameters from validity or reliability studies to estimate the impact of nondifferential biomarker measurement error on the risk ratio in an epidemiologic study that will use the biomarker. Under nondifferential error, the attenuation of the risk ratio is towards the null and is often quite substantial, even for reasonably accurate biomarker measures. Differential biomarker error between cases and controls can bias the risk ratio in any direction and completely invalidate an epidemiologic study.

Introduction

Importance of understanding the degree of measurement error in biomarkers

When a biomarker is being considered for use in an epidemiologic study, or has been selected, the researcher needs to become familiar with its measurement properties (i.e. how well the measure selected reflects the underlying exposure of interest). There are many sources of error in biomarkers when they are used in epidemiologic studies. These include not only laboratory error, but also errors due to variation in the specimen collection and processing methods, as well as a single measure of the biomarker not reflecting the longer time period during which the biomarker actually influences the disease. Validity and reliability studies that aim to assess the degree of biomarker error for use of a specific biomarker in epidemiologic studies must be designed to measure *all* of these sources of error; this differs from laboratory validation, which aims to assess only the laboratory component of error. If studies have not been published on these measurement issues, then a validity or reliability study of the biomarker should be conducted to determine its measurement error.

Once the measurement error in the biomarker has been quantified, the researcher can estimate the impact of that magnitude of error on the planned epidemiologic study in terms of the bias in the risk ratio of the relationship of the biomarker to the disease outcome. If there is a large degree of error, the researcher would need to improve the method or select a different one. If the biomarker measure is sufficiently valid to use in an epidemiologic study, then knowing the degree of measurement error will help in interpreting the results.

Definition of terms

The term parent epidemiologic study refers to the epidemiologic study that will use the biomarker. For simplicity, the assumption is made that the parent study is a case–control, cohort, or nested case–control study of the relationship between the biomarker and a binary outcome, such as incident disease or death. Measurement error in the biomarker leads to bias in the risk ratio for the association of the biomarker to disease in the parent study. This bias is called information bias or misclassification bias.

The measurement error for an individual can be defined as the difference between their measured biomarker and true exposure. The true exposure can be conceptualized as the underlying biologic or external factor that the biomarker is meant to measure (the causal factor for etiologic studies), without laboratory or other sources of error. If the biomarker measure can fluctuate over time, the true exposure would also be integrated over the time period of interest (e.g. the average of the true exposure over the etiologically important time period for etiologic studies). Nondifferential measurement error occurs when the measurement error does not differ between the disease and non-disease groups in the parent epidemiologic study. Differential measurement error occurs when the degree of biomarker error differs between those with and without the disease in the parent study. The sources and effects of both differential and nondifferential measurement error will be discussed in this chapter.

Validity is the relation of the biomarker measure to the true exposure in a population of interest. Measures of validity are parameters that describe the measurement error in the population. A validity study is defined here as one in which a sample of individuals is measured twice: once using the biomarker measure of interest and once using a perfect (or near-perfect) measure of the true exposure, and the values compared.

Often a perfect measure of the exposure does not exist or is not feasible to use in a validity study. In a reliability study, repeated measurements of the same biomarker are taken on a group of subjects and compared; they usually only measure part of the measurement error. However, certain designs of reliability studies can be used to measure the validity of a biomarker without having a perfect measure of the biomarker.

Overview of chapter

The first topics covered in this chapter are sources of measurement error in biomarkers and design issues in validity and reliability studies for biomarkers. The chapter then covers the parameters used in a validity study to measure the error in a binary biomarker and in a continuous biomarker. Equations are given for using these parameters to estimate the bias in the risk ratio in an epidemiologic study that will use the biomarker for both binary and continuous biomarkers. While these equations rely on simplifying assumptions, the purpose is to allow the researcher to easily estimate the impact of biomarker error on the parent epidemiologic study. Finally, these same concepts will be addressed for reliability studies.

Techniques to reduce biomarker measurement error, and therefore to reduce the bias in the results of the parent study caused by measurement error, are of great importance. Approaches to reduce measurement error are only briefly mentioned in this chapter, but are covered throughout the book.

Many related topics are beyond the scope of this chapter. The reader is referred to other sources for the effects of measurement error in a categorical measure with more than two categories (1–3), the design and analysis of more complex types of reliability studies (4), and the effect of measurement error when the parent study is the relationship between a biomarker and a continuous outcome (5). General reviews of measurement error effects in epidemiologic studies, and their correction for continuous and/or categorical exposures, are given in (6–12).

Sources of biomarker error and study designs to measure it

Sources of error in biomarkers

When used in an epidemiologic study, there are numerous sources of measurement error in a biomarker in comparison to the true exposure of interest (5,13). Examples have been discussed in previous chapters and are given here in Table 8.1. Measurement error can be introduced by errors in the choice of laboratory method selected to be a measure of the exposure of interest. The method selected may not measure all sources of the true exposure. For example, if the true exposure of interest is total carotenoids, then using only serum β-carotene as the biomarker would not capture all of the relevant exposure. Alternately, the measure selected may detect other related exposures beyond the etiologically significant one (i.e. it may not be 'specific' to the exposure of interest). For example, if the true exposure of etiologic importance is β-carotene, the choice of serum total carotenoids as the biomarker measure would include other exposures not relevant to the epidemiologic true exposure. Other sources of error that must be considered, especially in the selection of the biomarker method, are whether the method has a sufficiently long half-life in the tissue

Table 8.1. Sources of measurement error in biomarkers in epidemiologic studies

Errors in the choice of laboratory method or specimen (as a measure of the true exposure of interest)

- Method may not measure all sources of the true etiologic exposure of interest (e.g. use of serum beta-carotene when the disease is influenced by all carotenoids)
- Method may measure other related exposures that are not the true exposure of interest (e.g. use of serum total carotenoids, when the disease is only influenced by beta-carotene)
- Biomarker value in tissue sampled may not equal the value in the target tissue

Errors or omissions in the protocol

- Failure to specify the protocol in sufficient detail regarding timing and method of specimen collection, specimen handling and storage procedures
- Failure to specify the laboratory analytic procedures in sufficient detail
- Failure to include standardization of the instrument periodically throughout the data collection

Errors due to variations in execution of the protocol

- Variations in method of specimen collection
- Variations in specimen handling or preparation before reaching the laboratory or freezer
- Variations in length of specimen storage or freeze-thaw cycles (leading to possible analyte degradation)
- Contamination of specimen
- Variations in technique between laboratories
- Variations in technique between laboratory technicians
- Variations between batches (due to different batches of chemicals, drift in calibration of instrument)
- Any of the above that vary between disease and non-disease groups (e.g. unequal assignments of lab technicians to cases and controls)—*differential measurement error*
- Biases due to knowledge of lab technicians of disease status—*differential measurement error*
- Random variation within batch

Errors due to biomarker variability between and within subjects

- Biomarker may be influenced by the disease under study, its pre-clinical effects or its treatment or sequelae - *differential measurement error*[a]
- Short-term variability (hour-to-hour, day-to-day) in biomarker within-subjects due to diurnal variation, posture (sitting versus lying down), time since last meal, time since last exposure to agent of interest in relation to the half-life of the biomarker
- Medium-term variability (month-to-month) within subjects due to, for example, seasonal changes in diet, transient illness
- Long-term change (year-to-year) within subjects due to, for example, purposeful dietary changes over time, changes in occupational exposures
- Lack of variability in biomarker with changes in exposure to agent of interest, due to homeostasis

[a] This is a source of differential error for etiologic studies, not for studies of biomarkers for early detection.

selected to measure the exposure of interest, and whether homeostasis leads to the method not reflecting the actual level of external exposure. These sources of error can be minimized by careful selection of the biomarker measure to be used in the epidemiologic study.

Further sources of biomarker error are variations in the method of specimen collection and laboratory technique used between subjects. Due to variations in the collection and handling of the specimen by the study field staff, and by variations in the length of time or temperature the specimen was stored before analysis, errors often occur during the conduct of the epidemiologic study. Additional sources of error are the variations that occur between batches and between laboratory technicians even when the protocol is well specified. To reduce these sources of error, the protocol needs specific details in terms of subject instructions (e.g. fasting), method of specimen collection, handling of the specimen (e.g. maximum time at room temperature), methods of specimen processing before storage or analysis, and laboratory procedures. These procedures also must be carefully monitored throughout the study.

A final source of error that is common in molecular epidemiology is due to medium-term (e.g. month-to-month) variability or long-term change in the biomarker over years within-subjects. This type of error is often ignored when assessing laboratory measurement error, but can have great impact on an epidemiologic study. This is due to the fact that unless the biomarker is a fixed characteristic within individuals, the underlying true biomarker (that influences the disease of interest) is rarely an individual's measured biomarker on a single day, but rather the average over some much longer time period. Thus, even a perfect measure of the biomarker at a single point in time could be a poor measure of the true etiologic exposure. For example, even if an ideal laboratory method existed for serum estradiol in women, it could be a very poor measure of the true exposure (e.g. average serum estradiol) over the prior 15 years, which may influence breast cancer. This source of error can be controlled for by averaging multiple measures of exposure collected periodically over the time period of interest (12,14).

While it is essential to minimize the above sources of biomarker error, it is even more important to ensure that any errors are nondifferential between those with and without the disease (or other outcome) under study. Differential measurement error can invalidate a study, as discussed below, and should be avoided. A primary concern in case–control studies of biomarker-disease associations, when the specimens are obtained after diagnosis for the cases, is that the biologic effects of the disease or its treatment may affect the biomarker. In such situations, the biomarker does not measure the true (e.g. long-term, pre-disease) exposure for cases. (This concern is for etiologic studies, not for studies of biomarkers tested for early detection.) Differential measurement error can be avoided or reduced by selecting a cohort study rather than a case–control design when the disease or its treatment can affect the biomarker. The early cases occurring in a cohort study may also have their biomarker influenced by the preclinical phase of the disease under study. However, this can be tested by removing cases with diagnoses that occur within some time period (e.g. a year or two after the specimen collection) to see if this modifies the cohort study results.

Other potential sources of differential biomarker error are the laboratory technicians' knowledge of the disease status of the subjects and differences in the specimen collection, or other methods, between those with and without the disease. Thus, not only must laboratory personnel be blinded to disease status, but also the researcher must ensure that all procedures of specimen collection, processing, storage and analysis are identical for cases and controls. One cannot, for example, collect specimens in a clinic and immediately freeze them for cases, yet collect specimens in the field when freezing will be delayed for controls, if length of time at room temperature can have any impact on the biomarker. Since some variation is inevitable, it is important that the sources of error are matched by case–control status or adjusted for in the analysis. For example, one can control for systematic differences between laboratory technicians, between laboratory batches, or between specimens stored for different lengths of time, through matching controls to cases on these factors (15).

Design of validity and reliability studies

To design a validity or reliability study that measures the amount of measurement error that will occur in the parent epidemiologic study, several design issues must be considered. First, the subjects in the validity or reliability study should represent those in the parent epidemiologic study. The subjects could be a random sample from the parent study (e.g. 100–200 randomly selected individuals from the larger parent study). If that is not possible, the subjects in the validity/reliability study should be comparable to the subjects in the parent study in terms of age, sex and other parameters that could influence the distribution (variance) of the biomarker. Second, the biomarker should be collected, processed, stored and analysed in the validity/reliability study using the same procedures that will be used in the parent study.

A third issue is the selection of the comparison measure to be used in a validity study. Subjects need to be measured using a perfect measure of the true exposure in the validity study to compare to the imperfect biomarker. Sometimes a perfect or near-perfect measure exists that is too costly or not feasible for the parent epidemiologic study, but could be used for a validity study. This true measure must reflect the underlying true exposure without error, including without error due to variation in laboratory procedures or variations over time. The last issue is particularly problematic because the true biomarker of interest is often the average value over many years.

A reliability study can be conducted even when a perfect measure is not feasible or does not exist. For reliability studies, it is ideal if the two or more repeated biomarker measures taken on each subject vary in a way so as to capture all of the sources of error in the biomarker. Therefore, the repeated measures on a subject should be based on specimens taken years apart to reflect error due to year-to-year variation, and be analysed by different laboratory technicians if more than one will be used in the parent study, etc. This differs from a reliability study that aims to assess only the laboratory component of error; those studies might split a single specimen (e.g. blood from a single blood draw) from each subject and send the two samples per subject to the same laboratory for analysis. The importance and methods of measuring all sources of error will be discussed in more detail below.

In addition, the researcher should consider conducting the validity/reliability study on two groups: those with the disease and those without the disease. This is particularly important if there is the possibility that the biomarker error could differ between the disease and non-disease groups that will be used in the parent study. As noted earlier, this is a concern if the parent epidemiologic study is a case-control study in which the disease could influence the biomarker test. For validity/reliability studies to be able to assess differential measurement error in the biomarker measure between cases and controls, the comparison measure must be perfect (i.e. a validity study), or if an imperfect comparison measure is used, it must not have differential error. For example, the comparison measure could be based on specimens collected years before diagnosis for cases and during a similar period for controls. The design, analysis and interpretation of studies to measure differential error are only briefly discussed in this chapter (see (16) for a more detailed review).

Finally, a validity or reliability study should be analysed using parameters that provide information about the impact of biomarker measurement error on the parent epidemiologic study. These parameters, and their interpretations in terms of bias in the risk ratio in the parent study, are discussed in the remainder of this chapter.

Measuring the error in a binary biomarker: Sensitivity and specificity

Binary (dichotomous) biomarkers are those that classify an analyte or characteristic as present (positive) or absent (negative) for each study subject. Measurement error in a binary biomarker is usually referred to as misclassification. Binary biomarkers are subject to all of the sources of measurement error as described above and in Table 8.1.

The degree of misclassification in a binary biomarker is measured by its sensitivity and specificity. These can be measured in a validity study in which the biomarker under evaluation (the mismeasured biomarker) is compared to a perfect measure of the underlying true exposure among a sample of the population of interest. Individuals are then cross-classified by their results on each test:

		True Exposure	
		+	−
Classified by	+	A	B
Biomarker Test	−	C	D

The sensitivity (sens) of the biomarker under evaluation is the proportion of those who are true positives (positive on the criterion test) and are correctly classified as positive by the biomarker test:

$$\text{sens} = \frac{a}{a+c}.$$

(Note that the definition given here of sensitivity is different from the meaning in some laboratory contexts, i.e. the lowest level detectable by a measurement method.) The specificity (spec) is the proportion of those who are true negatives and are classified as negative by the biomarker test:

$$\text{spec} = \frac{d}{b+d}.$$

Even though both sensitivity and specificity can range from zero to one, it is assumed that sensitivity plus specificity is greater than or equal to one. In other words, for the biomarker test to be considered a measure of the true biomarker, the probability that the biomarker test classifies a truly positive person as positive (sensitivity) should be greater than the probability that it classifies a truly negative individual as positive (1 – specificity) (i.e. sens > 1 – spec, or sens + spec > 1). Thus, the parameter (sens + spec –1), called the Youden misclassification index (17), is a good measure of the total degree of misclassification. If the Youden index is close to 1, the biomarker test is close to perfect, and if the Youden index is close to zero, the biomarker has little association with the true exposure.

For a validity study to measure sensitivity and specificity of a biomarker, the study sample may be subjects selected independent of their biomarker status, or who are true positives and those who are true negatives by the perfect test. However, one cannot sample subjects based on the results of the mismeasured biomarker test and correctly compute sensitivity and specificity.

If the validity study is conducted on a sample of cases and a sample of controls, then sensitivity and specificity would be computed separately for the cases and for the controls.

Impact of error in a binary biomarker on epidemiologic studies

Effect of nondifferential misclassification on the odds ratio

The effects of misclassification of a binary biomarker on the results of the parent epidemiologic study of the relationship between the marker and a binary disease are straightforward (18–24). In an unmatched case–control study of a binary biomarker, under the assumption that the disease is correctly classified, the effect of misclassification of the biomarker is to rearrange individuals in the true 2x2 table into an observable 2x2 table. Individuals in the disease group remain in the disease group, but may be misclassified as to biomarker status, and the non-disease group is also rearranged as to biomarker status:

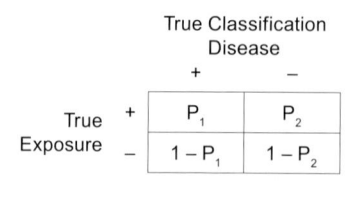

True Classification
Disease

True Exposure	+	–
+	P_1	P_2
–	$1 - P_1$	$1 - P_2$

Odds ratio : $OR_T = \dfrac{P_1(1 - P_2)}{P_2(1 - P_1)}$

P_1 and P_2 are the true proportions of subjects who are exposure-positive in the disease and non-disease groups respectively, and similarly p_1 and p_2 refer to the proportions that would be "observable" as positive by the biomarker test in the two groups. The term observable means what would be seen, on average, when the imperfect biomarker is used in the parent epidemiologic study.

There is nondifferential misclassification when the sensitivity of the biomarker test is the same for both disease and non-disease groups in the parent study, and the specificity is the same for both groups. The misclassification leads the observable p_1 and p_2 to be different from the true P_1 and P_2 (21):

$$\left. \begin{array}{l} p_1 = sens \cdot P_1 + (1 - spec) \cdot (1-P_1) \\ p_2 = sens \cdot P_2 + (1 - spec) \cdot (1-P_2) \end{array} \right\} (1)$$

The first equation states that the proportion of cases who will be classified by the biomarker test as positive (p_1) is made up of a proportion (sens) of those truly exposed (P_1) in the disease group, plus a proportion (1-spec) of those truly unexposed ($1-P_1$) in the disease group. The second equation expresses the same concept for the non-disease group.

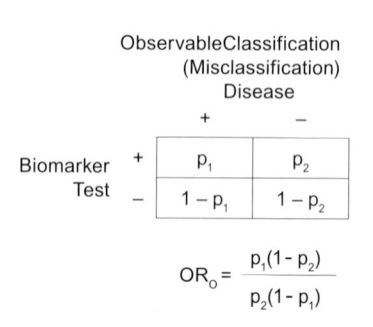

Observable Classification
(Misclassification)
Disease

Biomarker Test	+	–
+	p_1	p_2
–	$1 - p_1$	$1 - p_2$

$OR_O = \dfrac{p_1(1 - p_2)}{p_2(1 - p_1)}$

The association between the biomarker and disease in the parent study would be measured by the odds ratio in a case–control or nested case–control study. (The results presented here would be similar for the hazard ratio from a cohort study as well.) When there is measurement error, the observable odds ratio, OR_O, differs from the true odds ratio, OR_T, because OR_O is based on p_1 and p_2:

$$OR_O = \dfrac{p_1(1 - p_2)}{p_2(1 - p_1)}. \qquad (2)$$

The magnitude of the bias can be estimated by computing p_1 and p_2 from Equation 1 and OR_O from Equation 2 and comparing it to OR_T, using estimates of P_1 and P_2.

As an example, suppose current infection with *Helicobacter pylori* (*H. pylori*), an organism associated with several upper digestive tract diseases, is the true exposure of interest in a cohort study being planned. An ELISA test on serum, although imperfect, is the most feasible exposure measure to be used in the epidemiologic study. Information on the accuracy of the ELISA test comes from a validity study conducted in Taiwan, China on 170 patients undergoing gastroendoscopy (25). The serum ELISA test was compared to a highly accurate measure, assessed by either a positive culture or two positive tests among three others (histology, Campylobacter-Like Organism (CLO) test and 13C-urea breath test), with these results:

		True Exposure	
		H.pylori +	H.pylori −
ELISA Test	H.pylori +	103	16
	H.pylori −	4	47
	Total	107	63

Sensitivity and specificity of the ELISA test were:
sens = 103/107 = 0.96
spec = 47/63 = 0.75.

These estimates can be used to approximate the effect of the biomarker error on the results of the epidemiologic study. If one assumes that the measurement error in the future cohort study will be nondifferential, i.e. that the sensitivity and specificity are the same for cases and controls, then Equations 1 and 2 can be used to estimate the observed odds ratio, OR_O. If the estimated true *H. pylori* infection prevalence is 70% in cases (P_1) and 35% in controls (P_2), then the true odds ratio, OR_T, is:

$$OR_T = \frac{.70(.65)}{.35(.30)} = 4.3$$

Then by Equations 1 and 2:

$p_1 = 0.96 \cdot 0.70 + 0.25 \cdot 0.30 = 0.75$
$p_2 = 0.96 \cdot 0.35 + 0.25 \cdot 0.65 = 0.50$
$$OR_O = \frac{.75(.50)}{.50(.25)} = 3.0$$

This validity study shows that a study using the misclassified *H. pylori* test would find 75% of cases positive and 50% of controls positive (rather than 70% and 35%, respectively, as the true probabilities of exposure), and would yield an observed odds ratio of 3.0 rather than the true odds ratio of 4.3.

Nondifferential misclassification leads to an attenuation of the odds ratio towards the null hypothesis value of 1 (20). The observable odds ratio does not cross over the null value of 1, under the reasonable situation that sens + spec > 1 (see above). The degree of attenuation in the observable odds ratio depends on the sensitivity and specificity of the biomarker test and on P_1 and P_2, or equivalently, on the true odds ratio and on P_2, the proportion of the non-disease group who are truly exposed. Table 8.2 gives further examples of the effect of nondifferential misclassification on the odds ratio for reasonable values of sensitivity (0.5–0.9), specificity (0.8–0.99), P_2 (0.1, 0.5), and a true odds ratio of 2 and 4. As can be seen from the table, the attenuation in the odds ratio can be considerable. When the proportion who are truly exposed in the non-disease group is low (e.g. $P_2 = 0.1$ in upper half of Table 8.2), the attenuation of the odds ratio is severe except when the specificity is very high (e.g. spec = 0.99). When the proportion who are truly exposed is high (e.g. $P_2 = 0.5$ in lower half of Table 8.2), the observed OR is strongly attenuated from the true OR except when the sensitivity is very high (e.g. sens = 0.9). Even strong associations between the true biomarker and disease would be obscured by moderate values of sensitivity and specificity. For example, for sens = 0.7, spec = 0.8,

Table 8.2. Impact of nondifferential misclassification of a binary biomarker on the Observable Odds Ratio (OR_O)

Biomarker Test Sensitivity	Biomarker Test Specificity	True OR=2.0 OR_O[b]	True OR=4.0 OR_O[b]
$P_2 = 0.1$[a]			
0.5	0.80	1.14	1.38
0.7	0.80	1.23	1.64
0.9	0.80	1.32	1.92
0.5	0.90	1.28	1.76
0.7	0.90	1.39	2.09
0.9	0.90	1.48	2.41
0.5	0.99	1.75	3.06
0.7	0.99	1.83	3.33
0.9	0.99	1.89	3.61
$P_2 = 0.5$[a]			
0.5	0.80	1.24	1.46
0.7	0.80	1.40	1.83
0.9	0.80	1.64	2.59
0.5	0.90	1.35	1.69
0.7	0.90	1.50	2.07
0.9	0.90	1.73	2.85
0.5	0.99	1.48	1.96
0.7	0.99	1.61	2.33
0.9	0.99	1.82	3.11

[a] P_2 is the proportion with the true exposure in the non-disease group. By definition of OR_T, P_1, the proportion with the true exposure in the diseased group is: $P_1 = P_2 \cdot OR_T/(1 + P_2(OR_T - 1))$.
[b] OR_O from Equations 1 and 2.

and $OR_T = 4.0$, the observable odds ratio would be 1.64 for $P_2 = 0.01$ and 1.83 for $P_2 = 0.5$. These observable odds ratios would not be detectable as different from the null value of 1 unless the parent epidemiologic study sample size was large.

Effect of differential misclassification on the odds ratio

There is differential misclassification when the sensitivity of the biomarker test for the disease group differs from that for the non-disease group, and/or the specificity of the biomarker test for the disease group differs from that for the non-disease group. Differential misclassification can have any effect on the odds ratio: the observable odds ratio can be closer to the null hypothesis of OR = 1, be further from the null, or crossover the null compared with the true odds ratio. Thus, while the odds ratio under nondifferential measurement error can be assumed to be conservative (biased towards the null), the odds ratio when there is nondifferential error could be biased away from the null or even be in the wrong direction (e.g. it could make the biomarker appear to be a risk factor when it is, in fact, a protective factor). Equations 1 and 2 can be used to estimate the impact of differential measurement error, by using the estimates of sensitivity and specificity in the disease group for the equation for p_1, and estimates of sensitivity and specificity in the non-disease group for the equation for p_2.

Measuring the error in a continuous biomarker using a validity study

Often a biomarker assay yields quantitative information about the amount of an analyte in a biologic specimen; these measures can usually be considered to be continuous variables. This section covers the parameters used to assess measurement error in a continuous biomarker from a validity study in which each subject is measured twice: once using the mismeasured biomarker and once using a perfect measure of the true exposure of interest.

The theory of measurement error in continuous variables was developed in the fields of psychometrics, survey research and statistics (26–32). The effects of measurement error also have been derived in the context of epidemiologic studies of a continuous exposure variable and a binary disease outcome (3,33–35).

A model of measurement error

A simple model of measurement error in a continuous measure X is:

$$X_i = T_i + b + E_i,$$

where:
$\mu_E = 0$
$\rho_{TE} = 0.$

In this model for a given individual i, the measured biomarker, X_i, differs from the true value, T_i, by two types of measurement error. The first is systematic error, or bias, b, that would occur (on average) for all measured subjects. The second, E_i, is the additional error in X_i for subject i. E will be referred to as subject error to indicate that it varies from subject to subject. It does not refer just to error due to subject characteristics; rather it includes all of the sources of error outlined in Table 8.1.

For the population of potential study subjects, X, T and E are variables with distributions (e.g. the distribution of E is the distribution of subject measurement errors in the population of interest). X, T and E would have expected means in the population of interest denoted by μ_X, μ_T, and μ_E, respectively, and variances denoted by σ^2_X, σ^2_T, and σ^2_E. Because the average measurement error in X in the population is expressed as a constant, b, it follows that μ_E, the population mean of the subject error, is zero. The assumption of the model that the correlation coefficient of T with E, ρ_{TE}, is zero states that the true value of the biomarker is not correlated with the measurement error. In other words, individuals with high true values are assumed to not have systematically higher (or lower) errors than individuals with lower true values.

Measures of measurement error: Bias and validity coefficient

Two measures of measurement error are used to describe the relationship of X to T in the population of interest, based on the above model and assumptions. One is the bias (i.e. the average measurement error in the population):

$$b = \mu_X - \mu_T.$$

The bias in X can be estimated from a validity study as the difference between the mean of X and the mean of T: $b = \overline{X} - \overline{T}$. If b is positive, then X overestimates T on average; if b is negative, then X underestimates T on average.

The other measurement error is a measure of the precision of X (i.e. the variation of the measurement error in the population). One measure of precision is σ^2_E, the variance of E, which is often called the within-subject variance. (Note that b is a constant for all subjects and therefore does not contribute to the variance of the error, σ^2_E.) A more useful measure of precision is the correlation of T with X, ρ_{TX}, termed here as the validity coefficient of X. The measure ρ_{TX} is important because it relates the within-subject variance, σ^2_E, to the total variance, σ^2_X, and it is this ratio, along with the bias, that measures the impact of biomarker error on the parent epidemiologic study. ρ_{TX} can range between zero and one, with a value of one indicating that X is a perfectly precise measure of T. ρ_{TX} is assumed to be zero or greater (i.e. for X to be considered to be a measure of T, X must be at a minimum positively correlated to T).

The validity coefficient ρ_{TX} can be estimated in a validity study by the Pearson correlation coefficient of X with T. Thus, using the standard interpretation of a correlation coefficient, the correlation squared (ρ^2_{TX}) can be interpreted as the proportion of the variance of X explained by T. For example, if ρ_{TX} were 0.8, this would mean that only 64% of the variance in X is explained by T, with the remainder of the variance due to the error.

To further understand the concepts of bias and precision, consider a situation in which X has a systematic bias, but is perfectly precise (i.e. $E_i = 0$ for all subjects). Suppose that the only source of error in a measure of serum cholesterol, for example, were that it quantified each individual exactly 100 mg/dl too high. X would be biased (b = 100 mg/dl), but would have perfect precision ($\rho_{TX} = 1.0$). Then, in a population, the measure X, even though it has systematic measurement error, could be used to perfectly order each person in the population by their value of T.

There could also be situations in which there is no bias, yet there is lack of precision. Suppose that the measurement error, E_i, varied from person-to-person, but for some subjects their measured X was higher than their actual T, and for other subjects their measured X was lower than their actual T, but X on average in the population equaled the average T in the population. In this situation there is no bias (b = 0), but there is lack of precision ($\rho_{TX} < 1.0$). In this case, the ordering of subjects is lost. Of course, most likely a biomarker has both bias and lack of precision.

The degree of measurement error is not an inherent property of a biomarker test, but rather is a property of the test applied using a particular protocol to a specific population. Therefore, the error can vary for a biomarker test when applied using a different protocol or when applied to different population groups. In addition, ρ_{TX} is generally greater in populations with greater variance of the true exposure (36). Therefore, a validity study conducted on one population may not directly apply to another study population.

Finally, measurement error could differ between those with and without the disease, particularly in a case–control study. Separate assessment of the bias and precision in these two groups is needed to assess differential error (see below).

The terminology surrounding measurement error varies between fields. In this chapter, the terms validity, accuracy and measurement error are used as general terms reflecting the relationship of X to T, including both the concepts of bias and precision. (In laboratory quality control, the terms validity and accuracy are sometimes used to refer to unbiasedness only.)

Impact of error in a continuous biomarker on epidemiologic studies

When the bias and validity coefficient of the biomarker (X) are known, one can estimate the impact of the degree of measurement error in X on the parent epidemiologic study that will use X. Both nondifferential and differential measurement errors will be discussed, but first the effect of measurement error on a single study population will be considered.

Effect of measurement error on the observable mean and variance

In a single study population, both the mean and variance of the measured biomarker X would differ from the true mean and variance due to measurement error. Under the above model, the population mean of X would differ from the true mean (the population mean of T) by b:

$$\mu_X = \mu_T + b. \qquad (3)$$

The population variance of X, based on the above model, would be (30):

$$\sigma^2_X = \sigma^2_T / \rho^2_{TX}. \qquad (4)$$

The variance of X in the population is greater than the variance of T, due to the addition of the variance of the measurement error. For example, if the validity coefficient (ρ_{TX}) were 0.8, then the variance of X would be 56% greater than the variance of T ($\sigma^2_X = \sigma^2_T / .8^2 = 1.56\ \sigma^2_T$ by Equation 4).

Figure 8.1 demonstrates the effect of measurement error on the distribution of X in a population using a normally distributed biomarker and normally distributed error as an example. The bias in the measure X causes a shift in the distribution of

Figure 8.1. Distribution of true (T) and measured (X) biomarker

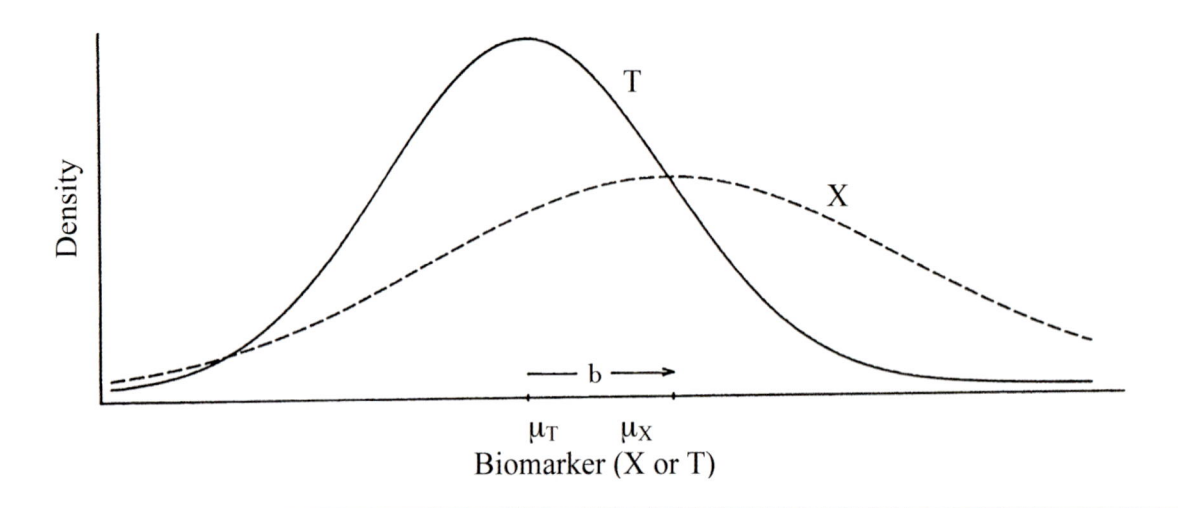

X compared with T. The increased variance of X compared with T (measured by ρ_{TX}) causes a flattening of the distribution of X. Even if a measure X were correct on average (b = 0), there could still be substantial measurement error due to lack of precision, which could lead to a greater dispersion in the measured exposures.

Effect of nondifferential measurement error on the odds ratio

While measurement error has an effect on the observable mean and variance of an exposure variable within a single population, a greater concern would be the impact of measurement error in an epidemiologic study comparing those who have the disease of interest to those who do not. In a case–control or nested case–control study, the common measure of association between a biomarker and disease is the odds ratio, which is often expressed as the odds ratio of disease for a *u* unit increase in the level of the biomarker. The results given here also approximately apply to estimates of the hazard ratio from data from a cohort study and the risk ratio from a matched case–control study (33). The results given in this section do not apply to odds ratios expressed as odds of disease for the upper quantile of the biomarker versus lowest quantile. They also do not apply when X and T are measured in different units in the validity study. For a discussion of these situations, see (12).

Errors in the measurement of the biomarker X would bias the odds ratio in the epidemiologic study. There is nondifferential misclassification when there is equal bias and equal precision (equal ρ_{TX}) in the biomarker test when applied to both the disease and non-disease groups in the parent epidemiologic study. Figure 8.2 illustrates the effects of nondifferential misclassification. Under nondifferential misclassification, the distribution of exposure in cases and in controls may shift, but because there is equal bias for the two groups, they are not shifted with respect to each other. However, the lack of precision flattens and leads to more overlap and less distinction between the distributions of X_D, the biomarker in the disease group, and of X_N, the biomarker in the non-disease group, compared with the distributions of the true exposure in the two groups (T_D and T_N).

The effect of nondifferential measurement error in X on the odds ratio can only be easily quantified when certain simplifying assumptions are made. Results can be derived for case–control studies under the following assumptions: a) X_D and X_N meet the assumptions of the simple model of measurement error given above; b) T_D and T_N are normally distributed with different true means in the disease and non-disease groups, but the same variance, σ^2_T; c) the bias in X is the same for the two groups; and d) the errors, E, are normally distributed with mean zero and common variance, σ^2_E, in the two groups. Assumption c and the second part of assumption d are the assumptions of nondifferential error (i.e. equal bias and equal precision of X_D and X_N).

The above assumptions imply a logistic regression model for the probability of disease (pr(d)) as a function of true biomarker T, with a true logistic regression coefficient β_T (37):

152

Figure 8.2. Effect of nondifferential measurement error (equal bias and precision) on distribution of true (T) versus measured (X) biomarker in an epidemiologic study comparing disease (D) and non-disease (N) groups

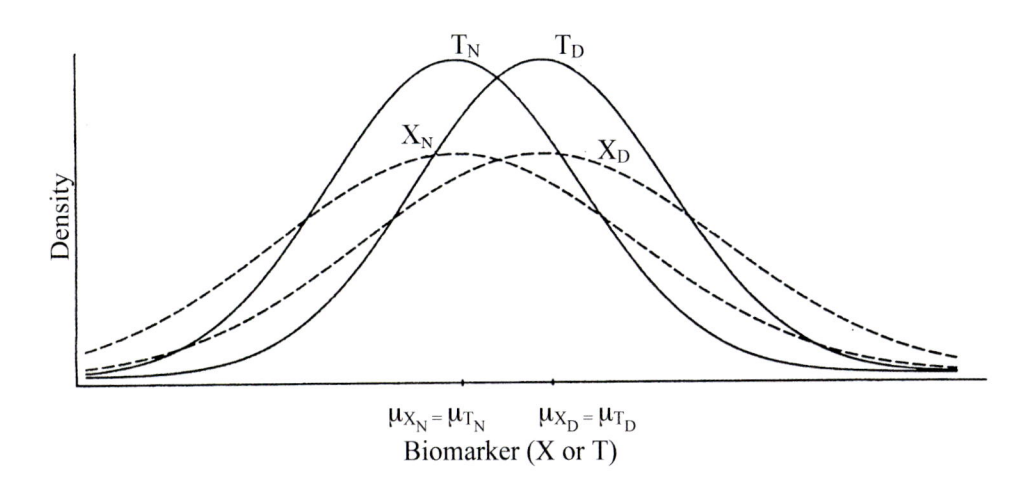

$$\log \frac{pr(d)}{1-pr(d)} = a_T + b_T T.$$

The true odds ratio for any u unit increase in T would be $OR_T = e^{b_T u}$.

With measurement error in the biomarker test X, the assumptions also lead to a logistic model:

$$\log \frac{pr(d)}{1-pr(d)} = a_O + b_O X.$$

$OR_O = e^{b_O u}$ is the observable odds ratio for a u unit increase in X.

The observable logistic regression coefficient, β_O, differs from β_T due to the measurement error in X. Under nondifferential misclassification, β_O is attenuated towards the null value of zero in comparison to β_T (34,37) by this equation:

$$\beta_O = \rho^2_{TX} \beta_T. \quad (5)$$

Equivalently, OR_O is attenuated towards the null value of 1 in comparison to OR_T by this equation:

$$OR_O = OR_T^{\rho^2_{TX}}. \quad (6)$$

This states that the observable odds ratio for any fixed difference in units of the biomarker is equal to the true odds ratio for the same fixed difference to the power ρ^2_{TX}. Since $0 \leq \rho^2_{TX} \leq 1$, the observable odds ratio will be closer to the null value of 1 (no association) than the true odds ratio. The observable odds ratio does not cross over the null value if X and T are at a minimum positively correlated.

Equation 6 shows that the attenuation in the odds ratio under nondifferential misclassification is a function of the precision of X (measured by ρ_{TX}), but *not* of the bias in X. Thus, even when X is correct on average for cases and correct on average for controls (bias = 0 for cases and for controls), the lack of precision of X can substantially bias the odds ratio. Examples of the effects of nondifferential measurement error in a biomarker on the odds ratio, based on the assumptions above and Equation 6, are given in Table 8.3. The table shows that biomarkers with a validity coefficient ρ_{TX} of 0.5 would obscure all but the strongest associations. For example, when $\rho_{TX} = 0.5$ and the true odds ratio for a u unit change in the biomarker was 4.0, this odds ratio would be attenuated to an observed odds ratio of 1.41. Furthermore, measures as precise as $\rho_{TX} = 0.9$ still lead to a modest attenuation (e.g. a true odds ratio of 4.0 would be attenuated to 3.07).

Effect of nondifferential measurement error on power and sample size

The examples above show that nondifferential measurement error in a biomarker leads to attenuation of the odds ratio for the association of the biomarker with the disease. This attenuation of the odds ratio would reduce the power of the epidemiologic study that uses the biomarker if the sample size were fixed. If the sample size was not fixed, it would lead to a need for a larger sample size to detect the attenuated odds ratio as different from the null value of 1.

When a continuous exposure with measurement error is used in an epidemiologic study, the sample size needed, N_X, is compared to the sample size needed in a study in which the exposure is measured without error, N_T. A simple formula shows this comparison (14,38):

$$N_X = N_T / \rho^2_{TX}.$$

This formula may be of theoretical interest only, since estimates of the parameters needed when calculating the required study sample size should be based on the mismeasured exposure (e.g. σ^2_X); as these estimates are usually available, the sample size calculations will yield the correct N. However, the above equation can be used to show the potentially dramatic effects of inaccurate biomarker measurement on the sample size required. For example, if the correlation between T and X is 0.7 ($\rho^2_{TX} = 0.49$), then the sample size required when the imperfect measure is used is twice that required if a perfect measure were available. This shows that the error in biomarkers, with what is considered to be a good validity coefficient, still leads to a large increase in required sample size for the epidemiologic study that will use the biomarker.

Effect of differential measurement error on the odds ratio

Differential measurement error occurs when the bias in the mismeasured biomarker in the disease group differs from the bias in the non-disease group, and/or the precision differs between groups. As noted above, differential measurement error should be a concern in a case–control study when the biomarker is measured within the preclinical disease phase before diagnosis or anytime after diagnosis, unless the marker is fixed (e.g. genotype). Differential bias has the most problematic effects: depending on the magnitude and the direction of the biases in X_D and X_N, the observable odds ratio for any u unit increase in X, $OR_o = e^{b_o u}$, could be towards the null value of one, away from the null, or cross over the null value compared with the true odds ratio.

Figure 8.3 presents a graphical example of differential measurement error, in particular, differential bias between cases and controls. In the figure, the true mean exposure level in the disease group, μ_{T_D}, is greater than the true mean level in the non-disease group, μ_{T_N}. This would lead to an odds ratio above 1 for any u unit increase in T. In this example, the bias for the non-disease group is positive, so the distribution of X_N is shifted to the right relative to T_N, and the bias among those with disease is negative, so the distribution of X_D is shifted to the left relative to T_D. Differential bias would cause the observable odds ratio to cross over the null value of one (i.e. the biomarker would appear to be a protective factor, rather than a risk factor, as the controls would appear to have higher mean exposure than the cases).

Differential bias could be assessed by comparing the bias ($\overline{X} - \overline{T}$) for cases with the bias ($\overline{X} - \overline{T}$) among controls. To assess differential bias, T does not need to be perfect; rather, T only needs to have nondifferential bias (e.g. T could be based on specimens drawn years before diagnosis).

An example of differential bias comes from several case–control studies which found that low serum cholesterol, measured at the time of diagnosis, was a risk factor for colon cancer (39). This could be an artefact if increased catabolism, or other effects of colon cancer, reduce serum cholesterol. In fact, it was found that serum cholesterol

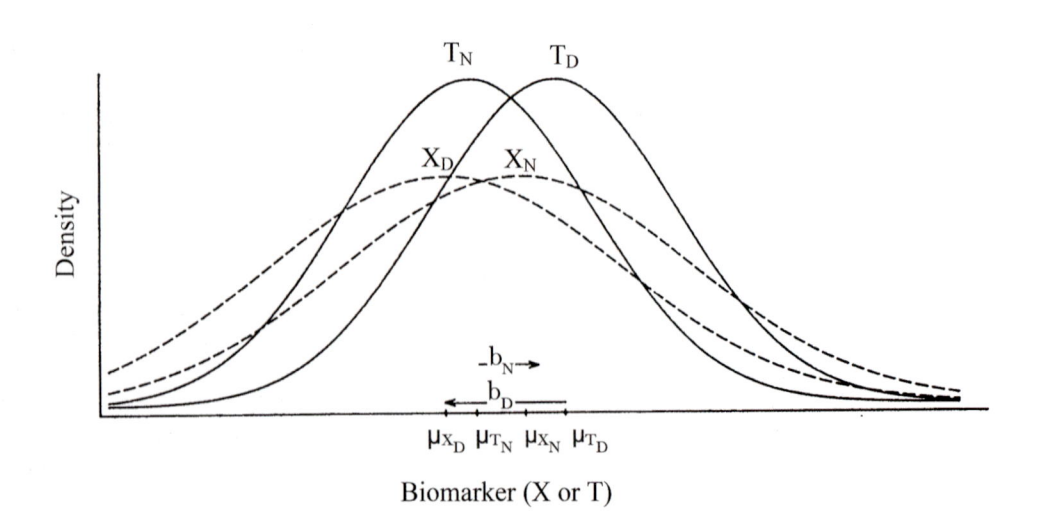

Figure 8.3. Effect of differential measurement error ($b_1 \neq b_2$) on distribution of true (T) versus measured (X) biomarker in an epidemiologic study comparing disease (D) and non-disease (N) groups

measured 10 years before diagnosis was higher in colon cancer cases than controls (40). This suggests that serum cholesterol measured at the time of diagnosis had differential bias; it likely underestimated the true etiologic exposure (say, true serum cholesterols a decade before) among cases due to the effects of the cancer, while it may have overestimated the true serum cholesterol a decade before among the controls (due to a tendency of cholesterol levels to increase with age).

Differential bias is a greater concern than differential precision because, as described above, differential bias can lead to a shift in the distribution of the biomarker in one group relative to the other. Differential measurement error will also occur if precision differs between groups. If there were no differential bias, but precision differed, the shape of the odds ratio function could change. For example, the observable odds ratio curve could be U-shaped when the true exposure–disease relationship is increasing (41).

More details about the design, analysis and interpretation of validity or reliability studies to assess differential measurement error are given in (16).

Measuring the error in a biomarker using a reliability study

The term reliability is used to refer to the reproducibility of a measure, that is, how consistently a measurement can be repeated on the same subjects. Reliability can be assessed in several ways, but only one type, intramethod reliability, will be covered in this chapter. Intramethod reliability studies measure the reproducibility of a measurement on the same subjects repeated two or more times using the same method, but often with some variation. For example, a comparison could be made of a biomarker from a single specimen from each subject analysed by the same laboratory technician twice, or analysed by two laboratory technicians, or from two specimens on each subject collected at two points in time. Reliability studies, in which two different analytic methods are compared, with one better than the other but neither perfect (intermethod reliability studies), are not covered here (see (12)). Measures of reliability are primarily important for what they reveal about the validity of a biomarker test, because as shown above, the bias in the odds ratio in the parent epidemiologic study is a function of the validity of the biomarker measure.

This section covers the parameters used to measure reliability, the interpretation of measures of reliability in terms of measures of validity, and the use of parameters from reliability studies to estimate the bias in the odds ratio in the parent epidemiologic study that will use the biomarker.

A model of reliability and measures of reliability for continuous biomarkers

Suppose each person in a sample of interest is measured two or more times using the same continuous biomarker test that will be used in the parent study. For a given subject i, two (or more) biomarker measurements, X_{i1} and X_{i2}, are obtained. The simple measurement error model described above applies to each measure:

$$X_{i1} = T_i + b_1 + E_{i1}$$

$$X_{i2} = T_i + b_2 + E_{i2}$$

Both X_{i1} and X_{i2} are measures of the subject's true biomarker T_i, but with different errors. In a reliability study, information is available on X_1 and X_2 for each subject, but not on T. A reliability study can yield estimates of the mean of X_1 and X_2 (μ_{X_1} and μ_{X_2}) and of the correlation between the two measures, ρ_X, termed the reliability coefficient.

The intraclass correlation coefficient (ICC) is generally used as the reliability coefficient for continuous biomarkers (see (12,14) for computational formulas). The ICC differs from the Pearson correlation coefficient in that it includes any systematic difference between X_1 and X_2 (i.e. any difference between b_1 and b_2) as part of the subject error E (the error that varies from subject-to-subject). The assumption is that in the parent epidemiologic study, each subject will be measured once, by either X_1 or X_2 (e.g. either by laboratory technician 1 or 2). Therefore, any systematic difference between X_1 and X_2 would not be a systematic bias affecting everyone in the parent study, but would vary between subjects because some are measured by X_1 and some by X_2. Thus, the ICC is equal to 1 only when there is exact agreement on all measures on each subject (which differs from the Pearson correlation coefficient, which is equal to 1 when X_1 is a linear function of X_2). Because X_1 and X_2 will be used as interchangeable measures of X in the parent study, and more than two replicates per subject can be used to compute the ICC, the reliability coefficient of X is written as ρ_X.

Two measures of the validity of a continuous biomarker measure X, the bias and the validity coefficient, were shown to be important in assessing the impact of measurement error on the parent epidemiologic study, which will use X. Unfortunately, reliability studies generally cannot provide information on the bias in X, because a similar bias often affects both X_1 and X_2. The inability of many reliability study designs to yield information on the bias, and on differential bias between cases and controls, is a major limitation. It should be recalled, however, that under nondifferential measurement error (and certain other assumptions), the attenuation of the odds ratio depends only on the validity coefficient and not on the bias. The reliability coefficient does provide information about the validity coefficient, and thus can be used to estimate the effects of measurement error on the parent study under the assumption of nondifferential measurement error.

Relation of reliability to validity under the parallel test model

When certain assumptions are met, reliability studies can yield estimates of the validity coefficient. One such set of assumptions is the model of parallel tests (27,29–31). The first assumption of the parallel test model is that the error variables, E_1 and E_2, are not correlated with the true value T. The second is that E_1 and E_2 have equal variance σ_E^2. This also implies that X_1 and X_2 have equal variance and that X_1 and X_2 are equally precise ($\rho_{TX_1} = \rho_{TX_2}$). This is usually a reasonable assumption in intramethod reliability studies, since X_1 and X_2 are measurements from the same method. Third, it is assumed that E_1 is not correlated with E_2. This important (and restrictive) assumption implies, for example, that an individual who has a positive error, E_1, on the first measurement is equally likely to have a positive or a negative error, E_2, on the second measurement. These assumptions are often summarized by saying that two measures are parallel measures of T if their errors are equal and uncorrelated.

Under the assumptions of parallel tests, it can be shown that (30):

$$\rho_{TX} = \sqrt{\rho_X}. \qquad (7)$$

This equation states that the validity coefficient of X, ρ_{TX}, can be estimated to be the square root of the reliability coefficient, ρ_X. This result is important because it shows that if the assumptions are correct, the reliability coefficient, which is a measure of the correlation between two imperfect measures, can be used to estimate the correlation between T and X without having a perfect measure of T. The correlation of the replicates of X is less than the correlation of X with T, as each replicate has measurement error.

A reliability study of a biomarker test can often be assumed to have equal and uncorrelated errors if the replicates within each person are sampled over the entire time period to which the true biomarker is intended to relate (if the biomarker can vary over time); the specimen handling, storage and analytic techniques vary in the reliability study as they will in the parent study; and the true exposure is defined as the mean measure over the relevant time period of repeated measures of the assay.

An example comes from a study which examined the reliability of serum hormone levels in premenopausal women (42). The goal was to understand whether a single blood draw was sufficiently accurate to be used in a large prospective study of serum hormones and cancer risk among premenopausal women. The reliability study included 113 women who had blood drawn once a year for three years during both the middle of the follicular and luteal phases of their menstrual cycles. The reliability coefficient (intraclass correlation coefficient) was 0.38 for total estradiol during the follicular phase and 0.45 during the luteal phase. The repeated measures in this study are close to the parallel test model: the errors on each of the repeated measures can be assumed to be equal because the same test procedure was repeated, and the errors are likely to be uncorrelated (i.e. a woman whose hormone measure was higher than her "true" three-year average on one measure is not more likely to be higher than her true average on another measure). This study also measured most sources of error, such as error due to variations in blood processing, storage and laboratory technique, and error due to long-term variation of plasma hormones within women. Therefore, the estimated validity coefficient (ρ_{TX}) for a single measure of total estradiol (X) as a measure of average estradiol over three years (T), based on Equation 7, is 0.62 if blood were drawn during the mid-follicular phase and 0.67 if drawn during the mid-luteal phase.

Based on Equation 7, the results in the last section on the effects of measurement error on the odds ratio can be expressed in terms of ρ_X rather than ρ_{TX}^2. When the parallel model holds, Equation 6 can be written:

$$OR_O = OR_T^{\rho_X}. \qquad (8)$$

From the example above (42), use of a single measure of total estradiol during the mid-luteal phase, with a reliability coefficient of 0.45 in a cohort study of total estradiol and breast cancer, would attenuate a true odds ratio of, for example, 4.0 to an observed odds ratio of 1.9 (from Equation 8). Other examples of the bias in the odds ratio (under the parallel test model)

Table 8.3. Impact of nondifferential measurement error in a normally distributed biomarker X on the Observable Odds Ratio (OR_O)

ρ_{TX}[a]	ρ_X[b]	True OR^c=2.0 OR_O[d]	True OR^c=4.0 OR_O[d]
.50	.25	1.19	1.41
.60	.36	1.28	1.65
.70	.49	1.40	1.97
.75	.56	1.48	2.18
.80	.64	1.56	2.42
.85	.72	1.65	2.72
.90	.81	1.75	3.07
.95	.90	1.87	3.49

[a] ρ_{TX} is the validity coefficient of X.
[b] ρ_X is the reliability coefficient of X under the parallel test model (see text):
$\rho_X = \rho^2_{TX}$
[c] The true OR = is the odds ratio for a *u* unit difference in T for comparison to OR_O for a *u* unit difference in X.
[d] OR_O from Equation 6 or 8. See text for model and assumptions.

from various degrees of unreliability are given in Table 8.3.

Relation of reliability to validity for correlated errors

In actual reliability studies, the assumptions of parallel tests are often incorrect. One assumption that is frequently violated is that of uncorrelated errors. Often the error in one measure is positively correlated with the error in the other ($\rho_{E_1E_2} > 0$). Correlated errors occur when the sources of error in the first measurement on a subject tend to repeat themselves in the second. If in a reliability study, for instance, blood was drawn only once on each subject and analysed twice, and the true marker of interest was the mean value of the biomarker over several years surrounding the time of measurement, there would be correlated error. For example, suppose the reliability of serum β-carotene was assessed by repeated laboratory analysis from a single blood draw. This would have correlated error, as an individual whose β-carotene level was higher on the first measure than the true long-term average (perhaps due to a seasonal variation in intake of β-carotene) would also likely be higher on the second measure than the true value, because the second measure used the same specimen. Because part of the error is repeated in both X_1 and X_2, the errors are correlated. This means the reliability study does not capture all sources of error in X, and therefore the reliability coefficient, ρ_X, is artificially too high.

When the errors of the measures used in a reliability study are positively correlated, then the reliability study can only yield an upper limit for the validity coefficient. Specifically, when X_1 and X_2 are equally precise, and the assumptions of the above model hold except that the errors are correlated, then the validity coefficient is less than the square root of the reliability coefficient (1):

$$\rho_{TX} < \sqrt{\rho_X}. \qquad (9)$$

Thus, a measure can appear to be reliable (repeatable) even if it has poor validity. While a low reliability coefficient implies poor validity, a high reliability does not necessarily imply a high validity coefficient. The high reliability may be due instead to correlated errors within subjects. The reliability coefficient is only diminished by part of the error in X (the part that is not repeated in X_1 and X_2), whereas the validity coefficient is a measure of all sources of error. When there is correlated error (i.e. when only part of the error is measured by a reliability study), then the attenuation of the odds ratio will be even greater than that predicted by Equation 8. Reliability studies should be designed, therefore, to capture all of the sources of error in the biomarker X, including error due to variations of specimen collection, variations between laboratory technicians, and within-person variations over time. (The concepts of correlated error, repeated within-person error, and failure of a reliability study to capture all sources of error, each describe the same phenomenon.)

Sometimes it is only possible or desirable to assess some components of error. To assess the laboratory error, a blinded test-retest reliability study on split samples from a single specimen from each subject in the reliability study, analysed in separate batches and by different laboratory technicians (if multiple laboratory technicians were going to be used in the parent epidemiologic study), would yield an intraclass correlation coefficient that measures the laboratory component of error only. Similarly, other reliability studies could be designed to test the effect of handling, storage, and short-, medium- and long-term biologic variation by only varying these components. When only some components of error are measured, there is correlation error and the resulting intraclass correlation just provides an upper estimate of the validity coefficient (see Equation 9). However, by estimating the components of error, the researcher can seek to improve those aspects having the most adverse effects. For example, enhanced laboratory quality control procedures could be used to reduce laboratory error, or multiple specimens (over time) per subject could be used to reduce the error due to medium- or long-term biologic variation. Finally, nested reliability study designs can be used to estimate the different components of error within one reliability study (4).

Coefficient of variation

One additional analytic technique for reliability studies of continuous biomarkers, the coefficient of variation (CV) deserves mention (43). For laboratory measures, reliability is often assessed by repeated analysis of a single reference material. For example, a single pooled blood sample might be analysed 10 times to yield measures of the biomarker X. In such studies, the mean and variance of X can be used to assess the reliability of X. A reliability coefficient cannot be estimated because there is only one sample. Instead a CV, defined as the standard deviation of X divided by the mean of X x 100, is often used:

$$CV\% = \frac{s.d. X}{\overline{X}} \times 100.$$

A small CV is considered to indicate a reliable measure.

The CV provides only limited information about measurement error for two reasons. First, this type of reliability study only assesses the laboratory error and excludes errors due to storage and handling of specimens, and to the variation in the measure over time within individuals, which are usually greater sources of error in epidemiologic studies than the laboratory error.

Second, the CV cannot be used to even assess the effect of laboratory error on the odds ratio. This is due to the fact that the CV is an estimate of the ratio of the standard deviation of X (which is an estimate of the standard deviation of the error (σ_E), as the true value of T is the same for each replicate) to $\overline{X}$, but it is ρ_{TX} which is a function of the ratio of the error variance to the total variance in X in the population of interest, that is needed to understand the impact of measurement error.

Reliability studies of binary biomarkers

Issues in the design and interpretation of reliability studies of binary biomarkers are similar to the issues discussed above for continuous biomarkers. However, the parameter used to measure the reliability of binary biomarkers is kappa (κ) rather than the intraclass correlation coefficient (44).

To compute κ for a binary marker, subjects are cross-classified by results on their first and second repeated measurements into a 2x2 table as follows:

	Measure 2 +	Measure 2 −	
Measure 1 +	p_{11}	p_{12}	r_1
Measure 1 −	p_{21}	p_{22}	r_2
	s_1	s_2	1

where p_{11} is the proportion of reliability study subjects classified as positive on both measures, p_{12} is the proportion classified as positive on measure 1 but negative on measure 2, etc. Note that the four proportions (p_{ij}) sum to 1. The overall (marginal) proportions of those who are positive and negative for measure 1 are r_1 and r_2 respectively, and the marginal proportions on the second measure are s_1 and s_2.

One measure of agreement is the observed proportion for whom there was agreement. The observed proportion of agreement, P_o, is the sum of the proportions on the diagonal:

$$P_o = p_{11} + p_{22}.$$

However, this simple measure does not take into consideration the agreement that would be expected by chance. For example, suppose the first reader of a stain on a slide accurately classified 10% of subjects as positive and 90% as negative, but the second reader simply classified all slides as negative. Then the percentage agreement would be 90%, which does not reflect the poor repeatability across readers.

Kappa is a measure of agreement that corrects for the agreement expected by chance. The expected agreement by chance (on the diagonal), P_e, is:

$$P_e = r_1 s_1 + r_2 s_2.$$

Kappa is the observed agreement beyond chance divided by the maximum possible agreement beyond chance, and is estimated as:

$$\hat{k} = \frac{P_o - P_e}{1 - P_e}.$$

Kappa ranges from zero (no agreement beyond chance) to 1 (perfect agreement), although it can be less than zero if agreement is less than expected by chance. (See (12,45) for the computation of confidence intervals for κ.)

Similar to the concepts discussed above for continuous biomarkers, the results of a reliability study of a binary biomarker can, in some situations, be used to estimate the impact of biomarker error in the parent epidemiologic study that will use the biomarker. If the reliability study meets the assumptions of equal and uncorrelated error of the parallel test model (described above), and of nondifferential measurement error, then κ can be used to estimate the bias in the odds ratio. Specifically, it has been shown that under these assumptions, this equation provides an approximation of the attenuation of the odds ratio (46):

$$OR_o = OR_T^{\sqrt{\kappa}}. \quad (10)$$

Similar to continuous measures, when the reliability study of a binary marker does not capture all sources of error (i.e. when some sources of error are repeated within-subjects (correlated)), κ will be artificially too high. Therefore, the attenuation of the odds ratio will be even greater than that predicted by Equation 10.

Review and conclusion

Before embarking on an epidemiologic study that uses a biomarker, it is important to research and understand the measurement error in that biomarker. This can be accomplished by reading previously published works on validity/reliability studies of the biomarker of interest, or conducting a new validity/reliability study. Measurement error in a biomarker refers to the error of a specific biomarker test, as applied in a specific way to a specific population, versus the true (etiologic) exposure. In epidemiologic studies, this error includes not only laboratory error, but also errors (variations) introduced during specimen collection, handling and storage, and due to month-to-month and year-to-year within-person variability of the biomarker.

Validity studies compare the biomarker to be used in an epidemiologic study to a perfect or near-perfect measure on a sample of subjects. The parameters used to quantify the error in a binary marker are sensitivity and specificity. For a continuous biomarker, X, the validity can be estimated by the bias ($\bar{X} - \bar{T}$) and by the validity coefficient ρ_{XT} (correlation coefficient of X with T), where T is the (continuous) measure of the true exposure. To assess whether the error is differential between those with and without the disease, separate analyses on a group of cases and a group of controls are needed.

Often a perfect measure of the exposure is not available, so reliability (repeatability) studies are conducted. For these, a sample of subjects is measured twice using the same marker to measure errors (variations) in the biomarker over time, between laboratory technicians, etc. The reliability study should be designed to capture all sources of error, so that the error in one measure is not repeated in (correlated with) the errors in the other measures. To design a reliability study without correlated error, the repeated specimens for each person must be collected at different times over the relevant etiologic time period and handled, stored, and analysed with the degree of variation (different specimen collectors/laboratory technicians/batches) as would occur in the parent epidemiologic study. Reliability studies are analysed using κ for binary biomarkers and the intraclass correlation coefficient for continuous biomarkers.

Equations 1, 2, 6, 8 and 10 can be used to interpret these parameters from validity or well-designed reliability studies to estimate the degree of bias in the risk ratio in an epidemiologic study that will use the biomarker. These equations assume nondifferential measurement error (i.e. equal biomarker error for those with and without the disease). Nondifferential measurement error in the biomarker attenuates the risk ratio in an epidemiologic study towards the null value of one. This attenuation is often quite substantial, even for reasonably accurate biomarker measures. For continuous markers, the impact of nondifferential measurement error depends only on the validity coefficient ρ_{XT}, and not on the bias (Equation 6).

Differential biomarker error between those with the disease and those without can bias the risk ratio in any direction, and even make a risk factor appear to be protective. Thus, differential error can completely invalidate an epidemiologic study and must be avoided. Differential measurement error is a particular concern in case–control studies (and among the early cases in cohort studies) when the biomarker is not a fixed marker (e.g. genotype), and, therefore could be influenced by the disease, its preclinical phase, or its treatment. Assessment of differential error requires specimens on a sample of cases years before diagnosis and comparable early specimens on controls. This measure can serve as the "true" marker, even if it is not perfect, as long as it does not have differential error. For continuous variables, differential bias has the most problematic effects on the risk ratio; it could be estimated by comparing bias ($\bar{X} - \bar{T}$) for cases with bias ($\bar{X} - \bar{T}$) among controls.

One goal of giving examples of the large effects that even moderate degrees of biomarker measurement error have on epidemiologic studies, is to motivate attention to reducing the biomarker error. The researcher should focus on reducing the errors through appropriate quality control techniques for specimen collection, storage and laboratory analyses, and, if needed, by use of multiple measures over time in the parent epidemiologic study to reduce the errors caused by biomarker variation over time. These important methods and additional approaches to reduce biomarker error in epidemiologic studies are covered in other chapters of this book and by (12,15).

References

1. Walker AM, Blettner M (1985). Comparing imperfect measures of exposure. *Am J Epidemiol*, 121:783–790. PMID:4014171

2. de Klerk NH, English DR, Armstrong BK (1989). A review of the effects of random measurement error on relative risk estimates in epidemiological studies. *Int J Epidemiol*, 18:705–712.doi:10.1093/ije/18.3.705 PMID:2807678

3. Armstrong BG, Whittemore AS, Howe GR (1989). Analysis of case-control data with covariate measurement error: application to diet and colon cancer. *Stat Med*, 8:1151–1163, discussion 1165–1166.doi:10.1002/sim.4780080916 PMID:2799135

4. Dunn G. Design and analysis of reliability studies. London: Edward Arnold and New York: Oxford University Press; 1989.

5. White E. Effects of biomarker measurement error on epidemiological studies. In: Toniolo P, Boffetta P, Shuker DEG et al., editors. Applications of biomarkers in cancer epidemiology. Lyon: IARC Scientific Publication; 1997. p. 73–94.

6. Chen TT (1989). A review of methods for misclassified categorical data in epidemiology. *Stat Med*, 8:1095–1106, discussion 1107–1108.doi:10.1002/sim.4780080908 PMID:2678350

7. Thomas D, Stram D, Dwyer J (1993). Exposure measurement error: influence on exposure-disease. Relationships and methods of correction. *Annu Rev Public Health*, 14:69–93.doi:10.1146/annurev.pu.14.050193.000441 PMID:8323607

8. Carroll RJ, Ruppert D, Stefanski LA, Crainiceanu CM. Measurement error in nonlinear models: a modern perspective. 2nd ed. London/Boca Raton: Chapman and Hall CRC Press; 2006.

9. Holford TR, Stack C (1995). Study design for epidemiologic studies with measurement error. *Stat Methods Med Res*, 4:339–358. doi:10.1177/096228029500400405 PMID:8745130

10. Bashir SA, Duffy SW (1997). The correction of risk estimates for measurement error. *Ann Epidemiol*, 7:154–164.doi:10.1016/S1047-2797(96)00149-4 PMID:9099403

11. Thürigen D, Spiegelman D, Blettner M et al. (2000). Measurement error correction using validation data: a review of methods and their applicability in case-control studies. *Stat Methods Med Res*, 9:447–474.doi:10.1191/096228000701555253 PMID:11191260

12. White E, Armstrong BK, Saracci R. Principles of exposure measurement in epidemiology. 2nd ed. Collecting, evaluating and improving measures of disease risk factors. Oxford: Oxford University Press; 2008.

13. Vineis P. Sources of variation in biomarkers. In: Toniolo P, Boffetta P, Shuker DEG et al., editors. Applications of biomarkers in cancer epidemiology. Lyon: IARC Scientific Publication; 1997. p. 59–72.

14. Fleiss JL. The design and analysis of clinical experiments. New York (NY): John Wiley and Sons; 1986.

15. Tworoger SS, Hankinson SE (2006). Use of biomarkers in epidemiologic studies: minimizing the influence of measurement error in the study design and analysis. *Cancer Causes Control*, 17:889–899.doi:10.1007/s10552-006-0035-5 PMID:16841256

16. White E (2003). Design and interpretation of studies of differential exposure measurement error. *Am J Epidemiol*, 157:380–387.doi:10.1093/aje/kwf203 PMID:12615602

17. Kotz S, Johnson NL, editors. Encyclopedia of statistical sciences, vol 8. New York (NY): John Wiley and Sons; 1988.

18. Bross I (1954). Misclassification in 2 x 2 tables. *Biometrics*, 10:478–486 doi:10.2307/3001619.

19. Newell DJ (1962). Errors in the interpretation of errors in epidemiology. *Am J Public Health Nations Health*, 52:1925–1928. doi: 10.2105/AJPH.52.11.1925 PMID:13938241

20. Gullen WH, Bearman JE, Johnson EA (1968). Effects of misclassification in epidemiologic studies. *Public Health Rep*, 83:914–918. PMID:4972198

21. Goldberg JD (1975). The effects of misclassification on the bias in the difference between two proportions and the relative odds in the fourfold table. *J Am Stat Assoc*, 70:561–567 doi:10.2307/2285933.

22. Copeland KT, Checkoway H, McMichael AJ, Holbrook RH (1977). Bias due to misclassification in the estimation of relative risk. *Am J Epidemiol*, 105:488–495. PMID:871121

23. Barron BA (1977). The effects of misclassification on the estimation of relative risk. *Biometrics*, 33:414–418.doi:10.2307/2529795 PMID:884199

24. Kleinbaum DG, Kupper LL, Morgenstern H. Epidemiologic research. Belmont (CA): Lifetime Learning Publications; 1982. p. 183–193, 220–241.

25. Wu IC, Wu DC, Lu CY et al. (2004). Comparison of serum and urine ELISA methods for the diagnosis of Helicobacter pylori–a prospective pilot study. *Hepatogastroenterology*, 51:1736–1741. PMID:15532816

26. Hansen MH, Hurwitz WN, Bershad M (1961). Measurement errors in censuses and surveys. *Bull Int Stat Inst*, 38:359–374.

27. Lord FM, Novick MR. Statistical theories of mental test scores. Reading (MA): Addison-Wesley; 1968.

28. Cochran WG (1968). Errors of measurement in statistics. *Technometrics*, 10:637–666 doi:10.2307/1267450.

29. Nunnally JC. Psychometric theory. 2nd ed. New York (NY): McGraw-Hill; 1978. p. 190–225.

30. Allen MJ, Yen WM. Introduction to measurement theory. Monterey (CA): Brooks/Cole; 1979. p. 1–117.

31. Bohrnstedt GW. Measurement. In Rossi PH, Wright JD, Anderson AB, editors. Handbook of survey research. Orlando (FL): Academic Press; 1983. p. 70–121.

32. Fuller WA. Measurement error models. New York (NY): John Wiley and Sons; 1987.

33. Prentice RL (1982). Covariate measurement errors and parameter estimation in a failure time regression model. *Biometrika*, 69:331–342 doi:10.1093/biomet/69.2.331.

34. Whittemore AS, Grosser S. Regression methods for data with incomplete covariates. In: Moolgavkar SH, Prentice RL, editors. Modern statistical methods in chronic disease. New York (NY): John Wiley and Sons; 1986. p. 19–34.

35. Rosner B, Willett WC, Spiegelman D (1989). Correction of logistic regression relative risk estimates and confidence intervals for systematic within-person measurement error. *Stat Med*, 8:1051–1069, discussion 1071–1073.doi:10.1002/sim.4780080905 PMID:2799131

36. White E, Kushi LH, Pepe MS (1994). The effect of exposure variance and exposure measurement error on study sample size: implications for the design of epidemiologic studies. *J Clin Epidemiol*, 47:873–880.doi:10.1016/0895-4356(94)90190-2 PMID:7730890

37. Wu ML, Whittemore AS, Jung DL (1986). Errors in reported dietary intakes. I. Short-term recall. *Am J Epidemiol*, 124:826–835. PMID:3766514

38. McKeown-Eyssen GE, Tibshirani R (1994). Implications of measurement error in exposure for the sample sizes of case-control studies. *Am J Epidemiol*, 139:415–421. PMID:8109576

39. Law MR, Thompson SG (1991). Low serum cholesterol and the risk of cancer: an analysis of the published prospective studies. *Cancer Causes Control*, 2:253–261. doi:10.1007/BF00052142 PMID:1831389

40. Winawer SJ, Flehinger BJ, Buchalter J et al. (1990). Declining serum cholesterol levels prior to diagnosis of colon cancer. A time-trend, case-control study. *JAMA*, 263:2083–2085.doi:10.1001/jama.263.15.2083 PMID:2319669

41. Gregorio DI, Marshall JR, Zielezny M (1985). Fluctuations in odds ratios due to variance differences in case-control studies. *Am J Epidemiol*, 121:767–774.doi:10.1093/aje/121.5.767 PMID:4014168

42. Missmer SA, Spiegelman D, Bertone-Johnson ER et al. (2006). Reproducibility of plasma steroid hormones, prolactin, and insulin-like growth factor levels among premenopausal women over a 2- to 3-year period. *Cancer Epidemiol Biomarkers Prev*, 15:972–978.doi:10.1158/1055-9965.EPI-05-0848 PMID:16702379

43. Garber CC, Carey RN. Laboratory statistics. In Kaplan L, Pesce A, editors. Clinical chemistry: theory, analysis, and correlation. St. Louis (MO): Mosby; 1984. p. 290–292.

44. Cohen J (1960). A coefficient of agreement for nominal scales. *Educ Psychol Meas*, 20:37–46 doi:10.1177/001316446002000104.

45. Fleiss JL, Levin B, Paik MC. Statistical methods for rates and proportions. 3rd ed. New York (NY): John Wiley and Sons; 2003.

46. Tavaré CJ, Sobel EL, Gilles FH (1995). Misclassification of a prognostic dichotomous variable: sample size and parameter estimate adjustment. *Stat Med*, 14:1307–1314.doi:10.1002/sim.4780141204 PMID:7569489

UNIT 3.
ASSESSING EXPOSURE TO THE ENVIRONMENT

CHAPTER 9.

Environmental and occupational toxicants

Frank de Vocht, Jelle Vlaanderen, Andrew C. Povey, Silvia Balbo, and Roel Vermeulen

Summary

Biological monitoring is the analysis of human biological materials for a substance of interest and/or its metabolites (biomarkers) or a biochemical change that occurs as a result of an exposure to provide a quantitative measure of exposure or dose. These measures can be used in epidemiological studies either directly as estimates of exposure or indirectly in the calibration of other exposure assessment methods, such as questionnaires. This chapter will discuss important methodological considerations for the implementation of biomarkers of exogenous exposure in epidemiology by focusing on biomarker characteristics (e.g. variability, half-life) and their application in different study designs.

Exposure assessment in environmental and occupational epidemiology

In general, the goal in environmental and occupational epidemiology is to estimate the association between levels of exposure and their impact on health in human populations in a valid and precise manner. (The analytical and technical aspects of measuring specific biomarkers of exposure will not be discussed here; see chapters 4 and 11 on biological monitoring of chemicals and nutrients, respectively). In these studies, 'exposure' is described as a substance or factor affecting human health, either adversely or beneficially, which in practice is usually regarded as an estimate of the 'true' exposure a subject under study might receive (3). Exposure might originate from environmental or occupational sources, which, within the context of this chapter, are included within environmental epidemiology. Exposure to humans can be considered a dose when a distinction is made between the available dose, which is the dose that is available for uptake in the human body; the administered dose (or intake); the absorbed dose, which actually enters the body (uptake); and the biologically effective dose, which reaches the target cells in the body. The objective of exposure measurements in any environmental or occupational epidemiological study is to provide an unbiased measure of the actual exposure or dose that an individual receives. To optimize the quantification of the association between exposure and health effects, these estimates of exposure should be accurate, precise, biologically relevant, apply to the etiological important exposure period, and show a range of exposure levels in the population under study (3).

Exposure is generally characterized by the physical and chemical properties of the agent, its intensity, and temporal variability (4,5). There can be considerable variability in all of these factors, temporal as well as between study subjects, which allow them to be used as metrics for exposure. Several distinct exposure metrics are used in epidemiological studies: cumulative exposure (total accumulated dose), average exposure (total accumulated dose divided by time), and peak exposure (highest exposure level experienced by a subject in a given time period). Each of these exposure metrics can be derived for the whole lifetime of each study subject or just for a particular etiologically relevant time period. Whereas cumulative exposure, average exposure, and peak exposure are basically interchangeable for short time periods, they might not be for long-term exposures due to complex exposure patterns over time (4).

Epidemiological studies generally deal with large population sizes. This makes estimating exposure for all individual study subjects difficult, as often not all subjects' exposure can be measured. Researchers therefore have to rely on some form of modelled or surrogate measure for true exposure. In general, there are two study types for exposure assignment: individual-based studies, in which exposure levels and health outcomes are measured for all persons; and group-based studies, in which samples of persons are measured in each of several groups and group-specific mean values of exposure levels are used to estimate the exposure-response association (6). In the group-based approach, it is important that measurements are made on a random selection of the population; often, however, they are based on convenience samples. In environmental epidemiological studies, these groups are generally defined on the basis of the presence or absence of an exposure source and the distance from it, while in occupational studies, exposure groups are often defined by factories, departments, or job titles (3). The underlying assumption when using this strategy for grouping is that subjects within each group are exposed to comparable exposure characteristics (e.g. intensity, cumulative exposure).

Environmental studies tend to have larger within-subject variability and smaller between-person and between-group variability than occupational studies. Therefore, group-based designs will generally be more appropriate to investigate exposure-response associations in the general population, but to a lesser extent for occupational studies (6). However, in both individual- and group-based designs, the relatively large within-subject variability in environmental and occupational exposures, emphasizes the importance of collecting multiple exposure measurements for each subject in the study (3,5). To assess the relative impact of temporal, between-subject, and between-group variability, studies using a repeated measures design should be conducted. This study design uses multiple exposure measurements for study subjects or groups in time to estimate these variance components by means of advanced statistical techniques, including (hierarchical) mixed effects models (7).

If the intensity or duration of exposure is poorly characterized, due to random measurement or misclassification error, the resulting estimated exposure-response associations will often underestimate the true risk for a given exposure level. This is known as attenuation bias. The expected attenuation in the risk estimate β in a common regression model ($Y_i \sim \alpha + \beta_1 x_1 + e_i$) to assess an exposure-response association, can, for group-based studies, be estimated by (8):

$$E(\hat{\beta}_1) = \frac{\beta_1}{1 + \frac{\sigma^2_{wg}}{kn\sigma^2_{bg} + n\sigma^2_{bh}}}$$

where $E(\hat{\beta}_1)$ is the expected risk estimate (β_1) adjusted for attenuation bias, σ^2_{wg} is the within-group variance (i.e. between-subject), σ^2_{bg} is the between-group variance, σ^2_{bh} is the within-subject variance, k is the number of randomly selected subjects, and n is the number of repeated measurements per subject.

While for individual-based studies this can be described by:

$$E(\hat{\beta}_1) = \frac{\beta_1}{\frac{\sigma^2_{ws}}{\sigma^2_{bs}}}$$

where σ^2_{ws} is the within-subject variance and σ^2_{bs} is the between-subject variance.

Non-random or differential measurement or misclassification, which can result from errors in the design of the study or the measurement technique, can both over- or underestimate an exposure-response association depending on the magnitude and direction of the bias.

Application of exposure markers in environmental epidemiology

Biomarkers of exposure generally aim at measuring the level of an external agent, or its metabolites, in either the free-state or bound to macromolecules. The range of biological samples that can be obtained and analysed includes: blood, urine, exhaled breath, hair, nails, milk, feces, sweat, saliva, semen, and cerebrospinal fluid. The choice of biological sample depends on the substance of interest, its characteristics (e.g. solubility, metabolism, transformation, and excretion), and how invasive the method to obtain it is. As such, several biomarkers can be available to represent the same exposure, including the parent compound itself, a metabolite, or a macromolecular DNA or protein adduct (9).

Whereas 'classical' methods of exposure assessment provide an estimate for exposure from one route of exposure only (e.g. inhalation through the respiratory system, ingestion through the gastrointestinal system, or absorption through the skin) (3), biological monitoring has the theoretical advantage of integrating exposures from all exposure routes. In addition, it also covers unexpected or accidental exposures and reflects interindividual differences in uptake, metabolism, genetic susceptibility, and excretion (10–12). However, some exposure biomarkers can also be formed endogenously and levels may then reflect both endogenous formation and exogenous exposures (13). Nonetheless, the use of exposure markers in epidemiology could potentially lead to a more accurate and/or more biologically relevant exposure estimate than 'classical' methods. For instance, biomarkers for tobacco specific N-nitrosamines, such as 4-(methylnitrosamino)-1-(3-pyridyl)-1-butanone (NNK), might be more relevant for certain research questions than self-reported smoking habits, as NNK is a known carcinogen and urinary levels of its reduction product 4-(methylnitrosamino)-1-(3-pyridyl)-1-butanol (NNAL) reflect differences in smoking habits, the type of tobacco, and individual metabolism (see the example below on tobacco smoke).

Biomarker characteristics

The choice of a biomarker will depend on several considerations, but the main issues are its kinetic parameters and the knowledge of the mechanistic basis of the adverse effects (9). Of these, the biological relevance (i.e. association with 'true' exposure at the site of action) is generally considered the most important selection criterion (14). However, although it is usually assumed that the biomarker is in some way associated with the exposure and the disease, limited information is often available on where the markers are located along the multistep pathway from exposure to human disease (14). Furthermore, to date, only a few biomarkers have been properly validated (15), which limits their application. The National Health and Nutrition Examination Survey (NHANES), conducted by the US Centers for Disease Control and Prevention (CDC), provides a good overview of exposure biomarkers and reference values in the normal population for many environmental exposures (http://www.cdc.gov/nchs/nhanes.htm).

In addition to the biological relevance of the biomarker, its biological half-life is a critical characteristic. 'Biological half-life' refers to the biological clearance of the biomarker from the target tissue. It can be derived from several sources, including empirical modeling of experimental data, compartment models incorporating experimentally determined rate constants, or from simulations based on physiologically and metabolically based parameters (16). Biological half-lives vary substantially between biomarkers. The half-lives of some compounds measured in the NHANES survey are presented in Figure 9.1.

The interpretation of a biomarker measurement depends on the sampling time, as each biomarker has a specific half-life. The analysis may reflect the amount of chemical absorbed shortly before the sample was taken, in the case of a biomarker with a short half-life (e.g. nicotine in blood); it may reflect exposure occurring during the preceding days for markers with intermediate half-lives (e.g. cotinine in blood); or it may reflect the dose integrated over a period of months for biomarkers with long half-lives (e.g. 3- and 4-aminobiphenyl-haemoglobin adducts). Additionally, some chemicals accumulate in specific tissues or organs; thus the biomarker value may reflect cumulative exposure over a period of years (16). However, most existing exposure markers have relatively short half-lives, with exceptions like some metals, and persistent organic pollutants like polychlorinated biphenyls and dioxins (Figure 9.1).

In general, biomarkers with relatively long half-lives are preferred, reflecting weeks, months, or even years of exposure when studying chronic health effects. This does not automatically mean that biomarkers with relatively short half-lives cannot be used in epidemiological studies; they are useful in studies of acute biological or health effects or where exposure is relatively constant over time.

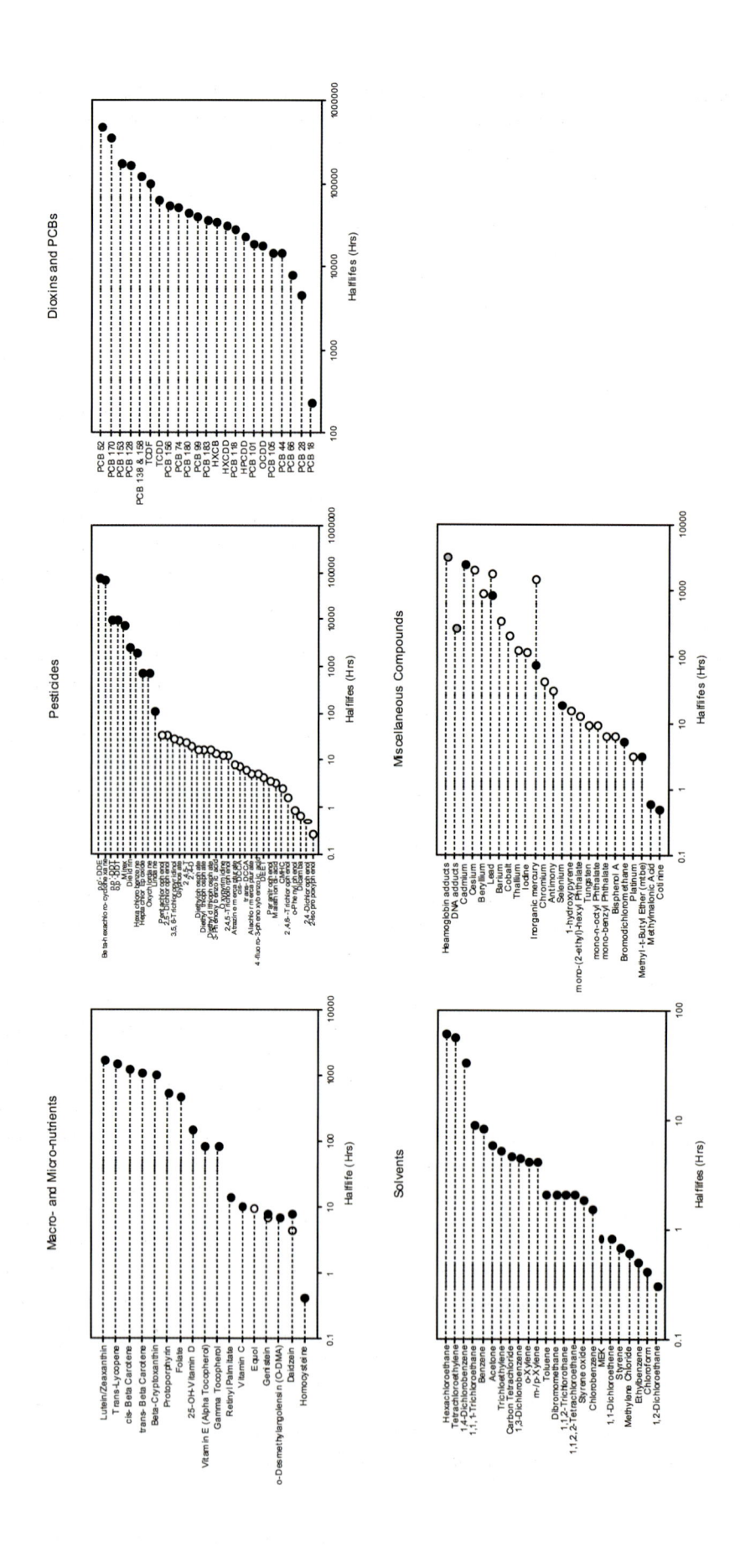

Figure 9.1. Examples of biological indicators and their half-lives measured as part of NHANES. Solid dots indicate measurements in blood. Open circles indicate measurements in urine

Analytical variability

One source of variability in biomarker studies is laboratory, or analytical, variability. Before a new, promising biomarker can be used in population studies, transitional studies should first be conducted to characterize the biomarker in terms of accuracy, reliability in the laboratory, and optimal conditions for use (17). These studies should make certain that the analytical results are sufficiently accurate to ensure correct interpretation of the biomarker results in population studies and that the results will be reproducible.

At present, the contribution of analytical variability to total variability is, in general, much lower than biological variability because of improved techniques and quality assurance procedures in biomarker assessment (9,18). This variability can be further reduced by sharing methods and techniques and exchange of reference materials between laboratories (9).

Individual and temporal variability

Variability in biomarker responses, for continuous, non-fixed biomarkers, has two dimensions: an individual dimension and time. Individual variability in biomarker responses will depend on external exposure variability and on interindividual differences as to how an individual metabolizes the agent of interest. The temporal variability in biomarker response depends primarily on the half-life and on the temporal variability in exposure. Driven by financial or logistic motivations, the assumption is often made in epidemiologic studies that biomarker levels (and other traditional measures of exposure) are a fixed attribute of an individual, rather than being time-dependent, and as such are measured at only one single point in time. However, for this to be valid, biological steady-state conditions are required. In practice, these are not likely to occur since they require stable biokinetics, a constant rate of exposure, dynamic equilibrium among body tissues, and a sufficiently long period of time for the biomarker to stabilize in all relevant tissues (19).

In general, ignoring the temporal variability in biomarker response leads to additional classic measurement error, which results in the attenuation of the biomarker-disease association (17). Biomarkers with relatively short half-lives generally display more temporal variability than biomarkers with relatively long half-lives, which is related to the dampening of the temporal variance in exposure over time (17,20). It has been shown that whereas less than 50% of the temporal variance in exposure is transmitted for many biological markers with a half-life of more than 40 hours, the dampening effect is negligible for markers with a half-life of less than five hours (21). In Figure 9.2, examples of constant and variable occupational and environmental exposure circumstances are given for biomarkers with different half-lives (i.e. five, 20, and 100 hours). These examples suggest that because biomarkers with a relatively short half-life are more sensitive to fluctuations of exposure from hour-to-hour and day-to-day, that timing of sample collection becomes increasingly important. The exception to this is when exposures are constant over time. Therefore, the use of biomarkers with relatively long half-lives is generally more appropriate for epidemiological studies, especially when they can be measured only at a single point in time and not necessarily in the optimal etiological time window. This also depends on the health effect under investigation, since, in principle, biomarkers with a short half-life are needed when (semi-)acute biological and health effects are studied.

Biomarker validity

Ideally, before starting a study involving biomarker measurements, information on the variation in exposure patterns between individuals, as well as over time, should be known to determine whether a specific biomarker of exposure will be appropriate for the particular study. If this information is not available at the start of the study and it is not feasible to conduct a pilot study to estimate the variability of exposure, the intraindividual variation in the biomarker response can also be evaluated. At the same time, all sources of unwanted variation (e.g. laboratory variation) should be taken into account, by conducting a repeated measures design in the main study.

Several methods are available to assess variability when using biomarkers with a continuous outcome. The coefficient of variation (CV), which is defined as the ratio of the standard deviation (σ) to the mean (μ), is generally used as a measure of the extent of variation between different batches, and/or duplicate samples within batches, and can be used to identify 'bad' sample batches. It does not, however, provide insight into the impact of the observed variance on the biomarker-disease association and it cannot be used to correct measures of association to account for measurement error (17,21). The intraclass correlation coefficient (ICC), described by:

$$ICC = \frac{\sigma_{bs}^2}{\sigma_{bs}^2 + \dfrac{\sigma_{ws}^2}{N}}$$

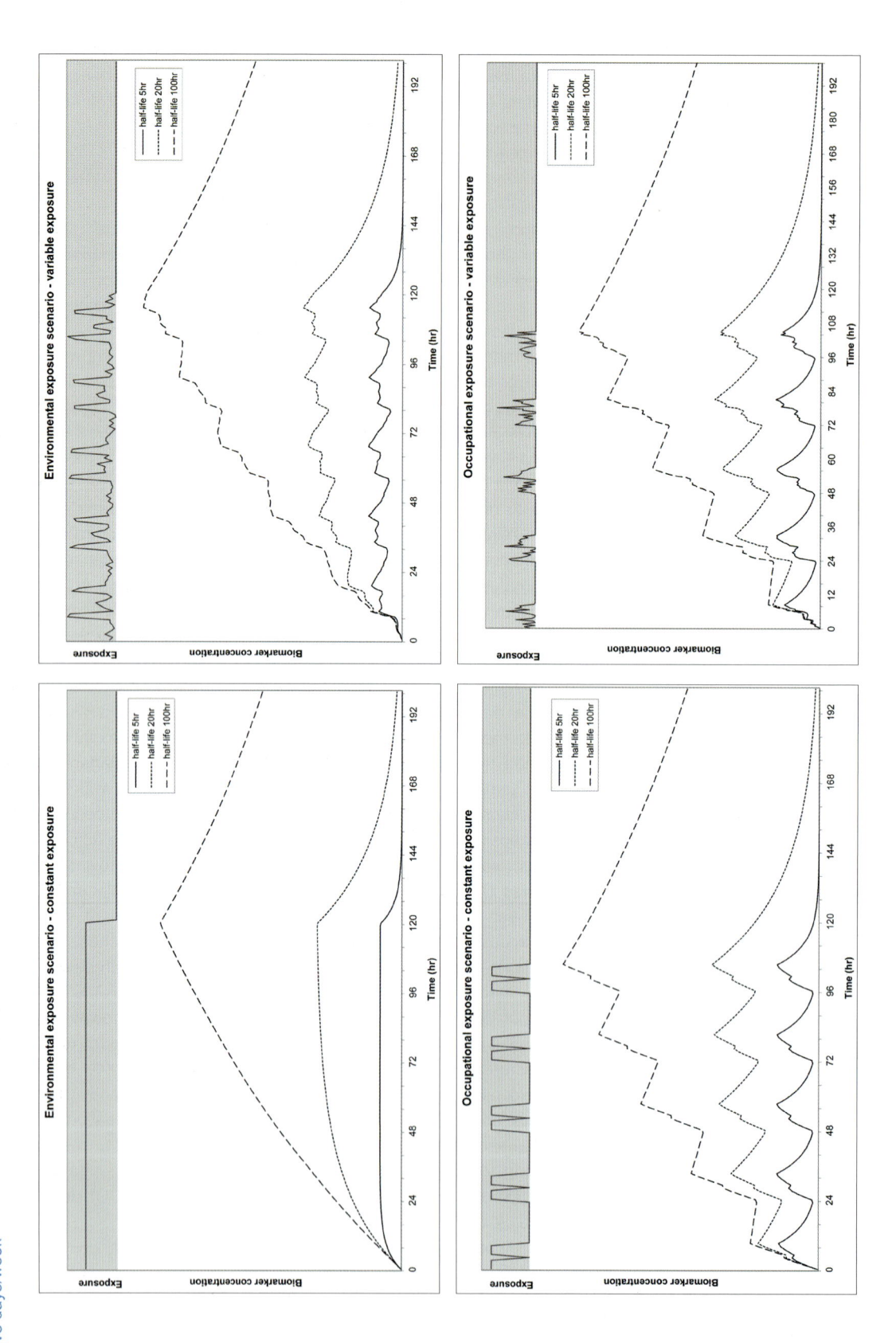

Figure 9.2. Effects of exposure variability on three biological indicators with half-lives of five, 20, and 100 hours. In all examples, the exposure scenario stopped after 120 hours. Ordinate is an arbitrary scale. Graph A1: Constant continuous environmental exposure. Graph A2: Constant occupational exposure. Graph B1: Variable environmental exposure with three high periods during morning (8–10), afternoon (12–14), and evening rush hour (17–19). Graph B2: Variable occupational exposure (eight hour day-1), one hour break, five days/week

(where σ_{bs}^2 is the between-subject variance, σ_{ws}^2 is the within-subject variance, and N is the number of repeated measures on an individual), is a more useful measure for evaluating the impact of the total measurement error (temporal plus analytical error).

In addition, the ICC can be used to adjust measures of association to account for measurement error. However, in the absence of a 'gold standard,' the results of such adjustments should not be interpreted as true associations, but instead as indicators for the degree of bias in the observed risk estimates (22; for more details, see Chapter 8).

Study designs

There is a spectrum of epidemiological study designs that make use of biological exposure markers. The choice of design depends on the specific research question and disease under study (e.g. rare versus common; acute versus chronic), and has implications for the use of biological exposure markers. The strengths and limitations of using exposure markers in relation to the major study designs are discussed below (for a more in depth discussion on study designs, see Chapters 14 and 15).

Cross-sectional studies

Cross-sectional studies are often initiated to assess whether a subset of a population has been exposed to a particular exposure, or to validate the exposure assessment from other sources, such as environmental monitoring or data obtained from questionnaires. For example, toenail nicotine levels, together with self-reported smoking habits and exposure to environmental tobacco smoke, were collected from 2485 women to assess the validity of toenail nicotine levels as a marker of tobacco smoke exposure, and to provide insight into its ability to capture non-reported exposure (23).

A distinct advantage of cross-sectional studies over alternative study designs is that detailed and accurate information can be collected on current exposure patterns and on determinants of exposure or potential confounders. To further improve the accuracy of the biomarker assessment, repeated measures should be considered, especially if the temporal variability is relatively large. However, one of the disadvantages of this study design is that current exposure patterns or determinants do not necessarily reflect historic levels, which might be more relevant to the exposure-disease pathway.

Case-control studies

The main goal of case-control studies is to compare exposure patterns in cases and in carefully matched controls during the etiologically relevant time period. One of the important advantages of case-control studies compared to prospective cohort studies, especially for biomarker studies, is their ability to enrol large numbers of cases relatively quickly and the potential to study uncommon diseases (17). A problem inherent to the way cases are recruited is that biological samples, exposure data, and other information is collected after diagnosis and even sometimes after commencing treatment of the disease. This makes these studies susceptible to differential misclassification and may lead to problems in the assessment of the temporal association between the disease and the biomarker under study (17).

For example, in a study on blood levels of organochlorines before and after chemotherapy among Non-Hodgkin lymphoma (NHL) patients, a marked decrease (25–30%) in serum levels of these compounds was found after treatment (24). This could lead to large exposure biases if cases are not enrolled before the start of chemotherapy, as blood levels of organochlorines among controls would not be influenced by therapy.

Prospective cohort studies

Prospective cohort studies are considered the only study design that allows researchers to look at biomarkers that are directly or indirectly affected by the exposure-disease mechanism, since biological specimens and exposure information are collected before disease diagnosis and, ideally, before the beginning of the disease process (25). It can be difficult to recruit enough subjects in the cohort and/or follow-up enough people for the duration of the study, therefore the study can be enriched with cases in subsequent nested case-control or case-cohort studies, which will improve the study efficiency (25). Unfortunately, larger prospective cohort studies have been able to collect only one biological sample at one point in time for individuals enrolled in the cohort. As discussed, this can cause problems for most types of biomarkers of exposure; especially short-term exposure markers which may vary substantially from day-to-day. It has further been discussed that although biomarkers can be collected in a variety of media, and that sometimes more media are available to assess exposure to the same chemical, most studies have only collected blood samples and only a few have collected urine.

An example of environmental exposure markers - tobacco smoke

Exposure to tobacco smoke represents one of the most prominent risk factors for cancer, cardiovascular diseases, and chronic respiratory diseases (26). Environmental tobacco smoke has also been implicated in the etiology of these diseases (27). The immense impact on public health of tobacco smoking and exposure to tobacco smoke has stimulated the development of tobacco-related biomarkers (Table 9.1).

Carbon monoxide and thiocyanate. Carbon monoxide (CO) and thiocyanate are considered the oldest biomarkers used as indicators of tobacco smoke exposure. CO is the product of incomplete combustion of organic materials. Inhaled CO is absorbed through the lungs and binds to haemoglobin (Hb) forming carboxyhaemoglobin (COHb). As the absorption is by the lung alveoli, levels of exhaled CO (COex) or COHb measured in blood are useful biomarkers of exposure, as CO does not undergo metabolic activation. CO has a short half-life

Table 9.1. Overview of tobacco exposure related biomarkers

Biomarker	Specimen	Reflected Exposure to Tobacco Smoke Product	Specificity	Half-Life	Detection Method
COex COHb	Breath Blood	Carbon monoxide	Low	2-3 hours	Infrared spectroscopy and GC
Thiocyanate	Saliva Blood Urine	Hydrogen cyanide	Low	1-2 weeks	Photometry, Ion exchange chromatography followed by UV detection, GC coupled with MS after derivatization
Nicotine	Saliva Blood Urine Toenail Hair	Nicotine	High	2 hours Several months	HPLC with UV detection
Cotinine	Saliva Blood Urine	Nicotine	High	3-4 days	HPLC with UV detection
NNAL and NNAL-Gluc	Urine	NNK uptake	High	Several months	GC
1-hydroxy-pyrene	Urine	Pyrene uptake	Low because of PAHs sources of exposure other than tobacco	Around 15 hours	HPLC
Benzo[a]pyrene diol epoxide DNA adducts	DNA	Benzo[a]pyrene biological effective dose	Low because of PAHs sources of exposure other than tobacco	In general, DNA adducts are considered to provide estimates of exposure for several half-lives of the adduct depending on adduct stability and repair capacity	
3- and 4- aminobiphenyl haemoglobin adducts	Blood	Aromatic amines uptake plus metabolic activation	Low because of aromatic amines sources of exposure other than tobacco	Around 120 days (haemoglobin life-span)	
Trans-trans-muconic acid	Urine	Benzene uptake	Low - influenced by food intake of sorbic acid from food	13 hours	LC/UV
S-Phenylmer-capturic acid	Urine	Benzene uptake	Low because of benzene sources of exposure other than tobacco	14 hours	LC/MS
Anabasine, anatabine and myosine	Saliva Urine	Tobacco products	High	Few hours	HPLC/MS GC/MS

(2–3 hours) making it a marker of recent exposure. However, COex and COHb levels can be affected by physical activity, sex, and the presence of lung or airway diseases.

Hydrogen cyanide. A chemical present in tobacco smoke, hydrogen cyanide (HCN) is formed in tobacco combustion mainly from proteins and nitrates. It is metabolized into thiocyanate (SCN) that can be measured in saliva, blood, and urine. Due to its relatively long half-life (1–2 weeks), SCN reflects at least several weeks of exposure (see section on *Temporal Variability*). However, both these biomarkers are considered non-specific. Levels of CO and SCN, can be affected by numerous sources other than tobacco smoke, such as air pollution and diet for CO and SCN, respectively (28).

Nicotine. Nicotine is a chemical found in all tobacco products and is the major addictive component. Levels of nicotine can be measured in blood, saliva, and urine, providing a specific biomarker of exposure. However, since this chemical has a short half-life (a few hours), the results are very dependent on time of sampling. Furthermore, urine levels are highly influenced by urine volume and pH, reducing the use of this biomarker. The development of methods for the detection of nicotine in hair and nails has recently been suggested as a promising marker for long-term exposure (29).

Cotinine. Cotinine is the major proximate metabolite of nicotine, but with a longer half-life in the blood (3–4 days) (30). The presence of cotinine in a biologic fluid indicates exposure to nicotine. There is some individual variation in the quantitative relationship between cotinine levels in blood, saliva, and urine and the intake of nicotine, due to the fact that people metabolize nicotine and cotinine differently. Still this metabolite has been widely used as a very specific biomarker of tobacco exposure. Cotinine is also of particular interest as a biomarker for the evaluation of exposure to environmental tobacco smoke (ETS). Cotinine concentrations in plasma, urine, and saliva of non-smokers have been used in assessing population exposure to ETS for developing risk estimates for ETS-related lung cancer (31).

N-nitrosamines. Tobacco smoke contains volatile N-nitrosamines, such as N-nitrosodimethylamine and N-nitrosopyrrolidine, as well as tobacco specific N-nitrosamines, such as N'-nitrosonornicotine (NNN) and 4-(methylnitrosamino)-1-(3-pyridyl)-1-butanone (NNK) (32). In particular, nitrosamines in tobacco are chemically related to nicotine, and other tobacco alkaloids, and therefore specific to tobacco products. For this reason, 4-(methylnitrosamino)-1-(3-pyridyl)-1-butanol (NNAL) (the major metabolite of NNK) together with its glucuronide derivative (NNAL-Glucs), which can be detected in urine, provide a particularly valuable biomarker due to their specificity. Moreover, both NNAL and NNAL-Glucs have a relatively long half-life compared to other measurable urinary metabolites. This biomarker has been used to quantify levels of NNK uptake in smokers and smokeless-tobacco users (33) to examine ethnic differences in NNK metabolism (34), and to study the effects of diet and potential cancer chemopreventive agents on NNK metabolism (35,36). There is a consistent correlation between levels of cotinine, NNAL, and NNAL-Glucs in urine (37). The measurement of NNAL and NNAL-Glucs in urine has been particularly useful in studies of ETS. Uptake of NNK by non-smokers exposed to ETS has been shown in several settings, including the detection of these biomarkers in amniotic fluid, indicating that NNK or NNAL are present in fetuses of mothers who smoke (38).

NNK and NNN can also lead to the formation of specific haemoglobin and DNA adducts, which can potentially measure uptake plus metabolic activation and the biological effective dose of these carcinogens, respectively. Methods for the detection of these biomarkers have been developed; however, their levels are frequently low and, in some cases, undetectable in many active smokers.

As for the N-nitrosamines, aromatic amines can undergo metabolic activation leading to the formation of DNA or protein adducts. 4-Aminobiphenyl (4-ABP) undergoes P450 catalysed N-oxidation to a hydroxylamine. O-Acetylation, catalysed by N-acetyltransferases (NATs), produces an O-acetoxy compound that reacts with DNA. Other esterification reactions of the hydroxylamine lead to related intermediates that can also react with DNA. However, since the levels of DNA adducts in humans are generally low (once every 106-108 normal bases), large amounts of DNA and sensitive methods are needed for the analysis. Moreover, little is known about their persistence in human tissue. Animal studies have shown a great variability in this respect depending on the different chemical structures formed and on the repairing systems, which might remove some adducts but not others (32). In general, studies on DNA adducts have reported higher levels in smokers compared to non-smokers and higher levels in tissue samples (from oral, lung, and bladder cancers) from cases than controls.

Aromatic amines, polycyclic aromatic hydrocarbons, and benzene. Aromatic amines (arylamines), such as o-toluidine,

2-aminonaphthalene, and 4-aminobiphenyl, occur in the environment and are constituents of tobacco smoke. A method for measuring these in cigarette smokers was developed, using the acid hydrolysis of the arylamine conjugates in urine. Urinary arylamine excretion in smokers was associated with the extent of smoking as assessed by daily cigarette consumption, urinary excretion of nicotine, cotinine in saliva, and carbon monoxide in exhaled breath. This analytical method is suitable for measuring short-term exposure to arylamines in urine of non-occupationally exposed smokers and non-smokers (39).

Haemoglobin adducts of aromatic amines are an informative type of carcinogen biomarker. Large amounts of haemoglobin are readily available in the blood and protein has a long half-life (120 days), which allows the adducts to accumulate and thus reflect a relatively long-term exposure. Levels of these adducts are consistently higher in smokers than in non-smokers (40). In a recent study, the relative risk of bladder cancer in women who smoked was found to be significantly higher than in men who smoked a comparable number of cigarettes. Consistent with this gender difference, levels of 3- and 4-aminobiphenyl-haemoglobin adducts, in relation to the number of cigarettes smoked per day, was statistically higher in women than in men (41).

Polycyclic aromatic hydrocarbons (PAHs), which cause lung cancer and other smoking-related cancers, are present in tobacco smoke. One of the main metabolites, 1-hydroxypyrene in urine, is the biomarker used to study the uptake of PAHs in smokers. Levels of 1-hydroxypyrene are 2–3 times higher in smokers than in non-smokers and decrease with smoking cessation (42). Benzo(a) pyrene, another main constituent of the PAHs mixture, is metabolized to Benzo(a)pyrene diol epoxides, which reacts with Hb and DNA forming adducts. However, since these adducts are difficult to detect even with highly sensitive methods, levels have been undetectable in many active smokers.

Benzene is another chemical present in tobacco smoke. Its metabolites trans-trans-muconic acid and S-phenylmercapturic acid can be detected in urine to measure benzene uptake; both biomarkers have been found elevated in smokers compared to non-smokers (42).

<u>Aromatic amines, PAHs, and benzene</u> are not exclusively contained in tobacco smoke: they also exist in environmental pollution, diesel exhaust, and as an outcome of many industrial productions. Thus their biomarkers are lacking in specificity towards exposure to tobacco smoke.

<u>Minor tobacco alkaloids.</u> Tobacco contains small amounts of minor alkaloids, such as anabasine, anatabine, and myosmine. As for nicotine, the main tobacco alkaloid, these minor alkaloids are absorbed systemically and can be measured in the urine of smokers and users of smokeless tobacco. The measurement of minor alkaloids is important as a way to quantitate tobacco use when a person is also receiving nicotine from other sources, such as nicotine medications or a non-tobacco nicotine delivery system, for instance, in smoking cessation studies (43).

The above example on tobacco smoke clearly demonstrates that a single environmental exposure can be represented by several biological exposure markers. Choosing the appropriate biomarker depends on several factors including chemical and biological characteristics of the biomarker itself, sources of variation (analytical, population, temporal), and the study design in which the biomarker is to be used.

The future of biomarkers of exposure – the exposome

The term 'exposome,' which encompasses all life-course environmental exposures, was coined to draw attention to the need for methodological developments in exposure assessment (44). It is known that environmental exposures play an important role in many common chronic diseases, yet the advances with regard to molecular epidemiology have been focused mostly on the genome. To some extent this can be explained by the complexity of measuring the exposome, as compared to the genome, due to its highly variable nature. However, more recently omics technologies, including transcriptomics, proteomics, metabolomics, and adductomics, are being applied to detect signatures of environmental exposures and to identify novel exposure markers. For instance, human metabolic phenotype diversity was found to be associated with dietary habits across different ethnic populations (45). This promising result suggests that in the future metabolomics might provide new leads to better individual exposure assessment. The development of adductomics, which measures the full complement of protein adducts, might, however, be more relevant for improving exposure assessment in epidemiological studies, as signals can be highly specific for certain (electrophilic) environmental exposures. Furthermore, given the relatively long half-lives of, for instance, haemoglobin adducts (~3000 hours if adducts are chemically stable), these markers would reflect months of exposure.

Conclusions

Given the potential issues associated with the use of biomarkers in epidemiological studies, it is certainly not a given that biomarkers of exposure always provide the most accurate and precise estimates of true exposure. Although the use of biological markers of exposure can improve the assessment of exposure in epidemiological studies, either by complementing other methods of assessment or by serving as the best method when other methods are absent or less valid, these are not always the most appropriate or valid assessment methods. As such, in addition to assessment of the use of biomarkers, it should be part of the design of any study to also consider 'classic' alternatives for exposure assessment, like personal external exposure measurements and advanced exposure modeling.

In summary, before deciding on a specific biological marker to assess exogenous exposures to investigate a specific hypothesis, there are several factors that should be considered. One should verify that the marker is indeed detectable in human populations and that its kinetics are known. A repeated measures design should be created to evaluate interindividual variation relative to intraindividual variation. In addition, duplicate samples should be included in the design to assess laboratory variation. Furthermore, the timing of sample collection in combination with the biological half-life of the biomarker should be optimized. The effect modifiers should be known and all major sources of variance quantified.

References

1. Lauwerys RR, Hoet P. Industrial chemical exposure: guideline for biological monitoring. 3rd ed. Boca Raton (FL): CRC Press, 2001.

2. Gropper SS, Smith JL, Groff JL. Advanced nutrition and human metabolism. 5th ed. Florence (KY): Wadsworth Publishing Company, 2008.

3. Nieuwenhuijsen MJ. Exposure assessment. In: Wild C, Vineis P, Garte S, editors. Molecular epidemiology of chronic diseases. Hoboken (NJ): John Wiley & Sons, Ltd.; 2008. p. 83–96.

4. Armstrong BK, White E, Saracci R. Principles of exposure measurements in epidemiology. Oxford: Oxford University Press; 1992.

5. White E, Armstrong BK, Saracci R. Principles of exposure measurement in epidemiology: collecting, evaluating and improving measures of disease risk factors. 2nd ed. Oxford: Oxford University Press; 2008.

6. Rappaport SM, Kupper LL. Some implications of random exposure measurement errors in occupational and environmental epidemiology. In: Wild C, Vineis P, Garte S, editors. Molecular epidemiology of chronic diseases. Hoboken (NJ): John Wiley & Sons, Ltd.; 2008. p. 223–231.

7. Snijders T, Bosker R. Multilevel analysis. London: SAGE Publications Ltd; 1999.

8. Tielemans E, Kupper LL, Kromhout H et al. (1998). Individual-based and group-based occupational exposure assessment: some equations to evaluate different strategies. Ann Occup Hyg, 42:115–119. PMID:9559571

9. Manini P, De Palma G, Mutti A (2007). Exposure assessment at the workplace: implications of biological variability. Toxicol Lett, 168:210–218.doi:10.1016/j.toxlet.2006.09.014 PMID:17157456

10. Bartsch H (2000). Studies on biomarkers in cancer etiology and prevention: a summary and challenge of 20 years of interdisciplinary research. Mutat Res, 462:255–279.doi:10.1016/S1383-5742(00)00008-9 PMID:10767637

11. Lin YS, Kupper LL, Rappaport SM (2005). Air samples versus biomarkers for epidemiology. Occup Environ Med, 62:750–760.doi:10.1136/oem.2004.013102 PMID:16234400

12. Rappaport SM, Symanski E, Yager JW, Kupper LL (1995). The relationship between environmental monitoring and biological markers in exposure assessment. Environ Health Perspect, 103 Suppl 3;49–53.doi:10.2307/3432560 PMID:7635112

13. De Bont R, van Larebeke N (2004). Endogenous DNA damage in humans: a review of quantitative data. Mutagenesis, 19:169–185.doi:10.1093/mutage/geh025 PMID:15123782

14. Kang D, Lee KH, Lee KM et al. (2005). Design issues in cross-sectional biomarkers studies: urinary biomarkers of PAH exposure and oxidative stress. Mutat Res, 592:138–146. PMID:16102785

15. Aitio A (2006). Guidance values for the biomonitoring of occupational exposure. State of the art. Med Lav, 97:324–331. PMID: 17017366

16. Watson WP, Mutti A (2004). Role of biomarkers in monitoring exposures to chemicals: present position, future prospects. Biomarkers, 9:211–242.doi:10.1080/13547500400015642 PMID:15764289

17. Garcia-Closas M, Vermeulen R, Sherman ME et al. Application of biomarkers in cancer epidemiology. In: Schottenfeld D, Fraumeni JF Jr, editors. Cancer epidemiology and prevention. 3rd ed. New York (NY): Oxford University Press; 2006. p. 70–88.

18. Jakubowski M, Trzcinka-Ochocka M (2005). Biological monitoring of exposure: trends and key developments. *J Occup Health*, 47:22–48.doi:10.1539/joh.47.22 PMID:15703450

19. Bartell SM, Griffith WC, Faustman EM (2004). Temporal error in biomarker-based mean exposure estimates for individuals. *J Expo Anal Environ Epidemiol*, 14:173–179. doi:10.1038/sj.jea.7500311 PMID:15014548

20. Symanski E, Greeson NM (2002). Assessment of variability in biomonitoring data using a large database of biological measures of exposure. *AIHA J (Fairfax, Va)*, 63:390–401.doi:10.1080/15428110208984727 PMID:12486772

21. Nieuwenhuijsen MJ, Droz P. Biological monitoring. In: Nieuwenhuijsen MJ, editor. Exposure assessment in occupational and environmental epidemiology. Oxford: Oxford University Press; 2008. p. 167–80.

22. White E. Effects of biomarker measurement error on epidemiological studies. In: Toniolo P, Boffetta P, Shuker DEG et al., editors. Application of biomarkers in cancer epidemiology. Lyon: IARC Scientific Publication; 1997. p. 73–93.

23. Al-Delaimy WK, Willett WC (2008). Measurement of tobacco smoke exposure: comparison of toenail nicotine biomarkers and self-reports. *Cancer Epidemiol Biomarkers Prev*, 17:1255–1261.doi:10.1158/1055-9965. EPI-07-2695 PMID:18483348

24. Baris D, Kwak LW, Rothman N et al. (2000). Blood levels of organochlorines before and after chemotherapy among non-Hodgkin's lymphoma patients. *Cancer Epidemiol Biomarkers Prev*, 9:193–197. PMID:10698481

25. Wacholder S (1991). Practical considerations in choosing between the case-cohort and nested case-control designs. *Epidemiology*, 2:155–158.doi:10.1097/00001648-199103000-00013 PMID:1932316

26. U.S. Department of Health and Human Services. The health consequences of smoking: a report of the Surgeon General. [Atlanta, GA]. U.S. Department of Health and Human Services, Centers for Disease Control and Prevention, National Center for Chronic Disease Prevention and Health Promotion, Office on Smoking and Health; 2004.

27. U.S. Department of Health and Human Services. The health consequences of involuntary exposure to tobacco smoke: a report of the Surgeon General. [Atlanta, GA]. U.S. Department of Health and Human Services, Centers for Disease Control and Prevention, National Center for Chronic Disease Prevention and Health Promotion, Office on Smoking and Health; 2006.

28. Scherer G (2006). Carboxyhemoglobin and thiocyanate as biomarkers of exposure to carbon monoxide and hydrogen cyanide in tobacco smoke. *Exp Toxicol Pathol*, 58:101–124.doi:10.1016/j.etp.2006.07.001 PMID:16973339

29. Stepanov I, Hecht SS, Lindgren B et al. (2007). Relationship of human toenail nicotine, cotinine, and 4-(methylnitrosamino)-1-(3-pyridyl)-1-butanol to levels of these biomarkers in plasma and urine. *Cancer Epidemiol Biomarkers Prev*, 16:1382–1386. doi:10.1158/1055-9965.EPI-07-0145 PMID:17627002

30. Benowitz NL (1999). Biomarkers of environmental tobacco smoke exposure. *Environ Health Perspect*, 107 Suppl 2;349–355.doi:10.2307/3434427 PMID:10350520

31. Boffetta P, Clark S, Shen M et al. (2006). Serum cotinine level as predictor of lung cancer risk. *Cancer Epidemiol Biomarkers Prev*, 15:1184–1188.doi:10.1158/1055-9965. EPI-06-0032 PMID:16775179

32. Hecht SS (2004). Tobacco carcinogens, their biomarkers and tobacco-induced cancer. Nat Rev Cancer 2003;3(10):733–44. Erratum: Nat Rev Cancer, 4:84.

33. Carmella SG, Akerkar S, Hecht SS (1993). Metabolites of the tobacco-specific nitrosamine 4-(methylnitrosamino)-1-(3-pyridyl)-1-butanone in smokers' urine. *Cancer Res*, 53:721–724. PMID:8428352

34. Richie JP Jr, Carmella SG, Muscat JE et al. (1997). Differences in the urinary metabolites of the tobacco-specific lung carcinogen 4-(methylnitrosamino)-1-(3-pyridyl)-1-butanone in black and white smokers. *Cancer Epidemiol Biomarkers Prev*, 6:783–790. PMID:9332760

35. Hecht SS, Chung FL, Richie JP Jr et al. (1995). Effects of watercress consumption on metabolism of a tobacco-specific lung carcinogen in smokers. *Cancer Epidemiol Biomarkers Prev*, 4:877–884. PMID:8634661

36. Taioli E, Garbers S (1997). Bradlow, Carmella SG, Akerkar S, Hecht SS. Effects of indole-3-carbinol on the metabolism of 4-(methylnitrosamino)-1-(3-pyridyl)-1-butanone (NNK) in smokers. *Cancer Epidemiol Biomarkers Prev*, 6:517–522. PMID:9232339

37. Hecht SS (2002). Human urinary carcinogen metabolites: biomarkers for investigating tobacco and cancer. *Carcinogenesis*, 23:907–922.doi:10.1093/carcin/23.6.907 PMID:12082012

38. Milunsky A, Carmella SG, Ye M, Hecht SS (2000). A tobacco-specific carcinogen in the fetus. *Prenat Diagn*, 20:307–310.doi:10.1002/(SICI)1097-0223(200004)20:4<307::AID-PD797>3.0.CO;2-M PMID:10740203

39. Riedel K, Scherer G, Engl J et al. (2006). Determination of three carcinogenic aromatic amines in urine of smokers and nonsmokers. *J Anal Toxicol*, 30:187–195. PMID:16803653

40. Phillips DH (2002). Smoking-related DNA and protein adducts in human tissues. *Carcinogenesis*, 23:1979–2004.doi:10.1093/carcin/23.12.1979 PMID:12507921

41. Castelao JE, Yuan JM, Skipper PL et al. (2001). Gender- and smoking-related bladder cancer risk. *J Natl Cancer Inst*, 93:538–545. doi:10.1093/jnci/93.7.538 PMID:11287448

42. Hatsukami DK, Hecht SS, Hennrikus DJ et al. (2003). Biomarkers of tobacco exposure or harm: application to clinical and epidemiological studies. 25–26 October 2001, Minneapolis, Minnesota. *Nicotine Tob Res*, 5:387–396.doi:10.1080/1462220031000094222 PMID:12791522

43. Benowitz NL, Hukkanen J, Jacob P 3rd (2009). Nicotine chemistry, metabolism, kinetics and biomarkers. *Handb Exp Pharmacol*, 192:29–60.doi:10.1007/978-3-540-69248-5_2 PMID:19184645

44. Wild CP (2005). Complementing the genome with an "exposome": the outstanding challenge of environmental exposure measurement in molecular epidemiology. *Cancer Epidemiol Biomarkers Prev*, 14:1847–1850. doi:10.1158/1055-9965.EPI-05-0456 PMID:16103423

45. Holmes E, Loo RL, Stamler J et al. (2008). Human metabolic phenotype diversity and its association with diet and blood pressure. *Nature*, 453:396–400.doi:10.1038/nature06882 PMID:18425110

UNIT 3.
ASSESSING EXPOSURE TO THE ENVIRONMENT

CHAPTER 10.

Infectious agents

François Coutlée and Eduardo L. Franco

Summary

The detection and characterization of microbial agents in biological specimens are essential for the investigation of disease outbreaks, for epidemiologic studies of the clinical course of infections, and for the assessment of the role of infectious agents in chronic diseases. Methodological approaches depend on the infectious agent, the specimens analysed and the target populations. Although the diagnosis of infectious diseases has traditionally relied on direct microscopic examination of samples and on the cultivation of microbial agents *in vitro*, novel techniques with increased sensitivity and specificity are now being used on samples that can be more easily collected and transported to microbiology laboratories (e.g. dried blood spots on filter paper for nucleic acid analysis). Direct detection techniques include the microscopic examination of specimens with special stains, antigen detection and nucleic acid detection by molecular assays. These assays are highly sensitive and provide rapid results for most agents. Genomic amplification assays greatly increase the sensitivity of nucleic acid-based tests by extensive amplification of specific nucleic acid sequences before detection. Real-time polymerase chain reaction (PCR) permits genomic amplification concurrently with detection of amplified products. Typing infectious agents requires additional investigation employing either serologic techniques to identify unique antigenic epitopes, or molecular techniques. These studies are important for epidemiologic purposes, as well as for the investigation of pathogenesis, disease progression, and to establish causality between a disease and a microbial agent. Much of bacteriology has relied on growth of organisms on artificial media, and on identification of bacterial growth with biochemical, serological, or more recently, nucleic acid-based tests. The detection of specific antibodies from the host directed against pathogens is another strategy to identify current or past infections.

Introduction

The detection and characterization of microbial agents in biological fluids are important for the prevention, control, and management of infectious diseases in clinical medicine, for the investigation of infectious outbreaks, for large-scale studies on the epidemiology of infectious agents, and for the assessment of the role of infectious agents in chronic diseases. Several approaches have been developed to attain these objectives depending on the nature of the infectious agent, the type of specimens available for analysis, the and populations evaluated (Table 10.1). Although the diagnosis of infectious diseases has traditionally relied on the direct microscopic examination of samples, and on the cultivation of microbial agents in *in vitro* systems, novel techniques with increased sensitivity and specificity are now being used on samples that can be more easily collected and transported to microbiology laboratories, such as the use of dried blood spots on filter paper for nucleic acid analysis. This chapter provides an overview of the approaches used to detect and identify infectious agents, investigate their relatedness, and characterize novel infectious agents in biological fluids. A comprehensive and detailed overview of available diagnostic molecular techniques by target agent is beyond the scope of this chapter, though the interested reader can find a wealth of detailed information on specific methods in several diagnostic microbiology and molecular epidemiology textbooks (1–3). Methods for the detection of microorganisms are classified into three categories: 1) direct detection techniques, 2) *in vitro* cultivation systems and 3) indirect detection based on serological methods that assess the host immune response against a putative infectious agent. This chapter also reviews the causal criteria for assessing the putative role of an infectious agent, and a chronic disease such as cancer, and the epidemiologic pitfalls due to measurement error inherent in diagnostic testing for an infectious agent.

Table 10.1. Methods for the detection and analysis of microbial agents in biological fluids

1.		**Direct detection of microbial agents**
	A.	Microscopic examination of specimens
	B.	Microbial antigen detection
	C.	Microbial nucleic acid detection
	D.	Promising techniques: real-time PCR and matrix hybridization
2.		*In vitro* **cultivation systems of microbial agents**
	A.	Non-cellular cultivation assays
	B.	Cell culture systems
3.		**Serological diagnosis of infectious diseases**
	A.	Screening assays
	B.	Confirmatory assays

Direct detection techniques

Microscopic examination

Microorganisms are often directly detected in biological fluids by special stains, such as the Gram stain or acridine orange for bacteria; mycobacterial stains, based on the ability of mycobacteria to retain dyes after treatment with alcohol-acid decoloriser; nocardia stains; and calcofluor white for fungi. Wet mounts are used for detection of fungi or parasites. Potassium hydroxide is often added to better visualize yeast or hyphal structures. Enteric parasitic infections can be diagnosed by detection of ova in stools. Classically, the viral agents responsible for gastroenteritis are not detectable by cell culture, but usually are with electron microscopy. This is especially the case for the investigation of outbreaks caused by noroviruses. There are several limitations to these direct techniques. Although rapid and easy to perform, they are often insensitive and non-specific. Electron microscopy is a costly procedure that requires the availability of an electron microscope and expertise in specimen processing and interpretation of results. Detection of a virus does not equate to active infection, as some individuals may simply shed a virus without active disease. Electron microscopy can, however, detect unsuspected or unknown agents (e.g. Severe Acute Respiratory Syndrome (SARS) coronavirus agent). Direct detection of microorganisms can also be accomplished histopathologically or cytologically by visualization of the pathogen itself with general-purpose stains, such as periodic acid-Schiff stain, or stains for substances produced by or contained in the pathogen, such as methenamine silver stain.

Detection of antigens from infectious agents

Simplified antigen detection assays are commonly used in diagnostic laboratories for a variety of microorganisms, including bacteria (e.g. group A streptococcus) or bacterial toxins (e.g. *Clostridium difficile*, enterohaemorrhagic *Escherichia coli*), viruses (e.g. varicella-zoster virus, influenza, rotavirus), fungi (e.g. *Pneumocystis jirovecii, Cryptococcus neoformans*) and parasites (e.g. *Toxoplasma gondii*). Immunoassays involve the specific non-covalent binding of a microbial antigen to an antibody that is detected by a labelled ligand. Since antigen and antibody can react under a wide range of conditions, these assays can be applied in most biological fluids, including cerebrospinal fluid, stools, serum, respiratory secretions or urine. Several assay formats have been used (Table 10.2). The most widespread assays use antibodies that are fixed to a solid phase to separate unbound from bound antigens.

Direct or indirect immunofluorescence assays use an antiserum conjugated to a fluorochrome dye, fluorescein isothiocyanate. Some of these assays have been proven to reach sensitivity endpoints that are clinically relevant. Fluorescent staining assays are always subject to technical concerns and require expertise. The quality of specimens tested and sensitivity of the fluorescent assays are improved by including a cytocentrifugation step. The use of multiple antisera, directed against various infectious agents, is also an advantage of direct immunofluorescence (antisera pools).

Membrane EIAs have the advantage of being simple assays that can be performed rapidly, do not require expertise or special equipment, and allow analysis of one specimen at a time. However, sensitivity is usually inferior to cell culture, and weak positive samples may be difficult to interpret. The same comments apply to latex agglutination techniques that have the added disadvantage of false-negative reactions due to prozone effects. The quality of these tests can be evaluated with periodic quality control panels.

The greatest drawback of the detection of microbial antigens is the limited sensitivity of these assays (e.g. *Legionella* fluorescent assays). Moreover, antigens can be degraded in clinical specimens, causing false-negative results. Intracellular antigens may not be detected as easily with these assays. Cross-reactivity, or high background staining in cellular material examined microscopically, can also generate false-positive results. The specificity of the antibody is influenced by the presence of non-microbial antigenic determinants co-purifying with microbial antigens during the steps of antigen production. Non-specific reactions can also be mediated by the antibody's F(c) portion, which can react with rheumatoid factor-like molecules in serum and some biological fluids. Finally, there may be cross-reactivity with other related organisms. Interestingly, direct immunofluorescence assays have proven to be more sensitive than cell culture (see below) for some enveloped viruses that are susceptible to adverse transportation conditions. Microbial antigens can also be detected in biopsies using immunofluorescence or immunoperoxidase reagents.

Detection of nucleic acid from infectious agents

The limitations of traditional direct (see above) and cultivation methods (see below) have provided the impetus for the development and validation of new methodologies based on the detection and analysis of nucleic acids contained in biological samples. Nowadays, these assays are highly sensitive, provide rapid results for most microbial agents, and can be partially automated. The unique sequence specificity of DNA from a given species permits the design of assays that are highly specific to a target agent. Performance of the test is not affected by death of the organism due to antimicrobial therapy, or to transportation or storage of specimens under suboptimal conditions (provided that extreme conditions that can affect DNA integrity can be avoided). DNA is one of the most stable and chemically resistant biological molecules in nature. Intracellular

Table 10.2. Most common assay formats for detection of microbial antigens in biological fluids

Assay Format	Time for completion	Potential for automation	Control of cellularity
Microtiterplate EIA	<4 hours	+	-
Particle agglutination	15 minutes	-	-
Membrane immunoassays	15 minutes	-	-
Direct or indirect IF	30-60 minutes	-	+

EIA, enzyme immunoassay; IF, immunofluorescence.

organisms can be detected after a DNA extraction treatment of samples. These techniques can also detect organisms involved in latent infections. Nucleic acid-based assays have been able to demonstrate integration into the human genome of viral DNA (e.g. hepatitis B virus (HBV) and human papillomavirus (HPV)). These assays can be quantitative or detect microbial mRNA, permitting the analysis of transcriptional activity of a microbial agent, such as for cytomegalovirus. Recombinant DNA technologies, as well as oligonucleotide synthesis strategies, have facilitated the synthesis of large quantities of reagents that can be standardized and quality-controlled more easily than for other diagnostic modalities. The cost of production of reagents has decreased substantially in recent years. Similar protocols for reagent synthesis can be used for different agents, since DNA is the target for all assays irrespective of the agents detected. This is in contrast to immunoassays, for which antibodies are produced by immunization of animals for polyclonal antibodies, and by hybridoma formation for monoclonal antibodies.

Although the use of direct nucleic acid assays was initially impeded by a lack of sensitivity, since the 1990s genomic amplification techniques have revolutionized diagnostic microbiology. The reliance on radiolabelled probes, initially to detect specific nucleic acid sequences (Figure 10.1), was also an important limitation for diagnostic laboratories. Radioactive probes had a short functional half-life and sometimes necessitated prolonged exposure to photographic plates to reach adequate sensitivity. The expense of the facilities required to manipulate radioactive material and dispose of it were significant drawbacks.

Fortunately, non-radioactive labels (i.e. enzyme-labelled probes, avidin-biotin systems, Europium, acridinium esters and others) have been successfully used with these technologies. Assays manufactured by Gen-Probe, with acridinium-labelled single-stranded DNA probes, have been used to detect *Chlamydia*, mycobacterial, fungal or Neisseriaceae rRNA by chemiluminescence. They are now mostly used to identify dimorphic fungi or mycobacterial isolates isolated by culture. Another format of nucleic acid detection tests, the signal amplification assays (branched DNA tests and Hybrid capture), have significantly increased the endpoint sensitivity of DNA-based assays (Figure 10.1). In these assays, the signal is amplified without amplification of the nucleic acid target (see below), thus avoiding carryover contamination and false-positive results. These assays are often less sensitive analytically than genomic amplification techniques, but reach useful clinical sensitivity endpoints (e.g. for human immunodeficiency virus (HIV), hepatitis C virus (HCV) or HPV), and are also highly specific.

Genomic amplification assays have greatly increased the sensitivity of nucleic acid-based tests by extensive amplification of the target nucleic acid sequence before detection (Figure 10.1). Several amplification technologies have been devised, as described in Table 10.3. Of note are those based on thermal cycling amplification, such as the polymerase chain reaction (PCR) (Roche Molecular Systems), and those based on isothermic amplification, such as transcription-mediated amplification (TMA) (Gen-Probe), Nucleic Acid Sequence-Based Amplification (NASBA) (Organon Technika) or strand displacement amplification (SDA) (Becton-Dickinson). PCR is the amplification format that has been the most widely used to develop assays available commercially and as in-house assays. When appropriately optimized and

Figure 10.1. Assay formats for the detection of nucleic acids in biological fluids. 1. Direct detection of nucleic acid (endpoint analytical sensitivity ± 105 copies per test). 2. Signal amplification tests (endpoint analytical sensitivity ± 103 copies per test). 3. Genomic amplification techniques (endpoint analytical sensitivity ± 1 copy per test)

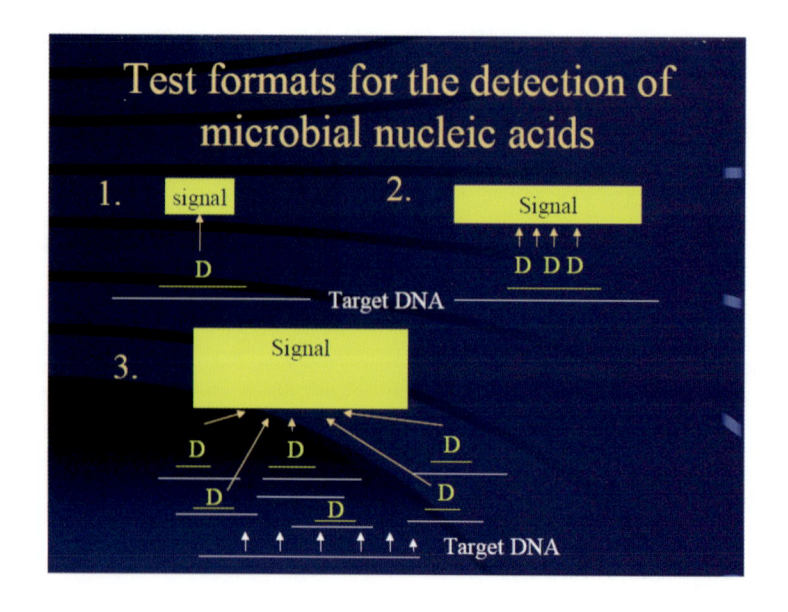

validated, they have consistently proven to be highly sensitive and specific. They can detect pathogens present in low quantities that are slow-growing or cannot be cultivated, and even infectious agents not yet discovered. Multiplex PCR assays can simultaneously detect several pathogens. By adding a reverse transcription step, RNA viruses can be detected with these molecular techniques. The complexity of some viral families (e.g. enteroviruses, HPVs) requires the use of consensus amplification assays to detect all relevant genotypes. These techniques can also be quantitative (Table 10.3). Measures of microbial loads are important information that can be predictive of existing disease, for deciding on initiation of treatment, or assisting the follow-up of treated individuals to assess response or resistance to therapy. For some pathogens, such as mycobacteria, PCR has been used to complement cultivation methods.

The exquisite sensitivity of amplification assays can cause problems in less experienced laboratories. Contamination of reagents by carryover of previously synthesized amplicons can generate false-positive results, but these mishaps can be prevented (Table 10.4). Good laboratory practices, and the use of separate working zones and plugged micropipette tips, effectively curtail the risk of contamination. These techniques are now widely employed without problem in accredited diagnostic laboratories. One limiting step of these assays is the extensive extraction procedures that are sometimes required to analyse samples. Automated extraction instruments resolve this issue in well-equipped laboratories. Finally, inhibitor substances that impede the amplification process can generate false-negative results, but can be screened for by the use of internal controls or amplification of human genes to assess specimen quality.

Table 10.3. Assay formats for amplification of nucleic acids

Assay Format	Manufacturer	Cycling temperature	Qualitative detection	Quantitation
PCR	Roche Molecular Systems	thermal cycling	yes	yes
TMA	Gen-Probe	isothermic	yes	yes
NASBA	Organon Technika/ BioMérieux	isothermic	yes	yes
SDA	Becton-Dickinson	isothermic	yes	no

PCR, polymerase chain reaction; TMA, transcription-mediated amplification; NASBA, nucleic acid sequence-based amplification; SDA, strand displacement amplification

Table 10.4. Most useful procedures to control false-positive results in nucleic acid amplification assays due to contamination

Good laboratory practices (supported by on-site manuals describing standard operating procedures) to prepare samples at perform testing

Separated working zones (pre- and post-PCR areas)

Dedicated instruments for each working zone

Aliquot reagents to avoid repeated use of reagents, especially the master mix

Use of plugged micropipette tips

Avoidance of strongly reactive positive controls

Chemical or enzymatic destruction of contaminating amplicons

New trends in nucleic acid detection tests

Real-time PCR is a new development in the science of genomic amplification. Amplification is performed concurrently with detection in a closed tube, significantly reducing the time to complete the assay and potential for cross-contamination by carryover. Probes are labelled with various fluorophors, and multiple targets can thus be detected. Quantitation of targets is typically done in the logarithmic phase of amplification, which provides for more reproducible measurements of the analyte. The detection of a positive signal is obtained during amplification, as shown in Figure 10.2, significantly shortening the time required to complete testing.

In the future, the use of microarrays will allow the detection of panels of infectious agents that will be selected depending on the disease screened and sample tested. These assays are based on the attachment to solid supports of up to thousands of oligonucleotide probes, generating a matrix of probes. Labelled amplified products are then hybridized with these fixed probes, and the specific signals generated by fixed labelled amplicons can be detected and analysed with a computer. One system utilizes photo-activation for the chemical synthesis of small DNA fragments directly onto a silicate solid support (silicon chip methodology), generating complex arrays of probes. After hybridization of labelled amplicons with these arrays, the silicon surface is screened with a scanning laser confocal fluorescence microscope. These assays can analyse complex mixtures of nucleic acids. The first systems developed using the chip technology successfully analysed HIV resistance to antiretroviral treatment. There are several drawbacks to these techniques, including the complexity of fabrication of the probe arrays, instruments required to perform these tests, and cost. These assays will still require extensive validation before application on populations or cohorts of individuals. These promising techniques are under investigation and could be applied in diagnostic and molecular epidemiologic laboratories in the next decade.

Figure 10.2. Amplification and detection of human DNA by real-time PCR. A titration curve of human DNA was tested and results were plotted in a titration curve (shown above). The rising curves indicate the presence of DNA amplified and detected in the assay. This information is provided online during the performance of the assay. The triplicates for each dilution show excellent reproducibility with curves being almost superimposed. The regression line obtained with these dilutions was excellent, with $r = -1.00$

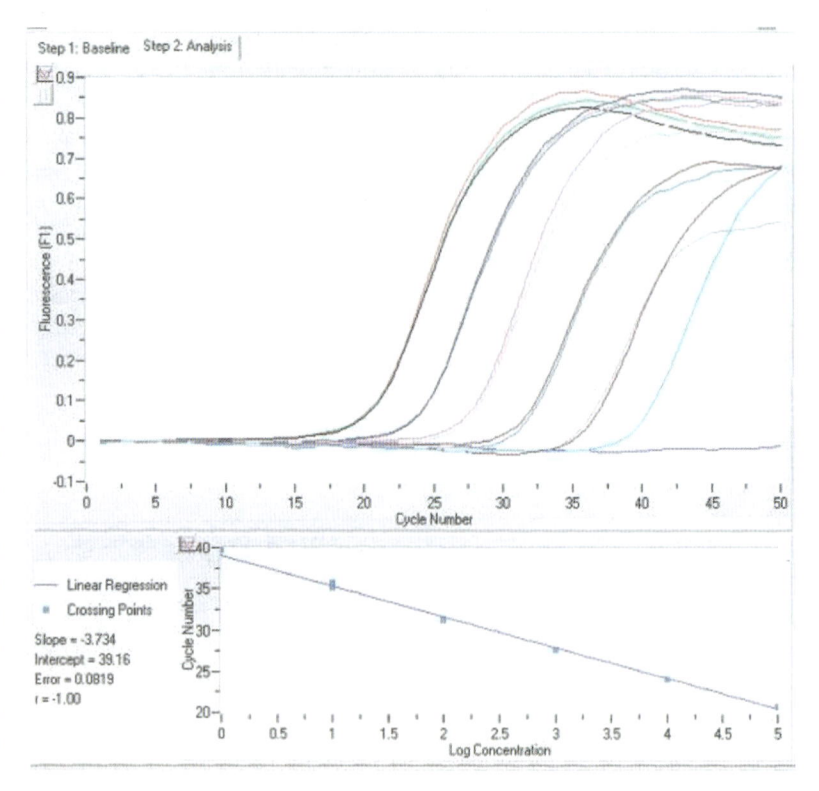

Molecular techniques for genotyping

Typing infectious agents requires additional investigations, which either employ serologic techniques to identify unique antigenic epitopes, or molecular techniques to analyse the microbial genome. These studies are important for epidemiologic purposes, and also for the investigation of pathogenesis, disease progression and the causal association between a disease and a microbial agent. Amplification methods are now the cornerstone for the molecular component of the epidemiologic investigation of infectious diseases, and are replacing phenotypic techniques (e.g. biotyping, susceptibility testing, serotyping, bacteriophage typing, and multilocus enzyme electrophoresis). A variety of DNA-based methods can be used to study the relatedness of different isolates of a species. Non-genomic amplification methods include: bacterial plasmid analyses, restriction endonuclease analysis of bacterial DNA or Southern blot analysis of restriction fragment length polymorphisms (RFLP), and pulsed-field gel electrophoresis (PGFE) of chromosomal DNA. The latter technique includes ribotyping for bacteria. Ribosomal sequences are highly conserved and could react with a wide range of bacterial species. All bacteria carry the ribosomal operons and are thus typeable. Ribotypes are stable, which facilitates the investigation of outbreaks. Variable regions of the microbial genome are ideal targets for these analyses. The study of insertion sequences (e.g. IS6110 DNA sequence for *Mycobacterium tuberculosis*) can also be used to investigate laboratory cross-contamination, identify sources of infection in outbreaks, and assess if a new recurrence is due to the initial organism or to reinfection, or if an infection is caused by multiple isolates.

Several genotyping methods have been adapted to PCR. The amplification step obviates the need for isolating the agents in culture and can be applied directly on samples. PCR-RFLP involves the digestion of PCR-generated amplicons with restriction enzymes, and depending on the various restriction patterns obtained, polymorphism can be studied. This low-cost technique is simple, easy to perform, and can accommodate testing of a large number of samples rapidly. However, only a limited number of DNA sites are analysed. PCR-single stranded conformation polymorphism (SSCP) is a technique in which radiolabelled amplicons are denatured and migrated in a non-denaturing polyacrylamide gel. The conformation of the migrating DNA strand is dependent on the nucleotide sequence of the amplicon, which ultimately affects the migration pattern of the latter. Single nucleotide changes can be detected with this technique. It has the advantage of analysing the complete amplicon, in contrast to PCR-RFLP, but it requires manipulation of radioactive reagents, it is more time-consuming, especially to optimize migration conditions, and it may miss some polymorphisms. Arbitrarily-primed PCR (AP-PCR), or randomly amplified polymorphic DNA (RAPD), is based on the observation that short non-specific primers of 10 nucleotides will hybridize and amplify random DNA sections of chromosomes that differ between genotypes. Since the number and locations of binding sites of short primers will vary, differences between genotypes can be established. Identification of suitable primers may require considerable effort. The technique has been described mainly for bacteria and fungi.

The heteroduplex mobility assay (HMA) is based on the hybridization of PCR amplicons generated from different isolates of a microbial agent. Duplexes containing bulges because of mismatches between amplicon strands from different genotypes will migrate differently during electrophoresis in neutral polyacrylamide gels. The relative retardation of migration is proportional to the DNA distance between genotypes analysed. This simple and rapid technique is limited by its capacity to detect genetic differences of at least 2%. The usefulness of this method was demonstrated in the analysis of viral quasi-species (as for HIV and HCV).

Automated sequencing facilities represent a significant, important improvement in nucleic acid-based tests for genotyping. PCR sequencing determines the nucleotide sequence of microbial DNA, thus permitting identification of the implicated microorganism. It is also essential for phylogenetic analysis and very useful for molecular epidemiology purposes, such as in the investigation of outbreaks or for examining the possible causal role of infectious agents in diseases. It is considered the gold standard method for genotyping. Although still a costly procedure, this has become less of a problem in recent years due to the availability of more affordable instrumentation. Results for each nucleotide position are generated. PCR sequencing does not require knowledge of the pathogen's complete DNA sequence. However, it does generate important quantities of data that must be systematically analysed. The analysis of RNA genomes is further complicated by the existence of several viral species of quasi-species, which add complexity to the genotyping process.

In vitro cultivation systems

Non-cellular cultivation assays

Much of diagnostic microbiology, especially bacteriology, has relied on the growth of organisms on artificial media. Bacterial growth has been identified using biochemical methods with antisera more often using agglutination tests (e.g. latex for β-haemolytic streptococci or whole-organism suspension for *Salmonella* and *Shigella* serogroups), or by nucleic acid-based tests. For most bacteria, cultivation on artificial media is the mainstay of diagnostic microbiology. Enriched all-purpose media, such as blood or chocolate agar, are used to grow common human pathogens. Selective media can also be used to screen for pathogens in the presence of normal microbial flora. Subculture in broth media increases the sensitivity of culture, but decreases its specificity; however, the microorganism is isolated and can be analysed more easily. Antimicrobial susceptibility testing can also be performed and used as an epidemiologic marker on isolated bacterial or fungal isolates. Cultivation techniques are often less sensitive for fastidious organisms, or when patients have started antimicrobial therapy before specimens were obtained for culture. Prolonged periods of incubation may be required for pathogens that grow slowly, such as several mycobacterial species and fungal dimorphic agents. Unfortunately, some key pathogenic bacteria cannot be readily cultivated *in vitro*, such as *Treponema pallidum*, *Mycobacterium leprae*, *Bartonella henselea* and *Tropheryma whippelii*. Recovery of bacterial pathogens in some specimens may be impeded by abundant normal bacterial flora competing for nutrients contained in artificial media. Moreover, pathogens may have similar phenotypes as bacterial agents from the normal flora. For instance, enterotoxin-producing strains of *Escherichia coli* (*E. coli*) that cause diarrhoea are undistinguishable from non-virulent *E. coli* strains.

Cell culture systems

Cell culture allows the detection of a wide range of viruses and the presence of mixed viral pathogens in specimens. After adding a specimen to a monolayer of cells obtained *in vitro*, the presence of a virus in cell cultures can be detected by the distinctive cytopathic effect on cells caused by viral replication (e.g. herpesviruses), by haemadsorption or haemagglutination (e.g. influenza viruses), or with virus-specific fluorescein-labelled antisera (e.g. cytomegalovirus). Viral isolates can be further characterized by molecular techniques for genotyping, antiviral susceptibility testing or immunoreagents for serotyping. However, the requirement for maintenance of several cell lines to support growth of most human viral agents limits cell culture to specialized laboratories. Moreover, propagation of some viruses, such as HIV, represents a significant biohazard for laboratory workers and requires Level-3 containment facilities. Likewise, some cell lines, such as Vero cells, can support the growth of the SARS agent, which represents a considerable biohazard for technologists. The viability of fragile viruses, mostly enveloped viruses, is adversely affected by inadequate transportation and storage conditions. For instance, the rate of positive cultures is lower in summer than in winter months for herpes simplex viruses. Also, Varicella-zoster virus is more frequently detected by direct immunofluorescent tests on samples than by cell culture. Some fastidious viruses do not grow well in cell culture. Furthermore, cell lines are not available for many key human pathogens, including rotavirus, norovirus, hepatitis A virus, HBV, HCV and Epstein-Barr virus (EBV). The delay before a cell culture turns positive is also a limitation of this procedure, as traditionally cell cultures are kept for 7 to 28 days. Shell vial spin amplification, most commonly used for cytomegalovirus and respiratory viruses but also for some fastidious bacteria (e.g. *Bartonella henselae* or *Francisella tularensis*), shortens this delay. In this procedure, specimens are added to a cell culture monolayer in a vial, centrifuged at low speed after inoculation, incubated, and reacted with a fluorescent antibody against viral antigens associated with a replicating virus. Detection of viral agents thus becomes possible before the development of a cytopathic effect.

Indirect detection via serological methods

The detection of specific host antibodies directed against pathogens is another strategy used to identify current or past infections. The detection of antibodies against infectious agents can be performed in serum, as well as cerebrospinal fluid (e.g. arboviruses). The diagnosis of acute infection is usually based on a four-fold increase, or more, of specific antibody titres in paired acute and convalescent sera obtained at two- to four-week intervals (e.g. respiratory viruses), or by the presence of specific immunoglobulin M (IgM) antibodies (e.g. Human parvovirus B19, *Toxoplasma gondii*). Detection of IgM antibodies is less sensitive in immunosuppressed individuals or newborns, however, and it can also be affected by heterologous responses and interference with rheumatoid factor-like molecules in the serum. Direct detection or cultivation methods provide faster results for diagnostic purposes. Chronic infections can be diagnosed by testing a serum to detect specific immunoglobulin G (IgG) antibodies against a preparation of the agent's antigens, such as HCV serology, or a panel of IgG antibodies directed against various microbial antigens that indicate current or resolved infection (HBV or EBV). For example, serologic methods are used for the diagnosis of acute primary EBV infection (by detecting IgG and IgM antibodies against viral capsid antigen, early antigen and Epstein-Barr nuclear antigen), and for the screening of EBV-associated nasopharyngeal carcinoma (mainly immunoglobulin A (IgA) against early and capsid antigens), while molecular techniques are most useful for the diagnosis of EBV-related lymphomas. Serology testing is very valuable for the diagnosis of chronic infections, such as HIV or viral hepatitis. In contrast with direct detection methods, serological testing provides information on past and current infection status of an infectious agent. Serological assays are also adequate tests to evaluate response to vaccination. Serology is most frequently used for the diagnosis of viral infections, but can be valuable in identifying individuals infected with protozoan or metazoan parasites and fungi.

Several techniques have been used to detect antibodies directed against infectious agents, including complement fixation, immunodiffusion, particle (e.g. latex) or erythrocyte agglutination, immunofluorescence, and enzyme immunoassays (also known as enzyme-linked immunosorbent assays (ELISAs)). In a diagnostic microbiology laboratory, most of today's serologic tests are performed with commercially available EIA or immunofluorescence formats. For several agents, screening is performed with EIA tests because of the ease, rapidity and low cost of this assay format. Better purification of viral antigens has resulted in improved sensitivity and specificity of EIAs for viral hepatitis diagnosis. Improved assays were thus designated as second- and third-generation assays. Positive results are then confirmed by a more specific technique that is often more cumbersome and costly. These techniques include recombinant immunoblot assays, radioimmunoprecipitation assays or Western blot assays (e.g. HIV and HCV).

Implicating infections as causes of cancer and other chronic diseases

The operational epidemiologic definition of a cause is a factor that alters the risk of disease occurrence. For infectious diseases, the definition has been more mechanistic: a cause is either a factor that must exist for disease to occur (i.e. is necessary) or always produces disease (i.e. is sufficient). A microbial agent is a necessary, and sometimes sufficient, cause of an infectious disease, depending on the interplay between agent, host and environmental factors. On the other hand, the situation is less clear for cancer: a group of diseases of multifactorial etiology, which ultimately result from the interaction between external environmental causes and the internal genetic makeup of the individual. Few of the accepted causes of human cancer are deemed necessary (e.g. HPV infection in cervical cancer) or sufficient (e.g. possibly some of the high penetrance cancer genes). Unlike most infectious diseases, cancer has a long latency period, which underscores the succession of time-dependent events that are necessary for normal tissue to develop into a malignant lesion and ultimately progress into invasive cancer. Carcinogenesis is a multistage process where final onset of disease is a function of the combined probabilities of relatively rare events occurring in each stage. These events depend on a myriad of factors related to carcinogen absorption and delivery to target cells, metabolic activation, binding with relevant gatekeeper or caretaker genes, and to the ability of the affected tissue to reverse these initiating processes. Also to be considered is the contribution of promoters, which will favour cell proliferation with consequent selection of clones with selective growth advantage within the surrounding tissue. Eventually, other factors will facilitate progression of a precancerous lesion to invasive cancer, and thus also contribute a causal role in carcinogenesis.

Historically, causal relationships in infectious diseases have been assessed using the mechanistically based Henle-Koch's postulates, which are based on the expectation that the microbial agent must be necessary, specific, and sufficient for the disease to occur. These postulates are only of indirect help in assessing cancer or chronic disease etiology, since they imply the causation of the immediate infectious disease or condition that originated from the agent, and not the final malignant process at the end of a lengthy chain of events triggered by the infection itself. A case-in-point is the causal pathway represented by the acquisition of infection with HBV in non-immune individuals, followed by the development of acute hepatitis, chronic hepatitis, and finally, many years later, the onset of hepatocellular carcinoma. Each step in succession affects smaller proportions of patients than the previous. Henle-Koch's postulates are useful up to the first or second steps of this pathway, but are of no guidance for the imputation of a causal link between the pathways' beginning and terminal events.

The reasoning into what constitutes the criteria for judging whether or not a given risk factor is a cause of cancer has primarily evolved from the so-called Bradford Hill criteria (4), a subset of which are referred to as the Surgeon General's criteria (5). These criteria were first proposed at the time of a vigorously debated health issue of the early 1960s, namely the interpretation of the accrued evidence on the role of tobacco smoking as a cause of lung cancer. Hill's nine criteria were: strength of the association, consistency, specificity, temporality, biological gradient, plausibility, coherence, experimental evidence and analogy. In his seminal paper (4), he downplayed the importance of specificity, plausibility and analogy, which are viewed today as non-essential and can even be considered counter-productive distractions to the discussion of any possible cause-effect relationship in cancer. Unfortunately, however, he also concluded that "...none of my nine viewpoints can bring indisputable evidence for or against the cause-and-effect hypothesis and none can be required as a *sine qua non*" (4). If published today, the second part of that statement would have been disputed immediately. Clearly, temporality is a necessary causal criterion, and biological gradient, consistency, and strength of the association are among the most frequently used in cancer risk assessment (reviewed in (6)).

Although highly persuasive in establishing causality, the availability of experimental evidence from randomized controlled trials is more the exception than the rule in public health. In the case of an infectious cause of cancer, one may include the results from vaccine trials of HPV and HBV, as well as post-deployment surveillance of these vaccines in different populations, which have provided strong evidence that these agents are unequivocally causal regarding their respective malignant diseases (i.e. cervical neoplasia) (7,8). (As of this writing, trials have shown a reduction in risk of precancerous lesions only, and not yet of cervical cancer and hepatocellular carcinoma (9)). Typically, the change in prevalence of a disease is observed after the prevalence of a causal determinant has been modified, subsequent to allowing for sufficient latency. More often, epidemiologists derive evidence from observational studies, such as case–control and cohort studies, which are prone to biases in interpretation because of confounding, measurement error (see below), and other issues that preclude isolating the effect of a single factor on causation.

Although useful for environmental, occupational and lifestyle determinants, Hill's criteria do not capture very well the evidential foundation of causal claims for microbial agents and their respective malignant diseases. Fortunately, useful guidelines for causal attributions involving infectious agents have been proposed (10–12). Summarized in Table 10.5, they are correlated with the original criteria formulated by Hill. These causal criteria take into account the knowledge about the timing, specificity and level of immune response against putative viruses, or the advances in nucleic acid detection methodology as used in modern molecular epidemiologic investigations.

In summary, what prevails today is an operational definition of cause, which incorporates the criteria required in different settings. Determining an exposure and intermediate endpoints related to an infectious agent depends on the type of mechanism being studied and its particular set of circumstances (13). Decisions concerning the etiologic role of specific infectious exposures must be a dynamic process that entertains both scientific and public health issues, and is constantly updated as new knowledge from more insightful and valid epidemiologic studies becomes available.

Epidemiologic pitfalls due to measurement error

Epidemiologic common sense has it that improper ascertainment of exposure variables will bias the relative risk (RR) estimates, generally towards the null hypothesis, if the misclassification is random and nondifferential with respect to the outcome (being a case of the disease or not). If the measurement error is not random or nondifferential with respect to the outcome, the direction and degree of the bias are difficult to predict. Although modern molecular methods to determine exposure to infectious agents have attained a substantial degree of accuracy, errors related to sampling, variations in viral load, and other mishaps all contribute to exposure misclassification. The following paragraphs describe the effects of misclassification in specific circumstances typical of epidemiologic studies, which attempt to examine the putative causal role of an infectious agent for a chronic disease such as cancer. In particular, the impact of measurement error on the prevalence of infection in field surveys, and on the magnitude of the association between the infectious agent and cancer in epidemiologic studies, is illustrated. Both of these issues are germane to our interpretation of the putative causal role of an infectious agent in a chronic disease that follows the exposure after a long latency period. (See Chapter 8 for additional discussion of misclassification and measurement error.)

Table 10.5. Criteria or guidelines used in attributing causality to candidate infectious, lifestyle and environmental risk factors* in the genesis of cancer and other chronic diseases of long latency

Role of Environment & Lifestyle		Role of Infections	
Hill (1965)	Evans (1976)	Evans and Mueller (1990)	Fredricks and Relman (1996)
Strength of the association: magnitude of the relative risk for the factor and incident disease or mortality		Presence of viral marker should be higher in cases than in controls	Nucleic acid from agent should be present in most cases and preferentially in affected organs
Consistency: findings are replicated by studies in different populations	IgG and IgM antibodies to the agent regularly appear during illness	Incidence of tumour should be higher in those with the viral marker than in those without it	Molecular or serological evidence should be reproducible
Specificity: Exposure to factor tends to be associated with only one outcome	Antibody to no other agent should be similarly associated with the disease unless a cofactor in its production	Geographic distributions of viral infection and tumour should coincide	Few or no copy numbers should occur in hosts or tissues without disease
Temporality: Onset of exposure should precede with sufficient latency to disease occurrence	Antibody to the agent is regularly absent prior to the disease and exposure to the agent	Appearance of viral marker should precede the tumour	Detection of DNA sequence should predate disease
Biological gradient: Dose–response relation between factor and rate of disease development			Copy number should decrease or become undetectable with disease regression (opposite with relapse or progression)
Plausibility: Does the association make sense in relation to the existing knowledge of likely mechanisms that could be affected by exposure?	Absence of antibody to the agent predicts susceptibility to both infection and the disease produced by the agent		Tissue-sequence correlates should be sought at the cellular level using *in situ* hybridization
	Antibody to the agent predicts immunity to the disease associated with infection by the agent		
Coherence: Does the association conflict with other knowledge about the disease?			
Experimental evidence: Data from randomized controlled trials that eliminate exposure or its consequences		Immunization with the virus should decrease the subsequent incidence of the tumour	
Analogy: Is there a comparable exposure–disease association that seems analogous?			Microorganism inferred by DNA sequence should be consistent with the biological characteristics of that group of organisms

* Or their respective exposure circumstances.

Bias in prevalence surveys

The effect of misclassification on the presumed prevalence of an infectious agent can be understood if the diagnostic performance of the chosen laboratory test is known, particularly its sensitivity (S) and specificity (W) with respect to the true exposure or infection status. The formula (14) to correct for the bias is as follows:

$Pc = (Pu + W - 1)/(S + W - 1)$

where Pc and Pu are the corrected and uncorrected prevalence rates, respectively.

Depending on the true prevalence rate that must be estimated via the test and its diagnostic performance, the estimated rate can be a gross overestimation of the true prevalence rate. For instance, for a rare infectious exposure prevalent among 2.5% of the individuals in the target population, a test with false-negative and false-positive rates of 10% (S = W = 90%) and 20% (S = W = 80%) will be positive 12% and 21.5% of the time in the survey, respectively, thus substantially overestimating the true rate. Under such conditions, the bias always results in overestimation of the prevalence rate and is more influenced by the specificity than by the sensitivity of the assay. Lowering sensitivity has only a moderate biasing effect on the presumed rate.

Bias in the magnitude of the association

As above, if the diagnostic properties of the assay that were used to ascertain exposure to an infectious agent are known, one can correct the estimated measure of the association for the relation between agent and disease. For instance, in a case–control or cross-sectional study, the formula (14) for correcting the odds ratio (OR) is as follows:

$$OR = \frac{(W_1 n_1 - b)(S_2 n_2 - c)}{(W_2 n_2 - d)(S_1 n_1 - a)}$$

where S = sensitivity, W = specificity, n is the number of subjects, and the subscripts 1 and 2 indicate that the information is for cases or controls, respectively. The frequencies a, b, c, and d are the study's 2x2 table frequencies as follows: a = exposed cases, b = unexposed cases, c = exposed controls and d = unexposed controls.

It is possible to simulate the impact of measurement error of an infectious exposure that causes a precursor cancerous lesion, affecting 2.5% of the population after a specified period of time (e.g. high grade cervical intraepithelial neoplasia (HGCIN)) (15). For illustration, assume that the prevalence of the putative agent (i.e. HPV) is 20%, and the underlying RR for the relation with the lesion outcome is 100. Under conditions of perfect measurement of lesion outcome (HGCIN and non-HGCIN), increasing misclassification of HPV status leads to biased estimates of RRs towards unity. For instance, at 10% misclassification (S = W = 90%), the original RR of 100 is erroneously measured as RR = 19. At 30% misclassification, the bias is so severe that the measured RR is just below 4.

In practice, study validity is further aggravated by concomitant misclassification of the outcome, which is a real concern in cohort studies, as for ethical and practical reasons they may have to rely on pre-invasive lesions as endpoints. On the other hand, case–control studies of invasive cancer are far less likely to be affected by outcome misclassification, but are prone to differential exposure misclassification, as detection of the infectious exposure may vary between cases and controls. In the case of HPV infection and cervical cancer, detection of the former is done in exfoliated cervical cells, which results in sampling differences between cases and controls. Moreover, the effects of fluctuation in viral load, transience of HPV infection, and other factors inherent to the dynamics of the infection make single testing for a virus, such as HPV, less likely to represent past exposure for controls than for invasive cancer cases. Capturing the actual exposure experience to HPV that led to cancer would have required sampling the cases' cervix earlier, when the infection was at a comparable state to that of the controls. The biasing effects of these two errors are in the same positive direction away from the null hypothesis (i.e. they produce RRs that are higher than the one truly underlying the relation between HPV and cervical cancer in the same population).

There is one important source of misclassification that cannot be corrected by knowledge of test parameters: it is caused by the biological variation in the ability to detect exposure to the agent over time. Again, the HPV–cervical cancer example is illustrative. Most instances of HPV infection are transient. It is clear, therefore, that collection of a single cervical specimen at the time of enrolment in a cohort study, or at the time of diagnosis of HGCIN, or of invasive cervical cancer in a case–control study, provides little assurance that the laboratory determination of the HPV positivity of that specimen accurately reflects the relevant past exposure to HPV infection that the subject may have had. Infections with low viral load may be labelled erroneously as HPV-negative. A subject with a mildly productive transient infection at the time of testing may be classified as HPV-positive in epidemiologic studies based on single-specimen assessment of exposure, regardless of whether the design is cohort or case–control. Such studies will also attribute exposure status to false-positive specimens resulting from contamination. The latter subjects' non-exposed status can be ascertained with greater validity if one determines a cumulative exposure status based on detection of HPV in multiple specimens collected over time in repeated measurement studies (16).

References

1. Murray PR, Baron EJ, Jorgensen JH et al., editors. Manual of clinical microbiology. 9th ed. Washington (DC): ASM Press; 2007.

2. Mahon CR, Lehman DC, Manuselis G. Textbook of diagnostic microbiology. 3rd ed. St. Louis (MO): Saunders-Elsevier; 2007.

3. Riley LW. Molecular epidemiology of infectious diseases: principles and practices. Washington (DC): ASM Press; 2004.

4. Hill AB (1965). The environment and disease: association or causation? *Proc R Soc Med*, 58:295–300. PMID:14283879

5. Surgeon General's Advisory Committee on Smoking and Health. Smoking and health. Washington (D.C.): U.S. Department of Health, Education and Welfare, Public Health Service, PHS Publication no. 1103; 1964.

6. Weed DL, Gorelic LS (1996). The practice of causal inference in cancer epidemiology. *Cancer Epidemiol Biomarkers Prev*, 5:303–311. PMID:8722223

7. Harper DM, Franco EL, Wheeler CM et al.; HPV Vaccine Study group (2006). Sustained efficacy up to 4.5 years of a bivalent L1 virus-like particle vaccine against human papillomavirus types 16 and 18: follow-up from a randomised control trial. *Lancet*, 367:1247–1255.doi:10.1016/S0140-6736(06)68439-0 PMID:16631880

8. FUTURE II Study Group (2007). Quadrivalent vaccine against human papillomavirus to prevent high-grade cervical lesions. *N Engl J Med*, 356:1915–1927.doi:10.1056/NEJMoa061741 PMID:17494925

9. Montesano R (2002). Hepatitis B immunization and hepatocellular carcinoma: The Gambia Hepatitis Intervention Study. *J Med Virol*, 67:444–446.doi:10.1002/jmv.10093 PMID:12116042

10. Evans AS (1976). Causation and disease: the Henle-Koch postulates revisited. *Yale J Biol Med*, 49:175–195. PMID:782050

11. Evans AS, Mueller NE (1990). Viruses and cancer. Causal associations. *Ann Epidemiol*, 1:71–92.doi:10.1016/1047-2797(90)90020-S PMID:1669491

12. Fredericks DN, Relman DA (1996). Sequence-based identification of microbial pathogens: a reconsideration of Koch's postulates. *Clin Microbiol Rev*, 9:18–33. PMID:8665474

13. Franco EL, Correa P, Santella RM et al. (2004). Role and limitations of epidemiology in establishing a causal association. *Semin Cancer Biol*, 14:413–426.doi:10.1016/j.semcancer.2004.06.004 PMID:15489134

14. Franco EL. Measurement errors in epidemiological studies of human papillomavirus and cervical cancer. In: Muñoz N, Bosch FX, Shah KV, Meheus A, editors. The epidemiology of human papillomavirus and cervical cancer. Oxford: Oxford University Press; 1992. p. 181–197.

15. Franco EL (2000). Statistical issues in human papillomavirus testing and screening. *Clin Lab Med*, 20:345–367. PMID:10863644

16. Franco EL, Rohan TE, Villa LL (1999). Epidemiologic evidence and human papillomavirus infection as a necessary cause of cervical cancer. *J Natl Cancer Inst*, 91:506–511.doi:10.1093/jnci/91.6.506 PMID:10088620

UNIT 3.
ASSESSING EXPOSURE TO THE ENVIRONMENT

CHAPTER 11.

Dietary intake and nutritional status

Jiyoung Ahn, Christian C. Abnet, Amanda J. Cross, and Rashmi Sinha

Summary

Though dietary factors are implicated in chronic disease risk, assessment of dietary intake has limitations, including problems with recall of complex food intake patterns over a long period of time. Diet and nutrient biomarkers may provide objective measures of dietary intake and nutritional status, as well as an integrated measure of intake, absorption and metabolism. Thus, the search for an unbiased biomarker of dietary intake and nutritional status is an important aspect of nutritional epidemiology. This chapter reviews types of biomarkers related to dietary intake and nutritional status, such as exposure biomarkers of diet and nutritional status, intermediate endpoints, and susceptibility. Novel biomarkers, such as biomarkers of physical fitness, oxidative DNA damage and tissue concentrations are also discussed.

Biomarkers of nutritional exposure and nutritional status: An overview

Food frequency questionnaires (FFQ), multiple food records, and 24-hour recalls are the most common methods to assess dietary intake in nutritional epidemiologic studies (1). The strengths and limitations of dietary assessment methods, as well as nutritional status biomarkers, are summarized in Table 11.1. Generally, the accuracy of the information collected depends on the ability to integrate complex eating patterns concisely and the subject's memory. Current dietary assessment methods may not completely capture nutrient interactions and metabolism, as food is a complex mixture; thus, the absorption and metabolism of any single nutrient is affected by the presence of another. For example, iron taken with vitamin C is absorbed more efficiently than by itself, but phytate can bind iron and make it unavailable. Cooking is another important factor that can change concentrations of nutrients or can form compounds not normally present in foods. Obtaining this level of detail using dietary assessment instruments is generally not feasible. Furthermore, food composition tables are not available for all nutrients, limiting the assessment of many of them. Finally, there are numerous nutrients, such as selenium and vitamin D, that cannot be measured adequately in the food source.

Table 11.1. Strengths and limitations of intake assessment methods and nutritional status biomarkers

	Assess by	Limitations	Strengths
Estimate of dietary intake	Questionnaire FFQ Food records 24-hour recall Diet history Food composition table	- Prone to different types of bias - Dependent on memory - May not capture variability in eating pattern - does not account for absorption or bioavailability when foods are cooked or eaten as complex mixtures - Not comprehensive especially for diaries and recalls - Focused on specific nutrient - Many newer dietary compounds of interest not covered	- Easier to administer in population-based studies - Long-term intake estimate
Nutritional status	Biomarker of: - absolute intake - correlated intake	- Collection - Storage - Sensitivity - Specificity - Laboratory variability - Single measure may not be representative of usual	- Objective measure - Error structure different than questionnaire-based information - Integrated measure

Diet and nutrient exposure biomarkers may be independent of the subject's memory or the capacity to describe foods consumed. Biomarkers may provide an integrated measure of intake, absorption and metabolism, which may improve the accuracy of the estimation of the association between the nutrient and disease, but limit the direct interpretation of the connection between intake and disease. However, since biomarkers may be affected by biospecimen collection methods, storage conditions and laboratory variations, these factors must be carefully considered in the study design.

Types of biomarkers

Exposure biomarkers

Biomarkers of absolute intake: Recovery biomarkers

Biomarkers of absolute intake, or recovery biomarkers, reflect a balance between intake and output over a defined period, with relatively high correlation between the absolute dietary intake and the biomarker (> 0.8) (2). The two well-studied recovery biomarkers are urinary nitrogen and doubly-labelled water. Urinary nitrogen is an example of a recovery biomarker of protein intake. A 24-hour urine collection is required, and subjects should take para-aminobenzoic acid (PABA) tablets with the three main meals of the day to validate the completeness of the collection (3). The amount of nitrogen recovered in a 24-hour urine collection can be converted to protein intake using estimates of the percent of nitrogen excreted in urine (~81%). Doubly-labelled water is another example of a recovery biomarker for energy expenditure, in which the average metabolic rate of a human is measured over a period of time. A dose of doubly-labelled water, in which both the hydrogen and the oxygen have been partly or completely replaced for tracking purposes (i.e. labelled) with an uncommon isotope of these elements, is administered to the individual. The loss of deuterium and O-18 is then measured over time by regular sampling of heavy isotope concentrations in the body water (by sampling saliva, urine or blood); the methods used to measure the recovered products are technically challenging (4).

Urinary nitrogen and doubly-labelled water are the only validated recovery biomarkers, but it must still be assumed that the testing period is representative of the subjects' usual habits. The relative complexity and high cost of these methods prevents these biomarkers from being applied to large cohort studies. Thus, these biomarkers are often used as gold standards for validating dietary questionnaires or developing correction factors to estimate measurement attenuation.

Biomarkers of correlated intake: Concentration biomarkers

Biomarkers of correlated intake are based on concentrations in the body (i.e. in blood, urine, saliva, hair, nails or tissue), reflecting current intake status. Concentration biomarkers are correlated with intake, such that higher concentrations of these biomarkers result from higher intake. The measured concentration is a consideration of intake, uptake, and metabolism. Concentration biomarkers can enhance dietary assessment, or in some cases be the primary method of assessment of nutrient exposure.

Nutrients

This type of biomarker may be used to enhance assessment and measurement of dietary components that are currently captured by dietary questionnaires. Vitamin C is thought to protect against oxidative stress, but assessment of intake is complicated by the varying concentration in foods and the widespread and episodic use of vitamin C supplements. Vitamin C is water soluble and responsive to short-term changes in intake; any single measure of vitamin C may not accurately rank subjects' typical exposure. Because the serum or plasma must be stored using metaphosphoric acid or other preservatives, few epidemiologic studies use vitamin C biomarkers (5).

Vitamin E, especially α-tocopherol, has been the focus of a great deal of scrutiny because of its potential benefits in reducing the risk of cancers and cardiovascular diseases (6,7). The correlation of estimates of vitamin E intake from questionnaires with serum concentrations is highly variable, since most dietary vitamin E is obtained from vegetable oils used in cooking (8) and intake of such oils is not estimated well by food frequency questionnaires (FFQs)(9). For example, the correlation between the FFQ-estimated vitamin E intake and serum α-tocopherol ranged from 0.47 in Dutch men to −0.08 in Italian men in the European Prospective Investigation into Cancer and Nutrition (EPIC) (10). Many studies have found an association between serum α-tocopherol levels and chronic disease risk, but not with dietary estimates of vitamin E. For example, in the EPIC study, high serum concentrations of α-tocopherol were associated with significantly lower risks of gastric cancer, but estimated dietary intake of vitamin E was not (11).

Biomarkers as the primary method of assessment. This type of biomarker may be used to measure intake for dietary components that are not currently captured by dietary questionnaires. The selenium content of foods is highly dependent on local soil concentrations, which range over several orders of magnitude. Wheat is an important selenium source in many populations, but the selenium content of wheat can vary considerably; therefore, wheat used to produce flour, bread, pasta and other noodles from different geographic areas can result in variable levels of selenium. Several well-established biomarkers of selenium have been developed, including serum and toenail selenium, which provide a valid estimate of selenium status. More than 20 studies have examined the association between serum, plasma or nail selenium and risk of prostate cancer; a meta-analysis concluded that serum and plasma selenium were consistently lower in cases compared with controls (12). Serum and toenail selenium are common validated biomarkers of selenium status (9).

Iron is another example of a concentration biomarker than may better reflect exposure and provide a more informative assessment of the association between iron and disease than intake estimates. Dietary iron is acquired from plant and animal sources, as well as fortified grain in some countries. There are large differences in the bioavailability and absorption pathways of heme and non-heme iron, suggesting that estimating total iron intake will not give a useful estimate of true exposure. In addition, because menstruation can lead to very different amounts of iron loss in women, the estimation of intake may not be biologically relevant. There are several biomarkers for iron, including serum iron and serum ferritin; both are subject to homeostatic control and influenced by inflammation, respectively.

Vitamin D is a third example of a nutrient that is not well measured by intake estimates. Liver, fatty fish, ergocalciferol in mushrooms, and fortified milk are major dietary sources of vitamin D; however, for most people, the primary source of vitamin D is produced internally upon exposure of the skin to ultraviolet B (UVB). This production depends on the melanin content of the skin and the amount of UVB exposure. Estimating sun exposure is complex, because of differences in time spent outside, amount of exposed skin, weather conditions and sunscreen use. Thus, circulating 25-hydroxy vitamin D is considered a more reliable indicator of vitamin D status, capturing both dietary intake and endogenous production. 25-hydroxy vitamin D has been used in several prospective epidemiologic studies to assess the role of vitamin D in chronic disease prevention (13).

Non-nutritional components

An important aspect of the connection between diet and chronic disease is the assessment of potentially hazardous dietary components. The human diet may contain inadvertent contaminants that are formed during food processing or cooking. Examples of contaminants formed during cooking are heterocyclic amines (HCAs) and polycyclic aromatic hydrocarbons (PAHs). Some developing countries lack an integrated food delivery system that affords the chance to regulate some undesirable food contaminants, such as mycotoxins or by-products of processing (e.g. silica from grinding grain or nitrosamines in salted fish).

Food-cooking by-products. HCAs and PAHs, both known carcinogens in animal models, are formed in the highest concentrations in meat cooked well-done using high-temperature cooking methods, such as pan-frying or grilling. The assessment of exposure to these compounds can be estimated using questionnaires, but may benefit from the use of biomarkers of exposure. Moreover, there is no national database available for food by cooking methods. A limited database, CHARRED, has been created that is based on the type of meat, cooking method and the degree of doneness (http://charred.cancer.gov/).

HCAs are formed from the reaction at high temperatures between creatine or creatinine (found in muscle meats), amino acids, and sugars (14–17). HCAs undergo extensive metabolism by phase I and II enzymes. Various biomarkers of HCAs have been investigated in urine, blood and hair, with each having advantages and limitations.

Urine is a useful biological fluid for the measurement of exposure to various classes of carcinogens, since large quantities may be obtained non-invasively. HCAs are rapidly absorbed from the gastrointestinal tract and eliminated in urine as multiple metabolites, with several percent of the dose present as the unmetabolized parent compound within 24 hours of consuming grilled meats. HCAs in urine have short half-lives, however, and may not be ideal measures of "usual" intake in etiologic studies, especially if there is substantial day-to-day variability.

Figure 11.1. Potential heterocyclic amines (HCAs) biomarkers (71)

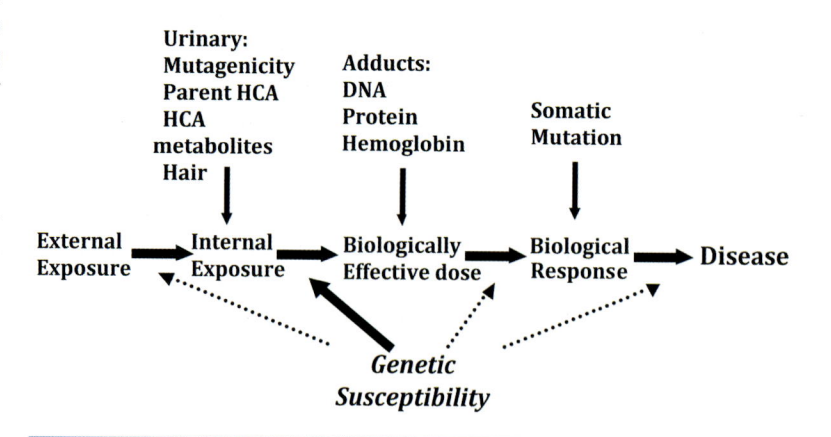

With a large sample size, though, urinary HCAs could still be used to validate intake of HCAs as estimated by questionnaires.

HCA-DNA adducts can be measured in lymphocytes and HCA metabolites bound to circulating blood proteins, such as haemoglobin (Hb) or serum albumin (SA). The measurement of these biomarkers can provide an estimate of exposure and the biologically effective dose, but they do not provide a measure of genetic damage directly in the target tissue. DNA and protein adducts of HCAs have been detected in experimental animal models by ^{32}P-postlabelling. There is a paucity of data on HCA biomarkers in humans, however, as their detection and quantification remains a challenging analytical problem: the concentration of HCAs in the diet is at the parts-per-billion level, and the quantity of HCA biomarkers formed in humans occurs at very low levels. Accumulation of HCAs in human hair, which may serve as a potential long-term biomarker to assess chronic exposure of HCAs, has been suggested but not yet validated (18,19). Similar, but larger, issues exist for PAHs, as these compounds are even more ubiquitous in the food source and environment.

Mycotoxins. Fungal carcinogens are another example of a food contaminant whose study may benefit from the use of an exposure biomarker. Aflatoxin (AFB1) is produced by Aspergillus flavus and other related species, and plays an important role in the high rates of hepatocellular carcinoma seen in southern China and parts of Africa (20). Assessment of AFB1 exposure by questionnaire is very limited, as the amount of infection in grain and the amount of toxin produced varies by locality, crop and storage conditions. In communities where most food is grown and stored at home, the ability to develop general exposure metrics applicable to questionnaire data is minimal. Therefore, the development of biomarkers of exposure to these dietary contaminants is critical.

Extensive work on the metabolism of AFB1 led to the identification of AFB1 adducts with DNA and albumin, including AFB1-DNA adducts in urine (21), as correlates of the effective dose of AFB1. Using urinary AFB1-DNA adducts as a biomarker, a nested case–control study demonstrated a 5-fold increased risk of liver cancer in subjects who had measurable levels of these adducts (22). In the same study, dietary aflatoxin intake was estimated by crossing the concentration of directly measured aflatoxin in food samples and typical intake by questionnaire for each of the contaminated food items. The intake estimates showed no association with risk of liver cancer (22).

Biomarkers of intermediate endpoints

Biomarkers of intermediate endpoints are defined as "...an exogenous substance or first metabolite or the product of an interaction between nutritional exposure and some target molecule or cell that is measured in a compartment within an organism." (23).

Biomarkers of intermediate endpoints have been used extensively to address the association between energy balance and chronic disease. There are multiple serologic indices able to reflect a state of obesity and/or physical activity, including circulating sex steroid and metabolic hormones, as well as inflammatory markers. Serum estradiol levels are higher in obese, compared to lean, post-menopausal women (24). Obesity is also associated with increased levels of adipokines (e.g. leptin and adiponectin), which can be related to insulin resistance (25), characterized by elevated insulin and glucose levels. Inflammatory markers, such as C-reactive protein and adiponectin, are suspected to be on the causal pathway between obesity and chronic disease. Obesity results in excessive production of storage lipids and high circulating levels of glucose, both of which create a proinflammatory oxidative environment (26,27). In addition, individuals who are physically active, after adjustment for body mass index (BMI), have decreased serum estradiol, estrone and androgens (28,29), and male athletes have low testosterone levels (30). Physical activity can improve insulin sensitivity, and thus decrease insulin levels (31). Proinsulin is enzymatically cleaved into insulin and C-peptide in the pancreas (32), and as C-peptide has a longer half-life than insulin (32), it is a better measure of insulin secretion (33–35). Increased physical activity has been associated with a reduction in inflammatory markers in many studies (36–39).

Other examples of biomarkers of intermediate endpoints include blood cholesterols, which are related to risk of cardiovascular disease and also related to saturate fat intake, and blood pressure, which is related to hypertension and also related to sodium intake.

Biomarkers of susceptibility

Humans have a myriad of enzymes that have evolved to maintain cellular homeostasis, including enzymes that metabolize exogenous environmental compounds and nutrients ingested in food. This metabolism allows the utilization of nutrients and the subsequent detoxification and excretion of potentially harmful compounds and metabolites. The genes encoding metabolic enzymes are polymorphically expressed in humans; molecular biology and enzymology studies have shown that there are many polymorphisms that have a functional consequence for the expressed protein. Therefore, the interaction of genetic polymorphisms with consumed nutrients or with foodborne promutagens could serve to modulate diet-influenced disease etiology. Including genetic heterogeneity may provide a better characterization of nutrient exposure and disease risk relationship. Several areas in which genetics may influence relationships between diet and disease risk are described below.

While there is limited evidence that fruit and vegetable intake is associated with risk of breast cancer (40), it is possible that this association may differ according to an individual's genetic profile. This is because fruits and vegetables contain compounds that serve to decrease oxidative load; reactive oxygen species are also endogenously generated or neutralized by numerous enzymes. Studies reported that a reduced breast cancer risk was particularly evident in women who had greater fruit and vegetable consumption and were among a subgroup with genetic variants in catalase (rs1001179) (41) and myeloperoxidase (rs2333227) (42), which is related to higher antioxidant capabilities.

A polymorphism of manganese superoxide dismutase (MnSOD) (rs1799725, Ex2+24T > C) in the mitochondrial targeting sequence results in a change of amino acids that is thought to alter antioxidant capacity. In the Physicians Health Study, investigators found that there was a significant interaction between prostate cancer risk, the MnSOD CC genotype, and low baseline plasma antioxidant levels; those with the CC genotype and low antioxidants had almost a four-fold increased risk of prostate cancer (43). These findings are also replicated in the Prostate, Lung, Colorectal and Ovarian (PLCO) study, where the MnSOD variant genotype was associated with increased risk of prostate cancer, particularly among men with lower intakes of dietary and supplemental vitamin E (44).

The Human Genome Project has opened unprecedented opportunities to comprehensively investigate inherited genetic variations. Ongoing work is exploiting these opportunities through the National Cancer Institute's Cancer Genetic Markers of Susceptibility (CGEMS) to characterize vitamin D/calcium-related pathway genetics and prostate cancer risk in the PLCO trial. Several single nucleotide polymorphisms that predicted serum 25(OH)D concentrations were identified (45). Studies of genetic variants that determine serum micronutrient concentrations, as well as adiposity and height, are ongoing with genome-wide scan data (46–52). Exploiting this information would provide a more coherent measure of chronic disease risk associated with dietary and nutrition exposures. Also, individualising dietary recommendations necessitates a detailed understanding of all genetic and physiological variables that influence the interaction of gene-diet and their relation to disease process. Thus, the potential benefits of understanding the interrelationships between genetic variation and nutrition are enormous.

Novel biomarkers

Physical activity as measured by markers of physical fitness

There has been a great deal of effort to expand studies of energy balance and its role in health and disease using both estimates of caloric intake and BMI. Recently, more attention has been paid to the other side of the energy balance equation, namely physical activity. Assessment of physical activity has primarily used questionnaire assessments of physical activity at work, during leisure time, and increasingly, activities of daily living (housework, etc.). Although physical fitness can be measured using factors such as resting pulse or aerobic capacity, this captures neither the amount of energy expended by a subject nor the amount of low-intensity activities, which may have health benefits (53). Epidemiologists are starting to use accelerometers to accurately capture activity over a test period. However, further work will be required to deploy these devices on a large scale; measurements will be restricted to a small number of days, and for etiologic studies this could only be meaningful in prospective studies.

Alternatively, biomarkers of effect have been examined to explore physical activity hypotheses. For example, one hypothesis for the protective effect of physical activity on breast cancer is the alteration of circulating hormone levels (54). Post-menopausal women with lower serum levels of sex steroid hormones have lower risk of breast cancer. Physical activity may lower these serum concentrations, but whether this is dependent on greater physical activity leading to lower BMI is unclear. Because adipose tissue is an important source of sex steroid hormones in post-menopausal women, determining whether the effect of physical activity on breast cancer risk is independent of BMI will require careful evaluation of serum sex steroid levels to assess the mechanism of action (54). Assessing the other potential mechanisms of action for the association of physical activity and cancer will require the use of molecular markers for inflammation and immunity (53).

Oxidative capacity of diet as measured by markers of oxidative DNA damage

The impact of dietary antioxidants on the incidence of cancer has been widely studied by assessing intake using FFQs or food records, as well as by status biomarkers, such as serum vitamin measures. An alternative biomarker strategy is to measure the amount of oxidative stress in an individual with a biomarker that integrates antioxidant intake, oxidative stress from exogenous and endogenous sources, and individual response to this stress (genetic and epigenetic factors). Oxidative stress can lead to modification of DNA nucleotides. The DNA adduct 7,8-dihydro-8-2'-deoxyguanosine (8-oxodG) has been widely used as a marker of oxidative stress (55). Direct measurement of this DNA adduct in peripheral blood mononuclear cells or urine reflects the sum of oxidative damage and the repair of this damage. Oxidative stress can also lead to the oxidation of thymine and the formation of 5-hydroxymethyl-2'deoxyuridine (HMdU). This adduct is immunogenic, and autoantibodies against it can be monitored as a marker of oxidative stress (56). Several studies have investigated the responsiveness of these markers to dietary intake and modification (57,58). Observational and intervention studies suggest that diet can significantly modify the amount of oxidative stress as measured by the DNA modification markers. Moreover, these markers can be employed directly, or alternatively used in conjunction with a diet/behaviour index. For example, the concentration of 8-oxodG could be measured in a subset of a cohort, and the dietary and other questionnaire data examined to build a predictive model for this oxidative stress marker. If a sufficiently powerful model can be built, it can then be used to examine the association between the index and the disease in the full cohort. This may be a more powerful technique for the integration of antioxidant and prooxidant exposures than is an index that arbitrarily assigns points based on median splits of intake for purportedly antioxidative foods or nutrients.

Exposure status as measured by tissue concentrations

Most previous biomarker studies of nutrient status or carcinogen exposure have used easily accessible biological compartments (e.g. serum, urine, hair or nails) to assess the association between exposure and cancer risk. Recently, the developing interest in molecular epidemiology has provided the impetus to use tissue banks to provide measures of exposure directly in the target tissue.

Large numbers of studies have examined the association between nutrient intake and the risk of disease. Some nutrients such as trace elements or minerals, however, are not amenable to intake estimation. For example, meaningful estimates of zinc intake are difficult because the bioavailability varies strongly with the other dietary constituents in the same meal; phytate from whole grain can almost completely block zinc absorption. Also, serum zinc may not be a sensitive indicator of status, because serum zinc concentration is under tight homeostatic control. Therefore, an alternative method has been devised whereby the concentration of zinc in the target tissue of interest is measured directly (59). This uses a sensitive technique to measure the zinc concentrations in the biopsy tissue directly, thus giving a clearer assessment of the importance of the element in the studied tissue.

An alternative use of tissue biomarkers is to directly assess the exposure to carcinogens that is derived from the diet. Antibodies against the adduct created when activated benzo[a]pyrene interacts with DNA have been used to assess the association between PAH exposure and breast cancer using breast biopsies. Using an immunohistochemical assay, an association was found between PAH-DNA adducts in breast tissue and the risk of breast cancer (60), but this work requires careful interpretation (61).

Studies using target tissue may be restricted to easily or routinely biopsied organs, such as those often biopsied during screening exams or positive exam work-ups (e.g. colon, prostate or breast). The direct assessment of nutrient status or carcinogen exposure in the target tissue may lead to a clearer understanding of the nutrient or exposure in the disease process.

Urinary mutagenicity

A urinary mutagenicity test using Salmonella typhimurium indicator strains (Ames test) has been used to monitor populations occupationally or environmentally exposed to genotoxic compounds (62). Genotoxic compounds in the diet may originate from contaminants in the food chain or from by-products of food preparation; for example, the urine of individuals who consumed well-done meat can be highly mutagenic (63). Mutagenic activity of the urine is substantially increased when the urine is acid-hydrolysed. Mutagenicity of unhydrolysed urine likely reflects excretion of unmetabolized mutagens, whereas the mutagenicity of hydrolysed urine reflects the excretion of both metabolized and unmetabolized mutagens. Other dietary components, such as cruciferous vegetables or parsley, may decrease urinary mutagenicity by enhancing the level of conjugation.

Current challenges and future directions

Dietary assessment

Food frequency questionnaires (FFQs) are the main dietary instrument used by nutritional epidemiologists, but in recent years this method has become controversial. Whether or not it is time to abandon the use of FFQs has been discussed (64,65). The inconsistencies in diet–disease associations observed in epidemiologic studies have been highlighted. Further emphasized were results from a methodologic study that used doubly labelled water as a gold standard for energy intake and urinary nitrogen for protein intake (4). Both energy and protein estimated by the FFQ were measured very poorly. The authors also argue that the associations observed using dietary biomarkers and food diaries are not detectable when FFQs are used (66,67). These assertions have been questioned, and it has been stated that some inconsistencies are to be expected in an area as complex as diet both due to chance and real biological interactions (68). The authors further assert that when large numbers of studies have been pooled the data are consistent, and the ability of the doubly-labelled water study to measure within-person variability has been questioned (68). The association between fat and breast cancer using food records has been seen in two cohort studies, one in the United Kingdom and the other in the Women's Health Initiative (69), in contrast to the null results using a FFQ (70). Further discussion must be undertaken to decide how to estimate dietary intake. For example, are all foods and nutrients substantially misclassified as observed for energy and protein intake? Should we use other forms of dietary instruments or possibly a combination of instruments? It is, however, crucial that dietary intake be estimated with less error if we are to correlate intake to a biomarker. An automated, web-based FFQ or 24-hour recall is currently being developed, and these new tools may help to improve dietary assessment.

Dietary biomarkers

There are several important issues that need consideration when deciding whether to use dietary biomarkers in an epidemiologic study. The application of dietary biomarkers is most likely to be useful in prospective cohort studies, as the biological samples will be collected and stored before the clinical manifestation of the disease.

In general, there are limited types of biospecimens that are easily available and can be used for measuring nutrients such as blood, urine, hair, nail, faeces, saliva, and tissue biopsies. These may not be specimens from the organ or site of interest; for example, fat soluble vitamins are stored in the liver, adipose tissue, cell membrane, and with smaller amounts in blood components.

Sample collection and storage of biospecimens in an appropriate manner is crucial for nutritional biomarkers. Certain nutrients must be collected under specific conditions (e.g. trace mineral-free tubes for zinc). Zinc contamination can be in dust, thus stringent laboratory conditions must be used to measure this mineral in biological samples. Other nutrients need to be stored with preservatives to maintain their integrity, such as vitamin C, which needs to be preserved with metaphosphoric acid. Such stringent control may not be a problem in smaller studies with targeted hypotheses, but it can become a constraint in large, prospective studies with competing interests and limited amounts of biological material.

Measurement of nutritional biomarkers in repeat samples is important for many dietary components that have short half-lives, such as water-soluble vitamins and meat-cooking carcinogens. Therefore, it is preferable to have biological specimens from multiple days to derive an estimate of usual nutritional status. A related issue to a short half-life is the need to collect fasting samples as certain nutrients respond with a postprandial spike for several hours. It is important to take into consideration the metabolism of the nutrient with the study aims and design.

The data generated from the Human Genome Project offer great opportunities to utilize genetic information. This rapidly expanding technology will provide valuable information to help understand disease etiology in a comprehensive way, but it will also provide a formidable challenge in design, analysis and implementation of molecular epidemiology studies. The application of molecular epidemiology to nutrition and disease prevention is in its infancy. Further investigations of diet, genetic variability, and disease risk will better elucidate the complex relationships between diet and disease risk, and support recommendations for healthful eating.

Conclusions

This chapter reviews types of biomarkers related to dietary intake and nutritional status. Exposure biomarkers include biomarkers of absolute intake and correlates of intake. Absolute intake biomarkers are often thought of as a gold standard to validate dietary questionnaires, with relatively high correlation with dietary intake. Concentration biomarkers reflect intake status, with moderate correlations in relation to nutritional (e.g. serum vitamin D, serum vitamin E, and toenail selenium) or non-nutritional components (e.g. food-cooking by-products, and mycotoxins). Intermediate biomarkers of biologic effect have been used extensively to address the association of surrogate endpoints of nutritional exposures. Biomarkers of susceptibility often include genetic heterogeneity, which may help better characterize nutrient exposure and disease–risk relationship. Novel biomarkers, such as biomarkers of physical fitness, oxidative DNA damage, and urinary mutagenicity, have been developed in this rapidly growing field. As analytic methods improve and more biochemical indicators are validated as measures of dietary intake, their use in nutritional epidemiology is likely to expand.

References

1. Willett W. Nutritional epidemiology. New York (NY): Oxford University Press; 1998.

2. Kipnis V, Midthune D, Freedman L et al. (2002). Bias in dietary-report instruments and its implications for nutritional epidemiology. *Public Health Nutr*, 5 6A;915–923.doi:10.1079/PHN2002383 PMID:12633516

3. Bingham SA (1994). The use of 24-h urine samples and energy expenditure to validate dietary assessments. *Am J Clin Nutr*, 59 Suppl;227S–231S. PMID:8279431

4. Subar AF, Kipnis V, Troiano RP et al. (2003). Using intake biomarkers to evaluate the extent of dietary misreporting in a large sample of adults: the OPEN study. *Am J Epidemiol*, 158:1–13.doi:10.1093/aje/kwg092 PMID:12835280

5. Galan P, Briançon S, Favier A et al. (2005). Antioxidant status and risk of cancer in the SU.VI.MAX study: is the effect of supplementation dependent on baseline levels? *Br J Nutr*, 94:125–132.doi:10.1079/BJN20051462 PMID:16115341

6. Traber MG, Frei B, Beckman JS (2008). Vitamin E revisited: do new data validate benefits for chronic disease prevention? *Curr Opin Lipidol*, 19:30–38. PMID:18196984

7. Huang HY, Caballero B, Chang S et al. (2006). The efficacy and safety of multivitamin and mineral supplement use to prevent cancer and chronic disease in adults: a systematic review for a National Institutes of Health state-of-the-science conference. *Ann Intern Med*, 145:372–385. PMID:16880453

8. McLaughlin PJ, Weihrauch JL (1979). Vitamin E content of foods. *J Am Diet Assoc*, 75:647–665. PMID:389993

9. Institute of Medicine. Dietary reference intakes for vitamin C, vitamin E, selenium, and carotenoids. Washington (DC): National Academy Press; 2000.

10. Kaaks R, Slimani N, Riboli E (1997). Pilot phase studies on the accuracy of dietary intake measurements in the EPIC project: overall evaluation of results. European Prospective Investigation into Cancer and Nutrition. *Int J Epidemiol*, 26 Suppl 1;S26–S36.doi:10.1093/ije/26.suppl_1.S26 PMID:9126531

11. Jenab M, Riboli E, Ferrari P et al. (2006). Plasma and dietary carotenoid, retinol and tocopherol levels and the risk of gastric adenocarcinomas in the European prospective investigation into cancer and nutrition. *Br J Cancer*, 95:406–415.doi:10.1038/sj.bjc.6603266 PMID:16832408

12. Brinkman M, Reulen RC, Kellen E et al. (2006). Are men with low selenium levels at increased risk of prostate cancer? *Eur J Cancer*, 42:2463–2471.doi:10.1016/j.ejca.2006.02.027 PMID:16945521

13. Institute of Medicine. Dietary reference intakes for calcium, phosphorus, magnesium, vitamin D, and fluoride. Washington (DC): National Academies Press; 1999.

14. Wakabayashi K, Nagao M, Esumi H, Sugimura T (1992). Food-derived mutagens and carcinogens. *Cancer Res*, 52 Suppl;2092s–2098s. PMID:1544146

15. Sugimura T, Wakabayashi K, Nakagama H, Nagao M (2004). Heterocyclic amines: Mutagens/carcinogens produced during cooking of meat and fish. *Cancer Sci*, 95:290–299.doi:10.1111/j.1349-7006.2004.tb03205.x PMID:15072585

16. Sugimura T, Wakabayashi K (1991). Heterocyclic amines: new mutagens and carcinogens in cooked foods. *Adv Exp Med Biol*, 283:569–578. PMID:2069025

17. Nagao M, Fujita Y, Wakabayashi K, Sugimura T (1983). Ultimate forms of mutagenic and carcinogenic heterocyclic amines produced by pyrolysis. *Biochem Biophys Res Commun*, 114:626–631.doi:10.1016/0006-291X(83)90826-4 PMID:6349633

18. Kobayashi M, Hanaoka T, Hashimoto H, Tsugane S (2005). 2-Amino-1-methyl-6-phenylimidazo[4,5-b]pyridine (PhIP) level in human hair as biomarkers for dietary grilled/stir-fried meat and fish intake. *Mutat Res*, 588:136–142. PMID:16289877

19. Alexander J, Reistad R, Hegstad S et al. (2002). Biomarkers of exposure to heterocyclic amines: approaches to improve the exposure assessment. *Food Chem Toxicol*, 40:1131–1137.doi:10.1016/S0278-6915(02)00053-4 PMID:12067575

20. Groopman JD, Kensler TW (1999). The light at the end of the tunnel for chemical-specific biomarkers: daylight or headlight? *Carcinogenesis*, 20:1–11.doi:10.1093/carcin/20.1.1 PMID:9934843

21. Groopman JD, Roebuck BD, Kensler TW (1992). Molecular dosimetry of aflatoxin DNA adducts in humans and experimental rat models. *Prog Clin Biol Res*, 374:139–155. PMID:1320272

22. Qian GS, Ross RK, Yu MC et al. (1994). A follow-up study of urinary markers of aflatoxin exposure and liver cancer risk in Shanghai, People's Republic of China. *Cancer Epidemiol Biomarkers Prev*, 3:3–10. PMID:8118382

23. IARC (1997). Proceedings of the workshop on application of biomarkers to cancer epidemiology. Lyon. IARC Scientific Publication; (142):1–318. PMID:9410826

24. Key TJ, Allen NE, Verkasalo PK, Banks E (2001). Energy balance and cancer: the role of sex hormones. *Proc Nutr Soc*, 60:81–89. doi:10.1079/PNS200068 PMID:11310427

25. Roden M, Price TB, Perseghin G et al. (1996). Mechanism of free fatty acid-induced insulin resistance in humans. *J Clin Invest*, 97:2859–2865.doi:10.1172/JCI118742 PMID:8675698

26. Esposito K, Nappo F, Marfella R et al. (2002). Inflammatory cytokine concentrations are acutely increased by hyperglycemia in humans: role of oxidative stress. *Circulation*, 106:2067–2072.doi:10.1161/01.CIR.0000034509.14906.AE PMID:12379575

27. Mohanty P, Hamouda W, Garg R et al. (2000). Glucose challenge stimulates reactive oxygen species (ROS) generation by leucocytes. *J Clin Endocrinol Metab*, 85:29 70–2973.doi:10.1210/jc.85.8.2970 PMID:1094 6914

28. Cauley JA, Gutai JP, Kuller LH et al. (1989). The epidemiology of serum sex hormones in postmenopausal women. *Am J Epidemiol*, 129:1120–1131. PMID:2729251

29. McTiernan A, Wu L, Chen C et al.; Women's Health Initiative Investigators (2006). Relation of BMI and physical activity to sex hormones in postmenopausal women. *Obesity (Silver Spring)*, 14:1662–1677.doi:10.1038/oby.2006.191 PMID:17030978

30. Hackney AC, Szczepanowska E, Viru AM (2003). Basal testicular testosterone production in endurance-trained men is suppressed. *Eur J Appl Physiol*, 89:198–201.doi:10.1007/s00421-003-0794-6 PMID:12665985

31. Regensteiner JG, Shetterly SM, Mayer EJ, et al. (1995). Relationship between habitual physical activity and insulin area among individuals with impaired glucose tolerance. The San Luis Valley Diabetes Study. *Diabetes Care*, 18 (4):0–497.

32. Polonsky K, Frank B, Pugh W et al. (1986). The limitations to and valid use of C-peptide as a marker of the secretion of insulin. *Diabetes*, 35:379–386.doi:10.2337/diabetes.35.4.379 PMID:3514322

33. Bonser AM, Garcia-Webb P, Harrison LC (1984). C-peptide measurement: methods and clinical utility. *Crit Rev Clin Lab Sci*, 19:297–352.doi:10.3109/10408368409165766 PMID:6373142

34. Clark PM (1999). Assays for insulin, proinsulin(s) and C-peptide. *Ann Clin Biochem*, 36:541–564. PMID:10505204

35. Hovorka R, Jones RH (1994). How to measure insulin secretion. *Diabetes Metab Rev*, 10:91–117.doi:10.1002/dmr.5610100204 PMID:7956679

36. Ford ES (2002). Does exercise reduce inflammation? Physical activity and C-reactive protein among U.S. adults. *Epidemiology*, 13:561–568.doi:10.1097/00001648-200209000-00012 PMID:12192226

37. Geffken DF, Cushman M, Burke GL et al. (2001). Association between physical activity and markers of inflammation in a healthy elderly population. Am J Epidemiol, 153:242–250. doi:10.1093/aje/153.3.242 PMID:11157411

38. King DE, Carek P, Mainous AG 3rd, Pearson WS (2003). Inflammatory markers and exercise: differences related to exercise type. Med Sci Sports Exerc, 35:575–581. doi:10.1249/01.MSS.0000058440.28108.CC PMID:12673139

39. Nassis GP, Papantakou K, Skenderi K et al. (2005). Aerobic exercise training improves insulin sensitivity without changes in body weight, body fat, adiponectin, and inflammatory markers in overweight and obese girls. Metabolism, 54:1472–1479. doi:10.1016/j.metabol.2005.05.013 PMID:16253636

40. van Gils CH, Peeters PH, Bueno-de-Mesquita HB et al. (2005). Consumption of vegetables and fruits and risk of breast cancer. JAMA, 293:183–193. doi:10.1001/jama.293.2.183 PMID:15644545

41. Ahn J, Gammon MD, Santella RM et al. (2005). Associations between breast cancer risk and the catalase genotype, fruit and vegetable consumption, and supplement use. Am J Epidemiol, 162:943–952. doi:10.1093/aje/kwi306 PMID:16192345

42. Ahn J, Gammon MD, Santella RM et al. (2004). Myeloperoxidase genotype, fruit and vegetable consumption, and breast cancer risk. Cancer Res, 64:7634–7639. doi:10.1158/0008-5472.CAN-04-1843 PMID:15492293

43. Li H, Kantoff PW, Giovannucci E, et al. (2004). Manganese superoxide dismutase (MnSOD) polymorphism, prediagnostic plasma antioxidants and prostate cancer risk. Proc Am Assoc Cancer Res, 45 (Abstract #2320).

44. Kang D, Lee KM, Park SK et al. (2007). Functional variant of manganese superoxide dismutase (SOD2 V16A) polymorphism is associated with prostate cancer risk in the prostate, lung, colorectal, and ovarian cancer study. Cancer Epidemiol Biomarkers Prev, 16:1581–1586. doi:10.1158/1055-9965.EPI-07-0160 PMID:17646272

45. Ahn J, Albanes D, Berndt SI et al.; Prostate, Lung, Colorectal and Ovarian Trial Project Team (2009). Vitamin D-related genes, serum vitamin D concentrations and prostate cancer risk. Carcinogenesis, 30:769–776. doi:10.1093/carcin/bgp055 PMID:19255064

46. Kathiresan S, Willer CJ, Peloso GM et al. (2009). Common variants at 30 loci contribute to polygenic dyslipidemia. Nat Genet, 41:56–65. doi:10.1038/ng.291 PMID:19060906

47. Tanaka T, Scheet P, Giusti B et al. (2009). Genome-wide association study of vitamin B6, vitamin B12, folate, and homocysteine blood concentrations. Am J Hum Genet, 84:477–482. doi:10.1016/j.ajhg.2009.02.011 PMID:19303062

48. Ahn J, Yu K, Stolzenberg-Solomon R et al. (2010). Genome-wide association study of circulating vitamin D levels. Hum Mol Genet, 19:2739–2745. doi:10.1093/hmg/ddq155 PMID:20418485

49. Weedon MN, Lango H, Lindgren CM et al.; Diabetes Genetics Initiative; Wellcome Trust Case Control Consortium; Cambridge GEM Consortium (2008). Genome-wide association analysis identifies 20 loci that influence adult height. Nat Genet, 40:575–583. doi:10.1038/ng.121 PMID:18391952

50. Gudbjartsson DF, Walters GB, Thorleifsson G et al. (2008). Many sequence variants affecting diversity of adult human height. Nat Genet, 40:609–615. doi:10.1038/ng.122 PMID:18391951

51. Lettre G, Jackson AU, Gieger C et al.; Diabetes Genetics Initiative; FUSION; KORA; Prostate, Lung Colorectal and Ovarian Cancer Screening Trial; Nurses' Health Study; SardiNIA (2008). Identification of ten loci associated with height highlights new biological pathways in human growth. Nat Genet, 40:584–591. doi:10.1038/ng.125 PMID:18391950

52. Frayling TM, Timpson NJ, Weedon MN et al. (2007). A common variant in the FTO gene is associated with body mass index and predisposes to childhood and adult obesity. Science, 316:889–894. doi:10.1126/science.1141634 PMID:17434869

53. Rundle A (2005). Molecular epidemiology of physical activity and cancer. Cancer Epidemiol Biomarkers Prev, 14:227–236. PMID:15668499

54. Campbell KL, McTiernan A (2007). Exercise and biomarkers for cancer prevention studies. J Nutr, 137 Suppl;161S–169S. PMID:17182820

55. Halliwell B (2000). Lipid peroxidation, antioxidants and cardiovascular disease: how should we move forward? Cardiovasc Res, 47:410–418. doi:10.1016/S0008-6363(00)00097-3 PMID:10963714

56. Ashok BT, Ali R (1998). Binding of human anti-DNA autoantibodies to reactive oxygen species modified-DNA and probing oxidative DNA damage in cancer using monoclonal antibody. Int J Cancer, 78:404–409. doi:10.1002/(SICI)1097-0215(19981109)78:4<404::AID-IJC2>3.0.CO;2-Y PMID:9797125

57. Hu JJ, Chi CX, Frenkel K et al. (1999). Alpha-tocopherol dietary supplement decreases titers of antibody against 5-hydroxymethyl-2'-deoxyuridine (HMdU). Cancer Epidemiol Biomarkers Prev, 8:693–698. PMID:10744129

58. Wallström P, Frenkel K, Wirfält E et al. (2003). Antibodies against 5-hydroxymethyl-2'-deoxyuridine are associated with lifestyle factors and GSTM1 genotype: a report from the Malmö Diet and Cancer cohort. Cancer Epidemiol Biomarkers Prev, 12:444–451. PMID:12750240

59. Abnet CC, Lai B, Qiao YL et al. (2005). Zinc concentration in esophageal biopsy specimens measured by x-ray fluorescence and esophageal cancer risk. J Natl Cancer Inst, 97:301–306. doi:10.1093/jnci/dji042 PMID:15713965

60. Rundle A, Tang D, Hibshoosh H et al. (2000). The relationship between genetic damage from polycyclic aromatic hydrocarbons in breast tissue and breast cancer. Carcinogenesis, 21:1281–1289. doi:10.1093/carcin/21.7.1281 PMID:10874004

61. Rundle A, Tang D, Hibshoosh H et al. (2002). Molecular epidemiologic studies of polycyclic aromatic hydrocarbon-DNA adducts and breast cancer. Environ Mol Mutagen, 39:201–207. doi:10.1002/em.10048 PMID:11921190

62. Cerná M, Pastorková A (2002). Bacterial urinary mutagenicity test for monitoring of exposure to genotoxic compounds: a review. Cent Eur J Public Health, 10:124–129. PMID:12298345

63. Peters U, DeMarini DM, Sinha R et al. (2003). Urinary mutagenicity and colorectal adenoma risk. Cancer Epidemiol Biomarkers Prev, 12:1253–1256. PMID:14652290

64. Kristal AR, Peters U, Potter JD (2005). Is it time to abandon the food frequency questionnaire? Cancer Epidemiol Biomarkers Prev, 14:2826–2828. doi:10.1158/1055-9965.EPI-12-ED1 PMID:16364996

65. Kristal AR, Potter JD (2006). Not the time to abandon the food frequency questionnaire: counterpoint. Cancer Epidemiol Biomarkers Prev, 15:1759–1760. doi:10.1158/1055-9965.EPI-06-0727 PMID:17021349

66. Midthune D, Kipnis V, Freedman LS, Carroll RJ (2008). Binary regression in truncated samples, with application to comparing dietary instruments in a large prospective study. Biometrics, 64:289–298. doi:10.1111/j.1541-0420.2007.00833.x PMID:17651458

67. Bingham SA, Luben R, Welch A et al. (2003). Are imprecise methods obscuring a relation between fat and breast cancer? Lancet, 362:212–214. doi:10.1016/S0140-6736(03)13913-X PMID:12885485

68. Willett WC, Hu FB (2006). Not the time to abandon the food frequency questionnaire: point. Cancer Epidemiol Biomarkers Prev, 15:1757–1758. doi:10.1158/1055-9965.EPI-06-0388 PMID:17021351

69. Bingham SA, Day N (2006). Commentary: fat and breast cancer: time to re-evaluate both methods and results? Int J Epidemiol, 35:1022–1024. doi:10.1093/ije/dyl142 PMID:16931532

70. Smith-Warner SA, Spiegelman D, Adami HO et al. (2001). Types of dietary fat and breast cancer: a pooled analysis of cohort studies. Int J Cancer, 92:767–774. doi:10.1002/1097-0215(20010601)92:5<767::AID-IJC1247>3.0.CO;2-0 PMID:11340585

71. Sinha R, Cross AJ, Turesky RJ. Biomarkers for dietary carcinogens: the example of heterocyclic amines in epidemiological studies. In: Wild C, Vineis P, Garte S, editors. Molecular epidemiology of chronic diseases. Chichester, UK: John Wiley & Sons Ltd; 2008.

UNIT 3.
ASSESSING EXPOSURE TO THE ENVIRONMENT

CHAPTER 12.

Assessment of the hormonal milieu

Susan E. Hankinson and Shelley S. Tworoger

Summary

The hormonal milieu has been hypothesized to play a role in a range of human diseases, and therefore has been a topic of much epidemiologic investigation. Hormones of particular interest include: sex steroids; growth hormones; insulin-like growth factors; stress hormones, such as cortisol; and hormones produced by the adipose tissue, termed adipokines. Depending on the hormone, levels may be measured in plasma or serum, urine, saliva, tissue, or by assessing genetic variation in the hormone or hormone metabolizing genes. Sample collection, processing, and storage requirements vary according to the type of sample collected (e.g. blood or urine) and the hormone of interest. Laboratory analysis of hormones is frequently complex, and the technology used to conduct the assays is constantly evolving.

For example, direct or indirect radioimmunoassay, bioassay or mass spectrometry can be used to measure sex steroids, each having advantages and disadvantages. Careful attention to laboratory issues, including close collaboration with laboratory colleagues and ongoing quality control assessments, is critical. Whether a single hormone measurement, as is frequently collected in epidemiologic studies, is sufficient to characterize the hormonal environment of interest (e.g. long-term adult hormone exposure) is also an important issue. While the assessment of hormones in epidemiologic studies is complex, these efforts have, and will continue to, add importantly to our knowledge of the role of hormones in human health.

Introduction

The study of hormones and their involvement in human health has been considered for many years, and their measurement has increasingly become an important part of many epidemiologic studies. Examining how various endogenous and exogenous hormones are related to disease increases our understanding of disease etiology, which may ultimately lead to improved prevention recommendations for both high-risk groups and the general population. Issues surrounding the appropriate use of hormone measures in epidemiologic studies are complex and require careful planning by study investigators. Many choices must be made, including the type of biospecimen to collect from participants, the timing and conduct of sample collection, the choice of hormones and assay modalities, and ultimately how to interpret the results. While several of these issues are dealt with in previous chapters (sample collection and processing in Chapter 3 and interpretation of assay results in Chapter 8), here the focus is on examples and concerns in measuring hormones. Because of the broad range of hormones found in humans, this chapter cannot cover every aspect of hormone measurement. However, general issues are addressed that should be considered when designing epidemiologic studies of hormones, such as the importance of hormones in medical research, their measurement in human samples, and issues regarding assay development and interpretation.

Context and public health significance

Hormones are chemicals produced by living cells that act as chemical messengers or signal molecules (1). The hormonal environment is a critical regulator of many physiologic processes, including growth, energy metabolism, fertility and the stress response. A wide range of well-documented diseases are linked to changes, either increases or decreases, in the hormonal milieu (Table 12.1). The role of hypoinsulinemia and insulin resistance in diabetes, excess insulin-like growth factor (IGF I in acromegaly and deficiency in dwarfism, excessive production of thyroid hormone in Graves disease, and overproduction of cortisol in Cushing syndrome are all examples. In addition to these well-established causal relationships, the hormonal milieu has been hypothesized to play a role in a range of other diseases, which have been the focus of many epidemiologic studies. Several examples, described further below, include the association of sex steroid hormones and breast cancer, the IGF system and cognitive function, and the role of adipokines such as adiponectin in both diabetes and heart disease.

This chapter focuses on the measurement of endogenous hormones (i.e. hormones produced by the body). The role of exogenous hormones (i.e. originating outside the body), particularly the use of oral contraceptives and postmenopausal hormones by millions of women worldwide, also has been the subject of substantial scientific study. The evaluation of exogenous hormones is not addressed here, largely because characterizing exposure to these agents is routinely accomplished via administration of questionnaires or tallying of pharmacy prescription records. However, any epidemiologic study of endogenous hormones must take into account sources of exogenous exposure that may influence endogenous hormone levels.

Table 12.1. Examples of diseases potentially caused (at least in part) by hormones that have been the subject of epidemiologic study

Hormone	Conditions known or hypothesized to be related
Sex steroids (e.g. estradiol and testosterone)	Infertility, osteoporosis, cancer (e.g. breast, endometrial and prostate cancers)
Vitamin D metabolites	Hypertension, osteoporosis, cancer (e.g. colon and breast cancer)
Insulin	Diabetes, heart disease, cancer (e.g. colon and endometrial cancers), cognitive function
Insulin-like growth factor/ Growth hormone axis	Cancer (e.g. colon, prostate and breast cancers), heart disease, cognitive function, osteoporosis
Prolactin	Immunologic diseases (e.g. rheumatoid arthritis and systemic lupus erythematosus), breast cancer
Adiponectin and other adipokines	Diabetes, heart disease, cancer (e.g. colon and breast cancers)
Stress hormones (e.g. cortisol)	Heart disease, cancer (e.g. breast cancer)
Melatonin	Breast cancer, sleep disorders

Examples/case studies

Sex steroids and breast cancer risk in postmenopausal women

Substantial data support a role of hormones, particularly sex steroids, in the etiology of breast cancer. There are consistent associations with reproductive factors, and increased risks associated with postmenopausal obesity and use of postmenopausal hormones (2). Further, drugs that either block estrogen binding to the estrogen receptor (selective estrogen receptor modulators, such as tamoxifen) or prevent the production of estradiol (aromatase inhibitors) are effective both in preventing breast cancer and improving survival of women with the disease (3–5). Considerable data assessing circulating sex steroids in postmenopausal women and breast cancer risk have accrued from prospective epidemiologic studies, where circulating levels of endogenous hormones are measured in study subjects before disease diagnosis (6). Overall, a strong positive association exists between breast cancer risk and circulating levels of both estrogens and androgens. Women in the top versus bottom 20% of estrogen levels have a two- to three-fold higher risk of breast cancer (7). The associations are similar for several forms of estrogen (e.g. estradiol, estrone, estrone sulfate). Although additional confirmation is required, the association appears strongest for estrogen receptor-positive breast tumours and is robust across groups of women at varying risk of breast cancer (e.g. defined by family history). Also, a single blood estrogen measure predicts subsequent breast cancer risk for at least 8–10 years. Generally, the more limited data available on urinary estrogens suggests similar predictive ability. For testosterone, a commonly measured androgen, the data are very consistent, with a significant positive association between circulating levels and postmenopausal breast cancer; the magnitude of the association is similar to that observed for estrogens (7). Most studies also noted a similar, although somewhat modest, positive association with other androgens, such as androstenedione, dehydroepiandrosterone (DHEA) and DHEA sulfate.

The insulin-like growth factor (IGF) axis and cognitive function

Insulin-like growth factor I (IGF-I) is a protein hormone that mediates many actions of growth hormone and plays a key regulatory role in cell growth and proliferation (1). Tissue bioavailability of IGF is regulated in large part by its binding to six known IGF binding proteins. Insulin-like growth factor binding protein 3 (IGFBP-3) is the most abundant of these, and it substantially prolongs the circulating half-life of IGF-I. Most circulating IGF is produced by the liver, although it can be produced locally in other body tissues. IGF is known to play a role in brain development and function (8); it is produced in the brain and can pass through the blood-brain barrier. In animal studies, IGF improves memory and learning (8), and raising IGF levels was found to decrease formation of amyloid β (9), a major constituent of the neural plaques that are a hallmark of Alzheimer's disease. In several recent cross-sectional and prospective studies, the association between circulating levels of IGF-I, IGFBP-3, or free (unbound) IGF-I have been assessed in relation to cognitive function. Although results have not been entirely consistent, it appears that older adults with higher levels of IGF, the IGF-1:IGFBP-3 ratio, or free IGF-1 tended to have better cognitive function as assessed by several cognitive tests (10–13).

Adiponectin and risk of diabetes and heart disease

In recent years, adipose tissue has been recognized as an active endocrine organ that secretes many biologically active substances, termed adipokines. Although adipokine research is relatively new, accruing laboratory and human data on adiponectin, one of the most abundant adipokines, support a role in diabetes and possibly heart disease (14–16). Adiponectin functions as an insulin-sensitizer, and also has important anti-inflammatory and anti-atherogenic actions. For example, adiponectin increases insulin sensitivity in animal models of insulin resistance, and reverses diet-induced insulin resistance in adiponectin knockout mice. The protein is inversely associated with body mass index (BMI) and insulin, and positively associated with serum lipids. Several cross-sectional, case–control, and prospective studies have reported either a strong significant inverse association for diabetes, or a modest inverse or no association for heart disease (17–21).

Predictors of hormone levels

In addition to the role of hormones in human health and disease, many studies have evaluated how the external environment influences the endogenous production of hormones in an effort to determine potential modes of action in causing (or preventing) disease. Examples include assessments of body size, physical activity and diet in relation to circulating hormone levels. The influence of alcohol intake on the hormonal milieu provides a good example. Both small randomized trials (22,23) and cross-sectional studies (24,25) have confirmed that alcohol intake increases estrogen levels in women, providing one potential mechanism for the positive association between alcohol intake and breast cancer risk. Further, alcohol intake has been found to increase insulin sensitivity (26) and HDL levels (27) and decrease fibrinogen levels (27), providing several mechanisms for the well-confirmed inverse association between alcohol intake and heart disease risk.

Strengths, limitations and lessons learned

Biologic samples for hormone evaluation

Many types of biologic specimens have been collected in epidemiologic studies where hormones are of interest (Table 12.2). In this section is a brief discussion of common sample types, and their advantages

Table 12.2. Collection and processing requirements and advantages and disadvantages of various biological specimen types collected in epidemiologic studies

Sample Type	Collection, processing and storage requirements	Advantages	Disadvantages
Blood, venipuncture	+Trained phlebotomist +Needles, tubes, biohazard waste, tourniquet, gloves, storage vials +Primarily clinic-based +Spin and aliquot plasma or serum +Delayed processing for 48 hours okay for some hormones +Mechanical or liquid nitrogen freezers	+Can measure broad range of hormones +Collect large volumes for future assays +For plasma tubes, can collect white blood cells for DNA and red blood cells +Can collect multiple types of samples (e.g. serum, plasma) simultaneously	+Less feasible for cost and logistic reasons for large studies, especially over a large geographic area +Invasive technique may lower participation rates +Requires laboratory and storage space +Biohazard potential
Blood, finger stick (filter paper)	+Finger lance, filter paper, instructions +Clinic or home-based +No processing +OK at room temperature for ~1 week for some hormones +Standard 4°C refrigerator or −20°C freezer	+Can measure some hormones +Feasible to collect on a large-scale +Allows serial sampling +Relatively non-invasive +Minimal storage space needed +Correlated with plasma levels for women; less clear for men +Not a biohazard after sample dries	+Low sample volume limits number of assays +Individual haematocrit levels can alter measured concentrations +Requires low humidity during transport and storage +Certain filter papers may inhibit some assays
Urine, cup	+Collection cup, antiseptic wipe, instructions, storage vials +For infants put a pad in the diaper +Clinic or home-based +May require addition of acid +Delayed processing/ freezing for 48 hours okay for some hormones +Mechanical or liquid nitrogen freezers	+Can measure broad range of hormone metabolites +Feasible to collect on a modest scale +Non-invasive and painless +Allows serial sampling +Generally correlated with plasma levels +Collect large volumes for future assays	+Assays often require larger volume than for blood +Must measure creatinine to determine urine concentration +Requires laboratory and storage space +Has biohazard potential +Can collect 24-hour, overnight, first-morning, or spot urine
Urine, filter paper	+Collection cup, antiseptic wipe, filter paper, instructions +Clinic or home-based +Okay at room temperature for ~1 week for some hormones +Standard −20°C freezer	+Can measure some hormone metabolites +Feasible to collect on a large scale +Non-invasive and painless +Minimal storage space needed +Allows serial sampling +Not a biohazard after sample dries	+Has been used only in a limited number of studies + Low sample volume limits number of assays +Certain filter papers may inhibit some assays
Saliva, spot collection	+Several collection types: Unstimulated, salivette, cotton rolls, gum-stimulated +Need instructions, storage vials +Clinic or home-based +Okay at room temperature for ~1 week feasible for some hormones +Standard −20°C freezer	+Can measure some hormones +Feasible to collect on a large scale +Non-invasive and painless +Allows serial sampling +May reflect tissue exposure, as hormone must diffuse across salivary gland cells +Correlated with plasma levels for women; less clear for men	+Can only detect free (not bound) hormones, which often have a low concentration +Can be contaminated with blood, which has high concentrations of total hormone +May not reflect some circulating hormones in men +Use of stimulation or salivettes increases some hormone values
Saliva, oral diffusion sink	+Diffusion sink (hormones diffuse across it), instructions, storage vials +Collect continuously for long periods +Home-based +Okay at room temperature for ~1 week feasible for some hormones +Standard −20°C freezer	+Measures integrated levels over time +May reflect tissue exposure of hormones under natural conditions +Separates bound and unbound hormones to reduce blood contamination +Participants find it acceptable	+Device cost can be high +Only available for a limited number of hormones +Not as feasible in a large, population-based study

Breast milk	+Breast pump, container, instructions, storage vial +Home-based +Little known about processing or acceptable amount of time until freezing	+May reflect breast tissue hormone exposure +Use for information on infant hormone consumption	+Few assays developed for this sample type +Only applies to lactating women
Breast nipple aspirate fluid	+Trained nurse or clinician +Syringe, applicator, storage vials +Little known about processing or acceptable amount of time until freezing	+May reflect breast tissue hormone exposure +May better measure intracrine production of hormones	+Few assays developed for this sample type +Only applies to women +Can only be obtained from about 50% of women +Only small volumes can be obtained
Other tissues	+Generally can only be obtained via surgery by a clinician +Fresh samples must be flash-frozen immediately +Processing into paraffin results in ability to store long-term at room temperature	+Can measure hormone levels at the tissue of interest +May be able to stain paraffin-embedded tissue to determine protein levels semi-quantitatively +Can extract DNA or RNA	+Few assays currently available for tissue +Paraffin embedding may limit assay types +Very invasive and difficult to obtain on a large-scale

and disadvantages, with respect to measuring the hormonal milieu.

Blood

Blood specimens, collected by venipuncture, are the most common and flexible sample type collected in epidemiologic studies. Levels of many hormones can be determined in blood, primarily because assay development traditionally has focused on this sample type. Various kinds of blood samples can be collected, including serum and plasma (e.g. EDTA, sodium heparin, citrate); each has advantages and disadvantages depending on the biomarker(s) of interest (28,29). For example, sex hormone levels generally are similar when comparing serum with EDTA or heparin plasma (30–33). However, most studies suggest slightly higher levels of sex hormones in plasma versus serum; despite this, both are acceptable (30,31,33).

The primary advantages of blood collected using venipuncture include the capability to measure many hormones, as well as the ability to collect multiple blood specimens simultaneously (e.g. plasma and serum) (34). Additionally, if collecting plasma, both red and white blood cells can be saved; the latter can be used to isolate DNA. Also, relatively large sample volumes can be collected at one time, allowing for many assays to be conducted per participant.

Despite these advantages, venipuncture is expensive, as it requires a trained phlebotomist and extensive equipment (e.g. needles, appropriate collection tubes, gloves, etc.) (34). Thus, it generally is not feasible to collect specimens using this method in very large studies, or studies where the population is geographically dispersed. One exception to this rule is studies of medical professionals who have training in phlebotomy, or are well connected to the medical care system, such as when nearly 33 000 blood samples from nurses who live across the United States were obtained in the Nurses' Health Study (NHS) cohort (24). After pilot testing for feasibility, women were mailed a blood collection kit with instructions, and asked to have someone draw the blood and ship it back, with a chill pack, to the study laboratory via overnight mail. Mobile clinics also have been used successfully, such as in the National Health and Nutrition Examination Survey (NHANES) (35). Another disadvantage of venipuncture is low participation rates because of its invasive nature. Specifically in the NHS, it was noted that among women who had declined giving blood, 50% agreed to give a cheek cell specimen (36). Further, liquid blood specimens have biohazard potential, requiring special training and storage conditions.

To overcome difficulties with venipuncture, several recent studies have focused on collecting blood on filter paper using a finger prick (34,37–43). This method has been used for years to test for uncommon, but treatable, genetic conditions in infants (38). Recently, extraction methods have improved such that many hormones can be assayed from blood spots, including thyroid hormones, prolactin, sex hormones, gonadotropins, growth factors, leptin and stress hormones (37,39–43). The primary advantages include that the method is relatively non-invasive, can be conducted serially, presents a reduced biohazard potential when dry, and is feasible to collect on a large scale (40–43). One study reported that women found repeated blood spot sampling to be less troublesome than venipuncture or saliva collection (40).

Unit 3 • Chapter 12. Assessment of the hormonal milieu 203

However, there are several disadvantages of blood collection via finger prick. The most important is that only a limited number of hormones can be assayed from one collection. In addition, differences in haematocrit between participants can introduce systemic or random measurement error in hormone levels, since the sample is whole blood (37,41). Some evidence suggests, however, that the filter paper can partially ameliorate this problem, as blood with high haematocrit tends to impregnate a smaller volume on the filter paper (37). Further, some studies have reported that while correlations for sex hormones between venipuncture and filter paper are high for women, the correlations appear lower in men (40,41); the mechanism behind this is unclear. When measuring estradiol, testosterone and progesterone in men and premenopausal women, blood spot hormone levels explained 89% of the variance of serum levels in women, but only 46% in men (41). Other issues to note are that certain filter paper types may inhibit some assays (38), clear participant instructions are important for obtaining reliable samples (34), and the filter paper must be kept at a relatively low humidity (38).

Urine

Urine specimens are another commonly collected biological specimen, particularly because their collection is non-invasive (34). It also can be easily collected in infants and toddlers by putting a pad in the child's diaper. In general, urine contains hormone metabolites, rather than primary hormones, and reflects excretion over the period of the collection. Therefore, it can be difficult to determine over what time period to collect the urine (e.g. 24 hours, overnight, first morning sample, or spot collection (34). The timing of urine collections depends on the hormone of interest. For example, the intraclass correlations (ICC) for a morning-spot versus 24-hour urine were 0.78 for estrone-3-glucuronide and 0.46 for pregnandiol-3-glucuronide; similar ICCs were observed for overnight versus 24-hour urines. This suggests that a morning spot urine was acceptable for the estrone metabolite, but neither the spot nor overnight urine appear to capture all the circadian variation for pregnandiol. Morning urines are acceptable for assessing nocturnal urinary melatonin production (44). Urine can be collected serially and in large volumes, allowing for multiple assays to be run. It can also be collected on filter paper to minimize storage needs and reduce the biohazard potential (34).

A primary disadvantage of urine is that its concentration varies substantially both between persons and within the same person over time (45). Creatinine is commonly used to measure urine concentration. Most studies have calculated the analyte/creatinine ratio to adjust for volume. However, it has been reported that creatinine levels should be included in the regression model as an independent variable. This adjusts for concentration while allowing one to assess the significance of other predictors in the model independently of creatinine levels (45). One other disadvantage of urine is that most hormones, or their metabolites, exist in low concentrations, often necessitating large volumes to conduct the assays.

Saliva

More recently, saliva has been used for measuring hormones. Generally, plasma and salivary hormone values were highly correlated, including for cortisol, androgens, estrogens and progesterones (46). These hormones enter the saliva by passive diffusion, thus the levels specifically reflect the free, unbound, circulating fraction (47,48). Melatonin levels also correlate well between plasma and saliva (49). However, for several protein hormones, such as thyroid hormones, prolactin, or IGF-1, the salivary level bears little relationship to plasma levels and is unlikely to be of any research value (47).

Advantages of saliva sampling include: it is non-invasive, painless, easily performed, relatively inexpensive, has higher rates of compliance, and can provide quantitative data of biologically active hormone levels in circulation (40,46). Salivary sampling also avoids stress sometimes associated with venipuncture, which can elevate some hormones, particularly cortisol (50). Another benefit is that saliva samples can be collected at home with minimal training (40,50); further serial collections are easily conducted.

Despite these advantages, a major disadvantage of saliva is that hormone concentrations are much lower than in plasma, because salivary hormone levels reflect the free levels, which is typically 1–10% of the total plasma level (40). Therefore, saliva assays often require large sample volumes and highly sensitive (and thus expensive) assays (50). Further, blood contamination in the oral mucosa can lead to substantial measurement error by increasing measured levels. Stimulated saliva collection methods, additionally, can bias hormone assays. For example, the use of cotton-based absorbent materials, or chewing gum, to stimulate saliva flow can artificially elevate assay results (40). Therefore, it is important to pilot sample collection techniques to ascertain whether such procedures interfere with the assays of interest.

One lesser known method of saliva sampling is a diffusion-sink device (51). This device is a small ring that the participant places orally for some set time period. The sink has a membrane that allows diffusion of free hormones, and thus measures the average freely diffusing concentration of an analyte over time. These devices reject artefacts arising from blood plasma contamination of saliva and provide a time-averaged sample, without requiring the subject to adhere to a frequent-sampling schedule (51).

Other sample types

Other specimen types, such as breast milk, breast nipple aspirate fluid and other tissues (e.g. tumour, adipose, colon polyps) can be obtained (34,52); however, they often are difficult to collect or require invasive procedures. The greatest disadvantage, though, is that it can be hard to conduct assays on these specimen types. Despite this, measuring hormones in these specimens may reflect true exposure at the tissue level better than from other sampling types.

Collection, processing and storage

As discussed in Chapter 3, the collection, processing, and storage of samples may affect the ability to accurately measure the hormonal milieu (36,53–56). Several factors for sample collection must be considered, including the study population, timing and location of the collection. Depending on these factors, the samples may need to be processed in a non-standard manner. A common issue in epidemiologic studies is that of delayed processing or delayed freezing. The effects of such protocols on the hormone of interest must be evaluated before assaying. Finally, storage of study specimens, particularly long-term storage in prospective studies, is an important and complex issue requiring appropriate acquisition of space and resources to maintain freezers and other related equipment.

Sample collection

Selection of the appropriate study population for any epidemiologic study is important. However, when studying hormone levels, careful consideration of the participants is often necessary to reduce bias. For example, in a study of predictors of estrogen levels, it would be inappropriate to combine men and women, premenopausal and postmenopausal women, or postmenopausal women taking hormones (PMH) versus not, as these groups have different mean estrogen levels. In this case, if gender, menopausal or PMH status was associated with the exposure, the observed association will be biased. Statistical adjustment alone generally cannot correct for this strong bias, particularly if the association varies across these subgroups.

The following example of the relationship between adiponectin and estradiol levels illustrates how combining inappropriate populations can alter study results. Experimental data suggest that adiponectin may, in part, regulate estradiol levels. But because adiponectin and estradiol are both derived primarily from adipose tissue in postmenopausal women, it is important to study this association on a population level, including adjustment for body mass index (BMI). One study observed no relationship after adjustment for BMI in postmenopausal women not using PMH (57). This contradicted two previous studies, which reported that additional adjustment for BMI did not attenuate the relationship (58,59). However, one study (59) combined PMH users and non-users, likely biasing the results, since PMH users had higher estrogen and lower adiponectin levels than non-users. The other study combined premenopausal and postmenopausal women (58). Given that premenopausal women have higher estradiol levels, and that the primary sources of estradiol are the ovaries in premenopausal women and body fat in postmenopausal women (1), the results of this study likely were biased as well.

Also important is the timing of the sample collection, since some hormones can fluctuate yearly, seasonally, monthly, daily, hourly and even from minute to minute (Figure 12.1). Other factors that can influence some hormones include fasting, alcohol intake, physical activity and medications. Understanding the underlying biology of the hormone(s) of interest is important to determining the optimal timing of sample collection. Three examples of this issue are elaborated upon.

Estrogen and progesterone are known to vary widely during the menstrual cycle in premenopausal women (1). Sample collection in this population, therefore, should either standardise the day(s) in the cycle on which samples are drawn, or collect detailed information about menstrual cycle start dates before and after the collection. In the NHSII, premenopausal women were asked to collect two blood samples, one in the early follicular phase and one in the mid-luteal phase, times when sex hormones are relatively stable from day to day (60). Women also returned a postcard with the

Figure 12.1. Four types of hormonal variation. Graph (a) shows how estradiol changes by day of the menstrual cycle with day 0 being the day of ovulation among 20–34 year olds ( ) and 35–46 year olds (•) cycling women (adapted from (116)). Graph (b) provides an example of how vitamin D concentrations vary over the course of one year (i.e. by season) among postmenopausal women living in New Zealand (adapted from (117)). Graph (c) demonstrates circadian variation across 3 days of a commonly measured urinary metabolite of melatonin, 6-sulphatoxymelatonin, in both young and elderly individuals (adapted from (118), copyright © 2005, Informa Healthcare. Reproduced with permission of Informa Healthcare). Graph (d) shows how IGF-1 levels vary by age in men and women, with a sharp increase early in life and a slow decline later in life (adapted from (119))

Figure 1. Types of hormonal variation

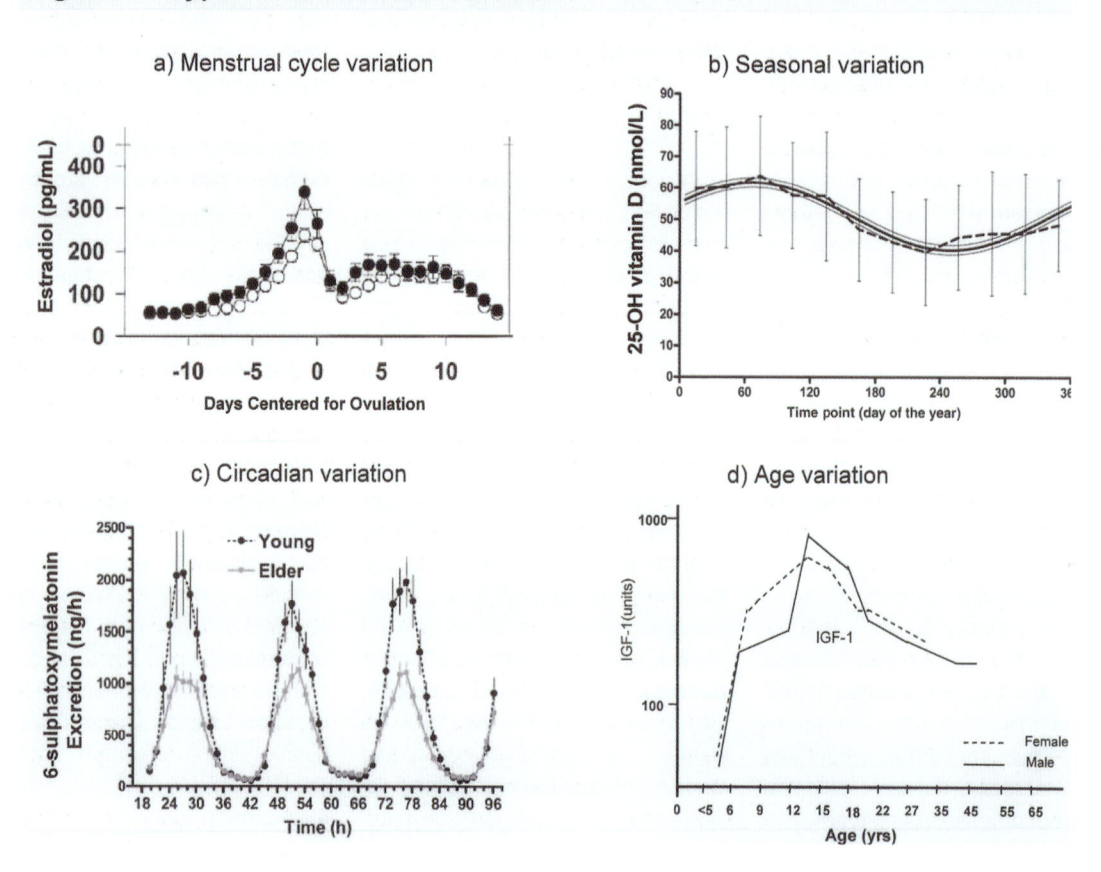

date their next menstrual cycle began. This information, along with the date of their previous cycle, allowed calculation of the cycle day on which the blood samples were drawn. One disadvantage of this method is that the investigator is reliant on women to remember when to collect the blood samples. Another method is to have women use home-based ovulation kits to time sample collection for a certain number of days after ovulation (61). The major disadvantages of this method are cost and the need to train women to use the kits appropriately. An alternative approach is to ask women to collect a sample on any day of their menstrual cycle and then provide specific dates of their cycle before (and if possible, after) collection and their average cycle length (62). This allows estimation of the cycle day. The main disadvantage of this method is that it reduces power to examine menstrual phase-specific associations. Ultimately, the collection method is dependent on the population and resources.

Several hormones have a circadian rhythm, with the most well characterized being melatonin. Levels are high at night while sleeping, and decrease during the daylight hours (44). In small, laboratory-based studies, the most common method of assessing melatonin has been to collect serial blood samples over a 24-hour period. However, this method is too labour-intensive for large epidemiologic studies. One alternative is to collect a blood sample at the same time of day for each participant. The utility of this method is limited, though, since the circadian pattern is not entrained to the same time of day for everyone (63). Another option is to collect urine to assay melatonin metabolites. Studies indicate that urinary levels, from either a 24-hour or first morning urine, or sequential saliva samples, are highly correlated with plasma melatonin (44,64,65).

A third example is that of plasma, or serum insulin, which is strongly affected by the number of hours since last eating (1). At minimum, time of last food consumption should be collected at the blood draw. However, if these hormones are important biomarkers, investigators should instruct participants to not consume any food or drink for at least 8 to 10 hours before the blood draw. Study staff should carefully ask participants about their food intake during that time and reschedule the collection if necessary. Of note is that for some diseases an alternative hypothesis exists: that the postprandial insulin response is most relevant to disease risk (66). If so, collecting a blood sample soon after eating would be preferred.

Sample processing

Sample collection and its processing should be conducted in a rigorous and standardized manner. In general, certain methods are preferred (e.g. immediate processing of samples); however this may not be feasible in some studies, particularly when participants are dispersed geographically. While this topic is covered in detail in Chapter 3, two issues are highlighted which are often faced when assessing hormones: delayed processing and delayed freezing.

Extensive pilot testing has shown that many hormones, including estrogens, androgens, prolactin, IGFs and gonadotropins, are not substantially affected by blood remaining unprocessed for 24 to 48 hours, while others, such as adrenocorticotropic hormone (ACTH), arginine vasopressin and free PSA, cannot be assessed with this protocol (31–33,67–70). Interestingly, several studies reported increasing testosterone levels with delayed processing, likely due to *ex vivo* conversion of precursor hormones (32,68,70). However, levels across delayed processing times remained highly correlated, suggesting that this approach is acceptable. Thus, each hormone of interest should be pilot tested for stability of the analyte over increasing time of delayed processing.

A delay in freezing can occur if there is delayed processing or if samples are processed immediately and frozen at a later time. For blood specimens, most hormones are stable if kept chilled (~4 °C) for up to 3 days before freezing; however, others, such as free PSA and ACTH, should be processed and frozen immediately (31,33,67,69). Urinary catecholamine levels appear to be stable when stored for 24 hours at room temperature or chilled, provided that samples are acidified at once (71). Saliva hormones tend to be stable, and can be stored at room temperature for at least one week without degradation (47). However, these samples can mould after 4–7 days; thus they should be frozen or refrigerated if possible (34). In general, filter paper collections need to be kept refrigerated at a low humidity to maintain hormone stability (38).

Storage options

Freezing and refrigeration are the most commonly used storage modalities; the merits of these options are enumerated in Chapter 3. In general, liquid nitrogen freezers (≤ 130 °C) are the best choice for long-term storage of samples, since temperatures in mechanical freezers can vary widely (72). Unfortunately it is difficult to directly assess the effects of long-term storage on hormone degradation. Two study designs can be used. One method is to collect samples at one time point and then measure the hormone(s) of interest several times over a period of years. Thus, baseline biomarker levels are the same for each person, but laboratory drift can make comparison of assay results over time difficult, especially if the assay changes. Interpretability strongly depends on the reliability of the assay. The second method is to collect samples from the same individuals, or population, over a period of years, storing them at each time point. Then, assay the samples together at the end of the study, reducing issues with assay variability. However, within-person changes in levels over time means that it is unclear whether the levels at each time are the same. Despite this, degradation of samples stored for long periods is an important issue.

There is some evidence that storage at −20 °C may not be acceptable for sex hormones (73,74). In particular, sex hormone binding globulin (SHBG) may dissociate from estradiol and testosterone, decreasing measurable non-bound levels of these hormones (73). However, long-term storage for blood at −70 °C or colder appears to be acceptable for estradiol, testosterone, DHEAS, prolactin, IGFs, TGF-β1 and urinary 6-sulfatoxymelatonin, among others (30,74–79). If degradation is at issue, samples should be transferred to a colder storage modality. Given that modest levels of degradation can be difficult to detect, another approach is to match the samples being compared (e.g. cases and controls) on storage time. This will reduce the effect of measurement error. Statistical modeling can also adjust for storage time or freezer temperature.

Overall issues surrounding sample collection, processing and storage are vital considerations when using hormone samples in epidemiologic studies. In particular, pilot studies that test sample collection and processing procedures are needed to determine feasibility and participant acceptability. Furthermore, if any non-standard protocols are used, it is important to test the effect of this on the hormones of interest before sending study samples for assay. In studies with long-term storage of samples, it is important to be aware of possible sample degradation over time and how that may affect the study design, analysis and interpretation.

Laboratory measurement issues

Three common sources of error are introduced when using biomarkers: issues related to specimen collection, processing, and storage (discussed in the previous section) (30,53,80); laboratory error and variability (36); and within-person variability over time (81). Since Chapter 8 discusses these latter two issues in detail, this chapter focuses specifically on hormone assays. In particular, there are often multiple methods for assaying a hormone, each with advantages and disadvantages. For example, one assay may require a large volume but have a higher sensitivity, while another uses a smaller volume but has lower sensitivity. Two examples are illustrated below.

The first example relates to measurement of sex hormones such as estradiol and testosterone. Three classes of assays are available to measure these hormones: mass spectrometry (MS); indirect radioimmunoassay (RIA), including a pre-extraction step; and direct immunoassays using chemiluminescent, colorimetric or fluorescent markers (82). When choosing which assay to use in a study, several factors should be taken into account, such as the amount of sample used, cost per sample, ease of assay, comparability with previous studies, and most importantly the assay reliability, validity and sensitivity in the hormone value range of the population under study (especially of concern if the values are low). Other issues are the abundance of structurally similar hormone metabolites that can cross-react with assay antibodies, and the binding of some hormones by SHBG, which can interfere with antibody binding (83,84). Differences in assays can be observed merely by noting the very different median levels of sex hormones measured in postmenopausal women across nine studies of breast cancer risk (7).

For sex hormones, the MS method obtains the highest marks for reliability, sensitivity and cross-reactivity (83). This method can measure multiple hormone metabolites simultaneously with a moderate amount of serum, plasma, or urine (~0.5mL). While MS is thought to be the gold standard for hormone measurement, recent analysis of inter-laboratory variation suggests that further standardization across laboratories is needed (83). This assay also is not widely available due to the expensive equipment and the need for highly-trained personnel to run the assays. The indirect assay methodology employs an extraction step before RIA to remove hormone metabolites that can cross-react with the antibody. This method has a high correlation with MS measures, although values tend to be slightly higher than those measured by MS, and uses a similar sample volume (85,86). The primary disadvantage of this method is that it cannot be easily automated and is thus labour-intensive and expensive (83). Direct assays, in general, do not have an extraction step before antibody binding. While these assays are high-throughput, easily available, inexpensive and use low volume, they may have only modest correlations with MS and indirect assays, and substantially overestimate hormone values, as well as a poor sensitivity for samples with low hormone concentrations (82–86). In general, these assays are not useful for clinical applications where precise levels must be determined, and are likely a major source of variability in epidemiologic study results (for a thorough review of this topic, see (83)).

The second example exemplifies the importance of understanding the biology of the hormone being measured, in this case prolactin. One limitation of the prolactin assay used in most epidemiologic studies to date (an immunoassay) is that it measures multiple forms of prolactin circulating in plasma (87). However, these forms likely have different biological activities (88,89). For example, glycosylated prolactin appears to have a higher metabolic clearance rate and lower biologic activity than the non-glycosylated form (89,90). Assays to specifically measure particular prolactin isoforms are difficult, time-intensive and require large amounts of plasma, and hence are not feasible in epidemiologic studies. The Nb2 lymphoma cell bioassay, however, is a sensitive measure of overall somatolactogenic activity in biological fluids. This assay measures the activity of both prolactin and growth hormone (91), although a modification of the assay, including anti-growth hormone antibodies, allows for specific evaluation of prolactin bioactivity. This measure and the ratio between the prolactin bioassay and immunoassay have been evaluated in several studies of systemic lupus erythematosus and found to be of importance (90,92,93). A breast cancer case–control study reported that prolactin levels measured by bioassay, but not immunoassay, were significantly higher in cases versus controls (94). The correlation between the immunoassay and bioassay is about 0.50, suggesting that the bioassay provides additional information beyond the immunoassay (94).

Another important issue arises when pooling data from multiple different studies, such as in the sex hormone-breast cancer study mentioned above (7). Since each study conducts assays at various laboratories and times, frequently the distribution of the analyte (e.g. hormone levels) differs across studies (95). Various analytic techniques are available to deal with this problem (96). One method is to use study-specific quantile cut-points to determine risk estimates comparing high versus low values for each study, which can then be pooled. The major drawbacks of this method are that it is difficult to evaluate what absolute hormone levels are related to disease, and to assess dose–response relationships. Another method used is to pool the risk estimates for a doubling (or tripling, etc.) of sex hormone concentrations within each study and then pool the risk estimates. However, the only way to assess how the absolute levels of a hormone compare across studies is to reassess a subset of samples from each laboratory used in each of the different studies (i.e. a calibration study) using a gold-standard assay.

Another approach to evaluating hormone levels is to assess genetic variation (e.g. single nucleotide polymorphisms (SNPs)), in the gene(s) associated with the hormone. Although estimates vary by hormone, a substantial component of the between-person variation observed in circulating hormone levels is genetically determined (e.g. ~40% for IGF-I (97)). For example, to assess the role of the steroid vitamin D in osteoporosis, several studies have evaluated SNPs in the vitamin D receptor (98). Furthermore, many studies have assessed variation in the sex steroid hormone metabolizing pathway in relation to cancer risk (99–101). Advantages to this approach include: retrospective case-control designs can be used without concern of the disease altering circulating hormone levels, genetic variation may provide information on the tissue hormonal environment, and genetic assays tend to be robust and with little to no variation in measures across studies. Disadvantages include not knowing the function (if any) of the SNPs measured, and small effect sizes. Ultimately, clearer answers likely will be obtained with approaches that evaluate multiple SNPs in a gene or across multiple genes in a pathway, but this requires extremely large sample sizes and complex statistical tools that are still in development. Although promising, this approach to evaluate the hormonal milieu has yet to provide substantial insight into the hormone-disease relationship (102). Yet with increasing sample sizes and method development, the potential of this approach should be realized soon.

Within-person stability over time

A particularly important source of measurement error in hormone studies is random variation in biomarker levels within an individual over time. Thus one measurement of the biomarker, as is common in many epidemiologic studies, may not accurately reflect an individual's long-term exposure. Measurement error correction, or inclusion of multiple samples per participant, can ameliorate the attenuating effects of biomarkers with a high intra-individual variability over time (103–105).

The intraclass correlation coefficient (ICC) can be used to measure the stability or reliability within individuals over time or across different assay platforms (103,106-108). It is the ratio of the between-person variance with the total variance (between- plus within-person variance), and ranges from 0 to 1.0 (80,109). The ICC is distinct from a Pearson or Spearman correlation coefficient in that a common mean is assumed between repeated measures. The ICC can be assessed on the natural log-transformed or untransformed scale, although if the data are skewed it is best to log-transform. An advantage of the ICC is that the impact of the within-assay variability is considered relative to the total variation. For example, a somewhat high assay coefficient of variation (CV) may not be acceptable if there is very limited between-person variation (resulting in a low ICC), as the additional laboratory variability could overshadow true differences between individuals.

Overall, most sex steroid hormones are reasonably stable within postmenopausal women over a 1–3 year period, with intraclass correlations ranging from 0.5 to 0.9

(103,110-112). Similarly, IGF-I and IGFBP–3 levels have correlations in the range of 0.6 to 0.8 over a 1–3 year period (106,113). Urinary melatonin levels also appear stable over several years (114). Thus, although using a single blood measure for these hormones will result in some misclassification and attenuation of relative risks, this level of reproducibility is similar to that observed for other biological variables, such as blood pressure and serum cholesterol measurements; all parameters that are considered well-measured and consistent predictors of disease in epidemiologic studies (115). In contrast, prolactin has a lower reproducibility, with correlations over time in the range of 0.45 or lower, indicating greater attenuation in the relative risk estimates when using a single blood measurement. In a setting such as this, where measurement error is higher (since the correlation of hormone levels within woman over time is lower), statistical methods that account for this error in the calculation of relative risks should be used (see Chapter 8), or collecting multiple samples per subject considered.

Future directions and challenges

Overall, the use of hormone measurements is becoming more common in epidemiologic studies. Several factors must be considered in the study design, sample collection, assay choice, and statistical analysis. Of greatest importance is a suitable choice of study population and sample type(s) (collected at an appropriate time), as well as proper storage facilities. Any non-standard methods should be pilot-tested before conducting the formal study. Pilot studies should also be conducted when considering the use of a new assay or laboratory. Choice of assay type can have a large impact on measurement error, and, ultimately, the interpretation of results. Although not discussed in this chapter, assessment of laboratory precision and reproducibility on an on-going basis is extremely important (36). Additionally, studies assessing hormone stability within an individual over time are important to conduct, particularly if the hormone is the exposure of interest. It is important to ultimately address all these issues to obtain results that are both reliable and valid. A better understanding of the role of hormones in human disease will benefit immensely from well-conducted epidemiologic studies.

References

1. Strauss JF, Barbieri RL. Yen and Jaffe's reproductive endocrinology. Philadelphia (PA): Elsevier Saunders; 2004.

2. Hankinson SE, Colditz GA, Willett WC (2004). Towards an integrated model for breast cancer etiology: the lifelong interplay of genes, lifestyle, and hormones. *Breast Cancer Res*, 6:213–218.doi:10.1186/bcr921 PMID:15318928

3. Cuzick J, Powles T, Veronesi U *et al.* (2003). Overview of the main outcomes in breast-cancer prevention trials. *Lancet*, 361:296–300.doi:10.1016/S0140-6736(03)12342-2 PMID:12559863

4. Lewis-Wambi JS, Jordan VC (2005-2006). Treatment of postmenopausal breast cancer with selective estrogen receptor modulators (SERMs). *Breast Dis*, 24:93–105. PMID:16917142

5. Lønning PE (2007). Adjuvant endocrine treatment of early breast cancer. *Hematol Oncol Clin North Am*, 21:223–238.doi:10.1016/j.hoc.2007.03.002 PMID:17512446

6. Hankinson SE (2005-2006). Endogenous hormones and risk of breast cancer in postmenopausal women. *Breast Dis*, 24:3–15. PMID:16917136

7. Key T, Appleby P, Barnes I, Reeves G; Endogenous Hormones and Breast Cancer Collaborative Group (2002). Endogenous sex hormones and breast cancer in postmenopausal women: reanalysis of nine prospective studies. *J Natl Cancer Inst*, 94:606–616. PMID:11959894

8. Sonntag WE, Ramsey M, Carter CS (2005). Growth hormone and insulin-like growth factor-1 (IGF-1) and their influence on cognitive aging. *Ageing Res Rev*, 4:195–212.doi:10.1016/j.arr.2005.02.001 PMID:16024298

9. Carro E, Trejo JL, Gomez-Isla T *et al.* (2002). Serum insulin-like growth factor I regulates brain amyloid-beta levels. *Nat Med*, 8:1390–1397.doi:10.1038/nm793 PMID:12415260

10. Dik MG, Pluijm SM, Jonker C *et al.* (2003). Insulin-like growth factor I (IGF-I) and cognitive decline in older persons. *Neurobiol Aging*, 24:573–581.doi:10.1016/S0197-4580(02)00136-7 PMID:12714114

11. Kalmijn S, Janssen JA, Pols HA *et al.* (2000). A prospective study on circulating insulin-like growth factor I (IGF-I), IGF-binding proteins, and cognitive function in the elderly. *J Clin Endocrinol Metab*, 85:4551–4555. doi:10.1210/jc.85.12.4551 PMID:11134107

12. Okereke O, Kang JH, Ma J *et al.* (2007). Plasma IGF-I levels and cognitive performance in older women. *Neurobiol Aging*, 28:135–142.doi:10.1016/j.neurobiolaging.2005.10.012 PMID:16337715

13. Okereke OI, Kang JH, Ma J et al. (2006). Midlife plasma insulin-like growth factor I and cognitive function in older men. J Clin Endocrinol Metab, 91:4306–4312.doi:10.1210/jc.2006-1325 PMID:16912125

14. Gable DR, Hurel SJ, Humphries SE (2006). Adiponectin and its gene variants as risk factors for insulin resistance, the metabolic syndrome and cardiovascular disease. Atherosclerosis, 188:231–244. doi:10.1016/j.atherosclerosis.2006.02.010 PMID:16581078

15. Menzaghi C, Trischitta V, Doria A (2007). Genetic influences of adiponectin on insulin resistance, type 2 diabetes, and cardiovascular disease. Diabetes, 56:1198–1209.doi:10.2337/db06-0506 PMID:17303804

16. Sattar N, Wannamethee G, Sarwar N et al. (2006). Adiponectin and coronary heart disease: a prospective study and meta-analysis. Circulation, 114:623–629.doi: 10.1161/CIRCULATIONAHA.106.618918 PMID:16894037

17. Choi KM, Lee J, Lee KW et al. (2004). Serum adiponectin concentrations predict the developments of type 2 diabetes and the metabolic syndrome in elderly Koreans. Clin Endocrinol (Oxf), 61:75–80.doi:10.1111/j.1365-2265.2004.02063.x PMID:15212647

18. Pischon T, Girman CJ, Hotamisligil GS et al. (2004). Plasma adiponectin levels and risk of myocardial infarction in men. JAMA, 291:1730–1737.doi:10.1001/jama.291.14.1730 PMID:15082700

19. Snijder MB, Heine RJ, Seidell JC et al. (2006). Associations of adiponectin levels with incident impaired glucose metabolism and type 2 diabetes in older men and women: the hoorn study. Diabetes Care, 29:2498–2503.doi:10.2337/dc06-0952 PMID:17065691

20. Vendramini MF, Ferreira SR, Gimeno SG et al.; Japanese-Brazilians Diabetes Study Group (2006). Plasma adiponectin levels and incident glucose intolerance in Japanese-Brazilians: a seven-year follow-up study. Diabetes Res Clin Pract, 73:304–309.doi:10.1016/j.diabres.2006.02.002 PMID:16546285

21. Wannamethee SG, Lowe GD, Rumley A et al. (2007). Adipokines and risk of type 2 diabetes in older men. Diabetes Care, 30:1200–1205.doi:10.2337/dc06-2416 PMID:17322479

22. Dorgan JF, Baer DJ, Albert PS et al. (2001). Serum hormones and the alcohol-breast cancer association in postmenopausal women. J Natl Cancer Inst, 93:710–715.doi: 10.1093/jnci/93.9.710 PMID:11333294

23. Reichman ME, Judd JT, Longcope C et al. (1993). Effects of alcohol consumption on plasma and urinary hormone concentrations in premenopausal women. J Natl Cancer Inst, 85:722–727.doi:10.1093/jnci/85.9.722 PMID:8478958

24. Hankinson SE, Willett WC, Manson JE et al. (1995). Alcohol, height, and adiposity in relation to estrogen and prolactin levels in postmenopausal women. J Natl Cancer Inst, 87:1297–1302.doi:10.1093/jnci/87.17.1297 PMID:7658481

25. Rinaldi S, Peeters PH, Bezemer ID et al. (2006). Relationship of alcohol intake and sex steroid concentrations in blood in pre- and post-menopausal women: the European Prospective Investigation into Cancer and Nutrition. Cancer Causes Control, 17:1033–1043.doi:10.1007/s10552-006-0041-7 PMID: 16933054

26. Davies MJ, Baer DJ, Judd JT et al. (2002). Effects of moderate alcohol intake on fasting insulin and glucose concentrations and insulin sensitivity in postmenopausal women: a randomized controlled trial. JAMA, 287:2559–2562.doi:10.1001/jama.287.19.2559 PMID:12020337

27. Rimm EB, Williams P, Fosher K et al. (1999). Moderate alcohol intake and lower risk of coronary heart disease: meta-analysis of effects on lipids and haemostatic factors. BMJ, 319:1523–1528. PMID:10591709

28. Holland NT, Smith MT, Eskenazi B, Bastaki M (2003). Biological sample collection and processing for molecular epidemiological studies. Mutat Res, 543:217–234.doi:10.1016/S1383-5742(02)00090-X PMID:12787814

29. Landi MT, Caporaso NE Sample collection, processing and storage. In: Toniolo P, Boffetta P, Shuker DEG et al., editors. Application of biomarkers in cancer epidemiology. Lyon: IARC Scientific Publication; 1997. p. 223–236.

30. Bolelli G, Muti P, Micheli A et al. (1995). Validity for epidemiological studies of long-term cryoconservation of steroid and protein hormones in serum and plasma. Cancer Epidemiol Biomarkers Prev, 4:509–513. PMID:7549807

31. Evans MJ, Livesey JH, Ellis MJ, Yandle TG (2001). Effect of anticoagulants and storage temperatures on stability of plasma and serum hormones. Clin Biochem, 34:107–112. doi:10.1016/S0009-9120(01)00196-5 PMID:11 311219

32. Key T, Oakes S, Davey G et al. (1996). Stability of vitamins A, C, and E, carotenoids, lipids, and testosterone in whole blood stored at 4 degrees C for 6 and 24 hours before separation of serum and plasma. Cancer Epidemiol Biomarkers Prev, 5:811–814. PMID: 8896892

33. Taieb J, Benattar C, Birr AS et al. (2000). Delayed assessment of serum and whole blood estradiol, progesterone, follicle-stimulating hormone, and luteinizing hormone kept at room temperature or refrigerated. Fertil Steril, 74:1053–1054.doi:10.1016/S0015-0282(00)01546-6 PMID:11056261

34. Rockett JC, Buck GM, Lynch CD, Perreault SD (2004). The value of home-based collection of biospecimens in reproductive epidemiology. Environ Health Perspect, 112:94–104.doi:10.1289/ehp.6264 PMID:14698937

35. National Center for Health Statistics. Plan and operation of the Third National Health and Nutrition Examination Survey, 1988–1994. Vital Health Stat 1. No. 32. US Department of Health and Human Services Publication PHS 94–1308. Hyattsville (MD): National Center for Health Statistics; 1994.

36. Tworoger SS, Hankinson SE (2006). Use of biomarkers in epidemiologic studies: minimizing the influence of measurement error in the study design and analysis. Cancer Causes Control, 17:889–899.doi:10.1007/s10552-006-0035-5 PMID:16841256

37. Diamandi A, Khosravi MJ, Mistry J et al. (1998). Filter paper blood spot assay of human insulin-like growth factor I (IGF-I) and IGF-binding protein-3 and preliminary application in the evaluation of growth hormone status. J Clin Endocrinol Metab, 83:2296–2301.doi: 10.1210/jc.83.7.2296 PMID:9661598

38. Howe CJ, Handelsman DJ (1997). Use of filter paper for sample collection and transport in steroid pharmacology. Clin Chem, 43:1408–1415. PMID:9267321

39. Miller AA, Sharrock KC, McDade TW (2006). Measurement of leptin in dried blood spot samples. Am J Hum Biol, 18:857–860. doi:10.1002/ajhb.20566 PMID:17039473

40. Shirtcliff EA, Granger DA, Schwartz EB et al. (2000). Assessing estradiol in biobehavioral studies using saliva and blood spots: simple radioimmunoassay protocols, reliability, and comparative validity. Horm Behav, 38:137–147. doi:10.1006/hbeh.2000. 1614 PMID:10964528

41. Shirtcliff EA, Reavis R, Overman WH, Granger DA (2001). Measurement of gonadal hormones in dried blood spots versus serum: verification of menstrual cycle phase. Horm Behav, 39:258–266.doi:10.1006/hbeh.2001. 1657 PMID:11374911

42. Worthman CM, Stallings JF (1994). Measurement of gonadotropins in dried blood spots. Clin Chem, 40:448–453. PMID:8131281

43. Worthman CM, Stallings JF (1997). Hormone measures in finger-prick blood spot samples: new field methods for reproductive endocrinology. Am J Phys Anthropol, 104:1–21.doi:10.1002/(SICI)1096-8644(199709)104:1<1::AID-AJPA1 >3.0.CO;2 -V PMID:9331450

44. Cook MR, Graham C, Kavet R et al. (2000). Morning urinary assessment of nocturnal melatonin secretion in older women. J Pineal Res, 28:41–47.doi:10.1034/j.1600-079x.2000.280106.x PMID:10626600

45. Barr DB, Wilder LC, Caudill SP et al. (2005). Urinary creatinine concentrations in the U.S. population: implications for urinary biologic monitoring measurements. Environ Health Perspect, 113:192–200.doi:10.1289/ehp.7337 PMID:15687057

46. King SL, Hegadoren KM (2002). Stress hormones: how do they measure up? Biol Res Nurs, 4:92–103.doi:10.1177/1099800402 238334 PMID:12408215

47. Chiappin S, Antonelli G, Gatti R, De Palo EF (2007). Saliva specimen: a new laboratory tool for diagnostic and basic investigation. Clin Chim Acta, 383:30–40.doi:10.1016/j.cca.2007.04.011 PMID:17512510

48. Lechner W, Marth C, Daxenbichler G (1985). Correlation of oestriol levels in saliva, plasma and urine of pregnant women. Acta Endocrinol (Copenh), 109:266–268. PMID:4013613

49. Voultsios A, Kennaway DJ, Dawson D (1997). Salivary melatonin as a circadian phase marker: validation and comparison to plasma melatonin. J Biol Rhythms, 12:457–466. PMID:9376644

50. Vining RF, McGinley RA, McGinley RA (1986). Hormones in saliva. Crit Rev Clin Lab Sci, 23:95–146. doi:10.3109/10408368609165797 PMID:3512171

51. Wade SE (1992). Less-invasive measurement of tissue availability of hormones and drugs: diffusion-sink sampling. Clin Chem, 38:1639–1644. PMID:1525992

52. Gann PH, Geiger AS, Helenowski IB et al. (2006). Estrogen and progesterone levels in nipple aspirate fluid of healthy premenopausal women: relationship to steroid precursors and response proteins. Cancer Epidemiol Biomarkers Prev, 15:39–44.doi:10.1158/1055-9965.EPI-05-0470 PMID:16434584

53. Tworoger SS, Hankinson SE (2006). Collection, processing, and storage of biological samples in epidemiologic studies: sex hormones, carotenoids, inflammatory markers, and proteomics as examples. Cancer Epidemiol Biomarkers Prev, 15:1578–1581.doi:10.1158/1055-9965.EPI-06-0629 PMID:16985015

54. Vaught JB (2006). Blood collection, shipment, processing, and storage. Cancer Epidemiol Biomarkers Prev, 15:1582–1584. doi:10.1158/1055-9965.EPI-06-0630 PMID:16985016

55. Vaught JB, Caboux E, Hainaut P (2010). International efforts to develop biospecimen best practices. Cancer Epidemiol Biomarkers Prev, 19:912–915.doi:10.1158/1055-9965.EPI-10-0058 PMID:20233852

56. Betsou F, Lehmann S, Ashton G et al.; International Society for Biological and Environmental Repositories (ISBER) Working Group on Biospecimen Science (2010). Standard preanalytical coding for biospecimens: defining the sample PREanalytical code. Cancer Epidemiol Biomarkers Prev, 19:1004–1011.doi:10.1158/1055-9965.EPI-09-1268 PMID:20332280

57. Tworoger SS, Mantzoros CM, Hankinson SE (2007). Relationship of plasma adiponectin with sex hormone and insulin-like growth factor levels. Obesity (Silver Spring), 15:2217–2224.doi:10.1038/oby.2007.263 PMID:17890489

58. Gavrila A, Chan JL, Yiannakouris N et al. (2003). Serum adiponectin levels are inversely associated with overall and central fat distribution but are not directly regulated by acute fasting or leptin administration in humans: cross-sectional and interventional studies. J Clin Endocrinol Metab, 88:4823–4831. doi:10.1210/jc.2003-030214 PMID:14557461

59. Im JA, Lee JW, Lee HR, Lee DC (2006). Plasma adiponectin levels in postmenopausal women with or without long-term hormone therapy. Maturitas, 54:65–71.doi:10.1016/j.maturitas.2005.08.008 PMID:16198517

60. Tworoger SS, Sluss P, Hankinson SE (2006). Association between plasma prolactin concentrations and risk of breast cancer among predominately premenopausal women. Cancer Res, 66:2476–2482. doi:10.1158/0008-5472.CAN-05-3369 PMID:16489055

61. Davis S, Mirick DK, Chen C, Stanczyk FZ (2006). Effects of 60-Hz magnetic field exposure on nocturnal 6-sulfatoxymelatonin, estrogens, luteinizing hormone, and follicle-stimulating hormone in healthy reproductive-age women: results of a crossover trial. Ann Epidemiol, 16:622–631.doi:10.1016/j.annepidem.2005.11.005 PMID:16458540

62. Kaaks R, Berrino F, Key T et al. (2005). Serum sex steroids in premenopausal women and breast cancer risk within the European Prospective Investigation into Cancer and Nutrition (EPIC). J Natl Cancer Inst, 97:755–765.doi:10.1093/jnci/dji132 PMID:15900045

63. Hsing AW, Meyer TE, Niwa S et al. (2010). Measuring serum melatonin in epidemiologic studies. Cancer Epidemiol Biomarkers Prev, 19:932–937.doi:10.1158/1055-9965.EPI-10-0004 PMID:20332275

64. Nowak R, McMillen IC, Redman J, Short RV (1987). The correlation between serum and salivary melatonin concentrations and urinary 6-hydroxymelatonin sulphate excretion rates: two non-invasive techniques for monitoring human circadian rhythmicity. Clin Endocrinol (Oxf), 27:445–452. doi:10.1111/j.1365-2265.1987.tb01172.x PMID:3436070

65. Pääkkönen T, Mäkinen TM, Leppäluoto J et al. (2006). Urinary melatonin: a noninvasive method to follow human pineal function as studied in three experimental conditions. J Pineal Res, 40:110–115.doi:10.1111/j.1600-079X.2005.00300.x PMID:16441547

66. Michaud DS, Wolpin B, Giovannucci E et al. (2007). Prediagnostic plasma C-peptide and pancreatic cancer risk in men and women. Cancer Epidemiol Biomarkers Prev, 16:2101–2109. PMID:17905943

67. Woodrum D, French C, Shamel LB (1996). Stability of free prostate-specific antigen in serum samples under a variety of sample collection and sample storage conditions. Urology, 48 Suppl;33–39.doi:10.1016/S0090-4295(96)00607-3 PMID:8973697

68. Hankinson SE, London SJ, Chute CG et al. (1989). Effect of transport conditions on the stability of biochemical markers in blood. Clin Chem, 35:2313–2316. PMID:2591049

69. Jane Ellis M, Livesey JH, Evans MJ (2003). Hormone stability in human whole blood. Clin Biochem, 36:109–112.doi:10.1016/S0009-9120(02)00440-X PMID:12633759

70. Kristal AR, King IB, Albanes D et al. (2005). Centralized blood processing for the selenium and vitamin E cancer prevention trial: effects of delayed processing on carotenoids, tocopherols, insulin-like growth factor-I, insulin-like growth factor binding protein 3, steroid hormones, and lymphocyte viability. Cancer Epidemiol Biomarkers Prev, 14:727–730.doi:10.1158/1055-9965.EPI-04-0596 PMID:15767358

71. Elfering A, Grebner S, Semmer NK et al. (2003). Two urinary catecholamine measurement indices for applied stress research: effects of time and temperature until freezing. Hum Factors, 45:563–574.doi:10.1518/hfes.45.4.563.27086 PMID:15055454

72. Su SC, Garbers S, Rieper TD, Toniolo P (1996). Temperature variations in upright mechanical freezers. Cancer Epidemiol Biomarkers Prev, 5:139–140. PMID:8850276

73. Langley MS, Hammond GL, Bardsley A et al. (1985). Serum steroid binding proteins and the bioavailability of estradiol in relation to breast diseases. J Natl Cancer Inst, 75:823–829. PMID:3863985

74. Phillips GB, Yano K, Stemmermann GN (1988). Serum sex hormone levels and myocardial infarction in the Honolulu Heart Program. Pitfalls in prospective studies on sex hormones. J Clin Epidemiol, 41:1151–1156.doi:10.1016/0895-4356(88)90018-2 PMID:3210063

75. Barba M, Cavalleri A, Schünemann HJ et al. (2006). Reliability of urinary 6-sulfatoxymelatonin as a biomarker in breast cancer. Int J Biol Markers, 21:242–245. PMID:17177163

76. Ito Y, Nakachi K, Imai K et al.; JACC Study Group (2005). Stability of frozen serum levels of insulin-like growth factor-I, insulin-like growth factor-II, insulin-like growth factor binding protein-3, transforming growth factor beta, soluble Fas, and superoxide dismutase activity for the JACC study. J Epidemiol, 15 Suppl 1;S67–S73.doi:10.2188/jea.15.S67 PMID:15881197

77. Cauley JA, Gutai JP, Kuller LH, Dai WS (1987). Usefulness of sex steroid hormone levels in predicting coronary artery disease in men. Am J Cardiol, 60:771–777.doi:10.1016/0002-9149(87)91021-6 PMID:3661391

78. Koenig KL, Toniolo P, Bruning PF et al. (1993). Reliability of serum prolactin measurements in women. Cancer Epidemiol Biomarkers Prev, 2:411–414. PMID:8220084

79. Orentreich N, Brind JL, Rizer RL, Vogelman JH (1984). Age changes and sex differences in serum dehydroepiandrosterone sulfate concentrations throughout adulthood. J Clin Endocrinol Metab, 59:551–555.doi: 10.1210/jcem-59-3-551 PMID:6235241

80. Armstrong BK, White E, Saracci R. Principles of exposure measurement in epidemiology. New York (NY): Oxford University Press; 1992.

81. Vineis P. Sources of variation in biomarkers. In: Toniolo P, Boffetta P, Shuker DEG et al., editors. Application of biomarkers in cancer epidemiology. Lyon: IARC Scientific Publication; 1997. p. 59–71.

82. Stanczyk FZ, Cho MM, Endres DB et al. (2003). Limitations of direct estradiol and testosterone immunoassay kits. Steroids, 68:1173–1178.doi:10.1016/j.steroids.2003.08.012 PMID:14643879

83. Stanczyk FZ, Lee JS, Santen RJ (2007). Standardization of steroid hormone assays: why, how, and when? Cancer Epidemiol Biomarkers Prev, 16:1713–1719. doi:10.1158/1055-9965.EPI-06-0765 PMID:17855686

84. Stanczyk FZ, Jurow J, Hsing AW (2010). Limitations of direct immunoassays for measuring circulating estradiol levels in postmenopausal women and men in epidemiologic studies. Cancer Epidemiol Biomarkers Prev, 19:903–906.doi:10.1158/1055-9965.EPI-10-0081 PMID:20332268

85. Hsing AW, Stanczyk FZ, Bélanger A et al. (2007). Reproducibility of serum sex steroid assays in men by RIA and mass spectrometry. Cancer Epidemiol Biomarkers Prev, 16:1004–1008.doi:10.1158/1055-9965.EPI-06-0792 PMID:17507629

86. Lee JS, Ettinger B, Stanczyk FZ et al. (2006). Comparison of methods to measure low serum estradiol levels in postmenopausal women. J Clin Endocrinol Metab, 91:3791–3797.doi:10.1210/jc.2005-2378 PMID:16882749

87. Haro LS, Lee DW, Singh RN et al. (1990). Glycosylated human prolactin: alterations in glycosylation pattern modify affinity for lactogen receptor and values in prolactin radioimmunoassay. J Clin Endocrinol Metab, 71:379–383.doi:10.1210/jcem-71-2-379 PMID:2380335

88. Sinha YN (1995). Structural variants of prolactin: occurrence and physiological significance. Endocr Rev, 16:354–369. PMID:7671851

89. Hoffmann T, Penel C, Ronin C (1993). Glycosylation of human prolactin regulates hormone bioactivity and metabolic clearance. J Endocrinol Invest, 16:807–816. PMID:8144855

90. García M, Colombani-Vidal ME, Zylbersztein CC et al. (2004). Analysis of molecular heterogeneity of prolactin in human systemic lupus erythematosus. Lupus, 13:575–583. doi:10.1191/0961203304lu1068oa PMID:15462486

91. Bernichtein S, Jeay S, Vaudry R et al. (2003). New homologous bioassays for human lactogens show that agonism or antagonism of various analogs is a function of assay sensitivity. Endocrine, 20:177–190.doi:10.1385/ENDO:20:1-2:177 PMID:12668884

92. Cruz J, Aviña-Zubieta A, Martínez de la EscaleraG et al. (2001). Molecular heterogeneity of prolactin in the plasma of patients with systemic lupus erythematosus. Arthritis Rheum, 44:1331–1335.doi:10.1002/1529-0131(200106)44:6<1331::AID-ART225>3.0.CO;2-Q PMID:11407692

93. Leaños-Miranda A, Chávez-Rueda KA, Blanco-Favela F (2001). Biologic activity and plasma clearance of prolactin-IgG complex in patients with systemic lupus erythematosus. Arthritis Rheum, 44:866–875.doi:10.1002/1529-0131(200104)44:4<866::AID-ANR143>3.0.CO;2-6 PMID:11315926

94. Maddox PR, Jones DL, Mansel RE (1992). Prolactin and total lactogenic hormone measured by microbioassay and immunoassay in breast cancer. Br J Cancer, 65:456–460.doi:10.1038/bjc.1992.92 PMID:1558804

95. Taioli E, Bonassi S (2002). Methodological issues in pooled analysis of biomarker studies. Mutat Res, 512:85–92.doi:10.1016/S1383-5742(02)00027-3 PMID:12220591

96. Key TJ, Appleby PN, Allen NE, Reeves GK (2010). Pooling biomarker data from different studies of disease risk, with a focus on endogenous hormones. Cancer Epidemiol Biomarkers Prev, 19:960–965. doi: 10.1158/1055-9965.EPI-10-0061 PMID: 20233851

97. Harrela M, Koistinen H, Kaprio J et al. (1996). Genetic and environmental components of interindividual variation in circulating levels of IGF-I, IGF-II, IGFBP-1, and IGFBP-3. J Clin Invest, 98:2612–2615. doi:10.1172/JCI119081 PMID:8958225

98. Uitterlinden AG, Ralston SH, Brandi ML et al.; APOSS Investigators; EPOS Investigators; EPOLOS Investigators; FAMOS Investigators; LASA Investigators; Rotterdam Study Investigators; GENOMOS Study (2006). The association between common vitamin D receptor gene variations and osteoporosis: a participant-level meta-analysis. Ann Intern Med, 145:255–264. PMID:16908916

99. Dunning AM, Dowsett M, Healey CS et al. (2004). Polymorphisms associated with circulating sex hormone levels in postmenopausal women. J Natl Cancer Inst, 96:936–945.doi:10.1093/jnci/djh167 PMID:15199113

100. Hunter DJ, Riboli E, Haiman CA et al.; National Cancer Institute Breast and Prostate Cancer Cohort Consortium (2005). A candidate gene approach to searching for low-penetrance breast and prostate cancer genes. Nat Rev Cancer, 5:977–985.doi:10.1038/nrc1754 PMID:16341085

101. Olson SH, Bandera EV, Orlow I (2007). Variants in estrogen biosynthesis genes, sex steroid hormone levels, and endometrial cancer: a HuGE review. Am J Epidemiol, 165:235–245.doi:10.1093/aje/kwk015 PMID:17110639

102. Rebbeck TR, Martínez ME, Sellers TA et al. (2004). Genetic variation and cancer: improving the environment for publication of association studies. Cancer Epidemiol Biomarkers Prev, 13:1985–1986. PMID:15598750

103. Hankinson SE, Manson JE, Spiegelman D et al. (1995). Reproducibility of plasma hormone levels in postmenopausal women over a 2–3-year period. Cancer Epidemiol Biomarkers Prev, 4:649–654. PMID:8547832

104. Michaud DS, Manson JE, Spiegelman D et al. (1999). Reproducibility of plasma and urinary sex hormone levels in premenopausal women over a one-year period. Cancer Epidemiol Biomarkers Prev, 8:1059–1064. PMID:10613337

105. White E. Effects of biomarker measurement error on epidemiological studies. In: Toniolo P, Boffetta P, Shuker DEG et al., editors. Application of biomarkers in cancer epidemiology. IARC Scientific Publication; 1997. p. 73–93.

106. Missmer SA, Spiegelman D, Bertone-Johnson ER et al. (2006). Reproducibility of plasma steroid hormones, prolactin, and insulin-like growth factor levels among premenopausal women over a 2- to 3-year period. Cancer Epidemiol Biomarkers Prev, 15:972–978.doi:10.1158/1055-9965.EPI-05-0848 PMID:16702379

107. Kotsopoulos J, Tworoger SS, Campos H et al. (2010). Reproducibility of plasma and urine biomarkers among premenopausal and postmenopausal women from the Nurses' Health Studies. Cancer Epidemiol Biomarkers Prev, 19:938–946.doi:10.1158/1055-9965.EPI-09-1318 PMID:20332276

108. Thomas CE, Sexton W, Benson K et al. (2010). Urine collection and processing for protein biomarker discovery and quantification. Cancer Epidemiol Biomarkers Prev, 19:953–959.doi:10.1158/1055-9965.EPI-10-0069 PMID:20332277

109. Rosner B. Fundamentals of biostatistics. Belmont (CA): Duxbury Press; 1995.

110. Micheli A, Muti P, Pisani P et al. (1991). Repeated serum and urinary androgen measurements in premenopausal and postmenopausal women. J Clin Epidemiol, 44:1055–1061.doi:10.1016/0895-4356(91)90007-V PMID:1940998

111. Muti P, Trevisan M, Micheli A et al. (1996). Reliability of serum hormones in premenopausal and postmenopausal women over a one-year period. Cancer Epidemiol Biomarkers Prev, 5:917–922. PMID:8922301

112. Toniolo P, Koenig KL, Pasternack BS et al. (1994). Reliability of measurements of total, protein-bound, and unbound estradiol in serum. Cancer Epidemiol Biomarkers Prev, 3:47–50. PMID:8118385

113. Muti P, Quattrin T, Grant BJ et al. (2002). Fasting glucose is a risk factor for breast cancer: a prospective study. Cancer Epidemiol Biomarkers Prev, 11:1361–1368. PMID:12433712

114. Schernhammer ES, Hankinson SE (2005). Urinary melatonin levels and breast cancer risk. *J Natl Cancer Inst*, 97:1084–1087. doi:10.1093/jnci/dji190 PMID:16030307

115. Willett WC (1998). Nutrition and chronic disease. *Public Health Rev*, 26:9–10. PMID:9775714

116. Welt CK, McNicholl DJ, Taylor AE, Hall JE (1999). Female reproductive aging is marked by decreased secretion of dimeric inhibin. *J Clin Endocrinol Metab*, 84:105–111. doi:10.1210/jc.84.1.105 PMID:9920069

117. Bolland MJ, Grey AB, Ames RW *et al.* (2007). The effects of seasonal variation of 25-hydroxyvitamin D and fat mass on a diagnosis of vitamin D sufficiency. *Am J Clin Nutr*, 86:959–964. PMID:17921371

118. Kripke DF, Youngstedt SD, Elliott JA *et al.* (2005). Circadian phase in adults of contrasting ages. *Chronobiol Int*, 22:695–709. doi:10.1080/07420520500180439 PMID:16147900

119. Yu H, Mistry J, Nicar MJ *et al.* (1999). Insulin-like growth factors (IGF-I, free IGF-I and IGF-II) and insulin-like growth factor binding proteins (IGFBP-2, IGFBP-3, IGFBP-6, and ALS) in blood circulation. *J Clin Lab Anal*, 13:166–172.doi:10.1002/(SICI)1098-2825(1999)13:4<166::AID-JCLA5>3.0.CO;2-X PMID:10414596

UNIT 3.
ASSESSING EXPOSURE TO THE ENVIRONMENT

CHAPTER 13.

Evaluation of immune responses

Robert Vogt and Paul A. Schulte

Summary

This chapter will present some general background material on the cellular, biochemical, and genetic mechanisms of the immune system, then focus on specific examples that illustrate the promise and pitfalls of using immune biomarkers as tools for molecular epidemiologic research and public health practice. Some of the most exciting frontiers in medical science will be discussed: early detection of cancer through autoimmunity; malignancies that arise from the immune system itself; newborn screening for lethal immune deficiencies and latent autoimmune disorders; and neurodevelopmental disabilities that could result from maternal immune responses, which protect the mother but harm the fetus. The chapter concludes with some thoughts about current challenges and future directions.

Introduction

Over the past 15 years, familiarity with the immune system has increased substantially among public health scientists, as well as the public at large. Since the use of immune biomarkers in molecular epidemiology was first addressed (1), the essential role of the immune system in maintaining health has been brought to public attention by the global HIV epidemic (2), the composite burden of autoimmune diseases (3,4,5), the genetic errors that lead to primary immune deficiencies (6), and the enigmatic relationship between immunity and malignancy (7,8). During this period, our understanding of the cellular and molecular processes that constitute the immune response has also increased in both scope and detail. These advances open new avenues for the use of immune biomarkers in epidemiologic field studies and public health applications. At the same time, the general principles advocated earlier remain fully relevant today, perhaps even more so, given that the pace of technological development often outstrips our ability to harness it in a meaningful fashion.

This chapter will first update some of the general background material presented before (1) with respect to the cellular, biochemical, and genetic mechanisms of the immune system. Thereafter, the focus will be on specific examples that illustrate the promise and pitfalls of using immune biomarkers as tools for translation research and public health practice. Some of the most exciting frontiers in medical science will be discussed: early detection of cancer through autoimmunity; environmental risk factors for malignancies that arise from the immune system itself; newborn screening for lethal immune deficiencies and latent autoimmune disorders; and neurodevelopmental disabilities that could result from maternal immune responses, which protect the mother but harm the fetus. The chapter concludes with some thoughts about current challenges and future directions.

Immune biomarkers as functional elements and sentinel indicators

The benefits and limitations of using immune markers in epidemiologic studies may be best appreciated by understanding their relationship to the basic biology of the host defence system: a complex network of cells and mediators with recognition and response functions that occur throughout most tissues of higher organisms (1). The primary functions of the host defence system are repairing injured tissue, identifying and removing foreign substances, destroying or containing infectious agents, and, in some cases, eradicating cancer cells.

Innate (non-specific) and acquired (specific) immunity

Host defence functions are carried out through non-specific mechanisms of innate immunity, and through specific mechanisms of acquired (adaptive) immunity, which develop as the organism encounters environmental agents (antigens). The term immune system is used in this chapter to refer to all components of both non-specific innate immunity and antigen-specific acquired immunity, as their components and activities are invariably intertwined (1). Nonetheless, the distinction between markers that are antigen-specific and those that are not is often important, especially in exposure-related studies. The ability of the immune system to recognize foreign molecules is so discerning that it has even been likened to a self-referential sensory organ (9).

Inflammation

Whether innate or acquired, the result of host defence activity is often inflammation. The cardinal signs of inflamed tissue were described by the ancient Greek physicians Celsus and Galen: *calor* (heat), *dolor* (pain), *rubor* (redness), *tumour* (swelling), and *functio laesa* (loss of function) (10). Our current knowledge of the cellular and molecular basis of inflammation is exhaustive, but the complexity of the *in situ* inflammatory response still lies beyond our complete understanding. Still, the cells and mediators of inflammation provide essential biomarkers for medicine, biomedical research, and, more recently, for epidemiologic studies.

Inflammation is essential for host defence, as it brings cells and mediators to the site of tissue injury and infection, sequestering the insult, destroying infectious agents or the cells they have infected, clearing the debris and promoting repair. However, it is a two-edged sword, and many of the symptoms following injury or infection come not from the insult but from the host response to it. Immunopathology is the study of how the immune system creates as well as prevents disease. From the classic animal models of viral meningitis (11) and tuberculosis (12) to the recent revelation that human cardiovascular disease and diabetes are largely inflammatory pathologies (13), biomarkers have shown that immunity and inflammation are inexorably linked.

Biological categories of immune biomarkers

The distinction between antigen-specific and non-specific biomarkers is fundamental and unique to the immune system. For convenience, this distinction can be overlaid onto three major types of intrinsic biological markers: cellular, biochemical and genetic (Table 13.1). These three categories are arbitrary and somewhat artificial, since biochemical and genetic markers originate in cells. In fact, many cells of the immune system are defined by the biochemical surface receptors they express or the unique gene rearrangements they contain.

Cellular biomarkers

Cellular immunology was for many years a phenomenologic area of study, often confusing and contradictory. Two technical breakthroughs combined to bring order to this area. Monoclonal antibodies allowed the development of specific probes without *a priori* knowledge of the properties of the cellular target. Flow cytometry allowed the cell-by-cell detection of these targets using fluorescent-labelled monoclonal antibodies and a dynamic streaming process that could analyse and sort thousands of cells per second (14,15). The sorted cells could then be characterized for their functional activities and other properties, and linked to their respective target. The cellular targets, which are all proteins (or glycoproteins) and are usually cell surface receptors, are most often identified by their cluster of differentiation (CD) number (16). International workshops are held periodically to update CD nomenclature (17); as of 2010, the list was up to CD363 (see http://www.hlda9.org/HLDA9Workshop/

Table 13.1. Examples of classifying immune biomarkers

	Innate	Acquired
Cellular	Granulocytes	Lymphocytes
	Macrophages	(T-cells and B-cells)
	Natural killer cells	
Biochemical	Complement	Immunoglobulin
	Cytokines	T-cell receptor
Genetic	MHC genes	V-genes of Ig and Tcr

tabid/60/Default.aspx). (Strictly speaking, the CD number refers to the monoclonal antibodies that recognize the cellular target and not to the target itself, but this convention is often ignored.) Many CD markers identify stages in the maturation from haematopoietic stem cells to the various mature forms. Targets that identify all the cells in a lineage and do not appear in any other types of cells are called lineage specific markers; they are relatively scarce compared to targets that are found in various states of differentiation among more than one type of cell.

It is important to note that the functional state of a cell depends on not simply the presence or absence of receptors, but rather on the quantitative extent to which a given receptor is expressed. Upregulation of certain receptors, such as CD69, is a typical consequence of lymphocyte activation, and this change in the degree of expression can be a highly informative biomarker (18,19). Neoplastic transformation is also often accompanied by altered receptor expression that may be correlated with genetic and clinical features (20).

Flow cytometry remains the customary method for using CD markers to identify lineages, sublineages, and functional states of lymphocytes and other blood cells. These methods have matured as they have become more commonly used (21), and our earlier cautions about methodological bias (1) may now be somewhat mitigated. Still, the less common tests, used mostly in research settings, may not have sufficient standardization to ensure comparability between methods and laboratories. In particular, quantitative measurements of cell receptor expression are often subject to considerable bias between methods (22–24).

Lymphocytes

Among cellular components, lymphocytes are the antigen-specific cells of the acquired immune response. B-cells are lymphocytes that differentiate into antibody-producing plasma cells. T-cells are lymphocytes that perform regulatory functions and differentiate into the cytotoxic "killer cells", which attack virus-infected and certain cancer cells. A third type of lymphocyte, the natural killer (NK) cell, does not show antigen specificity, but has a large role in innate immunity (recent studies have identified a hybrid form called the NK T-cell that has limited antigen specificity (25)). B-cells, T-cells and NK-cells are indistinguishable morphologically, but B-cells and T-cells have surface receptors and genomic mutations that are lineage-specific, and NK-cells may be identified by the presence of certain markers in the absence of the B-cell and T-cell markers.

Clonal expansion. Although resting lymphocytes are quiescent cells, when either B-cells or T-cells are stimulated in an appropriate fashion, they re-enter the mitotic cycle and multiply into a family of related cells called a clone. All the cells of a clone have the same antigenic specificity as the original lymphocyte. Clonal expansion accounts for the two fundamental properties of acquired immunity recognized since ancient times: memory and specificity. Memory comes from the expanded population of lymphocytes that persists after initial antigen exposure, which allows a more rapid and sizeable secondary response. Specificity comes from the fact that all the cells of a clone recognize the same antigen. However, clones do develop microheterogeneity as mutations occur in progeny cells. This process is important for maturation of the specific immune response, and it is also relevant to autoimmune disease and lymphoproliferative malignancies.

Lymphocyte subsets. Both T-cells and B-cells have subsets that may be identified by the presence of certain receptors or by functional assays.

T-cells. Mature T-cells can be identified by the lineage-specific CD3 receptor or by their antigen-specific receptor (Tcr). The Tcr is coded by one of two gene families: α-β or gamma-delta. Alpha-β T-cells are by far more common elements of the acquired immune response, while gamma-delta T-cells have a limited repertoire and are something of a bridge between innate and acquired immunity (26).

T-cells are further divided into those bearing the CD4 receptor, those bearing the CD8 receptor, and a small fraction of those that bear both. Most of the CD4-bearing cells are helper T-cells (T_h) that upregulate the immune response. The CD8-bearing T-cells were originally considered to be either cytotoxic (killer) cells (T_c) or suppressor cells that downregulate the immune response. CD8 cytotoxic T-cells are well characterized, but evidence for suppressor activity in this subset was never convincing. In 1995, the real suppressor population was identified among CD4 T-cells as a small proportion that also bears the CD25 receptor and contains a high concentration of the Foxp3 transcription factor (27). These CD4-CD25 T-cells are now called regulatory T-cells (T_{reg}); they are critical for preventing autoimmunity, preserving secondary immunity (immune memory), and protecting pregnancies (28).

Helper T-cells may also be characterized in terms of their regulatory roles. The original paradigm described a T_H1 response, which led to delayed hypersensitivity mediated by cellular responses, and a T_H2 response, which led to humoral immunity and allergy mediated by antibody production. The association of T_H2 responses with both parasitic infections and allergies has been well defined in laboratory and clinical studies (29–31). However, this simple picture has been replaced by a more complex interaction involving the cytokine IL-17, which mediates a third functional type called the T_H17 T-cell (32). Research on the T_H17 subset has progressed rapidly, and it is now seen as having a central role in immune regulation (33), autoimmunity (34), inflammation (35) and the link between innate and acquired immune activities (36,37). The T_H17 pathway may even explain the suspected immunotoxic effects of halogenated aryl hydrocarbons, such as PCB and dioxins (38,39).

The T_H1 response is associated with gamma interferon and tumour necrosis factor (TNF-α), the T_H2 response with interleukin-4 and interleukin-13, and the T_H17 response with the IL-17 family of six cytokines designated 17A-17F (35). The relative elevation of these cytokines in tissue or serum is generally taken as evidence for the respective type of *in vivo* response. However, the measurement of these factors (especially in serum) is not standardized, and their use in epidemiologic field studies should be approached cautiously. In particular, artefacts of the immunoassays used to measure cytokines may produce spurious differences (40,41); interestingly, such misleading artefacts may still have biologic and immunologic validity (42). In any case, the remarkable heterogeneity and plasticity of T helper cells (43) can make interpretation of relevant biomarkers enigmatic at best.

One other subclassification of T-cells deserves mention: the distinction between naive and memory T-cells. Naive (virgin) T-cells have not encountered antigen, while memory T-cells arose by clonal expansion caused by antigen-driven activation. The surface receptor CD45 exists in two isoforms: CD45RA is associated with naive T-cells, while CD45RO is associated with memory T-cells. While the two isoforms can be readily distinguished by flow cytometry, the categorization is probably oversimplified, especially for CD8 T-cells. However, the distinction may provide some insight into the pathogenesis of immune-mediated disorders (44) and environmental exposures (45,46).

B-cells. Mature B-cells may be identified by the lineage-specific CD19 receptor and by the presence of their surface immunoglobulin (sIg) molecules. All of the sIg molecules on a particular B-cell have the same antigen binding site, which gives B-cells their specificity. When B-cells are activated by antigen binding, they proliferate and redifferentiate into antibody-producing cells. The endpoint in this secondary differentiation is the plasma cell, which in essence is a cellular factory for making antibodies. Several other receptors are expressed during the various stages of progression towards plasma cells or diversion to memory cells (47); identification of these has long been a staple of diagnostic pathology for B-cell malignancies (48). B-cells do not have major functional subsets analogous to CD4 and CD8 in T-cells. However, the presence or absence of CD5 (a receptor found on all T-cells) appears to define distinct B-cell populations. CD5 B-cells are associated with chronic humoral responses, mucosal immunity, autoimmunity and possibly with an increased risk of transforming into a B-cell malignancy (49).

Non-lymphoid cells

In addition to lymphocytes, several other types of cells are important participants in immune function; most of them spend at least part of their life cycle in the bloodstream, where they (along with lymphocytes) are collectively referred to as leukocytes or white blood cells (WBC). The most numerous of the bloodstream leukocytes are granulocytes, end-stage cells with short lifetimes whose granules contain pre-formed mediators ready for immediate release. Most of them are neutrophils, which migrate into inflamed tissue where they ingest (phagocytise) and destroy bacteria. Eosinophils and basophils are normally present in much smaller numbers; they are involved in allergy and the host response to parasitic infections. Although granulocytes are endstage, they do have some limited ability to modify their functional status. For instance, activated neutrophils upregulate the expression of the CD64 receptor (50), a response that is now used clinically as a sign of occult infection, and activated eosinophils upregulate co-stimulatory and adhesion molecules in response to parasitic infection (51).

Resident cells in the connective tissue underlying the skin, mucosa and internal epithelium are also critical to immune function. Macrophages ingest, process and package antigens for presentation to T-cells, a process mediated by a transient intercellular macromolecular complex recently termed the immune synapse (52,53). Other accessory cells, such as dendritic cells (54), are also involved in antigen presentation. Mast cells contain histamine and other mediators of allergy in pre-formed granules ready for immediate release. They have surface receptors that bind very strongly to IgE antibodies, sensitizing them (and the individual they inhabit) to allergens recognized by the IgE. When allergens interact with their surface-bound IgE, activation and degranulation lead to immediate hypersensitivity (55). Mast cells also mediate signalling between the peripheral nerves and local immune activity, one reason that immediate hypersensitivity responses can be induced rather easily by Pavlovian conditioning (56).

Biochemical biomarkers

Biochemical biomarkers (excluding genomic DNA) include protein and RNA macromolecules as well as smaller molecules, such as steroid hormones and prostaglandins. Technical issues attend all the methods used to measure these markers, particularly the macromolecules. Proteins are often detected and quantified by antibody-binding methods, which may be subject to cross-reactivities or other interferences that cause spurious results (57). The use of mass spectroscopy for protein analysis has increased, particularly as a biomarker discovery tool (58–60). Some of the initial, promising results obtained this way have turned out to be disappointing (61); a careful approach to method evaluation and study design is required for meaningful results (62). RNA is generally detected and quantified by hybridization reactions, often in an expression microarray with thousands of targets. These methods are also subject to technical vagaries, but some standardization has been achieved (63). RNA microarrays have shown considerable promise in some clinical applications (64), but again, a careful approach to method evaluation and study design is required for meaningful results.

Antigen-specific biochemical markers

Among the wide range of biochemicals involved with immunity, only antibodies (also called immunoglobulins (Ig)) and T-cell receptors (Tcr) are antigen-specific.

Immunoglobulins. Ig molecules are rather large heterodimeric proteins that have a basic Y-shaped structure with two symmetrical antigen binding sites connected to a common stem through a "hinge" region (Figure 13.1). The basic Ig

Figure 13.1. This figure shows a space-filling molecular model of the human IgG1 antibody molecule. The constant region of the two heavy chains are shown in red and their variable region in yellow. The constant region of the two light chains is shown in blue and their variable region in grey. The two antigen binding sites are at the top of model, where the yellow and grey regions come together. The lower part of the model shows the region where various antibody functions, such as complement fixation, reside. Carbohydrate molecules that bind to this region are shown in violet, and areas specialized to the IgG1 subclass are shown in white. Copyright Mike Clark, adapted with permission from http://www.path.cam.ac.uk/~mrc7/mikeimages.html

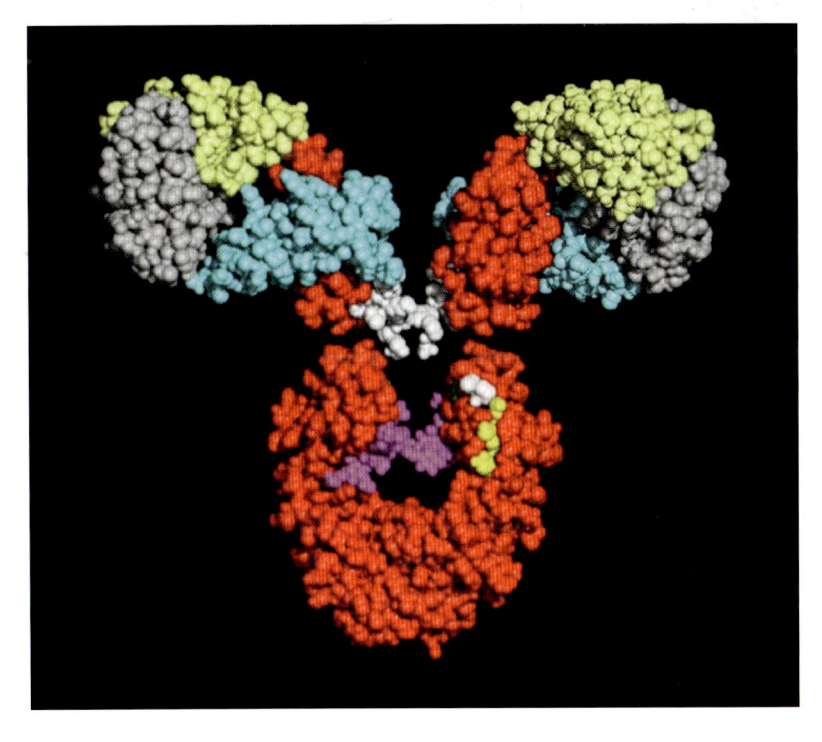

unit is composed of two identical copies of each of two peptide chains, the light chain and the heavy chain. Light chains come in two varieties, kappa and lambda, coded by different genes from different loci. Heavy chains are coded by only one genetic locus, but somatic recombination within the locus can produce different variants called isotypes through a process called class-switching.

Immunoglobulins have two functions in the immune response: membrane-bound Ig molecules are the antigen-specific receptors on B-cells, and secreted Ig molecules have a variety of effector functions critical to host defence including complement fixation, opsonization (which promotes phagocytosis) and viral inactivation. When antigen binds to Ig receptors on B-cells, they proliferate and secrete antibody of the same specificity as the original Ig receptor. Although the secreted antibody has the same antigen binding site as the B-cell receptor, it may differ in other portions of the molecule, which accounts for the different isotypes: IgM, IgG, IgA, IgD, and IgE. As an effector molecule, IgM is the first isotype to be produced in a primary immune response; it is especially good at binding antigen into complexes and activating complement. IgD functions only as a B-cell receptor and is not normally secreted. IgG is the most common isotype in serum, normally accounting for up to one-third of total serum protein. It exists in four subclasses, which differ in functions such as complement fixation and placental transfer. IgA is responsible for mucosal immunity (65,66) and is secreted from epithelial surfaces in the airways and the gut; a variant form is present in serum (67). IgE is (in humans) uniquely involved in immediate hypersensitivity, best known as the cause of common allergy and a major factor in the pathogenesis of asthma (68).

In terms of measurement, earlier difficulties in the standardization of assays to measure the major Ig isotypes in serum (1) have been largely resolved (69). However, measurement of isotype subclasses is not as well standardized. IgE concentrations in serum are much lower than the other secreted isotypes and must be measured by more sensitive methods. Measurement of antigen-specific IgE is an important marker for allergy in diagnostic, occupational, and research settings. Most of these assays are well standardized (70), but customized tests for IgE to novel allergens must be carefully characterized to assure sensitivity and specificity.

T-cell receptors (Tcr). Tcr proteins have a molecular structure and antigen specificity analogous to, but somewhat different than, that of antibodies. Tcr molecules are not secreted and have no effector function. There are two major types of Tcr: α-β and gamma-delta. These types may in turn be grouped into families that can be differentiated by monoclonal antibodies (71) or genetic analysis (see below). Like surface receptor antibodies on B-cells, the Tcr receptors on T-cells allow them to respond to antigen binding by proliferating and secreting effector molecules, in this case peptide regulators called lymphokines (13,72). Cytotoxic (killer) T-cells use their Tcr receptors to identify their cellular targets: viral-infected cells or, in some cases, cells that have undergone malignant transformation (73).

Non-specific biochemical markers

The immune system uses a variety of proteins, peptides, and smaller molecules to effect and regulate host defence.

Lymphokines, cytokines and interleukins. Lymphokines, which are produced by lymphocytes, are a subset of the localized cellular peptide mediators called cytokines elaborated by a variety of cells. The interleukins are cytokines particularly involved with signalling among leukocytes. The first interleukins (IL-1 and IL-2) were labouriously identified as functional activation and growth factors from cell culture supernatants (74). One of the most recent (IL-34) was uncovered by a systematic search of the extracellular proteome expressed by a set of recombinant secreted proteins, using a suite of assays that measured metabolic, growth or transcriptional responses in diverse cell types (75).

All cytokines are localized tissue mediators, and their concentration in peripheral blood is normally extremely low or undetectable. Increased concentrations due to spillover from tissue sites of inflammation may sometimes be detected in serum, but the assays used to measure them are not standardized and are subject to interferences, matrix effects and considerable bias between methods.

Genetic polymorphisms in cytokines and cytokine receptors have been shown to be useful biomarkers of susceptibility for lymphoid malignancies (76), other cancers (77–80), oral diseases (81), allergies (82) and autoimmune disorders (83,84). Soluble cytokine receptors, which are deliberately released from cells by a variety of specific mechanisms (85), are important mediators of inflammation (86). While serum concentrations of soluble receptors and receptor-cytokine complexes are good candidate biomarkers for inflammation-related disorders, the assays to measure them are not standardized, and results from different methods may not give concordant results.

Other non-antigenic-specific mediators. In addition to lymphokines, many other non-specific biochemical mediators are used in host defence activities. They often involve inflammation and include acute-phase reactant proteins, such as complement (a family of proteins that react in a cascading fashion) and C-reactive protein, as well as small molecules such as histamine, prostaglandins and endocrine hormones. Although these substances may serve as biomarkers of immune function, their non-specific nature and highly interactive functional pathways often make it difficult to distinguish cause from effect. The importance of neuropeptides and nerve growth factors in the immune response and inflammation has become increasingly apparent (87,88).

Non-antigen-specific cell receptors. Besides Ig and Tcr, several other cell surface receptor protein families are critical components of immune function even though they are not antigen-specific. Two of the most important are the receptors coded by the genes of the major histocompatibility complex (MHC), and the Toll-Like Receptors (TLR).

Major histocompatibility complex (MHC) (transplantation) antigens. The MHC receptors come in three primary classes; classes I and II comprise the human leukocyte antigens (HLA). (The class III MHC region contains a diverse set of genes, some of which code for certain immune-related proteins like cytokines and complement components.) Although they were discovered because they caused rejection of organ transplants, their normal biological function involves packaging antigens for presentation to T-cells. Class I proteins present antigens to CD8 cytotoxic T-cells, and class II proteins present antigens to CD4 helper T-cells. All of the HLA loci are highly polymorphic within a species; the four major loci in humans include over 2500 different alleles, leading to the difficulty in finding matches between organ donors and recipients. Different allotypes may confer relative susceptibility or resistance to autoimmune diseases and certain infectious agents (89). HLA protein allotypes can be identified by reactions with allo-specific antibody reagents, and alloantibodies in previously sensitized individuals can be detected by binding assays (90). Alloreactivity between tissues from two individuals may be detected by mixing lymphocytes from the two sources and measuring functional responses (e.g. proliferation, lymphokine secretion or mRNA production) (91). These types of assays are used clinically, but are highly specialized and should be performed by experienced histocompatibility laboratories. T-cell antigen recognition normally involves presentation by viral-infected epithelial cells or by accessory cells, such as macrophages that have packaged the antigen with HLA proteins. The detection of antigen-specific T-cells using *in vitro* stimulation assays is greatly enhanced by pre-packaging the antigen with a suitable HLA protein into a complex called a tetramer (92). These assays are highly specialized and used primarily in research settings, although some degree of standardization has been achieved (93). The use of tetramers to identify CD8 cytotoxic T-cells specific for viral or tumour antigens has been quite successful, but identification of antigen-specific CD4 helper T-cells remains problematic (94).

Toll-like receptors (TLRs). This class of receptor, named for its similarity to the Toll receptor of Drosophila, is largely responsible for initiating the inflammatory response to microbes and for the host perception of microbes in general (95). TLRs are evolutionarily ancient proteins, and most mammalian species have about a dozen different types; some shared across species and others unique. TLRs account for much of the protective effects of the host response to infection, such as the induction of lasting specific immunity, as well as its pathological effects (e.g. systemic inflammation and shock) (95). Their discovery illustrates the use of forward genetic methods in identifying genes for biomarkers that are constitutively expressed but conditionally functional (96). Originally considered as effectors of innate immunity, TLRs are now seen to have important roles in acquired immunity as well. They are involved in the class-switching maturation of B-cells, as the immune system transitions from the primary response dominated by IgM to the secondary responses dominated by IgG and IgA (97). They also appear to interact with the superantigen-mediated polyclonal activation of T-cells responsible for toxic shock syndrome (98), an often fatal condition associated with the use of tampons that reached epidemic proportions in the United States around 1980 (99).

TLRs belong to a larger group of molecules called the pattern recognition receptors (PRRs) that recognize conserved molecular motifs from pathogenic microbes: pathogen-associated molecular patterns (PAMP). A PAMP database has been established that contains about 500 patterns, including 177 recognized by TLRs (http://www.imtech.res.in/raghava/prrdb/) (100).

The biology of TLRs has just started to impact clinical medicine and public health (101–106), and aberrant variations in TLRs expression caused by genetic polymorphisms, mutations or dysregulation will become increasingly important biomarkers.

Genetic biomarkers

Immunoglobulin and T-cell receptor genes

Immunoglobulin and T-cell receptor genes code for the only antigen-specific proteins in the immune system. The three genes that produce Ig molecules are located on different chromosomes: the kappa light chain gene (chromosome 2), the lambda light chain gene (chromosome 22), and the heavy chain gene (chromosome 14). During differentiation, B-cells determine whether they will use the kappa or lambda gene, and that choice is maintained by all progeny of the clone. The two light chains are never expressed together in the same cell. The T-cell receptor, which has an analogous genetic basis, is coded for by either alpha and beta genes or by gamma and delta genes; like Ig light chains, the two pairs are never expressed in the same cell. The beta and gamma genes are on chromosome 7, while the alpha and delta genes are on chromosome 17.

All seven of the Ig and Tcr genetic loci contain two distinct regions separated by an intervening sequence of nucleotides. One region contains multiple sequences, referred to as V-genes, that code for the N-terminal portion of the peptide which will form the antigen binding site. The other region, referred to as the C-gene, contains sequences that code for the remainder of the peptide chain, which is not involved in antigen recognition. In a differentiation process unique to lymphocytes, somatic recombination removes the intervening sequence between the V-genes and C-genes to form a new gene that codes for the intact Ig or Tcr protein. This recombination persists as a genetic fingerprint in all the clonal progeny that subsequently arise from the lymphocyte undergoing the original recombination. The details differ for each locus, and Ig heavy chain genes have a special feature of multiple C-genes that sequentially recombine with the chosen V-gene to form different antibody isotypes with the same antigen specificity.

The V-genes that code for antigen binding sites might at first be considered precise markers for antigen specificity; however, immune specificity is a selective process. The antigen binding site in Ig or Tcr molecules is not designed to fit a particular antigen; rather, binding sites are created stochastically, and those that happen to react with antigens are selected for clonal expansion. In fact, each antigen binding site is polyspecific (107), and immune specificity to most antigens depends on the wide repertoire of specificities that reinforces common reactivities and dilutes out the others. The relationship between V-genes and antigen specificity is therefore indirect and ultimately must be determined by antigen binding, not gene sequences.

Because of the unique recombination of V-genes and the selective clonal expansion or diminution of the lymphocytes containing them, the repertoire of V-genes differs between individuals and within individuals over time. They can therefore serve as biomarkers of exposure and effect, as well as biomarkers of susceptibility.

Major histocompatibility complex (MHC) genes

The other major gene family involved in the immune response is the major histocompatibility complex (MHC), which comprises 26 different genes including those that code for the human HLA proteins (see http://www.ebi.ac.uk/imgt/hla/). These genes do not influence antigen specificity directly, but they do have notable effects on antigen-specific immune responses and are often used as biomarkers of susceptibility. In particular, the association of MHC polymorphisms has been a *sine qua non* for autoimmune disease since it was first uncovered in a mouse model of autoimmune thyroiditis (108,109).

The high degree of genetic polymorphism in the MHC presents difficulties for genetic analysis, which can be done at low resolution for modest costs or higher resolution for higher costs. The technical issues revolve first around the regions and primers selected for PCR amplification. Thereafter, the amplified product can be tested by probes, but they too may cross-react with different alleles. Selected regions associated with the allotypes (which may lie in intron-exon boundaries) can be sequenced, but even sequence-based typing cannot rule out a polymorphism that lies outside the sequenced region. As with HLA protein analysis, these analyses should be done by experienced laboratories, and epidemiologic investigators should understand the limitations of methods used.

Other genes

Many other genes may be biomarkers of susceptibility or effect modifiers for immune and inflammatory pathologies. Some of these genes code for other non-specific immune mediators, while others have no direct relationship to the immune system. Polymorphisms in cytokine genes, such as TNF-α and IL-8, have been associated with several clinical endpoints including severity of rheumatoid arthritis (110), incident cardiovascular disease (111), inflammatory bowel disease and cancer (112), type 2 diabetes (113) and thrombotic disease in children (114). They have also been associated with effect modification in chemical exposures (115) and nutritional biomarkers (116). However, such associations may not be apparent when tested in large-scale, longitudinal studies (117).

Genes that have no direct relationship to the immune system, such as those involved with metabolism, can also influence immunity and immunopathology. An association between oxidative metabolites of therapeutic drugs and the autoimmune disease systemic lupus erythematosis (118) was long attributed to a slow-acetylator polymorphism of the arylamine-N-acetyltransferase-2 gene. While subsequent epidemiological studies have cast doubt on the relationship with clinical disease (119,120), a relationship with autoimmunity may still exist. Observations in a mouse model of an association between expression of the aryl hydrocarbon receptor (AHR) and the T_H17 T-cell subset (38) suggest that exposure to aryl hydrocarbons, modified by AHR polymorphisms, may be associated with autoimmunity and perhaps with B-cell malignancies (121). Epigenetic changes could also account for differences between individuals in the way their gene–environment interactions lead to acquired susceptibility or resistance for autoimmunity (122) or lymphocyte malignancies (123). Genes, such as the autoimmune regulator that controls the expression of tissue-specific antigens, may have a profound impact on immune tolerance and autoimmunity (124). Genes concerned with the regulation of cell growth, such as *BCL2*, are especially important in lymphocytes, given their propensity for clonal expansion, and the non-coding regulatory microRNAs that are involved with cell growth pathways may be more important markers than many coding genes (125).

Special considerations for using immune biomarkers in epidemiologic studies

Immune biomarkers may be used to evaluate populations for disorders of the immune system itself, for immunogenic exposures, or for pathological conditions in other organ systems that provoke changes in immune status.

Disorders of the immune system

Three general types of disorders of the immune system may have adverse health consequences: immune deficiencies, inappropriate immune reactivities, and unregulated proliferation leading to lymphoid malignancies (1).

Immune deficiency disorders

Immune deficiency disorders are those in which the immune system fails to mount adequate protective responses against infection or certain forms of cancer. Deficiencies may be primary (caused by inherited genetic traits or spontaneous mutations) or secondary (caused by exposures or infections, such as HIV). Depending on the nature of the deficiency, the health consequences can range from almost unnoticeable, such as increases in the incidence of mild infections, to life-threatening, such as overwhelming sepsis. Immune deficiencies may be indicated by low or absent levels of serum immunoglobulins, low or absent numbers of immune cells, or decreased functional responses.

Immune reactive disorders

Immune reactive disorders are due to inappropriate or poorly regulated responses in which the ensuing inflammation damages host tissues. Autoimmune and allergic diseases are the major types of reactive disorders. Depending on their cause and nature, they can range from mild to severe.

Common allergies are caused by inappropriate responses to environmental antigens (usually referred to as allergens) that release histamine and lipid-derived mediators. These allergic reactions are often directed against airborne antigens and often contribute to the pathogenesis of asthma. Their severity ranges from mild localized symptoms, such as rhinitis, to life-threatening systemic anaphylaxis. Depending on the causative antigen, *in vitro* tests for allergen-specific immunoglobulin E (IgE) serum antibodies are often good markers for exposure to the antigens that evoke allergies (70).

Autoimmune disorders are often debilitating diseases in which the immune system reacts against its own host tissues. Autoimmune reactions can damage the skin, liver, kidneys, various glands, joints and other tissues, leading to diseases such as rheumatoid arthritis, ankylosing spondylitis, systemic lupus erythematosis, thyroiditis, multiple sclerosis, myasthenia gravis and type 1 diabetes. Autoimmune diseases are almost always associated with antibodies that react to self-proteins in particular tissues or cell components, and these autoantibodies can serve as predictive markers (108).

Immune proliferative disorders

Immune proliferative disorders include lymphoma, multiple myeloma, and chronic lymphocytic leukaemia. Like other forms of cancer, they involve the uncontrolled expansion of one family (clone) of cells. Immune proliferative disorders have unique clonal characteristics in both their receptor phenotypes and molecular genotypes, which can often serve as excellent biomarkers.

Immunogenic exposures

An acquired immune response can provide biomarkers of specific exposure to infectious agents or sensitizing chemicals. Antibodies are commonly employed in seroprevalence studies for viruses (126), bacteria (127) and parasites (128). IgE antibodies can reveal exposure to allergy-inducing antigens and their association with asthma. Antibodies to sensitizing chemicals, such as toluene diisocyanate, can serve as markers of exposure and susceptibility in occupational settings (129). Functional assays, such as T-cell proliferation, are more difficult to perform, but can be useful with particular exposures; for example, the lymphoproliferative test for beryllium has been carefully evaluated as a marker for sensitization (130). It should be emphasized that any biomarker of acquired immunity requires sensitization and therefore cannot rule out a non-sensitizing exposure.

Disorders of other organ systems

Infections

Infectious diseases that involve any tissues are likely to cause changes in the host defence system; in fact, many of the symptoms associated with infections are caused not by the infectious agents themselves, but by cellular and molecular activities of the host response. Antibodies and antigen-specific T-cells can provide markers for specific infectious agents, while elevations in acute phase serum proteins, some cytokines, and certain cell surface receptors (50) are non-specific markers that suggest infection.

Malignancies

Some solid tumours that release tumour-specific antigens may elicit immunogenic responses that serve as markers of the malignancy. These markers may manifest as tumour-associated autoantibodies or T-cell responses.

Other conditions

Malnutrition, chronic disease, stress, pregnancy and a variety of other factors can all influence and be influenced by the immune system. Immune markers could be used as indicators of these conditions; conversely, these effects can be confounding variables when immune markers are used in attempts to characterize the host defence system itself (1).

Immune biomarkers in animal models and epidemiologic studies

The type of samples used to test for immune components illustrates a major difference between the use of immune biomarkers in public health investigations compared to most basic research investigations (1). Animal models in general, and mice in particular, have been the mainstay of basic immunology research. Central lymphoid tissues, such as spleen, are readily harvested from mice; however, useful quantities of peripheral blood are difficult to obtain. Conversely, human studies are often limited to sampling peripheral blood. Although it does provide a convenient source of both cells and mediators, peripheral blood is by no means representative of the immune system as a whole. Host defence activities take place in the central lymphoid tissues (spleen, lymph nodes, epithelial-associated lymphoid tissues) and in interstitial tissue at local sites of injury and infection. Cell traffic and recirculation through the blood is carefully regulated: activated cells and molecules are quickly removed, while some cells and mediators persist outside the bloodstream for days and even years. With these points in mind, epidemiologists should appreciate the challenge of adapting findings from animal models of immune function to the design of epidemiologic studies.

Summary of basic concepts

Almost all markers used as tests of immune status are active participants in protective, regulatory or pathogenic processes of the immune system. This direct biological relevance provides special opportunities to learn about the mechanisms of host injury and response through tests for immune components. However, it can also make interpretation more difficult, since physiologic interactions among markers can mask changes or create internal confounders. Moreover, the continual changes that occur as the immune system senses and responds to environmental influences makes the normal ranges of variability for immune constituents very large between individuals, and even within individuals over time. Finally, the immune system of each individual continues to evolve throughout life, its course determined by a combination of inherited traits and acquired exposures. Since human beings are generally outbred and often exposed to a great variety of environmental stimuli, the diversity among individual immune systems is far greater than that among any organ system other than the neurobehavioural. The numerous confounding factors that can influence the immune response must also be taken into account (1).

Examples of using immune biomarkers in epidemiology and public health

Autoantibodies as pre-clinical markers of cancer

Cancer cells may express proteins that are not expressed in normal tissue or are expressed only at very low levels, and are therefore seen as foreign antigens by the immune system (Figure 13.2) (131). Such tumour-specific antigens (TSAs) were first uncovered in rodent tumours induced by coal tar dyes (132) and were later found in certain naturally-occurring human cancers (133). These discoveries augured two attractive concepts: first, that tumour defence was a primary function of the immune system that might be harnessed therapeutically; second, that testing for TSAs would allow early detection of cancer. For several decades, neither concept lived up to its presumed promise, but newly-uncovered biomarkers have revitalized efforts aimed at both therapy (131) and early detection. In part, the resurgence involved a more realistic perspective on the use of biomarkers. As studies showed that many so-called TSAs could be detected in persons without cancer, the term has largely been replaced by the more appropriate tumour-associated antigen (TAA).

The first human TAA identified was carcinoembryonic antigen (CEA) produced by colon cancer (133), and it remains useful as a marker for tumour recurrence and progression (134). While serum CEA levels did provide statistically significant predictive value when used in prospective population studies (135), neither the sensitivity nor specificity of the test justified its use in general screening (136). Prostate-specific antigen (PSA) has proven somewhat more serviceable,

Figure 13.2. Three ways for self antigens to become tumour antigens. Peptides from three normal self proteins (yellow, blue, and green) are presented on the cell surface as normal self peptides (yellow, blue, and green) in major histocompatibility complex (MHC) molecules. In cases of mutation (A), failure of the tumour cell to repair DNA damage can result in a mutation (red) in a normal protein and, consequently, presentation of mutated peptides (red) on the surface of tumour cells. Because of a mutation, or factors that regulate its expression, a normal protein (green) can be overexpressed in a tumour cell and its peptides presented on the cell surface at highly abnormal levels (B). In cases of post-translational modification (C), a normal protein can be abnormally processed (spliced, glycosylated, phosphorylated, or lipidated) post-translationally (green stripes), resulting in an abnormal repertoire of peptides on the surface of the tumour cell. Used directly from (131) by permission from the Massachusetts Medical Society.

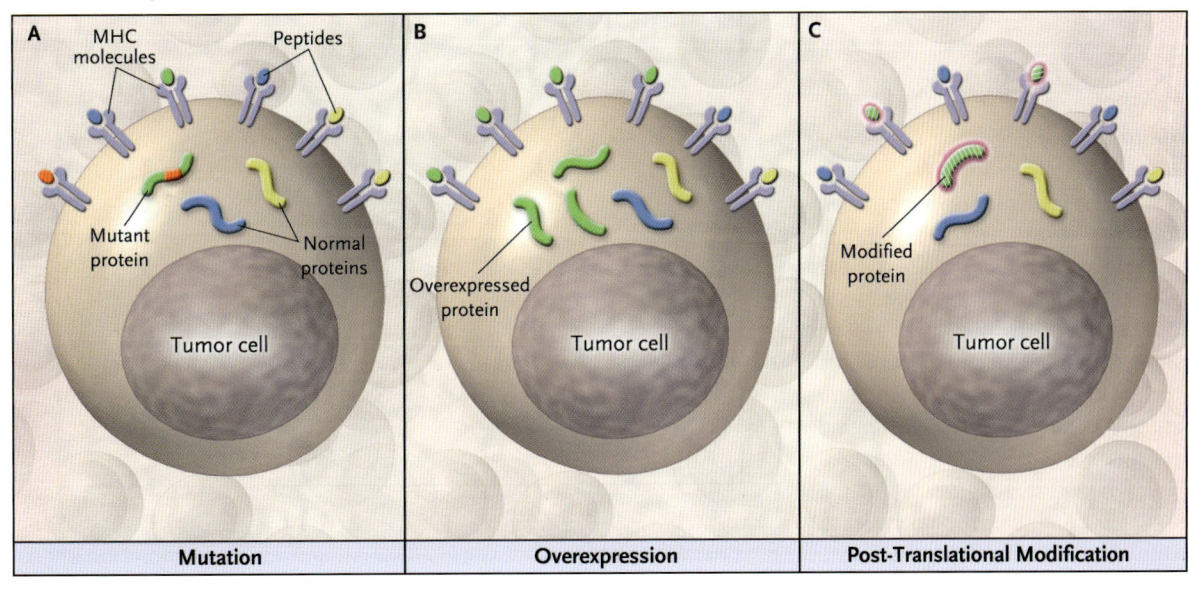

though still problematic, as a screen for prostate cancer (137), but in general blood screening for TAA has not improved the early detection.

The paradigm shift that has occurred over the last several years focused attention not on TAAs themselves, but rather on the immune responses to them. In essence, TAAs can act as autoantigens, producing a weak but detectable antibody or T-cell response. Although this is not a new idea (138,139), modern methods of molecular engineering allow TAA genes to be cloned and transfected, providing a ready supply of antigen to use in high-throughput multiplexed assays with attomolar sensitivity (140,141). Promising results have been obtained for the detection of autoantibodies in lung cancer (142), liver cancer (143), prostate cancer (141) and ovarian cancer (144). Assays for detecting TAA-specific T-cells have been applied mostly to studies of tumour vaccines or immunotherapy (145), but exploratory studies show that such T-cells can be detected in breast cancer patients naive to immunotherapy (146). Immune biomarkers may yet prove to be useful tools in the early detection of cancer.

Genetic and environmental risk factors for B-cell malignancies

While the role of the mammalian immune system in protection against cancer remains enigmatic, it is clear that lymphocytes themselves can lose control of their proliferative potential and expand uncontrollably into lymphoid malignancies. In the Eastern hemisphere, most lymphoid malignancies arise from T-cells, a phenomenon directly related to the endemic presence of human T-cell lymphocytotrophic viruses (HTLV). In the Western hemisphere, T-cell malignancies are rare, but B-cell malignancies represent a major proportion of cancers not related to the obvious risk factors of smoking and diet.

B-cells arise from haematopoietic stem cells and undergo multiple stages of differentiation, terminating as antibody-secreting plasma cells. Four major classes of cancer arise from these various stages: acute lymphoblastic leukaemia, non-Hodgkin lymphoma, chronic lymphocytic leukaemia, and multiple myeloma/plasmocytoma. B-cell acute lymphoblastic leukaemia, the most common form of childhood leukaemia, originates so early in the B-cell differentiation pathway that it is usually considered a stem cell malignancy. The remaining B-cell malignancies all arise from B-cells in later stages of differentiation (Figure 13.3) (147). The genetic and environmental risk factors for these B-cell malignancies remain surprisingly elusive, but information

Figure 13.3. A Venn Diagram illustrating the hypothetical relationship among the MBL, MGUS, and the malignant B-cell diseases to which they may progress. The diagram is based on the spectrum of progression endpoints for MBL and MGUS and the interrelationships of their respective biomarkers. The overlapping areas indicate some extent of shared biomarkers or clinical endpoints. Since the definition of MBL excludes any haematolymphoid disorder, areas of overlap between MBL and a malignancy is meant to convey shared biomarkers or progression from MBL. MBL and MGUS usually appear independently, but may appear together. Both conditions can remain quiescent over the lifespan of the individuals in whom they are found, or they can progress to clinical disease. CLL is shown as a complete subset of MBL, and MM/WM/SMM as nearly complete subsets of MGUS, under the presumption that all CLL is preceded by MBL and nearly all MM and WM is preceded by MGUS (although the precedent conditions may not be detected before the clinical disease endpoints are diagnosed). SMM frequently, but not always, progresses to MM. MGUS cases may infrequently develop CLL or NHL. At least one case with a combination of MBL and MGUS that developed into WM has been reported. MBL is detectable in a subset of already-diagnosed NHL cases (but to date there have been no reports of MBL developing into NHL). At least one type of NHL (small lymphocytic lymphoma) is considered to be a variant of CLL. AL may be associated with MGUS, SMM, MM, or WM and rarely with PC, CLL, or NHL. MBL progression directly to AL has not been reported to date. Used directly from (147) with permission

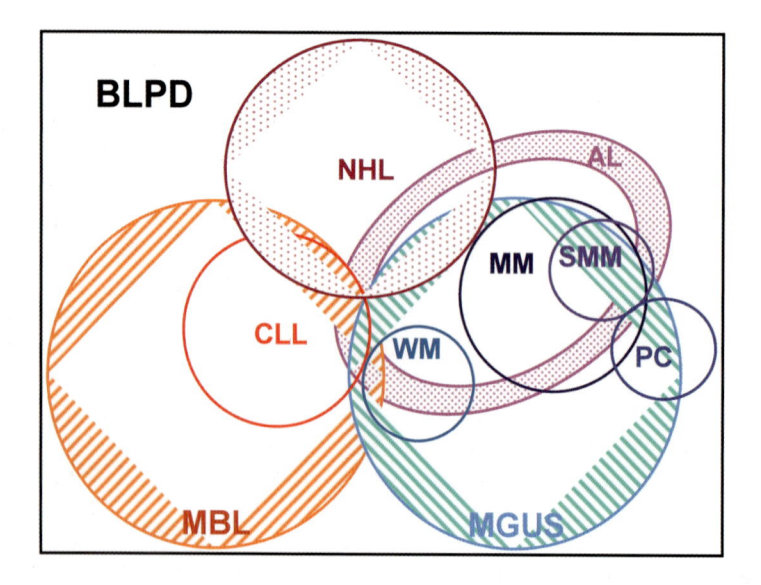

AL, immunoglobulin light chain amyloidosis; CLL, chronic lymphocytic leukaemia; MBL, monoclonal B-cell lymphocytosis; MGUS, monoclonal gammopathy of undetermined significance; MM, multiple myeloma; NHL, non-Hodgkin lymphoma; PC, plasmacytoma; SMM, smouldering multiple myeloma; WM, Waldenstrom macroglobulinemia.

revealed by population studies and immune biomarkers has allowed a much better understanding of their natural history.

The clonal expansion of lymphocytes must be carefully regulated to avoid overwhelming the host. Protective immunization generally involves the controlled proliferation of many different lymphocyte families, leading to a polyclonal response. When proliferation is dominated by a single clone, the result is a monoclonal response. Monoclonal proliferation is the first step in a progression that may lead to a lymphoid malignancy (148,149).

Chronic lymphocytic leukaemia (CLL) is a classic example of a B-cell malignancy arising from monoclonal expansion (150). Cellular, biochemical and genetic biomarkers have all contributed to our increased understanding of the natural history of CLL. The disease process probably begins with chronic immune stimulation by infectious agents, other external antigens, or autoantigens. As normal B-cell clones expand in response to antigen stimulation, the chance of individual cells acquiring genetic defects increases. Some of these defects cause the cell to escape regulatory control of proliferation: the best example to date is the loss of the microRNAs miR-15a and miR-16–1, critical regulatory factors in the bcl-2 pathway for apoptosis (151). Continued expansion of the deregulated clone promotes opportunity for other genetic lesions to accumulate, including epigenetic changes (123). At some point, the damaged clone exhibits phenotypic changes, typically an increased expression of CD5 and decreased expression of CD20. Eventually the clonal proliferation causes clinical disease by accumulating in lymphoid tissues and displacing normal haematopoietic cells in the bone marrow. While most cases of CLL are sporadic, a familial variant has been recognized for many years (152).

The transition from a pre-clinical B-cell proliferative disorder to CLL has been documented in a succession of studies made possible by the advent of flow cytometry (15). The term monoclonal B-cell lymphocytosis (MBL) is now used to describe the pre-clinical state (153). One of the first systematic studies of MBL originated from environmental public health studies in which 13 individuals with MBL were detected (prevalence of 0.9% among participants age 40 or above) (154). These individuals were followed for up to 12 years, along with other study participants who had high B-cell counts without MBL. The majority of MBL cases remained stable or died of unrelated causes, but progression to a B-cell malignancy was observed in two of the 13: one case of CLL and one case of Waldenstrom macroglobulinemia, a related disease. Interestingly, the high B-cell counts in individuals who did not develop MBL regressed to normal over the follow-up period.

The other seminal MBL study involved familial CLL, where 18% of first-degree family members without CLL were found to have MBL (155). This striking increase over the general population prevalence suggests the familial risk for CLL is reflected in the risk for MBL. Since these studies, other population surveys have shown that the prevalence of MBL increases with age, approaching 3–5% in older, otherwise healthy, adults (156,157).

Even with the power of multiparameter flow cytometry, the detection of MBL is not trivial, since it depends on subjective assessment and sequential selection ("gating") strategies that isolate B-cell subsets with distinct phenotypic characteristics. Once a distinct population has been identified, MBL can be identified by light chain restriction. Antibodies may have either kappa or lambda light chains, but all the cells of a particular clone must make the antibodies with the same light chain. A phenotypic cluster that shows only one type of light chain may therefore be considered monoclonal. The complexity of cell preparation, flow cytometry and data analysis make standardization of methods to detect clonality critical for epidemiological assessment. Fortunately, the raw data from flow cytometric analysis can be captured and re-analysed using different gating strategies, allowing retrospective analysis of existing data (158).

Long-term studies of MBL and CLL using established and newly-uncovered biomarkers, such as microRNAs, will be required to sort out environmental risk factors, innate susceptibility and biomarkers of progression.

Newborn screening for immune disorders

Newborn infants that appear healthy may actually have serious latent disorders that will cause future disease, disabilities or even premature death. Newborn bloodspot screening (NBS) is designed to identify such infants quickly so that medical intervention can begin before they fall victim to such disorders. A small amount of blood from a heel stick is collected on filter paper to form a dried blood spot (DBS). DBS samples are sent to central laboratories where they are analysed by various methods to detect biomarkers of latent disorders (159). The first conditions detected by NBS were metabolic or endocrine disorders, and they remain the dominant type screened for in current NBS programs. However, interest in screening for other types of disorders is growing rapidly (160), and the idea of screening for risk factors of future disorders, in addition to screening for established (though occult) conditions, has been gaining attention. Two immune disorders typify these two trends: severe combined immune deficiency (SCID), and type 1 (juvenile) diabetes.

Severe combined immune deficiency

Severe combined immune deficiency (SCID) is a lethal congenital failure of immune development (161). It is often called "bubble boy disease" because of early attempts to prevent infection through sequestering the child from the natural environment. Because of the persistence of placentally-transferred maternal antibodies, SCID remains concealed for several weeks after birth, but without a functional immune system, babies soon become infected and typically die in infancy. A series of landmark studies have shown that newborns with SCID can be rescued before they become symptomatic by transplanting bone marrow progenitor cells (162). However, such rescue is difficult or impossible after SCID babies become infected. SCID thus meets the ideal criteria for NBS: a lethal disorder with a latent onset that can be prevented by medical intervention. The birth prevalence of SCID is not certain, but is estimated to be in the range of 1 to 4 per 100 000.

The process of finding an immune biomarker for SCID that can be measured on a newborn DBS illustrates how far-reaching our knowledge of the immune system and our abilities to probe it have come. First came the understanding that, although SCID is expressed as a combined deficiency involving both humoral (B-cell) and cell-mediated (T-cell) immunity, it is actually a defect in T-cell development. B-cell counts in SCID babies are normal or even elevated, and they are fully functional. However, without functional T-cells to provide help, even humoral responses are deficient, giving the phenotype of a combined immune deficiency. The second realization was that mutations at any one of several unrelated genetic loci could result in failure of T-cell development: SCID was a single gene defect in each individual case, but with multiple genetic causes overall (Table 13.2). Other loci in which mutations could cause SCID may yet be uncovered (163). Moreover, the mutations at these various loci are widely scattered throughout the exons, so screening by conventional genetic tests is not feasible.

The third consideration came from the knowledge of T-cell development (Figure 13.4) (164), which led to a unique marker of T-cells that could be measured in DBS (165). When lymphocytes rearrange the genes that form their antigen receptors, a small segment of DNA is removed to juxtapose two previously separated segments. In T-cells, the excised segment is removed as a circular fragment called a T-cell receptor excision circle (TREC). TRECs are produced only in the original recombination event and are not duplicated in subsequent cell division, so fewer T-cells contain TRECs as the immune system develops. The newborn infant,

Figure 13.4. Defects in human T-cell development resulting in SCID phenotype. A simplified depiction of lymphocyte differentiation is shown. B-cells and NK cells mature in the bone marrow, whereas T-cells mature in the thymus. Normally, only the mature forms of these cells are released into the peripheral blood. Various stages in NK and T-cell development that are blocked by mutations in the genes known to cause SCID (IL2RG, JAK3, ADA, IL7R, RAG1, RAG1, ARTEMIS, and CD45) are indicated by X and dashed lines. Presence or absence of T-cell-specific antigenic markers (CD4,CD8TCR-α/β, and TCR-gamma/delta) is also indicated. Effects of these mutations on B-cell development is not shown. Used directly from (164) with permission

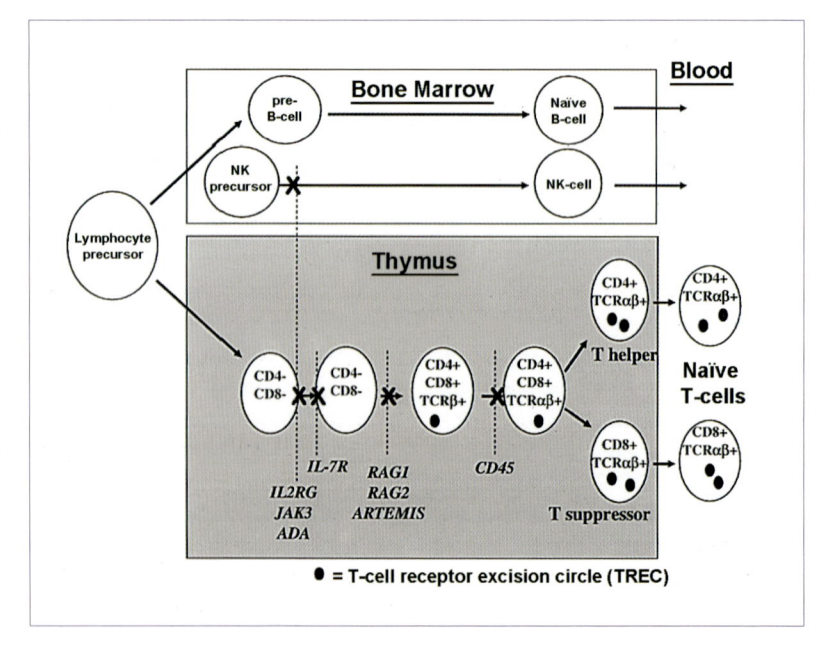

Table 13.2. Genetic loci associated with severe combined immune deficiency (SCID)

			Characteristics of SCID					
				Presence of				
Gene	Locus	Gene product/function	T-cell	B-cell	NK-cell	Mode of inheritance*	No. unique mutations identified	OMIM No.
IL7R	5p13	IL7 receptor. Needed for T-cell development. Activates JAK3 kinase	-	+	+	AR	5	146661
CD45	1q31-q32	Protein tyrosine phosphatase. Regulates Src kinases required for T-cell and B-cell antigen receptor signal transduction	-	+	+	AR	3	151460
IL2RG	Xq13.1	Gamma-c chain of IL2, 4, 7, 9, 15 cytokine receptors. Needed to activate JAK3 for intracellular signalling	-	+	-	XLR	169	308380
JAK3	19p13.1	Tyrosine kinase. Needed for differentiation of haematopoietic cells	-	+	-	AR	27	600173
RAG1	11p13	DNA recombinase. RAG1/RAG2 mediate DNA recombination during B-cell and T-cell development	-	-	+	AR	44	179615
RAG2	11p13	DNA recombinase. RAG1/RAG2 mediate DNA recombination during B-cell and T-cell development	-	-	+	AR	18	179616
ARTEMIS	10p	Involved in DNA repair during V(D)J recombination	-	-	+	AR	9	605988
ADA	20q13.11	Part of the purine salvage and methylation pathways. Needed for removal of toxic metabolites (e.g. ATP, S-adenosyl homocysteine) that inhibit lymphoid cells	-	-	-	AR	54	102700

*Mode of inheritance: AR, autosomal recessive; XLR, X-linked recessive. Adapted from (164).

however, is still producing new T-cells at a rapid pace, so about 10% of the peripheral blood T-cells contain TRECs. These cells may be measured by quantitative real-time polymerase chain reaction (Q-PCR), a technique in which each cycle of DNA amplification is monitored for the appearance of fluorescence from a probe released during the polymerase-mediated extension process (166). The greater the amount of DNA in the original sample, the more quickly the fluorescence signal increases. This allows construction of a calibration curve relating fluorescence to the number of TRECs.

The Q-PCR assay for TRECs has been tested on several thousand anonymized newborn DBS and some 18 DBS from newborns with SCID. About 99% of the anonymized DBS samples from newborns fall within the expected range of PCR amplification. In contrast, all of the DBS from newborns with SCID failed to show any amplification (165). The assay is therefore highly sensitive and specific, but given the rarity of SCID, its positive predictive value is still low (0.1–1%). Besides SCID, the newborn DBS that fail to amplify may be due to technical problems with the assay or to other T-cell deficiencies caused by genetic disorders, such as DiGeorge Syndrome (167), or acquired disorders, such as congenital HIV infection (168,169). Given the rarity of all of these disorders, large-scale population studies where NBS for SCID is performed under the aegis of translation research will be necessary to evaluate and refine testing protocols.

Type 1 (insulin-dependent) diabetes

Type 1 diabetes (T1D), formerly known as juvenile or type 1 diabetes, is the major cause of diabetes in children. T1D is generally caused by the autoimmune destruction of insulin-producing β cells of the pancreas (170). The autoimmune pathogenesis of T1D was revealed by two biomarkers, one genetic and one acquired. The genetic biomarker is linkage with certain alleles of the MHC genes that code for the human leukocyte antigens (HLA), a risk locus shared by all autoimmune disorders. About half of the attributable risk for T1D is genetic, and about half of that risk is contained in the HLA genes. The genetic risk for T1D is associated particularly with the class II MHC genes that code for the HLA-D antigens (171). Some alleles confer susceptibility, while others confer resistance. Interestingly, resistance is dominant, which allows more cost-effective screening approaches that identify protective alleles and eliminate them from further testing.

The acquired biomarker for T1D is a group of autoantibodies that react with pancreatic islet cell antigens (172). Autoantibodies are the other essential biomarker of autoimmune disorders. In the rheumatic disorders, such as systemic lupus erythematosis, they are an obvious part of the pathogenic process; in organ-specific disorders, such as T1D, they are thought to be largely paraphenomena, but still serve as useful markers. Originally discovered by immunofluorescence microscopy using pancreas tissue to visualize antibody binding to islet cells (173), most testing today is done biochemically using purified islet cell antigens produced by cloned genes. Autoantibodies to three major islet cell antigens have been the important determinants of T1D risk in epidemiologic and natural history studies, but antibodies to other islet cell antigens have been reported on the basis of distinct tissue binding patterns (172).

A series of prospective studies by research centres around the world has established a consensus paradigm (Figure 13.5) for the progression from innate risk to islet cell autoimmunity and ultimately to T1D. The major remaining puzzle is the role of environmental exposures in triggering or advancing the autoimmune process (174). The candidates for such exposures include bacterial and viral infections (particularly enteroviruses and rhinoviruses), food antigens, xenobiotic chemicals, allergens, ultraviolet light, and the immunomodulatory effects of stress. Clearly, the identification of environmental risk factors would open new possibilities for prevention and intervention.

With this goal in mind, a prospective multisite natural history study has been initiated to address comprehensively the role of environmental exposures in T1D. Called TEDDY (The Environmental Determinants of Diabetes in the Young), this study is recruiting infants at higher genetic risk for T1D (as well as controls without higher genetic risk) and assembling them into a long-term study cohort (175,176). To maximize the proportion of recruited children who will develop T1D, the highest genetic risk, defined as one of four MHC class II haplotypes (Table 13.3), is required for eligibility in the general population. However, since familial risk contributes independently, six additional MHC haplotypes are eligible in families where a first-degree relative of the prospective recruit already has T1D (Table 13.3).

Because risk from environmental exposures may begin very early, perhaps even *in utero* (177), TEDDY collects the first samples to look for environmental factors at three months of age. With such a short window to identify and recruit participants, TEDDY investigators seek informed consent for the initial genetic screen from the parents of newborns, making it a research application of newborn screening. By the close of the screening phase, some 300 000 newborns will have been screened, and about 8000 higher-risk infants enrolled. The

Figure 13.5. The stages in the natural history and pathogenesis of childhood type 1 diabetes (T1D). Genetic susceptibility creates an immunological environment that predisposes to pancreatic islet cell autoimmunity. An environmental trigger is suspected in most if not all cases of T1D. Autoimmunity then becomes evident by the presence of autoantibodies to islet cell antigens. The presence of autoantibodies to two or more antigens suggests a progressive condition and a significant risk (> 50%) of developing T1D. As islet cells are destroyed by immune-mediated inflammation, the pancreas loses the ability to produce insulin, ultimately resulting in type 1 (type 1) diabetes. Complications, largely related to inflammatory pathologies, cause morbidity and mortality. The progression from autoimmunity to frank diabetes is highly variable, but the strongest genetic risk factors tend to be associated with the earliest onset of disease

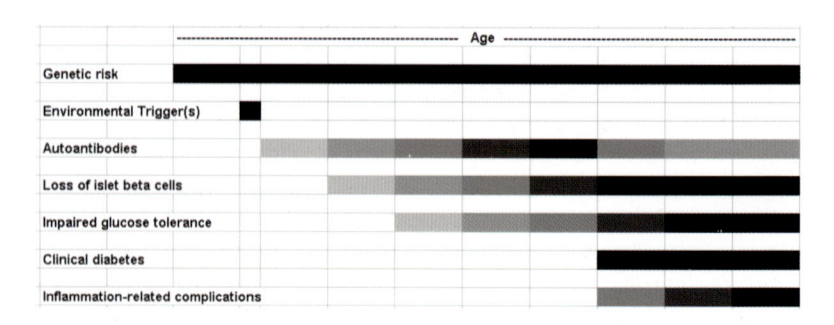

Table 13.3. Eligible MHC genotypes in the TEDDY study that confer higher risk for Type 1 diabetes

General Population Eligible Genotypes

DR4-DQA1*0301-DQB1*0302 /	DR3-DQA1*0501-DQB1*0201
DR4-DQA1*0301-DQB1*0302 /	DR4-DQA1*0301-DQB1*0302
DR4-DQA1*0301-DQB1*0302 /	DR8-DQA1*0401-DQB1*0402
DR3-DQA1*0501-DQB1*0201 /	DR3-DQA1*0501-DQB1*0201

First-Degree Relative Eligible Genotypes

DR4-DQA1*0301-DQB1*0302 /	DR4-DQA1*0301-DQB1*0201
DR4-DQA1*0301-DQB1*0302 /	DR1-DQA1*0101-DQB1*0501
DR4-DQA1*0301-DQB1*0302 /	DR13-DQA1*0102-DQB1*0604
DR4-DQA1*0301-DQB1*0302 /	DR4-DQA1*0301-DQB1*0304
DR4-DQA1*0301-DQB1*0302 /	DR9-DQA1*0301-DQB1*0303
DR3-DQA1*0501-DQB1*0201 /	DR9-DQA1*0301-DQB1*0303

participants will be followed with blood sampling every three months for islet autoantibody measurements until age four, and then every six months until the age of 15. These cohorts are to be followed over a period of 15 years for the appearance of islet cell autoantibodies and diabetes, with documentation of early childhood diet, reported and measured infections, vaccinations, and psychosocial stressors. The TEDDY Consortium will allow for a coordinated, multidisciplinary approach to this complex disease (see http://www.teddystudy.org). Collection of information and samples in a standardized manner will achieve greater statistical power than smaller independent investigations. Most importantly, the TEDDY study will establish a central repository of data and biologic samples for subsequent hypothesis-based research, applying immunologic and genetic biomarkers to samples collected in higher risk children.

Newborn screening for T1D risk is currently a research activity; since no intervention to prevent T1D onset currently exists, it is generally not considered a candidate for routine public health application. However, some diabetologists believe genetic risk for T1D should be part of routine newborn screening, since the knowledge can prevent morbidity, especially in young children where the acute onset of T1D can go unrecognized, occasionally with fatal consequences. Genetic assessment alone can increase the likelihood ratio several-fold, but the positive predictive value (PPV) is still quite low: 1–2%, depending on the population. The appearance of autoantibody to one of the islet cell antigens increases the risk substantially, and the presence of autoantibodies to two or more antigens raises the PPV to around 50% (178). Children with such positive serologies should be monitored for blood glucose levels, especially when they become acutely ill. This tiered approach to using biomarkers (in this case, metabolic as well as immune) may well be a model for future public health applications.

Several technical, operational and ethical caveats attend TEDDY, and other long-term population-based prospective studies. Technically, the high-throughput tests used for genetic screening do not identify HLA haplotypes with certainty, and indeed high resolution testing is neither necessary nor cost-efficient for this purpose (171). In TEDDY, screen-eligible haplotypes are independently confirmed by higher resolution methods after recruitment. In addition, the screening laboratories annually undergo a proficiency testing survey, which has repeatedly confirmed the validity of the different screening methods in use (176). The autoantibody tests have undergone rigorous standardization, but insulin autoantibody, the most important one for identifying the onset of autoimmunity, remains a technically challenging assay (179). Further concerns focus on the use of banked samples, since some of the tests to be done in the future may be affected by storage. This is particularly true for T-cell function tests (93) that could reveal the repertoire of lymphocyte specificities directed against islet cell antigens or environmental triggers. Operationally, long-term prospective studies are expensive and require dedicated management with committed field centres. Ethically, risk communication with the recruited families must be approached carefully, and the requirements of participation must not be unduly demanding to mitigate stress and foster retention in the long-term effort. Such effort is justified by the promise of primary prevention for T1D, eliminating life-long dependence on insulin, and the disabling morbidities that accompany it.

Immune biomarkers of neurodevelopmental disorders

Neuromental disorders (NMDs), such as autism, schizophrenia, attention deficit syndromes and epilepsy, represent a biological enigma and a public health imperative. While specific genetic mutations have been identified in a small proportion of NMDs (180), etiologic factors for the majority remain unclear. A growing body of scientific evidence suggests that many NMDs have an early etiological origin associated with aberrant brain development during gestation (181), and neuroimmunomodulatory factors have been implicated in the prenatal pathogenic process (182,183). Studies in animal models, as well as limited human studies and epidemiological data (184), suggest an etiologic role for autoimmune or cross-reactive antibodies maternally transferred across the placenta when the fetal blood-brain barrier is permeable to IgG. This concept has been developed into a full model termed the gestational neuroimmunopathology (GENIP) hypothesis (Figure 13.6) (181).

One line of evidence supporting the GENIP hypothesis comes from the consideration of the placental barrier and of the immune responses in very young children (181). The placenta restricts passage of maternal IgG2 antibodies, which are the subclass most enriched for reactivity to encapsulated bacteria, such as meningococcus B and *E. coli* K1. Similarly, very young children are unable to mount effective immune responses against bacterial polysaccharide antigens. The withholding of protective antibodies against common deadly pathogens in highly susceptible infants is evolutionarily difficult to explain, since the six-month-old fetus can readily make IgG antibodies to protein antigens of intrauterine infectious agents such as syphilis (185). However, polysaccharides, structurally identical to the bacterial α-2,8-linked capsular polysaccharides, are also synthesized by the mammalian central nervous system, where they regulate neuronal function in association with the neural cell adhesion molecule (186). If antibodies that react with these polysaccharides (or with other sialic acid epitopes, such as those of the many gangliosides in the developing nervous system) were present before the blood-brain barrier became fully impervious to IgG transfer, their effects on brain tissue could disrupt normal neurodevelopmental processes. The mechanism for such disruption could be directly upon neurons (187,188), or more subtly upon the regulation of axonal growth and connectivity (189,190). For instance, antibodies to ganglioside GM1, identified in cases of paediatric autoimmune neuropsychiatric disorders associated with streptococci, act by stimulating the enzyme calcium/calmodulin kinase II (191). In general, cross-reactive or polyreactive antibodies that effect neuronal development or function are seen as an emerging theme in neuroimmunology (192).

Another line of evidence comes from animal models that show

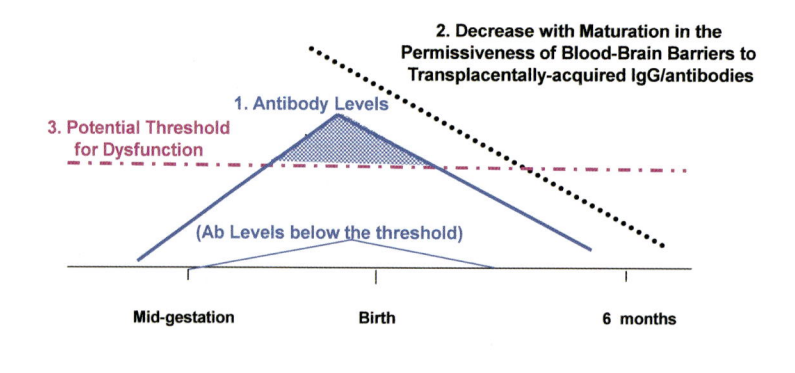

Figure 13.6. Factors influencing the risk that transplacental maternal IgG antibodies could cause developmental neuromental disorders, such as autism and schizophrenia. The relevant parameters are: (1) the concentration of maternal antibodies that could cross-react with brain antigens, (2) the permeability of the blood-brain barrier, and (3) the threshold at which neurodevelopmental damage may occur, presumably related to the developmental state of the fetus. The clinical manifestations for any particular type of neuromental dysfunction may be detected during infancy, or may only be recognized many years postnatally. This will depend on the degree of initial involvement in the function of the developing nervous system, and when the threshold for the neuropathological disorders to become clinically manifest, is lowered by genetic and/or postnatal environmental influences. The IgG antibody levels will also be affected by immune factors in the mother or progeny: the IgG subclass of antibodies to antigens common between an infectious agent and the brain; whether the antigens involved are polysaccharides or proteins; genetic influences on the immune responsiveness to these antigens of the mother; the time of her acquisition of the infectious agent, before, or during, pregnancy, as well as the agent's natural chronicity or reactivation potential, and the transfer by feto-fetal transfusion of IgG/antibodies from one member of a monozygotic twin pair to the other, due to a common chorionic placentation. The blood-brain barriers may become more permissive in case of trauma (e.g. a prolonged vaginal delivery), in which case lower antibody levels may cross to the nervous system. Adapted from (181) with permission

antibodies can influence behaviour if they reach the brain. Mice that were given autoantibodies associated with systemic lupus erythematosis developed behavioural anomalies, if, and only if, they also were given pharmacologic agents that opened the blood-brain barrier (193). Infant monkeys gestationally exposed to IgG antibodies purified from the serum of mothers with autistic children demonstrated the stereotypies characteristic of autistic behaviour (194).

A third line of evidence for GENIP is the identification of antibodies that react with brain tissue antigens in children with neurodevelopmental disorders (195,196). These studies have used research assays that are not widely employed or validated by inter-laboratory studies, but the methods are generally sound, and the differences with control samples are often striking. Although these studies cannot prove causality, they suggest an association that could at least provide useful markers for stratification within the complex spectrum of neurobehavioural diseases.

The key to providing direct evidence for the GENIP hypothesis is the newborn baby, whose blood contains large amounts of placentally-transferred IgG from the mother, as well as the low levels of antibodies made *in utero* by the fetus. Samples from newborns can therefore reveal the spectrum of antibody reactivities to which the developing brain has been exposed. Since blood samples from virtually every newborn in the United States are routinely collected as DBS for newborn screening, the samples required to test the GENIP hypothesis are readily available. Highly-multiplexed suspension arrays have been developed for detecting a wide variety of potentially relevant antibodies in DBS, and they are currently being applied to epidemiologic studies of autism, epilepsy, and other neuromental disorders. Perhaps the major concern in such studies lies not in the arcane realm of immune biomarkers, but rather with the classic epidemiologic dilemma of establishing a consistent case definition for these complex neurodevelopmental conditions.

Future opportunities and challenges

The continual advances of biomedical research in immunology offer an ever-widening opportunity to employ immune biomarkers in epidemiologic studies. In the foreseeable future biomarkers will lead to an increased understanding of the relationship between innate and acquired immunity; the influences of microRNAs on immune cell differentiation and function; the role of epigenetic mechanisms; the regulation of immune responses by the T_H1-T_H2-T_H17 network; the environmental factors that trigger allergy, autoimmunity and immune malignancy; and the interaction of the neurobehavioural and immune systems in neuromental disorders. Also envisioned is a more rapid pace of translation from basic and clinical research to epidemiologic field studies and public health applications, such as newborn screening and early cancer detection.

The translational process will have to confront challenges in the standardization of laboratory measurements, the establishment of biologic validity and the meaningful interpretation of results. A multidisciplinary approach that engages epidemiologists, laboratory scientists and clinicians offers the best chance for useful field studies and public health applications. All the scientists involved in such efforts should have some sense of the way immune markers are measured, of their functional role in health and disease, and of the statistical methods that will be used to analyse the data. While these aphorisms can be applied to any type of biomarker, the self-referential sensory nature of the immune system, its wide range of effector functions, and the narrow margin between protection and pathology make them especially pertinent to the use of immune biomarkers. When properly met, the challenges will open new routes to scientific discovery, disease prevention and public health practice.

Disclaimer: The findings and conclusions in this chapter are those of the author and do not necessarily represent the views of the Centers for Disease Control and Prevention.

References

1. Vogt RF, Schulte PA. Immune markers in epidemiologic field studies. In: Schulte PA, Perera FP, editors. Molecular epidemiology: principles and practices. New York (NY): Academic Press; 1993. p. 407–442.

2. Allen MA, Liang TS, La Salvia T et al. (2005). Assessing the attitudes, knowledge, and awareness of HIV vaccine research among adults in the United States. J Acquir Immune Defic Syndr, 40:617–624.doi:10.1097/01.qai.0000174655.63653.38 PMID:16284540

3. Jacobson DL, Gange SJ, Rose NR, Graham NM (1997). Epidemiology and estimated population burden of selected autoimmune diseases in the United States. Clin Immunol Immunopathol, 84:223–243.doi:10.1006/clin.1997.4412 PMID:9281381

4. Rose NR (2006). The significance of autoimmunity in myocarditis. Ernst Schering Res Found Workshop, 55:141–154.doi:10.1007/3-540-30822-9_9 PMID:16331858

5. Eaton WW, Rose NR, Kalaydjian A et al. (2007). Epidemiology of autoimmune diseases in Denmark. J Autoimmun, 29:1–9.doi:10.1016/j.jaut.2007.05.002 PMID:17582741

6. Lindegren ML, Kobrynski L, Rasmussen SA et al. (2004). Applying public health strategies to primary immunodeficiency diseases: a potential approach to genetic disorders. MMWR Recomm Rep, 53 RR-1;1–29. PMID:14724556

7. Beyer M, Schultze JL (2008). Immunoregulatory T cells: role and potential as a target in malignancy. Curr Oncol Rep, 10:130–136.doi:10.1007/s11912-008-0021-z PMID:18377826

8. Shim YK, Silver SR, Caporaso NE et al. (2007). B cells behaving badly. Br J Haematol, 139:658–662.doi:10.1111/j.1365-2141.2007.06842.x PMID:18021079

9. Davis MM, Krogsgaard M, Huse M et al. (2007). T cells as a self-referential, sensory organ. Annu Rev Immunol, 25:681–695. doi:10.1146/annurev.immunol.24.021605.090600 PMID:17291190

10. Scott A, Khan KM, Cook JL, Duronio V (2004). What is "inflammation"? Are we ready to move beyond Celsus? Br J Sports Med, 38:248–249.doi:10.1136/bjsm.2003.011221 PMID:15155418

11. Cole GA, Nathanson N, Prendergast RA (1972). Requirement for theta-bearing cells in lymphocytic choriomeningitis virus-induced central nervous system disease. Nature, 238:335–337.doi:10.1038/238335a0 PMID:4561841

12. Dannenberg AM. Pathogenesis of human pulmonary tuberculosis: insights from the rabbit model. Washington (DC): ASM Press; 2006.

13. Fisman EZ, Adler Y, Tenenbaum A (2008). Biomarkers in cardiovascular diabetology: interleukins and matrixins. Adv Cardiol, 45:44–64.doi:10.1159/000115187 PMID:18230955

14. Tung JW, Heydari K, Tirouvanziam R et al. (2007). Modern flow cytometry: a practical approach. Clin Lab Med, 27: 453–468, v.doi:10.1016/j.cll.2007.05.001 PMID:17658402

15. Shapiro HM (2007). Cytometry in monoclonal B-cell lymphocytosis and chronic lymphocytic leukaemia–the Hunting of the Snark? Br J Haematol, 139:772–773. doi:10.1111/j.1365-2141.2007.06855.x PMID:18021090

16. Ellmark P, Woolfson A, Belov L, Christopherson RI (2008). The applicability of a cluster of differentiation monoclonal antibody microarray to the diagnosis of human disease. Methods Mol Biol, 439:199–209.doi:10.1007/978-1-59745-188-8_14 PMID:18370105

17. Zola H, Swart B, Banham A et al. (2007). CD molecules 2006–human cell differentiation molecules. J Immunol Methods, 319:1–5. doi:10.1016/j.jim.2006.11.001 PMID:17174972

18. Beeler A, Zaccaria L, Kawabata T et al. (2008). CD69 upregulation on T cells as an in vitro marker for delayed-type drug hypersensitivity. Allergy, 63:181–188. PMID:18005225

19. Lindsey WB, Lowdell MW, Marti GE et al. (2007). CD69 expression as an index of T-cell function: assay standardization, validation and use in monitoring immune recovery. Cytotherapy, 9:123–132.doi:10.1080/14653240601182838 PMID:17453964

20. Wiestner A, Rosenwald A, Barry TS et al. (2003). ZAP-70 expression identifies a chronic lymphocytic leukemia subtype with unmutated immunoglobulin genes, inferior clinical outcome, and distinct gene expression profile. Blood, 101:4944–4951.doi:10.1182/blood-2002-10-3306 PMID:12595313

21. Clinical and Laboratory Standards Institute. Enumeration of immunologically defined cell populations by flow cytometry; approved guideline. 2nd ed. CLSI document H42-A2 [ISBN 1–56238–000–0]. Wayne (PA): Clinical and Laboratory Standards Institute; 2007.

22. Wang L, Abbasi F, Gaigalas AK et al. (2007). Discrepancy in measuring CD4 expression on T-lymphocytes using fluorescein conjugates in comparison with unimolar CD4-phycoerythrin conjugates. Cytometry B Clin Cytom, 72:442–449. PMID:17474131

23. Marti GE, Vogt RF Jr, Stetler-Stevenson M (2003). Clinical quantitative flow cytometry: "Identifying the optimal methods for clinical quantitative flow cytometry". Cytometry B Clin Cytom, 55:59.doi:10.1002/cyto.b.10053 PMID:12949961

24. Clinical and Laboratory Standards Institute. Fluorescence calibration and quantitative measurement of fluorescence intensity. NCCLS document I/LA24-A [ISBN 1–56238–000–0]. Wayne (PA): Clinical and Laboratory Standards Institute, 2004.

25. Elewaut D, Kronenberg M (2000). Molecular biology of NK T cell specificity and development. Semin Immunol, 12:561–568. doi:10.1006/smim.2000.0275 PMID:11145862

26. Hedges JF, Lubick KJ, Jutila MA (2005). Gamma delta T cells respond directly to pathogen-associated molecular patterns. J Immunol, 174:6045–6053. PMID:15879098

27. Tang Q, Bluestone JA (2008). The Foxp3+ regulatory T cell: a jack of all trades, master of regulation. Nat Immunol, 9:239–244.doi:10.1038/ni1572 PMID:18285775

28. Fehervari Z, Sakaguchi S (2006). Peacekeepers of the immune system. Sci Am, 295:56–63. PMID:16989481

29. Roumier T, Capron M, Dombrowicz D, Faveeuw C (2008). Pathogen induced regulatory cell populations preventing allergy through the Th1/Th2 paradigm point of view. Immunol Res, 40:1–17.doi:10.1007/s12026-007-0058-3 PMID:18193360

30. Levy DA (2004). Parasites and allergy: from IgE to Th1/Th2 and beyond. Clin Rev Allergy Immunol, 26:1–4.doi:10.1385/CRIAI:26:1:1 PMID:14755070

31. Kawamoto N, Kaneko H, Takemura M et al. (2006). Age-related changes in intracellular cytokine profiles and Th2 dominance in allergic children. Pediatr Allergy Immunol, 17:125–133.doi:10.1111/j.1399-3038.2005.00363.x PMID:16618362

32. Steinman L (2007). A brief history of T(H)17, the first major revision in the T(H)1/T(H)2 hypothesis of T cell-mediated tissue damage. Nat Med, 13:139–145.doi:10.1038/nm1551 PMID:17290272

33. Kaiko GE, Horvat JC, Beagley KW, Hansbro PM (2008). Immunological decision-making: how does the immune system decide to mount a helper T-cell response? *Immunology*, 123:326–338.doi:10.1111/j.1365-2567.2007.02719.x PMID:17983439

34. Weaver CT, Murphy KM (2007). The central role of the Th17 lineage in regulating the inflammatory/autoimmune axis. *Semin Immunol*, 19:351–352.doi:10.1016/j.smim.2008.01.001 PMID:18276155

35. Korn T, Oukka M, Kuchroo V, Bettelli E (2007). Th17 cells: effector T cells with inflammatory properties. *Semin Immunol*, 19:362–371.doi:10.1016/j.smim.2007.10.007 PMID:18035554

36. Stockinger B, Veldhoen M, Martin B (2007). Th17 T cells: linking innate and adaptive immunity. *Semin Immunol*, 19:353–361.doi:10.1016/j.smim.2007.10.008 PMID:18023589

37. Yu JJ, Gaffen SL (2008). Interleukin-17: a novel inflammatory cytokine that bridges innate and adaptive immunity. *Front Biosci*, 13:170–177.doi:10.2741/2667 PMID:17981535

38. Veldhoen M, Hirota K, Westendorf AM et al. (2008). The aryl hydrocarbon receptor links TH17-cell-mediated autoimmunity to environmental toxins. *Nature*, 453:106–109. doi:10.1038/nature06881 PMID:18362914

39. Stevens EA, Bradfield CA (2008). Immunology: T cells hang in the balance. *Nature*, 453:46–47.doi:10.1038/453046a PMID:18451850

40. Wilson SB, Kent SC, Patton KT et al. (1998). Extreme Th1 bias of invariant Valpha24JalphaQ T cells in type 1 diabetes. *Nature*, 391:177–181.doi:10.1038/34419 PMID:9428763

41. Wilson SB, Kent SC, Patton KT et al. (1999). Extreme Th1 bias of invariant V24JQ T cells in type 1 diabetes [Erratum]. *Nature*, 399:84 doi:10.1038/20007.

42. Redondo MJ, Gottlieb PA, Motheral T et al. (1999). Heterophile anti-mouse immunoglobulin antibodies may interfere with cytokine measurements in patients with HLA alleles protective for type 1A diabetes. *Diabetes*, 48:2166–2170.doi:10.2337/diabetes.48.11.2166 PMID:10535450

43. Zhu J, Paul WE (2010). Heterogeneity and plasticity of T helper cells. *Cell Res*, 20:4–12. doi:10.1038/cr.2009.138 PMID:20010916

44. Neidhart M, Fehr K, Pataki F, Michel BA (1996). The levels of memory (CD45RA-, RO+) CD4+ and CD8+ peripheral blood T-lymphocytes correlate with IgM rheumatoid factors in rheumatoid arthritis. *Rheumatol Int*, 15:201–209.doi:10.1007/BF00290522 PMID:8717104

45. Yamaoka M, Kusunoki Y, Kasagi F et al. (2004). Decreases in percentages of naïve CD4 and CD8 T cells and increases in percentages of memory CD8 T-cell subsets in the peripheral blood lymphocyte populations of A-bomb survivors. *Radiat Res*, 161:290–298.doi:10.1667/RR3143 PMID:14982485

46. Pawlik I, Mackiewicz U, Lacki JK et al. (1997). The differences in the expression of CD45 isoforms on peripheral blood lymphocytes derived from patients with seasonal or perennial atopic allergy. *Tohoku J Exp Med*, 182:1–8.doi:10.1620/tjem.182.1 PMID:9241767

47. Fairfax KA, Kallies A, Nutt SL, Tarlinton DM (2008). Plasma cell development: from B-cell subsets to long-term survival niches. *Semin Immunol*, 20:49–58.doi:10.1016/j.smim.2007.12.002 PMID:18222702

48. Stetler-Stevenson M, Braylan RC (2001). Flow cytometric analysis of lymphomas and lymphoproliferative disorders. *Semin Hematol*, 38:111–123.doi:10.1053/shem.2001.21923 PMID:11309693

49. Youinou P, Pers JO, Jamin C, Lydyard PM (2000). CD5-positive B cells at the crossroads of B cell malignancy and nonorgan-specific autoimmunity. *Pathol Biol (Paris)*, 48:574–576. PMID:10965537

50. Davis BH, Olsen SH, Ahmad E, Bigelow NC (2006). Neutrophil CD64 is an improved indicator of infection or sepsis in emergency department patients. *Arch Pathol Lab Med*, 130:654–661. PMID:16683883

51. Silveira-Lemos D, Teixeira-Carvalho A, Martins-Filho OA et al. (2006). High expression of co-stimulatory and adhesion molecules are observed on eosinophils during human Schistosoma mansoni infection. *Mem Inst Oswaldo Cruz*, 101 Suppl 1;345–351. doi:10.1590/S0074-02762006000900056 PMID:17308795

52. Davis DM (2006). Intrigue at the immune synapse. *Sci Am*, 294:48–55.doi:10.1038/scientificamerican0206-48 PMID:16478026

53. Wan S, Flower DR, Coveney PV (2008). Toward an atomistic understanding of the immune synapse: large-scale molecular dynamics simulation of a membrane-embedded TCR-pMHC-CD4 complex. *Mol Immunol*, 45:1221–1230.doi:10.1016/j.molimm.2007.09.022 PMID:17980430

54. Steinman RM (2007). Dendritic cells: understanding immunogenicity. *Eur J Immunol*, 37 Suppl 1;S53–S60.doi:10.1002/eji.200737400 PMID:17972346

55. Rivera J, Olivera A (2008). A current understanding of Fc epsilon RI-dependent mast cell activation. *Curr Allergy Asthma Rep*, 8:14–20.doi:10.1007/s11882-008-0004-z PMID:18377769

56. MacQueen G, Marshall J, Perdue M et al. (1989). Pavlovian conditioning of rat mucosal mast cells to secrete rat mast cell protease II. *Science*, 243:83–85.doi:10.1126/science.2911721 PMID:2911721

57. Clinical and Laboratory Standards Institute. Clinical evaluation of immunoassays; approved guideline. 2nd ed. CLSI document I/LA21-A2. Wayne (PA): Clinical and Laboratory Standards Institute; 2008.

58. Becker AM, Das S, Xia Z et al. (2008). Serum inflammatory protein profiles in patients with chronic rhinosinusitis undergoing sinus surgery: a preliminary analysis. *Am J Rhinol*, 22:139–143.doi:10.2500/ajr.2008.22.3151 PMID:18416969

59. Liu W, Li X, Ding F, Li Y (2008). Using SELDI-TOF MS to identify serum biomarkers of rheumatoid arthritis. *Scand J Rheumatol*, 37:94–102.doi:10.1080/03009740701747152 PMID:18415765

60. Langbein S (2008). Identification of disease biomarkers by profiling of serum proteins using SELDI-TOF mass spectrometry. *Methods Mol Biol*, 439:191–197.doi:10.1007/978-1-59745-188-8_13 PMID:18370104

61. Jacobs IJ, Menon U (2004). Progress and challenges in screening for early detection of ovarian cancer. *Mol Cell Proteomics*, 3:355–366.doi:10.1074/mcp.R400006-MCP200 PMID:14764655

62. McGuire JN, Overgaard J, Pociot F (2008). Mass spectrometry is only one piece of the puzzle in clinical proteomics. *Brief Funct Genomic Proteomic*, 7:74–83.doi:10.1093/bfgp/eln005 PMID:18308835

63. Clinical and Laboratory Standards Institute. Diagnostic nucleic acid microarrays; approved guideline. CLSI document MM12-A (ISBN 1–56238–608–5). Wayne (PA): Clinical and Laboratory Standards Institute; 2006.

64. Alizadeh AA, Eisen MB, Davis RE et al. (2000). Distinct types of diffuse large B-cell lymphoma identified by gene expression profiling. *Nature*, 403:503–511. doi:10.1038/35000501 PMID:10676951

65. Craig SW, Cebra JJ (2008). Peyer's patches: an enriched source of precursors for IgA-producing immunocytes in the rabbit. 1971. *J Immunol*, 180:1295–1307. PMID:18209023

66. Ogra PL, Welliver RC Sr (2008). Effects of early environment on mucosal immunologic homeostasis, subsequent immune responses and disease outcome. *Nestle Nutr Workshop Ser Pediatr Program*, 61:145–181. doi:10.1159/000113492 PMID:18196951

67. Otten MA, van Egmond M (2004). The Fc receptor for IgA (FcalphaRI, CD89). *Immunol Lett*, 92:23–31.doi:10.1016/j.imlet.2003.11.018 PMID:15081523

68. Gould HJ, Sutton BJ (2008). IgE in allergy and asthma today. *Nat Rev Immunol*, 8:205–217.doi:10.1038/nri2273 PMID:18301424

69. Ledue TB, Johnson AM, Cohen LA, Ritchie RF (1998). Evaluation of proficiency survey results for serum immunoglobulins following the introduction of a new international reference material for human serum proteins. *Clin Chem*, 44:878–879. PMID:9554502

70. Clinical and Laboratory Standards Institute. Analytical performance characteristics and clinical utility of immunological assays for human immunoglobulin E (IgE) antibodies and defined allergen specificities; approved guideline. 2nd ed. CLSI document I/LA20-A2 [ISBN 1–56238–000–0]. Wayne (PA): Clinical and Laboratory Standards Institute, 2008.

71. McLean-Tooke A, Barge D, Spickett GP, Gennery AR (2008). T cell receptor Vbeta repertoire of T lymphocytes and T regulatory cells by flow cytometric analysis in healthy children. *Clin Exp Immunol*, 151:190–198. doi:10.1111/j.1365-2249.2007.03536.x PMID:17983445

72. Perl A, Gergely P Jr, Puskas F, Banki K (2002). Metabolic switches of T-cell activation and apoptosis. *Antioxid Redox Signal*, 4:427–443.doi:10.1089/15230860260196227 PMID:12215210

73. Ng LG, Mrass P, Kinjyo I *et al.* (2008). Two-photon imaging of effector T-cell behavior: lessons from a tumor model. *Immunol Rev*, 221:147–162.doi:10.1111/j.1600-065X.2008.00596.x PMID:18275480

74. Smith KA, Lachman LB, Oppenheim JJ, Favata MF (1980). The functional relationship of the interleukins. *J Exp Med*, 151:1551–1556. doi:10.1084/jem.151.6.1551 PMID:6770028

75. Lin H, Lee E, Hestir K *et al.* (2008). Discovery of a cytokine and its receptor by functional screening of the extracellular proteome. *Science*, 320:807–811.doi:10.1126/science.1154370 PMID:18467591

76. Wang SS, Purdue MP, Cerhan JR *et al.* (2009). Common gene variants in the tumor necrosis factor (TNF) and TNF receptor superfamilies and NF-kB transcription factors and non-Hodgkin lymphoma risk. *PLoS One*, 4:e5360.doi:10.1371/journal.pone.0005360 PMID:19390683

77. Gunter MJ, Canzian F, Landi S *et al.* (2006). Inflammation-related gene polymorphisms and colorectal adenoma. *Cancer Epidemiol Biomarkers Prev*, 15:1126–1131.doi:10.1158/1055-9965.EPI-06-0042 PMID:16775170

78. Kong F, Liu J, Liu Y *et al.* (2010). Association of interleukin-10 gene polymorphisms with breast cancer in a Chinese population. *J Exp Clin Cancer Res*, 29:72.doi:10.1186/1756-9966-29-72 PMID:20553628

79. Sugimoto M, Yamaoka Y, Furuta T (2010). Influence of interleukin polymorphisms on development of gastric cancer and peptic ulcer. *World J Gastroenterol*, 16:1188–1200. doi:10.3748/wjg.v16.i10.1188 PMID:20222161

80. Han W, Kang SY, Kang D *et al.* (2010). Multiplex genotyping of 1107 SNPs from 232 candidate genes identified an association between IL1A polymorphism and breast cancer risk. *Oncol Rep*, 23:763–769. PMID:20127018

81. Dutra WO, Moreira PR, Souza PE *et al.* (2009). Implications of cytokine gene polymorphisms on the orchestration of the immune response: lessons learned from oral diseases. *Cytokine Growth Factor Rev*, 20:223–232.doi:10.1016/j.cytogfr.2009.05.005 PMID:19502097

82. Ryan JJ, Kashyap M, Bailey D *et al.* (2007). Mast cell homeostasis: a fundamental aspect of allergic disease. *Crit Rev Immunol*, 27:15–32. PMID:17430094

83. Aguillón JC, Cruzat A, Aravena O *et al.* (2006). Could single-nucleotide polymorphisms (SNPs) affecting the tumour necrosis factor promoter be considered as part of rheumatoid arthritis evolution? *Immunobiology*, 211:75–84.doi:10.1016/j.imbio.2005.09.005 PMID:16446172

84. Svejgaard A (2008). The immunogenetics of multiple sclerosis. *Immunogenetics*, 60:275–286.doi:10.1007/s00251-008-0295-1 PMID:18461312

85. Levine SJ (2008). Molecular mechanisms of soluble cytokine receptor generation. *J Biol Chem*, 283:14177–14181.doi:10.1074/jbc.R700052200 PMID:18385130

86. Mantovani A, Garlanda C, Locati M *et al.* (2007). Regulatory pathways in inflammation. *Autoimmun Rev*, 7:8–11.doi:10.1016/j.autrev.2007.03.002 PMID:17967718

87. Pavlovic S, Daniltchenko M, Tobin DJ *et al.* (2008). Further exploring the brain-skin connection: stress worsens dermatitis via substance P-dependent neurogenic inflammation in mice. *J Invest Dermatol*, 128:434–446.doi:10.1038/sj.jid.5701079 PMID:17914449

88. Elenkov IJ (2008). Neurohormonal-cytokine interactions: implications for inflammation, common human diseases and well-being. *Neurochem Int*, 52:40–51. doi:10.1016/j.neuint.2007.06.037 PMID:17716784

89. Leslie S, Donnelly P, McVean G (2008). A statistical method for predicting classical HLA alleles from SNP data. *Am J Hum Genet*, 82:48–56.doi:10.1016/j.ajhg.2007.09.001 PMID:18179884

90. Clinical and Laboratory Standards Institute. Detection of HLA-specific alloantibody by flow cytometry and solid phase assays; approved guideline. CLSI document I/LA29-A. Wayne (PA): Clinical and Laboratory Standards Institute; 2008.

91. Stordeur P (2007). Assays for alloreactive responses by PCR. *Methods Mol Biol*, 407:209–224.doi:10.1007/978-1-59745-536-7_15 PMID:18453258

92. Altman JD, Davis MM. MHC-peptide tetramers to visualize antigen-specific T cells. *Curr Protoc Immunol* 2003;17(Unit 17.3).

93. Clinical and Laboratory Standards Institute. Performance of single cell immune response assays; approved guideline. CLSI document I/LA26-A [ISBN 1–56238–546–1]. Wayne (PA): Clinical and Laboratory Standards Institute; 2004.

94. Vollers SS, Stern LJ (2008). Class II major histocompatibility complex tetramer staining: progress, problems, and prospects. *Immunology*, 123:305–313.doi:10.1111/j.1365-2567.2007.02801.x PMID:18251991

95. Hoebe K, Jiang Z, Tabeta K *et al.* (2006). Genetic analysis of innate immunity. *Adv Immunol*, 91:175–226.doi:10.1016/S0065-2776(06)91005-0 PMID:16938541

96. Beutler B (2005). The Toll-like receptors: analysis by forward genetic methods. *Immunogenetics*, 57:385–392.doi:10.1007/s00251-005-0011-3 PMID:16001129

97. Han JH, Akira S, Calame K *et al.* (2007). Class switch recombination and somatic hypermutation in early mouse B cells are mediated by B cell and Toll-like receptors. *Immunity*, 27:64–75.doi:10.1016/j.immuni.2007.05.018 PMID:17658280

98. Dalpke AH, Heeg K (2003). Synergistic and antagonistic interactions between LPS and superantigens. *J Endotoxin Res*, 9:51–54. PMID:12691619

99. Reingold AL. Epidemiology of toxic-shock syndrome, United States, 1960–1984. *MMWR CDC Surveill Summ* 1984;33(3):19SS-22SS.

100. Lata S, Raghava GP (2008). PRRDB: a comprehensive database of pattern-recognition receptors and their ligands. *BMC Genomics*, 9:180.doi:10.1186/1471-2164-9-180 PMID:18423032

101. Sheu JJ, Chang LT, Chiang CH *et al.* (2008). Prognostic value of activated toll-like receptor-4 in monocytes following acute myocardial infarction. *Int Heart J*, 49:1–11. doi:10.1536/ihj.49.1 PMID:18360060

102. Senthilselvan A, Rennie D, Chénard L *et al.* (2008). Association of polymorphisms of toll-like receptor 4 with a reduced prevalence of hay fever and atopy. *Ann Allergy Asthma Immunol*, 100:463–468.doi:10.1016/S1081-1206(10)60472-3 PMID:18517079

103. Ramanathan M Jr, Lee WK, Spannhake EW, Lane AP (2008). Th2 cytokines associated with chronic rhinosinusitis with polyps down-regulate the antimicrobial immune function of human sinonasal epithelial cells. *Am J Rhinol*, 22:115–121.doi:10.2500/ajr.2008.22.3136 PMID:18416964

104. Tversky JR, Le TV, Bieneman AP *et al.* (2008). Human blood dendritic cells from allergic subjects have impaired capacity to produce interferon-alpha via Toll-like receptor 9. *Clin Exp Allergy*, 38:781–788.doi:10.1111/j.1365-2222.2008.02954.x PMID:18318750

105. Mrabet-Dahbi S, Dalpke AH, Niebuhr M *et al.* (2008). The Toll-like receptor 2 R753Q mutation modifies cytokine production and Toll-like receptor expression in atopic dermatitis. *J Allergy Clin Immunol*, 121:1013–1019.doi:10.1016/j.jaci.2007.11.029 PMID:18234309

106. Zhao J, Kim KD, Yang X *et al.* (2008). Hyper innate responses in neonates lead to increased morbidity and mortality after infection. *Proc Natl Acad Sci USA*, 105:7528–7533.doi:10.1073/pnas.0800152105 PMID:18490660

107. Wucherpfennig KW, Allen PM, Celada F *et al.* (2007). Polyspecificity of T cell and B cell receptor recognition. *Semin Immunol*, 19:216–224.doi:10.1016/j.smim.2007.02.012 PMID:17398114

108. Rose NR (2007). Prediction and prevention of autoimmune disease: a personal perspective. *Ann N Y Acad Sci*, 1109:117–128. doi:10.1196/annals.1398.014 PMID:17785297

109. Vladutiu AO, Rose NR (1971). Autoimmune murine thyroiditis relation to histocompatibility (H-2) type. *Science*, 174:1137–1139.doi:10.1126/science.174.4014.1137 PMID:5133731

110. Nemec P, Pavkova-Goldbergova M, Stouracova M et al. (2008). Polymorphism in the tumor necrosis factor-alpha gene promoter is associated with severity of rheumatoid arthritis in the Czech population. *Clin Rheumatol*, 27:59–65.doi:10.1007/s10067-007-0653-7 PMID:17562093

111. Bis JC, Heckbert SR, Smith NL et al. (2008). Variation in inflammation-related genes and risk of incident nonfatal myocardial infarction or ischemic stroke. *Atherosclerosis*, 198:166–173.doi:10.1016/j.atherosclerosis.2007.09.031 PMID:17981284

112. Garrity-Park MM, Loftus EV Jr, Bryant SC et al. (2008). Tumor necrosis factor-alpha polymorphisms in ulcerative colitis-associated colorectal cancer. *Am J Gastroenterol*, 103:407–415.doi:10.1111/j.1572-0241.2007.01572.x PMID:18289203

113. Susa S, Daimon M, Sakabe J et al. (2008). A functional polymorphism of the TNF-alpha gene that is associated with type 2 DM. *Biochem Biophys Res Commun*, 369:943–947.doi:10.1016/j.bbrc.2008.02.121 PMID:18328809

114. Unal S, Gumruk F, Aytac S et al. (2008). Interleukin-6 (IL-6), tumor necrosis factor-alpha (TNF-alpha) levels and IL-6, TNF-polymorphisms in children with thrombosis. *J Pediatr Hematol Oncol*, 30:26–31.doi:10.1097/MPH.0b013e31815b1a89 PMID:18176176

115. Lv L, Kerzic P, Lin G et al. (2007). The TNF-alpha 238A polymorphism is associated with susceptibility to persistent bone marrow dysplasia following chronic exposure to benzene. *Leuk Res*, 31:1479–1485.doi:10.1016/j.leukres.2007.01.014 PMID:17367855

116. Fontaine-Bisson B, Wolever TM, Chiasson JL et al. (2007). Genetic polymorphisms of tumor necrosis factor-alpha modify the association between dietary polyunsaturated fatty acids and fasting HDL-cholesterol and apo A-I concentrations. *Am J Clin Nutr*, 86:768–774. PMID:17823444

117. Gallicchio L, Chang H, Christo DK et al. (2008). Single nucleotide polymorphisms in inflammation-related genes and mortality in a community-based cohort in Washington County, Maryland. *Am J Epidemiol*, 167:807–813.doi:10.1093/aje/kwm378 PMID:18263601

118. Rubin RL (2005). Drug-induced lupus. *Toxicology*, 209:135–147.doi:10.1016/j.tox.2004.12.025 PMID:15767026

119. Mongey AB, Sim E, Risch A, Hess E (1999). Acetylation status is associated with serological changes but not clinically significant disease in patients receiving procainamide. *J Rheumatol*, 26:1721–1726. PMID:10451068

120. Zschieschang P, Hiepe F, Gromnica-Ihle E et al. (2002). Lack of association between arylamine N-acetyltransferase 2 (NAT2) polymorphism and systemic lupus erythematosus. *Pharmacogenetics*, 12:559–563.doi:10.1097/00008571-200210000-00008 PMID:12360107

121. Vogt RF, Shim YK, Middleton DC et al. (2007). Monoclonal B-cell lymphocytosis as a biomarker in environmental health studies. *Br J Haematol*, 139:690–700.doi:10.1111/j.1365-2141.2007.06861.x PMID:18021083

122. Richardson B (2007). Primer: epigenetics of autoimmunity. *Nat Clin Pract Rheumatol*, 3:521–527.doi:10.1038/ncprheum0573 PMID:17762851

123. Plass C, Byrd JC, Raval A et al. (2007). Molecular profiling of chronic lymphocytic leukaemia: genetics meets epigenetics to identify predisposing genes. *Br J Haematol*, 139:744–752.doi:10.1111/j.1365-2141.2007.06875.x PMID:17961188

124. Ruan QG, Tung K, Eisenman D et al. (2007). The autoimmune regulator directly controls the expression of genes critical for thymic epithelial function. *J Immunol*, 178:7173–7180. PMID:17513766

125. Fabbri M, Garzon R, Andreeff M et al. (2008). MicroRNAs and noncoding RNAs in hematological malignancies: molecular, clinical and therapeutic implications. *Leukemia*, 22:1095–1105.doi:10.1038/leu.2008.30 PMID:18323801

126. Meyer TE, Bull LM, Cain Holmes K et al. (2007). West Nile virus infection among the homeless, Houston, Texas. *Emerg Infect Dis*, 13:1500–1503. PMID:18257995

127. Huang DB, Brown EL, DuPont HL et al. (2008). Seroprevalence of the enteroaggregative Escherichia coli virulence factor dispersin among USA travellers to Cuernavaca, Mexico: a pilot study. *J Med Microbiol*, 57:476–479.doi:10.1099/jmm.0.47495-0 PMID:18349368

128. Nissapatorn V, Lim YA, Jamaiah I et al. (2005). Parasitic infections in Malaysia: changing and challenges. *Southeast Asian J Trop Med Public Health*, 36 Suppl 4;50–59. PMID:16438180

129. Vogt RF, Whitfield WE, Jackson RL, Sampson EJ. Biomarkers of acquired immunity as indicators of prior environmental exposures. In: Mendelsohn ML, Mohr LC, Peeters JP, editors. Biomarkers: medical and workplace applications. Washington (DC): John Henry Press; 1998. p. 301–10.

130. Middleton DC, Fink J, Kowalski PJ et al. (2008). Optimizing BeLPT criteria for beryllium sensitization. *Am J Ind Med*, 51:166–172. doi:10.1002/ajim.20548 PMID:18181198

131. Finn OJ (2008). Cancer immunology. *N Engl J Med*, 358:2704–2715.doi:10.1056/NEJMra072739 PMID:18565863

132. Baldwin RW (1955). Immunity to methylcholanthrene-induced tumours in inbred rats following atrophy and regression of the implanted tumours. *Br J Cancer*, 9:652–657.doi:10.1038/bjc.1955.70 PMID:13304228

133. Gold P, Freedman SO (1965). Specific carcinoembryonic antigens of the human digestive system. *J Exp Med*, 122:467–481. doi:10.1084/jem.122.3.467 PMID:4953873

134. Jochmans I, Topal B, D'Hoore A et al. (2008). Yield of routine imaging after curative colorectal cancer treatment. *Acta Chir Belg*, 108:88–92. PMID:18411580

135. Cullen KJ, Stevens DP, Frost MA, Mackay IR (1976). Carcinoembryonic antigen (CEA), smoking, and cancer in a longitudinal population study. *Aust N Z J Med*, 6:279–283. PMID:1070982

136. Gold P, Shuster J, Freedman SO (1978). Carcinoembryonic antigen (CEA) in clinical medicine: historical perspectives, pitfalls and projections. *Cancer*, 42 Suppl;1399–1405.doi:10.1002/1097-0142(197809)42:3+<1399::AID-CNCR2820420803>3.0.CO;2-P PMID:361199

137. Hochreiter WW (2008). The issue of prostate cancer evaluation in men with elevated prostate-specific antigen and chronic prostatitis. *Andrologia*, 40:130–133. doi:10.1111/j.1439-0272.2007.00820.x PMID:18336465

138. Lejtenyi MC, Freedman SO, Gold P (1971). Response of lymphocytes from patients with gastrointestinal cancer to the carcinoembryonic antigen of the human digestive system. *Cancer*, 28:115–120.doi:10.1002/1097-0142(197107)28:1<115::AID-CNCR2820280121>3.0.CO;2-K PMID:5110615

139. Shuster J, Livingstone A, Banjo C et al. (1974). Immunologic diagnosis of human cancers. *Am J Clin Pathol*, 62:243–257. PMID:4135671

140. Shoshan SH, Admon A (2007). Novel technologies for cancer biomarker discovery: humoral proteomics. *Cancer Biomark*, 3:141–152. PMID:17611305

141. Casiano CA, Mediavilla-Varela M, Tan EM (2006). Tumor-associated antigen arrays for the serological diagnosis of cancer. *Mol Cell Proteomics*, 5:1745–1759.doi:10.1074/mcp.R600010-MCP200 PMID:16733262

142. Chapman CJ, Murray A, McElveen JE et al. (2008). Autoantibodies in lung cancer: possibilities for early detection and subsequent cure. *Thorax*, 63:228–233.doi:10.1136/thx.2007.083592 PMID:17932110

143. Zhang JY (2007). Mini-array of multiple tumor-associated antigens to enhance autoantibody detection for immunodiagnosis of hepatocellular carcinoma. *Autoimmun Rev*, 6:143–148.doi:10.1016/j.autrev.2006.09.009 PMID:17289549

144. Gagnon A, Kim JH, Schorge JO et al. (2008). Use of a combination of approaches to identify and validate relevant tumor-associated antigens and their corresponding autoantibodies in ovarian cancer patients. *Clin Cancer Res*, 14:764–771.doi:10.1158/1078-0432.CCR-07-0856 PMID:18245537

145. Baumgaertner P, Rufer N, Devevre E et al. (2006). Ex vivo detectable human CD8 T-cell responses to cancer-testis antigens. Cancer Res, 66:1912–1916.doi:10.1158/0008-5472.CAN-05-3793 PMID:16488988

146. Inokuma M, dela Rosa C, Schmitt C et al. (2007). Functional T cell responses to tumor antigens in breast cancer patients have a distinct phenotype and cytokine signature. J Immunol, 179:2627–2633. PMID:17675526

147. Vogt RF, Marti GE (2007). Overview of monoclonal gammopathies of undetermined significance. Br J Haematol, 139:687–689. doi:10.1111/j.1365-2141.2007.06860.x PMID:18021082

148. Ghia P, Caligaris-Cappio F (2006). The origin of B-cell chronic lymphocytic leukemia. Semin Oncol, 33:150–156.doi:10.1053/j.seminoncol.2006.01.009 PMID:16616061

149. Vogt RF Jr, Kyle RA (2009). The secret lives of monoclonal B cells. N Engl J Med, 360:722–723.doi:10.1056/NEJMe0810453 PMID:19213686

150. Landgren O, Albitar M, Ma W et al. (2009). B-cell clones as early markers for chronic lymphocytic leukemia. N Engl J Med, 360:659–667.doi:10.1056/NEJMoa0806122 PMID:19213679

151. Nicoloso MS, Kipps TJ, Croce CM, Calin GA (2007). MicroRNAs in the pathogeny of chronic lymphocytic leukaemia. Br J Haematol, 139:709–716.doi:10.1111/j.1365-2141.2007.06868.x PMID:18021085

152. Caporaso NE, Marti GE, Goldin L (2004). Perspectives on familial chronic lymphocytic leukemia: genes and the environment. Semin Hematol, 41:201–206.doi:10.1053/j.seminhematol.2004.05.002 PMID:15269880

153. Marti GE, Abbasi F, Raveche E et al. (2007). Overview of monoclonal B-cell lymphocytosis. Br J Haematol, 139:701–708.doi:10.1111/j.1365-2141.2007.06865.x PMID:18021084

154. Shim YK, Vogt RF, Middleton D et al. (2007). Prevalence and natural history of monoclonal and polyclonal B-cell lymphocytosis in a residential adult population. Cytometry B Clin Cytom, 72:344–353. PMID:17266153

155. Marti GE, Carter P, Abbasi F et al. (2003). B-cell monoclonal lymphocytosis and B-cell abnormalities in the setting of familial B-cell chronic lymphocytic leukemia. Cytometry B Clin Cytom, 52:1–12.doi:10.1002/cyto.b.10013 PMID:12599176

156. Rawstron AC, Green MJ, Kuzmicki A et al. (2002). Monoclonal B lymphocytes with the characteristics of "indolent" chronic lymphocytic leukemia are present in 3.5% of adults with normal blood counts. Blood, 100:635–639.doi:10.1182/blood.V100.2.635 PMID:12091358

157. Ghia P, Prato G, Scielzo C et al. (2004). Monoclonal CD5+ and CD5- B-lymphocyte expansions are frequent in the peripheral blood of the elderly. Blood, 103:2337–2342.doi:10.1182/blood-2003-09-3277 PMID:14630808

158. Vogt RF Jr, Henderson LO, Ethridge SF et al. (1993). Lymphocyte immunophenotyping with extended quantitative analysis of list-mode files for epidemiologic health studies. Ann N Y Acad Sci, 677 1 Clinical Flow;462–464.doi:10.1111/j.1749-6632.1993.tb38817.x PMID:8494243

159. Therrell BL, Hannon WH (2006). National evaluation of US newborn screening system components. Ment Retard Dev Disabil Res Rev, 12:236–245.doi:10.1002/mrdd.20124 PMID:17183567

160. Green NS, Rinaldo P, Brower A et al.; Advisory Committee on Heritable Disorders and Genetic Diseases in Newborns and Children (2007). Committee Report: advancing the current recommended panel of conditions for newborn screening. Genet Med, 9:792–796.doi:10.1097/GIM.0b013e318159a38e PMID:18007148

161. Buckley RH (2000). Advances in the understanding and treatment of human severe combined immunodeficiency. Immunol Res, 22:237–251.doi:10.1385/IR:22:2-3:237 PMID:11339359

162. Myers LA, Patel DD, Puck JM, Buckley RH (2002). Hematopoietic stem cell transplantation for severe combined immunodeficiency in the neonatal period leads to superior thymic output and improved survival. Blood, 99:872–878.doi:10.1182/blood.V99.3.872 PMID:11806989

163. Fischer A (2003). Have we seen the last variant of severe combined immunodeficiency? N Engl J Med, 349: 1789–1792.doi:10.1056/NEJMp038153 PMID:1460 2877

164. Kalman L, Lindegren ML, Kobrynski L et al. (2004). Mutations in selected genes required for T-cell development: IL7R, CD45, IL2R gamma chain, JAK3, RAG1, RAG2, ARTEMIS, and ADA and severe combined immunodeficiency. Genet Med, 6:16–26. doi:10.1097/01.GIM.0000105752.80592.A3 PMID:14726805

165. Chan K, Puck JM (2005). Development of population-based newborn screening for severe combined immunodeficiency. J Allergy Clin Immunol, 115:391–398.doi:10.1016/j.jaci.2004.10.012 PMID:15696101

166. Provenzano M, Mocellin S (2007). Complementary techniques: validation of gene expression data by quantitative real time PCR. Adv Exp Med Biol, 593:66–73.doi:10.1007/978-0-387-39978-2_7 PMID:17265717

167. Markert ML, Alexieff MJ, Li J et al. (2004). Complete DiGeorge syndrome: development of rash, lymphadenopathy, and oligoclonal T cells in 5 cases. J Allergy Clin Immunol, 113:734–741.doi:10.1016/j.jaci.2004.01.766 PMID:15100681

168. Brostowicz HM, Frazier ER, Harrison C (2007). Neonatal transmission of HIV: a persistent dilemma. J Ky Med Assoc, 105:541–544. PMID:18183805

169. Borkowsky W, Chen SH, Belitskaya-Levy I (2007). Distribution and evolution of T-cell receptor Vbeta repertoire on peripheral blood lymphocytes of newborn infants of human immunodeficiency virus (HIV)-infected mothers: differential display on CD4 and CD8 T cells and effect of HIV infection. Clin Vaccine Immunol, 14:1215–1222.doi:10.1128/CVI.00092-07 PMID:17652526

170. Eisenbarth GS (2007). Update in type 1 diabetes. J Clin Endocrinol Metab, 92:2403–2407.doi:10.1210/jc.2007-0339 PMID:17616634

171. Dantonio PD, Meredith-Molloy N, Hagopian WA et al. (2010). Proficiency testing of human leukocyte antigen-DR and human leukocyte antigen-DQ genetic risk assessment for type 1 diabetes using dried blood spots. J Diabetes Sci Technol, 4:929–941. PMID:20663459

172. Pietropaolo M, Yu S, Libman IM et al. (2005). Cytoplasmic islet cell antibodies remain valuable in defining risk of progression to type 1 diabetes in subjects with other islet autoantibodies. Pediatr Diabetes, 6:184–192.doi:10.1111/j.1399-543X.2005.00127.x PMID:16390386

173. Lendrum R, Walker G, Gamble DR (1975). Islet-cell antibodies in juvenile diabetes mellitus of recent onset. Lancet, 1:880–882.doi:10.1016/S0140-6736(75)91683-9 PMID:47533

174. Peng H, Hagopian W (2006). Environmental factors in the development of Type 1 diabetes. Rev Endocr Metab Disord, 7:149–162.doi:10.1007/s11154-006-9024-y PMID:17203405

175. Hagopian WA, Lernmark A, Rewers MJ et al. (2006). TEDDY–The Environmental Determinants of Diabetes in the Young: an observational clinical trial. Ann N Y Acad Sci, 1079:320–326.doi:10.1196/annals.1375.049 PMID:17130573

176. Kiviniemi M, Hermann R, Nurmi J et al.; TEDDY Study Group (2007). A high-throughput population screening system for the estimation of genetic risk for type 1 diabetes: an application for the TEDDY (the Environmental Determinants of Diabetes in the Young) study. Diabetes Technol Ther, 9:460–472.doi:10.1089/dia.2007.0229 PMID:17931054

177. Freiesleben De Blasio B, Bak P, Pociot F et al. (1999). Onset of type 1 diabetes: a dynamical instability. Diabetes, 48:1677–1685.doi:10.2337/diabetes.48.9.1677 PMID:10480594

178. LaGasse JM, Brantley MS, Leech NJ et al.; Washingtno State Diabetes Prediction Study (2002). Successful prospective prediction of type 1 diabetes in schoolchildren through multiple defined autoantibodies: an 8-year follow-up of the Washington State Diabetes Prediction Study. Diabetes Care, 25:505–511.doi:10.2337/diacare.25.3.505 PMID:11874938

179. Achenbach P, Schlosser M, Williams AJ et al. (2007). Combined testing of antibody titer and affinity improves insulin autoantibody measurement: Diabetes Antibody Standardization Program. Clin Immunol, 122:85–90.doi:10.1016/j.clim.2006.09.004 PMID:17059894

180. Weiss LA, Shen Y, Korn JM et al.; Autism Consortium (2008). Association between microdeletion and microduplication at 16p11.2 and autism. N Engl J Med, 358:667–675. doi:10.1056/NEJMoa075974 PMID:18184952

181. Nahmias AJ, Nahmias SB, Danielsson D (2006). The possible role of transplacentally-acquired antibodies to infectious agents, with molecular mimicry to nervous system sialic acid epitopes, as causes of neuromental disorders: prevention and vaccine implications. Clin Dev Immunol, 13:167–183.doi:10.1080/17402520600801745 PMID:17162360

182. Pickett J, London E (2005). The neuropathology of autism: a review. J Neuropathol Exp Neurol, 64:925–935. doi:10.1097/01.jnen.0000186921.42592.6c PMID:16254487

183. Braunschweig D, Ashwood P, Krakowiak P et al. (2008). Autism: maternally derived antibodies specific for fetal brain proteins. Neurotoxicology, 29:226–231. PMID:18078998

184. Wenner M (2008). Infected with insanity. Could microbes cause mental illness? Sci Am Mind, 19:40–47 doi:10.1038/scientificamericanmind0408-40.

185. Stoll BJ, Lee FK, Hale E et al. (1993). Immunoglobulin secretion by the normal and the infected newborn infant. J Pediatr, 122:780–786.doi:10.1016/S0022-3476(06)80026-0 PMID:8496761

186. Steenbergen SM, Vimr ER (2003). Functional relationships of the sialyltransferases involved in expression of the polysialic acid capsules of Escherichia coli K1 and K92 and Neisseria meningitidis groups B or C. J Biol Chem, 278:15349–15359.doi:10.1074/jbc.M208837200 PMID:12578835

187. Kowal C, Degiorgio LA, Lee JY et al. (2006). Human lupus autoantibodies against NMDA receptors mediate cognitive impairment. Proc Natl Acad Sci USA, 103:19854–19859.doi:10.1073/pnas.0608397104 PMID:17170137

188. Murphy TK, Snider LA, Mutch PJ et al. (2007). Relationship of movements and behaviors to Group A Streptococcus infections in elementary school children. Biol Psychiatry, 61:279–284.doi:10.1016/j.biopsych.2006.08.031 PMID:17126304

189. Lehmann HC, Lopez PH, Zhang G et al. (2007). Passive immunization with anti-ganglioside antibodies directly inhibits axon regeneration in an animal model. J Neurosci, 27:27–34.doi:10.1523/JNEUROSCI.4017-06.2007 PMID:17202469

190. El Maarouf A, Petridis AK, Rutishauser U (2006). Use of polysialic acid in repair of the central nervous system. Proc Natl Acad Sci USA, 103:16989–16994.doi:10.1073/pnas.0608036103 PMID:17075041

191. Kirvan CA, Swedo SE, Snider LA, Cunningham MW (2006). Antibody-mediated neuronal cell signaling in behavior and movement disorders. J Neuroimmunol, 179:173–179.doi:10.1016/j.jneuroim.2006.06.017 PMID:16875742

192. Kirvan CA, Cox CJ, Swedo SE, Cunningham MW (2007). Tubulin is a neuronal target of autoantibodies in Sydenham's chorea. J Immunol, 178:7412–7421. PMID:17513792

193. Huerta PT, Kowal C, DeGiorgio LA et al. (2006). Immunity and behavior: antibodies alter emotion. Proc Natl Acad Sci USA, 103:678–683.doi:10.1073/pnas.0510055103 PMID:16407105

194. Martin LA, Ashwood P, Braunschweig D et al. (2008). Stereotypies and hyperactivity in rhesus monkeys exposed to IgG from mothers of children with autism. Brain Behav Immun, 22:806–816.doi:10.1016/j.bbi.2007.12.007 PMID:18262386

195. Cabanlit M, Wills S, Goines P et al. (2007). Brain-specific autoantibodies in the plasma of subjects with autistic spectrum disorder. Ann NY Acad Sci, 1107:92–103. doi:10.1196/annals.1381.010 PMID:17804536

196. Connolly AM, Chez M, Streif EM et al. (2006). Brain-derived neurotrophic factor and autoantibodies to neural antigens in sera of children with autistic spectrum disorders, Landau-Kleffner syndrome, and epilepsy. Biol Psychiatry, 59:354–363.doi:10.1016/j.biopsych.2005.07.004 PMID:16181614

UNIT 4.

INTEGRATION OF BIOMARKERS INTO EPIDEMIOLOGY STUDY DESIGNS

CHAPTER 14.

Population-based study designs in molecular epidemiology

Montserrat García-Closas, Roel Vermeulen, David Cox, Qing Lan, Neil E. Caporaso, and Nathaniel Rothman

Summary

This chapter will discuss design considerations for epidemiological studies that use biomarkers in the framework of etiologic investigations. The main focus will be on describing the incorporation of biomarkers into the main epidemiologic study designs, including cross-sectional or short-term longitudinal designs to characterize biomarkers, and prospective cohort and case–control studies to evaluate biomarker-disease associations. The advantages and limitations of each design will be presented, and the impact of study design on the feasibility of different approaches to exposure assessment and biospecimen collection and processing will be discussed.

Introduction

There is a wealth of existing and emerging opportunities to apply a vast array of new biomarker discovery technologies, such as genome-wide scans of common genetic variants, mRNA and microRNA expression arrays, proteomics, metabolomics and adductomics, to further our understanding of the etiology of a broad range of diseases (1–9). These approaches are allowing investigators to explore biologic responses to exogenous and endogenous exposures, evaluate potential modification of those responses by variants in essentially the entire genome, and define disease processes at the chromosomal, DNA, RNA and protein levels. At the same time, most biomarkers analysed by these technologies can still be classified into the classic biomarker categories defined more than 20 years ago (Figure 14.1), which include biomarkers of exposure, intermediate endpoints (e.g. biomarkers of early biologic effect), disease and susceptibility (10–17). Biomarkers in epidemiological studies can also be used to evaluate behavioural characteristics that affect the likelihood of exposure, such as tobacco smoking, as well as clinical behaviour and progression of disease. The use of biomarkers associated with exposure, disease development, and clinical progression within the same overall design is of increasing interest and has recently been termed 'integrative epidemiology' (18,19).

Figure 14.1. A continuum of biomarker categories reflecting the carcinogenic process resulting from xenobiotic exposures

Exposure → Internal Dose → Biologically Effective Dose → Early Biologic Effect → Altered Structure/Function → Disease

SUSCEPTIBILITY (influences all stages)

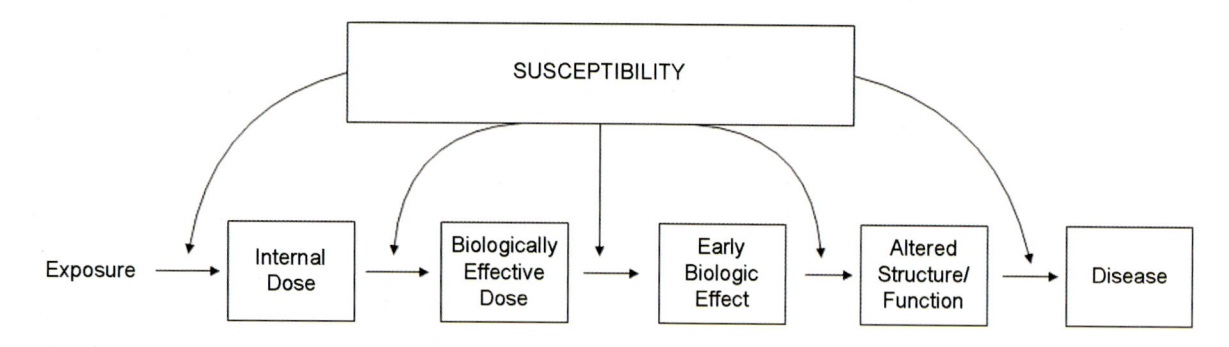

Figure compiled from (10) and (21, copyright © 2008, Informa Healthcare. Reproduced with permission of Informa Healthcare).

Regardless of the appellation used to describe the use of biomarkers in epidemiologic research, be it "molecular epidemiology," "integrative epidemiology," or the more limited "genetic epidemiology," the successful application of new and established biomarker technologies still depends on integrating them into the appropriate study design with careful attention to the time-tested principles of the epidemiologic method (16,20–24). Basic principles in vetting new biomarkers and technologies in pilot or transitional studies apply now more than ever (25–28). Understanding how to collect, process and store biologic samples (see Chapter 3), and the factors that influence biomarker levels, with particular attention to within- and between-person variation for non-fixed biomarkers (see Chapter 9), are key concerns. Testing for and optimizing laboratory accuracy and precision are also critical to the successful use of biomarkers in epidemiology studies (see Chapter 8). Finally, selecting the most appropriate, effective, and logistically feasible study design to use a given biomarker technology that answers a particular research question remains of paramount importance.

The focus of this chapter is on design considerations for epidemiological studies that use biomarkers, primarily in the context of etiologic research, including cross-sectional or short-term longitudinal designs to characterize biomarkers, and prospective cohort and case–control studies to evaluate biomarker–disease associations. A description of the general principles of study design (29–31) is outside the scope of this chapter. Instead, the focus is on describing the incorporation of biomarkers into the main epidemiologic study designs, pointing out the advantages and limitations of each, and showing how study design affects the feasibility of different approaches to both exposure assessment and biospecimen collection and processing.

Study designs in molecular epidemiology

Cross-sectional and short-term longitudinal studies with biomarker endpoints

In epidemiological terms, a cross-sectional study refers to a study design in which all of the information refers to the same point in time. As such, these studies provide a 'snapshot' of the population status with respect to exposure variables and intermediate endpoints, and, in some instances, disease at a specific point in time. Short-term longitudinal biomarker studies are studies in which subjects are prospectively followed for a short period of time (usually a few weeks to up to a year). Investigations are usually performed on healthy subjects exposed to particular exogenous or endogenous agents where the biomarker is treated as the outcome variable. These studies generally focus on exposure and intermediate endpoint biomarkers, and sometimes evaluate genetic and other modifiers of the exposure–endpoint relationship.

Questions addressed by cross-sectional and short-term longitudinal studies

Cross-sectional and/or short-term longitudinal studies are often used as follows:

1) To answer questions about whether or not a given population has been exposed to a particular compound, the level of exposure, the range of the exposure, and the external and internal determinants of the exposure. For instance, recent studies on haemoglobin adducts of acrylamide have shown that exposure to this toxic chemical

is widespread in the general population, due to dietary and lifestyle habits (32,33).

2) To evaluate intermediate biologic effects from a wide range of exposures in the diet and environment, as well as from lifestyle factors (e.g. obesity and reproductive status). This design can be used to provide mechanistic insight into well established exposure–disease relationships, and to supplement suggestive but inconclusive evidence on the possible adverse health effects of an exposure. For instance, studies have used haematological endpoints to investigate the effects of benzene on the blood forming system at low levels of exposure (34,35). These studies have found decreased levels in peripheral blood cell counts at exposures below 1 ppm, indicating that at low levels of exposure, perturbations in the blood forming system can be detected. These results hinted at the possibility of increased risk of leukaemia at low levels of benzene exposure, given the putative link between benzene poisoning (a severe form of haematotoxicity) and increased risk for leukaemia.

3) To evaluate whether or not there are early biologic perturbations caused by new exposures, or recent changes in lifestyle factors that have not been present long enough to have been evaluated for their association with disease. For example, there is considerable public health concern about the increased use of nanoparticles in both research and manufacturing operations (36). Various initial research studies and evaluations have demonstrated greater biological activity of nanoparticles compared with larger particles of the same material, and significant potential toxicity has been observed in laboratory animals exposed to some types of nanoparticles. However, given their recent introduction into commerce, the time between first exposure and the occurrence of any chronic health effect is most likely too short. In this particular example, the assessment of preclinical indicators of disease (e.g. markers of pulmonary inflammation) in asymptomatic individuals would be of importance to identify potential adverse health effects at an early stage.

4) To study changes in exposure and/or intermediate endpoints to determine the effectiveness of intervention studies. For instance, the effect of exercise and weight loss interventions on serum levels of four biomarkers related to knee osteoarthritis (cartilage oligometric matrix protein (COMP), hyaluronan, antigenic keratin sulfate, and transforming growth factor-β-1 (*TGF-β1*)), and clinical outcome measures (e.g. medial joint space, pain) were examined (37). Intervention programmes indeed resulted in changes in COMP (which was associated with decreased knee pain) and *TGF-β1*.

Cross-sectional and short-term longitudinal studies using exposure markers

Biomarkers of exposure measure the level of an external agent, its metabolic by-products in either the free state or bound to macromolecules, or the specific immunologic response it elicits. In addition, exposure biomarkers measure endogenously produced compounds, which may be influenced directly or indirectly by external factors (e.g. hormones), as well as by genetic factors. The first epidemiological evaluation of potential biomarkers of exposure generally occurs in cross-sectional studies in the general population, or in subgroups with specific, well characterized exposure and lifestyle patterns. Sometimes a biomarker of exposure can be used only in cross-sectional studies to determine if a population is exposed to an agent of concern, or used as an independent marker of exposure in studies evaluating intermediate biomarker endpoints.

The applicability of exposure biomarkers in cross-sectional studies depends on certain intrinsic features related to the marker itself (e.g. half-life, variability, and specificity of the marker) and the exposure pattern (see Chapter 9). The first requirements for successful application of an exposure marker are that the assay is reliable and accurate (see Chapter 8), the marker is detectable in human populations, and important effect modifiers (e.g. nutrition and demographic variables) and kinetics are known (20). Second, the timing of sample collection in combination with the biological half-life of a biomarker of exposure is key, as this determines the exposure time window that a marker of exposure reflects. The time of collection may be critical if, as is often the case in cross-sectional studies, only one sample per subject can be obtained on a given occasion, and if the exposure is of brief duration, highly variable in time, or has a distinct exposure pattern (e.g. diurnal variation in certain endogenous markers, such as hormones) (38). Chronic, near-constant exposures pose fewer problems. However, most biomarkers of internal dose generally provide information about recent exposures (hours to days), with the exception of markers of persistent pesticides, dioxins, polychlorobiphenyls, certain metals, and serological markers related to infectious agents, which can reflect exposures received many years before.

Cross-sectional and short-term longitudinal studies using intermediate endpoints

Intermediate biomarkers directly or indirectly represent events on the continuum between exposure and disease, and can provide important mechanistic insight into the pathogenesis of disease. As such, they complement classic epidemiological studies that use disease endpoints. For instance, the use of intermediate biomarkers in cross-sectional studies can provide initial clues about the disease potential of new exposures years before a disease develops (10,15,39–41).

One group of intermediate biomarkers, biomarkers of early biologic effect (10), generally measure early biologic changes that reflect early, and generally non-persistent, effects. Examples of early biologic effect biomarkers include: measures of cellular toxicity; chromosomal alterations; DNA, RNA and protein expression; and early non-neoplastic alterations in cell function (e.g. altered DNA repair, altered immune function). Generally, early biologic effect markers are measured in substances such as blood and blood components (red blood cells, white blood cells, DNA, RNA, plasma, sera, urine) because they are easily accessible, and, in some instances, it is reasonable to assume that they can serve as surrogates for other organs. Early biological effect markers also can be measured in other accessible tissues such as skin, cervical and colon biopsies, epithelial cells from surface tissue scrapings or sputum samples, exfoliated urothelial cells in urine, colonic cells in feces, and epithelial cells in breast nipple aspirates. Other early effect markers include measures of circulating biologically active compounds in plasma that may have epigenetic effects on disease development (e.g. hormones, growth factors, cytokines).

For maximum utility, an intermediate biomarker must be shown to be predictive of disease occurrence, preferably in prospective cohort studies (40) or potentially in carefully designed case–control studies. The criteria for validating intermediate biomarkers have focused on the calculation of the etiologic fraction of the intermediate endpoint, which varies from 0 to 1 (40,41). For intermediate endpoints with etiologic fractions that are close to 1.0, either positive or negative results in cross-sectional studies of an exposure–intermediate endpoint relationship are particularly informative. For intermediate endpoints linked to risk of developing disease but with a substantially lower etiologic fraction, the interpretation must be more circumspect. Specifically, a positive association between an exposure and an intermediate biomarker is informative, but a null association does not rule out that the exposure is associated with adverse health outcomes, as the exposure may act through a mechanism not reflected by the particular endpoint under study.

One of the most well known examples of a validated intermediate marker is low-density lipoprotein (LDL) cholesterol. Epidemiological studies have shown that elevated LDL cholesterol is one of the major causes of coronary heart disease (CHD). Given its high predictiveness, risk management/intervention programmes are focused on lowering and identifying factors that would reduce LDL levels and thus the risk for CHD (42). Unfortunately, there are very few examples like LDL. For instance, chromosomal aberrations in peripheral blood lymphocytes have been extensively used as the classic biomarker of early genotoxic effects in cross-sectional studies of populations exposed to a wide variety of potential carcinogens (43–45). Several cohort studies have reported that the prevalence of chromosomal aberrations in peripheral lymphocytes can predict subsequent risk of cancer (46–51). The predictive performance of this biomarker was shown to be similar irrespective of whether the subjects had been smokers or occupationally exposed to carcinogenic agents (52). In contrast, such associations were not observed for the sister chromatid exchange assay, another biomarker of genotoxicity also measured in peripheral lymphocytes (49–51).

Interpretation of results from cross-sectional studies using intermediate endpoints is, as indicated before, premised on the assumption that the intermediate endpoints reflect biological changes considered relevant to disease development. This may be based on *in vitro* and animal models or on previous observations that the biomarker is altered in human populations exposed to known toxicants. However, these studies are not capable, in and of themselves, of directly establishing or refuting a causal relationship between a given exposure or a given level of exposure and risk for developing diseases. Results of studies using most intermediate biomarkers as outcome measures are only suggestive; a biomarker may be overly sensitive (i.e. it may respond to low levels of chemical exposures that are below the disease threshold, if one exists), be insensitive, reflect phenomena that are irrelevant to the disease process, or fail to reflect important processes involved in the pathogenesis of disease.

Variance in biomarker response

The applicability of exposure and intermediate endpoint markers in cross-sectional and semi-longitudinal studies depends to a large extent on the variability in biomarker response between persons and over time (see Chapters 8 and 9). If a biomarker response is highly variable over time within a person, then it is clear that a single measurement of such a marker would be a poor estimate of the average marker level of a certain individual. However, even if the variance over time is small, and thus a reasonable estimate of the individual's average marker response, the applicability in epidemiological studies might be limited if the variance between individuals is small as well. In the end, the applicability of a marker in epidemiological research depends on the relative level between the interindividual and intraindividual variability in marker response. A useful measure in this regard is the intraclass correlation coefficient (ICC), which can be defined as the interindividual variance divided by the sum of the interindividual and intraindividual variance; in other words, it represents the fraction of the total variance that can be attributed to differences between individuals. Short-term longitudinal studies are ideal to collect information needed to estimate this key parameter. Chapter 9 provides a more detailed description of methods used to quantify biomarker variability and its impact on biomarker-disease associations.

Strengths and limitations

A distinct advantage of cross-sectional and short-term longitudinal studies is that detailed and accurate information can be collected on current exposure patterns, potential confounders, and effect modifiers. Furthermore, they can take advantage of a wide range of potential analytic (molecular) approaches, particularly those that require cell culturing and extensive processing within an often short period of time after collection (e.g. RNA, protein stabilization).

Cross-sectional and short-term longitudinal biomarker studies can collect very accurate information on the dose–response relationship between external or internal exposures and intermediate endpoints; these detailed exposure status data should be exploited to the fullest. As most biologic markers of exposure reflect exposures over the previous several days to months, this information must be collected over the etiologically relevant time period. For example, in a study on haematologic, cytogenetic and molecular endpoints among workers exposed to benzene, measurements were collected for over a year before determination of the biological endpoints to unequivocally assess individual exposures (53). Furthermore, given the increasing interest in identifying potential gene-environment interactions in chronic diseases, accurate measurement of the environment becomes very important. Simulation studies have shown that even a modest amount of nondifferential exposure misclassification can dramatically attenuate the estimate of the interaction parameter and increase sample size requirements (54). As such, cross-sectional and semi-longitudinal studies could have a distinct advantage in elucidating gene-environment interactions.

Summary and future directions

Cross-sectional and semi-longitudinal study designs have been successfully applied to: answer questions about whether or not a given population has been exposed to a particular compound; evaluate intermediate biologic effects from a wide range of exposures in the diet and environment; evaluate whether or not there are early biologic perturbations caused by new exposures or recent changes in lifestyle factors that have not been present long enough to have been evaluated for their association with disease; and to study changes in exposure and/or intermediate endpoints to determine the effectiveness of intervention studies. However, as indicated previously, the interpretation and therefore the usefulness of these studies depend heavily on the validity of the markers measured. The availability of numerous prospective cohort studies with stored blood specimens should enhance the ability to rapidly test the relationship between a wide variety of early biologic effect markers, using both standard and emerging technologies (55,56), and disease risk. Such studies could ultimately produce a novel endpoint to evaluate the disease potential and mechanisms of action of various risk factors.

Prospective cohort studies

In contrast to cross-sectional studies where biomarkers are the outcome variable, in prospective cohort and case–control studies the risk of disease is the outcome of interest. Prospective cohort studies collect exposure information and biological specimens from a group of healthy subjects who are then followed-up to identify those who develop disease. Establishing a cohort study is initially very costly and time-consuming, as large populations must be recruited and followed-up long enough to identify sufficient numbers of cases with the disease of interest.

Although power is limited by the overall cohort size and frequency of the outcome, in the long run the cohort design becomes more cost-efficient, since it can study multiple disease endpoints and provide a well defined population that can be easily sampled for efficiency (57–60). This section describes key features of cohort studies, particularly with regards to the use of biomarkers. Table 14.1 summarizes the strengths and limitations of this study design, as compared to the case–control designs described later in this chapter.

Participation in the study

Subjects in a cohort study often have distinct characteristics compared to their population of origin, by design or because of the motivation and level of commitment required to be included in such studies. Collection of biospecimens to measure biomarkers can have adverse effects on participation rates, even when collection procedures require minimally invasive procedures (e.g. buccal swab or oral rinse as opposed to blood collection). Cohort studies that collect questionnaires and biological samples at baseline or before disease onset can avoid selection biases, as long as specimens for each participant remain available and follow-up is complete. However, subjects with biological specimens collected after the cohort has been formed might have different characteristics from the rest of the cohort. Increasingly, concerns over privacy have also affected the willingness of participants to take part in some research studies.

Exposure assessment, timing of exposure, and misclassification

A major strength of the cohort design is that the sequence between exposure assessment and outcome is the same as the causal pathway: exposures are measured before disease diagnosis. This is particularly important for biomarkers that are directly or indirectly affected by the disease process (61), with the caveat that undiagnosed or preclinical disease may alter levels of specific biomarkers measured on specimens collected close to the date of diagnosis. Screening for disease at baseline (e.g. requiring recent colonoscopy for samples used in studies of colorectal cancer), or excluding cases diagnosed in the first few years after sample collection (lag analyses), can limit the effect of preclinical disease on biomarker levels.

Environmental data collected through questionnaires is less prone to recall biases (i.e. differential recall between cases and controls) than in case–control studies, thus facilitating the assessment of biomarker-environment interactions, such as gene-environment interactions (62–66). However, prospective studies often have a lower level of detail on specific exposures than case–control studies focusing on one or a few related diseases, due to the need to collect at least minimal data on the multiple exposures relevant to multiple outcomes. Therefore, although cohort studies can minimize the occurrence of differential misclassification, nondifferential misclassification of exposure might be larger than in alternative study designs.

Cohort studies with extended follow-up provide a wide range of time periods between biomarker collection and diagnosis of the outcome. This can be used to evaluate hypotheses relating to latent periods between the exposures of interest and the outcome. Theoretically, cohort studies have the advantage of collecting serial biological samples over time to evaluate biomarkers that vary in time. However, logistical and cost constrains often result in large studies collecting a single biological sample at one point in time. This results in diminishing the value of evaluating the relevant time window of exposure for disease causation, and studying markers with substantial seasonal or day-to-day variations, such as short-term exposure markers.

Chapter 9 describes important considerations in data analysis and inference related to the timing of exposure assessment. Below is a summary of these considerations in the context of cohort studies.

<u>Misclassification due to random within-person variation</u>. Most biomarkers vary from time to time within the same person. This variation could be due to multiple factors, including diurnal (e.g. melatonin) or monthly (e.g. estrogens) cycles, seasonal variation (e.g. vitamin D), recent dietary or supplement intake (e.g. vitamin C), as well as from the specificity of the assay used to measure the biomarker. If this variation is random and nondifferential between cases and controls, the bias will tend to attenuate measures of association.

When within-person variability for a particular biomarker is random, the correlation between single measurements in a population and the average of multiple measurements can be used to gauge the extent of attenuation in the association measure, or be used explicitly to correct the attenuation of relative risk estimates due to nondifferential misclassification (67). This information can be obtained from a representative subsample of the cohort in which the biomarkers are measured at two or more distinct time points. It is important that the subjects with repeated measurements represent the larger cohort, so that the correlation used to correct risk estimates can be generalized to the entire cohort. However, true random sampling is often difficult to perform in large cohort studies, due to geographic dispersion and lower-than-optimal participation rates in more burdensome subsampling studies.

Time integration. The most common conceptual timeframes for exposure data in the epidemiology of chronic disease in cohort studies are long-term average measurements, as the induction time of most chronic diseases (e.g. cardiovascular disease, diabetes or cancer) is thought to be in the order of years or decades. Therefore, biomarkers would optimally represent cumulative exposure over relatively long periods of time, such as months or years. Some biomarkers may be able to integrate exposure time, which might also depend on sampling, processing, and storage protocols. For example, concentrations of many nutrients are less susceptible to short-term fluctuations in erythrocytes than in plasma or serum. Concentrations in adipose tissue, which is often more difficult to acquire, reflect even more long-term exposure history. It is therefore important to balance feasibility of sample collection with implications for time-integration of the biomarkers of interest.

Multiple biomarker levels. Obtaining multiple samples over time can increase the time-integration of exposures and biomarker measurements. Multiple biomarker measurements can be used in several ways, including averaging measurements or comparing subjects with consistently high versus consistently low levels. If within-person variation in biomarker measurements is assumed to be random, methods exist to estimate the number of replicate measurements required to estimate the 'true' mean value of a biomarker within a specified range of error. On the other hand, if variation is not random, due to changes in behaviour or secular trends, multiple measurements can be used to estimate exposure error. From a practical point of view, collecting multiple biological specimens from large numbers of subjects in cohort studies increases the cost of sample collection and storage, as well as the burden on study subjects, and thus might not always be feasible or recommendable.

Inference from biomarker/disease associations. The association between a biomarker and disease may also be influenced by the point in time in which the biomarker was assayed with respect to where it influences disease on the causal pathway. In the case of cancer, early events on the causal pathway ('initiators') may need to be distinguished from later events ('promoters'). Therefore, it is important that any biomarker related to exposures that are either initiators or promoters be measured during a time period when the exposure is most likely to exert its influence on disease. For example, initiating exposures should most likely be measured many years, possibly decades, before cancer diagnosis. In contrast, exposures considered to be promoters should be measured more closely to the time of diagnosis. If a biomarker of exposure is not measured at the etiologically relevant time, the association between the exposure and disease could be attenuated or not observed at all. The problem is that the true latency between exposure and disease diagnosis is often not known. The within-person variability in the biomarker of exposure is also important to consider in respect to the optimum time point for measuring it with respect to disease risk. If the within-person variability is low, then careful timing of exposure measurement is not necessary. However, if there is large within-person variability in the biomarker, then measurements should be made as close to the time of predicted maximum effect as possible.

Considerations in biospecimen collection, processing and storage

The collection and storage of large numbers of samples needed for cohort studies using biological specimens is very complex (see Chapter 3 for considerations in sample collection, processing, and storage). Given that samples are often used years after collection, optimal biospecimen collection, processing and storage protocols that will allow the performance of a wide range of assays in the future are critical (68,69). Therefore, validation studies aimed at optimizing sample handling and storage protocols, according to the impact of these procedures on the stability of samples and biomarker measurements, are strongly recommended (69–72). Validation studies can assess considerations such as the influence of time of collection to arrival at the processing laboratory (e.g. blood collection tubes with different preservatives, anti-coagulants or clot accelerators, temperature during shipping, impact of time between collection to processing), processing protocols (e.g. isolation of serum for proteomic analyses (73)), and long-term storage (e.g. freezing temperature, impact of thaw/freeze cycles) on biomarker measurements.

Of paramount consideration is limiting the loss of information due to exhaustion of archived samples. The problem of sample exhaustion is most evident in prospective cohorts examining incident disease, as the amount of sample collected before diagnosis is by definition finite. Moreover, due to the advantages of the prospective cohort design, interest in the utilization of biological specimens could be great among the scientific community. Therefore, it is important for investigators to try to minimize the volume of sample used for measuring any one biomarker, either by reducing the volume of sample used or by maximizing the number of assays that can be made at any one time (multiplexing). Also beneficial is the formation of an advisory board to aid investigators in evaluating both internal use and external requests for access to precious prospective samples. While these considerations are obvious for those participants who are diagnosed with disease, biological samples from healthy or control subjects should also be carefully preserved, as these subjects may become cases in the future.

Despite advances in technology, such as whole-genome amplification to increase the amount of available DNA for assays, many biomarkers still require large amounts of biological samples. This limits the number of measurements that can be carried out on the limited resource of biological samples from cohort studies. Collecting additional specimens is often difficult in cohort studies, as members move, are lost to follow-up, or do not wish to go through the further inconvenience of providing an additional sample.

Statistical power

The major weakness of cohort studies is that even for common diseases the number of cases is limited by the cohort size and follow-up time. Even a very large cohort may not acquire enough cases of rare diseases to achieve adequate statistical power after long follow-up periods. Considering that cohorts using biological samples tend to be small or subsets of larger questionnaire-based cohorts, this is a particular problem for studies using biomarkers. Recently, a movement to form consortia of cohorts, such as the Cohort Consortium to study causes of cancer (http://epi.grants.cancer.gov/Consortia/cohort.html), has begun to address the problem of statistical power by coordinating biomarker measurements and analyses. Many consortia have been formed to support genome-wide association studies of many diseases (updated information on new publications from these efforts can be found at http://www.genome.gov/gwastudies/).

Sampling designs

When a cohort is chosen at random from the general population, the exposures in the cohort will be representative of the exposures in the general population. If the hypotheses to be tested rely on participants having either rare or extreme exposures in the general population, then oversampling these people or restricting the cohort to certain exposed groups would increase efficiency by increasing the prevalence of these exposures in the cohort.

For many large cohort studies, it is not feasible to assay all participants for a given biomarker. Therefore, with a few exceptions (such as assays that can only be performed on fresh samples), some selection of cases and controls will be necessary. This can be attained by using sampling designs in which only samples from cases and a random subset of non-cases are analysed, thus considerably reducing laboratory requirements and cost (60).

Nested case–control

The nested case–control study is an efficient sampling scheme that includes all cases identified in the cohort up to a particular point in time, and a random sample of subjects free of disease at the time of the case diagnosis. Increasing the case-to-control ratio to two or three controls per case can easily increase the efficiency of nested case–control studies. Optimally, controls should be selected for each case from the pool of participants that have not developed the disease at the time the case was diagnosed (risk set sampling). Alternatively, controls may be selected from all of the participants at baseline who were not diagnosed with disease throughout follow-up. Simulation studies have shown that as long as the proportion of the baseline cohort that acquires disease is low (e.g. less than 5%), the bias introduced by violating risk set sampling is minimal.

Case–cohort

A case–cohort design includes a random sample of the cohort population at the onset of the study and all cases identified in the cohort, up to a particular point in time (74). This design allows for the evaluation of several disease endpoints using the same comparison group (referred to as a subcohort). It may reduce the amount of laboratory work by assaying a subcohort of subjects at baseline, and then adding case information as cases accrue. While there are statistical considerations that must be taken into account when analysing case–cohort studies, these are now included in most statistical packages. Of greater concern in case–cohort studies are problems more unique to biomarker studies. For example, if the biomarker being assayed degrades over time or if there is substantial laboratory drift in measurement, then cases assayed at varying time periods after baseline (when controls were assayed) can lead to bias. Additionally, laboratory personal are less easily blinded to case or control status, which can also lead to bias. These factors limit the utility of the case–cohort design in biomarker studies. Another limitation of this design is that since the same disease-free subjects are repeatedly used as controls for different disease endpoints, depletion of samples from this group can become an issue.

Sample comparability

The methods by which biological samples are collected, handled, and stored can influence the measurement of many biomarkers of exposure (see Chapter 3). Therefore, to have valid biomarker studies, case and control samples must be handled in the same way. For prospective studies, it is also important to consider the length and type of storage, as some biomarkers may degrade over time even under ideal conditions. Thus, it is important to match cases and controls on the method of sample collection, duration of storage, as well as other factors that may be related to the biomarker of interest, such as fasting status or season of collection. Additionally, batch-to-batch variation in assay measurement should also be considered. This can and should be minimized by assaying matched cases and controls at the same time, regardless of the study design.

Screening cohorts

Prospective cohort studies are sometimes designed within screening cohorts. In this design, screening failures lead to missing prevalent cases among cohort participants that are misclassified as controls (75). Although repeated screening reduces misclassification of subjects, cases discovered in follow-up cannot be distinguished from prevalent cases missed by the initial screening or incident disease. However, the degree of misclassification of prevalent and incident cases can be assessed by analyses of time to diagnosis or pathological characteristics. Intensive screening may also uncover a reservoir of latent disease that would not otherwise become clinically relevant, and that might differ from disease detected through clinical symptoms (76,77).

Resources and infrastructure

The vastly greater size of cohorts compared to other designs, such as case–control studies, and the time period required for the cohort to mature, mean that a substantially greater initial investment is required to establish the cohort. For cohort studies that incorporate biological materials, the infrastructure to support biospecimens' databases, freezers, and processing require a correspondingly greater effort and cost. While all studies with biospecimens must consider the risk of untoward events (e.g. freezer failure), the anticipated long useful life of the samples from cohorts requires special emphasis on quality control and security issues (e.g. backup generators, monitoring, distributing samples among different freezers). In the next few years, however, the cost-per-case for studies fielded from a cohort will offer economies in comparison to fielding a new case–control study (57).

Summary and future directions

Prospective cohort studies provide invaluable resources to study biomarkers of risk, particularly those that can be affected by disease processes. Multiple prospective cohort studies are currently being followed-up for disease incidence with basic risk factor information from questionnaires and stored blood components, including white blood cells that can be used as a source of DNA. At the completion of ongoing collections, current studies will have stored DNA samples on over two million individuals (16). These studies will provide very large numbers of cases of the more common cancer sites (breast, lung, prostate, and colon) to evaluate genetic markers of susceptibility; biomarkers in serum or plasma, such as hormone levels; chemical carcinogen levels; and proteomic patterns. Most cohort studies do not have cryopreserved blood samples, as the procedure is very expensive and logistically challenging in large studies. Also, cohort studies often have a limited capability to collect detailed disease information or biological specimens to facilitate disease classification, as well as to follow-up cases for disease progression and survival studies. New cohort studies based on large institutions, such as health maintenance organizations (HMOs), could enable access to clinical records with more detailed disease information, archived biological specimens, and easier follow-up of cases for treatment response and survival. Caucasian populations in wealthier countries are overrepresented in studies of most diseases, and the recent establishment of consortia in other populations, such as the Asia Cohort Consortium (http://www.asiacohort.org/), will be critical to study disease across geographically and ethnically diverse populations that might have different exposures to environmental risk factors and frequencies of susceptibility alleles.

Case–control studies

Case-control studies are conceptualized as a retrospective sampling of cases and controls from an underlying prospective cohort, referred to as the source population (29,31). The case–control design has been a mainstay of molecular epidemiology studies due to its well known traditional strengths including depth and focus of questionnaire information, biologically intensive specimen collection, potential to enrol large numbers of cases rapidly, and ability to target rare diseases that occur in small numbers in prospective cohort studies.

Types of case–control designs

Case–control studies can be hospital- or population-based depending on how the cases and controls are identified (Table 14.1). A major concern of case–control studies is proper case and control selection. Proper controls are representative of the study base from which the cases arise (29). Identifying either a random sample from the general population or the source population for cases presenting at a particular hospital(s) may be difficult. Population-based studies attempt to identify all cases occurring in a pre-defined population during a specified period of time, and controls are a random sample of the source population where the cases came from. On the other hand, cases and controls in hospital-based studies are identified among subjects admitted to or seen in clinics associated with specific hospitals. As in the population-based design, the distribution of exposures in the control group should represent that from the source population of the cases. However, the source population is often more difficult to define in hospital-based studies.

Molecular epidemiology studies often use the hospital-based case–control design, as the hospital setting facilitates the enrolment of subjects, thus enhancing response rates, as well as the collection and processing of biological specimens. Enrolment of subjects is also made easier by having in-person contact with study participants by doctors, nurses or interviewers, which usually results in higher participation rates (78). Because study subjects are generally less geographically distributed than those in population-based or cohort studies, rapid shipment of specimens to central laboratories for more elaborate processing protocols, such as cyropreservation of lymphocytes, is made possible. Rapid ascertainment of cases through the hospitals also facilitates the collection of specimens from cases before treatment, thus avoiding the potential influence of treatment on some biomarker measurements.

Potential for selection bias is one of the most important limitations of case–control designs. The impact of selection bias in hospital-based studies is not only related to the reasons for non-participation, but also to diseases in the control population. An example of this is selection of controls admitted to the hospital for other diseases that might themselves introduce bias if they are related to the genetic or environmental exposures under study, particularly when evaluating gene–environment interactions or joint effects (79). Further potential for selection bias occurs if cases or controls are less likely to participate because of problems in the collection of biospecimens. Since the source population for cohorts is explicit, selection bias is less of a problem as long as follow-up rates are high (61). Low participation rates in case–control studies, and particularly refusals related to providing biological specimens, can bias results, especially when cases are less likely to participate than controls and selection is related to the biomarker of interest. Low participation rates additionally threaten the population-based nature of the study, undermining its use for estimating absolute and attributable risk (29). Use of non-intensive biospecimen collection protocols can increase participation rates—for instance, the collection of buccal cells as a source of DNA instead of the more invasive phlebotomy (80).

Single disease

Case–control studies are generally limited to one disease outcome (or a few related diseases), but are unconstrained by the rarity of the disease, while cohort studies (including full cohort, nested case–control, or case–cohort studies) may identify multiple disease endpoints. The focus of a case–control study on one disease entity permits more detailed documentation of disease information and detailed diagnostic procedures not routinely collected in clinical practice, such as specialized imaging and access to pathologic tissue and other

Table 14.1. Advantages and limitations of prospective cohort and case-control designs in molecular epidemiology relevant to the collection of biological specimens and data interpretation

	Advantages	Limitations
Prospective cohort	-Exposure measured and specimens collected prior to outcome - Allows study of multiple disease endpoints - Allows study of transient biomarkers and biomarkers affected by disease process - Selection bias and differential misclassification are avoided. Nondifferential misclassification may be reduced for some exposures - Nested case–control or case–cohort studies can be used to improve efficiency of the design - Opportunity to obtain repeated/serial biomarkers and questionnaires over time	-Power limited by overall cohort size and frequency of outcome - Detailed information on specific exposures often not available - Implementation of intense, specialized collection and processing protocols for the entire cohort can be logistically challenging and overly costly - Obtaining tissue samples and following-up cases for treatment response and survival can be challenging if hospitals are geographically dispersed - Loss to follow-up can cause a potential bias
Population-based case–control	-Facilitates enrolment of large numbers of cases (particularly for the less common diseases) in a relatively short period of time - Facilitates collection of more detailed exposure information by focusing on one or a few diseases - Less subject to biases (e.g. selection, exposure misclassification) than hospital-based studies	- Some biomarkers and responses to certain types of questions might be affected by disease process - May be harder to obtain high participation rates for biological collections than in hospital-based designs - Implementation of intense, specialized blood and tumor collection and processing protocols can be challenging - May be more difficult to carry out response to treatment and survival studies if cases are treated at many hospitals and clinics
Hospital-based case–control	- Facilitates enrolment of large numbers of cases (particularly for the less common diseases) in a relatively short period of time - Facilitates collection of more detailed exposure information by focusing on one or a few diseases - Facilitates intense collection and processing of specimens (e.g. freshly-frozen tumour samples, cryopreserved lymphocytes) - Participation rates for biological collections might be enhanced compared to other designs - Facilitates follow-up of cases for treatment response, recurrence and survival	- More prone to selection and differential misclassification biases than other designs - Some biomarkers and responses to certain types of questions might be affected by disease process or stay at the hospital

Adapted from (16).

biological specimens for application of novel biomarkers of disease. Obtaining disease-related data in cohorts entails mounting an effort that is generally less efficient and more costly. The advantage of cohorts' ability to examine multiple outcomes may be somewhat limited by resources and logistics, limited exposure information, the diverse approaches to documenting disease incidence or mortality, and the rarity of some outcomes.

Costs for a series of case–control studies of different diseases can sometimes be reduced by sharing a single control group. When different diseases require different exposures, the partial questionnaire design may offer reduction in the burden to respondents, thereby potentially increasing participation (81). Even if these options are not feasible, using the same infrastructure for control selection for repeated studies can reduce costs.

Exposure assessment and misclassification

Exposure assessment through questionnaires in case–control studies of a single disease or multiple diseases sharing risk factors (e.g. breast, ovarian, and endometrial cancer) can be more detailed and focused than prospective cohort studies that often study multiple unrelated diseases. However, studies that rely on retrospective exposure assessment may be affected by disease or its treatment. Also, questionnaire responses subject to rumination by respondents are susceptible to bias from differential misclassification. Biomarkers (except germ-line genetic variation) and responses to questionnaires may change as a consequence of the early disease process or diagnosis itself. Differential errors or recall bias from questionnaire information collected in case–control studies are possible, and their extent should be evaluated in the context of specific exposures and populations under study. Similarly, levels of certain biomarkers measured after diagnosis can be influenced by the disease process or treatment, and must be considered and evaluated to the extent possible for each biomarker of interest. Differences in biomarker levels among cases diagnosed at different stages of the disease can help evaluate whether differences in biomarker levels between cases and controls reflect an influence of the disease on the biomarker rather than the contrary.

The applicability of exposure biomarkers in case–control studies depends on certain intrinsic features related to the marker itself (e.g. half-life, variability, specificity) and the exposure time window that a marker of exposure reflects in relation to the biologically relevant time of exposure and timing of sample collection. Methods to evaluate these key biomarker features before their use in case–control and other epidemiological studies are described in the previous section and in Chapter 8. The time of collection may be critical if the exposure is of brief duration, is highly variable in time, or has a distinct exposure pattern (e.g. diurnal variation for certain endogenous markers, such as hormones). However, chronic, near-constant exposures pose fewer problems. Ideally, the biomarker should persist over time and not be affected by disease status in case–control studies. However, most biomarkers of internal dose generally provide information about recent exposures (hours to days), with the exception of markers such as persistent pesticides, dioxins, polychlorinated biphenyls, certain metals, and serological markers related to infectious agents, which can reflect exposures received many years before. If the pattern of exposure being measured is relatively continuous, short-term markers may be applicable in case–control studies of patients with early disease, so that disease bias would be less likely. However, short-term markers have generally limited use in case–control studies, as they are less likely to reflect usual patterns, and the disease or its treatment might influence its absorption, metabolism, storage, and excretion.

Biomarkers of susceptibility in case–control studies

The approaches to studying genetic susceptibility factors for disease have evolved very quickly over the last several years, owing to advances in genotyping technologies, substantial reductions in genotyping costs, and improvements in the annotation of common genetic variation, namely, the most common type of variant, the single nucleotide polymorphism (SNP). The principles and quality control approaches for the use of genetic makers in epidemiological studies are described in Chapter 6. Because inherited genetic markers measured at the DNA level are stable over time, the timing of measurement before disease diagnosis is irrelevant. In addition, it is highly likely that most genetic markers are not related to factors influencing the likelihood of participation in a study, and therefore selection bias in case–control studies is less of a concern for studying the main effect of genetic risk factors. Indeed, the robustness of genetic associations with disease for different study designs has been demonstrated in findings from consortia of studies that have shown remarkably consistent estimates of relative risk across studies of different design (82,83). Because genetic markers might influence disease progression, incomplete ascertainment of cases in case–control studies can introduce survival bias, particularly for cancers associated with high morbidity and mortality rates, such as pancreatic and ovarian cancers. This is a particular concern for population-based studies, unless a very rapid ascertainment system is implemented that enrols cases as close as possible to the time of diagnosis.

Susceptibility biomarkers can also be measured at the functional/phenotypic level (e.g. metabolic phenotypes, DNA repair capacity) (16). While genotypic measures are considerably easier to study than phenotypic measures, since they are stable over time and much less prone to analytical measurement error, phenotypic measures are likely to be closer to the disease process and can integrate the influences of multiple genetic and post-transcriptional influences on protein expression and function (84). Therefore, in spite of the advantages in measuring genotypic changes, when complex combinations of genetic variants and/or important post-transcriptional events determine a substantial portion of interindividual variation in a particular biologic process, phenotypic assays may be the only means to capture important variation in the population.

For example, several studies have assessed the role of DNA repair capacity (DRC) regarding cancer risk by using *in vitro* phenotypic assays mostly on circulating lymphocytes (e.g. mutagen sensitivity, host cell reactivation assay). These studies have shown differences in DRC between cases and controls; however, interpretation of these results must account for study design limitations, such as use of lymphocytes to infer DRC in target tissues, the possible impact of disease status on assay results, and confounding by unmeasured risk factors that influence the assay (85–87). The application of functional assays in multiple, large-scale epidemiological studies will require development of less costly and labour-intensive assays. In the future, assays that assess non-clonal mutations in DNA, through the analysis of DNA isolated from circulating white blood cells, may capture some of the same information as the above functional assays and have wider application because of greater logistic ease.

Considerations in biospecimen collection, processing and storage

A case–control study in a relatively small geographic region, or a defined set of hospitals, can permit efficient collection of medical records or specimens (e.g. blood, urine, surgical tissue and other pathologic material) along with supporting documentation. Hospital-based case–control studies or population-based studies served by a small number of hospitals can have direct contact with patients in a hospital setting, thus offering advantages for the collection of different types of specimens or elaborate processing protocols (e.g. cryopreserving lymphocytes and Epstein-Barr Virus transformation to ensure large quantities of DNA), since resources for collection, processing and storage are often available in diagnostic hospitals. This offers the potential for conducting functional assays that require live cells, such as mutagen sensitivity (85), which in general are not methodologically feasible in cohort studies. Pre-treatment specimens, critical for evaluation of biologic markers that could be affected by treatment, such as chemotherapy, can be obtained through rapid identification systems that recruit cases right at the time of diagnosis.

Biomarker measurements can be very sensitive to differences in handling of samples (e.g. fasting status at blood collection or time between collection and processing of specimens). Therefore, it is important that samples from cases and controls be collected during the same timeframe and use identical protocols to avoid differential biases. Ideally, the nursing and laboratory staff should be blinded with respect to the case–control status of the subjects. However, because the differences in handling samples between cases and controls are not always avoidable, it is important to record key information such as date and time of collection, processing and storage problems, time since last meal, current medication, and current tobacco and alcohol use to be able to account for the influence of these variables at the data analysis stage. This information can also be used to match cases and controls selected for specific biomarker measurements in a subset of the study population. This will ensure efficient adjustment for these extraneous factors during data analysis.

Biomarkers measured in samples collected from subjects during a hospital stay might not reflect measurements from samples collected outside the hospital, as habits and exposures change during hospitalization (e.g. dietary habits, medication used and physical activity). Therefore, even if cases and controls are selected through a hospital-based design, collection of specimens after the patients return home and are no longer taking medications for the conditions that brought them to the hospital should be considered, if feasible. On the other hand, specimens to measure biomarkers that are influenced by long-term effects of treatment should be collected before treatment is started at the hospital, within logistic limitations.

Case–control studies might also allow more detailed characterization of disease through the use of biomarkers, such as the presence of eosinophils in sputum to identify eosinophilic and non-eosinophilic asthma, typing of viruses in infectious diseases, or molecular characterization of tumours in cancer. This more detailed classification of disease permits the analysis of genetic and environmental risk factors and clinical outcomes by biologically important disease subtypes. These analyses can lead to improvements in risk assessment by identifying diseases with distinct risk profiles. In addition, identifying subclasses of disease of different etiology can aid in understanding the pathogenic pathways to disease, as well as developing targeted prevention programmes (e.g. use of hormonal chemoprevention for women at high risk of estrogen-receptor positive breast tumours). Review of medical records can be used to obtain information on disease characteristics determined for clinical practice, such as histological tumour type and tumour grade in cancer patients. However, more detailed characterization of disease might require large collections of biological specimens to determine disease biomarkers, which is facilitated in hospital-based studies.

Follow-up of cases to determine clinical outcomes

The prospective collection of clinical information from cases enrolled in case–control studies (e.g. treatment, recurrence of disease, and survival) greatly increases the value of these studies, since critical questions on the relationship between biomarkers and disease progression can be addressed in well characterized populations (see Chapter 4). Designing a survival study within a case–control study is easier to do at the beginning of the case–control study rather than later after subject enrolment is completed. Given the value that such studies have for carrying out translational research in a very efficient manner, consideration should be given to implementing this type of study whenever possible. The collection of clinical information is facilitated in hospital-based studies when cases are diagnosed in a relatively small number of hospitals, and in stable populations where patients are likely to be followed-up in the diagnostic hospitals or associated clinics.

Information on clinical outcomes can be obtained through active follow-up of the cases, in which patients are contacted individually through the course of their treatment and medical follow-up, or through passive follow-up by extracting information from medical records. Passive follow-up is less costly; however, it is often limited by difficulties in obtaining detailed information on treatment from medical records, or by loss to follow-up in populations where patients change cities or hospitals. Use of database resources, such as death registries in populations where cases are diagnosed, can be helpful in determining survival from cases lost to follow-up.

The case–control method in relation to other epidemiological designs

Existing cohort studies and their consortial groups that have or are collecting blood samples will accrue large numbers of cases with common diseases over the coming years. Appropriately, questions are being raised about the utility of carrying out new case–control studies, either population- or hospital-based, to study the main effects of common polymorphisms and their interaction with environmental exposures. Designers of a new case–control study will need to show that it offers benefits that cannot be obtained from existing cohorts. Below are some considerations when planning to carry out a new case–control study in contrast with performing nested studies within existing cohorts:

1) Disease incidence. A key advantage of case–control studies is the ability to enrol large numbers of cases with less common diseases in a relatively short period of time. Given the need for large sample sizes (up to several thousand cases and controls) to investigate weak to moderate associations, such as main effect for common susceptibility loci, as well as gene–environment interaction (88), and to explore data subsets, it is only feasible to collect enough cases of the more common diseases in most cohort studies. However, pooling efforts across cohorts or case–control studies, such as consortia of studies of specific tumour sites (http://epi.grants.cancer.gov/Consortia/tablelist.html), are critical to attain very large sample sizes.

2) Inclusion of diverse population groups. Case–control studies can focus on enrolling a narrow range of ethnic, racial, age or socioeconomic levels that are particularly interesting or important but not adequately represented in existing cohort studies.

3) Specialized specimen collection and processing protocols. Case–control studies can use labour- and technology-intensive biological collection, processing and storage protocols that would not be logistically feasible or cost-efficient in a large prospective cohort study.

4) Depth of exposure data. Case–control studies can collect more detailed and broader information about exposure from both interviews and records than is feasible in a cohort study. This is particularly important when there is concern about a specific type of exposure that is not generally assessed at all or in adequate detail in the typical cohort questionnaire (which usually focuses on diet and general lifestyle factors). Examples could include occupational and environmental exposures requiring complete occupational and residential histories, respectively. Cohorts have an inherent limitation in that their aim is to study multiple endpoints, and thus they collect less extensive data on exposures relevant to any one particular disease, although the opportunity to return to participants at later time points may partially ameliorate this point. Case–control studies can more readily focus on new exposures of concern for particular diseases, tailoring methods to optimally capture target data. In contrast, cohort studies will have instruments in place that will inevitably lack precision or entirely miss new exposures.

Summary and future directions

Case–control studies play a critical role in molecular epidemiologic research, particularly for biomarkers that are unlikely to have disease bias, such as DNA-based markers of genetic susceptibility. They can rapidly enrol large numbers of cases, even with rare conditions, in multicentre studies and by combining across studies in consortia. In addition, case–control studies can apply detailed diagnostic procedures, including specialized imaging approaches not routinely used in the usual healthcare setting, and state-of-the-art molecular analyses when tissue samples are collected. Given that a substantial number of rapidly developing new "omic" technologies can be readily applied to the case–control setting, this design should continue to be a core component of research programmes on the etiology of chronic diseases.

Case-only and other study designs

Case-only studies

Studies including subjects with the disease of interest without a control population (for instance, case series or clinical trials), are often used to evaluate questions related to disease treatment and progression, including secondary effects of treatment. These designs are also well suited to evaluate the influence of genetic and environmental risk factors on disease for disease progression and response to treatment, and can be very valuable to evaluate etiological questions, such as gene–gene and gene–environment interactions (89,90), and etiologic heterogeneity for different disease subtypes. An advantage of these designs is their ability to obtain extensive information on the disease to allow a more accurate definition of disease and refined classification of complex diseases, such as cancer, diabetes or hypertension, into entities more biologically or etiologically homogeneous among groups. By having direct access to patients, biological specimens, and clinical records, case series studies may be able to define diseases or preclinical conditions based on molecular events driving biological processes rather than clinical symptoms. For instance, cancers can be classified according to pathological and molecular characteristics of tumours, infectious diseases such as hepatitis can be classified according to the causal virus, and asthma can be more precisely defined according to pathophysiologic mechanisms (91).

The case-only design, however, has limitations when evaluating etiological questions, most notably related to the inability to directly estimate risk for disease. Although the case-only design can be used to estimate multiplicative interactions between risk factors under certain assumptions, it is susceptible to misinterpretation of the interaction parameter (92), is highly dependent on the assumption of independence between the exposure and the genotype under study (93), and it cannot be used to estimate additive interactions. The degree of etiologic heterogeneity in case-series studies can be quantified by the ratio of the relative risk for the effect of exposure on one disease subtype to the relative risk for another subtype. This parameter is equivalent to the relative risk for the association between exposure and disease subtype (94). However, case-only studies are limited to the estimation of the ratio of relative risk, and cannot be used to obtain estimates of the relative risk for different disease types. It should be noted that the relative risk from a case-only design would underestimate the relative risk derived in a case–control design when the exposure of interest is associated with more than one disease type.

Another potential limitation of the case-series design is the generalizability of findings, since this design can include highly selected cohorts of patients to address specific treatment protocols, such as in clinical trials. In etiological studies, it is always reassuring to observe associations between established factors and disease risk in a particular study population; however this cannot be observed in case series. Identification of cases through well characterized population-based registries, or evaluation of established associations between disease characteristics and clinical outcomes or risk factors, could address some of these limitations.

Other designs

Alternative study designs have been proposed to address some of the limitations of the classical epidemiological designs. For instance, the two-phase sampling design can be used to improve efficiency and reduce the cost of measuring biomarkers in large epidemiological studies (95). The first phase of this design could be a case–control or cohort study with basic exposure information and no biomarker measurements. In a second phase, more elaborate exposure information and/or determination of biomarkers (with collection of biological specimens if these were not collected in the first phase) is carried out in an informative sample of individuals defined by disease and exposure (e.g. subjects with extreme or uncommon exposures). Multiple statistical methods, such as simple conditional likelihood (96) or estimated-score (97), have been developed to analyse data from two-sampling designs. Another example is the use of the kin-cohort design as a more efficient alternative to case–control or cohort studies, when the goal is to estimate age-specific penetrance for rare inherited mutations in the general population (98,99). In this design, relatives of selected individuals with genetic testing form a retrospective cohort that is followed from birth to onset of disease or censoring.

Concluding remarks

The field of molecular epidemiology has undergone a transformational change with the incorporation of powerful genomic technology. Further, important advances are being made in the development of new approaches in exposure assessment (http://www.gei.nih.gov/exposurebiology). At the same time, large and high-quality case–control studies of many diseases have been established with detailed exposure data and stored biological specimens, previously established cohorts are being followed-up, and new cohort studies with biological samples are being established in developing as well as developed countries. The confluence of extraordinary technology and the availability of large epidemiologic studies should ultimately lead to new insights into the etiology of many important diseases and help to facilitate effective prevention, screening and treatment. However, this will only be achieved if molecular epidemiologists adhere to the fundamental epidemiologic principles of careful study design, vigilant quality control, thoughtful data analysis, cautious interpretation of results, and well powered replication of important findings.

Acknowledgements

This chapter has been adapted and updated from a book chapter on *Application of biomarkers in cancer epidemiology* (16) and a chapter on *Design considerations in molecular epidemiology* (21). We thank Elizabeth Azzato, David Hunter, Maria Teresa Landi, and Sholom Wacholder for their valuable comments to the chapter.

References

1. Aardema MJ, MacGregor JT (2002). Toxicology and genetic toxicology in the new era of "toxicogenomics": impact of "-omics" technologies. *Mutat Res*, 499:13–25. PMID:11804602

2. Wang W, Zhou H, Lin H et al. (2003). Quantification of proteins and metabolites by mass spectrometry without isotopic labeling or spiked standards. *Anal Chem*, 75:4818–4826. doi:10.1021/ac026468x PMID:14674459

3. Hanash S (2003). Disease proteomics. *Nature*, 422:226–232. doi:10.1038/nature01514 PMID:12634796

4. Baak JP, Path FR, Hermsen MA et al. (2003). Genomics and proteomics in cancer. *Eur J Cancer*, 39:1199–1215. doi:10.1016/S0959-8049(03)00265-X PMID:12763207

5. Sellers TA, Yates JR (2003). Review of proteomics with applications to genetic epidemiology. *Genet Epidemiol*, 24:83–98. doi:10.1002/gepi.10226 PMID:12548670

6. Staudt LM (2003). Molecular diagnosis of the hematologic cancers. *N Engl J Med*, 348:1777–1785. doi:10.1056/NEJMra020067 PMID:12724484

7. Strausberg RL, Simpson AJ, Wooster R (2003). Sequence-based cancer genomics: progress, lessons and opportunities. *Nat Rev Genet*, 4:409–418. doi:10.1038/nrg1085 PMID:12776211

8. Smith MT, Rappaport SM (2009). Building exposure biology centers to put the E into "G x E" interaction studies. *Environ Health Perspect*, 117:A334–A335. PMID:19672377

9. Schembri F, Sridhar S, Perdomo C et al. (2009). MicroRNAs as modulators of smoking-induced gene expression changes in human airway epithelium. *Proc Natl Acad Sci USA*, 106:2319–2324. doi:10.1073/pnas.0806383106 PMID:19168627

10. Committee on Biological Markers of the National Research Council (1987). Biological markers in environmental health research. *Environ Health Perspect*, 74:3–9. PMID:3691432

11. Rothman N, Wacholder S, Caporaso NE et al. (2001). The use of common genetic polymorphisms to enhance the epidemiologic study of environmental carcinogens. *Biochim Biophys Acta*, 1471:C1–C10. PMID:11342183

12. Schulte PA (1987). Methodologic issues in the use of biologic markers in epidemiologic research. *Am J Epidemiol*, 126:1006–1016. PMID:3318408

13. Perera FP (1987). Molecular cancer epidemiology: a new tool in cancer prevention. *J Natl Cancer Inst*, 78:887–898. PMID:3471998

14. Perera FP (2000). Molecular epidemiology: on the path to prevention? *J Natl Cancer Inst*, 92:602–612. doi:10.1093/jnci/92.8.602 PMID:10772677

15. Toniolo P, Boffetta P, Shuker DEG et al., editors. Application of biomarkers in cancer epidemiology. Lyon: IARC Scientific Publication; 1997.

16. García-Closas M, Vermeulen R, Sherman ME et al. Application of biomarkers in cancer epidemiology. In: Schottenfeld D, Fraumeni JF Jr, editors. Cancer epidemiology and prevention. 3rd ed. New York (NY): Oxford University Press; 2006. p. 70–88.

17. Perera FP, Weinstein IB (1982). Molecular epidemiology and carcinogen-DNA adduct detection: new approaches to studies of human cancer causation. *J Chronic Dis*, 35:581–600. doi:10.1016/0021-9681(82)90078-9 PMID:6282919

18. Spitz MR, Wu X, Mills G (2005). Integrative epidemiology: from risk assessment to outcome prediction. *J Clin Oncol*, 23:267–275. doi:10.1200/JCO.2005.05.122 PMID:15637390

19. Caporaso NE (2007). Integrative study designs–next step in the evolution of molecular epidemiology? *Cancer Epidemiol Biomarkers Prev*, 16:365–366. doi:10.1158/1055-9965.EPI-07-0142 PMID:17372231

20. Rothman N, Stewart WF, Schulte PA (1995). Incorporating biomarkers into cancer epidemiology: a matrix of biomarker and study design categories. *Cancer Epidemiol Biomarkers Prev*, 4:301–311. PMID:7655323

21. García-Closas M, Lan Q, Rothman N. Design considerations in molecular epidemiology. In: Rebbeck T, Ambrosone C, Shields P, editors. Molecular epidemiology: applications in cancer and other human diseases. New York (NY): Informa Healthcare; 2008. p. 1–18.

22. Ransohoff DF (2009). Promises and limitations of biomarkers. *Recent Results Cancer Res*, 181:55–59. doi:10.1007/978-3-540-69297-3_6 PMID:19213557

23. Pepe MS, Feng Z, Janes H et al. (2008). Pivotal evaluation of the accuracy of a biomarker used for classification or prediction: standards for study design. *J Natl Cancer Inst*, 100:1432–1438. doi:10.1093/jnci/djn326 PMID:18840817

24. Ransohoff DF (2007). How to improve reliability and efficiency of research about molecular markers: roles of phases, guidelines, and study design. *J Clin Epidemiol*, 60:1205–1219. doi:10.1016/j.jclinepi.2007.04.020 PMID:17998073

25. Schulte PA, Perera FP. Transitional studies. In: Toniolo P, Boffetta P, Shuker DEG et al., editors. Application of biomarkers in cancer epidemiology. Lyon: IARC Scientific Publications; 1997. p. 19–29.

26. Rothman N (1995). Genetic susceptibility biomarkers in studies of occupational and environmental cancer: methodologic issues. *Toxicol Lett*, 77:221–225. doi:10.1016/0378-4274(95)03298-3 PMID:7618141

27. Hulka BS, Margolin BH (1992). Methodological issues in epidemiologic studies using biologic markers. *Am J Epidemiol*, 135:200–209. PMID:1536135

28. Hulka BS (1991). ASPO Distinguished Achievement Award Lecture. Epidemiological studies using biological markers: issues for epidemiologists. *Cancer Epidemiol Biomarkers Prev*, 1:13–19. PMID:1845163

29. Wacholder S, McLaughlin JK, Silverman DT, Mandel JS (1992). Selection of controls in case-control studies. I. Principles. *Am J Epidemiol*, 135:1019–1028. PMID:1595688

30. Breslow NE, Day NE. Design considerations. In: Breslow NE, Day NE, editors. Statistical methods in cancer research. Vol 2. The design and analysis of cohort studies. Lyon: IARC Scientific Publication; 1987. p. 272–315.

31. Rothman KJ, Greenland S, editors. Modern epidemiology. Philadelphia (PA): Lippincott-Raven; 1998.

32. Paulsson B, Larsen KO, Törnqvist M (2006). Hemoglobin adducts in the assessment of potential occupational exposure to acrylamides – three case studies. *Scand J Work Environ Health*, 32:154–159. PMID:16680386

33. Wirfält E, Paulsson B, Törnqvist M et al. (2008). Associations between estimated acrylamide intakes, and hemoglobin AA adducts in a sample from the Malmö Diet and Cancer cohort. *Eur J Clin Nutr*, 62:314–323. doi:10.1038/sj.ejcn.1602704 PMID:17356560

34. Qu Q, Shore R, Li G et al. (2002). Hematological changes among Chinese workers with a broad range of benzene exposures. *Am J Ind Med*, 42:275–285. doi:10.1002/ajim.10121 PMID:12271475

35. Lan Q, Zhang L, Li G et al. (2004). Hematotoxicity in workers exposed to low levels of benzene. *Science*, 306:1774–1776. doi:10.1126/science.1102443 PMID:15576619

36. Schulte PA, Geraci C, Zumwalde R et al. (2008). Occupational risk management of engineered nanoparticles. *J Occup Environ Hyg*, 5:239–249. doi:10.1080/15459620801907840 PMID:18260001

37. Chua SD Jr, Messier SP, Legault C et al. (2008). Effect of an exercise and dietary intervention on serum biomarkers in overweight and obese adults with osteoarthritis of the knee. Osteoarthritis Cartilage, 16:1047–1053.doi:10.1016/j.joca. 2008.02.002 PMID:18359648

38. Rejnmark L, Vestergaard P, Heickendorff L et al. (2001). Loop diuretics alter the diurnal rhythm of endogenous parathyroid hormone secretion. A randomized-controlled study on the effects of loop- and thiazide-diuretics on the diurnal rhythms of calcitropic hormones and biochemical bone markers in postmenopausal women. Eur J Clin Invest, 31:764–772.doi:10.1046/j.1365-2362.2001. 00883.x PMID:11589718

39. Schulte PA, Rothman N, Schottenfeld D. Design considerations in molecular epidemiology. In: Schulte PA, Perera FP. Molecular epidemiology: principles and practices. San Diego (CA): Academic Press, Inc.;1993. p. 159–98.

40. Schatzkin A, Freedman LS, Schiffman MH, Dawsey SM (1990). Validation of intermediate end points in cancer research. J Natl Cancer Inst, 82:1746–1752.doi:10.1093/ jnci/82.22.1746 PMID:2231769

41. Schatzkin A, Gail M (2002). The promise and peril of surrogate end points in cancer research. Nat Rev Cancer, 2:19–27.doi:10. 1038/nrc702 PMID:11902582

42. Expert Panel on Detection, Evaluation, and Treatment of High Blood Cholesterol in Adults (2001). Executive summary of the third report of The National Cholesterol Education Program (NCEP) expert panel on detection, evaluation, and treatment of high blood cholesterol in adults (adult treatment panel III). JAMA, 285:2486–2497.doi:10.1001/ jama.285.19.2486 PMID:11368702

43. Tucker JD, Eastmond DA, Littlefield LG. Cytogenetic end-points as biological dosimeters and predictors of risk in epidemiological studies. In: Toniolo P, Boffetta P, Shuker DEG et al., editors. Application of biomarkers in cancer epidemiology. Lyon: IARC Scientific Publication; 1997. p. 185–200.

44. Zhang L, Eastmond DA, Smith MT (2002). The nature of chromosomal aberrations detected in humans exposed to benzene. Crit Rev Toxicol, 32:1–42.doi:10. 1080/20024091064165 PMID:11846214

45. Zhang L, Rothman N, Wang Y et al. (1999). Benzene increases aneuploidy in the lymphocytes of exposed workers: a comparison of data obtained by fluorescence in situ hybridization in interphase and metaphase cells. Environ Mol Mutagen, 34:260–268.doi:10.1002/(SICI)1098-2280(1999)34:4<260::AID-EM6>3.0.CO;2-P PMID:10618174

46. Boffetta P, van der Hel O, Norppa H et al. (2007). Chromosomal aberrations and cancer risk: results of a cohort study from Central Europe. Am J Epidemiol, 165:36–43. doi:10.1093/aje/kwj367 PMID:17071846

47. Bonassi S, Norppa H, Ceppi M et al. (2008). Chromosomal aberration frequency in lymphocytes predicts the risk of cancer: results from a pooled cohort study of 22 358 subjects in 11 countries. Carcinogenesis, 29:1178–1183.doi:10.1093/carcin/bgn075 PMID:18356148

48. Smerhovsky Z, Landa K, Rössner P et al. (2001). Risk of cancer in an occupationally exposed cohort with increased level of chromosomal aberrations. Environ Health Perspect, 109:41–45.doi:10.1289/ehp.01109 41 PMID:11171523

49. Liou SH, Lung JC, Chen YH et al. (1999). Increased chromosome-type chromosome aberration frequencies as biomarkers of cancer risk in a blackfoot endemic area. Cancer Res, 59:1481–1484. PMID:10197617

50. Bonassi S, Abbondandolo A, Camurri L et al. (1995). Are chromosome aberrations in circulating lymphocytes predictive of future cancer onset in humans? Preliminary results of an Italian cohort study. Cancer Genet Cytogenet, 79:133–135.doi:10.1016/0165-4608(94)00131-T PMID:7889505

51. Hagmar L, Brøgger A, Hansteen IL et al. (1994). Cancer risk in humans predicted by increased levels of chromosomal aberrations in lymphocytes: Nordic study group on the health risk of chromosome damage. Cancer Res, 54:2919–2922. PMID:8187078

52. Bonassi S, Hagmar L, Strömberg U et al.; European Study Group on Cytogenetic Biomarkers and Health (2000). Chromosomal aberrations in lymphocytes predict human cancer independently of exposure to carcinogens. Cancer Res, 60:1619–1625. PMID:10749131

53. Vermeulen R, Li G, Lan Q et al. (2004). Detailed exposure assessment for a molecular epidemiology study of benzene in two shoe factories in China. Ann Occup Hyg, 48:105–116.doi:10.1093/annhyg/meh005 PMID:14990432

54. García-Closas M, Rothman N, Lubin J (1999). Misclassification in case-control studies of gene-environment interactions: assessment of bias and sample size. Cancer Epidemiol Biomarkers Prev, 8:1043–1050. PMID:10613335

55. Nicholson JK, Wilson ID (2003). Opinion: understanding 'global' systems biology: metabonomics and the continuum of metabolism. Nat Rev Drug Discov, 2:668–676.doi:10.1038/nrd1157 PMID:12904817

56. Merrick BA, Tomer KB (2003). Toxicoproteomics: a parallel approach to identifying biomarkers. Environ Health Perspect, 111:A578–A579.doi:10.1289/ehp. 111-a578 PMID:12940285

57. Potter JD. Logistics and design issues in the use of biological specimens in observational epidemiology. In: Toniolo P, Boffetta P, Shuker DEG et al., editors. Application of biomarkers in cancer epidemiology. Lyon: IARC Scientific Publications; 1997. p. 31–37.

58. Rundle AG, Vineis P, Ahsan H (2005). Design options for molecular epidemiology research within cohort studies. Cancer Epidemiol Biomarkers Prev, 14:1899–1907. doi:10.1158/1055-9965.EPI-04-0860 PMID:16103435

59. Prentice RL (1995). Design issues in cohort studies. Stat Methods Med Res, 4:273–292.doi:10.1177/096228029500400402 PMID:8745127

60. Wacholder S (1991). Practical considerations in choosing between the case-cohort and nested case-control designs. Epidemiology, 2:155–158. doi:10.1097/00001648-199103000-00013 PMID:1932316

61. Hunter DJ. Methodological issues in the use of biological markers in cancer epidemiology: cohort studies. In: Toniolo P, Boffetta P, Shuker DEG et al., editors. Application of biomarkers in cancer epidemiology. Lyon: IARC Scientific Publications; 1997. p. 39–46.

62. Banks E, Meade T (2002). Study of genes and environmental factors in complex diseases. Lancet, 359:1156–1157, author reply 1157.doi:10.1016/S0140-6736(02)08140-0 PMID:11943294

63. Burton P, McCarthy M, Elliott P (2002). Study of genes and environmental factors in complex diseases. Lancet, 359:1155–1156, author reply 1157.doi:10.1016/S0140-6736 (02)08138-2 PMID:11943293

64. Clayton D, McKeigue PM (2001). Epidemiological methods for studying genes and environmental factors in complex diseases. Lancet, 358:1356–1360. doi:10.1016/S0140-6736(01)06418-2 PMID:11684236

65. Wacholder S, García-Closas M, Rothman N (2002). Study of genes and environmental factors in complex diseases. Lancet, 359:1155, author reply 1157.doi:10.1016/ S0140-6736(02)08137-0 PMID:11943292

66. García-Closas M, Thompson WD, Robins JM (1998). Differential misclassification and the assessment of gene-environment interactions in case-control studies. Am J Epidemiol, 147:426–433. PMID:9525528

67. Rosner B, Spiegelman D, Willett WC (1992). Correction of logistic regression relative risk estimates and confidence intervals for random within-person measurement error. Am J Epidemiol, 136:1400–1413. PMID:1488967

68. Holland NT, Smith MT, Eskenazi B, Bastaki M (2003). Biological sample collection and processing for molecular epidemiological studies. Mutat Res, 543:217–234.doi:10.1016/ S1383-5742(02)00090-X PMID:12787814

69. Tworoger SS, Hankinson SE (2006). Collection, processing, and storage of biological samples in epidemiologic studies: sex hormones, carotenoids, inflammatory markers, and proteomics as examples. Cancer Epidemiol Biomarkers Prev, 15:1578–1581.doi:10.1158/1055-9965.EPI-06-0629 PMID:16985015

70. Peakman TC, Elliott P (2008). The UK Biobank sample handling and storage validation studies. *Int J Epidemiol*, 37 Suppl 1;i2–i6.doi:10.1093/ije/dyn019 PMID:18381389

71. Jackson C, Best N, Elliott P (2008). UK Biobank Pilot Study: stability of haematological and clinical chemistry analytes. *Int J Epidemiol*, 37 Suppl 1;i16–i22. doi:10.1093/ije/dym280 PMID:18381388

72. Elliott P, Peakman TC; UK Biobank (2008). The UK Biobank sample handling and storage protocol for the collection, processing and archiving of human blood and urine. *Int J Epidemiol*, 37:234–244.doi:10.1093/ije/dym276 PMID:18381398

73. Fu Q, Garnham CP, Elliott ST *et al.* (2005). A robust, streamlined, and reproducible method for proteomic analysis of serum by delipidation, albumin and IgG depletion, and two-dimensional gel electrophoresis. *Proteomics*, 5:2656–2664.doi:10.1002/pmic.200402048 PMID:15924293

74. Prentice RL (1986). On the design of synthetic case-control studies. *Biometrics*, 42:301–310.doi:10.2307/2531051 PMID:3741972

75. Franco EL (2000). Statistical issues in human papillomavirus testing and screening. *Clin Lab Med*, 20:345–367. PMID:10863644

76. Welch HG, Black WC (1997). Using autopsy series to estimate the disease "reservoir" for ductal carcinoma in situ of the breast: how much more breast cancer can we find? *Ann Intern Med*, 127:1023–1028. PMID:9412284

77. Morrison AS. Screening. In: Rothman KJ, Greenland S, editors. Modern epidemiology. 2nd ed. Philadelphia (PA): Lippincott-Raven Publishers; 1998. p. 499–518.

78. Morton LM, Cahill J, Hartge P (2006). Reporting participation in epidemiologic studies: a survey of practice. *Am J Epidemiol*, 163:197–203.doi:10.1093/aje/kwj036 PMID:16339049

79. Wacholder S, Chatterjee N, Hartge P (2002). Joint effect of genes and environment distorted by selection biases: implications for hospital-based case-control studies. *Cancer Epidemiol Biomarkers Prev*, 11:885–889. PMID:12223433

80. Lum A, Le Marchand L (1998). A simple mouthwash method for obtaining genomic DNA in molecular epidemiological studies. *Cancer Epidemiol Biomarkers Prev*, 7:719–724. PMID:9718225

81. Wacholder S, Benichou J, Heineman EF *et al.* (1994). Attributable risk: advantages of a broad definition of exposure. *Am J Epidemiol*, 140:303–309. PMID:8059765

82. Cox A, Dunning AM, García-Closas M *et al.*; Kathleen Cunningham Foundation Consortium for Research into Familial Breast Cancer; Breast Cancer Association Consortium (2007). A common coding variant in CASP8 is associated with breast cancer risk. *Nat Genet*, 39:352–358.doi:10.1038/ng1981 PMID:17293864

83. Easton DF, Pooley KA, Dunning AM *et al.*; SEARCH collaborators; kConFab; AOCS Management Group (2007). Genome-wide association study identifies novel breast cancer susceptibility loci. *Nature*, 447:1087–1093.doi:10.1038/nature05887 PMID:17529967

84. Ahsan H, Rundle AG (2003). Measures of genotype versus gene products: promise and pitfalls in cancer prevention. *Carcinogenesis*, 24:1429–1434.doi:10.1093/carcin/bgg104 PMID:12819189

85. Wu X, Gu J, Spitz MR (2007). Mutagen sensitivity: a genetic predisposition factor for cancer. *Cancer Res*, 67:3493–3495. doi:10.1158/0008-5472.CAN-06-4137 PMID:17440053

86. Berwick M, Vineis P (2000). Markers of DNA repair and susceptibility to cancer in humans: an epidemiologic review. *J Natl Cancer Inst*, 92:874–897.doi:10.1093/jnci/92.11.874 PMID:10841823

87. Spitz MR, Wei Q, Dong Q *et al.* (2003). Genetic susceptibility to lung cancer: the role of DNA damage and repair. *Cancer Epidemiol Biomarkers Prev*, 12:689–698. PMID:12917198

88. García-Closas M, Lubin JH (1999). Power and sample size calculations in case-control studies of gene-environment interactions: comments on different approaches. *Am J Epidemiol*, 149:689–692. PMID:10206617

89. Yang Q, Khoury MJ, Sun F, Flanders WD (1999). Case-only design to measure gene-gene interaction. *Epidemiology*, 10:167–170. doi:10.1097/00001648-199903000-00014 PMID:10069253

90. Khoury MJ, Flanders WD (1996). Nontraditional epidemiologic approaches in the analysis of gene-environment interaction: case-control studies with no controls. *Am J Epidemiol*, 144:207–213. PMID:8686689

91. Bel EH (2004). Clinical phenotypes of asthma. *Curr Opin Pulm Med*, 10:44–50. doi:10.1097/00063198-200401000-00008 PMID:14749605

92. Schmidt S, Schaid DJ (1999). Potential misinterpretation of the case-only study to assess gene-environment interaction. *Am J Epidemiol*, 150:878–885. PMID:10522659

93. Albert PS, Ratnasinghe D, Tangrea J, Wacholder S (2001). Limitations of the case-only design for identifying gene-environment interactions. *Am J Epidemiol*, 154:687–693. doi:10.1093/aje/154.8.687 PMID:11590080

94. Begg CB, Zhang ZF (1994). Statistical analysis of molecular epidemiology studies employing case-series. *Cancer Epidemiol Biomarkers Prev*, 3:173–175. PMID:8049640

95. White JE (1982). A two stage design for the study of the relationship between a rare exposure and a rare disease. *Am J Epidemiol*, 115:119–128. PMID:7055123

96. Cain KC, Breslow NE (1988). Logistic regression analysis and efficient design for two-stage studies. *Am J Epidemiol*, 128:1198–1206. PMID:3195561

97. Chatterjee N, Chen Y, Breslow N (2003). A pseudoscore estimator for regression problems for two phase sampling. *J Am Stat Assoc*, 98:158–168 doi:10.1198/016214503388619184.

98. Wacholder S, Hartge P, Struewing JP *et al.* (1998). The kin-cohort study for estimating penetrance. *Am J Epidemiol*, 148:623–630. doi:10.1093/aje/148.7.623 PMID:9778168

99. Chatterjee N, Shih J, Hartge P *et al.* (2001). Association and aggregation analysis using kin-cohort designs with applications to genotype and family history data from the Washington Ashkenazi Study. *Genet Epidemiol*, 21:123–138.doi:10.1002/gepi.1022 PMID:11507721

UNIT 4.
INTEGRATION OF BIOMARKERS INTO EPIDEMIOLOGY STUDY DESIGNS

CHAPTER 15.

Family-based designs

Christopher I. Amos and Christoph Lange

Summary

Family-based designs are used for a variety of reasons in genetic epidemiology, including the initial estimation of the strength of genetic effects for a disease, genetic linkage analysis by which genetic causes can be sublocalized to chromosomal regions, as well as to perform association studies that are not confounded by ethnic background. This chapter describes some of the approaches that are followed in the initial characterizing of genetic components of disease and family-based designs for association analysis and linkage with genetic markers.

Family studies of phenotypes

To obtain an initial assessment of the genetic contributions to disease, and determine which subsequent approach is most likely to be effective, a variety of family-based designs are employed (Figure 15.1). The heritability of a disease indicates the proportion of the covariation in risk for disease that can be attributed to genetic factors. Heritability in the narrow sense excludes covariation due to gene–environment interactions, while in the broad sense includes all genetic contributions to disease. Heritability estimation can be performed using either data from population-based twin registries or from family studies that include different types of relatives. The study of twin registries allows investigators to contrast the similarity in disease among monozygotic twins, who share all their genetic material in common, versus dizygous twins. While the study of twin registries can provide important insights concerning the contribution of genetic factors to disease, twin studies have limitations. For the study of rare diseases, such as cancer, very large collections of twins must be followed for many years. Second, important assumptions that the monozygotic and dizygotic environments are similar are difficult to evaluate. Despite some concerns and weaknesses of this design, twin studies have indicated strong genetic components of risk for the most common cancers (1), as well as many autoimmune conditions (2). Since the development and maintenance of twin registries is beyond the scope of most epidemiologists, this design is not discussed here (see (3,4) for several comprehensive resources).

Figure 15.1. Designs for genetic epidemiological studies to identify genetic factors for diseases

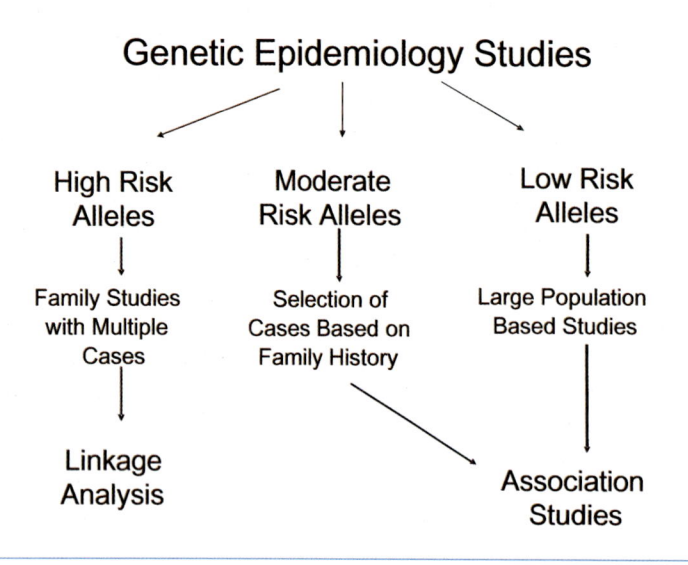

```
                    Genetic Epidemiology Studies
              ↙                    ↓                    ↘
        High Risk            Moderate              Low Risk
         Alleles            Risk Alleles            Alleles
            ↓                    ↓                     ↓
       Family Studies       Selection of         Large Population
       with Multiple       Cases Based on        Based Studies
          Cases            Family History
            ↓                    ↘                    ↓
         Linkage                            Association
         Analysis                             Studies
```

An alternate measure that is more often used to characterize the genetic contribution to disease is the recurrence risk to a class of relatives. For example, for genetically influenced diseases, a monozygous twin who shares all genes in common with their cotwin (a zeroth-degree relative) should have a higher risk of developing disease if their cotwin also has the disease, compared with dizygous twins, siblings or a parent or child who shares only half their genes in common. Each of these pairs of relatives is called a first-degree relative. Similarly, second-degree relatives (half-siblings, avuncular pairs, and grandchild-grandparent pairs) should show even lower risks for disease given that one of the pair members has the disease compared with first- or zeroth-degree relatives. Evidence that there are genetic contributions to disease is found by observing the relative risk for disease either among different classes of relatives, or if population-based estimates are available, by forming the relative recurrence risk by contrasting the risk to relatives of a certain type to the risk in the general population. The easiest such relative risk to estimate is the risk to cosiblings. Relative recurrence risks (RRR) to siblings for cancers range from 2–2.5 for most common epithelial cancers (5), but are much higher for selected cancers, such as non-medullary thyroid cancer (RRR = 15.6), Hodgkin's disease (RRR = 6.5), testicular cancer (RRR = 6.6), ovarian cancer (RRR = 4.9) and renal cancer (RRR = 4.7). Relative recurrence risks are much higher for some cancers if multiple relatives are affected and also higher for relatives of earlier onset cancers.

Contrasting recurrence risks for other types of relative combinations can provide initial insights into whether or not there are recessive or dominant effects for a disease, and the number of genetic factors that are likely to be important in disease causation (6,7). If there are recessive effects influencing disease causation, then risk to monozygous twins who share all their genetic material in common will be much greater than risks to dizygous pairs of siblings. In turn these risks will be higher than the risks to offspring or parents, because parent-offspring pairs never share two alleles in common, while siblings share on average one quarter of the time, and sibling pairs share both alleles in common. Thus, if a disease includes a recessive effect, then a co-sib of an affected individual is also more likely to have two deleterious alleles and be affected than a parent would be (see (8) for a detailed description concerning the estimation of allele frequencies in co-sibs and other relatives). The fall-off of the recurrence can also be used to provide insights into the number of loci that may influence a disease (6).

The probability that an individual becomes affected given that they carry a particular genotype is called the penetrance. The penetrance of disease can depend upon genotype(s) at one or more loci, as well as environmental factors. The genotypic risk ratio is the ratio of the penetrance given that an individual has one particular genotype compared to the risk for disease given another genotype. The relative recurrence risk depends upon the genotypic risk ratio and the population prevalence of the genetic factor (9). The genotypic recurrence risk determines the power of association studies, as described in more detail below.

Standard epidemiological approaches have been modified and further developed for characterizing evidence that genetic factors influence a disease. The most straightforward approach that epidemiologists will initially apply seeks to identify the odds ratio of a disease with family history for the disease in relatives. In this approach, the epidemiologist asks cases and controls to delineate the occurrence of disease among relatives of each type (e.g. siblings and parents). Then, usual epidemiological approaches such as logistic regression can be used to obtain an odds ratio that a case reports a family history of disease compared with a control. While this approach is similar to other analytical approaches commonly used in epidemiology, a historical cohort approach is preferred by genetic epidemiologists for many reasons. In the historical cohort approach, case and control participants are asked to provide medical history information on selected relatives, such as first-degree relatives. The data that must be collected for each relative includes either the age at disease onset(s), the current age if alive and unaffected, or the age at death if deceased without disease. The relatives of the cases and controls then form a historical cohort with the follow-up period extending from birth until either disease onset or last age (death or current age). There are many advantages of this approach over the case–control design (10). In the historical cohort design, both absolute and relative risks can be obtained. As this is a cohort design, multiple disease endpoints can be studied. In the case–control design, because cases and controls are typically selected not to have multiple diseases, it becomes impossible to evaluate whether one disease, such as rheumatoid arthritis, is associated with an increase in relatives for another disease, such as systemic lupus erythematosis. In addition, according to both the two-stage and multistage models of carcinogenesis, individuals with inherited susceptibility to disease should have higher hazards ratios for cancer(s) at earlier ages compared with older ages. Case–control designs that are typically matched on age have difficulty estimating differential risks for disease according to age, but for the historical cohort design, variation in disease risk according to age can be readily estimated. Because the relatives in a family are correlated, tests of the relative risks for disease are biased unless a variance correction is introduced to allow for this correlation. The Huber-White variance correction procedure can be applied and is readily available in standard analytical packages such as SAS, STATA or R. As an example, the occurrence of rheumatoid arthritis in relatives of cases compared to controls and the occurrence of other autoimmune conditions was studied (11). The results showed that aside from rheumatoid arthritis, which was more frequent in case relatives compared to control relatives, other autoimmune conditions occurred more frequently in relatives of controls.

When a candidate mutation has been studied in a family, an approach to estimate the penetrance specific to that mutation is the kin-cohort approach (12–14). This method takes advantage of the extensive data on family members that can be obtained using the historical cohort approach discussed above, but also allows the penetrance to be estimated specifically from the mutation. Among those probands who are found to have a rare mutation, about 50% carry the mutation, while nearly none of the relatives of probands not carrying the mutation are carriers. By contrasting the age-specific risk in relatives of carriers versus relatives of non-carriers, one can derive an estimate of the penetrance associated with the mutation being studied. An issue in applying this method is how to correct for the selection of probands based upon their being affected when there is risk for disease, not only from the mutation being studied, but also from other loci (14,15). Using these methods, the risk associated with carriage of breast cancer 1 (BRCA1) and BRCA2 mutations could be estimated from the population-based Washington Ashkenazi Study, since the prevalence of mutations in this population was sufficiently high. Using the kin-cohort approach, the risk for breast cancer due to carriage of either of the three common mutations in Ashkenazim was 56% to age 70, which is considerably less than had been estimated previously from the study of families ascertained through multiple affected relatives (16). This variation in risk according to the sampling design likely reflected the incomplete ascertainment correction provided by earlier studies of families that included many affected relatives. Previous approaches to ascertainment correction in family studies derived for linkage analysis conditioned only on the specific measured genetic factors (e.g. BRCA1 and BRCA2) and failed to allow for effects from unmeasured lower-penetrant loci. A more recent alternative approach to the kin-cohort method adapts segregation analysis to incorporate effects from a known measured genetic factor, such as BRCA1 and BRCA2, as well as residual risk from unmeasured genetic factors (17). Application of these methods has yielded penetrance estimates similar to those given by the kin-cohort approach.

Of concern when performing genetic epidemiological studies in which a case or control is interviewed about the occurrence of disease in relatives is the reliability of the reporting by such subjects. Numerous studies have shown that for some common epithelial cancers, such as breast, colon, prostate and lung, reporting of disease in relatives is acceptably accurate (18,19). For cancers of the internal organs or common metastatic sites, such as ovarian, liver and brain cancers, reliability of reporting is extremely poor (20). Studies of these cancers would entail obtaining medical records to verify reporting by the case or control. Reporting of autoimmune diseases also shows variable reliability, with rheumatoid arthritis, for example often being confused with other types of arthritis. Reports of rheumatoid arthritis in relatives were confirmed using medical records and reports from multiple relatives (11).

Reporting of disease in relatives can raise issues concerning the privacy of the relatives. The American Society of Human Genetics has issued a policy statement that indicates reporting by an individual about a relative is hearsay, and hence does not constitute a violation of privacy (21). However, an evaluation of risk associated with the collection of reported disease in relatives will require Internal Review Board review. Inadequate compliance with an approved protocol for the collection of reported medical data on relatives, led to temporary cessation of research at the University of Virginia, when a father complained that his child was being asked to report sensitive information about him as a part of a research study. In the USA, researchers involved in studies of diseases for which risk can accrue to either the patient or the researcher can obtain a certificate of confidentiality. This certificate protects the research from legal discovery.

Segregation analysis

To more precisely model the familial and genetic factors affecting disease expression, case–control studies have often been followed by segregation analyses. This is particularly useful when initial studies identify high risk associated with a family history of disease, and the disease is rare, suggesting the involvement of one or a few genetic factors having high penetrance. Segregation analyses seek to identify the relationship between an individual's genotype and the resulting phenotype. Inheritance of genetic factors results in a specific form of genotype dependence among family members. Although the genotypes at a disease locus cannot usually be determined, the inheritance of disease within families can be compared with that expected under specific genetic models. In segregation analyses, the model that most closely approximates the observed familial data is sought. The models that are evaluated by classic segregation analyses include a genetic factor, environmental effects which may be correlated among family members, and polygenic effects. These polygenic effects are a mathematical construct that corresponds to the inheritance of many independent genetic factors, each having small effects.

The classic paradigm of segregation analysis also requires scrupulous definition and attention to the ascertainment criteria. For most diseases, the occurrence of genetic susceptibility is sufficiently uncommon that random sampling would result in low power to detect genetic effects. However, most patterns of selection through affected individuals introduce biases into the genetic analyses. When the selection or ascertainment events are well characterized, these biases can often be controlled for appropriate mathematical conditioning (22). For segregation analysis, the units of observation are individuals within families, and although the modeling process is applied to individuals, it also requires information on their close relatives. Thus, the unit of sampling and analysis is the family. Summary statistics from segregation analytic studies include the gene frequency of the disease-causing locus, the penetrance for the susceptible genotypes, and the sporadic risk for the non-susceptible genotypes. During segregation analysis, the parameters describing the penetrance and the gene frequency are inferred using maximum likelihood methods. The parameters that most accurately describe the observed data are identified by computationally intensive numerical evaluations. To allow for the variable size and structure of human families, very general algorithms were developed, largely as a result of seminal works by R.C. Elston (23,24).

Genetic linkage analysis

Genetic linkage analysis has been an extremely powerful tool for identifying specific genetic factors for diseases. Linkage analysis has typically been applied for identifying novel genetic factors by using a genome-wide analysis of the co-inheritance of disease with genetic markers. Evidence in favour of linkage is typically expressed by the LOD score, which is the $\log_{10}$ ratio of the likelihood of the data assuming linkage between a modelled disease susceptibility locus and a genetic marker, to the likelihood of the data assuming no linkage of the disease susceptibility and genetic marker. To allow for the large number of tests that are indicated in a genome-wide analysis, several testing paradigms have been developed. If a Bayesian approach is adopted, a LOD score of about 3.0 leads to a 5% posterior probability of linkage assuming the existence of a single disease locus, even when many markers are genotyped over the entire genome. An approach for sequentially combining data from multiple studies by adding LOD scores across studies has been highly effective (25). From Bayesian and sequential analytical approaches, a LOD score of 3.0 was proposed as providing a meaningful critical value for declaring strong evidence for linkage. More recently, approaches to control the overall significance of genetic studies when studying multiple markers have been adopted (26). These criteria have been criticized for being excessively conservative (27), particularly when candidate regions are of primary interest (e.g. when prior studies indicated evidence for linkage to an area). The significance testing paradigm requires the slightly higher LOD score of 3.3 to declare that a significant result has been obtained while providing a genome-wide significance of 5%.

If a simple genetic mechanism explains inheritance of disease, then a genetic model can be specified and tested for co-inheritance of disease susceptibility with genetic markers. In order for linkage studies to be informative, the families chosen for study must be able to show inheritance of a genetic factor. For uncommon diseases for which the penetrance is reduced, the affected individuals provide the majority of information about the segregation or inheritance of genetic mutations predisposing to disease. For quantitative traits, sampling through individuals with extreme phenotypes can increase the probability of sampling a genetic variant influencing the trait of interest. Sampling through extreme individuals is an effective strategy for increasing the power of a linkage study, but may only be practical if the quantitative phenotype can be assayed inexpensively. Some studies of quantitative phenotypes look at many phenotypes. Sampling through extreme individuals only increases power for a single or a few correlated phenotypes.

Linkage analyses are mainly conducted using panels of single nucleotide polymorphisms (SNPs) with a density of at least 1 marker every 500 kilobases (usually at least 6K markers), but can also be performed using microsatellite panels with a density of at least 1 marker every 10 megabases (about 350 markers). SNPs are far less informative than microsatellites, so that a much denser mapping panel is required to obtain a comparable amount of information from a genetic study using SNPs compared with one using microsatellites. Evidence for genetic linkage in a region would often be followed by finer-scale mapping to improve the information for detecting linkage and to identify any recombinant individuals. Finer maps would be employed if a microsatellite panel or relatively sparse SNP panel was used, to search for associations between the disease or trait and particular marker alleles. Standard finer mapping panels for microsatellites provide a 0.5 to 0.2 megabase interval spacing (available from Decode Genetics (decodegenetics.com) or Invitrogen Genetics). Routine genotyping platforms for the purposes of genetic linkage analysis are available from Affymetrix and Illumina, and provide results from genotyping of between 6000 and 1 000 000 genome-wide SNPs, respectively. These much finer mapping panels can improve the power to detect linkages and may provide narrower intervals for positional cloning. However, the presence of strong linkage disequilibrium (LD) among the SNPs in these platforms raises many analytical complexities that must be dealt with for accurate inferences. In particular, biases occur when families are selected through multiple disease-affected relatives if LD is not precisely modelled (28).

A wide range of genetic linkage methods are available. The diversity of methods reflects, in part, the considerable success in identifying genetic causes of disease, and the consequent value and interest in using the methods by the scientific community. Computing statistics over a large number of genetic markers in families for diseases that do not show simple inheritance patterns is computationally demanding. There are three basic approaches that are used for analysis of the genetic marker data. The Elston-Stewart algorithm (23) summarizes information about haplotypes (the set of alleles on a chromosome) sequentially in a pedigree, and is therefore efficient for statistical analysis of large families, but limited in the number of markers that can be jointly modelled (usually fewer than five markers can be considered jointly). The Lander-Green-Kruglyak (LGK) method (29) adopts a different approach that facilitates the analysis of multiple markers. The LGK model first identifies the possible inheritance patterns of genotypes within families and stores this information as inheritance vectors. Because the number of inheritance vectors increases rapidly according to the number of individuals in a family, this approach is only suitable for small- or medium-sized families, usually allowing at most 25 individuals in a family to be studied. In addition, because the method stores all possible inheritance vectors in memory, the approach requires considerable RAM to be efficient. The major advantage of the LGK approach is that computational speed increases only linearly in the number of markers so that it is highly efficient for genome-wide analyses. In addition, the adaptations of the LGK algorithm allow haplotypes to be used as markers, thus allowing for the strong LD that can exist among tightly linked markers (30).

Analyses including many markers on large pedigrees, or analyses of pedigrees that include more than a few inbred individuals, may not be effectively performed using the Elston-Stewart or LGK algorithms. In this case, Monte-Carlo Markov Chain (MCMC) algorithms are used to approximate the likelihood of the data. MCMC methods provide tools for sampling the haplotype configurations in data (31,32). The MCMC procedure samples possible haplotypes according to the underlying probability distribution that generated the data and provides an accurate approximation to the likelihood. A major advantage of MCMC procedures is a decreased need for memory, since they do not require summing over all possible genotypes as in the Elston-Stewart algorithm, or over all possible inheritance vectors as in the LGK. One disadvantage is the complexity in storing output from analyses, since results from large numbers of realizations from the sampling of genotype configurations must be stored. MCMC methods infer the genotypes for all individuals that are specified as a part of the analytical file. Individuals with known genotypes have a limited number of potential haplotypes, but individuals who have not been genotyped can have a large number of potential genotypes and haplotypes. The probability distribution from which MCMC methods must sample can become quite large if many individuals who have not been genotyped are included in the analytical file.

Therefore, it is often beneficial to remove the ungenotyped individuals from MCMC analyses, particularly those who are not affected, since they contribute little in most linkage analyses.

An issue in performing genetic analysis is whether to use model-dependent or model-free methods for linkage analysis. Model-dependent methods have higher power for linkage analysis if an approximately valid genetic model can be specified to describe the manner in which disease susceptibility at a given locus is expressed. One approach for estimating penetrance to be used in a linkage study is to first perform a segregation analysis of families that have been ascertained according to a specified sampling scheme. The approach estimates parameters for models describing the inheritance of genetic and environmental factors that most closely fit the dependence in family data. For uncommon conditions, random sampling of families would not result in an informative family; a sampling scheme is usually followed in which relatives of cases with a disease are preferentially sampled. When the families are not randomly sampled, an ascertainment correction for non-random sampling is required to obtain parameter estimates that reflect the more general population of families. To correct for the non-random sampling approach usually used, a clearly defined sampling scheme must typically be followed. Using only a binary phenotype (e.g. affection or non-affection) one may not be able to estimate all the parameters that are necessary to describe the penetrance of the genotypes of the loci influencing disease susceptibility, unless restrictive assumptions about the interactions among the loci are made.

Sampling families and collecting information for segregation analysis can be an arduous task, and may not be fully informative about the parameters that describe the penetrance and disease allele frequencies. Therefore, investigators studying complex diseases may postulate genetic models from assumptions about the relative risks for disease that are observed from epidemiological studies. It has been shown that postulating an inaccurate genetic model for genetic linkage studies does not lead to false-positive results in a model-based linkage study. However, if multiple models are tested, there can be an inflation of the overall number of false-positive results from linkage studies because of the inherent multiple-testing problem that is introduced. A powerful approach for studying complex diseases is to evaluate the evidence for linkage, assuming simple recessive and dominant models of disease, and then to adjust the required critical value for the LOD score upwards by about 0.3 for the small multiple-testing problem so engendered (33).

If the genetic model influencing disease susceptibility cannot be inferred with any confidence, either because the genetic model appears too complex or because there is a lack of epidemiological data from which to postulate penetrance, then model-free methods are typically adopted. One approach is to set the penetrance to an artificially low level, thus restricting analysis to include only the affected subjects. With very low penetrance, unaffected individuals provide no information about their possible genotypes and so do not contribute in the linkage analysis, but this approach still makes some modelling assumptions about disease expression. An alternative approach is to evaluate the similarity in alleles that have been inherited by common parentage (identity by descent) and test whether or not there is evidence that affected relatives share more alleles than expected identical by descent. In some cases this approach may provide a more powerful test for linkage than a model-dependent approach, particularly when multiple independent loci additively increase disease risk. Because pedigrees are usually variable in size and contain different numbers of affected relatives, a variety of different tests have been proposed and are available for testing for linkage (34,35). These tests are optimal for varying disease penetrances (which are typically unknown). As a compromise, the pairs statistic is often used, which includes all affected relatives in a pedigree and gives only moderately higher weight to families that include multiple affected relatives (29).

The joint analysis of covariates, along with genetic markers in family studies, usually has limited utility. Typically, collecting covariate information in families is difficult because data cannot be directly collected from deceased or otherwise unavailable individuals. In addition, the genetic risks that are sought in linkage analyses are often large. Some non-genetic factors, such as smoking and reproductive behaviours, can be reliably collected through proxies (when needed), are inexpensive to collect, and may have a strong effect upon risk for some diseases.

For complex diseases, a large number of families may be needed to obtain adequate power to detect linkages. Meta-analyses combining multiple studies can assist in overcoming power limitations from a single study. However, in order for meaningful results to be obtained in meta-analyses, investigators must be studying comparable classifications of the same disease. Coordination of studies by using common definitions of disease outcomes, demographic measures and covariates is necessary for the study of complex diseases. Tools for meta-analysis of both linkage and association studies are available (36,37).

Association studies using families

While parametric and non-parametric linkage analysis approaches have proved successful for mapping many disease and trait genes, in some gene mapping investigations the limited number of meioses occurring within pedigrees limit one's ability to detect, by linkage recombination, events between closely spaced (< ~1 cM) loci (38). Association studies might be used instead to map more closely spaced disease genes. These studies generally have a case–control design, where cases are recruited from a disease registry or hospital-based populations. Controls can range from the cases' family members (e.g. parents or siblings), or unrelated individuals. Genetic variants observed in cases are contrasted with those observed among controls to determine if an association exists between genes and disease.

Association studies may permit one to get closer to the disease-causing gene than allowed by linkage studies (i.e. more recombinant events over evolutionary time). This type of study can also be used to directly examine genetic variants in known candidate genes. That is, association studies can be used either in an indirect manner, as a tool for mapping genes using linkage disequilibrium, or in a direct manner, for evaluating associations with postulated causal ("candidate") genes.

The growing use of association studies is driven in part by how quickly and easily they can be undertaken, and the availability of high-density SNP genotyping technology. The SNP consortium (39) has provided sequences for 1.8 million SNPs, and at least 250 000 of these have been confirmed as polymorphic by Perlegen alone, while polymorphisms in hundreds of thousands of additional SNPs have also been verified by the SNP consortium Applera, and by many investigators and companies.

The power to detect associations using unrelated cases and unrelated controls can be increased by selection of cases that are likely to have developed the disease because of increased genetic propensity. For rare or uncommon susceptibility factors, sampling unrelated cases on the basis that they have close relatives affected by the same disease can greatly increase the power to detect associations (40). Power to detect associations can also be accomplished by seeking a homogeneous genetic etiology for the disease, which entails selecting from isolated populations and cases that show a homogeneous clinical phenotype.

Linkage disequilibrium and haplotypes

The genetic variants that cause disease arise through, for example, novel mutations or immigration of mutation carriers into a population. When a mutation initially occurs, it has a particular chromosomal location and specific neighbouring marker alleles. At this incipient point in time, the mutation is completely associated with the adjacent alleles; it is only observed when the marker alleles are also present (41). Marker alleles that were in the neighbourhood of the disease gene when its mutation was introduced into the population will generally remain nearby over numerous generations, that is to say linkage disequilibrium. One can estimate whether particular marker alleles appear to be in disequilibrium, that is to say, are associated, with disease genes. In particular, if specific marker allele frequencies are higher among cases versus controls, this suggests linkage between the corresponding loci and a disease gene. The extent of this disequilibrium depends on the number of subsequent generations since the mutation was introduced into the population, the recombination between the disease and marker alleles, mutation rates, and selective values (e.g. epistatic components).

Alleles in linkage disequilibrium may be parts of haplotypes. Recent work indicates that there may exist discrete chromosomal regions with low haplotype diversity, termed haplotype blocks, that are separated by recombination hotspots. Information from some polymorphisms within each block may be redundant; in other words, having information on one SNP provides all the information about another if they are in strong linkage disequilibrium. The majority of the haplotypes within a block can thus be distinguished using a much smaller number of SNPs, known as haplotype tagging SNPs (htSNPs). Using such SNPs can drastically reduce the effort required to undertake large scale association studies. Instead of saturating an entire chromosomal region with genotypes in all study samples, an investigator can first screen for SNPs within a subsample of study subjects to determine the htSNPs. Then only these tagging SNPs (and possibly other promising SNPs) can be genotyped in the entire study population. Several approaches have been suggested for identifying optimal htSNPs. These include visual inspection of haplotypes, and analytic approaches that eliminate redundant markers (42–44).

Family-based association studies

The most common familial case–control designs use parents or siblings as controls. In the former, the parents themselves are not the controls, but the set of genotypes the parents could have transmitted to the case, given their own genotypes (the case's "pseudosibs"). For example, the Transmission/Disequilibrium Test (TDT) compares alleles transmitted from parents to diseased offspring with those alleles that are not transmitted (i.e. the non-diseased alleles) (45). The TDT provides a joint test of linkage and association (i.e. linkage in the presence of association or vice-versa). In doing so, when there is disequilibrium between marker and disease alleles, incorporating the additional information that the same alleles are associated across families with the TDT can provide increased power in comparison with linkage analysis. Furthermore, the use of pseudosib controls has better statistical efficiency than sibling or cousin controls (even more than population controls for a recessive gene), but the requirement that parents be available for genotyping limits its usefulness for late-onset diseases.

As with pseudosib controls, siblings are derived from the same gene pool as the cases, and thus provide another attractive source of controls for family-based studies. However, using siblings as controls can pose other difficulties. A major issue is that not every case will have an available sibling. If sibship size or other determinants of availability are associated with genotype, selection bias may result, possibly leading to false-negative or -positive results. Another issue is that controls should generally be selected from siblings who have already survived to the age at diagnosis of the case and be free of the disease. In practice, this will tend to limit control eligibility to older siblings, which can lead to confounding by factors related to year of birth, family size or birth order. Siblings are also more likely to have the same genotype as the case than are unrelated controls, thereby leading to some loss of statistical efficiency (i.e. larger sample sizes required to attain the same statistical precision).

The many successful applications of the TDT motivated the development of a large number of generalizations. The original TDT concept was extended to multiallelic marker data (45–47) and to different genetic models. In the framework of score tests for multivariate data, it has been shown that the TDT is the most powerful test under an additive mode of inheritance; alternative tests can be derived under a dominant and recessive mode of inheritance (48). (Extension to general pedigree designs and to scenarios in which parental genotypes are missing are discussed in (46,49–52)). Approaches to general pedigrees that are also valid under the null hypothesis of linkage, but no association has been developed are discussed in (53–55). Extensions to quantitative traits are described in (52,56–61).) The gamete competition model (62) provides one generalization of the TDT that can be applied to arbitrary pedigrees and extends to haplotype-based analyses. This approach has been integrated into the Mendel suite of programs (http://www.genetics.ucla.edu/software/mendel).

The family-based association tests (FBAT) approach

In this section is a review of a very general and adaptable approach to construct family-based association tests that are often referred to as the FBAT approach (60). FBATs can be applied under any mode of inheritance and in situations in which multiallelic data and/or general pedigrees are available. Various null hypotheses, and different phenotypic traits and arbitrary combinations of them (binary, quantitative, time-to-onset, repeated measurements, multivariate data, etc.), can be tested for association. FBAT can be computed for a single marker locus, haplotypes or multiple markers. The FBAT approach is built on the three key principles of the original TDT approach:

1. The FBAT statistic is a conditional test that conditions upon the parental genotype, or, as will be discussed later, equivalent information if parental data should be missing. By conditioning on the parental information, there is no need to estimate the genotype distribution of the data (e.g. the margins of the table in a case/control design) under the null hypothesis, and thereby eliminate the effects of population admixture. When parental information is missing, one can condition on the sufficient statistic for the genotype distribution in each family. For haplotype analysis, phase uncertainty will also be included in the conditioning.

2. The FBAT statistic is also computed conditional on the phenotype, which makes the approach robust against misspecification of the phenotypic assumptions that are used for the computation of the FBAT statistic.

3. Since the only random variable in the FBAT approach is the offspring genotype, whose distribution under the null hypothesis can be computed based on Mendelian transmission, Mendel's first law is the sole requirement for the validity of the approach.

The general FBAT statistic

The FBAT statistic assesses the association between the phenotype and the genetic locus by using a natural yardstick: the covariance between the phenotype and the Mendelian residuals. The covariance is defined by:

$$U = \Sigma \, T_{ij} (X_{ij} - E(X_{ij}|S_i)), \qquad (1)$$

where i indexes family and j indexes non-founders in the family. The summation is over all families i and all non-founders j. The parameter T_{ij} denotes the coded trait of interest in the jth non-founder of the ith family. The corresponding genotype is given by X_{ij} which is adjusted by its expected value $E(X_{ij}|S_i)$ under the null hypothesis. Using the assumption of Mendelian transmissions from the parents to the offspring, the expected marker score $E(X_{ij}|S_i)$ is computed conditional upon the parental genotypes Si of the ith family. If parental information is missing, S_i denotes the sufficient statistic of the genetic distribution in the ith family. The adjusted genotype, $(X_{ij} - E(X_{ij}|S_i))$, can be interpreted as an Mendelian residual, measuring a potential over- or undertransmission from the parents to the offspring. In this context, it is important to note that the Mendelian residuals for families with two homozygous parents will always be zero and that such families do not contribute to the FBAT statistic. The number of families that have at least one Mendelian residual $(X_{ij}-E(X_{ij}|S_i))$, which based on S_i can be different from zero, is typically referred to as 'number of informative families.'

As discussed below, the coded phenotype T_{ij} is either centred or unadjusted, depending on the absence or presence of a phenotypic ascertainment condition. By selecting appropriate coding functions, qualitative, quantitative, time-to-onset and multivariate phenotypes are incorporated into the FBAT approach.

The basic formula (1) is applicable in virtually any scenario; the appropriate selection of the phenotypic coding function and its adjustment, and the definition of the genotypes, reflecting the underlying genetic model.

Large sample distribution of the FBAT statistic under the null hypothesis

As outlined in the discussion of the key principles of the FBAT approach, the distribution of the FBAT statistic U is computed by treating the non-founder genotype as the only random variable and both the coded phenotype, T_{ij}, and the sufficient statistic, S_i, as deterministic variables by conditioning on them. The expected value of the FBAT statistic, U, is zero by definition (E(U) = 0), so to normalize U under the null hypothesis, all that is left to do is to compute the variance of U conditional upon the offspring phenotype and S_i. If the genotype and trait variable are both univariate, then

$Z = U/\sqrt{(var(U))}$, or equivalently,
$\chi^2_{FBAT} = U^2/var(U),$
where

$$Var(U) = \Sigma_i \Sigma_{jj'} T_{ij}T_{ij'} cov(X_{ij}, X_{ij'}|S_i, T_{ij}, T_{ij'}) \quad (2)$$

As for the expected marker score, the covariance $cov(X_{ij}, X_{ij'}|S_i, T_{ij}, T_{ij'})$ also conditions upon the traits and the sufficient statistics, assuming the null hypothesis is true. Under the null hypothesis of no association and no linkage, the covariance $cov(X_{ij}, X_{ij'} | S_i T_{ij}, T_{ij'})$ does not depend on the phenotype T_{ij}, and can be computed based on independent Mendelian transmissions within a family. However, when the null hypothesis of no association in the presence of linkage is selected, the transmissions to siblings within a family are correlated (55). In this situation, the derivation of the theoretical covariance is difficult, and an empirical variance can be used to estimate $var(U)$ (51).

Asymptotically, Z is normally distributed, $N(0,1)$, and χ^2_{FBAT} follows a χ^2 distribution with one degree of freedom. When multiple alleles and/or multiple traits are tested, U is the vector and $var(U)$ becomes a variance/covariance matrix. Then, the FBAT statistic is a quadratic form $U^T var(U)^- U$ and follows asymptotically a χ^2 distribution with degrees of freedom equal to the rank of $var(U)$ (60,63).

When the number of families is small (e.g. in linkage studies), it is recommended either to estimate the P-value of the FBAT statistic via Monte-Carlo simulations or to use an exact test (64).

Specifying the mode of inheritance in the FBAT statistic

In the FBAT statistic, the coding of the genotype reflects the specified mode of inheritance. When testing under an additive mode of inheritance is required, X_{ij} counts the number of target alleles (i.e. 0, 1 or 2). Under a recessive model, X_{ij} is defined to be 1 for subjects who carry 2 copies of the target allele, and 0 otherwise. For multiallelic markers or haplotypes, X_{ij} is a vector whose element reflects the coded genotype for each allele/haplotype.

Coding the phenotype: Testing binary phenotypes in the FBAT approach

When the phenotype of interest is affection status, an FBAT statistic that is equivalent to the classical TDT (61), and that only incorporates information on affected subjects, can be obtained by setting T_{ij} = 1 for affected subjects and 0 otherwise. Unaffected subjects can be included in the FBAT statistic by defining T_{ij} = $(Y_{ij} - \mu)$, where Y_{ij} is the original 1/0 phenotype and μ is a user-defined offset parameter in the range between 0 and 1. For example, by setting μ = 0, the original TDT statistic is obtained. Affected subjects (Y_{ij} = 1) then contributed $(1 - \mu)$ to the FBAT statistic and the unaffecteds $(1 - \mu)$. Here the FBAT statistic can be interpreted as a contrast between transmissions to affected offspring weighted by $(1 - \mu)$, and unaffected offspring weighted by μ.

In samples that have been recruited without a phenotypic ascertainment condition, e.g. population samples, the optimal offset choice is the prevalence of the disorder/trait $E(Y = 1)$ in the total population (63). Even for studies with phenotypic ascertainment conditions, this finding approximately holds (65,66). In many situations, the population prevalence of the disease/trait is unknown. Since most study designs over-sample affected subjects to maximize the genetic loading of the sample (e.g. trio-design), the population prevalence of the disease/trait cannot be estimated directly from the sample. Fortunately, the FBAT statistic achieves almost optimal power in a relatively large neighbourhood around the true population prevalence (66). In practice, rough estimates for the prevalence will be sufficient.

Handling general pedigrees and/or missing founders in the FBAT approach

The FBAT statistic is very general and can be applied to any complex pedigree as long the expected marker score, $E(X_{ij} | S)$, can be computed, as well as the corresponding variance/covariance structure, $cov(X_{ij}, X_{ij'} | S_i T_{ij}, T_{ij'})$, which requires the specification of the marker densities $p(X_{ij} | S_i T_{ij})$ and $p(X_{ij}, X_{ij'} | S_i T_{ij}, T_{ij'})$. For nuclear families in which both parents and one or multiple offspring are genotyped, the univariate density, $p(X_{ij} | S_i T_{ij})$, is completely defined by Mendel's law. Under the null hypothesis of no association and no linkage, the parental transmissions to all offspring are independent, $p(X_{ij}, X_{ij'} | S_i T_{ij}, T_{ij'}) = p(X_{ij} | S_i T_{ij}) * p(X_{ij'} | S_i T_{ij'})$, and computation of the expected marker score and its variance/covariance is straightforward. In the presence of linkage, the transmissions from the parents to the offspring are not independent anymore, but rather dependent on the recombination fraction which is known. Technically, it would be possible to remove the dependence on the unknown recombination fraction by conditioning on the identity-by-descent patterns among offspring (51); however, the inclusion of this additional condition would make many families uninformative for the computation of the test statistic, and would lead to a substantial drop in statistical power. It is therefore recommended to estimate the variance/covariance structure directly by using empirical variance estimators, as discussed above.

The same ideas for the computation of the expected marker scores and their variance/covariance structure are also applicable to extended pedigrees in which the genotypes of all founders are known (51,61). For the analysis of such data, the power of the FBAT statistic can be increased by computing the conditional marker distribution for the complete pedigree instead of splitting up the pedigree into nuclear families and analysing the data as such (51,61). For pedigrees in which founder genotypes are missing, the computation of the expected marker scores and its variance is more complex. Instead of conditioning on the parental genotypes, the distribution of the observed offspring genotypes is computed conditional on the sufficient statistics for the unobserved parental genotypes. The advantage of conditioning on the sufficient statistic here is that no assumptions about the unobserved parental genotypes are necessary. Such assumptions would make the FBAT statistic susceptible to the effects of population substructure and stratification. Although the concept of the conditioning on the sufficient statistic for the unobserved parental genotypes is very technical, the conditional distributed for the observed offspring genotypes can straightforwardly be computed using the algorithm by Rabinowitz and Laird (51). The details of the algorithm are not discussed here, and the interested reader is referred to the original paper.

Handling haplotypes and multiple markers in the FBAT approach

In candidate gene studies, and even in genome-wide association studies nowadays, closely spaced markers/SNPs are often available that characterize a gene or a well-defined region. In such scenarios, it might not be the optimal strategy to test each marker individually for association with the phenotype of interest for two reasons. First, in general, it is difficult to take the LD-structure/correlation structure between markers into account when adjusting for multiple comparisons. This often leads to adjustments for multiple comparisons that are too conservative. Second, by only testing one marker locus at a time, the available genetic information on the other marker loci is not used. Consequently, a more powerful strategy would be to test all markers that reside in a well-defined region simultaneously. Two approaches for this are available haplotype tests and multimarker tests.

Here a multiloci haplotype is defined as a set of alleles, one for each marker, that are located on the same copy of the chromosome and that are inherited from one generation to the next without recombination. There are several situations in which multiloci haplotype tests should be more powerful than single-marker tests. For example, consider the scenario in which a true disease susceptibility locus (DSL) is located in the region that is spanned by the markers, but the DSL has not been genotyped nor is in sufficiently high disequilibrium with one of the genotyped markers to be identified by a single-marker test. If the set of genotyped markers is able to capture the haplotype diversity in the region, a multiloci haplotype will exist that captures the variation at the DSL. Another scenario, in which a haplotype analysis will be more powerful than a single-marker approach, is when two or more of the observed markers have genetic effects on the phenotype of interest. On the other hand, if there is only a single DSL in the region, and its variation is sufficiently "tagged" by one of the genotyped markers, a haplotype analysis can be suboptimal.

If the phase of the haplotype (i.e. which alleles are located on the same copy of the chromosome and are inherited jointly) is known for each subject in the study, the set of markers defining the haplotypes can be interpreted as a single marker with multiple alleles and the FBAT statistic can be computed as outlined above. However, in most applications, the phase of the haplotypes will not be known and will have to be inferred. Despite the fact that family data is available here, for which it is generally easier to determine the phase of the haplotypes than for population-based data, resolving the phase in all subjects will not be possible, especially if parents' genotypes are missing.

However, an unresolved haplotype phase in a study subject does not prevent the computation of the FBAT statistic. The same trick can be applied here as in the case for missing parental genotypes. The haplotype distribution in offspring is computed conditional upon both the parental genotypes/ sufficient statistics and whether it is possible to infer the phase of the haplotypes (67). The FBAT statistic can then be calculated by assuming that the set of markers defines a multiallelic marker locus whose alleles are given by the phased haplotypes. Since this haplotype analysis approach does not make any assumptions about population parameters (e.g. haplotype frequencies, etc.), to infer haplotype phase, but conditions upon the ability/inability to reconstruct the phase, the approach maintains its robustness against population admixture and stratification. In the usual way, the FBAT statistic can either be computed for a specific target haplotype as a diallelic FBAT or as a global haplotype test based on a multiallelic FBAT. As discussed above, the presence of linkage can be accounted for by use of the empirical variance estimator.

As the numbers of markers increase, the advantages of a haplotype analysis are outweighed by characteristic disadvantages of the approach. Inferring the phase of a haplotype becomes increasingly difficult and numerically complex when the number of markers exceeds 5–10, particularly when parental information is missing or extended pedigrees are analysed. Furthermore, the assumption of non-recombination between the markers must be carefully considered. In this situation, which also applies to smaller numbers of markers, so-called multimarker FBATs can be an attractive alternative. Rather than trying to infer the underlying haplotype structure, multimarker FBATs account for the linkage disequilibrium between markers by directly estimating the variance/covariance structure between the markers. To construct a multimarker FBAT, in the FBAT statistic the univariate marker score X_{ij} is replaced by a vector X_{ij} whose elements are the genotypes for each individual marker. The vector of expected marker scores, $E(X_{ij}|S_j)$, is defined by the expected marker scores for each marker which are computed individually, conditioned upon the corresponding parental information/sufficient statistic. The linkage disequilibrium between the markers is incorporated by using the empirical estimator of $var(X_{ij})$ in the calculation of $var(U)$. The multimarker FBAT statistic is then a quadratic form which has an asymptotic χ_2 distribution, where the degrees of freedom are given by the number of markers that are linear independent. (A detailed discussion of multimarker FBATs is given in (68).) Alternative approaches are discussed in (69,70).

Complex trait analysis in the FBAT approach

Complex phenotypes are tested in the FBAT approach by selecting an appropriate coding function Tij that is selected by the user and that will depend on the trait type. The choice of the coding function should be motivated by an underlying phenotypic model, describing the phenotypes as a function of the genotypes. Since the FBAT approach conditions upon the parental genotypes and the offspring phenotype, the validity of the FBAT test will not depend upon the correct specification of the coding function, but a poor choice will affect the statistical power of the approach.

A more refined version of the phenotype affection status is the variable age-at-onset/time-to-onset. If the phenotype age-at-onset/time-to-onset contains more genetic information, such an analysis will result in greater statistical power (e.g. for childhood asthma). It can be assumed that an early onset is more related to genetic factors than is a late onset, which could be attributable to environmental factors. Various coding functions for an age-at-onset analysis are discussed in (71) and (72).

For quantitative phenotypes, standard phenotypic residuals are an obvious choice for the coding function, i.e. $T_{ij} = (Y_{ij} - \mu)$, where Y_{ij} is the original phenotype and μ is a user-defined offset parameter. For population samples (a study without any phenotypic ascertainment conditions), the optimal offset choice is the phenotypic sample mean. In such a situation, the FBAT statistic for the quantitative trait has higher statistical power than an FBAT statistic that is based on a dichotomized version of the same quantitative trait (73). To utilize this theoretical power advantage in a real data analysis, some additional work is usually required. By definition, quantitative traits contain more information and are therefore more powerful phenotypes in a statistical analysis, but they usually depend on other non-genetic factors (e.g. lung-volume measurements in asthma studies depend on age, gender and height). Such confounding variables can be probands characteristics, but they also include environmental/treatment information. For example, lung-volume measurements for asthmatics depend on smoking status/history and on treatment for asthma. An unadjusted, raw measurement of such a phenotype will be confounded by such factors and the genetic signal will be diluted, resulting in a potentially lower statistical power. For such phenotypes, it is recommended to regress the raw phenotypes on all known confounding variables and use the regression residuals as the coded phenotype in the computation of the FBAT statistic. Note that such an adjustment is study-specific and requires careful statistical model building; results might not be reproducible in other studies that do not have the same covariate information. The motivation for a within-study adjustment is to reduce the variability in the phenotype that is attributable to all other non-genetic factors. However, this requires knowledge and measurement of such variables, which are not necessarily known before the study. For such situations, efficient coding functions that do not require any covariate adjustment and that are able to achieve high power levels can be used (74).

For many complex diseases, the definition of affection status is based on a variety of phenotypes which describe and characterize the disease and its severity.

Consequently, when an association with affection status is tested for in such a situation, the aggregated and dichotomized information is assessed all at once. If now, to increase statistical power, the quantitative traits that define the disease and/or describe its severity are selected as target phenotypes instead of affection status, multiple FBATs have to be computed and the resulting multiple testing problem has to be addressed. Quantitative phenotypes for a complex disease typically are correlated and cluster together into groups (symptom groups). In asthma studies, it is standard practice to measure quantitative phenotypes that characterize such things as the lung-function of a proband (FEV1, FVC) and the atopy-reaction (number of positive skin tests, IGE-levels) (75). Depending on how well understood the disease is, symptom groups can be defined based on prior knowledge about the underlying biological pathways, clinical knowledge, or just the phenotypic correlation between the traits. A test strategy that does not incorporate this aspect of the data, but that tests all phenotypes individually and adjusts for multiple comparisons, would be optimal. Since the FBAT tests for the same symptom group will be correlated, standard adjustments for multiple testing will be too conservative here. Further, if the hypothesis is true that the phenotypes in the same symptom group are influenced by common genetic factors and/or share similar environmental confounding, it will be more powerful to assess the evidence for association for the entire symptom group at once. A multivariate method that tests all phenotypes jointly in a single test, without having to adjust for multiple comparisons, is the most desirable approach in this situation.

For the FBAT approach, such a multivariate test that examines all phenotypes simultaneously is the FBAT-GEE statistic (76). The FBAT-GEE statistic maintains the advantages of the original FBAT statistic. It is easy to compute and does not require any distributional assumptions about the phenotypes even if the selected phenotypes are of different trait types (e.g. normally distributed phenotypes, count variables, etc.).

FBAT-GEE

For each study subject it is assumed that m phenotypes have been recorded and are defined as a symptom group as described above. The vector containing all m observations for each proband is denoted by $Y_{ij} = (Y_{ij1},..., Y_{ijm})$, where Y_{ijk} is the kth phenotype for the jth offspring in the ith family. The multivariate FBAT-GEE statistic can then be obtained by defining the coding vector T_{ij},

$$T_{ij} = Y_{ij} - \hat{Y}_{ij1} = \begin{pmatrix} Y_{ij1} \\ Y_{ijk} \\ Y_{ijm} \end{pmatrix} - \begin{pmatrix} \hat{Y}_{ij1} \\ \hat{Y}_{ijk} \\ \hat{Y}_{ijm} \end{pmatrix}$$

where the parameter $\hat{Y}_{ijk}$ is given either by the observed sample means for the kth trait, or by the predicted trait value based on a regression $\hat{Y}_{ijk}$ on its known covariates/confounding variables. As discussed earlier, in the situation of a multivariate trait, the univariate coding variable T_{ij} in the FBAT statistic is replaced by the vector T_{ij} and the FBAT-GEE statistic given by

$$T_{FBAT-GEE} = C^T V^- C.$$

Under the null hypothesis, the FBAT-GEE statistic is asymptotically χ^2-distributed with m degrees of freedom. The name of the test statistic originates from its link with the generalized estimating equation approach (77). A generalized estimating equation model can be defined by modeling the m phenotypes as a function of the genotype, using appropriate trait-dependent link-functions and a predefined variance/covariance structure. When a family-based score test is derived for this estimating equation model, the link-functions and the assumptions for the variance/covariance structure cancel out and the model-free FBAT-GEE statistic is obtained, making the multivariate FBAT-GEE statistic invariant towards distributional assumptions for the phenotype.

Pedigree-based association tests (PBATS): Bypassing the multiple comparison problem in family-based association studies

To maintain the three key properties of the original TDT approach, the FBAT statistic conditions upon the phenotype and the parental genotypes, which comes at the price that not all information about linkage and association that is contained in the data can be used. While this ensures that the robustness and the model-free character of the original approach are maintained, FBATs are in general not the most efficient test statistic. However, this extra unutilized information can be brought into play in a screening step before the computation of the FBAT statistic. The information can be used to construct an optimally informed two-stage testing strategy, or an "optimal" FBAT statistic, which has been denoted as a pedigree-based association test (PBAT). This enhances the power of the FBAT approach substantially. FBAT, with a prior screening step, can achieve power levels that are comparable to power levels that would be obtained by a corresponding population-based analysis (78).

In particular, in large-scale association studies, with numerous genotyped markers and multiple complex traits, the screening step/extra information can be used to guide the testing strategy with respect to minimizing the effects of multiple comparisons, model-building and phenotype selection. Discussed here is a general approach that partitions family data into two independent components corresponding to the population information, and the within family information. The population information about association, which is susceptible to population substructure, is used for the screening step, or model development, and the within-family information is used for the construction of the confirmatory FBAT statistic. The idea is similar to cross validation, except that each subject contributes information to both parts of partitioning, minimizing the variability of the genetic effect in the two subsets. For simplicity it is assumed that offspring-parent trios are given.

The distribution of the complete data is the joint distribution of the offspring phenotype, Y, the offspring genotype, X, and the parental genotype, P (or more generally, the sufficient statistic, S). Using Bayes' rule, the joint distribution can be partitioned into two independent parts:

$$P(Y, X, S) = P(X|Y,S)P(S,Y). \quad (3)$$

If the screening step (e.g. model building, hypothesis generation) uses only information on S and Y, any subsequent hypothesis testing that is based on the FBAT statistic, whose distribution is given by $P(X|S,Y)$, will be independent of the prior screening step. There are various ways to model the variables S and Y so that information about a potential association between Y and X can be obtained. In general, the appropriate model for S and Y will depend on the specific design (e.g. ascertainment conditions, trait type, etc). For example, in the situation of an unascertained population sample with a quantitative target phenotype, the population-based information about the association between the offspring genotype and phenotype can be described by the conditional mean model (73,79):

$$E(Y) = m + a*E(X|S). \quad (4)$$

The genetic effect size, a, can be estimated by an ordinary regression of the phenotype, Y, on $E(X|S)$. Note that $E(X|S)$ is computed solely based on the parental genotypes. For the uninformative families (i.e. trios with doubly homozygous parents), the actual offspring genotype, X, is equal to $E(X|S)$. Otherwise, if parents are informative, the offspring genotype X can be thought of as missing and being imputed by $E(X|S)$. Since the conditional mean model is only based on information about Y and S, under the null hypothesis all its parameter estimates will be statistically independent of the FBAT statistic. Of course, the statistical independence of the screening step/conditional mean and the FBAT statistic does not hold under the alternative hypothesis. The conditional mean model (4) can therefore be fit repeatedly for any choice of genetic model, any number of phenotypes and any number of markers. Based on the parameter estimates for the conditional mean model, the Wald test for null hypothesis of no association, H_0: $a = 0$, can be computed. Alternatively, the parameter estimates can be used to compute the conditional, predicted power of the FBAT statistic. Such conditional power calculations will also depend upon the observed parental genotypes and phenotype (73,79). It is generally recommended to use the conditional power estimates to prioritize information for the subsequent FBAT testing step (80). This basic idea can be extended to handle longitudinal and repeated measurements (FBAT-PC) (81) and multivariate data (69). There the screening step can be used to compute optimal linear combinations of traits for subsequent testing. The approach has also been adapted to scenarios in which multiple markers are tested (69). A method has been proposed to estimate the genetically relevant age range for age-at-onset data (72). This extension is particularly useful for diseases in which an early onset suggests a strong genetic component, while a late onset is mostly attributable to non-genetic/environmental effects (e.g. Alzheimer's disease or childhood asthma).

Testing strategies for large-scale association studies

Genome-wide association studies offer great potential to the field of complex disease mapping, but to translate the dramatic increase in genetic information at a genome-wide level into the identification of new disease genes (a major statistical challenge) the multiple testing problem has to be tackled. For case–control studies, multistage designs have been proposed (82,83) as a cost-efficient way of handling this problem. In each stage of the design, the number of genotyped SNPs and genome-wide significance are achieved by a joint analysis of all stages.

For family-based association studies, the concept of partitioning the association information into two statistically independent components is well suited to efficiently address the multiple comparison problem within one study. By using the decomposition (3), a two-stage testing strategy can be constructed that consists of two statistically independent stages, the screening step and the testing step, which can be applied to the same data set (80). This approach is illustrated in Figure 15.2.

A two-stage approach is highly effective for screening and then testing results when family controls are available for application using FBAT. In step 1, association analysis based on the conditional mean model before the FBAT testing is used to minimize the multiple testing problem. In this example, one quantitative trait and M SNPs are analysed. In the first step, the marker data in the offspring is assumed to be missing and imputed by the expected markers scores conditional on the parental genotypes/sufficient statistic. Based on the imputed data, the conditional mean model is fitted, and its estimates are used to compute the power of the FBAT statistic for each SNP. The power is a function of the observed parental genotypes, their frequencies, and the genetic effect size estimated from the conditional mean model. In the final step, the K SNPs with the highest power estimates are selected to be subsequently tested for association with the FBAT statistic at a Bonferroni-adjusted significance level of $á/K$. Since only K SNPs are pushed forward to the testing step, it is only necessary to adjust for K comparisons instead of M. The markers that pass this first testing step are then validated in the second step, as depicted in Figure 15.2.

In family-based designs, the screening procedure utilizes information on all families, even the non-informative ones. Assuming moderate to small genetic effect sizes, simulation studies have shown that if a true DSL, or a SNP in LD with a DSL, is included in the data set, it is sufficient to select only the highest 10 or 20 SNPs for subsequent testing to achieve high power levels. The key advantage of this testing strategy for family-based designs is that the same data set is used twice; once for the genomic screening step and once for the testing step. Thereby the effects of study heterogeneity are minimized, which can cause, in a standard two-stage design that uses different samples in each step, the failure to discover an important association. Another advantage of this approach is that it is only necessary to recruit one sample to identify SNPs/associations that achieve genome-wide significance. Replications in other studies serve the sole purpose of generalizing a significant finding to other populations.

This testing strategy has been successfully applied to a 100 000-SNP scan for obesity in the family

Figure 15.2. Using the same data set for genomic screening and testing
Step 1: Screening SNPs using conditional power estimates for the FBAT statistics. The power estimates are based on genetic effect size estimates obtained from the conditional mean model.

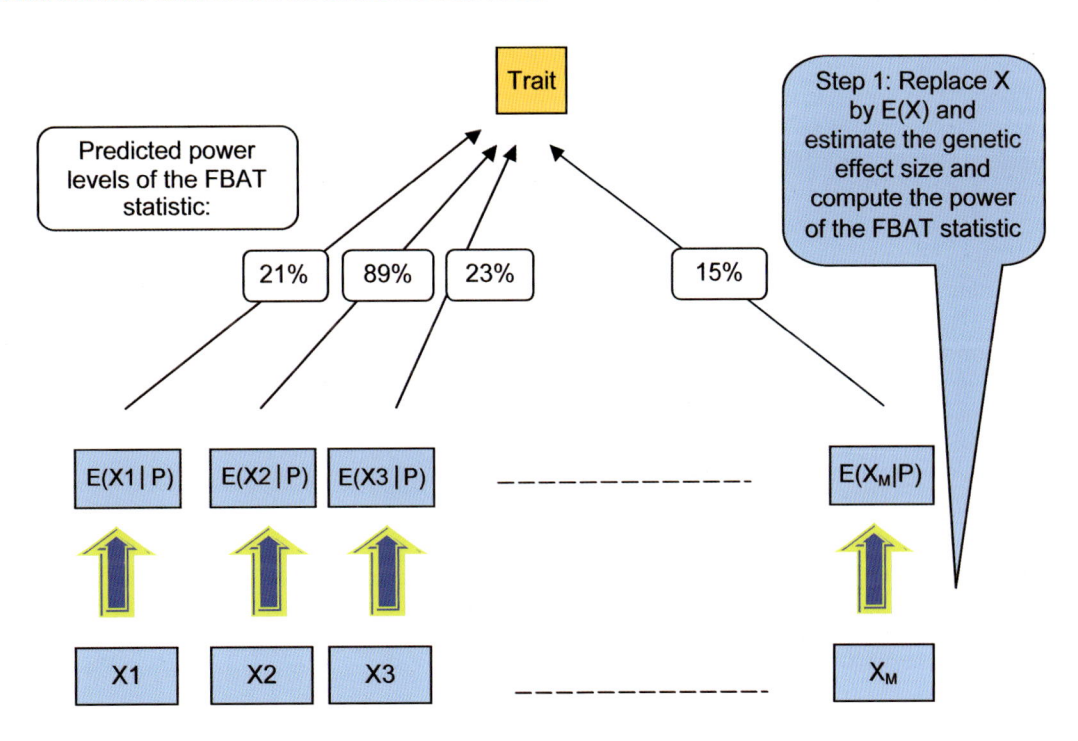

plates of the Framingham Heart Study. Among the top 10 SNPs from that study, as determined by estimated conditional power, there was a novel SNP whose association with body mass index (FBAT P-value = 0.0026) reached genome-wide significance, after having adjusted for 10 comparisons. If standard analysis methods would have been used (e.g. testing all SNPs for association and adjusting for multiple testing by the Bonferroni or Hochberg corrections), this association would have been missed. Using the same genetic model, the finding was replicated in four independent studies, including cohort, case–control and family-based samples of different ethnicities (84). Recently, the approach was extended so that all genotyped SNPs can be tested in the second stage of the testing strategy, making a decision on how many SNPs should be pushed forward to the test step redundant (78). Despite the larger number of tests in the second stage, this approach achieves power levels that are about 50% higher than in the original Van Steen approach, and that are comparable to the power levels of a population-based study with the same number of probands. The approach has been generalized so that phenotypic information on the parents can be incorporated as well (85). Extensions for case–control designs have been developed (86,87).

Other extensions of the FBAT approach include an extension to accommodate copy number variation calls (88), and an extension to allow covariate data from the parents to modify the weight assigned to transmissions of genetic information to the offspring, allowing the phenotype of the parents to influence the association analysis (89).

Figure 15.2, Step 2: The Testing Step. Select the top K SNPs with the highest power estimates for subsequent testing with the FBAT statistic. The P-value of the FBAT statistic must be smaller than á/K to achieve genome-wide/overall significance.

Power Rank	Estimated power of the FBAT statistic	SNP	P-value FBAT statistic
1	0.92	3	0.90
2	0.89	100	0.20
3	0.85	25	0.00001
....	...	...	...
K	0.70	53	0.20

Software

With family-based designs, there is generally a need for special software to analyse the data. For the FBAT approach, four software packages are available. Two packages were developed by the original authors of the methods and are home-grown (PBAT, P2BAT). Despite the lack of general support for such software packages in academia, the packages have proven to be reliable and user-friendly tools. Recently, a commercial package with professional user-support and documentation has become available that is particularly suited for less statistical-oriented users and for large-scale projects. Table 15.1 shows an overview of these packages and their functions.

Discussion

Studies of families have been instrumental for describing the genetic architecture of many Mendelian and complex diseases. For initial characterization of the strength of evidence for genetic factors influencing disease risk, twin studies and evaluations of the aggregation of disease within families provide key insights. For diseases that have strong influences from genetic factors, segregation analysis followed by linkage analysis has been a highly effective strategy. When the genetic and environmental factors influence disease risk in complex ways, linkage analysis using a model-free method is a preferred strategy. For diseases with weaker genetic influences, or that result from effects of many genetic factors with each individually having a weak effect on disease risk, association studies are more successful. Family-based association studies are robust to population stratification and can have power comparable to case–control studies with unrelated cases and controls.

The area of whole-genome association scans offers great promise for the field of genetic association mapping. Most predictions agree that studies with large sample sizes are needed to identify the "needles in the haystack," regardless of which design is used (80,83,90). It is much easier to achieve such sample sizes from existing cohorts, or from case–

Table 15.1. Software for the analysis of family-based association tests

Package	Genetic analysis capability	Phenotypic analysis capability	Special features
FBAT	Single marker, haplotype, multi-marker	Binary traits, quantitative/multivariate traits, ranked traits, time-to-onset	X-chromosome, permutation tests
PBAT	Single marker, haplotype, multi-marker	Binary traits, quantitative traits/multivariate, ranked traits, time-to-onset, gene-environment interaction	Covariate adjustment, Van Steen algorithm for multiple testing, X-chromosome, permutation tests
P2BAT R-implementation	Single marker, haplotype, multi-marker	Binary traits, quantitative traits/multivariate, ranked traits, time-to-onset, gene-environment interaction	Covariate adjustment, Van Steen algorithm for multiple testing, X-chromosome, permutation tests
PBAT GoldenHelix commercial package	Single marker, haplotype, multi-marker	Binary traits, quantitative traits/multivariate, ranked traits, time-to-onset, gene-environment interaction	Covariate adjustment, Van Steen algorithm for multiple testing, X-chromosome, permutation tests, active user-support and professional documentation

control studies, than from family samples. However the innovative use of the population information that is included in family-based data sets, combined with the robustness of the family-based association methods, can protect against both population substructures and misspecifications of the phenotypic model, creating a viable and powerful alternative to population-based studies. Further, with the ability to handle extended pedigrees with large numbers of subjects, the FBAT approach allows the continuing utilization of existing linkage studies. Recent developments to estimate and test gene–environment interaction in the FBAT approach, without any loss of robustness, are an additional advantage (91). For many complex diseases, genetic interactions with environmental exposure variables are thought to be crucial for the understanding of the disease (e.g. smoking status and/or smoking history in asthma and COPD studies) (92,93).

References

1. Lichtenstein P, Holm NV, Verkasalo PK et al. (2000). Environmental and heritable factors in the causation of cancer–analyses of cohorts of twins from Sweden, Denmark, and Finland. N Engl J Med, 343:78–85.doi:10.1056/NEJM200007133430201 PMID:10891514

2. Hemminki K, Li X, Sundquist K, Sundquist J (2008). Familial risks for common diseases: etiologic clues and guidance to gene identification. Mutat Res, 658:247–258.doi:10.1016/j.mrrev.2008.01.002 PMID:18282736

3. Neale MC, Cardon LR. Methodology for genetic studies of twins and families. Dordrecht (The Netherlands): Kluwer Academic Publishers; 1992 (vol 67).

4. Emery AEH. Methodology in medical genetics. An introduction to statistical methods. Edinburgh (UK): Churchill Livingstone; 1986.

5. Hemminki K, Sundquist J, Lorenzo Bermejo J (2008). Familial risks for cancer as the basis for evidence-based clinical referral and counseling. Oncologist, 13:239–247. doi:10.1634/theoncologist.2007-0242 PMID:18378534

6. Risch N (1990). Linkage strategies for genetically complex traits. II. The power of affected relative pairs. Am J Hum Genet, 46:229–241. PMID:2301393

7. Rigby AS, Voelm L, Silman AJ (1993). Epistatic modeling in rheumatoid arthritis: an application of the Risch theory. Genet Epidemiol, 10:311–320.doi:10.1002/gepi.1370100504 PMID:8224809

8. Khoury MJ, Beaty TH, Cohen BH. Fundamentals of genetic epidemiology. New York (NY): Oxford University Press; 1993.

9. Scott WK, Pericak-Vance MA, Haines JL (1997). Genetic analysis of complex diseases. Science, 275:1327–1330, author reply 1329–1330.doi:10.1126/science.275.5304.1327 PMID:9064788

10. Amos CI, Rubin LA (1995). Major gene analysis for diseases and disorders of complex etiology. Exp Clin Immunogenet, 12:141–155. PMID:8534501

11. Lin JP, Cash JM, Doyle SZ et al. (1998). Familial clustering of rheumatoid arthritis with other autoimmune diseases. Hum Genet, 103:475–482.doi:10.1007/s004390050853 PMID:9856493

12. Struewing JP, Hartge P, Wacholder S et al. (1997). The risk of cancer associated with specific mutations of BRCA1 and BRCA2 among Ashkenazi Jews. N Engl J Med, 336:1401–1408.doi:10.1056/NEJM199705153362001 PMID:9145676

13. Wacholder S, Hartge P, Struewing JP et al. (1998). The kin-cohort study for estimating penetrance. Am J Epidemiol, 148:623–630. doi:10.1093/aje/148.7.623 PMID:9778168

14. Gail MH, Pee D, Carroll R (1999). Kin-cohort designs for gene characterization. J Natl Cancer Inst Monogr, (26):55–60. PMID:10854487

15. Chatterjee N, Wacholder S (2001). A marginal likelihood approach for estimating penetrance from kin-cohort designs. Biometrics, 57:245–252. doi:10.1111/j.0006-341 X.2001.00245.x PMID:11252606

16. Ford D, Easton DF, Stratton M et al.; The Breast Cancer Linkage Consortium (1998). Genetic heterogeneity and penetrance analysis of the BRCA1 and BRCA2 genes in breast cancer families. Am J Hum Genet, 62:676–689.doi:10.1086/301749 PMID:9497246

17. Antoniou AC, Cunningham AP, Peto J et al. (2008). The BOADICEA model of genetic susceptibility to breast and ovarian cancers: updates and extensions. Br J Cancer, 98:1457–1466.doi:10.1038/sj.bjc.6604305 PMID:18349832

18. Ziogas A, Anton-Culver H (2003). Validation of family history data in cancer family registries. Am J Prev Med, 24:190–198.doi:10.1016/S0749-3797(02)00593-7 PMID:12568826

19. King TM, Tong L, Pack RJ et al. (2002). Accuracy of family history of cancer as reported by men with prostate cancer. Urology, 59:546–550.doi:10.1016/S0090-4295(01)01598-9 PMID:11927311

20. Bondy ML, Strom SS, Colopy MW et al. (1994). Accuracy of family history of cancer obtained through interviews with relatives of patients with childhood sarcoma. J Clin Epidemiol, 47:89–96. doi:10.1016/0895-4356(94)90037-X PMID:8283198

21. American Society of Human Genetics. Policy statement archives. Family history and privacy advisory; 2000. Available from URL: http://www.ashg.org/pages/statement_32000.shtml.

22. Cannings C, Thompson EA, Skolnick MH (1978). Probability functions on complex pedigrees. Adv Appl Prob, 10:26–61 doi:10.2307/1426718.

23. Elston RC, Stewart J (1971). A general model for the genetic analysis of pedigree data. Hum Hered, 21:523–542. doi:10.1159/000152448 PMID:5149961

24. Lange K, Elston RC (1975). Extensions to pedigree analysis I. Likehood calculations for simple and complex pedigrees. Hum Hered, 25:95–105.doi:10.1159/000152714 PMID:1150306

25. Morton NE (1955). Sequential tests for the detection of linkage. Am J Hum Genet, 7:277–318. PMID:13258560

26. Lander E, Kruglyak L (1995). Genetic dissection of complex traits: guidelines for interpreting and reporting linkage results. Nat Genet, 11:241–247.doi:10.1038/ng1195-241 PMID:7581446

27. Witte JS, Elston RC, Schork NJ (1996). Genetic dissection of complex traits. Nat Genet, 12:355–356, author reply 357–358. doi:10.1038/ng0496-355 PMID:8630483

28. Huang Q, Shete S, Amos CI (2004). Ignoring linkage disequilibrium among tightly linked markers induces false-positive evidence of linkage for affected sib pair analysis. Am J Hum Genet, 75:1106–1112. doi:10.1086/426000 PMID:15492927

29. Kruglyak L, Lander ES (1995). Complete multipoint sib-pair analysis of qualitative and quantitative traits. Am J Hum Genet, 57:439–454. PMID:7668271

30. Abecasis GR, Wigginton JE (2005). Handling marker-marker linkage disequilibrium: pedigree analysis with clustered markers. Am J Hum Genet, 77:754–767.doi:10.1086/497345 PMID:16252236

31. Sobel E, Lange K (1996). Descent graphs in pedigree analysis: applications to haplotyping, location scores, and marker-sharing statistics. Am J Hum Genet, 58:1323–1337. PMID:8651310

32. Heath SC (1997). Markov chain Monte Carlo segregation and linkage analysis for oligogenic models. Am J Hum Genet, 61:748–760.doi:10.1086/515506 PMID:9326339

33. Greenberg DA, Abreu PC (2001). Determining trait locus position from multipoint analysis: accuracy and power of three different statistics. Genet Epidemiol, 21:299–314.doi:10.1002/gepi.1036 PMID:11754466

34. McPeek MS (1999). Optimal allele-sharing statistics for genetic mapping using affected relatives. Genet Epidemiol, 16:225–249.doi:10.1002/(SICI)1098-2272(1999)16:3<225::AID-GEPI1>3.0.CO;2-# PMID:10096687

35. Cordell HJ (2004). Bias toward the null hypothesis in model-free linkage analysis is highly dependent on the test statistic used. Am J Hum Genet, 74:1294–1302. doi:10.1086/421476 PMID:15124101

36. Etzel CJ, Guerra R (2002). Meta-analysis of genetic-linkage analysis of quantitative-trait loci. *Am J Hum Genet*, 71:56–65. doi:10.1086/341126 PMID:12037716

37. Kavvoura FK, Ioannidis JP (2008). Methods for meta-analysis in genetic association studies: a review of their potential and pitfalls. *Hum Genet*, 123:1–14.doi:10.1007/s00439-007-0445-9 PMID:18026754

38. Hästbacka J, de la Chapelle A, Kaitila I et al. (1992). Linkage disequilibrium mapping in isolated founder populations: diastrophic dysplasia in Finland. *Nat Genet*, 2:204–211. doi:10.1038/ng1192-204 PMID:1345170

39. The SNP Consortium. Available from URL: http://snp.cshl.org/.

40. Amos CI (2007). Successful design and conduct of genome-wide association studies. *Hum Mol Genet*,16(Spec No. 2):R220–5.

41. Jorde LB (1995). Linkage disequilibrium as a gene-mapping tool. *Am J Hum Genet*, 56:11–14. PMID:7825565

42. Johnson GC, Esposito L, Barratt BJ et al. (2001). Haplotype tagging for the identification of common disease genes. *Nat Genet*, 29:233–237.doi:10.1038/ng1001-233 PMID:11586306

43. Meng Z, Zaykin DV, Xu C-F et al. (2003). Selection of genetic markers for association analyses, using linkage disequilibrium and haplotypes. *Am J Hum Genet*, 73:115–130. doi:10.1086/376561 PMID:12796855

44. Stram DO, Haiman CA, Hirschhorn JN et al. (2003). Choosing haplotype-tagging SNPS based on unphased genotype data using a preliminary sample of unrelated subjects with an example from the Multiethnic Cohort Study. *Hum Hered*, 55:27–36.doi:10.1159/000071807 PMID:12890923

45. Spielman RS, Ewens WJ (1996). The TDT and other family-based tests for linkage disequilibrium and association. *Am J Hum Genet*, 59:983–989. PMID:8900224

46. Sham PC, Curtis D (1995). An extended transmission/disequilibrium test (TDT) for multi-allele marker loci. *Ann Hum Genet*, 59:323–336.doi:10.1111/j.1469-1809.1995.tb00751.x PMID:7486838

47. Bickeböller H, Clerget-Darpoux F (1995). Statistical properties of the allelic and genotypic transmission/disequilibrium test for multiallelic markers. *Genet Epidemiol*, 12:865–870.doi:10.1002/gepi.1370120656 PMID:8788023

48. Schaid DJ (1996). General score tests for associations of genetic markers with disease using cases and their parents. *Genet Epidemiol*, 13:423–449.doi:10.1002/(SICI)1098-2272(1996)13:5<423::AID-GEPI1>3.0.CO;2-3 PMID:8905391

49. Spielman RS, Ewens WJ (1998). A sibship test for linkage in the presence of association: the sib transmission/disequilibrium test. *Am J Hum Genet*, 62:450–458.doi:10.1086/301714 PMID:9463321

50. Schaid DJ, Li HZ (1997). Genotype relative-risks and association tests for nuclear families with missing parental data. *Genet Epidemiol*, 14:1113–1118.doi:10.1002/(SICI)1098-2272(1997)14:6<1113::AID-GEPI92>3.0.CO;2-J PMID:9433633

51. Rabinowitz D, Laird N (2000). A unified approach to adjusting association tests for population admixture with arbitrary pedigree structure and arbitrary missing marker information. *Hum Hered*, 50:211–223. doi:10.1159/000022918 PMID:10782012

52. Fulker DW, Cherny SS, Sham PC, Hewitt JK (1999). Combined linkage and association sib-pair analysis for quantitative traits. *Am J Hum Genet*, 64:259–267.doi:10.1086/302193 PMID:9915965

53. Martin ER, Monks SA, Warren LL, Kaplan NL (2000). A test for linkage and association in general pedigrees: the pedigree disequilibrium test. *Am J Hum Genet*, 67:146–154.doi:10.1086/302957 PMID:10825280

54. Horvath S, Laird NM (1998). A discordant-sibship test for disequilibrium and linkage: no need for parental data. *Am J Hum Genet*, 63:1886–1897.doi:10.1086/302137 PMID:9837840

55. Lake SL, Blacker D, Laird NM (2000). Family-based tests of association in the presence of linkage. *Am J Hum Genet*, 67:1515–1525.doi:10.1086/316895 PMID:11058432

56. Allison DB (1997). Transmission-disequilibrium tests for quantitative traits. *Am J Hum Genet*, 60:676–690. PMID:9042929

57. Abecasis GR, Cardon LR, Cookson WOC (2000). A general test of association for quantitative traits in nuclear families. *Am J Hum Genet*, 66:279–292.doi:10.1086/302698 PMID:10631157

58. Rabinowitz D (1997). A transmission disequilibrium test for quantitative trait loci. *Hum Hered*, 47:342–350. doi:10.1159/000154433 PMID:9391826

59. Horvath S, Xu X, Laird NM (2001). The family based association test method: strategies for studying general genotype–phenotype associations. *Eur J Hum Genet*, 9:301–306.doi:10.1038/sj.ejhg.5200625 PMID:11313775

60. Laird NM, Horvath S, Xu X (2000). Implementing a unified approach to family-based tests of association. *Genet Epidemiol*, 19 Suppl 1;S36–S42.doi:10.1002/1098-2272(2000)19:1+<::AID-GEPI6>3.0.CO;2-M PMID:11055368

61. Laird NM, Lange C (2006). Family-based designs in the age of large-scale gene-association studies. *Nat Rev Genet*, 7:385–394.doi:10.1038/nrg1839 PMID:16619052

62. Sinsheimer JS, McKenzie CA, Keavney B, Lange K (2001). SNPs and snails and puppy dogs' tails: analysis of SNP haplotype data using the gamete competition model. *Ann Hum Genet*, 65:483–490.doi:10.1046/j.1469-1809.2001.6550483.x PMID:11806856

63. Lange C, Laird NM (2002). On a general class of conditional tests for family-based association studies in genetics: the asymptotic distribution, the conditional power, and optimality considerations. *Genet Epidemiol*, 23:165–180.doi:10.1002/gepi.209 PMID:12214309

64. Schneiter K, Laird N, Corcoran C (2005). Exact family-based association tests for biallelic data. *Genet Epidemiol*, 29:185–194. doi:10.1002/gepi.20088 PMID:16094642

65. Whittaker JC, Lewis CM (1998). The effect of family structure on linkage tests using allelic association. *Am J Hum Genet*, 63:889–897. doi:10.1086/302008 PMID:9718338

66. Lange C, Laird NM (2002). Power calculations for a general class of family-based association tests: dichotomous traits. *Am J Hum Genet*, 71:575–584.doi:10.1086/342406 PMID:12181775

67. Horvath S, Xu X, Lake SL et al. (2004). Family-based tests for associating haplotypes with general phenotype data: application to asthma genetics. *Genet Epidemiol*, 26:61–69. doi:10.1002/gepi.10295 PMID:14691957

68. Rakovski CS, Xu X, Lazarus R et al. (2007). A new multimarker test for family-based association studies. *Genet Epidemiol*, 31:9–17.doi:10.1002/gepi.20186 PMID:17086514

69. Xu X, Rakovski C, Xu XP, Laird N (2006). An efficient family-based association test using multiple markers. *Genet Epidemiol*, 30:620–626.doi:10.1002/gepi.20174 PMID:16868964

70. Rakovski CS, Weiss ST, Laird NM, Lange C (2008). FBAT-SNP-PC: an approach for multiple markers and single trait in family-based association tests. *Hum Hered*, 66:122–126.doi:10.1159/000119111 PMID:18382091

71. Lange C, Blacker D, Laird NM (2004). Family-based association tests for survival and times-to-onset analysis. *Stat Med*, 23:179–189.doi:10.1002/sim.1707 PMID:14716720

72. Jiang H, Harrington D, Raby BA et al. (2006). Family-based association test for time-to-onset data with time-dependent differences between the hazard functions. *Genet Epidemiol*, 30:124–132.doi:10.1002/gepi.20132 PMID:16374805

73. Lange C, DeMeo DL, Laird NM (2002). Power and design considerations for a general class of family-based association tests: quantitative traits. *Am J Hum Genet*, 71:1330–1341.doi:10.1086/344696 PMID:12454799

74. Fardo D, Celedón JC, Raby BA et al. (2007). On dichotomizing phenotypes in family-based association tests: quantitative phenotypes are not always the optimal choice. *Genet Epidemiol*, 31:376–382.doi:10.1002/gepi.20218 PMID:17342772

75. DeMeo DL, Silverman EK (2003). Genetics of chronic obstructive pulmonary disease. *Semin Respir Crit Care Med*, 24:151–160. doi:10.1055/s-2003-39014 PMID:16088534

76. Lange C, DeMeo D, Silverman EK *et al.* (2003). Using the noninformative families in family-based association tests: a powerful new testing strategy. *Am J Hum Genet,* 73:801–811.doi:10.1086/378591 PMID:14502464

77. Liang KY, Zeger S (1986). Longitudinal data analysis using generalized linear models. *Biometrika,* 73:13–22 doi:10.1093/biomet/73.1.13.

78. Ionita-Laza I, McQueen MB, Laird NM, Lange C (2007). Genomewide weighted hypothesis testing in family-based association studies, with an application to a 100K scan. *Am J Hum Genet,* 81:607–614. doi:10.1086/519748 PMID:17701906

79. Lange C, Silverman EK, Xu X *et al.* (2003). A multivariate family-based association test using generalized estimating equations: FBAT-GEE. *Biostatistics,* 4:195–206.doi:10.1093/biostatistics/4.2.195 PMID:12925516

80. Van Steen K, McQueen MB, Herbert A *et al.* (2005). Genomic screening and replication using the same data set in family-based association testing. *Nat Genet,* 37:683–691. doi:10.1038/ng1582 PMID:15937480

81. Lange C, van Steen K, Andrew T *et al.* (2004). A family-based association test for repeatedly measured quantitative traits adjusting for unknown environmental and/or polygenic effects. *Stat Appl Genet Mol Biol,* 3:e17. PMID:16646795

82. Thomas D, Xie R, Gebregziabher M (2004). Two-Stage sampling designs for gene association studies. *Genet Epidemiol,* 27:401–414.doi:10.1002/gepi.20047 PMID:15543639

83. Hirschhorn JN, Daly MJ (2005). Genome-wide association studies for common diseases and complex traits. *Nat Rev Genet,* 6:95–108. doi:10.1038/nrg1521 PMID:15716906

84. Herbert A, Gerry NP, McQueen MB *et al.* (2006). A common genetic variant is associated with adult and childhood obesity. *Science,* 312:279–283.doi:10.1126/science.1124779 PMID:16614226

85. Feng BJ, Goldgar DE, Corbex M (2007). Trend-TDT - a transmission/disequilibrium based association test on functional mini/microsatellites. *BMC Genet,* 8:75. doi:10.1186/1471-2156-8-75 PMID:17976242

86. Zheng G, Song K, Elston RC (2007). Adaptive two-stage analysis of genetic association in case-control designs. *Hum Hered,* 63:175–186.doi:10.1159/000099830 PMID:17310127

87. Won S, Elston RC (2008). The power of independent types of genetic information to detect association in a case-control study design. *Genet Epidemiol,* 32:731–756. doi:10.1002/gepi.20341 PMID:18481783

88. Ionita-Laza I, Perry GH, Raby BA *et al.* (2008). On the analysis of copy-number variations in genome-wide association studies: a translation of the family-based association test. *Genet Epidemiol,* 32:273–284.doi:10.1002/gepi.20302 PMID:18228561

89. Lu AT, Cantor RM (2007). Weighted variance FBAT: a powerful method for including covariates in FBAT analyses. *Genet Epidemiol,* 31:327–337.doi:10.1002/gepi.20213 PMID:17323371

90. Clayton DG, Walker NM, Smyth DJ *et al.* (2005). Population structure, differential bias and genomic control in a large-scale, case-control association study. *Nat Genet,* 37:1243–1246.doi:10.1038/ng1653 PMID:16228001

91. Vansteelandt S, Lange C. A unifying approach for haplotype analysis of quantitative traits in family-based association studies: Testing and estimating gene-environment interactions with complex exposure variables. COBRA Preprint Series 2006;(Article 11). Available at URL: http://biostats.bepress.com/cobra/ps/art11.

92. Celedón JC, Lange C, Raby BA *et al.* (2004). The transforming growth factor-beta1 (TGFB1) gene is associated with chronic obstructive pulmonary disease (COPD). *Hum Mol Genet,* 13:1649–1656.doi:10.1093/hmg/ddh171 PMID:15175276

93. Demeo DL, Mariani TJ, Lange C *et al.* (2006). The SERPINE2 gene is associated with chronic obstructive pulmonary disease. *Am J Hum Genet,* 78:253–264. doi:10.1086/499828 PMID:16358219

UNIT 4.
INTEGRATION OF BIOMARKERS INTO EPIDEMIOLOGY STUDY DESIGNS

CHAPTER 16.

Analysis of epidemiologic studies of genetic effects and gene-environment interactions

Montserrat García-Closas, Kevin Jacobs, Peter Kraft, and Nilanjan Chatterjee

Summary

This chapter describes basic principles in study design, data analysis, and interpretation of epidemiological studies of genetic polymorphisms and disease risk, including the assessment of gene–environment interactions. The case–control design (hospital-based, population-based or nested within a prospective cohort) is frequently used to study common genetic variants and disease risk. Because of their widespread use, the analysis of case–control data will be the focus of this chapter. Two key considerations in the study design will be addressed: the selection of genetic markers to be evaluated, and sample size considerations to ensure adequate power to detect associations with disease risk. Single nucleotide polymorphisms (SNPs) are the most frequent form of common genetic variation, thus the discussion on data analysis will be based on the evaluation of associations between SNPs and disease risk. This chapter will begin with the evaluation of quality control of genotyping data, which is a critical first step in the analysis of genetic data. A description of statistical methods will follow, aimed at the discovery of genetic susceptibility loci, including analysis of candidate SNPs and genome-wide association studies, haplotype analyses, and the evaluation of gene–gene and gene–environment interactions.

Introduction

The approaches to studying genetic susceptibility factors for disease have evolved very quickly over the last several years, due to advances in genotyping technologies, substantial reductions in genotyping costs, and improvements in the annotation of common genetic variation, particularly the most common type of variant, the single nucleotide polymorphism. These advances have enabled investigators to move beyond evaluating a few candidate variants in key genes, to conducting more comprehensive, as well as exploratory, evaluations of common genetic variation in candidate pathways/networks to disease, and performing genome-wide association studies (GWAS).

Over the last year, there has been an explosion of new discoveries of susceptibility loci for a wide range of diseases derived from GWAS (http://www.genome.gov/gwastudies/). This rapid trend of discoveries is likely to continue in the near future, as an increasing number of epidemiological studies use this approach to identify novel susceptibility loci. A major factor in the success of these breakthroughs has been the formation of very large collaborative efforts through consortia of studies that is creating unprecedented opportunities for discovery.

The discovery of disease susceptibility loci can bring about improvements in the understanding of disease etiology, and may ultimately lead to improvements in risk assessment, targeted preventive or screening strategies to reduce disease incidence and mortality, and improvements in therapy through the identification of drug targets. The aim of this chapter is to describe basic principles in study design and data analysis in studies on common genetic polymorphisms and disease. A discussion of biases, and other considerations in the interpretation of data analyses, is outside the scope of this chapter and can be found in previous publications (1–3).

Study design

The study designs used in molecular epidemiology studies, and a description of their advantages and disadvantages, can be found elsewhere in this book (see Chapter 14) and in previous publications (4). Discussed here are aspects of these epidemiologic study designs that are most relevant to studies of genetic susceptibility to disease. The hereditability of a disease, or the proportion of variation in disease susceptibility due to genetic factors, is directly related to the ability to identify susceptibility loci in epidemiological studies (5). Therefore, one of the first considerations is to evaluate the heritability, or *a priori* evidence, that the disease of interest is caused by genetic variation.

The case–control design (3,6), either nested in a prospective cohort or by retrospective sampling of a population, is by far the most commonly used design in genetic epidemiology studies of unrelated individuals. Hospital-based case–control studies are particularly popular, as the hospital setting facilitates the rapid enrolment of subjects, and the collection and processing of biological specimens with high participation rates. The case–control design is of particular importance for the study of uncommon diseases that occur in small numbers in the population or prospective cohort studies. Given that most members of a prospective cohort will not develop disease, these studies often use sampling strategies, such as nested case–control and (less commonly) case–cohort designs, to improve efficiency (7). In these designs, only samples from cases and a random subset of non-cases are analysed, reducing the DNA extraction and genotyping costs considerably. The case–cohort design allows for the evaluation of several disease endpoints using the same comparison group (referred to as a subcohort); however, since the same disease-free subjects are repeatedly used as controls for different disease endpoints, depletion of DNA samples from this group can be an issue. Until whole-genome sequencing is cost-effective and commonly available, whole-genome amplification of DNA, from cases and controls can be used to address the problem of limited DNA in epidemiologic studies; however, this amplified DNA might not be suitable for all genomic assays (8).

Biased sampling, or non-random selection of cases and/or controls, can be used to improve efficiency to discover genetic markers associated with disease. For instance, selection of cases with a family history of breast cancer can lead to gains in power to detect genetic susceptibility loci, assuming a polygenic model of inheritance with loci interacting multiplicatively (9). However, the generalizability of risk estimates and evaluations of gene-environment (G-E) interactions can be compromised.

Genotyping hundreds of thousands of genetic markers in thousands of individuals can be costly. Multistage designs are commonly used to reduce the cost of genotyping very large numbers of samples (10). In these designs, a proportion of samples are genotyped for a large number of markers (e.g. SNPs that represent genetic variation across the genome in GWAS). In subsequent stages, only those markers showing the most significant associations with disease are genotyped in additional samples (10). The reduced cost is offset by a reduction in power compared to a study genotyping all markers in all available samples. Since the majority of genetic association studies use some sort of case–control design, the description of methods for data analyses in the section *Analysis of Genetic Data*, which follows, will focus on case–control data.

Selection of genetic markers

This chapter focuses on single nucleotide polymorphisms (SNPs) – the most common form of variation in the human genome. A

SNP is a DNA sequence variation occurring when a single nucleotide base differs among members of a population. There are thought to be at least 10 million SNPs in the human genome, and the vast majorities are bi-alleleic, having only two alleles or nucleotide variant forms. SNPs occur throughout the genome and can be measured (genotyped) accurately. Although the genotyping costs have decreased dramatically in the last few years, it is still cost-prohibitive to genotype all known SNPs or sequence the entire human genome. Therefore, current studies must select subsets of markers to be evaluated.

SNP selection strategies take advantage of the correlation among genetic variants located close together on the same chromosome, or linkage disequilibrium (LD), to select a minimal set of tag SNPs that capture the majority of common genetic variation in human populations (11,12). The selection of tag SNPs has been aided by the International HapMap Project (http://www.hapmap.org/), a public resource that has genotyped millions of SNPs in 270 individuals from different ethnicities (30 Yoruba from Ibidan, Nigeria, 45 Japanese residents of Tokyo, 45 Han Chinese, and 30 Caucasian trios from Utah, USA) (13). Several methods have been proposed to use extensive data sets like the HapMap to select tag SNPs. Pairwise tagging is a method where tag SNPs are selected by examining the LD measures between pairs of SNPs using a squared correlation coefficient, r^2. A SNP is said to 'cover' another SNP if the r^2 value between them exceeds a given threshold (e.g. 0.80). The Carlson algorithm to select optimal tag SNPs is iterative and begins by considering all SNPs as potential tags. At each step, the SNP that covers the most correlated SNPs is chosen as a tag SNP. That SNP and all other SNPs that it covers (called a bin) are removed, and the algorithm begins again and continues until all SNPs are either taken as tags or are covered by a tag (14). Multimarker or aggressive tagging algorithms examine correlations among two or more SNPs using a generalized correlation coefficient to determine coverage (15,16). This approach typically reduces the total number of tag SNPs required; however, the selection algorithm is computationally more intense than pairwise methods, and statistically more complex, since an appropriate multimarker test should be used to test the associations with disease. Multimarker tagging approaches are also more affected by missing genotype data, since several SNPs are often required to perform tests. The current generation of genotyping arrays used to perform GWAS include about 300 000 to 1 million SNPs to capture common genetic variation. The proportion of SNPs in HapMap covered by SNPs in each of these genotyping assays depends on the ancestral origin of the underlying population.

Sample size considerations

As in any epidemiological study, sample size considerations are critical for the design of studies of genetic associations and G-E interactions (10,17,18). The main parameters that determine the required sample size to attain a specified statistical power are:
• Disease prevalence in the population
• Magnitude of association (often measured by the odds ratio)
• Alpha-level or P-value threshold to designate a 'statistically significant' finding
• Genotype or allele frequency in the population
• Mode of inheritance

Generally, hundreds to thousands of subjects are needed to evaluate genetic associations with risk of complex diseases, as the magnitude of association between individual genetic variants on disease risk tends to be small (see http://www.genome.gov/gwastudies/ for a catalogue of discoveries using GWAS in different diseases and traits).

Most studies measure genetic markers for disease rather than directly measuring the causal variant itself, as it is often unknown. The sample size needed to detect an association between a genetic marker and disease depends on the degree of linkage disequilibrium, or correlation due to physical proximity, between the marker and the causative variant. In approximate terms, the sample size requirements for studies using single SNPs as genetic markers are increased by a factor of $1/r^2$, where r^2 is the squared correlation between the marker and the causal unmeasured SNP (19). Sample size approximations are more complex when the disease susceptibility locus is in LD with multiple SNPs (10).

One limitation of standard power calculation methods is that they focus on the power for the detection of a single susceptibility locus with a given minor allele frequency (MAF) and disease odds ratio. In GWAS, however, there is likely to be a variety of susceptibility SNPs with a spectrum of MAF and disease odds ratios. The goal is to discover a certain number of underlying susceptibility loci, not some specific loci. A recent report has suggested novel approaches to power calculation that can provide realistic assessment for several probable discoveries in GWAS, accounting for the likely distribution of effect sizes for the underlying susceptibility loci (20).

An important challenge in large-scale evaluations of candidate genes/regions/pathways and GWAS, is to identify the few variants truly associated with disease among the large number being tested (21,22). Given the low probability of a true association (i.e. low prior probability) and the small expected magnitude of true associations (often resulting in low statistical power, particularly for less common variants), the standard threshold for statistical significance of an α-level of 0.05 results in the identification of a very high percentage of false-positive findings (23). Therefore, several authors have recommended reducing the P-value threshold to maintain a low probability that a statistically significant finding is a false-positive (i.e. false-positive report probability (FPRP)) (24,25). For instance, P-value thresholds of 10^{-4}–10^{-5} were estimated for variants in candidate genes, and 10^{-7} for random variants to reach high probabilities of true findings (~ >80%) (25). Inversely, FPRP also depends on the prior and statistical power to detect an association. Therefore since the priors are often low, to reach a desirably low FPRP, the sample size of the study should be large enough to attain adequate statistical power. For instance, a P-value of 0.0024 for a SNP with a prior probably of 0.001 in a study of 300 cases and 300 controls will correspond to an FPRP of 72%; however, increasing the sample size to 1500 cases and 1500 controls, and keeping everything else constant, would lower the FPRP to 20% (23). Figure 16.1 shows the sample size requirements to detect genetic associations with a per-allele OR ranging from 1.1–1.5 for a variety of frequencies of the at-risk allele (assuming a log-additive mode of inheritance) and P-value threshold of 10^{-5}. Sample size

Figure 16.1. Sample size requirements to attain 80% power to detect a range of per-allele odds ratios (OR) for an association between disease risk and a bi-allelic SNP

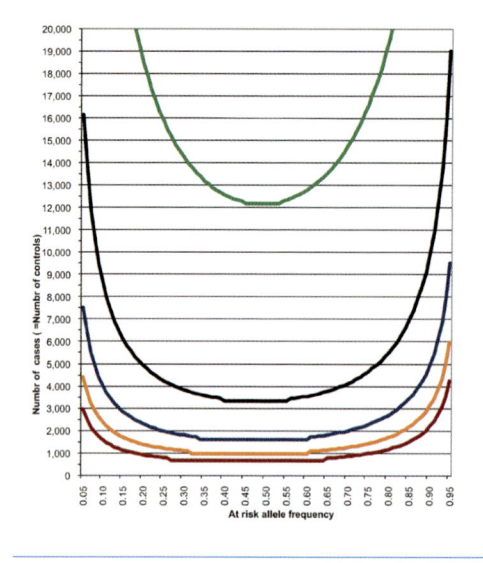

Assumptions: unmatched case-control (1:1) design; Type I error = 10^{-5}, two-sided; log-additive mode of inheritance; disease prevalence= 0.0001. Software: Quanto http://hydra.usc.edu/GxE/

needs increase dramatically for small changes in the OR when the magnitude of the OR is small, and for allele frequencies in the extremes (i.e. away from 0.50). The minimum sample size to detect a per-allele OR of 1.2 (i.e. assuming log-additive risk per allele and homozygous variant OR = 1.44) is 3300 cases and 3300 controls, whereas at least 12 000 cases and 12 000 controls are needed to detect a per-allele OR of 1.1 (i.e. homozygous variant OR = 1.21). Sample size needs would increase by a factor of 1.4 if a P-value threshold of 10^{-7} were to be used instead of 10^{-5}. These numbers illustrate that current studies of hundreds or a few thousand cases and controls have adequate power to detect an OR between 1.2–1.5 for common risk alleles (frequency > 10%); however, much larger studies are needed to detect ORs of 1.1 for less-common risk alleles.

The statistical power of multistage GWAS designs depends on several factors: total number of available samples, number of samples and markers genotyped in each stage, α-level, the size of the genetic effects to be detected, and type of analysis (10). The price of genotyping for different technologies used at each stage is also an important factor determining the optimal design of multistage studies. In general, joint analysis of data from the different stages is more powerful than replication analysis (26). As the cost of genotyping and sequencing methods continue to decrease, studies will be able to scan all individuals and eventually obtain a full genomic sequence, which will allow the evaluation of rarer variants as well as the mapping of causative variants. Sample size requirements for more complex analyses of genotype data, such as pathway-based, haplotype and novel high-dimensional analyses, are less well understood.

Evaluation of G-E interactions often requires large sample sizes that are further increased by the presence of errors measuring environmental and/or genetic exposures, even when the errors are small (17,27). Although multiplicative parameters for G-E interactions tend to be attenuated by differential misclassification of exposure (17), this type of bias could lead to overestimation of the main effects of the exposure, joint effects, and subgroup effects, or additive interactions. Thus, high-quality exposure assessment and almost perfect genotype determinations are required for the evaluation of G-E interactions. This highlights the importance of validating genotype assays and including quality control samples during genotype determinations. This will help assess the reproducibility of the assays to identify problematic ones for possible re-genotyping or assay optimization.

There are several free statistical software programs to carry out power calculations for genetic association studies. POWER (http://dceg.cancer.gov/bb/tools/power) can be used for binary outcome studies (case–control or cohort studies) based on a logistic-like regression model with one or two covariates (e.g. gene-exposure interactions) (18); POWER for Genetic Association Analyses (http://dceg.cancer.gov/bb/tools/pga) can be used in case–control studies, fine-mapping studies, and whole-genome scans, for power and sample size calculations under various genetic models and statistical constraints; QUANTO (http://hydra.usc.edu/gxe/) is useful in matched case–control, case–sibling, case–parent, and case-only designs to compute sample size or power calculations to evaluate genetic associations, G–E interaction, or gene–gene (G–G) interaction; the CaTS Power Calculator (http://www.sph.umich.edu/csg/abecasis/CaTS/) is a user-friendly interface for power calculations for large genetic association studies, including two-stage GWAS (26); a spreadsheet can be downloaded to calculate FPRP (http://jnci.oxfordjournals.org/cgi/content/full/96/6/434/DC1) (23).

Current case–control or cohort studies usually include between a few hundred to a few thousand cases and a similar numbers of controls. Therefore, to meet the larger sample size requirements to identify weak associations (Cf. Figure 16.1) and interactions, especially when considering disease subtypes, an increasing number of consortia of existing studies have been and continue to be formed (28). Consortia can achieve the large sample sizes necessary to confirm or refute associations by coordinating the analysis of pooled data from many studies, as well as evaluating consistency of findings across studies of different quality and with different sources of biases (29). However, comparability of data on environmental exposures across studies may be a limitation. Therefore, very large, well-designed studies with high-quality exposure data and tumour specimens might be needed. To date, there are very few examples of gene–environment interactions that have been demonstrated in large pooling efforts. One example is the demonstration of interactions between cigarette smoking and polymorphism in the *NAT2* and *GSTM1* genes in the context of a bladder cancer GWAS (30).

Analysis of genetic data

Quality control of DNA and genetic data

Quality control analyses are conducted both before and after genotyping of DNA samples (31). Ideally, DNA samples should be accurately quantified before genotyping (e.g. using fluorescence nucleic stains, such as PicoGreen® (Molecular Probes Inc.)), and profiled to obtain a "DNA fingerprint" using a panel of genetic markers that uniquely identify each sample (e.g. the Amp/STR® Identifiler® kit (Applied Biosystems) uses 15 SNPs and the Amelogenin marker for gender determination). This allows precise verification of duplicate DNA samples, identification of unexpected duplicates (e.g. due to sample collection, storage, labelling or plating errors), identification of gender mismatches between the DNA and self-reported gender, and identification of contaminated samples that should be excluded from further analyses. After genotyping assays have been performed, the quality of the resulting genotyping calls can be assessed by evaluating the scatter plots of allele-specific probe intensity values used for genotype determination. SNP genotype calls are made based on the clustering patterns of the probes, where clusters for each homozygote and heterozygote genotype state should be observed. High quality assays will demonstrate tight clusters with clear separation between them (Figure 16.2A versus 16.2B).

Genotyping completion rates can be calculated for DNA samples or loci:

• Overall completion rates—number of loci with genotype calls divided by the total number of genotyped loci

• Completion rates by sample—number of loci with genotype calls for a given sample divided by the total number of genotyped loci in that sample

• Genotype completion or call rates—number of samples with genotype calls for a given SNP divided by the total number of genotyped samples

Decreased completion rates often reflect poor assay performance, which may be due to chemical and physical properties of the assay or the quality of the input DNA. Completion rates should also be calculated separately by DNA source, processing laboratory, DNA extraction method, case–control status and genotyping plate to detect systematic variation in genotype quality. Low genotype call rates can help detect loci with problematic assays that require re-genotyping, a new assay, or selection of a surrogate SNP. Completion rates by sample or plate can detect problems with specific samples or plates that could result in exclusion of data from those samples or plates. Analyses of completion rates by case–control status can detect assay performance differences due to varying DNA quality for cases and controls, which would result in differential misclassification. When large numbers of SNPs are genotyped, such as in GWAS, it is useful to look at the distribution and plot the completion rates by sample or loci (genotype calls) to detect outliers. Figure 16.3A shows an example of such plots, which utilize data from a scan using the Illumina HumanHap 1M assay with good overall completion rates for most samples and loci.

Sample heterozygosity is the percentage of heterozygous genotypes in autosomal SNPs for a given sample. For instance, SNPs included in the Illumina HumanHap 1M genome-wide genotyping assay in populations of European origin have a mean heterozygosity of about 27%. Although samples from different racial origins will have different heterozygosity values, extreme outlier values can reflect sample quality or assay performance problems, which is reflected by a correlation between high (or low) sample heterozygosity and reduced sample completion rates. Plotting heterozygosity for all samples and against sample completion can help identify low performing samples (see Figure 16.3B for an example).

Analyses of data from duplicate quality control samples include calculation of percent agreement of informative genotypes (i.e. concordance of non-missing genotype calls for DNA samples from the same individual) among pairs of samples. As with completion rates, genotype concordance should also be evaluated by plate and sample, since this can give clues as to the source of error (e.g. systematic errors often reflect sample handling or plate labelling and orientation problems, whereas random errors reflect assays' reproducibility).

In very large, randomly mating populations with no selection, genotype frequencies are expected to be constant and 'in equilibrium' from generation to generation. This phenomenon is called Hardy–Weinberg equilibrium (HWE), and the expected genotype frequencies under HWE are called Hardy–Weinberg proportions (HWP). It should be noted that although HWE and HWP are often used interchangeably, HWE is a multigenerational phenomenon and cannot be directly assessed in standard epidemiological studies. Under random mating and no selection, HWP implies HWE; however, under selection and non-random mating, genotype frequencies can be in HWP but not HWE. For a bi-allelic SNP with A-allele frequency p and a-allele frequency $q = (1-p)$, the expected genotype proportions under HWP

Figure 16.2. Examples of genotype clustering plots used to make genotype calls (polar cluster plots of the normalized intensity and allelic intensity ratio)

A. Good assay with clear separation of genotype clusters

B. "Poor" assay with overlapping genotype clusters

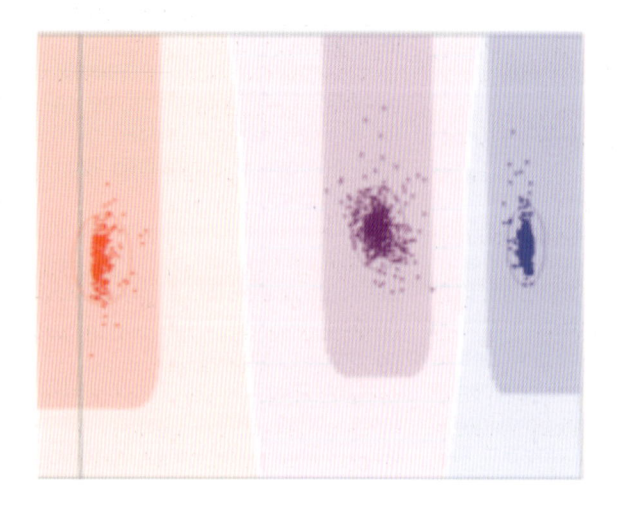

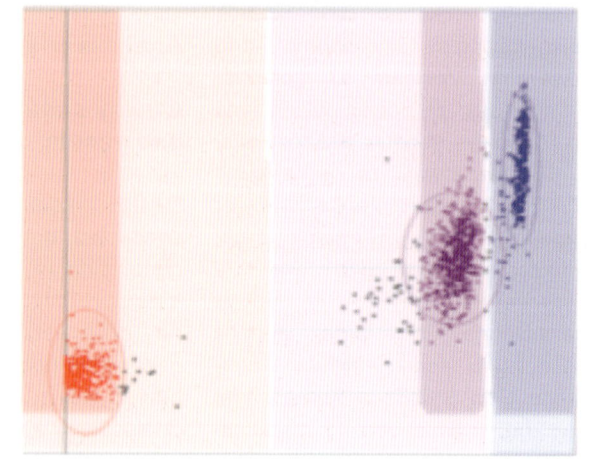

Figure 16.3. Examples of quality control plots using data from 1M Illumina platform

A. Sample and locus completion rates

B. Heterozygocity and completion rates

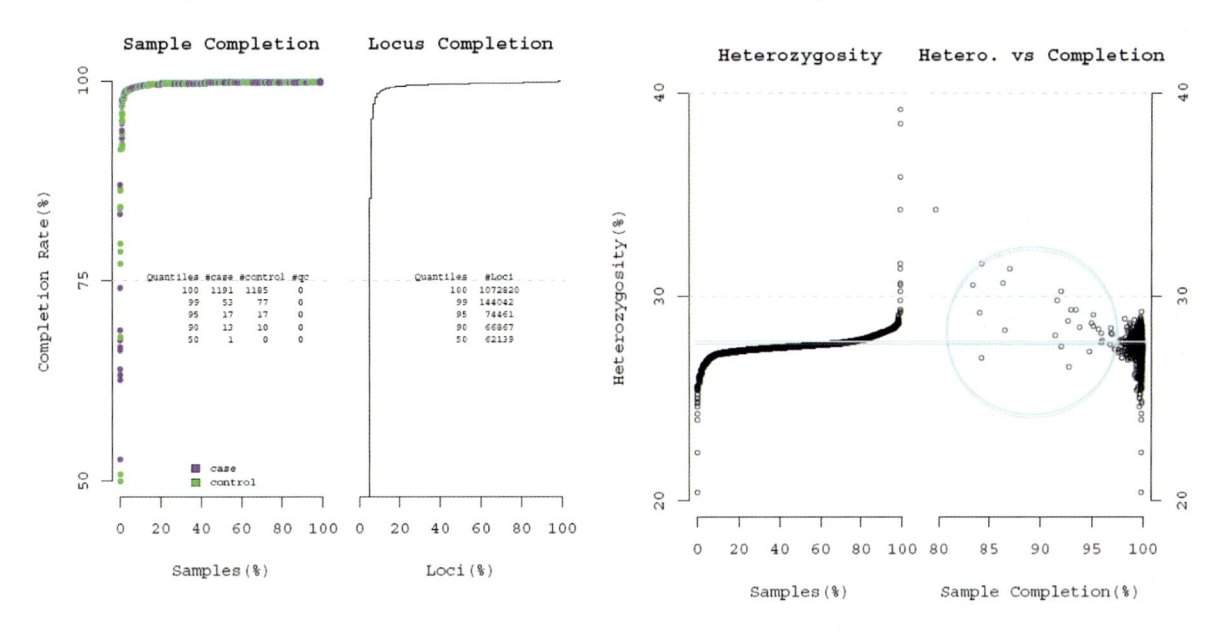

Circle: samples with low completion rates tend to have high heterozygocity values
Line: average heterozygocity in the population

---

are p^2 for genotype AA, $2pq$ for Aa and q^2 for aa. Extreme departures from HWP in the control population (departures in cases could be due to associations with disease) can reflect assay problems. Other reasons are also possible, such as non-random mating, selection, population admixture, and a genetically non-homogeneous control population. Therefore, a very careful evaluation of quality control measures should be performed when significant departures from HWP (e.g. using an exact test (32), Pearson's χ^2 test comparing observed and expected genotype frequencies) are observed for a specific SNP assay. Evaluation of HWP for all genotyped SNPs (i.e. comparison of expected and observed number of SNPs with significant HWP departures) can be helpful in determining if the observed departure reflects a problem with the controls, such as a problem during control selection, or the source population not being in HWE due to non-random mating. If there is no evidence of genotyping errors or control selection problems, the likely explanation for the observed departure is chance. In that case, methods of analyses for associations between the genotype and disease that assume HWP can be helpful in evaluating the impact of a chance departure on estimates of effect, such as the odds ratio (33).

In summary, a list of quality control checks before risk analysis of genotype data can include:

• Verifying duplicate samples and identifying unexpected duplicates using DNA profiling data
• Examining genotype clustering in scatter plots
• Identifying discrepancies between self-reported and genetically determined gender
• Completion rates by sample—excluding data from DNA samples with low completion rates
• Genotype call rates—excluding data from assays with low call rates
• Examining sample heterozygocity and excluding outlier samples
• Genotype concordance among verified duplicate samples (excluding assays with low genotype concordance)
• Genotype concordance among samples not from the same sample (excluding assays with unexpectedly high genotype concordance)
• Testing for deviations of Hardy–Weinberg proportions

Discovery of genetic susceptibility loci

Described here are statistical analyses of SNP data derived from candidate genes or regions, as well as genome-wide approaches in a case–control study (population, hospital-based or nested in a prospective cohort). Methods of analyses for prospective cohort data (e.g. Cox proportional-hazards regression analyses or analyses of quantitative traits) will not be addressed. The analyses described below can be implemented using widely available statistical packages, such as the commercial packages Stata (http://www.stata.com/), SAS (http://www.sas.com/) or the software R Project of Statistical Computing (http://www.r-project.org/). R is being used more frequently in analyses of genetic data, as many novel statistical methods are written and freely distributed as R add-on packages, providing a very flexible computing and graphical toolset.

Association between individual SNPs and disease risk: Genotype-based analyses

In genotype-based analyses, each individual SNP is evaluated in relation to disease risk by comparing the genotype distribution for cases and controls. The odds ratio (OR) approximates the ratio of disease incidence in exposed (or susceptible) and unexposed (or non-susceptible) individuals, and is often used as a measure of association in case–control studies, as it does not require estimates of the actual incidence rates (34). Table 16.1 illustrates a 2x3 table often used to display the number of cases and controls with the three possible genotypes in the population under study. This data can be used to estimate genotype ORs for subjects carrying the heterozygous and uncommon homozygous genotypes relative to subjects with the common homozygous genotype. Genotype-based ORs can be estimated using logistic regression models (34) with disease status as the outcome and the SNP as the explanatory variable coded as either indicator or dummy for each genotype. Although these analyses yield unbiased and efficient estimates of the OR, the estimate of the intercept parameter is biased due to the retrospective nature of the case–control design (35). Data from studies with cases individually matched to controls by variables such as age, hospital or region should be analysed by conditional logistic regression models to ensure unbiased and efficient estimates of the OR (34). However, when data on genotype or exposure information is missing for either the case or the control in a matched pair, information from both subjects is lost resulting in decreased efficiency. Therefore, individually matched studies are often analysed as unmatched studies, adjusting for categories of the matching factors using indicator variables. This can result in incomplete adjustment for the matched design, but the impact is generally minimal and compensated for by the gain in efficiency.

Below is the form of a logistic model for genotype (G) variables (additional variables can be added to adjust for potential confounders) and disease (D) outcome:

$$Pr(D|G) = \exp(\beta_0 + \beta_{Aa} Aa + \beta_{aa} aa) / (1 + \exp(\beta_0 + \beta_{Aa} Aa + \beta_{aa} aa))$$

or

$$\text{logit}(Pr(D|G)) = \beta_0 + \beta_{Aa} Aa + \beta_{aa} aa$$

where Aa, aa are 0,1 indicator variables for each genotype (AA is the reference).

The genotype-specific OR and 95% confidence intervals (CI) can be estimated from the logistic regression coefficients and its standard error (SE) as:

$OR(Aa) = \exp(\beta_{Aa})$;
95% CI = $\exp(\beta_{Aa} \pm 1.96\, SE(\beta_{Aa}))$
$OR(aa) = \exp(\beta_{aa})$;
95% CI = $\exp(\beta_{aa} \pm 1.96\, SE(\beta_{aa}))$

The null hypothesis (H_0) for a test of SNP-disease association (sometimes called co-dominant test) can be written as:

H_0: $OR(Aa) = 1.0$ and $OR(aa) = 1.0$ with 2 degrees of freedom (df)

Hypothesis testing can be carried out by conventional score-test, Wald test or likelihood ratio test.

Table 16.1. Genotype frequencies and odds ratio (OR) estimates for genotype-based analyses in a case–control Study

Genotype	Cases	Controls
AA	m	n
Aa	o	p
aa	q	r

m, n, o, p, q, r are cell counts for number of subjects

Genotype-specific estimates

$OR(Aa)$ vs. AA = $\dfrac{n * o}{m * p}$

$OR(aa)$ vs. AA = $\dfrac{n * q}{m * r}$

Variance for genotype-specific OR

$Var(LogOR(Aa)) = \dfrac{1}{m} + \dfrac{1}{n} + \dfrac{1}{o} + \dfrac{1}{p}$

$Var(LogOR(aa)) = \dfrac{1}{m} + \dfrac{1}{n} + \dfrac{1}{q} + \dfrac{1}{r}$

Dominant model estimate

$OR(Aa/aa)$ vs. AA = $\dfrac{n * (o + q)}{m * (p + r)}$

Recessive model Estimate

$OR(aa)$ vs. AA/Aa = $\dfrac{(n + p) * q}{(m + o) * r}$

The variance of the OR estimates for each genotype is inversely related to each of the cell counts (Table 16.1), which gives an intuitive sense of why the larger the cell counts are, the smaller the variance and the tighter the confidence intervals. This also shows that estimation of genotype-specific ORs can be unreliable for uncommon SNPs with only a few subjects carrying the homozygous variant genotype (aa).

Genotype ORs can also be estimated under the assumption of specific models of genetic inheritance, such as the log-additive (or multiplicative) model, which assumes a log linear trend for genotypes with an increasing number of variant alleles; the recessive model, which assumes the same risk for Aa and aa carriers; and the dominant model, which assumes the same risk for AA and Aa carriers. When using these approaches, keep in mind that these models of inheritance were originally developed for simple Mendelian diseases with near-complete penetrance, and thus might be over simplistic for complex diseases that are influenced by variants in multiple loci. Estimates of ORs and tests for genetic associations under different models for genetic risk can be obtained using logistic regression models with disease status as the outcome. The three possible genotypes for a given SNP are often coded as 0 for AA, 1 for Aa, and 2 for aa. ORs for Aa and aa (relative to AA) can be estimated by including two dummy variables for Aa and aa in a logistic regression model of the form:

Logit (P(D|G)) = $\beta_0 + \beta_G$ G

Log-additive trend model: G coded as 0 for AA, 1 for Aa, 2 for aa

$\exp(\beta_G)$ = per allele OR

Dominant model: G coded as 0 for AA and 1 for (Aa+aa)

$\exp(\beta_G)$ = OR(Aa/aa) vs AA

Recessive model: G coded as 0 for (AA+Aa) and 1 for aa

$\exp(\beta_G)$ = OR(aa) vs AA/Aa

The H_0 for a test of association under the models above can be written as:

H_0: OR = $\exp(\beta_G)$ = 1.0, 1 df.

The power to detect disease susceptibility loci is maximized when the assumed genetic model of inheritance is the true model. Thus, since the true mode of inheritance is often unknown, one might chose to test several models (with the caveat mentioned above that simple modes of inheritance might not hold for complex diseases). However, reporting the most significant finding after testing different models will result in an inflated type 1 error or an underestimate of the precision of the confidence intervals. Therefore, it is important to use appropriate statistical tests, such as permutation testing (36), to account for testing of multiple models. Multiple testing might result in no increases or even decreases in power, compared to testing only one pre-defined model with good performance under different alternative models, such as the 2 df genotype-based test or the trend test under the log-additive model (37). The advantage of the 1 df trend test, assuming a log-additive model, is that it uses one less df; it is generally more powerful than the 2 df genotype-based test, when the genetic effect is additive or log-additive; and it has good power to detect dominant effects. However, the trend test has poor power to detect recessive effects. Although the 2 df test for genotype effects has substantially better power to detect recessive effects, the actual power is often low. It should be noted that, even if the underlying model for the disease loci were recessive, the association with disease with a marker SNP (correlated with the causal SNP) would tend to look log-additive due to misclassification of subjects with respect to the true genotypes for the disease allele. Finally, a 2-sided test for trend is not affected by the sign of the LD between the minor allele of the marker SNPs and the causal SNP, whereas this can be affected for the other tests (38). No matter what model is used for testing for associations, genotype-specific estimates of the OR are often presented (unless the homozygous carriers are very uncommon), since they do not make any assumptions about the underlying mode of inheritance, and provide more information of the possible underlying models.

Association between individual SNPs and disease risk: Allele-based analyses

In genotype-based analyses, the unit of observation is the subject or genotype, whereas in allele-based analyses, the unit of observation is the allele. Since each subject has two alleles at any autosomal locus, the total number of observations in allele-based analyses is twice the number of subjects. Table 16.2 illustrates a 2x2 table for the allele frequencies for cases and controls, and the allele-based ORs and variance. Allele-based tests for association with disease assume independent distribution of alleles in the population, or HWP, for both cases and controls (39). HWP in controls implies HWP in cases only if there is no association with the disease (i.e. under the H_0), or if the alleles have multiplicative (log-additive) effects on disease risk. Therefore, if the control population is in HWP, the allele-based tests with variance estimates under Ho are valid, but estimation of confidence intervals will require the additional assumption that the alleles have multiplicative effects. It should be noted that under HWP, the allele-based test is asymptotically equivalent to the trend test in genotype-based analyses (39). The interpretation of the allele-based ORs is less intuitive than the genotype-based ORs, as individuals always carry a combination of two alleles, and thus might not have a useful risk interpretation. Because of the more restricted interpretation and set of assumptions of the allele-based analyses, genotype-based analyses are often preferable (39).

Association between other types of variants and disease risk

The previous section described analyses of SNP variants, which have two alleles for any autosomal loci (males are hemizygotes, i.e. they have only one allele for SNPs in X-linked genes). Analyses of other types of genetic variation, for instance multiallele variants in variable tandem repeats (VTR) or copy number variations (CNV), follow similar principles and will not be discussed here. Because of the increased number of categories, model-based analyses of multiallelic loci or haplotypes can be very helpful in reducing the number of parameters to be estimated (40).

Haplotype analyses

Haplotype analyses exploit the LD, or correlation among genetic markers that are physically close, to improve the statistical efficiency and interpretability of studies of genetic associations with disease risk (41). These analyses can be aimed at comprehensively scanning a candidate region for disease susceptibility loci, or used to detect associations with markers that act in cis (i.e. when two or more variants affect disease only if they are on the same chromosome). A methodological challenge in studies of unrelated individuals is the estimation of phased haplotypes using unphased genotype data. The estimation of haplotype frequencies among cases and controls is done iteratively using methods such as the expectation-maximization (EM) algorithm, which can be implemented using software packages such as PROC HAPLOTYPE in the SAS Genetics Package, SNPHAP (http://www-gene.cimr.cam.ac.uk/clayton/software/snphap.txt) and tagSNP (http://www-hsc.usc.edu/~stram/tagSNPs.html). There is a wide range of statistical methods to analyse haplotype associations with disease using regression models that allow for the adjustment of potential confounders (41,42). Single-imputation or "plug-in" methods model estimates of individual haplotypes as if they were observed, whereas marginal regression methods take into account phase ambiguity in the estimation of measures of association between haplotypes and disease risk (41,43). The main advantage of plug-in methods is that they are computationally simple; they can be implemented using standard statistical software, which uses estimates of posterior haplotype probabilities. More advanced methods use EM type algorithms for simultaneous estimation of haplotype frequencies and haplotype-disease odds ratio parameters. These methods produce more accurate variance estimates and confidence intervals, since they properly account for the fact that haplotype phases are not directly observed. These methods require specialized software, such as the R function haplo.glm in the haplo.stats package, Chaplin (http://www.genetics.emory.edu/labs/epstein/software/chaplin/) or HAPSTAT (http://www.bios.unc.edu/~lin/hapstat/).

Analysis of GWAS data

GWAS generate very large genotype data sets, often including billions of genotypes per study (e.g. 1 million SNPs in 4000 subjects) that require the development of tools to accommodate the demands for data storage, management, quality control and risk analyses. These tools are likely to expand and improve to meet the needs of

Table 16.2. Allele frequencies and odds ratio (OR) estimates for allele-based analysis in a case–control Study

Allele	Cases	Controls
A	2m+o	2n+p
a	2o+m	2p+m

Allele-based Estimates

$$OR(a) \text{ vs. } A = \frac{(2n+p) * (2o+m)}{(2m+o) * (2p+m)}$$

Variance for allele-specific OR

$$Var(LogOR(a)) = \frac{1}{2m+o} + \frac{1}{2n+p} + \frac{1}{2o+m} + \frac{1}{2p+m}$$

increasingly large data sets. Two free and available tools that can meet the needs of data management, quality control, population stratification and association analyses of GWAS data are PLINK (http://pngu.mgh.harvard.edu/~purcell/plink) (44) and the Genotype Library and Utilities (GLU) package (http://code.google.com/p/glu-genetics).

Analyses of GWAS data usually include: quality control analysis of genotype data, as described above; analysis of population structure and decisions on the need and method for adjustment for population stratification; definition of the analytical data set after exclusions of data from samples or loci based on quality control analyses; and analysis of the association between genotypes and disease risk.

Analysis of population structure. Epidemiological studies often collect information on self-reported ethnicity and geographical location from cases and controls. Self-reported ethnicity, race and geographical location are surrogates for a complex mixture of unmeasured factors, which reflect variation in genetic background, culture, language, religion and health-related behaviour (45). Because these unmeasured factors could introduce confounding bias when related to disease and exposures of interest, cases are usually matched to controls by race, ethnicity and geographical location to facilitate adjustment during the analyses. However, population substructure information (i.e. heterogeneous or admixed populations) not captured by self-reported ethnicity could lead to population stratification or confounding bias due to differences in allele frequency and disease risk across subpopulations (46–48). Population structure analyses use multilocus genetic data to assign individuals to populations of origin. This determines if there is population substructure not accounted for by variables measured in epidemiological studies. Population structure in GWAS can be analysed using a Bayesian clustering approach (49). This method uses information on linkage between a set of SNPs and Hardy–Weinberg disequilibrium to decompose a group of individuals (e.g. cases and controls) into genetically similar populations or clusters. Reference subjects of fixed populations (e.g. Asian, European and African from HapMap) can be used to guide the clustering process, to estimate the degree of admixture of each study sample. The SNPs for these analyses are selected from SNPs in the scan with high completion rates and low residual LD (e.g. r^2 < 0.1–0.01 for pairs of SNPs less than 200Kb apart). As an outcome of these analyses, each individual is assigned an admixture coefficient reflecting the estimated degree of membership with each population. The degree of membership in a structure analysis, assuming the use of the three HapMap populations as fixed reference populations, can be plotted in an equilateral triangle, also called an admixture plot. Membership estimates for each of the three populations are represented by the distance to each of the three corners of the triangle. The software STRUCTURE can be used to carry out these analyses and can be downloaded from http://pritch.bsd.uchicago.edu/software.html.

Genetically-determined race using admixture analyses can be compared to self-reported race and ethnicity to identify, and possibly exclude, outliers from subsequent risk analyses. Figure 16.4A shows an example of an admixture plot generated by STRUCTURE from a population self-identified as Caucasians. Red dots represent cases and green dots represent controls.

According to this plot, most subjects are estimated to be of more than 85% European descent (i.e. they are clustered in the European corner) with no evidence for substructure. The few subjects that are estimated to be of less than 85% European descent can be excluded from risk analyses to reduce population heterogeneity. However, if strong evidence for population structure were to be found, this method cannot be easily used to adjust for population structure in the risk analyses. EIGENSTRAT is a software that has been proposed to detect and adjust for population stratification in GWAS (50). This method uses principal component analyses (PCA) to reduce high-dimensional genotyping data to lower dimensions that can be used in the analyses. These analyses produce a set of continuous variables, called principal components (PC), that capture the maximum of the genetic variation across individuals in a data set (Figure 16.4B). Each PC is defined as the top eigenvector of a covariance matrix between samples, thus the name EIGENSTRAT. Inclusion of related individuals in PCA analyses can create problems because of the high genetic correlation between relatives. Epidemiological studies of unrelated individuals can occasionally unknowingly enrol family members. Therefore, analyses to determine the degree of relatedness between individuals, such as Pedigree Relationship Statistical Test (PREST) analyses (http://galton.uchicago.edu/~mcpeek/software/prest/) (51), should be carried out before the PCA analysis to identify and exclude relatives. Significant PC from these analyses can be used to model ancestry differences between

Figure 16.4A. Example of an triangular admixture plot generated by STRUCTURE for a population self-defined as Caucasians in the USA. The ancestry estimate is represented by the distance to each side of the triangle. Red dots represent cases and green dots represent controls. Reprinted by permission from Macmillan Publishers Ltd: Nature Genetics, copyright (2007).

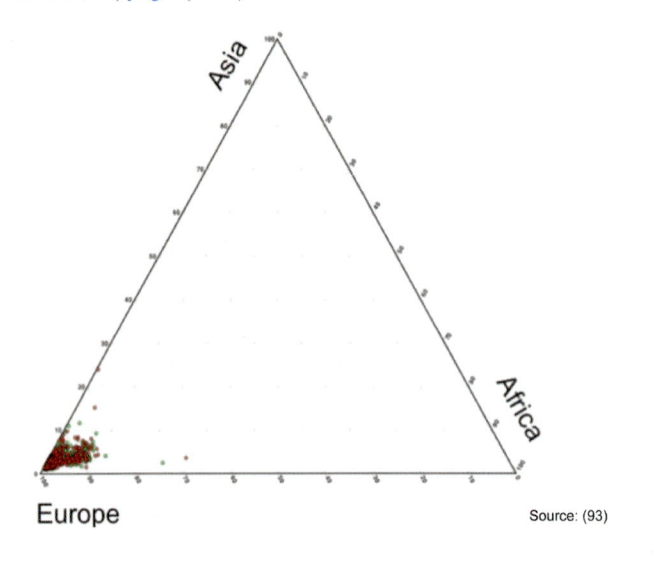

cases and controls and thus adjust for population stratification. These analyses can be performed using the EIGENSOFT package that includes population genetics methods (52) and the EIGENSRTAT stratification correction method (http://genepath.med.harvard.edu/~reich/Software.htm) (50).

SNP imputation methods can increase the power of studies of genetic regions or GWAS by filling in missing genotype data due to assay failures (e.g. 2% of samples with missing genotypes for a SNP with 98% completion rate), and increasing genetic coverage through imputation of SNPs that have not been genotyped. These methods use information from a reference panel, such as HapMap, to impute untyped SNPs. In addition to increasing power, SNP imputation can sometimes help localize signal for an association in a region, and facilitate the combination of data from studies using different genotyping chips or platforms, including overlapping, but not identical, sets of SNPs. Several methods and software are available to impute SNP data, such as IMPUTE (http://mathgen.stats.ox.ac.uk/impute/impute.html) (53), MACH (http://www.sph.umich.edu/csg/abecasis/MACH/download/) (54), and BimBam (http://stephenslab.uchicago.edu/software.html) (55). In all of these methods, it is important to keep in mind that imputation accuracy will depend

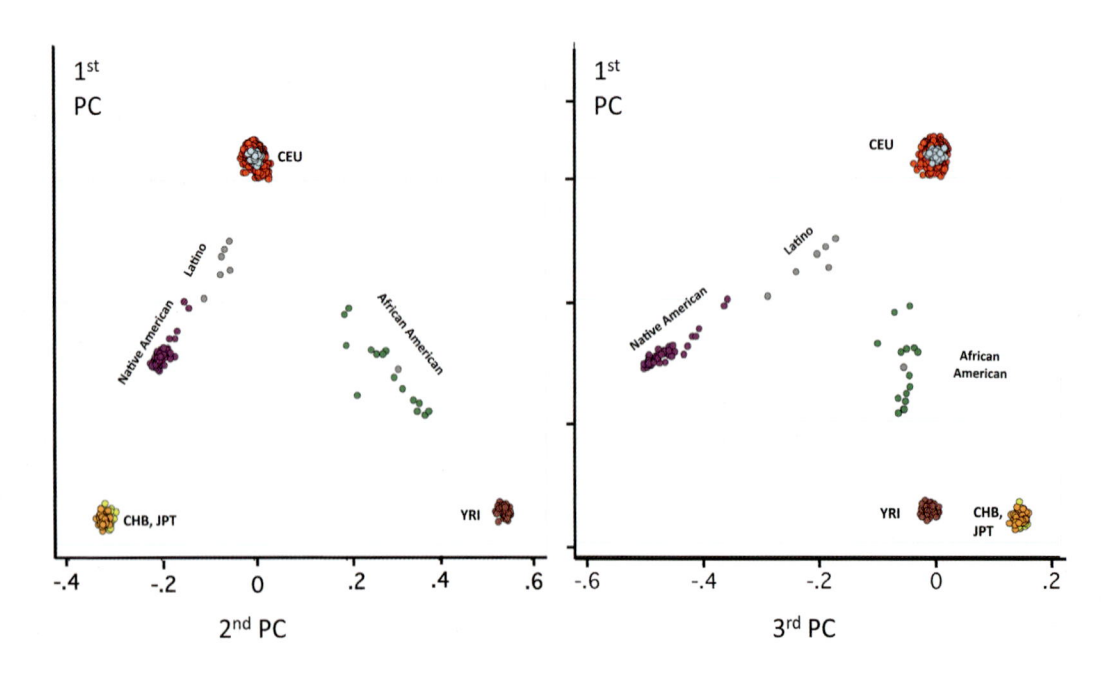

Figure 16.4B. Example of principal component analysis (PCA) to identify principal components (PC) that account for population structure

on many factors: completeness and accuracy to the reference SNP panel; the quality of the data (e.g. large amounts of missing data will decrease accuracy); density of SNPs and LD pattern in the region (i.e. areas with low coverage or low LD are more difficult to impute); the similarity of the LD structure between the reference population (e.g. HapMap) and the population under study (e.g. admixed or unique populations, such as Amish, might be difficult to impute); and allele frequency (i.e. rare SNPs are harder to impute). Thus, findings from analyses of data, including imputed SNPs, should be carefully interpreted taking into account these limitations.

Definition of the analytical data set. Analysis of quality control of the genotype data and population structure can be used to identify and exclude samples and loci from analyses (e.g. samples with low completion rates, samples from subjects with discrepancies between the self-reported race and ethnicity and genetically determined race, or loci with low call rates across samples or with discordant results in duplicated QC samples). Other criteria based on epidemiological data can also be used to define the analytical data set; for instance, exclusion of subjects with missing data in key variables, such as age, or exclusion of rare subtypes of disease, such as rare histological types of cancer, to decrease disease heterogeneity.

Analysis of association between genotypes and disease risk. The primary aim of analysis of GWAS data is to discover markers for genetic susceptibility loci. Initial analyses usually evaluate associations between each individual SNP and disease risk and follow the principles previously described. In multistage designs, an important consideration in the analysis is the criteria used in the first stage to select SNPs to be genotyped in subsequent stages (10). A subset of the most significant P-values for SNP-disease associations is often used to select SNPs to be carried forward to subsequent stages. Other approaches, such as hierarchical regression models incorporating prior knowledge on the SNP selection procedures, can also be used (56). In general, joint analysis of data from different stages is more powerful than replication analysis (26).

Graphical representation of results can be very helpful for summarizing the large amounts of GWAS data. For instance, quantile-quantile plots (Q-Q plots) for observed P-values for a test of the null hypothesis of no association between each SNP and disease risk against expected P-values under the null hypothesis can be useful. These plots summarize both systematic bias and evidence for association. Most SNPs in a GWAS will not be associated with disease risk; therefore, associated P-values will appear in the diagonal of the Q-Q plot. Small departures at the extreme of the Q-Q plot suggest associations with disease. Large departures from the diagonal can reflect systematic biases leading to increases in false-positive findings (e.g. due to different DNA quality for cases and controls or population stratification). Figure 16.5 shows an example of a Q-Q plot from an analysis of a GWAS of hair colour, before (back dots) and after (red dots) adjusting for population stratification using PC (57). This example shows how adjustment by PC was able to reduce a large deviation from the diagonal that reflected bias due to population stratification. Plots of the $-\log_{10}$ (P-value) for all SNP associations with disease sorted by chromosomal location, can also be helpful in showing the distribution of P-values and identifying the location of associations with genomic significance.

Analysis of additional outcomes. Data from GWAS can also be used in ancillary analyses to evaluate genetic association with secondary outcomes measured in cases and controls (e.g. other diseases or exposures, such as height or smoking habits). The original case–control sampling can affect measures of association with secondary outcomes (58,59); however, bias is only introduced when both the secondary outcome and the genetic loci under study are associated with the risk of the primary disease (58). Because most genetic loci in GWAS will not be associated with the primary disease, naïve analyses, ignoring the sampling design, will be valid for most loci.

Hierarchical-Bayesian methods

Prior information on the expected magnitude of genetic associations (e.g. the OR likely to vary from 1.1–1.5) can be used in hierarchical models to provide more constrained estimates than the conventional, frequentist analytical approaches mentioned above (60,61). Other advantages of this approach are that it can be used to address problems of multiple comparisons and to incorporate biological information from pathway in the analyses.

Hierarchical models can also be used in selecting SNPs to be followed-up in multistage GWAS (60,61). These models can increase the power to detect susceptibility loci by incorporating known information about the SNPs into the selection, rather than just relying on measure of association in the data set. Wider use of hierarchical models has been limited by the unfamiliarity of epidemiologists with software

Figure 16.5. Example of a Quartile-Quartile (Q-Q) plot of observed quartiles of log10(P-values) against the expected values under the null hypothesis. Black and red dots represent P values from analyses not adjusted and adjusted, respectively, for population stratification using principal components (57).

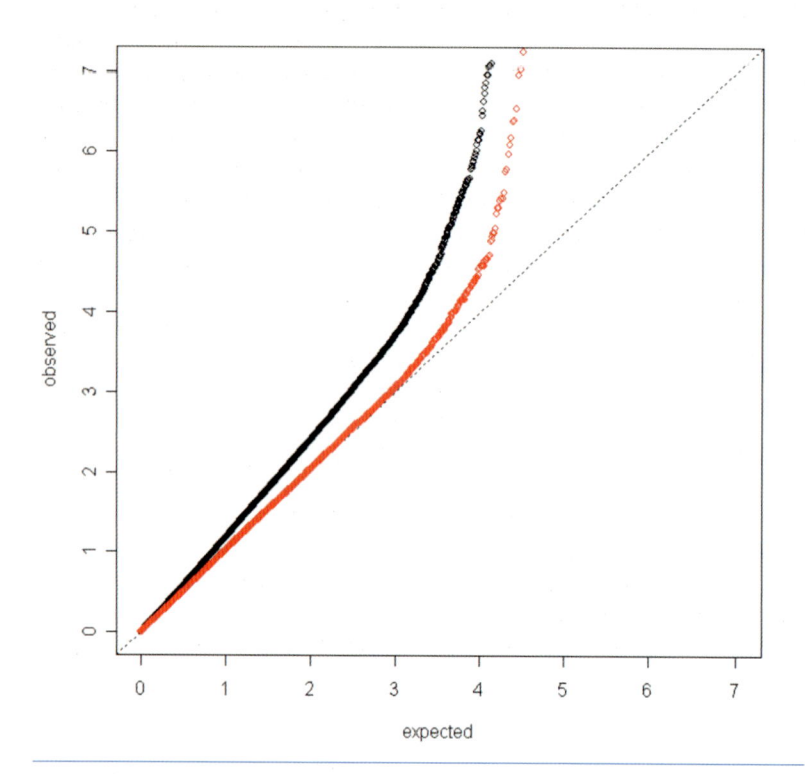

packages to fit these models. However, their use may increase now that SAS codes are available for analysing epidemiologic data with hierarchical models (60,61).

Evaluation of interactions

Complex diseases are likely to be caused by the interplay of multiple environmental exposures and genetic susceptibilities, hence the importance of evaluating G-G and G-E interactions (62). Specifically, evaluation of interactions can:
- Facilitate the identification of underlying risk factors for disease (e.g. improve power to detect a risk factor that varies according to the levels of another factor by stratifying on the modifier factor)
- Provide insights into the biological mechanisms of disease
- Provide public health benefits,

such as improved risk prediction models and strategies for disease prevention (e.g. benefits of targeting subjects susceptible to specific exposures)

In practice, however, evaluation of interactions can be quite challenging because it requires very large studies with high-quality exposure assessment and availability of biological specimens. Even in well-designed, well-powered epidemiological studies, exploring interactions can be a computationally daunting task, particularly in studies of a very large number of genetic markers, such as GWAS, evaluated in hundreds of thousands of SNPs.

Definition of interaction

In epidemiology, an interaction between two factors is usually defined as the statistical evaluation of whether the association between one factor (e.g. cigarette smoking) and disease risk varies according to the value of the other factor (e.g. *NAT2* genotype). A multiplicative interaction occurs if the association between the two factors is measured in a multiplicative scale by the relative risk (or odds ratio), and an additive interaction occurs if the association is measured in the additive scale by the risk difference (63). A multiplicative or additive interaction can also be described as a departure of the joint effect of the two factors from the expected effect under a multiplicative or additive model, respectively. Table 16.3 shows the definitions of measures of association between two dichotomous factors (an environmental exposure (E) and a genotype (G)) and disease risk, including joint effects and stratum-specific effects, as well as multiplicative and additive interactions (64,65). A set of three ORs characterize the E and G associations with disease: OR(G|E = 0), OR(E|G = 0), and OR(G,E). These ORs can be re-parameterized as the stratum-specific odds ratios and the interaction ORs (i.e. OR(G|E = 0), OR(G|E = 1), and ORint; or OR(E|G = 0), OR(E|G = 1), and ORint). The relationship between these parameters is shown in Table 16.3.

The biological implications of these two statistical forms of interactions have long been debated in the epidemiologic literature. The main problem in making biological inferences based on epidemiological interactions is that the presence or absence of interaction depends on the scale in which the association with disease is measured. The correspondence between statistical and biological modes for interaction can be defined under simple biological models (62). For instance, under models such as the single-

hit or the sufficient-component-cause, two factors with biologically independent actions on disease result in additive joint effects on the incident rate of the disease (3). However, relationships between biological actions and statistical models in complex diseases with multiple known and unknown causes cannot be easily made, except when the interaction is independent of the scale of measurement of association (66,67). These interactions occur when the effect of one or both factors exists only in the presence of the other, and can be referred to as non-removable interactions. Using notation from Table 16.3, 'non-removable interactions' can be defined as:
- OR(G|E = 0) = 1 and OR(E|G = 0) = 1 and OR(E,G)≠1
- OR(G|E = 0) = 1 and OR(E|G = 0)≠1 and OR(E,G)≠OR(E|G = 0)
- OR(G|E = 0)≠1 and OR(E|G = 0) = 1 and OR(E,G)≠OR(E|G = 0)

The interaction between the NAT2 genotype and smoking status in bladder cancer risk, where NAT2 slow acetylators are at increased risk of bladder cancer compared to rapid acetylators only among cigarette smokers, is an example of non-removable interactions (68). Crossover, or qualitative interactions, where the effect of one factor is reversed by the presence of the other, is an extreme form of non-removable interactions (69). There are only a few established examples of such interactions in the epidemiologic literature; for instance, the interaction between BMI and menopausal status, where BMI reduces the risk of breast cancer among pre-menopausal women, while it increases the risk among post-menopausal women (70). It is unclear how often G-E or G-G interactions are going to show crossover effects; however, biologically, this extreme type of interaction is generally believed to be rare.

Statistical evaluation of interaction

<u>Interactions between two factors</u>. Table 16.3 shows different definitions of interactions between two risk factors (either a G-E or G-G interaction) and the null hypotheses that can be tested using data from case–control studies of genetic associations. Logistic regression models, including interaction terms between two or more factors, are commonly used to test multiplicative interactions:

logit (Pr(D|G, E))
= $\beta_0 + \beta_G G + \beta_E E + \beta_{GE} G*E$

OR(G|E = 0) = $\exp(\beta_G)$

OR(E|G = 0) = $\exp(\beta_E)$

OR(G,E) = $\exp(\beta_G)*\exp(\beta_E)*\exp(\beta_{GE})$

OR(E|G = 1) = $\exp(\beta_E + \beta_{GE})$

OR(G|E = 1) = $\exp(\beta G + \beta_{GE})$

Table 16.3. Odds ratio (OR) estimates for the effects of two binary factors, exposure (E) and genotype (G)

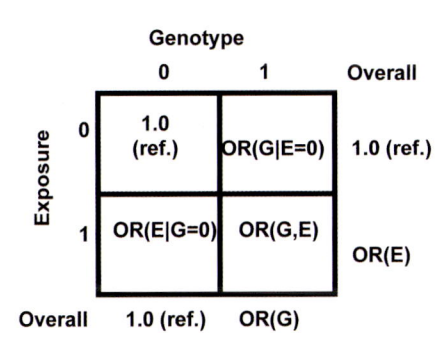

	Genotype 0	Genotype 1	Overall
Exposure 0	1.0 (ref.)	OR(G|E=0)	1.0 (ref.)
Exposure 1	OR(E|G=0)	OR(G,E)	OR(E)
Overall	1.0 (ref.)	OR(G)	

OR definitions	Symbol
Overall or marginal exposure effect	OR(E)
Exposure effect among non-susceptible	OR(E|G=0)
Exposure effect among susceptible	OR(E|G=1)
Overall or marginal genotype effect	OR(G)
Genotype effect among unexposed	OR(G|E=0)
Genotype effect among exposed	OR(G|E=1)
Joint genotype and exposure effect	OR(G,E)

Statistical model	No interaction	Interaction parameter
Multiplicative	OR(G,E) = OR(G|E=0)*OR(E|G=0) OR(E|G=0)= OR(E|G=1) OR(G|E=0)= OR(G|E=1)	ORint = OR(G,E) / [OR(G|E=0)*OR(E|G=0)] = OR(E|G=1)/OR(E|G=0) = OR(G|E=1)/OR(G|E=0)
Additive	OR(G,E) = OR(G|E=0)+OR(E|G=0)-1	OR*int= OR(G,E)/[OR(G|E=0)+OR(E|G=0)-1]

Re-parameterization to obtain stratum-specific ORs:

logit (Pr(D|G, E)) = $\beta_0 + \beta_{G|E=0} G_0 + \beta_{G|E=1} G_1 + \beta_E E$,

where G_0 and G_1 are two dummy variables defined as:

G_0 = G if E = 0; 0 otherwise

G_1 = G if E = 1; 0 otherwise

OR(G|E = 0) = exp($\beta_{G|E=0}$) = exp(βG)

OR(G|E = 1) = exp($\beta_{G|E=1}$)

Test for multiplicative interaction

$H_0: \beta_{GE} = 0$ or exp(β_{GE}) = ORint = 1

In addition to characterizing and testing differences in the relative risk of a factor across levels of another factor, interactions can also be used to increase the power to discover susceptibility loci. This is achieved by accounting for the underlying heterogeneity of the genetic risk due to G-G and G-E interactions (71,72). An omnibus test of the joint null hypothesis of no genetic main effects and interaction (e.g. Ho: $\beta_G = 0$ and $\beta_{GE} = 0$ in model above) can be used for this purpose. Thus, using notation from Table 16.3, one can specify three tests for detecting a genotype effect defined as:
- G-only test: H_0: OR(G) = 1, 1 df
- Subgroup-specific test:
H_0: OR(G|E = 1) = 1, 1 df
- Omnibus test: H_0: OR(G|E = 0) = 1 and OR(G|E = 1) = 1, 2 df

The power of the omnibus test to detect a genetic effect depends on the precision of both the main effect and the interaction parameter. Therefore, strategies that improve the efficiency of the interaction parameter can increase the power of the omnibus test. For instance, assuming independence between genetic factors, or between genetic and environmental factors, can lead to important gains in power; however, violation of these assumptions can lead to false-positive findings. Sampling strategies, such as oversampling for uncommon exposures, could interact with genetic markers and also increase the power of the omnibus test. The power advantage of the omnibus test, compared to testing for genetic main effects, is decreased by the presence of error in measuring the interacting exposure (2). The gain in power of the omnibus test with respect to the main effect test is robust to exposure measurement error. For poorly measured exposures, such as diet, there might not be much benefit in accounting for an underlying G-E interaction to detect genetic effects. The disadvantage of the omnibus test derived from the increase in degrees of freedom spent to account for the interaction and the performance of the test is that it can become poor when the degrees of freedom required to model the interaction becomes large. For example, when genetic variation is characterized by tag SNPs within a gene or region, the number of parameters in standard methods required to model interactions with other genes of exposures can become very large. Methods to address this limitation have been proposed (72). Another strategy is to perform multiple omnibus tests for a given genetic factor over a large number of other factors, such as potentially interacting SNPs or exposures. This approach can retain a gain in power, even after adjustment for multiple testing (15,72,73).

The odds ratio interaction parameter can be estimated using only data from cases, if the two interacting factors are independent in the source population of the cases and the disease is rare in the population (74). This can be easily shown if the ORint is expressed as: ORint = $OR_{EG|cases}/OR_{EG|controls}$ = 1.0, where $OR_{EG|cases}$ is the OR for the association between G and E among cases and $OR_{EG|controls}$ is the OR for the association between G and E among controls.
If the G-E independence assumption holds (i.e. $OR_{EG|controls}$ = 1.0), then ORint = $OR_{EG|cases}$.

An important limitation of this approach is that it does not allow the estimation of other important parameters estimable in case–control data, such as the stratum-specific effects and joint effects of G and E. However, when data from a case–control study is available, assuming independence between interacting factors can be used to increase the power to detect an interaction, without the limitation of the number of parameters that can be estimated (75,76). As in the case-only approach, these methods are subject to severe biases leading to detection of spurious interactions or masking of true interactions if the assumptions are violated. Two-step procedures first test for the G-E independence among the controls, and, based on the acceptance or rejection of the Ho, a second test uses the case-only (77) or case–control estimator. However, when the G-E association in the controls is modest or the sample size is small, the test in the first step might not have adequate power to reject Ho. Empirical Bayes methods have been proposed to address the trade-offs between bias and efficiency due to the independence assumption. A comparison of the different approaches mentioned above has been previously described (2,78,79).

Restricting evaluation of interactions only to loci that have previously shown some evidence of an overall association with disease, independent of the exposures of interest, substantially reduces the possible number of interactions to be evaluated. Of course, the cost of reducing complexity through this approach is that interactions that result in very weak or no overall associations with disease can easily be missed. Data mining techniques that attempt to address this problem are discussed in the next section. Variation in allele frequencies among interacting SNPs can have a strong impact on the power to detect their main effects, which can result in difficulties to replicate findings across populations (73). As larger numbers of epidemiological studies obtain comprehensive genetic data and the methods to evaluate interactions are further developed, the scientific community will be in a better position to characterize complex G-G interactions and evaluate their impact in risk characterization in the population.

Evaluation of high-order interactions

Many of the principles described above for studies on interactions between two factors, such as one G-E or G-G pair, also apply to studies of higher order interactions with three or more risk factors. One of the main difficulties in studying higher order interactions is the complexity of the models to capture relationships between many factors, since the number of possible combinations can be very large. Several data mining methods have been proposed to select models evaluating high order interactions in genetic studies. One method is the traditional stepwise regression approach, which uses statistical significance testing to decide whether higher or lower interaction terms should be kept in the model (80). Other methods include the Focused Interaction Testing Framework that uses a series of marginal and omnibus tests controlling for false discovery rates to detect susceptibly loci (81). Classification and Regression Tree (CART) is a data mining method that is increasingly being used to explore high order G-G and G-E interactions, and can be implemented using the Rpart package in R (82,83). This method uses a recursive-partitioning algorithm that splits a collection of subjects into groups based on the factor that results in the highest discrimination in the disease risk. The procedure starts with all the subjects in the study (root node) and ends with a set of final groups of subjects (nodes) with homogeneous disease risk. The problem of overfitting the data is minimized by cross-validation resulting in "pruning" or "trimming" of the tree. The main limitation of CART is that the resulting model can be very sensitive to peculiarities of the data set being used to generate it, and thus might not be replicated in independent data sets. The output models from CART can be stabilized by bagging, a procedure that combines results from a group or ensemble of trees generated by repeated bootstrap sampling of the data (84). The Random Forest procedure minimizes the correlation between the ensemble of trees by choosing a random subset of factors or exposures for growing the trees in each bootstrap replication (85). A useful feature of these ensemble approaches is that they can generate measures of variable importance of the contribution of each factor on risk, and these measures can be used as an omnibus test statistic capturing both the main effect of a factor and the interactions with other factors. P-values associated with the measures of variable importance can be generated using permutation-based resampling methods. The randomForest package available in R implements this procedure.

The main feature of logic regression (86), compared to logistic regression models and CART, is that it allows combinations of exposures using "and" and "or" operations rather than only "and." For instance, in a study evaluating the interaction between SNPs in three loci, a logic regression permits models to have similar risk of disease for subjects with the variant allele in locus 1 and variant alleles in either locus 2 or 3. This specifying "or" operator allows the flexibility of specifying biologically plausible models in which one variant resulting in disruption of a protein product only requires a variant in a class of genetic loci to determine the risk of a disease. In this model, the risk of carrying multiple variants in this class of loci is no higher than just carrying one variant. The optimal logic-tree is determined by cross-validation as in CART. Ensamples of logic trees can be generated by a Markov Chain Monte Carlo method that defines measure of variable importance (87). Logic regressions can be implemented using the LogicReg package in R.

The multifactorial dimension reduction (MDR) non-parametric method has also been proposed to evaluate high order G-G and G-E interactions (88). In contrast to tree-based methods that hierarchically build complex models, MDR reduces the dimensionality of multilocus genotype data by creating binary variables defining high-risk and low-risk groups. This method then evaluates the ability of the derived binary exposure variables to predict disease risk using cross-validation and permutation testing. The parsimony of this method is appealing; however, its performance depends on how well the simple dichotomization of high-risk and low-risk captures the underlying joint effects of multiple susceptibility loci (88,89). Information software to perform MDR analyses (90) can be found at http://chgr.mc.vanderbilt.edu/ritchielab/method.php?method=mdr, and an open-source version can be downloaded from http://www.epistasis.org/software.html.

The advantage of data mining methods is the flexibility to explore complex, high-order interactions without parametric constraints. However, this can also be a limitation since information on natural or highly plausible constraints is lost, which can result in decreased power and selection of implausible models of interaction. For instance, in studies of G-G interactions it might be reasonable to assume some sort of monotonic trend with increasing number of variant alleles on disease risk. In the case of SNP data, this would mean that the risk of carrying two variant alleles in a given locus is larger than carrying only one variant allele, irrespective of the genotype status of other loci. In logistic regression models, or other parametric models, this constrain is imposed by assuming additive or multiplicative (log-additive) effects of the variant on disease risk. Recent discoveries from GWAS studies provide support for additive or multiplicative effects of genetic markers on disease risk, although these studies might have been underpowered to detect other models, such as recessive mode of inheritance. When evaluating G-E interactions, on the other hand, it might also be reasonable to assume some sort of dose–response relationship between a continuous exposure, such as smoking dose, BMI, or dietary intake of vegetables, and disease risk. This limitation can be addressed by the FlexTree method, which allows imposing parametric constraints in binary tree-based regression models (91). An R-package to implement this method can be requested at http://www-stat.stanford.edu/~olshen/flexTree/.

In summary, data mining methods are promising tools for exploring higher-order G-G and G-E interactions. Their ability to identify reproducible interactions, however, has not yet been demonstrated. Different methods have complementary strengths, and thus the best analytical strategy might be to use a combination of methods and follow-up findings in independent data sets for replication.

Analyses of complex pathways or networks

Candidate genes are often selected from among genes involved in biochemical pathways that are known or thought to be related to the risk factors (e.g. carcinogen metabolizing genes in lung cancer, and other smoking-related cancers). However, the information on how the different genes act in the biological pathway is typically ignored in conventional analyses of the data. As information on the biochemical pathways and networks increases, thanks to the use of profiling or "omics" technologies, such as metabolomics, proteomics and transcriptomics, the interest in incorporating biochemical information in pathway/network analyses of epidemiological studies will grow. Hierarchical-Bayesian methods (92) have been proposed to integrate pathway information into the analyses, although the quantification and integration of biologic information from different sources can be very challenging and potentially limit the usefulness of these approaches. The need for methodologies for pathway analyses of complex data from molecular epidemiology studies is increasing, and novel methodologies to meet these requirements will likely be developed in the near future.

Concluding remarks

In the coming years, important advances in the understanding of the genetic contribution to complex diseases are likely to be made, facilitated by further advances in genotyping and sequencing technology. The initial discovery of markers of susceptibility in epidemiological studies is just the beginning of new areas of research. Others include:

• Identification of causal genetic variants through fine mapping and functional laboratory studies;

• Evaluation of differences in genotype frequencies and associations with disease in ethnic groups;

• Evaluation of complex interactions and joint effects of multiple susceptibility loci;

• Evaluation of G-E interactions that might facilitate the discovery and characterization of environmental risk factors for disease;

• Evaluation of heterogeneity of genetic associations by disease subtypes;

• Evaluation of the impact of susceptibility loci on individual risk prediction, and identification of population groups with low and high risk of disease; and

• Evaluation of associations between susceptibility loci with additional outcomes, such as disease recurrence, survival and response to therapy.

Therefore, this promising field of research is likely to lead to better understanding of disease etiology, enhancements in risk prediction at the individual and population levels, and improvements in treatment of disease.

Acknowledgements

Materials from the 2008 Core Genotyping Course on Genetic Analysis (National Cancer Institute, Rockville, MD), and a genetic analysis course at the 2008 Eastern North American Regional Meeting of the International Biometrics Society, were used for some sections in this chapter. We thank the course instructors for their contributions: Stephen Chanock, Meredith Yeager, Laufry Amundadottir, Sonja Berndt, Nianqing Xiao, Nick Orr, Belynda Hicks and Amy Hutchinson. We also thank Mark Sherman, Jonine Figueroa and Kelly Bolton for their comments on the chapter.

References

1. García-Closas M, Vermeulen R, Sherman ME et al. Application of biomarkers in cancer epidemiology. In: Schottenfeld D, Fraumeni JF Jr, editors. Cancer epidemiology and prevention. 3rd ed. New York (NY): Oxford University Press; 2006. p. 70–88.

2. Chatterjee N, Mukherjee B. Statistical approaches to studies of gene-gene and gene-environment interactions. In: Rebbeck TR, Ambrosone CB, Shields PG, editors. Molecular epidemiology: applications in cancer and other human diseases. New York (NY): Informa Healthcare; 2008. p. 145–168.

3. Rothman KJ, Greenland S. Modern epidemiology. Philadelphia (PA): Lippincott-Raven; 1998.

4. García-Closas M, Lan Q, Rothman N. Design considerations in molecular epidemiology. In: Rebbeck TR, Ambrosone CB, Shields PG, editors. Molecular epidemiology: applications in cancer and other human diseases. New York (NY): Informa Healthcare; 2008. p. 1–18.

5. Sham PC, Curtis D (1995). An extended transmission/disequilibrium test (TDT) for multi-allele marker loci. Ann Hum Genet, 59:323–336.doi:10.1111/j.1469-1809.1995.tb00751.x PMID:7486838

6. Wacholder S, McLaughlin JK, Silverman DT, Mandel JS (1992). Selection of controls in case-control studies. I. Principles. Am J Epidemiol, 135:1019–1028. PMID:1595688

7. Wacholder S (1991). Practical considerations in choosing between the case-cohort and nested case-control designs. Epidemiology, 2:155–158.doi:10.1097/00001648-199103000-00013 PMID:1932316

8. Bergen AW, Haque KA, Qi Y et al. (2005). Comparison of yield and genotyping performance of multiple displacement amplification and OmniPlex whole genome amplified DNA generated from multiple DNA sources. Hum Mutat, 26:262–270.doi:10.1002/humu.20213 PMID:16086324

9. Antoniou AC, Easton DF (2003). Polygenic inheritance of breast cancer: Implications for design of association studies. Genet Epidemiol, 25:190–202.doi:10.1002/gepi.10261 PMID:14557987

10. Kraft P, Cox D. Study designs for genome-wide association studies. In: Rao DC, Gu CC, editors. Genetic dissection of complex traits. 2nd ed. (Advances in genetics; vol 60). San Diego (CA): Academic Press; 2008. p. 465–504.

11. Johnson GC, Esposito L, Barratt BJ et al. (2001). Haplotype tagging for the identification of common disease genes. Nat Genet, 29:233–237.doi:10.1038/ng1001-233 PMID:11586306

12. Gabriel SB, Schaffner SF, Nguyen H et al. (2002). The structure of haplotype blocks in the human genome. Science, 296:2225–2229. doi:10.1126/science.1069424 PMID:12029063

13. Frazer KA, Ballinger DG, Cox DR et al.; International HapMap Consortium (2007). A second generation human haplotype map of over 3.1 million SNPs. Nature, 449:851–861. doi:10.1038/nature06258 PMID:17943122

14. Carlson CS, Eberle MA, Rieder MJ et al. (2004). Selecting a maximally informative set of single-nucleotide polymorphisms for association analyses using linkage disequilibrium. Am J Hum Genet, 74:106–120. doi:10.1086/381000 PMID:14681826

15. Chapman J, Clayton D (2007). One degree of freedom for dominance in indirect association studies. Genet Epidemiol, 31:261–271.doi:10.1002/gepi.20207 PMID:17266117

16. Stram DO, Haiman CA, Hirschhorn JN et al. (2003). Choosing haplotype-tagging SNPS based on unphased genotype data using a preliminary sample of unrelated subjects with an example from the Multiethnic Cohort Study. Hum Hered, 55:27–36.doi:10.1159/000071807 PMID:12890923

17. García-Closas M, Rothman N, Lubin J (1999). Misclassification in case-control studies of gene-environment interactions: assessment of bias and sample size. Cancer Epidemiol Biomarkers Prev, 8:1043–1050. PMID:10613335

18. García-Closas M, Lubin JH (1999). Power and sample size calculations in case-control studies of gene-environment interactions: comments on different approaches. Am J Epidemiol, 149:689–692. PMID:10206617

19. Pritchard JK, Przeworski M (2001). Linkage disequilibrium in humans: models and data. Am J Hum Genet, 69:1–14.doi:10.1086/321275 PMID:11410837

20. Park JH, Wacholder S, Gail MH et al. (2010). Estimation of effect size distribution from genome-wide association studies and implications for future discoveries. Nat Genet, 42:570–575.doi:10.1038/ng.610 PMID:20562874

21. Cardon LR, Bell JI (2001). Association study designs for complex diseases. Nat Rev Genet, 2:91–99.doi:10.1038/35052543 PMID:11253062

22. Gail MH, Pfeiffer RM, Wheeler W, Pee D (2008). Probability of detecting disease-associated single nucleotide polymorphisms in case-control genome-wide association studies. *Biostatistics*, 9:201–215.doi:10.1093/biostatistics/kxm032 PMID:17873152

23. Wacholder S, Chanock SJ, García-Closas M et al. (2004). Assessing the probability that a positive report is false: an approach for molecular epidemiology studies. *J Natl Cancer Inst*, 96:434–442.doi:10.1093/jnci/djh075 PMID:15026468

24. Colhoun HM, McKeigue PM, Davey Smith G (2003). Problems of reporting genetic associations with complex outcomes. *Lancet*, 361:865–872.doi:10.1016/S0140-6736(03)12715-8 PMID:12642066

25. Newton-Cheh C, Hirschhorn JN (2005). Genetic association studies of complex traits: design and analysis issues. *Mutat Res*, 573:54–69. PMID:15829237

26. Skol AD, Scott LJ, Abecasis GR, Boehnke M (2006). Joint analysis is more efficient than replication-based analysis for two-stage genome-wide association studies. *Nat Genet*, 38:209–213.doi:10.1038/ng1706 PMID:16415888

27. Kraft P, Hunter D (2005). Integrating epidemiology and genetic association: the challenge of gene-environment interaction. *Philos Trans R Soc Lond B Biol Sci*, 360:1609–1616.doi:10.1098/rstb.2005.1692 PMID:16096111

28. Ioannidis JP, Bernstein J, Boffetta P et al. (2005). A network of investigator networks in human genome epidemiology. *Am J Epidemiol*, 162:302–304.doi:10.1093/aje/kwi201 PMID:16014777

29. Ioannidis JP, Boffetta P, Little J et al. (2008). Assessment of cumulative evidence on genetic associations: interim guidelines. *Int J Epidemiol*, 37:120–132.doi:10.1093/ije/dym159 PMID:17898028

30. Rothman N, García-Closas M, Chatterjee N et al. (2010). A multi-stage genome-wide association study of bladder cancer identifies multiple susceptibility loci. *Nat Genet*, 42:978–984.doi:10.1038/ng.687 PMID:20972438

31. Chanock SJ. Principles of high-quality genotyping. In: Rebbeck TR, Ambrosone CB, Shields PG, editors. Molecular epidemiology: applications in cancer and other human diseases. New York (NY): Informa Healthcare; 2008. p. 63–80.

32. Wigginton JE, Cutler DJ, Abecasis GR (2005). A note on exact tests of Hardy-Weinberg equilibrium. *Am J Hum Genet*, 76:887–893. doi:10.1086/429864 PMID:15789306

33. Chen J, Chatterjee N (2007). Exploiting Hardy-Weinberg equilibrium for efficient screening of single SNP associations from case-control studies. *Hum Hered*, 63:196–204.doi:10.1159/000099996 PMID:17317968

34. Breslow NE, Day NE. Fundamental measure of disease occurrence and association. In: Breslow NE, Day NE, editors. The analysis of case-control studies. Lyon: IARC Scientific Publication; 1980. p. 43–81.

35. Prentice RL, Pyke R (1979). Logistic disease incidence models and case-control studies. *Biometrika*, 66:403–411 doi:10.1093/biomet/66.3.403.

36. Westfall P, Young S. Resampling based multiple testing. New York (NY): John Wiley & Sons, Inc; 1993.

37. Yen YC, Kraft P (2005). Model selection in genetic association studies. *Genet Epidemiol*, 29:289–289.

38. Pfeiffer RM, Gail MH (2003). Sample size calculations for population- and family-based case-control association studies on marker genotypes. *Genet Epidemiol*, 25:136–148. doi:10.1002/gepi.10245 PMID:12916022

39. Sasieni PD (1997). From genotypes to genes: doubling the sample size. *Biometrics*, 53:1253–1261.doi:10.2307/2533494 PMID:9423247

40. Wallenstein S, Hodge SE, Weston A (1998). Logistic regression model for analyzing extended haplotype data. *Genet Epidemiol*, 15:173–181.doi:10.1002/(SICI)1098-2272(1998)15:2<173::AID-GEPI5>3.0.CO;2-7 PMID:9554554

41. Kraft P, Chen J. Haplotype association analysis. In: Rebbeck TR, Ambrosone CB, Shields PG, editors. Molecular epidemiology: applications in cancer and other diseases. New York (NY): Informa Healthcare; 2008. p. 205–224.

42. Schaid DJ (2004). Evaluating associations of haplotypes with traits. *Genet Epidemiol*, 27:348–364.doi:10.1002/gepi.20037 PMID:15543638

43. Schaid DJ, Rowland CM, Tines DE et al. (2002). Score tests for association between traits and haplotypes when linkage phase is ambiguous. *Am J Hum Genet*, 70:425–434. doi:10.1086/338688 PMID:11791212

44. Purcell S, Neale B, Todd-Brown K et al. (2007). PLINK: a tool set for whole-genome association and population-based linkage analyses. *Am J Hum Genet*, 81:559–575. doi:10.1086/519795 PMID:17701901

45. Chaturvedi N (2001). Ethnicity as an epidemiological determinant–crudely racist or crucially important? *Int J Epidemiol*, 30:925–927.doi:10.1093/ije/30.5.925 PMID:11689494

46. Wacholder S, Rothman N, Caporaso NE (2000). Population stratification in epidemiologic studies of common genetic variants and cancer: quantification of bias. *J Natl Cancer Inst*, 92:1151–1158.doi:10.1093/jnci/92.14.1151 PMID:10904088

47. Thomas DC, Witte JS (2002). Point: population stratification: a problem for case-control studies of candidate-gene associations? *Cancer Epidemiol Biomarkers Prev*, 11:505–512. PMID:12050090

48. Wacholder S, Rothman N, Caporaso NE (2002). Counterpoint: bias from population stratification is not a major threat to the validity of conclusions from epidemiological studies of common polymorphisms and cancer. *Cancer Epidemiol Biomarkers Prev*, 11:513–520. PMID:12050091

49. Pritchard JK, Stephens M, Donnelly P (2000). Inference of population structure using multilocus genotype data. *Genetics*, 155:945–959. PMID:10835412

50. Price AL, Patterson NJ, Plenge RM et al. (2006). Principal components analysis corrects for stratification in genome-wide association studies. *Nat Genet*, 38:904–909.doi:10.1038/ng1847 PMID:16862161

51. Sun L, Wilder K, McPeek MS (2002). Enhanced pedigree error detection. *Hum Hered*, 54:99–110.doi:10.1159/000067666 PMID:12566741

52. Patterson N, Price AL, Reich D (2006). Population structure and eigenanalysis. *PLoS Genet*, 2:e190.doi:10.1371/journal.pgen.0020190 PMID:17194218

53. Marchini J, Howie B, Myers S et al. (2007). A new multipoint method for genome-wide association studies by imputation of genotypes. *Nat Genet*, 39:906–913.doi:10.1038/ng2088 PMID:17572673

54. Li Y, Mach Abecasis GR (2006). 1.0: Rapid haplotype reconstruction and missing genotype inference. *Am J Hum Genet*, S79:2290.

55. Servin B, Stephens M (2007). Imputation-based analysis of association studies: candidate regions and quantitative traits. *PLoS Genet*, 3:e114.doi:10.1371/journal.pgen.0030114 PMID:17676998

56. Lewinger JP, Conti DV, Baurley JW et al. (2007). Hierarchical Bayes prioritization of marker associations from a genome-wide association scan for further investigation. *Genet Epidemiol*, 31:871–882.doi:10.1002/gepi.20248 PMID:17654612

57. Han J, Kraft P, Nan H et al. (2008). A genome-wide association study identifies novel alleles associated with hair color and skin pigmentation. *PLoS Genet*, 4:e1000074.doi:10.1371/journal.pgen.1000074 PMID:18483556

58. Kraft P (2007). Analyses of genome-wide association scans for additional outcomes. *Epidemiology*, 18:838.doi:10.1097/EDE.0b013e318154c7e2 PMID:18049198

59. Richardson DB, Rzehak P, Klenk J, Weiland SK (2007). Analyses of case-control data for additional outcomes. *Epidemiology*, 18:441–445.doi:10.1097/EDE.0b013e318060d25c PMID:17473707

60. Witte JS, Greenland S, Kim LL (1998). Software for hierarchical modeling of epidemiologic data. *Epidemiology*, 9:563–566.doi:10.1097/00001648-199809000-00016 PMID:9730038

61. Chen GK, Witte JS (2007). Enriching the analysis of genomewide association studies with hierarchical modeling. *Am J Hum Genet*, 81:397–404.doi:10.1086/519794 PMID:17668389

62. Thompson WD (1991). Effect modification and the limits of biological inference from epidemiologic data. *J Clin Epidemiol*, 44:221–232.doi:10.1016/0895-4356(91)90033-6 PMID:1999681

63. Yang QH, Khoury MJ (1997). Evolving methods in genetic epidemiology. III. Gene-environment interaction in epidemiologic research. *Epidemiol Rev*, 19:33–43. PMID:9360900

64. Walter SD, Holford TR (1978). Additive, multiplicative, and other models for disease risks. *Am J Epidemiol*, 108:341–346. PMID:727202

65. Clayton D, McKeigue PM (2001). Epidemiological methods for studying genes and environmental factors in complex diseases. *Lancet*, 358:1356–1360.doi:10.1016/S0140-6736(01)06418-2 PMID:11684236

66. Ottman R (1996). Gene-environment interaction: definitions and study designs. *Prev Med*, 25:764–770.doi:10.1006/pmed.1996.0117 PMID:8936580

67. Khoury MJ, Beaty TH, Cohen BL, editors. Fundamentals of genetic epidemiology. New York (NY): Oxford University Press; 1993.

68. García-Closas M, Malats N, Silverman D et al. (2005). NAT2 slow acetylation, GSTM1 null genotype, and risk of bladder cancer: results from the Spanish Bladder Cancer Study and meta-analyses. *Lancet*, 366:649–659.doi:10.1016/S0140-6736(05)67137-1 PMID:16112301

69. Gail M, Simon R (1985). Testing for qualitative interactions between treatment effects and patient subsets. *Biometrics*, 41:361–372.doi:10.2307/2530862 PMID:4027319

70. Colditz GA, Baer HJ, Tamimi RM. Breast cancer. In: Schottenfeld D, Fraumeni JF Jr, editors. Cancer epidemiology and prevention. New York (NY): Oxford University Press; 2006. p. 995–1012.

71. Kraft P, Yen YC, Stram DO et al. (2007). Exploiting gene-environment interaction to detect genetic associations. *Hum Hered*, 63:111–119.doi:10.1159/000099183 PMID:17283440

72. Chatterjee N, Kalaylioglu Z, Moslehi R et al. (2006). Powerful multilocus tests of genetic association in the presence of gene-gene and gene-environment interactions. *Am J Hum Genet*, 79:1002–1016.doi:10.1086/509704 PMID:17186459

73. Marchini J, Donnelly P, Cardon LR (2005). Genome-wide strategies for detecting multiple loci that influence complex diseases. *Nat Genet*, 37:413–417.doi:10.1038/ng1537 PMID:15793588

74. Piegorsch WW (1994). Statistical models for genetic susceptibility in toxicological and epidemiological investigations. *Environ Health Perspect*, 102 Suppl 1;77–82. PMID:8187729

75. Umbach DM, Weinberg CR (1997). Designing and analysing case-control studies to exploit independence of genotype and exposure. *Stat Med*, 16:1731–1743.doi:10.1002/(SICI)1097-0258(19970815)16:15<1731::AID-SIM595>3.0.CO;2-S PMID:9265696

76. Chatterjee N, Carroll R (2005). Semiparametric maximum likelihood estimation exploiting gene-environment independence in case-control studies. *Biometrika*, 92:399–418 doi:10.1093/biomet/92.2.399.

77. García-Closas M, Thompson WD, Robins JM (1998). Differential misclassification and the assessment of gene-environment interactions in case-control studies. *Am J Epidemiol*, 147:426–433. PMID:9525528

78. Mukherjee B, Ahn J, Gruber SB et al. (2008). Tests for gene-environment interaction from case-control data: a novel study of type I error, power and designs. *Genet Epidemiol*, 32:615–626.doi:10.1002/gepi.20337 PMID:18473390

79. Mukherjee B, Chatterjee N (2008). Exploiting gene-environment independence for analysis of case-control studies: an empirical Bayes-type shrinkage estimator to trade-off between bias and efficiency. *Biometrics*, 64:685–694.doi:10.1111/j.1541-0420.2007.00953.x PMID:18162111

80. Millstein J, Conti DV, Gilliland FD, Gauderman WJ (2006). A testing framework for identifying susceptibility genes in the presence of epistasis. *Am J Hum Genet*, 78:15–27. doi:10.1086/498850 PMID:16385446

81. Storey J (2002). A direct approach to false discovery rates. *J Royal Stat Soc*, 64:479–498 doi:10.1111/1467-9868.00346.

82. Breiman L, Friedman JH, Olshen RA, Stone CJ. Classification and regression trees. Pacific Grove (CA): Wadsworth & Brooks/Cole Advanced Books & Software; 1984.

83. Zhang HP, Bonney G (2000). Use of classification trees for association studies. *Genet Epidemiol*, 19:323–332. doi:10.1002/1098-2272(200012)19:4<323::AID-GEPI4>3.0.CO;2-5 PMID:11108642

84. Breiman L (1996). Bagging predictors. *Mach Learn*, 24:123–140 doi:10.1007/BF00058655.

85. Breiman L (2001). Random forests. *Mach Learn*, 45:5–32 doi:10.1023/A:1010933404324.

86. Ruczinski I, Kooperberg C, LeBlanc M (2003). Logic regression. *J Graph Comput Stat*, 12:475–511 doi:10.1198/1061860032238.

87. Kooperberg C, Ruczinski I (2005). Identifying interacting SNPs using Monte Carlo logic regression. *Genet Epidemiol*, 28:157–170.doi:10.1002/gepi.20042 PMID:15532037

88. Ritchie MD, Hahn LW, Roodi N et al. (2001). Multifactor-dimensionality reduction reveals high-order interactions among estrogen-metabolism genes in sporadic breast cancer. *Am J Hum Genet*, 69:138–147. doi:10.1086/321276 PMID:11404819

89. Ritchie MD, Hahn LW, Moore JH (2003). Power of multifactor dimensionality reduction for detecting gene-gene interactions in the presence of genotyping error, missing data, phenocopy, and genetic heterogeneity. *Genet Epidemiol*, 24:150–157.doi:10.1002/gepi.10218 PMID:12548676

90. Hahn LW, Ritchie MD, Moore JH (2003). Multifactor dimensionality reduction software for detecting gene-gene and gene-environment interactions. *Bioinformatics*, 19:376–382.doi:10.1093/bioinformatics/btf869 PMID:12584123

91. Huang J, Lin A, Narasimhan B et al. (2004). Tree-structured supervised learning and the genetics of hypertension. *Proc Natl Acad Sci USA*, 101:10529–10534.doi:10.1073/pnas.0403794101 PMID:15249660

92. Conti DV, Cortessis V, Molitor J, Thomas DC (2003). Bayesian modeling of complex metabolic pathways. *Hum Hered*, 56:83–93. doi:10.1159/000073736 PMID:14614242

93. Hunter DJ, Kraft P, Jacobs KB et al. (2007). A genome-wide association study identifies alleles in FGFR2 associated with risk of sporadic postmenopausal breast cancer. *Nat Genet*, 39:870–874.doi:10.1038/ng2075 PMID:17529973

UNIT 4.
INTEGRATION OF BIOMARKERS INTO EPIDEMIOLOGY STUDY DESIGNS

CHAPTER 17.

Biomarkers in clinical medicine

Xiao-He Chen, Shuwen Huang, and David Kerr

Summary

Biomarkers have been used in clinical medicine for decades. With the rise of genomics and other advances in molecular biology, biomarker studies have entered a whole new era and hold promise for early diagnosis and effective treatment of many diseases. A biomarker is a characteristic that is objectively measured and evaluated as an indicator of normal biological processes, pathogenic processes or pharmacologic responses to a therapeutic intervention (1). They can be classified into five categories based on their application in different disease stages: 1) antecedent biomarkers to identify the risk of developing an illness, 2) screening biomarkers to screen for subclinical disease, 3) diagnostic biomarkers to recognize overt disease, 4) staging biomarkers to categorise disease severity, and 5) prognostic biomarkers to predict future disease course, including recurrence, response to therapy, and monitoring efficacy of therapy (1). Biomarkers can indicate a variety of health or disease characteristics, including the level or type of exposure to an environmental factor, genetic susceptibility, genetic responses to environmental exposures, markers of subclinical or clinical disease, or indicators of response to therapy. This chapter will focus on how these biomarkers have been used in preventive medicine, diagnostics, therapeutics and prognostics, as well as public health and their current status in clinical practice.

Introduction

Health sciences have been experiencing a shift from population-based approaches to individualized practice. The focus on individuals could make public health strategies more effective by allowing practitioners to direct resources towards those with the greatest need. However, the success of these efforts will largely depend on the continued identification of biomarkers that reflect the individual's health status and risk at key time points, and successful integration of these biomarkers into medical practice. To be clinically useful, tests for biomarkers must have high predictive accuracy, and be easily measurable and reproducible, minimally invasive, and acceptable to patients and physicians (2). Once a proposed biomarker has been validated, it can be used to assess disease risk in a general population, confirm diagnosis of disease in an individual patient, and tailor an individual's treatment (choice of drug treatment or administration regimes). In evaluating potential drug therapies, a biomarker may be used as a surrogate for a natural endpoint, such as survival or irreversible morbidity. If a treatment alters the biomarker, which has a direct connection to improved health, the biomarker serves as a surrogate endpoint for evaluating clinical benefit. More recently, with rapid advances in the molecular approaches to biology, genetics, biochemistry and medicine, and in particular with the rise of genomics, transcriptomics, proteomics and metabolomics, molecular biomarkers appear to hold the promise of transforming medical practice into personalized medicine—the right treatment at the right dose for the right person at the right time for the right outcome.

Context and public health significance

Clinical medicine covers disease prevention, diagnosis and treatment. Biomarkers play a critical role in all these aspects. There are three major types of biomarkers: biomarkers of exposure, effect and susceptibility. A biomarker of exposure is an exogenous chemical or its metabolite(s), or the product of an interaction between a xenobiotic agent and some target molecule or cell that is measured in a compartment within an organism. Specific markers of exposure include the presence of a xenobiotic compound or its metabolites in body tissues or fluids and in excretory products. For example, blood lead concentration has been used as a marker for lead exposure; saliva cotinine (a metabolite of nicotine) level has been used as a marker in investigating adolescents' cigarette consumption. A biomarker of effect is a measurable alteration of an endogenous factor that is shown to be linked with impairment or disease resulting from exposure to an exogenous agent. For example, the alteration in pulmonary function tests in children after exposure to environmental tobacco smoke is a biomarker of effect (3). Somatic mutations have been used as biomarkers of effect after exposure to carcinogens. A biomarker of susceptibility indicates individual factors that can affect response to environmental agents. These reflect variations between individuals in genetic structure, some of which make the individual more susceptible to health effects from environmental exposures (4). For example, skin cancer is related to excessive sun exposure, but not everyone develops skin cancer even with the same amount of exposure. Three recent studies revealed that genetic variants associated with three sections of genes were found to be linked with increased risk of skin cancer: 1) the variant of the TYR gene that encodes a R402Q amino acid substitution, previously shown to affect eye colour and tanning response, was associated with increased risk of developing cutaneous melanoma (CM) and basal cell carcinoma (BCC); 2) variations in a haplotype (set of closely associated genes) near the ASIP gene, known to affect pigmentation traits, conferred significant risk of CM and BCC; and 3) an eye colour variant in gene TYRP1 was also associated with risk of CM (5–7). The relationship between these biomarkers and their relationship with clinical medicine are illustrated in Figure 17.1.

There are two layers of exposure and effect biomarkers. The first represents hazardous exposures to a healthy human body that could cause negative biological effects (e.g. functional changes, somatic mutations) and eventually cause disease. Another layer indicates treatment exposures to a diseased human body that could induce positive biological effects and lead to the improvement of conditions or to the complete recovery from disease. Susceptibility biomarkers

Figure 17.1. Simplified flowchart of classes of biomarkers (indicated by boxes) representing a continuum of changes. Solid arrows indicate progression, if it occurs, to the next class of marker. Dashed arrows indicate that individual biomarker influences and/or indicates the rates of progression

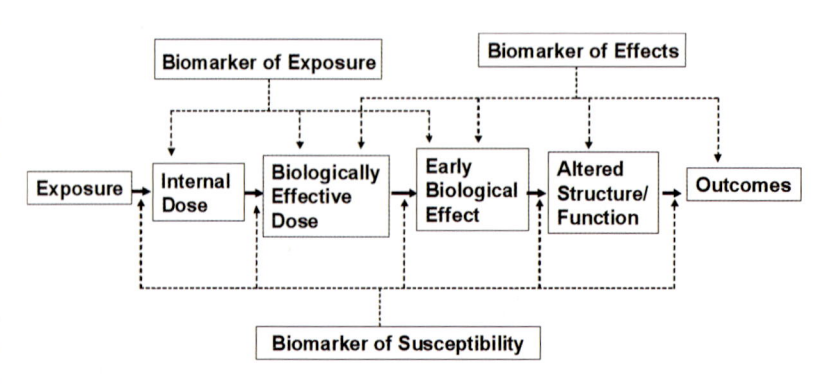

Source: Adapted from (106)

are present in every step of the process. For example, some individuals exposed to air pollutants show severe biological effects and manifest disease symptoms, while others experience little or no effect. The same discrepancy appears with drug treatment. While some patients benefit and are cured, others show no effect from treatment or develop severe side-effects or die. In clinical medicine, the first layer is more related to disease prevention and diagnosis, while the second layer is more relevant to disease treatment and recovery.

Biomarkers in preventive medicine

Preventive medicine aims to promote and preserve health and longevity in individuals and populations, use epidemiological approaches to define high-risk groups, prevent and limit disease and injury, facilitate early diagnosis through screening and education, enhance quality of the health care system and improve quality of life. To realize these aims, medical practitioners need the proper tools to facilitate decision-making and effect evaluation; biomarkers play an important role in these goals.

Exposure biomarkers have been used in the workplace for many years to identify exposed individuals. For example, macromolecule adducts and mutagenicity in urine have been successfully applied to identify workers exposed to carcinogens and as indicators of changes of exposure. Biomarkers of renal effects of cadmium, lead effects on haemoglobin synthesis and organophosphate effects on cholinesterase activities have been validated and are widely used in routine monitoring activities (8).

Antecedent and screening biomarkers have been used in preventive medicine for several decades to: screen before birth for genetic disorders, such as Down syndrome; screen newborn babies for genetic diseases, such as phenylketonuria (PKU) (9); check whether an individual is a carrier for a recessive disorder (where abnormal genes must be inherited from both parents to lead to the condition), such as cystic fibrosis; and indicate whether someone with a family history of a late-onset disease, such as Huntington's, is likely to develop the disease. These tests are aimed largely at single-gene disorders that have Mendelian patterns of inheritance. The identification of genetic variants responsible for diseas, in these single-gene disorders can lead directly to clinically helpful and reasonably accurate predictions and diagnosis of disease. Early diagnosis and proper treatment can make the difference between lifelong impairment and healthy development.

Common, complex diseases such as cancer, heart disease and diabetes contribute to the major disease burden and mortality both in developed and developing countries. These common diseases are caused by genetic and environmental factors (e.g. lifestyle and diet, and the interaction between them). Therefore, it is very difficult to define a single biomarker that could identify the risk of developing a particular disease. Although there are some rare subtypes of common diseases, such as breast and colorectal cancer, with a clear hereditary pattern, biomarkers for a single or several defective genes could indicate a lifetime risk of developing these cancers (e.g. overexpression of HER2/neu oncogene and loss of function mutations in *BRCA1* and *BRCA2* tumour suppressor genes for breast cancer (10–12), and activating mutation in Ras oncogene and loss of function mutations in APC and *p53* tumour suppressor genes for colon cancer) (12). Subtypes of these cancers and most other common diseases are less deterministic; even apparently simple Mendelian disorders may prove to have widely variable clinical phenotypes. For example, thalassaemia, an apparently simple genetic disease, has substantial complexities (13). Individuals with exactly the same globin mutations may suffer either from a severe life-threatening disorder or be relatively unaffected. Despite this, great efforts have been made towards simultaneous, systematic analysis of larger numbers of biomarkers for disease prediction, although these approaches are more suited to research than routine diagnostic activity. Biomarkers may help predict those individuals more susceptible to common disorders, so that specific attention can be directed towards them (e.g. enrolment in a screening programme). However, translating these biomarkers into clinical medicine to help prevent people having these common diseases still has a long way to go. As Kofi Annan, the former Secretary General of the United Nations, said, "We are under no illusion that preventive strategies will be easy to implement. For a start, the costs of prevention have to be paid in the present, while its benefits lie in the distant future. And the benefits are not tangible—when prevention succeeds, nothing happens. Taking such a political risk when there are few obvious rewards requires conviction and considerable vision." (14).

Biomarkers in diagnostics

Biomarkers have been used in disease diagnosis for over a century, beginning when the ABO blood group system was first discovered and used to detect ABO haemolytic disease of the newborn (HDN) and transfusion reactions. The four basic ABO phenotypes are O, A, B and AB. After it was found that blood group A's red blood cells (RBCs) reacted differently to a particular antibody (later called anti-A1), the blood group was divided into two phenotypes, A1 and A2. RBCs with the A1 phenotype react with anti-A1 and account for about 80% of blood type A. RBCs with the A2 phenotype do not react with anti-A1 and makeup about 20% of blood type A. HDN, caused by ABO antibodies, occurs almost exclusively in infants of blood group A or B who are born to group O mothers (15). This is because the anti-A and anti-B formed in group O individuals tends to be of the IgG type (and therefore can cross the placenta), whereas the anti-A and anti-B found in the serum of group B and A individuals, respectively, tends to be of the IgM type. Although uncommon, cases of HDN have been reported in infants born to mothers with blood group A2 (16) and blood group B (17). The most common cause of death from a blood transfusion is clerical error, in which an incompatible type of ABO blood is transfused. If a recipient who has blood group O is transfused with non-group O RBCs, the naturally occurring anti-A and anti-B in the recipient's serum binds to their corresponding antigens on the transfused RBCs. These antibodies fix complement and cause rapid intravascular haemolysis, triggering an acute haemolytic transfusion reaction that can cause disseminated intravascular coagulation, shock, acute renal failure and death. Routine biomarker tests can confirm the diagnosis.

Another important use of biomarkers in clinical medicine is the early detection and diagnosis of chromosome and single-gene disorders. Both cytogenetic and molecular genetic biomarkers have been used to accomplish this. Conditions caused by a change in the number (e.g. aneuploidy) or structure of chromosomes (e.g. translocation, inversion, deletion, and duplication) are known as chromosome disorders. Biomarkers used in the chromosome analysis developed in 1956 soon led to the discovery that several previously described conditions were due to an abnormality in chromosome number. For example, in Turner syndrome, only one intact X chromosome is present (45, X); all or part of the second X is deleted. Patients with Down syndrome have an extra chromosome 21 (47, XX/XY, +21). Patau syndrome is the result of trisomy 13, while trisomy 18 causes Edwards syndrome. The biomarker test in this case is assessment of chromosome numbers.

Microdeletion/microduplication syndromes are a group of chromosome disorders that could be detected by biomarker copy number variation (CNV). "Micro" represents submicroscopic, meaning that these deletions, normally smaller than 3Mb, cannot be detected using a microscope. New technologies, especially array comparative genomic hybridization (array-CGH), enabled many malformations and syndromes to be recognized. Figure 17.2 shows recently detected or confirmed microdeletions/duplications collected in DECIPHER (DatabasE of Chromosomal Imbalance and Phenotype in Humans using Ensembl Resources) (https://decipher.sanger.ac.uk/). Applications of new biomarkers in these disorders have generated particular interest. For example, most Angelman and Prader-Willi syndromes are related to microdeletion involving the proximal part of the long arm of chromosome 15q (15q11–12). It is now known that if the deletion occurs *de novo* on the paternally inherited number 15 chromosome, the child will have Prader-Willi syndrome; a deletion occurring at the same region on the maternally inherited number 15 chromosome causes Angelman syndrome. Non-deletion cases also exist and are often due to uniparental disomy (i.e. both homologues of a chromosome pair are inherited from only one of the parents), with both number 15 chromosomes being paternal in origin in Angelman syndrome and maternal in origin in Prader-Willi syndrome. This "parent of origin" effect is referred to as genomic imprinting and methylation of DNA. Here, CNV and mythelation biomarkers, coupled with clinical observations, have helped identify new underlying genetic mechanisms (18).

The most widely used biomarkers identified during the last few decades are for the diagnosis of single-gene disorders. More than 10 000 human diseases are believed to be caused by defects in single genes, affecting 1–2% of the population (18). The disease can be relatively trivial in its effects (e.g. colour blindness), or lethal like Tay-Sachs (a fatal inherited disease of the central nervous system; babies with Tay-Sachs lack an enzyme called hexosaminidase A (hex A) which is necessary for breaking down certain fatty substances in brain and nerve cells). Other disorders, though harmful to those afflicted with them, appear to offer some advantage to carriers. For example, carriers of sickle cell anaemia and thalassemia appear to have enhanced resistance to malaria. Some other examples of single-gene diseases are cystic fibrosis, Marfan syndrome,

Table 17.1. US FDA-published list of valid genomic biomarkers, approved drug labels, and test recommendation (1 = test required, 2 = test recommended, 3 = information only)

Biomarker	Representative label (Label context)	Test	Drug	Other drugs associated with this biomarker
Her2/neu over-expression	Over-expression of Her2/neu necessary for selection of patients appropriate for drug therapy (breast cancer)	1	Trastuzumab (Herceptin®)	
EGFR expression with alternate context	Epidermal growth factor receptor presence or absence (colorectal cancer)	1	Cetuximab (Erbitux®)	Gefitinib
UGT1A1 variants	UGT1A1 mutation patients, exposure to drug and hence their susceptibility to toxicity (colon-rectum cancer)	2	Irinotecan (Camptosar®)	
TPMT variants	Increased risk of myelotoxicity associated to thiopurine methyltransferase deficiency or lower activity	2	Azathioprine (Imuran®)	
Protein C deficiencies (hereditary or acquired)	Hereditary or acquired deficiencies of protein C or its cofactor protein S	2	Warfarin (Coumadin®)	
C-KIT expression	Gastrointestinal stromal tumour *c-Kit* expression	3	Imatinib mesylate (Glivec®)	
CYP2C19 variants	CYP2C19 variants (poor metabolizers PM and extensive metabolizers EM) with genetic defect leads to change in drug exposure	3	Voriconazole (Vfend®)	Omeprazole, Pantoprazole, Esomeprazole, Diazepam, Nelfinavir, Rabeprazole
CYP2C9 variants	CYP2C9 variants PM and EM genotypes and drug exposure	3	Celecoxib (Celebrex®)	Warfarin
CYP2D6 variants	CYP2D6 variants PM and EM genotypes and drug exposure	3	Atomoxetine (Strattera®)	Venlafaxine, Risperidone Tiotropium bromide inhalation, Tamoxifen, Timolol Maleate
CYP2D6 with alternate context	CYP2D6 PM and EM variants and drug exposure and risk	3	Fluoxetine HCl (Prozac®)	Fluoxetine HCl and Olanzapine, Cevimeline hydrochloride, Tolterodine, Terbinafine, Tramadol + Acetamophen, Clozapine Aaripipraxole, Metoprolol, Propanolol, Carvedilol, Propafenone, Thioridazine, Protriptyline
DPD deficiency	Severe toxicity (stomatitis, diarrhoea, neutropenia and neurotoxicity) associated to deficiency of dihydropyrimidine dehydrogenase	3	Capecitabine (Xeloda®)	Fluorouracil cream, Fluorouracil Topical Solution & Cream
EGFR expression	Epidermal growth factor receptor presence or absence (NSCLC, pancreas cancer)	3	Erlotinib (Tarceva®)	
EGFR expression with alternate context	Epidermal growth factor receptor presence or absence (SCCHN: squamous cell carcinoma of head and neck)	3	Cetuximab (Erbitux®)	Gefitinib
G6PD deficiency	G6PD deficiency and risk for haemolysis	3	Rasburicase (Elitek®)	Dapsone
G6PD deficiency with alternate context	G6PD deficiency (or NADH methemoglobin reductase deficiency) and risk for haemolytic reactions	3	Primaquine (Primaquine®)	Chloroquine
NAT variants	*N*-Acetyltransferase slow and fast acetylators and toxicity	3	Rifampin isoniazid (Rifater® and pyrazinamide)	Isosorbide dinitrate and hydralazine hydrochloride
Philadelphia chromosome deficiency	Philadelphia (Ph1) chromosome presence and efficacy-Busulfan is less effective in patients with CML lacking the Ph1 chromosome	3		Busulfan
UCD deficiency disorders	Valproate therapy and urea cycle disorders interaction	3	Valproic acid (Depakene®)	Sodium phenylacetate and sodium benzoate
VKORC1 variants	Polymorphisms of vitamin K epoxide reductase complex subunit identify warfarin-sensitive patients who require a lower dose of the drug	3	Warfarin (Coumadin®)	
PML/RAR alpha gene expression (retinoic acid receptor responder and non-responders)	PML/RAR (alpha) fusion gene presence	3	Tretinoin (Avita®, Renova®, Retin-A®)	Arsenic oxide

Source: (25)

rays. It measures glucose uptake by tumours using a radioactive form of fluorine incorporated in a sugar molecule. Tissues that accumulate radioactive glucose are visible through positron emission tomography. It is believed that FDG-PET could become a tool for gauging a cancer patient's response to chemotherapy or radiation by accurately measuring tumour metabolism. Physicians will quickly know whether or not the tumour is responding to therapy and when to switch therapies to provide the best chance for curing or managing the cancer. Cervical tumour uptake of F-18 FDG, measured as the maximal standardized uptake value (SUV_{max}) by PET, and its association with treatment response and prognosis in patients with cervical cancer were evaluated. It was found that a higher SUV_{max} was associated with an increased risk of lymph node metastasis at diagnosis (P = 0.0027) (36).

Biomarkers can help reduce adverse drug reactions. Studies estimate that over 2 million serious adverse drug reactions (ADRs) occur annually in the United States, and as many as 137 000 deaths are caused by ADRs (37). Using biomarkers to indicate if the patient is suitable for treatment with certain drugs, and what dose is appropriate for the patient, could prevent some of these deaths. Any given drug can be therapeutic to some individuals and ineffective to others, and likewise some individuals suffer from adverse drug effects whereas others experience drug resistance. Often, distinct molecular mechanisms underlie therapeutic and adverse effects. Studies have linked differences in drug responses to differences in genes that code for the production of drug-metabolizing enzymes, drug transporters or drug targets (38–40). These genetic variations could be used as biomarkers to direct a physician's drug choice for a patient and prevent adverse drug reactions. For example, the anticoagulant drug warfarin, marketed as Coumadin® by Bristol-Myers Squibb, is used to prevent potentially fatal clots in blood vessels. Approximately 2 million people start warfarin therapy each year to prevent blood clots, heart attacks and stroke. However, too much or too little of the drug can cause serious, life-threatening bleeding or blood clots. According to the FDA's adverse events reporting database, complications from warfarin are the second (just after that from insulin) most common reason for patients to go to the emergency department. Variability in the response to warfarin has been linked to mutations in two genes: CYP2C9 and VKORC1. Clinical studies have shown that patients with variations in these two genes may need a lower warfarin dose than patients without those variations. Recently, the FDA cleared the first test to detect gene variants in patients that are sensitive to the anticoagulant warfarin.

Thiopurine methyltransferase (TPMT) is another example of a biomarker being applied to drug treatment. TPMT is responsible for inactivating purine drugs used for treating acute lymphoblastic leukaemia (ALL) and other diseases (41). Variations in the *TPMT* gene can cause changes in TPMT enzymatic activity and thus drug metabolism. One in 300 patients has TPMT deficiency. In these patients, the normal dose of purine causes an accumulation of active compound, which may lead to a potentially fatal bone marrow reaction resulting in leucopenia, an abnormal lowering of the white blood cell counts. Therefore, if TPMT deficiency is detected, the dose is lowered by 10–15% to keep the systemic level of the drug comparable to that in patients with normal TPMT who have been given a standard dose of the drug (http://www.personalizedmedicinecoalition.org/communications/pmc_pub_11_06.php). (More information on genomic biomarkers for drug usage can be found at http://www.fda.gov/Drugs/ScienceResearch/ResearchAreas/Pharmacogenetics/ucm083378.htm.)

Biomarkers in clinical trials and drug discovery

With rapid advances in the molecular approaches to biology, genetics, biochemistry and medicine, a significant number of new drugs and treatments have been developed. But most of these discoveries still remain in the research field. Efficiently and effectively translating these discoveries into clinical practice is complex and involves the integration of scientific rationale and the regulatory process. Various models depict translational research as a process occurring in two stages (42–44). The first (sometimes referred to as type 1 (T1) translation) uses the findings from basic research, including preclinical studies, to inform the development and testing of an intervention in clinical trials, such as Phase I-III clinical trials. The second (type 2 (T2) translation) involves the translation of findings from clinical research into clinical and public health practice and policy (42,43). This section discusses how biomarkers have been used in clinical trial and drug development, and what changes can be brought about by biomarker application in these fields in clinical medicine.

A clinical trial is defined as a prospective study comparing the effect and value of intervention(s) against a control in human beings (45). Clinical trials are commonly classified into four phases. Phase I trials select drug dose, schedule and associated toxicities. Phase II trials determine the degree of efficacy and govern admission to Phase III testing. Phase III trials compare a new treatment against the existing standard treatment; if it gives better results, it may become the new gold standard. Phase IV trials are carried out after a drug has been licensed. Information is collected about side-effects, safety and the long-term risks and benefits of a drug (http://www.cancerhelp.org.uk/help/default.asp?page=52). For example, the conventional drug development process will normally proceed through all four stages over many years. A new drug is estimated to cost between US$800 and US$1700 million, and is expected to take anywhere between 7–12 years to be approved and launched. The complexity and duration of clinical trials are determined by the use of a long-term clinical endpoint (e.g. clinical progression, survival) to assess the clinical benefit of a new treatment or drug. Biomarkers have the potential to be used in clinical trials as validated surrogate endpoints to indicate drug efficacy or toxicity, or to make a "go/no-go" decision.

Biomarkers can be influential in every phase of drug development, from drug discovery and preclinical evaluations, through each phase of clinical trials and into post-marketing studies (Figure 17.3). Biomarkers have been used to identify and justify targets for therapy. For example, 95% of CML patients possess a mutation called Philadelphia chromosome, a translocation between chromosome 9 and chromosome 22 that produces a specific tyrosine kinase enzyme, BCR-ABL. Novartis' Gleevec® (imatinib mesylate) specifically targets this enzyme by attaching to the cancerous cells and stopping them from growing and spreading. But in some patients, the cancer cells mutate just enough to be resistant to imatinib. Bristol-Myers Squibb produced another drug, dasatinib (BMS-354 825), that inhibits five tyrosine kinase proteins, including BCR-ABL and SRC (a protein that may play a role in imatinib resistance). This new agent shows very good response in those who are resistant to imatinib.

Biomarkers play an important role in preclinical studies. Critical proof-of-concept studies typically involve appropriate animal models. In cancer studies, for example, the complexities of modeling human cancer in experimental systems are well known and have impeded cancer drug development over the years (46). Genetically engineered cancer models have improved the situation, but most current models have limited capability for predicting clinical effects. Models that feature biomarker properties, comparable with those seen in patient populations, will enhance their utility as predictive models. Specific effects on biomarkers in such models can, in turn, provide proof-of-concept for therapeutic

Figure 17.3. Roles of biomarkers (grey) and their associated technologies (blue) along the different phases of drug development and post-launch (yellow)

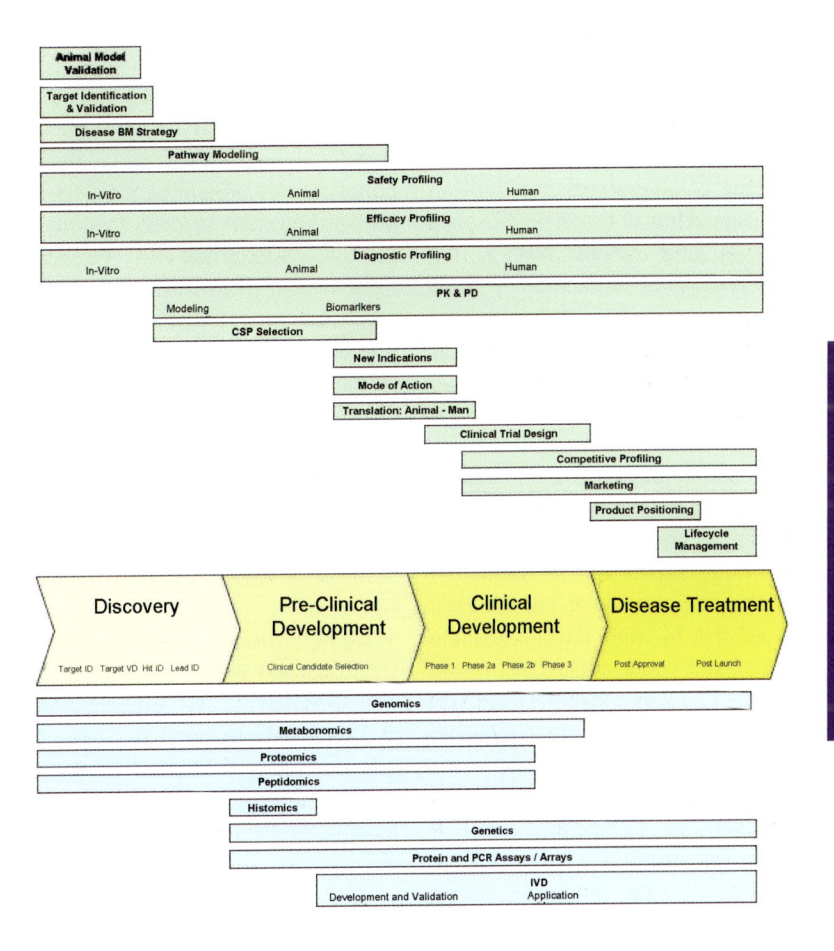

Source: (25)

approaches (46). For instance, in a preclinical study of dasatinib, biomarkers Phospho-BCR-ABL/phospho-CrkL were investigated in K562 human CML xenografts grown s.c. in severe combined immunodeficient mice. Results showed that following a single oral administration of dasatinib at a preclinical efficacious dose of 1.25 or 2.5 mg/kg, tumoural phospho-BCR-ABL/phospho-CrkL were maximally inhibited at 3 hours and recovered to basal levels by 24 hours. The time course and extent of inhibition correlated with the plasma levels of dasatinib in mice. Pharmacokinetic/pharmacodynamic modelling predicted that the plasma concentration of dasatinib required to inhibit 90% of phospho-BCR-ABL *in vivo* was 10.9 ng/mL in mice and 14.6 ng/mL in humans, which is within the range of concentrations achieved in CML patients who responded to dasatinib treatment in the clinic (47).

Use of biomarkers can shorten the clinical trial duration. In diseases with a long natural history, the final result of comparative trials with survival endpoints is often not known for 5–10 years after the study onset. If these clinical endpoints could be replaced by validated surrogate endpoints (biomarkers) that could be measured earlier, more conveniently or more frequently, then new drugs could be validated quicker and administered to patients. In addition, clinical trials could get by with smaller sample sizes, and costs could be lowered by using stratified patients based on molecular biomarkers. Traditional drug development relies on the random assignment of sufficient numbers of participants with a particular condition to investigational drug and control groups to enable detection of statistically significant drug responses. Some patients may be less genetically predisposed to respond to the investigational medication than others. As such, it is typical to enrol large numbers of patients to ensure sufficient power to detect with statistical certainty any true treatment effect among those who are responsive to the medication. In contrast, the application of biomarkers to clinical trials enables targeted selection of subjects and smaller trials by identifying subjects more likely to respond to a drug based on their genotype (48). The use of biomarkers may lead to more precise and effective inclusion and exclusion criteria in clinical trials and can be used at multiple points in the drug development process (49). Biomarkers will be most valuable when genotypes for adverse response and optimal efficacy for a given compound occur at a high frequency in the patient population (50). By applying biomarker-based stratification, based on these genotypes or protein biomarkers to predict and monitor drug response, specific subgroups of subjects examined in Phase III clinical trials would be expected to demonstrate greater response to and/or fewer adverse effects from the drug being studied. These trials would likely decrease drug development time, costs and potentially speed up the approval of drugs (51).

Examples/case studies

The goals of using biomarkers in drug treatment are to minimize toxicity and to maximize the effectiveness of therapy. Here are two cases of biomarker applications: UGT1A1 for minimizing toxicity and HER2 for maximizing drug efficiency.

UGT1A1 and irinotecan

Irinotecan (Camptosar®), a topoisomerase I poison, is approved for use in combination with 5-fluorouracil (5-FU) and leucovorin chemotherapy for first-line treatment of metastatic colorectal cancer, and also as a single agent in second-line salvage therapy of 5-FU refractory metastatic colorectal cancer. It is also commonly used to treat esophageal, non-small cell lung cancer, breast cancer and other solid tumours in a second- or third-line setting (52). Although it prolongs survival, it also causes severe diarrhoea and neutropenia in 20–35% of patients treated. Fatal events during single-agent irinotecan treatment have been reported (53,54). UDP-glucuronosyltransferase (UGT1A1) is responsible for the clearance by glucuronidation of drugs (e.g. irinotecan) and endogenous substances (e.g. bilirubin). As shown in Figure 17.4, the primary active and toxic metabolite of irinotecan, SN-38, is inactivated by UGT1A1 to form SN-38G, which is eliminated via the bile. It has been determined that variations of the TA repeat length in the UGT1A1 promoter TATA element may lead to decreased gene expression, accumulation of SN-38, and irinotecan-related toxicities.

The *UGT1A1* gene is located on chromosome 2q37. The polymorphic TA repeat in the 5′-promoter region of *UGT1A1* may consist of 5, 6, 7 or 8 repeats. The wild-type allele (*UGT1A1*1*) has six TA repeats, and the variant allele (*UGT1A1*28*) has seven TA repeats. Patients who are homozygous for the *UGT1A1*28* allele, glucuronidate SN-38, less efficiently metabolize than patients who have one or two wild-type alleles; therefore, homozygous patients are exposed to higher plasma concentrations of SN-38 (52,55). In a meta-analysis,

Figure 17.4. The irinotecan pathway shows the biotransformation of the chemotherapy prodrug, irinotecan, to form the active metabolite SN-38, an inhibitor of DNA topoisomerase I. SN-38 is primarily metabolized to the inactive SN-38 glucuronide by UGT1A1, the isoform catalysing bilirubin glucuronidation. Used with permission from PharmGKB and Stanford University; http://www.pharmgkb.org/do/serve?objId=PA2001&objCls=Pathway#

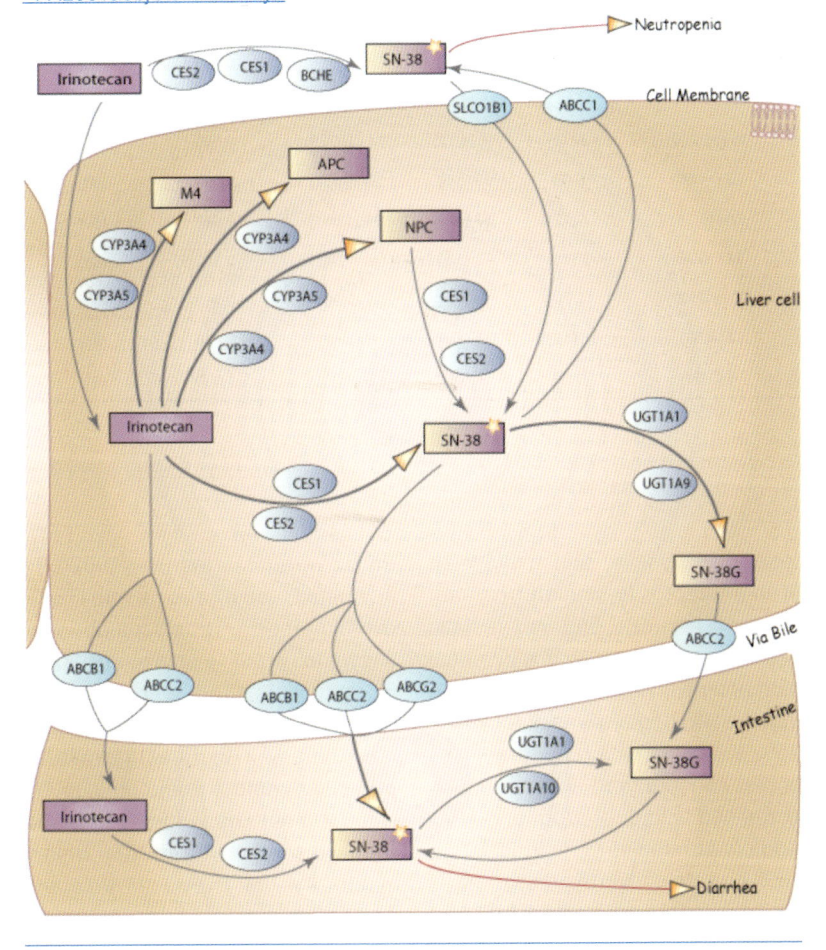

of neutropenia; however clinical results have been variable and such patients have been shown to tolerate normal starting doses." (http://www.fda.gov/Drugs/ScienceResearch/ResearchAreas/Pharmacogenetics/ucm083378.htm).

However, as shown in Figure 17.4, irinotecan interacts with multiple polymorphic drug metabolizing enzymes and transporters (54,56–62), being inactivated by CYP3A4 to APC and requiring conversion by carboxyesterases to the active metabolite SN38. The latter in turn is inactivated by *UGT1A1* glucuronidation as the main degradation pathway. In addition, irinotecan and its metabolites serve as substrates for transporters, including the ABC transporters (ATP-drive extrusion pumps) MDR1, MRP2, and BCRP. Each of these factors displays interindividual variability, with functional polymorphisms potentially contributing to variable irinotecan response. Haplotype analysis has provided additional insight into the regulation of gene transcription (54,56,57,62), but a quantitative assessment of all factors is lacking. As a result, use of TA repeat polymorphisms in predicting *in vivo* UGT activity and SN38 exposure after irinotecan administration has been only partially successful. The Clinical Pharmacology Subcommittee, Advisory Committee for Pharmaceutical Science, reviewing the product, further noted that "...although there is indication to start with a lower dosage, it is not necessarily an indication that sensitive patients will do well with this dosage" (54). This example illustrates that pharmacogenetic (PGx) testing can identify patients who are likely to respond differently to a particular drug and indicate the appropriate dosage, but that testing does not necessarily translate into dosing instructions. Hence, the

data presented in nine studies was reviewed that included 10 sets of patients (a total of 821 patients) and assessed the association of irinotecan dose with the risk of irinotecan-related haematologic toxicities (grade III–IV) for patients with a *UGT1A1*28/*28* genotype (52). As shown in Table 17.2, the risk of toxicity was higher among patients with a *UGT1A1*28/*28* genotype than among those with a *UGT1A1*1/*1* or *UGT1A1*1/*28* genotype at both medium and high doses of irinotecan; however, risk was similar at lower doses (52).

In 2005, the FDA approved the inclusion of *UGT1A1* genotype-associated risk of toxicity on the irinotecan package insert and cites that a clinical test (Invader *UGT1A1* Molecular Assay; Third Wave Technologies Inc.) to detect common *UGT1A1* alleles is available. The FDA-approved label for the test states that "Individuals who are homozygous for the *UGT1A*28* allele are at increased risk of neutropenia following initiation of Camptosar treatment. A reduced initial dose should be considered for patients known to be homozygous for the *UGT1A*28* allele. Heterozygous patients may be at increased risk

Table 17.2. Predictive value of UGT1A1*28 genotype upon irinotecan-induced Grade III-IV haematological toxicity

Irinotecan				Toxicity incidence					
Dose (mg/m^2)	Schedule (weeks)	Concomitant chemotherapy	No of Patients	Overall incidence of toxicity	No of *28/*28 patients (%)	UGT1A1 *28/*28	UGT1A1 *1/*1 or *1/*28	Two-sided Fisher's exact P	Reference
350	3	None	61	18 (11/61)	6 (10)	83 (5/6)	11 (6/55)	0.0004	(50)
300	3	None	20	10 (2/20)	4 (20)	50 (2/4)	0 (0/16)	0.030	(51)
200	3	OXA	103	17 (17/103)	11 (11)	55 (6/11)	12 (11/92)	0.002	(86)
180	2	5FU	250	15 (37/250)	22 (9)	18 (4/12)	14 (33/228)	0.550	(87)
180	2	5FU	56	25 (14/56)	5 (9)	60 (3/5)	22 (11/51)	0.090	(88)
180	2	None	58	28 (16/58)	7 (12)	57 (4/7)	24 (12/51)	0.080	(89)
180	2	5FU	46	33 (15/46)	5 (11)	60 (3/5)	29 (12/41)	0.310	(55)
100	1	5FU	109	10 (11/109)	11 (10)	18 (2/11)	9 (9/98)	0.310	(86)
80	1	RAL	56	7 (4/56)	7 (13)	14 (1/7)	6 (3/49)	0.420	(90)
100/125	1	CAP	64	5 (3/64)	6 (9)	0 (0/6)	5 (3/58)	1.000	(91)

OXA, oxaliplatin; 5FU, 5-fluorouracil; RAL, raltitrexed; CAP, capecitabine

Adapted from the summary table of analyses on 10 clinical trials that assessed the diagnostic value of the homozygous UGT1A1*28 genotype to predict irinotecan-induced grade III-IV hematologic toxicity (52, with permission of Oxford University Press). Other related references are (100–105).

value of prospective genotyping for UGT1A1 in irinotecan therapy must be determined empirically in the intended target populations (63).

HER2 and trastuzumab (Herceptin®)

Human epidermal growth factor receptor 2 (HER2), also known as ErbB2 and Neu, is a cell surface glycoprotein with intrinsic TK activity that is involved in cell growth and development (46,64). In normal quantities, HER2 promotes normal cell growth, but when a genetic mutation causes HER2 to be overexpressed on the cell surface, certain breast cancer cells are prompted to multiply uncontrollably and invade surrounding tissue (46,65). The cloned HER2, associated with a form of metastatic breast cancer, appeared as a potential monoclonal-antibody target in 1985. It has many of the properties required for such a target; it is overexpressed on the surface of tumour cells and not on normal cells, it has an extracellular domain that is readily accessible, and expression of the receptor is stable in primary tumour tissues and metastatic deposits. HER2 became a potential biomarker with the initial observation that the HER2 gene was amplified in 25% of axillary lymph node-positive breast cancers, and, when present, correlated with poor prognosis (66). Additional studies confirmed that HER2 protein overexpression was also a prognostic biomarker in breast cancer, correlating with decreased relapse-free and overall survival (66–68). Moreover, additional clinical data have shown that HER2 amplification/overexpression is a predictive biomarker for greater or lesser response to certain chemotherapies or hormonal therapies in breast cancer (69–73).

The role of HER2 as an oncogenic protein and clinically relevant biomarker led directly to the development of a specific targeted therapy: trastuzumab (Herceptin®; Genentech, South San Francisco, CA), a humanized IgG1 monoclonal antibody with high affinity and specificity for HER2. The clinical trials were started in 1992. In advanced breast cancers with HER2 overexpression, trastuzumab was shown to be active as a single agent in second- and third-line therapy (74,75), and subsequently as first-line therapy (76). Trastuzumab is particularly effective in combination with chemotherapy. In 1998, the drug was approved in the United States by the FDA as Herceptin. This drug was able to get a fast track approval status for two reasons: it demonstrated efficacy in patients previously resistant to more conventional treatments, and a diagnostic test was able to identify the patients that were expected to benefit from it. The HercepTest is the first example of a pharmacogenomic test that is marketed along with a drug. There are two tests to determine HER2 status and select patients for treatment with trastuzumab. The first approved was an immunohistochemistry (IHC) test, the HercepTest, which measures the level of expression of the HER2 protein. The possible outcomes of the test are reported as numbers from 0 to 3+, with 0 representing no overexpression and 3+ representing high overexpression. Only 3+ is defined as HER2 positive. The most recently approved method, fluorescence in situ hybridization (FISH), detects the underlying gene alteration in the patient's tumour cells. FISH makes the number of HER2/neu gene copies visible. In healthy cells, there are two copies of the HER2/neu gene per chromosome. If FISH detects more than two copies of the HER2 gene, it means that the cell is abnormal and is HER2-positive. This abnormality is also referred to as HER2 gene amplification. The results of the FISH test can be reported as positive or negative.

Recent comparison of FISH and IHC shows that FISH appears to be superior at providing prognostic information with respect to the detection of higher-risk breast cancers (77). Unfortunately, it is expensive and requires additional equipment and training beyond what is commonly found in most laboratories. For this reason, it is recommended that only IHC results of 2+ (which represents a little overexpression of HER2) should be retested with FISH to prevent false-negative outcomes (78), as shown in Figure 17.5.

The HER2 case is one of the most successful applications of biomarkers in drug development and disease treatment. The advantage of this case is the co-development of drugs and diagnostic tests, which greatly reduced the number of patients involved in the clinical trials and facilitated a fast-track approval status. It is known that women with HER2+ breast cancer do not respond well to standard therapy, and that patients whose breast cancers lack HER2 overexpression are highly unlikely to respond to trastuzumab alone (46). Moreover, HER2 positivity could predict the effect of adjuvant treatment of other drugs. For example, it was reported that HER2 positivity was associated with a significant benefit from the addition of paclitaxel to the treatment regimen (79). The interaction between HER2 positivity and the addition of paclitaxel was associated with a hazard ratio for recurrence of 0.59 ($P = 0.01$). Patients with a HER2-positive breast cancer benefited from paclitaxel regardless of estrogen-receptor status, but paclitaxel did not benefit patients with HER2-negative, estrogen-receptor-positive cancers.

The effect of adjuvant trastuzumab in the treatment of HER2-positive early breast cancers has been evaluated in randomized controlled trials and in a meta-analysis of published randomized trials. Results of a study on trastuzumab use after adjuvant chemotherapy in HER2-positive breast cancer patients found that one year of this treatment combination had a significant overall survival benefit after a median follow-up of two years (Figure 17.6) (80).

A meta-analysis of five randomized controlled trials was performed comparing adjuvant trastuzumab treatment for HER2-positive early breast cancer. Pooled results from the trials showed a significant reduction of mortality ($P < 0.00001$), recurrence ($P < 0.0001$), metastases rates ($P < 0.0001$) and second tumours other than breast cancer ($P = 0.007$), as compared to no-adjuvant-trastuzumab patients (81).

However, there are still questions about the HER2 biomarker and

Figure 17.5. Algorithm for HER2 testing

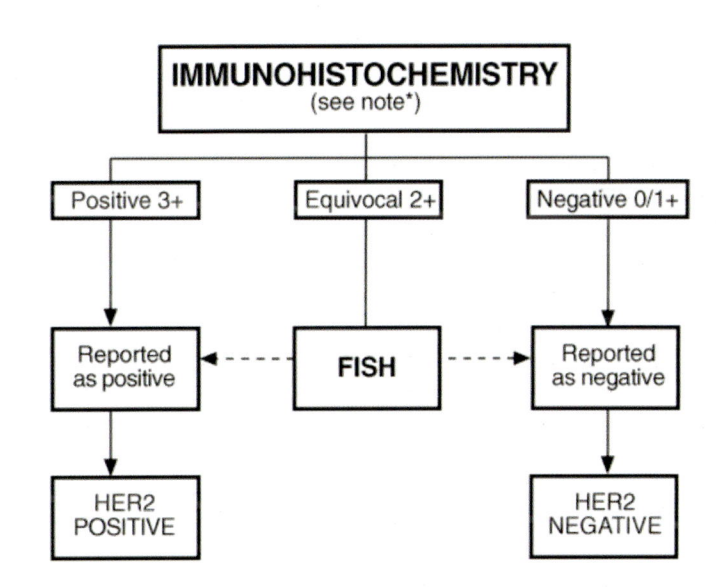

*All antibodies assessed by the modified HercepTest scoring system.

3+ is HER2 postive with > 10% of cancer cells showing strong, complete membrane staining without cytoplasmic staining and without staining of normal breast tissue.

2+ is equivocal
< 10% of cancer cells with strong complete membrane staining (rare).
> 10% weak to moderate complete membrane staining
Strong cytoplasmic staining making assesment of membrane staining difficult.

0 or 1+ is negative
no staining or <10% of cancer cells stained.

Adapted from http://www.iap-aus.org.au/2001no3.html.

Figure 17.6. Exploratory disease-free survival subgroup analysis for one year of trastuzumab versus observation

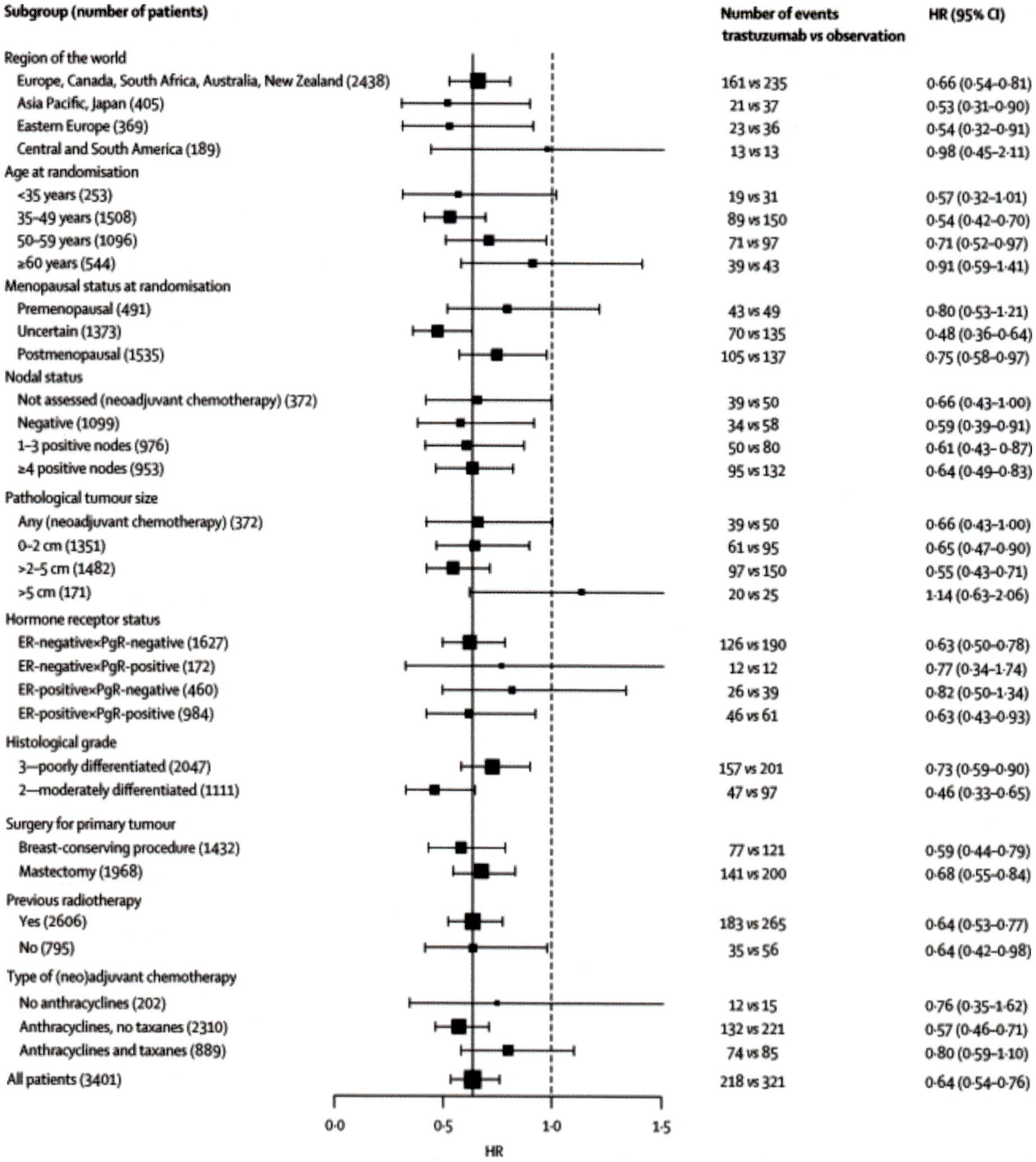

Subgroup (number of patients)	Number of events trastuzumab vs observation	HR (95% CI)
Region of the world		
Europe, Canada, South Africa, Australia, New Zealand (2438)	161 vs 235	0·66 (0·54–0·81)
Asia Pacific, Japan (405)	21 vs 37	0·53 (0·31–0·90)
Eastern Europe (369)	23 vs 36	0·54 (0·32–0·91)
Central and South America (189)	13 vs 13	0·98 (0·45–2·11)
Age at randomisation		
<35 years (253)	19 vs 31	0·57 (0·32–1·01)
35–49 years (1508)	89 vs 150	0·54 (0·42–0·70)
50–59 years (1096)	71 vs 97	0·71 (0·52–0·97)
≥60 years (544)	39 vs 43	0·91 (0·59–1·41)
Menopausal status at randomisation		
Premenopausal (491)	43 vs 49	0·80 (0·53–1·21)
Uncertain (1373)	70 vs 135	0·48 (0·36–0·64)
Postmenopausal (1535)	105 vs 137	0·75 (0·58–0·97)
Nodal status		
Not assessed (neoadjuvant chemotherapy) (372)	39 vs 50	0·66 (0·43–1·00)
Negative (1099)	34 vs 58	0·59 (0·39–0·91)
1–3 positive nodes (976)	50 vs 80	0·61 (0·43–0·87)
≥4 positive nodes (953)	95 vs 132	0·64 (0·49–0·83)
Pathological tumour size		
Any (neoadjuvant chemotherapy) (372)	39 vs 50	0·66 (0·43–1·00)
0–2 cm (1351)	61 vs 95	0·65 (0·47–0·90)
>2–5 cm (1482)	97 vs 150	0·55 (0·43–0·71)
>5 cm (171)	20 vs 25	1·14 (0·63–2·06)
Hormone receptor status		
ER-negative×PgR-negative (1627)	126 vs 190	0·63 (0·50–0·78)
ER-negative×PgR-positive (172)	12 vs 12	0·77 (0·34–1·74)
ER-positive×PgR-negative (460)	26 vs 39	0·82 (0·50–1·34)
ER-positive×PgR-positive (984)	46 vs 61	0·63 (0·43–0·93)
Histological grade		
3—poorly differentiated (2047)	157 vs 201	0·73 (0·59–0·90)
2—moderately differentiated (1111)	47 vs 97	0·46 (0·33–0·65)
Surgery for primary tumour		
Breast-conserving procedure (1432)	77 vs 121	0·59 (0·44–0·79)
Mastectomy (1968)	141 vs 200	0·68 (0·55–0·84)
Previous radiotherapy		
Yes (2606)	183 vs 265	0·64 (0·53–0·77)
No (795)	35 vs 56	0·64 (0·42–0·98)
Type of (neo)adjuvant chemotherapy		
No anthracyclines (202)	12 vs 15	0·76 (0·35–1·62)
Anthracyclines, no taxanes (2310)	132 vs 221	0·57 (0·46–0·71)
Anthracyclines and taxanes (889)	74 vs 85	0·80 (0·59–1·10)
All patients (3401)	218 vs 321	0·64 (0·54–0·76)

Source: (80). Reprinted from The Lancet, Copyright (2007), with permission from Elsevier.

trastuzumab treatment. For example, almost half of HER2-positive breast cancer patients are non-responsive to trastuzumab therapy or become drug resistant during treatment. Although other biomarkers have been investigated, and some drugs are in clinical trial, no breakthrough drug has been reported yet. Another unsolved issue is toxicity. In the meta-analysis, they reported more grade III or IV cardiac toxicity after trastuzumab (203/4555 = 4.5%) versus no trastuzumab patients (86/4562 = 1.8%); therefore, careful cardiac monitoring is warranted (81).

Strengths, limitations and lessons learned

Biomarkers have been used in disease prevention, diagnosis, treatment, prognosis and drug development for many years, but have only recently shown the potential to revolutionise the health paradigm into a new era. The successful completion of the human genome sequencing project laid the foundation for identifying mechanism-based biomarkers. Although US$1000 per individual for sequencing is still a ways off, BioNanomatrix and Complete Genomics Incorporated have formed a joint venture to develop a system capable of sequencing the entire human genome in eight hours at a cost of less than $100. By its completion, the proposed technology will have the potential to enable improvements in the diagnosis and personalized treatment of a wide variety of health conditions, as well as the ability to deliver individually tailored preventive medicine (http://nanotechwire.com/news.asp?nid=5087&ntid=130&pg=1).

Recently developed "omics" technologies, such as genomics, transcriptomics, proteomics, metabolomics and other high-throughput technologies, offer useful tools for biomarker discovery. Genomics studies organisms in terms of their genomes (i.e. their full DNA sequences) and the information they contain (an indication of what can happen). Transcriptomics is used to analyse gene expression (what appears to be happening). Proteomics is used to investigate proteins (compounds that make things happen). Metabolomics is used to measure metabolites (substances that indicate what has happened and is happening). It is widely known that early diagnosis and effective treatment of common diseases requires capturing and interpreting information at different levels and using a variety of novel techniques (as shown in Figure 17.3).

Computational technology and bioinformatics play a major role in the discovery of new biomarkers, the validation of potential biomarkers, and the analysis of disease states. For example, Figure 17.7 shows the detail of a subnetwork of the protooncogene MYC. Two types of technologies have made this work possible: the advent of a new wave of high-throughput biotechnology, with its sequencers, gene expression arrays, mass spectrometers and fluorescence microscopes; and information technology for qualitative changes in the way biological knowledge is stored, retrieved, processed and inferred.

Large and well-organized consortia and networks, as well as updated regulatory systems, guarantee the validation of biomarkers and their successful translation into clinical practice. A good example of this is the FDA consortium that includes members of the pharmaceutical industry and academia, and aims to observe how genetic biomarkers contribute to serious adverse events. The consortium launched two projects: to address drug-related liver toxicity; and to study a rare but serious drug-related skin condition called Stevens-Johnson syndrome. The Biomarkers Consortium has launched a web site to encourage researchers to submit biomarker project concepts (http://www.biomarkersconsortium.org/).

In clinical medicine, there are still many challenges that must be met before the full value of biomarkers, especially molecular biomarkers (e.g. cancer biomarkers), can be realized.

First, identification of highly prevalent targets that constitute key master promoters of oncogenesis in specific tumours is still very difficult. For instance, the oncogenetic process in malignant gliomas is driven by several signalling pathways that are differentially activated or silenced with both parallel and converging complex interactions. To date, no new molecule seems to be promising enough to justify a large Phase III trial (82).

Second, once a potential target is identified, it is not easy to discover new agents capable of restoring normal cell functions through interaction with the target. A major hurdle is that tumour cells acquire drug resistance. Certain cancers are effectively treated because the targeted drug is applied. But very often patients develop secondary mutations that recruit other kinases that are not affected by the inhibitor to substitute for the pharmacologically impaired kinase, and to restore downstream molecular signalling cascades that contribute to tumour growth (82–84).

Third, there are still many methodological issues to resolve. For example, how to define proper criteria for responsiveness, avoid measurement errors, interpret laboratory results, educate medical staff to accept and use biomarkers in

Figure 17.7. Detail of a subnetwork of the protooncogene MYC. Nodes are colour-coded according to their target status and available validation of direct MYC binding

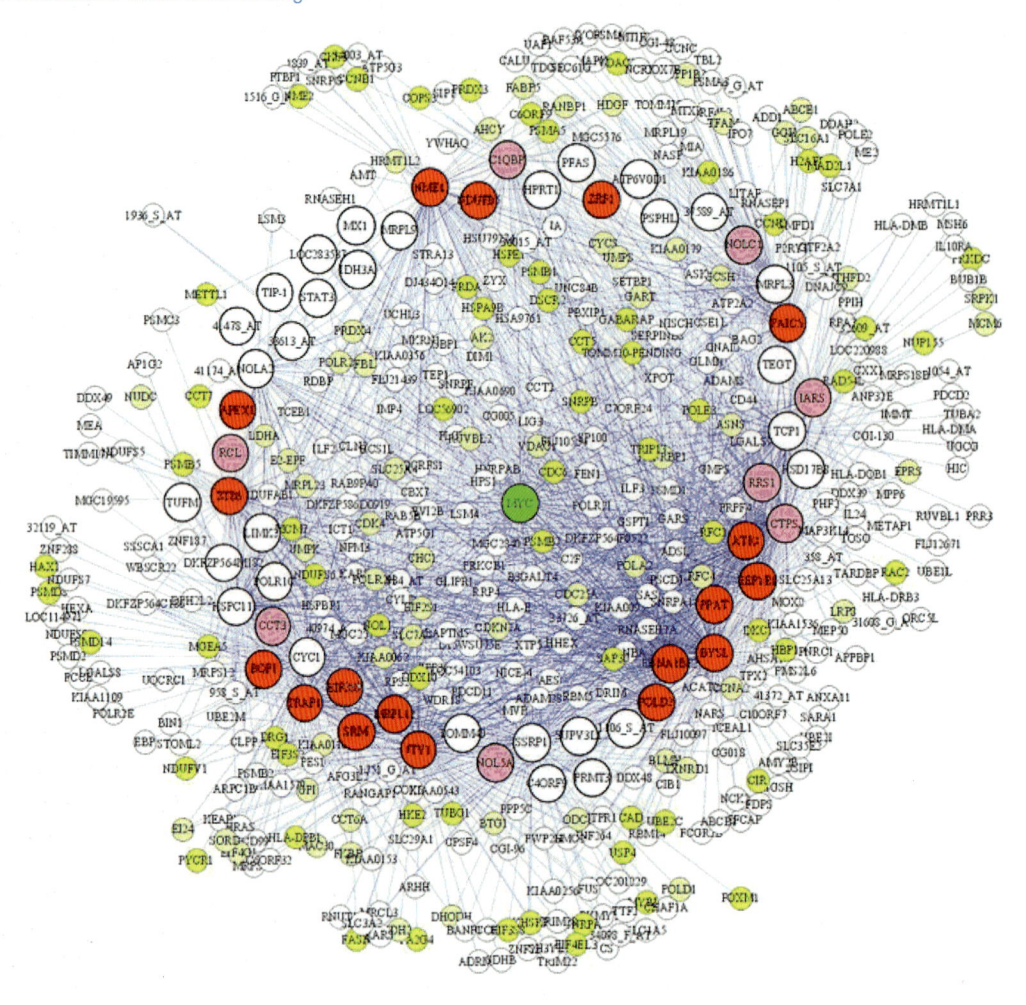

Source: (107). Reprinted by permission from Macmillan Publishers Ltd: Nature Genetics, copyright (2005).

their daily medical practice, and how to help the public better understand genetic tests (85).

Finally, ethical and social issues must be considered. Individual, family and societal goals may conflict with current health care practices and policies in regards to genetic testing. Current health care policies do not fully address these concerns. One major barrier is the potential loss of societal benefits, such as employment or insurability, based on one's genetic characteristics, which is referred to as genetic discrimination (85). Other issues include genetic testing on those who lack the capacity to consent, genetic testing on stored tissue samples and tissue banks, and ensuring appropriate monitoring of genetic tests. These concerns warrant the attention and action of society as a whole.

Future directions and challenges

Multiple targets, prevention and prediction, personalization and cooperation will be the future directions of biomarker applications in clinical medicine.

Multiple biomarkers will be more frequently applied in clinical tests, especially for common diseases. "Multiple" could represent many markers from the same profile, or markers from different profiles, such as DNA, mRNA, microRNA or protein and gene expression. In 2007, for example, the FDA approved a gene-based breast cancer test designed to determine the likelihood of early stage breast cancer recurrence within 5–10 years after treatment. The test called MammaPrintTM (Agendia) is a DNA microarray-based diagnostic kit that measures the level of transcription of 70 genes in breast cancer tumours. The profiles are scored to determine the risk or recurrence and with it the need for adjuvant therapy (86). There is currently a great deal of research being done on multitargeted therapies, which simultaneously target some of the many signalling pathways involved in tumour development and proliferation. "Mixing cocktails," as Charles L. Sawyers recently described it (84), will continue to grow, but should be under the appropriate molecular guidance.

Preventive and predictive biomarkers will play a key role in future health care. New agents, such as antiangiogenesis/vascular-targeting drugs, have moved from cancer therapy to cancer prevention. Molecular and epidemiologic studies of cancer risk and drug sensitivity and resistance began ushering in the era of personalized prevention (84,87). Development of new treatments has increased the need for markers that predict outcome and those that direct which treatment options are most likely to be effective for a particular patient with a particular tumour (88).

Personalized medicine is the use of detailed information about an individual's inherited and/or acquired characteristics and their phenotypic data to select a preventative measure or medication that is particularly suited to that person at the time of administration. This revolution in clinical care is predicated on the development and refinement of biomarkers, enabling disease prevention, and diagnosis and treatment of patients and populations (89). Biomarkers will be used before birth and throughout life. For example, a couple planning to have children could be tested for specific biomarkers to avoid haemolytic disease of the newborn (HDN) and some recessive diseases (carrier parents have a 25% chance of passing on the disease to the baby). Children with a family history of diabetes, heart disease or cancer may take a genetic test to adjust their lifestyle or consider preventive treatment. Therapeutic and prognostic biomarkers should be applied to all kinds of patients, especially cancer patients, to direct their treatment plans and predict the treatment outcomes. Within the foreseeable future, when the US$100 genome sequencer is developed, everybody would be able to have their whole-genome information on their ID card.

In the first decade of the 21st century, the fast-growing application of omics technologies in translational research and clinical medicine have been witnessed. It has accelerated biomarker development, improved the accuracy for diagnosis/treatment, and advanced personalized medicine. One example is the application of omics in reproductive medicine, in particular *in vitro* fertilization (IVF) treatment, an assisted reproduction. A key step in assisted reproduction is the assessment of oocyte and embryo viability to determine the embryo(s) most likely to result in a pregnancy. Although conventional systems such as morphological charaterization and cleavage rating have been successful in improving pregnancy rates, their precision is far from ideal (90,91). It was reported that two out of three IVF cycles fail to result in a pregnancy, and more than eight out of 10 embryos fail to implant (92). The presence of aneuploidy in embryos frequently causes failed implantation and pregnancy. In a recent study, CGH, a genomics approach, was used in assessment of embryo aneuploidy and achieved implantation and pregnancy rates of 68.9 and 82.2%, respectively (93). Alternatively, using microarray CGH (aCGH) and single nucleotide polymorphism microarray have the potential for further improvement in assessment of embryo aneuploidy at a higher resolution, as they can be used to detect more refined regions (less than megabases, or even less than kilobases of nucleotides) in any chromosome (94,95). Other omics have also been applied to assessing embryo viability, such as metabolomics (96,97), transcriptomics (98) and proteomics (99). These omics technologies present unique advantages as well as their own intrinsic limitations. However, a combined strategy of omics may enhance the thorough screening of gametes and embryos for their viability and reproductive potential. The applications of omics technologies in other medical fields are in different stages of development and ever expanding. It is envisioned that the biomarkers derived from those omics will realize their full potential before long in all fields of clinical medicine.

In summary, biomarkers have been widely used in clinical prevention, diagnostics, therapeutics, prognostics, clinical trials and drug development. With mapping of the human genome complete, rapid development of new technologies and the collaboration of different disciplines, biomarkers promise personalized medicine, though many challenges remain to be overcome.

References

1. Biomarkers Definitions Working Group. (2001). Biomarkers and surrogate endpoints: preferred definitions and conceptual framework. *Clin Pharmacol Ther*, 69:89–95. doi:10.1067/mcp.2001.113989 PMID:11240971

2. Srivastava S, Gopal-Srivastava R (2002). Biomarkers in cancer screening: a public health perspective. *J Nutr*, 132 Suppl;2471S–2475S. PMID:12163714

3. Young S, Le Souëf PN, Geelhoed GC et al. (1991). The influence of a family history of asthma and parental smoking on airway responsiveness in early infancy. *N Engl J Med*, 324:1168–1173. doi:10.1056/NEJM199104253241704 PMID:2011160

4. Bearer CF (1998). Biomarkers in pediatric environmental health: a cross-cutting issue. *Environ Health Perspect*, 106 Suppl 3;813–816. doi:10.2307/3434194 PMID:9646042

5. Brown KM, Macgregor S, Montgomery GW et al. (2008). Common sequence variants on 20q11.22 confer melanoma susceptibility. *Nat Genet*, 40:838–840. doi:10.1038/ng.163 PMID:18488026

6. Gudbjartsson DF, Sulem P, Stacey SN et al. (2008). ASIP and TYR pigmentation variants associate with cutaneous melanoma and basal cell carcinoma. *Nat Genet*, 40:886–891. doi:10.1038/ng.161 PMID:18488027

7. Sulem P, Gudbjartsson DF, Stacey SN et al. (2008). Two newly identified genetic determinants of pigmentation in Europeans. *Nat Genet*, 40:835–837. doi:10.1038/ng.160 PMID:18488028

8. Aitio A, Kallio A (1999). Exposure and effect monitoring: a critical appraisal of their practical application. *Toxicol Lett*, 108:137–147. doi:10.1016/S0378-4274(99)00082-X PMID:10511255

9. Secretary of State for Health. Our inheritance, our future: realising the potential of genetics in the NHS. London (UK): Stationary Office; 2003.

10. Martin AM, Weber BL (2000). Genetic and hormonal risk factors in breast cancer. *J Natl Cancer Inst*, 92:1126–1135. doi:10.1093/jnci/92.14.1126 PMID:10904085

11. Fabian CJ, Kimler BF, Elledge RM et al. (1998). Models for early chemoprevention trials in breast cancer. *Hematol Oncol Clin North Am*, 12:993–1017. doi:10.1016/S0889-8588(05)70038-1 PMID:9888018

12. Fearon ER, Vogelstein B (1990). A genetic model for colorectal tumorigenesis. *Cell*, 61:759–767. doi:10.1016/0092-8674(90)90186-I PMID:2188735

13. Weatherall DJ (2001). Phenotype-genotype relationships in monogenic disease: lessons from the thalassaemias. *Nat Rev Genet*, 2:245–255. doi:10.1038/35066048 PMID:11283697

14. Annan K (2000). The challenge of preventive medicine in the year 2000. *West J Med*, 172:408. doi:10.1136/ewjm.172.6.408 PMID:10854397

15. Ozolek JA, Watchko JF, Mimouni F (1994). Prevalence and lack of clinical significance of blood group incompatibility in mothers with blood type A or B. *J Pediatr*, 125:87–91. doi:10.1016/S0022-3476(94)70131-8 PMID:8021795

16. Jeon H, Calhoun B, Pothiawala M et al. (2000). Significant ABO hemolytic disease of the newborn in a group B infant with a group A2 mother. *Immunohematology*, 16:105–108. PMID:15373613

17. Haque KM, Rahman M (2000). An unusual case of ABO-haemolytic disease of the newborn. *Bangladesh Med Res Counc Bull*, 26:61–64. PMID:11508073

18. Turnpenny P, Ellard S, editors. Emery's elements of medical genetics. 12th ed. London: Elsevier Churchill Livingston; 2005.

19. Bradley J, Johnson D, Pober B. Medical genetics. 3rd ed. Oxford: Blackwell; 2007.

20. Peltonen L, McKusick VA (2001). Genomics and medicine. Dissecting human disease in the postgenomic era. *Science*, 291:1224–1229. doi:10.1126/science.291.5507.1224 PMID:11233446

21. Seo D, Ginsburg GS (2005). Genomic medicine: bringing biomarkers to clinical medicine. *Curr Opin Chem Biol*, 9:381–386. doi:10.1016/j.cbpa.2005.06.009 PMID:16006183

22. Hanash S (2003). Disease proteomics. *Nature*, 422:226–232. doi:10.1038/nature01514 PMID:12634796

23. Pittman J, Huang E, Dressman H et al. (2004). Integrated modeling of clinical and gene expression information for personalized prediction of disease outcomes. *Proc Natl Acad Sci USA*, 101:8431–8436. doi:10.1073/pnas.0401736101 PMID:15152076

24. Vasan RS (2006). Biomarkers of cardiovascular disease: molecular basis and practical considerations. *Circulation*, 113:2335–2362. doi:10.1161/CIRCULATIONAHA.104.482570 PMID:16702488

25. Marrer E, Dieterle F (2007). Promises of biomarkers in drug development–a reality check. *Chem Biol Drug Des*, 69:381–394. doi:10.1111/j.1747-0285.2007.00522.x PMID:17581232

26. Gold P, Freedman SO (1965). Specific carcinoembryonic antigens of the human digestive system. *J Exp Med*, 122:467–481. doi:10.1084/jem.122.3.467 PMID:4953873

27. Moertel CG, Fleming TR, Macdonald JS et al. (1993). An evaluation of the carcinoembryonic antigen (CEA) test for monitoring patients with resected colon cancer. *JAMA*, 270:943–947. doi:10.1001/jama.270.8.943 PMID:8141873

28. Wilson JF (2006). The rocky road to useful cancer biomarkers. *Ann Intern Med*, 144:945–948. PMID:16785487

29. Bangma CH, Grobbee DE, Schröder FH (1995). Volume adjustment for intermediate prostate-specific antigen values in a screening population. *Eur J Cancer*, 31A:12–14. doi:10.1016/0959-8049(94)00309-S PMID:7535074

30. Gann PH, Hennekens CH, Stampfer MJ (1995). A prospective evaluation of plasma prostate-specific antigen for detection of prostatic cancer. *JAMA*, 273:289–294. doi:10.1001/jama.273.4.289 PMID:7529341

31. Gillatt D, Reynard JM (1995). What is the 'normal range' for prostate-specific antigen? Use of a receiver operating characteristic curve to evaluate a serum marker. *Br J Urol*, 75:341–346. doi:10.1111/j.1464-410X.1995.tb07346.x PMID:7537603

32. Lepor H, Owens RS, Rogenes V, Kuhn E (1994). Detection of prostate cancer in males with prostatism. *Prostate*, 25:132–140. doi:10.1002/pros.2990250304 PMID:7520577

33. Manne U, Srivastava RG, Srivastava S (2005). Recent advances in biomarkers for cancer diagnosis and treatment. *Drug Discov Today*, 10:965–976. doi:10.1016/S1359-6446(05)03487-2 PMID:16023055

34. Johann DJ Jr, Veenstra TD (2007). Multiple biomarkers in molecular oncology. *Expert Rev Mol Diagn*, 7:223–225. doi:10.1586/14737159.7.3.223 PMID:17489728

35. Druker BJ (2003). Imatinib alone and in combination for chronic myeloid leukemia. *Semin Hematol*, 40:50–58. doi:10.1016/S0037-1963(03)70042-0 PMID:12563611

36. Kidd EA, Siegel BA, Dehdashti F, Grigsby PW (2007). The standardized uptake value for F-18 fluorodeoxyglucose is a sensitive predictive biomarker for cervical cancer treatment response and survival. *Cancer*, 110:1738–1744. doi:10.1002/cncr.22974 PMID:17786947

37. Lazarou J, Pomeranz BH, Corey PN (1998). Incidence of adverse drug reactions in hospitalized patients: a meta-analysis of prospective studies. *JAMA*, 279:1200–1205. doi:10.1001/jama.279.15.1200 PMID:9555760

38. Mangravite LM, Thorn CF, Krauss RM (2006). Clinical implications of pharmacogenomics of statin treatment. *Pharmacogenomics J*, 6:360–374.doi:10.1038/sj.tpj.6500384 PMID:16550210

39. Rieder MJ, Reiner AP, Gage BF et al. (2005). Effect of VKORC1 haplotypes on transcriptional regulation and warfarin dose. *N Engl J Med*, 352:2285–2293.doi:10.1056/NEJMoa044503 PMID:15930419

40. Terra SG, Hamilton KK, Pauly DF et al. (2005). Beta1-adrenergic receptor polymorphisms and left ventricular remodeling changes in response to beta-blocker therapy. *Pharmacogenet Genomics*, 15:227–234. doi:10.1097/01213011-200504000-00006 PMID:15864115

41. Wang L, Weinshilboum R (2006). Thiopurine S-methyltransferase pharmacogenetics: insights, challenges and future directions. *Oncogene*, 25:1629–1638.doi:10.1038/sj.onc.1209372 PMID:16550163

42. Sussman S, Valente TW, Rohrbach LA et al. (2006). Translation in the health professions: converting science into action. *Eval Health Prof*, 29:7–32.doi:10.1177/0163278705284441 PMID:16510878

43. Westfall JM, Mold J, Fagnan L (2007). Practice-based research–"Blue Highways" on the NIH roadmap. *JAMA*, 297:403–406. doi:10.1001/jama.297.4.403 PMID:17244837

44. Ozdemir V, Williams-Jones B, Cooper DM et al. (2007). Mapping translational research in personalized therapeutics: from molecular markers to health policy. *Pharmacogenomics*, 8:177–185.doi:10.2217/14622416.8.2.177 PMID:17286540

45. Friedman LM, Furberg CD, DeMets DL. Fundamentals of clinical trials. 3rd ed. New York (NY): Springer-Verlag; 1998.

46. Park JW, Kerbel RS, Kelloff GJ et al. (2004). Rationale for biomarkers and surrogate end points in mechanism-driven oncology drug development. *Clin Cancer Res*, 10:3885–3896.doi:10.1158/1078-0432.CCR-03-0785 PMID:15173098

47. Luo FR, Yang Z, Camuso A et al. (2006). Dasatinib (BMS-354825) pharmacokinetics and pharmacodynamic biomarkers in animal models predict optimal clinical exposure. *Clin Cancer Res*, 12:7180–7186.doi:10.1158/1078-0432.CCR-06-1112 PMID:17145844

48. Sadee W (2002). Pharmacogenomics: the implementation phase. *AAPS PharmSci*, 4:E5. doi:10.1208/ps040210 PMID:12141268

49. Emilien G, Ponchon M, Caldas C et al. (2000). Impact of genomics on drug discovery and clinical medicine. *QJM*, 93:391–423. doi:10.1093/qjmed/93.7.391 PMID:10874050

50. Shah J (2004). Criteria influencing the clinical uptake of pharmacogenomic strategies. *BMJ*, 328:1482–1486.doi:10.1136/bmj.328.7454.1482 PMID:15205293

51. Secretary's Advisory Committee on Genetics, Health and Society (2007). Realizing the promise of pharmacogenomics: opportunities and challenges. *Biotechnol Law Rep*, 26:261–291 doi:10.1089/blr.2007.9956.

52. Hoskins JM, Goldberg RM, Qu P et al. (2007). UGT1A1*28 genotype and irinotecan-induced neutropenia: dose matters. *J Natl Cancer Inst*, 99:1290–1295.doi:10.1093/jnci/djm115 PMID:17728214

53. Fuchs CS, Moore MR, Harker G et al. (2003). Phase III comparison of two irinotecan dosing regimens in second-line therapy of metastatic colorectal cancer. *J Clin Oncol*, 21:807–814.doi:10.1200/JCO.2003.08.058 PMID:12610178

54. Innocenti F, Undevia SD, Iyer L et al. (2004). Genetic variants in the UDP-glucuronosyltransferase 1A1 gene predict the risk of severe neutropenia of irinotecan. *J Clin Oncol*, 22:1382–1388.doi:10.1200/JCO.2004.07.173 PMID:15007088

55. Iyer L, Das S, Janisch L et al. (2002). UGT1A1*28 polymorphism as a determinant of irinotecan disposition and toxicity. *Pharmacogenomics J*, 2:43–47.doi:10.1038/sj.tpj.6500072 PMID:11990381

56. Kaniwa N, Kurose K, Jinno H et al. (2005). Racial variability in haplotype frequencies of UGT1A1 and glucuronidation activity of a novel single nucleotide polymorphism 686C>T (P229L) found in an African-American. *Drug Metab Dispos*, 33:458–465.doi:10.1124/dmd.104.001800 PMID:15572581

57. Ando Y, Saka H, Ando M et al. (2000). Polymorphisms of UDP-glucuronosyltransferase gene and irinotecan toxicity: a pharmacogenetic analysis. *Cancer Res*, 60:6921–6926. PMID:11156391

58. Mathijssen RH, Marsh S, Karlsson MO et al. (2003). Irinotecan pathway genotype analysis to predict pharmacokinetics. *Clin Cancer Res*, 9:3246–3253. PMID:12960109

59. Rouits E, Boisdron-Celle M, Dumont A et al. (2004). Relevance of different UGT1A1 polymorphisms in irinotecan-induced toxicity: a molecular and clinical study of 75 patients. *Clin Cancer Res*, 10:5151–5159.doi:10.1158/1078-0432.CCR-03-0548 PMID:15297419

60. Sai K, Saeki M, Saito Y et al. (2004). UGT1A1 haplotypes associated with reduced glucuronidation and increased serum bilirubin in irinotecan-administered Japanese patients with cancer. *Clin Pharmacol Ther*, 75:501–515. doi:10.1016/j.clpt.2004.01.010 PMID:15179405

61. Strassburg CP, Kneip S, Topp J et al. (2000). Polymorphic gene regulation and interindividual variation of UDP-glucuronosyltransferase activity in human small intestine. *J Biol Chem*, 275:36164–36171.doi:10.1074/jbc.M002180200 PMID:10748067

62. Innocenti F, Grimsley C, Das S et al. (2002). Haplotype structure of the UDP-glucuronosyltransferase 1A1 promoter in different ethnic groups. *Pharmacogenetics*, 12:725–733.doi:10.1097/00008571-200212000-00006 PMID:12464801

63. Sadée W, Dai Z (2005). Pharmacogenetics/genomics and personalized medicine. *Hum Mol Genet*, 14 Spec No. 2:R207–R214. doi:10.1093/hmg/ddi261 PMID:16244319

64. Yarden Y, Sliwkowski MX (2001). Untangling the ErbB signalling network. *Nat Rev Mol Cell Biol*, 2:127–137.doi:10.1038/35052073 PMID:11252954

65. Ménard S, Pupa SM, Campiglio M, Tagliabue E (2003). Biologic and therapeutic role of HER2 in cancer. *Oncogene*, 22:6570–6578.doi:10.1038/sj.onc.1206779 PMID:14528282

66. Slamon DJ, Clark GM, Wong SG et al. (1987). Human breast cancer: correlation of relapse and survival with amplification of the HER-2/neu oncogene. *Science*, 235:177–182. doi:10.1126/science.3798106 PMID:3798106

67. Slamon DJ, Godolphin W, Jones LA et al. (1989). Studies of the HER-2/neu proto-oncogene in human breast and ovarian cancer. *Science*, 244:707–712.doi:10.1126/science.2470152 PMID:2470152

68. Ravdin PM, Chamness GC (1995). The c-erbB-2 proto-oncogene as a prognostic and predictive marker in breast cancer: a paradigm for the development of other macromolecular markers–a review. *Gene*, 159:19–27.doi:10.1016/0378-1119(94)00866-Q PMID:7607568

69. Thor AD, Berry DA, Budman DR et al. (1998). erbB-2, p53, and efficacy of adjuvant therapy in lymph node-positive breast cancer. *J Natl Cancer Inst*, 90:1346–1360.doi:10.1093/jnci/90.18.1346 PMID:9747866

70. Stål O, Borg A, Fernö M et al.; South Sweden Breast Cancer Group. Southeast Sweden Breast Cancer Group (2000). ErbB2 status and the benefit from two or five years of adjuvant tamoxifen in postmenopausal early stage breast cancer. *Ann Oncol*, 11:1545–1550.doi:10.1023/A:1008313310474 PMID:11205461

71. Paik S, Bryant J, Tan-Chiu E et al. (2000). HER2 and choice of adjuvant chemotherapy for invasive breast cancer: National Surgical Adjuvant Breast and Bowel Project Protocol B-15. *J Natl Cancer Inst*, 92:1991–1998. doi:10.1093/jnci/92.24.1991 PMID:11121461

72. Lipton A, Ali SM, Leitzel K et al. (2002). Elevated serum Her-2/neu level predicts decreased response to hormone therapy in metastatic breast cancer. *J Clin Oncol*, 20:1467–1472.doi:10.1200/JCO.20.6.1467 PMID:11896093

73. De Placido S, Carlomagno C, De Laurentiis M, Bianco AR (1998). c-erbB2 expression predicts tamoxifen efficacy in breast cancer patients. *Breast Cancer Res Treat*, 52:55–64.doi:10.1023/A:1006159001039 PMID:10066072

74. Baselga J, Tripathy D, Mendelsohn J et al. (1996). Phase II study of weekly intravenous recombinant humanized anti-p185HER2 monoclonal antibody in patients with HER2/neu-overexpressing metastatic breast cancer. J Clin Oncol, 14:737–744. PMID:8622019

75. Cobleigh MA, Vogel CL, Tripathy D et al. (1999). Multinational study of the efficacy and safety of humanized anti-HER2 monoclonal antibody in women who have HER2-overexpressing metastatic breast cancer that has progressed after chemotherapy for metastatic disease. J Clin Oncol, 17:2639–2648. PMID:10561337

76. Vogel CL, Cobleigh MA, Tripathy D et al. (2002). Efficacy and safety of trastuzumab as a single agent in first-line treatment of HER2-overexpressing metastatic breast cancer. J Clin Oncol, 20:719–726.doi:10.1200/JCO.20.3.719 PMID:11821453

77. Mass RD, Press M, Anderson S et al. (2001). Improved survival benefit from Herceptin (Trastuzumab) in patients selected by fluorescence in situ hybridization (FISH). Proc Am Soc Clin Oncol;20 (abstr 85).

78. Ellis IO, Bartlett J, Dowsett M et al. (2004). Best practice No 176: Updated recommendations for HER2 testing in the UK. J Clin Pathol, 57:233–237.doi:10.1136/jcp.2003.007724 PMID:14990588

79. Hayes DF, Thor AD, Dressler LG et al.; Cancer and Leukemia Group B (CALGB) Investigators (2007). HER2 and response to paclitaxel in node-positive breast cancer. N Engl J Med, 357:1496–1506.doi:10.1056/NEJMoa071167 PMID:17928597

80. Smith I, Procter M, Gelber RD et al.; HERA study team (2007). 2-year follow-up of trastuzumab after adjuvant chemotherapy in HER2-positive breast cancer: a randomised controlled trial. Lancet, 369:29–36.doi:10.1016/S0140-6736(07)60028-2 PMID:17208639

81. Viani GA, Afonso SL, Stefano EJ et al. (2007). Adjuvant trastuzumab in the treatment of her-2-positive early breast cancer: a meta-analysis of published randomized trials. BMC Cancer, 7:153.doi:10.1186/1471-2407-7-153 PMID:17686164

82. Omuro AM, Faivre S, Raymond E (2007). Lessons learned in the development of targeted therapy for malignant gliomas. Mol Cancer Ther, 6:1909–1919.doi:10.1158/1535-7163.MCT-07-0047 PMID:17620423

83. Sawyers CL (2003). Opportunities and challenges in the development of kinase inhibitor therapy for cancer. Genes Dev, 17:2998–3010.doi:10.1101/gad.1152403 PMID:14701871

84. Sawyers CL (2007). Cancer: mixing cocktails. Nature, 449:993–996.doi:10.1038/449993a PMID:17960228

85. Williams JK, Skirton H, Masny A (2006). Ethics, policy, and educational issues in genetic testing. J Nurs Scholarsh, 38:119–125.doi:10.1111/j.1547-5069.2006.00088.x PMID:16773914

86. Glas AM, Floore A, Delahaye LJ et al. (2006). Converting a breast cancer microarray signature into a high-throughput diagnostic test. BMC Genomics, 7:278.doi:10.1186/1471-2164-7-278 PMID:17074082

87. Hoque A, Parnes HL, Stefanek ME et al. (2007). Meeting report: fifth annual AACR Frontiers in Cancer Prevention Research. Cancer Res, 67:8989–8993.doi:10.1158/0008-5472.CAN-07-3171 PMID:17895292

88. Chatterjee SK, Zetter BR (2005). Cancer biomarkers: knowing the present and predicting the future. Future Oncol, 1:37–50. doi:10.1517/14796694.1.1.37 PMID:16555974

89. Waldman S, Terzic A (2007). Targeted diagnostics and therapeutics for individualized patient management. Biomark Med, 1:3–8. doi:10.2217/17520363.1.1.3 PMID:20477454

90. Aydiner F, Yetkin CE, Seli E (2010). Perspectives on emerging biomarkers for non-invasive assessment of embryo viability in assisted reproduction. Curr Mol Med, 10:206–215.doi:10.2174/156652410790963349 PMID:20196727

91. Seli E, Robert C, Sirard MA (2010). OMICS in assisted reproduction: possibilities and pitfalls. Mol Hum Reprod, 16:513–530. doi:10.1093/molehr/gaq041 PMID:20538894

92. Bromer JG, Seli E (2008). Assessment of embryo viability in assisted reproductive technology: shortcomings of current approaches and the emerging role of metabolomics. Curr Opin Obstet Gynecol, 20:234–241.doi:10.1097/GCO.0b013e3282fe723d PMID:18460937

93. Schoolcraft WB, Katz-Jaffe MG, Stevens J et al. (2009). Preimplantation aneuploidy testing for infertile patients of advanced maternal age: a randomized prospective trial. Fertil Steril, 92:157–162.doi:10.1016/j.fertnstert.2008.05.029 PMID:18692827

94. Harper JC, Harton G (2010). The use of arrays in preimplantation genetic diagnosis and screening. Fertil Steril, 94:1173–1177.doi:10.1016/j.fertnstert.2010.04.064 PMID:20579641

95. Hellani A, Abu-Amero K, Azouri J, El-Akoum S (2008). Successful pregnancies after application of array-comparative genomic hybridization in PGS-aneuploidy screening. Reprod Biomed Online, 17:841–847.doi:10.1016/S1472-6483(10)60413-0 PMID:19079969

96. Botros L, Sakkas D, Seli E (2008). Metabolomics and its application for non-invasive embryo assessment in IVF. Mol Hum Reprod, 14:679–690.doi:10.1093/molehr/gan066 PMID:19129367

97. Seli E, Vergouw CG, Morita H et al. (2010). Noninvasive metabolomic profiling as an adjunct to morphology for noninvasive embryo assessment in women undergoing single embryo transfer. Fertil Steril, 94:535–542.doi:10.1016/j.fertnstert.2009.03.078 PMID:19589524

98. Assou S, Haouzi D, De Vos J, Hamamah S (2010). Human cumulus cells as biomarkers for embryo and pregnancy outcomes. Mol Hum Reprod, 16:531–538.doi:10.1093/molehr/gaq032 PMID:20435608

99. Estes SJ, Ye B, Qiu W et al. (2009). A proteomic analysis of IVF follicular fluid in women ≤ or =32 years old. Fertil Steril, 92:1569–1578.doi:10.1016/j.fertnstert.2008.08.120 PMID:18980758

100. McLeod HL, Parodi L, Sargent DJ et al. (2006). UGT1A1*28, toxicity and outcome in advanced colorectal cancer: results from Trial N9741. J Clin Oncol, 24 Suppl;18S.

101. Toffoli G, Cecchin E, Corona G et al. (2006). The role of UGT1A1*28 polymorphism in the pharmacodynamics and pharmacokinetics of irinotecan in patients with metastatic colorectal cancer. J Clin Oncol, 24:3061–3068.doi:10.1200/JCO.2005.05.5400 PMID:16809730

102. Marcuello E, Altés A, Menoyo A et al. (2004). UGT1A1 gene variations and irinotecan treatment in patients with metastatic colorectal cancer. Br J Cancer, 91:678–682. PMID:15280927

103. Chiara S, Serra M, Marroni P et al. (2005). UGT1A1 promoter genotype and toxicity in patients with advanced colorectal cancer treated with irinotecan-containing chemotherapy. J Clin Oncol, 23 Suppl;16S.

104. Massacesi C, Terrazzino S, Marcucci F et al. (2006). Uridine diphosphate glucuronosyl transferase 1A1 promoter polymorphism predicts the risk of gastrointestinal toxicity and fatigue induced by irinotecan-based chemotherapy. Cancer, 106:1007–1016.doi:10.1002/cncr.21722 PMID:16456808

105. Carlini LE, Meropol NJ, Bever J et al. (2005). UGT1A7 and UGT1A9 polymorphisms predict response and toxicity in colorectal cancer patients treated with capecitabine/irinotecan. Clin Cancer Res, 11:1226–1236. PMID:15709193

106. Biological Markers in Environmental Health Research (1987). Committee on biological markers of the National Research Council. Environ Health Perspect, 74:3–9. doi:10.1289/ehp.87743 PMID:3691432

107. Basso K, Margolin AA, Stolovitzky G et al. (2005). Reverse engineering of regulatory networks in human B cells. Nat Genet, 37:382–390.doi:10.1038/ng1532 PMID:15778709

UNIT 4.
INTEGRATION OF BIOMARKERS INTO EPIDEMIOLOGY STUDY DESIGNS

CHAPTER 18.

Combining molecular and genetic data from different sources

Evangelia E. Ntzani, Muin J. Khoury, and John P. A. Ioannidis

Summary

The rapidly growing number of molecular epidemiology studies is providing an enormous, often multidimensional, body of evidence on the association of various disease outcomes and biomarkers. The testing and validation of statistical hypotheses in genetic and molecular epidemiology presents a major challenge requiring methodological rigor and analytical power. The non-replication of many genetic and other biomarker association studies suggests that there may be an abundance of spurious findings in the field. This chapter will discuss ways of combining evidence from different sources using meta-analysis methods. Research synthesis not only aims at producing a summary effect estimate for a specific biomarker, but also offers a unique opportunity for a meticulous attempt to critically appraise a research field, identify substantial differences between or within studies, and detect sources of bias. Systematic reviews and meta-analyses in human genome epidemiology are specifically discussed, as they comprise the bulk of the available evidence in molecular epidemiology where these methods have been applied to date. Considered here are issues regarding validity and interpretation in genetic association studies, as well as strategies for developing and integrating evidence through international consortia. Finally, there is a brief look at how combining data through meta-analysis may be applied in other areas of molecular epidemiology.

Introduction

The number of molecular epidemiology studies is constantly growing, and this trend is expected to accelerate (1–4), especially with improvements in genotyping technology that allow massive testing of genetic variants in minimal time and at a decreasing cost on a genome-wide association study platform (5–8). The number of potentially identifiable genetic markers, and the multitude of clinical outcomes to which these may be associated, make the testing and validation of statistical hypotheses in genetic and molecular epidemiology a task of unprecedented scale. Currently, more than 6000 original articles on human genome epidemiology findings are published annually, and the numbers are increasing (9,10).

Yet, there has been considerable concern about non-replication in gene-disease association studies (11–19) and other areas of molecular epidemiology. The combination of high-throughput genotyping, selective reporting, and exploratory statistical analyses in studies with limited sample sizes could potentially generate a scientific literature replete with spurious findings and lead to wasted resources, unless mechanisms are put in place to promptly evaluate evidence as it accumulates (20,21). Related concerns also apply to other fields of molecular epidemiology where large amounts of data are produced and it is important to achieve unbiased integration of the evidence.

Combining evidence from different sources is discussed here. The goal of research synthesis is to estimate and explain between-study heterogeneity, arrive at summary effects, and appraise the quality and reliability of the evidence procured by many studies on the same research question. Specifically, systematic reviews and meta-analyses in human genome epidemiology are discussed, as they comprise the bulk of the available evidence in molecular epidemiology where these methods have been applied. Issues regarding validity and interpretation in genetic association studies are considered, as well as strategies for developing and integrating high-quality genomic evidence through international consortia. Finally, means for applying combined data through meta-analysis in other areas of molecular epidemiology are briefly examined.

Systematic reviews and meta-analyses: Definitions

Systematic reviews and meta-analyses provide valuable tools for summarizing genetic effects and for identifying and explaining the underlying differences and observed discrepancies between studies. The term systematic review has been used as a contrast to traditional review. Systematic reviews use a predefined, structured approach to the collection and integration of available evidence, whereas traditional reviews offer a non-structured, non-standardized appraisal of the current literature distorted in varying degrees by the reviewer's personal opinion and experience. The goal of this systematic approach is to guarantee the transparency and completeness of the review process. Meta-analyses use quantitative research synthesis methodology to derive summary estimates of the studied effects and to describe and explain the variability between and within studies (22). Systematic reviews and meta-analyses are well-established approaches to research synthesis in clinical trials, where their strengths and limitations have been widely assessed (23). Increasingly, they have also been applied to observational studies (24); meta-analyses of observational data are currently as common as those of clinical trials. Meta-analyses of gene–disease association studies have been accepted as a key method for establishing the genetic components of complex diseases (14,17). In 1998, The Human Genome Epidemiology Network (HuGENet) was launched as a global collaboration of individuals and organizations interested in accelerating the development of the knowledge base on genetic variation and common diseases. HuGENet has promoted the publication of HuGE reviews as a means of integrating evidence from human genome epidemiologic studies, that is, population-based studies of the impact of human genetic variation on health and disease (25). Initial efforts to apply quantitative methods were cautious, but there is now wide agreement that a meta-analysis of the evidence is almost always indicated and can provide more useful insights than a simple narrative review, provided the caveats of data synthesis are properly recognized. By the end of 2009, approximately 1200 systematic reviews and meta-analyses had been published on human genome epidemiology topics (http://www.cdc.gov/genomics/hugenet/default.htm); most of them, however, tried to integrate information on only one or a few specific gene–disease associations at a time. The unknown extent of unpublished data and the likelihood of biases inherent in single studies threaten the credibility of genetic findings.

While most meta-analyses in the past have been retrospective exercises, there is an increasing interest for prospective collaborative analyses that use the same statistical methods as traditional retrospective meta-analyses. Collaborative meta-analyses may be undertaken by consortia or networks of investigators working on the same disease and/or set of research questions. Participating teams may combine already-collected data, perform projects that use both retrospectively and prospectively collected information, or develop new collaborative projects on a completely prospective basis. With the advent of genome-wide association studies (GWAS), it is common practice to immediately seek replication of proposed discovered associations by other teams of investigators and publish the combined data in the same article. As a more recent alternative, meta-analyses have been implemented by combining multiple data sets at the discovery stage under a consortium umbrella (26–28). This is a prospective use of meta-analysis methods. Furthermore, for many diseases and research questions, numerous such coalitions of investigators may exist; bringing their data together presents a new field of application for meta-analysis methods.

Reviewing methods: Basic aspects

Recommendations have been developed regarding the conduct of systematic reviews and meta-analyses. In 2006, HuGENet posted online the first edition of a handbook for conducting HuGE reviews (29,30). The reporting of these studies may need further improvement and standardization in the literature and should become more evidence-based with increasing experience. Such standards may follow the examples of similar initiatives for genetic association studies (e.g. STREGA (31)), as well as other designs and disciplines (e.g. CONSORT (32,33), MOOSE (34), PRISMA (35), STARD (36) and TREND (37)).

First, typical retrospective systematic reviews and meta-analyses will be discussed. A typical systematic review includes the following stages: 1) formulation of the research question requiring appraisal of the available evidence, 2) identification of the eligible studies and data extraction, 3) synthesis of the available evidence, 4) assessing and addressing potential biases, and 5) interpreting the results.

Research questions

Formulating the research question is fundamental for systematic reviews, as for any other research endeavour. Decisions must be made upfront about which gene and variants and which disease and outcomes to assess, as well as the eligibility criteria for the study designs and the study and population characteristics. Different eligibility criteria may lead to different data being synthesized and possibly different conclusions.

Data

Identifying the studies eligible for inclusion in a systematic review requires comprehensive, systematic literature searches. One must specify which eligible databases to search, and decide whether to consider data without regard for their prior publication in peer-reviewed literature or the specific language(s) of publication (38). (For more details on issues pertaining to the eligibility and choice of sources of data, see (39).) Data extraction for published information is typically performed by two independent investigators with critical discussion of any discrepancies.

Data synthesis

Synthesizing the available evidence is best done in a quantitative way, producing summary estimates of the assessed effect and estimates of the between-study heterogeneity, as well as measures of the uncertainty thereof. A quantitative synthesis must be strongly encouraged, whenever feasible, as a means of producing a summary estimate, but most importantly for quantification of heterogeneity and identification of potential bias. Some key issues on methods for evaluation of between-data set heterogeneity and for obtaining summary effects will be touched on briefly; a discussion on issues of multivariate models and adjustments will follow.

Heterogeneity

One should distinguish between clinical, biological and statistical heterogeneity. Statistical heterogeneity can be tested in any quantitative synthesis. Its presence may signal genuine biological and clinical heterogeneity or bias and errors. Often it is difficult to pinpoint what the exact reasons are for heterogeneity, and inferences should be cautious. Conversely, the absence of demonstrable statistical heterogeneity cannot be interpreted as proof of clinical and biological homogeneity.

Several heterogeneity tests and metrics are traditionally used in meta-analyses. (For more details and mathematical formulas, see (39).) The Q statistic provides a χ^2-based test and is considered significant for $P < 0.10$, but it is still underpowered in most meta-analyses whenever there are few (roughly < 20) data sets combined (40).

The between-study variance, τ^2, is not commonly used as a metric of heterogeneity, because its magnitude depends on the respective effect size metric (e.g. standardized mean difference, odds ratio, hazard ratio) and it is not comparable among meta-analyses using different effect metrics (41). However, a useful metric often neglected is the ratio of τ over the effect size. This ratio can provide a measure of the extent of variability (between-study standard deviation) as compared with the effect size. Given that many molecular epidemiology effects are small, the relative magnitude of the uncertainty versus the effect is a useful measure to consider. The most popular metric for conveying between-study heterogeneity is nevertheless the I^2. This metric has the major advantage that it is independent of the number of studies, and thus can be standardized for use across different meta-analyses with different effect metrics and different numbers of studies (40). I^2 is directly interpreted as the percentage of total variation across studies due to heterogeneity rather than chance, and it takes values between 0 and 100% (42). Values over 50% indicate large heterogeneity. However, I^2 also becomes uncertain when only a few studies are combined, as in the large majority of current meta-analyses (41,43), and therefore presentation of 95% confidence intervals should be considered routinely. This will help avoid spuriously strong inferences regarding heterogeneity or lack thereof.

Summary effects

To date, most meta-analyses have used either fixed or random effects methods for combining the data across eligible studies and data sets. Fixed effects models assume a common effect estimate for all studies and attribute all observed between-study variability to chance. Fixed effects models include inverse-variance weighting, and Mantel-Haenszel and Peto methods (44), and seem inappropriate in the presence of demonstrable or anticipated heterogeneity if used as the single methodology for the effect estimate calculation. In the absence of demonstrable heterogeneity, keep in mind that failure to reject the null hypothesis of homogeneity does not prove homogeneity. Random effects assume that there is a different underlying effect size for each study. There are many different proposed estimators for the between-study variance; the most popular was suggested by DerSimonian and Laird (45). Random effects accommodate between-study heterogeneity and thus should be preferred in the presence or anticipation of heterogeneity. In the absence of any heterogeneity, fixed and random effects give similar results in any case.

Unfortunately, these issues are not yet well understood in the literature, as shown by empirical evaluations of candidate gene meta-analyses and also meta-analyses of GWAS (46,47). Until recently, the choice of model for combining results from candidate gene studies lay on the straightforward concept of underlying heterogeneity. Nevertheless, in a GWAS setting, the presence of heterogeneity may not necessarily correspond to replication failure, but can signal difficulty in extending the probed association in diverse populations (48). In light of the generally limited power to detect moderate signals of effect at the discovery stage, the exclusive use of random effects models, and the more conservative confidence intervals produced when heterogeneity is present would result in forbidding possibly true signals to pass the genome-wide significance threshold and seek further replication however large the discovery data sets might be (power desert phenomenon) (49). Thus, it would be more appropriate to report the results from both models and make critical decisions on the basis of the stage at which meta-analysis is performed.

Besides traditional fixed and random effects models, there is an increasing application of more fully Bayesian methods in meta-analysis. Their discussion is beyond the scope of this chapter, but the interested reader is referred to a reference textbook (50) and the WinBUGS software manual (51).

Adjustments for other variables

Both adjusted and unadjusted effect estimates from single studies may be combined in meta-analyses. Questionnaire-based data are used to some extent to adjust effects estimates, including minimum information, such as age and sex, or more complex data, such as clinical features of the disease under study defining a potentially differentiating risk profile, where genetic or other molecular information could add additional information (52). This is more likely to be the case in large multicenter clinical trials or cohort studies, where a "nested" genetic association study is performed.

An issue with adjusted estimates is to ensure that similar or at least comparable adjustments have been performed across different studies. For retrospective efforts there is usually large variability in the types of adjustments. Moreover, even data on the same variables may have been collected across different studies using different questionnaires or procedures, and standardization may be difficult or even impossible. Finally, caution should be used when differentiating between variables that are independent predictors and others that may be surrogates of the genetic/molecular effect under study.

Assessing and addressing potential biases

There are often considerable and justifiable concerns regarding the quality and validity of molecular epidemiology studies. Critical appraisal of the studies included in a systematic review is of paramount importance for identifying the sources of bias inherent in each study. The types of biases include selection bias, information bias and confounding. Moreover, issues such as multiple testing should be considered, as well as concerns pertinent to specific types of biomarkers and studies (e.g. Hardy–

Weinberg equilibrium violations for genetic association studies). Appraisal of potential biases is often hampered by poor reporting of the primary studies. Poor reporting of observational studies (53) is a common challenge in synthesizing evidence; statements about STrengthening the Reporting of OBservational studies in Epidemiology (STROBE) (54–56), and STrengthening the REporting of Genetic Associations (STREGA) (31), a similar effort in human genome epidemiology sponsored by the Human Genome Epidemiology Network (HuGENet), contribute to the transparency of reporting and the prompt identification of potential sources of study discrepancies and bias.

Detailed discussion of the specific biases that may be encountered in single studies is beyond the scope of this chapter. Some suggested references follow for the interested reader: selection bias (57–65); information bias involving biomarker measurement (e.g. genotyping), capture of environmental factors, or outcome assessment (8,66–69); and confounding which for genetic-association studies in particular manifests primarily through population stratification (70–75). For genome-wide investigations and other massive testing approaches, even minor biases on any of these fronts may create some highly statistically significant spurious signals among the many thousands being probed. Therefore, careful selection of cases and controls, high standards of genotyping and quality control, and routine use of principal component analysis, genomic control, family-based design, or other techniques that more rigorously control for stratification, are indispensable.

Interpreting the results

Interpreting the results of a systematic review and meta-analysis on an assessed biomarker includes consideration of the quantity and quality of the evidence and rigorous scrutiny for publication bias and selective reporting in the field at large. In terms of the quantity of accumulated evidence, it is unclear how much genetic information would be sufficient to validate a genetic association. Empirical evidence has demonstrated that initial research publications often fail to predict the subsequently established genetic effects and may even show substantial discrepancies with later research (14,76).

Publication and selective reporting bias

The tendency to publish studies with positive rather than negative results (preferring studies with large effects or statistically significant results) introduces publication bias (13,34). Publication bias is very difficult to address in a retrospective collection of published evidence. Tests such as funnel plots are notoriously unreliable and subjective, and they should be abandoned. Even formal statistical testing for funnel plot asymmetry cannot fully discriminate between publication bias and other sources of bias or genuine heterogeneity. In addition, the tests are generally underpowered (77–79) and subject to extensive limitations that make them useful only in a few meta-analyses (80). If these tests are employed, a suitable modified regression test should be selected that has appropriate type 1 error properties (81). Such tests would be more correctly called tests for small study effects, since they essentially evaluate whether small studies differ in their results from larger ones.

Another common issue that could have an increasingly important impact in molecular epidemiology is selective reporting of specific analyses and outcomes among the many that may be performed, often in pursuit of nominal statistical significance (82–85). Ideally, straightforward *a priori* hypotheses should be explicitly reported, and study objectives and future analyses should be documented at their outset under a collaborative initiative (20,31). However, this may not be as transparent as it should be, and lack of transparency is compounded by the exploratory nature of much molecular epidemiological research. A meta-analysis diagnostic that can be used to evaluate the presence of "significance-chasing" biases, including publication and other selective reporting biases, has been proposed (86). The test is most useful for application across many meta-analyses (e.g. evaluation of large research fields), while it is expected to be underpowered for meta-analyses with few studies.

Causal inference

An observed association may be spurious or real. Spurious associations may be due to chance, bias within studies, or bias across studies (reporting biases affecting the whole research field of interest). For genetics of common diseases, real associations, not attributable to confounding, may be due to a direct causal variant or to a variant in linkage disequilibrium (LD) with a direct causal variant (13,87,88). They can be a source of the heterogeneity found between studies of gene–disease associations.

Traditional epidemiological criteria for establishing causation include consistency, strength, biological plausibility (including analogy), dose–response, temporality, experimental support, and coherence (89,90). Nevertheless, rarely are all of these issues taken into account, and the last three are not really relevant to human genome epidemiology. In genetic epidemiology, replication as an expression of consistency has received the greatest attention (13,14,17). Strength would be difficult to assess, as genetic effects are generally modest with odds ratios below two or even below 1.5 (91). Furthermore, the size of an effect is a characteristic of the genetic association being studied rather than a biologically consistent feature, as it depends on the relative prevalence of other causes (92).

In theory, biological plausibility should be an important criterion for causation, bringing under the same denominator epidemiologic evidence and diverse forms of biological evidence (93–99). Biological data on gene function, and on the tissue(s) in which a gene is expressed, could contribute to making a causal inference about gene-disease associations. On the other hand, there is concern that a biological argument can be constructed for virtually any associated allele because of the "...relative paucity of current understanding of the mechanisms of action of complex trait loci." (11). Thus, some form of mechanistic evidence might be identified and (mis)used selectively to reinforce an assertion of causality. Empirical evidence suggests very low agreement between biological and epidemiological evidence for common genetic variants and complex diseases (100). While candidate gene studies are often based on some biological knowledge of the candidate gene, genome-wide linkage and association studies initially identify variants without consideration of their biological function. Yet, the absence of mechanistic evidence or evidence of high quality would not exclude inferring that an association is causal if other guidelines for causation are satisfied. As knowledge of the genome is incomplete, biological plausibility may not always be apparent (97,101-103).

Criteria for assessing cumulative evidence

For genetic associations, a consensus approach recently developed interim guidelines for grading of the cumulative epidemiological evidence (104). The grading considers three aspects (known as the Venice criteria): amount of evidence, consistency of replication, and protection from bias (Table 18.1, Figure 18.1). Particularly for retrospective meta-analyses, protection from bias cannot be assumed if: the effect size is small (odds ratio deviating less than 1.15 from the null), the summary results lose their formal statistical significance when the first study that proposed the association is removed or when Hardy–Weinberg equilibrium-violating studies are removed, there are strong signals of small-study effects (e.g. a significant modified regression test) or significance-chasing bias (as discussed above), or if there are other demonstrable major biases in any aspect. Additional, yet weaker, signals for potential bias would be unclear/misclassified phenotypes with possible differential misclassification against genotyping or vice versa, major concerns for population stratification, or any other reason (case-by-case basis) that would jeopardise the validity of the proposed association. For prospective consortium-endorsed meta-analyses, all of the above parameters must be taken into consideration with the exception of small effect size, small study effects, and significance-chasing bias based on the basic assumption that selective reporting bias is not operating in the field.

Figure 18.1. Categories for the credibility of cumulative epidemiological evidence. The three letters correspond (in order) to amount of evidence, replication and protection from bias. Evidence is categorized as strong when there is A for all three items, and is categorized as weak when there is a C for any of the three items. All other combinations are categorized as moderate

AAA	ABA	ACA
AAB	ABB	ACB
AAC	ABC	ACC

First letter = amount
Second letter = replication
Third letter = protection from bias

BAA	BBA	BCA
BAB	BBB	BCB
BAC	BBC	BCC

■ Strong evidence
■ Moderate evidence
□ Weak evidence

CAA	CBA	CCA
CAB	CBB	CCB
CAC	CBC	CCC

Source: (104). Reproduced with permission of Oxford University Press.

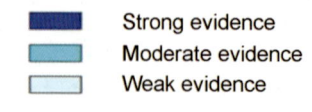

Table 18.1. Considerations for epidemiologic credibility in the assessment of cumulative evidence on genetic associations

Criteria	Categories	Proposed operationalization
Amount of evidence	A Large-scale evidence B Moderate amount of evidence C Little evidence	Thresholds may be defined based on sample size, power, or false discovery rate considerations. The frequency of the genetic variant of interest should be accounted for. As a simple rule, it is suggested that category A contains over 1000 subjects (total number of cases and controls assuming 1:1 ratio) evaluated in the least common genetic group of interest, B corresponds to 100–1000 subjects evaluated in this group, and C corresponds to <100 subjects evaluated in this group.*
Replication	A Extensive replication including at least one well-conducted meta-analysis with little between-study inconsistency B Well-conducted meta-analysis with some methodological limitations or moderate between-study inconsistency C No association, no independent replication, failed replication, scattered studies, flawed meta-analysis, or large inconsistency	Between-study inconsistency entails statistical considerations (e.g. defined by metrics such as I^2, where values of ≥50% are considered large, and values of 25–50% are considered moderate inconsistency) and also epidemiological considerations for the similarity/standardization, or at least harmonization, of phenotyping, genotyping, and analytical models across studies.
Protection from bias	A Bias, if at all present, could affect the magnitude but probably not the presence of the association B No obvious bias that may affect the presence of the association, but there is considerable missing information on the generation of evidence C Considerable potential for or demonstrable bias that can affect even the presence or not of the association	A prerequisite for A, is that the bias due to phenotype measurement, genotype measurement, confounding (population stratification), and selective reporting (for meta-analyses) can be appraised as not being high, plus there is no other demonstrable bias in any other aspect of the design, analysis, or accumulation of the evidence that could invalidate the presence of the proposed association. In category B, although no strong biases are visible, there is no such assurance that major sources of bias have been minimized or accounted for, as information is missing on how phenotyping, genotyping and confounding have been handled. Given that occult bias can never be ruled out completely, note that even in category A the qualifier "probably" is used.

* For example, if the association pertains to the presence of homozygosity for a common variant and if the frequency of homozygosity is 3%, then category A amount of evidence requires over 30 000 subjects, and category B between 3000 and 30 000.
Adapted from (104)

Networks in human genome epidemiology

Although meta-analyses of published data provide a mechanism for combining evidence from different sources, they cannot overcome methodological flaws originating from the primary studies. An alternative approach that may also help improve the quality of the primary data is a meta-analysis of individual participant data (MIPD), which involves collecting and analysing detailed data on individual subjects and, ideally, prospective meta-analysis of data collected from consortia of investigators (105).

Meta-analysis of individual participant data (MIPD)

The MIPD may offer some advantages over the meta-analysis of published data. In theory these advantages include: standardization of definitions of cases, molecular markers and other variables of interest, enhanced ability to contact meta-analysis of time-to-event outcomes, testing of the assumptions of time-to-event models, better control of confounding, standardized multivariable and adjusted analyses, consistent treatment of subpopulations, and assessment of sampling bias. Not every one of these advantages may be relevant in all MIPD applications and some may be impossible. For example, when studies have already been established with specific case definitions, it may not be possible to go back and achieve perfect standardization of definitions across all studies, or some adjusting variables may have been collected only in some of the studies but not others. Furthermore, an MIPD is far more labour-intensive and time-consuming than a meta-analysis of published data and may remain a retrospective effort (106).

Consortia and prospective collaborative efforts

An increasing number of consortia of investigators have been operating in molecular epidemiology. The value of such collaborative multicentre studies has long been recognized by epidemiologists for tackling important questions that are beyond the scope of a study at a single institution (107). Collaboration is of even greater significance in human genome epidemiology, due to the intrinsic characteristics of the field that can be better addressed through collaborative efforts (108): small sample sizes, weak expected genetic effects, genotype frequency variation in populations of different ethnic origin, and publication/selective reporting bias. Networks of scientists from multiple institutions can cooperate in research efforts involving, but not limited to, the conduct, analysis and synthesis of information from multiple population studies (3,20).

HuGENet has launched a global network of consortia working on human genome epidemiology, aimed at coordinating different research teams working on the same theme (109,110). The goal of the HuGENet Network of Investigator Networks is to create a resource to share information, offer methodological support, generate inclusive synopses of studies conducted in specific fields, and to facilitate rapid confirmation of findings. As of this writing, consortia in the Network of Investigator Networks comprise between five and more than 1000 teams, with accumulated sample sizes ranging from 3000 to over a half-million participants. Many other new consortia are continuously being developed. In particular for GWAS, it has become standard practice to try to replicate the derived associations across several other replicating teams as part of the first article to be published on a new proposed association (48). The replicating teams may already belong to an established consortium. Alternatively, their assembly may occur on an opportunity basis, but this may also form a nucleus for further collaborations. Besides choosing research targets based on agnostic massive testing approaches, other targets selected for study by consortia may be chosen based on *a priori* biological plausibility, supporting linkage evidence from genome-wide data, a perception of potentially high population risk (e.g. a common polymorphism), the number and consistency of published reports for a specific molecular marker, or a high-profile controversy in the literature (111,112). Also, consortia are increasingly being used to replicate associations derived from genome-wide association approaches independently from the first article that describes and partially replicates the associations (113).

Standardization issues

Members of consortia may share both prospective and retrospective features in the study design and accumulation of information. Standardization is one of the more significant benefits of consortia initiatives. Coordinating centres receive the incoming data, including both genotype and phenotype information, and guarantee adequate quality and transparency. Data standardization is best implemented at the beginning of a *de novo* collaborative study, while developing tools for data collection and definition of data items, and should achieve agreement on common data definitions to which all data layers must conform (114). Nevertheless, it may be difficult to achieve complete standardization if some data are already available. In this situation, consortia should still aim to maximize harmonization of data obtained from different sources.

Standardization or harmonization is crucial in order for a network to perform better than single studies. These processes increase the credibility of the derived evidence even when non-genetic measurements are difficult to standardise across teams. One criterion for the influence and success of a network may be its ability to adopt standards for phenotypes and covariates to establish the use of consistent definitions in subsequent studies. Standardization of genotypes, on the other hand, is usually achieved through central genotyping of all samples (115). Quality control of genotype results is typically straightforward, but additional checks are required in a multiteam collaboration. In the absence of central quality control, consortia may depend on post-hoc analyses, such as deviation from Hardy–Weinberg equilibrium (116) in the controls, to identify possible genotyping (or other) errors. Although large between-study heterogeneity in the final analyses may reflect measurement errors, these methods may still miss sizeable errors and their sensitivity and specificity are uncertain.

Meta-analyses of genome-wide association studies

As mentioned above, for many diseases several GWAS are performed and each may be accompanied by replication efforts by several other teams (117,118). These studies may have used different platforms, but it is still possible to combine data for markers that are in perfect or almost-perfect LD with a correlation coefficient r^2 approximating 1.00 (119,120). Examples of meta-analyses of several GWAS are available in the early literature (121,122). Meta-analysis is currently considered standard practice for a GWAS setting (123–126) (Figure 18.2). Apart from using meta-analysis in a sequential, multiple-stage manner to continuously update, refute or replicate association signals, it can also be implemented early on at the discovery stage by combining multiple data sets under a consortium umbrella, thus augmenting power to detect signals for subsequent replication (15,127).

Heterogeneity in the genome-wide association setting, where massive testing of agnostic (rather than candidate) markers takes place, has some special features. As previously mentioned, besides bias and errors, the possibility of genuine heterogeneity must be seriously considered, due to differential LD for the culprit gene variant and heterogeneity due to association with correlated phenotypes across the populations enrolled in different studies being combined (47).

Other applications of meta-analysis

Many fields of molecular epidemiology are characterized by large data sets that can be generated easily, due to the availability of sophisticated, low-cost technology. These data sets, derived from linkage scans, microarray-based gene expression profiling, mass spectra-based proteomics and many other massive testing platforms, usually capture information on hundreds of thousands of biological variables from a sometimes limited number of samples. To maximize the power to detect genuine signals requires combining data sets across different studies. However, combining data poses a further challenge, since the available data sets may have been obtained with different experimental conditions, platforms, analysis techniques or even sample types (e.g. different tissue, treatment conditions, or species). Meta-analysis could provide an appropriate framework for large data set synthesis. A few of these meta-analysis applications are mentioned here, but these are only indicative and the list is continuously expanding. Also discussed briefly are some issues that arise in the combination of information on other non-genetic biomarkers.

Figure 18.2. Typical work flow for conducting a meta-analysis of GWA data sets

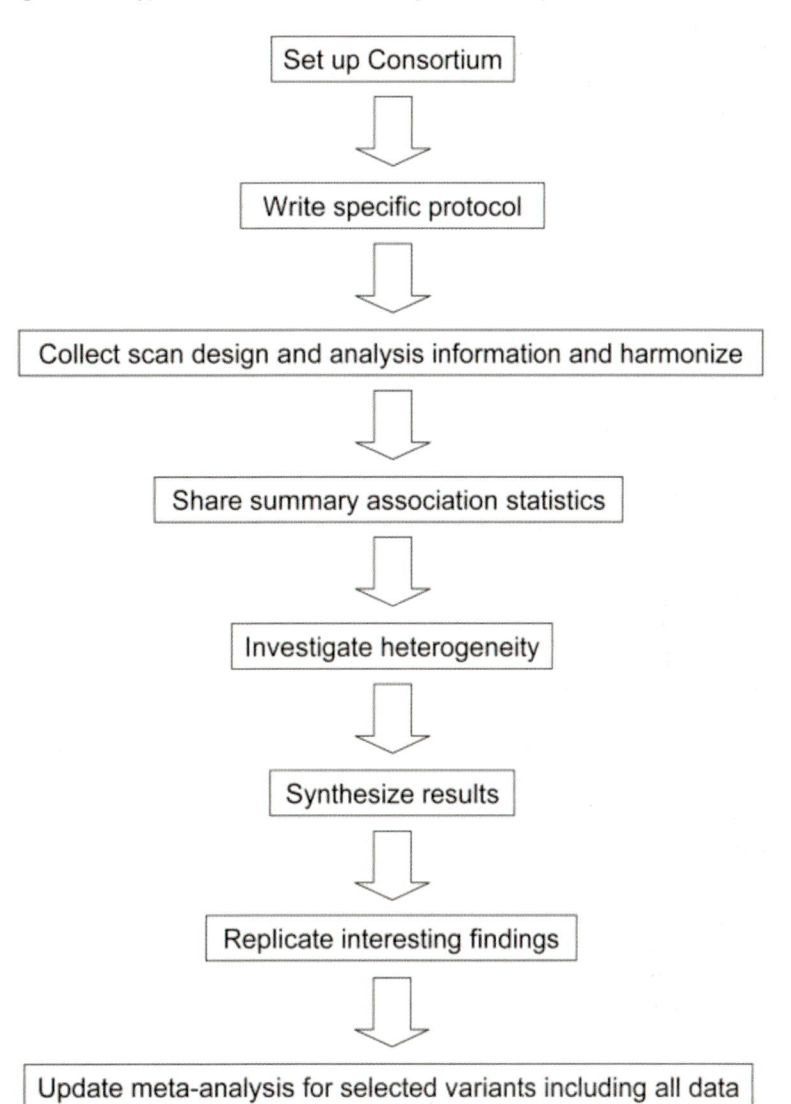

Set up Consortium
⬇
Write specific protocol
⬇
Collect scan design and analysis information and harmonize
⬇
Share summary association statistics
⬇
Investigate heterogeneity
⬇
Synthesize results
⬇
Replicate interesting findings
⬇
Update meta-analysis for selected variants including all data

Figure compiled from (104).

Meta-analysis of linkage signals in genome scans

Many teams of investigators have performed genome scans evaluating linkage between specific chromosomal loci and specific complex diseases (128–130). However, low linkage signals (131,132) and discrepancies in the findings of different teams often make the available evidence on a quantitative trait extremely difficult to summarize. Genome scan meta-analysis (GSMA) has been used as a method for summation of data from diverse genome scans through meta-analysis (131), and for formally testing whether the heterogeneity for specific chromosomal loci across genome scans (heterogeneity-based genome search meta-analysis (HEGESMA)) is large or small (133–134).

Microarrays and other multidimensional biology platforms

For various diseases, microarray platforms allow assessing differential expression of a large subset of genes. Research groups have approached the issue of synthesis across different platforms from different methodological perspectives (135–139). Significant computational power, multiple testing assumptions, and appropriate incorporation of heterogeneity estimates are only a few of the more challenging methodological issues. Given the small sample sizes of most microarray experiments and the complexity of the signals from single biological factors, meta-analysis may prove to be a very useful approach. Some non-parametric meta-analysis methods may allow synthesizing data from diverse platforms and different types of multidimensional data (140–142).

Meta-analyses of non-genetic prognostic markers

Besides the very large literature on genetic markers, there is also a burgeoning literature on non-genetic biomarkers. Single prognostic molecular markers, or combinations thereof, are still often considered in prognostic and predictive analyses for various clinical outcomes, such as mortality or other disease outcomes. Estrogen and other hormones, nutritional and related biochemical markers, and lipid or DNA adduct biomarkers are some of the commonly encountered examples in the literature (143–146).

Pertinent research synthesis methodology includes meta-analysis models as described above for genetic risk factors. Some of these predictors may be continuous variables, but the meta-analysis methods for combining information are very similar to the methods for combining data from binary markers (for details see (44)). Adjustment for covariates is more common in this literature, and may present problems related to the standardization of multivariate models and adjustments across the studies to be combined. Lack of standardization of biomarker measurements tends to be a more prominent problem than for genetic biomarkers, and error rates are expected to be larger and more variable across studies. Otherwise, heterogeneity testing and bias detection follow largely the same principles as described above for genetic markers.

Empirical evidence has shown that readily accessible published data can be misleading, producing a view of the literature that is distorted in a positive direction. An empirical evaluation has shown that almost all published prognostic marker studies on cancer report statistically significant results (147). Another empirical evaluation has shown that after standardising the definitions for the prognostic marker and the outcome under study, and, more importantly, after retrieving additional information that is unpublished or mentioned in only a cursory fashion in published articles, the statistical significance and predictive effect of a postulated prognostic/predictive factor may be abrogated (148).

In all, readily available information on prognostic factors may be the tip of the iceberg, and thus superficial perusal of the literature can lead to erroneous conclusions. This is yet another instance where selective reporting may spuriously inflate the importance of postulated prognostic factors unless retrieval of information and standardization of definitions in the literature are optimized. Meta-analyses of prognostic factors are likely to benefit from efforts to improve the conduct and reporting of primary studies, as exemplified by the REporting recommendations for tumour MARKer prognostic studies (REMARK) statement for tumour prognostic markers (149).

References

1. McPherson JD, Marra M, Hillier L et al.; International Human Genome Mapping Consortium (2001). A physical map of the human genome. *Nature*, 409:934–941.doi:10.1038/35057157 PMID:11237014

2. Guttmacher AE, Collins FS (2003). Welcome to the genomic era. *N Engl J Med*, 349:996–998.doi:10.1056/NEJMe038132 PMID:12954750

3. Little J, Khoury MJ, Bradley L et al. (2003). The human genome project is complete. How do we develop a handle for the pump? *Am J Epidemiol*, 157:667–673.doi:10.1093/aje/kwg048 PMID:12697570

4. Bell J (2004). Predicting disease using genomics. *Nature*, 429:453–456.doi:10.1038/nature02624 PMID:15164070

5. Lawrence RW, Evans DM, Cardon LR (2005). Prospects and pitfalls in whole genome association studies. *Philos Trans R Soc Lond B Biol Sci*, 360:1589–1595.doi:10.1098/rstb.2005.1689 PMID:16096108

6. Manolio TA, Collins FS, Cox NJ et al. (2009). Finding the missing heritability of complex diseases. *Nature*, 461:747–753.doi:10.1038/nature08494 PMID:19812666

7. Marchini J, Donnelly P, Cardon LR (2005). Genome-wide strategies for detecting multiple loci that influence complex diseases. *Nat Genet*, 37:413–417.doi:10.1038/ng1537 PMID:15793588

8. Thomas DC, Haile RW, Duggan D (2005). Recent developments in genomewide association scans: a workshop summary and review. *Am J Hum Genet*, 77:337–345. doi:10.1086/432962 PMID:16080110

9. Lin BK, Clyne M, Walsh M et al. (2006). Tracking the epidemiology of human genes in the literature: the HuGE Published Literature database. *Am J Epidemiol*, 164:1–4.doi:10.1093/aje/kwj175 PMID:16641305

10. Yu W, Wulf A, Yesupriya A et al. (2008). HuGE Watch: tracking trends and patterns of published studies of genetic association and human genome epidemiology in near-real time. *Eur J Hum Genet*, 16:1155–1158. doi:10.1038/ejhg.2008.95 PMID:18478035

11. Cardon LR, Bell JI (2001). Association study designs for complex diseases. *Nat Rev Genet*, 2:91–99.doi:10.1038/35052543 PMID:11253062

12. Gambaro G, Anglani F, D'Angelo A (2000). Association studies of genetic polymorphisms and complex disease. *Lancet*, 355:308–311.doi:10.1016/S0140-6736(99)07202-5 PMID:10675088

13. Hirschhorn JN, Lohmueller K, Byrne E, Hirschhorn K (2002). A comprehensive review of genetic association studies. *Genet Med*, 4:45–61.doi:10.1097/00125817-200203000-00002 PMID:11882781

14. Ioannidis JPA, Ntzani EE, Trikalinos TA, Contopoulos-Ioannidis DG (2001). Replication validity of genetic association studies. *Nat Genet*, 29:306–309.doi:10.1038/ng749 PMID:11600885

15. Ioannidis JP (2009). Population-wide generalizability of genome-wide discovered associations. *J Natl Cancer Inst*, 101:1297–1299.doi:10.1093/jnci/djp298 PMID:19726754

16. No authors listed. The Lancet (2003). In search of genetic precision. *Lancet*, 361:357.doi:10.1016/S0140-6736(03)12433-6 PMID:12573365

17. Lohmueller KE, Pearce CL, Pike M et al. (2003). Meta-analysis of genetic association studies supports a contribution of common variants to susceptibility to common disease. *Nat Genet*, 33:177–182.doi:10.1038/ng1071 PMID:12524541

18. Editorial (1999). Freely associating. *Nat Genet*, 22:1–2.doi:10.1038/8702 PMID:10319845

19. Tabor HK, Risch NJ, Myers RM (2002). Candidate-gene approaches for studying complex genetic traits: practical considerations. *Nat Rev Genet*, 3:391–397. doi:10.1038/nrg796 PMID:11988764

20. Ioannidis JP, Bernstein J, Boffetta P et al. (2005). A network of investigator networks in human genome epidemiology. *Am J Epidemiol*, 162:302–304.doi:10.1093/aje/kwi201 PMID:16014777

21. Ioannidis JP (2007). Non-replication and inconsistency in the genome-wide association setting. *Hum Hered*, 64:203–213. doi:10.1159/000103512 PMID:17551261

22. Lau J, Ioannidis JP, Schmid CH (1997). Quantitative synthesis in systematic reviews. *Ann Intern Med*, 127:820–826. PMID:9382404

23. Lau J, Ioannidis JP, Schmid CH (1998). Summing up evidence: one answer is not always enough. *Lancet*, 351:123–127.doi:10.1016/S0140-6736(97)08468-7 PMID:9439507

24. Stroup DF, Berlin JA, Morton SC et al. for the Meta-analysis of Observational Studies in Epidemiology (MOOSE) Group (2000). Meta-analysis of observational studies in epidemiology: a proposal for reporting. *JAMA*, 283:2008–2012.doi:10.1001/jama.283.15.2008 PMID:10789670

25. Khoury MJ, Dorman JS (1998). The Human Genome Epidemiology Network. *Am J Epidemiol*, 148:1–3. PMID:9663396

26. Easton DF, Pooley KA, Dunning AM et al.; SEARCH collaborators; kConFab; AOCS Management Group (2007). Genome-wide association study identifies novel breast cancer susceptibility loci. *Nature*, 447:1087–1093.doi:10.1038/nature05887 PMID:17529967

27. Ioannidis JP, Thomas G, Daly MJ (2009). Validating, augmenting and refining genome-wide association signals. *Nat Rev Genet*, 10:318–329.doi:10.1038/nrg2544 PMID:19373277

28. Khoury MJ, Bertram L, Boffetta P et al. (2009). Genome-wide association studies, field synopses, and the development of the knowledge base on genetic variation and human diseases. *Am J Epidemiol*, 170:269–279.doi:10.1093/aje/kwp119 PMID:19498075

29. Little J, Higgins J, editors. The HuGENet™ HuGE review handbook, version 1.0; 2006. Available from URL: http://www.medicine.uottawa.ca/public-health-genomics/web/assets/documents/HuGE_Review_Handbook_V1_0.pdf.

30. Higgins JP, Little J, Ioannidis JP et al. (2007). Turning the pump handle: evolving methods for integrating the evidence on gene-disease association. *Am J Epidemiol*, 166:863–866.doi:10.1093/aje/kwm248 PMID:17804859

31. Little J, Higgins JP, Ioannidis JP et al.; STrengthening the REporting of Genetic Association Studies (2009). STrengthening the REporting of Genetic Association Studies (STREGA): an extension of the STROBE statement. *PLoS Med*, 6:e22.doi:10.1371/journal.pmed.1000022 PMID:19192942

32. Altman DG, Schulz KF, Moher D et al.; CONSORT GROUP (Consolidated Standards of Reporting Trials) (2001). The revised CONSORT statement for reporting randomized trials: explanation and elaboration. *Ann Intern Med*, 134:663–694. PMID:11304107

33. Moher D, Schulz KF, Altman D; CONSORT Group (Consolidated Standards of Reporting Trials) (2001). The CONSORT statement: revised recommendations for improving the quality of reports of parallel-group randomized trials. *JAMA*, 285:1987–1991.doi:10.1001/jama.285.15.1987 PMID:11308435

34. Stroup DF, Thacker SB. Meta-analysis in epidemiology. In: Gail MH, Benichou J, editors. Encyclopedia of epidemiologic methods. Chichester (UK): Wiley & Sons Publishers; 2000. p. 557–570

35. Moher D, Liberati A, Tetzlaff J, Altman DG; PRISMA Group (2009). Preferred reporting items for systematic reviews and meta-analyses: the PRISMA statement. *BMJ*, 339 jul21 1;b2535.doi:10.1136/bmj.b2535 PMID:19622551

36. Bossuyt PM, Reitsma JB, Bruns DE et al.; Standards for Reporting of Diagnostic Accuracy (2003). Towards complete and accurate reporting of studies of diagnostic accuracy: The STARD Initiative. *Ann Intern Med*, 138:40–44. PMID:12513043

37. Des Jarlais DC, Lyles C, Crepaz N; TREND Group (2004). Improving the reporting quality of nonrandomized evaluations of behavioral and public health interventions: the TREND statement. *Am J Public Health*, 94:361–366. doi:10.2105/AJPH.94.3.361 PMID:14998794

38. Pan Z, Trikalinos TA, Kavvoura FK *et al.* (2005). Local literature bias in genetic epidemiology: an empirical evaluation of the Chinese literature. *PLoS Med*, 2:e334.doi:10.1371/journal.pmed.0020334 PMID:16285839

39. Kavvoura FK, Ioannidis JP (2008). Methods for meta-analysis in genetic association studies: a review of their potential and pitfalls. *Hum Genet*, 123:1–14.doi:10.1007/s00439-007-0445-9 PMID:18026754

40. Higgins JP, Thompson SG (2002). Quantifying heterogeneity in a meta-analysis. *Stat Med*, 21:1539–1558.doi:10.1002/sim.1186 PMID:12111919

41. Huedo-Medina TB, Sánchez-Meca J, Marín-Martínez F, Botella J (2006). Assessing heterogeneity in meta-analysis: Q statistic or I2 index? *Psychol Methods*, 11:193–206.doi:10.1037/1082-989X.11.2.193 PMID:16784338

42. Higgins JP, Thompson SG, Deeks JJ, Altman DG (2003). Measuring inconsistency in meta-analyses. *BMJ*, 327:557–560. doi:10.1136/bmj.327.7414.557 PMID:12958120

43. Ioannidis JP, Patsopoulos NA, Evangelou E (2007). Uncertainty in estimates of heterogeneity in meta-analysis. *BMJ*, 335:914–916. doi:10.1136/bmj.39343.408449.80 PMID:17974687

44. Sutton AJ, Abrams KR, Jones DR *et al.*, editors. Methods for meta-analysis in medical research. Chichester (UK): John Wiley & Sons; 2000.

45. DerSimonian R, Laird N (1986). Meta-analysis in clinical trials. *Control Clin Trials*, 7:177–188.doi:10.1016/0197-2456(86)90046-2 PMID:3802833

46. Attia J, Thakkinstian A, D'Este C (2003). Meta-analyses of molecular association studies: methodologic lessons for genetic epidemiology. *J Clin Epidemiol*, 56:297–303.doi:10.1016/S0895-4356(03)00011-8 PMID:12767405

47. Ioannidis JP, Patsopoulos NA, Evangelou E (2007). Heterogeneity in meta-analyses of genome-wide association investigations. *PLoS One*, 2:e841.doi:10.1371/journal.pone.0000841 PMID:17786212

48. Zeggini E, Ioannidis JP (2009). Meta-analysis in genome-wide association studies. *Pharmacogenomics*, 10:191–201.doi:10.2217/14622416.10.2.191 PMID:19207020

49. Pereira TV, Patsopoulos NA, Salanti G, Ioannidis JP (2009). Discovery properties of genome-wide association signals from cumulatively combined data sets. *Am J Epidemiol*,170:1197–1206.doi:10.1093/aje/kwp262 PMID:19808636

50. Spiegelhalter DJ, Abrams KR, Myles PJ, editors. Evidence synthesis. Chichester (UK): John Wiley & Sons Ltd; 2004.

51. Spiegelhalter DJ, Thomas A, Best NG, Lunn D. WinBUGS version 1.4 users manual; 2003.

52. Davis RL, Khoury MJ (2007). The emergence of biobanks: practical design considerations for large population-based studies of gene-environment interactions. *Community Genet*, 10:181–185.doi:10.1159/000101760 PMID:17575463

53. Pocock SJ, Collier TJ, Dandreo KJ *et al.* (2004). Issues in the reporting of epidemiological studies: a survey of recent practice. *BMJ*, 329:883.doi:10.1136/bmj.38250.571088.55 PMID:15469946

54. von Elm E, Altman DG, Egger M *et al.*; STROBE Initiative (2007). The Strengthening the Reporting of Observational Studies in Epidemiology (STROBE) statement: guidelines for reporting observational studies. *Lancet*, 370:1453–1457.doi:10.1016/S0140-6736(07) 61602-X PMID:18064739

55. STROBE Group. Strengthening the reporting of observational studies in epidemiology; 2005. Available from URL: http://www.strobe-statement.org/.

56. von Elm E, Egger M (2004). The scandal of poor epidemiological research. *BMJ*, 329:868–869.doi:10.1136/bmj.329.7471.868 PMID:15485939

57. Hill HA, Kleinbaum DG. Bias in observational studies. In: Gail MH, Benichou J, editors. Encyclopedia of epidemiologic methods. Chichester (UK): John Wiley & Sons, Ltd; 2000. p. 94–100.

58. Davey Smith G, Ebrahim S (2003). 'Mendelian randomization': can genetic epidemiology contribute to understanding environmental determinants of disease? *Int J Epidemiol*, 32:1–22.doi:10.1093/ije/dyg070 PMID:12689998

59. Little J, Khoury MJ (2003). Mendelian randomisation: a new spin or real progress? *Lancet*, 362:930–931.doi:10.1016/S0140-6736 (03)14396-6 PMID:14511923

60. Botto LD, Yang Q (2000). 5,10-Methylenetetrahydrofolate reductase gene variants and congenital anomalies: a HuGE review. *Am J Epidemiol*, 151:862–877. PMID:10791559

61. Brockton N, Little J, Sharp L, Cotton SC (2000). N-acetyltransferase polymorphisms and colorectal cancer: a HuGE review. *Am J Epidemiol*, 151:846–861. PMID:10791558

62. Cotton SC, Sharp L, Little J, Brockton N (2000). Glutathione S-transferase polymorphisms and colorectal cancer: a HuGE review. *Am J Epidemiol*, 151:7–32. PMID:10625170

63. Dorman JS, Bunker CH (2000). HLA-DQ and Type 1 Diabetes. *Epidemiol Rev*, 22:218–227. PMID:11218373

64. Madigan MP, Troisi R, Potischman N *et al.* (2000). Characteristics of respondents and non-respondents from a case-control study of breast cancer in younger women. *Int J Epidemiol*, 29:793–798.doi:10.1093/ije/29.5.793 PMID:11034958

65. Morton LM, Cahill J, Hartge P (2006). Reporting participation in epidemiologic studies: a survey of practice. *Am J Epidemiol*, 163:197–203.doi:10.1093/aje/kwj036 PMID:16339049

66. García-Closas M, Wacholder S, Caporaso NE, Rothman N. Inference issues in cohort and case-control studies of genetic effects and gene-environment interactions. In: Khoury MJ, Little J, Burke W, editors. Human genome epidemiology: a scientific foundation for using genetic information to improve health and prevent disease. New York (NY): Oxford University Press; 2004. p. 127–143.

67. Rothman N, Stewart WF, Caporaso NE, Hayes RB (1993). Misclassification of genetic susceptibility biomarkers: implications for case-control studies and cross-population comparisons. *Cancer Epidemiol Biomarkers Prev*, 2:299–303. PMID:8348052

68. Little J, Bradley L, Bray MS *et al.* (2002). Reporting, appraising, and integrating data on genotype prevalence and gene-disease associations. *Am J Epidemiol*, 156:300–310. PMID:12181099

69. Steinberg K, Gallagher M. Assessing genotypes in human genome epidemiology studies. In: Khoury MJ, Little J, Burke W, editors. Human genome epidemiology: a scientific foundation for using genetic information to improve health and prevent disease. New York (NY): Oxford University Press; 2004. P. 79–91.

70. Cardon LR, Palmer LJ (2003). Population stratification and spurious allelic association. *Lancet*, 361:598–604.doi:10.1016/S0140-6736 (03)12520-2 PMID:12598158

71. Thomas DC, Witte JS (2002). Point: population stratification: a problem for case-control studies of candidate-gene associations? *Cancer Epidemiol Biomarkers Prev*, 11:505–512. PMID:12050090

72. Wacholder S, Rothman N, Caporaso NE (2000). Population stratification in epidemiologic studies of common genetic variants and cancer: quantification of bias. *J Natl Cancer Inst*, 92:1151–1158.doi:10.1093/jnci/92.14.1151 PMID:10904088

73. Freedman ML, Reich D, Penney KL *et al.* (2004). Assessing the impact of population stratification on genetic association studies. *Nat Genet*, 36:388–393.doi:10.1038/ng1333 PMID:15052270

74. Ioannidis JP, Ntzani EE, Trikalinos TA (2004). 'Racial' differences in genetic effects for complex diseases. *Nat Genet*, 36:1312–1318.doi:10.1038/ng1474 PMID:15543147

75. Khlat M, Cazes MH, Génin E, Guiguet M (2004). Robustness of case-control studies of genetic factors to population stratification: magnitude of bias and type I error. *Cancer Epidemiol Biomarkers Prev*, 13:1660–1664. PMID:15466984

76. Trikalinos TA, Ntzani EE, Contopoulos-Ioannidis DG, Ioannidis JPA (2004). Establishment of genetic associations for complex diseases is independent of early study findings. *Eur J Hum Genet*, 12:762–769. doi:10.1038/sj.ejhg.5201227 PMID:15213707

77. Egger M, Davey Smith G, Schneider M, Minder C (1997). Bias in meta-analysis detected by a simple, graphical test. *BMJ*, 315:629–634. PMID:9310563

78. Lau J, Ioannidis JP, Terrin N et al. (2006). The case of the misleading funnel plot. *BMJ*, 333:597–600.doi:10.1136/bmj.333.7568.597 PMID:16974018

79. Terrin N, Schmid CH, Lau J (2005). In an empirical evaluation of the funnel plot, researchers could not visually identify publication bias. *J Clin Epidemiol*, 58:894–901.doi:10.1016/j.jclinepi.2005.01.006 PMID:16085192

80. Ioannidis JP, Trikalinos TA (2007). The appropriateness of asymmetry tests for publication bias in meta-analyses: a large survey. *Can Med Assoc J*, 176:1091–1096. doi:10.1503/cmaj.060410 PMID:17420491

81. Harbord RM, Egger M, Sterne JA (2006). A modified test for small-study effects in meta-analyses of controlled trials with binary endpoints. *Stat Med*, 25:3443–3457.doi:10.1002/sim.2380 PMID:16345038

82. Chan AW, Hróbjartsson A, Haahr MT et al. (2004). Empirical evidence for selective reporting of outcomes in randomized trials: comparison of protocols to published articles. *JAMA*, 291:2457–2465.doi:10.1001/jama.291.20.2457 PMID:15161896

83. Chan AW, Krleza-Jerić K, Schmid I, Altman DG (2004). Outcome reporting bias in randomized trials funded by the Canadian Institutes of Health Research. *Can Med Assoc J*, 171:735–740.doi:10.1503/cmaj.1041086 PMID:15451835

84. Chan AW, Altman DG (2005). Identifying outcome reporting bias in randomised trials on PubMed: review of publications and survey of authors. *BMJ*, 330:753.doi:10.1136/bmj.38356.424606.8F PMID:15681569

85. Contopoulos-Ioannidis DG, Alexiou GA, Gouvias TC, Ioannidis JP (2006). An empirical evaluation of multifarious outcomes in pharmacogenetics: beta-2 adrenoceptor gene polymorphisms in asthma treatment. *Pharmacogenet Genomics*, 16:705–711. doi:10.1097/01.fpc.0000236332.11304.8f PMID:17001289

86. Ioannidis JP, Trikalinos TA (2007). An exploratory test for an excess of significant findings. *Clin Trials*, 4:245–253.doi:10.1177/1740774507079441 PMID:17715249

87. Ardlie KG, Kruglyak L, Seielstad M (2002). Patterns of linkage disequilibrium in the human genome. *Nat Rev Genet*, 3:299–309. doi:10.1038/nrg777 PMID:11967554

88. Salanti G, Amountza G, Ntzani EE, Ioannidis JP (2005). Hardy-Weinberg equilibrium in genetic association studies: an empirical evaluation of reporting, deviations, and power. *Eur J Hum Genet*, 13:840–848. doi:10.1038/sj.ejhg.5201410 PMID:15827565

89. Hill AB (1965). The environment and disease: association or causation? *Proc R Soc Med*, 58:295–300. PMID:14283879

90. Surgeon General (Advisory Committee). Smoking and health. U.S. Department of Health, Education and Welfare. Washington (DC): U.S. Government Printing Office; 1964.

91. Ioannidis JP, Trikalinos TA, Khoury MJ (2006). Implications of small effect sizes of individual genetic variants on the design and interpretation of genetic association studies of complex diseases. *Am J Epidemiol*, 164:609–614.doi:10.1093/aje/kwj259 PMID:16893921

92. Rothman KJ, Greenland S, Lash TL, editors. Modern epidemiology. 3rd ed. Philadelphia (PA): Lippincott Williams & Wilkins; 2008.

93. Cloninger CR (2002). The discovery of susceptibility genes for mental disorders. *Proc Natl Acad Sci USA*, 99:13365–13367. doi:10.1073/pnas.222532599 PMID:12374853

94. Glazier AM, Nadeau JH, Aitman TJ (2002). Finding genes that underlie complex traits. *Science*, 298:2345–2349.doi:10.1126/science.1076641 PMID:12493905

95. Harrison PJ, Owen MJ (2003). Genes for schizophrenia? Recent findings and their pathophysiological implications. *Lancet*, 361:417–419.doi:10.1016/S0140-6736(03)12379-3 PMID:12573388

96. Editorial (2001). Challenges for the 21st century. *Nat Genet*, 29:353–354.doi:10.1038/ng1201-353 PMID:11726913

97. Page GP, George V, Go RC et al. (2003). "Are we there yet?": Deciding when one has demonstrated specific genetic causation in complex diseases and quantitative traits. *Am J Hum Genet*, 73:711–719.doi:10.1086/378900 PMID:13680525

98. Rebbeck TR, Martínez ME, Sellers TA et al. (2004). Genetic variation and cancer: improving the environment for publication of association studies. *Cancer Epidemiol Biomarkers Prev*, 13:1985–1986. PMID:15598750

99. Weiss KM, Terwilliger JD (2000). How many diseases does it take to map a gene with SNPs? *Nat Genet*, 26:151–157. doi:10.1038/79866 PMID:11017069

100. Ioannidis JP, Kavvoura FK (2006). Concordance of functional in vitro data and epidemiological associations in complex disease genetics. *Genet Med*, 8:583–593. doi:10.1097/01.gim.0000237775.93658.0c PMID:16980815

101. Begg CB (2005). Reflections on publication criteria for genetic association studies. *Cancer Epidemiol Biomarkers Prev*, 14:1364–1365.doi:10.1158/1055-9965.EPI-05-0407 PMID:15941941

102. Pharoah PD, Dunning AM, Ponder BA, Easton DF (2005). The reliable identification of disease-gene associations. *Cancer Epidemiol Biomarkers Prev*, 14:1362.doi:10.1158/1055-9965.EPI-05-0405 PMID:15941939

103. Ioannidis JP (2006). Common genetic variants for breast cancer: 32 largely refuted candidates and larger prospects. *J Natl Cancer Inst*, 98:1350–1353.doi:10.1093/jnci/djj392 PMID:17018776

104. Ioannidis JP, Boffetta P, Little J et al. (2008). Assessment of cumulative evidence on genetic associations: interim guidelines. *Int J Epidemiol*, 37:120–132.doi:10.1093/ije/dym159 PMID:17898028

105. Ioannidis JP, Rosenberg PS, Goedert JJ, O'Brien TR; International Meta-analysis of HIV Host Genetics (2002). Commentary: meta-analysis of individual participants' data in genetic epidemiology. *Am J Epidemiol*, 156:204–210.doi:10.1093/aje/kwf031 PMID:12142254

106. Steinberg KK, Smith SJ, Stroup DF et al. (1997). Comparison of effect estimates from a meta-analysis of summary data from published studies and from a meta-analysis using individual patient data for ovarian cancer studies. *Am J Epidemiol*, 145:917–925. PMID:9149663

107. Seminara D, Obrams GI (1994). Genetic epidemiology of cancer: a multidisciplinary approach. *Genet Epidemiol*, 11:235–254. doi:10.1002/gepi.1370110303 PMID:8088505

108. Kreeger K (2003). Consortia, 'big science' part of a paradigm shift for genetic epidemiology. *J Natl Cancer Inst*, 95:640–641. doi:10.1093/jnci/95.9.640 PMID:12734309

109. Ioannidis JP, Gwinn M, Little J et al.; Human Genome Epidemiology Network and the Network of Investigator Networks (2006). A road map for efficient and reliable human genome epidemiology. *Nat Genet*, 38:3–5. doi:10.1038/ng0106-3 PMID:16468121

110. Seminara D, Khoury MJ, O'Brien TR et al.; Human Genome Epidemiology Network; Network of Investigator Networks (2007). The emergence of networks in human genome epidemiology: challenges and opportunities. *Epidemiology*, 18:1–8.doi:10.1097/01.ede.0000249540.17855.b7 PMID:17179752

111. Colhoun HM, McKeigue PM, Davey Smith G (2003). Problems of reporting genetic associations with complex outcomes. *Lancet*, 361:865–872.doi:10.1016/S0140-6736(03)12715-8 PMID:12642066

112. Ioannidis JP (2005). Why most published research findings are false. *PLoS Med*, 2:e124.doi:10.1371/journal.pmed.0020124 PMID:16060722

113. Elbaz A, Nelson LM, Payami H et al. (2006). Lack of replication of thirteen single-nucleotide polymorphisms implicated in Parkinson's disease: a large-scale international study. *Lancet Neurol*, 5:917–923.doi:10.1016/S1474-4422(06)70579-8 PMID:17052658

114. John EM, Hopper JL, Beck JC et al.; Breast Cancer Family Registry (2004). The Breast Cancer Family Registry: an infrastructure for cooperative multinational, interdisciplinary and translational studies of the genetic epidemiology of breast cancer. *Breast Cancer Res*, 6:R375–R389.doi:10.1186/bcr801 PMID:15217505

115. Andrulis IL, Anton-Culver H, Beck J et al.; Cooperative Family Registry for Breast Cancer studies (2002). Comparison of DNA- and RNA-based methods for detection of truncating BRCA1 mutations. Hum Mutat, 20:65–73.doi:10.1002/humu.10097 PMID:12112659

116. Yonan AL, Palmer AA, Gilliam TC (2006). Hardy-Weinberg disequilibrium identified genotyping error of the serotonin transporter (SLC6A4) promoter polymorphism. Psychiatr Genet, 16:31–34.doi:10.1097/01.ypg.0000174393.79883.05 PMID:16395127

117. Hunter DJ, Kraft P (2007). Drinking from the fire hose–statistical issues in genomewide association studies. N Engl J Med, 357:436–439.doi:10.1056/NEJMp078120 PMID:17634446

118. Hindorff LA, Junkins HA, Mehta JP, Manolio TA. A catalog of published genome-wide association studies; 2009. Available from URL: http://genome.gov/gwastudies.

119. Voight BF, Scott LJ, Steinthorsdottir V et al.; MAGIC investigators; GIANT Consortium (2010). Twelve type 2 diabetes susceptibility loci identified through large-scale association analysis. Nat Genet, 42:579–589.doi:10.1038/ng.609 PMID:20581827

120. Elliott KS, Zeggini E, McCarthy MI et al.; Australian Melanoma Family Study Investigators; PanScan Consortium (2010). Evaluation of association of HNF1B variants with diverse cancers: collaborative analysis of data from 19 genome-wide association studies. PLoS One, 5:e10858.doi:10.1371/journal.pone.0010858 PMID:20526366

121. Zeggini E, Weedon MN, Lindgren CM et al.; Wellcome Trust Case Control Consortium (WTCCC) (2007). Replication of genome-wide association signals in UK samples reveals risk loci for type 2 diabetes. Science, 316:1336–1341.doi:10.1126/science.1142364 PMID:17463249

122. Evangelou E, Maraganore DM, Ioannidis JP (2007). Meta-analysis in genome-wide association datasets: strategies and application in Parkinson disease. PLoS One, 2:e196.doi:10.1371/journal.pone.0000196 PMID:17332845

123. Rivadeneira F, Styrkársdottir U, Estrada K et al.; Genetic Factors for Osteoporosis (GEFOS) Consortium (2009). Twenty bone-mineral-density loci identified by large-scale meta-analysis of genome-wide association studies. Nat Genet, 41:1199–1206.doi:10.1038/ng.446 PMID:19801982

124. Zeggini E, Scott LJ, Saxena R et al.; Wellcome Trust Case Control Consortium (2008). Meta-analysis of genome-wide association data and large-scale replication identifies additional susceptibility loci for type 2 diabetes. Nat Genet, 40:638–645. doi:10.1038/ng.120 PMID:18372903

125. McGovern DP, Gardet A, Törkvist L et al.; NIDDK IBD Genetics Consortium (2010). Genome-wide association identifies multiple ulcerative colitis susceptibility loci. Nat Genet, 42:332–337.doi:10.1038/ng.549 PMID:20228799

126. Houlston RS, Webb E, Broderick P et al.; Colorectal Cancer Association Study Consortium; CoRGI Consortium; International Colorectal Cancer Genetic Association Consortium (2008). Meta-analysis of genome-wide association data identifies four new susceptibility loci for colorectal cancer. Nat Genet, 40:1426–1435.doi:10.1038/ng.262 PMID:19011631

127. McCarthy MI, Abecasis GR, Cardon LR et al. (2008). Genome-wide association studies for complex traits: consensus, uncertainty and challenges. Nat Rev Genet, 9:356–369. doi:10.1038/nrg2344 PMID:18398418

128. Risch N (1990). Linkage strategies for genetically complex traits. III. The effect of marker polymorphism on analysis of affected relative pairs. Am J Hum Genet, 46:242–253. PMID:2301394

129. Becker KG, Simon RM, Bailey-Wilson JE et al. (1998). Clustering of non-major histocompatibility complex susceptibility candidate loci in human autoimmune diseases. Proc Natl Acad Sci USA, 95:9979–9984.doi:10.1073/pnas.95.17.9979 PMID:9707586

130. Cornélis F, Fauré S, Martinez M et al.; ECRAF (1998). New susceptibility locus for rheumatoid arthritis suggested by a genome-wide linkage study. Proc Natl Acad Sci USA, 95:10746–10750.doi:10.1073/pnas.95.18.10746 PMID:9724775

131. Wise LH, Lanchbury JS, Lewis CM (1999). Meta-analysis of genome searches. Ann Hum Genet, 63:263–272.doi:10.1046/j.1469-1809.1999.6330263.x PMID:10738538

132. Dempfle A, Loesgen S (2004). Meta-analysis of linkage studies for complex diseases: an overview of methods and a simulation study. Ann Hum Genet, 68:69–83.doi:10.1046/j.1529-8817.2003.00061.x PMID:14748832

133. Zintzaras E, Ioannidis JP (2005). HEGESMA: genome search meta-analysis and heterogeneity testing. Bioinformatics, 21:3672–3673.doi:10.1093/bioinformatics/bti536 PMID:15955784

134. Terwilliger JD, Ott J. Handbook of human genetic linkage. Baltimore (MD): Johns Hopkins University Press; 1994.

135. Conlon EM, Song JJ, Liu A (2007). Bayesian meta-analysis models for microarray data: a comparative study. BMC Bioinformatics, 8:80.doi:10.1186/1471-2105-8-80 PMID:17343745

136. Rhodes DR, Barrette TR, Rubin MA et al. (2002). Meta-analysis of microarrays: interstudy validation of gene expression profiles reveals pathway dysregulation in prostate cancer. Cancer Res, 62:4427–4433. PMID:12154050

137. Warnat P, Eils R, Brors B (2005). Cross-platform analysis of cancer microarray data improves gene expression based classification of phenotypes. BMC Bioinformatics, 6:265. doi:10.1186/1471-2105-6-265 PMID:16271137

138. Smid M, Dorssers LC, Jenster G (2003). Venn Mapping: clustering of heterologous microarray data based on the number of co-occurring differentially expressed genes. Bioinformatics, 19:2065–2071.doi:10.1093/bioinformatics/btg282 PMID:14594711

139. Grützmann R, Boriss H, Ammerpohl O et al. (2005). Meta-analysis of microarray data on pancreatic cancer defines a set of commonly dysregulated genes. Oncogene, 24:5079–5088.doi:10.1038/sj.onc.1208696 PMID:15897887

140. Zintzaras E, Ioannidis JP (2008). Meta-analysis for ranked discovery datasets: theoretical framework and empirical demonstration for microarrays. Comput Biol Chem, 32:38–46. PMID:17988949

141. Hsu YH, Zillikens MC, Wilson SG et al. (2010). An integration of genome-wide association study and gene expression profiling to prioritize the discovery of novel susceptibility Loci for osteoporosis-related traits. PLoS Genet, 6:e1000977.doi:10.1371/journal.pgen.1000977 PMID:20548944

142. Zhong H, Yang X, Kaplan LM et al. (2010). Integrating pathway analysis and genetics of gene expression for genome-wide association studies. Am J Hum Genet, 86:581–591.doi:10.1016/j.ajhg.2010.02.020 PMID:20346437

143. Friedenreich CM (1994). Methodologic issues for pooling dietary data. Am J Clin Nutr, 59 Suppl;251S–252S. PMID:8279435

144. Taioli E, Bonassi S (2002). Methodological issues in pooled analysis of biomarker studies. Mutat Res, 512:85–92.doi:10.1016/S1383-5742(02)00027-3 PMID:12220591

145. Veglia F, Matullo G, Vineis P (2003). Bulky DNA adducts and risk of cancer: a meta-analysis. Cancer Epidemiol Biomarkers Prev, 12:157–160. PMID:12582026

146. Flores-Mateo G, Navas-Acien A, Pastor-Barriuso R, Guallar E (2006). Selenium and coronary heart disease: a meta-analysis. Am J Clin Nutr, 84:762–773. PMID:17023702

147. Kyzas PA, Denaxa-Kyza D, Ioannidis JP (2007). Almost all articles on cancer prognostic markers report statistically significant results. Eur J Cancer, 43:2559–2579. PMID:17981458

148. Kyzas PA, Loizou KT, Ioannidis JP (2005). Selective reporting biases in cancer prognostic factor studies. J Natl Cancer Inst, 97:1043–1055.doi:10.1093/jnci/dji184 PMID:16030302

149. McShane LM, Altman DG, Sauerbrei W (2005). Identification of clinically useful cancer prognostic factors: what are we missing? J Natl Cancer Inst, 97:1023–1025.doi:10.1093/jnci/dji193 PMID:16030294

UNIT 5.
APPLICATION OF BIOMARKERS TO DISEASE

CHAPTER 19.

Cancer

Frederica P. Perera and Paolo Vineis

Summary

Molecular epidemiology was introduced in the study of cancer in the early 1980s, with the expectation that it would help overcome some important limitations of epidemiology and facilitate cancer prevention. The first generation of biomarkers has indeed contributed to our understanding of mechanisms, risk and susceptibility as they relate largely to genotoxic carcinogens, resulting in interventions and policy changes to reduce risk from several important environmental carcinogens. New and promising biomarkers are now becoming available for epidemiological studies, including alterations in gene methylation and gene expression, proteomics and metabolomics. However, most of these newer biomarkers have not been adequately validated, and their role in the causal paradigm is not clear. Systematic validation of these newer biomarkers is urgently needed and can take advantage of the principles and criteria established over the past several decades from experience with the first generation of biomarkers.

Prevention of only 20% of cancers in the United States alone would result in 300 000 fewer new cases annually, avoidance of incalculable suffering, and a savings in direct financial costs of over US$20 billion each year (1). Molecular epidemiology can play a valuable role in achieving this goal.

Introduction

In 1982, "molecular cancer epidemiology" was proposed as a new paradigm for cancer research that incorporated biomarkers into epidemiologic studies to reveal mechanisms and events occurring along the theoretical continuum between exposure and disease. Four categories of biomarkers were described: internal dose, biologically effective dose, early response/effect and susceptibility (2). In 1987, the United States National Academy of Sciences (NAS) convened a workshop on the use of biomarkers in environmental health research that adopted this concept and expanded it to include a fifth category: altered structure and function. Figure 19.1 summarizes the general paradigm proposed in 1982 and expanded in 1987 (3). The fundamental concept

Figure 19.1. Updated model for molecular epidemiology (figure compiled from (2,3,177))

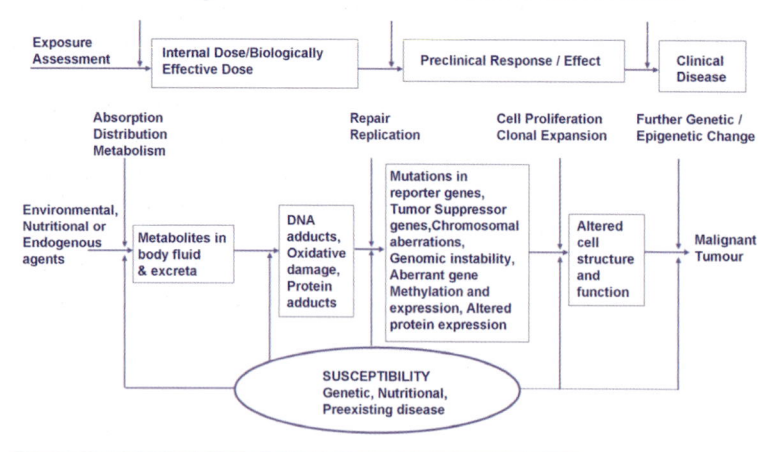

Primary Prevention and Clinical Interventions

of a continuum of molecular/genetic alterations leading to cancer that can be accessed using biomarkers remains valid.

Most of the focus thus far has been on biomarkers of genotoxicity. The field is now expanding rapidly to include high-throughput methods to detect alterations in the expression of genes, rather than structural changes. In this chapter, examples are provided of the accomplishments in molecular cancer epidemiology: studies that have provided evidence of causality and mechanisms, documented environment–susceptibility interactions and identified at-risk populations. The promise and challenge of new "omic" and epigenetic biomarkers (4–9), including their translational potential and need for validation, are then discussed. A discussion follows of the strengths, limitations and lessons learned from molecular epidemiologic research to date, and future directions for this field. Rather than an encyclopaedic review, presented are several paradigmatic examples of each area. Among promising biomarkers and technologies not included here are those related to inflammation and obesity (10,11), genome-wide scans (11), and tumour markers (12).

Context and public health significance

The context of this chapter on molecular cancer epidemiology is the need to prevent cancer, a disease that in the United States alone claims over half a million lives annually, with more than 1.5 million new cases diagnosed each year and attendant direct annual costs of US$107 billion (1). Many lines of evidence indicate, even more clearly than in 1982, that the great majority of cancers are in principle preventable, because the factors that determine their incidence are largely exogenous or environmental (5–8). These include exposures related to lifestyle (diet and smoking), occupation, and pollutants in the air, water and food supply. Genetic factors are largely important in terms of influencing individual susceptibility to carcinogens; only in some rare forms of human cancer do hereditary genetic factors play a decisive role. This awareness has lent greater urgency to the search for more powerful early-warning systems to identify causal environmental agents and flag risks well before the malignant process is entrenched.

Contributions of molecular epidemiology

The following sections refer to studies that have employed various study designs, the strengths and limitations of which are discussed in Chapters 14–18.

Providing evidence on causality and mechanisms: Examples

Polycyclic aromatic hydrocarbons/tobacco smoke and lung cancer

Most of the molecular epidemiologic research on lung cancer has targeted tobacco smoke as a model carcinogen. Polycyclic aromatic hydrocarbons (PAHs) such as benzo[a]pyrene (B[a]P) are one of 55 known carcinogens in tobacco smoke, are among the most studied, and often serve as a representative tobacco smoke carcinogen (13,14). Other tobacco carcinogens include 4-aminobiphenyl (4-ABP) and 4-(methylnitrosamino)-1-(3-pyridyl)-1-butanone (NNK) (14–16). PAHs are also found in outdoor air from fossil fuel combustion via automobile exhaust, emissions from coal-fired power plants, and other industrial sources; in indoor air from tobacco smoking, cooking and heating; and in the diet from consumption of smoked or grilled food (17,18). By several routes of exposure in adult animals, PAHs cause tumours including lung, liver and skin tumours (19) (see also (20) for review). PAHs are also transplacental carcinogens experimentally (21,22). PAHs such as B[a]P form adducts with DNA, a mechanism considered to be a critical early event in PAH-induced tumorigenesis, since adducts can lead to mutations and ultimately to cancer. As biomarkers, carcinogen-DNA adducts have the advantage of reflecting chemical-specific genetic damage that is mechanistically relevant to carcinogenesis (23,24).

In 1982, PAH-DNA adducts were detected in human subjects *in vivo*, specifically in white blood cells (WBCs) and lung tissue from lung cancer patients, most of whom were smokers (25). Using more sensitive laboratory methods to measure adducts, subsequent studies in healthy exposed populations (i.e. active smokers, coke-oven and foundry workers, and residents of Poland, the Czech Republic and China who were exposed to air pollution from coal burning) have found increased concentrations of PAH-DNA adduct levels in blood and other tissues compared to unexposed individuals, with no apparent threshold for DNA binding (26–29). These findings are consistent with traditional epidemiologic data showing elevated risk of lung cancer in PAH-exposed populations (see (20) for a review). Substantial interindividual variability has been observed in adduct levels among persons with similar exposure; about 30- to 70-fold for adducts in WBCs (29,30).

Although not all studies have been positive, since 1982 considerable evidence has mounted that PAH-DNA adducts in WBCs or lung tissue are risk markers for lung cancer (31–33). In one case–control study, higher PAH-DNA adduct levels were found in WBCs from 119 case subjects (compared with 98 control subjects), after adjusting for smoking, dietary PAH exposure and other potential confounders (32).

Caution is necessary in interpreting results from studies of DNA adduct levels and cancer risk. As discussed in Chapter 14, by their retrospective nature, case–control studies alone are unable to definitively establish causality. In addition, because the carcinogenic impact of adducts depends on the tissue and genes affected, one cannot assume *a priori* that adduct levels measured in blood are a valid surrogate for those in target tissue (34). The relationship between adduct concentration in blood and target tissue must be established for individual carcinogens. With respect to PAH-DNA, an experimental study (35) has shown ubiquitous binding of B[a]P metabolites to DNA and protein. Two other studies have found significant correlations between DNA adducts in WBCs and lung tissue from the same case subjects (35,36).

More recently, in a case–control study nested within the prospective Physicians' Health Study of over 14 000 men, it was evaluated whether DNA damage in blood samples collected at enrolment significantly predicted risk, consistent with the hypothesis that cases have greater biological susceptibility to PAHs and other aromatic tobacco carcinogens (37). The subjects in this nested case–control study were 89 cases of primary lung cancer and 173 controls, matched on smoking, age and duration of follow-up. Aromatic DNA adducts were measured in WBCs by the nuclease P1-enhanced ^{32}P-postlabelling method that primarily detects smoking-related adducts. Healthy current smokers who had elevated levels of aromatic DNA adducts in WBCs were approximately three times more likely to be diagnosed with lung cancer 1–13 years later than were current smokers with lower adduct concentrations (odds ratio (OR) = 2.98; 95% CI = 1.05–8.42; P = 0.04). The same relationship was not seen among former smokers and never smokers. The findings suggested that individuals who become cases have greater biological susceptibility to tobacco carcinogens, a biological difference that seems to manifest most clearly while exposure is still ongoing.

A second nested case–control study on lung cancer (newly diagnosed after recruitment) within the European Prospective Study Into Cancer and Nutrition (EPIC) cohort measured aromatic PAH-DNA adducts as markers of the biologically effective dose of PAHs, and mutations in the *ras* and *p53* genes in plasma DNA as markers of early preclinical effects. Cases included subjects with newly diagnosed lung cancer (n = 115) accrued after a median follow-up of seven years among the EPIC former smokers and never smokers. Unlike the prior nested case–control study, no current smokers were included. Adducts were associated with the subsequent risk of lung cancer among never smokers (OR = 4.04; 95% CI = 1.06–15.42) and among the younger age groups.

A meta-analysis of aromatic PAH-DNA adducts and lung cancer (38) concluded that current smokers with high levels of adducts have an increased risk of lung cancer, supporting a causal role of aromatic compounds in the etiology of lung cancer. While unmeasured variability in smoking, diet, or indoor/outdoor PAH concentrations may partially explain the finding of higher adduct levels in individuals with lung cancer, the results are also consistent with other evidence that some individuals are predisposed to genetic damage from PAHs and thereby to lung cancer (33,39,40). Taken together, the results of many studies support the theory that cumulative damage resulting from genotoxic chemicals that bind to DNA is a major cause of cancer (40).

Supporting molecular evidence that PAHs play an important role in lung cancer comes from observations that the *p53* tumour suppressor gene is mutated in 40–50% of lung tumours, and that the pattern of mutations in those tumours is consistent with the types of DNA adducts and mutations induced experimentally by B[a]P (41,42). Smokers with lung cancer show a pattern of mutations in *p53* that is different (with some exceptions) from that of non-smokers (43). Moreover, as discussed above, certain single nucleotide polymorphisms (SNPs) or genes involved in the metabolism or detoxification of PAHs or in the repair of PAH-DNA adducts have been implicated as effect modifiers in lung carcinogenesis.

In addition to genetic damage and gene mutations, epigenetic mechanisms are now emerging as important in lung cancer related to tobacco smoking (discussed in a later section).

In summary, studies using biomarkers of biologically effective dose, early preclinical effect/response, and individual susceptibility (SNPs) have been valuable in elucidating the steps that link tobacco smoke/PAH exposure to the onset of lung cancer.

AFB_1, HBV and liver cancer

During the past 30 years, research in experimental animals and humans has confirmed that the foodborne mutagen aflatoxin B1 (AFB_1) is a human hepatocarcinogen acting synergistically with the hepatitis B virus (HBV). AFB_1 is a fungal metabolite present in grains and cereals due to improper storage (44). Research has indicated that several biomarkers of the internal or biologically effective dose of AFB_1 (AFB_1 metabolites, AFB_1-albumin adducts, and AFB_1-N^7-guanine adducts in urine) and HBV surface antigen seropositivity are risk markers for liver cancer on a population level. In 1992, a prospective study in Shanghai, China found that among 18 244 men there were 22 incident cases of liver cancer (45). Analysis of urine samples banked 1–4 years before diagnosis from the case subjects and matched control subjects gave relative risks (RRs) of 2.4 (95% CI = 1.0–5.9) for any of the AFB_1 metabolites, and 4.9 (95% CI = 1.5–16.3) for detectable AFB_1-N^7-guanine adducts. There was a strong interaction between the serologic marker of HBV infection and the AFB_1 markers. Among individuals with chronic hepatitis infection who were also aflatoxin-positive, the RR was 60 (95% CI = 6.4–561.8). A subsequent follow-up study of 55 hepatocellular carcinoma (HCC) case subjects and 267 control subjects from the same cohort showed that the presence of any urinary AFB_1 biomarker significantly predicted liver cancer (RR = 5.0; 95% CI = 2.1–11.8) with an RR of 9.1 (95% CI = 2.9–29.2) for the presence of AFB_1-N^7-guanine adducts. A synergistic interaction between the presence of urinary AFB_1 biomarkers and HBV seropositivity resulted in a 59-fold (95% CI = 16.6–212.0) elevation in HCC risk (46). The implication for prevention is that both reduction in dietary levels of AFB_1 and wide-scale HBV vaccination are needed, since the benefits of the latter will not be manifest for many years (45). These biomarkers have subsequently been used as outcome measures in an intervention trial with the antischistosomal drug oltipraz (see further discussion below).

In Taiwan, China, subsequent studies of incident HCC case subjects and matched controls whose levels of AFB_1 metabolites, AFB_1-albumin and AFB_1-DNA adducts were measured in stored urine samples gave results consistent with the prior results from the PRC prospective study (47). In HBV-infected men with detectable AFB_1-albumin and AFB_1-DNA adduct levels, the risk of HCC was increased by 10-fold (RR = 10.0; 95% CI = 1.6–60.9) (48).

Other molecular data on the causal and mechanistic role of AFB_1 involve the *p53* gene. Early studies suggested a characteristic mutation spectrum in the human *p53* gene in HCC in South Africa and China, where it was observed that about 50% of the patients had a relatively rare mutation, a G to T transversion at codon 249 (49). This mutation was not previously observed in patients living in areas where food contamination by aflatoxins is not common; furthermore, the same mutation could be induced experimentally by AFB_1 *in vitro*. More recently, however, cells were incubated with AFB_1 and the types of DNA adducts induced in *p53* were studied (42). It was observed that adducts were mainly in sites different from codon 249, the one that the 'fingerprint' theory based on human data had implicated. In addition, the expected adducts in codon 249 were rapidly repaired (50% in seven hours). Therefore, the argument that aflatoxin exerts its carcinogenic activity by leaving a signature in a specific codon, and via a specific mechanism in *p53*, was considerably weakened. The apparent association of *p53*-specific mutations with aflatoxin now appears to be due to the selective advantage of mutated cells after exposure to HBV, rather than a causal event in the pathogenic process.

This example illustrates problems encountered in the use of human cancer gene fingerprints as definitive links between an exposure and a specific cancer. These difficulties include:

- the multifactorial nature of human cancers that hampers their attribution to single carcinogenic agents and/or the identification of a pathogenetic pathway common to several cancers;
- the high genetic instability of cancer cells that may increase the frequency of mutations in certain cancer genes regardless of exposure factors;
- the importance of DNA repair mechanisms and of the corresponding degree of population variation;
- tissue selection bias that may affect the results, although its extent is difficult to establish;
- the simultaneous presence of clinical (e.g. treatment) and biological factors (e.g. stage, grading) related to the exposure and to the frequency of mutations that may confound its association;
- the need for consideration of temporal sequences in the activation/deactivation of cancer genes;
- the fact that several different carcinogens may induce the same $p53$ mutation, and attribution to one of those carcinogens requires careful consideration of all relevant exposures.

For these reasons, the original hypothesis, that cancer fingerprints could be identified and used to recognize exposure-related tumours, has not been fully confirmed.

In summary, with this caveat in mind, studies using biomarkers of biologically effective dose, early preclinical response/effect and individual susceptibility (SNPs) have been valuable in elucidating the steps that link AFB_1 and HBV exposure to the risk of liver cancer.

Benzene and leukaemia

Benzene exposure occurs in the workplace and in the ambient environment largely because it is a component of gasoline (50). Another major source of public exposure to benzene is cigarette smoking. The example of benzene and haematological malignancies is paradigmatic, as it involves a single type of malignancy and combines biomarkers of several different classes that belong to the carcinogenic pathway shown in Figure 19.1. The various exposure markers include unmetabolized benzene in urine (UBz) and all major urinary metabolites (phenol (PH), E,E-muconic acid (MA), hydroquinone (HQ), and catechol (CA)), as well as the minor metabolite, S-phenylmercapturic acid (SPMA), all of which have been investigated among Chinese workers exposed to benzene (51). However, the most interesting results have come from investigations on early response/effect markers, specifically chromosomal aberrations.

Classical studies have shown that prospective data on chromosome aberrations are able to predict the onset of haematological malignancies. Combined analyses of data from Nordic and Italian prospective cohort studies, involving 3541 subjects, found that chromosomal aberrations were significant predictors of cancer (52). In the Nordic cohort, among subjects with high frequencies of chromosomal aberrations, the OR for all cancer deaths was 2.35 (95% CI = 1.31–4.23), compared with 2.66 (95% CI = 1.26–5.62) in the Italian cohort (53). In the Italian cohort, cancer predictivity of high chromosomal aberrations was greater for haematologic malignancies, with a standardized mortality ratio (SMR) of 5.49 (95% CI = 1.49–140.5) (54).

Specific chromosomal aberrations have been observed in both preleukemia and leukaemia patients exposed to benzene, as well as in otherwise healthy benzene-exposed workers (55). By use of fluorescent *in situ* hybridization (FISH) and the polymerase chain reaction (PCR), it was found that high occupational benzene exposure increased the frequencies of aberrations in chromosomes 5, 7, 9, 8 and 11—aberrations that are frequently seen in acute myeloid leukemias and in preleukemic myelodysplastic syndrome.

In the same studies on Chinese workers, protein-expression patterns were detected by surface-enhanced laser desorption/ionization time-of-flight mass spectrometry (SELDI-TOF MS). SELDI-TOF analysis of exposed and unexposed subjects revealed that lowered expression of PF4 and CTAP-III proteins is a potential biomarker of benzene's early biologic effects and may play a role in the immunosuppressive effects of benzene (56).

Finally, 20 candidate susceptibility genes were investigated in the same Chinese cohort (57). After accounting for multiple comparisons, SNPs in five genes were associated with a statistically significant decrease in total WBC counts among exposed workers (IL-1A (−889C > T), IL-4 (−1098T > G), IL-10 (−819T > C), IL-12A (8685G > A) and VCAM1 (−1591T > C)). This finding provides evidence that SNPs in genes that regulate haematopoiesis modify benzene-induced haematotoxicity. However, as is clarified later, much research on genetic variants and gene-environment interactions shows inconsistencies, and causal assessment is delicate, particularly when replication is lacking.

Molecular epidemiologic studies have also been conducted on acute lymphocytic leukaemia (ALL) in children, a disease that accounts for almost 25% of all childhood cancers. While more studies are needed, several have reported associations between parental or environmental exposure to benzene, or benzene-emitting sources, and childhood leukaemia, underscoring the potential importance of transplacental benzene exposures (58,59).

In summary, studies using biomarkers of internal dose, biologically effective dose, early preclinical effect/response and individual susceptibility (SNPs) have been valuable in elucidating the steps that link benzene exposure to the onset of leukaemia and other haematologic changes.

Nutritional factors and cancer

In the field of nutritional epidemiology, the investigation of biomarkers has shed some light on the role of obesity and metabolic syndrome in cancer. A high body mass index (BMI) has long been known to be associated with an increased risk of cancer at several sites, as the European Prospective Investigation into Cancer and Nutrition (EPIC) and other investigations have recently confirmed (60–63). The metabolic syndrome related to obesity is also suspected of a causal relationship with cancer (64). The metabolic syndrome is a constellation of central adiposity, impaired fasting glucose, elevated blood pressure and dyslipidemia (high triglyceride and low HDL cholesterol). The association of cancer with obesity and the metabolic syndrome has been unclear on biological grounds. Recently, however, several investigations have unveiled the role played by hormones and other intermediate markers related to key metabolic pathways. In particular, circulating insulin-like growth factor binding protein 1 (IGFBP-1), leptin, C-peptide and insulin are factors modified by obesity and have been associated with cancer. Higher circulating insulin levels may modulate cell proliferation and apoptosis, either directly or indirectly, by increasing the bioactivity of IGF-I, and decreasing the bioactivity of some of its binding proteins. Caloric restriction is a powerful way to reduce the occurrence of cancers, in particular lymphomas, induced by carcinogenic chemicals in *TP-53* deficient mice (65).

The evidence overall, however, is still incomplete. In a case–control study nested within the EPIC cohort involving 10 western European countries, serum C-peptide concentration was positively associated with an increased colorectal cancer risk for the highest versus the lowest quintile (OR = 1.56, 95% CI = 1.16–2.09, p for trend < 0.01). When stratified by anatomical site, the cancer risk was stronger in the colon (OR = 1.67, 95% CI = 1.14–2.46, p for trend < 0.01) than in the rectum (OR = 1.42, 95% CI = 0.90–2.25, p for trend = 0.35). No clear colorectal cancer risk associations were observed for IGFBP-1 or IGFBP-2. This large prospective study confirms that hyperinsulinemia, as determined by C-peptide levels, is associated with an increased colorectal cancer risk (66). In a nested case–control study in the prospective Prostate, Lung, Colorectal and Ovarian Cancer Screening Trial, which examined associations between IGF-1 and IGFBP-3 and risk of prostate cancer, a total of 727 incident prostate cancer cases and 887 matched controls were selected for a similar analysis. There was no clear overall association between IGF-1, IGFBP-3 and IGF-1:IGFBP-3 molar ratio (IGFmr) and prostate cancer risk; however, IGFmr was associated with risk in obese men (BMI > 30, p for trend = 0.04), with a greater than two-fold increased risk in the highest IGFmr quartile (OR = 2.34, 95% CI = 1.10–5.01). Risk was specifically increased for aggressive disease in obese men (OR = 2.80, 95% CI = 1.11–7.08) (67). However, in the EPIC study only a weak association was found between these factors (IGFmr not analysed) and prostate cancer (68).

Another associated line of research refers to the role of inflammation and immunity in obesity. The posited mechanism would imply immune impairment that accompanies obesity, and possibly a gene-environment interaction with leptin and other genes implicated in obesity (69).

While the relationships among the different factors involved in the relationship between cancer and obesity and the metabolic syndrome, as well as the precise causal pathways, are still far from clear (70), this is a promising area of research.

Arsenic and urothelial cancer

As in several studies mentioned above, tumour markers have been used to help identify causal environmental exposures in bladder cancer. A recent study of the differential expression of molecular markers in tissues of arsenic-related urothelial cancers (AsUC) (n = 33), non-AsUC (n = 20), and normal bladder urothelia from patients with benign diseases (n = 4) were examined for multiple selected molecular markers responsible for various cellular functions, including *Bcl-2*, *p53*, and *c-Fos* (71). The expression of *Bcl-2* and *c-Fos* in AsUC was significantly higher than in non-AsUC ($P = 0.004$ and $P = 0.02$, respectively), suggesting different carcinogenic pathways in the two etiologic groups. Such studies of the etiological heterogeneity of tumours at the molecular level may provide great insight into the mechanisms and causal pathways to carcinogenesis, which may lead to appropriate preventive strategies to reduce the incidence of cancer related to specific exposures (72).

Documenting environment-susceptibility interactions and identifying populations at greatest risk

To be effective, prevention strategies must target the most susceptible populations. This requires research to identify genetic and other susceptibility factors. Such research on exposure-susceptibility interactions must adhere to sound ethical principles, both in the conduct of research and in the communication of results and conclusions, in such a way as to discourage their inadvertent or intentional misuse (26,73–75). Although results from research on interactions have often been inconclusive and even conflicting, molecular epidemiologic studies indicate that some subgroups and individuals may have heightened susceptibility to environmental exposures. The categories of susceptibility factors that can modulate environmental risks, such as genetic predisposition, ethnicity, age, gender, and health and nutritional impairment, have been reviewed in detail elsewhere (26,73,74). With respect to the cancers and exposures discussed in this review, molecular epidemiologic studies have reported interactions between exposures to tobacco smoke, PAHs, AFB_1 or benzene and various susceptibility factors. These findings illustrate the complexities of interactions between environmental carcinogens and both genetic and non-genetic susceptibility factors. Susceptibility of the young has also been clearly demonstrated for several carcinogens.

Genetic susceptibility

Genes vary in their penetrance (the frequency, under given environmental conditions, with which a specific genotype is expressed by those individuals that possess it). Highly penetrant mutations in genes that are directly involved in carcinogenesis and confer a high risk of cancer in carriers represent the tail of a distribution of individual susceptibility to carcinogenesis (76). Less penetrant susceptibility may be conferred by common variants (SNPs) in genes that mediate the metabolism of carcinogens or DNA repair (77). For example, polymorphisms in certain cytochrome P450 (CYP) enzymes increase the oxidative metabolism of diverse endogenous and exogenous chemicals to their carcinogenic intermediates, while genetic variants in phase II (detoxifying) enzymes, such as glutathione S-transferase (*GST*), N-acetyltransferase (*NAT*), and epoxide hydrolase (*EH*) detoxify certain carcinogenic metabolites. Polymorphisms in DNA repair genes such as *XPD* or *XRCC1* can modulate risks from agents that directly or indirectly damage the DNA.

Rare and highly penetrant mutations in cancer genes may exert their effects without interacting with external exposures (usually by directly interfering with basic mechanisms of cell replication and differentiation), but gene–environment interactions are intrinsic to the mode of action of common, low-penetrance polymorphisms. The penetrance of a mutation is determined by other endogenous factors, including the importance of the function of the protein encoded by the gene (e.g. in key regulatory aspects of the cell cycle, as in the case of the *BRCA1* gene), the functional importance of the mutation (e.g. a total loss of function due to a truncating deletion versus a mild loss of function due to a point mutation), the existence of alternative pathways that can substitute for the loss of function, and interactions with other genes.

Most genes act in a sequence or in cascades. This is typical, for example, of metabolic and DNA repair genes. Genotyping according to pathways is likely to be much more rewarding then genotyping for single SNPs, in terms of both biological plausibility and statistical power (see discussion on the role of DNA repair genes (78)).

A large number of SNPs have been studied in molecular epidemiological investigations in recent decades, thanks to the development of quick and relatively inexpensive genetic techniques. However, only a few clear associations with cancer risk have been detected with reasonable certainty (i.e. consistently across different populations). Even with these SNPs most closely linked to cancer risk, caution is needed in extrapolating from one population and exposure scenario to another.

An example of a SNP consistently implicated in cancer is the methylenetetrahydrofolate reductase (*MTHFR*) gene, which plays an important role in the folate metabolism pathway (40,78). This enzyme provides the methyl group required for *de novo* methionine synthesis, and indirectly, for DNA methylation; therefore, it controls DNA stability and mutagenesis (79–81). Common *MTHFR* polymorphisms (C677T and A1298C) have been associated with reduced enzyme activity *in vitro* which, in the case of C677T, affects the metabolism of folate, consequently increasing homocysteine levels and (theoretically) the risk of colon cancer (82). According to a systematic review, in most studies *MTHFR* 677TT (10 studies, >4000 cases) and 1298CC (four studies, >1500 cases) were associated with moderately reduced colorectal cancer risk. In four of five genotype-diet interaction studies, 677TT subjects who had higher folate levels (or a high-methyl diet) had the lowest cancer risk (82).

An interaction of *MTHFR* SNPs with alcohol intake has also been reported, with high alcohol consumption levels decreasing DNA methylation, probably by hindering folate absorption, metabolism and excretion (83). Alcohol is thought to increase risk of cancer through its antagonist effects on folate. A study of health professionals examined folate, alcohol, *MTHFR* and alcohol dehydrogenase 3 (*ADH3*) polymorphisms in relation to risk of colorectal adenomas in 379 cases and 726 controls (84). *MTHFR* genotypes were not found to be appreciably related to risk of adenoma, but men who were TT homozygotes and who consumed 30+ g/day of alcohol had an OR of 3.52 (95% CI = 1.41–8.78) relative to drinkers of ≤ 5 g/day with the CC/CT genotypes (84).

Studies investigating the folate-*MTHFR*-cancer risk relationship have largely shown inverse associations of breast cancer risk with folate intake in all genotype groups, particularly among subjects with the 677TT genotype (85,86). Although the evidence is not conclusive, *MTHFR* provides a good example of how inherited gene variants can modify the cancer risk associated with dietary and other exposures.

With respect to lung cancer, various studies have implicated genetic polymorphisms involved in PAH metabolism (e.g. *CYP1A1* or *GST*) and DNA repair (e.g. *XRCC1*) as effect modifiers capable of increasing risk from PAHs (87–91). Increased risk of hepatocellular carcinoma has been associated with the *GSTM1* null/*GSTT1* null genotype in conjunction with smoking and drinking (92). The *GSTM1* null genotype, the low-activity epoxide hydrolase genotype, and a genetic polymorphism in *CYP2E1* also appear to confer greater risk of liver cancer (93,94).

Regarding leukaemia, the hepatic cytochrome P450 2E1 enzyme plays a key role in the activation of benzene to its ultimate haematotoxic and genotoxic benzoquinone metabolites (95). The NAD(P)H:quinone oxidoreductase (*NQO1*) and two subclasses of GSTs (M1 and T1) are involved in the detoxification of the ultimate benzoquinones and their reactive benzene oxide intermediates, respectively (50,95). A case–control study of occupational benzene poisoning in Shanghai showed that individuals homozygous for the *NQO1*[609] C→A mutation were at a 7.6-fold (95% CI = 1.8–31.2) greater risk of poisoning (96). Benzene poisoning was linked to risk of preleukemia and leukaemia. Theoretically, individuals with high activities of cytochrome P450 2E1 and homozygous mutations in the *NQO1*, *GSTT1* and *GSTM1* genes would be at highest risk of benzene haematotoxicity (50), but this inference has not been demonstrated conclusively. As noted earlier, polymorphisms in several IL and VCAM genes have been implicated in benzene haematoxicity (57).

DNA repair capacity is a particularly important source of variability in susceptibility to cancer. In addition to rare syndromes that involve faulty repair and genetic instability (e.g. ataxia-teleangectasia, Fanconi anaemia, Bloom syndrome, and xeroderma pigmentosum) (97), individuals commonly vary in their capability to repair DNA damage, at least in part due to genetics. The role of SNPs in three DNA repair genes (*XRCC1*-Arg399Gln, exon 10; *XRCC3*-Thr241Met, exon 7; and *XPD*-Lys751Gln, exon 23), and their combination in modulating the levels of DNA adducts in a population sample of healthy individuals has been investigated (98). The ^{32}P-postlabelling assay was used to measure aromatic DNA-adduct levels in WBCs from peripheral blood. A dose–response relationship between the number of at-risk alleles and the levels of adducts (*P* = 0.004) was observed, suggesting that the combination of multiple variant alleles may be more important than single SNPs in modulating cancer risk; hence the importance of focusing on gene pathways in the study of gene-environment interactions.

In addition to SNPs or polymorphisms in DNA repair genes, phenotypic tests have been widely used in recent years to measure DNA repair. The mutagen sensitivity assay, based on DNA damage (usually chromosome breaks) induced with chemical (bleomycin) or physical mutagens (radiation), unscheduled DNA synthesis, ^{3}H-thymidine incorporation, or count of pyrimidine dimers are examples of tests by which DNA repair is inferred from the different frequency of DNA damage induced in cancer cases and controls, without direct evidence of repair. Other phenotypic tests (e.g. the plasmid cat gene test, the ADPRT modulation test, or immunoassays based on antigenicity of thymidine) are based on some direct evidence of repair (77).

In contrast to genotype-based studies of DNA repair, for which the evidence is still largely inconsistent (78,99), most studies using phenotypic tests from which DNA repair is inferred have shown highly statistically significant results (100). When odds ratios were available, they were between 2.8 and 10.3, suggesting a strong association. However, the results are limited by potential confounding (i.e. the possibility that some exposure or characteristic of the patient is associated with DNA repair and is a risk factor for cancer, thus creating a spurious relationship between DNA repair and the disease). Repair enzymes can be induced in several ways, such as by stresses that damage DNA (e.g. pro-oxidative stress that could result from several endogenous and exogenous exposures). For example, in human studies, several tests of DNA repair were affected by characteristics such as age, sunlight, dietary habits, exposure to pro-oxidants, and cancer therapies (100). While age and therapies were usually controlled for in most studies, dietary habits might have acted as confounders, since both the intake and the plasma level of carotenoids and other antioxidants have been shown to be lower in cancer patients compared to healthy controls. The extent of such potential confounding is unknown, but could be substantial.

Another major limitation of many tests is that DNA repair is only indirectly inferred from DNA damage. To draw firm conclusions about a cause-effect relationship, more information about the biological meaning of tests is needed—for instance, whether they actually reflect DNA repair or a general or specific impairment of the DNA repair machinery.

Many investigations of gene–environment interactions (GEI) are underway in different parts of the world. Some ongoing studies are extremely large (e.g. EPIC, United Kingdom Biobank); all of them employ similar methods for genotyping (Taqman and high-throughput methods, such as Illumina). However, the quality of exposure assessment (e.g. diet, air pollution) is extremely variable. Ideally, understanding GEI requires determining, with equal resolution, both environmental exposures (e.g. to pesticides, air pollutants, ETS or dietary constituents) and genetic variants that are postulated to modulate the effects of the environmental exposures. However, there is an asymmetry between the two in that genotyping is much more accurate than most of methods used to measure environmental exposures. This implies a lower degree of genetic classification error, which in turn means an easier identification of associations between genes and disease than with environmental exposures and disease.

Suppose that classification error is expressed by the correlation coefficient between each exposure "assessor" and a reference standard (r = 1 means no error, r = 0.9 means a 10% classification error). For different expected relative risks that associate exposure with disease, one can compute the relative risks under different conditions of classification error. For example, a classification error of 10% implies the drop of a relative risk of 2.5 to 2.3 (i.e. little change). With an extreme classification error of 90%, however, even a relative risk of 2.5 becomes 1.1 (i.e. undetectable with common epidemiological methods). The lesson is that false-negative results are much more likely when analysing the role of environmental exposures than genetic variables (while in the latter case false-positives may be the main problem). In addition, very large numbers of subjects are needed if one wants to study interaction, for example, between a frequent exposure (prevalence 25%) and a frequent genotype (prevalence 50%). Presume that classification error is 20% for the environmental exposure (sensitivity = 80%) (in actuality, classification error for most exposures is likely to be much larger). Classification error could be around 7% for genotyping (sensitivity 93%). This is realistic, since genotyping techniques are currently validated and extremely accurate. The consequence of this situation is that approximately 1800 cases would be needed to observe main effects of genes if no classification error occurs, 2700 if exposure is incorrectly classified 20% of the time, and 3200 if the genotype is also mistaken 7% of the time. To study the effect of interactions, four times more subjects than those estimated above would be needed.

False-positives seem to be a common problem in genetic research, often due to small numbers and statistical instability. As pointed out by Ioannidis (the "Proteus phenomenon"), gene–disease associations that seem to be strong at first appear to be much weaker when larger studies are conducted (101). Publication bias contributes to this problem. For this reason, initiatives like the Venice criteria have been launched to provide sound systematic reviews of the genetic evidence (102).

In conclusion, the evidence concerning the role of low-penetrant genes in cancer is contradictory and difficult to interpret. Most observed associations between cancer and low-penetrant gene variants have been weak or very weak (with 20–50% increases in cancer risk). This, in fact, is inherent in the concept of low penetrance. However, the penetrance of a gene variant depends on interaction with external exposures, the internal environment, or other genes. Thus, the strength of association is a relative, not absolute, concept and requires the study of interactions. Nonetheless, interactions themselves are obviously difficult to investigate, as the study of a two-way interaction requires a sample size four times larger than the study of a main effect; therefore, little is known about the nature and strength of gene–environment interactions.

Genome-wide association studies (GWAS) and new methodological issues

Technical developments, with platforms such as Illumina or the Affymetrix microchips, offer the possibility of analysing up to 550 000 or even 1 million gene variants in one run. This revolution is giving rise to an unprecedented wave of new potential discoveries, as is testified by several papers in *Nature, Science*, and *Nature Genetics* in 2007, such as the Wellcome Trust Case-Control Study Consortium (103). Regarding cancer, a successful story is represented by the identification of chromosome 8q24 as the probable locus of a genetic risk factor for prostate cancer. Family-based linkage studies, association studies, and studies of tumours had already highlighted human chromosome 8q as a genomic region of interest for prostate cancer susceptibility loci. Recently, a locus at 8q24, characterized by both a SNP and a microsatellite marker, was shown to be associated with prostate cancer risk in Icelandic, Swedish and US samples (104). These data suggest that the locus on chromosome 8q24 harbours a genetic variant associated with prostate cancer, and that the microsatellite marker is a stronger risk factor for aggressive prostate cancers defined by poorly differentiated tumour morphology. Evidence has now been provided that colon cancer might also be associated with the same region. Using a multistage genetic association approach comprising 7480 affected individuals and 7779 controls, researchers have also identified markers in chromosomal region 8q24 associated with colorectal cancer (105). This example is interesting for two reasons: reverse genetics (the possibility that etiologic pathways for cancers that elude epidemiological research can be discovered starting from genetic susceptibility) and pleiotropy (the ability of certain gene variants to increase/modulate the risk for quite different diseases). (See Chapter 6 and (106,107) for a summary of recent GWAS findings.)

Apart from the 8q24 success story, many other contributions to the potential understanding of cancer and other diseases have come from GWAS. Exfoliation glaucoma is a striking example for which a potent signal has been identified, but this is an exception (in addition to being a non-cancer example). A cancer example is the KITLG gene and testicular carcinoma (see (108)). A summary of the locuses associated with cancer and other diseases after GWAS is available in the so-called GWAS catalogue of the National Human Genome Research Institute (http://www.genome.gov/GWAStudies).

However, genome-wide scans are clearly open to an even greater risk of false-positive findings related to multiple comparisons. Also, the interaction with external exposures is usually ignored. Design issues, including the investigation of traits that show strong familial aggregation, the selection of clinically homogeneous populations, and selection of cases that have a family history, are emerging as very influential on the success of genome-wide studies (109). (See also Chapter 6 for discussion of methodologic issues.)

Ethnicity, gender and nutritional factors

Ethnicity also appears to affect cancer risk. For example, higher rates of various smoking-related cancers in blacks may be partially explained by the finding that, in black smokers, urinary concentrations of NNK metabolites and serum concentrations of cotinine, a nicotine metabolite, exceeded those in white smokers (110). However, the effect of unmeasured differences in the exposure levels of the subjects cannot be ruled out.

Although studies have been inconsistent, there is evidence that women may be inherently more susceptible than men, on a dose-for-dose basis, to certain lung carcinogens. Several epidemiologic studies indicate that women smokers are 1.7- to 3-fold more likely to develop lung cancer than are male smokers with the same exposure (111,112). The level of PAH-DNA adducts and the frequency of G:C→T:A transversions in *p53* were elevated in lung tumours from female smokers compared with those from male smokers (112–114). Adduct levels in non-tumour lung tissue were also higher in women than in men, with a higher ratio of adduct levels to pack-years in women (115). The greater expression of the *CYP1A1* gene found in lung tissue of female smokers suggests a possible mechanism for this gender difference. In addition, a case–control study of lung cancer found that the effect of the *GSTM1* null genotype on lung cancer risk was significant among women, but not among men (116).

Nutritional deficits resulting in low levels of antioxidants can also heighten susceptibility to lung and other carcinogens by increasing DNA damage and subsequent mutation and carcinogenesis by oxygen radicals, PAHs and other chemical carcinogens (117). Heavy smokers with low plasma levels of micronutrients, such as retinol and the antioxidant α-tocopherol, appear to have reduced protection against carcinogen-induced DNA damage (118). In several studies, these effects were seen only in smokers with the *GSTM1* null genotype, illustrating the importance of interactions between multiple susceptibility factors (119,120). Sensitivity to mutagens, as measured by bleomycin-induced chromatid breaks, was also increased in cultured lymphocytes of healthy individuals with low plasma levels of antioxidants (121).

A special case of susceptibility: The fetus and young child

Compared with exposures occurring in adult life, exposures *in utero* and in the early years can disproportionately increase the risks of childhood cancer and many types of cancer later in life (122–124). Experimental and epidemiologic data indicate that because of differential exposure or physiologic immaturity, fetuses, infants and children experience greater risks than adults from a variety of environmental toxicants, including PAHs, nitrosamines, pesticides, tobacco smoke, air pollution and radiation. The underlying mechanisms may include increased exposure to toxicants, greater absorption or retention of toxicants, reduced detoxification and DNA repair, the higher rate of cell proliferation during early stages of development, and the fact that cancers initiated in the womb and in the early years have the opportunity to develop over many decades (for a review, see (125)) (126–129).

New evidence has emerged in recent years on the role played by *in utero* exposures in relation to the development of cancer in childhood and adult life. Fetuses and newborns seem to be particularly susceptible to diverse carcinogens (126–129). In a series of studies, PAH-DNA adducts were evaluated in mother-newborn pairs in central Europe, the USA and China (130,131). Consistently, levels of adducts in newborn cord blood were the same or higher than in the mothers' blood, although estimated transplacental exposure based on experimental studies was on the order of one tenth of maternal exposures. These observations across a gradient of exposure and in four different ethnic groups suggested that the fetus may be 10-fold more susceptible to DNA damage than the mother, and that *in utero* exposure to PAHs may disproportionately increase carcinogenic risk. Underscoring the potential risk of transplacental exposure to carcinogens, PAH/aromatic DNA adducts in cord blood were positively associated with hypoxanthine-guanine phosphoribosyltransferase (HPRT) mutant frequency in newborns (25). These studies provided molecular evidence of links between transplacental exposure to common air pollutants and somatic mutations indicative of increased cancer risk. In another study, airborne PAHs, measured by personal air monitoring during pregnancy, were significantly associated with stable aberration frequencies in cord blood (132). However, the epidemiologic evidence is still inconclusive on the role of transplacental exposure to PAHs and air pollution and childhood cancer (133).

Other investigators have reported that prenatal or postnatal exposure to tobacco smoke or its constituents were associated with increased frequencies of DNA and haemoglobin adducts, as well as chromosomal aberrations in newborns or children (134,135). A significant association between paternal smoking (without maternal smoking) and death from childhood cancer was found (136). A significant difference in the HPRT mutational spectrum was reported between newborns of mothers exposed to environmental tobacco smoke (ETS) and newborns of unexposed mothers. Their results suggested that V(D)J recombinase mutations, which are associated with leukaemia and lymphomas, are induced by ETS exposure (137). A meta-analysis of 11 studies of childhood exposure to maternal and paternal ETS found a very small excess risk for childhood cancer (RR = 1.10; CI = 1.03–1.19) (138). In addition, early-life exposure to ETS is suspected of playing a causal role in adult cancer. Three studies found that childhood exposure to ETS increased the risk of lung cancer in adults (139–141).

There is direct chromosomal evidence of a link between *in utero* exposures and cancer in infancy and childhood. Approximately 75% of infant acute leukemias have a reciprocal translocation between chromosome 11q23 and one of several partner chromosomes, including chromosome 4, which creates a fusion of the MLL gene at 11q23 and the AF4 gene at 4q21. Providing direct evidence of prenatal initiation of infant leukemias, the MLL-AF4 gene fusion sequence has been detected in neonatal blood spots of leukaemia patients subsequently diagnosed at ages five months to two years (141). Similarly, a signal mutation (TEL-AML1) observed in 25% of childhood acute lymphocytic leukaemia (ALL) cases was found to be present in neonatal bloodspots of children who subsequently developed ALL (142). The interpretation is that the TEL-AML1 fusion is acquired prenatally and constitutes the "first hit" in childhood leukaemia.

Adolescence and young adulthood are also viewed as sensitive stages of life because of greater proliferative activity in epithelial cells of certain tissues, as seen in radiation-induced breast cancer (143). Initiation of smoking at an early age confers a higher risk of lung, bladder and possibly breast cancer (144). Breast cancer risk associated with the *NAT2* slow acetylator genotype was higher in women who began smoking under the age of 16 years (145). In addition, aromatic DNA adduct levels were highest in lung tissue of former smokers who had smoked during adolescence, suggesting either that smoking at a young age induces more persistent adducts or that young smokers are more susceptible to DNA adduct formation (146).

In conclusion, molecular epidemiology has provided valuable data on the existence of complex interactions between environmental exposures and susceptibility factors, and has spurred researchers to investigate further differences in susceptibility among subsets of the population. Neither experimental nor conventional epidemiologic research alone could have done this. Although more research is needed before risk assessors can routinely develop quantitative estimates of the risks to sensitive subsets posed by specific environmental agents, the information obtained thus far has relevance to risk assessment and prevention. For example, government agencies are already beginning to require that regulatory policies explicitly protect children as a susceptible group.

The promise and challenge of new "omic" and epigenetic biomarkers

Types of new biomarkers

Several new and exciting biomarkers are becoming available for epidemiological studies thanks to the development of high-throughput technologies and theoretical advancements in biology. However, most of these markers have not yet been adequately validated, and their role in the causal paradigm is not clear. An exhaustive review is not possible here, and the reader is referred to Chapter 5 and other critical reviews, in particular for gene expression and toxicogenomics (147–149).

Toxicogenomics

Toxicogenomics refers to the study of the complex interaction between the cells' genome and chemicals in the environment or drugs, as they relate to disease. One method for genome-wide analysis, comparative genomic hybridization (CGH), provides a molecular cytogenetic approach for genome-wide scanning of differences in DNA sequence copy number (150). This technique has been attracting widespread interest among cancer researchers, as evidenced by the rapidly expanding database of CGH publications that already covers about 1500 tumours, and is beginning to reveal genetic abnormalities characteristic of certain tumour types or stages of tumour progression. In theory, such genomic differences could be exploited to gain insights into the risk factors involved (150).

Epigenetics and promoter methylation

Epigenetic mechanisms of carcinogenesis (i.e. mechanisms that do not depend on structural changes in DNA but on functional regulation, such as DNA methylation) are increasingly identified as key steps in the pathway from exposure to cancer. DNA methylation is an important epigenetic determinant of gene expression, since it determines the process by which the instructions in genes are converted to mRNA, directing protein synthesis (81). DNA methylation, that is, the covalent addition of methyl groups (CH3) to cytosine that precedes a guanine in the DNA sequence (the CpG dinucleotide), occurs naturally and plays a role in suppressing gene expression. CpG dinucleotides are enriched in the promoting regions of genes (CpG islands). Hypermethylation of promoter regions is associated with gene transcriptional silencing, and is a common mechanism for the inactivation of tumour suppressor genes in human cancer (151). DNA methylation is heritable; it passes from one generation of cells to the next.

Promoter methylation is a mechanism that regulates gene expression and is believed to play a crucial role in lung carcinogenesis. Several genes are commonly the target of promoter hypermethylation in lung cancer, including the *p16* gene (*p16^{INK4a}/CDKN2A*), *DAPK*, *RAR-β*, *RASSF1* and *O^{6}MGMT* (a DNA-repair gene) (152). Global hypomethylation has also been observed (153). Both current and former smoking have been associated with aberrant *p16*, *DAPK*, *RASSF1A* and *RAR-β* methylation (152). Recently, investigators have found that two alternative pathways can be detected in the biopsies of smoking and non-smoking lung cancer patients: one involving methylation and *K-ras* mutations, and the other *EGRF* mutations in the absence of gene methylation (154). In a prospective study, promoter hypermethylation of multiple genes (including those mentioned above) in the sputum was able to predict lung cancer onset with sensitivity and specificity of 64% (155). Notably, aberrant promoter methylation can be detected in the plasma of lung cancer patients (156); high frequencies of *ECAD* and *DAPK* methylation have been reported in lymphocytes of smokers versus non-smokers (157). The capacity of some airborne particulate carcinogens to induce hypermethylation in the regulatory regions of tumour suppressor genes has also been demonstrated in animal studies (158). Overall, the animal models support involvement of promoter methylation and other epigenetic mechanisms in carcinogen-induced lung carcinogenesis (159,160).

Acetylation is another key mechanism in epigenetic pathways, although it has been studied less extensively than methylation in cancer epidemiology (161).

Metabolomics

The study of the complete set of low-molecular weight metabolites present in a cell or organism at any time is metabolomics, sometimes referred to as metabolomics. With high-throughput techniques (NMR spectroscopy and LC-MS) it is possible to measure a large number of metabolites simultaneously, and to define individual metabolic profiles that can be used to predict the onset of common diseases (162). Use of data processing and chemometric models has already allowed the characterization of disease states and metabolic disorders (163). While several cross-sectional metabonomic studies investigating various cancers have been undertaken (164,165), no longitudinal study has yet been carried out, and few validation studies have been published. In one investigation of repeat samples from dietary studies (166), high-resolution ^{1}H NMR spectroscopy was used to characterize 24-hour urine specimens obtained from population samples in Japan (n = 259), Chicago, USA (n = 315), and China (n = 278). The authors investigated analytical reproducibility, urine specimen storage procedures, interinstrument variability, and split specimen detection. The multivariate analytical reproducibility of the NMR screening platform was > 98%, and most classification errors were due to heterogeneity in handling of urine specimens. In addition, cross-population differences in urinary metabolites could be related to genetic, dietary, and gut microbial factors.

Proteomics

The study of an organism's entire complement of proteins is known as proteomics. Proteomics has been used for the investigation of several types of cancer (167–170) and of physiological or pathological changes associated with external exposures. Proteomic studies have identified, for example, changes in proteins associated with oxidative stress (171). The investigation of proteomic patterns could be a powerful tool both for the identification of intermediate changes that relate environmental exposures to disease onset, and as an early marker of cancer. However, methodological issues need to be resolved before application in prospective studies. In a critique of early papers, Diamandis (2004) identified several methodologic problems: the lack of reproducibility in analytical methods; the lack of reproducibility of proteomic patterns in different series of patients and by different laboratories; unresolved effects of different protocols for sample collection and processing, freeze–thaw, and duration of storage; possible selection effects in cases and controls (bias, confounding), partly because of the opportunistic sampling that characterized the early studies; the possible effect of drugs/other treatments; and inappropriate or non-reproducible data analysis. Many of these concerns apply to other epigenetic and "omic" biomarkers and have been addressed in subsequent proteomic studies. In conclusion, for all the "omic" technologies, validation studies are urgently needed.

Incorporating new intermediate epigenetic or "omic" biomarkers into etiologic studies

Epigenetic and "omic" technologies can provide intermediate markers (either reflecting exposure/effective dose, early effects, or preclinical disease) for etiologic purposes (to investigate the causes and mechanisms of disease onset) or for clinical purposes (early diagnosis, prognosis, follow-up). This chapter refers to etiologic purposes, but many of the considerations apply to clinical purposes as well (see (172) and (173) for a review of biomarker-based tools for cancer screening, diagnosis and treatment). While past experience with earlier biomarkers is relevant, the current era is different and poses new challenges for the following reasons: "omic" and new epigenetic methods tend to be discovery-oriented, rather than oriented to testing specific hypotheses; the main feature of current technologies is the ability to perform massive testing of markers (i.e. thousands of markers at a time), potentially in thousands of subjects; and such new intermediate markers introduce increased potential for confounding. So, although our ability to measure new intermediate markers has considerably increased, making the current phase potentially very exciting, methodological challenges have expanded more than proportionally. In fact, much uncertainty surrounds the validity and applicability of new technologies (see (174, 175)).

Feasibility is also an issue. For example, it is currently prohibitively expensive and labour-intensive to perform expression array analysis for every subject in large studies. An alternative is to select a small subset of matched pairs of exposed and unexposed subjects (or subjects with and without preneoplastic lesions) and discover differentially exposed genes. Once several target genes are identified, real-time PCR analysis can be used to quantify expression of selected genes in all subjects (176).

Another important difference between the earlier and newer biomarkers is that the traditional cancer paradigm was very much centred around DNA damage and mutations, while recent research has uncovered several additional intermediate steps between genotype and phenotype, and has highlighted the importance of gene expression/modulation in carcinogenesis. Therefore, combinations of both types of biomarkers are expected to be informative, since pathways are not mutually exclusive.

Several critical steps in the putative causal pathway linking exposure to the onset of cancer can be explored with intermediate markers. Referring to the classical scheme (Cf. Figure 19.1), intermediate markers can play a role in each of the following steps: they can be related to exposure (e.g. metabolomics); related to early effects or changes in the causal pathways leading to disease (like promoter methylation, gene mutations, or changes in telomere length); or they can express epiphenomena of pre-clinical disease (e.g. mutations present in plasma DNA as a consequence of tumour cell apoptosis). It is very important that the biological significance of a marker be made explicit beforehand, because false expectations can arise as a consequence of an erroneous interpretation of a biomarker's role. For example, some markers (those on the right side of the scheme) have clinical relevance or can be useful for screening, others cannot.

Validating promising intermediate markers

A concept that is often unclear is the difference between technical and field validation. Technical validation has to do with intrinsic measurement error and analytical sensitivity. Field (or epidemiological) validation is related to how a certain marker behaves in the population, depending on biological variability within the population (177).

Biomarker validation requires several steps. A marker may be extremely powerful in increasing our understanding of the natural history and pathogenesis of a disease, but may still perform very poorly as a predictor for preventive or clinical purposes. One of the most important goals of validation is to characterize the ability of the marker to predict disease and, in intervention studies, reflect the modification of the natural course of disease.

One of the main summary measures of the contribution of a biomarker to the prediction of disease onset is the receiver operating characteristic (ROC) curve. The ROC curve is a measure of the overall capability of the marker to predict the disease, which is a function of sensitivity and specificity. An area under the curve (AUC) of 1 or close to 1 indicates perfect prediction, while an area close to 0.5 indicates random association between the marker measurement and the probability of disease onset. The maximum AUC for the prostate serum antigen (PSA) test (a serum tumour marker to predict the presence of prostate cancer) is only 0.77 (178). It is possible that gene expression microarrays or proteomics could perform better than the PSA test, but no candidate biomarker has yet been identified.

A major aim of biomarker validation is to characterize biomarker variability. The main components of biomarker variability that affect the design and interpretation of epidemiologic studies are: biologic variability related to the subject (i.e. variability between subjects (intersubject) and within subjects (intrasubject)); variability due to measurement error, including intralaboratory and interlaboratory variability; and random error. Methodological issues should be discussed within the context of specific biomarker categories. When epidemiologic studies employing biomarkers are designed and analysed, the goal is to minimize total intragroup variability to identify intergroup differences (e.g. between exposed and unexposed or between diseased and healthy subjects), if they exist. Total intragroup variation is the weighted sum of intersubject, intrasubject, sampling and laboratory variation, with weights that are inversely correlated to the numbers of subjects, number of measurements per subject, and analytical replicates used in the study design, respectively. Obviously, if detailed information is not available, intragroup variation cannot be adjusted for. Therefore, in epidemiologic studies employing biomarkers it is important to collect, whenever possible: repeat samples (day-to-day, month-to-month, or year-to-year variation may be relevant depending on the marker); information on subject characteristics that may influence intersubject variation; and information on conditions under which samples have been collected and laboratory analyses have been conducted (batch, assay, specific procedures). (For more about how the variability in laboratory measurements influences study design decisions, see (179).)

To increase power and improve the quality of studies, consortia like the NCI Cohort Consortium have recently been created. While these have been set up mainly to share questionnaire or GWAS data, consortia can also be extremely helpful for biomarker research including omics and biomarker validation. One recent example is the series of papers that examined the association between Vitamin D and several cancer sites (180).

Design issues

Study design issues identified with earlier biomarkers, such as mutation, oxidative damage, and adducts are particularly relevant for the use of newer intermediate markers, such as proteomic changes. Only prospective studies allow for a proper temporal evaluation of the role of intermediate biomarkers. The use of the cross-sectional design in the analysis of *p53* mutations has been an invaluable tool in the investigation of liver carcinogenesis. However, the cross-sectional design of the early studies did not allow researchers to exclude the possibility that mutations were a consequence of cell selection rather than of the original causal agent, such as aflatoxins (181). In other words, what was observed was the spectrum of mutations in liver cancers as the consequence of a long and complex process involving the effect of carcinogens, DNA repair, and the selection of cells carrying specific mutations conferring a selective advantage to cells. Therefore, in principle, prospective studies are better for the understanding of time relationships between exposure, intermediate biomarkers, and disease although they have the limitation of usually being based on a single spot biological sample, which does not allow the measurement of intraindividual variation.

Randomized clinical trials (RCTs) with biological samples have been repeatedly performed. For example, trials have used dietary changes as the intervention and oxidative damage or DNA adducts as the outcome. Though RCTs are probably the best design to conclude causality in epidemiology, they also have limitations, particularly the short half-life of most biomarkers and most interventions, compared to the long-term exposures that are needed to cause chronic diseases like cancer.

Another issue with respect to some biomarkers is that it is often difficult to understand whether the marker is intermediate in the pathway leading from exposure to disease, or it is just a consequence of exposure with no role in disease onset, or even an epiphenomenon of disease with no relationship to exposure. For example, micronuclei seem to originate from exposure to clastogens, but can lead to cell death and therefore are likely not to be intermediate in the causal pathway.

Types of bias that are common in other epidemiological studies may become dramatic when biological samples are collected and biomarkers are measured. For example, in a study on pancreas cancer, out of more than 1000 eligible patients, the investigators were only able to extract DNA from 46 biopsies (182). The patients with a biopsy available were more frequently white and the tumour size was on average 179 mm, versus 570 mm among the patients whose biopsy was not made available. This discrepancy is likely to introduce bias if one tries to correlate the prevalence of somatic mutations with exposure characteristics, such as occupation. As further example, in a case–case study, patients with pancreatic cancer seen at two general hospitals were retrospectively identified (183). Their clinical records were abstracted and paraffin-embedded samples retrieved from pathology records. DNA was amplified and mutations in codon 12 of the *K-ras* gene were detected. Results on the mutations were obtained for 51 of the 149 cases (34.2%). Mutation data were over five times more likely to be available from one of the hospitals. In particular, subjects with mutations were more likely to have received a treatment with curative intent (OR = 11.56; 95% CI = 2.88–46.36).

In addition, special forms of confounding may affect molecular epidemiology. An example is the levels of plasma DNA in cancer patients and controls, in the context of a multicentre cohort study. Researchers found that, although the level of plasma DNA seemed to predict the onset of cancer, it was also strongly associated with the recruitment centre. This was due to modalities of blood collection and storage, since a longer time elapsing between blood drawing and storage in liquid nitrogen was associated with higher DNA levels due to greater white blood cell death. Thus, the association between cancer and plasma DNA levels could be confounded by centre, since cancer rates also differed by centre in this multicentre study (184).

Translation of research into preventive programmes

Assessing risk

Risk assessment for low-level exposures

One of the main challenges for epidemiology in recent decades has been the need to characterize and quantify the effects of low-level exposures to carcinogens. Such exposures are widespread (e.g. traffic-related air pollution and ETS) but extremely difficult to study with conventional epidemiological tools. There has been a heated debate on the shape of dose–response relationships in carcinogenesis (i.e. on the extrapolation from high- to low-levels of dose), an issue of great public health significance. Epidemiological studies are often underpowered to study the carcinogenic effects of very low levels of exposure. Using biomarkers, molecular epidemiology can mitigate the problem that very large numbers of subjects are needed to detect small effects on cancer risk, by providing individual estimates of dose and intermediate markers of procarcinogenic damage that can be used *in lieu* of cancer as an outcome.

To illustrate these points, results are described from a series of analyses that have been carried out by investigators in the EPIC study on the effects of low-level exposure to ETS and air pollution on lung cancer. ETS and air pollution share several characteristics: they are widespread exposures in both developed and developing countries, they have chemical components in common, and they are associated with increased risks of lung cancer and other diseases (185,186). The lung cancer relative increase is around 20–30%, approximately of the same magnitude for both ETS and air pollution at the typical exposure levels in Western countries (158,187). In EPIC, relative risks were found in the order of 1.4–1.5 for exposure to ETS in adulthood and the risk of lung cancer, based on the prospective investigation of about 120 000 subjects with information on ETS and 117 newly diagnosed lung cancers in non-smokers (140). Biomarkers were used in several different ways. First, cotinine was used to validate the questionnaire information on ETS exposure, demonstrating a strong association with self-reported exposure ($P < 0.001$). Second, biomarkers of genetic susceptibility strengthened the epidemiological association between low-level exposures to carcinogens and cancer. The risk associated with ETS was higher in subjects with three or more at-risk alleles for genes involved in carcinogen metabolism (*GSTM1, GSTM3, GSTP1, GSTT1, CYP1A1, CYP1B1, NAT2, MnSOD, MPO,* and *NQO1*), with an odds ratio of 2.86 compared to 1.33 in those with less than three alleles (140). These results have implications for risk assessment in that they show a modest, but significant, increase in cancer risk at low levels of exposure to environmental carcinogens, and demonstrate that genetic factors can substantially increase risk to certain subsets of the population.

Developing dose–response models for assessing the risk of carcinogens

A major issue relevant to risk assessment is whether to view carcinogenesis as a linear process, involving the accumulation of several additive events, or as a nonlinear process. Molecular data on carcinogenic mechanisms have been instrumental in developing and validating different statistical models for carcinogen risk assessment. At the time of the initial development of the molecular epidemiology paradigm (2,3), the dominant model of carcinogenesis was the "multistage" model proposed by Armitage and Doll that postulated the existence of about six stages in cancer development (188). Armitage and Doll's model was consistent with the paradigm, implying several steps between exposure and cancer, and an important role for duration of exposure to carcinogenic stimuli. Steps were postulated to be heritable from one cell to the progeny, and critical genetic changes were hypothesized to be irreversible. In fact, after the multistage model was originally proposed, Vogelstein demonstrated, on the basis of molecular pathology, that the development of colon cancer was likely to require six mutations or chromosome aberrations (189).

In addition, Knudson had suggested, based on its age distribution, that retinoblastoma (Rb) in children was likely to be due to two mutations, one inherited and one acquired (190). The Knudson "two-hit" model for retinoblastoma was supported by the discovery of the first tumour suppressor gene (*Rb*) that in fact requires two mutations, one inherited and one acquired, to give rise to the tumour (191). Thus, both models were examples of the success of combining epidemiological observations, mathematical models, and molecular or biomarker evidence.

Yet another statistical model, based directly on molecular evidence, derives from the identification of hereditary syndromes that predispose to colon cancer (hereditary non-polyposis colon cancer) through mutations in the mismatch repair genes (77). The corresponding model postulates that the rate of mutations is too low to explain the incidence of cancer in human populations; therefore, a "mutator phenotype" (the inherited or acquired ability to develop frequent mutations, such as through a defect in DNA repair machinery) would be necessary (192,193). A cascade of mutations, originated by the inability to repair DNA damage, would better explain the high frequency of colon cancers in some families, than the simple accumulation of spontaneous or acquired mutations. The same could be true for "sporadic" cancers.

Several other models have been proposed in recent years that accommodate and reflect new molecular information on carcinogenic mechanisms. One recent model (194) is based on the concept that carcinogenesis is a Darwinian process in which transformed cells acquire a selective advantage over normal cells. The term "selectogen" has been proposed for carcinogens that act by increasing the ability of mutated cells to acquire selective advantage (in given environments) over normal cells. A biomarker that has been used to explore such a Darwinian concept of carcinogenesis is mutation in the *HPRT* reporter gene. The X-chromosomal gene for *HPRT* serves as a simple reporter gene (i.e. it indicates the induction of mutations) and is now finding use in studies of *in vivo* selection for mutations arising in either somatic or germinal cells (195). This line of research, however, is still in its infancy.

All these apparently diverse models are in fact generally compatible and consistent with molecular data. The picture that is emerging is that environmental stimuli can increase genomic instability (in addition to inherited variants of instability), which in turn leads to chromosome aberrations or mutations that increase the selective advantage of cells in stressful environments, and induces the carcinogenic process. However, the problem with mathematical models is that often they are compatible with different biological interpretations and do not easily accommodate certain aspects of carcinogenesis, such as epigenetics. The incorporation of non-genetic biomarkers into risk assessment models is still in a very early stage.

Developing new intervention strategies

Primary prevention encompasses a spectrum of measures that includes elimination or avoidance of exposure, prevention of carcinogen activation after it has entered the body, blocking interactions with the genome, and suppressing the propagation of premalignant changes. Several studies have used DNA damage as an intermediate biomarker or endpoint. An example is a study of smokers enrolled in a smoking cessation programme. Levels of biomarkers, PAH–DNA and 4-ABP–haemoglobin adducts, reflecting cessation were significantly reduced by eight weeks after quitting smoking (196). Similarly, following a reduction in air concentrations of PAHs in a Finnish iron foundry, both PAH–DNA and aromatic DNA adduct levels in workers' blood samples declined significantly (197).

Other prevention research has used biomarkers as intermediate endpoints in chemoprevention trials. Research studies have shown that isothiocyanates, which occur as conjugates in a wide variety of cruciferous vegetables, are involved in the inhibition of carcinogenesis (14). Isothiocyanates appear to selectively inhibit cytochrome P450 enzymes involved in carcinogen metabolic inactivation; they also induce Phase II enzymes and enhance apoptosis. Phenethyl isothiocyanate is a particularly effective inhibitor of lung tumour induction by the tobacco-specific nitrosamine 4-(methylnitrosamino)-1-(3-pyridyl)-1-butanone and is currently being developed as a chemopreventive agent against lung cancer (110).

Several dietary or vitamin supplementation randomized studies have used DNA adducts or oxidative damage markers as intermediate outcomes. Free radicals, which are produced naturally in the body, can cause oxidative damage of DNA, lipids, proteins and other cell constituents, contributing to the onset of cancers and other chronic diseases (198). Oxidative damage to DNA plays a major role in carcinogenesis, and all living cells have defence mechanisms in place to counter this damage. The simplest mechanism involves foods and nutrients with antioxidant properties, which work by intercepting free radicals and preventing cellular damage (198,199). To establish the potential chemopreventive properties of antioxidants, investigators have used markers such as 8-hydroxy-2'-deoxyguanosine (8-OHdG) and the comet assay as intermediate markers in interventions (198). A review of these intervention studies has concluded that most had extremely low statistical power (sample size usually ≤ 20) and that they led to modest changes in 8-OHdG (around 10%) (199,200). In conclusion, promising markers are available for intervention studies, but they await application in large-scale and well designed trials.

A randomized clinical trial of vitamins E and C in smokers found that among women, but not men, there was a significant decline in PAH-DNA adducts in the treatment group (201).

New biomarkers for clinical purposes

The field of biomarkers is rapidly expanding, particularly as far as biomarkers for clinical purposes are concerned. For example, microRNAs, which are very short stretches of RNA with regulatory functions, seem to be extremely promising for the understanding of cancer mechanisms, as well as for developing new therapies (202). In addition, microRNAs are also relevant to chemically-induced cancer (203). More about such new developments can be found in (204).

Policy changes

With regard to public health and environmental policy, molecular epidemiology has not yet led to broad policy changes to prevent or to reduce exposure to carcinogens. However, it has provided impetus for prevention of prolonged exposures to carcinogens, even at low levels, since they can result in DNA damage or epigenetic alterations that begin at a very early age, even *in utero*, and accumulate over a lifetime (41,205,206). In addition, molecular epidemiologic data on interindividual variation in susceptibility refute the default assumption in risk assessment that the population is biologically homogeneous in response to carcinogens. This default assumption can lead to substantial underestimates of risk to the population and to sensitive subgroups, leading to standards and policies that are not adequately health-protective or equitable (129,207).

The theoretical importance of focusing intervention strategies (regulations, public education programmes, health surveillance, behaviour modification, and chemoprevention programmes) on the subgroups at greatest risk as a result of genetic or acquired susceptibility (208,209) is illustrated in Figure 19.2 (210). The figure illustrates that while the distribution of susceptibility/risk is symmetrical on a log scale, it is right-skewed on the linear scale. Thus, for a hypothetical carcinogen with a linear low dose–response curve, the estimated risk would be 38-fold greater for a population of individuals with 99th-percentile sensitivity than for a population of median-sensitive individuals. (This number is the upper 95% confidence limit of risk with respect to uncertainty; the estimated increase in risk is similar if the arithmetic mean estimate of risk with respect to uncertainty is used.) Sensitivity due to genetic and nutritional factors can be compounded in the case of certain groups (e.g. children) who would be expected to have both more exposure and greater age-related susceptibility to certain carcinogens.

Figure 19.2. The theoretical distribution of cancer susceptibility and risk across a population that is heterogeneous with respect to sensitivity to a hypothetical non-threshold carcinogen [based on (200)]. The x-axis represents the percentile of sensitivity; the y-axis, the number of individuals. The numbers in parentheses are the estimated cancer risk for a population of individuals at the indicated percentile of sensitivity. They are derived by use of a Monte Carlo simulation using data on observed human variability in metabolic activation, detoxification and DNA repair, as well as uncertainty in cancer potencies for a set of genetically acting carcinogens. The numbers shown are the upper 95% confidence limit of risk with respect to uncertainty estimates and are similar if the arithmetic mean estimates are used. Panel [a] shows the distribution on a log scale; panel [b] shows the same distribution on a linear scale.

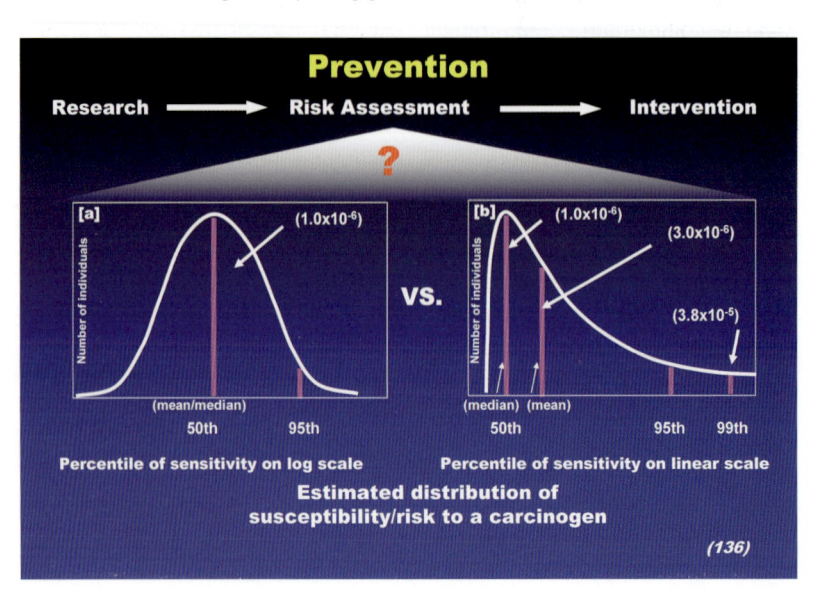

Table 19.1. Discoveries that support the original model of molecular epidemiology*

Marker linked to exposure or disease	Exposure	Reference
Metabolites in body fluids Urinary metabolites (NNK, NNN)	Nitroso compounds in tobacco	(211,212)
Exposure/biologically effective dose DNA adducts Albumin adducts	PAHs, aromatic compounds AFB$_1$	(37) (213,214)
Haemoglobin adducts	Acrylamide Styrene 1,3-butadiene	(215) (216) (217)
Preclinical effect Chromosome aberrations	**Exposure and/or Cancer** Lung Leukemia Benzene	(52) (218) (219)
HRPT	PAHs 1,3-butadiene	(220) (221)
Glycophorin A Gene expression	PAHs Cisplatin	(222) (223)
Genetic susceptibility Phenotypic markers	e.g. DNA repair capacity in head and neck cancer	(77,224)
SNPs: *NAT2, GSTM* *CYP1A1*	Bladder Lung	(225) (226)

*See (2) and (227).

Conclusions and future directions

The examples presented in this chapter show that molecular epidemiology has made extensive progress since the 1980s. It has contributed to prevention by providing new evidence that specific environmental agents pose human carcinogenic hazards, helping to establish their causal role, identifying subsets of the population at special risk, and using this information to develop new and more effective strategies to reduce risk. As a result, some interventions and policy changes have been mounted to reduce risk from several important environmental carcinogens.

As has been seen, recently developed epigenetic and "omic" biomarkers have considerable potential in molecular epidemiology, along with genotoxic markers, because they reflect another equally important mechanism of carcinogenicity: epigenetic alterations that affect the expression of genes and proteins. These can be measured by high-throughput methods, allowing large-scale studies that are discovery-oriented. However, a major challenge is the need for validation of these newer biomarkers so they may be applied in large-scale etiologic and intervention studies. An important development in molecular epidemiology has been the emergence of networks and consortia involving hundreds of researchers and multiple large studies. Examples include the Wellcome Trust Case-Control Consortium, CGEMS (Cancer Genetic Markers of Susceptibility), HuGE (Human Genome Epidemiology Network), ECNIS (Environmental Cancer Risk, Nutrition and Individual Susceptibility), NuGO (The European Nutrigenomics Organization linking genomics, nutrition and health research), and Interlymph in the field of lymphomas. Such initiatives allow coordinated efforts, avoid false-positives and publication bias from several small studies, and contribute to rapid dissemination and replication of new knowledge.

Another challenge and future direction is the timely translation of data from etiologic and intervention studies into risk assessment and public health policy, as well as focused research to fill gaps in scientific knowledge. Meeting these goals requires an infrastructure to promote a dialogue among scientists, policy-makers and other stakeholders, and a major investment in the second generation of molecular epidemiologic research, including large-scale collaborative studies incorporating validated biomarkers and automated technologies.

Acknowledgements

This work has been made possible by grants from the Compagnia di San Paolo to the ISI Foundation and from the European Union for the project ECNIS (PV), and by the National Institutes of Health, the National Institutes of Environmental Health Sciences (5P01ES09600, 5RO1ES08977, 1R01CA127532, R01CA69094), the US. Environmental Protection Agency (R827027, RD-832141), Bauman Family Foundation, and the New York Community Trust.

References

1. American Cancer Society. Cancer facts and figures 2006. Atlanta (GA): American Cancer Society; 2006.

2. Perera FP, Weinstein IB (1982). Molecular epidemiology and carcinogen-DNA adduct detection: new approaches to studies of human cancer causation. J Chronic Dis, 35:581–600.doi:10.1016/0021-9681(82)90078-9 PMID:6282919

3. National Academy of Sciences. Regulating pesticides in food: the Delaney Paradox. Washington (DC): National Academy Press; 1987.

4. Weinberg RA (2006). A lost generation. Cell, 126:9–10.doi:10.1016/j.cell.2006.06.022 PMID:16839866

5. Chiazze L Jr, Levin DL, Silverman DT. Recent changes in estimated cancer mortality. In: Hiatt HH, Watson JD, Winsten JA, editors. Origins of human cancer. Cold Spring Harbor conferences on cell proliferation, vol 4. Maine: Cold Spring Harbor Laboratory; 1977. p. 33–44.

6. Weinstein IB, Santella RM, Perera FP et al. The molecular biology and molecular epidemiology of cancer. The science and practice of cancer prevention and control. New York (NY): Marcel Dekker; 1993.

7. Tomatis L, Aitio A, Wilbourn J, Shukar L (1987). Human carcinogens so far identified. Jpn J Cancer Res, 78:887–898. PMID:2513295

8. Willett WC (2002). Balancing life-style and genomics research for disease prevention. Science, 296:695–698.doi:10.1126/science.1071055 PMID:11976443

9. U.S. Department of Health, Education and Welfare. Smoking and health. Report of the Advisory Committee to the Surgeon General of the Public Health Service. Washington (DC): Public Health Service; 1964.

10. Vineis P (2004). Individual susceptibility to carcinogens. Oncogene, 23:6477–6483. doi:10.1038/sj.onc.1207897 PMID:15322518

11. Suuriniemi M, Agalliu I, Schaid DJ et al. (2007). Confirmation of a positive association between prostate cancer risk and a locus at chromosome 8q24. Cancer Epidemiol Biomarkers Prev, 16:809–814.doi:10.1158/1055-9965.EPI-06-1049 PMID:17416775

12. Zhang ZF, Cordon-Cardo C, Rothman N et al. Methodological issues in the use of tumour markers in cancer epidemiology. In: Toniolo P, Boffetta P, Shuker DEG et al., editors. Application of biomarkers in cancer epidemiology. Lyon: IARC Scientific Publication; 1997(142). p. 201–213.

13. National Research Council. Environmental tobacco smoke: measuring exposures and assessing health effects. Washington (DC): National Academy Press; 1986.

14. Hecht SS (1999). Tobacco smoke carcinogens and lung cancer. J Natl Cancer Inst, 91:1194–1210.doi:10.1093/jnci/91.14.1194 PMID:10413421

15. Wogan GN, Hecht SS, Felton JS et al. (2004). Environmental and chemical carcinogenesis. Semin Cancer Biol, 14:473–486.doi:10.1016/j.semcancer.2004.06.010 PMID:15489140

16. Bryant MS, Skipper PL, Tannenbaum SR, Maclure M (1987). Hemoglobin adducts of 4-aminobiphenyl in smokers and nonsmokers. Cancer Res, 47:602–608. PMID:3791245

17. International Agency for Research on Cancer. Polynuclear aromatic compounds. Part I. Chemical, environmental, and experimental data. IARC monographs on the evaluation of the carcinogenic risk of chemicals to humans. Lyon: IARC Scientific Publication; 1983.

18. U.S. Environmental Protection Agency. Aerometric information retrieval system (AIRS), data for 1985–1990. Research Triangle Park (NC): U.S. Environmental Protection Agency; 1990.

19. Anderson LM, Jones AB, Miller MS, Chauhan DP. Metabolism of transplacental carcinogens. In Napalkov NP, Rice JM, Tomatis L, Yamasaki H, editors. Perinatal and multigeneration carcinogenesis. Lyon: IARC Scientific Publication; 1989(96). p. 155–188.

20. Boström CE, Gerde P, Hanberg A et al. (2002). Cancer risk assessment, indicators, and guidelines for polycyclic aromatic hydrocarbons in the ambient air. Environ Health Perspect, 110 Suppl 3;451–488. PMID:12060843

21. Anderson LM, Ruskie S, Carter J et al. (1995). Fetal mouse susceptibility to transplacental carcinogenesis: differential influence of Ah receptor phenotype on effects of 3-methylcholanthrene, 12-dimethylbenz[a]anthracene, and benzo[a]pyrene. Pharmacogenetics, 5:364–372. doi:10.1097/00008571-199512000-00005 PMID:8747408

22. Yu Z, Loehr CV, Fischer KA et al. (2006). In utero exposure of mice to dibenzo[a,l]pyrene produces lymphoma in the offspring: role of the aryl hydrocarbon receptor. Cancer Res, 66:755–762.doi:10.1158/0008-5472.CAN-05-3390 PMID:16424006

23. Miller EC, Miller JA (1981). Mechanisms of chemical carcinogenesis. Cancer, 47 Suppl;1055–1064.doi:10.1002/1097-0142(19810301)47:5+<1055::AID-CNCR2820471302>3.0.CO;2-3 PMID:7016297

24. Perera FP (1997). Environment and cancer: who are susceptible? Science, 278:1068–1073.doi:10.1126/science.278.5340.1068 PMID:9353182

25. Perera FP, Poirier MC, Yuspa SH et al. (1982). A pilot project in molecular cancer epidemiology: determination of benzo[a]pyrene–DNA adducts in animal and human tissues by immunoassays. Carcinogenesis, 3:1405–1410.doi:10.1093/carcin/3.12.1405 PMID:6295657

26. Perera FP, Santella RM, Brenner D et al. (1987). DNA adducts, protein adducts, and sister chromatid exchange in cigarette smokers and nonsmokers. J Natl Cancer Inst, 79:449–456. PMID:3114532

27. Mumford JL, Lee X, Lewtas J et al. (1993). DNA adducts as biomarkers for assessing exposure to polycyclic aromatic hydrocarbons in tissues from Xuan Wei women with high exposure to coal combustion emissions and high lung cancer mortality. Environ Health Perspect, 99:83–87.doi:10.1289/ehp.939983 PMID:8319664

28. Binková B, Lewtas J, Mísková I et al. (1996). Biomarker studies in northern Bohemia. Environ Health Perspect, 104 Suppl 3;591–597.doi:10.2307/3432828 PMID:8781388

29. Perera FP, Hemminki K, Gryzbowska E et al. (1992). Molecular and genetic damage in humans from environmental pollution in Poland. Nature, 360:256–258. doi:10.1038/360256a0 PMID:1436106

30. Dickey C, Santella RM, Hattis D et al. (1997). Variability in PAH-DNA adduct measurements in peripheral mononuclear cells: implications for quantitative cancer risk assessment. Risk Anal, 17:649–656. doi:10.1111/j.1539-6924.1997.tb00905.x PMID:9404054

31. Kriek E, Van Schooten FJ, Hillebrand MJ et al. (1993). DNA adducts as a measure of lung cancer risk in humans exposed to polycyclic aromatic hydrocarbons. Environ Health Perspect, 99:71–75.doi:10.1289/ehp.939971 PMID:8319662

32. Tang D, Santella RM, Blackwood AM et al. (1995). A molecular epidemiological case-control study of lung cancer. Cancer Epidemiol Biomarkers Prev, 4:341–346. PMID:7655328

33. Bartsch H, Rojas M, Alexandrov K, Risch A (1998). Impact of adduct determination on the assessment of cancer susceptibility. Recent Results Cancer Res, 154:86–96. PMID:10026994

34. Ambrosone CB, Kadlubar FF (1997). Toward an integrated approach to molecular epidemiology. Am J Epidemiol, 146:912–918. PMID:9400332

35. Stowers SJ, Anderson MW (1984). Ubiquitous binding of benzo[a]pyrene metabolites to DNA and protein in tissues of the mouse and rabbit. Chem Biol Interact, 51:151–166.doi:10.1016/0009-2797(84)90027-9 PMID:6088095

36. Wiencke JK, Thurston SW, Kelsey KT et al. (1999). Early age at smoking initiation and tobacco carcinogen DNA damage in the lung. J Natl Cancer Inst, 91:614–619.doi:10.1093/jnci/91.7.614 PMID:10203280

37. Tang D, Phillips DH, Stampfer M et al. (2001). Association between carcinogen-DNA adducts in white blood cells and lung cancer risk in the physicians health study. Cancer Res, 61:6708–6712. PMID:11559540

38. Veglia F, Matullo G, Vineis P (2003). Bulky DNA adducts and risk of cancer: a meta-analysis. Cancer Epidemiol Biomarkers Prev, 12:157–160. PMID:12582026

39. Kellermann G, Shaw CR, Luyten-Kellerman M (1973). Aryl hydrocarbon hydroxylase inducibility and bronchogenic carcinoma. N Engl J Med, 289:934–937. doi:10.1056/NEJM197311012891802 PMID:4126515

40. Vineis P, Malats N, Porta M, Real FX (1999). Human cancer, carcinogenic exposures and mutation spectra. Mutat Res, 436:185–194. PMID:10095140

41. Greenblatt MS, Bennett WP, Hollstein M, Harris CC (1994). Mutations in the p53 tumor suppressor gene: clues to cancer etiology and molecular pathogenesis. Cancer Res, 54:4855–4878. PMID:8069852

42. Denissenko MF, Pao A, Tang MS, Pfeifer GP (1996). Preferential formation of benzo[a]pyrene adducts at lung cancer mutational hotspots in P53. Science, 274:430–432. doi:10.1126/science.274.5286.430 PMID:8832894

43. Hainaut P, Hollstein M (2000). p53 and human cancer: the first ten thousand mutations. Adv Cancer Res, 77:81–137. doi:10.1016/S0065-230X(08)60785-X PMID:10549356

44. Wild CP, Hasegawa R, Barraud L et al. (1996). Aflatoxin-albumin adducts: a basis for comparative carcinogenesis between animals and humans. Cancer Epidemiol Biomarkers Prev, 5:179–189. PMID:8833618

45. Ross RK, Yuan JM, Yu MC et al. (1992). Urinary aflatoxin biomarkers and risk of hepatocellular carcinoma. Lancet, 339:943–946.doi:10.1016/0140-6736(92)91528-G PMID:1348796

46. Qian GS, Ross RK, Yu MC et al. (1994). A follow-up study of urinary markers of aflatoxin exposure and liver cancer risk in Shanghai, People's Republic of China. Cancer Epidemiol Biomarkers Prev, 3:3–10. PMID:8118382

47. Wang LY, Hatch M, Chen CJ et al. (1996). Aflatoxin exposure and risk of hepatocellular carcinoma in Taiwan. Int J Cancer, 67:620–625.doi:10.1002/(SICI)1097-0215(19960904)67:5<620::AID-IJC5>3.0.CO;2-W PMID:8782648

48. Yu MW, Lien JP, Chiu YH et al. (1997). Effect of aflatoxin metabolism and DNA adduct formation on hepatocellular carcinoma among chronic hepatitis B carriers in Taiwan. J Hepatol, 27:320–330.doi:10.1016/S0168-8278(97)80178-X PMID:9288607

49. Hsia CC, Kleiner DE Jr, Axiotis CA et al. (1992). Mutations of p53 gene in hepatocellular carcinoma: roles of hepatitis B virus and aflatoxin contamination in the diet. J Natl Cancer Inst, 84:1638–1641.doi:10.1093/jnci/84.21.1638 PMID:1279184

50. Smith MT, Zhang L (1998). Biomarkers of leukemia risk: benzene as a model. Environ Health Perspect, 106 Suppl 4;937–946. doi:10.2307/3434135 PMID:9703476

51. Kim S, Vermeulen R, Waidyanatha S et al. (2006). Modeling human metabolism of benzene following occupational and environmental exposures. Cancer Epidemiol Biomarkers Prev, 15:2246–2252. doi:10.1158/1055-9965.EPI-06-0262 PMID:17119053

52. Bonassi S, Znaor A, Norppa H, Hagmar L (2004). Chromosomal aberrations and risk of cancer in humans: an epidemiologic perspective. Cytogenet Genome Res, 104:376–382.doi:10.1159/000077519 PMID:15162068

53. Bonassi S, Hagmar L, Strömberg U et al.; European Study Group on Cytogenetic Biomarkers and Health (2000). Chromosomal aberrations in lymphocytes predict human cancer independently of exposure to carcinogens. Cancer Res, 60:1619–1625. PMID:10749131

54. Hagmar L, Bonassi S, Strömberg U et al. (1998). Chromosomal aberrations in lymphocytes predict human cancer: a report from the European Study Group on Cytogenetic Biomarkers and Health (ESCH). Cancer Res, 58:4117–4121. PMID:9751622

55. Zhang YJ, Chen Y, Ahsan H et al. (2005). Silencing of glutathione S-transferase P1 by promoter hypermethylation and its relationship to environmental chemical carcinogens in hepatocellular carcinoma. Cancer Lett, 221:135–143.doi:10.1016/j.canlet.2004.08.028 PMID:15808399

56. Vermeulen R, Lan Q, Zhang L et al. (2005). Decreased levels of CXC-chemokines in serum of benzene-exposed workers identified by array-based proteomics. Proc Natl Acad Sci USA, 102:17041–17046.doi:10.1073/pnas.0508573102 PMID:16286641

57. Lan Q, Zheng T, Rothman N et al. (2006). Cytokine polymorphisms in the Th1/Th2 pathway and susceptibility to non-Hodgkin lymphoma. Blood, 107:4101–4108.doi:10.1182/blood-2005-10-4160 PMID:16449530

58. Steffen C, Auclerc MF, Auvrignon A et al. (2004). Acute childhood leukaemia and environmental exposure to potential sources of benzene and other hydrocarbons; a case-control study. Occup Environ Med, 61:773–778.doi:10.1136/oem.2003.010868 PMID:15317919

59. Feingold L, Savitz DA, John EM (1992). Use of a job-exposure matrix to evaluate parental occupation and childhood cancer. Cancer Causes Control, 3:161–169.doi:10.1007/BF00 051656 PMID:1562706

60. Lahmann PH, Friedenreich C, Schuit AJ et al. (2007). Physical activity and breast cancer risk: the European Prospective Investigation into Cancer and Nutrition. Cancer Epidemiol Biomarkers Prev, 16:36–42.doi:10.1158/1055-9965.EPI-06-0582 PMID:17179488

61. Pischon T, Lahmann PH, Boeing H et al. (2006). Body size and risk of renal cell carcinoma in the European Prospective Investigation into Cancer and Nutrition (EPIC). Int J Cancer, 118:728–738.doi:10.1002/ijc. 21398 PMID:16094628

62. Pischon T, Lahmann PH, Boeing H et al. (2006). Body size and risk of colon and rectal cancer in the European Prospective Investigation Into Cancer and Nutrition (EPIC). J Natl Cancer Inst, 98:920–931.doi:10.1093/jnci/djj246 PMID:16818856

63. Friedenreich C, Cust A, Lahmann PH et al. (2007). Anthropometric factors and risk of endometrial cancer: the European prospective investigation into cancer and nutrition. Cancer Causes Control, 18:399–413.doi:10.1007/s105 52-006-0113-8 PMID:17297555

64. Russo A, Autelitano M, Bisanti L (2008). Metabolic syndrome and cancer risk. Eur J Cancer, 44:293–297.doi:10.1016/j.ejca.2007.11.005 PMID:18055193

65. Hursting SD, Lavigne JA, Berrigan D et al. (2004). Diet-gene interactions in p53-deficient mice: insulin-like growth factor-1 as a mechanistic target. J Nutr, 134:2482S–2486S. PMID:15333746

66. Jenab M, Riboli E, Cleveland RJ et al. (2007). Serum C-peptide, IGFBP-1 and IGFBP-2 and risk of colon and rectal cancers in the European Prospective Investigation into Cancer and Nutrition. Int J Cancer, 121:368–376.doi:10.1002/ijc.22697 PMID:17372899

67. Weiss JM, Huang WY, Rinaldi S et al. (2007). IGF-1 and IGFBP-3: risk of prostate cancer among men in the Prostate, Lung, Colorectal and Ovarian Cancer Screening Trial. Int J Cancer, 121:2267–2273.doi:10.1002/ijc. 22921 PMID:17597108

68. Allen NE, Key TJ, Appleby PN et al. (2007). Serum insulin-like growth factor (IGF)-I and IGF-binding protein-3 concentrations and prostate cancer risk: results from the European Prospective Investigation into Cancer and Nutrition. Cancer Epidemiol Biomarkers Prev, 16:1121–1127.doi:10.1158/1055-9965.EPI-06-1062 PMID:17548673

69. Skibola CF, Holly EA, Forrest MS et al. (2004). Body mass index, leptin and leptin receptor polymorphisms, and non-hodgkin lymphoma. Cancer Epidemiol Biomarkers Prev, 13:779–786. PMID:15159310

70. DeLellis Henderson K, Rinaldi S, Kaaks R et al. (2007). Lifestyle and dietary correlates of plasma insulin-like growth factor binding protein-1 (IGFBP-1), leptin, and C-peptide: the Multiethnic Cohort. Nutr Cancer, 58:136–145. PMID:17640159

71. Hour TC, Pu YS, Lin CC et al. (2006). Differential expression of molecular markers in arsenic- and non-arsenic-related urothelial cancer. Anticancer Res, 26 1A;375–378. PMID:16475721

72. Zhang ZF, Cordon-Cardo C, Rothman N et al. Methodological issues in the use of tumour markers in cancer epidemiology. In: Toniolo P, Boffetta P, Shuker DEG et al., editors. Application of biomarkers in cancer epidemiology.Lyon: IARC Scientific Publication; 1997(142). p. 201–213.

73. Garte S, Zocchetti C, Taioli E. Gene-environment interactions in the application of biomarkers of cancer susceptibility in epidemiology. In: Toniolo P, Boffetta P, Shuker DEG et al., editors. Application of biomarkers in cancer epidemiology. Lyon: IARC Scientific Publication; 1997. p. 251–264.

74. Caporaso NE, Goldstein A. Issues involving biomarkers in the study of the genetics of human cancer. In: Toniolo P, Boffetta P, Shuker DEG et al., editors. Application of biomarkers in cancer epidemiology. Lyon: IARC Scientific Publication; 1997. p. 237–250.

75. Schulte PA, Hunter D, Rothman N. Ethical and social issues in the use of biomarkers in epidemiological studies. In: Toniolo P, Boffetta P, Shuker DEG et al., editors. Application of biomarkers in cancer epidemiology. Lyon: IARC Scientific Publication; 1997. p. 313–318.

76. Vogelstein B, Kinzler KW, editors. The genetic basis of human cancer. New York (NY): McGraw-Hill; 1998.

77. Berwick M, Vineis P (2000). Markers of DNA repair and susceptibility to cancer in humans: an epidemiologic review. J Natl Cancer Inst, 92:874–897.doi:10.1093/jnci/92.11.874 PMID:10841823

78. Matullo G, Dunning AM, Guarrera S et al. (2006). DNA repair polymorphisms and cancer risk in non-smokers in a cohort study. Carcinogenesis, 27:997–1007.doi:10.1093/carcin/bgi280 PMID:16308313

79. Sharp L, Little J (2004). Polymorphisms in genes involved in folate metabolism and colorectal neoplasia: a HuGE review. Am J Epidemiol, 159:423–443.doi:10.1093/aje/kwh 066 PMID:14977639

80. Eichholzer M, Lüthy J, Moser U, Fowler B (2001). Folate and the risk of colorectal, breast and cervix cancer: the epidemiological evidence. Swiss Med Wkly, 131:539–549. PMID:11759174

81. Kim SJ, Kim TW, Lee SY et al. (2004). CpG methylation of the ERalpha and ERbeta genes in breast cancer. Int J Mol Med, 14:289–293. PMID:15254780

82. Little J, Sharp L, Duthie S, Narayanan S (2003). Colon cancer and genetic variation in folate metabolism: the clinical bottom line. J Nutr, 133 Suppl 1;3758S–3766S. PMID:14608111

83. Zhang SM, Willett WC, Selhub J et al. (2003). Plasma folate, vitamin B6, vitamin B12, homocysteine, and risk of breast cancer. J Natl Cancer Inst, 95:373–380.doi:10.1093/jnci/95.5.373 PMID:12618502

84. Giovannucci E, Chen J, Smith-Warner SA et al. (2003). Methylenetetrahydrofolate reductase, alcohol dehydrogenase, diet, and risk of colorectal adenomas. Cancer Epidemiol Biomarkers Prev, 12:970–979. PMID:14578131

85. Terry P, Jain M, Miller AB et al. (2002). Dietary intake of folic acid and colorectal cancer risk in a cohort of women. Int J Cancer, 97:864–867.doi:10.1002/ijc.10138 PMID:11857369

86. Harnack L, Jacobs DR Jr, Nicodemus K et al. (2002). Relationship of folate, vitamin B-6, vitamin B-12, and methionine intake to incidence of colorectal cancers. Nutr Cancer, 43:152–158.doi:10.1207/S15327914NC432_5 PMID:12588695

87. Zienolddiny S, Campa D, Lind H et al. (2006). Polymorphisms of DNA repair genes and risk of non-small cell lung cancer. Carcinogenesis, 27:560–567.doi:10.1093/carcin/bgi232 PMID:16195237

88. Leng S, Cheng J, Pan Z et al (2004). Associations between XRCC1 and ERCC2 polymorphisms and DNA damage in peripheral blood lymphocyte among coke oven workers. Biomarkers, 9:395–406.doi:10.1080/13547500400015618 PMID:15764301

89. Alexandrov K, Cascorbi I, Rojas M et al. (2002). CYP1A1 and GSTM1 genotypes affect benzo[a]pyrene DNA adducts in smokers' lung: comparison with aromatic/hydrophobic adduct formation. Carcinogenesis, 23:1969–1977.doi:10.1093/carcin/23.12.1969 PMID:12507920

90. Cascorbi I, Brockmöller J, Roots I (2002). Molecular-epidemiological aspects of carcinogenesis: the role of xenobiotic metabolizing enzymes. Int J Clin Pharmacol Ther, 40:562–563. PMID:12503814

91. Whyatt RM, Camann DE, Kinney PL et al. (2002). Residential pesticide use during pregnancy among a cohort of urban minority women. Environ Health Perspect, 110:507–514.doi:10.1289/ehp.02110507 PMID:12003754

92. Yu MW, Chiu YH, Chiang YC et al. (1999). Plasma carotenoids, glutathione S-transferase M1 and T1 genetic polymorphisms, and risk of hepatocellular carcinoma: independent and interactive effects. Am J Epidemiol, 149:621–629. PMID:10192309

93. McGlynn KA, Rosvold EA, Lustbader ED et al. (1995). Susceptibility to hepatocellular carcinoma is associated with genetic variation in the enzymatic detoxification of aflatoxin B1. Proc Natl Acad Sci USA, 92:2384–2387. doi:10.1073/pnas.92.6.2384 PMID:7892276

94. Yu MW, Gladek-Yarborough A, Chiamprasert S et al. (1995). Cytochrome P450 2E1 and glutathione S-transferase M1 polymorphisms and susceptibility to hepatocellular carcinoma. *Gastroenterology*, 109:1266–1273.doi:10.1016/0016-5085(95)90587-1 PMID:7557094

95. Ross D (1996). Metabolic basis of benzene toxicity. *Eur J Haematol Suppl,* 60:111–118. PMID:8987252

96. Rothman N, Smith MT, Hayes RB et al. (1997). Benzene poisoning, a risk factor for hematological malignancy, is associated with the NQO1 609C–>T mutation and rapid fractional excretion of chlorzoxazone. *Cancer Res*, 57:2839–2842. PMID:9230185

97. Vogelstein B, Kinzler KW (2004). Cancer genes and the pathways they control. *Nat Med,* 10:789–799.doi:10.1038/nm1087 PMID:15286780

98. Matullo G, Peluso M, Polidoro S et al. (2003). Combination of DNA repair gene single nucleotide polymorphisms and increased levels of DNA adducts in a population-based study. *Cancer Epidemiol Biomarkers Prev,* 12:674–677. PMID:12869411

99. Hung RJ, Hall J, Brennan P, Boffetta P (2005). Genetic polymorphisms in the base excision repair pathway and cancer risk: a HuGE review. *Am J Epidemiol,* 162:925–942. doi:10.1093/aje/kwi318 PMID:16221808

100. Berwick M, Vineis P (2005). Measuring DNA repair capacity: small steps. *J Natl Cancer Inst,* 97:84–85.doi:10.1093/jnci/dji038 PMID:15657333

101. Ioannidis JP, Trikalinos TA (2005). Early extreme contradictory estimates may appear in published research: the Proteus phenomenon in molecular genetics research and randomized trials. *J Clin Epidemiol,* 58:543–549.doi:10.1016/j.jclinepi.2004.10.019 PMID:15878467

102. Vineis P, Manuguerra M, Kavvoura FK et al. (2009). A field synopsis on low-penetrance variants in DNA repair genes and cancer susceptibility. *J Natl Cancer Inst,* 101:24–36. PMID:19116388

103. Zeggini E, Weedon MN, Lindgren CM et al.; Wellcome Trust Case Control Consortium (WTCCC) (2007). Replication of genome-wide association signals in UK samples reveals risk loci for type 2 diabetes. *Science,* 316:1336–1341.doi:10.1126/science.1142364 PMID:17463249

104. Haiman CA, Le Marchand L, Yamamato J et al. (2007). A common genetic risk factor for colorectal and prostate cancer. *Nat Genet,* 39:954–956.doi:10.1038/ng2098 PMID:17618282

105. Zanke BW, Greenwood CM, Rangrej J et al. (2007). Genome-wide association scan identifies a colorectal cancer susceptibility locus on chromosome 8q24. *Nat Genet,* 39:989–994.doi:10.1038/ng2089 PMID:17618283

106. Chung CC, Magalhaes WC, Gonzalez-Bosquet J, Chanock SJ (2010). Genome-wide association studies in cancer–current and future directions. *Carcinogenesis,* 31:111–120. doi:10.1093/carcin/bgp273 PMID:19906782

107. Hunter DJ, Chanock SJ (2010). Genome-wide association studies and "the art of the soluble". *J Natl Cancer Inst,* 102:836–837. doi:10.1093/jnci/djq197 PMID:20505151

108. Manolio TA (2010). Genomewide association studies and assessment of the risk of disease. *N Engl J Med,* 363:166–176.doi:10.1056/NEJMra0905980 PMID:20647212

109. Amos C. Successful design and conduct of genome-wide association studies. *Hum Mol Genet* 2007;16(Spec No. 2):R220–R225.

110. Hecht SS, Carmella SG, Akerkar S, Richie JP Jr (1994). 4-(Methylnitrosamino)-1-(3,pyridyl)-1-butanol (NNAL) and its glucuronide, metabolites of a tobacco-specific lung carcinogen, in the urine of black and white smokers. *Proc Am Assoc Cancer Res,* 35:286.

111. Zang EA, Wynder EL (1996). Differences in lung cancer risk between men and women: examination of the evidence. *J Natl Cancer Inst,* 88:183–192.doi:10.1093/jnci/88.3-4.183 PMID:8632492

112. Hennekens CH, Buring JE, Manson JE et al. (1996). Lack of effect of long-term supplementation with beta carotene on the incidence of malignant neoplasms and cardiovascular disease. *N Engl J Med,* 334:1145–1149.doi:10.1056/NEJM199605023341801 PMID:8602179

113. Ryberg D, Hewer A, Phillips DH, Haugen A (1994). Different susceptibility to smoking-induced DNA damage among male and female lung cancer patients. *Cancer Res,* 54:5801–5803. PMID:7954403

114. Kure EH, Ryberg D, Hewer A et al. (1996). p53 mutations in lung tumours: relationship to gender and lung DNA adduct levels. *Carcinogenesis,* 17:2201–2205.doi:10.1093/carcin/17.10.2201 PMID:8895489

115. Mollerup S, Ryberg D, Hewer A et al. (1999). Sex differences in lung CYP1A1 expression and DNA adduct levels among lung cancer patients. *Cancer Res,* 59:3317–3320. PMID:10416585

116. Tang DL, Rundle A, Warburton D et al. (1998). Associations between both genetic and environmental biomarkers and lung cancer: evidence of a greater risk of lung cancer in women smokers. *Carcinogenesis,* 19:1949–1953.doi:10.1093/carcin/19.11.1949 PMID:9855008

117. Block G (1993). Micronutrients and cancer: time for action? *J Natl Cancer Inst,* 85:846–848.doi:10.1093/jnci/85.11.846 PMID:8388060

118. Mooney LA, Perera FP (1996). Application of molecular epidemiology to lung cancer prevention. *J Cell Biochem,* 25 S25;63–68 doi:10.1002/(SICI)1097-4644(1996)25+<63::AID-JCB9>3.0.CO;2-0.

119. Grinberg-Funes RA, Singh VN, Perera FP et al. (1994). Polycyclic aromatic hydrocarbon-DNA adducts in smokers and their relationship to micronutrient levels and the glutathione-S-transferase M1 genotype. *Carcinogenesis,* 15:2449–2454.doi:10.1093/carcin/15.11.2449 PMID:7955090

120. Mooney LA, Bell DA, Santella RM et al. (1997). Contribution of genetic and nutritional factors to DNA damage in heavy smokers. *Carcinogenesis,* 18:503–509.doi:10.1093/carcin/18.3.503 PMID:9067549

121. Küçük O, Pung A, Franke AA et al. (1995). Correlations between mutagen sensitivity and plasma nutrient levels of healthy individuals. *Cancer Epidemiol Biomarkers Prev,* 4:217–221. PMID:7541679

122. National Research Council. Science and judgment in risk assessment. Washington (DC): National Academy Press; 1994.

123. Goldman LR (1995). Children–unique and vulnerable. Environmental risks facing children and recommendations for response. *Environ Health Perspect,* 103 Suppl 6;13–18. doi:10.2307/3432338 PMID:8549460

124. Perera FP (1996). Molecular epidemiology: insights into cancer susceptibility, risk assessment, and prevention. *J Natl Cancer Inst,* 88:496–509. doi:10.1093/jnci/88.8.496 PMID:8606378

125. Perera FP (2000). Molecular epidemiology: on the path to prevention? *J Natl Cancer Inst,* 92:602–612.doi:10.1093/jnci/92.8.602 PMID:10772677

126. Birnbaum LS, Fenton SE (2003). Cancer and developmental exposure to endocrine disruptors. *Environ Health Perspect,* 111:389–394.doi:10.1289/ehp.5686 PMID:12676588

127. Miller RL, Garfinkel R, Horton M et al. (2004). Polycyclic aromatic hydrocarbons, environmental tobacco smoke, and respiratory symptoms in an inner-city birth cohort. *Chest,* 126:1071–1078.doi:10.1378/chest.126.4.1071 PMID:15486366

128. Perera FP, Tang D, Tu YH et al. (2004). Biomarkers in maternal and newborn blood indicate heightened fetal susceptibility to procarcinogenic DNA damage. *Environ Health Perspect,* 112:1133–1136.doi:10.1289/ehp.6833 PMID:15238289

129. National Research Council. Pesticides in the diets of infants and children. Washington (DC): National Academy Press; 1993.

130. Perera FP, Tang D, Whyatt RM, et al. (2004). Comparison of PAH-DNA adducts in four populations of mothers and newborns in the US, Poland and China. *Proc Am Assoc Cancer Res,* 45 (abstract #1975).

131. Whyatt RM, Jedrychowski W, Hemminki K et al. (2001). Biomarkers of polycyclic aromatic hydrocarbon-DNA damage and cigarette smoke exposures in paired maternal and newborn blood samples as a measure of differential susceptibility. *Cancer Epidemiol Biomarkers Prev,* 10:581–588. PMID:11401906

132. Bocskay KA, Tang D, Orjuela MA et al. (2005). Chromosomal aberrations in cord blood are associated with prenatal exposure to carcinogenic polycyclic aromatic hydrocarbons. Cancer Epidemiol Biomarkers Prev, 14:506–511.doi:10.1158/1055-9965. EPI-04-0566 PMID:15734979

133. Buffler PA, Kwan ML, Reynolds P, Urayama KY (2005). Environmental and genetic risk factors for childhood leukemia: appraising the evidence. Cancer Invest, 23:60–75. PMID:15779869

134. Neri M, Ugolini D, Bonassi S et al. (2006). Children's exposure to environmental pollutants and biomarkers of genetic damage. II. Results of a comprehensive literature search and meta-analysis. Mutat Res, 612:14–39.doi:10.1016/j.mrrev.2005.04.003 PMID:16027031

135. Pluth JM, Ramsey MJ, Tucker JD (2000). Role of maternal exposures and newborn genotypes on newborn chromosome aberration frequencies. Mutat Res, 465:101–111. PMID:10708975

136. Sorahan T, Prior P, Lancashire RJ et al. (1997). Childhood cancer and parental use of tobacco: deaths from 1971 to 1976. Br J Cancer, 76:1525–1531.doi:10.1038/bjc.1997.589 PMID:9400953

137. Finette BA, O'Neill JP, Vacek PM, Albertini RJ (1998). Gene mutations with characteristic deletions in cord blood T lymphocytes associated with passive maternal exposure to tobacco smoke. Nat Med, 4:1144–1151.doi. 10.1038/2640 PMID:9771747

138. Boffetta P (2000). Molecular epidemiology. J Intern Med, 248:447–454. doi:10.1046/j.1365-2796.2000.00777.x PMID:11155137

139. Janerich DT, Thompson WD, Varela LR et al. (1990). Lung cancer and exposure to tobacco smoke in the household. N Engl J Med, 323:632–636.doi:10.1056/NEJM199009 063231003 PMID:2385268

140. Vineis P, Airoldi L, Veglia F et al. (2005). Environmental tobacco smoke and risk of respiratory cancer and chronic obstructive pulmonary disease in former smokers and never smokers in the EPIC prospective study. BMJ, 330:277–280.doi:10.1136/bmj.38327.648472.82 PMID:15681570

141. Gale KB, Ford AM, Repp R et al. (1997). Backtracking leukemia to birth: identification of clonotypic gene fusion sequences in neonatal blood spots. Proc Natl Acad Sci USA, 94:13950–13954.doi:10.1073/pnas.94.25. 13950 PMID:9391133

142. Wiemels JL, Greaves M (1999). Structure and possible mechanisms of TEL-AML1 gene fusions in childhood acute lymphoblastic leukemia. Cancer Res, 59:4075–4082. PMID:10463610

143. Shimizu Y, Kato H, Schull WJ (1991). Risk of cancer among atomic bomb survivors. J Radiat Res (Tokyo), 32 Suppl 2;54–63. doi:10.1269/jrr.32.SUPPLEMENT2_54 PMID:1823367

144. Hegmann KT, Fraser AM, Keaney RP et al. (1993). The effect of age at smoking initiation on lung cancer risk. Epidemiology, 4:444–448.doi:10.1097/00001648-199309000-00010 PMID:8399693

145. Moore CJ, Tricomi WA, Gould MN (1986). Interspecies comparison of polycyclic aromatic hydrocarbon metabolism in human and rat mammary epithelial cells. Cancer Res, 46:4946–4952. PMID:3093058

146. Palmer JR, Rosenberg L (1993). Cigarette smoking and the risk of breast cancer. Epidemiol Rev, 15:145–156. PMID:8405197

147. Potter JD (2003). Epidemiology, cancer genetics and microarrays: making correct inferences, using appropriate designs. Trends Genet, 19:690–695.doi:10.1016/j.tig.2003.10.005 PMID:14642749

148. Potter JD (2001). At the interfaces of epidemiology, genetics and genomics. Nat Rev Genet, 2:142–147.doi:10.1038/35052575 PMID:11253054

149. Kim YI (2004). Folate and DNA methylation: a mechanistic link between folate deficiency and colorectal cancer? Cancer Epidemiol Biomarkers Prev, 13:511–519. PMID:15066913

150. Forozan F, Karhu R, Kononen J et al. (1997). Genome screening by comparative genomic hybridization. Trends Genet, 13:405–409.doi:10.1016/S0168-9525(97)01244-4 PMID:9351342

151. Robertson KD, Wolffe AP (2000). DNA methylation in health and disease. Nat Rev Genet, 1:11–19.doi:10.1038/35049533 PMID:11262868

152. Alberg AJ, Brock MV, Samet JM (2005). Epidemiology of lung cancer: looking to the future. J Clin Oncol, 23:3175–3185.doi:10.1200/JCO.2005.10.462 PMID:15886304

153. Brena RM, Huang TH, Plass C (2006). Quantitative assessment of DNA methylation: Potential applications for disease diagnosis, classification, and prognosis in clinical settings. J Mol Med, 84:365–377.doi:10.1007/s00109-005-0034-0 PMID:16416310

154. Toyooka S, Tokumo M, Shigematsu H et al. (2006). Mutational and epigenetic evidence for independent pathways for lung adenocarcinomas arising in smokers and never smokers. Cancer Res, 66:1371–1375. doi:10.1158/0008-5472.CAN-05-2625 PMID:16452191

155. Belinsky SA (2005). Silencing of genes by promoter hypermethylation: key event in rodent and human lung cancer. Carcinogenesis, 26:1481–1487.doi:10.1093/carcin/bgi020 PMID:15661809

156. Bearzatto A, Conte D, Frattini M et al. (2002). p16(INK4A) Hypermethylation detected by fluorescent methylation-specific PCR in plasmas from non-small cell lung cancer. Clin Cancer Res, 8:3782–3787. PMID:12473590

157. Russo AL, Thiagalingam A, Pan H et al. (2005). Differential DNA hypermethylation of critical genes mediates the stage-specific tobacco smoke-induced neoplastic progression of lung cancer. Clin Cancer Res, 11:2466–2470.doi:10.1158/1078-0432.CCR-04-1962 PMID:15814621

158. Vineis P, Husgafvel-Pursiainen K (2005). Air pollution and cancer: biomarker studies in human populations. Carcinogenesis, 26:1846–1855.doi:10.1093/carcin/bgi216 PMID:16123121

159. Honorio S, Agathanggelou A, Schuermann M et al. (2003). Detection of RASSF1A aberrant promoter hypermethylation in sputum from chronic smokers and ductal carcinoma in situ from breast cancer patients. Oncogene, 22:147–150.doi:10.1038/sj.onc.1206057 PMID:12527916

160. Belinsky SA, Liechty KC, Gentry FD et al. (2006). Promoter hypermethylation of multiple genes in sputum precedes lung cancer incidence in a high-risk cohort. Cancer Res, 66:3338–3344.doi:10.1158/0008-5472.CAN-05-3408 PMID:16540689

161. Shen L, Issa JP (2002). Epigenetics in colorectal cancer. Curr Opin Gastroenterol, 18:68–73.doi:10.1097/00001574-200201000-00012 PMID:17031233

162. Lindon JC, Holmes E, Bollard ME et al. (2004). Metabonomics technologies and their applications in physiological monitoring, drug safety assessment and disease diagnosis. Biomarkers, 9:1–31.doi:10.1080/1354750041 0001668379 PMID:15204308

163. Lindon JC, Holmes E, Nicholson JK (2006). Metabonomics techniques and applications to pharmaceutical research & development. Pharm Res, 23:1075–1088.doi:10.1007/s11095-006-0025-z PMID:16715371

164. Yang SY, Xiao XY, Zhang WG et al. (2005). Application of serum SELDI proteomic patterns in diagnosis of lung cancer. BMC Cancer, 5:83.doi:10.1186/1471-2407-5-83 PMID:16029516

165. Odunsi K, Wollman RM, Ambrosone CB et al. (2005). Detection of epithelial ovarian cancer using 1H-NMR-based metabonomics. Int J Cancer, 113:782–788.doi:10.1002/ijc.20651 PMID:15499633

166. Dumas ME, Maibaum EC, Teague C et al. (2006). Assessment of analytical reproducibility of 1H NMR spectroscopy based metabonomics for large-scale epidemiological research: the INTERMAP Study. Anal Chem, 78:2199–2208.doi:10.1021/ac0517085 PMID:16579598

167. Esteller M (2006). The necessity of a human epigenome project. Carcinogenesis, 27:1121–1125.doi:10.1093/carcin/bgl033 PMID:16699174

168. Granville CA, Dennis PA (2005). An overview of lung cancer genomics and proteomics. Am J Respir Cell Mol Biol, 32:169–176.doi:10.1165/rcmb.F290 PMID:15713815

169. Chen J, Hunter DJ, Stampfer MJ et al. (2003). Polymorphism in the thymidylate synthase promoter enhancer region modifies the risk and survival of colorectal cancer. Cancer Epidemiol Biomarkers Prev, 12:958–962. PMID:14578129

170. Yanagisawa K, Shyr Y, Xu BJ et al. (2003). Proteomic patterns of tumour subsets in non-small-cell lung cancer. Lancet, 362:433–439.doi:10.1016/S0140-6736(03)14068-8 PMID:12927430

171. Dalle-Donne I, Scaloni A, Giustarini D et al. (2005). Proteins as biomarkers of oxidative/nitrosative stress in diseases: the contribution of redox proteomics. Mass Spectrom Rev, 24:55–99.doi:10.1002/mas.20006 PMID:15389864

172. Ambrosone CB, Rebbeck TR, Morgan GJ et al. (2006). New developments in the epidemiology of cancer prognosis: traditional and molecular predictors of treatment response and survival. Cancer Epidemiol Biomarkers Prev, 15:2042–2046. doi:10.1158/1055-9965.EPI-06-0827 PMID:17119026

173. Naas SJ, Moses HL. Cancer biomarkers: the promises and challenges of improving detection and treatment. Washington (DC): National Academies Press; 2007.

174. Ransohoff DF (2005). Lessons from controversy: ovarian cancer screening and serum proteomics. J Natl Cancer Inst, 97:315–319.doi:10.1093/jnci/dji054 PMID:15713968

175. García-Closas M, Vermeulen R, Sherman ME et al. Application of biomarkers in cancer epidemiology. In Schottenfeld D, Fraumeni JF Jr, editors. Cancer epidemiology and prevention. New York (NY): Oxford University Press; 2006.

176. Forrest MS, Lan Q, Hubbard AE et al. (2005). Discovery of novel biomarkers by microarray analysis of peripheral blood mononuclear cell gene expression in benzene-exposed workers. Environ Health Perspect, 113:801–807.doi:10.1289/ehp.7635 PMID:15929907

177. Schulte PA, Perera FP, editors. Molecular epidemiology: principles and practices. New York (NY): Academic Press; 1993.

178. Stephan C, Stroebel G, Heinau M et al. (2005). The ratio of prostate-specific antigen (PSA) to prostate volume (PSA density) as a parameter to improve the detection of prostate carcinoma in PSA values in the range of < 4 ng/mL. Cancer, 104:993–1003.doi:10.1002/cncr.21267 PMID:16007682

179. Rundle AG, Vineis P, Ahsan H (2005). Design options for molecular epidemiology research within cohort studies. Cancer Epidemiol Biomarkers Prev, 14:1899–1907. doi:10.1158/1055-9965.EPI-04-0860 PMID:16103435

180. Gallicchio L, Helzlsouer KJ, Chow WH et al. (2010). Circulating 25-hydroxyvitamin D and the risk of rarer cancers: Design and methods of the Cohort Consortium Vitamin D Pooling Project of Rarer Cancers. Am J Epidemiol, 172:10–20.doi:10.1093/aje/kwq116 PMID:20562188

181. Aguilar F, Harris CC, Sun T et al. (1994). Geographic variation of p53 mutational profile in nonmalignant human liver. Science, 264:1317–1319.doi:10.1126/science.8191284 PMID:8191284

182. Hoppin JA, Tolbert PE, Taylor JA et al. (2002). Potential for selection bias with tumor tissue retrieval in molecular epidemiology studies. Ann Epidemiol, 12:1–6.doi:10.1016/S1047-2797(01)00250-2 PMID:11750233

183. Porta M, Malats N, Corominas JM et al.; Pankras I Project Investigators (2002). Generalizing molecular results arising from incomplete biological samples: expected bias and unexpected findings. Ann Epidemiol, 12:7–14.doi:10.1016/S1047-2797(01)00267-8 PMID:11750234

184. Gormally E, Hainaut P, Caboux E et al. (2004). Amount of DNA in plasma and cancer risk: a prospective study. Int J Cancer, 111:746–749.doi:10.1002/ijc.20327 PMID:15252845

185. U.S. Department of Health and Human Services. The health consequences of involuntary exposure to tobacco smoke: a report of the Surgeon General. Atlanta, GA: U.S. Department of Health and Human Services, Centers for Disease Control and Prevention, Coordinating Center for Health Promotion, National Center for Chronic Disease Prevention and Health Promotion, Office on Smoking and Health. Washington (DC): U.S. Government Printing Office; 2006.

186. National Research Council. Issues in risk assessment. Washington (DC): National Academy Press; 1993.

187. Vineis P, Forastiere F, Hoek G, Lipsett M (2004). Outdoor air pollution and lung cancer: recent epidemiologic evidence. Int J Cancer, 111:647–652.doi:10.1002/ijc.20292 PMID:15252832

188. Armitage P, Doll R (1957). A two-stage theory of carcinogenesis in relation to the age distribution of human cancer. Br J Cancer, 11:161–169.doi:10.1038/bjc.1957.22 PMID:13460138

189. Vogelstein B, Fearon ER, Hamilton SR et al. (1988). Genetic alterations during colorectal-tumor development. N Engl J Med, 319:525–532.doi:10.1056/NEJM198809013190901 PMID:2841597

190. Knudson AG (1996). Hereditary cancer: two hits revisited. J Cancer Res Clin Oncol, 122:135–140.doi:10.1007/BF01366952 PMID:8601560

191. Hashimoto T, Takahashi R, Yandell DW et al. (1991). Characterization of intragenic deletions in two sporadic germinal mutation cases of retinoblastoma resulting in abnormal gene expression. Oncogene, 6:463–469. PMID:2011402

192. Bielas JH, Loeb KR, Rubin BP et al. (2006). Human cancers express a mutator phenotype. Proc Natl Acad Sci USA, 103:18238–18242.doi:10.1073/pnas.0607057103 PMID:17108085

193. Benedict WF, Banerjee A, Kangalingam KK et al. (1984). Increased sister-chromatid exchange in bone-marrow cells of mice exposed to whole cigarette smoke. Mutat Res, 136:73–80.doi:10.1016/0165-1218(84)90136-8 PMID:6717473

194. Gatenby RA, Frieden BR (2004). Information dynamics in carcinogenesis and tumor growth. Mutat Res, 568:259–273. PMID:15542113

195. Albertini RJ, Ardell SK, Judice SA et al. (2000). Hypoxanthine-guanine phosphoribosyltransferase reporter gene mutation for analysis of in vivo clonal amplification in patients with HTLV type 1-associated Myelopathy/Tropical spastic paraparesis. AIDS Res Hum Retroviruses, 16:1747–1752. doi:10.1089/08892220050193254 PMID:11080821

196. Mooney LA, Santella RM, Covey L et al. (1995). Decline of DNA damage and other biomarkers in peripheral blood following smoking cessation. Cancer Epidemiol Biomarkers Prev, 4:627–634. PMID:8547829

197. Hemminki K, Dickey CP, Karlsson S et al. (1997). Aromatic DNA adducts in foundry workers in relation to exposure, life style and CYP1A1 and glutathione transferase M1 genotype. Carcinogenesis, 18:345–350. doi:10.1093/carcin/18.2.345 PMID:9054627

198. Evans D, Sheares BJ, Vazquez TL (2004). Educating health professionals to improve quality of care for asthma. Paediatr Respir Rev, 5:304–310.doi:10.1016/j.prrv.2004.07.007 PMID:15531255

199. Møller P, Loft S (2002). Oxidative DNA damage in human white blood cells in dietary antioxidant intervention studies. Am J Clin Nutr, 76:303–310. PMID:12144999

200. Møller P, Loft S (2006). Dietary antioxidants and beneficial effect on oxidatively damaged DNA. Free Radic Biol Med, 41:388–415.doi:10.1016/j.freeradbiomed.2006.04.001 PMID:16843820

201. Mooney LA, Perera FP, Van Bennekum AM et al. (2001). Gender differences in autoantibodies to oxidative DNA base damage in cigarette smokers. Cancer Epidemiol Biomarkers Prev, 10:641–648. PMID:11401914

202. Nana-Sinkam P, Croce CM (2010). MicroRNAs in diagnosis and prognosis in cancer: what does the future hold? Pharmacogenomics, 11:667–669.doi:10.2217/pgs.10.57 PMID:20415558

203. Chen T (2010). The role of MicroRNA in chemical carcinogenesis. J Environ Sci Health C Environ Carcinog Ecotoxicol Rev, 28:89–124. PMID:20552498

204. García-Closas M, Vermeulen R, Sherman ME et al. Application of biomarkers in cancer epidemiology. In Schottenfeld D, Fraumeni JF Jr, editors. Cancer Epidemiology and Prevention. New York (NY): Oxford University Press; 2006. pp 70–88.

205. Perera FP, Whyatt RM, Jedrychowski W et al. (1998). Recent developments in molecular epidemiology: A study of the effects of environmental polycyclic aromatic hydrocarbons on birth outcomes in Poland. Am J Epidemiol, 147:309–314. PMID:9482506

206. Weinstein IB, Zhou P. Cell cycle control gene defects and human cancer. In Bertino JR, editor. Encyclopedia of cancer. Vol 1. New York (NY): Academic Press; 1997. p. 256–267.

207. Olden K, Poje J (1995). Environmental justice and environmental health. Bull Soc Occup Environ Health, 4:3–4.

208. Portier CJ, Bell DA (1998). Genetic susceptibility: significance in risk assessment. Toxicol Lett, 102-103:185–189.doi:10.1016/S0378-4274(98)00305-1 PMID:10022252

209. Hattis D (1996). Human interindividual variability in susceptibility to toxic effects: from annoying detail to a central determinant of risk. Toxicology, 111:5–14.doi:10.1016/0300-483X(96)03388-4 PMID:8711749

210. Hattis D, Barlow K (1996). Human interindividual variability in cancer risks. Technical and management challenges. Hum Ecol Risk Assess, 2:194–220 doi:10.1080/10807039.1996.10387468.

211. Hecht SS, Carmella SG, Murphy SE et al. (1993). A tobacco-specific lung carcinogen in the urine of men exposed to cigarette smoke. N Engl J Med, 329:1543–1546.doi:10.1056/NEJM199311183292105 PMID:8413477

212. Hecht SS (1997). Tobacco and cancer: approaches using carcinogen biomarkers and chemoprevention. Ann NY Acad Sci, 833:91–111.doi:10.1111/j.1749-6632.1997.tb48596.x PMID:9616743

213. Groopman JD, Kensler TW, Links JM (1995). Molecular epidemiology and human risk monitoring. Toxicol Lett, 82-83:763–769.doi:10.1016/0378-4274(95)03594-X PMID:8597140

214. Santella RM, Zhang YJ, Chen CJ et al. (1993). Immunohistochemical detection of aflatoxin B1-DNA adducts and hepatitis B virus antigens in hepatocellular carcinoma and nontumorous liver tissue. Environ Health Perspect, 99:199–202. PMID:8391434

215. Hagmar L, Wirfält E, Paulsson B, Törnqvist M (2005). Differences in hemoglobin adduct levels of acrylamide in the general population with respect to dietary intake, smoking habits and gender. Mutat Res, 580:157–165. PMID:15668117

216. Vodicka P, Koskinen M, Stetina R et al. (2003). The role of various biomarkers in the evaluation of styrene genotoxicity. Cancer Detect Prev, 27:275–284.doi:10.1016/S0361-090X(03)00096-5 PMID:12893075

217. Albertini RJ, Srám RJ, Vacek PM et al. (2003). Biomarkers in Czech workers exposed to 1,3-butadiene: a transitional epidemiologic study. Res Rep Health Eff Inst, (116):1–141, discussion 143–162. PMID:12931846

218. Smith MT, McHale CM, Wiemels JL et al. (2005). Molecular biomarkers for the study of childhood leukemia. Toxicol Appl Pharmacol, 206:237–245.doi:10.1016/j.taap.2004.11.026 PMID:15967214

219. Holecková B, Piesová E, Sivikova K (2004). Dianovský J. Chromosomal aberrations in humans induced by benzene. Ann Agric Environ Med, 11:175–179. PMID:15627321

220. Perera FP, Mooney LA, Stampfer M et al.; Physicians' Health Cohort Study (2002). Associations between carcinogen-DNA damage, glutathione S-transferase genotypes, and risk of lung cancer in the prospective Physicians' Health Cohort Study. Carcinogenesis, 23:1641–1646.doi:10.1093/carcin/23.10.1641 PMID:12376472

221. Ammenheuser MM, Bechtold WE, Abdel-Rahman SZ et al. (2001). Assessment of 1,3-butadiene exposure in polymer production workers using HPRT mutations in lymphocytes as a biomarker. Environ Health Perspect, 109:1249–1255.doi:10.1289/ehp.011091249 PMID:11748032

222. Lee KH, Lee J, Ha M et al. (2002). Influence of polymorphism of GSTM1 gene on association between glycophorin a mutant frequency and urinary PAH metabolites in incineration workers. J Toxicol Environ Health A, 65:355–363. doi:10.1080/15287390252808028 PMID:11936216

223. Gwosdz C, Balz V, Scheckenbach K, Bier H (2005). p53, p63 and p73 expression in squamous cell carcinomas of the head and neck and their response to cisplatin exposure. Adv Otorhinolaryngol, 62:58–71. PMID:15608418

224. Cheng L, Eicher SA, Guo Z et al. (1998). Reduced DNA repair capacity in head and neck cancer patients. Cancer Epidemiol Biomarkers Prev, 7:465–468. PMID:9641488

225. García-Closas M, Malats N, Silverman D et al. (2005). NAT2 slow acetylation, GSTM1 null genotype, and risk of bladder cancer: results from the Spanish Bladder Cancer Study and meta-analyses. Lancet, 366:649–659.doi:10.1016/S0140-6736(05)67137-1 PMID:16112301

226. Vineis P, Veglia F, Benhamou S et al. (2003). CYP1A1 T3801 C polymorphism and lung cancer: a pooled analysis of 2451 cases and 3358 controls. Int J Cancer, 104:650–657.doi:10.1002/ijc.10995 PMID:12594823

227. Committee on Biological Markers of the National Research Council (1987). Biological markers in environmental health research. Environ Health Perspect, 74:3–9. PMID:3691432

UNIT 5.
APPLICATION OF BIOMARKERS TO DISEASE

CHAPTER 20.

Coronary heart disease

Emanuele Di Angelantonio, Alexander Thompson, Frances Wensley, and John Danesh

Summary

Until recently, the potential relevance of genetic, biochemical and lifestyle factors to coronary heart disease have been studied in relative isolation from one another. Although this approach has yielded some major insights, it has resulted in a fragmented and incomplete understanding of the relative importance and interplay of nature and nurture in the development of coronary risk. New opportunities for more integrated, powerful and comprehensive approaches have been opened by major developments, including: establishment, collation and maturation of relevant population bioresources; emergence of technologies that enable rapid and accurate assessment of many genetic and biochemical factors, without necessitating assumptions about biological mechanisms; and advances in statistical analytical methods. This chapter provides a critical review of the strengths and limitations of established and emerging epidemiological approaches to the study of the separate and combined effects of genetic, biochemical and lifestyle factors in coronary heart disease.

Introduction

Coronary heart disease (CHD) remains a pre-eminent global public health concern. With over seven million deaths per year attributed to CHD, it is the leading cause of death worldwide, a major source of disability, and a considerable economic burden (1–3). Over the past half-century, several major modifiable coronary risk factors have been identified, such as smoking, diabetes, and elevated levels of blood pressure and low-density lipoprotein cholesterol (LDL-C) (4–7). These insights have led to improvements in primary and secondary prevention, prognosis and treatment strategies, and, ultimately, contributed to reductions in cardiovascular morbidity and mortality in many high-income countries (8–12). CHD remains, however, the leading killer in most high-income countries, and its incidence is increasing rapidly in many low- and middle-income countries, such as those in South Asia (13–15).

In parallel with greater efforts to control established risk factors, there is considerable interest in the discovery and evaluation of novel and emerging risk markers in CHD. By analogy with measurement and modification of LDL-C levels, it has been suggested that identification of usefully predictive and/or causal biomarkers in CHD should contribute to insights into disease pathophysiology that may translate into clinical benefits through identification of novel therapeutics, improved stratification of disease risk in vulnerable populations, more cost-effective targeting of existing interventions, and identification and understanding of joint gene–environment effects. The purpose of this chapter is to provide a critical survey of epidemiological approaches being used in the discovery and evaluation of genetic and molecular risk markers in CHD.

Studies of genetic sequence variation in coronary heart disease

Candidate gene approaches

The tendency for coronary heart disease (CHD) to cluster in families (coefficient of familial clustering [λs] estimated to be between 2 and 7) (16–18) suggests that genetic variation, through modulation of known or as-yet unidentified risk factors, importantly influences CHD risk (16). Until recently, genetic epidemiological studies in CHD tended to involve candidate variant or candidate gene studies involving focused investigation of relatively few genetic variants based on plausible biological hypotheses. Many of these studies had anticipated identification of common variants with large effects on CHD risk (e.g. odds ratios >2), and few were compatible with the reliable identification of variants with moderate effects or smaller (e.g. odds ratio <1.5). In retrospect, such expectations appear unrealistic because it now seems unlikely that the genetic architecture of CHD includes common variants of large effect, equivalent to HLA in type 1 diabetes (19,20) or *CFH* in age-related macular degeneration (21,22).

The combination of the low prior odds of the variants selected for study, inadequate power (i.e. small sample size) and over-liberal declarations of significance has resulted in the reporting of many "positive" findings that remain unreplicated or directly refuted, exemplified by studies of the insertion/deletion polymorphism of the angiotensin-converting enzyme gene (23,24) and of variants in the paraoxonase (25) and lymphotoxin-α genes (26,27). Indeed, a review of meta-analyses of about 50 candidate gene variants in CHD has indicated that available genetic association studies have typically been inconclusive (Figure 20.1) (28), with the notable exception of the apolipoprotein E gene, for which evidence of association is persuasive (Figure 20.2) (29). It is possible that some such candidate variants really are associated with CHD, but the available evidence is generally inadequate to reliably confirm or refute odds ratios of 0.8–1.2 per allele (which is the observed range for point estimates of odds ratios for the large majority of the variants listed in Figure 20.2).

Attempts to enhance statistical power by meta-analyses of the published literature can be helpful, but they are inherently limited by the scale of evidence available for review (e.g. only 15 variants listed in Figure 20.2 have been studied in a total of at least 10 000 CHD cases), and by potential reporting biases (e.g. preferential publication of striking findings) (30,31). As suggested by the power calculations in Table 20.1, analyses of about 20 000 myocardial infarction (MI) cases and a similar number of controls are generally required to provide excellent power to evaluate reliably common variants which may have odds ratios as low as 1.1, particularly when involving comparisons of many genotypes. So far, only a few studies have been established on this scale. The case–control study component of the International Study of Infarct

Table 20.1. Power to detect odds ratios of moderate size for the effect of common genetic variants on coronary disease outcomes in case–control studies with 2500 to 20 000 cases

		Odds ratio 1.1			Odds ratio 1.15			Odds ratio 1.2		
	MAF	0.1	0.2	0.3	0.1	0.2	0.3	0.1	0.2	0.3
No. of cases	2500	0	0	0	0	1	2	1	5	14
	5000	0	0	1	1	9	23	9	47	74
	10 000	1	7	18	15	64	87	62	98	100
	20 000	11	55	81	77	100	100	100	100	100

Assumptions include: $\alpha = 10^{-7}$, $r^2 = 0.8$, prevalence of coronary heart disease = 10%, multiplicative model, one control per case. Power is the ability to detect against type 2 error = 100*(1-β). MAF, minor allele frequency.

Figure 20.1. Summary estimates from meta-analyses of association studies of SNPs in various candidate genes and coronary disease (28)

Biological pathway	Author	Year	Gene	Variant	Allele	MAF*	Outcome	Cases	n[†]	Info[‡]	Per-allele OR & 95%CI	Adjusted[‡]	Consistency I²
Lipids													
	Bennet	2007	apoE	e2/e3/e4	e4	0.11	CHD	37 850	121	3335		1.13 (1.03, 1.25)	71 (65, 76)
	Bennet	2007	apoE	e2/e3/e4	e2	0.07	CHD	37 850	121	1818		0.88 (0.79, 1.00)	54 (44, 63)
	Sagoo	-	LPL	N291S	S	0.03	CHD	13 883	21	378		1.03 (0.88, 1.23)	5.0 (0, 37)
	Sagoo	-	LPL	S447X	X	0.11	CHD	11 047	26	893		0.88 (0.74, 1.00)	49 (20, 68)
	Sagoo	-	LPL	D9N	N	0.02	CHD	9807	21	279		1.24 (1.00, 1.56)	19 (0, 53)
	Sagoo	-	LPL	PvuII	P2	0.46	CHD	8440	18	1985		0.99 (0.86, 1.16)	32 (0, 62)
	Sagoo	-	LPL	T-93G	G	0.02	CHD	5045	7	107		1.17 (0.79, 1.65)	10 (0, 74)
	Sagoo	-	LPL	HindIII	H1	0.29	CHD	6226	23	634		0.94 (0.83, 1.07)	16 (0, 49)
	Sagoo	-	LPL	G188E	E	0.001	CHD	2524	3	3		1.41 (0.26, 6.99)	0 (-, -)
	Lawlor	2004	PON1	Q192R	R	0.32	CHD	10 816	38	2244		1.07 (0.96, 1.21)	61 (0, 64)
	Wheeler	2004	PON1	L55M	M	0.27	CHD	5989	20	944		1.00 (0.91, 1.10)	0 (0, 49)
	Wheeler	2004	PON1	T-107C	C	0.47	CHD	1366	4	334		1.01 (0.79, 1.44)	43 (0, 81)
	Wheeler	2004	PON2	S311C	C	0.22	CHD	1498	7	288		1.01 (0.72, 1.39)	67 (26, 85)
	Boekholdt	2005	CETP	TaqIB	B1	0.44	CHD	2857	7	767		0.88 (0.78, 1.03)	0 (0, 71)
	Chiodini	2003	apoB	SpIns/Del	Del	0.31	CHD	3777	19	859		1.10 (0.89, 1.45)	57 (28, 74)
	Chiodini	2003	apoB	XBaI	X-	0.63	CHD	2503	19	467		1.01 (0.79, 1.34)	59 (31, 75)
	Chiodini	2003	apoB	EcoRI	E-	0.14	CHD	1677	14	160		1.16 (0.89, 1.49)	19 (0, 56)
Inflammation													
	Sie	2006	IL6	G-174C	C	0.42	CHD	6927	8	1509		1.00 (0.92, 1.09)	54 (0, 79)
	Clarke	2006	LTA	909253	C	0.37	MI	9772	5	1534		1.04 (0.92, 1.28)	62 (0, 86)
	Koch	2008	THBS4	A387P	P	0.22	MI	6978	8	949		0.99 (0.84, 1.20)	63 (21, 83)
	Koch	2008	THBS1	T3949G	G	0.10	MI	6388	5	440		1.06 (0.81, 1.36)	35 (0, 76)
	Koch	2008	THBS2	N700S	S	0.25	MI	4930	4	592		1.00 (0.67, 1.40)	70 (14, 90)
	Pereira	2007	TNF-alpha	G-308A	A	0.15	CHD	6740	17	560		0.93 (0.80, 1.08)	49 (11, 71)
	Allen	2001	TNF-alpha	G-238A	A	0.05	CHD	822	2	45		1.02 (0.35, 3.16)	0 (-, -)
	Koch	2006	TLR4	D299G	G	0.07	MI	6143	7	213		0.91 (0.55, 1.36)	64 (18, 84)
	Abilleira	2006	MMP3	5A/6A	5A	0.31	MI	2549	7	269		1.23 (0.87, 1.80)	87 (76, 93)
	Abilleira	2006	MMP9	C-1562T	T	0.12	Stenosis	3909	5	211		1.05 (0.67, 1.78)	67 (15, 87)
Renin-angiotensin													
	Morgan	2003	ACE	I/D	Del	0.52	MI	13 506	40	1530		1.02 (0.96, 1.09)	60 (44, 72)
	Xu	2007	AGT	M235T	T	0.48	CHD	13 279	41	2714		1.00 (0.88, 1.12)	56 (37, 69)
	Xu	2007	AGT	T174M	M	0.11	CHD	8605	16	785		0.93 (0.81, 1.24)	54 (19, 74)
	Casas	2006	eNOS	E298D	D	0.26	CHD	13 298	40	1837		0.95 (0.85, 1.08)	69 (57, 77)
	Casas	2006	eNOS	T-786C	C	0.30	CHD	10 004	21	1414		1.09 (0.94, 1.28)	64 (42, 77)
	Casas	2006	eNOS	Intron 4	a	0.14	CHD	9704	30	945		1.04 (0.89, 1.20)	57 (35, 71)
	Ntzani	2007	AGT1R	A1166C	C	0.27	MI	9663	25	2232		1.05 (0.92, 1.24)	62 (41, 75)
Haemostasis													
	Ye	2006	Factor V	R506Q	Q	0.03	CHD	15 121	57	437		1.02 (0.90, 1.16)	15 (0, 40)
	Ye	2006	GPIIIa	P1A1/A2	A2	0.15	CHD	12 524	44	1440		0.94 (0.81, 1.09)	60 (45, 72)
	Ye	2006	Factor II	G20210A	A	0.01	CHD	11 309	32	134		1.00 (0.80, 1.24)	28 (0, 53)
	Ye	2006	PAI-1	4G/5G	4G	0.51	CHD	10 770	35	2658		0.99 (0.90, 1.07)	57 (37, 71)
	Ye	2006	Factor VII	R353Q	Q	0.12	CHD	6875	24	756		1.02 (0.92, 1.14)	0 (0, 45)
	Ye	2006	GPIa	C807T	T	0.38	CHD	5853	15	1311		1.01 (0.89, 1.16)	47 (2.4, 71)
	Ye	2006	GPIba	T-5C	C	0.13	CHD	4898	13	474		1.07 (0.87, 1.27)	39 (0, 68)
	Keavney	2006	FGB	G-455A	A	0.20	CHD	12 220	20	1915		0.96 (0.87, 1.03)	25 (0, 57)
	Voko	2007	Factor XIII	V34L	L	0.24	CHD	5751	16	990		1.02 (0.89, 1.14)	65 (40, 79)
Others													
	van der A	2006	HFE	C282Y	Y	0.06	CHD	8839	23	709		1.02 (0.91, 1.15)	22 (0, 53)
	van der A	2006	HFE	H63D	D	0.15	CHD	7239	13	1109		1.05 (0.96, 1.12)	14 (0, 53)
	Lewis	2005	MTHFR	C677T	T	0.34	CHD	22 196	80	1037		1.02 (0.96, 1.08)	43 (25, 56)
	Kjaergaard	2007	ESR1	T-397C	C	0.45	MI	4516	8	888		1.02 (0.91, 1.22)	47 (0, 76)
	Zafarmand	2008	ADRB3	W64R	R	0.09	CHD	4062	10	244		1.04 (0.82, 1.31)	42 (0, 72)

Ordered by biological category, review, then variant in descending order of total number of participants. Coronary heart disease (CHD) defined as myocardial infarction (MI), coronary stenosis, angina or fatal coronary event. * Frequency of allele of interest, not necessarily the rare allele, calculated under HWE where 3x2 data unavailable; † Number of studies included in meta-analysis; ‡ Pooled additive (per-allele) odds ratios from a random-effects meta-regression accounting for study sample size.

Survival (ISIS), for example, involves about 14 000 acute MI cases (about half of whom had a history of cardiovascular disease) and about 16 000 controls, all of whom were resident in the United Kingdom and > 90% of whom were of white ethnicity (4,32). The INTERHEART study involves about 15 000 first-ever MI cases and 15 000 controls from 52 countries (33). These studies have encouraged the initiation of similar research, such as the Pakistan Risk of Myocardial Infarction Study (PROMIS), which is recruiting about 20 000 patients with first-ever confirmed MI and 20 000 controls in urban Pakistan (http://www.phpc.cam.ac.uk/ MEU/PROMIS/).

Genomic approaches

While progress in identifying individual genetic variants associated with CHD risk has been relatively limited, recent successes in identifying susceptibility genes for CHD (e.g. chr9/CDKN2A: Figure 20.3 (29)) (34–37) and for lipid fractions (e.g. chr 1p13.3 in relation to

Figure 20.2. ApoE genotypes are significantly associated with coronary heart disease risk and levels of circulating low-density lipoprotein cholesterol from studies with ≥500 cases or ≥1000 healthy participants (29). Copyright © (2007) American Medical Association. All rights reserved.

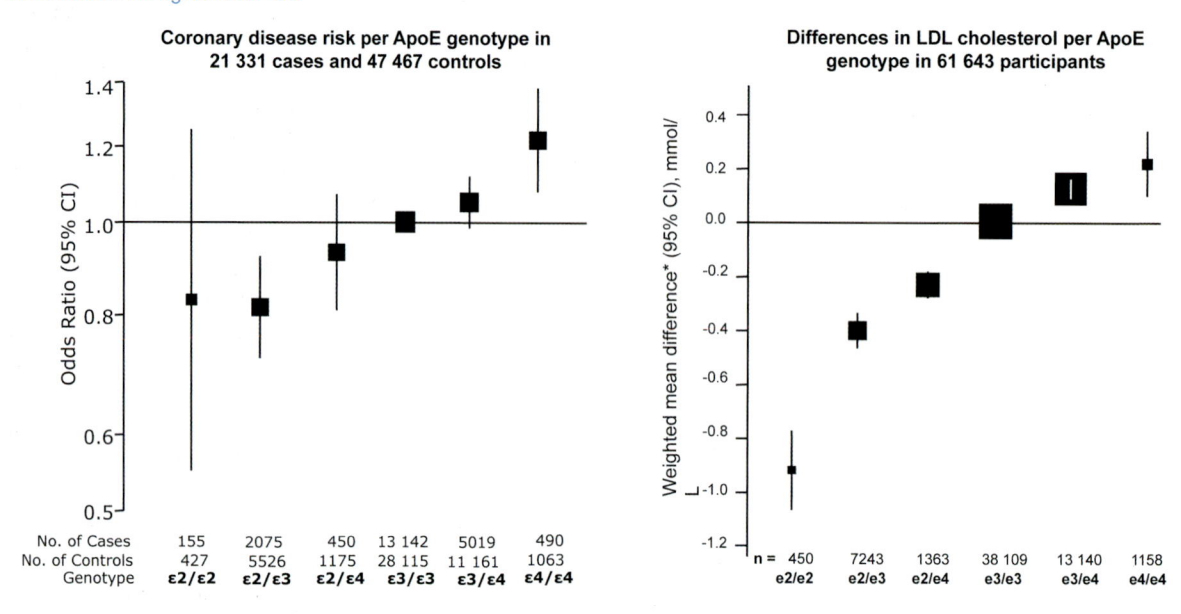

Sizes of data markers indicate the weight of each study in the analysis and the vertical lines represent 95% confidence intervals. The figure shows the weighted mean difference in LDL-C levels for each genotype compared with ε3/ε3.

LDL-C) (38–41) have demonstrated that novel, reproducible effects can be detected given appropriate study design and sample size. In particular, available genome-wide association studies (GWAS) have demonstrated that so-called hypothesis-free global-testing methods can advance discovery and can complement studies involving focused biological approaches to candidate identification (42–44). The nine GWAS of clinical coronary outcomes published by August 2010 have been generally conducted in populations of European ancestry (Table 20.2) (26,34–37,45-48). These studies have involved between 94 and 2967 coronary cases, some providing adequate power to detect per-allele odds ratios of 1.4 or larger, but not to detect more moderate odds ratios. Despite this relatively moderate power, initial GWAS have detected loci with odds ratios for CHD less than 1.4, suggesting that several (perhaps many) additional loci with effects of similar magnitude remain undetected and await discovery.

Similar considerations apply to the evaluation of less common genetic variants, as acknowledged in the Wellcome Trust Case Control Consortium (WTCCC) report: "Even with 2000 cases and 3000 controls, adequate power is restricted to common variants of relatively large effect," and "Given the likely distribution of effect sizes for most complex traits, there are strong grounds for prosecution of GWA studies on an even larger scale than ours." (35). Combined analysis of data from the WTCCC and the Cardiogenics Study (involving a total of 2801 coronary cases) suggested possible associations with four further loci not detected in either study separately, reinforcing how greater power can enhance discovery (34). Other variants, such as those seen at 16q23 and 6q25.1, have been replicated in at least one independent sample and require further investigation (34,35,45). As the numbers of single nucleotide polymorphisms (SNPs) assessed continues to increase, the sharing of findings and data sets by investigators will enable more rapid validation of results. For example, the lack of association at the 9p21.3 locus with CHD in the Japanese Osaka Acute Coronary Insufficiency Study (OACIS) study (26) contrasts notably with available evidence in predominantly Western populations. This may be an indication that the association varies between populations, although without access to full details of the SNPs genotyped it is not possible to determine if this region was adequately covered in the OACIS study. The prospect of increasingly powerful and detailed genomic studies of intermediate phenotypes (e.g. lipid levels) and cardiovascular outcomes is suggested by the initiation of further GWAS scans under the aegis of WTCCC-2 (http://www.wtccc.org.uk/), including 5000 MI cases and 5000 controls

Table 20.2. Examples of genome-wide association studies of coronary disease outcomes reported by 2010

Study (Ref)	Geographical location	No. of Cases / controls in the discovery stage	Case definition	Genotyping platform	No. of SNPs assessed	Loci declared significant
Celera (46)	USA	340 / 346	MI	Celera sequencing technology	11 053	4q32, 6q22, 12p13, 1q44, 19p13.2
DECODE (36)	Iceland	1607 / 6728	MI	Illumina Hap300	305 953	9p21.3
Framingham (45)	USA	118 / 1227	CHD	Affymetrix 100K	70 987	16q23, 2q32, 15q21, 17q24, 8q22, 4q22, 12p12 (not yet validated)
German MI Family Study (34)	Germany	875 / 1644	MI with family history of CAD	Affymetrix 500K	272 602	9p21.3, 6q25.1, 2q36.3
OACIS (26)	Japan	94 / 658	MI	PCR-Invader assay	65 671	6p21
Ottawa Heart Study (37)	Canada	322 / 312	Coronary revascularization	Oligonucleotide arrays	72 864	9p21.3
WTCCC (34,35)	UK	1926 / 2938	MI or coronary revascularization	Affymetrix 500K	469 557	9p21.3, 6q25.1
MIGen Consortium (47)	USA, Sweden, Finland, Spain, Italy	2967 / 3071	MI	Affymetrix 6.0	2 557 924	9p21.3, 1p13, 10q11, 1q41, 19p13, 1p32, 21q22, 6p24, 2q33
German MI Family Study (48)	Germany	1222 / 1298	MI with family history of CAD	Affymetrix 6.0	869 224	3q22.3, 12q24.31, 9p21.3, 1q41

CAD, coronary artery disease; CHD, coronary heart disease; MI, myocardial infarction; OACIS, Osaka Acute Coronary Insufficiency Study; PCR, polymerase chain reaction; SNPs, single nucleotide polymorphisms; WTCCC, Wellcome Trust Case Control Consortium

in PROMIS. The establishment of large international consortia, such as the EU-funded European Network of Genomic and Genetic Epidemiology (ENGAGE; http://www.euengage.org), which have pooled GWAS data in about 100 000 individuals, should also propel discovery and validation of novel loci in cardiovascular diseases and quantitative traits (49). The use of custom-designed microarrays, such as the Illumina MetaboChip of > 200 000 SNPs related to cardio-metabolic traits, should provide some of the advantages of GWAS at a considerably lower cost.

Studies of candidate plasma biomarkers in CHD

Approaches to prioritize hypotheses and enhance interpretation

Although technologies are emerging that enable rapid measurement of large numbers of many different blood-based molecules (biomarkers) (50–54), unlike GWAS for genetic markers, there are not as yet hypothesis-free global-testing methods that enable reliable quantitative assessment of concentrations of a large number of biomarkers in human blood samples. In the absence of such comprehensive tests, studies are needed to help prioritize the measurement of specific candidate biomarkers, assays for which may be costly and consume non-trivial quantities of limited blood samples that have been stored as part of long-term population studies. Moreover, in the absence of individual studies of very large size, appropriate synthesis of the available reports of such factors in CHD by meta-analysis should provide a better preliminary indication of their relevance to CHD

Figure 20.3. Meta-analysis summarising associations of chromosome 9 with coronary disease in 12 studies from populations of different ethnicity (28)

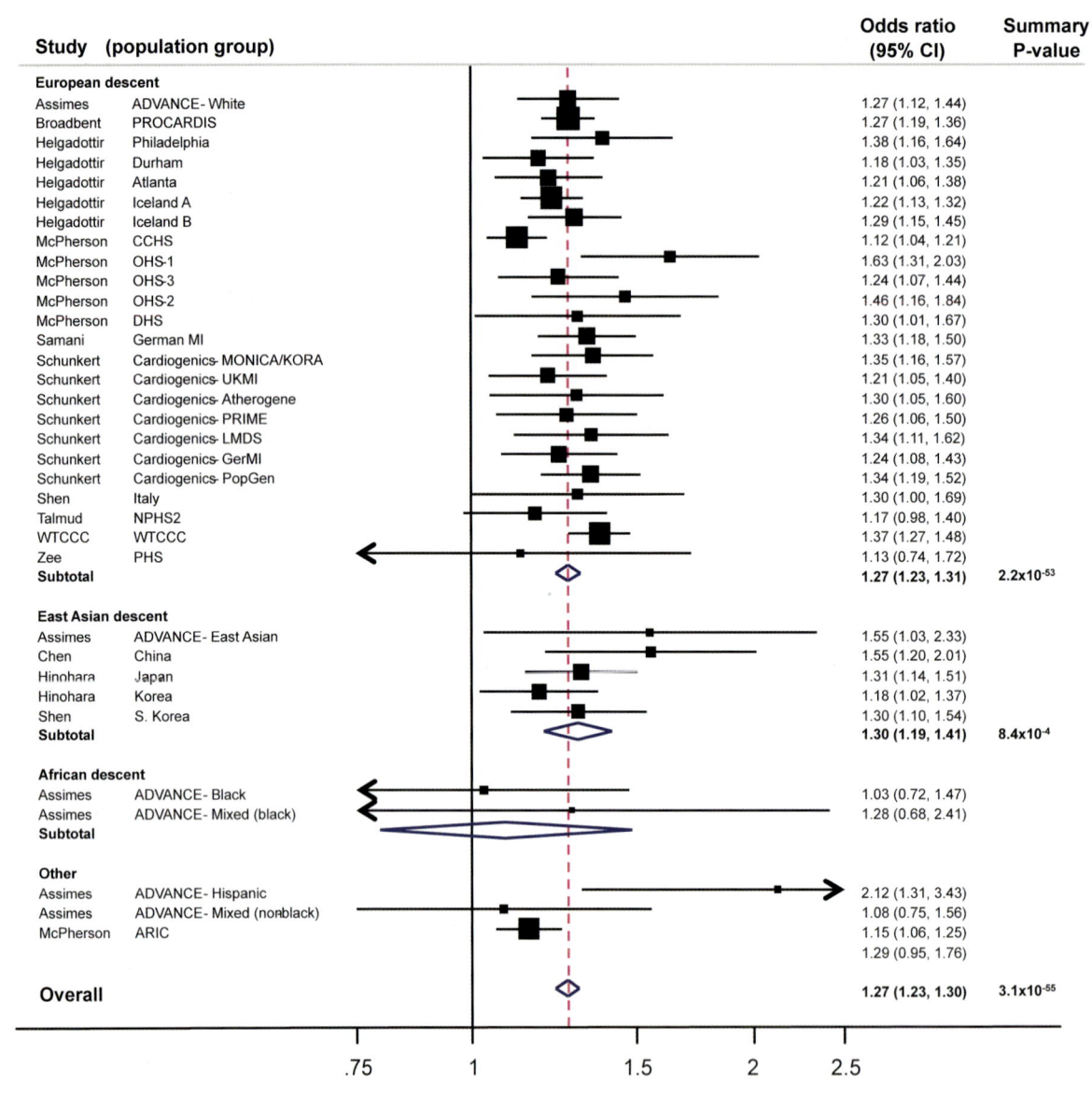

Study	(population group)	Odds ratio (95% CI)	Summary P-value
European descent			
Assimes	ADVANCE- White	1.27 (1.12, 1.44)	
Broadbent	PROCARDIS	1.27 (1.19, 1.36)	
Helgadottir	Philadelphia	1.38 (1.16, 1.64)	
Helgadottir	Durham	1.18 (1.03, 1.35)	
Helgadottir	Atlanta	1.21 (1.06, 1.38)	
Helgadottir	Iceland A	1.22 (1.13, 1.32)	
Helgadottir	Iceland B	1.29 (1.15, 1.45)	
McPherson	CCHS	1.12 (1.04, 1.21)	
McPherson	OHS-1	1.63 (1.31, 2.03)	
McPherson	OHS-3	1.24 (1.07, 1.44)	
McPherson	OHS-2	1.46 (1.16, 1.84)	
McPherson	DHS	1.30 (1.01, 1.67)	
Samani	German MI	1.33 (1.18, 1.50)	
Schunkert	Cardiogenics- MONICA/KORA	1.35 (1.16, 1.57)	
Schunkert	Cardiogenics- UKMI	1.21 (1.05, 1.40)	
Schunkert	Cardiogenics- Atherogene	1.30 (1.05, 1.60)	
Schunkert	Cardiogenics- PRIME	1.26 (1.06, 1.50)	
Schunkert	Cardiogenics- LMDS	1.34 (1.11, 1.62)	
Schunkert	Cardiogenics- GerMI	1.24 (1.08, 1.43)	
Schunkert	Cardiogenics- PopGen	1.34 (1.19, 1.52)	
Shen	Italy	1.30 (1.00, 1.69)	
Talmud	NPHS2	1.17 (0.98, 1.40)	
WTCCC	WTCCC	1.37 (1.27, 1.48)	
Zee	PHS	1.13 (0.74, 1.72)	
Subtotal		1.27 (1.23, 1.31)	2.2×10^{-53}
East Asian descent			
Assimes	ADVANCE- East Asian	1.55 (1.03, 2.33)	
Chen	China	1.55 (1.20, 2.01)	
Hinohara	Japan	1.31 (1.14, 1.51)	
Hinohara	Korea	1.18 (1.02, 1.37)	
Shen	S. Korea	1.30 (1.10, 1.54)	
Subtotal		1.30 (1.19, 1.41)	8.4×10^{-4}
African descent			
Assimes	ADVANCE- Black	1.03 (0.72, 1.47)	
Assimes	ADVANCE- Mixed (black)	1.28 (0.68, 2.41)	
Subtotal			
Other			
Assimes	ADVANCE- Hispanic	2.12 (1.31, 3.43)	
Assimes	ADVANCE- Mixed (nonblack)	1.08 (0.75, 1.56)	
McPherson	ARIC	1.15 (1.06, 1.25)	
		1.29 (0.95, 1.76)	
Overall		1.27 (1.23, 1.30)	3.1×10^{-55}

Odds ratio (95% confidence interval) for per-allele change in coronary heart disease risk

than can individual studies involving just a few hundred cases. This is because meta-analyses are less likely to be subject to random error than single studies, which due to their inherent statistical uncertainties may produce false-positive and false-negative results. The impact of random error in single studies can be compounded by unduly data-dependent analyses and selective reporting. Such situations arise when analytical cut-off values are chosen only after an exploration of the data has shown which values seemed to be most strongly related to CHD, prominence is given to extreme findings in selected subgroups based on sparse data, results are preferentially reported just for those few factors which show extreme associations (out of the many measured), and journals preferentially publish striking findings (55–60).

Consequently, to enhance appropriate interpretation and to prioritize hypotheses for further investigation, there is an increasing need for systematic reviews of publications on biomarkers in CHD (Table 20.3). Figure 20.4 suggests a schema for a staged approach to the evaluation of candidate biomarkers in CHD. This approach includes systematic reviews of published and unpublished data, measurement of emerging biomarkers in stored samples from existing large prospective studies, and the collaborative pooling of individual participant data from multiple studies.

Preliminary quantitative reviews (literature-based meta-analyses) have helped to prioritize research in CHD by

- identifying risk markers for which the available evidence is, in aggregate, comparatively unpromising, encouraging the study of other, potentially more fruitful hypotheses. For example, meta-analyses of *Chlamydia pneumoniae* infection (61), markers of iron status (62), or soluble adhesion molecules (63), have refuted inappropriate earlier claims of strongly positive associations;
- suggesting the need for new measurements in much larger studies than hitherto to achieve reliable results, exemplified by reviews of leptin and adiponectin (64), insulin and proinsulin (65), and lipoprotein(a) (66);
- indicating that existing data would, if properly brought together into a detailed synthesis, be sufficient to yield reliable results, encouraging the formation of collaborative groups to conduct individual participant meta-analyses based on the collation, harmonization and re-analysis of available worldwide data, as discussed below.

Collaborative analyses of primary data from prospective studies

Many long-term prospective studies of cardiovascular outcomes have reported on associations with established and emerging risk

Table 20.3. Examples of systematic reviews of studies of blood-based biomarkers and coronary disease outcomes

Type of factor	Examples (Ref)	No. of CHD cases	Risk ratio (top third vs. bottom third)*
Acute-phase reactants	Fibrinogen (146)	3000	1.8 (1.6-2.0)
	Albumin (146)	3800	1.5 (1.3-1.7)
	Leukocyte count (146)	6000	1.4 (1.3-1.5)
	Granulocyte count (153)	1500	1.3 (1.2-1.5)
	Neutrophil count (153)	1600	1.3 (1.2-1.5)
	Lymphocyte count (153)	1700	1.1 (1.0-1.3)
	Monocyte count (153)	1700	1.1 (1.0-1.2)
	Serum amyloid A protein (147)	600	1.6 (1.1-2.2)
	C-reactive protein (148)	7000	1.5 (1.4-1.6)
	Interleukin-6 (176)	5700	1.6 (1.4-1.8)
Haemostatic	von Willebrand factor (242)	1000	1.5 (1.1-2.0)
	tPA antigen (243)	2100	2.2 (1.8-2.7)
	Fibrin D-dimer (244)	1500	1.7 (1.3-2.2)
	PAI-I (243)	800	1.0 (0.5-1.8)
Lipids	Lipoprotein(a) (175)	9800	1.5 (1.3-1.6)
	Triglycerides (151)	10 000	1.7 (1.6-1.9)
	Apolipoprotein AI (152)	6300	1.6 (1.4-1.8)
	Apolipoprotein B (152)	6300	2.0 (1.7-2.4)
	Apolipoprotein B/AI ratio (152)	3700	1.9 (1.6-2.2)
Metabolic	Adiponectin (64)	1300	0.8 (0.7-1.0)
	Leptin (245)	1300	1.3 (0.8-2.0)
	Fasting insulin (65)	2600	1.1 (1.0-1.3)
	Random insulin (65)	2000	1.4 (1.1-1.6)
	Pro-insulin (65)	400	2.2 (1.7-3.0)
Renal function	eGFR (246)	4700	1.4 (1.2-1.7)
	Uric acid (247)	9400	1.1 (1.0-1.2)
Chronic infections	Cytomegalovirus (248)	700	0.9 (0.7-1.2)
	Mixed strains of *H. pylori* (249)	2300	1.2 (0.9-1.4)
	Cytotoxic strains of *H. pylori* (250)	600	1.3 (0.9-1.9)
	C pneumoniae IgG titres (154)	3000	1.2 (1.0-1.4)
	C pneumoniae IgA titres (61)	2300	1.2 (1.0-1.5)
Cell adhesion molecules	E-selectin (63)	800	1.2 (0.9-1.6)
	P-selectin (63)	600	1.2 (0.6-2.2)
	ICAM-1 (63)	1400	1.4 (1.1-1.7)
	VCAM-1 (63)	1300	1.0 (0.8-1.3)
Rheology	Viscosity (251)	1300	1.6 (1.3-1.9)
	Haematocrit (251)	8000	1.2 (1.1-1.3)
	ESR (250)	1700	1.3 (1.2-1.5)
Metalloproteins	Ferritin (62)	600	1.0 (0.8-1.3)
	Transferrin (62)	6000	0.9 (0.7-1.1)
Vitamin-related	Homocysteine (252)	1000	1.3 (1.1-1.5)

*Risk ratios presented are for a 1-sd increase for PAI-I, for a 1 mmol/L increase for fasting blood glucose and post load glucose, and for a comparison of <60 vs. ≥60 ml/min per 1.73 m² for eGFR. Albumin comparisons involve bottom third vs. top third.

Figure 20.4. Outline of a staged approach to prioritize and evaluate novel and emerging markers in cardiovascular diseases

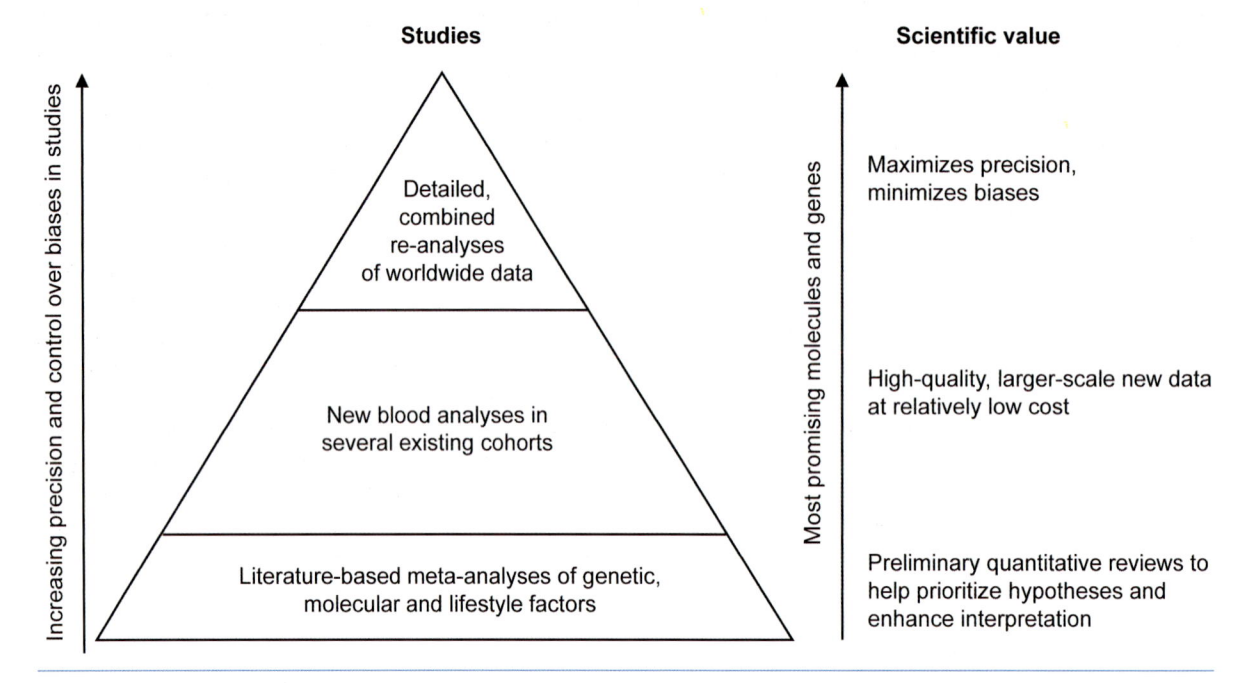

Studies — Scientific value

- Detailed, combined re-analyses of worldwide data — Maximizes precision, minimizes biases
- New blood analyses in several existing cohorts — High-quality, larger-scale new data at relatively low cost
- Literature-based meta-analyses of genetic, molecular and lifestyle factors — Preliminary quantitative reviews to help prioritize hypotheses and enhance interpretation

(Increasing precision and control over biases in studies ↑; Most promising molecules and genes ↑)

markers (67–144), but individually they have not generally been sufficiently powered to assess associations under different circumstances, or to correct for within-person variability and measurement error in the marker of interest. Although previous meta-analyses have attempted to summarize the evidence on such markers in CHD, they have typically been based on only published data (62–66,145–154). While such literature-based reviews can help to provide preliminary assessments, they cannot provide precise estimates of risk marker–disease associations under a range of different circumstances (including assessment of effect-modification), such as at different ages, in women and men, at different levels of established risk factors, nor reliable characterization of the shape of any dose–response relationships, nor consistent approaches to adjustment for possible confounding factors, or detailed investigation of potential sources of heterogeneity.

Moreover, most available assessments of emerging risk markers have related CHD risk solely to baseline measurements (which can lead to substantial underestimation of any associations due to regression dilution bias (155,156)), and have based statistical adjustment for possible confounding factors only on baseline values (which can lead to residual biases). But if a risk marker is of potential etiological relevance, it may also be important to characterize in detail its degree of within-person variability, both to understand the sources of this variability and to enable appropriate correction for regression dilution (156). It may also be informative to characterize in detail any lifestyle and biological correlates, thereby helping to identify possible determinants of the marker of interest (157).

Such uncertainties can be addressed by analyses of individual data from a comprehensive set of relevant prospective studies of cardiovascular outcomes (i.e. individual participant data meta-analysis). The value of this approach has been demonstrated by the Prospective Studies Collaboration (PSC) (158), an analysis of individual data on one million participants in 61 cohorts, including about 20 000 incident CHD deaths. The PSC has, for example, demonstrated approximately log-linear associations for each of blood pressure and total cholesterol with CHD mortality (Figures 20.5 & 20.6) (5,6). These findings are of considerable public health importance, refuting earlier suggestions of threshold levels at which these established risk factors cease to be relevant. They also demonstrated the importance of blood pressure and cholesterol to vascular outcomes under a wide range of circumstances, notably in the elderly for whom these risk factors were previously regarded by some authorities as unimportant. Individual participant meta-analysis is also being used in the 600 000 participant, 44-cohort Asia Pacific Cohort Studies Collaboration

(APCSC), which has recorded some lipid and other markers in relation to both cardiovascular morbidity and mortality (159). But, as the APCSC involves mostly East Asian participants, who tend to have a much lower incidence of CHD than Westerners, it has recorded less than one tenth of the numbers of incident CHD outcomes available in the PSC.

The Emerging Risk Factors Collaboration (ERFC) (160) and its related initiatives, such as the Fibrinogen Studies Collaboration (161,162) and the Lp-PLA$_2$ Studies Collaboration (163), are extending this approach to the study of several emerging risk markers (Table 20.4). The ERFC, for example, has collated and harmonized individual data on up to 500 characteristics in over 1.2 million participants in 110 long-term prospective studies in populations that are representative of the general population. During approximately 12 million person-years at risk, about 75 000 incident major cardiovascular outcomes have been recorded in the ERFC database. Over 300 000 of the participants in the ERFC have provided serial measurements of established or emerging risk markers (160). The ERFC complements and contrasts with the PSC and the APCSC by having a broader scope (investigating several lipid, inflammatory, and metabolic markers) (Table 20.5), recording a large panel of potentially relevant covariates (e.g. biochemical and lifestyle characteristics), and including both major cardiovascular morbidity and cause-specific mortality (whereas the PSC involves only cause-specific mortality). The establishment of the ERFC and related initiatives has also stimulated advancement of biostatistical methods to maximize the value of observational data from multiple studies (156,157,164-166). The emergence of findings from the ERFC over the next few years is likely to transform understanding of the relevance of several promising risk markers to CHD. A further

Figure 20.5. Age-specific associations of usual systolic blood pressure and coronary heart disease mortality in 34 283 cases among about 1 million participants from the Prospective Studies Collaboration (5). Reprinted from The Lancet, Copyright (2002), with permission from Elsevier.

Figure 20.6. Age-specific associations of usual total cholesterol levels and coronary heart disease mortality in 33 744 cases among about 1 million participants from the Prospective Studies Collaboration (6). Reprinted from The Lancet, Copyright (2007), with permission from Elsevier.

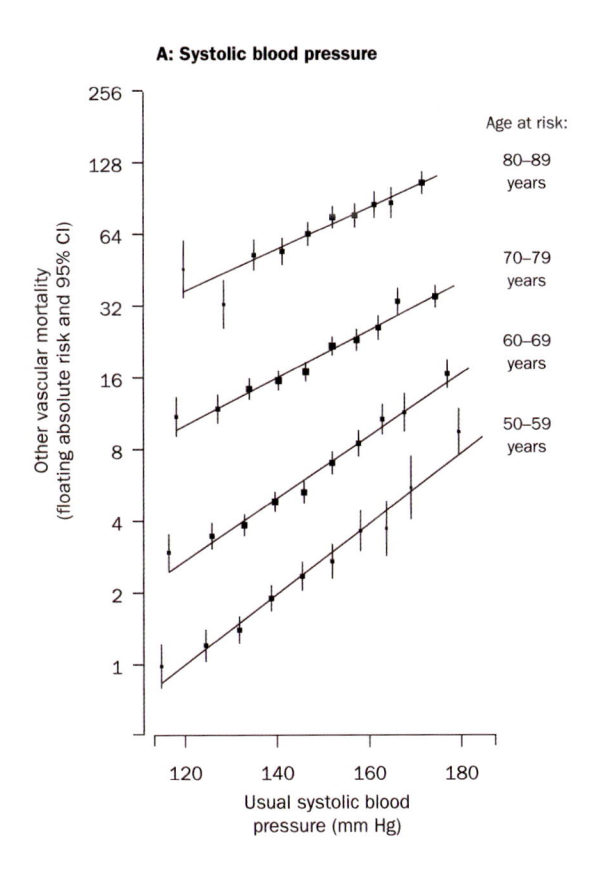

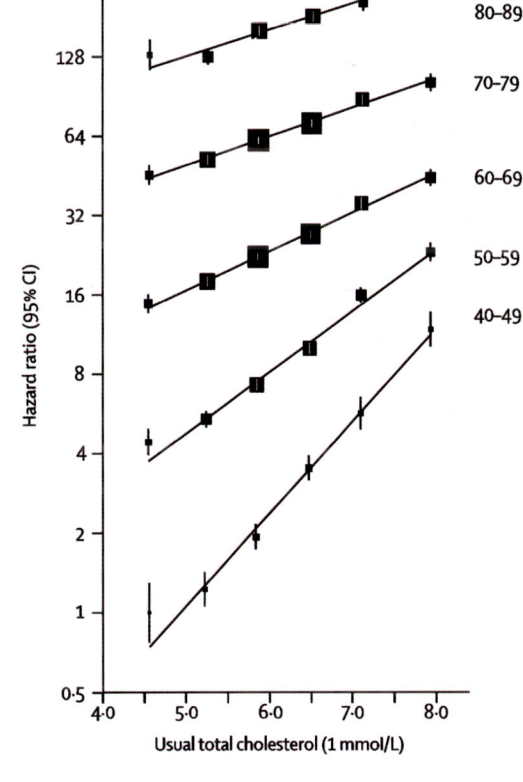

IHD, ischaemic heart disease

Table 20.4. Examples of collaborative groups conducting pooled analyses of individual participant data on established and emerging markers and major cardiovascular disease outcomes

	LSC	FSC	APCSC	PSC	ERFC
Biomarkers	Lp-PLA$_2$	Fibrinogen	Lipid and metabolic	Blood pressure and cholesterol	Lipid, inflammatory and metabolic
Studies	32	31	44	69	121
Participants	79K	154K	600K	1M	1.8M
Repeat measurements	3K	27K	50K	175K	300K
Person-years at risk	600K	1.4M	0.5M	12M	15M
Cardiovascular outcomes	15K	11K	10K	55K	90K

FSC, Fibrinogen Studies Collaboration; APCSC, Asia Pacific Cohort Studies Collaboration; CVD, cardiovascular disease; ERFC, Emerging Risk Factors Collaboration; LSC, Lp-PLA$_2$ (Lipoprotein-associated phospholipase A$_2$) Studies Collaboration; PSC, Prospective Studies Collaboration

Table 20.5. Preliminary summary of data available in the emerging risk factors collaboration on some lipid, inflammatory and metabolic markers

Marker	Participants with baseline measurements	No. with at least two measurements	Person-years at risk (million)	CHD outcomes	Stroke outcomes	Total mortality
Triglycerides	910K	150K	10	38K	17K	73K
HDL-C	638K	74K	6.5	23K	15K	48K
LDL-C	593K	63K	5	20K	13K	36K
Apolipoprotein-B	302K	9K	2.5	8K	8K	13K
Apolipoprotein-AI	295K	9K	2.5	8K	8K	13K
Leucocyte count	189K	39K	1	8K	3K	18K
Albumin	172K	9K	1.5	11K	4K	22K
Lipoprotein(a)	131K	0.5K	1	8K	3K	13K
C-reactive protein	125K	11K	1	12K	7K	16K
Diabetes	569K	93K	6	29K	16K	65K
Fasting glucose	544K	67K	6.5	25K	9K	63K
Post-load glucose	72K	23K	1	8K	2K	12K
Creatinine	154K	42K	1.5	13K	5K	28K

influence of the PSC, APCSC and the ERFC should be to facilitate the formation of further collaborative studies, as these initiatives have already brought together several hundred previously unconnected cardiovascular researchers to analyse and report data collaboratively.

Integration of information on genetic, biochemical and lifestyle factors in CHD

Several types of analyses require integration of data from different categories of exposures (e.g. genetic, biochemical and lifestyle factors). These include Mendelian randomization studies, optimization of risk stratification algorithms, and assessment of gene-lifestyle joint effects. Below, each is considered separately.

Mendelian randomization studies

Despite their advantages over individual studies of customary size, individual participant meta-analyses of several prospective studies of emerging risk markers may not distinguish reliably whether associations of particular biomarkers with CHD reflect a causal relationship, or mainly a marker of established cardiovascular risk factors to which the biomarker is correlated, or mainly a marker of subclinical disease, or some combination of these possibilities. For example, the Fibrinogen Studies Collaboration has reported approximately log-linear associations of fibrinogen with CHD risk under a wide range of different circumstances (Figure 20.7) (162). The magnitude of this association, however, reduced considerably following adjustment for several established cardiovascular risk factors (162), as could be expected given the large number of established and emerging risk factors to which plasma fibrinogen is correlated (Figure 20.8) (157). The existence of these many correlates makes it difficult, therefore, to determine to what extent the observed associations of fibrinogen with CHD risk are independent from these markers. Statistical adjustment for confounding factors is potentially limited, as not all relevant confounders have been (or can be) measured in a study. Moreover, even measured confounders may be incompletely adjusted for because allowances are typically not made for within-person variability or measurement error in levels of confounders (e.g. blood pressure, serum lipid concentrations). Alternatively, statistical overadjustment (the correction for markers in any causal pathway between fibrinogen levels and CHD risk) could, in principle, obscure a potentially important etiological relationship. In practice, however, it is difficult to judge the likelihood of overadjustment given that potential biological pathways are typically only partially understood (although they are probably better elucidated for fibrinogen than for most other candidate biomarkers in CHD).

Focused genetic studies may help to overcome some of these potential limitations of observational epidemiology (167–169). Mendelian randomization experiments attempt to minimize confounding and avoid reverse association bias by measurement of common polymorphisms or haplotypes in regulatory regions of genes that have been reliably associated with differences in circulating biomarker concentration (but not with any known change in biomarker function). According to Mendel's second law (170), the inheritance of genetic variants should be subject to

Figure 20.7. Age-specific associations of usual fibrinogen levels and coronary heart disease risk in 7118 cases among about 154 000 participants from the Fibrinogen Studies Collaboration (162). Copyright © (2005) American Medical Association. All rights reserved.

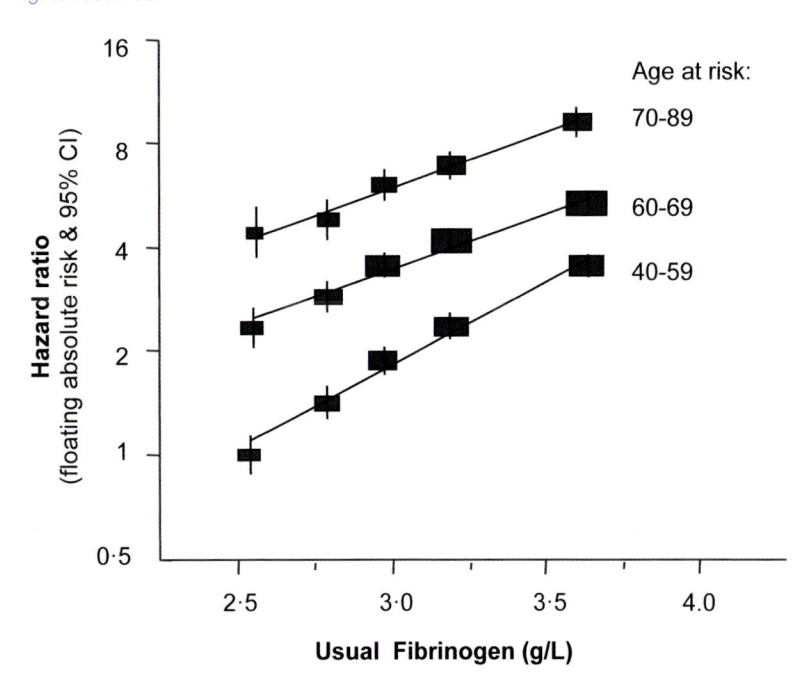

Figure 20.8. Sex-specific shape of cross-sectional associations of fibrinogen with some cardiovascular risk factors in about 154 000 individuals from the Fibrinogen Studies Collaboration (157). Reproduced with permission of Oxford University Press.

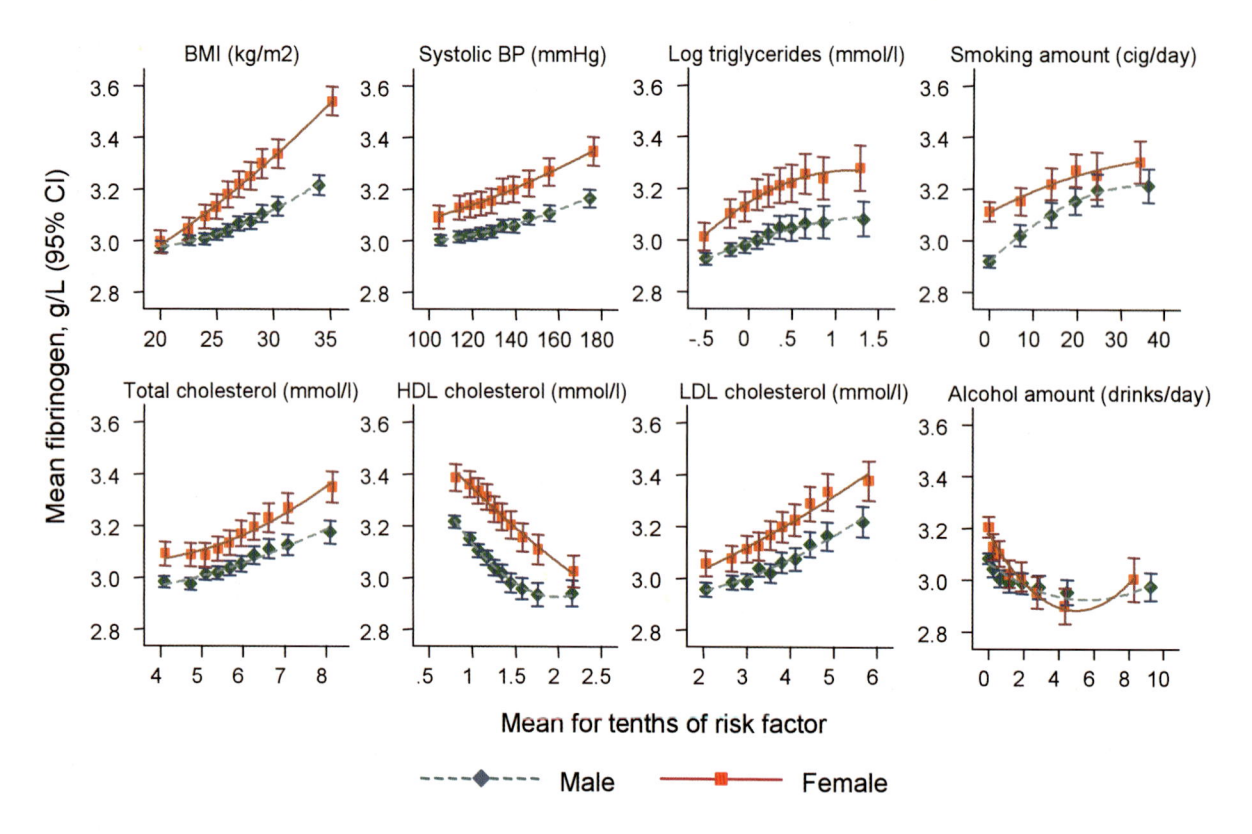

Note that the overall mean fibrinogen in each figure depends on which cohorts were included in the analysis having provided data for the relevant risk factor.

the random assortment of maternal and paternal alleles at the time of gamete formation. So, if the levels of a particular biomarker actually increase the risk of CHD, then carriage of alleles (or haplotypes) that expose individuals to a long-term elevation of that biomarker should confer an increased risk of CHD in proportion to the difference in biomarker levels attributable to the allele. Because of the randomized allocation of alleles from parents to offspring, potential confounders should be distributed evenly among the genotypic classes, and any bias due to reverse causation should be avoided because genotypes are fixed at conception and are unlikely to be modified by the onset of disease (171,172). Hence, by helping to judge the likelihood of any causal associations in CHD and estimating their magnitude, such focused genetic analyses should help to prioritize biomarkers for further study (e.g. as therapeutic targets) and elucidate disease pathways.

This approach has been applied to the study of plasma levels of fibrinogen (168,169). A report of a null association of fibrinogen genotypes with CHD risk, in a total of about 12 000 CHD cases and 18 000 controls, has decreased the likelihood of a major causal role for fibrinogen levels (Figure 20.9) (169), but even larger numbers would be needed to exclude the possibility of a modest but still potentially important effect. For example, it has been estimated that greater than 15 000 cases and greater than 15 000 controls would be needed to confirm or exclude 5–10% increases in CHD risk per 1 SD increase in blood levels of C-reactive protein (CRP) (173). The CRP CHD Genetics Collaboration is therefore generating data and conducting pooled analyses of known relevant CRP genetic variants in about 37 000 CHD cases and about 120 000 controls from 35 contributing studies (173). This approach is being extended to the study of several other candidate biomarkers, including high-density lipoprotein cholesterol (HDL-C) (6,39,174), lipoprotein(a) (175), and interleukin-6 (176).

The potential limitations of Mendelian randomization analyses include: the need for very large sample sizes, because most genotypes have only modest effects on concentrations of biomarkers; the scope for residual confounding by unrecognized pleiotropic effects of genotypes; and the potential obscuring of causal associations by processes related to developmental adaptation ("canalization") (171,172,177,178). Furthermore, ideal Mendelian randomization analyses should probably involve information on genotypes, biomarker levels, and CHD status derived from the same individuals in a single very large prospective study (which for clinical CHD outcomes may require upwards of 20 000 incident CHD cases). In the current absence of any such studies, however, it has been necessary to combine information from several different studies; only relatively few of which may involve concomitant assessment of genotype, biomarkers, and CHD status (indeed, in the case of fibrinogen, studies focusing on biomarker–CHD and gene–CHD associations have largely been non-overlapping). A possible limitation of such analyses is, of course, the increased scope for heterogeneity and the need for assumptions about similar effects across different populations and subgroups (171,172,177,178).

Risk prediction algorithms

Several risk prediction algorithms have been proposed to help stratify risk of cardiovascular disease in general Western populations, such as Framingham (179–181), PROCAM (69), SCORE (182), Reynolds (183) and QRISK (184,185) (Table 20.6). These algorithms each involve a core set of the same established risk factors (i.e. age, sex, smoking, blood pressure, total cholesterol), but differ in their inclusion of various other characteristics, such as HDL-C (in Framingham), triglycerides (in PROCAM only), CRP (in Reynolds

Table 20.6. Comparison of some features of selected risk scores in cardiovascular disease

Risk score (Ref)	Year	Population assessed	Outcome	Prediction period	Factors used in each risk score	Additional interview / physical measurements	Additional blood-based markers	Validation method*
FHS (179,180)	1991	USA	Fatal / non-fatal CHD	4-12 years	Age, sex, smoking, and blood pressure	Diabetes and ECG-LVH	HDL-C and LDL-C	External
FHS (181)	1998	USA	Fatal / non-fatal CHD	10 years		Diabetes	HDL-C and LDL-C	External
PROCAM (72)	2002	Germany	Fatal / non-fatal CHD	10 years		Diabetes and family history of CVD	HDL-C, LDL-C and triglycerides	Internal & external
SCORE (182)	2003	Multi-site Europe	Fatal CHD	10 years		None	Total cholesterol/HDL-C	External
Reynolds (183)	2007	USA	Fatal / non-fatal CHD	10 years		Family history of CVD	Total cholesterol, HDL-C, CRP and HbA1c (in diabetics)	Internal
QRISK (184)	2007	UK	Fatal / non-fatal CHD	10 years		Family history of CVD, SES, BMI and antihyper-tensive treatment	Total cholesterol/HDL-C	Internal
QRISK2 (185)	2008	UK	Fatal / non-fatal CHD, stroke or TIA	10 years		Diabetes, family history of CVD, BMI, ethnicity, deprivation, hyper-tension, rheumatoid arthritis, chronic renal disease, atrial fibrillation	Total cholesterol/HDL-C	Internal

BMI, body mass index; CHD, coronary heart disease; CRP, C-reactive protein; CVD, cardiovascular diseases; ECG-LVH, electrocardiogram left ventricular hypertrophy; FHS, Framingham Heart Study; HbA1c, glycosylated haemoglobin; HDL-C, high density lipoprotein cholesterol; LDL-C, low density lipoprotein cholesterol; PROCAM, Münster Heart Study; SES, socioeconomic status; TIA, transient ischaemic attack. *Internal validation refers to models derived and validated on data from the same study; external validation refers to models that are derived on one study and validated on data collected elsewhere.

Unit 5 • Chapter 20. Coronary heart disease

Figure 20.9. Meta-analysis of 20 studies of predominantly European descent showing an overall null association of fibrinogen genotypes with risk of coronary disease (169). Reproduced with permission of Oxford University Press.

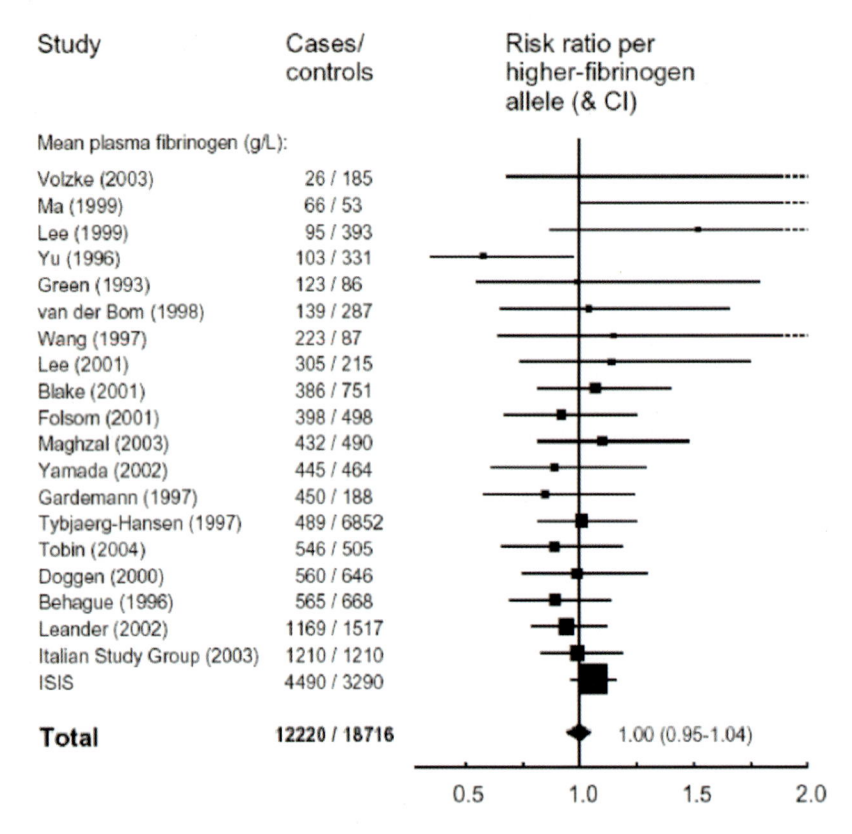

Study	Cases/controls	Risk ratio per higher-fibrinogen allele (& CI)
Mean plasma fibrinogen (g/L):		
Volzke (2003)	26 / 185	
Ma (1999)	66 / 53	
Lee (1999)	95 / 393	
Yu (1996)	103 / 331	
Green (1993)	123 / 86	
van der Bom (1998)	139 / 287	
Wang (1997)	223 / 87	
Lee (2001)	305 / 215	
Blake (2001)	386 / 751	
Folsom (2001)	398 / 498	
Maghzal (2003)	432 / 490	
Yamada (2002)	445 / 464	
Gardemann (1997)	450 / 188	
Tybjaerg-Hansen (1997)	489 / 6852	
Tobin (2004)	546 / 505	
Doggen (2000)	560 / 646	
Behague (1996)	565 / 668	
Leander (2002)	1169 / 1517	
Italian Study Group (2003)	1210 / 1210	
ISIS	4490 / 3290	
Total	**12220 / 18716**	1.00 (0.95-1.04)

Meta-analysis of studies of coronary disease and -148C/T or -455G/A polymorphisms in the beta-fibrinogen gene. These two polymorphisms are in complete linkage disequilibrium, so knowledge of genotype at one locus predicts genotype at the other locus with certainty. For each study, the risk ratio for coronary disease per higher-fibrinogen allele is represented by a square (area proportional to the information content of the study), with a horizontal line denoting the 99% confidence intervals (CI). The overall risk ratio and 95% CI is represented by a diamond, with values alongside.

only), and body mass index or markers of socioeconomic status (in QRISK only). Other authorities recommend measurement of markers of glycemic status (e.g. fasting or post-load glucose levels, glycosylated haemoglobin (186–190)), and novel biomarkers such as Lp-PLA$_2$, as adjuncts to established risk factors for the stratification of cardiovascular disease risk (191,192).

Such divergent recommendations by scientific and professional groups stem partly from differences in methodological approaches and partly from limitations in available epidemiological data. Although many published prospective studies have commented on the potential value of particular markers in risk prediction, they have often reported on measures of association only (e.g. odds ratios, hazard ratios), which do not directly address the issue of the utility of a marker in prediction or stratification. Furthermore, even studies that have involved statistics relevant to the assessment of risk prediction have emphasized different metrics, including measures of discrimination (e.g. the measure D (193) and the C index (194,195), with the latter related to the area under the receiver operating characteristic curve), and reclassification methods that aim to summarize the potential of a marker to reassign individuals into more appropriate risk groups (196). Each of these approaches may impart somewhat different information (197). As recommended by a 2006 workshop report by the US National Heart Lung and Blood Institute (http://www.nhlbi.nih.gov/meetings/workshops/crp/report.htm), further work is needed to compare and contrast the strengths and limitations of each of these approaches and to incorporate heath economic analyses to help judge the value of any such measurements in the light of potential additional costs and the consequences of any therapy (198).

Limitations in available data relate principally to the assessment of novel markers in comparative isolation from one another. For example, relatively few studies have assessed all of the risk markers named in the first paragraph of this section. This fragmentary approach has prevented direct comparisons of the relative merits of the different risk markers, a problem that has been compounded by development and evaluation of risk scores in studies of relatively moderate power. Advances in genetic epidemiology have encouraged recent suggestions that information on several genetic loci usefully add to conventional risk scores. But, as these analyses have so far been based on just several hundred CHD cases, much larger analyses in prospective studies are required to evaluate reliably any new risk scores that incorporate novel genetic loci (39,174) or lifestyle factors (199).

Joint effects of genetic and lifestyle factors

It has been proposed that reliable knowledge of the potential joint effect of genetic and lifestyle factors should contribute importantly to understanding the etiology of CHD and development of disease prevention strategies, such as optimum targeting of existing interventions (particularly if they are intensive or costly) and approaches for modifying the effects of deleterious genes by avoiding harmful lifestyle exposures (200–203). Although there is some evidence that the incidence of CHD is jointly determined by nature and nurture (200–203), the quantitative interplay of specific genetic and lifestyle components remains poorly understood. Assessment of genetic, biochemical and lifestyle factors has hitherto typically taken place in comparative isolation from one another, rather than in an integrated way, due to lack of sufficiently large prospective studies with appropriate and concomitant information on each of these exposure categories. Figure 20.10 indicates that at least 10 000 CHD cases and a similar number of controls may be required for reliable assessment of such joint effects in the presence of relatively common genetic variants. Data on apolipoprotein E (*apoE*) genotypes, which are among the best studied genetic variants in CHD, illustrate current limitations in the understanding of joint effects. Although it is now clear that there are approximately linear relationships of *apoE* genotypes with LDL-C concentrations and with CHD risk (Cf. Figure 20.1) (28), it remains unknown whether the impact of apoE genotypes differs considerably in different individuals, such as overweight people (204), those with higher lipid levels (205), or those who consume high quantities of fat (206). A prospective study involving a few hundred CHD cases has proposed that there are important interactions on CHD risk of the ε4 allele of the *apoE* gene and cigarette smoking (207), putatively mediated through a direct effect of LDL oxidation (208), but this was not confirmed by a large retrospective study (209). Data are even sparser in relation to proposed joint effects on CHD risk of apoE variants with

Figure 20.10. Sample size estimates for studies of joint effects between genetic and environmental factors and coronary risk (interaction effect, R_{ge})

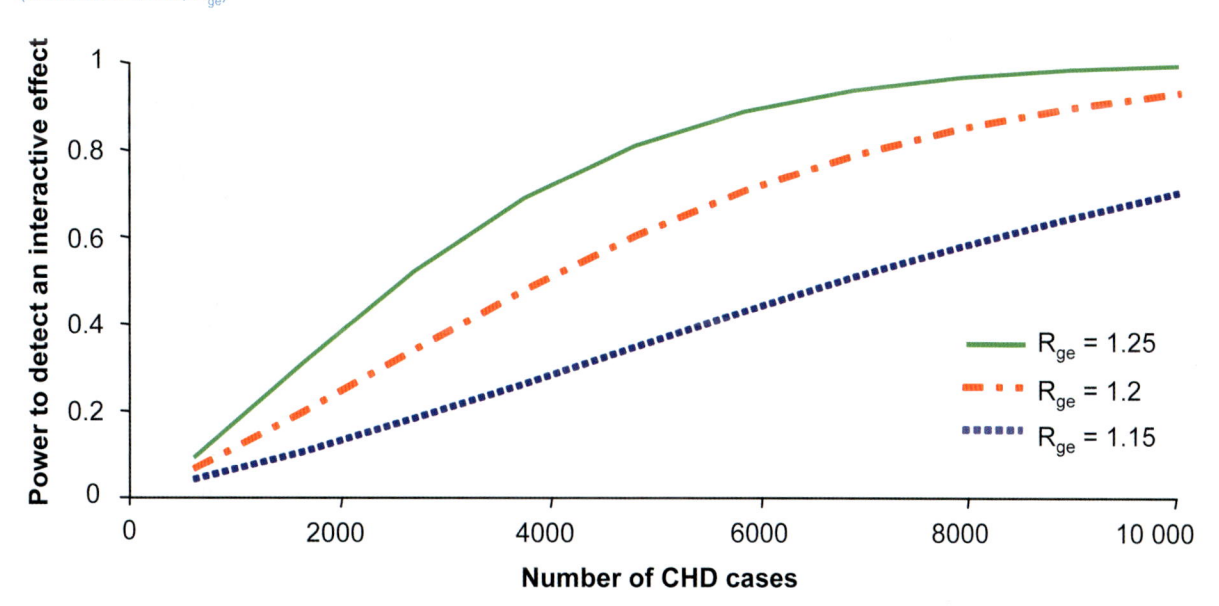

Assumptions include: population coronary heart disease (CHD) risk = 5%; additive genetic model (odds ratio = 1.2 per allele increase); minor allele frequency = 0.05; environmental exposure normally distributed (odds ratio = 1.25 per standard deviation increase); type 1 error = 0.01; 1 case per control. Source: Quanto version 1.2, 2006 dietary cholesterol (210), lipoprotein lipase gene variants and saturated fatty acid consumption (211), apolipoprotein AI gene variants and dietary fat consumption (203), and hepatic lipase gene variants and fat consumption (212).

Current nutritional guidelines, such as those of the Department of Health and the Food Standards Agency in the United Kingdom (213), encourage reduction in consumption of saturated fat, an increase in consumption of omega-3 fatty acids from fish oil or plant sources, and consumption of a diet high in fruits and vegetables. Yet evidence from prospective epidemiological studies of CHD (and dietary intervention trials) remains largely inconclusive (214,215). For example, one of the largest available studies, conducted in a cohort of American nurses, recently reported that diets higher in total and saturated fat were not significantly associated with CHD risk (216), and that there were only weak inverse associations of CHD risk with fruit and vegetable consumption (217). Interpretation of these findings has, however, been limited by relatively wide confidence intervals around estimates and by constraints of studying populations (such as health professionals) who may have comparatively homogeneous dietary habits. These limitations are compounded by likely measurement error in self-reported diet (218–220). Similar uncertainties apply to the emerging evidence on other dietary factors, such as foods (e.g. meat and dairy products), minerals (e.g. calcium) and nutrients (e.g. the optimum balance of fatty acids) (221–223). These uncertainties underscore the need for analyses of dietary factors in larger prospective studies with concomitant genetic and biomarker information and involving populations with considerable heterogeneity in dietary habits to enhance study generalizability and sensitivity (e.g. such as different populations across Europe), use of calibration studies to help optimize data from dietary questionnaires, and measurement of nutritional biomarkers to supplement self-reported diet. Similar considerations apply to studies of established lifestyle risk factors (e.g. physical activity and consumption of tobacco and alcohol), for which new evidence is needed to evaluate joint effects on CHD with genetic factors, to characterize important details of relationships (e.g. the shape of any dose–response relationships (224) and the magnitude of any associations in clinically relevant sungroups (225)), and to help better understand how lifestyle choices might mediate disease risk (226).

Maturation of prospective bioresources and discovery methods

The worldwide trend in recent decades towards the establishment of large epidemiological bioresources, notably those with prospective study designs and appropriate assessment of lifestyle factors, should facilitate the study of joint gene–lifestyle effects in CHD during the coming years. For example, the European Prospective Investigation of Diet in Cancer (EPIC) resource has recorded detailed lifestyle (notably, dietary) characteristics and stored biological samples for about 400 000 mostly middle-aged adults from 10 countries (227–229). By 2010, more than five million person-years at risk had accrued in this cohort, yielding over 15 000 incident CHD cases (228). EPIC-Heart, the cardiovascular component of EPIC, plans detailed studies of the separate and combined effects of genetic and lifestyle factors (such as on a case–cohort basis), including study of biomarkers in potentially causative intermediate pathways (229). Similar numbers of incident CHD cases will accrue from other large blood-based prospective studies as they mature and record several million-years of follow-up. For example, the Mexico City Prospective Study had by 2006 recruited about 150 000 middle-aged adults (230). The 500 000 participant Kadoorie prospective study in China, which involves assessment of many lifestyle characteristics and storage of biological samples, completed recruitment in 2008 (231). The 500 000 participant United Kingdom Biobank Study should be fully recruited by 2011 (232). Several further initiatives in CHD of comparable scale are planned, or have been started, in Australia, Canada, northern Europe and the USA (233).

The emergence of such bioresources has also encouraged the pursuit of large-scale "systems biology" studies in CHD (234–236). Such approaches aim to overlay and analyse multiple complementary layers of dense biological data (e.g. genomics (54,237), transcriptomics (238–240), and metabolomics (241)) from the same participants to help elucidate causal pathways. These methods are generally at relatively early stages in their development, but they should become increasingly valuable as laboratory and bioinformatics approaches mature.

Conclusions

Approaches that enable study of the separate and combined effects of genetic, biochemical and lifestyle factors should yield new scientific insights that contribute importantly to the prediction and prevention of CHD.

References

1. Leal J, Luengo-Fernández R, Gray A et al. (2006). Economic burden of cardiovascular diseases in the enlarged European Union. *Eur Heart J*, 27:1610–1619.doi:10.1093/eurheartj/ehi733 PMID:16495286

2. Murray CJ, Lopez AD (1997). Mortality by cause for eight regions of the world: Global Burden of Disease Study. *Lancet*, 349:1269–1276.doi:10.1016/S0140-6736(96)07493-4 PMID:9142060

3. Sanderson JE, Mayosi B, Yusuf S et al. (2007). Global burden of cardiovascular disease. *Heart*, 93:1175.doi:10.1136/hrt.2007.131060 PMID:17890692

4. Parish S, Collins R, Peto R et al. (1995). Cigarette smoking, tar yields, and non-fatal myocardial infarction: 14,000 cases and 32,000 controls in the United Kingdom. The International Studies of Infarct Survival (ISIS) Collaborators. *BMJ*, 311:471–477. PMID:7647641

5. Lewington S, Clarke R, Qizilbash N et al.; Prospective Studies Collaboration (2002). Age-specific relevance of usual blood pressure to vascular mortality: a meta-analysis of individual data for one million adults in 61 prospective studies. *Lancet*, 360:1903–1913.doi:10.1016/S0140-6736(02)11911-8 PMID:12493255

6. Lewington S, Whitlock G, Clarke R et al.; Prospective Studies Collaboration (2007). Blood cholesterol and vascular mortality by age, sex, and blood pressure: a meta-analysis of individual data from 61 prospective studies with 55,000 vascular deaths. *Lancet*, 370:1829–1839.doi:10.1016/S0140-6736(07)61778-4 PMID:18061058

7. Huxley R, Barzi F, Woodward M (2006). Excess risk of fatal coronary heart disease associated with diabetes in men and women: meta-analysis of 37 prospective cohort studies. *BMJ*, 332:73–78.doi:10.1136/bmj.38678.389583.7C PMID:16371403

8. Critchley JA, Capewell S (2003). Mortality risk reduction associated with smoking cessation in patients with coronary heart disease: a systematic review. *JAMA*, 290:86–97.doi:10.1001/jama.290.1.86 PMID:12837716

9. Ford ES, Ajani UA, Croft JB et al. (2007). Explaining the decrease in U.S. deaths from coronary disease, 1980–2000. *N Engl J Med*, 356:2388–2398.doi:10.1056/NEJMsa053935 PMID:17554120

10. Lewington S, MacMahon S; Prospective Studies Collaboration (1999). Blood pressure, cholesterol, and common causes of death: a review. *Am J Hypertens*, 12:96S–98S. doi:10.1016/S0895-7061(99)00162-4 PMID:10555608

11. O'Flaherty ME, Ford ES, Allender S et al. (2008). Coronary heart disease trends in England and Wales from 1984 to 2004: concealed levelling of mortality rates among young adults. *Heart*, 94:178–181.doi:10.1136/hrt.2007.118323 PMID:17641070

12. Prospective Studies Collaboration (1995). Cholesterol, diastolic blood pressure, and stroke: 13,000 strokes in 450,000 people in 45 prospective cohorts. Prospective studies collaboration. *Lancet*, 346:1647–1653.doi:10.1016/S0140-6736(95)92836-7 PMID:8551820

13. Ghaffar A, Reddy KS, Singhi M (2004). Burden of non-communicable diseases in South Asia. *BMJ*, 328:807–810.doi:10.1136/bmj.328.7443.807 PMID:15070638

14. McKeigue PM, Miller GJ, Marmot MG (1989). Coronary heart disease in south Asians overseas: a review. *J Clin Epidemiol*, 42:597–609.doi:10.1016/0895-4356(89)90002-4 PMID:2668448

15. Reddy KS (2004). Cardiovascular disease in non-Western countries. *N Engl J Med*, 350:2438–2440.doi:10.1056/NEJMp048024 PMID:15190135

16. Lusis AJ, Mar R, Pajukanta P (2004). Genetics of atherosclerosis. *Annu Rev Genomics Hum Genet*, 5:189–218.doi:10.1146/annurev.genom.5.061903.175930 PMID:15485348

17. Murabito JM, Pencina MJ, Nam BH et al. (2005). Sibling cardiovascular disease as a risk factor for cardiovascular disease in middle-aged adults. *JAMA*, 294:3117–3123.doi:10.1001/jama.294.24.3117 PMID:16380592

18. Lloyd-Jones DM, Nam BH, D'Agostino RB Sr et al. (2004). Parental cardiovascular disease as a risk factor for cardiovascular disease in middle-aged adults: a prospective study of parents and offspring. *JAMA*, 291:2204–2211.doi:10.1001/jama.291.18.2204 PMID:15138242

19. Noble JA, Valdes AM, Cook M et al. (1996). The role of HLA class II genes in insulin-dependent diabetes mellitus: molecular analysis of 180 Caucasian, multiplex families. *Am J Hum Genet*, 59:1134–1148. PMID:8900244

20. Dorman JS, Bunker CH (2000). HLA-DQ locus of the human leukocyte antigen complex and type 1 diabetes mellitus: a HuGE review. *Epidemiol Rev*, 22:218–227. PMID:11218373

21. Hughes AE, Orr N, Patterson C et al. (2007). Neovascular age-related macular degeneration risk based on CFH, LOC387715/HTRA1, and smoking. *PLoS Med*, 4:e355. doi:10.1371/journal.pmed.0040355 PMID:18162041

22. Seddon JM, Francis PJ, George S et al. (2007). Association of CFH Y402H and LOC387715 A69S with progression of age-related macular degeneration. *JAMA*, 297:1793–1800.doi:10.1001/jama.297.16.1793 PMID:17456821

23. Morgan TM, Coffey CS, Krumholz HM (2003). Overestimation of genetic risks owing to small sample sizes in cardiovascular studies. *Clin Genet*, 64:7–17.doi:10.1034/j.1399-0004.2003.00088.x PMID:12791034

24. Keavney B, McKenzie C, Parish S et al. (2000). Large-scale test of hypothesised associations between the angiotensin-converting-enzyme insertion/deletion polymorphism and myocardial infarction in about 5000 cases and 6000 controls. International Studies of Infarct Survival (ISIS) Collaborators. *Lancet*, 355:434–442. PMID:10841123

25. Wheeler JG, Keavney BD, Watkins H et al. (2004). Four paraoxonase gene polymorphisms in 11212 cases of coronary heart disease and 12786 controls: meta-analysis of 43 studies. *Lancet*, 363:689–695.doi:10.1016/S0140-6736(04)15642-0 PMID:15001326

26. Ozaki K, Ohnishi Y, Iida A et al. (2002). Functional SNPs in the lymphotoxin-alpha gene that are associated with susceptibility to myocardial infarction. *Nat Genet*, 32:650–654.doi:10.1038/ng1047 PMID:12426569

27. Clarke R, Xu P, Bennett D et al.; International Study of Infarct Survival (ISIS) Collaborators (2006). Lymphotoxin-alpha gene and risk of myocardial infarction in 6,928 cases and 2,712 controls in the ISIS case-control study. *PLoS Genet*, 2:e107.doi:10.1371/journal.pgen.0020107 PMID:16839190

28. Butterworth A, Higgins JPT, Sarwar N, Danesh J. Coronary heart disease. Chapter 19. In: Khoury MJ, Bedrosian SR, Gwinn M et al., editors. Human genome epidemiology. 2nd ed. Building the evidence for using genetic information to improve health and prevent disease. Oxford: Oxford University Press; 2010.

29. Bennet AM, Di Angelantonio E, Ye Z et al. (2007). Association of apolipoprotein E genotypes with lipid levels and coronary risk. *JAMA*, 298:1300–1311.doi:10.1001/jama.298.11.1300 PMID:17878422

30. Lohmueller KE, Pearce CL, Pike M et al. (2003). Meta-analysis of genetic association studies supports a contribution of common variants to susceptibility to common disease. Nat Genet, 33:177–182.doi:10.1038/ng1071 PMID:12524541

31. Wacholder S, Chanock SJ, García-Closas M et al. (2004). Assessing the probability that a positive report is false: an approach for molecular epidemiology studies. J Natl Cancer Inst, 96:434–442.doi:10.1093/jnci/djh075 PMID:15026468

32. Third International Study of Infarct Survival Collaborative Group (1992). ISIS-3: a randomised comparison of streptokinase vs tissue plasminogen activator vs anistreplase and of aspirin plus heparin vs aspirin alone among 41,299 cases of suspected acute myocardial infarction. ISIS-3 (Third International Study of Infarct Survival) Collaborative Group. Lancet, 339:753–770.doi:10.1016/0140-6736(92)91893-D PMID:1347801

33. Yusuf S, Hawken S, Ounpuu S et al.; INTERHEART Study Investigators (2004). Effect of potentially modifiable risk factors associated with myocardial infarction in 52 countries (the INTERHEART study): case-control study. Lancet, 364:937–952. doi:10.1016/S0140-6736(04)17018-9 PMID:15364185

34. Samani NJ, Erdmann J, Hall AS et al.; WTCCC and the Cardiogenics Consortium (2007). Genomewide association analysis of coronary artery disease. N Engl J Med, 357:443–453.doi:10.1056/NEJMoa072366 PMID:17634449

35. Burton PR, Clayton DG, Cardon LR et al.; Wellcome Trust Case Control Consortium (2007). Genome-wide association study of 14,000 cases of seven common diseases and 3,000 shared controls. Nature, 447:661–678. doi:10.1038/nature05911 PMID:17554300

36. Helgadottir A, Thorleifsson G, Manolescu A et al. (2007). A common variant on chromosome 9p21 affects the risk of myocardial infarction. Science, 316: 1491–1493.doi:10.1126/science.1142842 PMID:17478679

37. McPherson R, Pertsemlidis A, Kavaslar N et al. (2007). A common allele on chromosome 9 associated with coronary heart disease. Science, 316:1488–1491.doi:10.1126/science. 1142447 PMID:17478681

38. Kathiresan S, Manning AK, Demissie S et al. (2007). A genome-wide association study for blood lipid phenotypes in the Framingham Heart Study. BMC Med Genet, 8 Suppl 1;S17.doi:10.1186/1471-2350-8-S1-S17 PMID:17903299

39. Kathiresan S, Melander O, Guiducci C et al. (2008). Six new loci associated with blood low-density lipoprotein cholesterol, high-density lipoprotein cholesterol or triglycerides in humans. Nat Genet, 40:189–197.doi:10. 1038/ng.75 PMID:18193044

40. Sandhu MS, Waterworth DM, Debenham SL et al.; Wellcome Trust Case Control Consortium (2008). LDL-cholesterol concentrations: a genome-wide association study. Lancet, 371:483–491.doi:10.1016/S0140-6736(08)60208-1 PMID:18262040

41. Willer CJ, Sanna S, Jackson AU et al. (2008). Newly identified loci that influence lipid concentrations and risk of coronary artery disease. Nat Genet, 40:161–169.doi:10. 1038/ng.76 PMID:18193043

42. Casas JP, Cooper J, Miller GJ et al. (2006). Investigating the genetic determinants of cardiovascular disease using candidate genes and meta-analysis of association studies. Ann Hum Genet, 70:145–169.doi:10.1111/j.1469-1809.2005.00241.x PMID:16626327

43. Winslow RL, Gao Z (2005). Candidate gene discovery in cardiovascular disease. Circ Res, 96:605–606.doi:10.1161/01.RES. 0000162161.71447.3c PMID:15802616

44. Yagil C, Hubner N, Monti J et al. (2005). Identification of hypertension-related genes through an integrated genomic-transcriptomic approach. Circ Res, 96:617–625.doi:10. 1161/01.RES.0000160556.52369.61 PMID:15731461

45. Larson MG, Atwood LD, Benjamin EJ et al. (2007). Framingham Heart Study 100K project: genome-wide associations for cardiovascular disease outcomes. BMC Med Genet, 8 Suppl 1;S5.doi:10.1186/1471-2350-8-S1-S5 PMID:17903304

46. Shiffman D, Ellis SG, Rowland CM et al. (2005). Identification of four gene variants associated with myocardial infarction. Am J Hum Genet, 77:596–605.doi:10.1086/491674 PMID:16175505

47. Kathiresan S, Voight BF, Purcell S et al.; Myocardial Infarction Genetics Consortium; Wellcome Trust Case Control Consortium (2009). Genome-wide association of early-onset myocardial infarction with single nucleotide polymorphisms and copy number variants. Nat Genet, 41:334–341.doi:10.1038/ng.327 PMID:19198609

48. Erdmann J, Grosshennig A, Braund PS et al.; Italian Atherosclerosis, Thrombosis, and Vascular Biology Working Group; Myocardial Infarction Genetics Consortium; Wellcome Trust Case Control Consortium; Cardiogenics Consortium (2009). New susceptibility locus for coronary artery disease on chromosome 3q22.3. Nat Genet, 41:280–282.doi:10.1038/ng.307 PMID:19198612

49. Teslovich TM, Musunuru K, Smith AV et al. (2010). Biological, clinical and population relevance of 95 loci for blood lipids. Nature, 466:707–713.doi:10.1038/nature09270 PMID:20686565

50. German JB, Gillies LA, Smilowitz JT et al. (2007). Lipidomics and lipid profiling in metabolomics. Curr Opin Lipidol, 18:66–71. PMID:17218835

51. Gerszten RE, Wang TJ (2008). The search for new cardiovascular biomarkers. Nature, 451:949–952.doi:10.1038/nature06802 PMID:18288185

52. Pearson H (2008). Biologists initiate plan to map human proteome. Nature, 452:920–921.doi:10.1038/452920a PMID:18441544

53. Albert CJ, Anbukumar DS, Monda JK et al. (2007). Myocardial lipidomics. Developments in myocardial nuclear lipidomics. Front Biosci, 12:2750–2760.doi:10.2741/2269 PMID:17127277

54. Tyers M, Mann M (2003). From genomics to proteomics. Nature, 422:193–197. doi:10.1038/nature01510 PMID:12634792

55. Ioannidis JP, Bernstein J, Boffetta P et al. (2005). A network of investigator networks in human genome epidemiology. Am J Epidemiol, 162:302–304.doi:10.1093/aje/kwi201 PMID:16014777

56. Ioannidis JP, Gwinn M, Little J et al.; Human Genome Epidemiology Network and the Network of Investigator Networks (2006). A road map for efficient and reliable human genome epidemiology. Nat Genet, 38:3–5. doi:10.1038/ng0106-3 PMID:16468121

57. Seminara D, Khoury MJ, O'Brien TR et al.; Human Genome Epidemiology Network; Network of Investigator Networks (2007). The emergence of networks in human genome epidemiology: challenges and opportunities. Epidemiology, 18:1–8.doi:10.1097/01.ede.000 0249540.17855.b7 PMID:17179752

58. Dickersin K (1990). The existence of publication bias and risk factors for its occurrence. JAMA, 263:1385–1389.doi:10. 1001/jama.263.10.1385 PMID:2406472

59. Ioannidis JP (2005). Why most published research findings are false. PLoS Med, 2: e124.doi:10.1371/journal.pmed.0020124 PMID:16060722

60. Tang JL (2005). Selection bias in meta-analyses of gene-disease associations. PLoS Med, 2:e409.doi:10.1371/journal.pmed. 0020409 PMID:16363911

61. Danesh J, Whincup P, Lewington S et al. (2002). Chlamydia pneumoniae IgA titres and coronary heart disease; prospective study and meta-analysis. Eur Heart J, 23:371–375. doi:10.1053/euhj.2001.2801 PMID:11846494

62. Danesh J, Appleby P (1999). Coronary heart disease and iron status: meta-analyses of prospective studies. Circulation, 99:852–854. PMID:10027804

63. Malik I, Danesh J, Whincup P et al. (2001). Soluble adhesion molecules and prediction of coronary heart disease: a prospective study and meta-analysis. Lancet, 358:971–976.doi:10.1016/S0140-6736(01)06104-9 PMID:11583751

64. Sattar N, Wannamethee G, Sarwar N et al. (2006). Adiponectin and coronary heart disease: a prospective study and meta-analysis. Circulation, 114:623–629.doi: 10.1161/CIRCULATIONAHA.106.618918 PMID:16894037

65. Sarwar N, Sattar N, Gudnason V, Danesh J (2007). Circulating concentrations of insulin markers and coronary heart disease: a quantitative review of 19 Western prospective studies. *Eur Heart J*, 28:2491–2497.doi:10.1093/eurheartj/ehm115 PMID:17513304

66. Danesh J, Collins R, Peto R (2000). Lipoprotein(a) and coronary heart disease. Meta-analysis of prospective studies. *Circulation*, 102:1082–1085. PMID:10973834

67. Multiple Risk Factor Intervention Trial Research Group (1982). Multiple risk factor intervention trial. Risk factor changes and mortality results. *JAMA*, 248:1465–1477. doi:10.1001/jama.248.12.1465 PMID:7050440

68. The ARIC investigators (1989). The Atherosclerosis Risk in Communities (ARIC) Study: design and objectives. The ARIC investigators. *Am J Epidemiol*, 129:687–702. PMID:2646917

69. Steering Committee of the Physicians' Health Study Research Group (1989). Final report on the aspirin component of the ongoing Physicians' Health Study. *N Engl J Med*, 321:129–135.doi:10.1056/NEJM198907203210301 PMID:2664509

70. The RIFLE Research Group (1993). Presentation of the rifle project risk factors and life expectancy. *Eur J Epidemiol*, 9:459–476. PMID:8307130

71. ALLHAT Officers and Coordinators for the ALLHAT Collaborative Research Group (2002). Major outcomes in high-risk hypertensive patients randomized to angiotensin-converting enzyme inhibitor or calcium channel blocker vs diuretic: The Antihypertensive and Lipid-Lowering Treatment to Prevent Heart Attack Trial (ALLHAT). *JAMA*, 288:2981–2997.doi:10.1001/jama.288.23.2981 PMID:12479763

72. Assmann G, Cullen P, Schulte H (2002). Simple scoring scheme for calculating the risk of acute coronary events based on the 10-year follow-up of the prospective cardiovascular Münster (PROCAM) study. *Circulation*, 105:310–315.doi:10.1161/hc0302.102575 PMID:11804985

73. Bengtsson C, Blohmé G, Hallberg L et al. (1973). The study of women in Gothenburg 1968–1969–a population study. General design, purpose and sampling results. *Acta Med Scand*, 193:311–318.doi:10.1111/j.0954-6820.1973.tb10583.x PMID:4717311

74. Cantin B, Després JP, Lamarche B et al. (2002). Association of fibrinogen and lipoprotein(a) as a coronary heart disease risk factor in men (The Quebec Cardiovascular Study). *Am J Cardiol*, 89:662–666.doi:10.1016/S0002-9149(01)02336-0 PMID:11897206

75. Casiglia E, Palatini P (1998). Cardiovascular risk factors in the elderly. *J Hum Hypertens*, 12:575–581.doi:10.1038/sj.jhh.1000668 PMID:9783483

76. Cornoni-Huntley J, Ostfeld AM, Taylor JO et al. (1993). Established populations for epidemiologic studies of the elderly: study design and methodology. *Aging (Milano)*, 5:27–37. PMID:8481423

77. Cremer P, Nagel D, Labrot B et al. (1994). Lipoprotein Lp(a) as predictor of myocardial infarction in comparison to fibrinogen, LDL cholesterol and other risk factors: results from the prospective Göttingen Risk Incidence and Prevalence Study (GRIPS). *Eur J Clin Invest*, 24:444–453. PMID:7957500

78. Day N, Oakes S, Luben R et al. (1999). EPIC-Norfolk: study design and characteristics of the cohort. European Prospective Investigation of Cancer. *Br J Cancer*, 80 Suppl 1;95–103. PMID:10466767

79. Deeg DJ, van Tilburg T, Smit JH, de Leeuw ED (2002). Attrition in the Longitudinal Aging Study Amsterdam. The effect of differential inclusion in side studies. *J Clin Epidemiol*, 55:319–328.doi:10.1016/S0895-4356(01)00475-9 PMID:11927198

80. Ducimetiere P, Richard JL, Cambien F et al. (1980). Coronary heart disease in middle-aged Frenchmen. Comparisons between Paris Prospective Study, Seven Countries Study, and Pooling Project. *Lancet*, 1:1346–1350.doi:10.1016/S0140-6736(80)91796-1 PMID:6104139

81. Ducimetière P, Ruidavets JB, Montaye M et al.; PRIME Study Group (2001). Five-year incidence of angina pectoris and other forms of coronary heart disease in healthy men aged 50–59 in France and Northern Ireland: the Prospective Epidemiological Study of Myocardial Infarction (PRIME) Study. *Int J Epidemiol*, 30:1057–1062.doi:10.1093/ije/30.5.1057 PMID:11689522

82. Engström G, Stavenow L, Hedblad B et al. (2003). Inflammation-sensitive plasma proteins, diabetes, and mortality and incidence of myocardial infarction and stroke: a population-based study. *Diabetes*, 52:442–447. doi:10.2337/diabetes.52.2.442 PMID:12540619

83. Diez-Ewald M, Campos G, Rivero F et al. (2003). [Hemostatic coronary risk factors in a healthy population of Maracaibo, Venezuela]. *Invest Clin*, 44:21–30. PMID:12703180

84. Feinleib M, Kannel WB, Garrison RJ et al. (1975). The Framingham Offspring Study. Design and preliminary data. *Prev Med*, 4:518–525.doi:10.1016/0091-7435(75)90037-7 PMID:1208363

85. Fowkes FG, Housley E, Cawood EH et al. (1991). Edinburgh Artery Study: prevalence of asymptomatic and symptomatic peripheral arterial disease in the general population. *Int J Epidemiol*, 20:384–392.doi:10.1093/ije/20.2.384 PMID:1917239

86. Fried LP, Borhani NO, Enright P et al. (1991). The Cardiovascular Health Study: design and rationale. *Ann Epidemiol*, 1:263–276.doi:10.1016/1047-2797(91)90005-W PMID:1669507

87. García-Palmieri MR, Feliberti M, Costas R Jr et al. (2002). An epidemiological study on coronary heart disease in PR. The Puerto Rico Heart Health Program. 1969. *Bol Asoc Med PR*, 94(1–12):61–7.

88. Gardner CD, Winkleby MA, Fortmann SP (2000). Population frequency distribution of non-high-density lipoprotein cholesterol (Third National Health and Nutrition Examination Survey [NHANES III], 1988–1994). *Am J Cardiol*, 86:299–304.doi:10.1016/S0002-9149(00)00918-8 PMID:10922437

89. Giampaoli S, Palmieri L, Panico S et al. (2006). Favorable cardiovascular risk profile (low risk) and 10-year stroke incidence in women and men: findings from 12 Italian population samples. *Am J Epidemiol*, 163:893–902.doi:10.1093/aje/kwj110 PMID:16554350

90. Gillum RF, Makuc DM (1992). Serum albumin, coronary heart disease, and death. *Am Heart J*, 123:507–513.doi:10.1016/0002-8703(92)90667-K PMID:1736588

91. Goldbourt U, Medalie JH (1979). High density lipoprotein cholesterol and incidence of coronary heart disease–the Israeli Ischemic Heart Disease Study. *Am J Epidemiol*, 109:296–308. PMID:222135

92. Gram J, Bladbjerg EM, Møller L et al. (2000). Tissue-type plasminogen activator and C-reactive protein in acute coronary heart disease. A nested case-control study. *J Intern Med*, 247:205–212.doi:10.1046/j.1365-2796.2000.00604.x PMID:10692083

93. Håheim LL, Holme I, Hjermann I, Leren P (1993). The predictability of risk factors with respect to incidence and mortality of myocardial infarction and total mortality. A 12-year follow-up of the Oslo Study, Norway. *J Intern Med*, 234:17–24. doi:10.1111/j.1365-2796.1993.tb00699.x PMID:8326284

94. Iso H, Naito Y, Sato S et al. (2001). Serum triglycerides and risk of coronary heart disease among Japanese men and women. *Am J Epidemiol*, 153:490–499.doi:10.1093/aje/153.5.490 PMID:11226981

95. Kardys I, Kors JA, van der Meer IM et al. (2003). Spatial QRS-T angle predicts cardiac death in a general population. *Eur Heart J*, 24:1357–1364.doi:10.1016/S0195-668X(03)00203-3 PMID:12871693

96. Keil JE, Loadholt CB, Weinrich MC et al. (1984). Incidence of coronary heart disease in blacks in Charleston, South Carolina. *Am Heart J*, 108:779–786.doi:10.1016/0002-8703(84)90671-9 PMID:6475747

97. Knuiman MW, Jamrozik K, Welborn TA et al. (1995). Age and secular trends in risk factors for cardiovascular disease in Busselton. *Aust J Public Health*, 19:375–382. doi:10.1111/j.1753-6405.1995.tb00389.x PMID:7578538

98. Koenig W, Sund M, Fröhlich M et al. (1999). C-Reactive protein, a sensitive marker of inflammation, predicts future risk of coronary heart disease in initially healthy middle-aged men: results from the MONICA (Monitoring Trends and Determinants in Cardiovascular Disease) Augsburg Cohort Study, 1984 to 1992. *Circulation*, 99:237–242. PMID:9892589

99. Krobot K, Hense HW, Cremer P et al. (1992). Determinants of plasma fibrinogen: relation to body weight, waist-to-hip ratio, smoking, alcohol, age, and sex. Results from the second MONICA Augsburg survey 1989–1990. *Arterioscler Thromb*, 12:780–788. PMID:1616903

100. Laatikainen T, Critchley J, Vartiainen E et al. (2005). Explaining the decline in coronary heart disease mortality in Finland between 1982 and 1997. *Am J Epidemiol*, 162:764–773.doi:10.1093/aje/kwi274 PMID:16150890

101. Lakka HM, Lakka TA, Tuomilehto J et al. (2000). Hyperinsulinemia and the risk of cardiovascular death and acute coronary and cerebrovascular events in men: the Kuopio Ischaemic Heart Disease Risk Factor Study. *Arch Intern Med*, 160:1160–1168.doi:10.1001/archinte.160.8.1160 PMID:10789610

102. Lam TH, He Y, Li LS et al. (1997). Mortality attributable to cigarette smoking in China. *JAMA*, 278:1505–1508.doi:10.1001/jama.278.18.1505 PMID:9363970

103. Lawlor DA, Ebrahim S, Davey Smith G (2002). The association between components of adult height and Type II diabetes and insulin resistance: British Women's Heart and Health Study. *Diabetologia*, 45:1097–1106.doi:10.1007/s00125-002-0887-5 PMID:12189439

104. Lee ET, Welty TK, Fabsitz R et al. (1990). The Strong Heart Study. A study of cardiovascular disease in American Indians: design and methods. *Am J Epidemiol*, 132:1141–1155. PMID:2260546

105. Lindroos M, Kupari M, Heikkilä J, Tilvis R (1993). Prevalence of aortic valve abnormalities in the elderly: an echocardiographic study of a random population sample. *J Am Coll Cardiol*, 21:1220–1225.doi:10.1016/0735-1097(93)90249-Z PMID:8459080

106. Lubin F, Chetrit A, Lusky A, Modan M (1998). Methodology of a two-step quantified nutritional questionnaire and its effect on results. *Nutr Cancer*, 30:78–82.doi:10.1080/01635589809514645 PMID:9507518

107. Marín A, Medrano MJ, González J et al. (2006). Risk of ischaemic heart disease and acute myocardial infarction in a Spanish population: observational prospective study in a primary-care setting. *BMC Public Health*, 6:38. doi:10.1186/1471-2458-6-38 PMID:16503965

108. Marmot MG, Smith GD, Stansfeld S et al. (1991). Health inequalities among British civil servants: the Whitehall II study. *Lancet*, 337:1387–1393.doi:10.1016/0140-6736(91)93068-K PMID:1674771

109. Meade TW, Mellows S, Brozovic M et al. (1986). Haemostatic function and ischaemic heart disease: principal results of the Northwick Park Heart Study. *Lancet*, 2:533–537.doi:10.1016/S0140-6736(86)90111-X PMID:2875280

110. Meade TW, Roderick PJ, Brennan PJ et al. (1992). Extra-cranial bleeding and other symptoms due to low dose aspirin and low intensity oral anticoagulation. *Thromb Haemost*, 68:1–6. PMID:1514166

111. Meade TW; For the British Medical Research Council General Practice Research Framework and participating vascular clinics (2001). Design and intermediate results of the Lower Extremity Arterial Disease Event Reduction (LEADER)* trial of bezafibrate in men with lower extremity arterial disease. *Curr Control Trials Cardiovasc Med*, 2:195–204. doi:10.1186/CVM-2-4-195 PMID:11806795

112. Meisinger C, Loewel H, Mraz W, Koenig W (2005). Prognostic value of apolipoprotein B and A-I in the prediction of myocardial infarction in middle-aged men and women: results from the MONICA/KORA Augsburg cohort study. *Eur Heart J*, 26:271–278. doi:10.1093/eurheartj/ehi003 PMID:15618061

113. Miller GJ, Bauer KA, Barzegar S et al. (1996). Increased activation of the haemostatic system in men at high risk of fatal coronary heart disease. *Thromb Haemost*, 75:767–771. PMID:8725721

114. Mooy JM, Grootenhuis PA, de Vries H et al. (1995). Prevalence and determinants of glucose intolerance in a Dutch caucasian population. The Hoorn Study. *Diabetes Care*, 18:1270–1273.doi:10.2337/diacare.18.9.1270 PMID:8612442

115. Onat A (2001). Risk factors and cardiovascular disease in Turkey. *Atherosclerosis*, 156:1–10.doi:10.1016/S0021-9150(01)00500-7 PMID:11368991

116. Pai JK, Pischon T, Ma J et al. (2004). Inflammatory markers and the risk of coronary heart disease in men and women. *N Engl J Med*, 351:2599–2610.doi:10.1056/NEJMoa040967 PMID:15602020

117. Palmieri L, Donfrancesco C, Giampaoli S et al. (2006). Favorable cardiovascular risk profile and 10-year coronary heart disease incidence in women and men: results from the Progetto CUORE. *Eur J Cardiovasc Prev Rehabil*, 13:562–570.doi:10.1097/01.hjr.0000221866.27039.4b PMID:16874146

118. Ridker PM, Cook NR, Lee IM et al. (2005). A randomized trial of low-dose aspirin in the primary prevention of cardiovascular disease in women. *N Engl J Med*, 352:1293–1304. doi:10.1056/NEJMoa050613 PMID:15753114

119. Rodeghiero F, Tosetto A (1996). The VITA Project: population-based distributions of protein C, antithrombin III, heparin-cofactor II and plasminogen–relationship with physiological variables and establishment of reference ranges. *Thromb Haemost*, 76:226–233. PMID:8865536

120. Rosengren A, Wilhelmsen L, Welin L et al. (1990). Social influences and cardiovascular risk factors as determinants of plasma fibrinogen concentration in a general population sample of middle aged men. *BMJ*, 300:634–638.doi:10.1136/bmj.300.6725.634 PMID:2322698

121. Rosengren A, Eriksson H, Larsson B et al. (2000). Secular changes in cardiovascular risk factors over 30 years in Swedish men aged 50: the study of men born in 1913, 1923, 1933 and 1943. *J Intern Med*, 247:111–118. doi:10.1046/j.1365-2796.2000.00589.x PMID:10672138

122. Salomaa VV, Rasi VP, Vahtera EM et al. (1994). Haemostatic factors and lipoprotein (a) in three geographical areas in Finland: the Finrisk Haemostasis Study. *J Cardiovasc Risk*, 1:241–248. PMID:7621304

123. Sato S, Nakamura M, Iida M et al. (2000). Plasma fibrinogen and coronary heart disease in urban Japanese. *Am J Epidemiol*, 152:420–423.doi:10.1093/aje/152.5.420 PMID:10981454

124. Schnohr P, Jensen G, Nyboe J, Eybjaerg Hansen A (1977). [The Copenhagen City Heart Study. A prospective cardiovascular population study of 20,000 men and women]. *Ugeskr Laeger*, 139:1921–1923. PMID:906112

125. Sempos CT, Cleeman JI, Carroll MD et al. (1993). Prevalence of high blood cholesterol among US adults. An update based on guidelines from the second report of the National Cholesterol Education Program Adult Treatment Panel. *JAMA*, 269:3009–3014.doi:10.1001/jama.269.23.3009 PMID:8501843

126. Shaper AG, Pocock SJ, Walker M et al. (1981). British Regional Heart Study: cardiovascular risk factors in middle-aged men in 24 towns. *Br Med J (Clin Res Ed)*, 283:179–186.doi:10.1136/bmj.283.6285.179 PMID:6789956

127. Shepherd J, Blauw GJ, Murphy MB et al. (1999). The design of a prospective study of Pravastatin in the Elderly at Risk (PROSPER). PROSPER Study Group. PROspective Study of Pravastatin in the Elderly at Risk. *Am J Cardiol*, 84:1192–1197.doi:10.1016/S0002-9149(99)00533-0 PMID:10569329

128. Shepherd J, Cobbe SM, Ford I et al.; West of Scotland Coronary Prevention Study Group (2004). Prevention of coronary heart disease with pravastatin in men with hypercholesterolemia. 1995. *Atheroscler Suppl*, 5:91–97. doi:10.1016/j.atherosclerosissup.2004.08.029 PMID:15531281

129. Sigurdsson G, Baldursdottir A, Sigvaldason H et al. (1992). Predictive value of apolipoproteins in a prospective survey of coronary artery disease in men. *Am J Cardiol*, 69:1251–1254.doi:10.1016/0002-9149(92)91215-P PMID:1585854

130. Simons LA, McCallum J, Simons J et al. (1990). The Dubbo study: an Australian prospective community study of the health of elderly. *Aust N Z J Med*, 20:783–789. PMID:2291727

131. Smith WC, Crombie IK, Tavendale R et al. (1987). The Scottish Heart Health Study: objectives and development of methods. *Health Bull (Edinb),* 45:211–217. PMID:3497906

132. Soyama Y, Miura K, Morikawa Y et al.; Oyabe Study (2003). High-density lipoprotein cholesterol and risk of stroke in Japanese men and women: the Oyabe Study. *Stroke,* 34:863–868.doi:10.1161/01.STR.0000060 869.34009.38 PMID:12637692

133. Stehouwer CD, Weijenberg MP, van den Berg M et al. (1998). Serum homocysteine and risk of coronary heart disease and cerebrovascular disease in elderly men: a 10-year follow-up. *Arterioscler Thromb Vasc Biol,* 18:1895–1901. PMID:9848881

134. Thelle DS, Førde OH, Try K, Lehmann EH (1976). The Tromsø heart study. Methods and main results of the cross-sectional study. *Acta Med Scand,* 200:107–118.doi:10.1111/j.0954-6820.1976.tb08204.x PMID:785953

135. Thøgersen AM, Söderberg S, Jansson JH et al. (2004). Interactions between fibrinolysis, lipoproteins and leptin related to a first myocardial infarction. *Eur J Cardiovasc Prev Rehabil,* 11:33–40.doi:10.1097/01.hjr.0000116824.84388.a2 PMID:15167204

136. Tverdal A, Foss OP, Leren P et al. (1989). Serum triglycerides as an independent risk factor for death from coronary heart disease in middle-aged Norwegian men. *Am J Epidemiol,* 129:458–465. PMID:2916539

137. Ulmer H, Kelleher C, Diem G, Concin H (2003). Long-term tracking of cardiovascular risk factors among men and women in a large population-based health system: the Vorarlberg Health Monitoring & Promotion Programme. *Eur Heart J,* 24:1004–1013. doi:10.1016/S0195-668X(03)00170-2 PMID:12788300

138. Vartiainen E, Jousilahti P, Alfthan G et al. (2000). Cardiovascular risk factor changes in Finland, 1972–1997. *Int J Epidemiol,* 29:49–56.doi:10.1093/ije/29.1.49 PMID:10750603

139. Wald NJ, Law M, Watt HC et al. (1994). Apolipoproteins and ischaemic heart disease: implications for screening. *Lancet,* 343:75–79.doi:10.1016/S0140-6736(94)90814-1 PMID:7903777

140. Walldius G, Jungner I, Holme I et al. (2001). High apolipoprotein B, low apolipoprotein A-I, and improvement in the prediction of fatal myocardial infarction (AMORIS study): a prospective study. *Lancet,* 358:2026–2033.doi:10.1016/S0140-6736(01)07098-2 PMID:11755609

141. Wilhelmsen L, Johansson S, Rosengren A et al. (1997). Risk factors for cardiovascular disease during the period 1985–1995 in Göteborg, Sweden. The GOT-MONICA Project. *J Intern Med,* 242:199–211.doi:10.1046/j.1365-2796.1997.00163.x PMID:9350164

142. Willeit J, Kiechl S (1993). Prevalence and risk factors of asymptomatic extracranial carotid artery atherosclerosis. A population-based study. *Arterioscler Thromb,* 13:661–668. PMID:8485116

143. Wingard DL, Barrett-Connor EL, Scheidt-Nave C, McPhillips JB (1993). Prevalence of cardiovascular and renal complications in older adults with normal or impaired glucose tolerance or NIDDM. A population-based study. *Diabetes Care,* 16:1022–1025. doi:10.2337/diacare.16.7.1022 PMID:8359095

144. Yarnell JW, Baker IA, Sweetnam PM et al. (1991). Fibrinogen, viscosity, and white blood cell count are major risk factors for ischemic heart disease. The Caerphilly and Speedwell collaborative heart disease studies. *Circulation,* 83:836–844. PMID:1999035

145. Craig WY, Neveux LM, Palomaki GE et al. (1998). Lipoprotein(a) as a risk factor for ischemic heart disease: metaanalysis of prospective studies. *Clin Chem,* 44:2301–2306. PMID:9799757

146. Danesh J, Collins R, Appleby P, Peto R (1998). Association of fibrinogen, C-reactive protein, albumin, or leukocyte count with coronary heart disease: meta-analyses of prospective studies. *JAMA,* 279:1477–1482. doi:10.1001/jama.279.18.1477 PMID:9600484

147. Danesh J, Whincup P, Walker M et al. (2000). Low grade inflammation and coronary heart disease: prospective study and updated meta-analyses. *BMJ,* 321:199–204.doi:10.1136/bmj.321.7255.199 PMID:10903648

148. Danesh J, Wheeler JG, Hirschfield GM et al. (2004). C-reactive protein and other circulating markers of inflammation in the prediction of coronary heart disease. *N Engl J Med,* 350:1387–1397.doi:10.1056/NEJMoa032804 PMID:15070788

149. Gordon DJ, Probstfield JL, Garrison RJ et al. (1989). High-density lipoprotein cholesterol and cardiovascular disease. Four prospective American studies. *Circulation,* 79:8–15. PMID:2642759

150. Hokanson JE, Austin MA (1996). Plasma triglyceride level is a risk factor for cardiovascular disease independent of high-density lipoprotein cholesterol level: a meta-analysis of population-based prospective studies. *J Cardiovasc Risk,* 3:213–219. doi:10.1097/00043798-199604000-00014 PMID:8836866

151. Sarwar N, Danesh J, Eiriksdottir G et al. (2007). Triglycerides and the risk of coronary heart disease: 10,158 incident cases among 262,525 participants in 29 Western prospective studies. *Circulation,* 115:450–458. doi:10.1161/CIRCULATIONAHA.106.637793 PMID:17190864

152. Thompson A, Danesh J (2006). Associations between apolipoprotein B, apolipoprotein AI, the apolipoprotein B/AI ratio and coronary heart disease: a literature-based meta-analysis of prospective studies. *J Intern Med,* 259:481–492.doi:10.1111/j.1365-2796.2006.01644.x PMID:16629854

153. Wheeler JG, Mussolino ME, Gillum RF, Danesh J (2004). Associations between differential leucocyte count and incident coronary heart disease: 1764 incident cases from seven prospective studies of 30,374 individuals. *Eur Heart J,* 25:1287–1292. doi:10.1016/j.ehj.2004.05.002 PMID:15288155

154. Danesh J, Whincup P, Walker M et al. (2000). Chlamydia pneumoniae IgG titres and coronary heart disease: prospective study and meta-analysis. *BMJ,* 321:208–213.doi:10.1136/bmj.321.7255.208 PMID:10903653

155. Clarke R, Shipley M, Lewington S et al. (1999). Underestimation of risk associations due to regression dilution in long-term follow-up of prospective studies. *Am J Epidemiol,* 150:341–353. PMID:10453810

156. Wood AM, White I, Thompson SG et al.; Fibrinogen Studies Collaboration (2006). Regression dilution methods for meta-analysis: assessing long-term variability in plasma fibrinogen among 27,247 adults in 15 prospective studies. *Int J Epidemiol,* 35:1570–1578.doi:10.1093/ije/dyl233 PMID:17148467

157. Kaptoge S, White IR, Thompson SG et al.; Fibrinogen Studies Collaboration (2007). Associations of plasma fibrinogen levels with established cardiovascular disease risk factors, inflammatory markers, and other characteristics: individual participant meta-analysis of 154,211 adults in 31 prospective studies: the fibrinogen studies collaboration. *Am J Epidemiol,* 166:867–879.doi:10.1093/aje/kwm191 PMID:17785713

158. Prospective Studies Collaboration (1999). Collaborative overview ('meta-analysis') of prospective observational studies of the associations of usual blood pressure and usual cholesterol levels with common causes of death: protocol for the second cycle of the Prospective Studies Collaboration. *J Cardiovasc Risk,* 6:315–320. PMID:10534135

159. Woodward M, Barzi F, Martiniuk A et al.; Asia Pacific Cohort Studies Collaboration (2006). Cohort profile: the Asia Pacific Cohort Studies Collaboration. *Int J Epidemiol,* 35:1412–1416.doi:10.1093/ije/dyl222 PMID:17060333

160. The Emerging Risk Factors Collaboration (2007). Danesh J, Erqou S, Walker M, Thompson SG. Analysis of individual data on lipid, inflammatory and other markers in over 1.1 million participants in 104 prospective studies of cardiovascular diseases. *Eur J Epidemiol,* 22:839–869. doi:10.1007/s10654-007-9165-7 PMID:17876711

161. Fibrinogen Studies Collaboration (2004). Collaborative meta-analysis of prospective studies of plasma fibrinogen and cardiovascular disease. *Eur J Cardiovasc Prev Rehabil,* 11:9–17.doi:10.1097/01.hjr.0000114968.39211.01 PMID:15167201

162. Danesh J, Lewington S, Thompson SG et al.; Fibrinogen Studies Collaboration (2005). Plasma fibrinogen level and the risk of major cardiovascular diseases and nonvascular mortality: an individual participant meta-analysis. *JAMA,* 294:1799–1809.doi:10.1001/jama.294.14.1799 PMID:16219884

163. Ballantyne C, Cushman M, Psaty B et al.; Lp-PLA$_2$ Studies Collaboration (2007). Collaborative meta-analysis of individual participant data from observational studies of Lp-PLA$_2$ and cardiovascular diseases. Eur J Cardiovasc Prev Rehabil, 14:3–11. doi:10.1097/01.hjr.0000239464.18509.f1 PMID:17301621

164. Thompson S, Kaptoge S, White I et al.; Emerging Risk Factors Collaboration (2010). Statistical methods for the time-to-event analysis of individual participant data from multiple epidemiological studies. Int J Epidemiol, 39:1345–1359.doi:10.1093/ije/dyq063 PMID:20439481

165. Fibrinogen Studies Collaboration (2009). Measures to assess the prognostic ability of the stratified Cox proportional hazards model. Stat Med, 28:389–411.doi:10.1002/sim.3378 PMID:18833567

166. Fibrinogen Studies Collaboration (2009). Correcting for multivariate measurement error by regression calibration in meta-analyses of epidemiological studies. Stat Med, 28:1067–1092.doi:10.1002/sim.3530 PMID:19222086

167. Casas JP, Bautista LE, Smeeth L et al. (2005). Homocysteine and stroke: evidence on a causal link from mendelian randomisation. Lancet, 365:224–232. PMID:15652605

168. Smith GD, Harbord R, Milton J et al. (2005). Does elevated plasma fibrinogen increase the risk of coronary heart disease? Evidence from a meta-analysis of genetic association studies. Arterioscler Thromb Vasc Biol, 25:2228–2233.doi:10.1161/01.ATV.0000183937.65887.9c PMID:16123313

169. Keavney B, Danesh J, Parish S et al. (2006). Fibrinogen and coronary heart disease: test of causality by 'Mendelian randomization'. Int J Epidemiol, 35:935–943. doi:10.1093/ije/dyl114 PMID:16870675

170. Mendel G. Experiments in plant hybridisation. Edinburgh and London: Oliver & Boyd; 1965.

171. Davey Smith G, Ebrahim S (2003). 'Mendelian randomization': can genetic epidemiology contribute to understanding environmental determinants of disease? Int J Epidemiol, 32:1–22.doi:10.1093/ije/dyg070 PMID:12689998

172. Keavney B (2002). Genetic epidemiological studies of coronary heart disease. Int J Epidemiol, 31:730–736.doi:10.1093/ije/31.4.730 PMID:12177010

173. CRP CHD Genetics Collaboration (2008). Collaborative pooled analysis of data on C-reactive protein gene variants and coronary disease: judging causality by Mendelian randomisation. Eur J Epidemiol, 23:531–540.doi:10.1007/s10654-008-9249-z PMID:18425592

174. Kathiresan S, Melander O, Anevski D et al. (2008). Polymorphisms associated with cholesterol and risk of cardiovascular events. N Engl J Med, 358:1240–1249.doi:10.1056/NEJMoa0706728 PMID:18354102

175. Bennet A, Di Angelantonio E, Erqou S et al. (2008). Lipoprotein(a) levels and risk of future coronary heart disease: large-scale prospective data. Arch Intern Med, 168:598–608.doi:10.1001/archinte.168.6.598 PMID:18362252

176. Danesh J, Kaptoge S, Mann AG et al. (2008). Long-term interleukin-6 levels and subsequent risk of coronary heart disease: two new prospective studies and a systematic review. PLoS Med, 5:e78.doi:10.1371/journal.pmed.0050078 PMID:18399716

177. Hingorani A, Humphries S (2005). Nature's randomised trials. Lancet, 366:1906–1908.doi:10.1016/S0140-6736(05)67767-7 PMID:16325682

178. Hingorani AD, Shah T, Casas JP (2006). Linking observational and genetic approaches to determine the role of C-reactive protein in heart disease risk. Eur Heart J, 27:1261–1263. doi:10.1093/eurheartj/ehi852 PMID:16684780

179. Anderson KM, Odell PM, Wilson PW, Kannel WB (1991). Cardiovascular disease risk profiles. Am Heart J, 121:293–298. doi:10.1016/0002-8703(91)90861-B PMID:1985385

180. Anderson KM, Wilson PW, Odell PM, Kannel WB (1991). An updated coronary risk profile. A statement for health professionals. Circulation, 83:356–362. PMID:1984895

181. Wilson PW, D'Agostino RB, Levy D et al. (1998). Prediction of coronary heart disease using risk factor categories. Circulation, 97:1837–1847. PMID:9603539

182. Conroy RM, Pyörälä K, Fitzgerald AP et al.; SCORE project group (2003). Estimation of ten-year risk of fatal cardiovascular disease in Europe: the SCORE project. Eur Heart J, 24:987–1003.doi:10.1016/S0195-668X(03)00114-3 PMID:12788299

183. Ridker PM, Buring JE, Rifai N, Cook NR (2007). Development and validation of improved algorithms for the assessment of global cardiovascular risk in women: the Reynolds Risk Score. JAMA, 297:611–619. doi:10.1001/jama.297.6.611 PMID:17299196

184. Hippisley-Cox J, Coupland C, Vinogradova Y et al. (2007). Derivation and validation of QRISK, a new cardiovascular disease risk score for the United Kingdom: prospective open cohort study. BMJ, 335:136.doi:10.1136/bmj.39261.471806.55 PMID:17615182

185. Hippisley-Cox J, Coupland C, Vinogradova Y et al. (2008). Predicting cardiovascular risk in England and Wales: prospective derivation and validation of QRISK2. BMJ, 336:1475–1482.doi:10.1136/bmj.39609.449676.25 PMID:18573856

186. Beckman JA, Creager MA, Libby P (2002). Diabetes and atherosclerosis: epidemiology, pathophysiology, and management. JAMA, 287:2570–2581.doi:10.1001/jama.287.19.2570 PMID:12020339

187. Danaei G, Lawes CM, Vander Hoorn S et al. (2006). Global and regional mortality from ischaemic heart disease and stroke attributable to higher-than-optimum blood glucose concentration: comparative risk assessment. Lancet, 368:1651–1659.doi:10.1016/S0140-6736(06)69700-6 PMID:17098083

188. DECODE Study Group, European Diabetes Epidemiology Group (2003). Is the current definition for diabetes relevant to mortality risk from all causes and cardiovascular and noncardiovascular diseases? Diabetes Care, 26:688–696.doi:10.2337/diacare.26.3.688 PMID:12610023

189. Khaw KT, Wareham N (2006). Glycated hemoglobin as a marker of cardiovascular risk. Curr Opin Lipidol, 17:637–643.doi:10.1097/MOL.0b013e3280106b95 PMID:17095908

190. Rydén L, Standl E, Bartnik M et al.; Task Force on Diabetes and Cardiovascular Diseases of the European Society of Cardiology (ESC); European Association for the Study of Diabetes (EASD) (2007). Guidelines on diabetes, pre-diabetes, and cardiovascular diseases: executive summary. Eur Heart J, 28:88–136. PMID:17220161

191. Davidson MH, Corson MA, Alberts MJ et al. (2008). Consensus panel recommendation for incorporating lipoprotein-associated phospholipase A2 testing into cardiovascular disease risk assessment guidelines. Am J Cardiol, 101 12A;51F–57F.doi:10.1016/j.amjcard.2008.04.019 PMID:18549872

192. Lerman A, McConnell JP (2008). Lipoprotein-associated phospholipase A2: a risk marker or a risk factor? Am J Cardiol, 101 12A;11F–22F.doi:10.1016/j.amjcard.2008.04.014 PMID:18549867

193. Royston P, Sauerbrei W (2004). A new measure of prognostic separation in survival data. Stat Med, 23:723–748.doi:10.1002/sim.1621 PMID:14981672

194. Harrell FE Jr, Califf RM, Pryor DB et al. (1982). Evaluating the yield of medical tests. JAMA, 247:2543–2546.doi:10.1001/jama.247.18.2543 PMID:7069920

195. Harrell FE Jr, Lee KL, Mark DB (1996). Multivariable prognostic models: issues in developing models, evaluating assumptions and adequacy, and measuring and reducing errors. Stat Med, 15:361–387.doi:10.1002/(SICI)1097-0258(19960229)15:4<361::AID-SIM168>3.0.CO;2-4 PMID:8668867

196. Pencina MJ, D'Agostino RB Sr, D'Agostino RB Jr, Vasan RS (2008). Evaluating the added predictive ability of a new marker: from area under the ROC curve to reclassification and beyond. Stat Med, 27:157–172, discussion 207–212.doi:10.1002/sim.2929 PMID:17569110

197. Cook NR (2007). Use and misuse of the receiver operating characteristic curve in risk prediction. Circulation, 20;115(7):928–35.

198. Vickers AJ, Elkin EB (2006). Decision curve analysis: a novel method for evaluating prediction models. Med Decis Making, 26:565–574.doi:10.1177/0272989X06295361 PMID:17099194

199. Khaw KT, Wareham N, Bingham S et al. (2008). Combined impact of health behaviours and mortality in men and women: the EPIC-Norfolk prospective population study. PLoS Med, 5:e12.doi:10.1371/journal.pmed.0050012 PMID:18184033

200. Hunter DJ (2005). Gene-environment interactions in human diseases. Nat Rev Genet, 6:287–298.doi:10.1038/nrg1578 PMID:15803198

201. Loktionov A (2003). Common gene polymorphisms and nutrition: emerging links with pathogenesis of multifactorial chronic diseases (review). J Nutr Biochem, 14:426–451.doi:10.1016/S0955-2863(03) 00032-9 PMID:12948874

202. Manolio TA, Bailey-Wilson JE, Collins FS (2006). Genes, environment and the value of prospective cohort studies. Nat Rev Genet, 7:812–820.doi:10.1038/nrg1919 PMID:16983377

203. Ordovas JM (2006). Genetic interactions with diet influence the risk of cardiovascular disease. Am J Clin Nutr, 83:443S–446S. PMID:16470010

204. Elosua R, Demissie S, Cupples LA et al. (2003). Obesity modulates the association among APOE genotype, insulin, and glucose in men. Obes Res, 11:1502–1508.doi:10.1038/oby.2003.201 PMID:14694215

205. Moreno JA, Pérez-Jiménez F, Marín C et al. (2004). Apolipoprotein E gene promoter -219G->T polymorphism increases LDL-cholesterol concentrations and susceptibility to oxidation in response to a diet rich in saturated fat. Am J Clin Nutr, 80:1404–1409. PMID:15531693

206. Campos H, D'Agostino M, Ordovás JM (2001). Gene-diet interactions and plasma lipoproteins: role of apolipoprotein E and habitual saturated fat intake. Genet Epidemiol, 20:117–128.doi:10.1002/1098-2272(200101)20:1<117::AID-GEPI10>3.0.CO;2-C PMID:11119301

207. Humphries SE, Talmud PJ, Hawe E et al. (2001). Apolipoprotein E4 and coronary heart disease in middle-aged men who smoke: a prospective study. Lancet, 358:115–119.doi:10.1016/S0140-6736(01)05330-2 PMID:11463413

208. Talmud PJ, Stephens JW, Hawe E et al. (2005). The significant increase in cardiovascular disease risk in APOEepsilon4 carriers is evident only in men who smoke: potential relationship between reduced antioxidant status and ApoE4. Ann Hum Genet, 69:613–622.doi:10.1111/j.1529-8817.2005.00205.x PMID:16266401

209. Keavney B, Parish S, Palmer A et al.; International Studies of Infarct Survival (ISIS) Collaborators (2003). Large-scale evidence that the cardiotoxicity of smoking is not significantly modified by the apolipoprotein E epsilon2/epsilon3/epsilon4 genotype. Lancet, 361:396–398.doi:10.1016/S0140-6736(03)12386-0 PMID:12573381

210. Rubin J, Berglund L (2002). Apolipoprotein E and diets: a case of gene-nutrient interaction? Curr Opin Lipidol, 13:25–32.doi:10.1097/00041433-200202000-00005 PMID:11790960

211. Talmud PJ, Stephens JW (2004). Lipoprotein lipase gene variants and the effect of environmental factors on cardiovascular disease risk. Diabetes Obes Metab, 6:1–7. doi:10.1111/j.1463-1326.2004.00304.x PMID:14686956

212. Ordovas JM, Corella D, Demissie S et al. (2002). Dietary fat intake determines the effect of a common polymorphism in the hepatic lipase gene promoter on high-density lipoprotein metabolism: evidence of a strong dose effect in this gene-nutrient interaction in the Framingham Study. Circulation, 106:2315–2321.doi:10.1161/01.CIR.0000036597.52291.C9 PMID:12403660

213. UK Department of Health. Food and health action plan. July 2003. Available at URL: http://www.dh.gov.uk/en/Publichealth/Healthimprovement/Healthyliving/Foodandhealthactionplan/DH_4065832.

214. Bucher HC, Hengstler P, Schindler C, Meier G (2002). N-3 polyunsaturated fatty acids in coronary heart disease: a meta-analysis of randomized controlled trials. Am J Med, 112:298–304.doi:10.1016/S0002-9343(01)01114-7 PMID:11893369

215. Howard BV, Van Horn L, Hsia J et al. (2006). Low-fat dietary pattern and risk of cardiovascular disease: the Women's Health Initiative Randomized Controlled Dietary Modification Trial. JAMA, 295:655–666. doi:10.1001/jama.295.6.655 PMID:16467234

216. Halton TL, Willett WC, Liu S et al. (2006). Low-carbohydrate-diet score and the risk of coronary heart disease in women. N Engl J Med, 355:1991–2002.doi:10.1056/NEJMoa055317 PMID:17093250

217. Hung HC, Joshipura KJ, Jiang R et al. (2004). Fruit and vegetable intake and risk of major chronic disease. J Natl Cancer Inst, 96:1577–1584.doi:10.1093/jnci/djh296 PMID:15523086

218. Kristal AR, Potter JD (2006). Not the time to abandon the food frequency questionnaire: counterpoint. Cancer Epidemiol Biomarkers Prev, 15:1759–1760.doi:10.1158/1055-9965.EPI-06-0727 PMID:17021349

219. Schatzkin A, Kipnis V (2004). Could exposure assessment problems give us wrong answers to nutrition and cancer questions? J Natl Cancer Inst, 96:1564–1565.doi:10.1093/jnci/djh329 PMID:15523078

220. Willett WC, Hu FB (2006). Not the time to abandon the food frequency questionnaire: point. Cancer Epidemiol Biomarkers Prev, 15:1757–1758.doi:10.1158/1055-9965.EPI-06-0388 PMID:17021351

221. Elwood PC, Pickering JE, Hughes J et al. (2004). Milk drinking, ischaemic heart disease and ischaemic stroke II. Evidence from cohort studies. Eur J Clin Nutr, 58:718–724. doi:10.1038/sj.ejcn.1601869 PMID:15116074

222. Hu FB, Manson JE, Willett WC (2001). Types of dietary fat and risk of coronary heart disease: a critical review. J Am Coll Nutr, 20:5–19. PMID:11293467

223. Tholstrup T (2006). Dairy products and cardiovascular disease. Curr Opin Lipidol, 17:1–10. PMID:16407709

224. Boekholdt SM, Sandhu MS, Day NE et al. (2006). Physical activity, C-reactive protein levels and the risk of future coronary artery disease in apparently healthy men and women: the EPIC-Norfolk prospective population study. Eur J Cardiovasc Prev Rehabil, 13:970–976.doi:10.1097/01.hjr.0000209811.97948.07 PMID:17143130

225. Jensen MK, Mukamal KJ, Overvad K, Rimm EB (2008). Alcohol consumption, TaqIB polymorphism of cholesteryl ester transfer protein, high-density lipoprotein cholesterol, and risk of coronary heart disease in men and women. Eur Heart J, 29:104–112.doi:10.1093/eurheartj/ehm517 PMID:18063597

226. Kloner RA, Rezkalla SH (2007). To drink or not to drink? That is the question. Circulation, 116:1306–1317.doi:10.1161/CIRCULATIONAHA.106.678375 PMID:17846 344

227. Bingham S, Riboli E (2004). Diet and cancer–the European Prospective Investigation into Cancer and Nutrition. Nat Rev Cancer, 4:206–215.doi:10.1038/nrc1298 PMID:14993902

228. Danesh J, Saracci R, Berglund G et al.; EPIC-Heart (2007). EPIC-Heart: the cardiovascular component of a prospective study of nutritional, lifestyle and biological factors in 520,000 middle-aged participants from 10 European countries. Eur J Epidemiol, 22:129–141.doi:10.1007/s10654-006-9096-8 PMID:17295097

229. Riboli E, Hunt KJ, Slimani N et al. (2002). European Prospective Investigation into Cancer and Nutrition (EPIC): study populations and data collection. Public Health Nutr, 5 6B;1113–1124.doi:10.1079/PHN2002394 PMID:12639222

230. Tapia-Conyer R, Kuri-Morales P, Alegre-Díaz J et al. (2006). Cohort profile: the Mexico City Prospective Study. Int J Epidemiol, 35:243–249.doi:10.1093/ije/dyl042 PMID:16556648

231. Chen Z, Lee L, Chen J et al. (2005). Cohort profile: the Kadoorie Study of Chronic Disease in China (KSCDC). Int J Epidemiol, 34:1243–1249.doi:10.1093/ije/dyi174 PMID:16131516

232. Peakman TC, Elliott P (2008). The UK Biobank sample handling and storage validation studies. Int J Epidemiol, 37 Suppl 1;i2–i6.doi:10.1093/ije/dyn019 PMID:18381389

233. Davey Smith G, Ebrahim S, Lewis S et al. (2005). Genetic epidemiology and public health: hope, hype, and future prospects. Lancet, 366:1484–1498.doi:10.1016/S0140-6736(05)67601-5 PMID:16243094

234. Bennett MR, Hasty J (2008). Systems biology: genome rewired. Nature, 452:824–825.doi:10.1038/452824a PMID:18421342

235. Newman JR, Weissman JS (2006). Systems biology: many things from one. *Nature*, 444:561–562.doi:10.1038/nature05407 PMID:17122769

236. Santos SD, Ferrell JE (2008). Systems biology: On the cell cycle and its switches. *Nature*, 454:288–289.doi:10.1038/454288a PMID:18633407

237. Baker M (2008). Genome studies: genetics by numbers. *Nature*, 451:516–518. doi:10.1038/451516a PMID:18235474

238. No Authors (1999). Proteomics, transcriptomics: what's in a name? *Nature*, 402:715. PMID:10617184

239. Abbott A (1999). A post-genomic challenge: learning to read patterns of protein synthesis. *Nature*, 402:715–720.doi:10.1038/45350 PMID:10617183

240. Dalton R, Abbott A (1999). Can researchers find recipe for proteins and chips? *Nature*, 402:718–719.doi:10.1038/45359 PMID:10617186

241. Brindle JT, Antti H, Holmes E *et al.* (2002). Rapid and noninvasive diagnosis of the presence and severity of coronary heart disease using 1H-NMR-based metabonomics. *Nat Med*, 8:1439–1445.doi:10.1038/nm802 PMID:12447357

242. Whincup PH, Danesh J, Walker M *et al.* (2002). von Willebrand factor and coronary heart disease: prospective study and meta-analysis. *Eur Heart J*, 23:1764–1770. doi:10.1053/euhj.2001.3237 PMID:12419296

243. Lowe GD, Danesh J, Lewington S *et al.* (2004). Tissue plasminogen activator antigen and coronary heart disease. Prospective study and meta-analysis. *Eur Heart J*, 25: 252–259.doi:10.1016/j.ehj.2003.11.004 PMID:14972427

244. Danesh J, Whincup P, Walker M *et al.* (2001). Fibrin D-dimer and coronary heart disease: prospective study and meta-analysis. *Circulation*, 103:2323–2327. PMID:11352877

245. Sattar N, Wannamethee G, Sarwar N *et al.* (2009). Leptin and coronary heart disease: prospective study and systematic review. *J Am Coll Cardiol*, 53:167–175.doi:10.1016/j.jacc.2008.09.035 PMID:19130985

246. Di Angelantonio E, Danesh J, Eiriksdottir G, Gudnason V (2007). Renal function and risk of coronary heart disease in general populations: new prospective study and systematic review. *PLoS Med*, 4:e270.doi:10.1371/journal.pmed.0040270 PMID:17803353

247. Wheeler JG, Juzwishin KD, Eiriksdottir G *et al.* (2005). Serum uric acid and coronary heart disease in 9,458 incident cases and 155,084 controls: prospective study and meta-analysis. *PLoS Med*, 2:e76.doi:10.1371/journal.pmed.0020076 PMID:15783260

248. Danesh J, Collins R, Peto R (1997). Chronic infections and coronary heart disease: is there a link? *Lancet*, 350:430–436.doi:10.1016/S0140-6736(97)03079-1 PMID:9259669

249. Whincup P, Danesh J, Walker M *et al.* (2000). Prospective study of potentially virulent strains of Helicobacter pylori and coronary heart disease in middle-aged men. *Circulation*, 101:1647–1652. PMID:10758045

250. Danesh J, Youngman L, Clark S *et al.* (1999). Helicobacter pylori infection and early onset myocardial infarction: case-control and sibling pairs study. *BMJ*, 319:1157–1162. PMID:10541503

251. Danesh J, Collins R, Peto R, Lowe GD (2000). Haematocrit, viscosity, erythrocyte sedimentation rate: meta-analyses of prospective studies of coronary heart disease. *Eur Heart J*, 21:515–520.doi:10.1053/euhj.1999.1699 PMID:10775006

252. Danesh J, Lewington S (1998). Plasma homocysteine and coronary heart disease: systematic review of published epidemiological studies. *J Cardiovasc Risk*, 5:229–232. doi:10.1097/00043798-199808000-00004 PMID:9919470

UNIT 5.
APPLICATION OF BIOMARKERS TO DISEASE

CHAPTER 21.

Work-related lung diseases

Ainsley Weston

Summary

Work-related respiratory diseases affect people in every industrial sector, constituting approximately 60% of all disease and injury mortality and 70% of all occupational disease mortality. There are two basic types: interstitial lung diseases, that is the pneumoconioses (asbestosis, byssinosis, chronic beryllium disease, coal workers' pneumoconiosis (CWP), silicosis, flock workers' lung, and farmers' lung disease), and airways diseases, such as work-related or exacerbated asthma, chronic obstructive pulmonary disease and bronchiolitis obliterans (a disease that was recognized in the production of certain foods only 10 years ago). Common factors in the development of these diseases are exposures to dusts, metals, allergens and other toxins, which frequently cause oxidative damage. In response, the body reacts by activating primary immune response genes (i.e. cytokines that often lead to further oxidative damage), growth factors and tissue remodelling proteins. Frequently, complex imbalances in these processes contribute to the development of disease. For example, tissue matrix metalloproteases can cause the degradation of tissue, as in the development of CWP small profusions, but usually overexpression of matrix metalloproteases is controlled by serum protein inhibitors. Thus, disruption of such a balance can lead to adverse tissue damage. Susceptibility to these types of lung disease has been investigated largely through candidate gene studies, which have been characteristically small, often providing findings that have been difficult to corroborate. An important exception to this has been the finding that the *HLA-DPB1^{E69}* allele is closely associated with chronic beryllium disease and beryllium sensitivity. Although chronic beryllium disease is only caused by exposure to beryllium, inheritance of *HLA-DPB1^{E69}* carries an increased risk of between two- and 30-fold in beryllium exposed workers. Most, if not all, of these occupationally related diseases are preventable; therefore, it is disturbing that rates of CWP, for example, are again increasing in the United States in the 21st century.

Introduction

Excluding lung cancer, which is thought to account for 10 000–12 000 occupationally-related deaths annually in the United States (1), and infectious diseases like tuberculosis and histoplasmosis which may be work-related, several work-related lung diseases have been identified. These have been broadly divided into two types: interstitial lung diseases that are typified by the pneumoconioses (asbestosis, byssinosis, chronic beryllium disease, coal workers' pneumoconiosis, silicosis, flock workers' lung and farmers' lung disease), and airways diseases like asthma, chronic obstructive pulmonary disease (COPD) and bronchiolitis obliterans. Work-related respiratory diseases are a problem of major magnitude. They cut across all industrial sectors, constituting ~60% of all disease and injury mortality and ~70% of all occupational disease mortality (2).

Even though the capability has existed for many years to prevent pneumoconioses (e.g. silicosis, coal workers' pneumoconiosis (CWP) and asbestosis), they still cause or contribute to more than 2500 deaths per year in the United States (3). The threat of other interstitial lung diseases, such as chronic beryllium disease in beryllium metal extraction, production and processing, or hypersensitivity pneumonitis in those exposed to metal working fluids, are also important concerns in specific industries (4,5).

Airways diseases, such as asthma and COPD, are important occupational problems. In 2004, 11.4 million adults (aged ≥ 18) in the USA were estimated to have COPD (6). In the interval from 1997–1999, an estimated 7.4 million people in the United States (aged ≥ 15) reported an episode of asthma or asthma attack in the previous 12 months (7). A 2003 statement by the American Thoracic Society estimated that 15% of COPD and adult asthma cases were work-related, with a conservative annual estimated cost of nearly $7 billion in the USA alone (8).

An emerging area that thus far has not been explored in terms of molecular epidemiology is that of engineered nanotechnology. Nanoparticles and nanomaterials have diverse applications (e.g. drug delivery, electronics and cosmetics); however, their large surface area to volume and respirable nature suggest that they may pose a risk of lung disease. Studies in rodents have shown the potential of nanomaterials to cause oxidative stress, inflammation, and fibrosis (9).

In the last three decades, with the expansion of the emerging field of molecular epidemiology, several genetic susceptibility factors for work-related lung diseases and biomarkers of exposure and effect have been recognized. The majority of these findings took clues from physiological or pathobiological observations, and in some cases genetic linkage analysis, and applied them to candidate gene investigations in molecular epidemiological association studies. Though these types of studies may help to identify high risk subpopulations, their current utility is most valuable in understanding disease mechanisms and developing better laboratory models of disease.

Interstitial lung diseases

Asbestosis, asbestos-related lung cancer and mesothelioma

Several mineral fibres, including chrysotile, amosite, crocidolite, tremolite, actinolite and anthophyllite, are collectively known as asbestos. Asbestos mineral fibres are flame- and heat resistant, pliable, strong, refractory to corrosive chemicals, and provide insulation. Therefore asbestos has been used as a building material to insulate buildings from heat and protect against fire (it has been especially important in the shipbuilding industry), in fabric to make protective suits, as a brake liner (e.g. in automobiles and railroad rolling stock) and for engine gaskets, and in making filters (e.g. in the chemical industry).

Although known and used for its fire resistant properties as early as 3000 B.C., asbestos started to become widely used in the mid- to late-nineteenth century (10). Asbestos-associated fibrosis (asbestosis) was described in the 1920s, and mesothelioma (a very rare cancer of the mesothelium, an epithelial lining of the serous cavities: thorax and peritoneum) and lung cancer were linked to asbestos exposure in the 1960s (11). Thus, asbestosis, mesothelioma (almost exclusively associated with asbestos exposure) and asbestos-associated lung cancer are diseases frequently found in workers employed or formerly employed in construction, shipbuilding, mining, manufacturing and heat and frost insulation.

Fibrous particles generally have a large length to diameter aspect; asbestos fibres are generally considered to have a length to diameter ratio of at least 3:1. Respirable fibrous particles have an effective aerodynamic diameter that more closely resembles particle diameter than length. Thus, long, narrow fibres can reach the alveoli. Fibrous asbestos particles can exert their biological effects in several ways. Physiological attempts by the body to remove asbestos fibres from the deep lung may result in "frustrated phagocytosis" by macrophages that engulf long, narrow fibres. These macrophages then disgorge digestive enzymes and other cytological materials potentially leading to inflammation, fibrosis and malignancy. It has also been proposed that the mineral fibres themselves can promote oxidative damage provoked by Fenton chemistry and the release of iron in the form of Fe^{3+} (12).

Several approaches have been taken to assess potential biomarkers of asbestosis. A major pathway is thought to be mediated through macromolecular and chromosomal damage resulting from reactive oxygen species (ROS) (e.g. O_2^-, $HO^·$, $ONOO$, NO_2, NO_3) formed in the processes described above (13). Because fibrosis and inflammation are major components of the pathobiology of asbestosis, various procollagen genes and cytokine genes have been suggested as potential disease susceptibility markers. In addition, because asbestos exposure is a risk factor for lung cancer and mesothelioma, various tumour markers have been investigated.

Carboxyterminal propeptide of type 1 procollagen (PICP) is a marker for collagen synthesis; it is also associated with tissue and organ fibrosis (14). In this context it has been investigated as a marker for asbestosis. Levels of PICP in bronchoalveolar lavage fluid (BALF) and epithelial lining fluid (ELF) were found to be highest among asbestosis patients (n = 5), with ranges of greater than 7 µl/L to approximately 12 µl/L (mean = 9.8 ± 1.8 µl/L) and approximately 300–800 µl/L (mean = 489 ± 209) in BALF and ELF, respectively. Among 25 asbestos-exposed patients, pleural plaques levels were in the range of zero to less than 5 µl/L (mean = 0.6 ± 1.3 µl/L), and zero to 200 µl/L (mean = 51 ± 23 µl/L) in BALF and ELF, respectively. Among 12 persons with no X-ray evidence of abnormalities, only two were positive, and both of these had levels of PICP of less than 3 µl/L and 200 µl/L in BALF and ELF, respectively. Data for N-terminal propeptide of type 3 procollagen did not support it as a marker of asbestosis. These results are supportive of PICP as a biomarker for asbestosis; however, PICP has been associated with several other fibrotic and chronic inflammatory conditions (e.g. idiopathic fibrosing alveolitis (15), sarcoidosis (16) and myocardial fibrosis (17)). PICP has also been implicated in bone growth and bone metastasis (18). Thus, whereas PICP appears to be a good biomarker of asbestosis, it is not entirely specific.

Leukocyte glycoproteins (cluster of differentiation) CD66b and CD69 are antigens that signify leukocyte activation or hypersensitivity. Elevated levels of interleukins indicate increased inflammatory activity. Asbestos-exposed workers (n = 61 asbestos cement factory) and two groups of non-asbestos-exposed control workers (n = 48 "town" and n = 21 "factory") were evaluated for expression of multiple eosinophilic leukocyte cluster of differentiation marker expression by flow cytometry, as well as serum interleukin (IL) levels by immunoassay (19). A statistically significantly increased expression of markers CD69 and CD66b on eosinophils was found in blood samples collected from asbestos exposed workers. In addition, serum levels of the proinflammatory cytokines IL6 and IL8 were statistically significantly elevated (20). Although these findings reached statistical significance, they did not support the use of these biomarkers as robust screening tests. Furthermore, others have shown that CD69 can be induced in human peripheral blood mononuclear cells *in vivo* by silica, but not by chrysotile asbestos (20).

Asbestosis progression has been monitored by X-ray analysis; the radiographic changes (International Labour Office (ILO) classified) over 2–10 years were correlated with a large series of biomarkers: adenosine deaminase, α-1-antitrypsin, angiotensin-converting enzyme (ACE), β-2-microglobulin, β-*N*-acetylglucosaminidase, carcinoembryonic antigen (CEA), complement components (C3 and C4), erythrocyte sedimentation rate (ESR), ferritin, fibronectin, and lysozyme (21). Radiographic changes, which ranged from ILO 1/1 to ILO 2/2 (at an average of 0.4 minor ILO categories per year), were seen in 32 of 85 patients (OR = 1.54; 95% CI = 0.96–2.47). The only biomarkers that correlated with radiographic changes were fibronectin, ESR and ACE. The ranges of biomarker levels displayed overlap between the patient groups, and while the differences were statistically significant between those measured in patients who progressed compared to those who did not, they were relatively unimpressive (fibronectin OR = 1.01; 95% CI = 1.00–1.02; ESR OR = 1.05; 95% CI = 1.00–1.10; ACE OR = 1.10; 95% CI = 1.00–1.20) (21).

An important tumour marker that has been investigated in asbestos exposed groups is *p53*. Altered expression or overexpression of p53 can be detected in various ways: *p53* mutations can be detected in DNA from tumour tissue (22) or as exfoliated material in blood before a tumour is clinically detected (23), *p53* protein can be detected in blood if it is expressed at high enough levels, and *p53* autoantibodies can be detected. In a study of 115 compensable asbestosis cases, blood samples were drawn from 103 cases between 1980 and

1988. Autoantibodies for *p53* were assayed using an enzyme-linked immunosorbant assay (ELISA); 17 individuals were found to be positive. This cohort was followed for 20 years, and cancers developed in 49 people, among whom 13 were seropositive for *p53* autoantibodies (11 lung cancers, one mesothelioma and one lymphoma). The hazard ratio (HR) for cancer development in seropositive *p53* autoantibody asbestosis patients was determined to be statistically significant (HR = 5.5; 95% CI = 2.8–10.9) (24). Similar results have been obtained by others (25). These results, plus data that showed that both tumour and histologically normal tissue may test positive for *p53* expression, support the idea that *p53* changes are an early event in asbestos-associated lung cancer (25). Several reports have attempted to establish links between *p53* expression as measured in tumour tissue or serum, and *p53* mutations in DNA and autoantibodies (24,26,27). However, caution is recommended in consideration of such associations, as *p53* is both a tumour suppressor gene, and when mutated, an oncogene. Mechanisms that lead to detectable expression of *p53* can result from mutation or stabilization of wild-type *p53*. Mechanisms that lead to absence of detectable *p53* are normal expression of wild-type *p53*, and deletion of chromosome p17.13, which may be in the presence or absence of a *p53* mutation.

A panel of markers was evaluated as a "fuzzy classifier" in both lung cancer patients (n = 216) and asbestosis patients (n = 76). This panel consisted of CEA, neuron specific enolase, squamous cell carcinoma antigen, cytokeratin fragment and C-reactive protein. This panel of markers had 95% specificity in distinguishing cancer cases from asbestosis patients; they were present in 70–98% (overall 92%) of cancer patients, but only 1.3% (1/76) of asbestosis cases (28,29). Other studies of asbestosis cases have found expression of CEA, but this appears to be a preclinical marker of asbestos-related lung cancer and mesothelioma (30,31). Similarly, soluble mesothelin-related protein was found to be higher in mesothelioma patients (n = 24) than asbestosis patients (n = 33) or healthy controls (n = 109; $P < 0.05$) (32).

Osteopontin is a glycoprotein expressed in several malignancies (e.g. lung, gastric, colorectal, breast and ovarian, as well as mesothelioma and melanoma) (33,34). Osteopontin interacts with the integrin receptor and the CD44 receptor to mediate cell matrix interactions and cell signalling. Although it has been identified as a potentially valuable serum marker for mesothelioma, its expression appears to be associated with asbestos exposure. An ELISA test was used to determine serum osteopontin levels in 76 mesothelioma patients, 69 patients with asbestos-related non-malignant pulmonary disease, and 45 controls (no known asbestos exposure). The lowest serum osteopontin levels were found in the control group (20 ± 4 ng/ml) and the highest levels in mesothelioma patients (133 ± 10 ng/ml); the levels in the asbestos-related non-malignant pulmonary disease patient group were 30 ± 3 ng/ml. Interestingly, osteopontin levels in this last group increased with the onset of fibrosis. In addition, levels of osteopontin were higher in those study participants with greater duration of asbestos exposure (0–9 years, 16 ng/ml versus ≥10 years 34 ng/ml; $P = 0.02$) (33).

In summary, since asbestosis itself is a risk factor for lung cancer and pleural mesothelioma it is difficult to disentangle specific biomarkers of asbestosis from biomarkers of asbestos-related lung cancer and mesothelioma. In addition, more robust biomarkers of asbestosis tend to be biomarkers of other conditions where the underlying pathobiology involves chronic inflammation and fibrosis.

Berylliosis

The elemental metal beryllium was discovered in 1798, isolated in 1828, and became an important strategic commodity in 1923 when a patent for a copper-aluminum-beryllium alloy was filed (35). Beryllium has a wide range of interesting properties that have made this metal important in the manufacture of a host of products. It is light, with an atomic weight of 9.012, strong, and has a high melting point (1560°K). It is a neutron moderator and is X-ray transparent. It is non-sparking, corrosion resistant, and acts as an anti-galling agent. It has excellent heat and electrical conductivity, formability, castability and dimensional stability. With these properties it is invaluable in the aerospace, telecommunications, biomedical, defence and automotive industries (36).[1]

In the 1940s, exposure to beryllium in the fluorescent lamp industry was recognized as a respiratory hazard with the emergence of acute chemical pneumonitis (acute beryllium disease (ABD)) (37,38). In addition, extraction and primary production of beryllium metal was also associated with dermatitis, reversible pneumonitis and lung granulomas. In 1949, the Atomic

---

[1] This reference contains a more detailed listing of specific applications. See also: http://www.berylliumproducts.com/

Energy Commission introduced an occupational exposure limit for beryllium of 2 µg/m³ and ABD disappeared. However, chronic beryllium disease (CBD), which is characterized by a cell-mediated immunologic (type 2) hypersensitivity and lung granulomas, remains problematic today (4).

Immunological sensitization to beryllium, which is generally considered to precede CBD, was originally recognized in the 1950s when beryllium salts were applied to the skin with a patch (39). Patch testing is not considered to be a viable procedure for diagnosis of beryllium sensitization, since it requires beryllium exposure itself, albeit through the skin (40,41). In 1987, an *in vitro* test for beryllium sensitization (BeS) was developed in which peripheral blood lymphocytes from beryllium sensitized individuals displayed beryllium specific proliferation (42). This beryllium lymphocyte proliferation test (BeLPT), though not perfect (43), has proved to be an important tool for occupational health screening and medical surveillance in the beryllium industry (44).

Latency in CBD is obscure; workers who are found to be positive for BeS are referred for bronchoalveolar lavage, to seek evidence of sensitized T-lymphocytes in the lung, and/ or lung biopsy, to seek evidence of granulomas formation (4). Workers found to be BeS, through medical surveillance or screening, often have asymptomatic CBD. In other cases of BeS, clinical CBD has only developed decades later (4). These issues concerning latency have provoked debate over the value of using the BeLPT in medical surveillance, because early diagnosis provides no information on which to base treatment options. Moreover, there is no evidence to support the notion that a BeS worker can avoid CBD by leaving the industry, and having a positive BeLPT absent CBD might be an unwelcome source of anxiety.

The benefits of medical surveillance using the BeLPT are that evidence of BeS can support claims under the Energy Employees Occupational Illness Compensation Program Act of 2000 (20 CFR Part 30), help set priorities for disease prevention, and provide confirmation of the efficacy of intervention (4,45).

Together with the BeLPT, a genetic marker of BeS and CBD risk have also been described. In 1989, the BeLPT was used to show that the proliferative response in peripheral blood lymphocytes from a BeS individual could be inhibited in the presence of antibodies elicited against the major histocompatibility complex two molecule, HLA-DPβ1. This finding led to seven molecular epidemiologic association studies that unequivocally demonstrated that the genetic marker *HLA-DPβ1^{E69}* (a DNA sequence that codes for a glutamic acid residue at position 69 of the β chain of the HLA-DP molecule, an antigen presenting entity located on the surface of T-cells, macrophages, and Langerhans cells) is a risk factor for BeS and CBD (46–53).

The identification of a genetic marker closely associated with risk/ susceptibility to CBD in the presence of occupational exposure raises serious ethical, legal and social issues. Indeed, a major United States beryllium producer briefly used an anonymous toll-free telephone line to introduce prospective employees to the possibility of undergoing an industry-sponsored genetic test for *HLA-DPβ1^{E69}* and pre-employment counselling. This programme was discontinued because of a hiring freeze and was not revived. However, it is reasonable to note that it has been shown that the positive predictive value (PPV) of *HLA-DPβ1^{E69}* is poor (around 10%), because the frequency of this marker in the population is high (~0.2 for the allele and 0.3–0.5 for carrier frequency) (54).

More recent refinements to these studies have provided evidence that not all *HLA-DPβ1^{E69}* alleles are equal with respect to CBD susceptibility. The *HLA-DPβ1* gene represents a family of at least 150 alleles having more than 40 single nucleotide polymorphisms (SNPs) in the hypervariable region (55). Consequently, there are 50 *HLA-DPβ1^{E69}* alleles, 5 *HLA-DPβ1^{R69}* alleles and 95 *HLA-DPβ1^{K69}* alleles. Among *HLA-DPβ1^{E69}* alleles, there appears to be a hierarchy of risk which ranges from approximately two- to 20-fold (36,56–58). Most recently these data have been used to shape the design of a transgenic mouse model. Moreover, scrutiny of specific genotypes is likely to reveal genetic biomarkers that have PPVs close to unity.

Coal workers' pneumoconiosis (black lung disease)

Coal workers' pneumoconiosis (CWP) is an interstitial lung disease that is caused by over-exposure to coal mine dust. In the United States, before the Coal Mine Health and Safety Act of 1969 (42 CFR Part 37), coal mine dust levels were as high as six to eight milligrams per cubic metre. The Act dictated that dust levels be capped at two milligrams per cubic metre. At that time, between 30 and 35% of miners developed CWP. As coal mine dust levels dropped to reported levels in the range of one milligram per cubic metre, the percentage of miners developing CWP dropped to approximately 5%. Diagnosis of

CWP is made by the observation of radiographic changes according to the ILO's classification system. In simple pneumoconiosis these changes are described as small opacities (graded, with increasing progression, as 1/0, 1/1, 1/2, 2/1, 2/2, 2/3, 3/2, 3/3; where 0/0 or 0/- reflects a normal x-radiogram, and 0/1 is no disease but stage 1 was considered), and in progressive massive fibrosis (PMF or macular CWP) these are described as large opacities (graded, with increasing progression, as A, B, C). CWP, a chronic inflammatory and fibrotic disease, is characterized by shortness of breath, cough, and deterioration of pulmonary function, all of which become progressively worse with increasing radiographic stage (59).

There is some blurring of distinction between CWP and silicosis in that both show characteristic small opacities on X-ray examination, and coal mine dust is often contaminated with crystalline silica, which is the more toxic component. It appears that oxygen free radical damage can be attributed to coal mine dust exposure from both ferrous iron, in the absence of silica, and silica itself (60,61). Apart from drawing a distinction between these two diseases, another challenge that faces the epidemiology of CWP is exposure assessment. One study that considered five strategies for exposure assessment found that using job and mine led to the most homogeneous exposure categories and most contrast between groups, although that method was the least precise (62).

It has been possible to determine measures of inflammatory response among miners (e.g. alveolar macrophages), polymorphonuclear leukocytes (PMNs), and the antioxidant superoxide dismutase (SOD). One small study of 20 coal miners and 16 control subjects (non-miners) was able to demonstrate a correlation between cumulative exposure to quartz, estimated from work histories and mine air sampling data, and PMNs in bronchoalveolar lavage ($P < 0.0001$), SOD ($P < 0.01$), and radiographic category ($P < 0.0001$) (63). However, a SOD promoter region polymorphism (SOD$^{9Val/Ala}$) was not associated with progression to PMF (n = 700 National Coal Workers Autopsy Study (NCWAS)) (64).

It has been shown that $TNF\text{-}α$, pulmonary surfactant protein A and phospholipids are increased in bronchoalveolar lavage fluids in response to coal mine dust, that $TNF\text{-}α$ levels fall in response to cessation of exposure, and that these biomarkers increase with increasing radiographic evidence of disease progression (65). However, here as in most molecular epidemiologic studies of biomarkers of exposure and effect of coal mine dust exposures, the number of participants was small (n = 48).

Remodeling of extracellular matrix is also a critical event in the progression of fibrotic diseases. A small study of coal miners from Zonguldak, an old coal port on the Turkish Black Sea coast, found that serum pro-matrix metalloproteinase-3 (proMMP-3, also known as Stromelysin 1) was elevated in CWP (n = 44 CWP, 24 ILO 0/0, 0/1, and 17 surface worker controls) (61). In addition, among the CWP group, increasing serum proMMP-3 levels were detected with disease progression or severity measured x-radiographically ($P < 0.01$).

Observations that coal mine dust exposure can induce macrophages and monocytes to secrete cytokines, chemokines, and growth factors *in vivo* and *in vitro*, has led to the development of hypotheses implicating polymorphisms in members of these gene types in susceptibility to CWP and disease progression (66). The promoter region $TNF\text{-}α$ G/A transversion polymorphism at positions −238 and −308, with respect to the ATG translation signal, has been investigated in numerous studies of diseases that involve inflammation and fibrosis (67). In a study of 78 coal miners and 56 controls (healthy members of a non-mining Belgian population), evidence of an association between the minor variant (A) of the −308 polymorphism and development of CWP was obtained by polymerase chain reaction–restriction fragment length polymorphism (PCR-RFLP) (*NcoI*) (68). Miners inheriting a $TNF\text{-}α\text{-}308$ A-variant (or 2) allele were three times more likely to develop CWP (OR = 3.0; 95% CI = 1.0–9.0, $χ^2$ = 4.1, $P < 0.05$). In this study there was no association between inheritance of the minor variant (A or 2-allele) of the $TNF\text{-}α\text{-}238$ polymorphism and CWP. When peripheral blood monocytes from 66 retired miners were exposed to coal mine dust *in vitro*, levels of $TNF\text{-}α$ release were stimulated five- to 10-fold irrespective of genotype.

Eighty CWP patients and 54 healthy volunteers were recruited at a hospital in the Republic of Korea. Peripheral blood mononuclear cells were harvested to provide DNA, and a segment of the TNF-α promoter region −331 to +14 was amplified to determine the identity of the $TNF\text{-}α\text{-}308$ G/A polymorphism (NcoI digestion) (69). The data showed that the frequency of the minor variant (A or *TNF2*) was over-represented in CWP patients by more than two-fold (F = 0.102 versus 0.206; $χ^2$ = 5.121, $P = 0.024$). Moreover, when simple CWP (n = 41) was compared to cases of PMF (n = 39), the frequency of the minor variant was higher in PMF than in simple CWP (F = 0.282 versus 0.134; χ^2 = 5.517, P = 0.019).

A study of 259 unrelated coal miners in France investigated an association between inheritance of the *TNF-α-308* A-variant and CWP. There were 99 cases of CWP (80 active and 19 retired), and 152 without x-radiographic abnormalities for which genotyping data were presented (total n = 212), but no direct association was found (70). However, an interaction was observed in coal mine dust overexposed miners with disease between the *TNF-α-308* A-variant and erythrocyte glutathione peroxidase levels (OR = 2.5; 95% CI = 0.7–9.3; n = 61). In the same population, genotypes (n = 210) were also obtained for the biallelic A/G transversion polymorphism at nucleotide +252 (intron 1) of the lymphotoxin α gene (*LTα*, formerly known as TNF-β). In this case, again there was no difference in allelic distributions by disease status at the inception of the study (*LTα* A-allele frequencies of 0.277 and 0.367, and *LTα* A-homozygosity of 7% and 13% for radiologically normal and CWP groups respectively). However, after five years of follow-up, the CWP group constituted 33.6% of the remaining study population (n = 202), an increase of 5%. At that time, the *LTα* A-allele frequencies were 0.254 and 0.433, and *LTα* A-homozygosity was 4% and 16% for radiologically normal and CWP groups, respectively, which were borderline significant (P = 0.07).

Among 246 Chinese (124 CWP patients and 122 controls), the frequency of the *TNF-α-308* A-variant was found to be 0.0635 and 0.0205, respectively (P = 0.036). However, when a similar analysis was performed for the *TNF-α-238* or the *TNF-α-376* polymorphisms, no associations were found (71). Another study of 674 Chinese (234 CWP patients and 450 coal worker controls) was less conclusive, finding no difference between CWP and controls (F = 0.1034 versus 0.1091, respectively), but finding an elevated frequency of the *TNF-α-308* A-variant among workers with advanced disease (0.2000) (72).

Polymorphisms in the chemokine receptor genes *CCR5* and *CX3CR1* and interleukin 6 and 18 (*IL6, IL18*) have been implicated in the development of CWP (73–75), as has the urokinase-plasminogen activator *PLAU (P141L)* (76). Elevated levels of serum, urine and bronchioalveolar lavage fluid neoptrin, a marker of cell mediated immune activation, have been reported in both simple CWP and PMF (77). In addition, proMMP-3 was found to be higher in miners with more advanced disease (61). In the NCWAS, multiple polymorphisms in a variety of cytokines, growth factors and matrix metalloproteinase genes were evaluated for associations with PMF (78), but only the polygenotype $VEGF^{+405C}/ICAM-1^{+241A}/IL-6^{-174G}$ appeared to have a positive relationship with disease (OR = 3.4; 95% CI = 1.3–8.8, n = 700) (78).

Silicosis

Silicosis is a problematic occupational lung disease; exposure to silica (quartz and cristobalite) causes an inflammatory fibrosing response that can result in interstitial disease (silicosis) or lung cancer (79). The primary origin of the tissue damage leading to these conditions is oxidative, thus fresh fractured silica is much more potent than aged materials (80). Therefore, silicosis has some commonality with both asbestosis and CWP. Occupations that incur prodigious risk are silica sand blasting and coal mining – especially roof-bolters. Indeed, in recent years, as coal seams have become thinner, there is need to cut more siliceous rock to extract the coal, which involves greater hazard of silicosis. An emerging area of concern is roadway repair and demolition, which generates airborne silica dust. Despite these problems, deaths from silicosis in the USA have fallen from more than 1060 in 1968 to less than 170 in 2005 (81).

Dosimetric methods for the assessment of silica are generally problematic. Methods have been developed that can detect silica in blood, urine, lung tissue, lymph nodes, and bronchoalveolar lavage cells, and range from chemical staining to a variety of electron microscopy techniques. However, measures of crystalline silica have not proved to be useful in establishing any kind of dose–response relationship with silicosis, and these methods are not recommended for routine laboratory use (79).

Exposure to silica, like asbestos and coal mine dust, results in oxidative damage (80). This primary damage, mediated by *MIP-2, TNF-α,* IL-β, and *TGF-β*, is central to our current basic understanding of the pathobiology of silicosis (82–84). Because various environmental and occupational exposures, as well as infections and chronic conditions, trigger oxidative stress, using measures of oxidative damage would be too non-specific to be a useful biomarker of silica exposure; indeed, few studies have assessed this possibility. However, 8-hydroxydeoxy-guanosine (8-OHdG) has been measured in leukocyte DNA and urine of quartz exposed workers (n = 42) and silicotics (n = 63) (85). The data from this study showed no difference in either 8-OHdG in leukocyte DNA and 8-OHdG exfoliated in urine between healthy workers and silicotics. There was, however, an inverse relationship between urinary 8-OHdG and DNA-adducts in silicotics, suggesting impaired nucleotide and/or base excision DNA-repair of 8-OHdG, which may be a factor associated with lung cancer susceptibility in silicosis patients (86). In another study of silicosis patients (n = 46, with 27 controls), serum heme oxygenase-1 (heme-HO-1) levels were found to be elevated in silicosis patients compared to controls; serum heme-HO-1 was inversely correlated with serum 8-OHdG levels, but positively correlated with measures of pulmonary function (87). Taken together, the results of these studies (85,87) suggest that both nucleotide/base excision repair activity and antioxidant activity may play a role in protection against the adverse lung function effects in silicosis.

In addition to oxyradical damage, several potential biomarkers associated with oxidative stress have been investigated. A comprehensive review concluded that several factors may potentially be reliable biomarkers of early effects of exposure to crystalline silica (79). These include generation of reactive oxygen species from alveolar macrophages, activation of NFκB, total radical trapping antioxidant capacity, serum isoprostane and glutathione levels, antioxidant enzyme activities (glutathione peroxidase and superoxide dismutase), DNA damage in lymphocytes (measured by the comet assay), neoptrin (2-amino-6-[1, 2, 3-trihydroxypropyl]-1H-pteridin-4-one, a purine nucleotide derivative) (88), and clara cell 16 (CC16) (a protein secreted by non-ciliated cells unique to bronchioles).

More recent studies have investigated these markers further. Increased lipid peroxidation, resulting in isoprostane production, has been measured in urine and exhaled breath, and has been found to be elevated in silicosis patients (P = 0.0001, n = 85) (89,90); however, this marker of oxidative stress is not specific for silica exposure (91). Plasma erythrocyte glutathione levels were decreased among cement manufacturing workers (n = 48) compared to controls (n = 28); conversely, plasma malondialdehyde levels were elevated (92). These data indicate an adverse shift in oxidative balance in cement workers that is likely associated with exposure to silica. In addition, all objective measures of pulmonary function were depressed in the cement worker group.

Among 90 silica-exposed workers (3 groups of 30 each; silicotics phase I, silicosis phase 0+, and non-silicotics phase 0) compared with healthy controls, serum CC16 levels were reduced in all silica exposed workers (P < 0.0001) (93). In the same study, surfactant protein D was increased in silicotics (phase I). In an autopsy study of 29 Canadian hard rock miners, there was a correlation between the amount of silica in the lungs and lymph nodes, the X-ray classification (ILO), and the amount of hydroxyvaline in the lung tissues (94).

Just as in CWP, *TNF-α* promoter region SNPs have been implicated in silicosis. In 2001, it was reported that among 489 study subjects (325 silicotics and 164 controls) silicotics were one and a half- to two-fold more likely to have inherited the minor *TNF-α-238* A-variant (OR = 1.56; 95% CI = 1.0–2.5) and the minor *TNF-α-308* A-variant (OR = 2.35; 95% CI = 1.4–3.6) than controls (95,96). The same study also implicated the minor *IL-1RA+2018* allele (OR = 2.12; 95% CI = 1.3–3.5); however, there were no associations with IL1α and IL1β polymorphisms that were investigated. The association of silicosis with the minor *IL-1RA+2018* allele was confirmed in 212 Chinese silica-exposed workers (75 cases and 137 controls) (97). The association was confirmed between the minor A-variants of *TNF-α-308* and *TNF-α-238* and silicosis in 241 South African miners (121 silicosis cases and 120 controls) (98). This study further implicated the minor A-variant of the *TNF-α-376* promoter region polymorphism. Other proinflammatory cytokines that have been linked to silicosis include CD25+ and CD69+ (99).

The tumour suppressor and prooncogene *p53* has an important role in programmed cell death (apoptosis) and DNA-repair mechanisms (100). Silica has been shown to cause *p53* transactivation through both induction of *p53* protein expression and *p53* protein phosphorylation *in vitro* and *in vivo* (101). It was observed that most apoptotic cells in mice instilled with fresh fractured silica were macrophages. Although it was not investigated in this study, different polymorphic variants of *p53* have been implicated in carcinogenesis (102).

Silicosis patients frequently have associated autoimmune disease disorders (103). These appear to be mediated through the Fas or CD95 pathway. Fas is an important component of the TNF receptor pathway that triggers apoptosis upon ligand binding. Numerous studies have reported elevated Fas levels and variant Fas transcripts in bronchoalveolar lavage fluid and peripheral blood mononuclear cells of silicosis patients (79,104,105). Moreover, serum soluble Fas ligand (sFas) is elevated in silicosis patients and in systemic lupus erythematosus patients (106).

Airways diseases

Asthma

Occupational asthma, or work-exacerbated asthma, is a widespread constriction or obstruction of the airways due to exposure to an irritant present in the workplace that may occur through an allergic or non-allergic mechanism. Work-related asthma was recognized by Hippocrates (460–370 BCE) and associated with occupations involving work with metals, textiles and animals, including fish (107). Today work-related asthma is commonly encountered in isocyanate production, in healthcare workers who use natural rubber latex gloves (although this is becoming less of a problem due to the substitution of other materials), and among office workers due to poor indoor environmental quality (108–110). It is estimated that between 15 and 30% of asthmatics have new-onset adult asthma or work exacerbated asthma. Thus, over two million workers in the United States suffer from work-related or work exacerbated asthma (7). Despite these facts and statistics that suggest a major occupational disease that has been known for more than 2000 years, asthmagens remain difficult to identify, and the connection of asthma with materials or conditions in the workplace may be hard to establish.

Asthma has long been recognized to have both an environmental and a genetic component in addition to being a recognized multigenic disease. A large number of genetic linkage studies, molecular genetic studies, and molecular epidemiology association studies of asthma have been conducted. Examples of fifteen molecular epidemiology association studies or candidate gene studies are given in Table 21.1 (111–125). These studies have focused on: major histocompatibility genes (*HLA-DR, HLA-DQ, HLA-DP*), chemical detoxication genes (*GSTM1, GSTT1, GSTP1, GSTM3*), cytokines (*CD13, CD14, IL4, IL10, IL12b, IL13, IL18, TNF-α*), oxyradical associated pathways (*PTGS2*), proteinase inhibitors (*PAI* or *SERPINE2*), growth factors (*TGF-β*), chemokines (*RANTES*) and related receptors (*CCR3, FCER1B*).

In addition to these studies, linkage studies have implicated genes on chromosomes 5q and 11q. These regions of the genome code are for atopy-related genes, cytokine genes, and the β-2-adrenoceptor gene (or β-2-adrenergic receptor *ADRB2*) (126). These studies have led to the conclusion that asthma is a multigenic disease with an environmental component.

Multiple studies have implicated the *ADRB2*; the product of this gene is present on smooth muscle cells in pulmonary airways. Polymorphisms in this receptor may dispose individuals to be susceptible to nocturnal asthma (127). A meta-analysis suggests that the *ADRB2*^{G16} adrenoceptor glycine 16 allele is associated with nocturnal asthma (OR = 2.2; 95% CI = 1.6–3.1), and that β-2-adrenoceptor glutamic acid 27 (*ADRB2*^{E16}) is not an asthma risk factor (OR = 1.0; 95% CI = 0.7–1.4).

A transmembrane protein, *ADAM33* (also known as MMP33), is a disintegrin and metalloprotease (endopeptidase) that has also been implicated in bronchial hyperresponsiveness. Matrix metalloproteases are normally involved with the structural modeling of tissues, like the lung, therefore disruption of their normal function, either through lack of proteinase inhibition or chronic inflammatory processes, may result in adverse pathology. In a study of 652 nuclear families, a haplotype of 16 *ADAM33* SNPs was associated with susceptibility to asthma ($P < 0.006$); however, no single polymorphism alone was found to have a statistically significant association (128). All of these data contribute to asthma—a complex multigenic disease that has an environmental trigger.

With the advent of the HapMap, a collection of millions of SNP markers arrayed across the genome, genome-wide association studies (GWAS) have become popular. These studies are unfettered by formal hypotheses, and multiplex SNP analysis is used to interrogate the entire genome simultaneously. For asthma, the following chromosomal regions have been found to contain markers that have P-values for association as low as 0.0000000001. They are: 1q32, 2q12, 5q12, 5q22, 5q33, 6q23, 8p21, 9q21, 17q21 and 20pter-p12 (129,130). These GWAS studies have confirmed the involvement of various genes in asthma, while others have suggested new candidates. Examples of genes that have been confirmed by GWAS include: *IL4, IL5, IL13, CD14, ADRB2, HLA-DQB1* and *HLA-DRB1* (131). New candidate genes that have been suggested by GWAS include: *ORMDL3* (a transmembrane protein of unknown function that is associated with the endoplasmic reticulum) (132), *ADRA1B* (an adrenergic receptor distinct from *ADRB2*), *PRNP* (a prion related protein found on chromosome 20p), *DPP10* (adipeptidyl peptidase) (130), *PDE4D* (a protein involved in the regulation of smooth muscle) (133), *IL3, TLE4* (a transcription corepressor that in part regulates *PAX5*, a transcription factor), *IL1R1, IL33, WDR36* (a gene involved in the synthesis of ribosomes), *MYB* (a transcription factor) and *CHI3L1* (a chitinase-3-like protein) (129).

Table 21.1. Genetic epidemiology association studies of asthma

Study and Subjects (n)	Allele(s)	Association[†]	Reference
Paris, France	HLA-DR4	$P<0.0004$	(111)
Cases (56, 62% ♀)	HLA-DR7	$P<0.05$	
Controls (39, 62% ♀)	HLA-DQB1*0103	$P<0.002$	
	HLA-DQB1*0302	$P<0.01$	
Helsinki, Finland[‡]	NAT1[§]	OR=2.5 (1.3-4.9)	(112)
Cases (109, 22% ♀)	GSTM1 + NAT1	OR=4.5 (1.8-11.6)	
Controls (73, 12% ♀)	GSTM1 + NAT2	OR=3.1 (1.1-8.8)	
	NAT1 + NAT2	OR=4.2 (1.5-11.6)	
Cincinnati, OH, USA	CD14^{159T}	$P=0.03$	(113)
Cases (175)	CD14^{159TT}	OR=2.3 (0.9-5.8)	
Controls (61)	CD14^{159TT**}	OR=3.1 (1.1-9.1)	
Taichung, Taiwan, China	IL10^{627AA}	OR=3.6 (1.2-10.4)	(114)
Cases (117, 48% ♀)	IL10^{627AC}	OR=4.8 (1.7-13.9)	
Controls (47, 64% ♀)			
SE Anatolia, Turkey			(115)
Cases (210, 74% ♀)	GSTP1^{105val}	OR=0.3 (0.1-0.6)	
Controls (265, 69% ♀)			
Tokyo, Japan			(116)
Japanese (210)	CCR3^{51C}	OR=1.4 (0.7-2.7)	
Controls (181)			
British (142)		OR=2.4 (1.3-4.3)	
Controls (92)			
San Diego, CA, USA			(117)
Cases (236)	TNF-α-308 A	OR=1.9 (1.0-3.3)	
Controls (275)		OR=1.7 (1.0-2.9)[††]	
Osaka, Japan			(118)
Cases (479)	IL18^{105A}	$P<0.01$	
Controls (85)			
Sapporo, Japan			(119)
Cases (298)	RANTES^{-28G}	OR=2.0 (1.4-3.0)	
Controls (311)			
Boston, MA, USA			(120)
Cases (527, 51% ♀)	TGF-β^{509TT}	OR=2.5 (1.3-5.1)	
Controls (170, 36% ♀)	TGF-β^{509TC}	OR=1.3 (0.9-1.8)	
Vancouver, Canada	HLA-DRB1*0101	OR=0.3 (0.1-0.8)	(121)
Cases (56, 2% ♀)	HLA-DQB1*0603	OR=2.9 (1.0-8.2)	
Controls (63, 0% ♀)	HLA-DQB1*0302	OR=4.9 (1.3-18.6)	
Helsinki, Finland			(122)
Cases (42)	GSTM1^{null}	OR=1.9 (1.0-3.5)	
Controls (56)	GSTM3^{MnI+}, GSTP1^{313val}, GSTT1^{null}	Not significant	

Amsterdam, Netherlands			(123)
Cases (101)	IL13$^{-1055TT}$	P<0.002	
Controls (107)			
Hong Kong SAR, China			(124)
Cases (299)	PTGS2^{8473C}	OR=1.5 (1.0-2.3)	
Controls (175)			
Sapporo, Japan			(125)
Cases (374)	PAI-1^{5G}/ FCER1B$^{109T/654C}$	OR=0.2 (0.1-0.5)	
Controls (374)			

†Statistics given as either P-values or odds ratios (OR) with 95% confidence intervals in parentheses
‡Isocyanate workers
§ Slow acetylator phenotype. Risk of NAT2 alone not significant (OR=1.4; 95% CI=0.7-2.6)
** Nonatopy only (n=47)
††European-Americans only (n=169 cases, 170 controls)
‡‡Statistics given as either P-values or odds ratios (OR) with 95% confidence intervals in parentheses

To address the multigenic nature of asthma, a statistical modeling attempt has been made to elucidate asthma risk. Sixteen alleles, most conveying susceptibility, but some with evidence of protection, were used as a basis of the model (134). A similar model has been used to predict overall risk of breast cancer (135). The model revealed a broad spectrum of potential risk and may help to more clearly identify susceptible populations; however, it will be challenging to integrate an environmental component. As noted in the section on berylliosis, this may be accomplished through an understanding of gene–environment interaction at the molecular level using the tools of computational chemistry (58).

Bronchiolitis obliterans

Bronchiolitis obliterans syndrome (BOS) is a fibroproliferative process that causes intraluminal obstruction of the smallest airways, the bronchioles. This condition can be caused by exposure to toxic chemicals (e.g. diacetyl in artificial butter flavoring, responsible for popcorn workers' lung), it can occur following transplant surgery (notably bone marrow, lung, or heart and lung) and as the result of infection (136–139). Only a few studies exist that have looked for biomarkers of susceptibility, exposure and effect.

The first study of six lung transplant recipients evaluated transcripts of platelet-derived growth factor (PDGF)-β and *TGF-β1* in bronchoalveolar lavage cells. Slightly elevated levels of both growth factors were found in BOS patients compared to controls, and the PDGF-β increase was associated with lung function decrement (140). Another study of 93 lung transplant recipients evaluated SNPs in *TNF-α, TGF-β, IL-6, INF-γ,* and *IL-10.* Both of the high expression variants of *IL-6*$^{-174G}$ and *INF-γ*$^{+874T}$ were found to be correlated with BOS ($P < 0.05$ and 0.04 respectively). In addition, onset of BOS was more rapid in patients carrying these variants (141). A third study extended these data by examining the frequency of the same alleles in a cohort of 78 lung transplant recipients. This study was able to confirm that *IL-6*$^{-174G}$ was associated with earlier onset BOS ($P < 0.04$) and a decreased overall survival ($P < 0.05$) (142).

A novel receptor gene, *NOD2/CARD15*, can interact with *NFκB* to trigger an inflammatory response. Three SNPs in this gene (*Arg702Trp, Gly908Arg,* and *Leu1007finsC*) were investigated in a cohort of 427 donor-recipient pairs involved in allogenic stem cell transplantation. The cumulative incidence of BOS rose in donor recipient pairs with a minor variant of this gene (F = 0.187 versus F = 0.013 (those without mutation), $P < 0.001$); donor variants alone were significantly associated with the complication of BOS (F = 0.132, $P < 0.04$) (143).

Chronic obstructive pulmonary disease

Chronic obstructive pulmonary disease (COPD) results in shortness of breath (dyspnea) due to thickening of the airways of the lung. This is an inflammatory condition, which in contrast to asthma is irreversible, and is caused by toxic exposure to tobacco smoke, dust and/or gases. COPD may be an occupational hazard caused by exposures to dusts and gases in the textile industry, coal and other mining industries, construction industry (silica), services industry (secondhand smoke), and damp non-industrial indoor environments (volatile organic compounds) (6).

COPD is a leading cause of morbidity and mortality in the United States and worldwide (6). In 2003, 10.7 million United States adults were estimated to have COPD, although close to 24 million adults had evidence of impaired lung function, indicating underdiagnosis of COPD in the United States (144). The economic burden in the United States is approximately US$37.2 billion, which includes health care expenditures of US$20.9 billion in direct costs, US$7.4 billion in indirect morbidity costs, and US$8.9 billion in indirect mortality costs (145). Although smoking accounts for the majority of COPD cases, occupational factors associated with many industries are estimated to account for 19% of all cases and 31% among never smokers (144).

COPD is a complex, mutagenic disease that only affects a fraction of smokers (15–20%), therefore it has been reasoned that genetic predisposition and environmental factors are important in its development. A genetic factor that was implicated about 40 years ago was α-1-antitrypsin (α1AT), or rather its deficiency (146). Alpha-1-antitrypsin is a serum protease inhibitor (SERPIN). This family of glycoproteins prevents massive tissue damage from proteases released by host cells during inflammation.

Deficiency of *SERPINA1*, also known as α1AT (PiZ homozygotes), accounts for approximately 2% of COPD patients. Six *SERPINA1* 5 SNP haplotypes were shown to increase risk of COPD by six- to 50-fold (147). In contrast, there was no such association with *SERPINA3* even after an initial study had yielded positive results (141,147). *SERPIN1A* deficiency has also been implicated in liver disease. Another serum protease inhibitor, *SERPINE2*, was implicated in COPD by linkage analysis of 127 probands and 949 total individuals in a family-based study (148).

Several matrix metalloprotease molecules have been implicated in COPD using a linkage strategy. They are: MMP1 or interstitial collagenase, MMP2 or gelatinase-A, MMP8 or neutrophil collagenase, MMP9 or gelatinase-B, and macrophage mellatoelastase. The allele $MMP1^{-1607G}$ was found to be associated with lung function decline (P = 0.02 for allele frequency between 284 patients with rapid decline and 306 with no decline) (149). In addition, this group found evidence that the $MMP12^{357Ser}$ allele was also associated with lung function decline. In several other epidemiological association studies, $MMP9^{-1562T}$ was found to be associated with COPD diagnosed with conventional computed tomography (CT) scans (150), spirometry (151) or high-resolution CT scans (152). Two further studies also implicated *MMP9* alleles, $MMP9^{279Arg}$ that modifies substrate binding (153), and a promoter region polymorphism $MMP9-82G$ (154). A large study using Boston, USA early-onset COPD study subjects set out to confirm COPD associations with SNPs of 12 genes, including MMP1, MMP9 (short tandem repeats, not −1562T), and *TIMP2* (155). The association between $TIMP2^{853A}$ and COPD (P < 0.0001), originally reported in Japanese subjects (85 cases, 40 controls), was found to be of marginal significance in the Boston population (P = 0.08) (155,156). Associations previously reported for $MMP1^{-1607G}$ and the short tandem repeats in *MMP9* were not confirmed. A contemporary study has also implicated multiple SNPs in *ADAM33* (157).

As with asthma and pneumoconioses, which are driven to some extent by oxidative damage, cytokines have been implicated in COPD. Several studies have examined the influence of SNPs in *TNF-α* (158–165). Most of these studies were null, and a meta-analysis that included several of them confirmed this. Other cytokine genes that were investigated for COPD-associated SNPs include: *LTα* (159,164), *IL6* and *IL10* (159), and *IL13* (162); of these the *IL10–1082G* was associated with COPD (OR = 2.6; 95% CI = 1.5–4.4) (159). In a recent study of 374 active firefighters with at least five serial lung function tests, $TNF\text{-}\alpha^{-238}$ was found to be associated with a more rapid rate of FEV1 decline (166).

Several polymorphisms in xenobiotic metabolizing genes have received some attention. It is reasonable to assume that some of these genes could at least contribute to oxidative damage since induction of, for example, cytochrome P450s leads to redox cycling and the formation of oxygen free radicals (167). The isoleucine/valine polymorphism in residue 462 of *CYP1A1*, previously considered to be involved in gene induction (168), was investigated in patients recruited at the University of Edinburgh Medical School, Scotland (36 cases, 281 controls). An association was found between inheritance of the $CYP1A1^{462val}$ and COPD (OR = 2.3; 95% CI = 1.0–5.2) (169). Other xenobiotic metabolism genes that have been investigated are *GSTM1*, *GSTP1*, *GSTT1* and *EPHX1* (165,170-172). With the exceptions of epoxide hydrolase (EPHX1) and *GSTP1*, none have shown a positive association that could be confirmed (155). In the case of *EPHX1*, there is an histidine/arginine polymorphism in residue 139, $EPHX1^{139Arg}$, which was found to be associated with COPD (P = 0.02) (155). In the case of *GSTP1*, there is an isoleucine/valine polymorphism in residue 105; $GSTP1^{105Val}$ was found to be associated with COPD (P = 0.05) (155).

More recently, GWAS technology has also been applied to analysis of genetic factors in COPD. Using this strategy, involvement of several of the above implicated genes has been confirmed, including *SERPINE2* (at 2q33–2q37), *EPHX1* (at 1q42) and *GSTP1* (at 1p13) (173,174). These and other GWAS have implicated additional genes: *SFTPB* (a pulmonary surfactant protein at 2p11), *ADRB1* (at 5q32), *TGF-β* (at 19q13) (175), and *FAM13A* (involved in hypoxia response through signal transduction in human lung epithelial cells at 4q22) (176). In addition, GWAS studies of COPD have also identified an association with *CHRNA* sub-units 3 and 5 (an α-nicotinic acetylcholine receptor, located at chromosome 15q25) (177) and *ADAM33* (the metalloprotease located at chromosome 20p13) (178). For both asthma and COPD, it can be seen from the GWAS approach that there is some genetic overlap in these airways diseases.

Summary

The interstitial lung diseases asbestosis, silicosis and CWP have in common exposure to dusts and fibres that induce oxygen free radical damage. These exposures tend to stimulate inflammation and fibrosis, at least in part mediated through the *TNF-α* pathway. In silicosis and CWP this probably influenced the choice of SNP biomarkers that have been examined, and there is a preponderance of evidence to suggest that the promoter region polymorphism of *TNF-α* is implicated in susceptibility and severity of these diseases; this has not been the case for CBD. While most molecular epidemiology has focused on the major histocompatibility complex type 2 molecules, and especially the *HLA-DPB1* gene, there are several studies concerning the *TNF-α* promoter regions in CBD, but none of them have provided support for implication of this gene (49,178,179).

The studies on berylliosis provide an interesting example of a susceptibility marker for several reasons. First, the *HLA-DPB1*[E69] allele has been shown to be associated with CBD and beryllium sensitization in at least three sufficiently-sized, well-characterized study populations (51–53) and several smaller studies, and essentially all of the studies agree. Second, it is a marker that could be used for pre-employment screening, but the positive predictive value is only about 7–14% (54). (This is a cautionary note: despite the strong and uncontested association with disease, it would not make good economic or ethical sense to use beryllium for testing, as exposure to it is what drives disease.) Third, if similar markers could be found for asthma, it may be possible to learn about asthmagens through computational chemical modelling (57,58).

In the case of the airways diseases, asthma and COPD, it is clear that aberrant tissue remodeling is a major contributory factor to pathology (180). Imbalances in matrix metalloproteases and serum protease inhibitors (SERPINs) in the presence of inflammation, which are associated to some extent with genetic polymorphisms, appear to be critical factors. These findings have prompted therapeutic targeting of matrix metalloproteases through the use of inhibitors for the treatment of COPD (181).

In terms of occupational diseases, molecular epidemiological studies of bronchiolitis obliterans, byssinosis and flock workers' lung have not yet been developed. Byssinosis, or brown lung disease, was highly prevalent in the United States in the early 1970s, but numbers have declined due to implementation of the Cotton Dust standard (29 CFR Part 1910) and migrations of textile work to Asia. Thus, research in this area would now be confined to populations in India, China and other parts of Asia. A similar situation is evolving for flock workers.

Many of the molecular epidemiological association studies reported on here are small, and the variation in the quality of participant characterization is considerable. Many of the control populations are convenience samples, and less-than-appropriate samples that come from expired units from blood banks. This has led to considerable disparity across the field of molecular epidemiology with respect to the soundness of specific associations. One study, using a well-characterized molecular epidemiologic case–control population to attempt to verify previous reports for 15 alleles in COPD, is a model and an approach that should be adopted if meaningful associations are to be established (155).

Disclaimer: The findings and conclusions in this chapter are those of the author and do not necessarily represent the views of the National Institute for Occupational Safety and Health.

References

1. Steenland K, Loomis D, Shy C, Simonsen N (1996). Review of occupational lung carcinogens. *Am J Ind Med*, 29:474–490.doi:10.1002/(SICI)1097-0274(199605)29:5<474::AID-AJIM6>3.0.CO;2-M PMID:8732921

2. Steenland K, Burnett C, Lalich N et al. (2003). Dying for work: The magnitude of US mortality from selected causes of death associated with occupation. *Am J Ind Med*, 43:461–482.doi:10.1002/ajim.10216 PMID:12704620

3. National Institute for Occupational Safety and Health. All pneumoconioses and related exposures. Section 6. In: Work-related lung disease surveillance report 2002. U.S. Department of Health and Human Services. Public Health Service. Centers for Disease Control and Prevention. Number 2003–111; Cincinnati (OH); 2003 p. 125–56.

4. Kreiss K, Day GA, Schuler CR (2007). Beryllium: a modern industrial hazard. *Annu Rev Public Health*, 28:259–277.doi:10.1146/annurev.publhealth.28.021406.144011 PMID:17094767

5. U.S. Department of Health and Human Services. Criteria for a recommended standard: occupational exposure to metalworking fluids. Public Health Service. Centers for Disease Control and Prevention. National Institute for Occupational Safety and Health Number, 98–102; Cincinnati (OH);1998.

6. American Lung Association. Chronic obstructive pulmonary disease (COPD) fact sheet. 2008. Available from URL: http://www.lungusa.org/site/apps/nlnet/content3.aspx?c=dvLUK9O0E&b=2058829&content_id={EE451F66-996B-4C23-874D-BF66586196FF}&notoc=1.

7. Sama SR, Milton DK, Hunt PR et al. (2006). Case-by-case assessment of adult-onset asthma attributable to occupational exposures among members of a health maintenance organization. *J Occup Environ Med*, 48:400–407.doi:10.1097/01.jom.0000199437.33100.cf PMID:16607195

8. Balmes J, Becklake M, Blanc P et al.; Environmental and Occupational Health Assembly, American Thoracic Society (2003). American Thoracic Society Statement: Occupational contribution to the burden of airway disease. *Am J Respir Crit Care Med*, 167:787–797.doi:10.1164/rccm.167.5.787 PMID:12598220

9. Shvedova AA, Kisin ER, Mercer R et al. (2005). Unusual inflammatory and fibrogenic pulmonary responses to single-walled carbon nanotubes in mice. *Am J Physiol Lung Cell Mol Physiol*, 289:L698–L708.doi:10.1152/ajplung.00084.2005 PMID:15951334

10. Barbalace RC. A brief history of asbestos use and associated health risks. 2004. Available from URL: http://environmentalchemistry.com/yogi/environmental/asbestoshistory2004.html.

11. Williams PRD, Phelka AD, Paustenbach DJ (2007). A review of historical exposures to asbestos among skilled craftsmen (1940–2006). *J Toxicol Environ Health B Crit Rev*, 10:319–377. PMID:17687724

12. Luster MI, Simeonova PP (1998). Asbestos induces inflammatory cytokines in the lung through redox sensitive transcription factors. *Toxicol Lett*, 102-103:271–275.doi:10.1016/S0378-4274(98)00321-X PMID:10022265

13. Bhattacharya K, Dopp E, Kakkar P et al. (2005). Biomarkers in risk assessment of asbestos exposure. *Mutat Res*, 579:6–21. PMID:16112146

14. Lammi L, Ryhänen L, Lakari E et al. (1999). Carboxyterminal propeptide of type I procollagen in ELF: elevation in asbestosis, but not in pleural plaque disease. *Eur Respir J*, 14:560–564.doi:10.1034/j.1399-3003.1999.14c13.x PMID:10543275

15. Lammi L, Ryhänen L, Lakari E et al. (1999). Type III and type I procollagen markers in fibrosing alveolitis. *Am J Respir Crit Care Med*, 159:818–823. PMID:10051256

16. Lammi L, Kinnula V, Lähde S et al. (1997). Propeptide levels of type III and type I procollagen in the serum and bronchoalveolar lavage fluid of patients with pulmonary sarcoidosis. *Eur Respir J*, 10:2725–2730.doi:10.1183/09031936.97.10122725 PMID:9493651

17. Querejeta R, Varo N, López B et al. (2000). Serum carboxy-terminal propeptide of procollagen type I is a marker of myocardial fibrosis in hypertensive heart disease. *Circulation*, 101:1729–1735. PMID:10758057

18. Wallace JD, Cuneo RC, Lundberg PA et al. (2000). Responses of markers of bone and collagen turnover to exercise, growth hormone (GH) administration, and GH withdrawal in trained adult males. *J Clin Endocrinol Metab*, 85:124–133.doi:10.1210/jc.85.1.124 PMID:10634375

19. Ilavská S, Jahnová E, Tulinská J et al. (2005). Immunological monitoring in workers occupationally exposed to asbestos. *Toxicology*, 206:299–308.doi:10.1016/j.tox.2004.09.004 PMID:15588921

20. Wu P, Hyodoh F, Hatayama T et al. (2005). Induction of CD69 antigen expression in peripheral blood mononuclear cells on exposure to silica, but not by asbestos/chrysotile-A. *Immunol Lett*, 98:145–152.doi:10.1016/j.imlet.2004.11.005 PMID:15790520

21. Oksa P, Huuskonen MS, Järvisalo J et al. (1998). Follow-up of asbestosis patients and predictors for radiographic progression. *Int Arch Occup Environ Health*, 71:465–471. doi:10.1007/s004200050307 PMID:9826079

22. Greenblatt MS, Bennett WP, Hollstein M, Harris CC (1994). Mutations in the p53 tumor suppressor gene: clues to cancer etiology and molecular pathogenesis. *Cancer Res*, 54:4855–4878. PMID:8069852

23. Chen JG, Kuang SY, Egner PA et al. (2007). Acceleration to death from liver cancer in people with hepatitis B viral mutations detected in plasma by mass spectrometry. *Cancer Epidemiol Biomarkers Prev*, 16:1213–1218.doi:10.1158/1055-9965.EPI-06-0905 PMID:17548687

24. Husgafvel-Pursiainen K, Kannio A, Oksa P et al. (1997). Mutations, tissue accumulations, and serum levels of p53 in patients with occupational cancers from asbestos and silica exposure. *Environ Mol Mutagen*, 30:224–230.doi:10.1002/(SICI)1098-2280(1997)30:2<224::AID-EM15>3.0.CO;2-F PMID:9329647

25. Li Y, Karjalainen A, Koskinen H et al. (2005). p53 autoantibodies predict subsequent development of cancer. *Int J Cancer*, 114:157–160.doi:10.1002/ijc.20715 PMID:15523685

26. Rusin M, Butkiewicz D, Malusecka E et al. (1999). Molecular epidemiological study of non-small-cell lung cancer from an environmentally polluted region of Poland. *Br J Cancer*, 80:1445–1452.doi:10.1038/sj.bjc.6690542 PMID:10424749

27. Trivers GE, De Benedetti VM, Cawley HL et al. (1996). Anti-p53 antibodies in sera from patients with chronic obstructive pulmonary disease can predate a diagnosis of cancer. *Clin Cancer Res*, 2:1767–1775. PMID:9816128

28. Schneider J, Bitterlich N, Kotschy-Lang N et al. (2007). A fuzzy-classifier using a marker panel for the detection of lung cancers in asbestosis patients. *Anticancer Res*, 27 4A;1869–1877. PMID:17649786

29. Shitrit D, Zingerman B, Shitrit AB et al. (2005). Diagnostic value of CYFRA 21–1, CEA, CA 19–9, CA 15–3, and CA 125 assays in pleural effusions: analysis of 116 cases and review of the literature. Oncologist, 10:501–507.doi:10.1634/theoncologist.10-7-501 PMID:16079317

30. Krajewska B, Lutz W, Piłacik B (1996). [Exposure to asbestos and levels of selected tumor biomarkers]. Med Pr, 47:89–96. PMID:8657007

31. Eyden BP, Banik S, Harris M (1996). Malignant epithelial mesothelioma of the peritoneum: observations on a problem case. Ultrastruct Pathol, 20:337–344.doi: 10.3109/01913129609016334 PMID:8837340

32. Di Serio F, Fontana A, Loizzi M et al. (2007). Mesothelin family proteins and diagnosis of mesothelioma: analytical evaluation of an automated immunoassay and preliminary clinical results. Clin Chem Lab Med, 45:634–638.doi:10.1515/CCLM. 2007.112 PMID:17484626

33. Pass HI, Lott D, Lonardo F et al. (2005). Asbestos exposure, pleural mesothelioma, and serum osteopontin levels. N Engl J Med, 353:1564–1573.doi:10.1056/NEJMoa051185 PMID:16221779

34. Cullen MR (2005). Serum osteopontin levels–is it time to screen asbestos-exposed workers for pleural mesothelioma? N Engl J Med, 353:1617–1618.doi:10.1056/NEJMe 058176 PMID:16221786

35. Kolanz ME (2001). Introduction to beryllium: uses, regulatory history, and disease. Appl Occup Environ Hyg, 16:559–567.doi:10.1080/ 10473220119088 PMID:11370935

36. Weston A, Snyder J, McCanlies EC et al. (2005). Immunogenetic factors in beryllium sensitization and chronic beryllium disease. Mutat Res, 592:68–78. PMID:16054169

37. No authors listed (1984). Archives of the Cleveland Clinic Quarterly 1943: Chemical pneumonia in workers extracting beryllium oxide. Report of three cases. By H.S. VanOrdstrand, Robert Hughes and Morris G. Carmody. Cleve Clin Q, 51:431–439. PMID:6380823

38. DeNARDI JM, Van Ordstrand HS, Carmody MG (1949). Acute dermatitis and pneumonitis in beryllium workers; review of 406 cases in 8-year period with follow-up on recoveries. Ohio Med, 45:567–575. PMID:18127900

39. Curtis GH (1951). Cutaneous hypersensitivity due to beryllium; a study of thirteen cases. AMA Arch Derm Syphilol, 64:470–482. PMID:14867858

40. Tinkle SS, Antonini JM, Rich BA et al. (2003). Skin as a route of exposure and sensitization in chronic beryllium disease. Environ Health Perspect, 111:1202–1208. doi:10.1289/ehp.5999 PMID:12842774

41. Day GA, Stefaniak AB, Weston A, Tinkle SS (2006). Beryllium exposure: dermal and immunological considerations. Int Arch Occup Environ Health, 79:161–164.doi:10.1007/s00420-005-0024-0 PMID:16231190

42. Kreiss K, Newman LS, Mroz MM, Campbell PA (1989). Screening blood test identifies subclinical beryllium disease. J Occup Med, 31:603–608.doi:10.1097/00043764-19890 7000-00011 PMID:2788726

43. Deubner DC, Goodman M, Iannuzzi J (2001). Variability, predictive value, and uses of the beryllium blood lymphocyte proliferation test (BLPT): preliminary analysis of the ongoing workforce survey. Appl Occup Environ Hyg, 16:521–526.doi:10. 1080/104732201750169598 PMID:11370932

44. Kreiss K, Wasserman S, Mroz MM, Newman LS (1993). Beryllium disease screening in the ceramics industry. Blood lymphocyte test performance and exposure-disease relations. J Occup Med, 35:267–274. PMID:8455096

45. Cummings KJ, Deubner DC, Day GA et al. (2007). Enhanced preventive programme at a beryllium oxide ceramics facility reduces beryllium sensitisation among new workers. Occup Environ Med, 64:134–140.doi:10.1136/oem.2006.027987 PMID:17043076

46. Richeldi L, Sorrentino R, Saltini C (1993). HLA-DPB1 glutamate 69: a genetic marker of beryllium disease. Science, 262:242–244. doi:10.1126/science.8105536 PMID:8105536

47. Richeldi L, Kreiss K, Mroz MM et al. (1997). Interaction of genetic and exposure factors in the prevalence of berylliosis. Am J Ind Med, 32:337–340.doi:10.1002/(SICI)1097-0274(199710)32:4<337::AID-AJIM3>3.0.CO;2-R PMID:9258386

48. Wang Z, White PS, Petrovic M et al. (1999). Differential susceptibilities to chronic beryllium disease contributed by different Glu69 HLA-DPB1 and -DPA1 alleles. J Immunol, 163:1647–1653. PMID:10415070

49. Saltini C, Richeldi L, Losi M et al. (2001). Major histocompatibility locus genetic markers of beryllium sensitization and disease. Eur Respir J, 18:677–684.doi:10.1183/09031936. 01.00106201 PMID:11716174

50. Wang Z, Farris GM, Newman LS et al. (2001). Beryllium sensitivity is linked to HLA-DP genotype. Toxicology, 165:27–38. doi:10.1016/S0300-483X(01)00410-3 PMID:11551429

51. Rossman MD, Stubbs J, Lee CW et al. (2002). Human leukocyte antigen Class II amino acid epitopes: susceptibility and progression markers for beryllium hypersensitivity. Am J Respir Crit Care Med, 165:788–794. PMID:11897645

52. Maier LA, McGrath DS, Sato H et al. (2003). Influence of MHC class II in susceptibility to beryllium sensitization and chronic beryllium disease. J Immunol, 171: 6910–6918. PMID:14662898

53. McCanlies EC, Ensey JS, Schuler CR et al. (2004). The association between HLA-DPB1Glu69 and chronic beryllium disease and beryllium sensitization. Am J Ind Med, 46:95–103.doi:10.1002/ajim.20045 PMID:15273960

54. Weston A, Ensey J, Kreiss K et al. (2002). Racial differences in prevalence of a supratypic HLA-genetic marker immaterial to pre-employment testing for susceptibility to chronic beryllium disease. Am J Ind Med, 41:457–465.doi:10.1002/ajim.10072 PMID:12173370

55. European Bioinformatics Institute. IMGT/HLA database. 2005. Available from URL: http://www.ebi.ac.uk/imgt/hla/allele.html.

56. McCanlies EC, Kreiss K, Andrew M, Weston A (2003). HLA-DPB1 and chronic beryllium disease: a HuGE review. Am J Epidemiol, 157:388–398.doi:10.1093/aje/kwg 001 PMID:12615603

57. Snyder JA, Weston A, Tinkle SS, Demchuk E (2003). Electrostatic potential on human leukocyte antigen: implications for putative mechanism of chronic beryllium disease. Environ Health Perspect, 111:1827–1834.doi: 10.1289/ehp.6327 PMID:14630515

58. Snyder JA, Demchuk E, McCanlies EC et al. (2008). Impact of negatively charged patches on the surface of MHC class II antigen-presenting proteins on risk of chronic beryllium disease. J R Soc Interface, 5:749–758.doi:10.1098/rsif.2007.1223 PMID:17956852

59. Wang XR, Christiani DC (2000). Respiratory symptoms and functional status in workers exposed to silica, asbestos, and coal mine dusts. J Occup Environ Med, 42:1076–1084.doi:10.1097/00043764-200011000-00009 PMID:11094786

60. Huang C, Li J, Zhang Q, Huang X (2002). Role of bioavailable iron in coal dust-induced activation of activator protein-1 and nuclear factor of activated T cells: difference between Pennsylvania and Utah coal dusts. Am J Respir Cell Mol Biol, 27:568–574. PMID:12397016

61. Altin R, Kart L, Tekin I et al. (2004). The presence of promatrix metalloproteinase-3 and its relation with different categories of coal workers' pneumoconiosis. Mediators Inflamm, 13:105–109.doi:10.1080/096293504 10001688549 PMID:15203551

62. Heederik D, Attfield M (2000). Characterization of dust exposure for the study of chronic occupational lung disease: a comparison of different exposure assessment strategies. Am J Epidemiol, 151:982–990. PMID:10853637

63. Kuempel ED, Attfield MD, Vallyathan V et al. (2003). Pulmonary inflammation and crystalline silica in respirable coal mine dust: dose-response. J Biosci, 28:61–69. doi:10.1007/BF02970133 PMID:12682426

64. Yucesoy B, Johnson VJ, Kashon ML et al. (2005). Lack of association between antioxidant gene polymorphisms and progressive massive fibrosis in coal miners. Thorax, 60:492–495. doi:10.1136/thx.2004.029090 PMID:15923250

65. Xing JC, Chen WH, Han WH et al. (2006). Changes of tumor necrosis factor, surfactant protein A, and phospholipids in bronchoalveolar lavage fluid in the development and progression of coal workers' pneumoconiosis. Biomed Environ Sci, 19:124–129. PMID:16827183

66. Borm PJ, Schins RP (2001). Genotype and phenotype in susceptibility to coal workers' pneumoconiosis. the use of cytokines in perspective. Eur Respir J Suppl, 32:127s–133s. PMID:11816820

67. Wilson AG, di Giovine FS, Duff GW (1995). Genetics of tumour necrosis factor-alpha in autoimmune, infectious, and neoplastic diseases. J Inflamm, 45:1–12. PMID:7583349

68. Zhai R, Jetten M, Schins RP et al. (1998). Polymorphisms in the promoter of the tumor necrosis factor-alpha gene in coal miners. Am J Ind Med, 34:318–324.doi:10.1002/(SICI)1097-0274(199810)34:4<318::AID-AJIM4>3.0.CO;2-O PMID:9750937

69. Kim KA, Cho YY, Cho JS et al. (2002). Tumor necrosis factor-alpha gene promoter polymorphism in coal workers' pneumoconiosis. Mol Cell Biochem, 234-235:205–209.doi:10.1023/A:1015914409661 PMID:12162435

70. Nadif R, Jedlicka A, Mintz M et al. (2003). Effect of TNF and LTA polymorphisms on biological markers of response to oxidative stimuli in coal miners: a model of gene-environment interaction. Tumour necrosis factor and lymphotoxin alpha. J Med Genet, 40:96–103.doi:10.1136/jmg.40.2.96 PMID:12566517

71. Wang XT, Ohtsuka Y, Kimura K et al. (2005). Antithetical effect of tumor necrosis factor-alpha gene polymorphism on coal workers' pneumoconiosis (CWP). Am J Ind Med, 48:24–29.doi:10.1002/ajim.20180 PMID:15940715

72. Li L, Yu C, Qi F et al. (2004). [Potential effect of tumor necrosis factor-alpha and its receptor II gene polymorphisms on the pathogenesis of coal worker's pneumoconiosis]. Zhonghua Lao Dong Wei Sheng Zhi Ye Bing Za Zhi, 22:241–244. PMID:15355698

73. Zhai R, Liu G, Ge X et al. (2002). Serum levels of tumor necrosis factor-alpha (TNF-alpha), interleukin 6 (IL-6), and their soluble receptors in coal workers' pneumoconiosis. Respir Med, 96:829–834.doi:10.1053/rmed.2002.1367 PMID:12412984

74. Nadif R, Mintz M, Rivas-Fuentes S et al. (2006). Polymorphisms in chemokine and chemokine receptor genes and the development of coal workers' pneumoconiosis. Cytokine, 33:171–178.doi:10.1016/j.cyto.2006.01.001 PMID:16524739

75. Nadif R, Mintz M, Marzec J et al. (2006). IL18 and IL18R1 polymorphisms, lung CT and fibrosis: A longitudinal study in coal miners. Eur Respir J, 28:1100–1105.doi:10.1183/09031936.00031506 PMID:16971411

76. Chang LC, Tseng JC, Hua CC et al. (2006). Gene polymorphisms of fibrinolytic enzymes in coal workers' pneumoconiosis. Arch Environ Occup Health, 61:61–66.doi:10.3200/AEOH.61.2.61-66 PMID:17649957

77. Ulker OC, Yucesoy B, Durucu M, Karakaya A (2007). Neopterin as a marker for immune system activation in coal workers' pneumoconiosis. Toxicol Ind Health, 23:155–160.doi:10.1177/0748233707083527 PMID:18220157

78. Yucesoy B, Johnson VJ, Kissling GE et al. (2008). Genetic susceptibility to progressive massive fibrosis in coal miners. Eur Respir J, 31:1177–1182.doi:10.1183/09031936.00075107 PMID:18256065

79. Gulumian M, Borm PJ, Vallyathan V et al. (2006). Mechanistically identified suitable biomarkers of exposure, effect, and susceptibility for silicosis and coal-worker's pneumoconiosis: a comprehensive review. J Toxicol Environ Health B Crit Rev, 9:357–395.doi:10.1080/15287390500196537 PMID:16990219

80. Castranova V (2004). Signaling pathways controlling the production of inflammatory mediators in response to crystalline silica exposure: role of reactive oxygen/nitrogen species. Free Radic Biol Med, 37:916–925. doi:10.1016/j.freeradbiomed.2004.05.032 PMID:15336307

81. U.S. Department of Health and Human Services. Work-related lung disease surveillance report: silicosis and related exposures. Public Health Service. National Institute for Occupational Safety and Health. Centers for Disease Control and Prevention. DHHS (NIOSH) Number 2003-111. Cincinnati (OH); 2002. p. 51–86.

82. Driscoll KE (2000). TNFalpha and MIP-2: role in particle-induced inflammation and regulation by oxidative stress. Toxicol Lett, 112-113:177–183.doi:10.1016/S0378-4274(99)00282-9 PMID:10720729

83. Rimal B, Greenberg AK, Rom WN (2005). Basic pathogenetic mechanisms in silicosis: current understanding. Curr Opin Pulm Med, 11:169–173.doi:10.1097/01.mcp.0000152998.11335.24 PMID:15699791

84. Barrett EG, Johnston C, Oberdörster G, Finkelstein JN (1999). Silica-induced chemokine expression in alveolar type II cells is mediated by TNF-alpha-induced oxidant stress. Am J Physiol, 276:L979–L988. PMID:10362723

85. Pilger A, Germadnik D, Schaffer A et al. (2000). 8-Hydroxydeoxyguanosine in leukocyte DNA and urine of quartz-exposed workers and patients with silicosis. Int Arch Occup Environ Health, 73:305–310. doi:10.1007/s004200000117 PMID:10963413

86. Dusinská M, Dzupinková Z, Wsólová L et al. (2006). Possible involvement of XPA in repair of oxidative DNA damage deduced from analysis of damage, repair and genotype in a human population study. Mutagenesis, 21:205–211.doi:10.1093/mutage/gel016 PMID:16613913

87. Sato T, Takeno M, Honma K et al. (2006). Heme oxygenase-1, a potential biomarker of chronic silicosis, attenuates silica-induced lung injury. Am J Respir Crit Care Med, 174:906–914.doi:10.1164/rccm.200508-1237OC PMID:16858012

88. Altindag ZZ, Baydar T, Isimer A, Sahin G (2003). Neopterin as a new biomarker for the evaluation of occupational exposure to silica. Int Arch Occup Environ Health, 76:318–322. PMID:12768284

89. Pelclová D, Fenclová Z, Kacer P et al. (2007). 8-isoprostane and leukotrienes in exhaled breath condensate in Czech subjects with silicosis. Ind Health, 45:766–774. doi:10.2486/indhealth.45.766 PMID:18212471

90. Saenger AK, Laha TJ, Edenfield MJ, Sadrzadeh SM (2007). Quantification of urinary 8-iso-PGF2alpha using liquid chromatography-tandem mass spectrometry and association with elevated troponin levels. Clin Biochem, 40:1297–1304.doi:10.1016/j.clinbiochem.2007.07.023 PMID:17854792

91. Nourooz-Zadeh J, Cooper MB, Ziegler D, Betteridge DJ (2005). Urinary 8-epi-PGF2alpha and its endogenous beta-oxidation products (2,3-dinor and 2,3-dinor-5,6-dihydro) as biomarkers of total body oxidative stress. Biochem Biophys Res Commun, 330:731–736.doi:10.1016/j.bbrc.2005.03.024 PMID:15809058

92. Orman A, Kahraman A, Cakar H et al. (2005). Plasma malondialdehyde and erythrocyte glutathione levels in workers with cement dust-exposure [corrected]. Toxicology, 207:15–20.doi:10.1016/j.tox.2004.07.021 PMID:15590118

93. Wang SX, Liu P, Wei MT et al. (2007). Roles of serum clara cell protein 16 and surfactant protein-D in the early diagnosis and progression of silicosis. J Occup Environ Med, 49:834–839.doi:10.1097/JOM.0b013e318124a927 PMID:17693780

94. Verma DK, Ritchie AC, Muir DC (2008). Dust content of lungs and its relationships to pathology, radiology and occupational exposure in Ontario hardrock miners. Am J Ind Med, 51:524–531.doi:10.1002/ajim.20589 PMID:18459150

95. Yucesoy B, Vallyathan V, Landsittel DP et al. (2001). Association of tumor necrosis factor-alpha and interleukin-1 gene polymorphisms with silicosis. Toxicol Appl Pharmacol, 172:75–82.doi:10.1006/taap.2001.9124 PMID:11264025

96. Yucesoy B, Vallyathan V, Landsittel DP et al. (2001). Polymorphisms of the IL-1 gene complex in coal miners with silicosis. Am J Ind Med, 39:286–291.doi:10.1002/1097-0274(200103)39:3<286::AID-AJIM1016>3.0.CO;2-7 PMID:11241561

97. Wang DJ, Yang YL, Xia QJ et al. (2006). [Study on association of interleukin-1 receptor antagonist(RA) gene + 2018 locus mutation with silicosis]. Wei Sheng Yan Jiu, 35:693–696. PMID:17290743

98. Corbett EL, Mozzato-Chamay N, Butterworth AE et al. (2002). Polymorphisms in the tumor necrosis factor-alpha gene promoter may predispose to severe silicosis in black South African miners. Am J Respir Crit Care Med, 165:690–693. PMID:11874815

99. Carlsten C, de Roos AJ, Kaufman JD et al. (2007). Cell markers, cytokines, and immune parameters in cement mason apprentices. Arthritis Rheum, 57:147–153.doi:10.1002/art.22483 PMID:17266079

100. Hussain SP, Harris CC (2006). p53 biological network: at the crossroads of the cellular-stress response pathway and molecular carcinogenesis. J Nippon Med Sch, 73:54–64. doi:10.1272/jnms.73.54 PMID:16641528

101. Wang L, Bowman L, Lu Y et al. (2005). Essential role of p53 in silica-induced apoptosis. Am J Physiol Lung Cell Mol Physiol, 288:L488–L496.doi:10.1152/ajplung.00123.2003 PMID:15557088

102. Weston A, Wolff MS, Morabia A (1998). True extended haplotypes of p53: indicators of breast cancer risk. Cancer Genet Cytogenet, 102:153–154. PMID:9546072

103. Otsuki T, Maeda M, Murakami S et al. (2007). Immunological effects of silica and asbestos. Cell Mol Immunol, 4:261–268. PMID:17764616

104. Otsuki T, Miura Y, Nishimura Y et al. (2006). Alterations of Fas and Fas-related molecules in patients with silicosis. Exp Biol Med (Maywood), 231:522–533. PMID:16636300

105. Hamzaoui A, Ammar J, Graïri H, Hamzaoui K (2003). Expression of Fas antigen and Fas ligand in bronchoalveolar lavage from silicosis patients. Mediators Inflamm, 12:209–214.doi:10.1080/09629350310001599648 PMID:14514471

106. Tomokuni A, Otsuki T, Isozaki Y et al. (1999). Serum levels of soluble Fas ligand in patients with silicosis. Clin Exp Immunol, 118:441–444.doi:10.1046/j.1365-2249.1999.01083.x PMID:10594565

107. Pepys J, Bernstein IL. Historical aspects of occupational asthma. In: Bernstein IL, Chan-Yeung M, Malo J-L, Bernstein DI, editors. Asthma in the workplace. 3rd ed. New York (NY): Taylor & Francis; 2006. p. 9–35.

108. Bello D, Herrick CA, Smith TJ et al. (2007). Skin exposure to isocyanates: reasons for concern. Environ Health Perspect, 115:328–335.doi:10.1289/ehp.9557 PMID:17431479

109. Straus DC, Cooley JD, Wong WC, Jumper CA (2003). Studies on the role of fungi in Sick Building Syndrome. Arch Environ Health, 58:475–478.doi:10.3200/AEOH.58.8.475-478 PMID:15259426

110. Mirabelli MC, Zock JP, Plana E et al. (2007). Occupational risk factors for asthma among nurses and related healthcare professionals in an international study. Occup Environ Med, 64:474–479.doi:10.1136/oem.2006.031203 PMID:17332135

111. Aron Y, Desmazes-Dufeu N, Matran R et al. (1996). Evidence of a strong, positive association between atopy and the HLA class II alleles DR4 and DR7. Clin Exp Allergy, 26:821–828.doi:10.1111/j.1365-2222.1996.tb00614.x PMID:8842557

112. Wikman H, Piirilä P, Rosenberg C et al. (2002). N-Acetyltransferase genotypes as modifiers of diisocyanate exposure-associated asthma risk. Pharmacogenetics, 12:227–233. doi:10.1097/00008571-200204000-00007 PMID:11927838

113. Woo JG, Assa'ad A, Heizer AB et al. (2003). The -159 C−>T polymorphism of CD14 is associated with nonatopic asthma and food allergy. J Allergy Clin Immunol, 112:438–444. doi:10.1067/mai.2003.1634 PMID:12897754

114. Hang LW, Hsia TC, Chen WC et al. (2003). Interleukin-10 gene -627 allele variants, not interleukin-I beta gene and receptor antagonist gene polymorphisms, are associated with atopic bronchial asthma. J Clin Lab Anal, 17:168–173.doi:10.1002/jcla.10088 PMID:12938145

115. Aynacioglu AS, Nacak M, Filiz A et al. (2004). Protective role of glutathione S-transferase P1 (GSTP1) Val105Val genotype in patients with bronchial asthma. Br J Clin Pharmacol, 57:213–217.doi:10.1046/j.1365-2125.2003.01975.x PMID:14748821

116. Fukunaga K, Asano K, Mao XQ et al. (2001). Genetic polymorphisms of CC chemokine receptor 3 in Japanese and British asthmatics. Eur Respir J, 17:59–63.doi:10.1183/09031936.01.17100590 PMID:11307756

117. Witte JS, Palmer LJ, O'Connor RD et al. (2002). Relation between tumour necrosis factor polymorphism TNFalpha-308 and risk of asthma. Eur J Hum Genet, 10:82–85. doi:10.1038/sj.ejhg.5200746 PMID:11896460

118. Higa S, Hirano T, Mayumi M et al. (2003). Association between interleukin-18 gene polymorphism 105A/C and asthma. Clin Exp Allergy, 33:1097–1102.doi:10.1046/j.1365-2222.2003.01739.x PMID:12911784

119. Hizawa N, Yamaguchi E, Konno S et al. (2002). A functional polymorphism in the RANTES gene promoter is associated with the development of late-onset asthma. Am J Respir Crit Care Med, 166:686–690.doi:10.1164/rccm.200202-090OC PMID:12204866

120. Silverman ES, Palmer LJ, Subramaniam V et al. (2004). Transforming growth factor-beta1 promoter polymorphism C-509T is associated with asthma. Am J Respir Crit Care Med, 169:214–219.doi:10.1164/rccm.200307-973OC PMID:14597484

121. Horne C, Quintana PJ, Keown PA et al. (2000). Distribution of DRB1 and DQB1 HLA class II alleles in occupational asthma due to western red cedar. Eur Respir J, 15:911–914.doi:10.1034/j.1399-3003.2000.15e17.x PMID:10853858

122. Piirilä P, Wikman H, Luukkonen R et al. (2001). Glutathione S-transferase genotypes and allergic responses to diisocyanate exposure. Pharmacogenetics, 11:437–445. doi:10.1097/00008571-200107000-00007 PMID:11470996

123. van der Pouw Kraan TC, van Veen A, Boeije LC et al. (1999). An IL-13 promoter polymorphism associated with increased risk of allergic asthma. Genes Immun, 1:61–65. doi:10.1038/sj.gene.6363630 PMID:11197307

124. Chan IH, Tang NL, Leung TF et al. (2007). Association of prostaglandin-endoperoxide synthase 2 gene polymorphisms with asthma and atopy in Chinese children. Allergy, 62:802–809.doi:10.1111/j.1398-9995.2007.01400.x PMID:17573729

125. Hizawa N, Maeda Y, Konno S et al. (2006). Genetic polymorphisms at FCER1B and PAI-1 and asthma susceptibility. Clin Exp Allergy, 36:872–876.doi:10.1111/j.1365-2222.2006.02413.x PMID:16839401

126. Thakkinstian A, McEvoy M, Minelli C et al. (2005). Systematic review and meta-analysis of the association between β2-adrenoceptor polymorphisms and asthma: a HuGE review. Am J Epidemiol, 162:201–211.doi:10.1093/aje/kwi184 PMID:15987731

127. Contopoulos-Ioannidis DG, Manoli EN, Ioannidis JP (2005). Meta-analysis of the association of beta2-adrenergic receptor polymorphisms with asthma phenotypes. J Allergy Clin Immunol, 115:963–972.doi:10.1016/j.jaci.2004.12.1119 PMID:15867853

128. Raby BA, Silverman EK, Kwiatkowski DJ et al. (2004). ADAM33 polymorphisms and phenotype associations in childhood asthma. J Allergy Clin Immunol, 113:1071–1078.doi:10.1016/j.jaci.2004.03.035 PMID:15208587

129. Kabesch M (2010). Novel asthma-associated genes from genome-wide association studies: what is their significance? Chest, 137:909–915.doi:10.1378/chest.09-1554 PMID:20371526

130. Mathias RA, Grant AV, Rafaels N et al. (2010). A genome-wide association study on African-ancestry populations for asthma. J Allergy Clin Immunol, 125:336–346,e4. doi:10.1016/j.jaci.2009.08.031 PMID:19910028

131. Li X, Howard TD, Zheng SL et al. (2010). Genome-wide association study of asthma identifies RAD50-IL13 and HLA-DR/DQ regions. J Allergy Clin Immunol, 125:328–335, e11.doi:10.1016/j.jaci.2009.11.018 PMID: 20159242

132. Willis-Owen SA, Cookson WO, Moffatt MF (2009). Genome-wide association studies in the genetics of asthma. Curr Allergy Asthma Rep, 9:3–9.doi:10.1007/s11882-009-0001-x PMID:19063818

133. Himes BE, Hunninghake GM, Baurley JW et al. (2009). Genome-wide association analysis identifies PDE4D as an asthma-susceptibility gene. Am J Hum Genet, 84:581–593.doi:10.1016/j.ajhg.2009.04.006 PMID:19426955

134. Demchuk E, Yucesoy B, Johnson VJ et al. (2007). A statistical model for assessing genetic susceptibility as a risk factor in multifactorial diseases: lessons from occupational asthma. Environ Health Perspect, 115:231–234.doi:10.1289/ehp.8870 PMID:17384770

135. Pharoah PD, Antoniou A, Bobrow M et al. (2002). Polygenic susceptibility to breast cancer and implications for prevention. Nat Genet, 31:33–36.doi:10.1038/ng853 PMID:11984562

136. Sharples LD, Tamm M, McNeil K et al. (1996). Development of bronchiolitis obliterans syndrome in recipients of heart-lung transplantation–early risk factors. Transplantation, 61:560–566.doi:10.1097/00007890-199602270-00008 PMID:8610381

137. Kanwal R (2008). Bronchiolitis obliterans in workers exposed to flavoring chemicals. Curr Opin Pulm Med, 14:141–146.doi:10.1097/MCP.0b013e3282f52478 PMID:18303424

138. Hubbs AF, Goldsmith WT, Kashon ML et al. (2008). Respiratory toxicologic pathology of inhaled diacetyl in sprague-dawley rats. Toxicol Pathol, 36:330–344.doi:10.1177/0192623307312694 PMID:18474946

139. Epler GR (2001). Bronchiolitis obliterans organizing pneumonia. Arch Intern Med, 161:158–164.doi:10.1001/archinte.161.2.158 PMID:11176728

140. Bergmann M, Tiroke A, Schäfer H et al.; (1998). Gene expression of profibrotic mediators in bronchiolitis obliterans syndrome after lung transplantation. Scand Cardiovasc J, 32:97–103.doi:10.1080/14017439850140247 PMID:9636965

141. Lu KC, Jaramillo A, Lecha RL et al. (2002). Interleukin-6 and interferon-gamma gene polymorphisms in the development of bronchiolitis obliterans syndrome after lung transplantation. Transplantation, 74:1297–1302.doi:10.1097/00007890-200211150-00017 PMID:12451269

142. Snyder LD, Hartwig MG, Ganous T et al. (2006). Cytokine gene polymorphisms are not associated with bronchiolitis obliterans syndrome or survival after lung transplant. J Heart Lung Transplant, 25:1330–1335.doi:10.1016/j.healun.2006.07.001 PMID:17097497

143. Hildebrandt GC, Granell M, Urbano-Ispizua A et al. (2008). Recipient NOD2/CARD15 variants: a novel independent risk factor for the development of bronchiolitis obliterans after allogeneic stem cell transplantation. Biol Blood Marrow Transplant, 14:67–74.doi:10.1016/j.bbmt.2007.09.009 PMID:18158963

144. Mannino DM, Buist AS (2007). Global burden of COPD: risk factors, prevalence, and future trends. Lancet, 370:765–773. doi:10.1016/S0140-6736(07)61380-4 PMID:17765526

145. Hnizdo E, Sullivan PA, Bang KM, Wagner G (2002). Association between chronic obstructive pulmonary disease and employment by industry and occupation in the US population: a study of data from the Third National Health and Nutrition Examination Survey. Am J Epidemiol, 156:738–746.doi:10.1093/aje/kwf105 PMID:12370162

146. Sampsonas F, Karkoulias K, Kaparianos A, Spiropoulos K (2006). Genetics of chronic obstructive pulmonary disease, beyond a1-antitrypsin deficiency. Curr Med Chem, 13:2857–2873.doi:10.2174/092986706778521922 PMID:17073633

147. Chappell S, Daly L, Morgan K et al. (2006). Cryptic haplotypes of SERPINA1 confer susceptibility to chronic obstructive pulmonary disease. Hum Mutat, 27:103–109. doi:10.1002/humu.20275 PMID:16278826

148. Demeo DL, Mariani TJ, Lange C et al. (2006). The SERPINE2 gene is associated with chronic obstructive pulmonary disease. Am J Hum Genet, 78:253–264. doi:10.1086/499828 PMID:16358219

149. Joos L, He JQ, Shepherdson MB et al. (2002). The role of matrix metalloproteinase polymorphisms in the rate of decline in lung function. Hum Mol Genet, 11:569–576. doi:10.1093/hmg/11.5.569 PMID:11875051

150. Minematsu N, Nakamura H, Tateno H et al. (2001). Genetic polymorphism in matrix metalloproteinase-9 and pulmonary emphysema. Biochem Biophys Res Commun, 289:116–119.doi:10.1006/bbrc.2001.5936 PMID:11708786

151. Zhou M, Huang SG, Wan HY et al. (2004). Genetic polymorphism in matrix metalloproteinase-9 and the susceptibility to chronic obstructive pulmonary disease in Han population of south China. Chin Med J (Engl), 117:1481–1484. PMID:15498369

152. Ito I, Nagai S, Handa T et al. (2005). Matrix metalloproteinase-9 promoter polymorphism associated with upper lung dominant emphysema. Am J Respir Crit Care Med, 172:1378–1382.doi:10.1164/rccm.200506-953OC PMID:16126934

153. Tesfaigzi Y, Myers OB, Stidley CA et al. (2006). Genotypes in matrix metalloproteinase 9 are a risk factor for COPD. Int J Chron Obstruct Pulmon Dis, 1:267–278. PMID:18046864

154. Korytina GF, Akhmadishina LZ, Ianbaeva DG, Viktorova TV (2008). [Polymorphism in promoter regions of matrix metalloproteinases (MMP1, MMP9, and MMP12) in chronic obstructive pulmonary disease patients]. Genetika, 44:242–249. PMID:18619044

155. Hersh CP, Demeo DL, Lange C et al. (2005). Attempted replication of reported chronic obstructive pulmonary disease candidate gene associations. Am J Respir Cell Mol Biol, 33:71–78.doi:10.1165/rcmb.2005-0073OC PMID:15817713

156. Hirano K, Sakamoto T, Uchida Y et al. (2001). Tissue inhibitor of metalloproteinases-2 gene polymorphisms in chronic obstructive pulmonary disease. Eur Respir J, 18:748–752.doi:10.1183/09031936.01.00102101 PMID:11757622

157. van Diemen CC, Postma DS, Vonk JM et al. (2005). A disintegrin and metalloprotease 33 polymorphisms and lung function decline in the general population. Am J Respir Crit Care Med, 172:329–333.doi:10.1164/rccm.200411-1486OC PMID:15879414

158. Küçükaycan M, Van Krugten M, Pennings HJ et al. (2002). Tumor necrosis factor-alpha +489G/A gene polymorphism is associated with chronic obstructive pulmonary disease. Respir Res, 3:29.doi:10.1186/rr194 PMID:12537602

159. Seifart C, Dempfle A, Plagens A et al. (2005). TNF-alpha-, TNF-beta-, IL-6-, and IL-10-promoter polymorphisms in patients with chronic obstructive pulmonary disease. Tissue Antigens, 65:93–100.doi:10.1111/j.1399-0039.2005.00343.x PMID:15663746

160. Sakao S, Tatsumi K, Igari H et al. (2002). Association of tumor necrosis factor-alpha gene promoter polymorphism with low attenuation areas on high-resolution CT in patients with COPD. Chest, 122:416–420. doi:10.1378/chest.122.2.416 PMID:12171811

161. Chierakul N, Wongwisutikul P, Vejbaesya S, Chotvilaiwan K (2005). Tumor necrosis factor-alpha gene promoter polymorphism is not associated with smoking-related COPD in Thailand. Respirology, 10:36–39.doi:10.1111/j.1440-1843.2005.00626.x PMID:15691236

162. Jiang L, He B, Zhao MW et al. (2005). Association of gene polymorphisms of tumour necrosis factor-alpha and interleukin-13 with chronic obstructive pulmonary disease in Han nationality in Beijing. Chin Med J (Engl), 118:541–547. PMID:15820084

163. Gingo MR, Silveira LJ, Miller YE et al. (2008). Tumour necrosis factor gene polymorphisms are associated with COPD. Eur Respir J, 31:1005–1012.doi:10.1183/09031936.00100307 PMID:18256059

164. Ruse CE, Hill MC, Tobin M et al. (2007). Tumour necrosis factor gene complex polymorphisms in chronic obstructive pulmonary disease. Respir Med, 101:340–344.doi:10.1016/j.rmed.2006.05.017 PMID:16867312

165. Brøgger J, Steen VM, Eiken HG et al. (2006). Genetic association between COPD and polymorphisms in TNF, ADRB2 and EPHX1. Eur Respir J, 27:682–688.doi:10.1183/09031936.06.00057005 PMID:16585076

166. Yucesoy B, Kurzius-Spencer M, Johnson VJ et al. (2008). Association of cytokine gene polymorphisms with rate of decline in lung function. *J Occup Environ Med*, 50:642–648.doi:10.1097/JOM.0b013e31816515e1 PMID:18545091

167. Weston A, Harris CC. Chemical Carcinogenesis In: Holland JF, Frei E, Bast R et al., editors. Cancer Medicine, 7th ed. Ontario, Canada: B.C. Decker Inc.; 2006. p. 1–13.

168. Hayashi SI, Watanabe J, Nakachi K, Kawajiri K (1991). PCR detection of an A/G polymorphism within exon 7 of the CYP1A1 gene. *Nucleic Acids Res*, 19:4797.doi:10.1093/nar/19.17.4797 PMID:1891387

169. Cantlay AM, Lamb D, Gillooly M et al. (1995). Association between the CYP1A1 gene polymorphism and susceptibility to emphysema and lung cancer. *Clin Mol Pathol*, 48:M210–M214.doi:10.1136/mp.48.4.M210 PMID:16696009

170. Ishii T, Matsuse T, Teramoto S et al. (1999). Glutathione S-transferase P1 (GSTP1) polymorphism in patients with chronic obstructive pulmonary disease. *Thorax*, 54:693–696.doi:10.1136/thx.54.8.693 PMID:10413721

171. Cheng SL, Yu CJ, Chen CJ, Yang PC (2004). Genetic polymorphism of epoxide hydrolase and glutathione S-transferase in COPD. *Eur Respir J*, 23:818–824.doi:10.1183/09031936.04.00104904 PMID:15218992

172. Vibhuti A, Arif E, Deepak D et al. (2007). Genetic polymorphisms of GSTP1 and mEPHX correlate with oxidative stress markers and lung function in COPD. *Biochem Biophys Res Commun*, 359:136–142.doi:10.1016/j.bbrc.2007.05.076 PMID:17532303

173. Cha SI, Kang HG, Choi JE et al. (2009). SERPINE2 polymorphisms and chronic obstructive pulmonary disease. *J Korean Med Sci*, 24:1119–1125.doi:10.3346/jkms.2009.24.6.1119 PMID:19949669

174. Kim WJ, Hersh CP, DeMeo DL et al. (2009). Genetic association analysis of COPD candidate genes with bronchodilator responsiveness. *Respir Med*, 103:552–557.doi:10.1016/j.rmed.2008.10.025 PMID:19111454

175. Cho MH, Boutaoui N, Klanderman BJ et al. (2010). Variants in FAM13A are associated with chronic obstructive pulmonary disease. *Nat Genet*, 42:200–202.doi:10.1038/ng.535 PMID:20173748

176. Boezen HM (2009). Genome-wide association studies: what do they teach us about asthma and chronic obstructive pulmonary disease? *Proc Am Thorac Soc*, 6:701–703.doi:10.1513/pats.200907-058DP PMID:20008879

177. Wang X, Li L, Xiao J et al. (2009). Association of ADAM33 gene polymorphisms with COPD in a northeastern Chinese population. *BMC Med Genet*, 10:132.doi:10.1186/1471-2350-10-132 PMID:20003279

178. Sato H, Silveira L, Fingerlin T et al. (2007). TNF polymorphism and bronchoalveolar lavage cell TNF-alpha levels in chronic beryllium disease and beryllium sensitization. *J Allergy Clin Immunol*, 119:687–696.doi:10.1016/j.jaci.2006.10.028 PMID:17208287

179. McCanlies EC, Schuler CR, Kreiss K et al. (2007). TNF-alpha polymorphisms in chronic beryllium disease and beryllium sensitization. *J Occup Environ Med*, 49:446–452.doi:10.1097/JOM.0b013e31803b9499 PMID:17426528

180. Warburton D, Gauldie J, Bellusci S, Shi W (2006). Lung development and susceptibility to chronic obstructive pulmonary disease. *Proc Am Thorac Soc*, 3:668–672.doi:10.1513/pats.200605-122SF PMID:17065371

181. Daheshia M (2005). Therapeutic inhibition of matrix metalloproteinases for the treatment of chronic obstructive pulmonary disease (COPD). *Curr Med Res Opin*, 21:587–593.doi:10.1185/030079905X41417 PMID:15899108

UNIT 5.
APPLICATION OF BIOMARKERS TO DISEASE

CHAPTER 22.

Neurodegenerative diseases

Harvey Checkoway, Jessica I. Lundin, and Samir N. Kelada

Summary

Degenerative diseases of the nervous system impose substantial medical and public health burdens on populations throughout the world. Alzheimer's disease (AD), Parkinson's disease (PD), and amyotrophic lateral sclerosis (ALS) are three of the major neurodegenerative diseases. The prevalence and incidence of these diseases rise dramatically with age; thus the number of cases is expected to increase for the foreseeable future as life spans in many countries continue to increase. Causal contributions from genetic and environmental factors are, with some exceptions, poorly understood. Nonetheless, molecular epidemiology approaches have proven valuable for improving disease diagnoses, characterizing disease prognostic factors, identifying high-risk genes for familial neurodegenerative diseases, investigating common genetic variants that may predict susceptibility for the non-familial forms of these diseases, and for quantifying environmental exposures. Incorporation of molecular techniques, including genomics, proteomics, and measurements of environmental toxicant body burdens into epidemiologic research, offer considerable promise for enhancing progress on characterizing pathogenesis mechanisms and identifying specific risk factors, especially for the non-familial forms of these diseases. In this chapter, brief overviews are provided of the epidemiologic features of PD, AD, and ALS, as well as illustrative examples in which molecular epidemiologic approaches have advanced knowledge on underlying disease mechanisms and risk factors that might lead to improved medical management and ultimately disease prevention. The chapter concludes with some recommendations for future molecular epidemiology research.

Introduction

Increasingly, epidemiologic research on neurodegenerative diseases has applied molecular techniques to identify host susceptibility factors to elucidate more clearly pathogenesis mechanisms, and to characterize exposures to potential environmental risk factors. Advances in molecular genetics and exposure measurement have facilitated expanded use of these techniques. Largely due to the ease and availability of genotyping assays, studies of candidate gene variants have been the most common applications. In this chapter, illustrations of the contributions of molecular epidemiology related primarily to elucidating disease pathogenesis processes and identifying etiologic factors will be presented. The focus will be on Alzheimer's disease (AD), Parkinson's disease (PD) and amyotrophic lateral sclerosis (ALS), as they share some common clinical, pathological and epidemiologic features. Other chronic neurological disorders, such as multiple sclerosis and Huntington's disease, are also significant public health concerns, but will not be discussed in the interest of brevity.

As background, brief descriptions of the clinical and pathological features of the three disorders will be provided, as well as summaries of epidemiologic aspects, including the relative contributions of genetics and the environment. No attempt to provide comprehensive reviews of these topics will be made, as they would be far beyond the scope of this chapter. Examples of various types of molecular epidemiologic approaches applied to investigations of AD, PD and ALS will be presented in the second section of this chapter.

Context and public health significance

Alzheimer's disease

Alzheimer's disease (AD) is the most common neurodegenerative disease. It also represents the most frequent cause of dementia, accounting for roughly half of all cases. The prevalence of AD is roughly 30% among people 85 years and older. Incidence rates climb steeply from 0.5% per year for ages 65 to 75 to 6–8% per year for ages 85 and up. AD onset is rare before the age of 50, except in cases of familial AD, which comprise roughly 5–10% of cases (1).

The primary clinical manifestation of AD is dementia, which is an accelerated loss of cognitive function beyond that due to normal aging. Alterations in mood and behaviour often accompany the onset of dementia, followed by memory loss, disorientation and aphasia. The hippocampus and cerebral cortex are preferentially and severely affected in AD. Pathologically, senile or neuritic plaques and neurofibrillary tangles (NFTs) are the two characteristic lesions in affected tissues (2). Neuritic plaques in blood vessels and neurons of the hippocampus are primarily composed of amyloid β (Aβ) peptide aggregates. The second pathological hallmark, NFTs, are filamentous bundles comprised of abnormal (hyper-phosphorylated) tau proteins that accumulate in the cytoplasm of affected neurons. Tau protein is normally involved in nutrient transport along neuronal axons. Various lines of evidence indicate that AD develops primarily as a result of an "amyloid cascade" (i.e. an imbalance in the production and clearance of Aβ is the central mechanism) (3). Aggregation of hyperphosphorylated tau proteins leading to tangles may also contribute to this cascade mechanism. Other potentially relevant disease mechanisms include: microvascular damage, leading to diminished blood flow and nutrient deficiency to brain cells; oxidative stress; inflammation; and mitochondrial dysfunction (2).

Family studies have established that genetic factors play a substantial role in AD, especially in younger-onset cases (<65 years). Familial AD has an autosomal dominant inheritance pattern. Three mutations in genes encoding proteins involved in amyloid plaque formation, the amyloid precursor protein (APP), presenilin-1, and presenilin-2 genes, have been identified as causal genes for early-onset AD (4–6). Non-familial AD, typically defined as having an onset at age 65 years or older, accounts for most of cases. Non-familial AD has been associated most consistently with the ε4 allele of the *apolipoprotein* gene (*ApoE-ε4*), which is a very low-density lipoprotein carrier that is required for Aβ deposition (7). Carriers of the *ApoE-ε4* allele have reduced AD ages at onset, with 3-fold and 15-fold risk excesses observed in heterozygotes and homozygotes, respectively (8,9). Numerous other candidate genes have been investigated as AD susceptibility factors, such as sortilin-related receptor-1 gene (10), but no strong or consistent findings have emerged.

Increasing age is a clear risk factor for non-familial AD, and rates are generally higher in women than in men (6,10). Other factors that have been investigated in relation to AD risk include: cardiovascular diseases (largely motivated by the link of lipid metabolism with *ApoE-ε4*) (11), head trauma (12), smoking (13), dietary antioxidants and fats (14), alcohol (15), occupational exposures to solvents (16,17), electromagnetic fields (18), educational status (19), and occupational exposures to pesticides (20,21). Epidemiologic evidence has been mixed thus far, as exemplified by contradictory findings for cigarette smoking (22). It is possible, yet remains to be established conclusively, whether genetic factors account for the majority of the population attributable risk for AD.

Parkinson's disease

Parkinson's disease (PD), the second most common neurodegenerative disease, is a movement disorder whose cardinal clinical features are rest tremor, rigidity, bradykinesia and postural instability (23). PD is relatively rare before age 50, after which incidence and prevalence rise sharply through the eighth decade of life. Epidemiologic surveys, mainly in western countries, indicate a small (20–30%) excess risk in men. Annual incidence rates of 10–15 per 100 000 have been noted in most surveys worldwide. Prevalence may reach 2% in persons aged 65 years and older (24).

The underlying cause of PD is a loss of dopamine-producing neurons of the mid-brain substantia nigra (SN). PD pathogenesis involves complex interactions among several mechanisms, including abnormal protein aggregation and deficient clearance of aggregates, altered dopamine metabolism, impaired mitochondrial function, oxidative stress, inflammation, necrosis and accelerated apoptosis (25). Intracellular deposits of aggregated α synuclein, ubiquitin, and other proteins (known as Lewy bodies) found in many surviving neuronal populations are considered to be the pathologic characteristic of PD (26,27). Whether Lewy bodies are themselves neurotoxic, or represent the end product of cellular defence mechanisms to sequester toxic abnormal proteins, remains to be determined.

Similar to Alzheimer's disease, epidemiologic differences in early-onset (< 50 years) and late-onset PD have been described. Genetic factors, especially specific causal mutations, appear to be more prominent in early-onset PD, although the distinctions are by no means absolute. Kindred studies of heavily affected families have identified at least five genetic loci for PD (7,28). The initial discoveries were mutations of the gene encoding the α-synuclein protein that have been related to autosomal dominant early-onset PD, typified by rapid disease onset and progression (29). The functional consequences of mutations in these genes are incompletely understood, although abnormal brain protein aggregation and clearance appears to be a common feature. Mutations in the leucine-rich repeat kinase 2 (*LRRK2* or *PARK8* gene), first identified from kindred studies in Japan (30) and subsequently confirmed in Europe (31) and North America (32), have also been associated with typical late-onset PD, and thus may also contribute to risk for non-familial PD (28). Identified mutations in other genes include: parkin (*PARK2*), PTEN-induced putative kinase I (*PINK1* or *PARK6*) and DJ-1 (*DJ-1* or *PARK7*), all of which follow a recessive inheritance mode (7).

Candidate gene studies for late-onset non-familial PD have explored associations with the same genes related to familial PD. In general, the rare causal mutations observed for familial PD have not been associated consistently with non-familial disease. Extensive efforts have also been undertaken to identify common variants of biologically-based candidate genes that may confer PD susceptibility, either independently or in combination with host or environmental factors. These include variants of genes related to the metabolism of dopamine and toxic environmental chemicals, and to presumed PD pathogenesis mechanisms (e.g. oxidative stress). Perhaps not surprisingly, numerous associations have been observed, yet attempts at replication have been largely disappointing. An illustration is the inconsistent pattern of results for the gene encoding the enzyme monoamine oxidase B (MAO-B) that catabolizes dopamine (33–35).

Apart from older age, the most consistent epidemiologic observation has been an inverse relation between cigarette smoking and PD, with smokers having approximately half the rate as never smokers, and strong evidence for an inverse dose–response ("protective") effect with duration and pack-years smoked (36–39). The reduced risk among smokers does not appear to be due to selective survival bias. A biochemical basis may be the lowering of MAO-B enzyme activity in the brain, and consequent reduced dopamine catabolism (40,41). Alternatively, aversion to novelty-seeking behaviour, such as smoking, by persons who ultimately develop PD may explain the relation with smoking. Inverse PD risk associations have also been reported for caffeine (37,42) and non-steroidal anti-inflammatory medications (43,44), although the evidence is less consistent than for smoking. Additionally, family history of PD (45) and history of severe head trauma (46,47) have been related to elevated PD risks.

The discovery in the early 1980s of PD among intravenous drug users who had injected a synthetic heroin contaminated with 1-methyl-4-phenyl-1,2,3,6-tetrahydropyridine (MPTP), prompted great interest in the possibility that there are important etiologic roles of environmental toxicants (48). Induction of PD by MPTP in experimental animals and recognition of the chemical structural similarity of MPTP provided strong impetus for a focus on pesticides (49). Occupational exposures to pesticides have been associated with elevated risk in some studies (50–54), although consistent associations with specific pesticides have not been identified. Metals, especially manganese, have been implicated as risk factors in several epidemiologic studies (55,56). Epidemiologic findings for PD risk among welders, whose jobs entail chronic exposures to various metal mixtures including manganese, have been inconsistent (57–60). There is only limited evidence supporting associations with solvents and other environmental chemicals (61–63).

Amyotrophic lateral sclerosis (ALS)

Amyotrophic lateral sclerosis (ALS) is a disease of the motor neurons of the anterior horns of the spinal cord and motor neurons in the cerebral cortex. Similar to AD and PD, there are both familial and non-familial forms of ALS, with the familial ALS accounting for about 10% of cases. The incidence of non-familial ALS is approximately 1–2 cases per 100 000 per year, and appears to be slightly more common in men (64). ALS onset usually occurs in the middle to later years of life, and the incidence rises with increasing age. ALS is generally a rapidly fatal condition within two to three years of onset (65).

Excitotoxicity mediated by glutamate and elevated calcium ion (Ca^{2+}) is considered to be a major pathogenesis mechanism in the neuronal death that occurs in ALS (66). As a consequence of neuronal cell death, neuronal muscle atrophy occurs, resulting in diminished muscle strength and bulk, fasciculations, and hyperreflexia. Effects on respiratory muscles can lead to pulmonary infection, and eventually amyotrophy leads to paralysis and death. The histopathology of ALS is characterized by intracytoplasmic inclusion bodies composed of neurofilaments and spheroids containing ubiquitinated copper and zinc superoxide dismutase (CuZnSOD) or SOD1, an enzyme that catalyses the conversion of the superoxide free radical to hydrogen peroxide.

Mutations in the *SOD1* gene are present in roughly 20% of familial cases and perhaps as much as 10% of non-familial ALS (7,67). *SOD1* mutations may cause a reduced capacity to counteract oxidative stress. Additionally, mutations may result in mis-folded *SOD1* proteins that aggregate and form toxic inclusion bodies, reminiscent of the presumed mechanisms involved in AD and PD pathogenesis (68). Mutations in a second gene associated with familial ALS, *alsin* (*ALS2*), has been identified in juvenile-onset recessive PD (69). Investigations of other genetic variants in non-familial ALS have not yielded consistent findings, although several potentially promising candidate gene loci have been identified by genetic linkage studies (70).

Potentially important etiologic roles of environmental factors are indicated, at least by default, by the absence of convincing support for ALS being a predominantly genetically-determined disease. Various environmental risk factors have been investigated, including smoking (71); pesticides and other agricultural chemicals (72–73); heavy metals, especially lead (74); electric shock; and electromagnetic fields (75). Reasonably consistent yet modest excess risks have been observed among cigarette smokers (71,76). In addition, reports of apparent case clusters of ALS among United States military personnel deployed in the first Gulf War prompted epidemiologic studies that are suggestive of associations. Exposures to pesticides, petroleum combustion products and mycotoxins have been speculated as the causative agents among Gulf War veterans, but none have been established (77).

Examples

Application of molecular epidemiology methods will be illustrated with some selected examples from the literature on neurodegenerative diseases. These examples span a range of applications, including molecular methods for biomonitoring of environmental neurotoxicants, candidate susceptibility investigation, gene–gene and gene–environment interaction analyses, and the more recently developed method of proteomics profiling to identify early disease markers.

Example 1: Occupational lead exposure and risk of ALS

An etiologic association between occupational lead exposure and ALS has been suggested primarily from case–control studies in which exposures have been based on the classification of self-reported jobs. To investigate whether lead is causally related to ALS with more precise exposure characterization, a population-based case-control study was conducted in the New England region of the USA (78). Study subjects were 109 ALS cases and age/sex/region-matched controls identified by random digit dialling. Exposures to lead were assessed as lifetime number of days worked in lead-exposed occupations, blood lead levels, and bone lead concentrations determined by X-ray fluorescence. Blood and bone lead measurements were obtained for 107 cases, but only for 41 controls. Blood lead levels represent recent exposure (within three months, or the lifespan of red blood cells that store the majority of blood lead). Lead concentrations were measured in patella and tibia bones, representing the shorter (3–5 years) and longer-term (10–15 years) body storage compartments.

The most striking finding from this study was a monotonically increasing exposure–response relation with cumulative lifetime lead-exposure work days, with a 2.3 relative risk estimate found for the highest exposure category (≥ 2000 days) compared to 0 days. Comparisons of cases' and controls' blood and bone lead levels (Table 22.1) yielded slightly larger lead body burdens among cases. Overall, the study findings offer some support to the hypothesis that lead is a risk factor for ALS.

Several features of this example warrant comment. Measurement of exposure biomarkers, as opposed to reliance strictly on questionnaire response data, has a theoretical advantage of improved precision. However, it should also be realized that biological measurements may not necessarily be more valid than questionnaire data, even in situations where the precision of measurement techniques is well established, such as for blood and bone lead. Biological monitoring of exposure does have the advantage of taking into account multiple sources of exposures (i.e. occupational and non-occupational), which is both a strength and a limitation. The strength is that it provides a more complete picture of exposure levels than does, e.g., an occupational history. The limitation is that it can be difficult to identify specific exposure sources from biomonitoring if the goal of the epidemiologic study is intervention to minimize or eliminate exposure. Additionally, the low participation rate among controls in this study was perhaps not surprising, given that presumably healthy controls would have less motivation to undergo biological sampling, albeit relatively non-invasive.

Table 22.1. Blood and bone lead levels in ALS cases and controls

Lead measurement (units)	Cases	Controls
Blood (µg/dl)	5.2 ± 0.4[†]	3.4 ± 0.4
Patella (µg/g)	20.5 ± 2.1	16.7 + 2.0
Tibia (µg/g)	14.9 ± 1.6	11.1 ± 1.6

[†] Mean (± standard error). Adapted from (78).

Example 2: Alpha-synuclein (SNCA) promoter region variants in PD

As mentioned earlier, mutations in the alpha synuclein (*SNCA*) gene have been associated with increased risks of PD in familial, and to a lesser extent non-familial PD. Epidemiologic studies of an apparently functionally important dinucleotide repeat in the *SNCA* promoter region (Rep1) have provided mixed evidence for an association with PD risk (79,80). A pooled analysis of Rep1 variability was performed that combined data for 2692 PD cases and 2652 unrelated controls from 11 study centres (in six western European countries, the USA and Australia) (81). Common genotyping protocols and quality control assessments, including selective re-genotyping, were incorporated into the study to minimize laboratory bias.

Analyses were performed for the three most common Rep1 base pair repeat lengths: 259, 261 and 263. The results of analysis comparing the Rep1 263 base pair genotype versus all others are summarized in Table 22.2. Overall, there was a modest yet statistically significant association (OR = 1.43; 95% CI = 1.22–1.69).

Notably, PD risk was elevated among carriers of 263 base pair length in each of the 11 studies. Moreover, the findings showed positive associations for both dominant (OR = 1.44; 95% CI = 1.21–1.70) and recessive (OR = 2.46; 95% CI

Table 22.2. Associations of Parkinson's disease with the Rep1 263 base pair repeat of the alpha synuclein gene promoter region

Group	No. cases/controls	OR (95% CI)[†] P-value
All subjects	2686/2454	1.43 (1.22–1.69) <0.001
Negative family history	2241/676	1.33 (1.03–1.72) 0.03
Positive family history	413/38	1.67 (0.51–5.50) 0.40
Age ≤68	1361/1317	1.47 (1.17–1.84) 0.001
Age >68	1325/1137	1.31 (1.03–1.66) 0.03
Women	1083/1205	1.33 (1.06–1.67) 0.01
Men	1603/1249	1.54 (1.22–1.95) <0.001

[†] Odds ratio (95% confidence interval) for 263 versus other base pair lengths
Adapted from (81). Copyright © (2006) American Medical Association. All rights reserved.

= 0.95–6.37) inheritance models, and varied little with respect to age, gender or family history of PD. There was a slight, although less consistently noted, reduced risk (OR = 0.86; 95% CI = 0.79–0.94) related to the 259 base pair repeat length. The 261 base pair repeat length was unrelated to PD risk.

This study exemplifies the approach of focusing on a single candidate gene that has a plausible relation to the phenotype of interest. By combining data from multiple studies and following standardized laboratory protocols, the investigators were able to achieve greater statistical precision than was possible in any previous study while maintaining a high level of validity. As with all studies of single gene associations with complex diseases, this study could not address the interactions among genes or with environmental factors. In fact, the investigators estimated that variability of the *SNCA* Rep1 promoter region may account for a population attributable risk of only 3%, but might be a component cause of a constellation of genetic and environmental factors that confer substantially larger population effects.

Example 3: Interaction of estrogen receptor and ApoE genetic polymorphisms in Alzheimer's disease

A study was conducted in Italy of the single and combined associations of two candidate genes, *ApoE-ε4* and the estrogen receptor-α (ER-α) gene, in a case–control study of 131 non-familial AD cases and 109 age-matched controls, comprised mostly of cases' spouses (82). The rationale for selecting these candidate genes was provided by previous studies demonstrating strong risks related to *ApoE-ε4*, and suggestions from the literature, albeit controversial, that estrogen may protect against dementia (82,83). Estrogenic activity is known to be mediated by α and β estrogen receptors; reduced AD risks among users of estrogen replacement therapy has been reported previously (83). Also, previous research indicated variable associations between two *ER-α* intronic single nucleotide polymorphisms (SNPs) in intron 1, rs2234693 [-397 T→C] and rs9340799 [-351 A→G], and AD (84,85).

Consistent with previous literature, *ApoE-ε4* carrier status was strongly associated with AD in both women and men, as indicated by observed relative risk estimates OR = 6.48 (95% CI = 2.99–14.0) and 4.67 (95% CI = 1.98–11.0), respectively. No associations with AD were detected for either of the ER-α SNPs individually or in combination. In contrast, analysis of the joint effects of *ApoE-ε4* and the *ER-α* intronic alleles revealed evidence for interactive effects, as shown in Table 22.3. The strongest associations were observed in women for the combinations of *ApoE-ε4/*-397 T allele (OR = 7.24; 95% CI = 2.22–23.6) and *ApoE-ε4/*-351 A allele (OR = 8.33; 95% CI = 1.73–40.1). Evidence of combined gene effects was considerably weaker in men. Notwithstanding the relatively small sample size of the study, this pattern of results could be interpreted as a gender-specific interaction between an established high risk allele, *ApoE-ε4*, and *ER-α* gene intronic variants, where the presence of the latter enhances the effects of *ApoE-ε4*. Moreover, the results from this study are strengthened insofar as they replicate findings from a previous study (85).

As this example illustrates, investigation of associations with combinations of gene variants, rather than a focus on a single gene, can provide further etiologic insight.

Table 22.3. Alzheimer's disease risk in relation to combinations of ApoE-ε4 and estrogen receptor α (ER-α) intronic alleles

	Women	Men
ApoE allele	OR (95%CI)[†] P-value	OR (95%CI) P-value
ε4+	6.48 (2.99–14.0) <0.001	4.67 (1.98–11.0) <0.001
ε4-	Reference	Reference
ApoE/*ER*-α intron 1 -397[‡]		
ε4+/TT or TC	7.24 (2.22–23.6) 0.001	3.47 (0.71–16.9) 0.125
ε4+/CC	2.00 (0.43–9.26) 0.375	0.80 (0.10–6.35) 0.833
ε4-/TT or TC	0.76 (0.27–2.12) 0.603	0.51 (0.12–2.07) 0.343
ε4-/CC	Reference	Reference
ApoE/*ER*-α intron 1 -351[‡]		
ε4+/AA or AG	8.33 (1.73–40.0) 0.008	2.31 (0.44–12.1) 0.320
ε4+/GG	1.25 (0.18–8.44) 0.819	0.60 (0.05–6.80) 0.680
ε4-/AA or AG	0.73 (0.18–2.96) 0.656	0.37 (0.08–1.69) 0.199
ε4-/GG	Reference	Reference

[†] Odds ratio (95% confidence interval)
[‡] *ER*-α intron 1 -397 T/C (rs2234693); *ER*-α intron 1 -351 A/G allele (rs9340799)
Adapted from (82).

This approach can be especially informative when one of the genes under study bears a predictable relation to disease risk, as is the case for *ApoE* and AD.

Example 4: Interaction of pesticides and CYP2D6 genetic polymorphism in Parkinson's disease

Potential interactions between environmental and genetic risk factors for neurodegenerative diseases have become an increasingly prominent research focus, with the growing recognition that some persons may be especially susceptible to environmental toxicants, as illustrated by the following example.

The interaction between genetic polymorphisms of the cytochrome P4502D6 (*CYP2D6*) gene and pesticide exposure were investigated in a case–control study of PD in France (86). Pesticides have been regarded as plausible causes of non-familial PD, as reviewed earlier in this chapter. The *CYP2D6* enzyme is known to metabolize MPTP and various toxic environmental chemicals, including some pesticides (87,88). The *CYP2D6* gene is polymorphic, with carriers of the *4 (minor) allele, which contains a SNP at an intron/exon junction, having diminished metabolic capacity proportional to the number of variant alleles. Pesticide exposures, determined by exposure assessment experts, and *CYP2D6* genotypes were compared between 190 PD cases identified from the French health insurance organization for workers in agricultural occupations (Mutualite Sociale Agricole), and 419 age/gender/regionally-matched controls who were also members of this insurance organization (86). A qualitative exposure gradient was defined as "no use," "gardening use," and "professional use," assuming that the last category would represent the heaviest exposures. Analyses were adjusted for cigarette smoking, in addition to the matching variables.

For the entire study population, there was a modest gradient of PD associated with the *CYP2D6*4* genotypes: OR = 1.02; 95% CI = 0.69–1.51 and OR = 1.56; 95% CI = 0.67–3.65 for carriers of one and two *4 alleles, respectively. The joint effects of pesticides and *CYP2D6*4* (Table 22.4) suggest synergism, whereby the most pronounced exposure-response trend was found among carriers of two *4 alleles, who would be classified as 'poor metabolizers.'

The notable strengths of the study were the selection of a study population with a relatively high prevalence of the environmental exposure of interest, pesticides, and the choice of a candidate gene variant whose functional consequences are well understood and plausibly related to pesticide metabolism. As with most case–control studies, this study was prone to exposure assessment uncertainties, particularly insofar as quantification of exposure levels to specific pesticides was not possible.

Example 5: Protein analysis of cerebrospinal fluid in Alzheimer's disease

Protein biomarker profiles in biological tissues have promise as early markers of disease onset and progression that may ultimately have diagnostic and medical management benefits. In addition, protein measurements may reveal characteristic patterns of response to toxic endogenous or exogenous agents that predict disease occurrence. Thus, from a neuro-epidemiologic standpoint, protein profiles offer several potential advantages by serving as early or surrogate disease markers, improving diagnostic accuracy, and suggesting host susceptibility factors. Because of its intimate anatomical and biochemical relations to the brain, cerebrospinal fluid (CSF) is the most relevant biological

Table 22.4. Joint effects on Parkinson's disease risk of CYP2D6*4 genotype and pesticide exposure

	Pesticide exposure		
CYP2D6*4	None	Gardening use	Professional use
Alleles	OR (95%CI)‡	OR (95%CI)	OR (95%CI)
0	1.00 [reference] --	1.73 (0.86–3.48) 0.12	1.85 (0.96–3.55) 0.06
1	1.39 (0.70–2.76) 0.35	1.17 (0.49–2.77) 0.72	1.83 (0.84–3.95) 0.13
2	0.41 (0.04–3.99) 0.44	2.75 (0.55–13.7) 0.22	4.74 (1.29–17.5) 0.02

‡ Odds ratio (95% confidence interval). Adapted from (86).

medium that can be accessed ante-mortem for epidemiologic studies of neurodegenerative disorders. In contrast, brain tissue can only be examined directly post-mortem, and blood or urine protein levels may reflect biological processes that are not-specific to the brain.

CSF levels of Aβ-amyloid$_{1-42}$ (Aβ$_{1-42}$) have been consistently associated with AD. Specifically, Aβ$_{1-42}$ levels are lower, whereas tau protein levels are higher in AD cases compared to controls (89). This pattern probably reflects impaired brain clearance (via CSF) of Aβ$_{1-42}$ and overexpression of tau protein in AD. A study was conducted in Sweden to determine whether these proteins were predictive of conversion to AD among persons with mild cognitive impairment (MCI) (90). These groups were compared at baseline: 93 AD cases, 52 MCI cases, and 10 healthy controls. AD and MCI were diagnosed according to established criteria. During a follow-up period of 3–15 months, 29 MCI cases were determined to have converted to probable AD. As summarized in Table 22.5, Aβ$_{1-42}$ levels decreased consistently from lowest to highest among AD cases, MCI converters, MCI non-converters and healthy controls. For tau protein, a similar pattern in the opposite direction was observed. Relative to reference values established in an earlier study of 231 healthy Swedish subjects (91), abnormally low Aβ$_{1-42}$ levels were associated with an MCI conversion sensitivity of 59% and specificity of 100%. Sensitivity and specificity for MCI conversion to AD associated with abnormally high tau protein levels were 83% and 90%, respectively.

This study provides a vivid illustration of the utility of measuring well-established biomarkers of a defined clinical outcome, AD, to assess clinical progression from earlier symptomatic states. Serial measurements in addition to baseline assessments of Aβ$_{1-42}$ and tau among the MCI and control groups would have been valuable, although the requirement of multiple lumbar punctures would certainly have posed a logistical hurdle. The relatively small sample size, especially of healthy controls, is another limitation, partly offset by the availability of normative data obtained previously in a larger sample.

Strengths, limitations and lessons learned

There are some formidable and characteristic challenges that epidemiologists confront when investigating the causes and prognostic factors for AD, PD and ALS. Each is a complex disorder with varying phenotypes that may in fact represent different clinical entities. Subdivisions of disease phenotypes into familial and non-familial forms, or with respect to age at onset, is a convenient approach, although may be fraught with considerable uncertainty. For example, age 50 is often cited as the demarcation of early- versus late-onset PD, largely based on the onset ages of familial cases, yet the age distinction is arbitrary. From an epidemiologic perspective, the relative homogeneity or heterogeneity of any disease rubric is especially important for identifying risk and prognostic factors. The generally slow rate of disease progression among the majority of cases (non-familial) of the neurodegenerative diseases complicates establishing precise disease onset times. The net result is often inclusion of prevalent rather than incident disease cases in epidemiologic studies, and attendant biases due to differential survival associated with risk factors of interest. Other challenges, which are not unique to research on neurodegenerative disorders, include uncertainties of diagnoses that are based solely on clinical examination; reliance on questionnaire responses, sometimes by proxies, such as with AD, to determine exposure status; availability of very few population-based neurologic disease registries (in contrast to cancer registries, for example); and typically low

Table 22.5. Cerebrospinal fluid levels of β-amyloid and tau protein in relation to conversion from mild cognitive impairment (MCI) to Alzheimer's disease

CSF protein (ng/l)	Group[†]			
	AD (n=93)	MCI converters (n=29)	MCI non-converters (n=23)	Healthy controls (n=10)
Aβ-amyloid$_{1-42}$	545 (± 230)	577 (± 197)	805 (± 368)	962 (± 182)
Tau	725 (± 266)	640 (± 162)	576 (± 275)	341 (± 118)

[†] Mean (± standard deviation). Table compiled from (90).

response rates among controls when biological sampling is included in a study.

Due to the rarity of AD, PD and ALS, population-based case–control studies have been the predominant study design. There have been some cohort studies in which neurodegenerative and other diseases have been investigated. Large cohort sizes and thorough exposure assessments are needed. The Nurses' Health Study cohort in the USA is a good example of a valuable study population for investigating associations with common exposures, such as smoking (37). Cohorts with well-characterized environmental exposures can also be investigated for associations with specific agents, such as a study of neurodegenerative disease mortality among US workers exposed to polychlorinated biphenyls (92). Occupational cohorts, however, generally are limited by relatively small numbers of cases and reliance on death certificates for case identification. Investigations of disease incidence or clinical indicators of neurologic disease, such as symptoms determined from standardized clinical exams, may be desirable where there are clear *a priori* hypotheses regarding risk in relation to specific exposures. Studies of PD-related signs and symptoms among cohorts of career orchardists exposed to pesticides (93) and welders exposed to manganese and other metals (58) typify this approach.

With respect to study size, large samples are generally required to detect low to modest risk associations, such as those usually observed in case–control studies of candidate genes. Collaborative pooled studies following similar protocols, as illustrated in the example of *SNCA* and PD (81), are thus highly desirable in that they offer the opportunity to examine consistency of associations among various populations, with attendant increased statistical power.

The issue of sample size is especially relevant for genome-wide association studies (GWAS), which have become increasingly prominent. Typically, GWAS include extremely large sample sizes (thousands of cases and controls) assembled across multiple collaborating studies to achieve adequate statistical power to detect modest associations. A particular advantage of this method is the opportunity to replicate findings in heterogeneous study populations worldwide with high levels of statistical power. This is illustrated by two large independent GWAS of PD among persons of European ancestry (94) and of Japanese ancestry (95), both of which identified *SNCA* and *LRRK2* as important disease-related genetic loci. Similarly, several GWAS for AD have consistently replicated findings for *ApoE*, but differences were noted for other loci (96–98). For a comprehensive review of GWAS results, the reader is referred to the National Human Genome Research Institute's GWAS catalogue, at http://www.genome.gov/26525384.

Future directions and challenges

Molecular epidemiology cannot eliminate or mitigate all of the previously mentioned research shortcomings. Nevertheless, there are some distinct advantages to incorporating molecular methods into neuroepidemiologic studies. As demonstrated with the examples presented in this chapter, the range of potential benefits include: improved diagnoses and phenotypic characterization, identification of genetic susceptibility factors, and, at least in theory, more precise exposure assessments in some instances.

A particular challenge that arises in molecular epidemiologic studies of neuro-degenerative disorders is that the target tissues of the central nervous system are not directly accessible except in post-mortem studies, which typically have epidemiologic shortcomings (e.g. convenience sampling). Consequently, surrogate measurements of toxicants, metabolites and other biomarkers are necessitated. The exception is DNA that can be assayed for genotyped validity from multiple tissue sources. Figure 22.1 summarizes the inter-relations between tissue sources for molecular biomarker assessment.

Molecular methods to date have mainly been applied to address relatively narrowly defined hypotheses, such as associations with a small number of genetic polymorphisms or exposure biomarkers. Nevertheless, as molecular technology becomes increasingly affordable and flexible, epidemiologists will be able to capitalize on technological advances to broaden the scope of research. GWAS and extensive linkage studies of PD (99) and ALS (100), and broad-based proteomic assessments of CSF in AD (101), indicate that this trend is underway. GWAS approaches are well suited to identifying common variants associated with disease, but other types of gene variation, namely rare variants and/or copy number variation, are not amenable to the current genotyping platforms. Hence, newer, more advanced approaches (whose development is in progress) will be necessary to query the genome fully, and in some cases this may necessitate even larger sample sizes. Molecular methods should also be particularly

Figure 22.1. Inter-relations between brain, cerebrospinal fluid and blood for molecular biomarker assessment

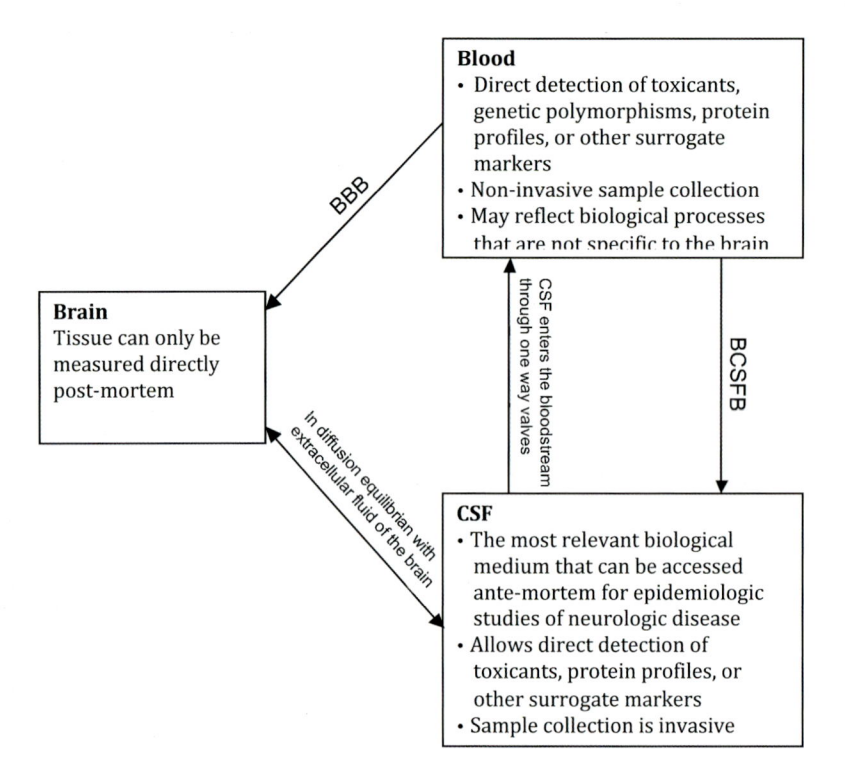

Blood
- Direct detection of toxicants, genetic polymorphisms, protein profiles, or other surrogate markers
- Non-invasive sample collection
- May reflect biological processes that are not specific to the brain

Brain
Tissue can only be measured directly post-mortem

CSF
- The most relevant biological medium that can be accessed ante-mortem for epidemiologic studies of neurologic disease
- Allows direct detection of toxicants, protein profiles, or other surrogate markers
- Sample collection is invasive

Arrows: BBB; In diffusion equilibrium with extracellular fluid of the brain; CSF enters the bloodstream through one way valves; BCSFB

CSF, cerebrospinal fluid; CNS, central nervous system; BBB, blood-brain barrier; BCSFB, blood-cerebrospinal fluid barrier.

---

advantageous for investigating risk and prognostic factors for pre-clinical neurodegenerative disease outcomes, such as neuroimaging abnormalities or proteomic profiles, for which there are demonstrated high predictive values for late-stage disease.

Ultimately, consistent findings from epidemiologic studies, focused on narrow hypotheses that are corroborated by results from broader-based molecular epidemiology investigations, will be important for the prevention and management of the neurodegenerative diseases.

Acknowledgement

Support for this work comes from the US National Institute of Environmental Health Sciences (grant P42ES04696) and the International Agency for Research on Cancer, where Dr Checkoway was a Visiting Scientist in 2006.

References

1. Ferri CP, Prince M, Brayne C et al.; Alzheimer's Disease International (2005). Global prevalence of dementia: a Delphi consensus study. Lancet, 366:2112–2117. doi:10.1016/S0140-6736(05)67889-0 PMID:16360788

2. Blennow K, de Leon MJ, Zetterberg H (2006). Alzheimer's disease. Lancet, 368:387–403.doi:10.1016/S0140-6736(06)69113-7 PMID:16876668

3. Hardy J, Selkoe DJ (2002). The amyloid hypothesis of Alzheimer's disease: progress and problems on the road to therapeutics. Science, 297:353–356.doi:10.1126/science.1072994 PMID:12130773

4. Goate A, Chartier-Harlin MC, Mullan M et al. (1991). Segregation of a missense mutation in the amyloid precursor protein gene with familial Alzheimer's disease. Nature, 349:704–706.doi:10.1038/349704a0 PMID:1671712

5. Patterson C, Feightner JW, Garcia A et al. (2008). Diagnosis and treatment of dementia: 1. Risk assessment and primary prevention of Alzheimer disease. CMAJ, 178:548–556. PMID:18299540

6. Turner RS (2006). Alzheimer's disease. Semin Neurol, 26:499–506.doi:10.1055/s-2006-951622 PMID:17048151

7. Bertram L, Tanzi RE (2005). The genetic epidemiology of neurodegenerative disease. J Clin Invest, 115:1449–1457.doi:10.1172/JCI24761 PMID:15931380

8. Farrer LA, Cupples LA, Haines JL et al.; APOE and Alzheimer Disease Meta Analysis Consortium (1997). Effects of age, sex, and ethnicity on the association between apolipoprotein E genotype and Alzheimer disease. A meta-analysis. JAMA, 278:1349–. doi:10.1001/jama.278.16.1349 PMID:9343467

9. Rubinsztein DC, Easton DF (1999). Apolipoprotein E genetic variation and Alzheimer's disease. a meta-analysis. Dement Geriatr Cogn Disord, 10:199–209.doi:10.1159/000017120 PMID:10325447

10. Webster JA, Myers AJ, Pearson JV et al. (2008). Sorl1 as an Alzheimer's disease predisposition gene? Neurodegener Dis, 5:60–64.doi:10.1159/000110789 PMID:17975299

11. Mayeux R (2003). Epidemiology of neurodegeneration. Annu Rev Neurosci, 26:81–104.doi:10.1146/annurev.neuro.26.043002.094919 PMID:12574495

12. Jellinger KA (2004). Head injury and dementia. Curr Opin Neurol, 17:719–723. doi:10.1097/00019052-200412000-00012 PMID:15542981

13. Aggarwal NT, Bienias JL, Bennett DA et al. (2006). The relation of cigarette smoking to incident Alzheimer's disease in a biracial urban community population. Neuroepidemiology, 26:140–146.doi:10.1159/000091654 PMID:16493200

14. Staehelin HB (2005). Micronutrients and Alzheimer's disease. Proc Nutr Soc, 64:565–570.doi:10.1079/PNS2005459 PMID:16313699

15. Ruitenberg A, van Swieten JC, Witteman JC et al. (2002). Alcohol consumption and risk of dementia: the Rotterdam Study. Lancet, 359:281–286.doi:10.1016/S0140-6736(02)07493-7 PMID:11830193

16. Kukull WA, Larson EB, Bowen JD et al. (1995). Solvent exposure as a risk factor for Alzheimer's disease: a case-control study. Am J Epidemiol, 141:1059–1071, discussion 1072–1079. PMID:7771442

17. Graves AB, Rosner D, Echeverria D et al. (1998). Occupational exposures to solvents and aluminium and estimated risk of Alzheimer's disease. Occup Environ Med, 55:627–633.doi:10.1136/oem.55.9.627 PMID:9861186

18. Qiu C, Fratiglioni L, Karp A et al. (2004). Occupational exposure to electromagnetic fields and risk of Alzheimer's disease. Epidemiology, 15:687–694.doi:10.1097/01.ede.0000142147.49297.9d PMID:15475717

19. Bennett DA, Schneider JA, Wilson RS et al. (2005). Education modifies the association of amyloid but not tangles with cognitive function. Neurology, 65:953–955. doi:10.1212/01.wnl.0000176286.17192.69 PMID:16186546

20. Tyas SL, Manfreda J, Strain LA, Montgomery PR (2001). Risk factors for Alzheimer's disease: a population-based, longitudinal study in Manitoba, Canada. Int J Epidemiol, 30:590–597.doi:10.1093/ije/30.3.590 PMID:11416089

21. Baldi I, Lebailly P, Mohammed-Brahim B et al. (2003b). Neurodegenerative diseases and exposure to pesticides in the elderly. Am J Epidemiol, 157:409–414.doi:10.1093/aje/kwf216 PMID:12615605

22. Kukull WA (2001). The association between smoking and Alzheimer's disease: effects of study design and bias. Biol Psychiatry, 49:194–199.doi:10.1016/S0006-3223(00)01077-5 PMID:11230870

23. Samii A, Nutt JG, Ransom BR (2004). Parkinson's disease. Lancet, 363:1783–1793.doi:10.1016/S0140-6736(04)16305-8 PMID:15172778

24. Elbaz A, Bower JH, Maraganore DM et al. (2002). Risk tables for parkinsonism and Parkinson's disease. J Clin Epidemiol, 55:25–31.doi:10.1016/S0895-4356(01)00425-5 PMID:11781119

25. Dawson TM, Dawson VL (2003). Molecular pathways of neurodegeneration in Parkinson's disease. Science, 302:819–822. doi:10.1126/science.1087753 PMID:14593166

26. Agid Y (1991). Parkinson's disease: pathophysiology. Lancet, 337:1321–1324. doi:10.1016/0140-6736(91)92989-F PMID:1674304

27. Wakabayashi K, Tanji K, Mori F, Takahashi H (2007). The Lewy body in Parkinson's disease: molecules implicated in the formation and degradation of alpha-synuclein aggregates. Neuropathology, 27:494–506. doi:10.1111/j.1440-1789.2007.00803.x PMID:18018486

28. Gosal D, Ross OA, Toft M (2006). Parkinson's disease: the genetics of a heterogeneous disorder. Eur J Neurol, 13:616–627.doi:10.1111/j.1468-1331.2006.01336.x PMID:16796586

29. Gwinn-Hardy K (2002). Genetics of parkinsonism. Mov Disord, 17:645–656.doi: 10.1002/mds.10173 PMID:12210852

30. Funayama M, Hasegawa K, Kowa H et al. (2002). A new locus for Parkinson's disease (PARK8) maps to chromosome 12p11.2-q13.1. Ann Neurol, 51:296–301.doi:10.1002/ana.10113 PMID:11891824

31. Paisàn-Ruìz C, Sàenz A, Lòpez de Munain A et al. (2005). Familial Parkinson's disease: clinical and genetic analysis of four Basque families. Ann Neurol, 57:365–372. doi:10.1002/ana.20391 PMID:15732106

32. Nichols WC, Pankratz N, Hernandez D et al.; Parkinson Study Group-PROGENI investigators (2005). Genetic screening for a single common LRRK2 mutation in familial Parkinson's disease. Lancet, 365:410–412. PMID:15680455

33. Checkoway H, Franklin GM, Costa-Mallen P et al. (1998). A genetic polymorphism of MAO-B modifies the association of cigarette smoking and Parkinson's disease. Neurology, 50:1458–1461. PMID:9596006

34. Mellick GD, McCann SJ, Le Couter DG (1999). Parkinson's disease, MAOB, and smoking. Neurology, 53:658. PMID:10449149

35. Hernán MA, Checkoway H, O'Brien R et al. (2002). Monoamine oxidase B intron 13 and catechol-O-methyltransferase codon 158 polymorphisms, cigarette smoking, and the risk of Parkinson's disease. Neurology, 58:1381–1387. PMID:12011284

36. Hellenbrand W, Seidler A, Robra BP et al. (1997). Smoking and Parkinson's disease: a case-control study in Germany. Int J Epidemiol, 26:328–339.doi:10.1093/ije/26.2.328 PMID:9169168

37. Hernán MA, Zhang SM, Rueda-deCastro AM et al. (2001). Cigarette smoking and the incidence of Parkinson's disease in two prospective studies. Ann Neurol, 50:780–786. doi:10.1002/ana.10028 PMID:11761476

38. Checkoway H, Powers K, Smith-Weller T et al. (2002). Parkinson's disease risks associated with cigarette smoking, alcohol consumption, and caffeine intake. Am J Epidemiol, 155:732–738.doi:10.1093/aje/155.8.732 PMID:11943691

39. Ritz B, Ascherio A, Checkoway H et al. (2007). Pooled analysis of tobacco use and risk of Parkinson disease. Arch Neurol, 64: 990–997.doi:10.1001/archneur.64.7.990 PMID:17620489

40. Fowler JS, Volkow ND, Wang GJ et al. (1996). Inhibition of monoamine oxidase B in the brains of smokers. Nature, 379:733–736. doi:10.1038/379733a0 PMID:8602220

41. Castagnoli K, Murugesan T (2004). Tobacco leaf, smoke and smoking, MAO inhibitors, Parkinson's disease and neuroprotection; are there links? Neurotoxicology, 25:279–291.doi:10.1016/S0161-813X(03)00107-4 PMID:14697903

42. Ross GW, Abbott RD, Petrovitch H et al. (2000). Association of coffee and caffeine intake with the risk of Parkinson disease. JAMA, 283:2674–2679.doi:10.1001/jama.283.20.2674 PMID:10819950

43. Chen H, Jacobs E, Schwarzschild MA et al. (2005). Nonsteroidal antiinflammatory drug use and the risk for Parkinson's disease. Ann Neurol, 58:963–967.doi:10.1002/ana.20682 PMID:16240369

44. Ton TG, Heckbert SR, Longstreth WT Jr et al. (2006). Nonsteroidal anti-inflammatory drugs and risk of Parkinson's disease. Mov Disord, 21:964–969.doi:10.1002/mds.20856 PMID:16550541

45. Sveinbjörnsdottir S, Hicks AA, Jonsson T et al. (2000). Familial aggregation of Parkinson's disease in Iceland. N Engl J Med, 343:1765–1770.doi:10.1056/NEJM200012143432404 PMID:11114315

46. Bower JH, Maraganore DM, Peterson BJ et al. (2003). Head trauma preceding PD: a case-control study. Neurology, 60:1610–1615. PMID:12771250

47. Goldman SM, Tanner CM, Oakes D et al. (2006). Head injury and Parkinson's disease risk in twins. Ann Neurol, 60:65–72. doi:10.1002/ana.20882 PMID:16718702

48. Langston JW, Ballard P, Tetrud JW, Irwin I (1983). Chronic Parkinsonism in humans due to a product of meperidine-analog synthesis. Science, 219:979–980.doi:10.1126/science.6823561 PMID:6823561

49. Langston JW, Irwin I, Ricaurte GA (1987). Neurotoxins, parkinsonism and Parkinson's disease. Pharmacol Ther, 32:19–49. PMID:3295897

50. Liou HH, Tsai MC, Chen CJ et al. (1997). Environmental risk factors and Parkinson's disease: a case-control study in Taiwan. Neurology, 48:1583–1588. PMID:9191770

51. Petrovitch H, Ross GW, Abbott RD et al. (2002). Plantation work and risk of Parkinson disease in a population-based longitudinal study. Arch Neurol, 59:1787–1792.doi:10.1001/archneur.59.11.1787 PMID:12433267

52. Baldi I, Lebailly P, Mohammed-Brahim B et al. (2003a). Neurodegenerative diseases and exposure to pesticides in the elderly. Am J Epidemiol, 157:409–414.doi:10.1093/aje/kwf216 PMID:12615605

53. Kamel F, Tanner C, Umbach D et al. (2007). Pesticide exposure and self-reported Parkinson's disease in the agricultural health study. Am J Epidemiol, 165:364–374. doi:10.1093/aje/kwk024 PMID:17116648

54. Brighina L, Frigerio R, Schneider NK et al. (2008). Alpha-synuclein, pesticides, and Parkinson disease: a case-control study. Neurology, 70(16):1461–1469.doi:10.1212/01.wnl.0000304049.31377.f2 PMID:18322262

55. Wang JD, Huang CC, Hwang YH et al. (1989). Manganese induced parkinsonism: an outbreak due to an unrepaired ventilation control system in a ferromanganese smelter. Br J Ind Med, 46:856–859. PMID:2611159

56. Gorell JM, Johnson CC, Rybicki BA et al. (1997). Occupational exposures to metals as risk factors for Parkinson's disease. Neurology, 48:650–658. PMID:9065542

57. Fryzek JP, Hansen J, Cohen S et al. (2005). A cohort study of Parkinson's disease and other neurodegenerative disorders in Danish welders. J Occup Environ Med, 47:466–472. doi:10.1097/01.jom.0000161730.25913.bf PMID:15891525

58. Racette BA, Tabbal SD, Jennings D et al. (2005). Prevalence of parkinsonism and relationship to exposure in a large sample of Alabama welders. Neurology, 64:230–235. PMID:15668418

59. Fored CM, Fryzek JP, Brandt L et al. (2006). Parkinson's disease and other basal ganglia or movement disorders in a large nationwide cohort of Swedish welders. Occup Environ Med, 63:135–140.doi:10.1136/oem.2005.022921 PMID:16421393

60. Marsh GM, Gula MJ (2006). Employment as a welder and Parkinson disease among heavy equipment manufacturing workers. J Occup Environ Med, 48:1031–1046.doi:10.1097/01.jom.0000232547.74802.d8 PMID:17033503

61. McDonnell L, Maginnis C, Lewis S et al. (2003). Occupational exposure to solvents and metals and Parkinson's disease. Neurology, 61:716–717. PMID:12963777

62. Frigerio R, Sanft KR, Grossardt BR et al. (2006). Chemical exposures and Parkinson's disease: a population-based case-control study. Mov Disord, 21:1688–1692.doi:10.1002/mds.21009 PMID:16773614

63. Gash DM, Rutland K, Hudson NL et al. (2008). Trichloroethylene: Parkinsonism and complex 1 mitochondrial neurotoxicity. Ann Neurol, 63:184–192.doi:10.1002/ana.21288 PMID:18157908

64. Logroscino G, Beghi E, Zoccolella S et al.; SLAP Registry (2005). Incidence of amyotrophic lateral sclerosis in southern Italy: a population based study. J Neurol Neurosurg Psychiatry, 76:1094–1098.doi:10.1136/jnnp.2004.039180 PMID:16024886

65. del Aguila MA, Longstreth WT Jr, McGuire V et al. (2003). Prognosis in amyotrophic lateral sclerosis: a population-based study. Neurology, 60:813–819. PMID:12629239

66. Sen I, Nalini A, Joshi NB, Joshi PG (2005). Cerebrospinal fluid from amyotrophic lateral sclerosis patients preferentially elevates intracellular calcium and toxicity in motor neurons via AMPA/kainate receptor. J Neurol Sci, 235:45–54.doi:10.1016/j.jns.2005.03.049 PMID:15936037

67. Rosen DR, Siddique T, Patterson D et al. (1993). Mutations in Cu/Zn superoxide dismutase gene are associated with familial amyotrophic lateral sclerosis. Nature, 362:59–62.doi:10.1038/362059a0 PMID:8446170

68. Nordlund A, Oliveberg M (2006). Folding of Cu/Zn superoxide dismutase suggests structural hotspots for gain of neurotoxic function in ALS: parallels to precursors in amyloid disease. Proc Natl Acad Sci USA, 103:10218–10223.doi:10.1073/pnas.0601696103 PMID:16798882

69. Chandran J, Ding J, Cai H (2007). Alsin and the molecular pathways of amyotrophic lateral sclerosis. Mol Neurobiol, 36:224–231.doi:10.1007/s12035-007-0034-x PMID:17955197

70. Pasinelli P, Brown RH (2006). Molecular biology of amyotrophic lateral sclerosis: insights from genetics. Nat Rev Neurosci, 7:710–723. doi:10.1038/nrn1971 PMID:16924260

71. Weisskopf MG, McCullough ML, Calle EE et al. (2004). Prospective study of cigarette smoking and amyotrophic lateral sclerosis. Am J Epidemiol, 160:26–33.doi:10.1093/aje/kwh179 PMID:15229114

72. McGuire V, Longstreth WT Jr, Nelson LM et al. (1997). Occupational exposures and amyotrophic lateral sclerosis. A population-based case-control study. Am J Epidemiol, 145:1076–1088. PMID:9199537

73. Morahan JM, Pamphlett R (2006). Amyotrophic lateral sclerosis and exposure to environmental toxins: an Australian case-control study. Neuroepidemiology, 27:130–135.doi:10.1159/000095552 PMID:16946624

74. Kamel F, Umbach DM, Hu H et al. (2005). Lead exposure as a risk factor for amyotrophic lateral sclerosis. Neurodegener Dis, 2:195–201.doi:10.1159/000089625 PMID:16909025

75. Li CY, Sung FC (2003). Association between occupational exposure to power frequency electromagnetic fields and amyotrophic lateral sclerosis: a review. *Am J Ind Med*, 43:212–220.doi:10.1002/ajim.10148 PMID:12541277

76. Nelson LM, McGuire V, Longstreth WT Jr, Matkin C (2000). Population-based case-control study of amyotrophic lateral sclerosis in western Washington State. I. Cigarette smoking and alcohol consumption. *Am J Epidemiol*, 151:156–163. PMID:10645818

77. Weisskopf MG, O'Reilly EJ, McCullough ML et al. (2005). Prospective study of military service and mortality from ALS. *Neurology*, 64:32–37. PMID:15642900

78. Kamel F, Umbach DM, Munsat TL et al. (2002). Lead exposure and amyotrophic lateral sclerosis. *Epidemiology*, 13:311–319. doi:10.1097/00001648-200205000-00012 PMID:11964933

79. Spadafora P, Annesi G, Pasqua AA et al. (2003). NACP-REP1 polymorphism is not involved in Parkinson's disease: a case-control study in a population sample from southern Italy. *Neurosci Lett*, 351:75–78. doi:10.1016/S0304-3940(03)00859-0 PMID:14583385

80. Mellick GD, Maraganore DM, Silburn PA (2005). Australian data and meta-analysis lend support for alpha-synuclein (NACP-Rep1) as a risk factor for Parkinson's disease. *Neurosci Lett*, 375:112–116.doi:10.1016/j.neulet.2004.10.078 PMID:15670652

81. Maraganore DM, de Andrade M, Elbaz A et al.; Genetic Epidemiology of Parkinson's Disease (GEO-PD) Consortium (2006). Collaborative analysis of alpha-synuclein gene promoter variability and Parkinson disease. *JAMA*, 296:661–670.doi:10.1001/jama.296.6.661 PMID:16896109

82. Porrello E, Monti MC, Sinforiani E et al. (2006). Estrogen receptor alpha and APOEepsilon4 polymorphisms interact to increase risk for sporadic AD in Italian females. *Eur J Neurol*, 13:639–644.doi:10.1111/j.1468-1331.2006.01333.x PMID:16796589

83. Shumaker SA, Legault C, Rapp SR et al.; WHIMS Investigators (2003). Estrogen plus progestin and the incidence of dementia and mild cognitive impairment in postmenopausal women: the Women's Health Initiative Memory Study: a randomized controlled trial. *JAMA*, 289:2651–2662.doi:10.1001/jama.289.20.2651 PMID:12771112

84. Isoe-Wada K, Maeda M, Yong J et al. (1999). Positive association between an estrogen receptor gene polymorphism and Parkinson's disease with dementia. *Eur J Neurol*, 6:431–435.doi:10.1046/j.1468-1331.1999.640431.x PMID:10362895

85. Mattila KM, Axelman K, Rinne JO et al. (2000). Interaction between estrogen receptor 1 and the epsilon4 allele of apolipoprotein E increases the risk of familial Alzheimer's disease in women. *Neurosci Lett*, 282:45–48.doi:10.1016/S0304-3940(00)00849-1 PMID:10713392

86. Elbaz A, Levecque C, Clavel J et al. (2004). CYP2D6 polymorphism, pesticide exposure, and Parkinson's disease. *Ann Neurol*, 55:430–434.doi:10.1002/ana.20051 PMID:14991823

87. Gilham DE, Cairns W, Paine MJ et al. (1997). Metabolism of MPTP by cytochrome P4502D6 and the demonstration of 2D6 mRNA in human foetal and adult brain by in situ hybridization. *Xenobiotica*, 27:111–125. doi:10.1080/004982597240802 PMID:9041683

88. Sams C, Mason HJ, Rawbone R (2000). Evidence for the activation of organophosphate pesticides by cytochromes P450 3A4 and 2D6 in human liver microsomes. *Toxicol Lett*, 116:217–221.doi:10.1016/S0378-4274(00)00221-6 PMID:10996483

89. Sunderland T, Linker G, Mirza N et al. (2003). Decreased beta-amyloid1–42 and increased tau levels in cerebrospinal fluid of patients with Alzheimer disease. *JAMA*, 289:2094–2103.doi:10.1001/jama.289.16.2094 PMID:12709467

90. Hampel H, Teipel SJ, Fuchsberger T et al. (2004). Value of CSF beta-amyloid1–42 and tau as predictors of Alzheimer's disease in patients with mild cognitive impairment. *Mol Psychiatry*, 9:705–710. PMID:14699432

91. Sjögren M, Vanderstichele H, Agren H et al. (2001). Tau and Abeta42 in cerebrospinal fluid from healthy adults 21–93 years of age: establishment of reference values. *Clin Chem*, 47:1776–1781. PMID:11568086

92. Steenland K, Hein MJ, Cassinelli RT 2nd et al. (2006). Polychlorinated biphenyls and neurodegenerative disease mortality in an occupational cohort. *Epidemiology*, 17:8–13. doi:10.1097/01.ede.0000190707.51536.2b PMID:16357589

93. Engel LS, Checkoway H, Keifer MC et al. (2001). Parkinsonism and occupational exposure to pesticides. *Occup Environ Med*, 58:582–589.doi:10.1136/oem.58.9.582 PMID:11511745

94. Simón-Sánchez J, Schulte C, Bras JM et al. (2009). Genome-wide association study reveals genetic risk underlying Parkinson's disease. *Nat Genet*, 41:1308–1312.doi:10.1038/ng.487 PMID:19915575

95. Satake W, Nakabayashi Y, Mizuta I et al. (2009). Genome-wide association study identifies common variants at four loci as genetic risk factors for Parkinson's disease. *Nat Genet*, 41:1303–1307.doi:10.1038/ng.485 PMID:19915576

96. Coon KD, Myers AJ, Craig DW et al. (2007). A high-density whole-genome association study reveals that APOE is the major susceptibility gene for sporadic late-onset Alzheimer's disease. *J Clin Psychiatry*, 68:613–618.doi:10.4088/JCP.v68n0419 PMID:17474819

97. Harold D, Abraham R, Hollingworth P et al. (2009). Genome-wide association study identifies variants at CLU and PICALM associated with Alzheimer's disease. *Nat Genet*, 41:1088–1093.doi:10.1038/ng.440 PMID:19734902

98. Seshadri S, Fitzpatrick AL, Ikram MA et al.; CHARGE Consortium; GERAD1 Consortium; EADI1 Consortium (2010). Genome-wide analysis of genetic loci associated with Alzheimer disease. *JAMA*, 303:1832–1840. doi:10.1001/jama.2010.574 PMID:20460622

99. Maraganore DM, de Andrade M, Lesnick TG et al. (2005). High-resolution whole-genome association study of Parkinson disease. *Am J Hum Genet*, 77:685–693. doi:10.1086/496902 PMID:16252231

100. Vance C, Al-Chalabi A, Ruddy D et al. (2006). Familial amyotrophic lateral sclerosis with frontotemporal dementia is linked to a locus on chromosome 9p13.2–21.3. *Brain*, 129:868–876.doi:10.1093/brain/awl030 PMID:16495328

101. Zhang J, Goodlett DR, Quinn JF et al. (2005). Quantitative proteomics of cerebrospinal fluid from patients with Alzheimer disease. *J Alzheimers Dis*, 7:125–133, discussion 173–180. PMID:15851850

UNIT 5.
APPLICATION OF BIOMARKERS TO DISEASE

CHAPTER 23.

Infectious diseases

Betsy Foxman

Summary

Molecular tools have enhanced our understanding of the epidemiology of infectious diseases by describing the transmission system, including identifying novel transmission modes and reservoirs, identifying characteristics of the infectious agent that lead to transmission and pathogenesis, identifying potential vaccine candidates and targets for therapeutics, and recognizing new infectious agents. Applications of molecular fingerprinting to public health practice have enhanced outbreak investigation by objectively confirming epidemiologic evidence, and distinguishing between time-space clusters and sporadic cases. Clinically, molecular tools are used to rapidly detect infectious agents and predict disease course. Integration of molecular tools into etiologic studies has identified infectious causes of chronic diseases, and characteristics of the agent and host that modify disease risk. The combination of molecular tools with epidemiologic methods provides essential information to guide clinical treatment, and to design and implement programmes to prevent and control infectious diseases. However, incorporating molecular tools into epidemiologic studies of infectious diseases impacts study design, conduct, and analysis.

Historical perspectives

The development of epidemiology as a discipline was roughly simultaneous with the development of microbiology. As the presence of infectious agents was linked to disease, laboratory methods were incorporated into epidemiologic studies. One epidemiologic hero is John Snow, who identified a strong epidemiologic association between sewage-contaminated water and cholera. Despite extremely well-documented evidence supporting thoroughly researched and reasoned arguments, his findings remained in doubt for some time. Max Von Pettenkofer, 1818–1901, a contemporary of Snow and also an early epidemiologist, is related (in a perhaps apocryphal story) to have drunk a glass of the stool of someone with cholera to test the hypothesis; Pettenkofer remained disease-free. Snow's conclusions were not generally accepted until 25 years after his death, when the cholera vibrio was discovered by Joseph Koch, who definitively demonstrated the causal relationship between the vibrio and cholera (1). The strategy of isolating an organism from an ill individual, showing it can cause disease in a disease-naïve individual, and then be re-isolated as described in the landmark postulates of Henle and Koch reflects how incorporating laboratory methods enhances our ability to make causal inferences about disease transmission and pathogenesis from even the most carefully researched epidemiologic evidence.

Early epidemiologists made tremendous strides with what are now relatively simple molecular tools: using microscopy for identification, which showed that agents not visible by microscope caused disease ("filterable"); and detection of protective antibodies with haemagglutination assays. For example, Charles Nicolle and Alphonse Laveran showed that a protozoan caused malaria (2,3), and Charles Nicolle demonstrated, by injecting a monkey with small amounts of infected louse, that lice transmitted typhus. He also observed that some animals carry infection asymptomatically (3). Wade Hampton Frost used the presence of protective antibodies in the serum of polio patients to explain the emergence of polio epidemics (4). These early, dramatic successes combined with the successful development and implementation of vaccines against major childhood diseases, including smallpox, measles, diphtheria, whooping cough and polio, and the identification of antibiotics, led to a rather simplistic view of infectious disease, and ultimately to the incorrect impression that we might "close the book" on infectious disease during the 20th century. This assertion was quickly undermined in the last quarter of the 20th century by the emergence of new infectious agents such as human immunodeficiency virus (HIV), Ebola and Hantavirus, the re-emergence of tuberculosis and malaria, and the transcontinental transmission of agents such as West Nile Virus and Dengue.

Infectious disease epidemiologists were early adapters of modern molecular biologic techniques to epidemiology, such as those used in genomics. Indeed, the term molecular epidemiology came from infectious disease work (5). Modern molecular techniques have fundamentally changed our understanding of the epidemiology of infectious agents. Characterizing the genetics of human pathogens has revealed the tremendous heterogeneity of various infectious agents, and the rapidity with which they evolve. This heterogeneity and rapid evolution helps explain our difficulties in creating successful vaccines for the more heterogeneous organisms, such as *Neisseria gonorrhoeae*. With the increased ability to detect host immunologic response to infectious agents, we increase our ability to test and refine our theories about the extent and duration of immunity, a key parameter in disease spread. Molecular analysis has also revealed the role of infectious agents in the initiation and promotion of previously classified chronic diseases. Further, molecular tools have enhanced our understanding of the epidemiology of infectious diseases by describing the transmission systems, identifying novel transmission modes and reservoirs, identifying characteristics of the infectious agent that lead to transmission and pathogenesis, revealing potential targets for vaccines and therapeutics, and recognizing new infectious agents. The combination of molecular tools with epidemiologic methods thus provides essential information to design and implement programmes to prevent and control infectious diseases.

Outbreak investigations

Molecular tools have substantially improved inferences from outbreak investigations. We can more rapidly provide laboratory confirmation of disease diagnosis, and detect the presence of difficult-to-culture or uncultivable agents, enhancing the sensitivity and specificity of case definitions. Molecular tools make it possible to determine if the epidemiologically identified outbreak source, such as a food item, contains the infecting organism, and if the identified organism has the same genotype causing the outbreak, enhancing causal inference. Molecular typing can also be used to determine order of transmission (Table 23.1). It is not surprising that molecular typing has become a standard tool in outbreak investigation.

Case definition is an essential component of successful outbreak investigation. The detection of the infectious agent from all cases more accurately classifies cases than clinical diagnosis alone; molecular typing further minimizes misclassification. For point-source outbreaks, all individuals are expected to be infected with the same strain of a particular genus and species; for propagated outbreaks similar molecular fingerprints are expected that vary only in the

Table 23.1. Applications of molecular tools in outbreak investigations

- Enhance case definitions
- Determine whether cases occurring in the same time frame are part of the same outbreak
- Confirm or refute epidemiologic inferences regarding etiologic pathways
- Determine the order of transmission

mutations accrued over repeated transmission events. The resulting reduction of misclassification of disease status increases the study power and the validity of inferences.

The first and most typical application of molecular tools in an outbreak situation is to confirm or refute epidemiologic information. For example, in a foodborne outbreak, isolates might be collected from all those with disease, the person suspected to have introduced the infected agent into a food item, and the suspected food item. Figure 23.1 shows the molecular fingerprints, determined using pulsed-field gel eletrophoresis (PFGE), of a foodborne *Staphylococcus aureus* (*S. aureus*) outbreak isolated from the suspected food that had the same molecular type as *S. aureus* found in the food handler, and in those with disease.

In most foodborne outbreaks, neither specimens from all individuals that meet the case definition, nor the putative food is available for testing by the time it is identified. With clear epidemiologic evidence, the demonstration of genetic relatedness between cases and the putative item is solely confirmatory, and adds little unless the food vehicle has not been previously identified. For example, epidemiologic evidence linked consumption of toasted oats cereal with a multistate outbreak of Salmonella in the USA. A culture of the cereal from one opened and two unopened boxes found *Salmonella agona* (*S. agona*) with the same PFGE pattern as the *S. agona* causing the outbreak (7). This was the first time a commercial cereal product was implicated in a Salmonella outbreak, so the PFGE evidence was particularly compelling.

A second application in an outbreak investigation is determining whether cases occurring in the same time frame are part of the same outbreak. With disseminated outbreaks, there can be apparent sporadic cases that are actually linked. For example, in 1997 a large (n = 126) foodborne outbreak of hepatitis A occurred in Michigan (8). Epidemiologic evidence implicated frozen strawberries from a single processor. During the same time period, a much smaller outbreak (n = 19) occurred in Maine, and sporadic cases occurred in three other states among individuals suspected to have consumed frozen strawberries from the same processor. The genetic sequences of the virus from all tested individuals were the same, confirming a common source of infection for all cases. Without molecular fingerprinting, it would have been very difficult, if not impossible, to link these apparently disparate cases to one common source using solely epidemiologic methods. This example also highlights the importance of using molecular tools with surveillance, discussed in detail in the next section.

A third application is to determine the order of transmission. This application has been particularly useful in forensic cases. Understanding the evolution of an organism and the ability to trace the order of that evolution has made it possible to detect cases where an individual deliberately infected others. For example, a gastroenterologist was convicted of infecting a former girlfriend with the blood of an HIV patient. A variety

Figure 23.1. Example of molecular typing using pulsed-field gel electrophoresis (PFGE). In this foodborne outbreak of methicillin sensitive *Staphylococcus aureus*, isolates from food handler (lane 6), cases of food poisoning (lanes 7-9, 11) and infected food (lane 12) had the same PFGE type (6).

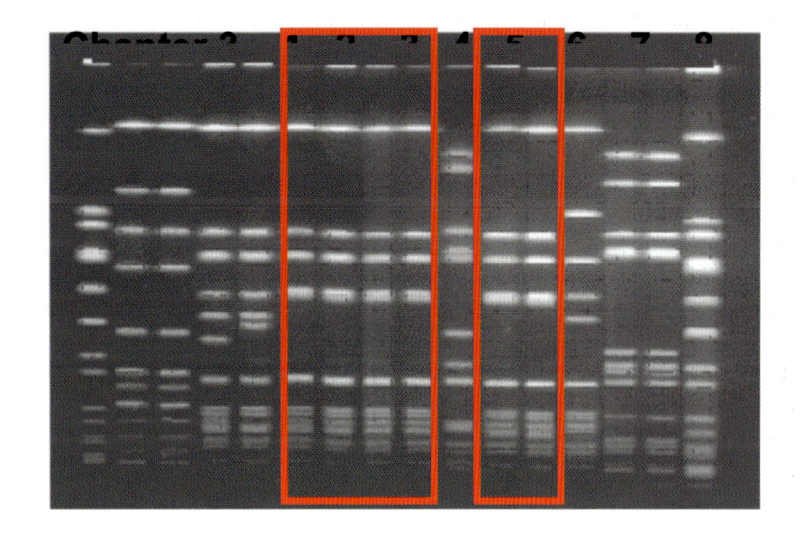

Unit 5 • Chapter 23. Infectious diseases   423

of molecular analyses, including phylogenetic analyses of HIV-1 reverse transcriptase and *env* DNA sequences isolated from the victim, the patient, and a local population sample of HIV-1-positive individuals, strongly supported not only that there was transmission between the two individuals, but who had infected whom (9).

Surveillance

Surveillance is an essential component of a successful public health infrastructure. Laboratories are key components of many surveillance systems; hospital laboratories may be part of regional surveillance networks, as well as part of a local surveillance system. Monitoring of infectious disease isolates identifies time-space clusters of infection; molecular typing distinguishes between infectious agents of the same species, allowing differentiation among clusters of disease occurring by chance and true outbreaks. True outbreaks and clusters of the same strain can be traced back to a common source and presumably are amenable to public health intervention. Spurious clusters cannot, and their investigation wastes time and resources. Applying molecular tools to surveillance isolates can also identify new strains with increased virulence or changing patterns of resistance (Table 23.2).

Hospitals have high endemic rates of bacterial infection, but the infections are often due to a bacterial strain that was colonising an individual before entering the hospital, for example, *S. aureus*. The prevalence of *S. aureus* colonization among the general population is 32% in the nares (10), but much higher in patients and personnel in hospitals and long-term care facilities. By typing strains causing infection among patients, a distinction can be made between a strain from the community and one circulating endemically or causing an outbreak within the hospital. The prevention and control strategies are different in each case, and thus it is important to make a distinction between them.

A second application of molecular tools in surveillance is to identify clusters requiring further investigation. By monitoring isolates from time-space clusters for the presence of a common molecular type, one can distinguish between common-source outbreaks that are local and those that are widely disseminated. Processed foods may be distributed widely, as demonstrated in the earlier example of the Michigan hepatitis A outbreak caused by frozen strawberries (8), so adding molecular typing to laboratory monitoring of specimens is essential. The US Centers for Disease Control and Prevention's PulseNet, a molecular subtyping surveillance system for foodborne bacterial disease, monitors *Escherichia coli (E. coli)* O157:H7, *Salmonella, Shigella*, and *Listeria monocytogenes*, and other bacterial pathogens (11) causing disease throughout the United States. In 2006, clusters of a common *E. coli* O157:H7 pulsed-field type were observed at several monitoring sites. An investigation revealed the source of the outbreak to be washed, pre-packaged, fresh spinach. Once the epidemiologic investigation identified spinach, the public was notified and *E. coli* O157:H7 with the putative pulsed-field type was isolated from an unopened package of spinach from an individual's home. Molecular typing enabled rapid linkage of cases occurring across several states, the identification of the disease source, and facilitated quick public health intervention.

A third application is the detection of infectious agents resistant or insensitive to prevailing therapies. A cluster of drug-resistant agents is often the first indication of an outbreak, particularly in a hospital setting. Mobile genetic elements that confer resistance can be exchanged between bacteria, even across species, complicating outbreak investigation. Molecular tools can distinguish between a common strain of a single bacterial species or a mobile genetic element, conferring antibiotic resistance across strains of the same or even different species. Outbreak control must take into account whether a mobile genetic element is being exchanged between species or if there is clonal spread of a single organism.

In the United States there is selective culture of organisms. In outpatient settings the most common bacterial infections, urinary tract infection and pneumonia, are generally treated empirically. Only if treatment fails is a culture taken. Thus surveillance for antimicrobial resistance reflects a biased sample, suggesting the subset most likely to have a resistant infection. The inherent selection biases should be taken into consideration when

Table 23.2. Applications of molecular tools in surveillance

- Distinguish between time-space clusters and sporadic cases of the same infection
- Identify clusters requiring further investigation
- Detect the emergence of strains with new resistance profiles
- Estimate prevalence of infection and observe trends over time

suggesting policy changes in therapies based on surveillance data.

Molecular tools have also been applied to screen biological specimens collected as part of ongoing national databases for the presence of known and newly discovered infectious agents. For example, blood samples are collected as part of the National Health and Nutrition Examination Survey, a multistage probability sample of the United States conducted every 10 years. This has enabled the estimation of the prevalence of various infectious agents, including hepatitis B and C viruses, human herpes virus 8 (which causes Kaposi sarcoma) and herpes simplex viruses 1 and 2. These studies provide insight into the frequency of new agents, and the distributions of agents by spatial-temporal and host characteristics. Such studies are extremely useful for generating hypotheses about transmission systems, potential prevention and control strategies, evaluating the effectiveness of ongoing prevention and control programmes, and observing time trends.

Describe the transmission system

The transmission system of an infectious agent determines how infectious agents are circulated within a population, and includes the transmission mode, interactions between the infectious agent and the host, the natural history of the infection, and interactions between hosts that lead to infection. The emergence and re-emergence of a variety of infectious agents highlights the utility of understanding the various transmission systems, as this understanding is central to identifying effective prevention and control strategies. Combining molecular typing methods with questionnaire data can confirm self-reported behaviours, especially important when the validity of self-report may be in doubt, such as contact tracing of sexually transmitted diseases. As described in detail below, molecular tools facilitate estimating parameters key to understanding the transmission system, including the incidence, prevalence, transmission probability, duration of carriage, effective dose, and probability of effective contact.

Estimation of key parameters

When using simple transmission models to estimate R_0, the average number of new cases generated from each infectious case in a fully susceptible population, the transmission probability per effective contact is needed, as well as the duration of infectivity and the rate of effective contact. Molecular tools can usefully be applied to estimate each of these parameters. Prior to the availability of modern molecular tools, our ability to empirically estimate transmission probabilities was limited. For example, the transmission of a sexually transmitted infection can be estimated by following couples where one is infected and the other is susceptible; however, without molecular tools it is difficult to ensure that the transmission event is not attributable to a person outside the partnership. For respiratory infections, such as pulmonary tuberculosis, our estimates of the transmission probability and natural history have been based on careful documentation of outbreaks. However, as we have been able to type individual strains, it has been determined that tuberculosis cases that previously were considered sporadic, and not part of apparent time-space clusters (because the exposure to the index case was very limited), were indeed part of the same outbreak (12).

Key transmission system parameters are incidence, prevalence, duration of infection, and transmission probabilities. The estimation of these parameters assumes the accurate measure of identical strains or subtypes of an infectious agent. As we have increased our ability to type infectious agents, we have been forced to re-evaluate many of our previous assumptions. One key assumption is that pathogens are clonal; that is, during active infection all infecting organisms are the same. A second, parallel assumption is that during an infectious process the pathogenic organism will be the one most frequently isolated from the infected site. For many diseases, we now know these assumptions

Table 23.3. Components of the transmission system and associated parameters

Component	Parameter
Occurrence in a population	Incidence Prevalence
Transmission mode	Probability of transmission given contact
Natural history of infection	Duration of infection
Interactions between agent and host	Effective dose
Interactions between hosts leading to transmission	Probability of contacting an infected individual

are false. For example, individuals can be infected with different strains of human papillomavirus (HPV), gonorrhea and even tuberculosis. During a diarrheal episode, the predominant organism isolated from the stool may not be the one causing the symptoms: a toxin-secreting organism occurring at low frequency may be the culprit. For infectious agents that also are human commensals, such as Streptococcus agalactiae, strains causing disease may be different from normal inhabitants and different strains may have different transmission systems.

These observations have profound impacts on the conduct of future studies. If the population genetic structure of pathogens is not clonal and the pathogen is not readily isolated, this must be reflected in the sampling of isolates for study. Multiple isolates must be sampled and tested from an individual. For example, if there is a second strain only 5% of the time and the pathogen is uniformly distributed in the sample, 28 different isolates must be sampled from an individual to reliably detect the second strain. Further, if an infectious agent mutates rapidly within a host, such as HIV, determining the mutation rate will be essential for accurately estimating transmission probabilities and following transmission chains.

As molecular tools can detect the presence of the organism or host response to a specific organism, in some cases, for a particular strain (13), studies can detect both incidence and prevalence of asymptomatic infection and clinical disease. Understanding the full extent of the circulation of a particular infectious agent is essential for making accurate predictions and determining appropriate prevention and control strategies.

The duration of carriage can be estimated from the prevalence and incidence, presuming that the average duration across strain type is of interest. However, if duration is short but incidence is high, an individual might become re-infected with a different strain type, suggesting a longer duration if strain types are not determined. By contrast, if a strain mutates rapidly within the human host, duration of carriage might be underestimated. Thus, strain-specific estimates of prevalence and incidence are essential to our understanding of disease etiology, especially if different strains have different propensities to cause diseases.

Using molecular tools to estimate contact patterns

Molecular tools can also assist in the estimation of contact patterns by identifying asymptomatic and low levels of infections. Asymptomatic infection is often a key component in maintaining disease transmission. For example, in a study of intra-family transmission of *Shigella*, asymptomatic carriage increased risk of a symptomatic episode within 10 days by nine-fold (14). Molecular typing can also be used to enrich and validate contact tracing information. The addition of molecular typing to epidemiologic information on gonorrhea cases in Amsterdam identified large clusters of individuals with related strains, individuals infected with different strains at different anatomical sites, and persons with high rates of re-infection (15). The results suggested that the transmission networks for men who have sex with men and for heterosexuals were essentially separate—a key public health insight for planning interventions.

Increase understanding of the epidemiology of infectious diseases

While the contributions of molecular tools to outbreak investigation and surveillance have been substantial, there have also been significant contributions to our understanding of the epidemiology of infectious diseases. Molecular tools enable us to trace the dissemination of a particular subtype across time and space, and thus develop theories of transmission and dissemination; determine the origin of an epidemic, and therefore test theories about reservoirs and evolution of a particular agent; follow the emergence of new infections as they cross species, testing our hypotheses about the apparent transmissibility and rate of evolution; and follow mobile genetic elements conferring antimicrobial resistance or virulence between strains within a species or between species, and so develop theories about evolution and transmission within the populations of infectious agents.

Tracing the dissemination of infectious agents across time and space

Infectious agents are constantly emerging and re-emerging. Some agents, like influenza, have a well-understood pattern, where new strains generally emerge from southeastern Asia. This allows not only set up of sentinel surveillance points, but prediction, with some accuracy, of which influenza strain type(s) are most likely to cause the next epidemic. As the genetics of influenza is fairly well-understood, appropriate vaccines for known variants can be prepared. The difficultly is when the virus undergoes an antigenic shift. At this writing, an influenza A strain

Table 23.4. Ways molecular tools increase understanding of the epidemiology of infectious diseases

- Tracing the dissemination of infectious agents across time and space
- Determine the origin of an epidemic
- Follow emergence of new infections
- Follow mobile genetic elements conferring virulence of antimicrobial resistance

Table 23.5. Applications of evolutionary theory to infectious disease epidemiology

- Identify genetic lineages
- Estimate rate of evolution
- Generate theories about the emergence and maintenance of specific lineages of infectious agents

(H5N1, also known as avian or bird influenza) has repeatedly caused human infection with very high case fatality rates (~50%), although the chains of transmission have been relatively short and the total number of cases is relatively small. However, there is an ongoing widespread epidemic among wild birds, and there have been several outbreaks among domestic birds, resulting in large-scale culling of birds and considerable adverse economic impact.

Other infectious agents have hit by surprise, such as the emergence of HIV, and the migration of West Nile virus to the United States. Further, some infectious agents have mutated in unpredicted ways, such as the emergence of multidrug-resistant tuberculosis, penicillin-resistant *Streptococcus pneumoniae*, and community-acquired methicillin resistant *S. aureus*. In addition to understanding the transmission system, the origin and source of entry of infectious agents into the population must be traceable to prevent and control the spread of infection. By comparing strains, it can be determined if there has been single or multiple points of entry, and if emerging resistance was from multiple spontaneous mutations or from dissemination of a single clone. For example, until 2004, only occasional isolates of gonorrhea found in Sweden were resistant to azithromycin, and these cases were attributed to acquisition elsewhere (16). However, in 2004, epidemiologic evidence suggested that domestic transmission might have occurred; this was confirmed by molecular typing. The ongoing transmission of the azithromycin-resistant strain in Sweden has short-term implications for surveillance and long-term implications for treatment recommendations.

Streptococcus pneumoniae (*S. pneumonia*) is a major cause of pneumonia, but also causes meningitis and otitis media. A major human pathogen, it is one of the most common indications for antibiotic use. Resistance to penicillin emerged relatively slowly, but once it emerged it was widely disseminated in relatively few clones as defined by multilocus sequence typing. By contrast, the recent emergence of *S. pneumonia* resistant to fluoroquinolones has been due to a diverse set of genetic mutations (17), suggesting spontaneous emergence following treatment. As *S. pneumonia* resistant to fluoroquinolones rapidly followed the introduction of fluoroquinolones, alternative antibiotics will be needed in relatively short order to treat S. pneumonia infections.

Determine the origin of an epidemic

Molecular tools enable us to trace an outbreak or epidemic back in time to its origin, and back in space to its reservoir. Knowing the origin in time is essential for predicting future spread and identifying the reservoir for infection is central for controlling disease spread. The use of molecular techniques has solved long-standing mysteries, such as cholera's reservoir between cholera epidemics. The same strains of cholera that infect humans also thrive in aquatic environments (19). While the importance of pigs and fowl as the origin of antigenic shifts in the genetics of influenza is understood, molecular tools have clarified that avian influenza need not first pass through the pig before jumping to humans, and that direct transmission from birds to humans is often more virulent (20). Molecular tools can also provide insight into the origins of infection in highly endemic populations, such as hospitals. The prevalence of methicillin resistant *Staphylococcus aureus* (MRSA) has been steadily increasing in hospitals in the United States; in 2004 the prevalence among some intensive care units was as high as 68% (21). However, in the early 2000s, new strains of MRSA emerged among individuals in the community that could not be traced back to hospitals. Genetic typing of the strains confirmed that strains isolated from those who had no epidemiologic linkage with hospitals had genotypically different strains (Figure 23.2) (18). More recently, community-acquired MRSA has joined hospital-acquired strains in causing infection in hospital settings (22).

Emergence of new infectious agents

Surveillance, outbreak investigation, sentinel networks, and the astute healthcare worker are keystones for identifying the presence of new disease syndromes. While a clearly defined clinical syndrome facilitates epidemiologic investigation, the potential for misclassification and associated bias can be high for non-specific syndromes. Molecular tools, such as non-culture techniques, have dramatically improved our ability to rapidly identify the etiologic agent and develop diagnostic tools. In addition, detection of the agent improves our ability to predict transmission routes, and identify potential therapies and prevention strategies by analogy to similar organisms.

Severe acute respiratory syndrome (SARS) was the first emerging disease identified this century. The story of the rapid isolation, identification, and sequencing of the coronavirus causing SARS, is illustrative of the synergistic effects of the marriage of molecular methods with epidemiology. SARS was first reported in southern China in 2002 and rapidly spread worldwide (Figure 23.3). Basic epidemiologic methods were essential for tracking the outbreak; a carefully collected epidemiologic case definition was sufficient for case ascertainment, clinical management, infection control, and identifying chains of transmission (23). However, key to characterizing and ultimately preventing and controlling the outbreak was the ability to detect mild cases and confirm that widely disseminated cases were caused by the same agent, which required a validated antibody test (24). Early in the epidemic there were many possible candidates identified as the causative agent, but these agents were not found in all SARS patients. A variety of state-of-the-art and standard molecular techniques were used to identify the viral agent, a new coronavirus. Molecular techniques established that the genetic sequences were the same throughout the world, and a rapidly developed test demonstrated that SARS patients had antibodies to the new coronavirus. Further, healthy controls not having SARS had no evidence of either past or present infection (25).

Trace mobile genetic elements

Mobile genetic elements are sequences of genetic material that can change places on a chromosome, and be exchanged between chromosomes, between bacteria, and even between species. A type of mobile genetic element, known as a plasmid, can integrate directly into the chromosome or extra-chromosomal in the cytoplasm of bacteria and still code for proteins. The recognition of mobile genetic elements, and the ability to trace these genetic elements as they move within and between species, has caused a re-thinking of the rate of, and potential for, evolution of infectious agents. For example, Shiga toxin-producing *E. coli* probably emerged from the transfer of genes coding for Shiga toxin from *Shigella* into *E. coli*.

Antibiotic resistance is often spread via a mobile genetic element. These elements tend to code for genes providing resistance against multiple antibiotics. This explains

Figure 23.2. Pulsed-field gel electrophoresis pattern relatedness of community-associated and health care-associated methicillin resistant Staphylococcus aureus (MRSA) isolates to a reference strain. The reference strain was MR14, which was the most commonly identified pattern among Minnesota MRSA isolates with a community-associated case definition (18). Copyright © (2003) American Medical Association. All rights reserved.

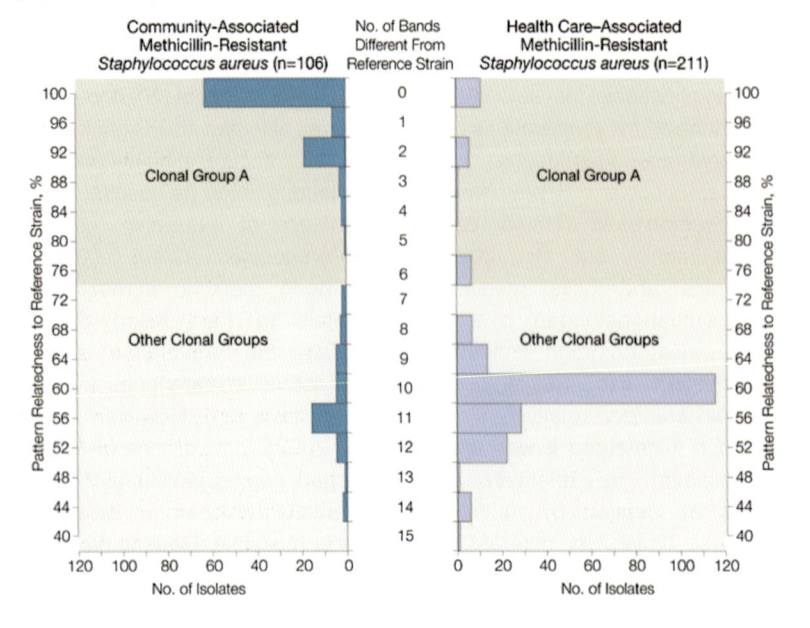

Figure 23.3. The rapid dissemination of severe acute respiratory syndrome (SARS) (http://yaleglobal.yale.edu/reports/images/SARS_MAP1.jpg, permission given by Yale Center for the Study of Globalization and YaleGlobal Online).

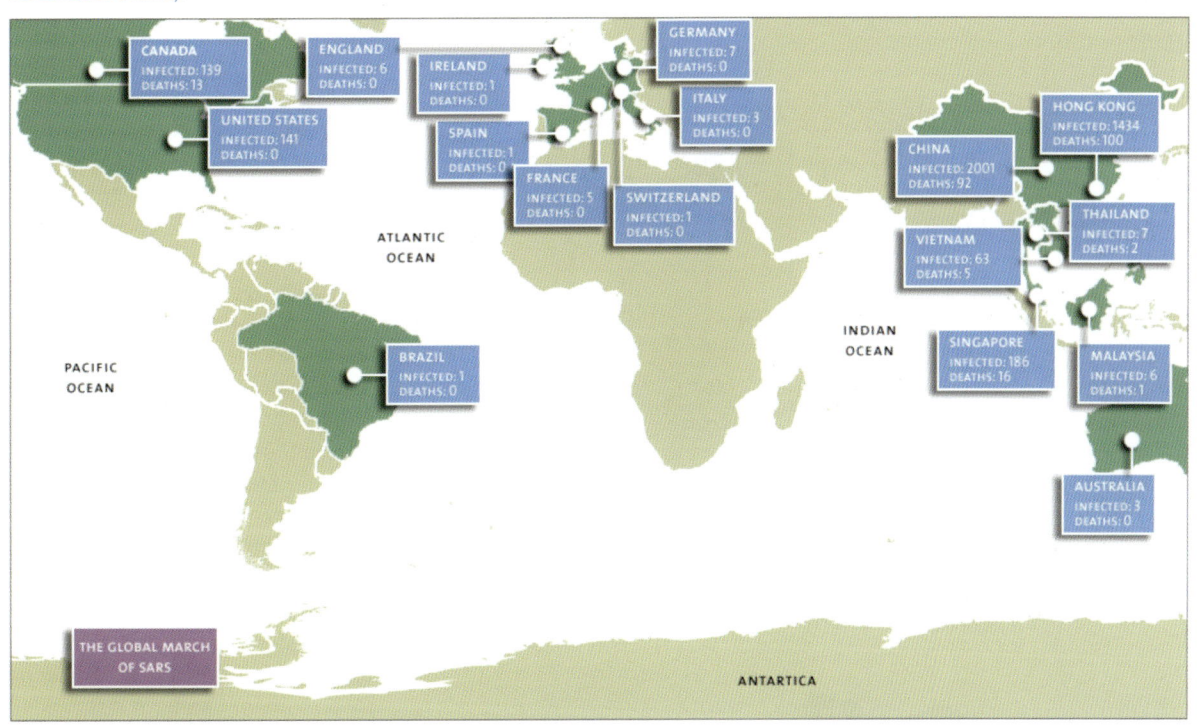

several apparent mysteries, such as the spread across several bacterial species within a hospital of the same antibiotic resistance profile, and why treating an individual with one antibiotic can result in resistance to multiple unrelated antibiotics.

Determine phylogenetic relationships

Genetic sequence and other molecular typing methods enable the construction of phylogenetic trees. Phylogenetics enables the use of evolutionary theory to explain epidemiologic phenomena, particularly emergence and transmission of more (or less) virulent strains, strains resistant to antimicrobials, simply to trace the transmission of a rapidly evolving species, or, in an outbreak situation, determine order of transmission. Separate phylogenies can be constructed for mobile genetic elements, or conserved elements on the chromosome.

Human immunodeficiency virus (HIV), which causes acquired immunodeficiency syndrome (AIDS), evolves quite rapidly even with a single host. Thus, the strain that infects an individual is not genetically identical to the strains that the individual might transmit to others. This property of HIV has made it possible to confirm the deliberate infection of one individual by another using a single blood sample from an individual (9) and to gain insight into the origin and spread of HIV worldwide.

There are three primary applications of phylogenic analyses in an epidemiologic context. First, phylogenic analysis enables us to determine genetic lineages. Using this type of analysis, researchers traced the introduction and spread of HIV in the Ukraine (Figure 23.4) (26). They were able to demonstrate that two subtypes introduced into drug networks in the 1990s still contributed to the epidemic in 2001 and 2002, and that one subtype spread widely throughout the Ukraine and into the Russian Federation, the Republic of Moldova, Georgia, Uzbekistan and Kyrgyzstan. Further studies to determine the biologic and social contributions to the success of the one subtype over another will provide important insights into how to control HIV.

A second application is to determine the rate of evolution. This is a standard application in biology, but understanding how fast infectious agents evolve has profound implications for choosing a molecular typing technique and interpreting epidemiologic data. For example, some agents change very rapidly, so that the agent infecting an individual is different from the agent that is transmitted to another, for

Figure 23.4. Phylogenetic analysis of strains from different cities in Ukraine. Phylogenetic trees of strains from Kiev (a), Crimea (b), Nikolayev (c), Odessa (d), Donetsk (e), and Poltava (f). A, B, C, D (capital letters) shown on tree branches are the HIV-1 subtypes. A scale bar of 0.01 substitutions per site is shown under each tree (26). The publisher for this copyrighted material is Mary Ann Liebert, Inc. publishers.

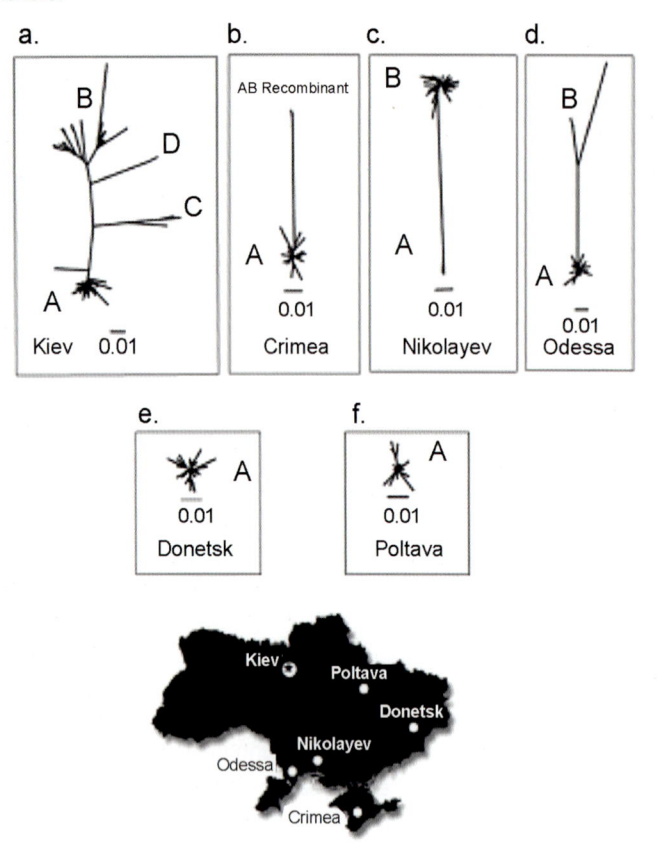

Test hypotheses about transmission systems

Applying molecular typing to ongoing or endemic disease transmission increases our understanding of how contact patterns produce observed patterns of disease, revealing novel prevention and control strategies. In addition to characterizing ongoing chains of transmission, molecular typing can clarify who had contact with whom, and who was the source of infection, and thus identify a transmission network. Identifying transmission networks provides essential information for targeting intervention programmes, particularly when designing and implementing vaccine programmes.

Using polymerase chain reaction (PCR)-restriction length polymorphism typing of the porin and opacity genes of *Neisseria gonorrhoreae* and questionnaire data, a study of successive gonorrhea cases in Amsterdam identified several ongoing transmission chains. The epidemiologic characteristics, including number of sexual partners and choice of same or opposite partners of patients with different molecular types differed, suggesting that the transmission chains represented different transmission networks (15). Molecular typing has also improved our understanding of tuberculosis transmission. Until confirmed by molecular typing, tuberculosis was not believed to be transmitted by short-term casual contact. Several investigations have demonstrated that this assumption is incorrect, such as clusters associated with use of services at day shelters (28), and even linked to only a few brief visits to an infected individual's worksite (12). Molecular typing has also demonstrated linkage between apparently sporadic tuberculosis cases, and determined that at least some recurrent tuberculosis

example HIV. Other agents change very slowly, such as tuberculosis. Thus, the appropriate typing technique must be chosen, so that phylogeny can be used to determine if rapidly changing agents evolved from a common ancestor, and to be able to distinguish between slowly evolving isolates. A typing technique for a rapidly evolving agent might focus on a region of the genome that evolves relatively slowly; a typing technique for a slowly evolving agent might focus on a genetic region that evolves fairly quickly, so that the investigator can distinguish between outbreak and non-outbreak strains.

A final application is to generate theories about the emergence and maintenance of specific genetic lineages. This application is a by-product of studies of genetic lineages and the rate of evolution. A key insight from the study in the Ukraine was the differential spread of different HIV subtypes (26). A study of the molecular epidemiology of norovirus outbreaks in Norway demonstrated the emergence of a new variant that accounted for a change in both the seasonal distribution and common transmission mode (27). A next step for furthering our understanding of the epidemiology of HIV and norovirus would be to generate theories to explain these phenomena.

is attributable to exogenous re-infection (reviewed by (29)).

Identify agent characteristics that lead to transmission and pathogenesis

The Microbial Genome Program of the US Department of Energy has sequenced more than 500 microbial genomes (http://microbialgenomics.energy.gov/brochure.pdf); the genetic sequence of numerous human pathogens have already been published, and many more are ongoing, as well as experiments to compare the sequences of other strains to a sequenced strain. A great deal can be learned from sequence data; of particular interest here is the identification of new open reading frames (ORF) which correspond to gene sequences. Although inferences can be made about a particular ORF based on the genetic sequence by comparing it to other gene sequences of known function, we cannot be certain of the gene's function or its importance to disease transmission or pathogenesis. However, conducting epidemiologic studies on appropriately collected samples can be done to increase understanding of the potential function of the genes and their relative prevalence using a molecular epidemiologic strategy (Table 23.6) (30). Many bacterial species are found in both diseased and healthy individuals and have highly diverse genomes. For example, *E. coli*, the most common cause of urinary tract infection and diarrhoea, is found in the normal bowel flora of virtually all humans and animals. When disease and commensal isolates are compared, the genome of *E. coli* is quite diverse, even when limited to human isolates. *E. coli* O157:H7, which causes diarrhoea and haemolytic urea syndrome, is substantially different in genetic content from the well-studied *E. coli* K12 strain, which was originally isolated from human feces in 1922 (http://www.sgm.ac.uk/pubs/micro_today/pdf/080402.pdf). Epidemiologic studies can take advantage of this variation in genetic content to compare the frequency of a putative virulence factor present among strains isolated from individuals with a specific pathology with the frequency among commensal isolates. For infectious agents with less diverse genomes, studies can determine differences in genetic alleles or in gene expression.

Studies using a molecular epidemiologic strategy can be done using high-throughput methods, such as multiplex PCR or microarrays. For example, Library on a Slide is a novel microarray platform that enables the screening of thousands of bacterial isolates for the presence of a putative virulence gene in a single experiment. The genomes of up to 5000 bacterial isolates can be arrayed, in duplicate, on a single array, and screened for the presence or absence of a single gene using dot blot hybridization (Figure 23.5) (31). Library on a Slide has been created for *Mycobacterium tuberculosis, Haemaphilis influenza, E. coli, S. pneumoniae,* Group B Streptococcus and *Streptococcus mutans.*

The molecular epidemiologic strategy has been productively applied to many bacteria species. For example, a study screening collections of middle ear and throat non-typeable Haemaphilis influenzae isolates, a major cause of

Table 23.6. Molecular epidemiologic strategy for gene discovery

- Identify candidate genes by combining bioinformatics information with molecular data
- Screen well-characterized representative samples of isolates causing different pathologies and asymptomatic infection
- Analyse to determine relative frequency of selected characteristics in various populations

Figure 23.5. Library on a slide microarray platform compared with a United States quarter. Each spot on the slide contains the total genomic DNA of a strain of *E. coli*. Photograph and slide by Dr. Lixin Zhang. Reprinted from (32), Copyright (2007), with permission from Elsevier.

otitis media, identified a gene found significantly more frequently among middle ear isolates, lic2B. lic2B was found 3.7 times more frequently among middle ear isolates than in throat isolates from children attending day care (33,34).

Identify infectious agents that are adapted to particular human hosts

Humans and infectious agents are extremely well adapted to each other. Some infectious agents are essential to human development, such as for digestion of foods and production of nutrients such as vitamins (35). Thus, it is not surprising that recent evidence suggests that normal microbiota co-evolved with their human hosts (36). Further, several studies suggest certain pathogenic species are adapted to certain human populations, and that infectious agents may select for mutations in human lineages.

Evidence that a particular genetic lineage of a pathogen may be better adapted to certain human populations, has been gathered by comparing the epidemiology of specific lineages in human populations of mixed genetic origin. For example, HPV variants of African origin persist longer among African American women than among white women, and European variants persist longer in white women than among African American women (37). A similar association between the infected host's region of origin and the infecting strain has been observed for tuberculosis (38). Social and behavioural factors have been associated with acquisition and persistence of each of these infectious agents, but tangible evidence of host susceptibility tied to specific agent characteristics implies that strategies based on the identification and development of more effective and specific therapies and prevention strategies may be more successful than attempts at changing human behaviour.

Infectious agents may also contribute to the evolution of humans. A well-documented case is the impact of malaria on the human host: there are several different genetic variations that protect against malaria. These variations are found in countries that either currently or in the past had endemic malaria. The effectiveness of the human genetic variant at reducing malarial disease varies with *Plasmodium* species. The most well-known variant is the sickle cell trait, but there are others, such as the Duffy blood group. When the Duffy blood group is absent, *Plasmodium vivax* is unable to enter the red blood cells (39). Using different molecular techniques, other human adaptations can be identified. These may provide insight into potential therapeutics, such as understanding the role of CCR5 in blocking HIV, or generate theories to explain observed human variation.

Identify new infectious agents causing disease

Epidemiologic studies have often suggested a possible infectious origin for a clinical syndrome. However, our ability to detect the etiologic agent has been hampered by our inability to culture most infectious agents. The development of the polymerase chain reaction (PCR) has dramatically increased our ability to detect infectious agents, both those cultivable and uncultivable. PCR has been essential to such public health triumphs as the development of a vaccine against HPV (the primary cause of cervical cancer), the identification of human herpes virus 8 as the infectious cause of Kaposi sarcoma, and the rapid identification of the coronavirus as the cause of SARS.

The development of a vaccine against cervical cancer was a direct result of our ability to use molecular tools in an epidemiologic context. Cervical cancer has an epidemiology that strongly suggests a sexually transmitted infection. The disease is associated with a greater lifetime number of sex partners, early age at first intercourse, and history of a sexually transmitted infection. The precursors of cervical cancer (cervical dysplasia detectable via PAP smear), were studied extensively, but widespread misclassification obscured the results. Cervical infection with different HPV types have different propensities to progress to cervical cancer, but the clinical presentation at the initial stages of infection, cervical dysplasia, is indistinguishable among types. It was not until the tools were available to identify HPV and to determine the different HPV types that the epidemiology was truly understood and an effective vaccine developed (40).

The Kaposi sarcoma (KS) story demonstrates the potential of combining an exquisitely sensitive molecular detection technique with epidemiologic study design. Using representational difference analysis to identify DNA sequences present in KS lesions but absent or present in low copy number in non-diseased tissue obtained from the same patient, researchers identified non-human DNA sequences in KS lesions of HIV patients (41). The sequences were determined to be herpes-like, subsequently designated human herpes virus 8. A randomized, blind, evaluation of tissue from patients with KS of different origin was then performed: AIDS-associated, classic, and among homosexual men who were HIV-seronegative (42). This confirmed that the DNA sequences were present in all types of KS, suggesting that the sequences were not found only among AIDS patients. Seroepidemiology studies confirmed that seroprevalence was correlated with risk of KS, and that seroconversion and seropositivity predicted development of KS (43).

The ability to detect non-culturable infectious agents provides new strategies to more rapidly identify emerging infections. A surveillance system has been established in the United States to identify the infectious components of unexplained deaths and critical illnesses possibly due to infectious causes (http://www.cdc.gov/ncidod/eid/vol8no2/01-0165.htm). This system enabled the rapid detection of West Nile Virus encephalitis when it first appeared in the United States (44).

Identify infectious agents involved in the initiation and promotion of chronic disease

Infectious agents are popularly associated with acute disease processes, but many well-studied infectious processes lead to chronic diseases, such as tuberculosis and AIDS. However, other diseases that were previously attributed to genetics, behavioural, or lifestyle factors are now known to have an infectious component, including stomach ulcers, chronic liver disease, and arthritis (Table 23.7). Pathogenesis occurs at the infectious-chronic axis complex (Figure 23.6), and includes interactions between the agent, host and environment. Some of these diseases result from molecular mimicry, that is, the infectious agent has epitopes so similar to the host that the host response attacks itself, such as in reactive arthritis or rheumatic fever following infection.

Table 23.7. Selected infectious causes of chronic diseases

Chronic disease	Infectious agent
Arthritis	*Borrelia burgdorferi*, Epstein-Barr Virus, *Salmonella spp*, *Campylobacter spp*, *Yersinia spp*, *Chlamydia spp*
Bladder cancer	*Schistosoma spp.*
Cervical, anal, penile, head and neck cancers	Human papillomavirus
Chronic liver diseases, hepatocellular carcinoma	Hepatitis B, Hepatitis C
Creutzfeldt-Jakob disease	Variant Creutzfeldt-Jakob disease
Gastric cancer	*Helicobacter pylori*
Heart disease	*Chlamydia pneumoniae*
Kaposi sarcoma	Human herpes virus 8
Leukemia	Human T-lymphotropic virus type 1
Lymphoma	Epstein-Barr virus, Human T-lymphotropic virus type 1
Peptic ulcer disease, chronic gastritis	*Helicobacter pylori*
Whipple disease	*Tropheryma whipplei*

Modified and expanded from (45).

Figure 23.6. Schematic showing how multiple factors interact leading to chronic sequelae of infectious diseases (45).

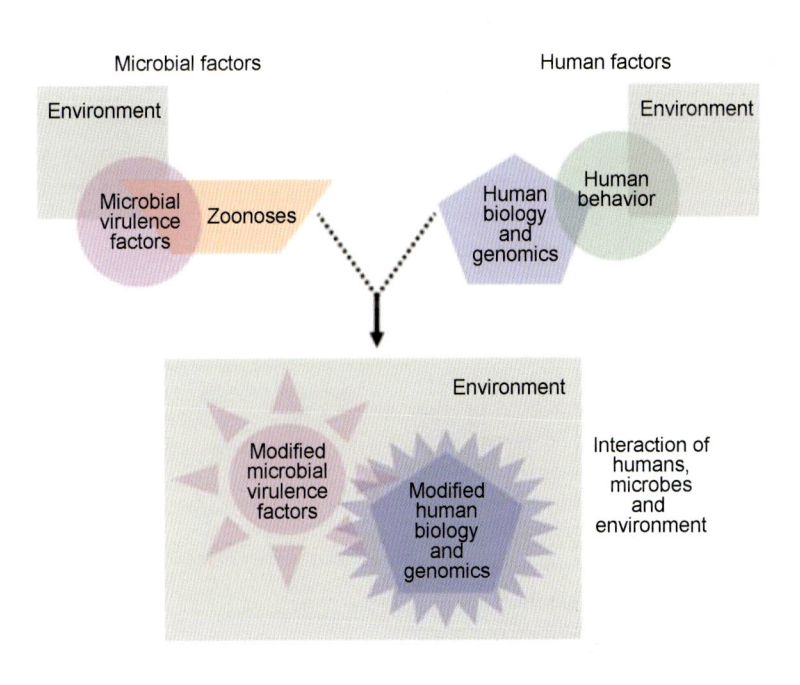

Alternatively, there may be disease that results from a host primed for infection that does not happen.

Molecular tools have definitively demonstrated that there are infectious causes of cancer: hepatitis B and C can cause liver cancer; serotypes of human papillomavirus cause cervical, anal, penile, and head and neck cancers; and herpes virus 8 causes KS (45). Infectious diseases, such as *Chlamydophila pneumoniae (C. pneumoniae)*, are hypothesized to lead to the promotion of atherosclerotic plaques and thus coronary artery disease. Numerous studies have demonstrated C. pneumoniae in atherosclerotic tissue, and severity of disease has been positively associated with antibodies to *C. pneumoniae* after adjustment for other known risk factors. However, results of antibiotic therapy in preventing progression of cardiovascular disease among those who already have disease have been disappointing. The totality of the evidence suggests that C. pneumoniae is neither a necessary nor sufficient cause of coronary artery disease, but is likely a modifiable risk factor (46).

Determining an infectious cause of a chronic disease is difficult: active infection often has ceased by the time the chronic disease is manifest. Disease clusters may be quite informative: Lyme disease was identified as a cause of juvenile arthritis because of careful epidemiologic investigation of a disease cluster (47). As the disease process is not solely a function of the presence of the infectious agent, but an interaction of the agent with the host, it is as likely to detect the presence of specific genes associated with the disease as the infectious agent. Moreover, there is evidence that infectious agents can incorporate their genetic material into the human genome: human endogenous retroviruses are the remnants of ancient germ cell infections (48). However, as we increase our ability to detect specific host response to infectious agents and detect traces of infection within the host, we will likely increasingly detect infectious causes of many diseases of unknown etiology.

Guide clinical treatment and intervention strategies

Knowledge of the molecular genetics of infectious agents and the interaction of the infectious agent with the host gained from molecular epidemiologic studies can be used to more rapidly detect infectious agents and thus improve patient diagnosis, predict disease course and identify potential vaccine candidates. Molecular techniques can also be applied to characterize the ecology of normal human flora, detect disruptions in the flora that lead to disease, and to detect the presence of biofilms (microbial structures which often contain multiple species that can initiate or promote disease or protect the host from disease) (Table 23.8).

Rapid detection of infectious agents

Increasingly, there are rapid methods for the detection of infectious agents. These methods have profound implications for public health practice, clinical diagnosis and epidemiologic studies. Rapid detection methods are based on either PCR, the amplification of genetic sequences of the infectious agent that can identify the agent present, or an antigen-antibody reaction that detects the host response to the infectious agent or a metabolite of the agent. Rapid detection means faster and more accurate diagnosis. For example, intrapartum prophylaxis with antibiotics has reduced by 50% the incidence of neonatal Group B streptococcal (GBS) disease. However, GBS colonization is often transient, so many women may be treated unnecessarily. Detection of GBS colonization at time of labour and delivery would minimize inappropriate antibiotic use, decreasing unnecessary pressure for the development of antibiotic resistance. Further, GBS is increasingly resistant to the second-line antibiotics used for women sensitive to the first-choice antibiotic. Rapid detection of antibiotic resistance, based on the detection of resistance genes and the ability to discriminate between more virulent GBS strains, will potentially improve medical care.

Rapid techniques are often extremely sensitive, so when applied in an epidemiologic context

Table 23.8. Using molecular tools in a clinical epidemiologic context

- Rapidly detect infectious agents
- Distinguish between pathogens and commensals
- Distinguish between relapse and re-infection
- Predict disease course
- Evaluate potential vaccine candidates
- Guide intervention

may be cost saving. For example, epidemiologic studies of colonization with MRSA, which has emerged as a community-acquired pathogen of some significance, can be screened for using rapid techniques; only specimens screening positive by rapid methods might be cultured. Culture is not only more time-consuming, but costly in terms of reagents and personnel. However, the ability to propagate an infectious agent is highly desirable, as it facilitates more detailed studies at the molecular level.

Distinguish between commensals and pathogens

Increasingly, it is recognized that many species of infectious agents previously thought to be harmless commensals can, under certain circumstances, cause disease (49). For example, fungal infections are only a problem among immune-suppressed patients. Acquired immunosuppression can result either from medical therapy, such as chemotherapy for cancer or immunosuppressive drugs prescribed to transplant patients, or as a result of infection, such as HIV. It has been discovered that all strains within a species do not have equal disease potential; indeed, an opportunistic infection may arise because of special characteristics of the infectious agent itself. This makes the identification of the causal agent in the laboratory difficult, as there are cases where basic clues such as quantity of the agent or agent type may be insufficient to identify the cause. Molecular tools can be used to identify and characterize the specific virulence potential, as well as identify humans particularly susceptible. This ability should eventually translate into improved laboratory tests, making it much easier for the laboratorian to determine the causal agent and the physician to prescribe appropriate therapy.

Distinguish between relapse and re-infection

Some infections have a chronic, recurring nature. In this situation, it is extremely useful to distinguish between a relapse, which implies treatment failure, and a new infection. Strain typing can be extremely useful in this situation, as typing allows us to distinguish between strains of the same species. For example, molecular typing demonstrated that individuals can be infected with more than one strain of tuberculosis (50) and of HIV (51).

Predict disease course

Disease course is a function of both host and agent factors. While an individual who receives a larger infectious dose of an infectious agent is, on average, more likely to become ill and to manifest symptoms more rapidly, this may not always be the case. The virulence of the infectious agent, whether the host has had previous exposure to the same or a similar strain of the infectious agent, or if the host has an underlying genetic predisposition or presence of predisposing factors, such as co-morbidities, all influence the infectious course. For example, initial viral load in HIV patients has been demonstrated as a good predictor of disease prognosis, as well as potential to transmit to others (52). Similar predictors for other infections are sure to follow.

Identify potential vaccine candidates

Early microbiology led to the development of several vaccines that have dramatically improved public health worldwide: smallpox has been eradicated, and measles and polio are largely under control in developed countries. Tetanus, influenza, diphtheria, mumps, rubella, chickenpox, hepatitis A and B, and yellow fever can be prevented, and the Bacille Calmette–Guérin vaccine for tuberculosis minimizes the most adverse manifestations of tuberculosis in children. Most recently, a vaccine against the HPV serotypes 16 and 18, which are most likely to cause cancer, was licensed. Nonetheless, many other infectious diseases that cause significant morbidity and mortality have remained intractable to prevention via vaccine using conventional strategies, but not from lack of trying. Some of these infectious agents, such as the bacteria that cause gonorrhea and bacterial meningitis, *Neisseria gonorrhoeae* and *Neisseria menigitidis*, can rapidly vary their surface antigens (53,54) making it difficult to identify an appropriate target.

The human pathogen sequencing project has resulted in a new strategy for the identification of vaccine candidates based on the predicted protein products based on the genetic sequence (Figure 23.7) (55). *In silico* analyses using bioinformatics enable the detection of potential epitopes. Many species are very diverse at the genetic level, thus not only must appropriate epitopes be identified, but epitopes that are found across the range of potentially diverse members of a particular species. Thus, epidemiologic screening of population-based samples of the species of interest for the presence of candidate epitopes can assist in selecting between potential candidates by ruling out those with limited geographic distribution (56).

Figure 23.7. Reverse vaccinology for identification of novel vaccine antigens (55). Reprinted by permission from Macmillan Publishers Ltd: Nature Biotechnology, copyright (2006).

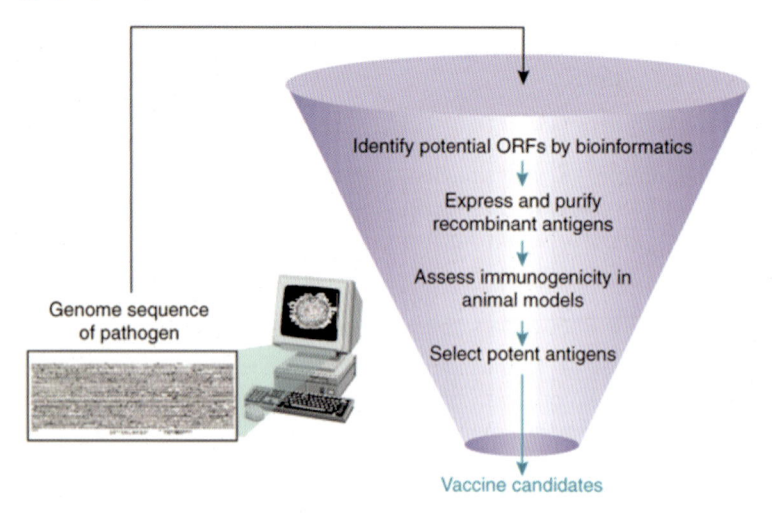

Guide intervention strategies

The ability of molecular tools to detect asymptomatic infection not only increases our understanding of the transmission system, but also has implications for clinical practice. For example, using culture, the protozoan parasite *Trichomonas vaginalis (T. vaginalis)* is detected in only 8–20% of male partners of women infected with *T. vaginalis*; testing urine with PCR detects *T. vaginalis* in up to 70% of partners (57). This has profound implications for preventing the spread of this common infection, suggesting that routine PCR testing of sex partners is in order.

Detecting the emergence and spread of resistance to therapy should lead to changes in clinical practice. Resistance genes are often carried on the same mobile genetic elements, so that resistance to one drug often implies resistance to others. Further, not only does treating an individual with antibiotics select for resistant organisms within that host, it also increases risk of acquiring resistant organisms in their contacts (58). A better understanding of the transmission of resistance genes between bacteria, and of resistant bacteria between individuals, will aid in designing effective policies. New therapies or combination therapies may be introduced or steps taken to minimize the spread of resistance.

Polymicrobial infections and interactions

Many diseases result from infection not with a single infectious agent, but as a result of a change in the microbial community that results from microbial activities (35). Microbes have certain nutrient and other requirements, but as they grow and die, they themselves serve as a source of nutrients for additional microorganisms. For example, an upper respiratory infection caused by a virus can enhance the environment for bacterial growth. Thus, upper respiratory infections are often a precursor to otitis media or bacterial pneumonia; influenza vaccination of children reduces rates of otitis media. Further, by-products of microbial growth can create local variations in pH, and in the presence of oxygen result in the growth of biofilms that enable colonization by other infectious agents. Biofilms are complex, often polymicrobial structures formed on a variety of surfaces in the environment and human body. The scum that forms on the insides of water pipes is a biofilm, but so is plaque on teeth, and the slimy surface on the tongue, inside the nose and throat, in the vaginal cavity and on other surfaces. There are also natural synergies or anergies between organisms of the same or different species; for example, the anergy between *S. pneumoniae* and *S. aureus* colonization in the nasal cavity (59). There are also synergies, such as is observed in the vaginal flora with lactobacillus modifying the pH and enabling the growth of other species.

PCR and high-throughput sequencing have enabled the description of the complex microbiota found on and in the human body. These studies use the fact that all cells have a ribosome, and that the sequence of genes that code for the ribosome can be used for taxonomy (61). These techniques have a great advantage over detection by culture, as culture requires at least a rough idea of what organisms might be present and their growth requirements. For example, applications of non-culture techniques to microorganisms in the human gut suggest that as many as 93% of the rRNA sequences identified are from uncultured organisms (62). Most non-culture techniques are based on some type of PCR and detect highly-conserved genes, such as those coding for 16sRNA, which vary at the species level and are semiquantitative. Others involve *in situ* hybridization techniques, enabling both detection

Figure 23.8. Vaginal epithelium from a healthy premenopausal woman hybridized with a universal probe (x400) and lactobacillus probe (inset x1000). Only a small number of bacteria are scattered over the surface of intact epithelium (A). Long rods can be seen with high magnification (inset). Bacteria are found in similar concentrations on the subepithelial surface of the biopsy that was exposed by mechanical trauma of the tissue (B) (60).

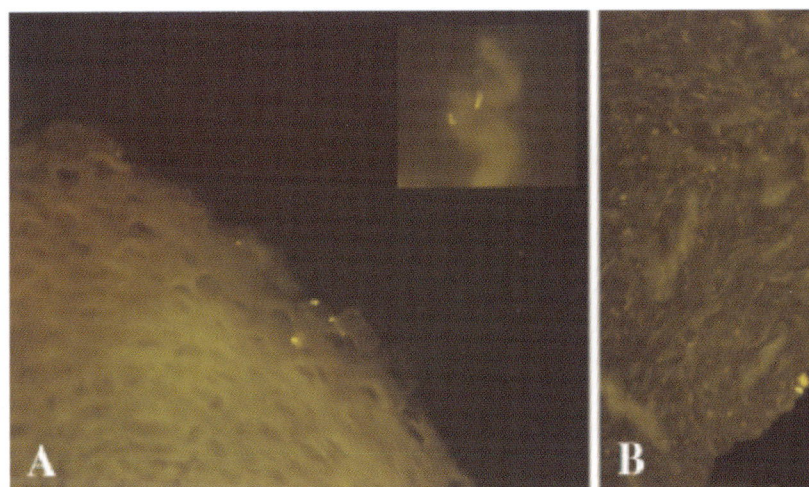

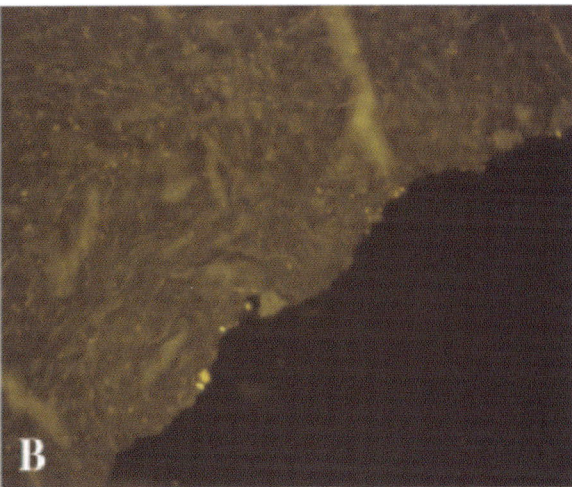

of organism presence and visualization of structure, such as vaginal epithelium shown in Figure 23.8 (60). There is much to learn about what constitutes normal flora; the dynamics of colonization; how colonization varies with normal biovariations such as the menstrual cycle, pregnancy and aging; and in the face of antibiotic therapy and disease.

Implications of using molecular tools for the design, conduct and analysis of epidemiologic studies

Modern molecular techniques combined with epidemiologic methods allow us to identify novel methods of disease prevention and control, markers of disease diagnosis and prognosis, and fertile research areas for potential new therapeutics and/or vaccines. However, the success of these studies depends not only on the molecular measure chosen, but also on whether the strengths and limitations of the chosen measure are considered in the design, conduct, analysis, and interpretation of the study results (Table 23.9).

Some molecular measures are relatively invariate with time, such as human genes. Studies associating genetic susceptibility to an infectious agent might be conducted using genetic material collected long before or after the disease occurred. By contrast, studies of host response to infection must be collected within a fairly tight time frame. Antibodies to an infectious agent may not appear until a defined period after infection, such as for HIV, or the infectious agent may be present only for a short duration, such as for *Streptococcus agalactiae*. Thus, some studies might be nested in large cohort studies or conducted using a case–control technique, while others require a prospective design.

The requirements of the molecular tool also affect sampling. For example, if a test must be conducted on fresh samples, the sampling of cases and controls in a case–control study should be done so that the groups are sampled and tested in similar time periods to minimize potential biases resulting from assay drift, which is where a method gives increasingly higher or lower results with time. For nested case–control studies, how specimens are collected may determine whether controls can be sampled from the base population (case-based, also called case–cohort, sampling) or at time

Table 23.9. Impact of using molecular tools on the design, conduct and interpretation of epidemiologic studies

- Study design: design choice, sampling
- Conduct: specimen collection and handling
- Analysis: translating laboratory measures to interpretable variables
- Interpretation: limitations of measures

of incidence disease (incidence density sampling) (63).

The conduct of the study must take into account the requirements of molecular testing. Some tests are sensitive to freezing and thawing and must be tested immediately, while others degrade over time, even if stored properly. Multiple different strains of a single infectious agent might be isolated from one individual, such as from different body sites. Labelling should make it possible to identify the appropriate strain and link it back to the appropriate individual and isolation site. Further, infectious agents may be grown in different media, passed in multiple cultures that might change phenotypic characteristics, or a plasmid might be lost. Thus, noting the number of times an isolate is cultured and on what media is important.

Epidemiologic analyses often use simple cut-offs: e.g. diseased versus not diseased. Laboratory measures often are continuous, but the scales may be ordinal, that is, the differences between values are not consistent. The interpretation of the measures might vary with the study population or the presence of other ancillary information. For example, >100 000 of a single bacterial species in the urine of an asymptomatic, healthy, non-pregnant individual has no clinical significance. If the individual has symptoms referable to the urinary tract, such as urgency and frequency, the individual probably has a urinary tract infection (64). By contrast, >100 000 of a single bacterial species in the urine of an asymptomatic, healthy, pregnant woman, is a treatable condition, because of the increased risk of pyelonephritis due to physiologic changes that occur during pregnancy.

Conclusions and future challenges

The applications of molecular tools to the study of infectious disease are varied, including applications to public health practice, diagnostics, and understanding of the transmission, evolution and pathogenesis. To date, the major potentials have been explored using genomics, but applying the power of proteomics and transcriptomics to the understanding of disease transmission and pathogenesis and host-agent interactions will open new avenues to understanding.

Much remains to be learned about infectious agents. Some future challenges are listed in Table 23.10. One area that is particularly amenable to study using molecular techniques is the normal human flora or microbiota. Extremely little is known about normal human microbiota, its response to invasion by pathogens, and its response to therapeutic treatment. Disruptions of normal microbiota are associated with a variety of pathogenic syndromes, such as bacterial vaginosis, that put the affected host at increased risk of acquiring other, often more serious, infectious agents. Interactions between disrupted normal microbiota and the host may also be important in explaining chronic recurring infections. Relatively little is understood about the structures formed by microbes within the human body; there are also structures that microbes stimulate the human host to form, such as pedestals on which *E. coli* O157:H7 sit.

Another area to explore is the interaction between the host and the agent. With molecular tools it was possible to identify why some individuals are repeatedly exposed to HIV but do not develop disease: these individuals have a variant in their CCR5 receptor that makes it difficult for HIV to invade the cell (65). It is also possible to identify human genes that explain why some individuals infected with HIV do not progress to AIDS. For example, a genome-wide association study (GWAS) identified the *HCP5* gene of the HLA region in chromosome 6 (66). GWAS have been applied to hepatitis C virus to identify why the infection spontaneously resolves in some individuals and treatment of chronic disease only eradicates infection in 40% of cases (67). Other human genetic variants likely modify risk of infection and response to infection, both positively and negatively, for many other infectious agents.

We have only begun to explore these interactions, and many challenges, both technological and methodological, remain (68). By combining modern molecular tools with epidemiologic methods, we have a powerful means to understand host agent interactions—an understanding essential for us to learn to live peacefully with microbes within our bodies, which, after all, outnumber the human cells that comprise us.

Table 23.10. Future challenges in the study of infectious disease

- Normal flora
- Biofilm formation
- Host agent interactions
- Successive infection

References

1. Kaufmann SHE, Schaible UE (2005). 100th anniversary of Robert Koch's Nobel Prize for the discovery of the tubercle bacillus. Trends Microbiol, 13:469–475.doi:10.1016/j.tim.2005.08.003 PMID:16112578

2. Laveran A. The Nobel Prize in physiology or medicine 1907. In: Nobel lectures, physiology or medicine 1901–1921. Amsterdam: Elsevier Publishing Company; 1967 (http://nobelprize.org/nobel_prizes/medicine/laureates/1907/laveran-bio.html).

3. Nicolle C. The Nobel Prize in physiology or medicine 1928. In: Nobel lectures. Physiology or medicine 1922–1941. Amsterdam: Elsevier Publishing Company; 1965 (http://nobelprize.org/nobel_prizes/medicine/laureates/1928/nicolle-lecture.html).

4. Daniel TM, editor. Wade Hampton Frost, pioneer epidemiologist 1880–1938: up to the mountain. Rochester (NY): University of Rochester Press; 2004.

5. Kilbourne ED (1973). The molecular epidemiology of influenza. J Infect Dis, 127:478–487.doi:10.1093/infdis/127.4.478 PMID:4121053

6. Jones TF, Kellum ME, Porter SS et al. (2002). An outbreak of community-acquired foodborne illness caused by methicillin-resistant Staphylococcus aureus. Emerg Infect Dis, 8:82–84.doi:10.3201/eid0801.010174 PMID:11749755

7. Centers for Disease Control and Prevention (CDC) (1998). Multistate outbreak of Salmonella serotype Agona infections linked to toasted oats cereal–United States, April–May, 1998. MMWR Morb Mortal Wkly Rep, 47:462–464. PMID:9639368

8. Hutin YJF, Pool V, Cramer EH et al.; National Hepatitis A Investigation Team (1999). A multistate, foodborne outbreak of hepatitis A. N Engl J Med, 340:595–602.doi:10.1056/NEJM199902253400802 PMID:10029643

9. Metzker ML, Mindell DP, Liu XM et al. (2002). Molecular evidence of HIV-1 transmission in a criminal case. Proc Natl Acad Sci USA, 99:14292–14297.doi:10.1073/pnas.222522599 PMID:12388776

10. Mainous AG 3rd, Hueston WJ, Everett CJ, Diaz VA (2006). Nasal carriage of Staphylococcus aureus and methicillin-resistant S aureus in the United States, 2001–2002. Ann Fam Med, 4:132–137.doi:10.1370/afm.526 PMID:16569716

11. Swaminathan B, Barrett TJ, Hunter SB, Tauxe RV; CDC PulseNet Task Force (2001). PulseNet: the molecular subtyping network for foodborne bacterial disease surveillance, United States. Emerg Infect Dis, 7:382–389. PMID:11384513

12. Golub JE, Cronin WA, Obasanjo OO et al. (2001). Transmission of Mycobacterium tuberculosis through casual contact with an infectious case. Arch Intern Med, 161:2254–2258.doi:10.1001/archinte.161.18.2254 PMID:11575983

13. Gieseler S, König B, König W, Backert S (2005). Strain-specific expression profiles of virulence genes in Helicobacter pylori during infection of gastric epithelial cells and granulocytes. Microbes Infect, 7:437–447.doi:10.1016/j.micinf.2004.11.018 PMID:15788154

14. Khan AI, Talukder KA, Huq S et al. (2006). Detection of intra-familial transmission of shigella infection using conventional serotyping and pulsed-field gel electrophoresis. Epidemiol Infect, 134:605–611.doi:10.1017/S0950268805005534 PMID:16288683

15. Kolader ME, Dukers NH, van der Bij AK et al. (2006). Molecular epidemiology of Neisseria gonorrhoeae in Amsterdam, The Netherlands, shows distinct heterosexual and homosexual networks. J Clin Microbiol, 44:2689–2697.doi:10.1128/JCM.02311-05 PMID:16891479

16. Lundbäck D, Fredlund H, Berglund T et al. (2006). Molecular epidemiology of Neisseria gonorrhoeae- identification of the first presumed Swedish transmission chain of an azithromycin-resistant strain. J Clin Microbiol, 114:67–71.doi:10.1111/j.1600-0463.2006.apm_332.x PMID:16499664

17. Doern GV, Richter SS, Miller A et al. (2005). Antimicrobial resistance among Streptococcus pneumoniae in the United States: have we begun to turn the corner on resistance to certain antimicrobial classes? Clin Infect Dis, 41:139–148.doi:10.1086/430906 PMID:15983908

18. Naimi TS, LeDell KH, Como-Sabetti K et al. (2003). Comparison of community- and health care-associated methicillin-resistant Staphylococcus aureus infection. JAMA, 290:2976–2984.doi:10.1001/jama.290.22.2976 PMID:14665659

19. Reidl J, Klose KE (2002). Vibrio cholerae and cholera: out of the water and into the host. FEMS Microbiol Rev, 26:125–139. doi:10.1111/j.1574-6976.2002.tb00605.x PMID:12069878

20. Van Reeth K (2007). Avian and swine influenza viruses: our current understanding of the zoonotic risk. Vet Res, 38:243–260. doi:10.1051/vetres:2006062 PMID:17257572

21. National Nosocomial Infections Surveillance System (2004). National Nosocomial Infections Surveillance (NNIS) System Report, data summary from January 1992 through June 2004, issued October 2004. Am J Infect Control, 32:470–485.doi:10.1016/j.ajic.2004.10.001 PMID:15573054

22. Sabol KE, Echevarria KL, Lewis JS 2nd (2006). Community-associated methicillin-resistant Staphylococcus aureus: new bug, old drugs. Ann Pharmacother, 40:1125–1133. doi:10.1345/aph.1G404 PMID:16735661

23. Weinstein RA (2004). Planning for epidemics–the lessons of SARS. N Engl J Med, 350:2332–2334.doi:10.1056/NEJMp048082 PMID:15175434

24. Berger A, Drosten Ch, Doerr HW et al. (2004). Severe acute respiratory syndrome (SARS)–paradigm of an emerging viral infection. J Clin Virol, 29:13–22.doi:10.1016/j.jcv.2003.09.011 PMID:14675864

25. Ksiazek TG, Erdman D, Goldsmith CS et al.; SARS Working Group (2003). A novel coronavirus associated with severe acute respiratory syndrome. N Engl J Med, 348:1953–1966.doi:10.1056/NEJMoa030781 PMID:12690092

26. Saad MD, Shcherbinskaya AM, Nadai Y et al. (2006). Molecular epidemiology of HIV Type 1 in Ukraine: birthplace of an epidemic. AIDS Res Hum Retroviruses, 22:709–714. doi:10.1089/aid.2006.22.709 PMID:16910825

27. Vainio K, Myrmel M (2006). Molecular epidemiology of norovirus outbreaks in Norway during 2000 to 2005 and comparison of four norovirus real-time reverse transcriptase PCR assays. J Clin Microbiol, 44:3695–3702. doi:10.1128/JCM.00023-06 PMID:17021099

28. DeRiemer K, Daley CL (2004). Tuberculosis transmission based on molecular epidemiologic research. Semin Respir Crit Care Med, 25:297–306.doi:10.1055/s-2004-829502 PMID:16088471

29. Mathema B, Kurepina NE, Bifani PJ, Kreiswirth BN (2006). Molecular epidemiology of tuberculosis: current insights. Clin Microbiol Rev, 19:658–685.doi:10.1128/CMR.00061-05 PMID:17041139

30. Zhang L, Foxman B, Manning SD et al. (2000). Molecular epidemiologic approaches to urinary tract infection gene discovery in uropathogenic Escherichia coli. Infect Immun, 68:2009–2015.doi:10.1128/IAI.68.4.2009-2015.2000 PMID:10722596

31. Zhang L, Srinivasan U, Marrs CF et al. (2004). Library on a slide for bacterial comparative genomics. BMC Microbiol, 4:12. doi:10.1186/1471-2180-4-12 PMID:15035675

32. Foxman B (2007). Contributions of molecular epidemiology to the understanding of infectious disease transmission, pathogenesis, and evolution. *Ann Epidemiol,* 17:148–156.doi:10.1016/j.annepidem.2006.09.004 PMID:17175168

33. Pettigrew MM, Foxman B, Marrs CF, Gilsdorf JR (2002). Identification of the lipooligosaccharide biosynthesis gene lic2B as a putative virulence factor in strains of nontypeable Haemophilus influenzae that cause otitis media. *Infect Immun,* 70:3551–3556.doi:10.1128/IAI.70.7.3551-3556.2002 PMID:12065495

34. Xie J, Juliao PC, Gilsdorf JR et al. (2006). Identification of new genetic regions more prevalent in nontypeable Haemophilus influenzae otitis media strains than in throat strains. *J Clin Microbiol,* 44:4316–4325. doi:10.1128/JCM.01331-06 PMID:17005745

35. Wilson M, editor. Microbial inhabitants of humans: their ecology and role in health and disease. New York (NY): Cambridge University Press; 2005.

36. Oh PL, Benson AK, Peterson DA et al. (2010). Diversification of the gut symbiont Lactobacillus reuteri as a result of host-driven evolution. *ISME J,* 4:377–387.doi:10.1038/ismej.2009.123 PMID:19924154

37. Xi LF, Kiviat NB, Hildesheim A et al. (2006). Human papillomavirus type 16 and 18 variants: race-related distribution and persistence. *J Natl Cancer Inst,* 98:1045–1052.doi:10.1093/jnci/djj297 PMID:16882941

38. Hirsh AE, Tsolaki AG, DeRiemer K et al. (2004). Stable association between strains of Mycobacterium tuberculosis and their human host populations. *Proc Natl Acad Sci USA,* 101:4871–4876.doi:10.1073/pnas.0305627101 PMID:15041743

39. Kwiatkowski DP (2005). How malaria has affected the human genome and what human genetics can teach us about malaria. *Am J Hum Genet,* 77:171–192.doi:10.1086/432519 PMID:16001361

40. Trottier H, Franco EL (2006). The epidemiology of genital human papillomavirus infection. *Vaccine,* 24 Suppl 1;S1–S15.doi:10.1016/j.vaccine.2005.09.054 PMID:16406226

41. Chang Y, Cesarman E, Pessin MS et al. (1994). Identification of herpesvirus-like DNA sequences in AIDS-associated Kaposi's sarcoma. *Science,* 266:1865–1869. doi:10.1126/science.7997879 PMID:7997879

42. Moore PS, Chang Y (1995). Detection of herpesvirus-like DNA sequences in Kaposi's sarcoma in patients with and without HIV infection. *N Engl J Med,* 332:1181–1185. doi:10.1056/NEJM199505043321801 PMID:7700310

43. Cathomas G (2000). Human herpes virus 8: a new virus discloses its face. *Virchows Arch,* 436:195–206.doi:10.1007/s004280050031 PMID:10782877

44. Hajjeh RA, Relman D, Cieslak PR et al. (2002). Surveillance for unexplained deaths and critical illnesses due to possibly infectious causes, United States, 1995–1998. *Emerg Infect Dis,* 8:145–153.doi:10.3201/eid0802.010165 PMID:11897065

45. O'Connor SM, Taylor CE, Hughes JM (2006). Emerging infectious determinants of chronic diseases. *Emerg Infect Dis,* 12:1051–1057. PMID:16836820

46. Watson C, Alp NJ (2008). Role of Chlamydia pneumoniae in atherosclerosis. *Clin Sci (Lond),* 114:509–531.doi:10.1042/CS20070298 PMID:18336368

47. Steere AC, Malawista SE, Snydman DR et al. (1977). Lyme arthritis: an epidemic of oligoarticular arthritis in children and adults in three connecticut communities. *Arthritis Rheum,* 20:7–17.doi:10.1002/art.1780200102 PMID:836338

48. Bannert N, Kurth R (2004). Retroelements and the human genome: new perspectives on an old relation. *Proc Natl Acad Sci USA,* 101 Suppl 2;14572–14579.doi:10.1073/pnas.0404838101 PMID:15310846

49. Casadevall A, Pirofski LA (2000). Host-pathogen interactions: basic concepts of microbial commensalism, colonization, infection, and disease. *Infect Immun,* 68:6511–6518.doi:10.1128/IAI.68.12.6511-6518.2000 PMID:11083759

50. Warren RM, Victor TC, Streicher EM et al. (2004). Patients with active tuberculosis often have different strains in the same sputum specimen. *Am J Respir Crit Care Med,* 169:610–614.doi:10.1164/rccm.200305-714OC PMID:14701710

51. Janini LM, Tanuri A, Schechter M et al. (1998). Horizontal and vertical transmission of human immunodeficiency virus type 1 dual infections caused by viruses of subtypes B and C. *J Infect Dis,* 177:227–231. doi:10.1086/517360 PMID:9419195

52. Pilcher CD, Tien HC, Eron JJ Jr et al.; Quest Study; Duke-UNC-Emory Acute HIV Consortium (2004). Brief but efficient: acute HIV infection and the sexual transmission of HIV. *J Infect Dis,* 189:1785–1792.doi:10.1086/386333 PMID:15122514

53. Seifert HS (1996). Questions about gonococcal pilus phase- and antigenic variation. *Mol Microbiol,* 21:433–440. doi:10.1111/j.1365-2958.1996.tb02552.x PMID:8866467

54. Davidsen T, Tønjum T (2006). Meningococcal genome dynamics. *Nat Rev Microbiol,* 4:11–22.doi:10.1038/nrmicro1324 PMID:16357857

55. Ulmer JB, Valley U, Rappuoli R (2006). Vaccine manufacturing: challenges and solutions. *Nat Biotechnol,* 24:1377–1383. doi:10.1038/nbt1261 PMID:17093488

56. Hebert AM, Talarico S, Yang D et al. (2007). DNA polymorphisms of Mycobacterium tuberculosis pepA and PPE18 genes among clinical strains: implications for vaccine efficacy. *Infect Immun,* 75:5798–5805. doi:10.1128/IAI.00335-07 PMID:17893137

57. Seña AC, Miller WC, Hobbs MM et al. (2007). Trichomonas vaginalis infection in male sexual partners: implications for diagnosis, treatment, and prevention. *Clin Infect Dis,* 44:13–22.doi:10.1086/511144 PMID:17143809

58. Samore MH, Lipsitch M, Alder SC et al. (2006). Mechanisms by which antibiotics promote dissemination of resistant pneumococci in human populations. *Am J Epidemiol,* 163:160–170.doi:10.1093/aje/kwj021 PMID:16319292

59. Teti G, Tomasello F, Chiofalo MS et al. (1987). Adherence of group B streptococci to adult and neonatal epithelial cells mediated by lipoteichoic acid. *Infect Immun,* 55:3057–3064. PMID:3316030

60. Swidsinski A, Mendling W, Loening-Baucke V et al. (2005). Adherent biofilms in bacterial vaginosis. *Obstet Gynecol,* 106:1013–1023. doi:10.1097/01.AOG.0000183594.45524.d2 PMID:16260520

61. Clarridge JE 3rd (2004). Impact of 16S rRNA gene sequence analysis for identification of bacteria on clinical microbiology and infectious diseases. *Clin Microbiol Rev,* 17:840–862.doi:10.1128/CMR.17.4.840-862.2004 PMID:15489351

62. Bäckhed F, Ley RE, Sonnenburg JL et al. (2005). Host-bacterial mutualism in the human intestine. *Science,* 307:1915–1920. doi:10.1126/science.1104816 PMID:15790844

63. Rundle AG, Vineis P, Ahsan H (2005). Design options for molecular epidemiology research within cohort studies. *Cancer Epidemiol Biomarkers Prev,* 14:1899–1907. doi:10.1158/1055-9965.EPI-04-0860 PMID:16103435

64. Warren JW, Abrutyn E, Hebel JR et al.; Infectious Diseases Society of America (IDSA) (1999). Guidelines for antimicrobial treatment of uncomplicated acute bacterial cystitis and acute pyelonephritis in women. *Clin Infect Dis,* 29:745–758.doi:10.1086/520427 PMID:10589881

UNIT 5.
APPLICATION OF BIOMARKERS TO DISEASE

CHAPTER 24.

Obesity

Salma Musaad and Erin Haynes

Summary

The adverse effects of obesity support the use of biomarkers to help elucidate disease mechanism, therapeutic interventions, and preventive strategies. Emerging biomarkers for obesity-associated cardiovascular disease (CVD), type 2 diabetes and cancer play diverse roles in biological pathways including immune modulation and fat metabolism. Animal and *in vitro* data support the association of these biomarkers with obesity-associated diseases, but evidence in humans is still lacking. In humans, plasma levels of biomarkers are widely used to determine risk, but many studies are limited by ethnicity/race, gender or sample size. In this chapter, the use of biomarkers in obesity research and in the context of CVD, type 2 diabetes and cancer will be discussed. Markers of exposure (adipokines), effect (resulting metabolic abnormalities), and susceptibility (genetic determinants for obesity and related disorders) are covered for each of the three diseases.

Introduction

Obesity epidemiology has typically relied upon long-established markers, such as blood cholesterol, triglycerides and blood pressure. It is now recognized that novel, non-traditional biomarkers have the potential to augment the utility of traditional markers. Emerging as a more formative tool in obesity epidemiology, the majority of non-traditional biomarkers act in relation to fat cells, or adipocytes. Acting as endocrine organs, adipocytes produce a variety of peptides and metabolites that result in a cascade of events leading to inflammation and oxidative stress. These products are being explored as biomarkers for the prevention, diagnosis, risk stratification and control of obesity co-morbidities, such as cardiovascular disease (CVD) (1,2), type 2 diabetes (3,4) and certain cancers (5,6). Despite the dangerous health effects of obesity, little is known regarding the clinical utility of adipokines in modifying disease risk, especially cancer.

This chapter focuses on the use of non-traditional biomarkers in obesity research and more specifically, in the context of CVD, type 2 diabetes and cancer. Adipokines will be referred to as markers of exposure and the resulting metabolic abnormalities as markers of effect. Genetic determinants for obesity and related disorders are referred to as markers of susceptibility. Markers of exposure, effect and susceptibility are discussed for each of the three co-morbidities.

Obesity is multifactorial

Generally defined, obesity is a state of excess weight gain and increased body fat that is disproportionate to the individual's height. Obesity is a prevalent disorder adversely impacting quality of life (7,8) and life expectancy (9). In the USA it is estimated that 32% of adults and 17% of children and adolescents aged 2–19 years were obese in 2003–2004, a dramatic increase from the previous two decades (10), justifying the need for prevention of obesity and related disorders. Caused by a combination of genetic, metabolic and environmental factors, obesity is characterized by an imbalance between energy intake and energy expenditure. This imbalance is closely regulated by signals emanating from and controlled by the central nervous system.

The central melanocortin system is integral in regulating food intake and peripheral lipid metabolism (11). Peripheral signalling molecules, such as ghrelin (12) and cholecystokinin (13), communicate with this system to control energy metabolism (11). Other signalling networks, such as the endocannabinoids (14), are believed to be involved in the development of obesity. In addition to the central nervous system, the adipose tissue is integral in the development of obesity because its expansion signifies the obese state (Figure 24.1).

Adipose tissue is an active organ innervated by the sympathetic nervous system (15), communicating with the hypothalamo-pituitary axis and the adrenal glands (16) to influence food intake, hunger, energy expenditure and adipose tissue mass. Adipose tissue secretes a variety of signals that influence energy balance, including leptin and adiponectin (17). Leptin reduces food intake, and resistance to its activities is often found in obesity. Adiponectin promotes insulin sensitivity and exhibits anti-inflammatory actions (17). In addition to the influence of adipose tissue, obesity can be caused by inherited defects.

Rare, monogenic forms of obesity are caused by single genetic mutations in genes, such as leptin, leptin receptor and pro-opiomelanocortin, as well as chromosomal rearrangements (18). These defects are among the few direct causes of obesity and are not the focus of this chapter.

Non-genetic causes of obesity are commonly implicated in the current rise in obesity prevalence in the USA and worldwide. Modifiable behavioural risk factors, including energy intake and a sedentary lifestyle, are principal components in the development of obesity (19,20). As a result, improving dietary habits by increasing the intake of fruits and vegetables and decreasing fat-laden foods, together with an active lifestyle, are vital techniques for the prevention of obesity and have been

Figure 24.1. The central melanocortin system, endocannabinoid system, and autonomic system are among the most important central nervous system (CNS) control areas regulating energy intake and expenditure. Short-term signals control food intake, while long-term signals chronically regulate lipid storage and metabolism, as well as glucose homeostasis. The signals simultaneously exert their peripheral effects to control energy metabolism via multiple tissues including adipose tissues. Fat deposition modifies adipose tissue function, leading to abnormal production of various molecules. The resulting potential biomarkers occur at various stages in the course of obesity and may interact with environmental, behavioural, and genetic factors and possibly with traditional risk factors, leading to disease manifestations.

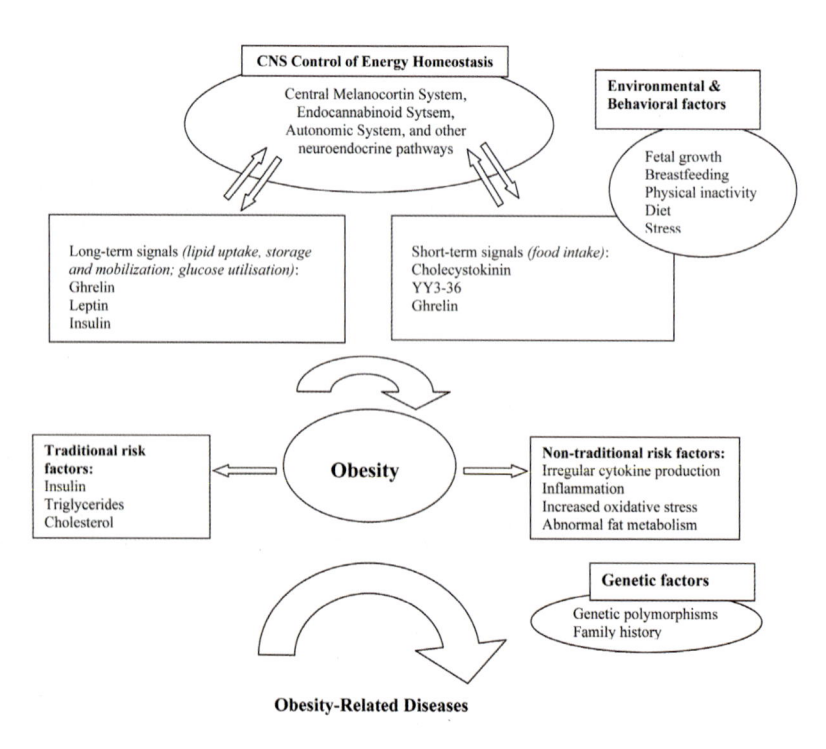

the focus of large obesity prevention trials (21).

Just as obesity is a complex disorder, molecular biomarkers that occur in obesity and that help identify individuals most at risk for subsequent cardiovascular disease, diabetes or cancer are complex. There is a significant degree of functional overlap among the biomarkers, and in some cases it is hard to discern the order in which they first appear in the body. For simplicity, however, the biomarkers have been broadly classified into exposure, effect or susceptibility (Table 24.1).

Overview of inflammation and oxidative stress

Abnormal adipokine production with consequent inflammation and oxidative stress may play an instrumental role in obesity-related disorders. Increased accumulation of adipose tissue, particularly visceral obesity, causes abnormal cytokine release and macrophage recruitment, all of which induce systemic inflammation (22).

Accumulation of adipose tissue also leads to increased oxidative stress partly via the oxidant effects of free fatty acids (23). Leukocytes derived from obese individuals and healthy individuals infused with free fatty acids generate reactive oxygen species (24,25). Consequently, reactive oxygen species induce a pro-inflammatory state and promote adverse metabolic complications, such as insulin resistance (23). Despite the association of oxidative stress with CVD independent of traditional risk factors (26), prospective clinical trials of antioxidant supplementation for reducing CVD risk provide conflicting evidence (27,28), possibly due to lack of a strong effect of oxidative stress on atherosclerosis, choice of antioxidant therapy, confounding by other dietary and non-dietary factors, or unsuitable choice of the biomarker. Nevertheless, oxidative stress is increasingly recognized as a potential mechanism for obesity-related disorders, hence it is included in this chapter.

Owing to the clear link between fat accumulation, inflammation and oxidative stress, the following three sections focus on these mechanisms in the context of CVD, diabetes and cancer. The sections are not intended to be all-inclusive, but aim to introduce the reader to major biomarkers of potential benefit in disease prevention and early detection. A brief introduction to the major epidemiologic study designs used to evaluate the biomarkers and common methodological issues then follows.

Biomarkers of obesity and subsequent cardiovascular disease

For the purposes of this section, cardiovascular disease (CVD) outcomes comprise cerebrovascular disease (cerebral embolism, thrombosis and haemorrhage), peripheral arterial disease, coronary heart disease, and atherosclerosis. Different mechanisms are implicated in the link between obesity and CVD. For example, the fetal origin of metabolic risk (29), epigenetic gene regulation (30), and the "pup in a cup" model (31) are potential causes of increased CVD risk in obesity. In addition, cardiovascular injury is promoted by adipokines, cytokines and other molecules that affect multiple pathways, such as lipid metabolism and immune modulation, eventually leading to inflammation or oxidative stress (32).

Inflammatory biomarkers that are elevated in obesity include leptin, plasminogen activator inhibitor 1, and adiponectin. Beyond regulation of energy expenditure, leptin induces a myriad of inflammatory mediators (33) and alters myocardial structure (34). Leptin has extensive regulatory functions and has been

Table 24.1. Classification of molecular biomarkers for obesity and subsequent cardiovascular disease, diabetes or cancer

Class	Molecular biomarkers	Examples of studies
Exposure	Adipokines: surrogates for adipose tissue deposition	Characterise adipose tissue type and activity; association with disease outcome; assess change with weight loss
Effect	Markers of inflammation and oxidative stress: mechanisms by which obesity may exert its toxic effects	Monitor disease progression; association with disease outcome; disease prognosis; disease intervention
Genetic susceptibility	Gene polymorphisms: account for variation in susceptibility to obesity-related disorders	Gene-phenotype relations; heritable variations in the quantity of systemic biomarkers; risk stratification

implicated in multiple adverse CVD endpoints independent of traditional risk factors. The National Health and Nutrition Examination Survey (NHANES) III was used to conduct a retrospective analysis of leptin concentrations and history of myocardial infarction and stroke (35). Plasminogen activator inhibitor 1 is an indirect correlate of abdominal obesity (36), though the molecular mechanism is still uncertain (37). Increased plasminogen activator inhibitor 1 contributes to CVD by impairing fibrinolysis and promoting cardiovascular tissue remodeling and the formation of blood clots. Its involvement in CVD is still controversial, as shown, for example, in a case–control study and meta-analysis on stroke (38,39); nevertheless, data suggest a role in patients with a prior history of CVD. In these patients, genetic variants in plasminogen activator inhibitor 1 have been implicated in disease recurrence such as recurrent myocardial infarction (40,41). Plasminogen activator inhibitor 1 is also associated with traditional risk factors, as well as other CVD risk indicators including low adiponectin (42). Adiponectin is a protective molecule inversely related to obesity and the associated metabolic abnormalities (43), and some studies, including prospective evaluations, strongly suggest that low plasma concentrations are implicated in increased CVD risk (43–45). Adiponectin's protective effects arise from its versatile immune functions and its ability to protect the vascular endothelium by antagonising inflammation and oxidative stress (43).

A hallmark of oxidative stress, reactive oxygen species mount their effects by reacting with diverse biological molecules including lipids and lipid derivatives. For example, oxidized low density lipoproteins are readily taken up by macrophages, thus promoting atherosclerosis (46). F_2-isoprostanes are prostaglandin-like compounds formed from the oxidation of cell membrane-derived fatty acids and have been implicated in atherosclerosis (47). The Framingham Heart Study was accessed to examine the utility of using F_2-isoprostane as a biomarker of oxidative stress. After adjustment for age and sex, it was found that the biomarker was increased in obese individuals (48).

In addition to oxidation products, enzymatic manipulation of reactive oxygen species further determines the effects of oxidative stress. One of the most studied oxidative stress-related enzymes in the field of cardiovascular medicine is glutathione peroxidase 1, an antioxidant that has been shown to drop in obese individuals in both prospective (49) and cross-sectional (50) studies. It is also expressed in the endothelium and protects blood vessels against oxidative stress, not just by counteracting reactive oxygen species, but by inhibiting oxidative enzymes that contribute to atherosclerosis (51).

Other potential markers of CVD risk in obese individuals include monocyte chemoattractant protein 1 (52). In addition to macrophages and endothelial cells in the vascular wall, monocyte chemoattractant protein 1 is synthesized by adipose tissues. Microarray gene expression profiles of subcutaneous adipose tissues demonstrate increased expression in obese compared to non-obese individuals (53). By recruiting macrophages to blood vessel walls, this chemokine contributes to atherosclerosis, but its link to CVD risk is still under investigation (54), as some studies do not find an association with subsequent cardiovascular events. Other studies provide evidence, however, for a role in long-term CVD prognosis. For example, prospective evaluation of patients with acute coronary syndromes indicates that monocyte chemoattractant protein 1 is independently associated with long-term mortality and cardiovascular events (55). Another biomarker that is associated with obesity and cardiovascular events is the angiotensin converting enzyme (56,57). In addition to blood pressure regulation, this enzyme exerts local pro-inflammatory effects in several tissues, including cardiac myocytes, and has long been employed for the management of heart failure. A recent meta-analysis maintains that angiotensin converting enzyme inhibitors reduce the risk of cardiovascular mortality, as well as specific endpoints, such as myocardial infarction and stroke, but the mechanism is yet to be explained (58). Table 24.2 summarizes the effects of the above biomarkers, as well as other molecules that can potentially be used for disease prevention or intervention. Also, see Chapter 20 for additional discussion of biomarkers associated with CVD.

Biomarkers of obesity and type 2 diabetes

Insulin resistance is a precursor of type 2 diabetes, an inflammatory condition (59,60) characterized by glucose intolerance (61). Due to the central role of glucose metabolism in the pathogenesis of type 2 diabetes, molecular biomarkers that function in this pathway may prove instrumental in targeting individuals with the highest risk for disease, designing tailored interventions, and improving risk stratification. One gene involved in glucose synthesis is PCK1, encoding the enzyme phosphoenolpyruvate carboxykinase. PCK1 is expressed in adipocytes, intestinal epithelia,

Table 24.2. Biomarkers associated with obesity and cardiovascular disease outcomes

Class	Biomarker	Disease/risk factor (Reference)
Exposure	Interleukin 18	Coronary heart disease (113)
	Leptin	Haemorrhagic stroke (114) Acute myocardial infarction (115)
	Plasminogen activator inhibitor 1	Recurrent myocardial infarction (116)
Effect	Low adiponectin	Coronary artery disease (43-45)
	Oxidized low density lipoproteins	Ischemic damage in cortical lesions (115) Coronary heart disease (117)
	Glutathione peroxidase	Reduced risk of death from cardiovascular events or non-fatal myocardial infarction (118)
	F_2-Isoprostanes	Coronary artery calcification (119) Coronary artery stenosis (120)
	Monocyte chemoattractant protein 1	Long-term mortality and cardiovascular events (55)
Genetic susceptibility	Angiotensin-converting enzyme	Coronary heart disease (121,122)
	Plasminogen activator inhibitor 1	Recurrent myocardial infarction (40,41)

and hepatocytes. Hepatic overexpression is associated with a diabetic phenotype in mice and overexpression in adipose tissues causes obesity. Among the few human studies performed to date, some show a link between PCK1 variants with type 2 diabetes (62). Another gene, ectonucleotide pyrophosphatase/phosphodiesterase 1, has shown conflicting associations with type 2 diabetes and obesity, but its role as an inhibitor of insulin signalling warrants further examination as a potential candidate gene for obesity-associated type 2 diabetes. In some populations, polymorphisms in this gene are associated with childhood and adult obesity, as well as type 2 diabetes in obese individuals (63–65) (see Chapter 7). Further population-based studies are needed to validate its use as a predictive biomarker. Glucose homeostasis is also regulated by several obesity-associated adipokines that control multiple immune pathways. Low adiponectin concentrations (66,67) and high tumour necrosis factor-α levels (68–70) have often been associated with insulin resistance. Despite conflicting evidence, promoter polymorphisms in the tumour necrosis factor-α gene have been implicated in increased insulin resistance, particularly in obese adults with type 2 diabetes (71–73). Further, one large (n = 809) population-based cross-sectional study of unrelated Caucasians showed that this gene interacts with adiponectin resulting in lower adiponectin levels and higher glucose and insulin concentrations two hours after glucose administration (74).

In addition to glucose, perturbed fatty acid metabolism, uptake and transport has been implicated in insulin resistance and diabetes. Uptake of fatty acids is partly controlled by fatty acid translocase (CD36), which regulates long chain fatty acid transport in skeletal muscles and adipose tissue (75). Subcutaneous adipose tissue expression of this binding protein was increased in obese individuals and further increased in those with type 2 diabetes (75). A promoter polymorphism in CD36 was also linked with insulin resistance and type 2 diabetes (76). Interestingly, CD36 is linked to oxidative stress, because it is a scavenger receptor for oxidized lipoproteins on the surface of macrophages, rendering them insulin-resistant (77).

Other genetic candidates include stearoyl-coenzyme A desaturase type 1 (SCD1) and 11β-hydroxysteroid dehydrogenase type 1 (11HSD1), a glucocorticoid-amplifying enzyme. Genetic variants in the fatty acid metabolizing enzyme SCD1 have been linked with decreased waist circumference and improved insulin sensitivity in adults (78). The glucocorticoid-amplifying enzyme 11HSD1 is an intriguing molecule with varying roles ranging from regulation of adipocyte differentiation to possible amplification of macrophage-driven adipose tissue inflammation in obesity (79). It stimulates lipid synthesis in the intra-abdominal fat depots of diet-induced obese mice (80). Several studies find

Unit 5 • Chapter 24. Obesity   445

dysregulated adipose tissue activity in obesity (81) and type 2 diabetes (82), including one prospective study (83). Lipid storage and adipocyte differentiation is partially regulated by peroxisome proliferator-activated receptor gamma (PPAR-γ). This nuclear receptor is highly expressed in adipose tissues and favourably controls the release of several adipokines, such as adiponectin, leptin, resistin, interleukin 6 and monocyte chemoattractant protein 1, mounting anti-inflammatory and anticoagulant actions that intensely counteract the adipose tissue dysfunction plaguing obesity (84,85). Notably, PPAR-γ improves glucose uptake and insulin sensitivity, as evidenced by the actions of receptor agonists for treatment of type 2 diabetes. According to multiple investigations, a Pro12Ala polymorphism that decreases receptor activity, protects against hyperinsulinemia, type 2 diabetes (86), and high free fatty acid concentrations (87). The Ala12 carriers show up to a 19% risk reduction, and the protective effect is greatest at a lower body mass index (BMI) (88). It is important to note that, as in the case with any gene, the effects of this variant can be modified by other genetic influences, such as variants within the same or other genes, and by environmental factors, such as diet, BMI or physical activity. Indeed, physical activity was found to modify its effect in one follow-up study (89). LDL receptor-related protein 1 is another vital regulator of systemic lipid transport and absorption in liver, muscle, heart and adipocytes. This receptor plays a role in the uptake and hydrolysis of triglyceride-rich lipoproteins. Adipose-specific knockout mice are protected from diet-induced obesity and exhibit increased metabolic rate and glucose tolerance (90). By understanding the potential role of LDL receptor-related protein 1 in humans, this receptor can potentially serve as a valuable marker for conferring susceptibility to obesity and type 2 diabetes in individuals at risk. In addition to the above biomarkers, other candidates for obesity-associated insulin resistance and type 2 diabetes are highlighted in Table 24.3.

Biomarkers of obesity and cancer

Cancer is a condition of uncontrolled cell growth triggered by a variety of factors. It is estimated that one third of all cancer deaths in 2006 were related to physical inactivity, nutrition and obesity (91). Compared to other known genetic or environmental risk factors for cancer, obesity may play a minor role. Nevertheless, its effect can be magnified in susceptible individuals, such as those with a family history of cancer or who belong to a particular race or gender.

Table 24.3. Biomarkers associated with obesity and type 2 diabetes risk

Class	Biomarker	Disease/Risk Factor (Reference)
Exposure	Lipocalin 2	Insulin sensitivity (123)
	Cideb	Insulin sensitivity (124)
	Monocyte chemoattractant protein 1	Insulin resistance (125)
	Interleukin 8	Insulin resistance (126)
	Low adiponectin	Insulin resistance (66,67)
Genetic Susceptibility	SCD1	Insulin sensitivity (78)
	PCK1	Blood glucose and triglyceride synthesis (62)
	11HSD1	Fasting glucose and insulin resistance (82,83) Increased lipid synthesis and adipose tissue mass in mice (80)
	CD36	Insulin resistance and type 2 diabetes (75,76)
	Peroxisome proliferator-activated receptor gamma	Type 2 diabetes (86)
	Ectonucleotide pyrophosphatase/ phosphodiesterase 1	Type 2 diabetes (64,65)
	Tumour necrosis factor-α	Increased postprandial free fatty acid concentrations and insulin resistance (71,72) associated with lower adiponectin concentrations (74)

Defined using BMI or a high upper body (central) fat distribution, obesity modified risk for the development and progression of cancers affecting multiple target organs, including the gastrointestinal system (92–94), ovaries (95), breasts (96) and prostate (97). For example, women with a high BMI exhibited cytological abnormalities in the breast that may predispose to cancer (98). A few of the many possible mechanisms implicated in these findings is detection bias, hormonal imbalance (e.g. sex hormones or insulin) or genetic predisposition. Notably, abnormal adipokine regulation is another potential mechanism for this predisposition.

In the obese environment, insulin stimulates leptin activity in breast cancer cells (99), and an imbalance in leptin and adiponectin secretion is highly implicated as one mechanism for breast cancer development (100). Leptin exerts mitogentic and antiangiogenic effects that appear to be counteracted by adiponectin, which is decreased in obesity and protects against breast cancer (101). High leptin or low adiponectin levels are also implicated in other malignancies, such as non-Hodgkin lymphoma (102) and endometrial cancer (103).

In addition to adipokines, genetic variants in lipid metabolizing genes are associated with breast cancer, such as the leptin receptor and the paraoxonase gene (*PON1*), which prevents low density lipoprotein oxidation. Polymorphisms in these genes are protective against breast cancer development in postmenopausal Caucasian women with benign breast disease (104) (these findings should be interpreted with caution due to the small number of cases (61 cases out of a total of 994)). The aforementioned molecules and other molecules putatively implicated in obesity-linked cancer are summarized in Table 24.4.

Common epidemiological study designs and methodological issues

Multiple epidemiologic study designs have been used to examine putative biomarkers in human populations (see Chapter 14). Case–control and cross-sectional studies are among the most common designs. Relatively cheap and rapid, the designs are a sound stepping stone for collecting background information on the desired criteria for any biomarker, including average plasma concentrations, inter-individual and intra-individual variability, effect size in cases compared to controls, stability, half-life, circadian variation and ethnic/racial differences, as well as age and gender effects. Several biomarkers illustrated in this chapter have been preliminarily identified and repeatedly investigated using these designs to justify further study in more demanding prospective evaluations. More difficult to conduct, population-based prospective, longitudinal studies and randomized controlled trials greatly help strengthen the predictive role of the biomarker and support its predictive and clinical utility.

Conclusions and future directions

The adverse effects of obesity propagate through many human generations, begging the use of biomarkers that help elucidate disease mechanism, therapeutic interventions and preventive strategies. Emerging biomarkers

Table 24.4. Biomarkers associated with obesity and cancer risk

Class	Biomarker	Disease/Risk factor (Reference)
Exposure	Leptin	Cellular proliferation, anti-apoptosis (126) Differentiation of breast cancer (99)
	Adiponectin	Antiangiogenic (127) Low levels implicated in breast cancer (128) and prostate cancer (129)
	Interleukin 6	Prostate cancer (129) Breast cancer (130)
Effect	Vascular endothelial growth factor	Regulates angiogenesis and cell migration, implicated in prostate cancer (129)
Genetic susceptibility	Leptin receptor	Polymorphism associated with lower risk of breast cancer in benign breast disease (104)
	PON1	Polymorphism associated with lower risk of breast cancer in benign breast disease (104)

for obesity-associated CVD, type 2 diabetes or cancer play diverse roles in biological pathways including immune modulation and fat metabolism. Support for the association of these biomarkers with obesity-associated diseases stems from animal and *in vitro* data, but evidence in humans is still lacking. In humans, plasma levels of biomarkers or genetic polymorphisms are widely used to determine risk, but many studies are limited by ethnicity/race, gender or sample size. These deficits are perhaps the most challenging to overcome in future studies, but it is the only way by which a biomarker can be validated for reliable use in human populations. Along these lines, there has recently been a series of large-scale genome-wide association studies of BMI that are uncovering a substantial number of new loci associated with obesity (105–111).

In the field of chronic inflammatory disorders exist acute indicators, such as C-reactive protein, an acute phase protein that rises in any inflammatory condition, and chronic prognostic indicators, such as monocyte chemoattractant protein 1 in the case of CVD. Management of obesity-associated disorders may benefit from the use of acute indicators augmented with chronic markers to better predict disease progression. In fact, multiple biomarkers may be required to complement standard traditional risk factors to enhance risk stratification and to develop measurable therapeutic targets.

Additional measures of obesity aside from the typical BMI are required to better characterize obesity in the context of other chronic disorders. Other non-invasive measures (e.g. waist circumference, waist-to-height ratio and the conicity index) are surrogates of abdominal (central) obesity. Central obesity is often found to predict disease outcome better than BMI (112). One major cause of central obesity is a large visceral adipose tissue distribution. Visceral adipose tissue actively expresses and secretes a myriad of adipokines and other agents that act locally and systemically to promote obesity-associated disorders. Therefore, central obesity should be fully investigated in the context of relevant biomarkers, for it may add to their predictive utility and clinical validity.

References

1. Wilson PW, D'Agostino RB, Sullivan L et al. (2002). Overweight and obesity as determinants of cardiovascular risk: the Framingham experience. *Arch Intern Med*, 162:1867–1872.doi:10.1001/archinte.162.16.1867 PMID:12196085

2. Fox CS, Massaro JM, Hoffmann U et al. (2007). Abdominal visceral and subcutaneous adipose tissue compartments: association with metabolic risk factors in the Framingham Heart Study. *Circulation*, 116:39–48.doi:10.1161/CIRCULATIONAHA.106.675355 PMID:17576866

3. Jiang Y, Chen Y, Mao Y; CCDPC Obesity Working Group (2008). The contribution of excess weight to prevalent diabetes in Canadian adults. *Public Health*, 122:271–276.doi:10.1016/j.puhe.2007.06.002 PMID:17931673

4. Krishnan S, Rosenberg L, Djoussé L et al. (2007). Overall and central obesity and risk of type 2 diabetes in U.S. black women. *Obesity (Silver Spring)*, 15:1860–1866.doi:10.1038/oby.2007.220 PMID:17636105

5. Stoll BA (2002). Upper abdominal obesity, insulin resistance and breast cancer risk. *Int J Obes Relat Metab Disord*, 26:747–753. PMID:12037643

6. Takahashi H, Yoneda K, Tomimoto A et al. (2007). Life style-related diseases of the digestive system: colorectal cancer as a life style-related disease: from carcinogenesis to medical treatment. *J Pharmacol Sci*, 105:129–132.doi:10.1254/jphs.FM0070022 PMID:17928742

7. Schwimmer JB, Burwinkle TM, Varni JW (2003). Health-related quality of life of severely obese children and adolescents. *JAMA*, 289:1813–1819.doi:10.1001/jama.289.14.1813 PMID:12684360

8. Varni JW, Limbers CA, Burwinkle TM (2007). Impaired health-related quality of life in children and adolescents with chronic conditions: a comparative analysis of 10 disease clusters and 33 disease categories/severities utilizing the PedsQL 4.0 Generic Core Scales. *Health Qual Life Outcomes*, 5:43. doi:10.1186/1477-7525-5-43 PMID:17634123

9. Fontaine KR, Redden DT, Wang C et al. (2003). Years of life lost due to obesity. *JAMA*, 289:187–193.doi:10.1001/jama.289.2.187 PMID:12517229

10. Ogden CL, Carroll MD, Curtin LR et al. (2006). Prevalence of overweight and obesity in the United States, 1999-2004. *JAMA*, 295:1549-1555.doi:10.1001/jama.295.13.1549 PMID:16595758

11. Nogueiras R, Wiedmer P, Perez-Tilve D et al. (2007). The central melanocortin system directly controls peripheral lipid metabolism. *J Clin Invest*, 117:3475-3488.doi:10.1172/JCI31743 PMID:17885689

12. Gil-Campos M, Aguilera CM, Cañete R, Gil A (2006). Ghrelin: a hormone regulating food intake and energy homeostasis. *Br J Nutr*, 96:201-226.doi:10.1079/BJN20061787 PMID:16923214

13. Date Y, Toshinai K, Koda S et al. (2005). Peripheral interaction of ghrelin with cholecystokinin on feeding regulation. *Endocrinology*, 146:3518-3525.doi:10.1210/en.2004-1240 PMID:15890776

14. Matias I, Di Marzo V (2007). Endocannabinoids and the control of energy balance. *Trends Endocrinol Metab*, 18:27-37.doi:10.1016/j.tem.2006.11.006 PMID:17141520

15. Bartness TJ, Song CK (2007). Brain-adipose tissue neural crosstalk. *Physiol Behav*, 91:343-351.doi:10.1016/j.physbeh.2007.04.002 PMID:17521684

16. Roberge C, Carpentier AC, Langlois MF et al. (2007). Adrenocortical dysregulation as a major player in insulin resistance and onset of obesity. *Am J Physiol Endocrinol Metab*, 293:E1465-E1478.doi:10.1152/ajpendo.00516.2007 PMID:17911338

17. Trayhurn P, Bing C, Wood IS (2006). Adipose tissue and adipokines—energy regulation from the human perspective. *J Nutr*, 136 Suppl;1935S-1939S. PMID:16772463

18. Farooqi IS (2005). Genetic and hereditary aspects of childhood obesity. *Best Pract Res Clin Endocrinol Metab*, 19:359-374.doi:10.1016/j.beem.2005.04.004 PMID:16150380

19. Hill JO (2006). Understanding and addressing the epidemic of obesity: an energy balance perspective. *Endocr Rev*, 27:750-761.PMID:17122359

20. Sallis JF, Glanz K (2006). The role of built environments in physical activity, eating, and obesity in childhood. *Future Child*, 16:89-108. doi:10.1353/foc.2006.0009 PMID:16532660

21. Flodmark CE, Marcus C, Britton M (2006). Interventions to prevent obesity in children and adolescents: a systematic literature review. *Int J Obes (Lond)*, 30:579-589.doi:10.1038/sj.ijo.0803290 PMID:16570086

22. Fantuzzi G (2005). Adipose tissue, adipokines, and inflammation. *J Allergy Clin Immunol*, 115:911-919, quiz 920.doi:10.1016/j.jaci.2005.02.023 PMID:15867843

23. Furukawa S, Fujita T, Shimabukuro M et al. (2004). Increased oxidative stress in obesity and its impact on metabolic syndrome. *J Clin Invest*, 114:1752-1761. PMID:15599400

24. Dandona P, Mohanty P, Ghanim H et al. (2001). The suppressive effect of dietary restriction and weight loss in the obese on the generation of reactive oxygen species by leukocytes, lipid peroxidation, and protein carbonylation. *J Clin Endocrinol Metab*, 86:355-362.doi:10.1210/jc.86.1.355 PMID:11232024

25. Tripathy D, Mohanty P, Dhindsa S et al. (2003). Elevation of free fatty acids induces inflammation and impairs vascular reactivity in healthy subjects. *Diabetes*, 52:2882-2887.doi:10.2337/diabetes.52.12.2882 PMID:14633847

26. Stephens JW, Gable DR, Hurel SJ et al. (2006). Increased plasma markers of oxidative stress are associated with coronary heart disease in males with diabetes mellitus and with 10-year risk in a prospective sample of males. *Clin Chem*, 52:446-452.doi:10.1373/clinchem.2005.060194 PMID:16384883

27. Yusuf S, Dagenais G, Pogue J et al.; The Heart Outcomes Prevention Evaluation Study Investigators (2000). Vitamin E supplementation and cardiovascular events in high-risk patients. *N Engl J Med*, 342:154-160. PMID:10639540

28. Stephens NG, Parsons A, Schofield PM et al. (1996). Randomised controlled trial of vitamin E in patients with coronary disease: Cambridge Heart Antioxidant Study (CHAOS). *Lancet*, 347:781-786.doi:10.1016/S0140-6736(96)90866-1 PMID:8622332

29. Barker DJ, Eriksson JG, Forsén T, Osmond C (2002). Fetal origins of adult disease: strength of effects and biological basis. *Int J Epidemiol*, 31:1235-1239.doi:10.1093/ije/31.6.1235 PMID:12540728

30. Waterland RA, Jirtle RL (2003). Transposable elements: targets for early nutritional effects on epigenetic gene regulation. *Mol Cell Biol*, 23:5293-5300.doi:10.1128/MCB.23.15.5293-5300.2003 PMID:12861015

31. Patel MS, Srinivasan M (2002). Metabolic programming: causes and consequences. *J Biol Chem*, 277:1629-1632.doi:10.1074/jbc.R100017200 PMID:11698417

32. Musaad S, Haynes EN (2007). Biomarkers of obesity and subsequent cardiovascular events. *Epidemiol Rev*, 29:98-114.doi:10.1093/epirev/mxm005 PMID:17494057

33. Hekerman P, Zeidler J, Korfmacher S et al. (2007). Leptin induces inflammation-related genes in RINm5F insulinoma cells. *BMC Mol Biol*, 8:41.doi:10.1186/1471-2199-8-41 PMID:17521427

34. Karmazyn M, Purdham DM, Rajapurohitam V, Zeidan A (2007). Leptin as a cardiac hypertrophic factor: a potential target for therapeutics. *Trends Cardiovasc Med*, 17:206-211.doi:10.1016/j.tcm.2007.06.001 PMID:17662916

35. Sierra-Johnson J, Romero-Corral A, Lopez-Jimenez F et al. (2007). Relation of increased leptin concentrations to history of myocardial infarction and stroke in the United States population. *Am J Cardiol*, 100:234-239.doi:10.1016/j.amjcard.2007.02.088 PMID:17631076

36. Appel SJ, Harrell JS, Davenport ML (2005). Central obesity, the metabolic syndrome, and plasminogen activator inhibitor-1 in young adults. *J Am Acad Nurse Pract*, 17:535-541. doi:10.1111/j.1745-7599.2005.00083.x PMID:16293162

37. Lindeman JH, Pijl H, Toet K et al. (2007). Human visceral adipose tissue and the plasminogen activator inhibitor type 1. *Int J Obes (Lond)*, 31:1671-1679.doi:10.1038/sj.ijo.0803650 PMID:17471294

38. Saidi S, Slamia LB, Mahjoub T et al. (2007). Association of PAI-1 4G/5G and -844G/A gene polymorphism and changes in PAI-1/tPA levels in stroke: a case-control study. *J Stroke Cerebrovasc Dis*, 16:153-159.doi:10.1016/j.jstrokecerebrovasdis.2007.02.002 PMID:17689411

39. Tsantes AE, Nikolopoulos GK, Bagos PG et al. (2007). Plasminogen activator inhibitor-1 4G/5G polymorphism and risk of ischemic stroke: a meta-analysis. *Blood Coagul Fibrinolysis*, 18:497-504.doi:10.1097/MBC.0b013e3281ec4eee PMID:17581326

40. Corsetti JP, Ryan D, Moss AJ et al. (2008). Plasminogen activator inhibitor-1 polymorphism (4G/5G) predicts recurrence in nonhyperlipidemic postinfarction patients. *Arterioscler Thromb Vasc Biol*, 28:548-554. doi:10.1161/ATVBAHA.107.155556 PMID:18096824

41. Morange PE, Saut N, Alessi MC et al. (2007). Association of plasminogen activator inhibitor (PAI)-1 (SERPINE1) SNPs with myocardial infarction, plasma PAI-1, and metabolic parameters: the HIFMECH study. *Arterioscler Thromb Vasc Biol*, 27:2250-2257.doi:10.1161/ATVBAHA.107.149468 PMID:17656673

42. Mertens I, Ballaux D, Funahashi T et al. (2005). Inverse relationship between plasminogen activator inhibitor-I activity and adiponectin in overweight and obese women. Interrelationship with visceral adipose tissue, insulin resistance, HDL-chol and inflammation. *Thromb Haemost*, 94:1190-1195. PMID:16411393

43. Goldstein BJ, Scalia R (2004). Adiponectin: A novel adipokine linking adipocytes and vascular function. *J Clin Endocrinol Metab*, 89:2563-2568.doi:10.1210/jc.2004-0518 PMID:15181024

44. Cavusoglu E, Ruwende C, Chopra V et al. (2006). Adiponectin is an independent predictor of all-cause mortality, cardiac mortality, and myocardial infarction in patients presenting with chest pain. *Eur Heart J*, 27:2300-2309.doi:10.1093/eurheartj/ehl153 PMID:16864609

45. Wolk R, Berger P, Lennon RJ et al. (2007). Association between plasma adiponectin levels and unstable coronary syndromes. Eur Heart J, 28:292–298.doi:10.1093/eurheartj/ehl361 PMID:17090613

46. Weinbrenner T, Schröder H, Escurriol V et al. (2006). Circulating oxidized LDL is associated with increased waist circumference independent of body mass index in men and women. Am J Clin Nutr, 83:30–35, quiz 181–182. PMID:16400046

47. Morrow JD (2005). Quantification of isoprostanes as indices of oxidant stress and the risk of atherosclerosis in humans. Arterioscler Thromb Vasc Biol, 25:279–286. doi:10.1161/01.ATV.0000152605.64964.c0 PMID:15591226

48. Keaney JF Jr, Larson MG, Vasan RS et al.; Framingham Study (2003). Obesity and systemic oxidative stress: clinical correlates of oxidative stress in the Framingham Study. Arterioscler Thromb Vasc Biol, 23:434–439. doi:10.1161/01.ATV.0000058402.34138.11 PMID:12615693

49. Bougoulia M, Triantos A, Koliakos G (2006). Plasma interleukin-6 levels, glutathione peroxidase and isoprostane in obese women before and after weight loss. Association with cardiovascular risk factors. Hormones (Athens), 5:192–199. PMID:16950753

50. Ustundag B, Gungor S, Aygün AD et al. (2007). Oxidative status and serum leptin levels in obese prepubortal children. Cell Biochem Funct, 25:479–483.doi:10.1002/cbf.1334 PMID:16874844

51. Espinola-Klein C, Rupprecht HJ, Bickel C et al.; AtheroGene Investigators (2007). Glutathione peroxidase-1 activity, atherosclerotic burden, and cardiovascular prognosis. Am J Cardiol, 99:808–812.doi:10.1016/j.amjcard.2006.10.041 PMID:17350371

52. Christiansen T, Richelsen B, Bruun JM (2005). Monocyte chemoattractant protein-1 is produced in isolated adipocytes, associated with adiposity and reduced after weight loss in morbid obese subjects. Int J Obes (Lond), 29:146–150.doi:10.1038/sj.ijo.0802839 PMID:15520826

53. Dahlman I, Kaaman M, Olsson T et al. (2005). A unique role of monocyte chemoattractant protein 1 among chemokines in adipose tissue of obese subjects. J Clin Endocrinol Metab, 90:5834–5840.doi:10.1210/jc.2005-0369 PMID:16091493

54. Frangogiannis NG (2007). The prognostic value of monocyte chemoattractant protein-1/CCL2 in acute coronary syndromes. J Am Coll Cardiol, 50:2125–2127.doi:10.1016/j.jacc.2007.08.027 PMID:18036448

55. de Lemos JA, Morrow DA, Blazing MA et al. (2007). Serial measurement of monocyte chemoattractant protein-1 after acute coronary syndromes: results from the A to Z trial. J Am Coll Cardiol, 50:2117–2124.doi:10.1016/j.jacc.2007.06.057 PMID:18036447

56. Kramer H, Wu X, Kan D et al. (2005). Angiotensin-converting enzyme gene polymorphisms and obesity: an examination of three black populations. Obes Res, 13:823–828.doi:10.1038/oby.2005.94 PMID: 15919834

57. Moran CN, Vassilopoulos C, Tsiokanos A et al. (2005). Effects of interaction between angiotensin I-converting enzyme polymorphisms and lifestyle on adiposity in adolescent Greeks. Obes Res, 13:1499–1504. doi:10.1038/oby.2005.181 PMID:16222048

58. Saha SA, Molnar J, Arora RR (2007). Tissue ACE inhibitors for secondary prevention of cardiovascular disease in patients with preserved left ventricular function: a pooled meta-analysis of randomized placebo-controlled trials. J Cardiovasc Pharmacol Ther, 12:192–204.doi:10.1177/1074248407304791 PMID:17875946

59. Homo-Delarche F, Calderari S, Irminger JC et al. (2006). Islet inflammation and fibrosis in a spontaneous model of type 2 diabetes, the GK rat. Diabetes, 55:1625–1633.doi:10.2337/db05-1526 PMID:16731824

60. Herder C, Peltonen M, Koenig W et al. (2006). Systemic immune mediators and lifestyle changes in the prevention of type 2 diabetes: results from the Finnish Diabetes Prevention Study. Diabetes, 55:2340–2346. doi:10.2337/db05-1320 PMID:16873699

61. McLaughlin T, Abbasi F, Cheal K et al. (2003). Use of metabolic markers to identify overweight individuals who are insulin resistant. Ann Intern Med, 139:802–809. PMID:14623617

62. Beale EG, Harvey BJ, Forest C (2007). PCK1 and PCK2 as candidate diabetes and obesity genes. Cell Biochem Biophys, 48:89–95.doi:10.1007/s12013-007-0025-6 PMID:17709878

63. Matsuoka N, Patki A, Tiwari HK et al. (2006). Association of K121Q polymorphism in ENPP1 (PC-1) with BMI in Caucasian and African-American adults. Int J Obes (Lond), 30:233–237.doi:10.1038/sj.ijo.0803132 PMID:16231022

64. Bochenski J, Placha G, Wanic K et al. (2006). New polymorphism of ENPP1 (PC-1) is associated with increased risk of type 2 diabetes among obese individuals. Diabetes, 55:2626–2630.doi:10.2337/db06-0191 PMID:16936213

65. Böttcher Y, Körner A, Reinehr T et al. (2006). ENPP1 variants and haplotypes predispose to early onset obesity and impaired glucose and insulin metabolism in German obese children. J Clin Endocrinol Metab, 91:4948–4952.doi:10.1210/jc.2006-0540 PMID:16968801

66. Blüher M, Fasshauer M, Tönjes A et al. (2005). Association of interleukin-6, C-reactive protein, interleukin-10 and adiponectin plasma concentrations with measures of obesity, insulin sensitivity and glucose metabolism. Exp Clin Endocrinol Diabetes, 113:534–537. doi:10.1055/s-2005-872851 PMID:16235156

67. Singhal A, Jamieson N, Fewtrell M et al. (2005). Adiponectin predicts insulin resistance but not endothelial function in young, healthy adolescents. J Clin Endocrinol Metab, 90:4615–4621.doi:10.1210/jc.2005-0131 PMID:15886241

68. Kempf K, Hector J, Strate T et al. (2007). Immune-mediated activation of the endocannabinoid system in visceral adipose tissue in obesity. Horm Metab Res, 39:596–600.doi:10.1055/s-2007-984459 PMID:17712725

69. Serino M, Menghini R, Fiorentino L et al. (2007). Mice heterozygous for tumor necrosis factor-alpha converting enzyme are protected from obesity-induced insulin resistance and diabetes. Diabetes, 56:2541–2546.doi:10.2337/db07-0360 PMID:17646208

70. Krogh-Madsen R, Plomgaard P, Møller K et al. (2006). Influence of TNF-alpha and IL-6 infusions on insulin sensitivity and expression of IL-18 in humans. Am J Physiol Endocrinol Metab, 291:E108–E114.doi:10.1152/ajpendo.00471.2005 PMID:16464907

71. Fontaine-Bisson B, Wolever TM, Chiasson JL et al. (2007). Tumor necrosis factor alpha -238G>A genotype alters postprandial plasma levels of free fatty acids in obese individuals with type 2 diabetes mellitus. Metabolism, 56:649–655.doi:10.1016/j.metabol.2006.12.013 PMID:17445540

72. Sookoian SC, González C, Pirola CJ (2005). Meta-analysis on the G-308A tumor necrosis factor alpha gene variant and phenotypes associated with the metabolic syndrome. Obes Res, 13:2122–2131. doi:10.1038/oby.2005.263 PMID:16421346

73. Zeggini E, Groves CJ, Parkinson JR et al. (2005). Large-scale studies of the association between variation at the TNF/LTA locus and susceptibility to type 2 diabetes. Diabetologia, 48:2013–2017.doi:10.1007/s00125-005-1902-4 PMID:16132956

74. González-Sánchez JL, Martínez-Calatrava MJ, Martínez-Larrad MT et al. (2006). Interaction of the -308G/A promoter polymorphism of the tumor necrosis factor-alpha gene with single-nucleotide polymorphism 45 of the adiponectin gene: effect on serum adiponectin concentrations in a Spanish population. Clin Chem, 52:97–103.doi:10.1373/clinchem.2005.049452 PMID:16254197

75. Bonen A, Tandon NN, Glatz JF et al. (2006). The fatty acid transporter FAT/CD36 is upregulated in subcutaneous and visceral adipose tissues in human obesity and type 2 diabetes. Int J Obes (Lond), 30:877–883. doi:10.1038/sj.ijo.0803212 PMID:16418758

76. Corpeleijn E, van der Kallen CJ, Kruijshoop M et al. (2006). Direct association of a promoter polymorphism in the CD36/FAT fatty acid transporter gene with Type 2 diabetes mellitus and insulin resistance. Diabet Med, 23:907–911.doi:10.1111/j.1464-5491.2006.01888.x PMID:16911630

77. Handberg A, Levin K, Højlund K, Beck-Nielsen H (2006). Identification of the oxidized low-density lipoprotein scavenger receptor CD36 in plasma: a novel marker of insulin resistance. *Circulation*, 114:1169–1176. doi:10.1161/CIRCULATIONAHA.106.626135 PMID:16952981

78. Warensjö E, Ingelsson E, Lundmark P et al. (2007). Polymorphisms in the SCD1 gene: associations with body fat distribution and insulin sensitivity. *Obesity (Silver Spring)*, 15:1732–1740.doi:10.1038/oby.2007.206 PMID:17636091

79. Ishii T, Masuzaki H, Tanaka T et al. (2007). Augmentation of 11beta-hydroxysteroid dehydrogenase type 1 in LPS-activated J774.1 macrophages–role of 11beta-HSD1 in pro-inflammatory properties in macrophages. *FEBS Lett*, 581:349–354.doi:10.1016/j.febslet.2006.11.032 PMID:17239856

80. Berthiaume M, Laplante M, Festuccia W et al. (2007). Depot-specific modulation of rat intraabdominal adipose tissue lipid metabolism by pharmacological inhibition of 11beta-hydroxysteroid dehydrogenase type 1. *Endocrinology*, 148:2391–2397.doi:10.1210/en.2006-1199 PMID:17272400

81. Desbriere R, Vuaroqueaux V, Achard V et al. (2006). 11beta-hydroxysteroid dehydrogenase type 1 mRNA is increased in both visceral and subcutaneous adipose tissue of obese patients. *Obesity (Silver Spring)*, 14:794–798. doi:10.1038/oby.2006.92 PMID:16855188

82. Alberti L, Girola A, Gilardini L et al. (2007). Type 2 diabetes and metabolic syndrome are associated with increased expression of 11beta-hydroxysteroid dehydrogenase 1 in obese subjects. *Int J Obes (Lond)*, 31:1826–1831. doi:10.1038/sj.ijo.0803677 PMID:17593901

83. Koska J, de Courten B, Wake DJ et al. (2006). 11beta-hydroxysteroid dehydrogenase type 1 in adipose tissue and prospective changes in body weight and insulin resistance. *Obesity (Silver Spring)*, 14:1515–1522.doi:10.1038/oby.2006.175 PMID:17030962

84. Sharma AM, Staels B (2007). Review: Peroxisome proliferator-activated receptor γ and adipose tissue–understanding obesity-related changes in regulation of lipid and glucose metabolism. *J Clin Endocrinol Metab*, 92:386–395.doi:10.1210/jc.2006-1268 PMID:17148564

85. Cole SA, Mitchell BD, Hsueh WC et al. (2000). The Pro12Ala variant of peroxisome proliferator-activated receptor-γ2 (PPAR-γ2) is associated with measures of obesity in Mexican Americans. *Int J Obes Relat Metab Disord*, 24:522–524.doi:10.1038/sj.ijo.0801210 PMID:10805513

86. Tönjes A, Stumvoll M (2007). The role of the Pro12Ala polymorphism in peroxisome proliferator-activated receptor gamma in diabetes risk. *Curr Opin Clin Nutr Metab Care*, 10:410–414.doi:10.1097/MCO.0b013e3281e389d9 PMID:17563457

87. Tan GD, Neville MJ, Liverani E et al. (2006). The in vivo effects of the Pro12Ala PPARgamma2 polymorphism on adipose tissue NEFA metabolism: the first use of the Oxford Biobank. *Diabetologia*, 49:158–168.doi:10.1007/s00125-005-0044-z PMID:16362285

88. Ludovico O, Pellegrini F, Di Paola R et al. (2007). Heterogeneous effect of peroxisome proliferator-activated receptor γ2 Ala12 variant on type 2 diabetes risk. *Obesity (Silver Spring)*, 15:1076–1081.doi:10.1038/oby.2007.617 PMID:17495182

89. Kilpeläinen TO, Lakka TA, Laaksonen DE et al. (2008). SNPs in PPARG associate with type 2 diabetes and interact with physical activity. *Med Sci Sports Exerc*, 40:25–33. PMID:18091023

90. Hofmann SM, Zhou L, Perez-Tilve D et al. (2007). Adipocyte LDL receptor-related protein-1 expression modulates postprandial lipid transport and glucose homeostasis in mice. *J Clin Invest*, 117:3271–3282.doi:10.1172/JCI31929 PMID:17948131

91. American Cancer Society. Cancer facts and figures 2006. Atlanta (GA): American Cancer Society; 2006.

92. Dai Z, Xu YC, Niu L (2007). Obesity and colorectal cancer risk: a meta-analysis of cohort studies. *World J Gastroenterol*, 13:4199–4206. PMID:17696248

93. Larsson SC, Wolk A (2007). Obesity and colon and rectal cancer risk: a meta-analysis of prospective studies. *Am J Clin Nutr*, 86:556–565. PMID:17823417

94. Merry AH, Schouten LJ, Goldbohm RA, van den Brandt PA (2007). Body mass index, height and risk of adenocarcinoma of the oesophagus and gastric cardia: a prospective cohort study. *Gut*, 56:1503–1511.doi:10.1136/gut.2006.116665 PMID:17337464

95. Olsen CM, Green AC, Whiteman DC et al. (2007). Obesity and the risk of epithelial ovarian cancer: a systematic review and meta-analysis. *Eur J Cancer*, 43:690–709.doi:10.1016/j.ejca.2006.11.010 PMID:17223544

96. Daling JR, Malone KE, Doody DR et al. (2001). Relation of body mass index to tumor markers and survival among young women with invasive ductal breast carcinoma. *Cancer*, 92:720–729.doi:10.1002/10970142(20010815)92:4<720::AID-CNCR1375>3.0.CO;2-T PMID:11550140

97. Freedland SJ, Platz EA (2007). Obesity and prostate cancer: making sense out of apparently conflicting data. *Epidemiol Rev*, 29:88–97.doi:10.1093/epirev/mxm006 PMID:17478439

98. Seewaldt VL, Goldenberg V, Jones LW et al. (2007). Overweight and obese perimenopausal and postmenopausal women exhibit increased abnormal mammary epithelial cytology. *Cancer Epidemiol Biomarkers Prev*, 16:613–616.doi:10.1158/1055-9965.EPI-06-0878 PMID: 17372261

99. Garofalo C, Koda M, Cascio S et al. (2006). Increased expression of leptin and the leptin receptor as a marker of breast cancer progression: possible role of obesity-related stimuli. *Clin Cancer Res*, 12:1447–1453.doi:10.1158/1078-0432.CCR-05-1913 PMID:16533767

100. Vona-Davis L, Rose DP (2007). Adipokines as endocrine, paracrine, and autocrine factors in breast cancer risk and progression. *Endocr Relat Cancer*, 14:189–206.doi:10.1677/ERC-06-0068 PMID:17639037

101. Vona-Davis L, Howard-McNatt M, Rose DP (2007). Adiposity, type 2 diabetes and the metabolic syndrome in breast cancer. *Obes Rev*, 8:395–408.doi:10.1111/j.1467-789X.2007.00396.x PMID:17716297

102. Skibola CF, Holly EA, Forrest MS et al. (2004). Body mass index, leptin and leptin receptor polymorphisms, and non-hodgkin lymphoma. *Cancer Epidemiol Biomarkers Prev*, 13:779–786. PMID:15159310

103. Soliman PT, Wu D, Tortolero-Luna G et al. (2006). Association between adiponectin, insulin resistance, and endometrial cancer. *Cancer*, 106:2376–2381.doi:10.1002/cncr.21866 PMID:16639730

104. Gallicchio L, McSorley MA, Newschaffer CJ et al. (2007). Body mass, polymorphisms in obesity-related genes, and the risk of developing breast cancer among women with benign breast disease. *Cancer Detect Prev*, 31:95–101.doi:10.1016/j.cdp.2007.02.004 PMID:17428620

105. Frayling TM, Timpson NJ, Weedon MN et al. (2007). A common variant in the FTO gene is associated with body mass index and predisposes to childhood and adult obesity. *Science*, 316:889–894.doi:10.1126/science.1141634 PMID:17434869

106. Dina C, Meyre D, Gallina S et al. (2007). Variation in FTO contributes to childhood obesity and severe adult obesity. *Nat Genet*, 39:724–726.doi:10.1038/ng2048 PMID:17496892

107. Scuteri A, Sanna S, Chen WM et al. (2007). Genome-wide association scan shows genetic variants in the FTO gene are associated with obesity-related traits. *PLoS Genet*, 3:e115.doi:10.1371/journal.pgen.0030115 PMID:17658951

108. Loos RJ, Lindgren CM, Li S et al.; Prostate, Lung, Colorectal, and Ovarian (PLCO) Cancer Screening Trial; KORA; Nurses' Health Study; Diabetes Genetics Initiative; SardiNIA Study; Wellcome Trust Case Control Consortium; FUSION (2008). Common variants near MC4R are associated with fat mass, weight and risk of obesity. *Nat Genet*, 40:768–775.doi:10.1038/ng.140 PMID:18454148

109. Willer CJ, Speliotes EK, Loos RJ et al.; Wellcome Trust Case Control Consortium; Genetic Investigation of ANthropometric Traits Consortium (2009). Six new loci associated with body mass index highlight a neuronal influence on body weight regulation. *Nat Genet*, 41:25–34.doi:10.1038/ng.287 PMID:19079261

110. Thorleifsson G, Walters GB, Gudbjartsson DF et al. (2009). Genome-wide association yields new sequence variants at seven loci that associate with measures of obesity. *Nat Genet,* 41:18–24.doi:10.1038/ng.274 PMID:19079260

111. Speliotes EK, Willer CJ, Berndt SI et al.; MAGIC; Procardis Consortium (2010). Association analyses of 249,796 individuals reveal 18 new loci associated with body mass index. *Nat Genet,* 42:937–948.doi:10.1038/ng.686 PMID:20935630

112. Musaad SM, Patterson T, Ericksen M et al. (2009). Comparison of anthropometric measures of obesity in childhood allergic asthma: central obesity is most relevant. *J Allergy Clin Immunol,* 123:1321–1327, e12.doi:10.1016/j.jaci.2009.03.023 PMID:19439348

113. Blankenberg S, Luc G, Ducimetière P et al.; PRIME Study Group (2003). Interleukin-18 and the risk of coronary heart disease in European men: the Prospective Epidemiological Study of Myocardial Infarction (PRIME). *Circulation,* 108:2453–2459.doi:10.1161/01.CIR.0000099509.76044.A2 PMID:14581397

114. Söderberg S, Ahrén B, Stegmayr B et al. (1999). Leptin is a risk marker for first-ever hemorrhagic stroke in a population-based cohort. *Stroke,* 30:328–337. PMID:9933268

115. Söderberg S, Ahrén B, Jansson JH et al. (1999). Leptin is associated with increased risk of myocardial infarction. *J Intern Med,* 246:409–418.doi:10.1046/j.1365-2796.1999.00571.x PMID:10583712

116. Wiman B, Andersson T, Hallqvist J et al. (2000). Plasma levels of tissue plasminogen activator/plasminogen activator inhibitor-1 complex and von Willebrand factor are significant risk markers for recurrent myocardial infarction in the Stockholm Heart Epidemiology Program (SHEEP) study. *Arterioscler Thromb Vasc Biol,* 20:2019–2023. PMID:10938026

117. Meisinger C, Baumert J, Khuseyinova N et al. (2005). Plasma oxidized low-density lipoprotein, a strong predictor for acute coronary heart disease events in apparently healthy, middle-aged men from the general population. *Circulation,* 112:651–657.doi:10.1161/CIRCULATIONAHA.104.529297 PMID:16043640

118. Blankenberg S, Rupprecht HJ, Bickel C et al.; AtheroGene Investigators (2003). Glutathione peroxidase 1 activity and cardiovascular events in patients with coronary artery disease. *N Engl J Med,* 349:1605–1613.doi:10.1056/NEJMoa030535 PMID:14573732

119. Gross M, Steffes M, Jacobs DR Jr et al. (2005). Plasma F2-isoprostanes and coronary artery calcification: the CARDIA Study. *Clin Chem,* 51:125–131.doi:10.1373/clinchem.2004.037630 PMID:15514100

120. Shishehbor MH, Zhang R, Medina H et al. (2006). Systemic elevations of free radical oxidation products of arachidonic acid are associated with angiographic evidence of coronary artery disease. *Free Radic Biol Med,* 41:1678–1683.doi:10.1016/j.freeradbiomed.2006.09.001 PMID:17145556

121. Riera-Fortuny C, Real JT, Chaves FJ et al. (2005). The relation between obesity, abdominal fat deposit and the angiotensin-converting enzyme gene I/D polymorphism and its association with coronary heart disease. *Int J Obes (Lond),* 29:78–84.doi:10.1038/sj.ijo.0802829 PMID:15520830

122. Sekuri C, Cam FS, Ercan E et al. (2005). Renin-angiotensin system gene polymorphisms and premature coronary heart disease. *J Renin Angiotensin Aldosterone Syst,* 6:38–42.doi:10.3317/jraas.2005.005 PMID:16088850

123. Yan QW, Yang Q, Mody N et al. (2007). The adipokine lipocalin 2 is regulated by obesity and promotes insulin resistance. *Diabetes,* 56:2533–2540.doi:10.2337/db07-0007 PMID:17639021

124. Li JZ, Ye J, Xue B et al. (2007). Cideb regulates diet-induced obesity, liver steatosis, and insulin sensitivity by controlling lipogenesis and fatty acid oxidation. *Diabetes,* 56:2523–2532.doi:10.2337/db07-0040 PMID:17646209

125. Kim CS, Park HS, Kawada T et al. (2006). Circulating levels of MCP-1 and IL-8 are elevated in human obese subjects and associated with obesity-related parameters. *Int J Obes (Lond),* 30:1347–1355.doi:10.1038/sj.ijo.0803259 PMID:16534530

126. Chen C, Chang YC, Liu CL et al. (2007). Leptin induces proliferation and anti-apoptosis in human hepatocarcinoma cells by up-regulating cyclin D1 and down-regulating Bax via a Janus kinase 2-linked pathway. *Endocr Relat Cancer,* 14:513–529.doi:10.1677/ERC-06-0027 PMID:17639064

127. Bråkenhielm E, Veitonmäki N, Cao R et al. (2004). Adiponectin-induced antiangiogenesis and antitumor activity involve caspase-mediated endothelial cell apoptosis. *Proc Natl Acad Sci USA,* 101:2476–2481.doi:10.1073/pnas.0308671100 PMID:14983034

128. Miyoshi Y, Funahashi T, Kihara S et al. (2003). Association of serum adiponectin levels with breast cancer risk. *Clin Cancer Res,* 9:5699–5704. PMID:14654554

129. Mistry T, Digby JE, Desai KM, Randeva HS (2007). Obesity and prostate cancer: a role for adipokines. *Eur Urol,* 52:46–53.doi:10.1016/j.eururo.2007.03.054 PMID:17399889

130. Lorincz AM, Sukumar S (2006). Molecular links between obesity and breast cancer. *Endocr Relat Cancer,* 13:279–292. doi:10.1677/erc.1.00729 PMID:16728564

UNIT 5.

APPLICATION OF BIOMARKERS TO DISEASE

CHAPTER 25.

Disorders of reproduction

Anne Sweeney and Deborah del Junco

Summary

This chapter focuses on biomarkers of reproductive health and disease that have been developed in the past 15 years. Due to the gender- and age-dependency of most of the advances in measuring reproductive health status and outcomes, these biomarkers have been categorized with respect to the unique member of the reproductive triad of interest (i.e. mother, father, conceptus). Biomarkers of female and male puberty, female reproductive function, fetal and infant development, and male reproductive function are discussed. The strengths and limitations of developing and implementing biomarkers in reproductive health studies over the past decade are explored.

Introduction

The utilization of biomarkers in reproductive and perinatal health research has greatly enhanced our understanding of these critical areas of public health. There has been increasing emphasis on these time periods in early development as vulnerable windows over the life course, during which humans are most highly susceptible to the effects of exposure to toxic agents in the environment. The periconceptional, prenatal, perinatal and peripubertal time periods are considered to be the most susceptible intervals for adverse health events (1–7). However, methodologic issues unique to this area of research render the identification of appropriate biomarkers a daunting challenge.

Reproductive epidemiology studies often consist of a triad, including the mother, father and conceptus, which constitutes the unit of both observation and analysis. The ability to obtain biomarkers for all three of these subjects varies greatly, compounded by the differing time intervals of concern for each subject in terms of biomarkers of exposure, susceptibility and effect. Other challenges include the interrelatedness of reproductive outcomes across the spectrum of time-dependent endpoints of interest and the accuracy and reliability of the markers available for evaluation. Examination of effects that occur at later time points in gestation (e.g. recognized spontaneous abortions or preterm delivery) are restricted to those conceptions that have survived long enough to be identified for evaluation. This reinforces the urgent need to develop methodologies, including biomarkers, that enable us to examine the earliest outcomes along this spectrum (8).

Context and public health significance

In developing biomarkers that would be appropriate for use in large-scale epidemiological studies of reproductive outcomes, one must consider not only sensitivity, specificity, predictive value, within-subject reliability (low coefficients of variation) and cost, but also acceptability and ease of use by study participants (9). Several studies have reported an increase in participation rates when study subjects are taught how to collect and ship biological specimens from the privacy of their own homes, as opposed to having the samples collected in clinics or field offices (10).

In this chapter, the earlier review by Lemasters and Schulte (11) is updated, focusing on biomarkers of reproductive health and disease that have been developed in the 15 years since that publication. Due to the gender- and age-dependency of most of these advances in measuring reproductive health status and outcomes, the biomarkers have been categorized with respect to the unique member of the reproductive triad of interest (i.e. mother, father, conceptus). Detailed discussions of advances in molecular biomarker technologies to measure exposure to environmental and infectious agents in reproductive epidemiology studies (beyond the scope of this chapter) are covered in Chapters 9–13 of this text and in comprehensive reviews devoted to these topics (12–20). Likewise, readers are referred to Chapter 7 and the wealth of resources described in Perera and Herbstman (14), Burke et al. (21), Seminara et al. (22), Field & Sansone (23) and Ho & Tang (24) for more in-depth information on developments in genomic, transcriptomic, proteomic, metabolomic and epigenomic technologies to examine disease susceptibility and etiopathogenetic pathways in reproductive epidemiology research. An overview (25) describes how the combination of bioengineering and bioinformatics has evolved to help reveal integrated, dynamic molecular networks underlying complex functions in biological systems like human reproduction and early development. Also beyond the scope of this chapter, but covered well in several recent publications, are genome-wide association studies of reproductive health outcomes (26–33).

Figure 25.1 provides an illustration of the spectrum of reproductive outcomes (although not exhaustive) that are available for investigation. This figure attempts to present these topics in a chronological fashion, from the earliest sentinel of potential adverse reproductive function among males and females, to early or delayed onset of puberty, and extending to childhood cancers in their offspring that may be linked to prenatal exposures. Again, the earliest events along this chronological spectrum represent the target areas of greatest focus, as these outcomes enable the examination of a representative cohort at risk, and may serve as early sentinels of exposure to toxic agents in the environment, a major and unresolved public health concern.

Figure 25.1. Selection of reproductive outcomes available for study

Age at pubertal onset
Decreased libido (males, females)
Menstrual cycle function
Sperm/semen abnormalities
Infertility
Time-to-pregnancy
Early pregnancy loss
Recognized spontaneous abortion
Stillbirths
Neonatal/infant mortality
Congenital anomalies
Intrauterine growth restriction
Low birth weight
Preterm birth
Sudden Infant Death Syndrome
Neuropsychological/cognitive disorders
Developmental disorders
Childhood cancer

Biomarkers of female and male puberty

There has been persistent, increasing concern over the past several years regarding the observation that children in the USA are entering puberty at younger ages. While advancing age at puberty may reflect inadequate nutritional or socioeconomic conditions, younger ages may be indicative of other adverse scenarios, including obesity and exposure to endocrine active compounds in the environment (34). Most expert panelists assembled by the US Environmental Protection Agency (EPA), the National Institute of Environmental Health Sciences (NIEHS) and Serono Inc. to evaluate secular trends in the timing of puberty concluded that there is sufficient evidence of earlier breast development onset and menarche in girls (35). On the other hand, almost all the panelists agreed that there is insufficient evidence regarding trends in male pubertal development.

The Tanner scales have been widely used by clinicians for several years to examine onset of puberty in girls and boys. The scales assess stages of breast development, pubic hair growth, genitalia changes, and age at menarche (36). A method of self-assessment using photographs and written descriptions of the various stages of development was developed and evaluated by researchers in different populations (37–39). A recent review of biological markers for assessing puberty status, however, indicated that many study participants are reluctant to undergo this examination by a clinician, and several studies have indicated a range of correlations between self-reported and physician Tanner scores, depending upon many factors including race/ethnicity, age and certain psychological disorders (40). Moreover, many young individuals are reluctant to perform this self-assessment, even in the privacy of their own homes; one study had a 61% response rate when participants were asked to complete the procedure at home and mail in their information (41).

There are several biomarkers currently undergoing evaluation for use in ascertaining pubertal status. Most have limited feasibility for use in large, population-based studies, as the components of interest have short serum half-lives and would require the collection of serial blood samples. These include leptin, an adipocyte hormone involved in energy homeostasis that also interacts with the reproductive axis, and Müllerian inhibiting substance (MIS), a glycoprotein hormone produced by the Sertoli cells of the male during fetal development that causes regression and atrophy of the Müllerian ducts (42–44). Leptin is a critical regulator of body fat stores, which may underlie its role as a possible biomarker of approaching puberty. Given the well-known changes in body fat mass and percent body fat associated with puberty, it is hypothesized that leptin may serve as a biomarker of peripuberty and pubertal advancements (45). In girls, serum leptin levels rise markedly as they approach puberty, and this increase is correlated with body fat mass. The levels continue to increase throughout puberty, whereas among boys, there is an initial increase during peripuberty, followed by a return to prepubertal levels as they advance through puberty (43). The ability to measure leptin in urine would greatly enhance the utility of this biomarker in studies of peripubertal events. A recent cross-sectional study of 188 children, aged 5–19 years, reported a correlation of $r = 0.65$ ($P < 0.01$) between serum and urinary leptin levels (46). Of note, urinary leptin levels corresponded to serum levels and patterns by gender during puberty.

The gonadotropins, luteinizing hormone (LH) and follicle stimulating hormone (FSH), and the sex steroid hormones, estrogen and testosterone, also increase in early puberty, and can be measured in urine (47,48). A small longitudinal study recently evaluated the relationship between urinary leptin and gonadotropin levels in 13 boys and seven girls over a six-month period as they were expected to approach puberty (49). Three consecutive first morning urine samples were collected each month. These results indicated significant correlations between urinary leptin and LH levels ($r = 0.43$, $P < 0.001$) and FSH levels ($r = 0.32$, $P < 0.001$). Moreover, urinary leptin levels were higher among the girls, and were increased among both girls and boys nearing puberty compared with those remaining prepubertal over the course of the study.

Both MIS and inhibins, peptides that suppress FSH levels, have been characterized as potential biomarkers of pubertal onset in males. MIS is detected at high levels during late infancy in males, then declines gradually until the presence of primary spermatocytes are detected, which appear to inhibit MIS (50). In contrast, MIS is only synthesized postnatally by granulosa cells in pubertal girls, and is measured in serum (51,52). Serum concentrations are similar in both sexes after puberty (53). Inhibin B is a gonadal polypeptide hormone that regulates, via a negative feedback loop, the synthesis and secretion of FSH (54). Similar to MIS, among males there is a peak concentration of serum inihibin B in infancy, followed by a rapid decline until onset of puberty, when another rise in serum levels is noted. Inhibin B is also associated with FSH, LH, testosterone and testicular volume in varying patterns throughout puberty (54,55). In girls, serum inhibin B is positively correlated with age and FSH levels during childhood, and an increase during early breast development stages (56). However, because these biomarkers currently require drawing blood samples for measurement and biological variability (e.g. diurnal fluctuations) is unknown, research is needed to develop valid and sensitive urinary biomarkers that will also be feasible in terms of cost and acceptance by study participants in longitudinal studies (57).

Molecular epidemiology studies associating environmental exposures with puberty onset

The few epidemiologic studies that have associated pubertal development with exposure to endocrine disrupting chemicals (e.g. polychlorinated and polybrominated biphenyls (PCBs and PBBs) and phthalate esters) have been the topic of several recent reviews (58–60). Conflicting findings and uncertainties regarding critical windows of susceptibility and the possibility of exposure to complex mixtures of chemicals that may have antagonistic effects highlight the need for further research with objective biomarkers. A recent epidemiologic investigation of pubertal stages in nine-year-old inner-city girls in New York City was unique in associating delayed breast development with high levels of the hormonally active agents phytoestrogens and isoflavones measured in urine (61). An expert panel recently convened to review the association between endocrine-active chemicals in the environment and altered timing of pubertal onset concluded that the evidence available appears suggestive (62). Future epidemiologic research to elucidate gene-environment interactions and the molecular pathways mediating the influence of environmental exposures on pubertal development is eagerly awaited.

Biomarkers of female reproductive function

Female libido

Changes in usual patterns of sexual desire can also serve as an early sentinel for exposures that may adversely affect reproductive health. Most of the concerns regarding a relationship between exposure to endocrine active compounds and decreased libido have focused on males, although some have questioned whether this may also be a problem among women (63). The complex interactions between sex steroid hormones and the hypothalamic-pituitary-gonadal axis at varying times over the menstrual cycle also add to the challenges in measuring these hormones as biomarkers of decreased libido (64). There have been several recent studies of diminished sexual desire in females, but the majority have examined this condition among postmenopausal women (65–68). A related outcome, hypoactive sexual desire disorder (HSDD), is defined as low sexual desire accompanied by personal distress caused by this decrease in sexual desire (69,70). The prevalence of low sexual desire among younger and middle-aged women that is not ascribed to menopausal effects ranges from 24–31% (71,72). While questionnaires to ascertain HSDD have been tested for validity and reliability, a biomarker for HSDD in women remains elusive. Although there is a growing body of evidence supporting the role of testosterone and sexual desire in women, the association between decreased serum androgen levels and women reporting low libido remains unclear (73,74). The development of assays to measure testosterone levels in saliva samples would greatly facilitate the investigation into the potential relationship between exposure to endocrine active compounds in the environment and decreased libido (75,76).

Menstrual cycle characteristics

Alterations in menstrual cycle characteristics may also serve as early sentinels of exposure to potentially harmful environmental contaminants. Furthermore, changes in menstrual cycle parameters have been associated with adverse reproductive outcomes, including infertility and spontaneous abortion (77–80). One small study (n = 14) determined that urinary FSH was significantly lower in the periovulatory period in cycles that did not result in conception compared with those that did, rendering urinary FSH a potentially useful predictor of cycle fecundity.

Numerous studies have been conducted using questionnaires to obtain information on cycle length, days and severity of menstrual flow, and dysmenorrhea; however, these methods often do not yield valid information. One study described the use of urinary biomarkers to determine sex steroid hormone levels throughout the menstrual cycle among healthy premenopausal women (n = 403) (81). The women were required to collect and freeze daily morning urine samples that were analysed for pregnanediol-3-glucuronide, estrone sulfate and estrone glucuronide (combined and referred to as E1C) by enzyme-linked immunoassay. Using computer-generated algorithms to define menstrual cycle events, the researchers were able to utilize the hormone data to describe menstrual cycle length, the length of both the follicular and luteal phases, as well as occurrence and timing of ovulation.

Using the same sampling frame as the study above, the Kaiser Permanente Medical Care Program in California evaluated variability in estrogen and progesterone metabolites according to these menstrual cycle characteristics, as well as demographic variables and reproductive history (82). With an average length of participation of 141 days, urine samples were collected on over 95% of the study days in this sample of mostly white, highly-educated women. They reported urinary estrogen metabolite levels 10–13% higher in the baseline interval (days 1–5) and during the follicular phase among women who experienced shorter menstrual cycles. There was also an association between increased urinary progesterone metabolites and a longer luteal phase. Associations between estrogen and progesterone metabolites and race/ethnicity, prior reproductive experiences, age, BMI and educational level were also noted.

A large study was conducted in 1989–1991 to determine the relationship between occupational exposures among women employed in the semiconductor industry and fertility and early pregnancy losses (83). Investigators evaluated the influence of demographic and lifestyle factors on menstrual cycle characteristics among 309 women from this cohort. Duration of follicular and luteal phase segments, and occurrence and timing of ovulation were determined using urinary estrogen and progesterone metabolite levels that were analysed in daily morning urine samples. These efforts confirmed earlier reports that menstrual cycle characteristics vary by age, race/ethnicity and lifestyle factors (e.g. alcohol consumption) (84–86).

Taken together, these studies provide evidence that urinary biomarkers can be used in large, population-based studies of menstrual cycle function, as well as in selected fertility and pregnancy outcomes. However, it must be noted that the majority of these protocols required daily specimen collection over varying lengths of time. While compliance rates with urine specimen collection in prospective pregnancy studies is generally quite high (8), it must also be noted that these studies typically involve women (or couples) who are planning to conceive. There may therefore be considerable variability with compliance, depending upon the goal of the study and the pregnancy intentions of the sample.

Additional biomarkers that would be feasible for use in large, population-based studies focusing on the detection of ovulation include cervical mucus monitoring, which has been available and used for several years, and salivary sex steroid hormones levels. However, studies examining the validity and feasibility of analysing salivary samples for various biomarkers of reproductive function have included relatively small sample sizes, and reported considerable intra- as well as interindividual variability for the sex steroid hormones (87–91).

Infertility/early pregnancy loss

The only approach that allows researchers to distinguish between women having infertility and those experiencing subclinical or early pregnancy loss is the prospective pregnancy study design. Though beyond the scope of this chapter because of limited application in large-scale epidemiologic research, diagnostic advances in assisted reproductive technology (ART) have shed light on this distinction, and readers are referred to several recent and thorough reviews (92–94). As illustrated in Figure 25.2, there is increased emphasis on the periconceptional environment that may impact fertilization as well as implantation. Scientists have made great strides in the ability to detect early pregnancies once implantation takes place. The measurement of urinary human chorionic gonadotropin (hCG) as a marker of early pregnancy loss has been used since the late 1980s (95–97). The algorithm for determining the level of hCG that exceeds normal background levels and indicates that conception has occurred was developed by examining hCG levels throughout the menstrual cycle in women who had been surgically sterilized (96). Concerns were later raised suggesting that urinary hCG alone may not be a sensitive indicator of early pregnancy among those conceptions terminating in subclinical losses (98–100). However, when patterns of serum and urinary hCG levels were compared between successful pregnancies and those terminating in early pregnancy loss and clinical spontaneous abortion, there were no differences, thus validating the solitary measure of urinary hCG as a biomarker of pregnancies around the time of implantation (101).

Figure 25.2. The sensitive periods in human development (230). Copyright Elsevier (1998).

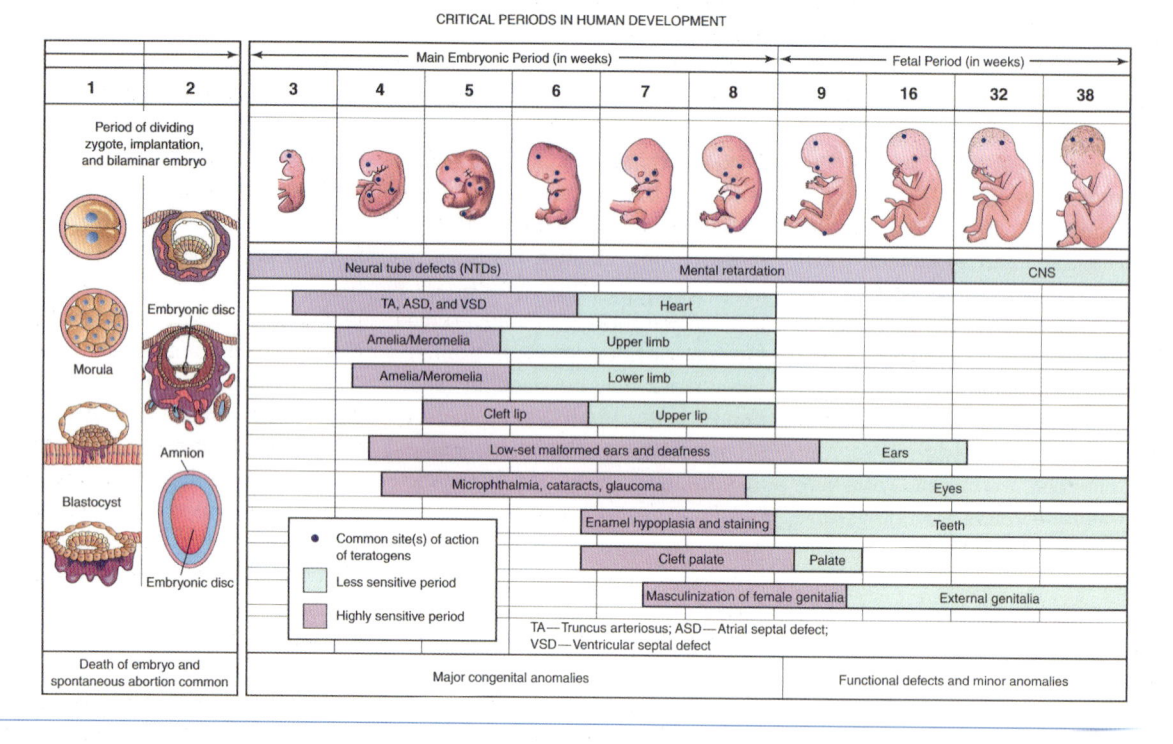

A prospective, longitudinal study of early pregnancy loss (conceptions ending within five weeks of ovulation) among indigenous women of rural Bolivia correlated salivary progesterone measurements with urinary hCG measurements (102). Among the 191 study women who were eligible (taking no active steps to either prevent or achieve conception) and visited a clinic every other day (to collect samples and record the first day of menstrual bleeding), eight early pregnancy losses were detected and 32 pregnancies were sustained past the five-week cut point. Overcoming some of the limitations of salivary progesterone measurements described above, the investigators were able to detect ovulation as a sudden, steep rise in salivary progesterone, and they reported a significant association between elevated follicular phase (pre-ovulatory) progesterone and subsequent early pregnancy loss.

The development of fertility monitors and sensitive urinary hCG assays that can easily be used by women in the home to detect early pregnancy has greatly enhanced the ability to examine fecundability, infertility, time-to-pregnancy and early pregnancy loss (103,104). The Oxford Conception Study is a randomized clinical trial to determine if knowledge of the timing of peak fertility increases conception rates among couples trying to achieve pregnancy (103). The study is using the Clearblue® Easy fertility monitor, which measures levels of urinary estrone-3-glucuronide (E3G) and LH. The monitor screen displays bars to indicate high and peak (which displays an egg symbol) fertile days in the cycle. There are two intervention arms to the study: one third of women receive feedback only on the early fertile time, defined as the first rise in E3G until the LH surge is detected, and another one third receive information only about the late fertile time, or the onset of the LH surge and the subsequent two days. The control group does not receive any fertility monitor feedback concerning fertile windows. This same fertility monitor is also currently being used in the Longitudinal Investigation of Fertility and the Environment (LIFE) study, a prospective examination of the impact of environmental exposures and lifestyle factors on fecundability and fertility, which will be described later.

There are additional advantages to using the fertility monitors. Women are requested to collect first morning urine samples beginning on day six of their cycle, and continue for 10 or 20 days depending upon their cycle length; thus the need for daily urine sample collection is reduced somewhat. Moreover, the monitors store data on the estrogen and progesterone metabolite levels, which may be downloaded using a data card and processed for data analysis. These hormone data are also used to generate graphs for study participants, which provide a clear illustration of their fertile windows during each cycle, and reinforce the importance of engaging in sexual intercourse on those days to increase likelihood of conception.

The utilization of fertility monitors and home pregnancy test kits as described above has greatly enhanced the information that can be obtained in prospective pregnancy studies. However, the home pregnancy tests cannot detect fertilization, and only begin to measure hCG around the time of implantation. Thus, the interval between fertilization and implantation remains a 'black box' in the investigation of factors affecting fertility. There have been some attempts to measure a substance known as early pregnancy factor (EPF) (105–107). EPF is believed to be a substance secreted by the ovary in response to a trigger from the zygote (107). This 'ovum factor' is secreted at the time that the ovum is fertilized, and can be measured in the maternal serum within 2–6 days post ovulation (106). Despite the fact that several earlier studies demonstrated the ability of EPF to detect fertilization in humans before implantation, this biomarker has not been evaluated for use in prospective studies of fecundability and fertility. Again, the need to draw blood samples during each menstrual cycle to detect EPF, renders this less than optimal in large-scale population-based studies, but a recent study reported the presence of EPF in the cervical mucus of pregnant women (108). Mean EPF activity, measured in rosette inhibition titres (RIT), was significantly higher among 53 pregnant women during their first trimester of gestation, and seven women in the second trimester, compared with 25 non-pregnant women (6.58, 5.71, and 3.44 RTI, respectively ($P < 0.001$). Moreover, there was a significant correlation between serum and cervical mucus RTI values, r = 0.611 ($P < 0.0005$). Additional research on EPF activity in cervical mucus around the time of ovulation is needed to determine the feasibility of this biomarker in identifying fertilization and preimplantation events in fecundability and fertility studies.

The challenges inherent in establishing and maintaining prospective pregnancy study cohorts underscore the importance of developing biomarkers that are sensitive, specific and cost-effective but also acceptable and relatively easy to use. The LIFE study, designed to examine the relationship between environmental and lifestyle factors and fecundability and fertility, is funded by the US National Institute for Child Health and Human Development and is currently enrolling couples in Texas and Michigan (http://www.lifestudy.us). This study methodology is the only approach that allows for ascertainment and examination of the early critical events in the reproductive process in humans (i.e. to distinguish between failures of fertilization versus subclinical pregnancy losses), and hopefully soon, implantation failures versus these other two outcomes. The LIFE study recruitment efforts have determined that willingness to comply with protocol requirements, including collection and testing of biological samples and completion of diaries during the interval of attempting to conceive, as well as changes in pregnancy intentions due to varied life events while enrolled in the study, illustrate the need both for the development of sensitive and acceptable biomarkers, as well as very large sampling frames in population-based prospective pregnancy studies.

Biomarkers of fetal and infant development

Adverse fetal or infant outcomes

The major focus of this review is the identification of new biomarkers of reproductive health that can be employed in large-scale epidemiologic studies of reproduction. Although the many promising biomarkers developed recently for use in clinical obstetrics and reproductive endocrinology are beyond the scope of this chapter, readers are referred to in-depth discussions of advances in the detection of pregnancy complications (e.g. pre-eclampsia and intrauterine growth retardation) and other adverse prenatal events provided elsewhere (109–112). Several resources are also available for updates on screening and diagnostic tests for aneuploidy, including Down syndrome (113–117), other congenital anomalies (118), fetal lung maturation (119) and haematologic disorders and complications (120) during gestation. Epidemiologic studies of perinatal outcomes and long-term health in children conceived by ARTs have been an important source of prospective and high quality data. Outcomes following the more recently developed ARTs (e.g. intracytoplasmic sperm injection) are currently being compared with outcomes following the more established technologies, such as *in vitro* fertilization (121–123).

There have been recent advances in the utilization of dried blood spots (DBS) that hold great promise for large-scale epidemiologic studies aiming to identify valuable biomarkers of susceptibility, exposure and effect. Blood spots, obtained via heel stick, are collected within 24–48 hours after birth on nearly all newborns in the USA. The spots are stored on special filter paper Guthrie cards and used by states in screening programmes to diagnose a variety of disorders among newborns. The number of different disorders assessed in the screening programmes is determined by each state's department of health. There has been a surge of interest recently in the analyses of DBS to investigate diverse disorders in large population-based studies, including environmental exposures and their relationship to congenital anomalies and developmental disorders, the prevalence of infectious diseases among newborns, and genetic disorders and potential biomarkers of susceptibility (124–129).

A recent meeting discussed the issues and approaches to using DBS in studies investigating environmental exposures and infant health outcomes (130). Preliminary work has been conducted to determine the feasibility of analysing DBS for several exposures of concern, including persistent bioaccumulative toxics (PBTs is the term used by the US EPA, whereas the United Nations Environment Programme uses the term persistent organic pollutants, or POPs), metals, infectious agents, immune factors and genetic disorders. The scientists concluded that DBS represent a very valuable source of biomarkers of exposure, effect and susceptibility, but there are limitations that must be resolved before they can be used to the maximum potential. The limitations include inadequate sample volume, as many laboratory techniques to measure environmental contaminants require relatively large volumes; development of reference values for elements with large variability in whole blood matrices; and issues related to stability, recoverability, half-life and storage over time. Great care must be taken in the collection, drying, storage and transport of the spots if they are to be of future use in research studies (131). There is also the critical need to develop policies regarding the ethics and human subjects research challenges presented by these specimens (132,133).

Molecular epidemiology studies associating environmental exposures with fetal and infant development

An extensive review of the epidemiologic literature (including original studies, expert panel reports, meta-analyses, and pooled analyses) published between 1970 and 2006 examined adverse reproductive and developmental outcomes in relation to preconceptual and prenatal exposures to environmental compounds (18). The pregnancy outcomes examined included fetal loss, intrauterine growth retardation, preterm birth, birth weight, congenital anomalies and childhood cancers among others. The environmental exposures included metals, pesticides and hormonally active agents (e.g. methyl mercury and PCBs). The question of whether *in utero* exposure of male offspring to hormonally active agents in the environment (e.g. the plasticisers, phthalates) (134) increases their risk of testicular dysgenesis syndrome (TDS) (i.e. impaired spermatogenesis, hypospadias, cryptorchidism and testicular cancer) in a manner similar to *in utero* exposure to diethylstilbestrol (DES), was evaluated in a recent meta-analysis (135). Although the meta-analysis confirmed the association of DES with TDS, there is as yet no compelling evidence that any other hormonally active agents increase the risk of TDS.

An enormous amount of valuable information on the reproductive and developmental effects of the atomic bombing of Hiroshima and Nagasaki has been compiled and summarized (136). The unique and precedent-setting contributions to the field of reproductive epidemiology, as a result of the investigators' careful and extensive use of biomarkers throughout the extended course of this research, are remarkable. Recent studies reporting increased risks of preterm birth, low birth weight, and small for gestational age among the births to female survivors of childhood cancer further illustrate the advantages of accurate, carefully documented biomarkers in revealing the reproductive health effects of exposures at time points early in the mother's development (137,138).

Gene–environment interaction studies in pregnancy outcomes

Molecular epidemiology studies examining the influence of gene–environment interactions on reproductive health have focused most often on the relatively common birth defects, such as neural tube defects, oral clefts, hypospadias and gastroschisis (139–147). Several these studies have linked either folate metabolism genes with maternal nutritional factors (139–141) or metabolic/detoxification pathway genes with maternal smoking (143,144,146). Maternal exposure has been assessed by questionnaire or interview (without biomarkers) and genotyping has been performed on DNA from the mother (146), infant (142,145,147), or rarely, the entire triad including the father (143,144).

Researchers have emphasized the utility of examining Mendelian randomization, or the random transmission of genes that occurs between parent and offspring, in drawing inferences from studies that examine the developmental health effects of *in utero* exposures in combination with candidate susceptibility genes (148). Studies that genotype each member of the triad separately may confer an important advantage to the interpretation of study results. Specifically, if the genetic variant under study either influences exposure to the etiologic factor of interest or modifies the exposure–disease relationship, then the expected relative risk for the adverse pregnancy outcome would be farthest from the null when presence of the risk gene is measured in mothers' DNA, closer to the null when the presence of the risk gene is measured in fathers' DNA, and somewhere in-between when presence of the risk gene is measured in infants' DNA.

Research interest in the interactions between metabolic/detoxification genes and maternal/*in utero* exposure to environmental agents has begun to extend across the spectrum of pregnancy outcomes including preterm delivery (149), infant birth weight (150), pervasive developmental disorders (e.g. autism) (151) and childhood-onset attention deficit disorder (152). These studies further illustrate the methodologic challenges often confronted by molecular epidemiologists: constraints on the study design (e.g. statistical power issues and the need to rely on case-control or case-only designs); source populations of varying race, ethnicity and genetic backgrounds (e.g. case and control groups may not be comparable); participant recruitment (e.g. requiring consent for sensitive, invasive or complicated sample collection procedures); exposure assessment (that may include idiosyncratic sample collection, processing and storage requirements and expensive analytical techniques); and data analysis strategies that must take into account important interrelationships (e.g. confounding and effect-measure modification) across large numbers of measured independent and dependent variables.

Childhood cancer

The extent to which the etiology of childhood cancer involves *in utero* or preconceptional parental exposure to environmental agents remains a major public health concern and an important research question for both reproductive and cancer epidemiologists. Biomarkers of childhood cancer and a review of the epidemiologic investigations that have incorporated molecular markers of exposure and susceptibility are discussed in detail in Chapter 26. A recent comprehensive review of pesticides and childhood cancer (153) is discussed only briefly here to emphasize the critical need for improvements in exposure assessment (e.g. objective biomarkers of exposure to environmental chemicals) and for the detection of specific gene–environment interactions that are likely contributors to the complex etiology of childhood cancer. Among the 77 studies included in the review (153), parental exposure to pesticides (including herbicides and insecticides) was measured indirectly (e.g. by proximity of residential address to chemical production plants or by classification of usual occupation on birth certificate) by responses to interviews or self-administered questionnaires, review of employment records, or by environmental monitoring techniques. None of the studies had measured exposure to pesticides using biological samples (e.g. urine or blood). This is understandable given that over time, study participants may have been exposed to several different compounds (either one at a time or in complex mixtures), that many pesticides have short half-lives in the biological samples commonly used for analysis, and that the limits of detection for many of the standard chemical analyses are relatively high.

Transgenerational health effects and epigenetic mechanisms

Several recent studies have advanced yet another challenging but essential area of research to determine adverse reproductive effects of environmental exposures—the conduct of second-generation studies in humans. The critical importance of this effort is illustrated by the studies of the offspring of women who took DES during pregnancy to prevent spontaneous abortions. Earlier animal studies suggested that the carcinogenic effects of prenatal DES exposure may be transgenerational, reporting an increase in reproductive tract tumours among the offspring of mice with prenatal exposure to DES (154–156). After noting the occurrence of hypospadias among two boys born to mothers who had been exposed to DES prenatally, researchers in the Netherlands conducted a cohort study among women experiencing fertility problems to examine this association (157). Among 205 women who reported having been exposed to DES *in utero*, four gave birth to sons with hypospadias, compared with eight cases reported among the 8729 sons born to non-exposed women (prevalence ratio = 21.3; 95% CI = 6.5–70.1). This finding generated several studies in more representative populations, all indicating an increase in hypospadias among sons of prenatally exposed DES mothers. However, the magnitude was much lower, with prevalence ratios of 5.0 (95% CI = 1.2–16.8) among a French population and 1.7 (95% CI = 0.4–6.8) among women in the USA, and a case–control study of both maternal and paternal *in utero* DES exposures yielded an adjusted odds ratio (OR) of 4.9 (95% CI = 1.1–22.3) for maternal exposures, but no increase among males whose fathers were prenatally exposed to DES (OR = 0.9; 95% CI = 0.1–6.7) (158–160). In addition, there were case reports of other congenital anomalies among second-generation offspring of DES-exposed women, including limb reduction defects, deafness and ovarian carcinoma (161,162).

There have been several recent reports on approaches to examine the transgenerational adverse reproductive effects of *in utero* DES exposure in animal models. Experiments in mice exposed to DES within 1–5 days after birth have supported the hypothesis that epigenetic dysregulation (e.g. hypomethylation of multiple CpG sites of the proto-oncogene *c-fos*) could be a causal mechanism underlying the adverse effects of DES on uterine tissue (163). As described in greater detail in Chapter 26 and recent reviews (164,165) of the 'developmental origins' or Barker hypothesis (166–168) (i.e. that environmental exposures during the earliest and most plastic stages of human development may be among the most significant causal factors underlying many chronic diseases in children and adults), epigenetic mechanisms regulate gene expression through DNA methylation, histone modification of chromatin structure, and autoregulatory DNA binding proteins.

Although epigenetic mechanisms can cause phenotypic discordance between monozygotic twins, epigenetic influences can also be inherited (e.g. an imprinted gene in which the only allele expressed in the offspring is the one inherited from either the mother or the father, never both). Epigenetic alterations can occur at the level of transcription, translation or post-translation, and appear to mediate the development of adverse reproductive health effects following experimental exposure to several hormonally active environmental toxicants, at least in animal models (e.g. dioxins and PBBs) (169). In rodents, for example, *in utero* exposure to the endocrine-active compounds bisphenol A (170) and vinclozolin (171) produced epigenetically-mediated changes in coat colour and in reproductive organs including testicular defects and prostate tumours, respectively. In the latter study of rats, the epigenetic changes and adverse reproductive health effects appeared to be transmitted through a paternal allele in three consecutive generations (i.e. transgenerational), despite the lack of any vinclozolin exposure beyond the first generation of pups.

As is so often the case for findings from novel and intriguing animal experiments, they await confirmation in other experiments, since replication is the hallmark of good science. The validity and reliability (e.g. coefficient of variation) of the intricate molecular techniques to measure epigenetic mechanisms in different human biological matrices with varying amounts of each individual sample are highly uncertain at this time, and gold standard methodologies have yet to emerge (24). In humans, molecular epidemiology studies able to examine biomarkers of epigenetic mechanisms and reveal the ways in which such mechanisms mediate the relation between exposure to environmental chemicals and reproductive outcomes may lag only a few years behind the groundbreaking studies in animal models. However, it may take many years before molecular epidemiologists can measure epigenetic markers and their ultimate effects on reproductive health across multiple generations. There is hope that answers will come from the National Children's Study (http://www.nationalchildrensstudy.gov), which plans to examine mothers and fathers before and during pregnancy, and to follow their children for decades thereafter (172).

Biomarkers of male reproductive function

Male libido

Biomarkers of male reproductive function have been reviewed (173), as well as a comprehensive and in-depth look at the current array of diagnostic tests available for male sexual dysfunction (174). Libido is the biological need for sexual activity (i.e. the sex drive), and male sexual desire is regulated by past sexual activity, psychosocial factors, activation of brain and spinal cord dopamine receptors, and gonadal hormones. Little is known of the physiologic basis of libido, and assessments of libido, erection, ejaculation, orgasm and detumescence would be difficult to make in large-scale epidemiologic studies. There are electronic devices adapted for home use that monitor nocturnal penile tumescence (the penile erections that occur spontaneously during rapid eye movement stages of sleep) (174,175). Though relatively little is known about the effect of occupational or environmental exposures on sexual desire in men, it has been suggested that lead, carbon disulfide, stilbene or cadmium exposure may have adverse effects (176). The few studies that have associated erectile dysfunction with exposure to hazardous environmental and occupational chemicals, and the fewer still that have incorporated biomarkers of exposure, have been carefully reviewed (175). Clearly, male libido and erectile dysfunction are important reproductive health outcomes requiring further epidemiologic research and biomarker development.

Hormones

Chemical analyses for male hormones are usually performed on serum or urine samples, and the latter are readily available for use in large-scale epidemiologic studies. Abnormal levels of male hormones in serum or urine are indicators of problems in the hypothalamic-pituitary-gonadal axis that underlie abnormalities observed in semen analysis (e.g. azoospermia and oligospermia) (177). As a modestly invasive procedure, requiring relatively little formal training, blood collection in males is fairly well tolerated and inexpensive. The National Institute for Occupational Safety and Health recommends a profile including FSH, LH, testosterone and prolactin to evaluate endocrine dysfunction in the male. Although LH and FSH can be measured in urine, prolactin is currently measured only in serum (173).

As high rates of refusal among study candidates has remained a major barrier to the assessment of semen quality in population-based epidemiologic investigations, the identification of alternative reliable biomarkers is a pressing need (20). In a recent comparison with the gold standard measurement of sperm concentration in semen, serum levels of inhibin B, the peptide hormone produced in Sertoli cells, looked promising as a potential surrogate biomarker for large-scale epidemiologic research (178). Although serum FSH has also been used as a surrogate biomarker for semen quality, FSH levels may be less desirable on the biological grounds that they are affected by gonadotropin releasing hormone, estradiol and testosterone, and unlike inhibin B, FSH is not produced in the testes (178). In the future, serum biomarkers for other male hormones, such as activin and follistatin, may be explored for their utility in epidemiologic studies of male reproductive function (173).

Molecular epidemiology studies associating environmental exposures with male hormone levels

A recent, unique molecular epidemiology study compared serum prolactin and inhibin B levels in male welders with corresponding levels in an age-matched comparison group. The investigators reported significantly positive associations after adjusting for smoking and alcohol consumption (179). Whole blood manganese concentration was also positively associated with serum prolactin level. As the higher serum inhibin B concentrations in welders compared with the referent group were contrary to expectation, the findings of this novel study await confirmation (179). Several recent epidemiologic studies have examined the effects of exposure to environmental agents on both male hormones and semen quality, allowing for a more thorough assessment of the potentially complex environmental effects on male reproduction (180–182).

Semen characteristics

Despite conflicting reports and substantial geographic variation, the question of whether declines in semen quality and sperm counts over the past several decades have resulted from exposure to post-industrial age environmental toxicants remains a major unresolved public health concern. While the number of reports of declining sperm counts continues to grow, there is as yet no compelling evidence of decreased fertility in the human populations studied (173,183). The assessment of reproductive function in males usually begins with semen analysis (177). In addition to the challenges of recruiting participants willing to submit semen samples for large-scale epidemiologic studies, the samples must be collected in appropriate containers, analysed within one hour of collection, and kept warm during transportation (177).

The systematic analysis of semen includes macroscopic and microscopic evaluations and chemical assays. Samples are examined for liquefaction, viscosity, colour, pH (normal = 7.2–7.8), volume (normal = 2–5 millilitres), sperm concentration (normal = 20–50 million per millilitre), morphology (e.g. size of the acrosomal cap and length of the tail, normal ≥ 50% of sperm have a typical acrosomal shape and size and a tail around 45 μm in length), motility and velocity/progression (normal ≥ 50% of observed sperm are motile and move forward rapidly in a straight line with little lateral movement), agglutination (clumping) and the presence of other cellular elements (e.g. immature germ cells and leukocytes) (20,177).

Although automated semen analysis systems have been developed, visualization and interpretation of important subtleties in human sperm are difficult. Evaluation by manual methods remains the standard practice (177). Nevertheless, researchers should be aware that evaluations of sperm concentration and motility have demonstrated acceptable interlaboratory reliability and low coefficients of variation, but the assessment of other sperm characteristics (e.g. progression) has been more difficult to standardise (184). Semen contains a vast array of antigenically diverse proteins, including those carried on the surface of sperm, and the largely unknown influence of specific male reproductive proteins on human conception and pregnancy outcomes will be an important focus for future molecular epidemiology research (185).

Although semen has been used for biomonitoring of exposure to metals and xenobiotics (20), concerns have been raised that routine semen analysis may be an insensitive measure of many important reproductive health effects resulting from environmental exposures (19,20). A more comprehensive approach includes assessments for cytogenetic sperm abnormalities and DNA damage (19,20). The fluorescent *in situ* hybridization (FISH) technique has been widely used to detect aneuploidy, chromosomal breaks, and rearrangements in sperm cells. Although the FISH technique is efficient for large-scale use, an immense number of each subject's sperm cells must be evaluated (up to 10 000) (20).

Sperm chromatin is extremely compact and stable relative to chromatin from somatic cells (20), and this property has led to the development of novel biomarkers of sperm DNA integrity. Damage to sperm from reactive oxygen species (e.g. oxygen ions, free radicals and peroxides) has been shown to contribute to reductions in male fertility (20,186). In fact, the production of free radicals due to oxidative stress was first reported in sperm cells (186). Oxidative DNA damage refers to the functional or structural alteration of DNA that contributes to many degenerative diseases of aging including cancer (20). Oxidative stress to sperm DNA integrity can arise endogenously or from exposure to environmental toxicants including xenobiotics (19). Oxidative stress leads to impaired sperm motility, reduced fertilization, and DNA damage.

Although there are over 30 assays of oxidative stress available for sperm assessment, the cost and complexity, combined with difficulties in standardization across laboratories, limit their use especially for large-scale epidemiologic research (186). The level of the oxidative DNA adduct 8-hydroxy-2'-deoxyguanosine (8-OHdG) in sperm is considered a sensitive and precise biomarker of oxidative DNA damage (19,20). High levels of 8-OHdG are positively correlated with abnormal sperm morphology and negatively correlated with sperm concentration, number and motility (19). In addition to the 8-OHdG assay, a variety of methods to measure sperm DNA strand breaks have developed over the past 25 years, leading to the four major tests of sperm DNA fragmentation in current use today (mostly in ART laboratories): the Comet, Tunel, sperm chromatin structure assay (SCSA) and the acridine orange test (AOT) (186,187). The Comet, Tunel, and AOT techniques use light microscopy; however, the Tunel technique can also be performed with flow cytometry. The SCSA method requires flow cytometry (187). Challenges remain in standardization, establishing thresholds and reference ranges, and achieving acceptable levels of interlaboratory reliability for these assays (187).

Molecular epidemiology studies associating environmental exposures with semen quality

The literature on epidemiologic studies relating semen quality to exposure to pesticides, and other endocrine disrupting chemicals has been thoroughly reviewed (188–190). The overwhelming consensus is that the studies have varied considerably in methods, exposures and outcomes, and the results have been equivocal. The need for further research in this area is compelling. Several molecular epidemiologic studies have begun to incorporate biomarkers of DNA damage to sperm (e.g. SCSA and the DNA fragmentation index) as more sensitive measures examine the effects of exposure to potentially toxic environmental compounds (191–194). Molecular epidemiology studies have also begun to examine the influence of gene–environment interactions on sperm DNA integrity (195,196).

The evidence for male-mediated reproductive and developmental toxicity has been reviewed (197). There is some evidence that irradiation and exposure to certain chemical compounds can be genotoxic to sperm in experimental animals, leading to the development of malformations and tumours in their offspring. On the other hand, paternal exposure to low levels of non-mutagenic compounds, such as lead, has altered learning and mating behaviour in offspring, but has not led to obvious malformations or tumours (197). Evidence of male-mediated reproductive and developmental effects in humans has derived mostly from studies of children born to men exposed to environmental toxicants (e.g. methyl mercury, anaesthetic gases, lead, solvents and pesticides) through their occupations. These studies have been limited by the lack of objective and precise measures of exposure and by the failure to adequately account for exposures in the mother. The mechanisms proposed for male-mediated reproductive toxicity in humans involve direct effects from contaminated seminal fluid, as well as both genetic and epigenetic pathways. These pathways could involve germ cell mutation, sperm DNA instability, suppression of germ cell apoptosis, or interference with genomic imprinting (197). As available animal models may have reproductive systems and exposure regimens that poorly approximate conditions in humans, evidence for or against male-mediated reproductive toxicity and the hypothesized underlying molecular mechanisms will largely depend on the validity and precision of future epidemiologic investigations.

Strengths, limitations and lessons learned

There have been tremendous strides in the development and implementation of biomarkers in reproductive health studies over the past decade. These advances have greatly enhanced our ability to explore potential etiologies of and increased susceptibility to adverse reproductive outcomes. In addition, these new techniques have enabled us to identify precursor events and more subtle manifestations of these disorders, as well as to elucidate underlying mechanisms. This is of great benefit to public health, as we now are better positioned to identify early sentinels of exposure to toxicants and to intervene to reduce human exposures.

The capability of measuring the dose of exposure to a toxic agent in individuals greatly reduces exposure misclassification (e.g. recall bias), especially notable in scenarios where the individuals have little knowledge of their exposure. Biomarkers of exposure also enable more precise determinations of dose–response relationships, which can vary across age and gender. The recent focus on critical windows of susceptibility has also affected the risk assessment process, as evidenced by the use of age-dependent adjustment factors (ADAFs) by the EPA in risk assessment regarding early life exposures to carcinogens (198). For ages 0–2 years, the ADAF is 10, indicating a 10-fold increase in carcinogenic potency during this period. For ages 2–16 years, the ADAF is 3, and for ages ≥ 16 years, the ADAF decreases to 1. At present, there is no ADAF for the prenatal period, which is a limitation.

Studies in reproductive epidemiology that incorporate biochemical markers of exposure are only just beginning to address the "windows of susceptibility" issue. It is important to consider the independent and joint influences of age at initial exposure and intensity of exposure, as an environmental agent may have irreversible, long-term reproductive health effects in addition to acute or immediate effects. Perhaps because the timing of exposure is an emergent area of inquiry for epidemiologic studies of environmental chemicals, this issue is not without controversy. Recent commentaries (199,200) following our report of a significant association between infant birth weight and mother's age at initial exposure to PBBs, independent of the association with maternal serum PBB levels (201), highlight the need to consider the time-dependency of maternal exposures and the potential for differential effects on pregnancy outcomes. For example, the effects of *in utero* exposure to a biological agent that remains in maternal circulation (due either to a single exposure event or long-standing cumulative exposure) during critical periods of embryogenesis and organogenesis may be very different from *in utero* effects mediated by the mother's initial exposure to the agent during critical periods of her own childhood and reproductive development (202). Such complexities pose a significant challenge to occupational and environmental epidemiology, as well as to molecular epidemiology (203). As researchers further explore the underlying biological relationships among the growing number of measurable biomarkers, they must take care to operationally define all relevant variables, allow for time-dependencies, and be prepared to reach beyond conventional statistical modeling strategies that may be oversimplified or poorly specified (204,205).

At the same time, epidemiologists have become more alert to the periods of heightened susceptibility during early human development; there is growing awareness of the uncertainties and complexities relating to the toxicokinetics of different environmental chemicals. There is greater appreciation that nonlinear dynamics may be involved and that many interacting factors can influence the rates of uptake, biotransformation, metabolism and elimination. An especially important development is the recognition among toxicologists that the methodologic challenges to valid exposure assessment are not unique to epidemiologic research. As underscored in a recent thought-provoking review (206), these challenges extend to experimental animal models as well. There is room for substantial improvement in biomonitoring that incorporates sensitive, repeated measures to better reveal biological variability both within and between subjects (whether human or animal).

Accurate exposure assessment for some of the most toxic environmental chemicals (e.g. dioxins and the most toxic congener, 2,3,7,9 tetrachlorodibenzo-p-dioxin (TCDD)), requires potentially uncomfortable or invasive specimen collection and highly expensive, technically complicated sample preparation and analysis (207,208). This may partially explain why, for example, specific reproductive health effects of dioxin congeners, including TCDD, have been studied extensively with biomarkers in animal models, but infrequently in population-based epidemiologic studies (209). The rarity of many specific adverse reproductive health outcomes in the general population (e.g. narrowing the broad class of reportable birth defects to diagnostic categories, such as neural tube defects, or to single, clinically-defined entities, such as spina bifida) has made conclusive findings difficult to compile from the few cohorts of highly exposed individuals that have been tested for serum TCDD levels and closely monitored for adverse reproductive outcomes (209). The expected number of cases of many of the single, well-defined reproductive health outcomes of interest in these exposed cohorts is relatively small. Results from well-designed epidemiologic studies in larger populations of individuals that are expected to have a gradient of dioxin exposure (due to the environmentally ubiquitous and persistent nature of these compounds) are eagerly awaited despite the methodologic and political challenges (210–212). One strategy that has been proposed for biomarker studies of TCDD and other environmental chemicals that require either relatively large aliquots for each individual sample, expensive analytical methodologies, or both, is to design statistically powerful and cost-conserving protocols for the pooling of individual blood or serum samples (213–218).

The ability to utilize easily obtained biological specimens (e.g. urine samples and buccal swabs), rather than relying on collection of blood samples (often serially), will greatly improve the acceptability of biomarker studies in large population-based studies. A major limitation of biomarker studies is the reluctance of study participants to undergo serial blood draws, but as discussed above, there is considerable research focused on the development of alternative biological media that would be of huge benefit for researchers in this area. Use of home-based collection and storage protocols also improves compliance and reduces the costs associated with the transport and processing of the samples.

Despite the allure of inexpensive, high-throughput technologies, the appropriate application of biomarker assays to measure exposure, susceptibility, and reproductive health effects in large-scale epidemiologic research will require painstaking validation, comprehensive quality control procedures, and active participation of collaborating laboratories in regular programmes of proficiency testing. Detailed results from a comprehensive quality control programme should be thoroughly reviewed before selecting a collaborating laboratory. If at all possible, studies applying novel biomarkers or high-throughput technologies should incorporate a validation study comparing results using the new test (for at least a reasonably large random sample of participants) with results using a suitable gold standard methodology.

A validation study of a lower cost screening test for dioxin-like compounds in serum (i.e. the chemically activated luciferase reporter gene expression (CALUX) assay) in a case–control study of neural tube defects in the children of US veterans of the Viet Nam War provides a striking example (219). Of interest was the potential use of the CALUX assay to reduce costs in large-scale molecular epidemiology studies that seek to examine the association between parental exposure to dioxin congeners, especially TCDD, and adverse pregnancy outcomes. To assess the validity of the CALUX assay, results were compared with results from the gold standard method of dioxin analysis—high resolution gas chromatography/high resolution mass spectrometry (HRGC/HRMS). Figure 25.3 shows a lack of correlation (and in the negative direction) between the CALUX results for dioxin-like activity in serum (in TEQs) and the serum TCDD levels measured by the gold standard, HRGC/HRMS, on paired serum samples. A recent epidemiologic study of the Seveso, Italy cohort of residents exposed to TCDD in 1976 (from an explosion in a plant that manufactured 2,4,5-trichlorophenol) reported a similar lack of correlation in results from serum analyses comparing the CALUX assay with the gold standard, HRGC/HRMS (220). Had

Figure 25.3. Correlation between log-transformed serum values of 2,3,7,8-tetrachlorodibenzo-p-dioxin (TCDD), measured by high-resolution gas chromatography/high-resolution mass spectrometry (HRGC/HRMS), and serum values of dioxin-like activity measured by the chemically activated luciferase gene expression (CALUX) assay

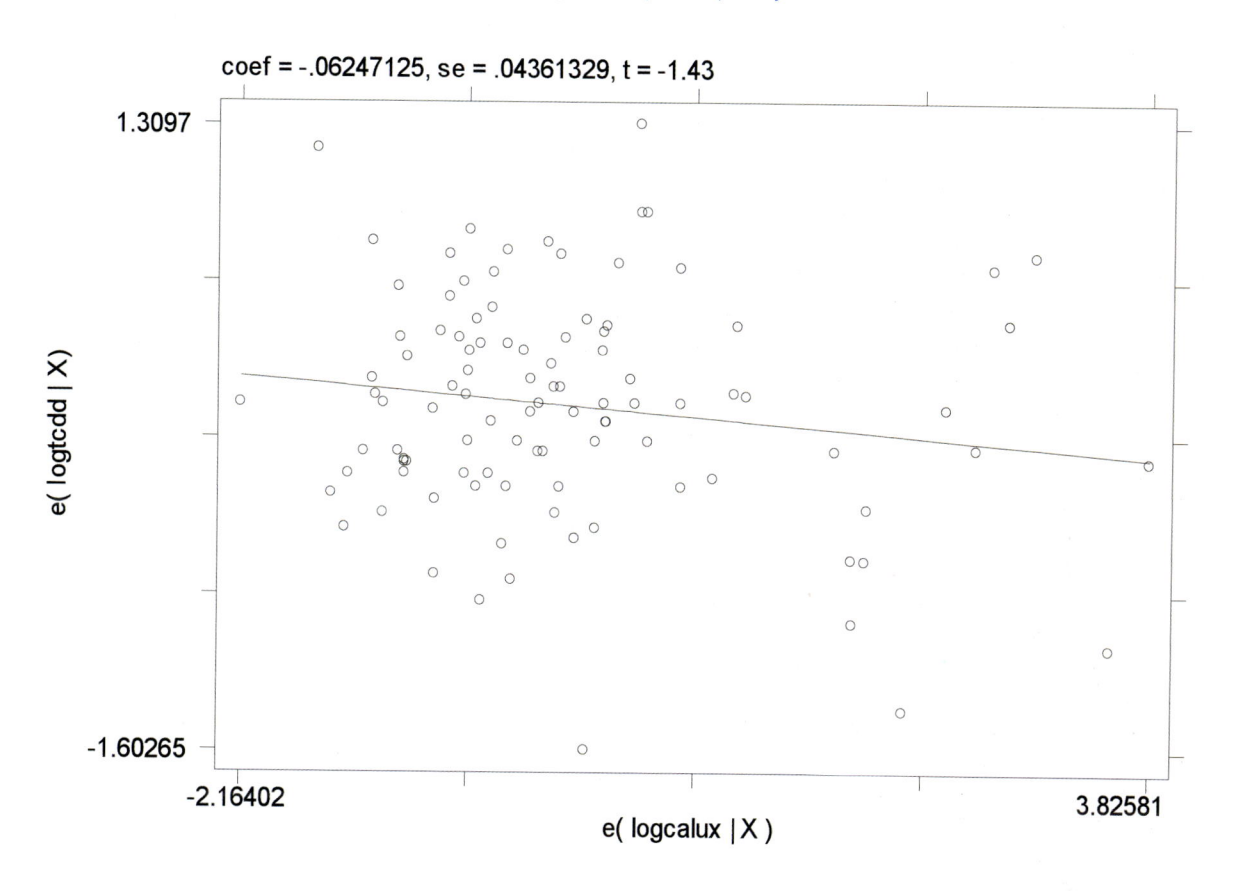

Y axis = log TCDD values; X axis = log CALUX values; P=0.389.

the validity of this CALUX assay been assumed (e.g. on the basis of reports of its validity in matrices other than human blood (221)), the significant three-fold association between paternal serum TCDD level (measured by the gold standard methodology) and the occurrence of neural tube defects in the children of US Viet Nam War veterans (219) would have been missed.

The number of studies examining biomarkers of genetic susceptibility for adverse reproductive outcomes is growing rapidly, with important implications for prenatal screening, as well as for the detection of gene–environment interactions. In addition to new challenges regarding the acquisition and ethical utilization of biological samples, there are longstanding constraints posed by the conventional approaches to data analysis that measure gene–environment interactions on a multiplicative rather than additive scale (e.g. relative risk versus risk difference) and require extremely large sample sizes (222–226).

Future directions and challenges

In addition to the development of biomarkers using easily obtained biological samples, research into the identification of biomarkers that would enable detection when fertilization occurs would enhance our understanding of fecundability and fertility as related to environmental and lifestyle factors. Identification of additional genetic susceptibility markers will open multiple avenues of research in both the genetic screening area, as well as research into gene–environment interactions in the etiology of adverse reproductive outcomes. All of the advances will result in the increasing necessity to develop safeguards for the confidentiality and ethical use of the data obtained. It is also critical that effective means of communication of study results to participants be developed, so that important information regarding their health status is provided to them while also taking into consideration situations in which the interpretation of the data and their relevance to reproductive health may not yet be understood.

To be sure, the vast and growing array of biomarkers and measurements that can be made at the molecular level fuels hope and raises expectations for breakthrough discoveries in reproductive epidemiology, with the potential for rapid translation and significant health benefits. At the same time, exciting developments in molecular epidemiology must be balanced by advances in research design and data analysis that foster methodologic rigor, replication and sustained scientific vigilance. Recent rulings by the US federal Vaccine Injury Compensation Program, highlighted by the Hannah Poling case alleging vaccine-induced autism, raise concern for the temptation to accept biologically plausible molecular mechanisms on the perceived elegance of the argument over the weight of the empirical evidence (227).

Future molecular epidemiology studies of reproductive and developmental health will be shaped also by the increasing pressures to register clinical research (228) and to share data within a limited time frame (229). A socioeconomic exigency of modern, multidisciplinary, multicentre epidemiologic studies (e.g. genome-wide association studies (26–33)) data sharing has broad and complex ethical implications for study participants, study investigators and other stakeholders ranging from corporate interests to the scientific community and the population at large. For future studies aiming to advance the fields of reproductive and molecular epidemiology, the ethical challenges of data sharing must be weighed against the growing demand for scientific synergies and public health benefit (229).

References

1. Selevan SG, Kimmel CA, Mendola P (2000). Identifying critical windows of exposure for children's health. *Environ Health Perspect,* 108 Suppl 3;451–455.doi:10.2307/3454536 PMID:10852844

2. Lemasters GK, Perreault SD, Hales BF *et al.* (2000). Workshop to identify critical windows of exposure for children's health: reproductive health in children and adolescents work group summary. *Environ Health Perspect,* 108 Suppl 3;505–509.doi:10.2307/3454542 PMID:10852850

3. Pryor JL, Hughes C, Foster W *et al.* (2000). Critical windows of exposure for children's health: the reproductive system in animals and humans. *Environ Health Perspect,* 108 Suppl 3;491–503.doi:10.2307/3454541 PMID: 10852849

4. Olshan AF, Anderson L, Roman E *et al.* (2000). Workshop to identify critical windows of exposure for children's health: cancer work group summary. *Environ Health Perspect,* 108 Suppl 3;595–597.doi:10.2307/3454550 PMID:10852858

5. Barr M Jr, DeSesso JM, Lau CS *et al.* (2000). Workshop to identify critical windows of exposure for children's health: cardiovascular and endocrine work group summary. *Environ Health Perspect,* 108 Suppl 3;569–571. doi:10.2307/3454548 PMID:10852856

6. Adams J, Barone S Jr, LaMantia A *et al.* (2000). Workshop to identify critical windows of exposure for children's health: neurobehavioral work group summary. *Environ Health Perspect,* 108 Suppl 3;535–544.doi:10.2307/3454544 PMID:10852852

7. Dietert RR, Etzel RA, Chen D *et al.* (2000). Workshop to identify critical windows of exposure for children's health: immune and respiratory systems work group summary. *Environ Health Perspect,* 108 Suppl 3;483–490.doi:10.2307/3454540 PMID:10852848

8. Buck GM, Lynch CD, Stanford JB *et al.* (2004). Prospective pregnancy study designs for assessing reproductive and developmental toxicants. *Environ Health Perspect,* 112:79–86.doi:10.1289/ehp.6262 PMID:14698935

9. Lasley BL, Overstreet JW (1998). Biomarkers for assessing human female reproductive health, an interdisciplinary approach. *Environ Health Perspect,* 106 Suppl 4;955–960. PMID:9703478

10. Rockett JC, Buck GM, Lynch CD, Perreault SD (2004). The value of home-based collection of biospecimens in reproductive epidemiology. *Environ Health Perspect,* 112:94–104. doi:10.1289/ehp.6264 PMID:14698937

11. Lemasters GK, Schulte PA. Biologic markers in the epidemiology of reproduction. In: Schulte PA, Perera FP, editors. Molecular epidemiology: principles and practices; 1993. p. 385–406.

12. Wilson SH, Suk WA. Biomarkers of environmentally associated disease: technologies, concepts and perspectives. Boca Raton (FL): Lewis Publishers; 2002.

13. National Research Council. Hormonally active agents in the environment. Washington (DC): National Academy Press; 1999.

14. Perera FP, Herbstman JB (2008). Emerging technology in molecular epidemiology: what epidemiologists need to know. *Epidemiology,* 19:350–352.doi:10.1097/EDE.0b013e318162 a920 PMID:18277170

15. Thompson RCA. Molecular epidemiology of infectious diseases. New York (NY): Arnold (Co-published by Oxford University Press); 2000.

16. Riley LA. Molecular epidemiology of infectious diseases: principles and practices. Washington (DC): ASM Press; 2004.

17. Foxman B (2007). Contributions of molecular epidemiology to the understanding of infectious disease transmission, pathogenesis, and evolution. *Ann Epidemiol,* 17:148–156.doi:10.1016/j.annepidem.2006. 09.004 PMID:17175168

18. Wigle DT, Arbuckle TE, Turner MC et al. (2008). Epidemiologic evidence of relationships between reproductive and child health outcomes and environmental chemical contaminants. *J Toxicol Environ Health B Crit Rev*, 11:373–517. PMID:18470797

19. Rockett JC, Kim SJ (2005). Biomarkers of reproductive toxicity. *Cancer Biomark*, 1:93–108. PMID:17192035

20. Ong CN, Shen HM, Chia SE (2002). Biomarkers for male reproductive health hazards: are they available? *Toxicol Lett*, 134:17–30.doi:10.1016/S0378-4274(02)00159-5 PMID:12191857

21. Burke W, Khoury MJ, Stewart A, Zimmern RL; Bellagio Group (2006). The path from genome-based research to population health: development of an international public health genomics network. *Genet Med*, 8:451–458. doi:10.1097/01.gim.0000228213.72256.8c PMID:16845279

22. Seminara D, Khoury MJ, O'Brien TR et al.; Human Genome Epidemiology Network; Network of Investigator Networks (2007). The emergence of networks in human genome epidemiology: challenges and opportunities. *Epidemiology*, 18:1–8.doi:10.1097/01.ede.0000249540.17855.b7 PMID:17179752

23. Field D, Sansone SA (2006). A special issue on data standards. *Omics: a journal of integrative biology*, 10(2):84–93.

24. Ho SM, Tang WY (2007). Techniques used in studies of epigenome dysregulation due to aberrant DNA methylation: an emphasis on fetal-based adult diseases. *Reprod Toxicol*, 23:267–282.doi:10.1016/j.reprotox.2007.01.004 PMID:17317097

25. Beyer A, Bandyopadhyay S, Ideker T (2007). Integrating physical and genetic maps: from genomes to interaction networks. *Nat Rev Genet*, 8:699–710.doi:10.1038/nrg2144 PMID:17703239

26. Aston KI, Carrell DT (2009). Genome-wide study of single-nucleotide polymorphisms associated with azoospermia and severe oligozoospermia. *J Androl*, 30:711–725.doi:10.2164/jandrol.109.007971 PMID:19478329

27. Beaty TH, Murray JC, Marazita ML et al. (2010). A genome-wide association study of cleft lip with and without cleft palate identifies risk variants near MAFB and ABCA4. *Nat Genet*, 42:525–529.doi:10.1038/ng.580 PMID:20436469

28. Freathy RM, Mook-Kanamori DO, Sovio U et al.; Genetic Investigation of ANthropometric Traits (GIANT) Consortium; Meta-Analyses of Glucose and Insulin-related traits Consortium; Wellcome Trust Case Control Consortium; Early Growth Genetics (EGG) Consortium (2010). Variants in ADCY5 and near CCNL1 are associated with fetal growth and birth weight. *Nat Genet*, 42:430–435.doi:10.1038/ng.567 PMID:20372150

29. Ingersoll RG, Hetmanski J, Park JW et al. (2010). Association between genes on chromosome 4p16 and non-syndromic oral clefts in four populations. *Eur J Hum Genet*, 18:726–732.doi:10.1038/ejhg.2009.228 PMID:20087401

30. Liu YZ, Guo YF, Wang L et al. (2009). Genome-wide association analyses identify SPOCK as a key novel gene underlying age at menarche. *PLoS Genet*, 5:e1000420. doi:10.1371/journal.pgen.1000420 PMID:19282985

31. Loos RJ, Lindgren CM, Li S et al.; Prostate, Lung, Colorectal, and Ovarian (PLCO) Cancer Screening Trial; KORA; Nurses' Health Study; Diabetes Genetics Initiative; SardiNIA Study; Wellcome Trust Case Control Consortium; FUSION (2008). Common variants near MC4R are associated with fat mass, weight and risk of obesity. *Nat Genet*, 40:768–775. doi:10.1038/ng.140 PMID:18454148

32. Mangold E, Ludwig KU, Birnbaum S et al. (2010). Genome-wide association study identifies two susceptibility loci for nonsyndromic cleft lip with or without cleft palate. *Nat Genet*, 42:24–26.doi:10.1038/ng.506 PMID:20023658

33. Voorhuis M, Onland-Moret NC, van der Schouw YT et al. (2010). Human studies on genetics of the age at natural menopause: a systematic review. *Hum Reprod Update*, 16:364–377.doi:10.1093/humupd/dmp055 PMID:20071357

34. Herman-Giddens ME (2006). Recent data on pubertal milestones in United States children: the secular trend toward earlier development. *Int J Androl*, 29:241–246, discussion 286–290.doi:10.1111/j.1365-2605.2005.00575.x PMID:16466545

35. Euling SY, Herman-Giddens ME, Lee PA et al. (2008). Examination of US puberty-timing data from 1940 to 1994 for secular trends: panel findings. *Pediatrics*, 121 Suppl 3;S172–S191.doi:10.1542/peds.2007-1813D PMID:18245511

36. Tanner JM. Growth at adolescence. Oxford: Blackwell Scientific; 1962.

37. Duke PM, Litt IF, Gross RT (1980). Adolescents' self-assessment of sexual maturation. *Pediatrics*, 66:918–920. PMID:7454482

38. Neinstein LS (1982). Adolescent self-assessment of sexual maturation: reassessment and evaluation in a mixed ethnic urban population. *Clin Pediatr (Phila)*, 21:482–484.doi:10.1177/000992288202100806 PMID:7083719

39. Wu T, Mendola P, Buck GM (2002). Ethnic differences in the presence of secondary sex characteristics and menarche among US girls: the Third National Health and Nutrition Examination Survey, 1988–1994. *Pediatrics*, 110:752–757.doi:10.1542/peds.110.4.752 PMID:12359790

40. Rockett JC, Lynch CD, Buck GM (2004). Biomarkers for assessing reproductive development and health: Part 1–Pubertal development. *Environ Health Perspect*, 112:105–112.doi:10.1289/ehp.6265 PMID:14698938

41. Blanck HM, Marcus M, Tolbert PE et al. (2000). Age at menarche and tanner stage in girls exposed in utero and postnatally to polybrominated biphenyl. *Epidemiology*, 11:641–647.doi:10.1097/00001648-2000110 00-00005 PMID:11055623

42. Mantzoros CS, Flier JS, Rogol AD (1997). A longitudinal assessment of hormonal and physical alterations during normal puberty in boys. V. Rising leptin levels may signal the onset of puberty. *J Clin Endocrinol Metab*, 82:1066–1070.doi:10.1210/jc.82.4.1066 PMID:9100574

43. Roemmich JN, Rogol AD (1999). Role of leptin during childhood growth and development. [viii.]. *Endocrinol Metab Clin North Am*, 28:749–764, viii.doi:10.1016/S0889-8529(05)70100-6 PMID:10609118

44. Lee MM, Donahoe PK, Hasegawa T et al. (1996). Mullerian inhibiting substance in humans: normal levels from infancy to adulthood. *J Clin Endocrinol Metab*, 81:571–576.doi:10.1210/jc.81.2.571 PMID:8636269

45. Kiess W, Reich A, Meyer K et al. (1999). A role for leptin in sexual maturation and puberty? *Horm Res*, 51 Suppl 3;55–63. doi:10.1159/000053163 PMID:10592445

46. Zaman N, Hall CM, Gill MS et al. (2003). Leptin measurement in urine in children and its relationship to other growth peptides in serum and urine. *Clin Endocrinol (Oxf)*, 58:78–85. doi:10.1046/j.1365-2265.2003.01677.x PMID:12519416

47. Wu FC, Brown DC, Butler GE et al. (1993). Early morning plasma testosterone is an accurate predictor of imminent pubertal development in prepubertal boys. *J Clin Endocrinol Metab*, 76:26–31.doi:10.1210/jc.76.1.26 PMID:8421096

48. Demir A, Dunkel L, Stenman UH, Voutilainen R (1995). Age-related course of urinary gonadotropins in children. *J Clin Endocrinol Metab*, 80:1457–1460.doi:10.1210/jc.80.4.1457 PMID:7714124

49. Maqsood AR, Trueman JA, Whatmore AJ et al. (2007). The relationship between nocturnal urinary leptin and gonadotrophins as children progress towards puberty. *Horm Res*, 68:225–230.doi:10.1159/000101335 PMID:17389812

50. Rajpert-De Meyts E, Jørgensen N, Graem N et al. (1999). Expression of anti-Müllerian hormone during normal and pathological gonadal development: association with differentiation of Sertoli and granulosa cells. *J Clin Endocrinol Metab*, 84:3836–3844.doi:10.1210/jc.84.10.3836 PMID:10523039

51. Josso N, Cate RL, Picard JY et al. (1993). Anti-müllerian hormone: the Jost factor. *Recent Prog Horm Res*, 48:1–59. PMID:8441845

52. Lee MM, Donahoe PK (1993). Mullerian inhibiting substance: a gonadal hormone with multiple functions. *Endocr Rev*, 14:152–164. PMID:8325249

53. Hudson PL, Dougas I, Donahoe PK *et al.* (1990). An immunoassay to detect human müllerian inhibiting substance in males and females during normal development. *J Clin Endocrinol Metab*, 70:16–22.doi:10.1210/jcem-70-1-16 PMID:2294129

54. Chada M, Průsa R, Bronský J *et al.* (2003). Inhibin B, follicle stimulating hormone, luteinizing hormone and testosterone during childhood and puberty in males: changes in serum concentrations in relation to age and stage of puberty. *Physiol Res*, 52:45–51. PMID:12625806

55. Radicioni AF, Anzuini A, De Marco E *et al.* (2005). Changes in serum inhibin B during normal male puberty. *Eur J Endocrinol*, 152:403–409.doi:10.1530/eje.1.01855 PMID:15757857

56. Crofton PM, Evans AE, Groome NP *et al.* (2002). Dimeric inhibins in girls from birth to adulthood: relationship with age, pubertal stage, FSH and oestradiol. *Clin Endocrinol (Oxf)*, 56:223–230.doi:10.1046/j.0300-0664.2001.01449.x PMID:11874414

57. Foster CM, Olton PR, Padmanabhan V (2005). Diurnal changes in FSH-regulatory peptides and their relationship to gonadotrophins in pubertal girls. *Hum Reprod*, 20:543–548.doi:10.1093/humrep/deh607 PMID:15550493

58. Rogan WJ, Ragan NB (2007). Some evidence of effects of environmental chemicals on the endocrine system in children. *Int J Hyg Environ Health*, 210:659–667.doi:10.1016/j.ijheh.2007.07.005 PMID:17870664

59. Den Hond E, Schoeters G (2006). Endocrine disrupters and human puberty. *Int J Androl*, 29:264–271, discussion 286–290.doi:10.1111/j.1365-2605.2005.00561.x PMID:16466548

60. Schoeters G, Den Hond E, Dhooge W *et al.* (2008). Endocrine disruptors and abnormalities of pubertal development. *Basic Clin Pharmacol Toxicol*, 102:168–175.doi:10.1111/j.1742-7843.2007.00180.x PMID:18226071

61. Wolff MS, Britton JA, Boguski L *et al.* (2008). Environmental exposures and puberty in inner-city girls. *Environ Res*, 107:393–400. doi:10.1016/j.envres.2008.03.006 PMID:18479682

62. Buck Louis GM, Gray LE Jr, Marcus M *et al.* (2008). Environmental factors and puberty timing: expert panel research needs. *Pediatrics*, 121 Suppl 3;S192–S207. doi:10.1542/peds.1813E PMID:18245512

63. Kavlock RJ, Daston GP, DeRosa C *et al.* (1996). Research needs for the risk assessment of health and environmental effects of endocrine disruptors: a report of the U.S. EPA-sponsored workshop. *Environ Health Perspect*, 104 Suppl 4;715–740. PMID:8880000

64. Graziottin A (2000). Libido: the biologic scenario. *Maturitas*, 34 Suppl 1;S9–S16. doi:10.1016/S0378-5122(99)00072-9 PMID:10759059

65. McHorney CA, Rust J, Golombok S *et al.* (2004). Profile of Female Sexual Function: a patient-based, international, psychometric instrument for the assessment of hypoactive sexual desire in oophorectomized women. *Menopause*, 11:474–483.doi:10.1097/01.GME.0000109316.11228.77 PMID:15243286

66. Bancroft J, Cawood EH (1996). Androgens and the menopause; a study of 40–60-year-old women. *Clin Endocrinol (Oxf)*, 45:577–587.doi:10.1046/j.1365-2265.1996.00846.x PMID:8977755

67. Avis NE, Stellato R, Crawford S *et al.* (2000). Is there an association between menopause status and sexual functioning? *Menopause*, 7:297–309.doi:10.1097/00042192-200007050-00004 PMID:10993029

68. Dennerstein L, Dudley E, Burger H (2001). Are changes in sexual functioning during midlife due to aging or menopause? *Fertil Steril*, 76:456–460.doi:10.1016/S0015-0282(01)01978-1 PMID:11532464

69. Basson R, Berman J, Burnett A *et al.* (2000). Report of the international consensus development conference on female sexual dysfunction: definitions and classifications. *J Urol*, 163:888–893.doi:10.1016/S0022-5347(05)67828-7 PMID:10688001

70. Bancroft J (2002). The medicalization of female sexual dysfunction: the need for caution. *Arch Sex Behav*, 31:451–455. doi:10.1023/A:1019800426980 PMID:12238614

71. Leiblum SR, Koochaki PE, Rodenberg CA *et al.* (2006). Hypoactive sexual desire disorder in postmenopausal women: US results from the Women's International Study of Health and Sexuality (WISHeS). *Menopause*, 13:46–56.doi:10.1097/01.gme.0000172596.76272.06 PMID:16607098

72. Laumann EO, Paik A, Rosen RC (1999). Sexual dysfunction in the United States: prevalence and predictors. *JAMA*, 281:537–544.doi:10.1001/jama.281.6.537 PMID:10022110

73. Turna B, Apaydin E, Semerci B *et al.* (2005). Women with low libido: correlation of decreased androgen levels with female sexual function index. *Int J Impot Res*, 17:148–153. doi:10.1038/sj.ijir.3901294 PMID:15592425

74. Nyunt A, Stephen G, Gibbin J *et al.* (2005). Androgen status in healthy premenopausal women with loss of libido. *J Sex Marital Ther*, 31:73–80.doi:10.1080/00926230590475314 PMID:15841707

75. Granger DA, Shirtcliff EA, Booth A *et al.* (2004). The "trouble" with salivary testosterone. *Psychoneuroendocrinology*, 29:1229–1240. doi:10.1016/j.psyneuen.2004.02.005 PMID:15288702

76. Whembolua GL, Granger DA, Singer S *et al.* (2006). Bacteria in the oral mucosa and its effects on the measurement of cortisol, dehydroepiandrosterone, and testosterone in saliva. *Horm Behav*, 49:478–483.doi:10.1016/j.yhbeh.2005.10.005 PMID:16309679

77. Baird DD (1999). Characteristics of fertile menstrual cycles. *Scand J Work Environ Health*, 25 Suppl 1;20–22, discussion 76–78. PMID:10235401

78. van Zonneveld P, Scheffer GJ, Broekmans FJ *et al.* (2003). Do cycle disturbances explain the age-related decline of female fertility? Cycle characteristics of women aged over 40 years compared with a reference population of young women. *Hum Reprod*, 18:495–501. doi:10.1093/humrep/deg138 PMID:12615813

79. Li H, Nakajima ST, Chen J *et al.* (2001). Differences in hormonal characteristics of conceptive versus nonconceptive menstrual cycles. *Fertil Steril*, 75:549–553.doi:10.1016/S0015-0282(00)01765-9 PMID:11239540

80. Small CM, Manatunga AK, Klein M *et al.* (2006). Menstrual cycle characteristics: associations with fertility and spontaneous abortion. *Epidemiology*, 17:52–60.doi:10.1097/01.ede.0000190540.95748.e6 PMID:16357595

81. Waller K, Swan SH, Windham GC *et al.* (1998). Use of urine biomarkers to evaluate menstrual function in healthy premenopausal women. *Am J Epidemiol*, 147:1071–1080. PMID:9620051

82. Windham GC, Elkin E, Fenster L *et al.* (2002). Ovarian hormones in premenopausal women: variation by demographic, reproductive and menstrual cycle characteristics. *Epidemiology*, 13:675–684. doi:10.1097/00001648-200211000-00012 PMID:12410009

83. Liu Y, Gold EB, Lasley BL, Johnson WO (2004). Factors affecting menstrual cycle characteristics. *Am J Epidemiol*, 160:131–140.doi:10.1093/aje/kwh188 PMID:15234934

84. Harlow SD, Campbell B (1996). Ethnic differences in the duration and amount of menstrual bleeding during the postmenarcheal period. *Am J Epidemiol*, 144:980–988. PMID:8916509

85. Harlow SD, Campbell B, Lin X, Raz J (1997). Ethnic differences in the length of the menstrual cycle during the postmenarcheal period. *Am J Epidemiol*, 146:572–580. PMID:9326435

86. Rowland AS, Baird DD, Long S *et al.* (2002). Influence of medical conditions and lifestyle factors on the menstrual cycle. *Epidemiology*, 13:668–674.doi:10.1097/00001648-20021 1000-00011 PMID:12410008

87. Chatterton RT Jr, Mateo ET, Hou N *et al.* (2005). Characteristics of salivary profiles of oestradiol and progesterone in premenopausal women. *J Endocrinol*, 186:77–84.doi:10.1677/joe.1.06025 PMID:16002538

88. Zinaman MJ (2006). Using cervical mucus and other easily observed biomarkers to identify ovulation in prospective pregnancy trials. *Paediatr Perinat Epidemiol*, 20 Suppl 1;26–29.doi:10.1111/j.1365-3016.2006.00767.x PMID:17061970

89. Gann PH, Giovanazzi S, Van Horn L et al. (2001). Saliva as a medium for investigating intra- and interindividual differences in sex hormone levels in premenopausal women. *Cancer Epidemiol Biomarkers Prev*, 10:59–64. PMID:11205490

90. Lu Y, Bentley GR, Gann PH et al. (1999). Salivary estradiol and progesterone levels in conception and nonconception cycles in women: evaluation of a new assay for salivary estradiol. *Fertil Steril*, 71:863–868. doi:10.1016/S0015-0282(99)00093-X PMID:10231047

91. Lipson SF, Ellison PT (1996). Comparison of salivary steroid profiles in naturally occurring conception and non-conception cycles. *Hum Reprod*, 11:2090–2096. PMID:8943508

92. Hardy K, Wright C, Rice S et al. (2002). Future developments in assisted reproduction in humans. *Reproduction*, 123:171–183.doi:10.1530/rep.0.1230171 PMID:11866685

93. Goldberg JM, Falcone T, Attaran M (2007). In vitro fertilization update. *Cleve Clin J Med*, 74:329–338.doi:10.3949/ccjm.74.5.329 PMID:17506238

94. Gleicher N, Weghofer A, Barad D (2008). Preimplantation genetic screening: "established" and ready for prime time? *Fertil Steril*, 89:780–788.doi:10.1016/j.fertnstert.2008.01.072 PMID:18353323

95. Canfield RE, O'Connor JF, Birken S et al. (1987). Development of an assay for a biomarker of pregnancy and early fetal loss. *Environ Health Perspect*, 74:57–66. doi:10.1289/ehp.877457 PMID:3319556

96. Wilcox AJ, Weinberg CR, O'Connor JF et al. (1988). Incidence of early loss of pregnancy. *N Engl J Med*, 319:189–194.doi:10.1056/NEJM198807283190401 PMID:3393170

97. Sweeney AM, Meyer MR, Aarons JH et al. (1988). Evaluation of methods for the prospective identification of early fetal losses in environmental epidemiology studies. *Am J Epidemiol*, 127:843–850. PMID:3354549

98. Lasley BL, Lohstroh P, Kuo A et al. (1995). Laboratory methods for evaluating early pregnancy loss in an industry-based population. *Am J Ind Med*, 28:771–781.doi:10.1002/ajim.4700280611 PMID:8588563

99. Stewart DR, Overstreet JW, Nakajima ST, Lasley BL (1993). Enhanced ovarian steroid secretion before implantation in early human pregnancy. *J Clin Endocrinol Metab*, 76:1470–1476.doi:10.1210/jc.76.6.1470 PMID:8501152

100. Qiu Q, Overstreet JW, Todd H et al. (1997). Total urinary follicle stimulating hormone as a biomarker for detection of early pregnancy and periimplantation spontaneous abortion. *Environ Health Perspect*, 105:862–866.doi:10.1289/ehp.97105862 PMID:9347902

101. Lohstroh PN, Overstreet JW, Stewart DR et al. (2005). Secretion and excretion of human chorionic gonadotropin during early pregnancy. *Fertil Steril*, 83:1000–1011.doi:10.1016/j.fertnstert.2004.10.038 PMID:15820813

102. Vitzthum VJ, Spielvogel H, Thornburg J, West B (2006). A prospective study of early pregnancy loss in humans. *Fertil Steril*, 86:373–379.doi:10.1016/j.fertnstert.2006.01.021 PMID:16806213

103. Pyper C, Bromhall L, Dummett S et al. (2006). The Oxford Conception Study design and recruitment experience. *Paediatr Perinat Epidemiol*, 20 Suppl 1;51–59.doi:10.1111/j.1365-3016.2006.00771.x PMID:17061974

104. Lynch CD, Jackson LW, Buck Louis GM (2006). Estimation of the day-specific probabilities of conception: current state of the knowledge and the relevance for epidemiological research. *Paediatr Perinat Epidemiol*, 20 Suppl 1;3–12.doi:10.1111/j.1365-3016.2006.00765.x PMID:17061968

105. Morton H, Rolfe B, Clunie GJ (1977). An early pregnancy factor detected in human serum by the rosette inhibition test. *Lancet*, 1:394–397.doi:10.1016/S0140-6736(77)92605-8 PMID:65512

106. Fan XG, Zheng ZQ (1997). A study of early pregnancy factor activity in preimplantation. *Am J Reprod Immunol*, 37:359–364. PMID:9196793

107. Cavanagh AC, Morton H, Rolfe BE, Gidley-Baird AA (1982). Ovum factor: a first signal of pregnancy? *Am J Reprod Immunol*, 2:97–101. PMID:7102890

108. Cheng SJ, Zheng ZQ (2004). Early pregnancy factor in cervical mucus of pregnant women. *Am J Reprod Immunol*, 51:102–105. doi:10.1046/j.8755-8920.2003.00136.x PMID:14748834

109. Olsen J, Basso O. Reproductive epidemiology. In: Ahrens W, Pigeot I, editors. Handbook of epidemiology. Berlin: Springer; 2005. p. 1043–109.

110. Baumann MU, Bersinger NA, Surbek DV (2007). Serum markers for predicting pre-eclampsia. *Mol Aspects Med*, 28:227–244. doi:10.1016/j.mam.2007.04.002 PMID:17532461

111. Papageorghiou AT (2008). Predicting and preventing pre-eclampsia-where to next? *Ultrasound Obstet Gynecol*, 31:367–370. doi:10.1002/uog.5320 PMID:18383482

112. Murphy DJ, Fowlie PW, McGuire W (2004). Obstetric issues in preterm birth. *BMJ*, 329:783–786.doi:10.1136/bmj.329.7469.783 PMID:15459053

113. Breathnach FM, Malone FD, Lambert-Messerlian G et al.; First and Second Trimester Evaluation of Risk (FASTER) Research Consortium (2007). First- and second-trimester screening: detection of aneuploidies other than Down syndrome. *Obstet Gynecol*, 110:651–657.doi:10.1097/01.AOG.0000278570.76392.a6 PMID:17766613

114. Spencer K (2007). Aneuploidy screening in the first trimester. *Am J Med Genet C Semin Med Genet*, 145C:18–32. PMID:17290444

115. Wald NJ, Rudnicka AR, Bestwick JP (2006). Sequential and contingent prenatal screening for Down syndrome. *Prenat Diagn*, 26:769–777.doi:10.1002/pd.1498 PMID:16821246

116. Rosen T, D'Alton ME, Platt LD, Wapner R; Nuchal Translucency Oversight Committee, Maternal Fetal Medicine Foundation (2007). First-trimester ultrasound assessment of the nasal bone to screen for aneuploidy. *Obstet Gynecol*, 110:399–404.doi:10.1097/01.AOG.0000275281.19344.66 PMID:17666617

117. Scott A (2007). Nuchal translucency measurement in first trimester Down syndrome screening. *Issues Emerg Health Technol*, (100):1–6. PMID:17595751

118. Sekizawa A, Purwosunu Y, Matsuoka R et al. (2007). Recent advances in non-invasive prenatal DNA diagnosis through analysis of maternal blood. *J Obstet Gynaecol Res*, 33:747–764.doi:10.1111/j.1447-0756.2007.00652.x PMID:18001438

119. Torday JS, Rehan VK (2003). Testing for fetal lung maturation: a biochemical "window" to the developing fetus. *Clin Lab Med*, 23:361–383.doi:10.1016/S0272-2712(03)00030-1 PMID:12848449

120. Rubin LP, Hansen K (2003). Testing for hematologic disorders and complications. *Clin Lab Med*, 23:317–343.doi:10.1016/S0272-2712(03)00031-3 PMID:12848447

121. Lightfoot T, Bunch K, Ansell P, Murphy M (2005). Ovulation induction, assisted conception and childhood cancer. *Eur J Cancer*, 41:715–724, discussion 725–726. doi:10.1016/j.ejca.2004.07.032 PMID:15763647

122. Cheung AP (2006). Assisted reproductive technology: both sides now. *J Reprod Med*, 51:283–292. PMID:16739266

123. Knoester M, Helmerhorst FM, Vandenbroucke JP et al.; Leiden Artificial Reproductive Techniques Follow-up Project (L-art-FUP) (2008). Perinatal outcome, health, growth, and medical care utilization of 5- to 8-year-old intracytoplasmic sperm injection singletons. *Fertil Steril*, 89:1133–1146.doi:10.1016/j.fertnstert.2007.04.049 PMID:18177652

124. Parker SP, Khan HI, Cubitt WD (1999). Detection of antibodies to hepatitis C virus in dried blood spot samples from mothers and their offspring in Lahore, Pakistan. *J Clin Microbiol*, 37:2061–2063. PMID:10325381

125. Wiemels JL, Cazzaniga G, Daniotti M et al. (1999). Prenatal origin of acute lymphoblastic leukaemia in children. *Lancet*, 354:1499–1503.doi:10.1016/S0140-6736(99)09403-9 PMID:10551495

126. Crawford DC, Caggana M, Harris KB et al. (2002). Characterization of beta-globin haplotypes using blood spots from a population-based cohort of newborns with homozygous HbS. Genet Med, 4:328–335. doi:10.1097/00125817-200209000-00003 PMID:12394345

127. Klotz J, Bryant P, Wilcox HB et al. (2006). Population-based retrieval of newborn dried blood spots for researching paediatric cancer susceptibility genes. Paediatr Perinat Epidemiol, 20:449–452.doi:10.1111/j.1365-3016.2006.00749.x PMID:16911024

128. Spector LG, Hecht SS, Ognjanovic S et al. (2007). Detection of cotinine in newborn dried blood spots. Cancer Epidemiol Biomarkers Prev, 16:1902–1905.doi:10.1158/1055-9965.EPI-07-0230 PMID:17855712

129. Patton JC, Akkers E, Coovadia AH et al. (2007). Evaluation of dried whole blood spots obtained by heel or finger stick as an alternative to venous blood for diagnosis of human immunodeficiency virus type 1 infection in vertically exposed infants in the routine diagnostic laboratory. Clin Vaccine Immunol, 14:201–203.doi:10.1128/CVI.00223-06 PMID:17167036

130. Olshan AF (2007). Meeting report: the use of newborn blood spots in environmental research: opportunities and challenges. Environ Health Perspect, 115:1767–1779.doi:10.1289/ehp.10511 PMID:18087597

131. Mei JV, Alexander JR, Adam BW, Hannon WH (2001). Use of filter paper for the collection and analysis of human whole blood specimens. J Nutr, 131:1631S–1636S. PMID:11340130

132. Mandl KD, Feit S, Larson C, Kohane IS (2002). Newborn screening program practices in the United States: notification, research, and consent. Pediatrics, 109:269–273.doi:10.1542/peds.109.2.269 PMID:11826206

133. Olney RS, Moore CA, Ojodu JA et al. (2006). Storage and use of residual dried blood spots from state newborn screening programs. J Pediatr, 148:618–622.doi:10.1016/j.jpeds.2005.12.053 PMID:16737872

134. Ge RS, Chen GR, Tanrikut C, Hardy MP (2007). Phthalate ester toxicity in Leydig cells: developmental timing and dosage considerations. Reprod Toxicol, 23:366–373. doi:10.1016/j.reprotox.2006.12.006 PMID:17258888

135. Martin OV, Shialis T, Lester JN et al. (2008). Testicular dysgenesis syndrome and the estrogen hypothesis: a quantitative meta-analysis. Environ Health Perspect, 116:149–157.doi:10.1289/ehp.10545 PMID:18288311

136. Schull WJ (2003). The children of atomic bomb survivors: a synopsis. J Radiol Prot, 23:369–384.doi:10.1088/0952-4746/23/4/R302 PMID:14750686

137. Signorello LB, Cohen SS, Bosetti C et al. (2006). Female survivors of childhood cancer: preterm birth and low birth weight among their children. J Natl Cancer Inst, 98:1453–1461. doi:10.1093/jnci/djj394 PMID:17047194

138. Nagarajan R, Robison LL (2005). Pregnancy outcomes in survivors of childhood cancer. J Natl Cancer Inst Monogr, 2005:72–76.doi:10.1093/jncimonographs/lgi020 PMID:15784829

139. Botto LD, Yang Q (2000). 5,10-Methylenetetrahydrofolate reductase gene variants and congenital anomalies: a HuGE review. Am J Epidemiol, 151:862–877. PMID:10791559

140. Finnell RH, Shaw GM, Lammer EJ et al. (2004). Gene-nutrient interactions: importance of folates and retinoids during early embryogenesis. Toxicol Appl Pharmacol, 198:75–85.doi:10.1016/j.taap.2003.09.031 PMID:15236946

141. Blom HJ, Shaw GM, den Heijer M, Finnell RH (2006). Neural tube defects and folate: case far from closed. Nat Rev Neurosci, 7:724–731.doi:10.1038/nrn1986 PMID:16924 261

142. Zeiger JS, Beaty TH, Liang KY (2005). Oral clefts, maternal smoking, and TGFA: a meta-analysis of gene-environment interaction. Cleft Palate Craniofac J, 42:58–63.doi:10.1597/02-128.1 PMID:15643916

143. Krapels IP, Raijmakers-Eichhorn J, Peters WH et al.; Eurocran Gene-Environment Interaction Group (2008). The I,105V polymorphism in glutathione S-transferase P1, parental smoking and the risk for nonsyndromic cleft lip with or without cleft palate. Eur J Hum Genet, 16:358–366. doi:10.1038/sj.ejhg.5201973 PMID:18159215

144. Lie RT, Wilcox AJ, Taylor J et al. (2008). Maternal smoking and oral clefts: the role of detoxification pathway genes. Epidemiology, 19:606–615.doi:10.1097/EDE.0b013e3181690731 PMID:18449058

145. Torfs CP, Christianson RE, Iovannisci DM et al. (2006). Selected gene polymorphisms and their interaction with maternal smoking, as risk factors for gastroschisis. Birth Defects Res A Clin Mol Teratol, 76:723–730. doi:10.1002/bdra.20310 PMID:17051589

146. Kishi R, Sata F, Yoshioka E et al. (2008). Exploiting gene-environment interaction to detect adverse health effects of environmental chemicals on the next generation. Basic Clin Pharmacol Toxicol, 102:191–203.doi:10.1111/j.1742-7843.2007.00201.x PMID:18226074

147. Etheredge AJ, Christensen K, Del Junco D et al. (2005). Evaluation of two methods for assessing gene-environment interactions using data from the Danish case-control study of facial clefts. Birth Defects Res A Clin Mol Teratol, 73:541–546.doi:10.1002/bdra.20167 PMID:15965987

148. Smith GD (2008). Assessing intrauterine influences on offspring health outcomes: can epidemiological studies yield robust findings? Basic Clin Pharmacol Toxicol, 102:245–256.doi:10.1111/j.1742-7843.2007.00191.x PMID:18226080

149. Tsai HJ, Liu X, Mestan K et al. (2008). Maternal cigarette smoking, metabolic gene polymorphisms, and preterm delivery: new insights on GxE interactions and pathogenic pathways. Hum Genet, 123:359–369.doi:10.1007/s00439-008-0485-9 PMID:18320 229

150. Wang X, Zuckerman B, Pearson C et al. (2002). Maternal cigarette smoking, metabolic gene polymorphism, and infant birth weight. JAMA, 287:195–202.doi:10.1001/jama.287.2.195 PMID:11779261

151. Williams TA, Mars AE, Buyske SG et al. (2007). Risk of autistic disorder in affected offspring of mothers with a glutathione S-transferase P1 haplotype. Arch Pediatr Adolesc Med, 161:356–361.doi:10.1001/archpedi.161.4.356 PMID:17404132

152. Todd RD, Neuman RJ (2007). Gene-environment interactions in the development of combined type ADHD: evidence for a synapse-based model. Am J Med Genet B Neuropsychiatr Genet, 144B:971–975. PMID:17955458

153. Infante-Rivard C, Weichenthal S (2007). Pesticides and childhood cancer: an update of Zahm and Ward's 1998 review. J Toxicol Environ Health B Crit Rev, 10:81–99. PMID:18074305

154. Newbold RR, Hanson RB, Jefferson WN et al. (1998). Increased tumors but uncompromised fertility in the female descendants of mice exposed developmentally to diethylstilbestrol. Carcinogenesis, 19:1655–1663.doi:10.1093/carcin/19.9.1655 PMID:9771938

155. Newbold RR, Hanson RB, Jefferson WN et al. (2000). Proliferative lesions and reproductive tract tumors in male descendants of mice exposed developmentally to diethylstilbestrol. Carcinogenesis, 21:1355–1363.doi:10.1093/carcin/21.7.1355 PMID:10874014

156. Tomatis L (1994). Transgeneration carcinogenesis: a review of the experimental and epidemiological evidence. Jpn J Cancer Res, 85:443–454. PMID:8014100

157. Klip H, Verloop J, van Gool JD et al.; OMEGA Project Group (2002). Hypospadias in sons of women exposed to diethylstilbestrol in utero: a cohort study. Lancet, 359:1102–1107.doi:10.1016/S0140-6736(02)08152-7 PMID:11943257

158. Palmer JR, Wise LA, Robboy SJ et al. (2005). Hypospadias in sons of women exposed to diethylstilbestrol in utero. Epidemiology, 16:583–586.doi:10.1097/01.ede.0000164789.59728.6d PMID:15951681

159. Pons JC, Papiernik E, Billon A et al. (2005). Hypospadias in sons of women exposed to diethylstilbestrol in utero. Prenat Diagn, 25:418–419.doi:10.1002/pd.1136 PMID:15906411

160. Brouwers MM, Feitz WF, Roelofs LA et al. (2006). Hypospadias: a transgenerational effect of diethylstilbestrol? Hum Reprod, 21:666–669.doi:10.1093/humrep/dei398 PMID:16293648

161. Stoll C, Alembik Y, Dott B (2003). Limb reduction defects in the first generation and deafness in the second generation of intrauterine exposed fetuses to diethylstilbestrol. Ann Genet, 46:459–465. PMID:14659782

162. Blatt J, Van Le L, Weiner T, Sailer S (2003). Ovarian carcinoma in an adolescent with transgenerational exposure to diethylstilbestrol. J Pediatr Hematol Oncol, 25:635–636.doi:10.1097/00043426-200308000-00009 PMID:12902917

163. Li S, Hansman R, Newbold R et al. (2003). Neonatal diethylstilbestrol exposure induces persistent elevation of c-fos expression and hypomethylation in its exon-4 in mouse uterus. Mol Carcinog, 38:78–84.doi:10.1002/mc.10147 PMID:14502647

164. Waterland RA, Michels KB (2007). Epigenetic epidemiology of the developmental origins hypothesis. Annu Rev Nutr, 27:363–388.doi:10.1146/annurev.nutr.27.061406.093705 PMID:17465856

165. Nijland MJ, Ford SP, Nathanielsz PW (2008). Prenatal origins of adult disease. Curr Opin Obstet Gynecol, 20:132–138. doi:10.1097/GCO.0b013e3282f76753 PMID:18388812

166. Barker DJ (2004). The developmental origins of adult disease. J Am Coll Nutr, 23 Suppl;588S–595S. PMID:15640511

167. Gillman MW, Barker D, Bier D et al. (2007). Meeting report on the 3rd International Congress on Developmental Origins of Health and Disease (DOHaD). Pediatr Res, 61:625–629. PMID:17413866

168. Barker DJ (2007). The origins of the developmental origins theory. J Intern Med, 261:412–417.doi:10.1111/j.1365-2796.2007.01809.x PMID:17444880

169. Trosko JE, Chang CC, Upham B. Modulation of GAP junctional communication by "epigenetic" toxicants: A shared mechanism in teratogenesis, carcinogenesis, atherogenesis, immunomodulation, reproductive and neurotxicities. In: Wilson SH, Suk WA, editors. Biomarkers of environmentally associated disease: technologies, concepts and perspectives. Boca Raton (FL): Lewis Publishers; 2002. p. 445–54.

170. Dolinoy DC, Huang D, Jirtle RL (2007). Maternal nutrient supplementation counteracts bisphenol A-induced DNA hypomethylation in early development. Proc Natl Acad Sci USA, 104:13056–13061.doi:10.1073/pnas.0703739104 PMID:17670942

171. Chang HS, Anway MD, Rekow SS, Skinner MK (2006). Transgenerational epigenetic imprinting of the male germline by endocrine disruptor exposure during gonadal sex determination. Endocrinology, 147:5524–5541.doi:10.1210/en.2006-0987 PMID:16973722

172. Ness RB, Catov J (2007). Invited commentary: Timing and types of cardiovascular risk factors in relation to offspring birth weight. Am J Epidemiol, 166:1365–1367.doi:10.1093/aje/kwm314 PMID:17977895

173. Golden AL. Biomarkers of male reproductive health. In: Wilson SH, Suk WA, editors. Biomarkers of environmentally associated disease: technologies, concepts, and perspectives. Boca Raton (FL): Lewis Publishers; 2002. p. 387–410.

174. Kandeel FR, Koussa VK, Swerdloff RS (2001). Male sexual function and its disorders: physiology, pathophysiology, clinical investigation, and treatment. Endocr Rev, 22:342–388.doi:10.1210/er.22.3.342 PMID:11399748

175. Burnett AL (2008). Environmental erectile dysfunction: can the environment really be hazardous to your erectile health? J Androl, 29:229–236.doi:10.2164/jandrol.107.004200 PMID:18187396

176. Schrader SM. Male reproductive toxicity. In: Massaro EJ, editor. Handbook of human toxicology. Boca Raton (FL): CRC Press; 1997. p. 962–80.

177. Webster RA. Reproductive function and pregnancy. In: McPherson RA, Pincus MR, editors. Henry's clinical diagnosis and management by laboratory methods. 21st ed. Philadelphia (PA): Saunders Elsevier; 2007. p. 364–78.

178. Mabeck LM, Jensen MS, Toft G et al. (2005). Fecundability according to male serum inhibin B–a prospective study among first pregnancy planners. Hum Reprod, 20:2909–2915.doi:10.1093/humrep/dei141 PMID:16024538

179. Ellingsen DG, Chashchin V, Haug E et al. (2007). An epidemiological study of reproductive function biomarkers in male welders. Biomarkers, 12:497–509.doi:10.1080/13547500701366496 PMID:17701748

180. Mocarelli P, Gerthoux PM, Patterson DG Jr et al. (2008). Dioxin exposure, from infancy through puberty, produces endocrine disruption and affects human semen quality. Environ Health Perspect, 116:70–77.doi:10.1289/ehp.10399 PMID:18197302

181. Dhooge W, van Larebeke N, Koppen G et al.; Flemish Environment and Health Study Group (2006). Serum dioxin-like activity is associated with reproductive parameters in young men from the general Flemish population. Environ Health Perspect, 114:1670–1676. PMID:17107851

182. Telisman S, Colak B, Pizent A et al. (2007). Reproductive toxicity of low-level lead exposure in men. Environ Res, 105:256–266. doi:10.1016/j.envres.2007.05.011 PMID:17632096

183. Sripada S, Fonseca S, Lee A et al. (2007). Trends in semen parameters in the northeast of Scotland. J Androl, 28:313– 319.doi:10.2164/jandrol.106.000729 PMID:17079743

184. Brazil C, Swan SH, Tollner CR et al.; Study for Future Families Research Group (2004). Quality control of laboratory methods for semen evaluation in a multicenter research study. J Androl, 25:645–656. PMID:15223854

185. Ness RB, Grainger DA (2008). Male reproductive proteins and reproductive outcomes. Am J Obstet Gynecol, 198:620–624, e1–e4.doi:10.1016/j.ajog.2007.09.017 PMID:18191798

186. Tremellen K (2008). Oxidative stress and male infertility–a clinical perspective. Hum Reprod Update, 14:243–258.doi:10.1093/humupd/dmn004 PMID:18281241

187. Evenson DP, Wixon R (2006). Clinical aspects of sperm DNA fragmentation detection and male infertility. Theriogenology, 65:979–991.doi:10.1016/j.theriogenology.2005.09.011 PMID:16242181

188. Roeleveld N, Bretveld R (2008). The impact of pesticides on male fertility. Curr Opin Obstet Gynecol, 20:229–233.doi:10.1097/GCO.0b013e3282fcc334 PMID:18460936

189. Phillips KP, Tanphaichitr N (2008). Human exposure to endocrine disrupters and semen quality. J Toxicol Environ Health B Crit Rev, 11:188–220. PMID:18368553

190. Perry MJ (2008). Effects of environmental and occupational pesticide exposure on human sperm: a systematic review. Hum Reprod Update, 14:233–242.doi:10.1093/humupd/dmm039 PMID:18281240

191. Lemasters GK, Olsen DM, Yiin JH et al. (1999). Male reproductive effects of solvent and fuel exposure during aircraft maintenance. Reprod Toxicol, 13:155–166.doi:10.1016/S0890-6238(99)00012-X PMID:10378465

192. Rignell-Hydbom A, Axmon A, Lundh T et al. (2007). Dietary exposure to methyl mercury and PCB and the associations with semen parameters among Swedish fishermen. Environ Health, 6:14.doi:10.1186/1476-069X-6-14 PMID:17488503

193. Krüger T, Spanò M, Long M et al. (2008). Xenobiotic activity in serum and sperm chromatin integrity in European and inuit populations. Mol Reprod Dev, 75:669–680. doi:10.1002/mrd.20747 PMID:18076054

194. Hernández-Ochoa I, García-Vargas G, López-Carrillo L et al. (2005). Low lead environmental exposure alters semen quality and sperm chromatin condensation in northern Mexico. Reprod Toxicol, 20:221–228. PMID:15907657

195. Rubes J, Selevan SG, Sram RJ et al. (2007). GSTM1 genotype influences the susceptibility of men to sperm DNA damage associated with exposure to air pollution. Mutat Res, 625:20–28. PMID:17714740

196. Giwercman A, Rylander L, Rignell-Hydbom A et al.; INUENDO (2007). Androgen receptor gene CAG repeat length as a modifier of the association between persistent organohalogen pollutant exposure markers and semen characteristics. Pharmacogenet Genomics, 17:391–401.doi:10.1097/01.fpc.0000236329.26551.78 PMID:17502831

197. Cordier S (2008). Evidence for a role of paternal exposures in developmental toxicity. Basic Clin Pharmacol Toxicol, 102:176–181.doi:10.1111/j.1742-7843.2007.00162.x PMID:18226072

198. Risk Assessment Forum. Supplemental guidance for assessing susceptibility from early-life exposure to carcinogens. Report No.: EPA/630/R-03/003F. Washington (DC): U.S. Environmental Protection Agency; 2005.

199. Terrell ML, Small CM, Cameron LL, Marcus M (2008). Comment on "The influence of age at exposure to PBBs on birth outcomes". *Environ Res,* 108:117–120, discussion 121–126.doi:10.1016/j.envres.2008.04.007 PMID:18555987

200. Sweeney AM, Symanski E, del Junco D (2008). Rebuttal to Comment on: "The influence of age at exposure to PBBs on birth outcomes [Environ Res 2007;105(3):370–9]. *Environ Res,* 108:121–126. doi:10.1016/j.envres.2008.04.008

201. Sweeney AM, Symanski E (2007). The influence of age at exposure to PBBs on birth outcomes. *Environ Res,* 105:370–379.doi:10.1016/j.envres.2007.03.006 PMID:17485077

202. Hertz-Picciotto I. Environmental epidemiology. In: Rothman KJ, Greenland S, Lash TL, editors. Modern epidemiology. 3rd ed. Philadelphia (PA): Lippincott Williams & Wilkins; 2008. p. 598–619.

203. Lawson CC, Grajewski B, Daston GP *et al.* (2006). Workgroup report: Implementing a national occupational reproductive research agenda–decade one and beyond. *Environ Health Perspect,* 114:435–441.doi:10.1289/ehp.8458 PMID:16507468

204. Greenland S. Applications of stratified analysis methods. In: Rothman KJ, Greenland S, Lash TL, editors. Modern epidemiology. 3rd ed. Philadelphia (PA): Lippincott Williams & Wilkins; 2008. p. 283–302.

205. Greenland S. Introduction to regression modeling. In: Rothman KJ, Greenland S, Lash TL, editors. Modern epidemiology. 3rd ed. Philadelphia (PA): Lippincott Williams & Wilkins; 2008. p. 418–55.

206. Ritter L, Arbuckle TE (2007). Can exposure characterization explain concurrence or discordance between toxicology and epidemiology? *Toxicol Sci,* 97:241–252. doi:10.1093/toxsci/kfm005 PMID:17234646

207. Needham LL, Barr DB, Caudill SP *et al.* (2005). Concentrations of environmental chemicals associated with neurodevelopmental effects in U.S. population. *Neurotoxicology,* 26:531–545.doi:10.1016/j.neuro.2004.09.005 PMID:16112319

208. Kang HK, Dalager NA, Needham LL *et al.* (2006). Health status of Army Chemical Corps Vietnam veterans who sprayed defoliant in Vietnam. *Am J Ind Med,* 49:875–884.doi:10.1002/ajim.20385 PMID:17006952

209. Committee to Review the Health Effects in Vietnam Veterans of Exposure to Herbicides, editor. Reproductive and developmental effects. In: Veterans and Agent Orange: Update 2006.Washington (DC): National Academies Press; 2007. p. 517–65.

210. Butler D (2008). Further delays to full Agent Orange study. *Nature,* 452:786–787. doi:10.1038/452786a PMID:18431821

211. Butler D (2005). US abandons health study on Agent Orange. *Nature,* 434:687. doi:10.1038/434687a PMID:15815597

212. Anonymous (2008). A ghost of battles past. *Nature,* 452:781–782 doi:10.1038/452781b.

213. Weinberg CR, Umbach DM (1999). Using pooled exposure assessment to improve efficiency in case-control studies. *Biometrics,* 55:718–726.doi:10.1111/j.0006-341X.1999.00718.x PMID:11314998

214. Faraggi D, Reiser B, Schisterman EF (2003). ROC curve analysis for biomarkers based on pooled assessments. *Stat Med,* 22:2515–2527.doi:10.1002/sim.1418 PMID:12872306

215. Mumford SL, Schisterman EF, Vexler A, Liu A (2006). Pooling biospecimens and limits of detection: effects on ROC curve analysis. *Biostatistics,* 7:585–598.doi:10.1093/biostatistics/kxj027 PMID:16531470

216. Schisterman EF, Perkins NJ, Liu A, Bondell H (2005). Optimal cut-point and its corresponding Youden Index to discriminate individuals using pooled blood samples. *Epidemiology,* 16:73–81.doi:10.1097/01.ede.0000147512.81966.ba PMID:15613948

217. Vexler A, Schisterman EF, Liu A (2008). Estimation of ROC curves based on stably distributed biomarkers subject to measurement error and pooling mixtures. *Stat Med,* 27:280–296.doi:10.1002/sim.3035 PMID:17721905

218. Vexler A, Liu A, Schisterman EF (2006). Efficient design and analysis of biospecimens with measurements subject to detection limit. *Biomed J,* 48:780–791.doi:10.1002/bimj.200610266 PMID:17094343

219. del Junco DJ, Sweeney AM, Paepke O (2006). Paternal blood dioxin, CYP1A1 and neural tube defects in children of Vietnam veterans. *Am J Epidemiol,* 163 Suppl 11;S117–S467.

220. Warner M, Eskenazi B, Patterson DG *et al.* (2005). Dioxin-Like TEQ of women from the Seveso, Italy area by ID-HRGC/HRMS and CALUX. *J Expo Anal Environ Epidemiol,* 15:310–318.doi:10.1038/sj.jea.7500407 PMID:15383834

221. Tsutsumi T, Amakura Y, Nakamura M *et al.* (2003). Validation of the CALUX bioassay for the screening of PCDD/Fs and dioxin-like PCBs in retail fish. *Analyst,* 128:486–492. doi:10.1039/b300339f PMID:12790202

222. Rothman KJ, Greenland S, Poole C, Lash TL. Causation and causal inference. In: Rothman KJ, Greenland S, Lash TL, editors. Modern epidemiology. 3rd ed. Philadelphia (PA): Lippincott Williams & Wilkins; 2008. p. 5–31.

223. Greenland S, Lask TL, Rothman KJ. Concepts of interaction. In: Rothman KJ, Greenland S, Lash TL, editors. Modern epidemiology. 3rd ed. Philadelphia (PA): Lippincott Williams & Wilkins; 2008. p. 71–83.

224. Cox C (2006). Model-based estimation of the attributable risk in case-control and cohort studies. *Stat Methods Med Res,* 15:611–625. doi:10.1177/0962280206071930 PMID:17260927

225. Assmann SF, Hosmer DW, Lemeshow S, Mundt KA (1996). Confidence intervals for measures of interaction. *Epidemiology,* 7: 286–290.doi:10.1097/00001648-199605000-00012 PMID:8728443

226. Skrondal A (2003). Interaction as departure from additivity in case-control studies: a cautionary note. *Am J Epidemiol,* 158:251–258.doi:10.1093/aje/kwg113 PMID: 12882947

227. Offit PA (2008). Vaccines and autism revisited–the Hannah Poling case. *N Engl J Med,* 358:2089–2091.doi:10.1056/NEJMp0802904 PMID:18480200

228. Scherer M, Trelle S (2008). Opinions on registering trial details: a survey of academic researchers. *BMC Health Serv Res,* 8:18. doi:10.1186/1472-6963-8-18 PMID:18215264

229. Foster MW, Sharp RR (2007). Share and share alike: deciding how to distribute the scientific and social benefits of genomic data. *Nat Rev Genet,* 8:633–639.doi:10.1038/nrg2124 PMID:17607307

230. Moore KL, Persaud TVN. The Developing Human: Clinically Oriented Embryology, 6th ed. Philadelphia: W.B. Saunders, 1998, p. 548

UNIT 5.
APPLICATION OF BIOMARKERS TO DISEASE

CHAPTER 26.

Studies in children

Frederica P. Perera and Susan C. Edwards

Summary

This chapter first discusses the urgent need for prevention of childhood diseases that impose a huge and growing burden on families and society. It provides a review of recent research in this area to illustrate both the strengths and limitations of molecular epidemiology in drawing needed links between environmental exposures and illness in children. For illustration, three of the major diseases in children are discussed: asthma, cancer and developmental disorders. All three impose significant difficulties, have increased in recent decades, and are thought to be caused in substantial part by environmental factors, such as toxic exposures due to lifestyle choices (i.e. smoking and diet), pollutants in the workplace, ambient air, water and the food supply. These exogenous exposures can interact with "host" factors, such as genetic susceptibility and nutritional deficits, to cause disease. Molecular epidemiology has provided valuable new insights into the magnitude and diversity of exposures beginning *in utero*, the unique susceptibility of the young, and the adverse preclinical and clinical effects resulting from the interactions between these factors. However, molecular epidemiology also faces certain constraints and challenges that are specific to studies of the very young, including ethical issues, technical issues due to the limited amount of biological specimens that can be obtained, and communication of results to parents and communities. These challenges are particularly apparent when incorporating the newer epigenetic and "omic" techniques and biomarkers into studies of children's diseases.

Introduction

Molecular epidemiology, which combines epidemiologic methods and molecular/genetic techniques to measure biomarkers, has been a valuable tool in the study of environmental causes of diseases and disorders in children. Over the past 25 years, the field has made many notable contributions to intervention and prevention efforts that have significantly improved the health of children. These contributions include the phase-out of lead in gas in the 1970s, which was a result of studies on the negative effects of low-levels of lead on child neurodevelopment, and documentation of the benefits to fetal growth of a 2001 regulation which restricted the use of the pesticide chlorpyrifos (CPF). These positive changes provided the impetus for US federal policies that require agencies to explicitly address risks to children (e.g. the US Federal Insecticide, Fungicide, and Rodenticide Act (FIFRA)) and the revised US Environmental Protection Agency's (EPA) Cancer Guidelines (1990s–2000s)).

In the context of studies of disease in children, molecular epidemiology has enhanced the power and capabilities of researchers to better delineate mechanisms and causal pathways involved in the exposure-outcome pathway. It has also improved estimates of dose-biologic dosimetry; reduced misclassification of exposure and disease status; augmented the understanding of the variability in individual responses and risk, especially interactions between environment and genes or other susceptibility factors; identified preclinical cases for intervention, thereby enabling earlier interventions; and improved quantitative risk assessment and public policy.

However, molecular epidemiology also faces certain generic constraints and challenges, as well as others that are specific to studies of the very young, including ethical issues, technical issues due to the limited amount of biological specimens that can be obtained, and communication of results to parents and communities. These challenges are apparent in thinking about future directions, such as the incorporation of newer epigenetic and "omics" into studies of children's diseases.

Context and public health significance: The need for prevention

Between 1980 and 1995, the percentage of children with asthma has doubled in the USA (from 3.6% in 1980 to 7.5% in 1995), and has also increased in other countries (1). An estimated 9 million (12.5%) children aged <18 years in the USA have had asthma diagnosed at some time in their lives (2); an estimated 8.7% (6.3 million) of children had asthma in 2001 alone (1). Rates vary by geographic area and ethnic group; a recent study found that over 25% of elementary schoolchildren in Harlem, New York had asthma (3).

The overall cancer incidence rate increased from the mid-1970s in the USA, but rates in the past decade have been fairly stable (4). Leukaemia is the most common diagnosis for those < 15 years of age, but the relative proportion of it decreases with age: from 36% for those < 5 years of age, to 22% for 10–14-year-olds, and 12% for adolescents 15–19 years of age. Incidence rates in Europe have shown an increase over time since the middle of the last century: the yearly increase averages 1.1% for the 1978–1997 period and ranges from 0.6% for the leukemias to 1.8% for soft-tissue sarcomas (5). According to the databases of population-based cancer registries, which joined forces in cooperative projects such as Automated Childhood Cancer Information System (ACCIS) and EUROCARE, leukemias (34%), brain tumours (23%) and lymphomas (12%) represent the largest diagnostic groups among the < 15 year olds in Europe.

Developmental disabilities, the name given to a broad group of conditions caused by learning or physical impairments, affect an estimated 17% of children in the USA under age 18 (6). The high rates of these childhood disorders have significant medical-related costs and social impact on individual families and the country as a whole.

Rising rates of asthma and certain cancers, the high rates of developmental disabilities, and the growing evidence that the risk of certain adult diseases is influenced by *in utero* and childhood exposures, indicate that maintaining an "early focus" can have a significant impact on the overall burden of disease (1,7).

Exposures of concern

Environmental factors, such as toxic exposures due to lifestyle choices (i.e. smoking and diet), pollutants in the workplace, ambient air, water, and the food supply, can interact with 'host' factors, such as genetic susceptibility and nutritional deficits, to cause disease. Therefore, there is a need to understand the role of both environmental and susceptibility factors in childhood disease and neurodevelopmental disorders, and to identify the primary environmental toxins affecting them so that preventive measures can be taken.

A focus of this research must be early exposure to toxic chemicals, which has risen exponentially in the past 50 years. Over 80 000 synthetic chemical compounds have been created and registered for commercial use with the US Environmental Protection Agency (EPA) (8), and 2000–3000 new chemicals are submitted for review by the EPA every year (9). Nearly 3000 registered substances are produced in quantities of almost 500 000 kg every year, yet no basic toxicity information is publicly available for 43% of the high volume chemicals manufactured in the USA; a full set of basic toxicity information is available for only 7% of these chemicals, and there is no information about developmental or paediatric toxicity for 80% (8).

The exposures of concern include genotoxic and non-genotoxic chemicals, as well as chemicals that exert both types of effects. Recent research suggests that endocrine-related cancers, or susceptibility to cancer, may be a result of developmental exposures (10). There is differential exposure of the young to diverse toxic chemicals. During pregnancy and lactation, certain toxicants stored in the bodies of mothers can become bioavailable, exposing the fetus and child. In addition, the fetus and child clear toxicants less readily than an adult (11–15). Young children breathe air closer to the ground, exposing them to particles and vapours present in carpets and soil. While playing and crawling on the floor, children can inhale or dermally absorb toxicants, which are subsequently absorbed more efficiently in children than in adults (16). Compounding the effects of these behaviours is the fact that infants have twice the breathing rate of the average adult. Hand-to-mouth behaviour and thumb sucking habits can also increase exposure.

Dietary habits of children also cause increased exposure to foodborne toxicants. In the USA, children under five years of age eat 3–4 times more food per unit of body weight than the average adult; the average one-year-old drinks 10–20 times more juice than the average adult (17). Dermal exposures may also be higher, as a typical newborn has more than double the surface area of skin per unit of body weight than an adult (18).

Susceptibility of the young

The biological susceptibility of the young is another important research area. Experimental and human data indicate that the fetus and young child are especially vulnerable to the toxic effects of environmental tobacco smoke (ETS), polycyclic aromatic hydrocarbons (PAHs), particulate matter, nitrosamines, pesticides, polychlorinated biphenyls (PCBs), metals and radiation (11). There is mounting evidence, much of it from molecular epidemiologic studies, that the fetus, infant and child are biologically more sensitive to a variety of environmental toxicants than adults (7,11,19). Specifically, the *in utero* and childhood periods are characterized by rapid physical and mental growth, and gradual maturation of major organ systems. In fact, typical newborns double their weight within six months of birth, while integral parts of the nervous and immune systems are formed during the first six years of life (20). Additionally, sex organ development, myelination, and alveoli formation begin late in pregnancy and continue until adolescence (16). Since cells are proliferating rapidly and organ systems are immature, they are sensitive to the potentially harmful effects of environmental toxins.

Absorption, metabolism and excretion pathways in infants and children differ from those in adults. These pathways dictate the amount of a toxicant, in its various forms, that is present in the body. Epidemiological studies with biomarkers have demonstrated placental transfer of toxicants, and in some cases slower fetal clearance of chemicals such as PAHs, PCBs and mercury (21–23). An infant's kidney filtration rate is lower than an adult's, thus increasing potential susceptibility (16). DNA repair systems are also immature in the fetus and young child, leading to higher levels of genetic damage per unit of exposure in cord blood leukocytes compared to maternal blood leukocytes (24). Studies have also shown a 65-fold range of variability in sensitivity to the pesticide chlorpyrifos between the most sensitive newborn and the least-sensitive mother (based on paraoxonase 1 (*PON1*) status) (25).

Finally, infants and children have more years of future life than most adults. Thus, there is more time for early exposures to trigger diseases that have long latency periods. For example, early exposure to carcinogens will more likely lead to cancer than the same exposure experienced later in life. In addition, it has been hypothesized that fetal growth restriction due to nutritional deprivation in early life is an important cause of some of the most common, costly and disabling medical disorders of adult life, including coronary heart disease and the related disorders hypertension, stroke and type 2 diabetes (26). It has been proposed that individuals with a 'thrifty phenotype' will have "…a smaller body size, a lowered metabolic rate and a reduced level of behavioural activity… adaptations to an environment that is chronically short of food" (27). This hypothesis, now widely (though not universally) accepted, is known as the Barker Early Origins Hypothesis or thrifty phenotype hypothesis; it is a source of concern for societies undergoing a transition from poor to better nutrition (28). All in all, there is increasing evidence that exposures in early life strongly influence risk of chronic diseases in adulthood, including heart disease and cancer (29).

Additional susceptibility factors

The differential susceptibility of the fetus, infant and child can be influenced by nutritional deficits, genetic predispositions, and psychosocial stressors that can modify (i.e. reduce or increase) the toxic effect of exposures.

Nutritional factors

Micronutrients are known to have major effects on child health and development, and there is evidence that they interact with environmental exposures. Certain micronutrient deficiencies have been associated with childhood asthma, adverse birth outcomes, child development and childhood cancer. For example, essential fatty acid deficiencies are associated with low birth weight, smaller head circumference, and reduced cognitive and motor function (30–32). Antioxidants modulate inflammatory response to air pollution and its effects on childhood asthma (33–35). By removing free radicals and oxidant intermediates, antioxidants protect DNA from the genotoxic, procarcinogenic effects of chemicals that bind to DNA (36,37). Nutritional deficiencies are often closely related to lower socioeconomic status, although variations exist within each socioeconomic bracket.

Genetics

Genetic susceptibility can take the form of common polymorphisms or haplotypes that modulate the individual response to a toxic exposure. For example, two genes have been identified that can increase an individual's vulnerability to organophosphates (OPs), such as chlorpyrifos (CPF), by reducing the reservoir of functioning protective enzymes (38). The first gene has a prevalence of 4% and results in a poorly functioning form of the enzyme acetylcholinesterase. The second gene results in a relatively inactive form of the enzyme paraoxonase (PON1) (prevalence of 30–38%), an enzyme that detoxifies CPF before the toxin can inhibit acetylcholinesterase (39–41). The effect of CPF on head circumference at birth was significant only among women with low PON1 activity, which could be evidence of an interaction between the *PON1* genotype and OP pesticides (42). However, there are still limited data relating *PON1* to clinical outcomes in individuals exposed to OPs. Other examples of gene–environment interactions of interest include the gene coding for the d-ALA enzyme that affects lead metabolism and storage (39,43), and a genetic polymorphism in the dopamine transporter that is associated with increased behavioural problems in children prenatally exposed to tobacco smoke (44). Other research has found that the P450 and glutathione-S-transferase gene families play a role in the activation and detoxification of various xenobiotics. They are involved in the metabolism of PAHs and can influence the level of PAH–DNA damage. PAH-DNA adduct levels in human placenta were significantly higher in infants with the *CYP1A1* MspI restriction site, a genetic marker associated with lung cancer risk, than in infants without the restriction site (45). The *GSTT1* genotype was also shown to be a susceptibility marker for lower birth weight and pre-term birth among babies of pregnant women who actively used tobacco (37,46). Children with the *GSTM1* genotype who were exposed *in utero* to tobacco smoke had increased risk for persistent asthma and wheezing (47,48). The *GSTM1* genotype in asthmatic children is also associated with increased susceptibility to the harmful effects of ozone, such as reduced forced expiratory flow (49).

Individual- and community-level psychosocial stressors

The notion that individual- and community-level conditions can produce profound effects on host susceptibility to disease is derived from the long-standing existence of strong social class gradients in health (50). Recent studies have shown that women who live in violent, crime-ridden, physically decayed neighbourhoods are more likely to experience pregnancy complications and adverse birth outcomes, after adjusting for a range of individual-level sociodemographic attributes and health behaviours (51,52). Other studies have suggested that the stresses of racism and community segregation are associated with lower birth weight (53). The effects of individual poverty on birth outcomes have been shown to be exacerbated by residence in a disadvantaged neighbourhood (54). In one of the few studies that has measured interactions between physical toxicants and individual psychosocial stressors, a prospective cohort study of Northern Manhattan (New York, USA) mothers and toddlers (by the Columbia Center for Children's Environmental Health (CCCEH) cohort) found that the risk of developmental delay among children exposed prenatally to maternal ETS was significantly greater among those whose mothers experienced material hardship during pregnancy (55).

Examples/case studies: Molecular epidemiology and children's diseases related to environmental factors

The research reviewed in this article is based on the paradigm of a continuum of molecular/genetic alterations between exposure to an external agent and an adverse outcome that can be accessed using biomarkers to provide links in the chain of causality (Figure 26.1). The following is not an encyclopaedic review; rather, examples are provided for illustration of the methodology (Table 26.1).

Based on the growing understanding that children may be more susceptible to toxicants than adults, the last few decades of molecular epidemiologic research have provided a new perspective on studying environmental risks in paediatric populations. As such, the emphasis in conducting studies of disease in children has been placed on identifying less-invasive methods of biological specimen collection; specific approaches to interpretation and validation of biomarkers; methods for translating biomarker results into intervention strategies, and for integrating them with environmental monitoring and health data; optimal ways to obtain consent and provide information to children and/or their parents participating in the studies; and techniques for the effective communication with policy-makers and the public (56).

Respiratory disease/asthma

Air pollution and allergens are among the best-studied environmental risk factors, and have been established as important triggers for asthma and respiratory disease in childhood. There appears to be a critical window in both prenatal and postnatal development during which exposure to irritants and other toxicants can modify the formation and maturation of the lung, which occurs through years six to eight of life (57). Most of the research on air pollution has focused on the postnatal window of exposure and has not used biomarkers. For example, an association of poorer air quality with increased prevalence of respiratory symptoms in children has been documented in the Netherlands and in Indonesia (58,59). Moreover, increased levels of fine ambient particulate matter have been associated with decreased peak expiratory flow rates among inner-city children with asthma (60). These studies seem to implicate vehicle exhaust emissions and/or ambient particulate matter, especially diesel exhaust particles (DEP) and PAHs, in the exacerbation of asthma. Moreover, a community study of exposure to traffic evaluated 5–7 year old schoolchildren in southern California, USA, and found that residing near a major road was associated with asthma (61). Experimental studies have shown that DEP was associated with a greater risk of becoming sensitized to allergens (62,63). Importantly, a prospective study in southern California detected significant declines in lung function (FEV1) in association with exposure to nitrogen dioxide, acid vapour, PM$_{2.5}$, and elemental carbon (EC) among children ages 10–18 (64).

The very early causal role of air pollutants in childhood asthma has been less well understood, but has recently benefited from prospective molecular epidemiologic studies that have enrolled pregnant women and assessed *in utero* exposures resulting from the maternal environment. Recent results from such a study in New York City (the CCCEH cohort study) highlight the importance of the prenatal period of development, showing that adaptive immune responses may begin *in utero*, as evidenced by the occurrence of cord blood T-cell proliferation in response to specific allergens, independent of maternal sensitization (65). Moreover, high prenatal exposure to certain airborne PAHs (e.g. pyrene) was found to increase the likelihood of children's allergic response to cockroach, mouse and dust mite allergens as measured by elevated IgE (a biomarker of preclinical effect and a known asthma predictor) at two years of age (66).

Figure 26.1. Molecular epidemiology paradigm. Figure compiled from 136–138

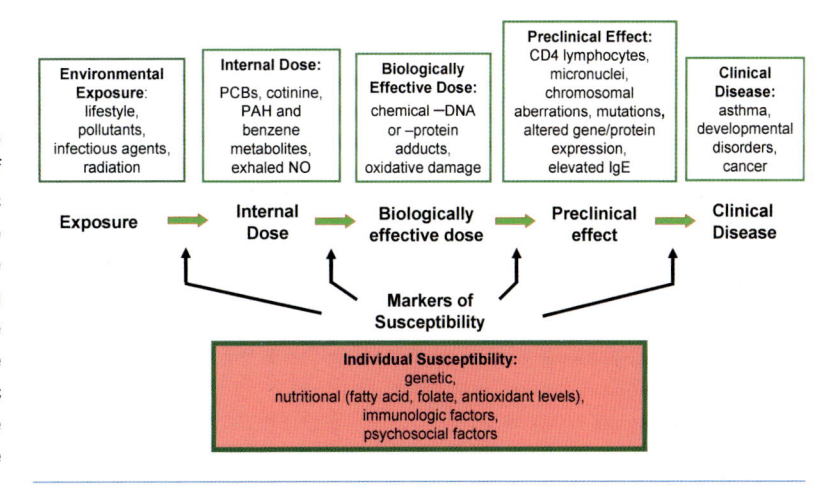

Table 26.1. Examples of biomarkers used in children's studies

Type of marker	Biomarker	Sample	Indicator of exposure to	Outcome of concern	Ref
Marker of exposure	S-Phenylmercapturic acid (S-PMA)	Urine	Metabolite of benzene, thought to be derived from the condensation product of benzene oxide with glutathione	Cardiovascular risk (as determined by prevalence of arterial hypertension (AH) and pathologic changes in electrocardiography (ECG))	(128,129)
Internal dose	Levels of 2,2', 4,4', 5,5'-hexachloro-biphenyl (CB-153), as a proxy of the total PCB burden, and of p,p'-DDE	Serum	Persistent organochlorine pollutants, such as polychlorinated biphenyls (PCBs) and dichlorodiphenyldichloroethylene (p,p'-DDE)	Altered sperm DNA/chromatin integrity (i.e. male fertility)	(130)
Internal dose	Newborn cotinine levels (nanograms per milliliter)	Plasma	Active and passive cigarette smoke exposure of the mother	Birth weight and birth length	(131)
Biologically effective dose	PAH-DNA adducts	White blood cells (WBC)	Ambient air pollution (polycyclic aromatic hydrocarbons (PAH), particulate matter, and environmental tobacco smoke (ETS)), which is significantly associated with the amount of PAH bound to DNA	Birth weight, birth length, and head circumference (ultimately, reduction of head circumference at birth correlates with lower intelligence quotient, as well as poorer cognitive functioning and school performance in childhood)	(132)
Preclinical effect	CD4 count	Lymphocytes	HIV virus	HIV-related mortality and time to death, psychological resources (positive affect, positive expectancy regarding health outcomes, finding meaning in challenging circumstances) may be protective	(133)
Clinical disease	Soluble Nogo-A (a development-related molecule inhibiting axonal regeneration) is a major component of central nervous system (CNS) myelin	Cerebrospinal fluid (CSF) and central nervous system (CNS) tissue		Multiple sclerosis (MS) (the etiology of non-reversible neurologic dysfunctions is thought to have something to do with failure of damaged axons to regenerate)	(134)
Individual susceptibility	Polymorphism of enzymes paraoxonase (PON1) and glutathione S-transferase (GSTM1 and GSTT1), which are involved in the detoxification of pesticides	Erythrocytes	Pesticides, including organophosphates and paraquat (exposures to which were evaluated thanks to erythrocyte delta-aminolevulinic acid dehydratase (ALA-D), an important biological indicator of pesticide exposure)	PON1 and GSTT1 are relevant determinants of susceptibility to chronic pesticide poisoning and pesticide-related symptomatology	(135)

Laboratory studies have suggested possible mechanisms for the effects observed in children. *In vitro* studies have shown that nasal challenges with DEP, combined with ragweed allergen, heightened the production of the Th2 cytokine IL-4 and isotype class switching to IgE (67–69). In addition, DEP has been shown to upregulate the Th2 chemokines, including I-309 and PARC, even among non-atopic subjects (70,71). Combined, these studies suggest that DEP may promote asthma by upregulating IL-4-mediated IgE pathways in response to allergen exposure. DEP exposure has also been shown to increase procollagen gene expression and tissue hydroxyproline levels in explanted rat tracheas (72), which suggests that it may trigger airway remodeling, a pathological phenotype associated with severe asthma (73).

While it is clearly established that ETS exposure is associated with respiratory infections, reduced lung function, and asthma in children (74,75), recent studies suggest that ETS exposure may modulate the respiratory response to other toxic exposures, such as PAHs. There is evidence that cigarette smoke may delay pulmonary clearance of inhaled insoluble particles (76,77). ETS exposure has been shown to exacerbate the IgE-promoting effects of ragweed exposure (78), providing another mechanism whereby ETS can worsen airway disease. In the CCCEH cohort, more cough and wheeze were reported by 12 months of age among children exposed to prenatal PAH in concert with ETS postnatally. By 24 months, difficulty breathing and probable asthma were reported more frequently among children exposed to prenatal PAH and ETS postnatally (66). Most recently, a parallel cohort study in Poland has assessed PAH exposure during pregnancy by personal air monitors worn by women (n = 333) for 48 hours during the second or third trimester of pregnancy. After delivery, the mothers were interviewed every three months over the course of a year. Prenatal PAH exposure was associated independently with an increased risk for respiratory symptoms including wheezing without cold (RR = 3.8; 95% CI = 1.2–12.4) during the course of the infants first year of life (79). The data were adjusted for confounders including ETS, which was verified by cotinine, a biomarker of ETS exposure. These results suggest an independent effect of urban PAH exposure on respiratory outcomes in children (79).

Cancer risk/genetic damage

Environmental exposures are recognized as potentially important risk factors for childhood cancer (80), and again biomarkers are proving useful in assessing causal relationships. For example, carcinogen-DNA adducts are considered a biomarker of the biologically effective dose of PAHs and increased cancer risk (81). Experimental evidence shows that the amount of PAHs crossing the placenta and reaching the fetus is on the order of one-tenth of the dose to the mother (12,13), yet the levels of PAH–DNA adducts measured in rodent fetal tissue are higher than expected based on transplacental dose (14). Similarly, research in mothers and newborns has consistently shown that PAH–DNA adduct levels in the white blood cells (WBC) of newborns were similar to or exceeded those in paired maternal samples, despite the estimated 10-fold lower dose of the parent compound to the fetus (82,83). This research indicates that the differential effect of exposure to PAHs in the fetus is not limited to a particular ethnic or geographic group (84). Increased adducts in the fetus relative to the adult could result from lower levels of phase II (detoxification) enzymes and decreased DNA repair efficiency in the fetus (19,82,85,86). In addition, fetal plasma cotinine levels were higher than in paired maternal samples, suggesting reduced ability of the fetus to clear carcinogenic cigarette smoke constituents (82,83).

Chromosomal aberrations have been associated with increased risk of cancer in multiple studies, and are a well-validated biomarker of the preclinical effect of carcinogens (87,88). In New York City newborns, maternal exposure to airborne PAHs during pregnancy was associated with increased frequency of chromosomal aberrations in WBCs, suggesting that risk of cancer can be increased by exposure *in utero* (89). Studies have also linked maternal tobacco smoking to increased chromosomal aberrations in the WBCs of newborns (90). Other research has shown an approximately 10-fold higher risk of infant acute myeloid leukaemia (AML) with increasing maternal consumption of DNA topo 2 inhibitor-containing foods, raising concerns about benzene, a topo 2 inhibitor (91).

There is a growing body of evidence from studies of adults that polymorphisms of the DNA repair genes X-ray repair cross-complementing group 1 (*XRCC1*) and xeroderma pigmentosum group D (*XPD*) may constitute susceptibility factors to cancer. The results of a case–control study in a Chinese population suggested that *XRCC1* 194 Trp/Trp and *XPD* 751 Lys allele might be risk genotypes for lung cancer in this population

(92). *XPD* polymorphisms at codons 312 and 751 were both significantly associated with elevated levels of PAH–DNA adducts in tumour tissue from breast cancer cases, suggesting that by increasing DNA damage that may lead to further mutations and contribute to genetic instability in the tumour, *XPD* may play a role in tumour progression (93). While no studies are available in children, it is possible that decreased ability to repair early genetic damage from environmental carcinogens may render children more susceptible to cancer later in life.

Recent reports have established the prenatal origin of leukaemia translocations and resultant fusion genes in some patients, including *MLL-AF4* translocations in infants and *TEL-AML1* translocations in children. Using twins with concordant leukaemia, it was found that a hallmark genetic event in these acute leukemias (i.e. chromosomal translocation) can have a prenatal origin (94–98). More explicit evidence is provided by the finding of clonotypic chromosomal fusion sequences in archived neonatal (Guthrie) heel-prick spots matched to children who later contracted leukaemia (98–100). Additional indirect support for prenatal origin of leukaemia clones is derived from the demonstration of the presence of clonotypic rearrangements at the *IGH* and *TCR* loci in Guthrie spots (101,102).

Furthermore, new evidence for the prenatal origin of a translocation in childhood AML was reported (100). The t(8;21) *AML1-ETO* translocations in childhood AML can arise *in utero*, possibly as an initiating event, and may establish a long-lived or stable parental clone that requires additional secondary genetic alterations to cause leukaemia.

While progress in this field is substantial, a basic understanding of the timing of the origin of the crucial molecular abnormalities, or the natural history of leukaemic clones, is incomplete. The recognition of crucial temporal and developmental windows for the formation of leukemogenic genetic alterations will help to focus epidemiologic analysis as well as to prompt preventive strategies (103).

Neurobehavioural disorders

The exquisitely sensitive process of the development of the human central nervous system involves the production of 100 billion nerve cells and 1 trillion glial cells, which then must follow a precise stepwise choreography involving migration, synaptogenesis, selective cell loss, and myelination (104). A mistake at any point in the process can have permanent consequences. Experimental studies of prenatal and neonatal exposure to the organophosphate CPF have demonstrated, for example, neurochemical and behavioural effects, as well as selected brain cell loss (105–109). The behavioural and morphologic effects of developmental toxicants are highly dependent on the timing as well as on the dose and duration of exposure. This is illustrated by both rodent and human studies showing that the effect of irradiation on brain malformation is heightened during the window of susceptibility throughout fetal development (104). Adverse neurological development, including lowered intelligence, diminished school performance, and increased rates of behavioural problems have been associated with exposure to low-levels of several environmental toxicants and pollutants.

Cohort studies have demonstrated that low-level exposure to lead (even below 10ug/dL in blood) during early childhood is inversely associated with neuropsychological development through the first 10 years of life (110–114). Prenatal exposure to PCBs and methylmercury, predominantly from maternal seafood consumption, has been associated with neurocognitive deficits (115). In these studies, biomarkers (including levels of blood lead, tooth lead concentrations, blood mercury levels and cord tissue PCB) have been instrumental in quantifying the internal dose of the pollutants. Taking the recent example of pesticides, New York City children in the CCCEH cohort who were prenatally exposed to high levels of CPF, as measured by high cord plasma CPF levels, were significantly more likely than children with low cord levels to experience delay in both psychomotor and cognitive development at three years of age (116). In addition, the highly-exposed children were significantly more likely than less-exposed children to manifest symptoms of attention disorders, attention-deficit/hyperactivity disorder (ADHD) and pervasive personality disorder at age three. Although the EPA banned residential use of CPF in 2001, this pesticide is still widely used in agriculture. In addition, children with high prenatal exposure to airborne PAHs also had significantly lower test scores at age three on the Bayley test for cognitive development, after controlling for pesticide exposure (plasma CPF) (117), and at age 5 had significantly lower IQ (118). Moreover, in the same study, children prenatally exposed to ETS (cotinine-verified), especially children whose mothers experienced material hardship (unmet basic food, clothing and housing needs) had significantly reduced scores on tests of cognitive development at two years of age (55). Finally, cohort children with high prenatal exposure to PAHs were more likely to experience developmental delays in childhood, after controlling for ETS and CPF (119).

Table 26.1 lists a few hand-picked examples from the literature of biomarkers that have been used to study relationships between exposure and disease in children.

Strengths, limitations and lessons learned: Special considerations in mounting children's studies

Long-term studies that follow participants from the prenatal period into adolescence and early adulthood are considered essential to assess the full range of developmental consequences of exposure to environmental chemicals. (The advantages of prospective cohort studies and their logistical, ethical, and financial challenges are discussed in Chapter 17.)

To address some of the important lessons from prior research regarding the logistics, ethics, and the financing of these long-term studies in children, the platform in which many of these challenges were encountered, tackled and, for the most part, overcome is introduced.

National centers for children's environmental health

In 1998, the US National Institute of Environmental Health Sciences (NIEHS) and the EPA collaborated to develop a research programme (the Children's Centers) that would coordinate efforts to better understand toxic exposures to infants and young children, study the health effects of such exposures to clarify the mechanisms by which they work, and explore intervention strategies for reducing such exposures in a way that would provide evidence for practice. Each centre is designed around a central theme focusing on important questions regarding the role of exposures in one of the following health outcome areas: respiratory disease, childhood learning, and growth and development, including developmental disabilities. The purpose of the Children's Centers programme was to create local research environments that promote multidisciplinary interactions among basic, clinical and behavioural scientists through university/community partnering, to accelerate translation of basic research findings into clinical prevention or intervention strategies. Additionally, it was designed to support a coordinated, nationwide network of scientists and community advocacy groups synergistically sharing their experiences to address relevant questions related to the role of environmental exposures in the health of children, to enhance community-level capacity to identify and address environmental threats and prevention opportunities (120). (A full description of the Children's Centers can be found on the NIEHS web site (121), and a summary of the first eight centres has been more fully discussed by (122) and (123).)

The Children's Centers have addressed and overcome many hurdles in their efforts to understand the link between environmental exposures and health outcomes, as well as interactions between exposures, and a variety of social and cultural factors. Out of their enterprise, several lessons have been learned on the practicalities of conducting longitudinal birth cohorts, such as the critical importance of long-term studies for assessing the full range of developmental consequences of environmental exposures, recognition of the unique challenges presented at different life stages for both outcome and exposure measurement, and the importance of ethical issues that must be dealt with in a changing medical and legal environment (120). In the following section, some of the more specific shared experiences are paraphrased as they pertain to the methodological, logistical, ethical and communication issues (124).

Successes, challenges and lessons learned

Logistical issues in children's studies

Barriers to recruitment. The most common and important barriers to recruitment into prospective cohort studies, especially for working women, is the time required for each visit and the length of the follow-up period. Many members of a population are also distrustful of Western medicine and research. Solutions to these obstacles include hiring study staff familiar with or from the target population, recruitment by or at clinics known by the community to respect patient confidentiality, and allowing time for potential participants to discuss the study with their families before enrolment.

Staffing issues. Building trusting relationships with participants in the cohort is best accomplished by hiring bilingual, bicultural staff from the local community, who are assigned to follow particular families ideally from pregnancy through the child's assessments. Although this is helpful, it can introduce systematic bias. Often, more in-depth training on data collection techniques is needed than when hiring from within the academic community. In addition, the number of staff required to maintain a birth cohort, which includes conducting weekday, weeknight and weekend assessments, as well as completing quality control tasks, is often much larger than projected. Gaps in funding are extremely detrimental.

Retention. Besides the decrease in cohort size, one of the main issues with retention of participants is that loss to follow-up often differs from continuing participants with some demographic characteristics, such as age, marital status, medical insurance status, race and ethnicity. The most common reason for attrition is the inability to locate participants due to disconnected phones and/or frequent moves, regularly missed appointments leading to exclusion from the study, refusal to continue, or in a few cases, infant deaths. Different incentives that have been successful in improving retention rates include payments, often times incremental over the course of enrolment, gift certificates (e.g. to grocery stores), and bonus incentives for certain activities (e.g. calling study staff when in labour, returning on a separate day to finish an assessment or provide an additional sample, or providing new contact information upon moving). Incentives remain a major budget item to be considered in the planning stages of such a study. Maintaining contact every few months with birthday cards, brief telephone interviews about the child's health, or simple check-ins with the family to remind them of the next phases of the study, is also critical.

Environmental assessments. Home inspections to assess housing quality are time-consuming and require extensive training. It might be necessary to visit homes multiple times to reassess household exposures, which may vary by season or change when families relocate (125). Collecting environmental measurements often requires the purchase of expensive, specialized collection equipment (e.g. air monitors) and a delay between home assessments to allow for cleaning of equipment. In addition, standard practices for interpreting ambient measurements are not yet fully developed; for example, it is unclear for most contaminants whether house dust concentration (μg per gram of dust) or loading (μg of surface area) is a better predictor of children's exposure or body burden.

Delivery events. Shortened post-delivery hospital stays in the United States leave a limited window of opportunity to collect information and samples from mothers and neonates in the hospital. Because of the slow notification of participants' admission for delivery, a large proportion of women fail to be tracked at the time of delivery. For efficient notification, researchers must rely on both participating women and delivery ward staff. Some of the solutions developed by researchers include: distributing cell phones to enable mothers without home phones to call the research team, or alternatively, distributing t-shirts or socks to wear to the hospital, which will alert the delivery staff; providing lists of participants approaching their due date to medical stations; and checking delivery logbooks daily. Cord blood samples are particularly difficult to obtain. Most missed collections occur when women's delivery admissions are not reported to research staff; additional samples are missed from high-risk children with emergency deliveries. The greatest collection rate tends to be reported by the research teams that involve physicians in collecting the samples. Conducting neonatal assessments is also a difficult task. Few tests are available to assess newborn behaviour, and their predictive validity is not high. Many assessment tests require trained evaluators, who are not easily replaced when they leave projects, which can create gaps in cohort assessment. Within the context of shortened post-delivery hospital stays, post-delivery assessments have to be scheduled both after the effects of delivery medications wear off, and between the child's sleeping and eating schedule. Finding a quiet assessment room in the hospital can also be a challenge. Due to these various impediments, assessments intended for the neonatal period are in many cases conducted several weeks after delivery. Early-morning assessments and assessment of the child both soon after delivery and again one month later tend to increase success with hospital assessments.

Child assessments. Conducting assessments on small children in the home is nearly impossible, thus the provision of a standardized testing facility may be essential. Minimizing distractions to children during neurobehavioural assessment is particularly challenging. For children > 12 months of age, it is desirable to assess the child separately from the mother to reduce interference; this requires additional time for the tester to build a rapport with the child. Siblings can also be a source of distraction during assessments. On-site childcare, giving reimbursements for off-site childcare, and/or using videos or games to busy these children are some handy options; however, the ideal arrangement is on-site childcare with dedicated space and personnel.

Problems in sample collection. Blood collection from children is a challenge. Collecting research blood samples at the same time as clinical samples helps to avoid participant concerns about taking blood from children and pregnant women, especially in certain cultural groups. Consulting with community physicians to determine the amount of blood collection that is both clinically and culturally acceptable to the target population is also helpful. Collecting breast milk samples soon after delivery, although most convenient for the research team, can be daunting for mothers, as the milk supply may not yet be fully developed. Later collection of breast milk avoids some of these problems, but timing issues may arise for other sample types as well. Studies conducted in rural areas face additional barriers to successful collection and processing of samples, such as limited laboratory facilities that are not adequately equipped to process samples (e.g. to separate whole blood into blood products). In this case, it is necessary to transport samples over long distances, which increases costs. In locations where necessary goods and services (e.g. dry ice or courier services) are in short supply, it can also be difficult to ensure the prompt stabilization of samples. Finally, some rural areas may lack skilled paediatric phlebotomists.

Participant fatigue. Longitudinal studies are demanding for families. To minimize participant fatigue, researchers should aim to optimize contact frequency such that attrition is prevented, but participants are not overly burdened. It is important to design contact between researchers and study participants to be as brief and efficient as possible; respect for participants' time may require that the focus of research be narrowed down. Strategies employed to minimize the impact of participant fatigue include using multiple workers to simultaneously collect information at each visit. This requires that each research worker is trained in multiple aspects of the study protocol (sometimes though, multiple short visits were preferred to one long visit, both for convenience and to prevent child fatigue). Other approaches were to provide snacks in case of lengthy and demanding assessments; develop qualitative assessments that allowed study staff to document participants' level of fatigue, cooperation, and attention; and to record any changes made to the usual study protocol.

Quality control of assessments and interviews. Proper staff training is critical. Adequate pilot testing is important, but often hindered by time, cost and the need for prior Institutional Review Board (IRB) approval. Insufficient time and resources are the main reasons why clear quality control protocols (e.g. direct observation, review of videotapes by the other evaluators and lead psychologists) fall short. Even after extensive staff training, inter-rater differences and reliability issues also remain a concern.

Lack of transportation. Transportation can be a barrier to successful completion of assessments. Paying for taxi services, reimbursing participants for alternate travel costs, transporting participants to the office for an assessment after completing a home visit, purchasing and outfitting an RV that could be driven to participants' homes and used as a roving assessment room, or simply purchasing a car for the study to reduce mileage reimbursement costs and wear and tear on staff cars, are a few of the solutions to which Children's Centers have turned to address this problem.

Issues of literacy, language and culture. The wording and phrasing of all study documents, including consent forms, must be simple; in addition, most study instruments, including those designed for self-administration, must be administered orally to attend to the issue of participants with limited education and low literacy. Potentially embarrassing topics that evade translation (e.g. specific birth control methods) may have to be described graphically. Other culturally sensitive issues to be attentive to include participants not knowing or not sharing their exact date of birth, being hesitant to provide biologic samples, and reporting pregnancy relatively late in gestation. Understanding these types of issues and planning the research accordingly relies heavily on focus groups with community members. Because young children cannot precisely answer questionnaires regarding behaviour or habits, researchers must often ask the child's mother for a specific answer in a follow-up.

In terms of the logistics of a longitudinal birth cohort study, it is critical that funding be adequate for the start-up period, continuous without gaps, and extend for the long-term. Costs are often underestimated for labour intensive activities, such as tracking and maintaining study participants.

Ethical issues

The ethical issues in a longitudinal birth cohort study are likely to become increasingly more complex in the changing medical and legal environment, and must be carefully considered in designing research protocols and following the cohort. It is particularly necessary to develop clear plans of referral when children with disease, developmental difficulties, or adverse social situations emerge.

Increasing ethical complexity. Since the implementation of the Health Insurance Portability and Accountability Act (HIPAA), it has become more time-consuming to obtain participants' informed consent for studies in the USA. Concerns about potential lawsuits have increased, and conflicting ethical issues are routine (e.g. deciding when the health and safety of a child takes precedence over a promise of confidentiality).

Consent and assent. Because longitudinal studies demand lengthy and complex consent forms, ensuring that participants are well informed is challenging and requires the allocation of adequate time and resources. For studies using medical records in the USA, the completion of HIPAA subject authorization forms adds time to the consent process. It has been found to be important to inculcate in staff an understanding that consent is an ongoing process; instead of training staff to simply procure participant signatures, centres have trained staff to solicit and answer participants' questions so that they can make informed decisions. Clearly, writing consent forms at a reading level understandable to all is critical. In cases where the level is suspected to be too high to assure comprehension, research workers have the option of reading consent forms aloud to participants to ensure that everyone, including participants who are embarrassed to admit their low literacy level, fully understands the information. Solicited feedback from community partners, community board members and community-based staff (in addition to the IRB) also helps ensure that appropriate language is used. Additional measures to enhance understanding of consent forms include providing study participants with timetables and schedules to communicate study procedures, or lists outlining the important items on the consent. Providing short checklists to verify that participants understand the key aspects of the study is also effective. All solutions that decrease the amount of complex information that participants have to digest at each visit and give them an opportunity to re-evaluate their participation at a midway point are helpful. While studies often operate with uncertainty about funding and the future direction in the long run, continuing requests for participation can be a source of frustration; full disclosure of the protocol up-front is thus preferable. Careful thought must be given to *who* must consent to participate at each stage of the research. In all cases, a pregnant woman or mother should be asked to consent to her own participation and that of her child. However, once children reach a certain age (generally 5–9 years), child assent is typically also required by the IRB, leading to new challenges.

Banked samples and informed consent. The process of banking samples for future studies requires special consideration, as participants must be informed about and consent to future uses of these samples. Consent forms may allow participants the option of either not having samples banked and/or not allowing future analysis of samples for unrelated studies. IRB re-approvals may additionally be required for each new analysis of banked samples.

Confidentiality and consideration of children at risk. Protecting the identity and personal information of all participants can be difficult in small or close-knit communities, especially when the research staff was hired from the local community.

Confidentiality within computerized databases also requires particular attention. All computerized files should be password-protected with knowledge of passwords restricted to a small number of staff. The number of computer or paper files containing both the participant study number and identifying information (e.g. name) should be limited. In complex studies with multiple contacts, it can be necessary to work with both the IRB and the research staff to identify the types of linked information necessary for day-to-day operations, and to provide that information with the least possible risk to participants.

Protocols on intervening in cases of clear developmental delays or undiagnosed physical health problems are compulsory. Most protocols include timely screening of developmental assessments and questionnaires to ensure prompt referral or treatment. Reporting some exposure measures, such as lead results, to public health authorities when they exceed certain action levels is also essential. Lastly, certificates of confidentiality, which protect identifiable research information from forced disclosure, including in the case of legal action, are an important component in protecting participant confidentiality. However, an investigator might need to break the promise of confidentiality without participant consent, for example, in cases where child abuse, severe depression, drug use or traffic in the home, and other potentially dangerous conditions are observed. Disclosure of such requirements (e.g. the need to report child abuse) is typically incorporated into consent forms, despite concern that it would deter some participants.

Communication issues

Communicating study results is a key step in any research project. In addition to publishing results in scientific journals, centres should seek to share findings with participants and community members. Researchers may benefit from soliciting the guidance of community collaborators to decide when and how to disseminate results, including how to clearly craft messages so they would be understood and of interest to the community. In some cases, communities expect interventions and actions that are outside the scope of the research; therefore it is important to concisely communicate the purposes and limitations of the research beforehand to prevent false expectations.

Timing of the results communication. It is advisable to disclose findings to participants and/or community advisory boards before their publication in journal or newspaper articles. This disclosure is an important step in building trust between researchers, participants, and communities. Community members resent hearing findings for the first time in the media.

Communication tools. Strategies to disseminate information, developed in collaboration with community advisory boards, have included newsletters, fact sheets, pamphlets, press releases in local papers, pay-stub inserts, radio programmes (particularly useful in rural areas), town hall meetings, and internet sites. Ideally, researchers concerned with children's health would also like to communicate results to children themselves. Based on results from their study of pesticide exposure in children, the University of Washington's Center for Children's Environmental Health (Center for Child Environmental Health Risk Research) created colouring books and curricula to educate preschool and school-age children on how to prevent exposure to pesticides. Specialized tools are sometimes needed for studies that target low-literacy or non-majority-language-speaking communities. Publishing information in more than one language is essential, and successful attempts to develop pictorial rather than verbal messages have also been made (126) (http://www.checnet.org/healthehouse/education/articles-detail.asp?Main_ID=644).

Group- versus individual-level results. Perhaps the biggest communication issue has to do with whether to provide individual-level results, particularly for measures of exposure or internal dose. The argument in favour of providing such results is that participants have the right to know; the counterargument is that participants may be unnecessarily alarmed by results with no interpretable meaning. Generally, results with a clear clinical implication (e.g. blood lead levels) have been reported to participants, whereas results without clear clinical implications (e.g. urinary pesticide metabolite levels) have not been shared. One centre, however, on the basis of community advisory board input, decided to offer participants the option of requesting their individual pesticide levels. That centre is currently in the process of developing materials to provide these results and will work closely with community health care providers when clinical questions arise (124). Regardless of whether group- or individual-level results are returned, it is important to provide participants a context for these results. Providing a comparison, either to other study participants or nationwide data, has been particularly helpful. In communicating results, centres aim to clearly describe their implications for health and well-being; when these implications are not known (as in the case of pesticides), centres state this honestly (104).

Active and meaningful participation of the community is essential for determining the relevant research questions, enrolling and retaining the cohort in an intensive investigation over the long term, and contributing to translation of scientific principles and research results for communities and the public at large. This requires establishing trust and respecting differences in culture and knowledge of the community. Sufficient time and resources are necessary to develop community partnerships.

Future directions and challenges: Looking ahead

Summary

Recent molecular epidemiologic research has helped identify etiologic factors in childhood diseases. Exposures of particular concern for the fetus and young child are environmental tobacco smoke (ETS); polycyclic aromatic hydrocarbons (PAHs); particulate matter; nitrosamines; pesticides; polychlorinated biphenyls (PCBs); metals and radiation, which may be involved in respiratory disease and asthma; cancer risk and genetic damage; and neurobehavioural disorders. Molecular epidemiology has also been a useful tool in identifying the interactions between exposures and certain "host" factors, such as genetic susceptibility, nutritional deficits, and psychosocial stressors that can lead to disease. It has also provided compelling new evidence that early exposure to environmental factors is a likely contributor to disease in later years of life.

The powerful need for prevention, as directed by current trends in disease incidence, requires that strong, collaborative research be performed, especially on populations most at risk. The logistics involved in designing and conducting such observational research, adequate communication of the research findings to populations enrolled in the study, as well as the ethics at stake, present real challenges. Strategies for conducting cutting-edge yet safe and responsible research must therefore be widely shared and understood.

Future directions: Suggestions

New epigenetic and "omic" biomarkers

The previous generation of biomarkers has contributed to our understanding of risk and susceptibility related to genotoxic carcinogens. As a result, interventions and policy changes have been mounted to reduce risk to children from several important environmental carcinogens. More recently developed biomarkers have considerable potential in molecular epidemiology, as they reflect another equally important mechanism of carcinogenicity: epigenetic alterations that affect the expression of genes and proteins. These can be measured by high-throughput methods, allowing large-scale studies that are 'discovery-oriented.' Research using these techniques is needed to study the effects of multiple exposures and their interactions both with each other and in combination with different types of susceptibility factors. Gene–environment interactions should be a major focus, but interactions with nutritional and psychosocial factors also deserve emphasis in future research. Studying the low-level exposure effects of global pollutants and their interactions with susceptibility factors in different geographic locations, exposure scenarios, and ethnic/racial groups will help in understanding etiology and confirm findings. However, most of these markers have not yet been validated, and their role in the causal paradigm is not clear. There is an urgent need for validation of these newer biomarkers so that they can be combined with the more traditional ones in hypothesis-testing studies. Large-scale, long-term prospective studies will be the key to achieving this.

National Children's Study

In the USA, studies under way at the National Centers for Children's Environmental Health were among the first in line to meet these needs. The National Children's Study (NCS) (a programme that will follow 100 000 children across the US from before birth until age 21) will examine the effects of natural and man-made environment factors, biological and chemical factors, physical surroundings, social factors, behavioural influences and outcomes, genetics, cultural and family differences, and geographic location on the health and development of children in the cohort (127). (A summary of the participating cohorts and links to their web pages, as well as links to updates on the NCS, can be found at: http://www.nationalchildrensstudy.gov/studylocations/Pages/websites.aspx.)

Conclusion

Molecular epidemiology has contributed much to our knowledge of risk factors for environmental health-related diseases in children. The high rates of asthma, certain cancers and developmental disabilities, and the growing evidence that risk of certain adult diseases is associated with *in utero* and childhood exposures, indicate that maintaining an early focus in molecular epidemiology can have a significant impact on the overall burden of disease. When preventive measures have been enacted based on this knowledge in the past, children's health has benefited. Incorporating strategic principles to translate existing and future data into public health policy will ensure benefits in children's environmental health.

Acknowledgements

This work has been made possible by the following NIEHS grants: 5P01ES09600, R01ES09089, 5RO1 ES101654, and by the Harriman Foundation and the New York City Community Trust.

References

1. Woodruff TJ, Axelrad DA, Kyle AD et al. America's children and the environment: measures of contaminants, body burdens, and illnesses. Washington (DC): U.S. Environmental Protection Agency; 2003.

2. Dey AN, Bloom B. Summary health statistics for U.S. children: National Health Interview Survey, 2003. Vital Health Stat 2005;223(Series 10):1–78.

3. Nicholas SW, Jean-Louis B, Ortiz B et al. (2005). Addressing the childhood asthma crisis in Harlem: the Harlem Children's Zone Asthma Initiative. Am J Public Health, 95:245–249.doi:10.2105/AJPH.2004.042705 PMID:15671459

4. Ries LAG, Smith MA, Gurney JG et al., editors. Cancer incidence and survival among children and adolescents: United States SEER program 1975–1995, National Cancer Institute. NIH Publication No. 99–4649. Bethesda, MD, 1999.

5. Kaatsch P (2010). Epidemiology of childhood cancer. Cancer Treat Rev, 36: 277–285.doi:10.1016/j.ctrv.2010.02.003 PMID:20231056

6. Boyle CA, Decouflé P, Yeargin-Allsopp M (1994). Prevalence and health impact of developmental disabilities in US children. Pediatrics, 93:399–403. PMID:7509480

7. Grandjean P, Landrigan PJ (2006). Developmental neurotoxicity of industrial chemicals. Lancet, 368:2167–2178.doi:10.1016/S0140-6736(06)69665-7 PMID:17174709

8. U.S. Environmental Protection Agency. Chemical hazard data availability study. Washington (DC): U.S. Environmental Protection Agency; 1998.

9. U.S. Environmental Protection Agency. Proposed guidelines for carcinogen risk assessment. Washington (DC): Office of Research and Development; 1996.

10. Birnbaum LS, Fenton SE (2003). Cancer and developmental exposure to endocrine disruptors. Environ Health Perspect, 111:389–394.doi:10.1289/ehp.5686 PMID:12676588

11. Perera FP, Illman SM, Kinney PL et al. (2002). The challenge of preventing environmentally related disease in young children: community-based research in New York City. Environ Health Perspect, 110:197–204.doi:10.1289/ehp.02110197 PMID:11836150

12. Srivastava VK, Chauhan SS, Srivastava PK et al. (1986). Fetal translocation and metabolism of PAH obtained from coal fly ash given intratracheally to pregnant rats. J Toxicol Environ Health, 18:459–469.doi:10.1080/15287398609530885 PMID:3712502

13. Neubert D, Tapken S (1988). Transfer of benzo(a)pyrene into mouse embryos and fetuses. Arch Toxicol, 62:236–239. doi:10.1007/BF00570149 PMID:3196162

14. Lu LJ, Disher RM, Reddy MV, Randerath K (1986). 32P-postlabeling assay in mice of transplacental DNA damage induced by the environmental carcinogens safrole, 4-aminobiphenyl, and benzo(a)pyrene. Cancer Res, 46:3046–3054. PMID:3698023

15. Lu LJ, Wang MY (1990). Modulation of benzo[a]pyrene-induced covalent DNA modifications in adult and fetal mouse tissues by gestation stage. Carcinogenesis, 11:1367–1372.doi:10.1093/carcin/11.8.1367 PMID:2387022

16. Bearer CF (1995). Environmental health hazards: how children are different from adults. Future Child, 5:11–26.doi:10.2307/1602354 PMID:8528683

17. Wiles R, Campbell C. Pesticides in children's food. Washington (DC): Environmental Working Group; 1993.

18. International Programme on Chemical Safety Commission of the European Communities. Principles for evaluating health risks from chemicals during infancy and early childhood: the need for a special approach. Environmental Health Criteria. 59th ed. Geneva (Switzerland): World Health Organization; 1986.

19. National Research Council. Pesticides in the diets of infants and children. Washington (DC): National Academy Press, 1993.

20. Sonawane B, Beliles R. The susceptibility of children to immunotoxic and neurotoxic agents. Poster Abstract, 1st National Research Conference on Children's Environmental Health, Children's Environmental Health Network, February 21–23, 1997, Washington (DC); 1997.

21. Perera FP (1996). Molecular epidemiology: insights into cancer susceptibility, risk assessment, and prevention. J Natl Cancer Inst, 88:496–509.doi:10.1093/jnci/88.8.496 PMID:8606378

22. National Research Council. Toxicological effects of methylmercury. Washington (DC): National Academy Press; 2000.

23. Ramirez GB, Cruz MC, Pagulayan O et al. (2000). The Tagum study I: analysis and clinical correlates of mercury in maternal and cord blood, breast milk, meconium, and infants' hair. Pediatrics, 106:774–781. doi:10.1542/peds.106.4.774 PMID:11015522

24. Perera FP, Tang D, Whyatt R et al. (2005). DNA damage from polycyclic aromatic hydrocarbons measured by benzo[a]pyrene-DNA adducts in mothers and newborns from Northern Manhattan, the World Trade Center Area, Poland, and China. Cancer Epidemiol Biomarkers Prev, 14:709–714.doi:10.1158/1055-9965.EPI-04-0457 PMID:15767354

25. Furlong CE, Holland N, Richter RJ et al. (2006). PON1 status of farmworker mothers and children as a predictor of organophosphate sensitivity. Pharmacogenet Genomics, 16:183–190. PMID:16495777

26. Hales CN, Barker DJ (1992). Type 2 (non-insulin-dependent) diabetes mellitus: the thrifty phenotype hypothesis. Diabetologia, 35:595–601.doi:10.1007/BF00400248 PMID:1644236

27. Bateson P, Martin P. Design for a life: how behaviour develops. London (U.K.): Jonathan Cape; 1999.

28. Robinson R (2001). The fetal origins of adult disease. BMJ, 322:375–376.doi:10.1136/bmj.322.7283.375 PMID:11179140

29. Barker DJP. Mothers, babies and health in later life. 2nd ed. Edinburgh (U.K.): Churchill Livingston; 1998.

30. Crawford MA, Doyle W, Drury P et al. (1989). n-6 and n-3 fatty acids during early human development. J Intern Med Suppl, 731 Suppl 1;159–169. PMID:2706039

31. Voigt RG, Jensen CL, Fraley JK et al. (2002). Relationship between omega3 long-chain polyunsaturated fatty acid status during early infancy and neurodevelopmental status at 1 year of age. J Hum Nutr Diet, 15:111–120.doi:10.1046/j.1365-277X.2002.00341.x PMID:11972740

32. Makrides M, Neumann M, Simmer K et al. (1995). Are long-chain polyunsaturated fatty acids essential nutrients in infancy? Lancet, 345:1463–1468.doi:10.1016/S0140-6736(95)91035-2 PMID:7769900

33. Hatch GE (1995). Asthma, inhaled oxidants, and dietary antioxidants. Am J Clin Nutr, 61 Suppl;625S–630S. PMID:7879729

34. Greene LS (1995). Asthma and oxidant stress: nutritional, environmental, and genetic risk factors. J Am Coll Nutr, 14:317–324. PMID:8568107

35. Peat JK, Britton WJ, Salome CM, Woolcock AJ (1987). Bronchial hyperresponsiveness in two populations of Australian schoolchildren. III. Effect of exposure to environmental allergens. Clin Allergy, 17:291–300. doi:10.1111/j.1365-2222.1987.tb02017.x PMID:3621548

36. Perera FP, Whyatt RM, Jedrychowski W et al. (1998). Recent developments in molecular epidemiology: A study of the effects of environmental polycyclic aromatic hydrocarbons on birth outcomes in Poland. Am J Epidemiol, 147:309–314. PMID:9482506

37. Wang X, Zuckerman B, Pearson C et al. (2002). Maternal cigarette smoking, metabolic gene polymorphism, and infant birth weight. JAMA, 287:195–202.doi:10.1001/jama.287.2.195 PMID:11779261

38. Mutch E, Blain PG, Williams FM (1992). Interindividual variations in enzymes controlling organophosphate toxicity in man. Hum Exp Toxicol, 11:109–116.doi:10. 1177/096032719201100209 PMID:1349216

39. Greater Boston Physicians for Social Responsibility. In harm's way: toxic threats to child development. Cambridge (MA): Greater Boston Physicians for Social Responsibility; 2000.

40. Costa LG, Cole TB, Vitalone A, Furlong CE (2005). Measurement of paraoxonase (PON1) status as a potential biomarker of susceptibility to organophosphate toxicity. Clin Chim Acta, 352:37–47.doi:10.1016/j.cccn.2004.09.019 PMID:15653099

41. Costa LG (2006). Current issues in organophosphate toxicology. Clin Chim Acta, 366:1–13.doi:10.1016/j.cca.2005.10.008 PMID:16337171

42. Berkowitz GS, Wetmur JG, Birman-Deych E et al. (2004). In utero pesticide exposure, maternal paraoxonase activity, and head circumference. Environ Health Perspect, 112:388–391.doi:10.1289/ehp.6414 PMID:14998758

43. Shen XM, Wu SH, Yan CH et al. (2001). Delta-aminolevulinate dehydratase polymorphism and blood lead levels in Chinese children. Environ Res, 85:185–190. doi:10.1006/enrs.2000.4230 PMID:11237505

44. Kahn RS, Khoury J, Nichols WC, Lanphear BP (2003). Role of dopamine transporter genotype and maternal prenatal smoking in childhood hyperactive-impulsive, inattentive, and oppositional behaviors. J Pediatr, 143: 104–110.doi:10.1016/S0022-3476(03)002087 PMID:12915833

45. Whyatt RM, Bell DA, Jedrychowski W et al. (1998). Polycyclic aromatic hydrocarbon-DNA adducts in human placenta and modulation by CYP1A1 induction and genotype. Carcinogenesis, 19:1389–1392.doi:10.1093/carcin/19.8.1389 PMID:9744534

46. Delpisheh A, Brabin L, Topping J et al. (2009). A case-control study of CYP1A1, GSTT1 and GSTM1 gene polymorphisms, pregnancy smoking and fetal growth restriction. Eur J Obstet Gynecol Reprod Biol, 143:38–42.doi:10.1016/j.ejogrb.2008.11.006 PMID:19147266

47. Gilliland FD, Li YF, Gong H Jr, Diaz-Sanchez D (2006). Glutathione s-transferases M1 and P1 prevent aggravation of allergic responses by secondhand smoke. Am J Respir Crit Care Med, 174:1335–1341.doi:10.1164/rccm.200509-1424OC PMID:17023730

48. Rogers AJ, Brasch-Andersen C, Ionita-Laza I et al. (2009). The interaction of glutathione S-transferase M1-null variants with tobacco smoke exposure and the development of childhood asthma. Clin Exp Allergy, 39:1721–1729.doi:10.1111/j.1365-2222.2009.03372.x PMID:19860819

49. Romieu I, Sienra-Monge JJ, Ramírez-Aguilar M et al. (2004). Genetic polymorphism of GSTM1 and antioxidant supplementation influence lung function in relation to ozone exposure in asthmatic children in Mexico City. Thorax, 59:8–10. PMID:14694237

50. Cassel J (1976). The contribution of the social environment to host resistance: the Fourth Wade Hampton Frost Lecture. Am J Epidemiol, 104:107–123. PMID:782233

51. Zapata BC, Rebolledo A, Atalah E et al. (1992). The influence of social and political violence on the risk of pregnancy complications. Am J Public Health, 82:68 5–690.doi:10.2105/AJPH.82.5.685 PMID:15 66947

52. Kliegman RM (1992). Perpetual poverty: child health and the underclass. Pediatrics, 89:710–713. PMID:1557266

53. David RJ, Collins JW Jr (1997). Differing birth weight among infants of U.S.-born blacks, African-born blacks, and U.S.-born whites. N Engl J Med, 337:1209–1214.doi:10.1056/NEJM199710233371706 PMID:9337381

54. Wise PH (1993). Confronting racial disparities in infant mortality: reconciling science and politics. Am J Prev Med, 9 Suppl;7–16. PMID:8123287

55. Rauh VA, Whyatt RM, Garfinkel R et al. (2004). Developmental effects of exposure to environmental tobacco smoke and material hardship among inner-city children. Neurotoxicol Teratol, 26:373–385. doi:10.1016/j.ntt.2004.01.002 PMID:15113599

56. Neri M, Ugolini D, Bonassi S et al. (2006). Children's exposure to environmental pollutants and biomarkers of genetic damage. II. Results of a comprehensive literature search and meta-analysis. Mutat Res, 612: 14–39.doi:10.1016/j.mrrev.2005.04.003 PMID:16027031

57. Plopper CG, Fanucchi MV (2000). Do urban environmental pollutants exacerbate childhood lung diseases? Environ Health Perspect, 108:A252–A253.doi:10.1289/ehp.108-a252 PMID:10856036

58. Hong CY, Chia SE, Widjaja D et al. (2004). Prevalence of respiratory symptoms in children and air quality by village in rural Indonesia. J Occup Environ Med, 46:1174–1179.doi:10.1097/01.jom.0000141666.21758.86 PMID:15534505

59. Janssen NA, Brunekreef B, van Vliet P et al. (2003). The relationship between air pollution from heavy traffic and allergic sensitization, bronchial hyperresponsiveness, and respiratory symptoms in Dutch schoolchildren. Environ Health Perspect, 111:1512–1518. doi:10.1289/ehp.6243 PMID:12 948892

60. Mortimer KM, Neas LM, Dockery DW et al. (2002). The effect of air pollution on inner-city children with asthma. Eur Respir J, 19:699–705.doi:10.1183/09031936.02.00247102 PMID:11999000

61. McConnell R, Berhane K, Yao L et al. (2006). Traffic, susceptibility, and childhood asthma. Environ Health Perspect, 114:766–772.doi:10.1289/ehp.8594 PMID:16675435

62. Diaz-Sanchez D, Tsien A, Fleming J, Saxon A (1997). Combined diesel exhaust particulate and ragweed allergen challenge markedly enhances human in vivo nasal ragweed-specific IgE and skews cytokine production to a T helper cell 2-type pattern. J Immunol, 158:2406–2413. PMID:9036991

63. Wyler C, Braun-Fahrländer C, Künzli N et al.; The Swiss Study on Air Pollution and Lung Diseases in Adults (SAPALDIA) Team (2000). Exposure to motor vehicle traffic and allergic sensitization. Epidemiology, 11:450–456. doi:10.1097/00001648-200007000-00015 PMID:10874554

64. Gauderman WJ, Avol E, Gilliland F et al. (2004). The effect of air pollution on lung development from 10 to 18 years of age. N Engl J Med, 351:1057–1067.doi:10.1056/NEJMoa040610 PMID:15356303

65. Miller RL, Chew GL, Bell CA et al. (2001). Prenatal exposure, maternal sensitization, and sensitization in utero to indoor allergens in an inner-city cohort. Am J Respir Crit Care Med, 164:995–1001. PMID:11587985

66. Miller RL, Garfinkel R, Horton M et al. (2004). Polycyclic aromatic hydrocarbons, environmental tobacco smoke, and respiratory symptoms in an inner-city birth cohort. Chest, 126:1071–1078.doi:10.1378/chest.126.4.1071 PMID:15486366

67. U.S. Environmental Protection Agency. Health assessment document for diesel engine exhaust. Available from URL: http://www.epa.gov/fedrgstr/EPA-AIR/2002/September/Day-03/a22368.htm 2002;67(170).

68. Devouassoux G, Saxon A, Metcalfe DD et al. (2002). Chemical constituents of diesel exhaust particles induce IL-4 production and histamine release by human basophils. J Allergy Clin Immunol, 109:847–853.doi:10.1067/mai.2002.122843 PMID:11994710

69. Fujieda S, Diaz-Sanchez D, Saxon A (1998). Combined nasal challenge with diesel exhaust particles and allergen induces In vivo IgE isotype switching. Am J Respir Cell Mol Biol, 19:507–512. PMID:9730879

70. Chang Y, Sénéchal S, de Nadai P et al. (2006). Diesel exhaust exposure favors TH2 cell recruitment in nonatopic subjects by differentially regulating chemokine production. J Allergy Clin Immunol, 118:354–360.doi: 10.1016/j.jaci.2006.04.050 PMID:16890758

71. Sénéchal S, de Nadai P, Ralainirina N et al. (2003). Effect of diesel on chemokines and chemokine receptors involved in helper T cell type 1/type 2 recruitment in patients with asthma. Am J Respir Crit Care Med, 168:215–221.doi:10.1164/rccm.200211-1289OC PMID:12724126

72. Dai J, Xie C, Vincent R, Churg A (2003). Air pollution particles produce airway wall remodeling in rat tracheal explants. Am J Respir Cell Mol Biol, 29:352–358.doi:10.1165/rcmb.2002-0318OC PMID:12649123

73. Ohta K, Yamashita N, Tajima M et al. (1999). Diesel exhaust particulate induces airway hyperresponsiveness in a murine model: essential role of GM-CSF. J Allergy Clin Immunol, 104:1024–1030.doi:10.1016/S0091-6749(99)70084-9 PMID:10550748

74. U.S. Environmental Protection Agency. Respiratory health effects of passive smoking: lung cancer and other disorders. Washington (DC): Office of Health and Environmental Assessment, Office of Research and Development; 1992.

75. Pinkerton KE, Joad JP (2000). The mammalian respiratory system and critical windows of exposure for children's health. Environ Health Perspect, 108 Suppl 3;457–462.doi:10.2307/3454537 PMID:10852845

76. Bohning DE, Atkins HL, Cohn SH (1982). Long-term particle clearance in man: normal and impaired. Ann Occup Hyg, 26:259–271. doi:10.1093/annhyg/26.2.259 PMID:7181269

77. Finkelstein JN, Johnston CJ (2004). Enhanced sensitivity of the postnatal lung to environmental insults and oxidant stress. Pediatrics, 113 Suppl;1092–1096. PMID:15060204

78. Diaz-Sanchez D, Rumold R, Gong H Jr (2006). Challenge with environmental tobacco smoke exacerbates allergic airway disease in human beings. J Allergy Clin Immunol, 118:441–446.doi:10.1016/j.jaci.2006.04.047 PMID:16890770

79. Jedrychowski W, Galas A, Pac A et al. (2005). Prenatal ambient air exposure to polycyclic aromatic hydrocarbons and the occurrence of respiratory symptoms over the first year of life. Eur J Epidemiol, 20:775–782. doi:10.1007/s10654-005-1048-1 PMID:16170661

80. Van Larebeke NA, Birnbaum LS, Boogaerts MA et al. (2005). Unrecognized or potential risk factors for childhood cancer. Int J Occup Environ Health, 11:199–201. PMID:15875896

81. Perera FP (2000). Molecular epidemiology: on the path to prevention? J Natl Cancer Inst, 92:602–612.doi:10.1093/jnci/92.8.602 PMID:10772677

82. Whyatt RM, Jedrychowski W, Hemminki K et al. (2001). Biomarkers of polycyclic aromatic hydrocarbon-DNA damage and cigarette smoke exposures in paired maternal and newborn blood samples as a measure of differential susceptibility. Cancer Epidemiol Biomarkers Prev, 10:581–588. PMID:11401906

83. Perera FP, Tang D, Tu YH et al. (2004). Biomarkers in maternal and newborn blood indicate heightened fetal susceptibility to procarcinogenic DNA damage. Environ Health Perspect, 112:1133–1136.doi:10.1289/ehp.6833 PMID:15238289

84. Perera FP, Tang D, Whyatt RM et al. (2004). Comparison of PAH-DNA adducts in four populations of mothers and newborns in the US, Poland, and China [Abstract #1975]. Proc Am Assoc Cancer Res, 45.

85. Calabrese EJ. Age and susceptibility to toxic substances. New York (NY): John Wiley and Sons; 1986.

86. Laib RJ, Klein KP, Bolt HM (1985). The rat liver foci bioassay: I. Age-dependence of induction by vinyl chloride of ATPase-deficient foci. Carcinogenesis, 6:65–68.doi:10.1093/carcin/6.1.65 PMID:3155670

87. Hagmar L, Brøgger A, Hansteen IL et al. (1994). Cancer risk in humans predicted by increased levels of chromosomal aberrations in lymphocytes: Nordic study group on the health risk of chromosome damage. Cancer Res, 54:2919–2922. PMID:8187078

88. Bonassi S, Hagmar L, Strömberg U et al.; European Study Group on Cytogenetic Biomarkers and Health (2000). Chromosomal aberrations in lymphocytes predict human cancer independently of exposure to carcinogens. Cancer Res, 60:1619–1625. PMID:10749131

89. Bocskay KA, Tang D, Orjuela MA et al. (2005). Chromosomal aberrations in cord blood are associated with prenatal exposure to carcinogenic polycyclic aromatic hydrocarbons. Cancer Epidemiol Biomarkers Prev, 14:506–511.doi:10.1158/1055-9965.EPI-04-0566 PMID:15734979

90. Pluth JM, Ramsey MJ, Tucker JD (2000). Role of maternal exposures and newborn genotypes on newborn chromosome aberration frequencies. Mutat Res, 465:101–111.PMID:10708975

91. Ross JA (1998). Maternal diet and infant leukemia: a role for DNA topoisomerase II inhibitors? Int J Cancer Suppl, 11 S11;26–28. doi:10.1002/(SICI)1097-0215(1998)78:11+<26::AID-IJC8>3.0.CO;2-M PMID:9876473

92. Chen S, Tang D, Xue K et al. (2002). DNA repair gene XRCC1 and XPD polymorphisms and risk of lung cancer in a Chinese population. Carcinogenesis, 23:1321–1325. doi:10.1093/carcin/23.8.1321 PMID:12151350

93. Tang D, Cho S, Rundle A et al. (2002). Polymorphisms in the DNA repair enzyme XPD are associated with increased levels of PAH-DNA adducts in a case-control study of breast cancer. Breast Cancer Res Treat, 75:159–166.doi:10.1023/A:1019693504183 PMID:12243508

94. Ford AM, Ridge SA, Cabrera ME et al. (1993). In utero rearrangements in the trithorax-related oncogene in infant leukaemias. Nature, 363:358–360.doi:10.1038/363358a0 PMID:8497319

95. Gill Super HJ, Rothberg PG, Kobayashi H et al. (1994). Clonal, nonconstitutional rearrangements of the MLL gene in infant twins with acute lymphoblastic leukemia: in utero chromosome rearrangement of 11q23. Blood, 83:641–644. PMID:8298125

96. Ford AM, Pombo-de-Oliveira MS, McCarthy KP et al. (1997). Monoclonal origin of concordant T-cell malignancy in identical twins. Blood, 89:281–285. PMID:8978302

97. Ford AM, Bennett CA, Price CM et al. (1998). Fetal origins of the TEL-AML1 fusion gene in identical twins with leukemia. Proc Natl Acad Sci USA, 95:4584–4588. doi:10.1073/pnas.95.8.4584 PMID:9539781

98. Wiemels JL, Cazzaniga G, Daniotti M et al. (1999). Prenatal origin of acute lymphoblastic leukaemia in children. Lancet, 354:1499–1503.doi:10.1016/S0140-6736(99)09403-9 PMID:10551495

99. Gale KB, Ford AM, Repp R et al. (1997). Backtracking leukemia to birth: identification of clonotypic gene fusion sequences in neonatal blood spots. Proc Natl Acad Sci USA, 94:13950–13954.doi:10.1073/pnas.94.25.13950 PMID:9391133

100. Maia AT, Ford AM, Jalali GR et al. (2001). Molecular tracking of leukemogenesis in a triplet pregnancy. Blood, 98:478–482. doi:10.1182/blood.V98.2.478 PMID:11435320

101. Yagi T, Hibi S, Tabata Y et al. (2000). Detection of clonotypic IGH and TCR rearrangements in the neonatal blood spots of infants and children with B-cell precursor acute lymphoblastic leukemia. Blood, 96:264–268.PMID:10891460

102. Fasching K, Panzer S, Haas OA et al. (2000). Presence of clone-specific antigen receptor gene rearrangements at birth indicates an in utero origin of diverse types of early childhood acute lymphoblastic leukemia. Blood, 95:2722–2724. PMID:10753857

103. Wiemels JL, Xiao Z, Buffler PA et al. (2002). In utero origin of t(8;21) AML1-ETO translocations in childhood acute myeloid leukemia. Blood, 99:3801–3805.doi:10.1182/blood.V99.10.3801 PMID:11986239

104. Faustman EM, Silbernagel SM, Fenske RA et al. (2000). Mechanisms underlying Children's susceptibility to environmental toxicants. Environ Health Perspect, 108 Suppl 1;13–21.doi:10.2307/3454629 PMID:10698720

105. Kreider ML, Aldridge JE, Cousins MM et al. (2005). Disruption of rat forebrain development by glucocorticoids: critical perinatal periods for effects on neural cell acquisition and on cell signaling cascades mediating noradrenergic and cholinergic neurotransmitter/neurotrophic responses. Neuropsychopharmacology, 30:1841–1855. doi:10.1038/sj.npp.1300743 PMID:15841102

106. Slotkin TA (2004). Guidelines for developmental neurotoxicity and their impact on organophosphate pesticides: a personal view from an academic perspective. Neurotoxicology, 25:631–640.doi:10.1016/S0161-813X(03)00050-0 PMID:15183016

107. Slotkin TA, Tate CA, Ryde IT et al. (2006). Organophosphate insecticides target the serotonergic system in developing rat brain regions: disparate effects of diazinon and parathion at doses spanning the threshold for cholinesterase inhibition. *Environ Health Perspect*, 114:1542–1546.doi:10.1289/ehp.9337 PMID:17035140

108. Chanda SM, Pope CN (1996). Neurochemical and neurobehavioral effects of repeated gestational exposure to chlorpyrifos in maternal and developing rats. *Pharmacol Biochem Behav*, 53:771–776. doi:10.1016/0091-3057(95)02105-1 PMID:8801577

109. Campbell CG, Seidler FJ, Slotkin TA (1997). Chlorpyrifos interferes with cell development in rat brain regions. *Brain Res Bull*, 43:179–189.doi:10.1016/S0361-9230(96)00436-4 PMID:9222531

110. Baghurst PA, McMichael AJ, Wigg NR et al. (1992). Environmental exposure to lead and children's intelligence at the age of seven years. The Port Pirie Cohort Study. *N Engl J Med*, 327:1279–1284.doi:10.1056/NEJM 199210293271805 PMID:1383818

111. Bellinger DC, Stiles KM, Needleman HL (1992). Low-level lead exposure, intelligence and academic achievement: a long-term follow-up study. *Pediatrics*, 90:855–861. PMID:1437425

112. Needleman HL, Gatsonis CA (1990). Low-level lead exposure and the IQ of children. A meta-analysis of modern studies. *JAMA*, 263:673–678.doi:10.1001/jama.263.5.673 PMID:2136923

113. Lanphear BP, Dietrich K, Auinger P, Cox C (2000). Cognitive deficits associated with blood lead concentrations <10 microg/dL in US children and adolescents. *Public Health Rep*, 115:521–529.doi:10.1093/phr/115.6.521 PMID:11354334

114. Canfield RL, Henderson CR Jr, Cory-Slechta DA et al. (2003). Intellectual impairment in children with blood lead concentrations below 10 microg per deciliter. *N Engl J Med*, 348:1517–1526.doi:10.1056/NEJMoa022848 PMID:12700371

115. Grandjean P, Weihe P, Burse VW et al. (2001). Neurobehavioral deficits associated with PCB in 7-year-old children prenatally exposed to seafood neurotoxicants. *Neurotoxicol Teratol*, 23:305–317.doi:10.1016/S0892-0362(01)00155-6 PMID:11485834

116. Rauh VA, Garfinkel R, Perera FP et al. (2006). Impact of prenatal chlorpyrifos exposure on neurodevelopment in the first three years of life among inner-city children. *Pediatrics*, 118:1845–1859.doi:10.1542/peds.2006-0338.

117. Perera FP, Rauh V, Whyatt RM et al. (2005). A summary of recent findings on birth outcomes and developmental effects of prenatal ETS, PAH, and pesticide exposures. *Neurotoxicology*, 26:573–587.doi:10.1016/j.neuro.2004.07.007 PMID:16112323

118. Perera FP, Li Z, Whyatt R et al. (2009). Prenatal airborne polycyclic aromatic hydrocarbon exposure and child IQ at age 5 years. *Pediatrics*, 124:e195–e202.doi:10.1542/peds.2008-3506 PMID:19620194

119. Perera FP, Rauh V, Whyatt RM et al. (2006). Effect of prenatal exposure to airborne polycyclic aromatic hydrocarbons on neurodevelopment in the first 3 years of life among inner-city children. *Environ Health Perspect*, 114:1287–1292.doi:10.1289/ehp.9084 PMID:16882541

120. Kimmel CA, Collman GW, Fields N, Eskenazi B (2005). Lessons learned for the National Children's Study from the National Institute of Environmental Health Sciences/U.S. Environmental Protection Agency Centers for Children's Environmental Health and Disease Prevention Research. *Environ Health Perspect*, 113:1414–1418.doi:10.1289/ehp.7669 PMID:16203257

121. National Institute of Environmental Health Sciences. Centers for children's environmental health and disease prevention research. 2005. Available from URL: http://www.niehs.nih.gov/research/supported/centers/prevention/.

122. Dearry AD, Collman GW, Saint C et al. (1999). Building a network of research in children's environmental health. *Environ Health Perspect*, 107 Suppl 3;391–392.doi:10.1289/ehp.99107s3391 PMID:10346987

123. O'Fallon LR, Collman GW, Dearry A (2000). The National Institute of Environmental Health Sciences' research program on children's environmental health. *J Expo Anal Environ Epidemiol*, 10:630–637.doi:10.1038/sj.jea.7500117 PMID:11138655

124. Eskenazi B, Gladstone EA, Berkowitz GS et al. (2005). Methodologic and logistic issues in conducting longitudinal birth cohort studies: lessons learned from the Centers for Children's Environmental Health and Disease Prevention Research. *Environ Health Perspect*, 113:1419–1429.doi:10.1289/ehp.7670 PMID:16203258

125. Yiin LM, Rhoads GG, Lioy PJ (2000). Seasonal influences on childhood lead exposure. *Environ Health Perspect*, 108:177–182.doi:10.1289/ehp.00108177 PMID:10656860

126. Thompson B, Coronado GD, Vigoren EM et al. (2008). Para niños saludables: a community intervention trial to reduce organophosphate pesticide exposure in children of farmworkers. *Environ Health Perspect*, 116:687–694.doi:10.1289/ehp.10882 PMID:18470300

127. National Children's Study. 2010. Available from URL: http://www.nationalchildrensstudy.gov/.

128. Farmer PB, Kaur B, Roach J et al. (2005). The use of S-phenylmercapturic acid as a biomarker in molecular epidemiology studies of benzene. *Chem Biol Interact*, 153-154:97–102.doi:10.1016/j.cbi.2005.03.013 PMID:15935804

129. Kotseva K, Popov T (1998). Study of the cardiovascular effects of occupational exposure to organic solvents. *Int Arch Occup Environ Health*, 71 Suppl;S87–S91. PMID:9827890

130. Spanò M, Toft G, Hagmar L et al.; INUENDO (2005). Exposure to PCB and p,p'-DDE in European and Inuit populations: impact on human sperm chromatin integrity. *Hum Reprod*, 20:3488–3499.doi:10.1093/humrep/dei297 PMID:16223788

131. Perera FP, Jedrychowski W, Rauh V, Whyatt RM (1999). Molecular epidemiologic research on the effects of environmental pollutants on the fetus. *Environ Health Perspect*, 107 Suppl 3;451–460.doi:10.1289/ehp.99107s3451 PMID:10346993

132. Albertini RJ, Srám RJ, Vacek PM, Lynch J, Nicklas JA, van Sittert NJ, et al. Biomarkers in Czech workers exposed to 1,3-butadiene: a transitional epidemiologic study. *Res Rep Health Eff Inst* 2003;116:1–141 (discussion 143–162).

133. Ickovics JR, Milan S, Boland R et al.; HIV Epidemiology Research Study (HERS) Group (2006). Psychological resources protect health: 5-year survival and immune function among HIV-infected women from four US cities. *AIDS*, 20:1851–1860.doi:10.1097/01.aids.0000244204.95758.15 PMID:16954726

134. Jurewicz A, Matysiak M, Raine CS, Selmaj K (2007). Soluble Nogo-A, an inhibitor of axonal regeneration, as a biomarker for multiple sclerosis. *Neurology*, 68:283–287. doi:10.1212/01.wnl.0000252357.30287.1d PMID:17242333

135. Hernández AF, López O, Rodrigo L et al. (2005). Changes in erythrocyte enzymes in humans long-term exposed to pesticides: influence of several markers of individual susceptibility. *Toxicol Lett*, 159: 13–21.doi:10.1016/j.toxlet.2005.04.008 PMID: 15922524

136. Perera FP, Weinstein IB (1982). Molecular epidemiology and carcinogen-DNA adduct detection: new approaches to studies of human cancer causation. *J Chronic Dis*, 35:581–600.doi:10.1016/0021-9681(82)90078-9 PMID:6282919

137. Perera FP, Santella RM. Carcinogenesis. In: Schulte PA, Perera FP, editors. Molecular epidemiology: principles and practices. New York (NY): Academic Press, Inc; 1993. p. 277–300.

138. National Research Council. Regulating pesticides in food: The Delaney Paradox. Washington (DC): National Academy of Sciences, National Academy Press; 1987.

UNIT 5.

APPLICATION OF BIOMARKERS TO DISEASE

CHAPTER 27.

Future perspectives on molecular epidemiology

Martyn T. Smith, Pierre Hainaut, Frederica Perera, Paul A. Schulte, Paolo Boffetta, Stephen J. Chanock, and Nathaniel Rothman

The current status of molecular epidemiology

As witnessed in the previous chapters, since the development of molecular epidemiology in the early 1980s (1), the field has advanced so that large-scale, in-depth studies have been performed or are ongoing. These studies have not only monitored the external environment and ascertained clinical disease status, but have also collected data on biomarkers of exposure, biologically effective dose, preclinical effects, and susceptibility within population studies. Links have been drawn between various environmental and nutritional factors and diseases as diverse as childhood asthma, cardiovascular disease, cancer, developmental disorders, obesity and metabolic disorders. In some cases, educational or regulatory interventions have been mounted as a result of these studies.

As described throughout this book, non-genetic environmental factors, broadly defined to include diet, lifestyle, infections, stress, ionizing radiation, and chemical pollutants in the air, water, food supply and workplace, are important contributors to chronic disease. Adverse gene–environment interactions (GxE) probably influence most chronic diseases, including neurological disorders and cancer. The genetic (G) contribution to different diseases varies, but several lines of evidence, including classic studies of migrant populations in which the genetics remain essentially the same but the incidence of disease changes because of the new environment, clearly show that non-genetic factors have high attributable risks (2). For some diseases, incidence rates increase or decrease dramatically within the first or second generation of immigrants, with disease patterns becoming more similar to the adoptive country and less similar to the country of origin. This highlights the fact that environmental factors (E) can contribute to a large portion of at least some chronic diseases (3,4).

Genomic tools arising from the Human Genome Project combined with bioinformatics have allowed scientists to begin to examine the genetic component of many chronic diseases. Initially, variations in candidate genes were examined in great detail, most notably in xenobiotic metabolizing and DNA repair genes (5–7). More recently, genome-wide association studies (GWAS) have been increasing in number and scope and have provided important insights into the roles that particular genes, gene regions, regulatory elements, and other parts of the genome with function yet to be defined play in disease development (8). Thus, a key focus of most current molecular epidemiology studies is on the genome and genetic variation. The reason for this focus on genetics, even though the environment may be more critical, is simply that we have extremely precise, accurate, and global tools to examine the genome, either measured as external factors or biologically as reflected by the "exposome". Such tools are not available to examine the environment. At the same time, by examining the G component of GxE we may find clues as to where to look for E factors (9,10), although success in this regard is very limited to date.

The most productive approach to assessing the environmental contribution to disease may be to examine environmental exposures agnostically (11). Unfortunately, compared to genomics, the tools for assessment of exposures, based upon measurements of chemicals in air, water, food and the human body, have undergone a more gradual evolution in the past 30 years and have not experienced the same exponential gains. This is due to both lack of technological progress in the tools available for exposure assessment, as well as the more challenging task of obtaining data on, or estimates of, non-fixed exogenous and endogenous individual exposures. These exposures can vary day-to-day as well as over time, as individuals age and secular changes occur in a given population. The use of questionnaires has been the core approach for assessing exposure in studies of chronic diseases in the general population that arise, in part, from exposure patterns present over many years. This approach relies on self-reports, which can be imprecise and inaccurate. However, they have been successfully used to identify consistent patterns of chronic disease risk for several exposures such as tobacco, alcohol, obesity, components of the reproductive history, air pollution, and some aspects of diet (e.g. intake of cruciferous vegetables). Also, the increased ability to obtain objective occupational and residential histories from study subjects, linked by sophisticated methods to comprehensive exposure databases, has allowed advances in identifying associations between certain chemical exposures and disease risk. At the same time, methods to measure chemicals in biologic samples have steadily evolved to measure a wider array of compounds in smaller amounts of samples. Nevertheless, these advances are not comparable to the quantum leap that has occurred in genomics.

The Human Genome Project and at least 20 years of investment in genetics are very helpful to molecular epidemiologists in understanding the genetic determinants of diseases, but we remain much more limited when it comes to quantifying human exposures. This disparity in current knowledge between genetic contributions and environmental exposures was recognized by Wild, who defined the exposome, representing all environmental exposures and lifestyle factors from conception onwards, as a quantity of critical interest to disease etiology (12). If we expect to have any success at identifying the effects of G, E and GxE on chronic diseases, we must develop 21st century tools to measure exposure levels in large human populations (11). That is, we need to quantify the exposome, a topic we will return to later.

Many lifestyle factors such as exercise levels, dietary choices and stress levels also contribute to the environmental component of disease, but are hard to quantify retrospectively and prospectively. Modern tools to capture, store and use information about physical activity, diet and stress levels are needed for epidemiological studies. Such tools are being developed. For example, it soon may be possible to perform population-scale, longitudinal measurement of physical activity using common cell phones that include internal accelerometers and low-power wireless communication capabilities (13). Dietary assessment methods suitable for use in large epidemiologic studies (e.g. dietary recall, food diaries and food frequency questionnaires) are subject to considerable inaccuracy. More accurate methods (e.g. metabolic ward studies and doubly-labelled water) are prohibitively costly and/or labour-intensive for use in population-based studies. Several research groups are assessing methods that use cell phones to capture both voice recordings and photographs of dietary intake in real time that, along with computerized analysis, may revolutionise nutritional epidemiology studies (14).

Accumulating evidence is also consistent with the role of psychosocial stress in moderating the effects of genetic and other environmental factors on health outcomes. Further advances in this area will require the development of standardized, psychometrically sound instruments for quantifying exposures to psychosocial stress. Again, progress is being made in this area using, for example, colorimetric test strips to rapidly detect and quantify salivary α-amylase, a biomarker of the body's adrenergic stress response (15,16). The measurement of telomere length is another stress biomarker that is gaining acceptance (17,18).

The potential contributions of genomics to molecular epidemiology

Genomics is the study of all of the nucleotide sequences, including structural genes, regulatory sequences, and noncoding DNA segments, in the chromosomes of an organism. Because of the tools provided by the Human Genome Project, the current focus of many molecular epidemiological studies is on genomic variation. GWAS and, most recently, examinations of copy number variation have revealed many surprising insights. Results from these studies show that many common causal variants, each of small, additive effect, probably contribute to complex disease risk.

As of 2010, GWAS had identified over 750 regions in the genome strongly associated with more than 125 traits and diseases (http://www.genome.gov/gwastudies) (19). In chronic complex diseases, such as type 2 diabetes and Crohn's disease, over 40 genetic regions have been associated with each disease. For certain heritable traits such as height, recent studies have identified several hundred regions, each of which contributes to their heredity (20). In cancer alone there are over 135 regions associated with 21 cancers (21). However, the early GWAS have not sufficiently explained the heredity of any given common disease. This is not surprising since GWAS interrogate only common variants, which represent only a proportion of genetic variation in the human genome. For example, despite the addition of 10 positively-associated SNPs, the performance of breast cancer risk models only improved modestly; the area under the curve of the receiver operating curve increased from 58% to a mere 61.8% (22). Thus far, GWAS have been most successful in identifying regions that harbour genetic variants that are directly associated with risk for a complex disease, such as cancer or inflammatory bowel disease. For the latter, GWAS have pointed towards a region on chromosome 2q37.1 and identified a novel mechanism of autophagy previously not well described in the pathogenesis of inflammatory bowel disease, specially as it relates to the genes in this pathway (23). Fine mapping of regions together with functional work is required to elucidate the biological underpinnings of the direct association of common variants with complex diseases such as cancer (24). Certainly, the advent of new technologies, in conjunction with computational resources, will enable investigators to use next generation sequencing to explore the contribution of uncommon and rare variants in the near future.

The potential contributions of molecular epidemiology in the near future

In the near future, there will also likely be a maturing of omic applications and the incorporation of systems biology into molecular epidemiology, which will produce what some have called systems epidemiology (25). Studies of epigenetic changes are already coming to the forefront of molecular epidemiology, and studies of changes in DNA methylation, histone modifications and microRNA (miRNA) expression, both in cells and body fluids, have recently been published (26–28). ChIP-on-Chip (chromatin immunoprecipitation with microarray analysis) and ChIP-seq (chromatin immunoprecipitation with sequencing) will help in understanding epigenetic effects on gene–protein interactions. Advances in mass spectrometry will soon make it possible to measure post-translation modifications of proteins such as histones in small volumes of biological sample, adding to our repertoire of epigenetic changes that can be studied in human populations.

Advances in mass spectrometry and in laboratory-on-a-chip devices that use nanotechnology may also soon permit us to profile all the major protein and DNA adducts in humans using adductomics. This will allow for the examination of multiple biomarkers in very small sample volumes, such as a few microlitres of serum, a drop of blood, or a dried blood spot.

These tools are expected to have great application in molecular epidemiology studies in the near future. There are emerging opportunities to apply these technologies in molecular epidemiologic studies with banked biological samples, including cross-sectional, case–control, and, in particular, prospective cohort studies, to study a wide range of diseases.

This should advance the ability of molecular epidemiology to more broadly explore exposure–disease relationships, to study effect modifiers, and to obtain insight into the fundamental underlying pathogenesis of these conditions. Further, beyond providing etiologic insights, it is expected that molecular epidemiology will be newly positioned to make important contributions to translating these findings into primary, secondary and tertiary prevention strategies. This would begin with broad public health practices that could include removal or substantial reduction of exposure to hazardous environmental compounds, making available healthier food in schools and better education on lifestyle risk factors.

At the same time, molecular epidemiology is likely to play an important role in the upcoming revolution in personalized medicine (29). At present, identifying individual genetic risk is at the forefront of this personalization of health care. But given the limited role for genetics in comparison to the environment in causing disease, the focus must eventually shift to include individual environmental risk factors, again broadly defined to include toxic exposures, lifestyle, diet, drugs, etc. This could help bring about not only lifestyle modification to prevent disease and improve drug treatment, but it could also help individuals gain an understanding of their prior and current chemical exposures and other risk factors, leading to personalized risk assessment. Molecular epidemiologists may be able to identify not only broad subgroups of the population with a higher probability of developing disease given genetic and other risk factors, but also move to further develop predictive models that can be applied to individuals by preventive and clinical medicine practitioners. An example of this is the Gail Model for predicting breast cancer, which is based on all known risk factors including *BRCA1* and *BRCA2* mutations (30). Genetics is now poised to augment this model and provide even greater sensitivity and specificity, but as mentioned previously, success to date using GWAS data is limited.

Additional profoundly important steps taking place in molecular epidemiology are the increased size of studies and the formation of dozens of international consortia, including those that focus on specific diseases as well as those that are based on study design (e.g. various cohort consortia). There are now a large number of prospective cohort studies in North America, Europe, Asia and Australia that have enrolled or are continuing to enrol several million study subjects. These cohort studies have millions of samples of DNA, serum, blood cells, and other biological material stored at low temperatures. Some studies are tracking individuals *in utero* through adolescence, providing an opportunity to assess the earliest determinants of disease. These samples are precious as well as numerous. Efficient, high-throughput methods that work on minute amounts of sample are needed to analyse nested case–control or case–cohort studies carried out within them. The combination of new nanotechnology/laboratory-on-a-chip methodologies with large prospective cohort studies holds great promise for new research findings. At the same time, there will still be a need for focused, hypothesis-testing studies carried out within these cohorts, in addition to the application of discovery technologies. Such studies can often be carried out on smaller sample sizes, as they generally do not need to contend with the low prior hypothesis/multiple testing problem.

In addition, there is still a need for focused, cross-sectional studies of well-defined populations with particular exposure or lifestyle patterns of interest, such as new exposures recently introduced into the environment, high or low levels of exposures, etc. (e.g. (31,32)). These studies can use extensive and complicated protocols as sample sizes are generally small (a few hundred subjects). Often they can have very detailed assessment of the target exposure, evaluate potential confounders, modifiers, and other contributors to endpoints under study in greater detail (e.g. nutritional, genetic, psychosocial factors) (33), and arrange for samples to be transported and processed very quickly, allowing specialty assays to be carried out. They can incorporate both state-of-the-art omics platforms as well as in-depth hypothesis-testing (34).

Future use of omic technologies in molecular epidemiology

The field of molecular epidemiology is entering an exciting new phase in which the innovative tools of omics, such as microarrays and metabolic and peptide profiling, are being applied along with novel laboratory-on-a-chip microdevices that can act as biosensors of everything from glucose levels to protein adducts (Figure 27.1). The term omics has come to mean any field of study in biology in which the totality of something is studied, beginning with genomics which surveys across the genome. The tools of genomics, developed as a consequence of the Human Genome Project, include microarrays allowing the examination of gene variation and expression and high-throughput sequencing. The latter is now being used not only to sequence DNA but the RNA transcriptome to give a more complete picture of gene and siRNA expression. Transcriptomics is the study of all forms of RNA that are transcribed from the DNA and includes mRNA and miRNA expression.

Figure 27.1

Omics in Molecular Epidemiology

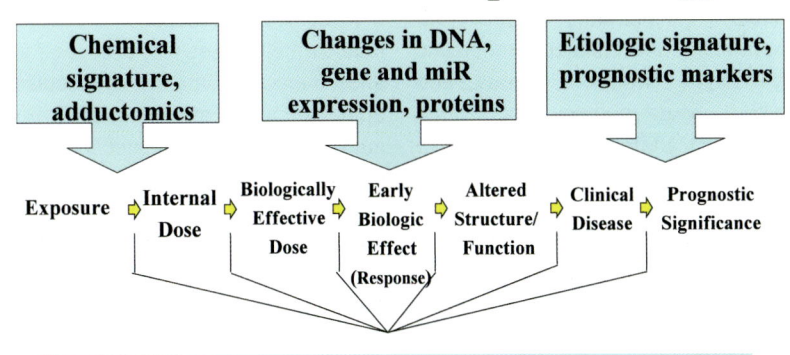

| Chemical signature, adductomics | Changes in DNA, gene and miR expression, proteins | Etiologic signature, prognostic markers |

Exposure ⇨ Internal Dose ⇨ Biologically Effective Dose ⇨ Early Biologic Effect (Response) ⇨ Altered Structure/Function ⇨ Clinical Disease ⇨ Prognostic Significance

Markers of Susceptibility
Genomics identifies all genetic variants of importance
Functional role?

Recent and near-future contributions of transcriptomics

Distinctive blood transcriptional profiles have been demonstrated for over 35 human diseases (35). As more data become available on global gene expression in the blood of humans following exposure, it will become easier to identify molecular mechanisms by which environmental chemicals promote/cause human disease. Initiatives such as the Comparative Toxicogenomics Database (http://ctd.mdibl.org/) (36) have been developed towards this goal.

A broad array of environmental exposures including pharmaceuticals, pesticides, air pollutants, industrial chemicals, heavy metals, hormones, nutrition and behaviour can change gene expression through several gene regulatory mechanisms (37). The potential of toxicogenomics in the discovery of biomarkers of complex environmental exposures was illustrated by a study in which gene expression profiling of leukocytes was shown to distinguish individuals exposed to cigarette smoke (CS) from unexposed individuals (38). An association between CS-induced gene expression and DNA adduct formation was later shown in a study of monozygotic twin pairs (39). The impact of air pollution on children at the transcriptional level in blood cells was investigated by comparing children from urban and rural regions of the Czech Republic (40). Several genes were differentially expressed and a correlation with micronuclei frequencies was shown. Further, the effects on children and adults at the transcriptional level differed (41). A small study of children in New York City found that a gene-specific methylation change in umbilical cord white blood cell DNA was associated both with prenatal exposure to PAH air pollutants and with reported asthma in the children by age 5 years (42).

More recently, many groups have begun profiling miRNA expression. A role for miRNAs in mediating the response to environmental exposures has been demonstrated by a study showing that smoking induces gene expression changes in the human airway epithelium (43) with some genes modulated by miRNA (44). Expression profiling analyses have revealed characteristic miRNA signatures in certain human cancers (45) and other diseases. The study of miRNA in molecular epidemiology will likely explode in the near future as new tools become available and the biology is better understood.

Applications of proteomics

While toxicogenomics studies using global transcriptional analysis have enormous potential, the transcriptome does not always reflect the functional proteome. Further, proteins may be subject to post-translational modification and translocation. However, proteomics, the analysis of the total protein output encoded by the genome using techniques such as mass spectrometry and antibody arrays, is more challenging and less amenable to application in a high-throughput capacity due to differences in protein properties, location and abundance. Recently, a multilaboratory study has attempted to dispel some of the notions of the irreproducibility of mass spectrometry-based proteomics by pinpointing where the methodological problems are and where challenges remain (46). By addressing these methodological issues researchers hope to bring proteomics to the forefront of biomarker research.

Applications of metabolomics

Metabolomics is defined as the study of metabolic profiles in easily collected biological samples such as urine, saliva or plasma. The metabolome is highly variable and time-dependent, and consists of several thousand chemical structures. Since it is sensitive to

Unit 5 • Chapter 27. Future perspectives on molecular epidemiology age, gut microbial composition, and lifestyle, metabolomics is ideal for the characterization of dietary and therapeutic interventions, metabolism and metabolism-related disorders (47). While successfully established in the screening of inborn errors in neonates, metabolomics is being increasingly applied to several diseases. For many years specific metabolites have been measured in body fluids to diagnose particular diseases such as diabetes, by measurement of glucose, and vascular diseases, by determination of cholesterol. Metabolomics, with its ever-increasing coverage of endogenous compounds and its high-throughput capacity, now provides a much more comprehensive assessment of health status and can be used in the identification, qualification, and development of biomarkers.

An important challenge in metabolomics is the acquisition of qualitative and quantitative information concerning the metabolites that occur under normal circumstances to be able to detect perturbations in the complement of metabolites as a result of changes in environmental factors. Technologies that rely on UPLC-MS/MS, FT-ICR-MS, Orbitrap, and asymmetric waveform ion mobility analysers are emerging as dominant analytical methods for metabolomic studies because of the accuracy, high throughput and coverage (>1000 unique metabolites) that can be achieved (48). However, even though these methods provide accurate mass values that may reduce the number of potential molecular formulas down to a few candidates, further development is needed to provide complete structural information. The exchange of chemical and analytical information must be encouraged for metabolomics to expand.

Importantly for epidemiologists, metabolomics is relatively easy to apply in large-scale human studies. For example, a large-scale exploratory analytical approach investigated metabolic phenotype variation across and within four human populations using 1H NMR spectroscopy (49). Metabolites discriminating across populations were then linked to data for individuals on blood pressure. Spectra were analysed from two 24-hour urine specimens for each of 4630 participants from the INTERMAP epidemiological study, which involved 17 population samples in China, Japan, the United Kingdom, and the USA. It was shown that urinary metabolite excretion patterns for East Asian and western population samples, with contrasting diets, diet-related major risk factors, and coronary heart disease/stroke rates, were significantly differentiated ($P < 10(-16)$). Among discriminatory metabolites, four were quantified and showed associations with blood pressure.

Potential impact of molecular epidemiology on public health and regulatory policy

A bioinformatics database could be built of the human response to different chemical exposures and associated chronic diseases. This database may well be useful in many ways for risk assessment. For example, by comparing the molecular effects of newly tested chemicals to those of established carcinogens, we could identify potential carcinogens (hazard identification) and establish modes of action by studying the effects of the same chemicals in experimental animals and on human cells *in vitro*. This would allow for better prediction of human carcinogenicity and assessment of carcinogenic mechanisms (50). Given the sensitivity of omic analyses, low-dose adverse effects can also be observed and distinguished from high-dose phenomena, if exposure is accurately assessed, allowing for dose–response data from molecular epidemiology studies to be incorporated into risk assessments.

For additional public health impact, molecular epidemiology must continue to expand its contributions to surveillance, mechanistic research, efficacy trials, translational research and health policy. We must assemble and communicate information to decision-makers, medical and health professionals, and the public. If molecular epidemiology is to make a major impact on population health, it must be preventive and have a global as well as a local focus. A life-course approach is also important in establishing the earliest causes of diseases both in children and adults. We must expand our horizons to develop affordable population-wide tools for combating common diseases.

Serving as the linking hub for laboratory and population, problem and solution, molecular epidemiology can help translate research to practice. To do this, there will be a need to continue current trends in the discipline and establish new ones. Continuation of the trend towards large-scale consortia and biobanks, use of bioinformatics, and attention to individual and collective ethical issues will serve to move the field forward, as will in-depth hypothesis-driven studies of at-risk populations. Powerful impacts will be achieved by incorporating epigenetic and biological systems theory in research and by expanding skill sets and professional knowledge to complement translation research and risk communication and to foster public health perspectives. A broad population-wide vision for using biological markers is required to leverage the power of molecular scale insight to give beneficial macro-scale impacts on public health.

Future challenges: Dealing with complexity and lack of resources

A major challenge to many of the novel approaches described above is the size and complexity of the data generated. Currently, it is a major biostatistical undertaking to analyse terabytes of data, and the emerging results require extensive further analysis by bioinformatics. Efforts must be made to simplify the analysis and reduce the data. New statistical approaches and computer programs are urgently needed to assist in the analysis.

Exposure assessment must also be able to address low-level exposure to complex mixtures. The current cost of analysis for most chemicals in blood and other fluids is prohibitive if one wishes to assess multiple compounds. New analytical chemical approaches are needed to assess the thousands of chemicals and their metabolites to which we are exposed.

One method to overcome resource difficulties may be to pool samples. Recently, this approach has been used with considerable success in GWAS and in studies of the plasma proteome (51,52).

Conclusion

Molecular epidemiology is poised to make ever-greater contributions to understanding the genetic and environmental causes of human disease. Both agnostic and hypothesis-driven approaches to both categories of risk factors could lead to leaps in our understanding. Investment in new methods and approaches will be needed, however. Strong links between population scientists, bench scientists, bioinformaticians and engineers must also be forged if progress is to be made.

Disclaimer: The findings and conclusions in this chapter are those of the author and do not necessarily represent the views of the Centers for Disease Control and Prevention.

References

1. Perera FP, Weinstein IB (1982). Molecular epidemiology and carcinogen-DNA adduct detection: new approaches to studies of human cancer causation. *J Chronic Dis*, 35:581–600.doi:10.1016/0021-9681(82)90078-9 PMID:6282919

2. Willett WC (2002). Balancing life-style and genomics research for disease prevention. *Science*, 296:695–698.doi:10.1126/science.1071055 PMID:11976443

3. Muir CS (1996). Epidemiology of cancer in ethnic groups. *Br J Cancer Suppl*, 29:S12–S16. PMID:8782793

4. Sasco AJ (2003). Breast cancer and the environment. *Horm Res*, 60 Suppl 3;50. doi:10.1159/000074500 PMID:14671396

5. García-Closas M, Malats N, Real FX et al. (2007). Large-scale evaluation of candidate genes identifies associations between VEGF polymorphisms and bladder cancer risk. *PLoS Genet*, 3:e29.doi:10.1371/journal.pgen.0030029 PMID:17319747

6. Lan Q, Zhang L, Shen M et al. (2009). Large-scale evaluation of candidate genes identifies associations between DNA repair and genomic maintenance and development of benzene hematotoxicity. *Carcinogenesis*, 30:50–58.doi:10.1093/carcin/bgn249 PMID: 18978339

7. Wang SS, Purdue MP, Cerhan JR et al. (2009). Common gene variants in the tumor necrosis factor (TNF) and TNF receptor superfamilies and NF-kB transcription factors and non-Hodgkin lymphoma risk. *PLoS One*, 4:e5360.doi:10.1371/journal.pone.0005360 PMID:19390683

8. Khoury MJ, Bertram L, Boffetta P et al. (2009). Genome-wide association studies, field synopses, and the development of the knowledge base on genetic variation and human diseases. *Am J Epidemiol*, 170:269–279.doi:10.1093/aje/kwp119 PMID:19498075

9. Rothman N, Wacholder S, Caporaso NE et al. (2001). The use of common genetic polymorphisms to enhance the epidemiologic study of environmental carcinogens. *Biochim Biophys Acta*, 1471:C1–C10. PMID:11342183

10. Vineis P, Marinelli D, Autrup H et al. (2001). Current smoking, occupation, N-acetyltransferase-2 and bladder cancer: a pooled analysis of genotype-based studies. *Cancer Epidemiol Biomarkers Prev*, 10:1249–1252. PMID:11751441

11. Rappaport SM, Smith MT (2010). Epidemiology. Environment and disease risks. *Science*, 330:460–461.doi:10.1126/science.1192603 PMID:20966241

12. Wild CP (2005). Complementing the genome with an "exposome": the outstanding challenge of environmental exposure measurement in molecular epidemiology. *Cancer Epidemiol Biomarkers Prev*, 14:1847–1850.doi:10.1158/1055-9965.EPI-05-0456 PMID:16103423

13. Bexelius C, Löf M, Sandin S et al. (2010). Measures of physical activity using cell phones: validation using criterion methods. *J Med Internet Res*, 12:e2. PMID:20118036

14. Arab L, Winter A (2010). Automated camera-phone experience with the frequency of imaging necessary to capture diet. *J Am Diet Assoc*, 110:1238–1241.doi:10.1016/j.jada.2010.05.010 PMID:20656101

15. Groer M, Murphy R, Bunnell W et al. (2010). Salivary measures of stress and immunity in police officers engaged in simulated critical incident scenarios. *J Occup Environ Med*, 52:595–602.doi:10.1097/JOM.0b013e3181e129da PMID:20523239

16. Klein LC, Bennett JM, Whetzel CA et al. (2010). Caffeine and stress alter salivary alpha-amylase activity in young men. *Hum Psychopharmacol*, 25:359–367.doi:10.1002/hup.1126 PMID:20589924

17. Epel ES, Blackburn EH, Lin J et al. (2004). Accelerated telomere shortening in response to life stress. Proc Natl Acad Sci USA, 101: 17312–17315.doi:10.1073/pnas.0407162101 PMID:15574496

18. Puterman E, Lin J, Blackburn E et al. (2010). The power of exercise: buffering the effect of chronic stress on telomere length. PLoS One, 5:e10837.doi:10.1371/journal.pone.0010837 PMID:20520771

19. Hindorff LA, Sethupathy P, Junkins HA et al. (2009). Potential etiologic and functional implications of genome-wide association loci for human diseases and traits. Proc Natl Acad Sci USA, 106:9362–9367.doi:10.1073/pnas.0903103106 PMID:19474294

20. Lango Allen H, Estrada K, Lettre G et al. (2010). Hundreds of variants clustered in genomic loci and biological pathways affect human height. Nature, 467:832–838. doi:10.1038/nature09410 PMID:20881960

21. Chung CC, Magalhaes WCS, Gonzalez-Bosquet J, Chanock SJ (2010). Genome-wide association studies in cancer–current and future directions. Carcinogenesis, 31:111–120. doi:10.1093/carcin/bgp273 PMID:19906782

22. Wacholder S, Hartge P, Prentice R et al. (2010). Performance of common genetic variants in breast-cancer risk models. N Engl J Med, 362:986–993.doi:10.1056/NEJMoa0907727 PMID:20237344

23. Rioux JD, Xavier RJ, Taylor KD et al. (2007). Genome-wide association study identifies new susceptibility loci for Crohn disease and implicates autophagy in disease pathogenesis. Nat Genet, 39:596–604.doi:10.1038/ng2032 PMID:17435756

24. Donnelly P (2008). Progress and challenges in genome-wide association studies in humans. Nature, 456:728–731.doi: 10.1038/nature07631 PMID:19079049

25. Lund E, Dumeaux V (2008). Systems epidemiology in cancer. Cancer Epidemiol Biomarkers Prev, 17:2954–2957.doi:10.1158/1055-9965.EPI-08-0519 PMID:18990736

26. Perera F, Herbstman J (2011). Prenatal environmental exposures, epigenetics, and disease. Reprod Toxicol, 31:363–373 doi: 10.1016/j.reprotox.2010.12.055 PMID:21256208

27. Smeester L, Rager JE, Bailey KA et al. (2011). Epigenetic changes in individuals with arsenicosis. Chem Res Toxicol, 24:165–167. doi:10.1021/tx1004419 PMID:21291286

28. Ren X, McHale CM, Skibola CF et al. (2011). An emerging role for epigenetic dysregulation in arsenic toxicity and carcinogenesis. Environ Health Perspect, 119:11–19.doi:10.1289/ehp.1002114 PMID:20682481

29. Aebersold R, Auffray C, Baney E et al. (2009). Report on EU-USA workshop: how systems biology can advance cancer research (27 October 2008). Mol Oncol, 3:9–17.doi:10.1016/j.molonc.2008.11.003 PMID:19383362

30. Gail MH (2009). Value of adding single-nucleotide polymorphism genotypes to a breast cancer risk model. J Natl Cancer Inst, 101:959–963.doi:10.1093/jnci/djp130 PMID: 19535781

31. Lan Q, Zhang L, Li G et al. (2004). Hematotoxicity in workers exposed to low levels of benzene. Science, 306:1774–1776. doi:10.1126/science.1102443 PMID:15576619

32. Turner PC, Collinson AC, Cheung YB et al. (2007). Aflatoxin exposure in utero causes growth faltering in Gambian infants. Int J Epidemiol, 36:1119–1125.doi:10.1093/ije/dym122 PMID:17576701

33. Vineis P, Perera F (2007). Molecular epidemiology and biomarkers in etiologic cancer research: the new in light of the old. Cancer Epidemiol Biomarkers Prev, 16:1954–1965.doi:10.1158/1055-9965.EPI-07-0457 PMID:17932342

34. Zhang L, McHale CM, Rothman N et al. (2010). Systems biology of human benzene exposure. Chem Biol Interact, 184:86–93.doi:10.1016/j.cbi.20 09.12.011 PMID:20026094

35. Mohr S, Liew CC (2007). The peripheral-blood transcriptome: new insights into disease and risk assessment. Trends Mol Med, 13:422–432.doi:10.1016/j.molmed.2007.08.003 PMID:17919976

36. Mattingly CJ, Rosenstein MC, Colby GT et al. (2006). The Comparative Toxicogenomics Database (CTD): a resource for comparative toxicological studies. J Exp Zool A Comp Exp Biol, 305:689–692.doi:10.1002/jez.a.307 PMID:16902965

37. Edwards TM, Myers JP (2008). Environmental exposures and gene regulation in disease etiology. Cien Saude Colet, 13:269–281.doi:10.1590/S1413-81232008000100030 PMID:18813540

38. Lampe JW, Stepaniants SB, Mao M et al. (2004). Signatures of environmental exposures using peripheral leukocyte gene expression: tobacco smoke. Cancer Epidemiol Biomarkers Prev, 13:445–453. PMID:15006922

39. van Leeuwen DM, Gottschalk RW, van Herwijnen MH et al. (2005). Differential gene expression in human peripheral blood mononuclear cells induced by cigarette smoke and its constituents. Toxicol Sci, 86:200–210. doi:10.1093/toxsci/kfi168 PMID:15829617

40. van Leeuwen DM, van Herwijnen MH, Pedersen M et al. (2006). Genome-wide differential gene expression in children exposed to air pollution in the Czech Republic. Mutat Res, 600:12–22. PMID:16814814

41. van Leeuwen DM, Pedersen M, Hendriksen PJ et al. (2008). Genomic analysis suggests higher susceptibility of children to air pollution. Carcinogenesis, 29:977–983. doi:10.1093/carcin/bgn065 PMID:18332047

42. Perera F, Tang WY, Herbstman J et al. (2009). Relation of DNA methylation of 5'-CpG island of ACSL3 to transplacental exposure to airborne polycyclic aromatic hydrocarbons and childhood asthma. PLoS One, 4:e4488.doi:10.1371/journal.pone.0004488 PMID: 19221603

43. Sridhar S, Schembri F, Zeskind J et al. (2008). Smoking-induced gene expression changes in the bronchial airway are reflected in nasal and buccal epithelium. BMC Genomics, 9:259.doi:10.1186/1471-2164-9-259 PMID:18513428

44. Schembri F, Sridhar S, Perdomo C et al. (2009). MicroRNAs as modulators of smoking-induced gene expression changes in human airway epithelium. Proc Natl Acad Sci USA, 106:2319–2324.doi:10.1073/pnas.0806383106 PMID:19168627

45. Calin GA, Croce CM (2006). MicroRNA signatures in human cancers. Nat Rev Cancer, 6:857–866.doi:10.1038/nrc1997 PMID:17060945

46. Aebersold R (2009). A stress test for mass spectrometry-based proteomics. Nat Methods, 6:423–430 doi:10.1038/nmeth.f.255. PMID:19448641

47. Bictash M, Ebbels TM, Chan Q et al. (2010). Opening up the "Black Box": metabolic phenotyping and metabolome-wide association studies in epidemiology. J Clin Epidemiol, 63:970–979.doi:10.1016/j.jclinepi.2009.10.001 PMID:20056386

48. Want EJ, Wilson ID, Gika H et al. (2010). Global metabolic profiling procedures for urine using UPLC-MS. Nat Protoc, 5:1005–1018. doi:10.1038/nprot.2010.50 PMID:20448546

49. Holmes E, Loo RL, Stamler J et al. (2008). Human metabolic phenotype diversity and its association with diet and blood pressure. Nature, 453:396–400.doi:10.1038/nature06882 PMID:18425110

50. Guyton KZ, Kyle AD, Aubrecht J et al. (2009). Improving prediction of chemical carcinogenicity by considering multiple mechanisms and applying toxicogenomic approaches. Mutat Res, 681:230–240. doi:10.1016/j.mrrev.2008.10.001 PMID:19010444

51. Skibola CF, Bracci PM, Halperin E et al. (2009). Genetic variants at 6p21.33 are associated with susceptibility to follicular lymphoma. Nat Genet, 41:873–875.doi:10.1038/ng.419 PMID:19620980

52. Prentice RL, Paczesny S, Aragaki A et al. (2010). Novel proteins associated with risk for coronary heart disease or stroke among postmenopausal women identified by in-depth plasma proteome profiling. Genome Med, 2:48.doi:10.1186/gm169 PMID:20667078

Authors

Christian C. Abnet
National Cancer Institute
Bethesda, MD
abnetc@mail.nih.gov

Jiyoung Ahn
NYU School of Medicine
New York, NY
jiyoung.ahn@nyumc.org

Christopher I. Amos
MD Anderson Cancer Center
Houston, TX
camos@mdanderson.org

Silvia Balbo
University of Minnesota
Minneapolis, MN
balbo006@umn.edu

Paolo Boffetta
International Prevention Research
Institute Lyon, France and
Mount Sinai School of Medicine
New York, NY
paolo.boffetta@i-pri.org

Jesus Gonzalez Bosquet
H. Lee Moffitt Cancer Center &
Research Institute
Tampa, FL
jesus.gonzalezbosquet@moffitt.org

Pierre R. Bushel
National Institute of Environmental
Health Sciences
Research Triangle Park, NC
bushel@niehs.nih.gov

Neil Caporaso
National Cancer Institute
Bethesda, MD
caporasn@mail.nih.gov

Stephen Chanock
National Cancer Institute
Bethesda, MD
chanocks@mail.nih.gov

Nilanjan Chatterjee
National Cancer Institute
Bethesda, MD
chattern@mail.nih.gov

Harvey Checkoway
University of Washington
Seattle, WA
checko@u.washington.edu

Xiao-He Chen
University of Oxford
Oxford, UK
xiao-he.chen@ndcls.ox.ac.uk

Francois Coutlee
Centre Hospitalier de l'Universite
de Montreal
Montreal, Canada
francois.coutlee@ssss.gouv.qc.ca

David G. Cox
Harvard School of Public Health
Boston, MA
dcox@hsph.harvard.edu

Amanda Cross
National Cancer Institute
Bethesda, MD
crossa@mail.nih.gov

John Danesh
University of Cambridge
Cambridge, UK
john.danesh@phpc.cam.ac.uk

Frank de Vocht
University of Manchester
Manchester, UK
frank.devocht@manchester.ac.uk

Deborah J. del Junco
University of Texas Health Science
Center at Houston
Houston, TX
deborah.j.deljunco@uth.tmc.edu

Emanuele Di Angelantonio
University of Cambridge
Cambridge, UK
ed303@medschl.cam.ac.uk

Susan Edwards
Columbia University
New York, NY
se2171@columbia.edu

Betsy Foxman
University of Michigan School of
Public Health
Ann Arbor, Michigan
bfoxman@umich.edu

Eduardo L. Franco
McGill University
Montreal, Canada
eduardo.franco@mcgill.ca

Montserrat Garcia-Closas
Institute of Cancer Research
Surrey, UK
montse.garciaclosas@icr.ac.uk

Sherry F. Grissom
National Institute of Environmental
Health Sciences
Research Triangle Park, NC
sgrissom@illumina.com

John Groopman
Johns Hopkins Bloomberg School
of Public Health
Baltimore, MD
jgroopma@jhsph.edu

Susan E. Hankinson
Brigham and Women's Hospital and
Harvard Medical School
Boston, MA and
Harvard School of Public Health
Boston, MA
sue.hankinson@channing.harvard.edu

Pierre Hainaut
International Agency for Research
on Cancer
Lyon, France
hainaut@iarc.fr

Erin Haynes
University of Cincinnati College of
Medicine
Cincinnati, OH
erin.haynes@uc.edu

Marianne K. Henderson
National Cancer Institute
Bethesda, MD
hendersm@mail.nih.gov

Shuwen Huang
National Genetics Research
Laboratory (Wessex)
Salisbury, UK
shuwen@soton.ac.uk

David J. Hunter
Harvard School of Public Health
Boston, MA
dhunter@hsph.harvard.edu

John P.A. Ioannidis
University of Ioannina, School of
Medicine
Ioannina, Greece
jioannid@cc.uoi.gr

Kevin Jacobs
National Cancer Institute
Bethesda, MD
jacobske@mail.nih.gov

Samir N. Kelada
National Human Genome Research
Institute
Bethesda, MD
keladas@mail.nih.gov

David Kerr
University of Oxford
Oxford, UK
david.kerr@clinpharm.ox.ac.uk

Muin J. Khoury
Centers for Disease Control and
Prevention
Atlanta, GA
muk1@cdc.gov

Peter Kraft
Harvard School of Public Health
Boston, MA
pkraft@hsph.harvard.edu

Qing Lan
National Cancer Institute
Bethesda, MD
qingl@mail.nih.gov

Christoph Lange
Harvard School of Public Health
Boston, MA
clange@hsph.harvard.edu

Robert E. London
National Institute of Environmental
Health Sciences
Research Triangle Park, NC
london@niehs.nih.gov

Jessica I. Lundin
University of Washington
Seattle, WA
jlundin2@u.washington.edu

B. Alex Merrick
National Institute for Environmental
Health Sciences
Research Triangle Park, NC
merrick@niehs.nih.gov

Salma Musaad
University of Cincinnati
Cincinnati, OH
salma.musaad@gmail.com

Evangelia E. Ntzani
University of Ioannina, School of
Medicine
Ioannina, Greece
entzani@hotmail.com

Richard S. Paules
National Institute of Environmental
Health Sciences
Research Triangle Park, NC
paules@niehs.nih.gov

Frederica P. Perera
Columbia University
New York, NY
fpp1@columbia.edu

Andrew Povey
University of Manchester
Manchester, UK
andy.povey@manchester.ac.uk

Nathaniel Rothman
National Cancer Institute
Bethesda, MD
rothmann@exchange.nih.gov

Paul A. Schulte
National Institute for Occupational
Safety and Health
Cincinnati, OH
pas4@cdc.gov

Joe Shuga
University of California
Berkeley, CA
joe.shuga@gmail.com

Rashmi Sinha
National Cancer Institute
Bethesda, MD
sinhar@mail.nih.gov

Andrea Smith
National Institute for Occupational
Safety and Health,
Cincinnati, OH
andrea.smith@mcmaster.ca

Martyn T. Smith
University of California
Berkeley, CA
martynts@berkeley.edu

Anne Sweeney
Texas A&M Health Science Center
College Station, TX
sweeney@srph.tamhsc.edu

Alexander Thompson
University of Cambridge
Cambridge, UK
alex.thompson@roche.com

Shelley S. Tworoger
Brigham and Women's Hospital and
Harvard Medical School
Boston, MA and
Harvard School of Public Health
Boston, MA
nhsst@channing.harvard.edu

Jelle Vlaanderen
Institute for Risk Assessment
Sciences
The Netherlands
j.j.Vlaanderen@uu.nl

Jimmie B. Vaught
National Cancer Institute
Bethesda, MD
vaughtj@mail.nih.gov

Roel Vermeulen
Institute for Risk Assessment
Sciences
The Netherlands
r.c.h.vermeulen@uu.nl

Paolo Vineis
Imperial College
London, UK
p.vineis@imperial.ac.uk

Robert F. Vogt
Centers for Disease Control and
Prevention
Atlanta, GA
rfv1@cdc.gov

Jia-Sheng Wang
University of Georgia
Athens, GE
jswang@uga.edu

Frances Wensley
University of Cambridge
Cambridge, UK
frances.wensley@gmail.com

Ainsley Weston
National Institute for Occupational
Safety and Health
Morgantown, WV
agw8@cdc.gov

Emily White
University of Washington and
Fred Hutchinson Cancer Research
Center
Seattle, WA
ewhite@fhcrc.org

Index

1000 genomes project
    112–116
12(S)-hydroxyeicosaintetraenoic acid (12(S)-HETE)
    132
2,3,7,9 tetrachlorodibenzo-p-dioxin (TCDD)
    466–467 (See also TCDD and Dioxin)
24-hour recall
    190
32P-post-labeling assay
    49–57
8-hydroxy-2'-deoxyguanosine (8-OHdG)
    354, 393–394, 464
8-oxodG
    194
Accelerator mass spectrometry (AMS)
    49, 53
Acquired (adaptive) immunity
    216–233
Additive interaction
    294–298
Adductomics
    495
Adipokines
    193, 199–201
Adiponectin
    200–205, 369, 442–447
Adipose tissue
    199, 201, 205, 442–448
Aflatoxin (AFB1)
    52, 55–56, 68, 192, 340
Airways diseases
    395–399, 481
Albumin adducts
    55–56
Aliquoting
    30–32

Allele
    -based analyses, 289–290
    -specific PCR, 78 (See also Polymerase chain reaction)
Alzheimer's disease (AD)
    407–419
Amyotrophic lateral sclerosis (ALS)
    410–411
Analysis of GWAS data
    19, 290–293, 351, 367
Analytical variability
    31, 167
Androgens
    193, 201
Aneuploidy
    68, 82, 104, 319, 489, 464
Antibody arrays
    128–129
Antigen detection
    175–176
ApoE genotypes with LDL-C
    377–378
Arbitrarily-primed PCR (AP-PCR)
    181
Aromatic amines
    53–55, 171–172
Arrayed primer extension (APEX)
    75, 77, 80–81
Arsenic and urothelial cancer
    342
Asbestos-associated fibrosis (asbestosis)
    388
Asia cohort consortium
    250
Association studies
    106, 116, 261–280
    using families (See Family-based association studies)
Asthma
    220, 224, 233, 272–279, 395–397, 475–488

Atherosclerosis
    443–444
Atomic absorption (AA)
    46
Attenuation bias
    164
Autoantibodies
    225–226, 230–233
Autoimmune disorders
    223
Automated systems
    32–34
Bacterial plasmid analyses
    181
Barker hypothesis
    462 (See also Developmental origins)
Banking specimens
    12
Bayesian approach
    265, 291, 326
B-cells
    217–218
    malignancies, 226
Bead-array technology
    109
BeadArray platforms
    123
BEAMing
    80
Benzene
    83, 84, 125, 129, 170, 171, 172, 243, 245
    and leukaemia, 341–342
Berylliosis
    390
Bias
    143–161
    in prevalence surveys, 186
    in the magnitude of the association, 186
Biobanks
    90, 498

Biochemical biomarkers
  219
Bioinformatics
  37, 115, 121–141, 498–499
Biological plausibility
  327–328
Biomarker
  characteristics, 165
  validation, 350–351
  validity, 167
  of effect, 44, 304
  of exposure, 44, 190–196, 304
  of susceptibility, 44, 304
Biomarkers
  analysis of genetic variation, 99–120
  analytical methods, 43–62
  assessment of genetic damage, 63–98
  biosample management, 23–42
  cellular –, 216
  Consortium, 317
  measurement errors, 143–162
  of absolute intake, 190–196
  of correlated intake, 190–196
  of female and male puberty, 453–457
  of female reproductive function, 456–459
  of fetal and infant development, 459–460
  of intermediate endpoints, 193
  of internal dose, 43–57
  of male reproductive function, 462–465
  of obesity and cancer, 446–447
  platforms for analysis, 121–142
Biorepository
  24–39, 113–114
Biosafety
  38
Biospecimen
  collection, processing, and storage, 34, 40, 190, 241–242, 247–253
  resource, 23
Bisphenol A
  462
Blood
  26, 27–33, 125, 132, 166, 170, 202, 203–204, 243, 306
  pressure, 363, 367–379
Body mass index (BMI)
  201, 276, 342, 376

Breast cancer
  51–53, 126, 132, 133, 146, 156, 193, 194, 200–201, 263, 308, 314–319, 447, 496
Bronchiolitis obliterans syndrome (BOS)
  387, 388, 397–399
C-reactive protein (CRP)
  374, 375
Canalization
  375 (*See also* Developmental adaptation)
Cancer
  337–362
Candidate
  plasma biomarkers in CHD, 367
  susceptibility genes and investigation, 341, 410, 461
Capillary electrophoresis
  49, 88, 110, 123, 130
Carcinoembryonic antigen (CEA)
  225, 308, 389
Carcinogen–DNA adducts
  49–57, 481
Case–cohort design
  248–249, 282
Case–control design
  250–251, 282
Case-only and other study designs
  255–256, 285, 296, 461
Causality
  184–185, 338–339, 351
CCR5 receptor
  393, 438
CD3
  217
CD4
  217–221
CD8
  218, 221
CD19
  218
CD36
  445–446
CD45
  218, 228, 229
Cell culture systems
  182
Cellular biomarkers (*See* Biomarkers)

Centers for Disease Control and Prevention (US)
  and blood spot cards, 30
  and integration of genetic variation in population-based research, 13
  information management at the –, 37
  NHANES, 165
  PulseNet, 424
Central melanocortin system
  442
Cerebrovascular disease
  443
Cervical cancer
  183, 184, 186, 310, 432
CFSUM1 and 2
  133
Chemotherapy
  52, 85, 169, 308, 310–315, 435
Childhood cancer
  347, 454, 461, 475, 481
Children
  478–492
Chlamydophila pneumoniae
  434
Cholecystokinin
  442
Cholesterol
  102, 155, 441 (*See also* High-density lipoprotein cholesterol, Low-density lipoprotein cholesterol, Total cholesterol)
  and colon cancer, 154
  and CHD mortality, 370, 371
  as a marker of intermediate endpoints, 193
  dietary –, 378
  LDL –, 244, 363, 366
  HDL –, 342
  total –, 375
Chromatin immunoprecipitation (ChIP)
  113, 126, 495
Chromosome aberrations
  73, 81–85, 353, 355
Chronic
  beryllium disease (CBD), 387–391
  lymphocytic leukemia (CLL), 226–227
  obstructive pulmonary disease (COPD), 388, 397–399
Circadian rhythm
  206
Classification and regression tree (CART)
  297

Clinical
　endpoint, 308, 311
　medicine, 303–319
　treatment, 421–434
　trials, 310–312
Clonal expansion
　217, 218, 222, 223, 227
Cluster of differentiation (CD)
number
　216, 389
Clustering
　Bayesian –, 291
　familial –, 364
　genotype –, 285, 286, 287
　behaviour, 131
　for classification, 135
Coal workers' pneumoconiosis (CWP)
　387, 391–393, 394, 399
Coefficient of variation
　158, 167, 209, 462
Cohort
　consortium, 248
　　Asia –, 250
　　NCI –, 351
　design, 246, 248, 256, 263
　studies, 245
Collaboration
　330
Comet assay
　354, 394, 464
Commensals and pathogens
　426, 434–435
Common rule
　13, 14
Communicating test and study results
　10, 14, 15–17, 486–487
Comparative genome hybridization (CGH)
　123, 306, 319, 348
Complement fixation
　219–220
Complex phenotypes
　272
Concentration biomarkers
　190, 196
Confidentiality of data
　10, 14, 15, 19, 116, 264, 467, 483, 486
Confounding
　11, 16, 184, 225, 253, 268, 272, 273, 291, 326, 327, 329, 345, 349, 350, 351, 370, 373, 375, 443, 461
Conicity index
　448
Consortia
　323, 330

Copy number variations (CNVs)
　103–104
Coronary heart disease (CHD)
　363–379
Correlated error
　157–159
C-reactive protein
　193, 197, 221, 369, 372, 374, 375, 448
Cross-sectional studies
　169, 242–245, 351, 447, 495, 496
CRP (See C-reactive protein)
Cryopreservation
　31, 33, 249, 251, 253
Cultivation
　175, 176, 177, 179
　assays, 181–182
Cysteine
　131
Cytokine IL17
　218
Cytokines
　220, 443, 481
Data
　-base for SNPs, 102
　mining methods, 297–298
　normalisation, 134
　standardization, 330
dbSNP (See Database for SNPs)
Dementia
　408, 412
Dermal exposures
　477
DES
　460
Design issues
　146, 346, 351
Detection of mutations
　74–78
Developmental adaptation
　375 (See also Canalization)
　defect, 123, 476
　disorders, 232, 454, 460, 461, 475, 493
　exposures, 477
　origins, 462 (See also Barker hypothesis)
　studies, 25
　toxicity, 464, 477, 482, 483
Diabetes
　200, 201, 247, 319 (See also Type 1 –, Type 2–, TEDDY)
Diagnostics
　130, 303, 305, 308, 438
Diagnostics prenatal
　82

Diesel exhaust
　172, 479
Dietary
　antioxidants, 194, 408
　biomarkers, 195, 196
　factors, 107, 189, 378
Difference gel electrophoresis (DIGE)
　127
Differential
　measurement error, 144, 145, 146, 147, 150, 154, 155, 159
　misclassification, 150, 169, 246, 251, 252, 285, 286, 328
Diffusion-sink device
　205
Dimension reduction
　134, 297
Dioxins
　165, 218, 243, 252, 462, 466-467 (See also TCDD)
Disease course
　45, 303, 350, 421, 434, 435
Disease susceptibility locus (DSL)
　264, 271, 283
Disorders of the immune system
　223
DNA
　adducts in excretion, 52
　and protein adducts, 48, 56, 57, 192
　damage, 52, 63, 65, 66–71, 86, 194–196, 339, 344, 345, 347, 350, 353, 354, 464, 482
　diagnostics, 126
　extraction, 31–32, 75, 76, 178, 282, 286
　microarrays, 124, 126, 128, 129
　repair, 341, 343, 344, 345, 353, 355, 477, 481, 494
　sequencing, 74, 89, 111–112, 121–124
Dose-response models
　352
Doubly labelled water
　195
Dried blood spot (DBS)
　175, 176, 227, 459
Drug
　development, 130, 310, 311–312, 315, 317, 319
　discovery, 310–311
Duplicates
　105, 108, 110, 114, 285, 287
Elston-Stewart algorithm
　265–266

Endocannabinoids
442
Endogenous
DNA damage, 52 (*See also* DNA damage)
hormones, 27, 33, 190, 200 (*See also* Hormones)
Environmental
163–174, 337
exposures, 172, 230, 254, 343, 345, 456, 460, 475, 483, 494, 497
exposure markers, 170
neurotoxicants, 410
tobacco smoke (ETS), 54, 170–171, 347, 352, 477, 481
Enzyme-linked immunosorbent assay (ELISA)
49, 53, 55, 57, 148, 149, 183, 390
EPIC-Heart
378
Epidemiology of infectious diseases
422, 426, 427
Epigenetic mechanisms
462
Epigenetics
348–353
Estrogens
53, 200, 201, 205, 412, 457
Ethical issues
9–22, 10, 460, 483, 488
Ethnicity
291, 293, 346, 365, 368, 441, 457
ETS exposure (*See* Environmental tobacco smoke)
European Prospective Study Into Cancer and Nutrition (EPIC)
191, 339, 342, 345, 352, 378
Evaluation of interactions
295, 296
Exogenous hormones
199, 200
Exposome
172, 494
Exposure
assessment, 11, 163, 165, 169, 172, 241, 246, 251–252, 256, 285, 294, 345, 392, 413, 461, 466, 494, 499
assessment and misclassification, 246
biomarkers, 22, 165, 190, 196, 243, 252, 305, 411, 415, 465
in utero, 230, 347, 354, 460, 461, 462, 465, 475, 476, 478, 481, 488
variability, 167–168
Expression profiling
113, 124, 125, 126, 331, 497
F2-isoprostane
444, 445
Family-based
association studies, 268, 273, 274, 276
association tests, 269, 277
designs, 261–280, 327
FBAT approach
269–277
FBAT-GEE statistic
273
Federal Drug Administration (US FDA)
310, 313, 317, 318
biomarkers approved by the –, 308
list of valid genomic biomarkers, 309
Female libido
456
Fetal and infant development
459–460
Fetal lung maturation
459
FEV1
273, 398, 479
Fibrinogen studies collaboration
371, 372, 373, 374
Field validation
350
Finger prick
203, 204
Fingerprint
carcinogen –, 67, 72, 340
gene –, 114, 222, 285, 340, 422, 423
Fixed effects models
326 (*See also* random effects)
FlexTree method
298
Flow cytometry
80, 89, 216, 217, 218, 227, 389, 464
Fluorescence in situ hybridization (FISH)
66, 82, 83, 124, 315, 341, 464
Fold change
125, 127, 134
Follicle stimulating hormone (FSH)
455
Follow-up of cases to determine clinical outcomes
250, 251, 254
Food frequency questionnaires (FFQ)
189, 190, 191, 194, 195–196
Foodborne outbreak
423
Forward phase arrays
128
Fourier transform infrared (FTIR)
130
Framingham
276, 367, 375, 444
Fungal carcinogens
68, 192, 340
Future use of specimens
12, 13, 31, 37, 460
Gail model
496
Gas chromatography (GC)
48, 130, 133, 466, 467
Gender
346–347
Gene
expression profiling, 113, 124, 125, 126, 331, 497
fingerprint (*See* Fingerprint)
methylation, 66
-environment interactions, 277, 281–302, 342, 397, 410, 460
-gene interaction, 255, 281, 285, 299, 301, 410
Genetic
algorithm (GA), 135, 105, 265, 266
and lifestyle factors, 377–378
biomarkers, 222–223, 231, 306, 317, 332, 391
lineages, 427, 429, 430, 432
linkage analysis, 261, 264–266, 388, 395, 410
polymorphisms, 102, 112, 281, 282, 344, 399, 412, 413, 415, 448
susceptibility, 230, 252, 255, 281, 282, 287, 303, 343, 352, 355, 443, 445-447, 467, 478
susceptibility loci, 264, 271, 281, 282, 283, 287, 293, 336, 359
testing, 256, 307, 318, 319, 391
variation, 99–120, 194, 199, 209, 252, 281–302, 324, 364, 494, 495

Genome
 99–120, 122, 123, 136, 137, 172, 193, 196, 248, 265, 282, 283, 319, 332, 346, 348, 430, 431, 494, 496, 497
  scan meta-analysis (GSMA), 332
  -wide association studies (GWAS), 19–20, 99–120, 194, 264–280, 324–330, 338, 346, 348, 367, 438, 448, 454, 494 (See also Meta-analyses)
  -wide association studies of BM, 448
Genomic
  amplification, 124, 175, 178, 180
  approaches, 365
  imprinting, 306, 465
Genomics
 4, 28, 88, 113, 121–142, 317, 319, 324, 333, 355, 495, 496
Genotoxic compounds
 50, 52, 74, 86, 195, 244, 317, 339, 344, 464, 477, 478, 488
Genotype
  analysis, 108–109
  -based analyses, 288–290
Genotypic risk ratio
 262
Genotyping
 88, 102–123, 180–182, 252, 265–268, 281–287, 323, 324, 327, 329, 330, 343, 345, 367, 411, 460
Gestational neuroimmunopathology (GENIP) hypothesis
 232–233
Ghrelin
 442
Glutathione
 68, 131, 343, 394, 478, 480
  peroxidase 1, 393–394, 444, 445
Glycemic status
 376
Half-lives
 165–168, 192, 455, 461
Haplotype
 analyses, 269–272, 281, 290, 313
 blocks, 103, 268
 mapping, 122–123
 phase, 102, 271
 tagging SNPs (htSNPs), 268

Haplotypes
 103, 122, 230, 231, 265, 266, 267–271, 290
Hardy-Weinberg equilibrium
 105, 286
HDL-C (See High-density lipoprotein cholesterol) (See also Cholesterol, Low-density lipoprotein cholesterol, Total cholesterol)
Health Insurance Portability and Accountability Act (HIPAA)
 10, 13, 486
Helper T-cells (See T-cells)
Haemoglobin adducts
 54, 165, 170, 172, 173, 242, 347, 355
Henle-Koch's postulates
 183, 422
Hepatitis A
 182, 423, 424, 435
Hepatitis B
 56, 178, 182, 340, 425, 433, 434, 435
Hepatitis C
 178, 182, 425, 433, 434
Herceptin
 314
Heritability
 261, 282
Herpes virus 8
 425, 432, 433, 434
Heterocyclic amines (HCAs)
 191, 192
Heteroduplex mobility assay (HMA)
 181
Heterogeneity, 325–327
 disease –, 293, 414
 genetic –, 196
 etiologic –, 343, 255
Heterogeneity-based genome search meta-analysis (HEGESMA)
 332
Hierarchical-Bayesian methods
 293, 298
High performance liquid chromatography (HPLC) (See Liquid chromatography)
High-density
 lipoprotein cholesterol, 372, 374, 375
 SNP detection, 109–110, 267
High-order interactions
 297–298
High-throughput sequencing
 71, 80–81, 124, 126, 436, 496, 112, 121
HIPAA (See Health Insurance)
Historical cohort approach
 263

HLA-DPβ1
 387, 391, 399
Hormonal variation
 206
Hormone assays
 204, 208
Hormones
 199–213
Hospital-based studies
 250, 251, 253, 254, 282, 287
Host defence system
 176, 182, 216, 219, 220, 221, 224
HPV serotypes 16 and 18
 435
Huber-White variance correction procedure
 263
HuGENet (See Human genome epidemiology network)
Human
 chorionic gonadotropin (hCG), 457, 458, 459
 epidermal growth factor receptor 2 (HER2), 305, 312, 314, 315, 317
 genome epidemiology network (HuGENet), 324, 325, 327, 330
 immunodeficiency virus (HIV), 215, 429, 430, 432, 435, 438, 480
 leukocyte antigens (HLA) (See also HLA-Pβ1), 221, 222, 230, 231, 438
 microbiome project (HMP), 113
 papillomavirus (HPV), 39, 178, 183, 184, 186, 187, 426, 432, 435
I2
 325, 326, 329
Identifiability of biological specimens
 14
IgA
 220, 221
IgD
 220
IgE
 219, 220–224, 479, 481
IgG
 182, 185, 220, 221, 232, 233, 306, 369
IgM
 182, 185, 220, 221, 306

Immune
 biomarkers in animal models, 224
 biomarkers of neurodevelopmental disorders, 232
 deficiency disorders, 223–229
 proliferative disorders, 224
 reactive disorders, 223–224
 system, 215–240
Immunoaffinity chromatography (IAC)
 48
Immunoassays
 48, 49, 53, 57, 177, 178, 183, 208, 218, 345
Immunodiffusion
 183
Immunogenic exposures
 223–224
Immunoglobulin
 219–226
Immunohistochemistry (IHC) test
 49, 57, 314, 315
Immunopathology
 216, 223, 232
In utero
 230, 233, 347, 460, 461, 462, 465, 475–488
 and childhood periods, 476, 477, 488
Indirect detection (*See* Serological methods)
Individual variability
 167
Inductively coupled plasma-optical emission spectrometer (ICP-OES)
 46
Infectious
 causes of cancer, 175–188, 421, 433, 434
 diseases, 175–188, 421–440
Infertility/early pregnancy loss
 200, 454, 456, 457–488
Inflammation
 216, 441–445
Inflammatory biomarkers
 44, 193
Informatics
 34–37
 system security, 35–37
Information
 bias, 144, 326–327
 management (*See also* Laboratory information management), 23–42
Informed consent
 12–13, 486

Innate immunity
 216, 221
Institutional review board (IRB)
 10–20, 486
Insulin
 resistance, 193, 200–201, 444–446
 -like growth factor I (IGF-I), 201, 209–210, 342
 -like growth factor binding protein 3 (IGFBP-3), 201, 210, 342
Intake assessment methods
 190
Interleukins
 218–220, 374, 389, 393
Intermediate
 biologic effects, 243–245
 endpoints, 242–245
International HapMap project
 102, 104, 108, 110, 111, 116, 122, 283, 291
International Agency for Research on Cancer (IARC)
 biological resource centre guidelines, 27, 28, 29, 31
 classification of carcinogens, 71
 p53 database, 70, 73
International Organization for Standardization (ISO)
 38
International Society for Biological and environmental repositories (ISBER)
 best practices, 27, 28, 29, 31, 34, 37, 38
Interpretation of biomarker data
 16
Interpreting the results
 15, 327
Interstitial lung diseases
 388, 391, 399
Intervention strategies
 353–354, 434, 436, 479, 483
Intraclass correlation coefficient
 143, 155–157, 158, 159, 167, 209, 245
Intramethod reliability
 155–156
Intrauterine growth retardation
 459–460
Inventory control
 36
Irinotecan
 312–314
Iron
 191, 369, 389, 392

Kadoorie prospective study in China
 378
Kaposi's sarcoma (KS)
 425, 432, 433
Kappa in reliability studies
 143, 158
Kin-cohort approach
 263
K-nearest neighbors (KNN)
 135
Laboratory information management system (LIMS)
 110, 115
Lab-on-a-chip
 88–89, 96
Lander-Green-Kruglyak (LGK) method
 265–266
Leptin
 203, 342, 369, 442, 443, 444, 445, 447, 455
Libido
 454, 456, 462, 463
Lifestyle exposures
 373–378
Linkage
 analyses, 264–268, 276, 388, 398
 disequilibrium (LD), 102–103, 106, 265, 267, 268, 272, 283, 327, 376
Liquid chromatography (LC)
 48, 75, 77, 79, 127, 130, 132
 High performance – (HPLC), 48, 49, 52, 53, 54, 55, 57, 79, 132, 170
 Ultra-performance – (UPLC), 130, 498
Liver cancer
 45, 47, 52, 56, 79, 192, 193, 226, 340, 341, 344, 434
LOD score
 264–266
Logistical issues in children's studies
 483
Longitudinal investigation of fertility and the environment (LIFE) study
 458
Low-density lipoprotein (LDL-C) cholesterol
 244, 363, 364, 366, 372, 375, 377, 408 (*See also* Cholesterol, High-density lipoprotein cholesterol, Total cholesterol)

Low-level exposures
    352, 482, 499
Lp-PLA2
    371, 372, 376
    studies collaboration, 371
Lung cancer
    56, 67, 86, 132, 171, 172, 226,
        338–340, 347, 348, 349,
        352, 353, 388, 390, 478,
        481
Luteinizing hormone (LH)
    455
Lyme disease
    434
Lymphocytes
    86, 217, 244, 253, 391, 394,
        480
    subsets, 217
Lymphokines
    220, 221
Major histocompatibility complex
(MHC)
    (transplantation) antigens, 221
    genes, 222
Malaria
    422, 432
Male
    hormones, 463
    libido, 462–463
Manganese superoxide dismutase
(MnSOD)
    193
Mapping by admixture of linkage
disequilibrium (MALD) (See also
Linkage disequilibrium)
    106
Mass spectrometry (MS)
    48, 79, 124, 127–128, 495, 497
    (See also Tandem – (MS/
        MS), 79, 127
Mast cells
    219
Matrix-assisted laser desorption/
ionization time-of-flight (MALDI-
TOF)
    87, 109, 128 (See also Surface-
        enhanced laser)
Measurement error
    143–162, 184
Measuring DNA adducts
    49–57
Mechanisms of mutagenesis
    64–72
Membrane EIAs
    177, 183
Mendelian randomization
    373–375, 461
Menstrual cycle
    205–206, 454, 456–457

Mesothelioma
    388–390
Meta-analyses
    323
    of genome-wide association
        studies, 330–332 (See
        also Genome-wide
        association studies)
    of individual participant data
        (MIPD), 329–330
Metabolome
    122–123, 130, 497
Metabolomics
    121, 129–133, 337, 497
Methicillin resistant Staphylococcus
aureus (MRSA)
    427, 428, 435
Methyl mercury
    46, 460, 465
Methylation
    86–88, 126, 348–349
Methylenetetrahydrofolate
reductase (MTHFR) gene
    88, 343–344
Mexico City Prospective Study
    378
Microarray
    platforms, 104, 125, 129, 332,
        431
    -based gene expression
        profiling, 331
Microarrays
    83, 124, 125, 126, 128, 180,
        219, 332–333, 350, 367,
        431, 496
Microbial genome
    180, 181
    Program, 431
Microbiota
    432, 436, 438
Microfabricated genetic analysis
systems
    88
Micronuclei
    85–86, 351, 497
Microsatellite
    instability, 76, 86
    panels, 265
Microscopic examination
    82, 175–176, 463
Minor allele frequency (MAF)
    101, 110, 283, 364, 377, 413
Misclassification (See also
Measurement error)
    147, 148, 150, 186, 246, 251
    bias, 144
    due to random within-person
        variation, 246

Mitochondrial genome
    63, 73
Mixtures of carcinogen exposure
    49, 190, 410, 456, 461, 499
Mobile genetic elements
    424, 426, 427, 428–429, 436
Model
    of parallel tests, 156
    -dependent methods, 266, 267
    -free methods, 266, 273
Molecular
    cancer epidemiology, 337–338
    epidemiology, 1, 1–7, 9–22,
        323–324, 242
    fingerprints, 421–423
Monoclonal B-cell lymphocytosis
(MBL)
    226–227
Monte-Carlo Markov chain (MCMC)
algorithms
    266
Müllerian inhibiting substance (MIS)
    455
Multidimensional protein
identification technology (MuDPIT)
    124, 127
Multi-factorial dimension reduction
(MDR)
    297–298
Multilocus haplotype
    271
Multimarker
    FBATs, 272
    tagging, 283
Multiple food records
    189
Multiplicative interaction
    294–298, 467
Multistage designs
    263, 274, 282, 284, 293, 352
Multistep carcinogenesis
    72–74
Mutation
    patterns, 70–71
    -enriched PCR, 77–78 (See
        also Polymerase chain
        reaction)
Mutations
    63–98
    in mitochondrial DNA, 73
Mycotoxins
    192, 410
Myocardial infarction
    129, 364–367, 444, 445
Nanotechnology
    388, 495, 496

National Cancer Institute (US)
	iv, 23, 27, 37, 194, 299, 308
	informatics at the –, 37
	Cancer genetic markers of susceptibility (CGEMS), 194
	FDG-PET, 308
National Centers for Children's Environmental Health
	483, 488
National children's study
	462, 488
National health and nutrition examination Survey (NHANES)
	47, 165, 166, 203, 444
National Institute of Environmental Health Sciences (NIEHS)
	73, 137, 455, 483, 488
Natural killer (NK) cell
	217
NBS (See Newborn bloodspot screening)
NCI cohort consortium
	248, 351
Nested case-control study
	56, 148, 169, 251, 282, 339, 437
Networks
	pathways and –, 281, 298
	consortia and –, 317, 324, 329–330, 355
	transmission –, 426, 430
Neurobehavioural disorders
	233, 482–487
Neurodegenerative disease
	407–419
Neuromental disorders (NMDs)
	232, 233
New infectious agents
	176, 182, 421, 422, 428, 429, 432–433
Newborn bloodspot screening (NBS)
	227
Newborn screening
	233, 307
	for immune disorders, 216, 227
	for Type 1 diabetes, 231
NHANES (See National Health and Nutrition)
NMDs (See Neuromental disorders)
Nondifferential misclassification (See also Misclassification)
	148–154, 246, 251
Non-lymphoid cells
	218
Non-specific markers
	171, 216, 220, 224

Nuclear magnetic resonance
	49, 57, 124, 130, 132, 133, 349, 498
Nucleic acid detection
	177–185
Nurses health study
	203, 415
Nutrient intake
	195
Nutritional
	factors, 342, 346–347, 354, 460, 478
	status biomarkers, 189, 190
Obesity
	193, 200, 441–452
Occupational carcinogen exposure
	51–52
"Omics"
	122, 134–136, 319, 496–498
"Omic" technologies
	348–350, 497
Open reading frames (ORFs)
	431
Opportunistic infection
	435
ORFs (See Open reading frames)
Origin of an epidemic
	426–427
Outbreak investigations
	422–426
Oxidative
	damage 353, 387, 393, 398
	stress 194, 349, 388, 393, 394, 409, 410, 443–444, 464
	– biomarker 8-oxoguanosine, 131
p53
	aflatoxin and –, 340
	amplichip – microarray, 47
	and apoptosis, 394
	and asbestos exposure, 389–390
	autoantibodies, 389–390
	mutations, 71, 76, 305, 339, 341, 351
	smoking and –, 51
PAH-DNA adducts
	50, 51, 56, 195, 339, 340, 346, 347, 353, 354, 478, 480, 482
Pairs statistic
	267
Pairwise tagging
	283
Parent compounds and metabolites
	46, 192, 481
Parkinson's disease (PD)
	409–413

PCR (See Polymerase chain reaction (PCR))
Pedigree-based association test (PBAT)
	276, 277
Penetrance
	262–264, 266, 343, 346
Persistent
	bioaccumulative toxics, 460
	organic pollutants, 165, 460
Personalized medicine
	100, 123, 126, 304, 319, 496
Pesticide exposure
	413, 480, 482, 487
Phenotypic sample mean
	272
Phthalates
	460
Phylogenetics
	181, 424, 429, 430
Physical activity
	193, 194
Physicians' health study
	193, 399
Plasmodium
	432
Pneumoconioses
	387, 388, 398
Policy
	40, 337, 356, 476, 488, 498
	public health and environmental –, 354
Polycyclic aromatic hydrocarbons (PAHs)
	67, 72, 171–172, 191, 338, 477, 480, 487
Polymerase chain reaction (PCR)
	66, 74, 432
	allele-specific –, 78
	measurement of rearrangements by –, 83–84
	multiplex–, 179
	mutation-enriched –, 77–78
	real-time–, 83, 180
	quantitative (Q-PCR) –, 229
	single-template –, 89
	sensitivity and specificity, 75–76
	-RFLP, 181, 392
	-single stranded conformation polymorphism (SSCP), 181
Polymicrobial infections
	436

Population
  genetics, 99, 100, 105–107, 292
  stratification, 105–106, 276, 291–294, 327–329
  structure, 291–293
  substructure, 106, 271, 274, 291
    –based case-control studies, 250–255
    –based studies, 250
Precautionary principle
  18
Pre-eclampsia
  459
Pregnancy outcomes
  460–461
Preventive medicine
  303, 305, 317
Principal component analyses
  105–106, 131, 291, 327
Privacy of subjects
  10, 13, 14, 15, 19, 25, 40, 246, 264
PROCAM risk calculator
  375
Progesterone
  204–205, 457–458
Prognosis and prognostics
  308
Promoter methylation
  86, 348–349
Prospective
  cohort, 241
  cohort studies, 245–254, 282, 483, 495–496
  studies collaboration, 370–373
Prostate, lung, colorectal and ovarian (PLCO) study
  193–194
Protein
  biomarker profiles, 413
  microarrays, 128
  quantitation, 127–128
Proteome
  122–126, 130, 135, 499
  human – organization, 27
Proteomics
  28, 121–142, 317, 331, 349–351, 407, 438,
  applications, 497
Protocol development
  11
Puberty onset
  453–456
Pubertal status
  455
Public health ethics
  18–19

Publication bias
  327, 355
Q statistic
  325
QRISK
  375–376
Quality
  assurance, 37, 25, 115, 167
  control, 25, 37, 90, 104, 110, 113–116, 285–287, 291, 466
    bioinformatics –, 136
    genetic data –, 285–287
    microarray –, 134
Quantitative phenotypes
  265, 272–273
Quartile-quartile plots (Q-Q plots)
  293–294
Radioimmunoassay (RIA)
  48, 129, 183, 199, 208
Random
  effects models (*See also* Fixed effects), 326
  Forest procedure, 297
  genetic drift, 107
Randomly amplified polymorphic DNA (RAPD)
  181
Rapid ascertainment of cases
  250, 252
Rapid detection of infectious agents
  434
Recovery biomarkers
  190
Recruiting participants
  12, 230, 365, 444, 461, 463
Recurrence risk
  262
  relative –, 262
Reliability
  coefficient, 155–158
  study, 144, 146–147, 155–159
Restriction endonuclease analysis
  181
Restriction fragment length polymorphism (RFLP)
  75, 77, 108, 181
Retentate chromatography-mass spectrometry (RC-MS)
  124, 128
Reticulocytes
  85
Reverse phase arrays
  128
Reynolds algorithm
  375

Risk assessment
  46, 184
  individual –, 16, 496
  and distinct risk profiles, 253
  and early-life exposures, 465
  for cancer-causing agent, 57
  for low level exposures, 352–353, 184
  of complex diseases, 116
  process, 44
Risk prediction algorithms
  294, 375–376
Rolling circle amplification (RCA)
  114
Saliva
  30–31, 190, 204–205
  /buccal cell collection, 30–31
Sample
  collection, 23–42, 205–208
    cost of –, 247
    feasability of –, 247
    invasive or complicated –, 461
    method of –, 249
    problems in –, 484, 184
    protocols of –, 349
    timing of –, 167, 173, 199, 252
  heterozygosity, 286–287
  processing, 26, 110, 207
  size, 153, 154, 283–285
    in GWAS, 415 (*See also* under Genome-wide*)
Screening cohorts
  249
Security systems for biospecimen facilities
  38–39
Segregation analysis
  264–266
Selection bias
  268, 326, 327
  – in case-control studies, 250, 251, 251
Selective reporting bias
  327, 328, 329
Selenium
  191

Sensitivity
- beryllium –, 387
- in error quantification –, 143, 147-150, 159
- insulin –, 193, 201, 442, 446
- PCR –, 75-76, 78, 83
- of analytical techniques, 25, 45, 46, 47
- of DNA assays, 49, 53, 57, 90, 177, 178, 186
- of mutation detection, 74-75, 76, 80, 82, 88
- of omic analyses, 498

Sequence analysis
- 104, 110, 113, 115

Serological methods
- 176, 182

Severe acute respiratory syndrome (SARS) (See New infectious agents)

Severe combined immune deficiency (SCID)
- 228, 229

Sex steroids
- 199, 200

Short
- oligonucleotide mass analysis (SOMA), 75, 79
- tandem repeats (STRs), 105, 285
- -term longitudinal biomarker studies, 242

Silicosis
- 393, 394
- characteristics of –, 392
- and coal workers' pneumoconiosis, 388, 392, 399

Single
- gene disorders, 307
- nucleotide polymorphisms (SNPs), 101–103, 104, 106, 109, 110, 111
- the – consortium, 267
- in GWASs, 123,
- and microsatellites, 265, (See also dbSNP, tag SNPs)

Sources of measurement error in biomarkers
- 144, 145, 147

Specificity
- antibody –, 220
- antigen –, 217, 220, 222
- assay – and deviation, 106
- carcinoembryonic antigen test –, 308
- in error quantification, 143, 147–150, 159
- method, 87, 88
- PCR – and sensitivity, 75
- sequence – of DNA, 177
- and Hill's criteria, 184

Specimen
- collection, 25, 26, 27–39
- and errors, 145–146, 157, 159, 208
- and molecular tools, 437
- and TCDD exposure assessment, 466
- less-invasive methods of – for children studies, 479
- processing, 31–32, 146, 176
- tracking, 35, 37

Spectral karyotyping (SKY)
- 83

Sperm chromatin structure assay (SCSA)
- 464

Spontaneous mutations
- 69, 223, 427

Spot urine (See Urine)

Spotted microarrays
- 124

Standard operating procedures
- 33, 35, 36, 37, 106, 114, 179

Standardization
- data, 330
- of terms, 37
- International Organization for –, 38
- method –, 46

Statistical
- evaluation of interaction, 294–295
- heterogeneity, 325
- power, 248, 272, 273, 283, 284, 415

Storage
- 23–42,
- alternate technologies for –, 39
- at the NCI, 37
- conditions, 34
- options, 207–208
- requirements, 202, 203
- system maintenance, 34

STROBE-ME
- 327

Study design
- 11, 169
- basic principles in –, 281–302
- common epidemiological –s, 447
- family-based –, 261–280
- issues in –, 351
- population-based –, 241–260

Study results
- interpreting and communicating –, 15–17

Subject error
- 150, 155

Surface-enhanced laser desorption ionization time-of-flight mass spectrometry (SELDI-TOF-MS)
- 128, 133, 341 (See also Matrix-assisted laser)

Surrogate endpoint
- 304, 308

Surveillance
- 2, 424–425
- in outbreak investigation, 428
- medical –, 391
- of vaccines, 184
- population-based – systems, 19

Survival study
- 249, 251, 254, 316

Susceptibility
- 2, 16
- biomarkers of –, 44, 45, 193–194, 220, 222
- in case-control studies, 252
- loci, 282–283

System interoperability
- 36

Systematic reviews
- 323, 324–325, 326, 327
- of published and unpublished studies, 369

Systematic variation normalization (SVN)
- 134

Systems biology
- 122, 135
- studies, 378

Tag SNPs
- 283, 296

TagSNPs
- 103, 104, 109, 290

Tandem mass spectrometry (MS/MS)
- 79, 127–129, 498 (See also Mass spectrometry)

Target tissue
- 145, 192, 195, 339

TCDD (*See also* Dioxin)
   466–467
   half-life, 165
T-cells
   217–218, 223, 224
   development, 228
   malignancies, 226
   helper –, 218, 221
T-cell receptor (TR)
   217, 220, 222
      excision circle (TREC), 228
      genes, 222
Technical validation
   350
TEDDY (The Environmental Determinants of Diabetes in the Young)
   230–231
Temporal variability
   167
Testicular dysgenesis syndrome (TDS)
   460
Testosterone
   201, 204, 207, 208, 455, 456, 463
TH1 response
   218
TH2 response
   218
TH17 pathway
   218
The emerging risk factors collaboration (ERFC)
   371–373
Therapeutics
   308–310
Time integration
   247
Tissue
   banks, 195, 318
   collections, 28–29, 32
Tobacco smoke
   as an exposure marker, 170–172
   and lung cancer, 338–340
Toll-like receptors (TLRs)
   221
Total cholesterol (*See also* Cholesterol, High-density lipoprotein cholesterol, Low-density lipoprotein cholesterol)
   370, 371, 375
Toxic exposures
   476, 481, 483, 496
Toxicogenomics
   136, 348, 497
Transcript profiling
   125

Transcriptome
   113, 122, 123, 496, 497
Transcriptomics
   121–142, 317, 319, 438, 496, 497
Transgenerational health effects
   461
Translational research
   310
Transmission system
   425, 427, 436
Transmission/disequilibrium test (TDT)
   268
Tricarboxylic acid cycle (TCA)
   132
Triglycerides
   as a marker of obesity, 441
   in literature reviews, 369, 372
   in risk prediction algorithms, 375
   PUFA in –, 53
Tumour-associated antigen (TAA)
   225–226
Tumour-specific antigens (TSAs)
   225
Twin registries
   261
Two-dimensional polyacrylamide gel electrophoresis (2D PAGE)
   127
Two-phase sampling design
   255
Two-stage testing strategy
   273, 275
Type 1 diabetes (T1D)
   229, 230, 231
Type 2 diabetes (T2D)
   132, 223, 441–452, 477 (*See also* Diabetes, TEDDY)
Ultra-performance liquid chromatography (UPLC) (*See* Liquid chromatography)
United Kingdom Biobank
   26, 28, 345
   Ethics and Governance Framework, 12
   study, 378
U.S. Environmental Protection Agency (EPA)
   455
   acknowledgement, 356
   method standardization, 46
   –'s cancer guidelines, 476
Urinary
   mutagenicity, 195, 196
   nitrogen, 190, 195

Urine
   collection, 29–30, 190, 457
   specimens, 349
   24-hour –, 204
   morning –, 30, 204, 207
   spot –, 202, 204
Vaccine
   422, 426
   candidates, 421, 434, 435
   induced autism, 468
   HPV –, 432
   HPV – trial, 39
   HBV – trials, 184
   programmes, 430
   tumour –s, 226
   US federal – Injury Compensation Program, 468
Validity
   coefficient, 143, 150–151
   of a biomarker test ,155
   studies, 11, 143, 144, 150, 154
      design of – and reliability studies, 146–147, 155
      clinical –, 16
      relation of reliability to –, 156-157
Variation
   coefficient, 158
Venice criteria
   328, 345
Vitamin C
   191
Vitamin D
   191
   –/calcium-related pathway genetics, 194
   metabolites, 200
   receptor, 209
   association of – and several cancer sites, 351
   variation of – concentrations, 206
Vitamin E
   191
Waist
   circumference, 445, 448
   to-height ratio, 448
West Nile virus
   encephalitis, 433
   and dengue, 422
   in the United States, 427, 433
White blood cells (WBC)
   50, 202, 203, 218, 249, 253, 339, 480, 481

Whole-genome
    amplification (WGA), 30, 109, 114, 248,
    research, 13, 19-20, 110, 123, 124, 276, 285
Windows of susceptibility
    456, 465
Within-person variability
    195
    biomarker –, 159, 246, 247, 370
    over time, 208
Xenobiotic
    131, 464
    chemicals, 230
    metabolism, 398, 494
    activation and detoxification of –s, 478
    exposure to a –, 16, 43, 44, 242, 464
    markers of –, 304
X-ray fluorescence (XRF)
    46, 410
Zinc
    195, 196
    superoxide dismutase, 410